DYNAMICS AND CONTROL OF CHEMICAL REACTORS, DISTILLATION COLUMNS AND BATCH PROCESSES

(DYCORD+ '95)

A Postprint volume from the 4th IFAC Symposium, Helsingør, Denmark, 7 - 9 June 1995

Edited by

J.B. RAWLINGS
University of Wisconsin-Madison, Wisconsin, USA

Co-edited by

W. MARQUARDT *(Germany)*

D. BONVIN *(Switzerland)*

S. SKOGESTAD *(Norway)*

Published for the

INTERNATIONAL FEDERATION OF AUTOMATIC CONTROL

by

PERGAMON
An Imprint of Elsevier Science

UK Elsevier Science Ltd, The Boulevard, Langford Lane, Kidlington, Oxford, OX5 1GB, UK

USA Elsevier Science Inc., 660 White Plains Road, Tarrytown, New York 10591-5153, USA

JAPAN Elsevier Science Japan, Tsunashima Building Annex, 3-20-12 Yushima, Bunkyo-ku, Tokyo 113, Japan

---

First edition 1995

Library of Congress Cataloging in Publication Data

A catalogue record for this book is available from the Library of Congress

British Library Cataloguing in Publication Data

A catalogue record for this book is available from the British Library

ISBN 0-08-042368 X

This volume was reproduced by means of the photo-offset process using the manuscripts supplied by the authors of the different papers. The manuscripts have been typed using different typewriters and typefaces. The lay-out, figures and tables of some papers did not agree completely with the standard requirements: consequently the reproduction does not display complete uniformity. To ensure rapid publication this discrepancy could not be changed: nor could the English be checked completely. Therefore, the readers are asked to excuse any deficiencies of this publication which may be due to the above mentioned reasons.

The Editors

Printed in Great Britain

4th IFAC SYMPOSIUM ON DYNAMICS AND CONTROL OF CHEMICAL REACTORS, DISTILLATION COLUMNS AND BATCH PROCESSES (DYCORD+ '95)

Sponsored by

IFAC - The International Federation of Automatic Control

International Program Committee
James B. Rawlings (USA) - Chairman and editor

Subcommittee on Chemical Reactors

W. Marquardt (D)
- co-chairman and co-editor
D. Gay (CH)
E. D. Gilles (D)
I. Hashimoto (J)
J. F. Mac Gregor (CAN)
P. Nielsen (DK)
B. A. Ogunnaike (USA)
L. Zullo (NL)

Subcommittee on Distillation Columns

S. Skogestad (N)
- co-chairman and co-editor
A. Brambilla (I)
E. W. Jacobson (S)
L. Lang (D)
W. L. Luyben (USA)
J. Mandler (USA)
T. J. McAvoy (USA)
K. V. Waller (FIN)
G. Wozny (D)

Subcommittee on Batch Processes

D. Bonvin (CH)
- co-chairman and co-editor
C. Georgakis (USA)
S. Hasebe (J)
N. Jerome (USA)
K. Levien (USA)
D. R. Lewin (ISR)
S. Macchietto (UK)
H. Preisig (AUS)
H. Schuler (D)

National Organizing Committee

S. B. Jørgensen
- Chairman
M. Y. Andersen
L. v. Gersdorff
O. Hansen
N. Jensen
J. Knudsen
M. Kümmel
P. Martin Larsen
J. Saietz

Organized by the
Danish Automation Society
Copenhagen

Acknowledgements

The organizers greatly appreciate support and deficit guarantees obtained from the following foundations, companies and institutions

Carlsbergs Mindelegat
Danisco Distillers
Danisco Ingredients
Fabrikant Mads Clausens Fond
FLS Automation A/S
i/s Sjællandske Kraftværker
Kuwait Petroleum Danmark A/S
Novo Nordisk A/S
Otto Mønsteds Fond
Siemens A/S
Thomas B. Thriges Fond

CONTENTS

PLENARY PAPER I

A Contemporary Industrial Perspective on Process Control Theory and Practice 1
B.A. OGUNNAIKE

CHEMICAL REACTORS I

Control System Improvement of an Industrial Chemical Reactor 9
S. MURAKAMI, T. YOSHINARI, T. FUJIWARA, H. NISHITANI

Computational Fluid Dynamics as a Tool for Chemical Reactor Optimization 15
H.-J. PREISIGKE

Control of Forced Cyclic Process 21
J.E. HANSEN, S.B. JORGENSEN

Fuzzy Logic Controller Design for pH-Control in a CSTR 27
T. HECKENTHALER, S. ENGELL

Reliable and Efficient Optimization Strategies for Nonlinear Model Predictive Control 33
L.O. SANTOS, N.M.C. DE OLIVEIRA, L.T. BIEGLER

CHEMICAL REACTORS II

Optimal Estimation of States in Differential-Algebraic Equation Systems 39
Y.S. CHENG, T. MONGKHONSI, L.S. KERSHENBAUM

Use of Iterative Dynamic Programming in Optimization of Reactor Systems 45
R. LUUS, V. DONG

Hard Constraints in Control and State Variables of Multivariable Nonlinear Processes Resolved by Elementary Nonlinear Decoupling 51
J.G. BALCHEN, B. SANDRIB

Time-Variable Models for Mixing Processes Under Unsteady Flow and Volume 57
K. ZENGER

General Methods of Kinetic Model Reduction 63
K. EDWARDS, T.F. EDGAR

Robust Decentralized Control of A CSTR with Complex Reaction Scheme 69
R. MULLER, J.O. TRIERWEILER

DISTILLATION COLUMNS I

Input Multiplicity and Right Half Plane Zeros in Ideal Two-Product Distillation 75
E.W. JACOBSEN

Heterogeneous Azeotropic Distillation Involving an Embedded Two-Liquid Phase Region 81
H.S. MUMTAZ, G. CORBETT, M. REAGAN, W.D. SEIDER, L.A. FABIANO, S. WIDAGDO

Hierarchical Sub-Optimal Control of Batch Distillation with Slop Recycling 87
B.H.L. BETLEM, H.C. KRIJNSEN, Th.H. TIMMER

High Performance Distillation Column Control Through Novel Control Configurations 93
J. ZHANG, M.T. THAM, A.J. MORRIS

A DAE Framework for Modeling and Control of Reactive Distillation Columns 99
A. KUMAR, P. DAOUTIDIS

Identification of a Pilot Plant Reactive Distillation Process using RBF Networks 105
P. SAYNATJOKI, L. HAMMARSTROM

BATCH PROCESSES I

Reduced Modeling of Complex Fluid Mixtures by the Wavelet-Galerkin Method 111
R.v. WATZDORF, W. MARQUARDT

An Object-Oriented Method for Process Modelling 117
S.O. WASBO, B.A. FOSS

Hybrid Simulation of Flexible Batch Plants 123
S. ENGELL, M. FRITZ, C. SCHULZ, K. WOLLHAF

Neural Network Based Estimators for a Batch Polymerization Reactor 129
J. ZHANG, X. YANG, A.J. MORRIS, C. KIPARISSIDES

ALAMBIC - A Software Package for Optimising Design and Operation of Batch Distillation Columns 135
I.M. MUJTABA, G. STUART, S. MACCHIETTO

Comparison of Inverted and Regular Batch Distillation 141
E. SORENSEN, S. SKOGESTAD

POSTER PAPERS

Experimental Evaluation of a Fault Detection and Identification Scheme for Chemical Processes 147
P.A.A. AFONSO, J.L. FERREIRA, J.A.A.M. CASTRO, N.M.C. OLIVEIRA

Steady State and Dynamic Simulation of FCC Reactor Units 153
R. MAYA-YESCAS, D. BOGLE, F. LOPEZ ISUNZA

Dynamics of the Jarosite Conversion Process 159
G. HOLCROFT, L. ROSATO, M. PERRIER, D. BERK

Model-Based Temperature Control of Fed-Batch Reactors 165
F. SZEIFERT, L. NAGY, T. CHOVAN

Practical Control Methods for Distillation Columns Using Neural Networks 171
M.S. BASUALDO, H.A. CECCATTO

Fine Tuning of Multiloop PI Controllers Using Fuzzy Set Theory 177
H. LEE, S. PARK

Multivariable Control of Distillation Column 185
J. MACHACEK, J. KOTYK

Simulation of Multicomponent Batch Distillation with Chemical Reactions Processes 191
M.S. BASUALDO, C.A. RUIZ, N.J. SCENNA

Physical Implications of Periodic Solutions in a Fixed Bed Reactor with Recycle 197
B. RECKE, S.B. JORGENSEN

Analysis of Process Dynamics Using Steady State Flowsheeting Tools 203
I. NAOT, D.R. LEWIN

Neural Network Identification of an In-Line pH Process 209
S.K. DOHERTY, J.B. GOMM, D. WILLIAMS

PLENARY PAPER II

Some Aspects of Integrated Process Operation 215
E.D. GILLES, G. LAUSCHKE, A. KIENLE, M. STORZ

PLENARY PAPER III

Optimization in Model Based Control 229
D.Q. MAYNE

DISTILLATION COLUMNS II

Multiple Steady States in Azeotropic Distillation and Their Effect on Column Operation and Design 243
N. BEKIARIS, G.A. MESKI, M. MORARI

Comparison of the Separation Performances of a Multi-Effect Batch Distillation System and a Continuous Distillation System 249
S. HASEBE, T. KUROOKA, I. HASHIMOTO

Analysis and Design of CASCADE Control Schemes for Distillation 255
G. LORENZI, D. SEMINO, C. SCALI, A. BRAMBILLA

Evaluation of Dynamic Models of Distillation Columns with Emphasis on the Initial Response 261
B. WITTGENS, S. SKOGESTAD

Plantwide Control Problem for a Process with Three Distillation Columns and Two Recycle Streams 269
M.L. LUYBEN, W.L. LUYBEN

CHEMICAL REACTORS III

Wave-Net Based On-Line Quality Inference System for Polymerization Processes 275
M. OHSHIMA, A. KOULOURIS, S. TOMITA, G. STEPHANOPOULOS

State and Parameter Estimation with Augmented Kalman Filter Applied to an Industrial Polymer Reactor 281
H. DIDRIKSEN, M. HILLESTAD, K.S. ANDERSEN, J.G. BALCHEN

Kalman Filter and Neural Network for On-Line Estimation of Polymer Chain Characteristics 287
S. GUIOCHON, G. DEFAYE, C. VIDAL, L. CARALP

Interactions Between Production Rate Optimization and Temperature Control in Gas Phase Polyethylene Reactors 293
S.A. DADEBO, K.B. McAULEY, P.J. McLELLAN

Controllability Issues Concerning Particle Size in Emulsion Polymerization 299
V. LIOTTA, C. GEORGAKIS, M. EL-AASSER

Multivariable Control of a Continuous Crystallization Process Using Experimental Input-Output Models 305
R.A. EEK, O.H. BOSGRA

DISTILLATION COLUMNS III

A Dynamic Model of a Superfractionator: A Test Case for Comparing Distillation Control Techniques 311
V. GOKHALE, S.E. HUROWITZ, J.B. RIGGS

Identification for Control of High-Purity Distillation Columns - A Benchmark Problem 317
E.W. JACOBSEN

Distillation Column Control Benchmarks: Four Typical Problems 323
A. KOGGERSBOL, S.B. JORGENSEN

The Impact of Process Directionality on Robust Control in Non-Ideal Distillation 327
M.F. SAGFORS, K.V. WALLER

Robust Analysis of Distillation Column Dynamics 333
E.I. VARGA, G.Z. KOVACS, S.B. JORGENSEN, K.M. HANGOS

Investigation of Operability and Controllability Properties of a Pilot-Scale Distillation Column 339
K.E. HAGGBLOM, K. LEHTINEN

BATCH PROCESSES II

Control of Chemical Reactors Using Multiple-Model Adaptive Control (MMAC) 345
K.D. SCHOTT, B.W. BEQUETTE

Flexibility Analysis of Dynamic Chemical Processes 351
V.D. DIMITRIADIS, E.N. PISTIKOPOULOS

Control of a Batch Reactor Using a Multivariate Statistical Controller Design 357
M.J. PIOVOSO, K.S. DAHL, K.A. KOSANOVICH

Integration of Process and Control Designs by Nonlinear Geometric Methods 363
J. ALVAREZ, F. ZALDO, S. PADILLA

Dynamic Heat Balance Model for Industrial Batch Reactors and its Applications 369
K.S. LEE, C.M. JEONG, S. YI, J.S. SON, S.C. YOON

Methods of Duality in Problems about Expediency of Usage and Optimization of Cyclic Regimes of Technological Processes 375
A. TSIRLIN

PLENARY PAPER IV

Status and Challenges of Intelligent Plant Control 381
M. LIND

BATCH PROCESSES III

Risk-Conscious Operation of Batch Processes 401
P. TERWIESCH, D. RAVEMARK, D.W.T. RIPPIN

Synthesis of Procedural Controllers for Batch Chemical Processes 407
A. SANCHEZ, G.E. ROTSTEIN, S. MACCHIETTO

Temperature Control of a Batch Esterification Reactor Overcoming Non-Linearities Using Adaptive Gain 413
P. DJAVDAN

Temperature Control of a Batch Reactor by On-Line Identification of a Vapor Pressure Law 419
J. CEZERAC, V. GARCIA, M. CABASSUD, M.V. LE LANN, G. CASAMATTA

Effect of Model Uncertainty on the Tendency Modeling, Optimization and Control of Batch Reactors 425
J. FOTOPOULOS, C. GEORGAKIS, H.G. STENGER, Jr.

CHEMICAL REACTORS IV

Nonlinear Inferential Parallel Cascade Control 433
T.J. McAVOY, NAN YE

Nonlinear Observer Design with Application to Chemical Reactors 439
M. SOROUSH

pH Control in the Presence of Precipitation Equilibria 445
R.A. WRIGHT, C. KRAVARIS

Partitioning-Space Approach for Nonlinear Process Control 451
Z.H. LIU, S. MACCHIETTO

Limitations to Input-Output Analysis of Cascade Control of Unstable Chemical Reactors 457
L.P. RUSSO, B.W. BEQUETTE

Automatic Control of Chemical Reactors at Unstable Steady-States 463
D. SEMINO, M. GORI

DISTILLATION COLUMNS IV

Multivariable Quality Control of a Crude Oil Fractionator 469
M.V. DE OLIVEIRA MAGALHAES, D. ODLOAK

Composition Estimators for a Distillation Column: Theory and Experiments 475
R. BARATTI, A. BERTUCCO, A. DA ROLD, M. MORBIDELLI

Prediction and Analysis of the Effect of Reboiler and Condenser Holdup on the Dynamic Behaviour of Distillation Columns 481
M. MONTESI, A. BRAMBILLA

Development and Experimental Verification of a Simulation Tool for Column Startup 489
Ch. KRUSE, G. FIEG, G. WOZNY, L. JEROMIN

Verification of a Distillation Column Benchmark 495
A. KOGGERSBOL, S.B. JORGENSEN

Author Index 501

A CONTEMPORARY INDUSTRIAL PERSPECTIVE ON PROCESS CONTROL THEORY AND PRACTICE

Babatunde A. Ogunnaike

E.I. DuPont de Nemours and Company, Wilmington, Delaware, 19880-0101

Abstract. The chemical process industry, a highly successful global enterprise, currently faces certain challenges having significant implications for Process Control theory and practice. This paper takes the perspective that these challenges actually represent a unique opportunity for Process Technology in general, and Process Control in particular, to make an impact. An overview of some of the most important of these challenges is provided along with some suggestions regarding what the process control community needs to do in order to translate this *potential* opportunity into actual success.

Keywords. Industrial Processes, Process Control, Plant-Wide Control, Academic research, Industrial Practice

1 INTRODUCTION

The Chemical Process Industry occupies a unique position in the global economic landscape in general, and of the U.S. in particular (cf. Miller, 1994):

1. It is currently assessed as a one trillion dollar giant worldwide, accounting for about 10% of all U.S. manufacturing, with annual sales exceeding $100 Billion;
2. It is the only unsubsidized industrial sector in the U.S. to have sustained a long-term positive balance of trade.

But this industry, which, by any yardstick — global or local — must be assessed as one of the most successful enterprises in the world, is facing unprecedented challenges with implications of historic magnitude for the theory and practice of Process Technology in general, and Process Control in particular. A U.S. perspective of these challenges have been succintly captured in Miller, 1994:

> "In the last fourteen years, our net returns have not been sufficient to allow the industry to attract the capital needed for sustained growth. We seem trapped in a box defined both by major sociopolitical forces and by the way we develop and implement technology.
>
> "Our future depends on finding a way of breaking out of this box, and such a breakout I believe will require a renewed commitment not only to the development of new technology, but also to implementing it more quickly and effectively than we do now. I am confident that if we in industry, academia and government work together to grow and sustain fundamental research and develop the tools to facilitate the commercialization process, we will find a way of breaking out. The chemical industry is truly a national resource but it will need such a commitment if we are to sustain and grow it.
>
> "... Aggressive global competition and increased customer expectations combined with rising societal demands for environmentally benign plants will require new and better chemical processes for current and future products. With costs increasing and prices decreasing, productivity must increase to justify the capital investments needed. *These productivity improvements can only be obtained with continuing advances in [process] technology and through their more effective implementation.*"

The essence of these challenges is therefore that of meeting the demands of increased productivity through *increased production rate of top quality products, at the lowest cost, employing the most "flexible" process, all with the barest minimum amount of new investment.* The general consensus in industrial circles currently is that no group of people is better equipped to confront these challenges successfully than those trained in Process Synthesis and Design, Process Dynamics, Process Modeling and Simulation, Process Optimization, and Process Control. This is the main premise of this paper: that at no time in the history of the chemical process industry has there been such an opportunity for Process Technology in general, and Process Control in particular, to make such an impact. However, in translating this potential opportunity into actual success the current state of Process Control Theory and Practice (among other things) will have to change, and the time to begin the change is *now*.

The discussion in this paper is limited to the role that Process Control specifically — not Process Technology in general — must play as the Chemical Process Industry attempts to deal successfully with the contemporary challenges it faces. To the extent possible within the confines of the alloted number of pages, we will attempt to

- elaborate on what these contemporary challenges are from a process control perspective, and place them in historical context; and
- suggest what the Process Control community needs to do in order to rise up to these challenges.

To do this, the paper is organized as follows: In Section 2 we re-examine "Process Control" within the context of the Process Industry, as a subject matter and as an enabling technology; this is followed by a brief historical perspective of the evolution of industrial process control in Section 3. Some of the contemporary issues are identified specifically and discussed in Section 4, emphasizing that the truly important issues will not be resolved without true collaboration between theory and practice. To the extent that any aspect of the future can be anticipated, an attempt to do so is presented in Section 5 before providing a summary and some conclusions in Section 6.

2 RE-EXAMINING "PROCESS CONTROL"

2.1 The Essence of "Process Control"

If, as thought-provokingly stated by Alistair MacFarlane (MacFarlane, 1993), "control is effective interaction," then, as a special case of this wide-ranging concept, "Process Control" *is* the industrial process, as one object, interacting effectively with its environment. From a different but related perspective, control as a *technology* is what enables the achievement of this "effective interaction."

Within the context of MacFarlane's paradigm therefore (of control as a concept as well as an enabling technology) we are able to see the obvious and the not-so-obvious connection between a perfectly suspended hovering humming bird and an aircraft landing; and also between a tightrope walker and a crude-oil refinery producing various grades of useful fuel, of consistent quality. These are all examples of diverse "objects" interacting effectively with their environments in a controlled way, with the animate providing a subtle metaphor for the inanimate.

One of the benefits of this somewhat general perspective is that it permits us to reduce the essence of process control to the following three elements:

1. **The primary object:** The industrial process itself, not the mathematical abstraction of it called the process model; the latter might be a convenient means to an end, but it should never be mistaken for an end in itself. *Thus Process Control as a discipline will ultimately be meaningless if the focus on industrial process itself is not steadfastly maintained.*
2. **The "raison d'être:"** The environment (physical, economic and social) which influences, and is influenced by, (i.e. interacts with) the industrial process.
3. **The main task:** enabling the primary object (the industrial process) to interact with the "environment" in a manner that is considered "effective" by currently acceptable norms. (Let us therefore recognize that what may be effective today may no longer be so tommorow.)

One of the objectives of this paper is to attempt to examine the evolution of the problems and challenges of industrial process control from this perspective. And then by noting the synergistic partnership between theory and practice in the development of useful solutions, it is possible to place the contemporary issues into context and hence anticipate what future problems are likely to be and how theory and practice must combine to solve them effectively.

2.2 Why "Process Control"

In the chemical process industry, the primary objective is to combine chemical processing units (chemical reactors, distillation columns, heat exchangers, etc) in a rational fashion into a "chemical process" (or a "plant") in order to transform raw materials and input energy into finished product. The basic principles guiding the operation of these chemical processes (from the dawn of the process industries until now) may be stated as follows:

- They must be operated *safely*;
- Specified *production rates* must be maintained;
- *Product quality* specifications must be met.

Because chemical processes are by nature dynamic and are always subject to "disturbances," the "process control system" has the responsibility of monitoring, and inducing change in, the appropriate variables related to safety, production rate, and product quality. In discharging its responsibility, there are, in essence, two main aspects to the process control system's task:

- Information gathering and transmission;
- Control action computation and implementation.

This rationalization is important in understanding the evolution of industrial process control practice as discussed in the next section.

As already introduced in Section 1, the universal drive for more consistent attainment of high product quality, more efficient use of energy, and tighter safety and environmental regulations, has forced industrial processes to evolve over the past few decades into very complex and highly integrated systems. Such stringent demands have created more difficult and challenging control problems for today's industrial process control engineer — problems requiring more sophisticated solutions than can be provided by traditional techniques alone.

We may now observe that the current status of process control derives largely from the following facts:

1. **The primary object:** The industrial process itself, has become more complex;

2. **The raison d'être:** has become more compelling. The economic and social (if not so much the physical) environment now introduces more frequent and more significant "disturbances;"

3. **The main task:** has now become more difficult since not only has the "environment" become more "disruptive," the standards by which the effectiveness of process operation is measured (in terms of product quality, energy utilization, emmisions, etc) are now stricter. What was previously considered acceptable process operation is now no longer acceptable.

With this as background, let us now examine somewhat more closely the historical evolution of industrial process control.

3 A BRIEF HISTORICAL PERSPECTIVE

Prior to the 1940's most chemical processes were operated manually. This is because the processes themselves were not terribly complex, the product and process operation requirements were not very strict, and thus the process was able to "interact effectively with its environment" by employing manual control alone.

As production volumes increased, however, and the interconnections between processing units became more complex, the need for automation became more compelling. The application of basic feedback control theory to solve important chemical process automation problems therefore became popular in the 1950's. During this period, significant advances continued to be made in control theory (see for example, Kalman, 1958), but hardware limitations primarily prevented immediate application.

In most of the earliest implementations of automatic control systems, information flow — the first half of the control system's task — was accomplished by pneumatic transmission, and computation — the second half of the control system's task — was done by mechanical devices using bellows, springs, etc. Because the medium of transmission was air, there was a fundamental limitation to the the speed of information flow. The mechanical nature of the computational devices meant that the controllers could only perform limited and very simple operations; in addition, mechanical devices tend to wear out quickly. The dual problem of speed of transmission and longevity of computational devices led to the evolution of electronic analog implementation. Analog signals travel much faster than pneumatic ones, and analog computational devices are not subject to the same degree of wear as their mechanical counterparts. However, the computations possible with analog devices are still limited, and because analog signals are easily susceptible to contamination, signal quality tends to degrade over long transmission lines.

Over the ensuing decades, however, there has been a steady increase in the awareness of environmental responsibility, in addition to a universal drive for more consistent attainment of high product quality, and for more efficient use of energy. These and other such issues have combined to impose far stricter demands on the control system than can be met by traditional feedback techniques alone. These circumstances (which still exist to this day) created the need for the application of more sophisticated control theories to industrial processes — control theories whose development had been mostly driven by perceived practical need, but whose implementation had been limited by the existing hardware.

The complex interconnection of processes partly contributed to the need to preserve process signal integrity over long transmission lines; it also led to the need for more sophisticated controllers, which in turn led to the need for more versatile computational devices with which to implement these controllers. The implementation of such advanced control systems would not have been possible but for the major developments in digital computer technology which, fortuitously, have occured during this same time span.

The transmission medium for digital signals is electrical, but they are transmitted as binary numbers so that they are far less susceptible to noise. Also, the digital computer as the computational device is more flexible because it is programmable, and more versatile because there is virtually no limitation to the complexity of the computations it can carry out. (See Ogunnaike and Ray, 1994, Chapter 2 for a fuller discussion.)

It is important to observe that it is nothing more than an attempt to enable the process to interact effectively with its environment that has driven the evolution of process control summarized briefly here: as the social, economic, and physical "environment" changed, more demands were placed on the industrial process, and effective interaction therefore meant an elevation of the level of process control sophistication.

What this means for the present and the future will be considered in the next sections.

4 CONTEMPORARY ISSUES

4.1 Overview and An Illustration

When the challenges of increased customer expectations and rising societal demands for environmentally benign process operation are combined with the reality of "aggressive global competition," one is confronted with a situation in which exisiting industrial processes must be operated much more efficiently than ever before. Quite often this leads to industrial process control problems that are far more complex than can be solved by traditional techniques alone.

To illustrate, consider the process for manufacturing a certain polymer product represented as it stands today in the highly simplified schematic diagram shown in Figure 1. Without regard to (i) environmetal regulations; (ii) production costs; and (iii) product quality, it is easy to operate the process as follows: The basic reactor operating conditions are set *once* at startup and need not be closely monitored or regulated over the entire production period; the unreacted monomer may simply be "boiled off" the solid product; and the resulting product may be dried and sent directly to customer without monitoring the final product quality too closely. This of course is not acceptable today.

In an attempt to meet environmental regulations and also reduce production cost, the indicated separations process is introduced as a means of recovering and reusing the (somewhat toxic) unreacted monomer. However, in thus solving one set of problems, another one is created: the recycle loop thus introduced increases the final product quality variability significantly. With increasing customer pressure for quality improvements, basic regulatory control of key process variables is imperative, but this alone is not sufficient: on the one hand, the process interactions are too strong, and on the other, the final product quality itself still cannot be measured on-line. The indicated blenders constitute an expensive, and not-very-effective, means of coping with the product quality variability issue: such a strategy relies solely on the "capacitance" of the blenders for "damping out" the product quality variability. This is clearly not an effective way to achieve acceptable quality control; it is an admission of the fact that the traditional control techniques are unable to provide acceptable solutions, but yet *something* must be done simply to stay in business.

To expand into additional markets for the polymer product, different grades of the product must be manufactured to different customer specifications. This process — difficult enough to operate, as currently configured, to produce a single grade to specification — must now operate efficiently through several startups and shutdowns, (or transitions,) while maintaining desired product quality, and minimizing the production of "off-spec" material in the interim period.

The means by which commercially successful operation of this example process can be achieved summarizes the essence of the contemporary challenges facing industrial chemical process control. But in its broadest sense, there are currently two aspects to industrial chemical process control:

1. **Plant-wide control:** which, as implied by the name, involves the control of the entire operating plant, considered as a single entity whose components are the various individual, interconnected processing units;

2. **Control of Individual Processing Units** such as chemical reactors, distillation columns, heat exchangers, etc.

Even though inextricably interrelated, these two aspects are still considered as distinct entities; we shall therefore examine each separately.

4.2 Plant-Wide Control

The plant-wide control problem clearly involves the control of individual units, but it involves much more. In plant-wide control, the problem of control system design for individual units is considered at the *second* stage, after the more fundamental issue of appropriate basic *overall control structure selection* has been resolved. This basic, "first stage" control structure is primarily to ensure the "smoothest possible" flow of material and/or energy from one processing unit to another; easy containment of disturbance propagation from one unit to the other; and stable overall process operation. Such terms as "process operability," "process controllability," and "process stability" are used to define the objectives at this stage. Some metrics for quantifying these properties, as well as other important issues concerning plant-wide control, may be found, for example, in Morari *et al.*, 1980; Govind and Powers, 1982; Douglas, 1982.

It is now clear that effective plant-wide control *must* play a significant role in meeting the contemporary challenges facing the process industries. However, there is as yet no unified theory for the design of plant-wide control systems. The basic principles employed in Industry to date still have not deviated far from those laid down in Buckley, 1964 — one of the pioneering systematic treatments of the subject matter. And at the foundation of all such techniques is the traditional segregation of the process design and process control tasks, with the assumption that acceptable process operation ("effective interaction with the environment") can always be obtained by employing control strategies developed after the process has been designed (cf. Downs *et al*, 1994). However, it is also now clear that effective plant-wide control *must* involve the interaction of process design and process control. This interface between design and control remains an active area of research, even though many important concepts and results are already available (see for example, Morari, 1982; Stephanopoulos, 1982; Fisher *et al.*, 1988; Morari, 1992).

4.3 Control of Individual Processing Units

A significant portion of the industrial control engineer's efforts is still directed towards the control of individual processing units.

Virtually all processing units of commercial importance are multivariable, with significant interactions between the input and output variables; they typically exhibit nonlinear characteristics, and often suffer from time delays either as a result of material transport through pipes, or as a result of measurement and analysis delay.

Control Techniques Currently Applied in Industry. The control techniques most commonly applied on industrial processes (listed in order of total number of estimated applications) are as follows: (see also Yamamoto and Hashimoto, 1991):

1. Classical PID Control:
   - Both analog and digital versions; sometimes augmented with *feedforward, cascade, ratio, Smith Predictor* schemes for enhanced performance, as needed etc.;
   - For multivariable systems, the relative gain array (RGA) methodology is used to determine appropriate loop pairing; sometimes augmented with a *Decoupling* scheme, as needed.
2. Statistical Quality Control:
   - Quality control techniques (such as Shewart Charts, CUSUM, etc.);
   - PLS/PCA techniques have recently been introduced for multivariate process data analysis, with potential for control decisions still being explored.
3. Model Predictive Control (in all flavors)
4. Others
   - Special purpose, custom techniques;
   - Industry-specific "trade secret" techniques;
   - Neural networks for modeling (and sometimes for control); etc.

These techniques have been employed in a variety of cases to solve a significant number of industrial control problems. Nevertheless, to meet the current challenges, many important problems are still in need of effective and widely applicable solutions.

4.4 Some Current Problems and Needs

Effective control of industrial processes is still being hindered by some important problems and needs; a representative few of these are listed below.

1. On-Line Measurements:

   There are many processes for which critical process variables cannot be measured on-line. The most common examples are polymer processes for which most critical product end-use quality variables are not easily measured on-line. This complicates otherwise "solvable" problems since it is significantly more difficult to control (with confidence) what cannot be measured.

   When overall product quality was not so critical, the "on-line" availability of these measurements was not an issue; in the current environment, it is impossible to meet the stringent quality demands without confronting this issue head on.

2. Severely Nonlinear Processes:

   One of the lasting legacies of increased environmental and economic demands is that many chemical processes are now forced to operate in regions in which inherent nonlinear characteristics are exhibited more prominently. Since most standard control techniques applied in industry are for linear systems, controlling chemical processes effectively when they exhibit severely nonlinear behavior remains a major problem. Standard linear control theory is inadequate; linear model predictive control (MPC) provides poor performance; adaptive control has had only limited success; differential geometric methods, (linear) robust control theory, and nonlinear MPC are still largely impractical so far.

3. Modeling and Identification for control system design:

   The current "conventional wisdom" is that high performance control of a process is not possible without the aid of a model of the process in question. As such, a significant number of control schemes are, in fact, model-based. It is also widely recognized, however, that obtaining the process model is the single most time consuming task in the application of model-based control (cf. Andersen *et al.*, 1991). Surprisingly, many issues regarding effective modeling and identification of industrial processes are still unresolved. For example, the current, almost universally adopted philosophy of treating model identification like statisticians treat the problem of "fitting equations to data" may be inadequate. For statisticians, modeling for the purpose of "explaining the data" is an end; not so for the control engineer whose objective (in this context) is to use the model for controller design. There is abundant evidence in industrial practice that when modeling for control is *not* based on criteria related to the actual end use, the results can sometimes be quite disappointing.

4. Modeling for Simulation and Operator Training:

   The development of high fidelity, low maintenance models that can be used in simulators for control system evaluation as well as operator training is crucial if more sophisticated control systems are to become a permanent part of the chemical process of today and the future. Issues regarding what modeling paradigm to employ, and on what computing platforms to deploy these simulators remain to be resolved.

5. Process Monitoring and Diagnosis:

   The development of effective paradigms and tools for *automatic* monitoring and diagnosis of the increasingly complex process operations is important for continuous assessment of overall process and control system performance. The early detection of sensor and/or actuator failure (or other process operating faults), and an effective system for devising and implementing corrective actions rapidly and efficiently, is inevitable if the current and projected future *safety* and environmental regulations are to be met without jeopardizing *economic viability* in the attempt.

Since it was not meant to be exhaustive, clearly there are many more items worthy of inclusion in this list; such items as, for example, (i) Production Scheduling and Process Optimization (off-line as well as on-line); (ii) Control of Large-scale systems; (iii) Conceptual, computational and algorithmic issues for on-line data analysis for "Process Understanding" (as distinct from data analysis for process monitoring and diagnostics in which the result of the analysis is restricted to ascertaining *first* whether or not the process is operating "normally," and *then* recommending the source of any detected faults); these are all important, but have not been discussed in the interest of space.

4.5 What must we do?

Being the one entrusted with the responsibility of making the actual industrial process operate under the current stringent conditions, the typical process control practitioner in industry is confronted daily with a bewildering array of individual, specialized, industry-specific catalog of problems; he/she is therefore often unable to "see the tree for the forest." Without the "organizing power of abstraction" inherent in *theory*, it will be difficult to make truly real and lasting progress in dealing with such an array of problems. Ironically, the very nature of the industrial practitioner's responsibilities is what prevents him/her from being able to afford the reflection required for successfully collating related but isolated problems into a coherent formulation, and developing the appropriate theory needed to provide widely applicable solutions. On the other hand, the academic researcher, by the very nature of his/her job — and training — is best suited to provide this "organizing power of abstraction" and from there provide the appropriate theory.

The indicated complementary nature of the roles of the process control practitioner and the process control theoretician is, of course, quite obvious. *Practice* without guidance from *theory* is inefficient, potentially dangerous, and ultimately counterproductive; and theory without contact with practice quickly becomes a hollow exercise in futility with ultimately no real, or lasting impact. Unfortunately, the current environment is characterized essentially by the fact that *a significant number* of academic researchers have lost sight of the "primary object" of process control, having been seduced by the "elegance" of mathematics and so-called control theory into formulating and solving abstract and largely irrelevant problems with no prospect of making any real impact. On the other hand, a significant number of industrial practitioners have in turn developed such a cynical attitude towards results of academic research to the point that any proposed *new* control technology (i.e. anything more complicated than the PID controller) is often viewed *first* with a degree of irrational prejudice.

Thus the more fundamental issue to be addressed is the perennial disconnect between theory and practice, the perpetual mutual distrust between the theoretician and the practitioner. This issue has been around for so long and has been analyzed and discussed to the point that it has become almost trite. Nevertheless, it is still the contention of this author that a positive resolution of this issue is critical to the timely solution of the serious problems currently facing the chemical process industry. And the time for such resolution is *now*.

The Theory-and-Practice Synergism. Properly employed, theory and practice are supposed to be synergistic: the one incomplete without the other; and the whole being greater than the sum of its parts. This is the way in which chemical engineering as a whole made progress; and this is the only way process control will make progress.

This point is well illustrated by the relatively unknown 1958 paper by R.E. Kalman entitled "Design of a Self-Optimizing Control System" based on work he did while with the Engineering Research Lab at the DuPont Company. The paper anticipates modern model-predictive control deployed on a digital computer, complete with the required model identification technique as well as procedures for updating model parameters on-line. The abstract stated, among other things, that *"... it should find widespread application in the automation of complex systems such as aircrafts or chemical processes, where present methods would be too expensive or time-consuming to apply."* And the acknowledgment noted that the paper grew out of "several discussions with Dr. J. R. Ragazzini," then of Columbia University, now recognized widely as one of the founders of Digital Control theory. This paper illustrates two facts at once: (i) the more obvious and immediate fact that the solutions presented in this particular paper as well as its underlying theory came about as a result of the Kalman-Ragazzini industry/academic collaboration; (ii) the more remote fact that Kalman's strong connection to real industrial problems early in his career, no doubt influenced the results for which he would later become most famous.

The main point is that:

- to *formulate* current and future problems properly;

- to *develop* appropriate and timely solutions;
- to *transfer* the technology effectively and in a timely fashion,

stronger collaboration must be established between theoreticians and practitioners: i.e. between academia and industry, and within industry, between R & D, and manufacturing groups.

Specific Recommendations.

1. *Industry must become more open.* Without compromising its proprietary information, industry needs to provide interested academic researchers more access to problems of real significance, and must be more willing to implement new technology. The DuPont company, Weyerhaeuser, 3M, are some examples of companies in the U.S. that have begun to take steps in this direction.

2. *Academic Researchers must refocus their efforts.* It is important not to lose sight of the fact that familiarity with the peculiarities of the industrial process is the Process Control researcher's competitive advantage over other control theorists. The Process Control researcher is more likely to appear as being no better than an *average* mathematician if all his/her efforts are directed towards the same goals as those of the pure mathematical control theorist.

3. *The academic/industrial partnership should be strengthened.* There are now enough people in industrial R & D organizations who understand the language of the most abstruse academic theoretician and who can also communicate with the most unrepentant practitioner (the one who is naturally suspicious of anything more complex than a PI controller). A strong and fruitful collaboration between industry and academia will necessarily speed up the development *and implementation* of appropriate technology. An arrangement by which the government contributes to funds for joint industrial/academic collaborative research could further strengthen this partnership.

5 ANTICIPATING THE FUTURE

To the extent that anyone can truly anticipate the future with any degree of accuracy, the following are some thoughts regarding what to expect for industrial process control.

1. *The "Primary Object"* The chemical process is likely to become more complex; current "non-traditional" processes (e.g. bioprocesses, microelectronics processes) will become more mainstream; some manufacturing processes will become smaller in order to be more responsive (or "agile") but not neccesarily less complex; it is likely that more processes will combine biological steps with traditional chemical reactions, leading to more interesting process dynamics; more importantly from a process control perspective, processes will more likely be viewed more "holistically" in the sense that controllers, sensors, and actuators will be considered more integral to, rather than distinct from, the process.

2. *The "Environment"* will more likely call for production practices characterizable as "faster," "better," "greener," "cheaper," "less wasteful," etc. i.e. the challenges will be tougher versions of what exists today.

3. *The "Task"* will more likely become more formidale because of increased "complexity" of both the chemical process and the environment with which it is to interact effectively.

The implications of these issues are that:

1. *Process Controllers* will be required to have higher functionality, greater flexibility and some element of "built-in intelligence."

2. The complexity of the environment will force the development of a Theory (or Theories) of much more fundamental nature than is currently supported by today's fragmentation. One could imagine for example, development of:

   (a) a theory of process characterization (dealing with the issue of what model structure is required for certain purposes, and how best to obtain such a model);

   (b) a theory of sufficiency of control functionality (given a control objective, what is the minimum controller functionality sufficient to the task, and how best to obtain such a controller);

   (c) a theory of information and knowledge (how much information, and of what quality, is required to meet or exceed process operation objective)

   These developments will be motivated (and aided) by a strong *practical* pull towards timely translation into actual design and implementation.

3. *Control System Implementation Hardware and Software* could conceivably incorporate "smart sensors and actuators" — with built-in data handling capabilities, and with limited delegated autonomous control responsibilities; more easily reconfigurable, more flexible man/machine interfaces.

6 SUMMARY/CONCLUSION

In this paper, we have presented a brief assessment of the current problems and challenges facing the chemical process industry, stressing the fact that people trained in various aspects of process technology in general, and process control in particular are ideally equipped to confront these challenges successfully. While some specific problems in immediate need of effective solutions have been presented, enough emphasis was also placed on the more fundamental issue: the basic framework for evolving truly lasting solutions must be based on true collaboration between theory and practice — between academia and industry. Some suggestions and specific recommendations for the process control community have also been included.

It is important to conclude with the view expressed in the introduction that at no time in the history of the chemical process industry has there been such an opportunity for Process Technology in general, and Process Control in particular, to make such an impact. It is time to *do* what is needed to translate this potential opportunity into actual success.

REFERENCES

[1] Andersen, H. W., K. H. Rasmussen, and S. B. Jorgensen; "Advances in Process Identification" in *Chemical Process Control – IV*, Y. Arkun and W. H. Ray (Eds) American Institute of Chemical Engineers, New York, (1991).

[2] Buckley, P. S.; *Techniques of Process Control*, John Wiley, New York, (1964).

[3] Downs, J. J., A. C. Hiester, S. M. Miller, and K. B. Yount; "Industrial Viewpoint on Design/Control Tradeoffs", Presented at IFAC Workshop on Integration of Design and Control, Baltimore, Maryland, June 27-28 (1994).

[4] Douglas, J. M.; "Preliminary Design Procedure for Steady-State Control Systems for Complete Chemical Plants," in *Chemical Process Control – II*, D. E. Seborg, and T. F. Edgar (Eds) *American Institute of Chemical Engineers*, New York, (1982).

[5] Fisher, W. R., M. F. Doherty, J. M. Douglas; "The Interface between Design and Control. 3. Selecting a Set of Controlled Variables," *I&EC Research*, 27, 611 (1988).

[6] Govind, R., and G. J. Powers; " Control System Synthesis Strategies," *A.I.Ch.E. Journal*, 28(1), 60 (1982).

[7] Kalman, R. E.; "Design of a self-optimizing control system" *Trans. ASME*, 80, 468-478, (1958).

[8] MacFarlane, A.G.J.; "Information, Knowledge and Control" in *Essays on Control: Perspectives in the Theory and its Applications*, H. L. Trentelman and J. C. Williams (Eds), Birkhäuser, 1-28, (1993).

[9] Miller, J. A.; "Impact of Global Economy on New Directions for the Competitiveness of the Chemical Industry" Plenary paper, FOCAPD '94 Conference, Snowmass, Colorado, July (1994)

[10] Morari, M.; "Integrated Plant Control: A Solution at Hand or a Research Topic for the Next Decade?" in *Chemical Process Control – II*, D. E. Seborg, and T. F. Edgar (Eds), American Institute of Chemical Engineers, New York, (1982).

[11] Morari, M., Y. Arkun, and G. Stephanopoulos; "Studies in the Synthesis of Control Structures for Chemical Processes," Parts I, II, III, *A.I.Ch.E. Journal*, 26, 220 (1980).

[12] Morari, M.; "Effect of Design on the Controllability of Chemical Plants" Presented at IFAC Workshop on Interactions between Process Design and Process Control, London, September 6-8, (1992).

[13] Ogunnaike, B. A. and W. H. Ray; *Process Dynamics, Modeling, and Control*, Oxford University Press, New York, (1994).

[14] Stephanopoulos, G.; "Synthesis of Control Systems for Chemical Plants. A Challenge for Creativity" in *Proceedings of the International Symposium on Process Systems Engineering*, Kyoto, Japan, (1982).

[15] Yamamoto, S. and Hashimoto, I.; "Present Status and Future Needs: The View from Japanese Industry" in *Chemical Process Control – IV*, Y. Arkun and W. H. Ray (Eds) American Institute of Chemical Engineers, New York, (1991).

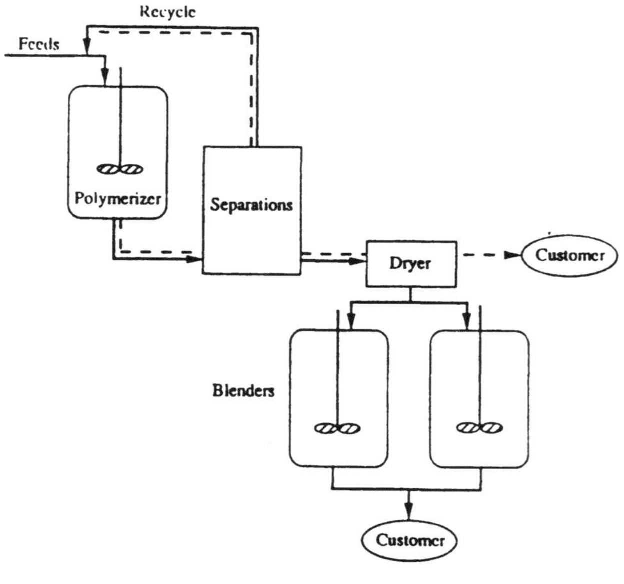

Fig 1. An Illustrative Chemical Process

CONTROL SYSTEM IMPROVEMENT OF AN INDUSTRIAL CHEMICAL REACTOR

SATORU MURAKAMI* , TOHRU YOSHINARI,*
TAKESHI FUJIWARA** and HIROKAZU NISHITANI**

* *Kaneka Corporation, 1-8 Miyamaemachi, Takasagocho, Takasago, Hyogo, 676, Japan*

** *Nara Institute of Science and Technology, 8916-5, Takayama, Ikoma, Nara, 630-01, Japan (e-mail:nisitani@is.aist-nara.ac.jp)*

Abstract. The current control system of an industrial reactor, which was designed under constant throughput at first, creates undesirable behavior under recent large throughput changes due to material and energy conservation. To cope with this situation, intervention by the operator is necessary. Data analysis of the existing plant and modeling of the system enabled us to find strategies to solve the problem. As a result, remarkable improvement was obtained by retrofit of the control system using these strategies.

Key Words. Continuous Flow Reactor, Control System Retrofit, Process Identification, Data Analysis, Simulation, Model Predictive Control

1. INTRODUCTION

The flow sheet of our plant is highly integrated from the viewpoint of material and energy conservation to increase the process efficiency. These factors cause a difficulty in regulatory control under large throughput(load) changes based on cost-minimum operation. In other words, although it is necessary to follow up throughput changes smoothly, there often arises undesirable behavior during transient conditions under the current control system. To cope with this situation, supervisory control based in the operator's knowledge is required in plant operation.

In this paper, time series data of a chemical reactor in the plant are analyzed carefully to identify essential problems and the current control system is reconsidered to improve performance in actual operation.

2. OUTLINE OF EXISTING PLANT

In the reactor, the following addition reaction occurs in the liquid phase:

$$A + B = C \qquad (1)$$

Reactant A is supplied directly from an upstream plant. On the other hand, reactant B is fed to the reactor in the following three ways:

1. Main flow by ratio control
2. Subsidiary flow by feedback control
3. Recovery flow from other processes

The main flow rate, u_1 is determined by flow ratio control of reactant A. The subsidiary flow rate, u_2 is controlled to maintain the concentration of A in the vapor phase at the top of the reactor. Its set-point is determined by the operator optimally from the economical viewpoint. The flow capacity of the subsidiary flow is about one tenth of the capacity of

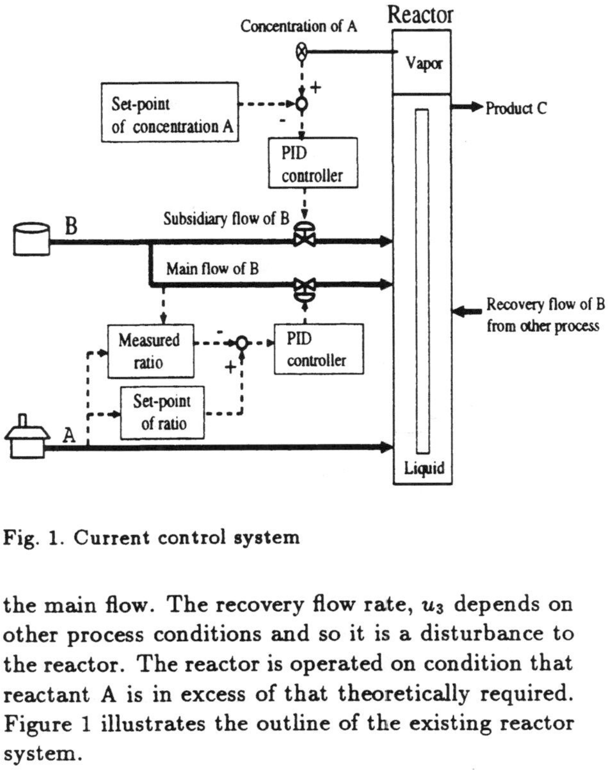

Fig. 1. Current control system

the main flow. The recovery flow rate, u_3 depends on other process conditions and so it is a disturbance to the reactor. The reactor is operated on condition that reactant A is in excess of that theoretically required. Figure 1 illustrates the outline of the existing reactor system.

In this reaction process, the feed rate of reactant A is changed deliberately by taking account of energy cost. There often occurs large fluctuation in the concentration of A under large load changes of A. This is due to the current control system designed under constant throughput. So far, when the concentration of A fluctuates largely, the operator intervenes in ratio control. The target of this study is to realize more stable control performance and to minimize operator intervention in the reactor. For this purpose, running data on the existing reactor system were analyzed in detail, and process models were built to reason out the phenomena.

3. MODELING OF EXISTING PLANT

3.1. *Reactor statics*

So far the ratio of the main flow rate to the load A has been given by a linear curve. But the value was often changed by the operator in a critical situation such as control saturation of the subsidiary flow. First, process data were analyzed in steady state operation where the concentration of A in the vapor phase is around the set-point. Figure 2 illustrates the stoichiometry of the reaction (dotted line) and the main flow rate of B(chain line) under ratio control and a constant recovery flow rate for B. In this paper, every flow rate is represented by the percentage of the maximum flow rate of A.

The discrepancy between the stoichiometry and the main flow rate of B is the sum of the subsidiary flow rate of B and the excess amount of A, if any exists. The excess amount of A, v was defined to measure the discrepancy with the stoichiometry under steady state as follows:

$$v = x - (u_1 + u_2 + u_3) \tag{2}$$

where x is the flow rate of A (load A) and u_1, u_2, and u_3 are the flow rates of the main, subsidiary and recovery flows. Steady state data showed that the artificial variable v is correlated well with a quadratic curve to x, as shown in Figure 3.

$$v^*(x) = ax^2 + bx + c \tag{3}$$

where v^* is the estimated value of the excess amount of A. This relation provides useful information to determine a better value for the main flow rate. The set-point value of the main flow rate (u_1^{sp}) was chosen as follows:

$$u_1^{sp} = \{x - v^*\} - u_2^0 - u_3 \tag{4}$$

where u_2^0 is a nominal value for the subsidiary flow rate and its value is chosen around half of the flow capacity. u_3 is the estimated value of the recovery flow rate. This formula helps avoid saturation of the subsidiary flow. Figure 4 illustrates the determination of the main flow rate of B. It is calculated from the stoichiometry of the reaction, estimated value of the excess amount of A, nominal value of subsidiary flow rate, and recovery flow rate .

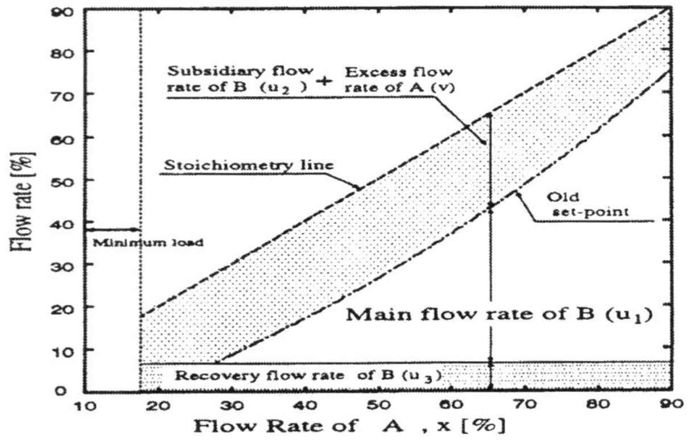

Fig. 2. Quantitative relation between A and B

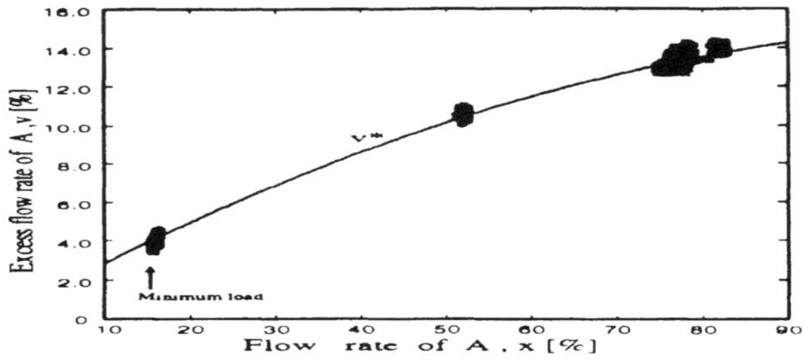

Fig. 3. Static relationship between flow rate of A and excess flow rate of A

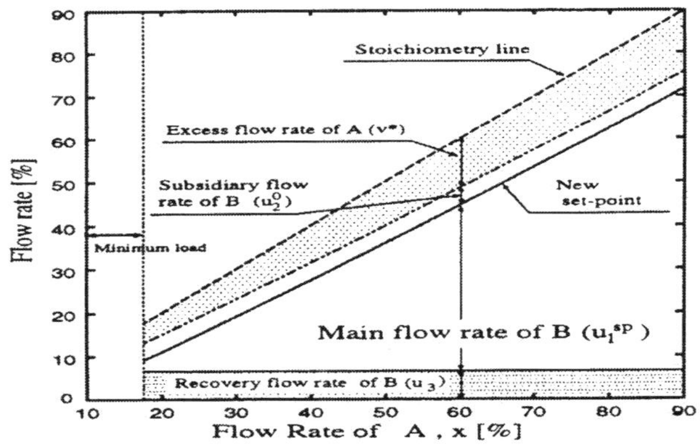

Fig. 4. Quantitative relation between A and B

3.2. *Reactor dynamics*

As the system input and output to represent the reactor dynamics, the excess amount of A, v and the concentration of A in the vapor phase in the reactor, y were chosen. The excess amount of A at the ith sampling time is defined as follows:

$$v(i) = x(i) - \{u_1(i) + u_2(i) + u_3(i)\} \tag{5}$$

For system identification the feedback loop of subsidiary flow was opened and its flow rate was changed manually under a constant load of A. Figure 5 shows a set of time series data on system input and output. The output is represented by the percentage of a specified value of concentration of A.

Through time series data analysis of this unsteady state operation, a parametric model called the ARX model was developed for the reactor dynamics as follows:

$$y'(k) = \sum_{i=1}^{m} a_i y'(k-i) + \sum_{i=1}^{n} b_i v'(k-i) \tag{6}$$

where,

$$\begin{aligned} y'(k) &= y(k) - \bar{y} \\ v'(k) &= v(k) - \bar{v} \end{aligned} \tag{7}$$

and, $\bar{y}$ and $\bar{v}$ are the averaged values of y and v, respectively. a_i and b_i are model parameters. The num-

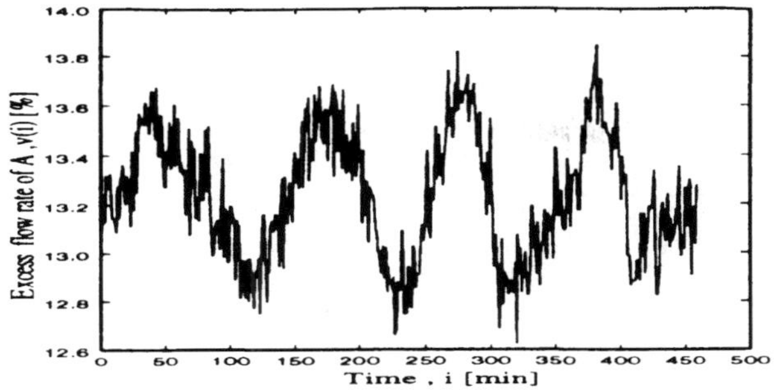

(a) Excess flow rate of A

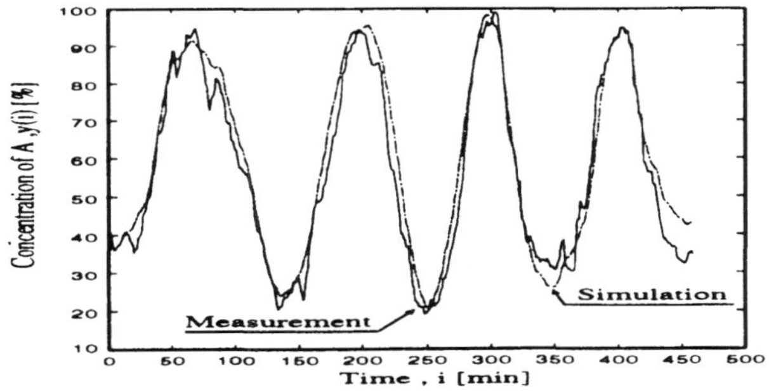

(b) Concentration of A
(measurement and simulation)

Fig. 5. System input and output

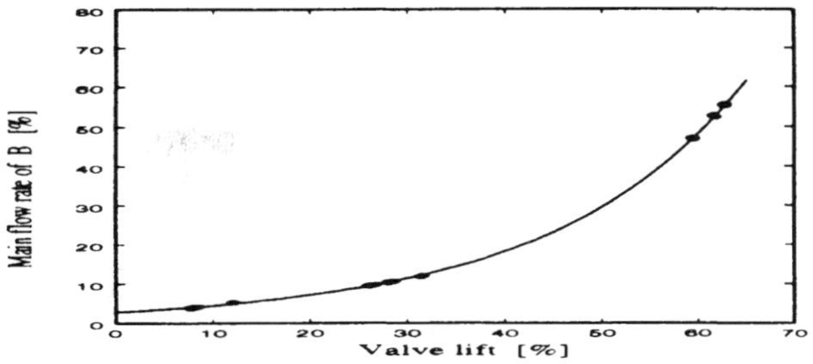

(a) Main flow control valve

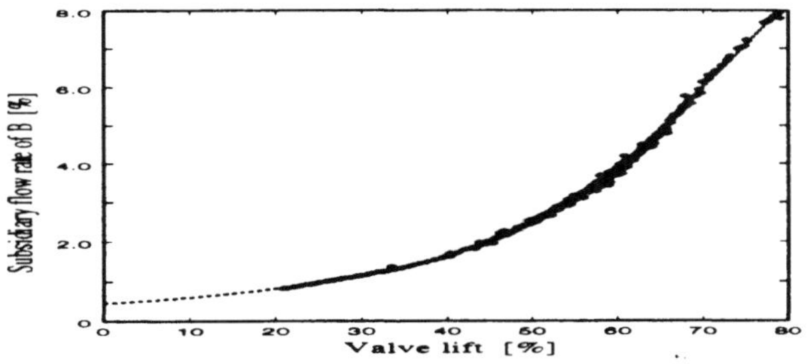

(b) Subsidiary flow control valve

Fig. 6. Static flow characteristic curve of control valves

ber of m and n can be determined optimally by using AIC(Akaike Information Criterion)(?). Figure 5 also shows a comparison between calculated output and measured output. In the reactor system under constant value control, $y'(k)$ and $v'(k)$ are replaced as follows:

$$
\begin{aligned}
y'(k) &= y(k) - y^{sp} \\
v'(k) &= v(k) - v^{*}(x(k))
\end{aligned} \tag{8}
$$

where, y^{sp} is the set-point of y in the feedback control loop.

3.3. *Control valve statics and dynamics*

The static flow characteristic curve shows the relationship between the flow rate and valve lift. The curves of the two control valves were examined under steady state operation. Figure 6 shows nonlinearity in the flow-characteristics for two control valves. The characteristic curve of the main flow can be approximated by a quadratic curve, but that of the subsidiary flow has an inflection point. The dynamics of the subsidiary flow control valve can be neglected because of its small capacity. But the dynamics of the main flow control valve were approximated by the 1-st order lag model (K.Fukuda et al.,1989).

3.4. *Block diagram of total system*

Figure 7 is the block diagram of the total reactor system including current control loops. G(x) is a function that gives the set-point of ratio control. Dynamic simulation of the system reproduced well the measured data. This means that the set of models were

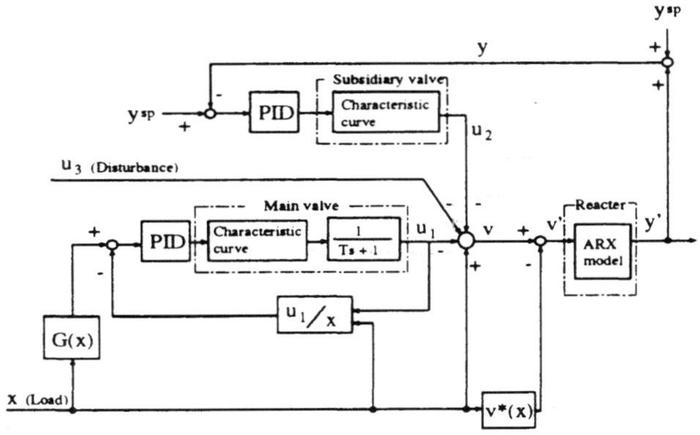

Fig. 7. Block diagram of current control system

validated by simulation. Therefore, simulation study can be used to evaluate proposed retrofit alternatives modified from the current control system.

4. BASIC STRATEGY FOR IMPROVEMENT

4.1. *Strategy #1: Improvement of ratio control*

The correlation curve between load A and the excess flow rate of reactant A can help solve the saturation problem of the subsidiary flow control under large throughput changes. Equation(4) is used to determine the main flow rate for varying load. Figure 8 shows the block diagram after the modification of ratio control. This modified control system will decrease changes in the subsidiary flow rate considerably.

4.2. *Strategy #2: Improvement of nonlinearity of main flow control valve*

The characteristic curve of the main flow is convex as shown in Figure 6(a). This means that the differential coefficient changes greatly at different values of valve lift. This causes variations in the static gain of

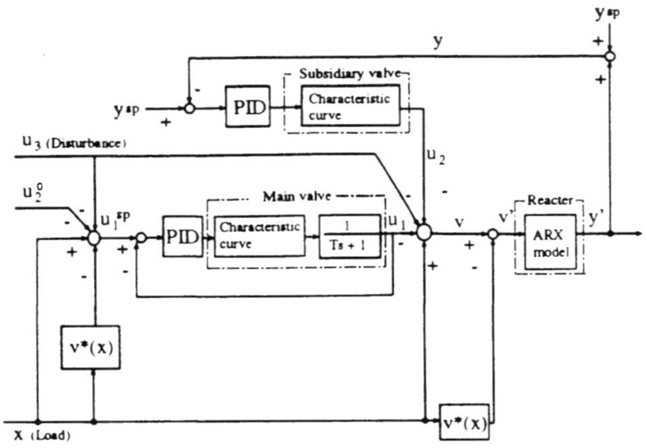

Fig. 8. Block diagram of modified control system

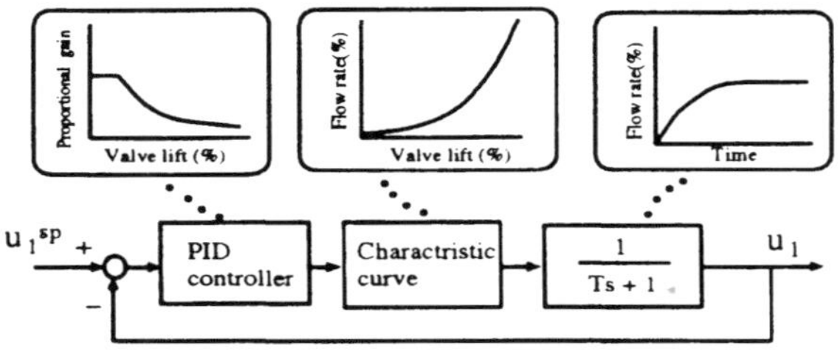

Fig. 9. Modified main flow control loop

the main flow control valve. The gain becomes small at a small value of valve lift, i.e., at a lower load. This fact accounts for decrease in the loop gain of the main flow control and the slow response of the main flow to load changes. This enforces excessive changes in the subsidiary flow. The gain scheduling method was introduced to keep the loop gain around a constant value (H.Nishitani et al.,1989) . Figure 9 illustrates scheduling of the proportional gain in the PI controller. Consequently, changes in the loop gain of the main flow control become small. This will realize fast response of the main flow rate at a lower load.

4.3. *Effect of strategy #1 and #2*

Figure 10 shows actual control performance of the existing plant before improvement. The output deviates largely from the set-point. Figure 11 shows the result after the improvement from following strategy #1 and #2. Change in the subsidiary flow was suppressed by the appropriate determination of the main flow rate and its fast response.

5. ADVANCED STRATEGY FOR IMPROVEMENT

5.1. *Strategy #3: Application of MPC*

Feedback control is the only way to cope with uncertainties such as inputs and changes in dynamics. However, practically some information on inputs and process dynamics is available as follows:

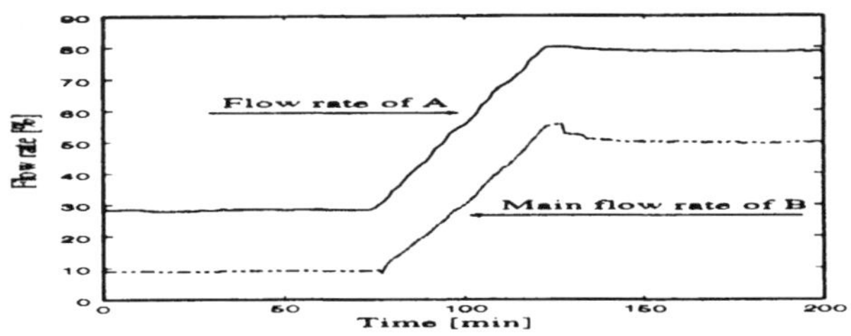

(a) Flow rate of A and main flow rate of B

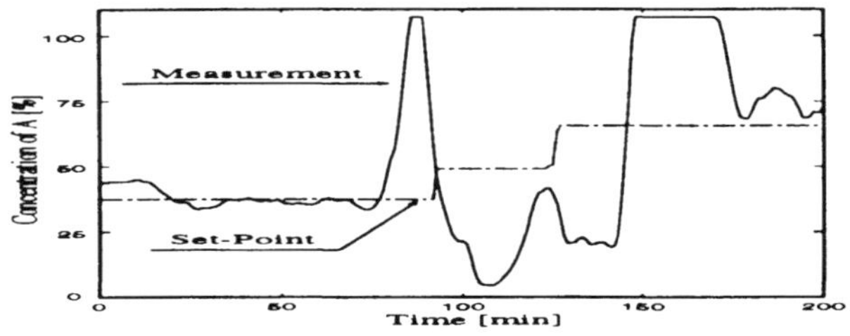

(b) Concentration of A

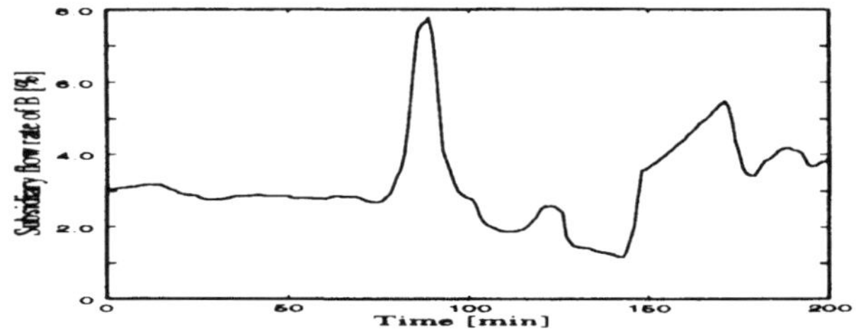

(c) Subsidiary flow rate of B

Fig. 10. Data before improvement by strategy #1 and #2

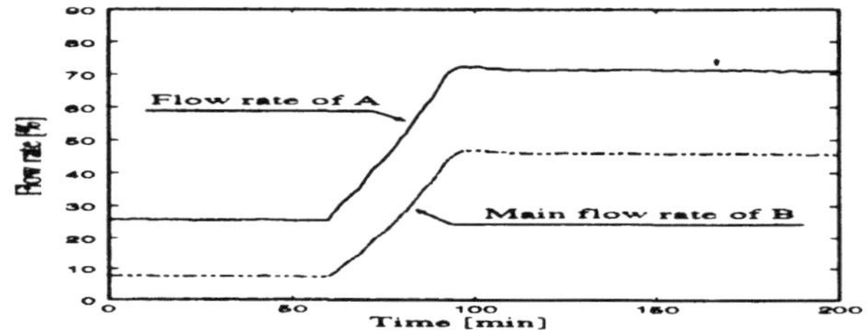

(a) Flow rate of A and main flow rate of B

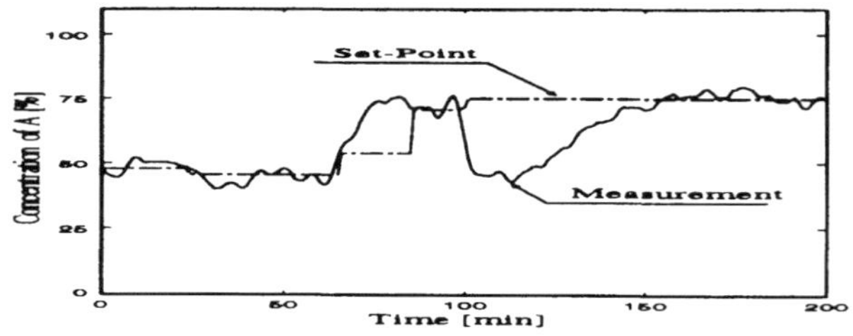

(b) Concentration of A

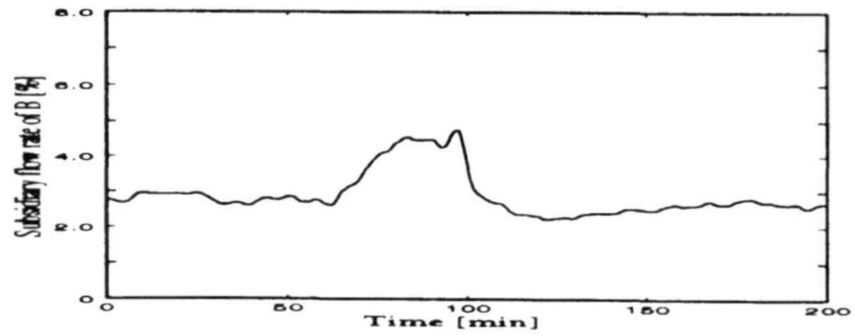

(c) Subsidiary flow rate of B

Fig. 11. Data after improvement by strategy #1 and #2

1. Load changes are given by the upstream plant, where an optimal production rate is planned and executed. Therefore, load changes in the shot term can be predicted.
2. The rate of recovery flow from other process can be estimated with some accuracy.
3. The dynamics of the reactor and control valves are obtained from operating data on the existing plant.

The above information can be used effectively to improve the control properties. But it has not been used in the current control system. A simple method for using such information is the framework of model predictive control. It can include the future reference and process dynamics in the control algorithm (?).

The block diagram of incorporating model predictive control in the reactor system is presented in Figure 12. The augmented process, which is the controlled object for model predictive control, is a three-input, one-output process as shown in Figure 13. Load A, u_1 and the rate of recovery flow, u_3 are treated as measurable inputs and the set-point of subsidiary flow rate, u_2^{sp} is a manipulated variable. The output is the concentration of A, y in the vapor phase. Responses of y to a unit step change in u_1, u_3 and u_2^{sp} were obtained by simulation of the developed system model. These responses are shown in Figure 14. Future output prediction until the Pth sampling time at the kth sampling period is represented as follows:

$$y_p(k+j) = y(k) + y_m(k+j) - y_m(k) \tag{9}$$

where,

$$\begin{aligned} y_m(k+j) &= \sum_{i=1}^{N} A_i \Delta U(k+j-i) \\ &\quad + A_N U(k+j-N-1) \end{aligned} \tag{10}$$

and $j = 1, 2, ..., P$. A_i(i=1,..,N) are coefficient vectors of the step response model of x, u_3 and u_2^{sp}.

The objective function is given as a quadratic form of the discrepancy between the reference, $y_s(k+j)$,and the predicted output, $y_p(k+j)$, and change in the manipulated variable, $u_2^{sp}(k+j-i)$ on the horizon.

$$\| Q(Y_s - Y_p)\|^2 + \|R\Delta U_2\|^2 \rightarrow Min \tag{11}$$

$$\begin{aligned} Y_s &= [y_s(k+1), .., y_s(k+P)]^t \\ Y_p &= [y_p(k+1), .., y_p(k+P)]^t \\ \Delta U_2 &= [\Delta u_2^{sp}(k), .., \Delta u_2^{sp}(k+P-1)]^t \\ \|X\|^2 &= X^t X \end{aligned}$$

where Q and R are weighting vectors. They are given as follows in this study:

$$\begin{aligned} Q &= [1, 1, .., 1]^t \\ R &= r[1, 1, .., 1]^t \end{aligned}$$

where r is a scalar parameter. Minimization of Equation(11) subject to Equation(9) is executed under the following assumptions:

$$\begin{aligned} \Delta u_2^{sp}(k+j) &= 0 \\ j &= M, M+1, ..., P-1 \end{aligned} \tag{12}$$

Under no inequality condition, the control action to be placed at present is determined analytically and its value is implemented.

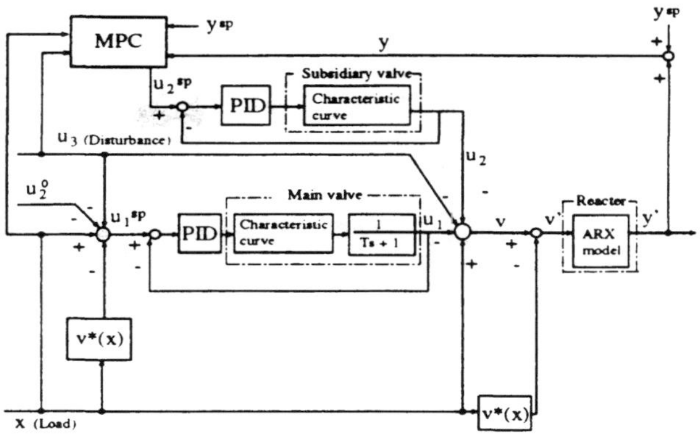

Fig. 12. Block diagram of MPC control system

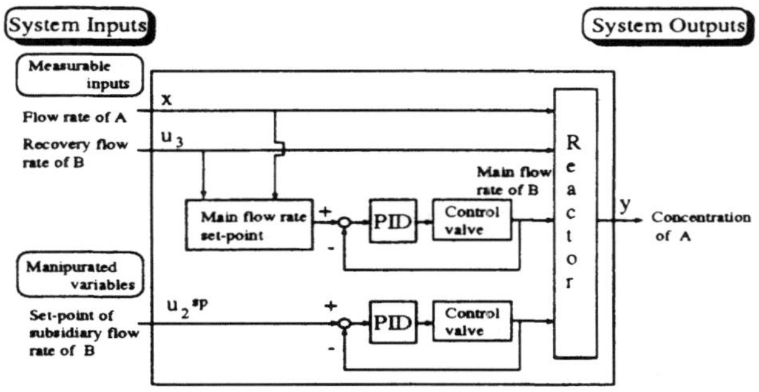

Fig. 13. Augmented process

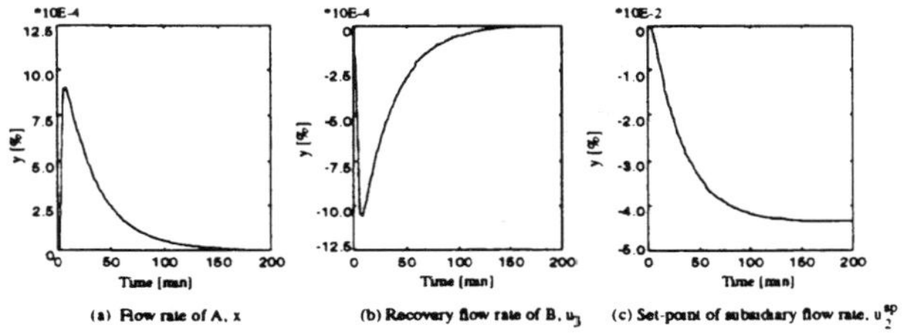

Fig. 14. Step responses for three system inputs

5.2. *Simulation*

A set of parameters such as length of horizon, P and weighting factor, r must be determined appropriately. For this purpose, simulations were repeated under the following conditions:

- Command responses under various values of P with a fixed value of r (r=1) are shown in Figure 15(a).
- Command responses under various values of r with a fixed value of P (P=7) are shown in Figure 15(b).
- Disturbance attenuation under various values of P with a fixed value of r (r=1) are shown in Fig-

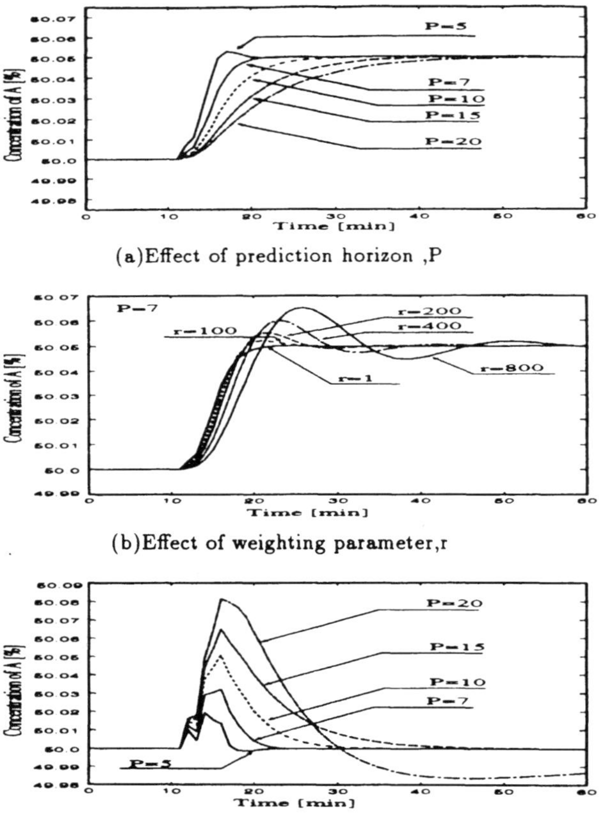

(a)Effect of prediction horizon ,P

(b)Effect of weighting parameter,r

(c)Effect of prediction horizon ,P

Fig. 15. Simulation for MPC parameter tuning

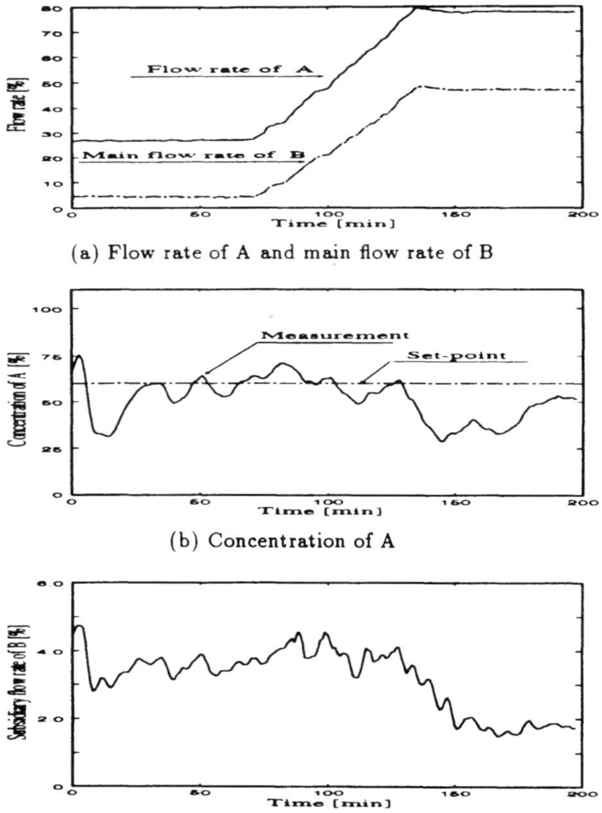

(a) Flow rate of A and main flow rate of B

(b) Concentration of A

(c) Subsidiary flow rate of B

Fig. 16. Data after improvement by three strategies

ure 15(c).

These figures supported some observations on parameter tuning.

1. The length of horizon affects both stability and speed of response. The appropriate value of P is around the peak of impulse response(?).
2. The weighting factor is usually increased to avoid rapid changes in manipulated variables. However, an excessive value brings oscillation.

5.3. *Effect of strategy #3*

Model predictive control was implemented for the augmented process shown in Figure 13. The MPC algorithm is included in a process computer (workstation) connected to the DCS. Figure 16 shows an actual result under a sharp load change. The output of concentration A is kept around the desired value as shown in Figure 16(b).

6. CONCLUSION AND FUTURE DEVELOPMENT

Control performances of an existing continuous flow reactor have been improved under large load changes. Firstly, essential factors were identified by analyzing process data on the existing plant. Secondly, dynamic models were built to consider strategies for control system improvement. These strategies brought no intervention by operators under large load changes. However, the set-point value of concentration A in subsidiary flow control is adjusted by the operator to satisfy the upper level goal, i.e., economy. Automatic determination of the set-point value is a topic for future discussion.

In the plant, load changes are executed deliberately based on production planning. This means that future load values can be estimated. This information should be used effectively in the framework of model predictive control.

7. REFERENCES

Fukuda K., K.Okamoto, E.Fukui and H.Nishitani (1989). Modeling and control system design of natural gas supply system. *Computers and Chem. Eng.* **13**(11/12), 1299–1308.

Nakamizo, T. (1988). *Signal Analysis and System Identification (in Japanese).* Corona Pub. Co., Ltd.

Nishitani, H. (1987). Model predictive control of chemical processes (in Japanese). In: *Proc. SICE Symposium on Industrial Process Modeling and Control System Design.* Osaka. pp. T1–T8.

Nishitani, H. (1989). Application of model predictive control(in Japanese). *Jr. of Society of Instrument and Control Eng.,Japan* **28**(11), 54–61.

Nishitani, H., E.Kunugita and X.W.Tao (1989). Design of gain scheduling control system with required tolerance. *Jr. of Chem. Eng., Japan* **22**(5), 454–459.

COMPUTATIONAL FLUID DYNAMICS AS A TOOL FOR CHEMICAL REACTOR OPTIMIZATION

Hans-Jörg Preisigke

Hans-Jörg Preisigke, Bayer AG, Central Research Department, D-51368 Leverkusen, Germany

Abstract: The minimization of side products in chemical reactors and the associated reduction of energy, raw materials and waste disposal costs are still ambitious goals for the chemical engineer. Computational Fluid Dynamics (CFD) is a highly useful tool in this field. If the reaction kinetics are known, the reactor geometry and inlet conditions can be optimized, giving an optimum yield of desired main product. Four examples which demonstrate the capabilities of CFD are presented.

Keywords: models, computational methods, numerical solutions, finite difference method, nozzles, chemical industry.

1. INTRODUCTION

The minimization of side products in chemical reactors and the associated reduction of energy, raw materials and waste disposal costs are still ambitious goals for the chemical engineer. Experimental means often enable one to find reaction conditions near the optimum by changing the inlet temperatures and the concentration profiles. However, experimental testing, especially of changes in geometry, is often very expensive. Computational Fluid Dynamics (CFD) is a highly useful tool in this application. The continuously improving hardware performance and advances in software capabilities make it possible to model more complicated reaction processes. Commercial fluid flow simulation packages enable the user to model fluid flow, heat transfer and chemical reactions to obtain information about the interaction between fluid flow, turbulence, temperature field, concentration profiles and residence time distribution.

At Bayer AG the simulation package FLUENT is used for flow simulation in reaction engineering. Four examples which demonstrate the possibilities of Computational Fluid Dynamics are presented.

2. REACTOR MODEL

A simplified scheme of the interaction between fluid flow, temperature field and concentration field is shown in fig.1. The velocity field is determined by the reactor geometry and the boundary conditions via forced convection. In most of the technical processes the fluid flow is turbulent, therefore the interaction between the flow field and the structure of turbulence has to be taken into account. Due to the dependence on physical properties, the velocity field is influenced by the temperature and the concentration field. All of these phenomena influence the chemical reaction which leads to the formation of main products and side products. The goal of experimental or numerical optimizations is to obtain a maximum yield of main products.

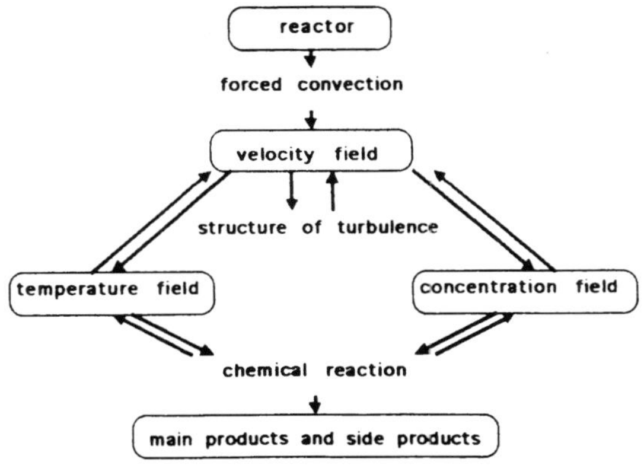

Fig.1 Interacting phenomena inside a chemical reactor.

There are still many problems in modeling all of these physical phenomena in a real reaction process. The influence of turbulence on the reaction rate is difficult to describe. In most of the commercial fluid simulation packages the eddy dissipation model (Magnussen and Hjertager, 1976) is implemented. This model relates the reaction rate to the rate of dissipation of turbulent eddies. The rate of reaction, which is limited by turbulence, is expressed by the concentration of the reacting species, the turbulent kinetic energy and the rate of dissipation of this energy.

Moreover, the modeling of the turbulence structure often still poses difficulties. Flow separation problems, highly rotating flows and laminar-turbulent transitions are especially difficult to simulate and the results are often rather inaccurate. Nevertheless, the possibilities and the accuracy of the models involving all these critical phenomena are continuously being improved. In many applications CFD and other simulation methods are not used as a tool to calculate an absolute temperature value with an accuracy of one Kelvin, but to give quick information about the effect of changes in the reactor configuration.

3. EXAMPLES OF FLUID FLOW SIMULATION

3.1 Influence of swirl on mixing time and flow recirculation

This example demonstrates the use of CFD in receiving helpful information to contribute to the improvement of reaction behaviour, even if the reaction kinetics are unknown.

The simulated reactor is a gas-phase reactor. Component A enters through a central nozzle. The inlet area of component B is a concentric annulus. Fig.2 shows the flow field which is visualized via velocity vectors. The length of the vectors

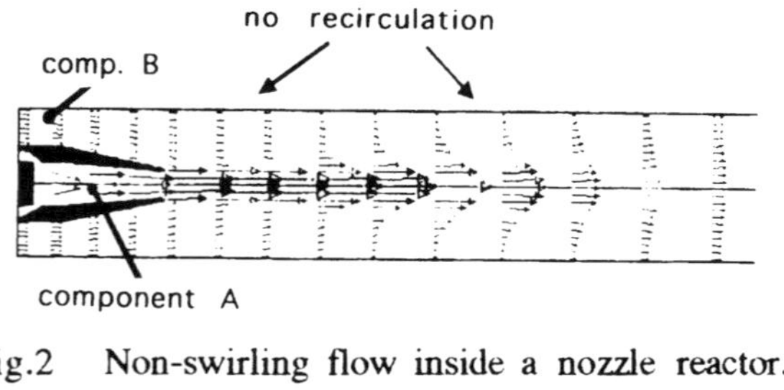

Fig.2 Non-swirling flow inside a nozzle reactor.

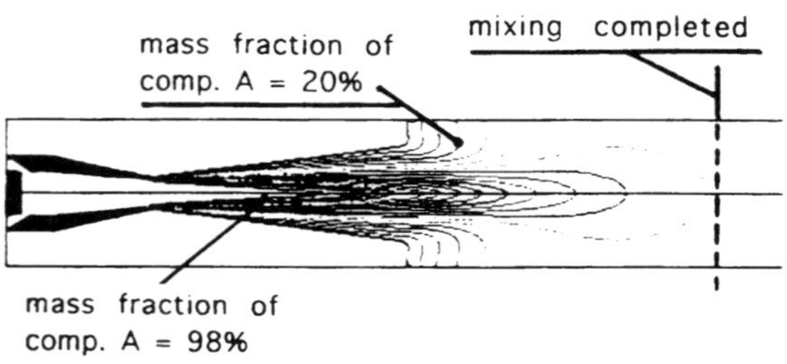

Fig.3 Contours of constant mass fraction for the case of non-swirling flow.

corresponds to the velocity magnitudes. For modeling highly turbulent flow the k-ε turbulence model is used. This model is based on the solution of transport equations for the turbulence kinetic energy, k, and the dissipation rate, ε.

For this non-swirling flow a separation free flow field over the total reactor can be observed. Because consecutive reactions take place inside the reactor, a fast mixing of the reaction species is desired. Fig.3 visualizes the mixing behavior via contours of constant concentration. Mixing is finished after about four reactor diameters. Beyond this length there are no concentration gradients to be seen across the reactor.

One possibility to achieve a faster mixing which has been discussed as an improvement of the existing reactor is a swirling flow condition of educt B at the inlet. For modeling this flow and mixing process, the Reynolds stress turbulence model has been used. This model takes into account non-isotropic turbulence effects. Due to this fact, the simulation requires more computational time.

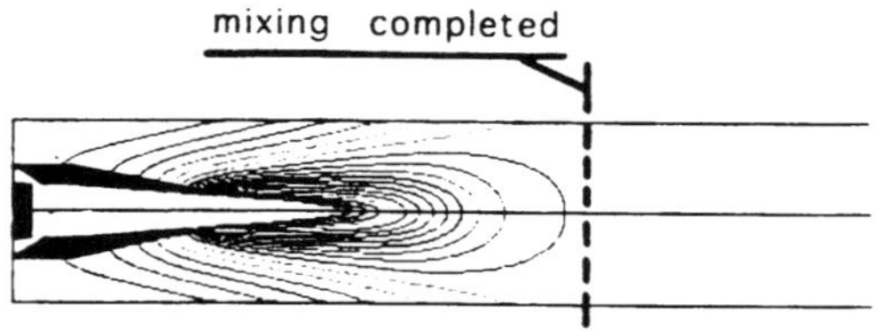

Fig.4 Contours of constant mass fraction for the case of swirling flow.

The circumferential velocity of component B entering the reactor is ten times the axial velocity. For this configuration a considerable faster mixing can be achieved in comparison to the non-swirling flow case (fig.4). The penalty, however, is a recirculating flow zone at the inlet nozzle (fig.5) which would lead to side product forming and to side product agglomerations at the nozzle.

Other possibilities for faster mixing without such a strong recirculation are being investigated. For example, component A can be introduced through a small concentric annulus or a multitude of small jets.

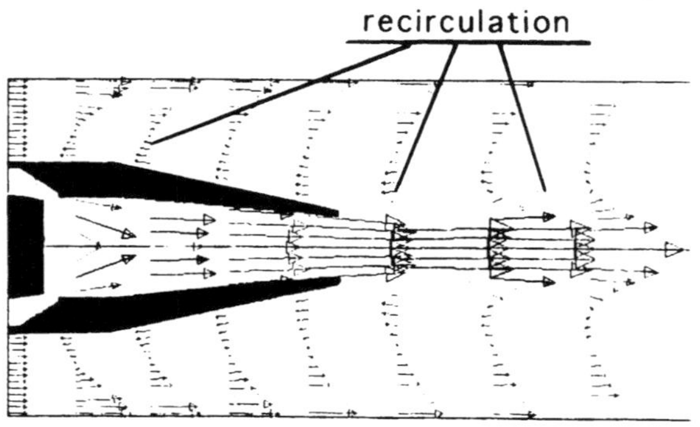

Fig.5 Swirling flow inside the nozzle reactor. Only the axial and radial velocity components are visualized.

In this case, the users of CFD at Bayer work closely together with the experts in the production process. The numerical results are compared to experimental data received at the process reactor.

3.2 Achieving a homogeneous velocity profile with an optimized pressure drop

The flow of component B into the reactor shown in fig.2 has been homogenized as a result of numerical calculations. The velocity profile of component B into the reactor has a strong non-uniformity. The flow enters the reactor through a concentric annulus. This annulus together with axial velocity profiles and streaklines of massless tracer particles are shown in fig.6.

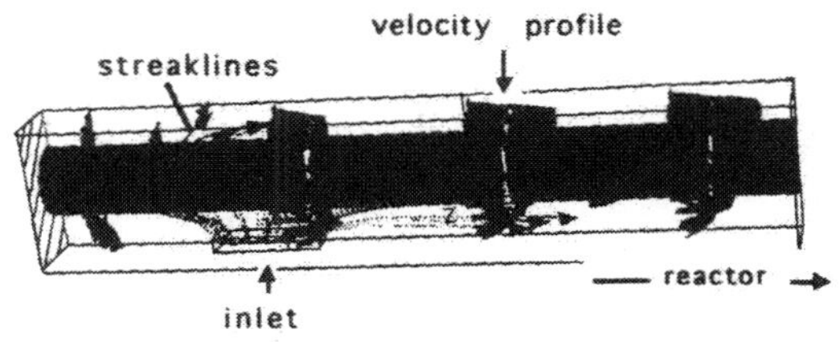

Fig.6 Inhomogeneous velocity profiles inside a concentric annulus.

Due to symmetry, only one half of the annulus is evaluated. From an attached pipe, the gas enters radially and then flows along both sides in the circumferential direction. On the side opposite to the entrance, the streams impinge on each other and are redirected in the axial direction.

The contours of the axial velocity show an undesirable non-uniformity which does not level-off on the way to the reactor. In the flow shown, circumferential distribution of the axial velocity up to the addition of component A still varies from 6.5 to 12.5 meters per second.

By adding a structured packing, a higher flow uniformity can be achieved. The relevant parameter here is pressure drop. In the model the pressure drop inside the packing across a unit length is related to the fluid velocity and the squared fluid velocity in a manner analogous to the Ergun equation.

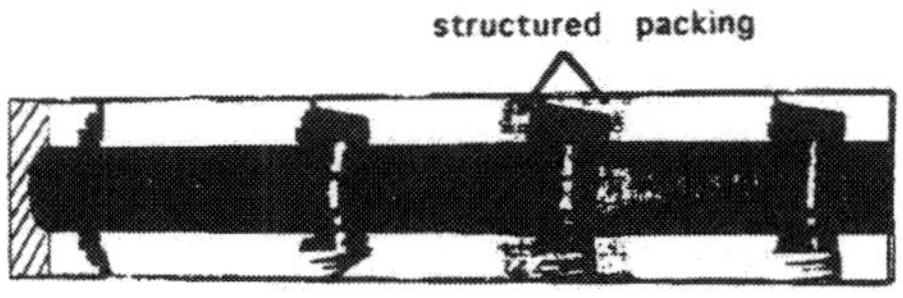

Fig.7 Homogeneous velocity profile due to the installation of a structured packing.

The tested packing provides a pressure drop of about 1300 Pa. The axial velocity presentation in fig.7 clearly shows the effect on flow uniformity. The circumferential variation of the axial velocity is reduced to between 9.3 and 10.2 m/s. For practical purposes, the flow can be regarded as having a uniform velocity. Due to the homogeneous inflow of component B, non-uniform product agglomerations at the reactor wall were reduced.

3.3 Optimization of a reactor geometry

This example demonstrates the capabilities of CFD for optimizing a reactor geometry if the reaction

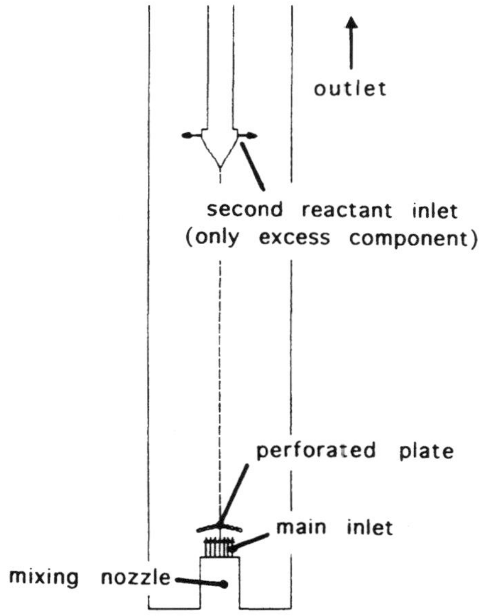

Fig.8 Vertical gas-phase reactor.

kinetics are known. The reaction system encompasses one main reaction and five side reactions. The reactor is a vertical gas-phase reactor, shown in fig.8. Through the mixing nozzle at the bottom of the reactor, the total amount of the limiting reactant enters together with the excess reaction component. The second reactant stream located inside the reactor contains only a further amount of excess component for

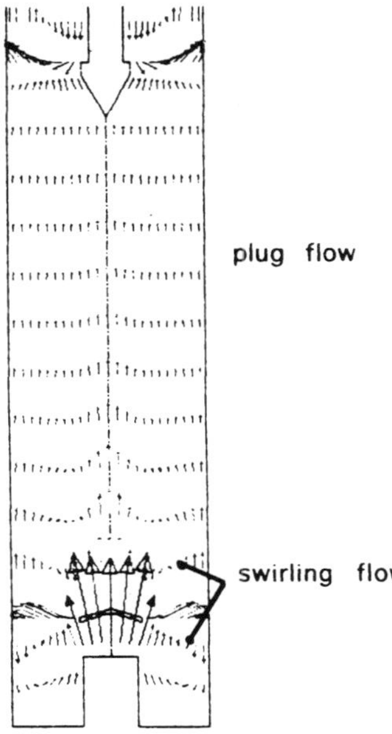

Fig.9 Velocity vectors inside the gas-phase reactor.

cooling the gas flow. At the top of the educt mixing nozzle a perforated plate is located. Due to this configuration a highly swirling flow field in the vicinity of the nozzle is generated (fig.9). Above this region the velocity profile is plug flow-like. The goal of this simulation is to find a reactor condition that maximizes the yield of main product. The yield of main product depends on the maximum temperature within the reactor, the size of the hot-spot region, the intensity of backflow and the location of the recirculation region.

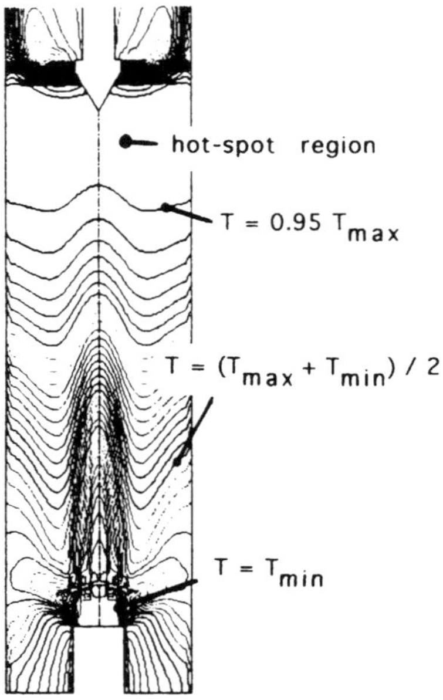

Fig.10 Temperature field inside the gas-phase reactor.

The temperature field, shown in fig.10, visualizes the locations of main product formation due to the source term inside the energy equation based on the reaction enthalpy. It corresponds with the concentration field of the limiting educt. Due to the plug flow behaviour

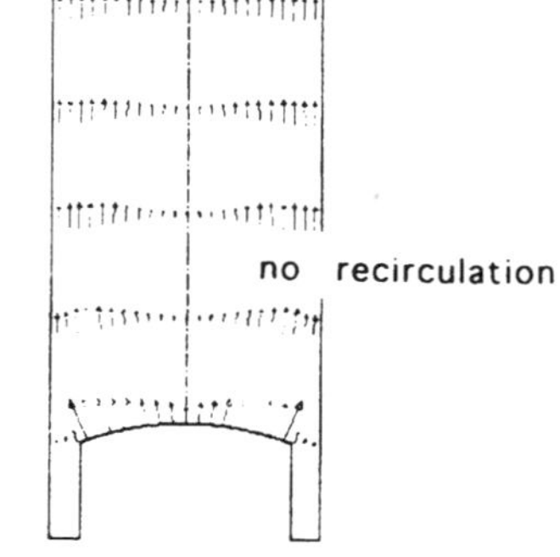

Fig.11 Optimum reactor configuration and velocity field.

in the high temperature zone, the yield of main product is already quite good for this configuration; it is approximately 95%.

The optimum reactor configuration and velocity profile is shown in fig.11. The mixing nozzle has to be enlarged to occupy almost the total cross-section area of the reactor. The annulus between the nozzle and the reactor wall is necessary because undesired granular side products collect at this part of the reactor bottom wall. The nozzle outlet velocities have to increase linearly from the reactor axes to the reactor wall. This construction provides the best temperature field (fig.12) for an optimum reactor yield (95.5%). This value is near the theoretical

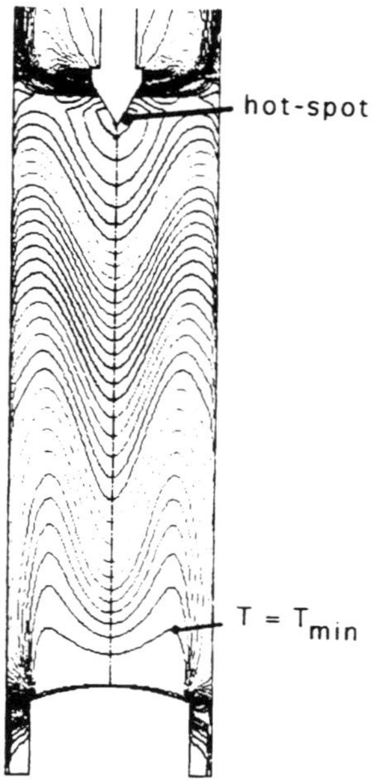

Fig.12 Temperature field inside the optimized reactor.

optimum which has been found by kinetic experiments. The hot spot region is very small. Plug flow behaviour occurs over almost the total reactor up to the addition of the second educt charge. In the high temperature zone, there is no recirculation.

The linearly increasing velocity magnitude at the mixing nozzle has to be realized by a variable pressure drop across the nozzle.

3.4 Assessment of new reactor types

In the process of finding new reactor types for a given reaction system, numerical simulations may be used before costly and time consuming experiments are carried out.

The competitive-consecutive liquid-phase reaction system measured by Bourne et al. (1985) was investigated by numerical calculations:

main reaction: A + B -> R,

consecutive reaction: B + R -> S.

The kinetic reaction rate of the main reaction is much higher than the kinetic reaction rate of the side reaction.

In the first stage, the simulation is used to model a nozzle reactor (fig.13). The results for the selectivity

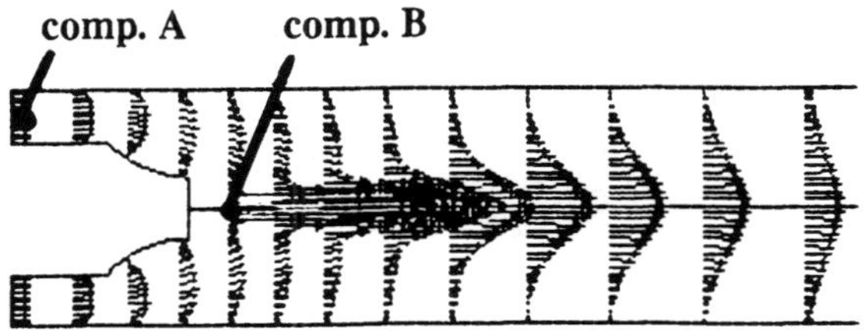

Fig.13 Velocity field inside a nozzle reactor. A competitive-consecutive liquid-phase reaction system is investigated.

of the side product for different specific power inputs and mass flow rates are compared to experimental data (fig.14). The agreement between simulation and experiment is very good.

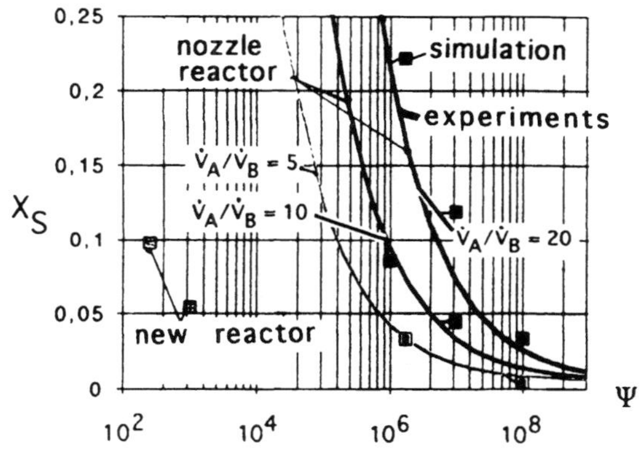

Fig.14 Selectivity X_S of undesirable product S as a function of dimensionless specific power input Ψ. Experimental data are approximated by the curves for different mass flow rates. The points denote simulation results.

In the second stage, this verified model is used to investigate new reactor designs for the same reaction. A reactor type with a very small selectivity of the undesirable product S has been identified. This new reactor design is now tested by measurements.

4. CONCLUSIONS

This paper has demonstrated the possibilities of fluid flow simulations for the improvement of existing production processes and for testing new reactor conditions or new reactor designs.

The examples provide only a glimpse of the various possibilities of CFD in reaction engineering. Advances in software capabilities are continuously being made available to the users; the models are being improved, especially in the area of two-phase flow. The Eulerian formulation of the conservation equations can be used for both phases (e.g. Sokolichin, et al., 1994). Another example of the new simulation capabilities is the modeling of transient flow, mixing and reaction in stirred tank reactors using sliding coordinate grids (e.g. Luo, et al., 1993).

REFERENCES

Bourne, J. R, C. Hilber and G. Tovstiga (1985). Kinetics of the azo coupling reactions between 1-naphthol and diazotized sulphanilic acid. *Chem. Eng. Commun.*, **37**, 293-314.

Luo, J. Y., A. D. Gosman, R. I. Issa, J. C. Middleton and M. K. Fitzgerald (1993). Full flow field computation of mixing in baffled stirred vessels. *Trans. of the Inst. of Chem. Eng.*, **71**, 342-344.

Magnussen, B. F. and B. H. Hjertager (1976). On mathematical modeling of turbulent combustion with special emphasis on soot formation and combustion. *16th. symp. on combustion, Cambridge, Ma*, 719-729.

Sokolichin, A., S. Becker and G. Eigenberger (1994). Modellierung und numerische Simulation von G/L-Blasen-Strömungen (modeling and numerical simulation of gas/liquid bubble flows). *Chem.-Ing.-Tech.*, **66**, 505-510.

Control of Forced Cyclic Process

Jens Erik Hansen * and Sten Bay Jørgensen **

**Haldor Topsøe A/S, Nymøllevej 55, 2800 Lyngby, Denmark*

***Technical Univercity of Denmark, Department of Chemical Engineering, DTU, 2800 Lyngby, Denmark, Email: SBJ@ktuserver.kt.dtu.dk*

Abstract: Forced cycled processes are increasingly being used in practice due to their potential high material and energy efficiencies. In this paper the quasi static and dynamic properties of a regenerative oxidation reactor using periodic flow reversal are investigated. A linear model which depicts the behaviour from cycle to cycle is developed to be used for model based control design for this type of process. Three possible actuators are investigated. It is shown that one of the more likely actuators to be used in practice exhibits a slow inverse response. A control configuration with a number of temperature sensors and three possible actuators is proposed. The control problems considered are that of avoiding extinction of the reactor in case of a low concentration inlet gas reaching the reactor and that of avoiding overheating in case of a high concentration disturbance.

Keywords: Periodic cycling process, Fixed bed reactor, Regenrative reactor Model based control, Dynamic modelling

1. INTRODUCTION

Periodic Cycling processes have received increasing attention during the last few years, due to their potential economical and ecological benefits. One process example is pressure swing absorption another is packed bed reactors operated with flow reversal to combust low concentration volatiles in exhaust air. These reactors have received significant practical interest following the pioneering work of Matros (1985 and 1989). Allthough many authors have considered computational aspects of simulating these transient processes very few have shown interest in evaluating the static characteristics and even fewer have shown interest in the dynamic aspects ranging over several periods of operation of the reactor. Since the transients in these types of processes are rather slow it is the period to period dynamics which is relevant for evaluating the dynamic charateristics. Control of these reactor processes is a prctically relevant subject since the reactors may exhibit runaway or extinction depeding upon the disturbances in the inlet volatiles concentration.

The purpose of the present paper is to present analysis of the quasi static and dynamic behaviour and to develop a linear model for the cycle to cycle behaviour. The paper formalise first a model for a periodic cycling process. The model is solved both in the a quasi steady state case and also in the dynamic case. Thereafter a a linear state space model is developed which can be used to predict the cycle to cycle behaviour. The quasi static and dynamic behaviours are analysed. It is shown that the period to period linear model predicts the behaviour fairly well over a limited number of (up to 25) periods. Finally design of a model based controller is proposed.

1.1. *Process Description*

The REGENOX unit is used for combustion of organic combustibles in off-gases. The unit serves as a combined reactor and heatexchanger. In principle it consists of a fixed bed reactor connected with fixed beds of inert material at each end. The *active* fixed bed is divided into two sections which enables heating of the gas to prevent extinction or purging a fraction of gas to avoid overheating. (see fig. 1).

The unit is operated with switch flow, in this way the inert section at the *present* exit is heated by the hot product stream, and afterwards when the flow is reversed the heated inert material will preheat the product stream to the ignition temperature. Then by continuously reversing the flow, autothermal operation of the unit can be achieved. Obviously at startup of the unit either the reactant flow has to be preheated before entering the unit or the reactor must have a sufficiently high

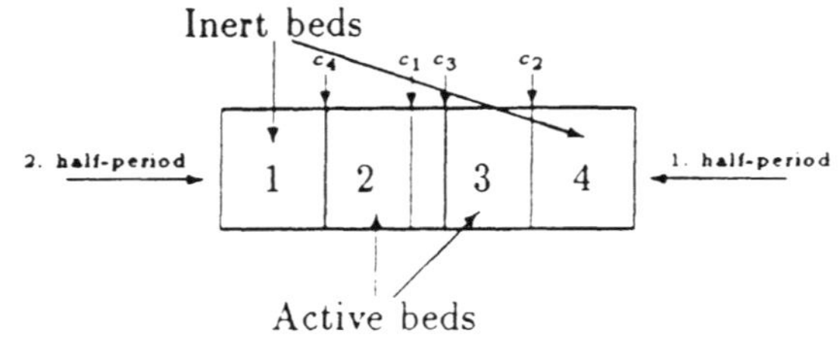

Fig. 1. Draft of the REGENOX unit

temperature to ignite the reactive gas. Alternatively heating in the midzone during start up may be applied.

The change of flow mechanismn called HAPS is mounted on the reactor. This method increases the overall conversion since unconverted reactant in the *dead volume* of the inert beds, just before flow reversal, is recycled through the reactor.

Because of the reversed flow operating procedure moving temperature and concentration profiles are observed. The flow is reversed every 5-10 minutes which is approximately one fourth of the total thermal residence time of the reactor. At a steady state cycle in autothermal operation the amount of energy which is lost at the exit during a half-cycle equals the liberated reaction energy.

2. METHODS

2.1. *Mathematical Modelling*

To describe the cyclic behaviour of the reactor a *complete* model which is used to determine the movements of the temperature and concentration profiles during a half-period is needed together with a procedure for handling the swiching of flow direction. Then the same model is used for both flow directions and instead of switching the flow direction one (mathematically) turns the reactor around.

The complete model of the unit is assembled by models for each of the four reactor sections. The same type of model is applied for each section - however the reaction terms in the inert sections are neglected. The most important assumptions are:

- The combustion reactions can be describe by one fictive reaction undergoing one reaction
- The reaction takes place in the solid phase of the active beds.
- The HAPS system is ignored.
- Plug flow (radial gradients are to some extent accounted for by
- Constant pressure
- Constant density and superficial velocity
- Isothermal catalyst pellets
- 1. order reaction, Arrhenius type temperature dependence: $R = a_r \exp(-\frac{E}{R_G T_s})c_s$
- Quasi-steady state assumption on gas phase energy and mass balance

These assumptions result in a pseudo heterogeneous model consisting of three partial differential equations:

Gas phase mass balance : (1)

$$0 = -\frac{\partial \theta_g^p}{\partial(\zeta^p)} + \frac{1}{Pe_H^p}\frac{\partial^2 \theta_g^p}{\partial(\zeta^p)^2} + H^p(\theta_s^p - \theta_g^p) \quad (2)$$

Gas phase mass balance : (3)

$$0 = -\frac{\partial x_g^p}{\partial(\zeta^p)} + \frac{1}{Pe_M^p}\frac{\partial^2 x_g^p}{\partial(\zeta^p)^2} - \frac{1}{\beta^p} g^p(\theta_s^p) x_g^p \quad (4)$$

Solid phase energy balance : (5)

$$t_s^p \frac{\partial \theta_s^p}{\partial t} = H^p(\theta_g^p - \theta_s^p) + g^p(\theta_s^p) x_g^p \quad (6)$$

where the function $g^p(\theta_s^p)$ is defined by:

$$g^p(\theta_s^p) = \begin{cases} \beta^p Da^p \exp[\gamma^p(1 - \frac{1}{\theta_s^p})] & p = 2, 3 \\ 0 & p = 1, 4 \end{cases} \quad (7)$$

2.2. *Numerical procedure*

The simulation strategy is to use the same model in both flow directions with the initial condition of one half-period being the reversed end-condition of the pervious half-period. Hereby the numerical problem has been reduced to the simulation of only one half-period. A differential algebraic equation (DAE) system is obtained by discritization of the complete model in the axial position by an orthogonal collocation procedure. Since solid to solid heat transfer is neglected only the gas phase balances contains partial derivatives with respect to the position. Then the gas phase temperature and the concentration in section p are approximatated by $N+1$ order polynomials given by the linear combinations:

$$\theta_g^p(t, \zeta^p) = \sum_{j=1}^{N_p+2} l_j^p(\zeta^p)\theta_g^p(t, \zeta_j^p) \quad p = 1, 2, 3, 4 \quad (8)$$

$$x_g^p(t, \zeta^p) = \sum_{j=1}^{N_p+2} l_j^p(\zeta^p) x_g^p(t, \zeta_j^p) \quad p = 2, 3 \quad (9)$$

where $l_j^p(\zeta^p)$ is a $N_p + 1$ order Lagrange polynomial constructed from the Jacoby polynomial $p_{N_p+2}(\zeta) = \zeta(\zeta - 1)p_{N_p}^{(0,0)}(\zeta)$ All the roots of the $(N_p - 1)$'th order polynomial $p_{N_p}(\zeta)$ are located in the interval]0; 1[and are the collocation points of the section p. The Lagrange polynomials are constructed to posess the following property:

$$l_j^p(\zeta_i^p) = \delta_{ij} = \begin{cases} 1 & i = j \\ 0 & i \neq j \end{cases}$$

where δ_{ij} are Kronecker's delta. This gives a DAE system to find the movements during one half-period given an *initial* condition. The solid phase energy balances are the only equations which contain derivatives with respect to time, then this is the only dependent variable that must be supplied with an initial condition. The profile (represented by the temperature at the internal collocation point of each of the four sections) can be found in the beginning of a half-period by the profile at the end of the previous using the equations:

$$\theta_i^1(0,\varsigma_i^1) = \theta_i^1(t_{sw},1-\varsigma_i^1) \quad i=2,\ldots,N_1+1 \quad (10)$$

$$\theta_i^2(0,\varsigma_i^2) = \theta_i^2(t_{sw},1-\varsigma_i^2) \quad i=2,\ldots,N_2+1 \quad (11)$$

$$\theta_i^3(0,\varsigma_i^3) = \theta_i^3(t_{sw},1-\varsigma_i^3) \quad i=2,\ldots,N_3+1 \quad (12)$$

$$\theta_i^4(0,\varsigma_i^4) = \theta_i^4(t_{sw},1-\varsigma_i^4) \quad i=2,\ldots,N_4+1 \quad (13)$$

where the right hand sides are the end-conditions. t_{sw} is the switch-time. A fourth order Runge-Kutta method is used integrate the DAE-system. The method has proven to be capable to describe even in very extreme situations involving steep concentration and tempeture profiles.

Since the model is quasi-stationary the solid phase profile at the end of a half-cycle is exactly the reversed of the profile in the beginning of that particular half-cyle, if the system is in a steady state cycle. This provides a condition to find the steady state cycle as the solution to a boundary value problem in time. By a double orthogonal collocation procedure an nonlinear equation system is obtained, which is the solved by a Newton-Rapson procedure. This method reduces the time to find the steady state by approximately a factor 50 over the traditional approach by integrating in time untill a steady cycle is reached.

2.3. *Discrete Time Model*

In this paper a MPC strategy is proposed for control the reactor. In order to apply such a method it is convenient to use a state space model. Since the reactor is a distributed parameter system and furthermore the dynamic behaviour is periodic there is no obvious choice of state variables.

In this work the states are determined as the solid phase temperature at fixed locations in the reactor system (corresponding to the internal collocation points of the model) at a specific time instant once per cycle. It seems reasonable to choose this time instant as the end of every cycle, since then one can by knowledge of the state variables determine whether full conversion has been achieved in that particular cycle. This is computed by using the quasi-stationary mass balance. Actually only the question of full conversion within the last half-period of a cycle can be answered.

By this particular choice of *discrete* state model, the states are constant when the system is in a steady state cycle

Using the the solid phase temperature at the internal collocation points is very favorable since the nonlinear model can then easily be applied for control purposes, e.g. for nonlinear estimation of the states and updating the linearized model to be used in the MPC algoritm. Updating of the linearized model will be nessecary.

Let x_i be the states at some time $i(2t_{sw})$ and let the actuators and disturbances on the system during the following cycle be represented by the vectors u_i and d_i. By integrating the nonlinear model the next end-of-cycle profile (and the new *discrete* states) one cycle later can be found exclusively from knowledge of x_i, u_i, and d_i. This implies:

$$x_{i+1} = F(x_i, u_i, d_i) \quad (14)$$

In section 5 it will be discussed which inputs are concidered as actuators and as disturbances. The function in equation (14) can be linearized around the reference $(x^{ref}, u^{ref}, d^{ref})$ to give:

$$x_{i+1} = Ax_i + Bu_i + Ed_i + K \quad (15)$$

where $K = F(x^{ref}, u^{ref}, d^{ref}) - Ax^{ref} - Bu^{ref} - Ed^{ref}$. The reference does not neccesarily correspond to a steady state. As a consequence of the quasi-stationary gas phase mass balance the gas phase temperature at any point can be obtained by multiplying the state vector by a constant vector. This means that the output vector of the linear state model can be chosen to consist of the end-of-cycle, gas phase temperature at distinct points in the reactor.

It must be mentioned that a somewhat more artificial state space model can be determined, in which the discrete states exist at every half-cycle instead of every full cycle. This implies that the locations at which the solid phase temperature constitutes the states dependends on the flow direction. Such a state model could be used to alter the actuators of the system every half-period instead of every full cycle uptaining a better control. Unfortunately this aproach is very complicated due to the asymmetries in the system, when it is not at steady state.

3. STATIC BEHAVIOUR

This section includes discussion of the steady state behaviour As a point of reference a set of realistic parameter values have been selected. The parameters correspond to a complete autothermal operation (this means no heating or purging from the midzone) in which the maximum temperature inside the reactor is approximately twice the inlet temperature (measured in Kelvin). The switch time is chosen to be 5 *min*, which is about 1/8 of the thermal residence time of one sandwich bed (one inert plus one active bed).

Fig. 2 illustrates how the solid phase temperature profile moves through the reactor during one half period (the flow direction is from left to right). Notice that even at the end of the cycle full con-

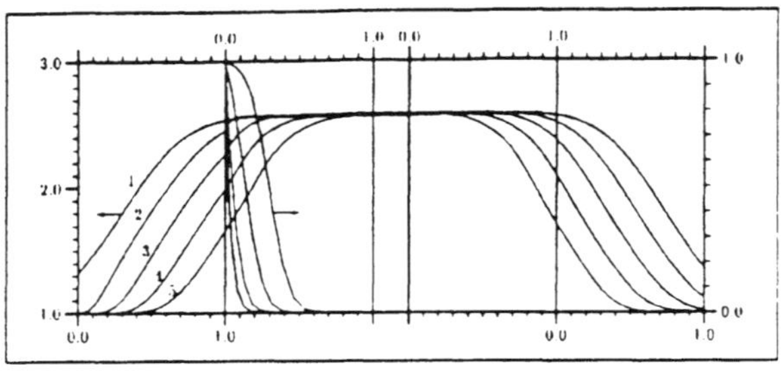

Fig. 2. Movements of the solid phase temperature profile during one half period

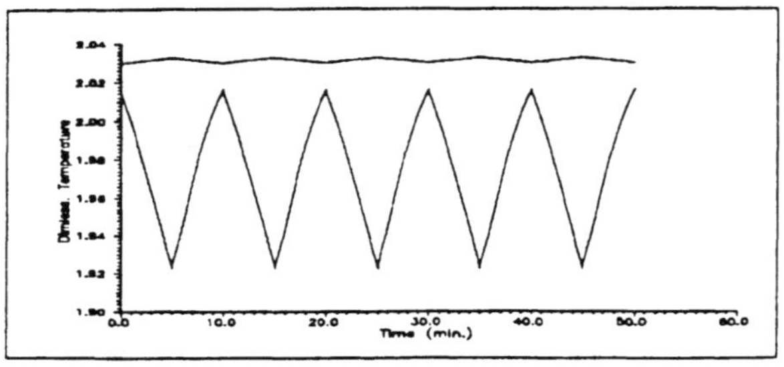

Fig. 3. Dimensionless temperatures in the reactor a fixed points. The upper curve shows the solid phase temperature at the location c_1, and the lower curve at the location c_2

version is achived in the *first* active bed, this is a favourably situation and the reactor must be designed to reach this performance. The reason being that if a purge stream from the midzone is applied in order to cool the reactor one must be assured that no unconverted gas is vented.

Due to the symmetry of the system (and the quasi-stationary model) the solid phase temperature profile at the end of the half cycle is seen to be the exact mirror image of the profile at the beginning of the half period. Furthermore the time average solid temperature profile within a cycle will be completely symmetric around the midzone.

The solid phase temperature at fixed points in the reactor are smooth functions of time within a half-period this is shown on figure 3 for the location between the first inactive and the active bed, together with the tempreature in the midzone. Note that the smooth behaviour during the half cycles to some extent justifies the usage of a double, othogonal collocation procedure to establish the steady state cycle.

It can be shown that the length of the horizontal movement of the solid phase temperature profile during one half-period relative to the length of one inert section ($L^{1,4}$) is approximately equal to the thermal residence time of the of an inert section ($t_s^{1,4}$) relative to the switch-time. This may be expressed as

$$\Delta\zeta \approx \frac{t_{sw}}{t_s^{1,4}} \tag{16}$$

where $\Delta\zeta$ is the relative movement of the temperature profile. The simulations demonstrates good

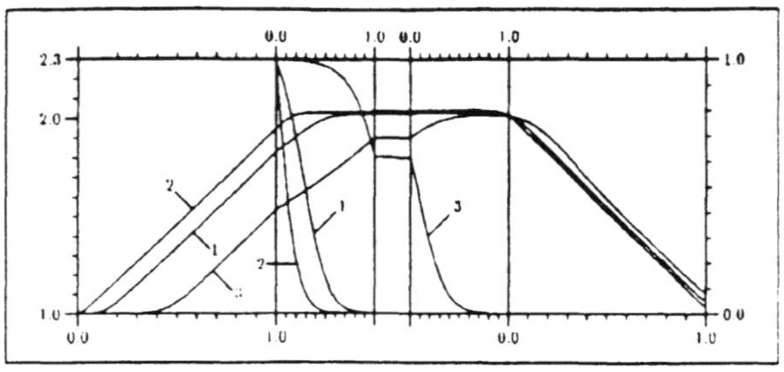

Fig. 4. Effects of heating in the midzone. 1) $c_0 = 0.0800$, $\Delta T_h = 0.000\ K$ 2) $c_0 = 0.0400$, $\Delta T_h = 2.876\ K$ 3) $c_0 = 0.0072$, $\Delta T_h = 5.104\ K$ ΔT_h is the gas temperature rise past the midzone.

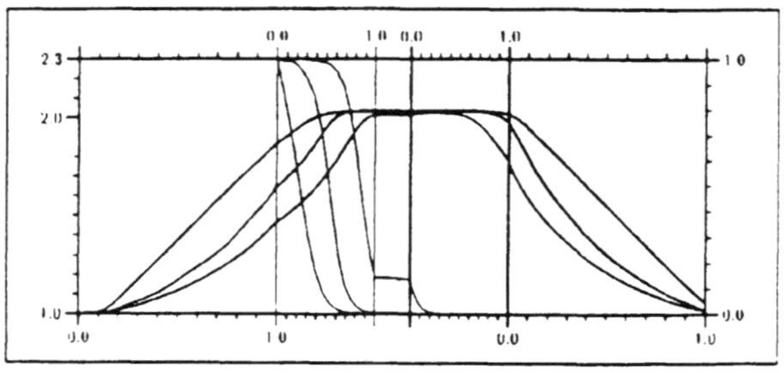

Fig. 5. Effects of venting from the midzone 1) $c_0 = 0.0800$, $\kappa_p = 0.00$, $t_p = 0$ 2) $c_0 = 0.1600$, $\kappa_p = 0.0381$, $t_p = t_{sw}$ 3) $c_0 = 0.1600$, $\kappa_p = 0.0761$, $t_p = t_{sw}/2$, κ and t_p is the purge fraction and time respectively. c_0 is the inlet concetration.

agreement with this result.

The effects of heating in the midzone is illustrated in fig. 4, where the concentration and the solid phase temperature profiles at the end of the half cycle are plotted in three different situations. The amount of energy added to the gas is adjusted by a reduced inlet concentration (c_0) to give the same outlet temperature as in the original autothermal situation. It is seen that the solid phase temperature profiles parrallels with the original profile in the inert beds, but in the active bed the profiles are no longer flat, instead a significant change in the shape of steady state profile is observed.

If the reactor is operated by applying venting a fraction of the gas flow from the midzone, then the solid phase temperature profiles becomes more narrow, the reason is that the amount of cold gas, blown through the inert bed is larger that the amount of hot gas blown in the other direction one half-period later.

The way to find the amount energy release by venting is to determine the fraction of gas to be vented and the duration. If the venting time is less than the swicth time, venting will be applied at the beginning of the half period since purging a fraction of the gas in the middle or in the end of the half-period will only increase the risk of purging unconverted gas, without improving the cooling effect. A short duration of venting time and large vent fraction gives a narrower tempera-

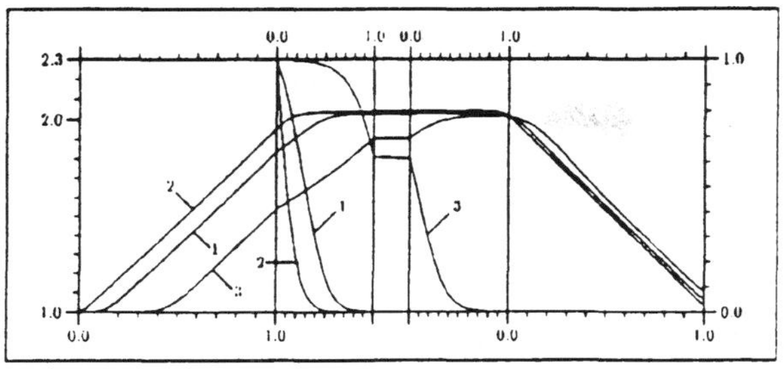

Fig. 6. Effects of changes in the switch-time. 1) $t_{sw} = 5.0\ min$ 2) $t_{sw} = 1.0\ min$ 3) $t_{sw} = 15.0\ min$

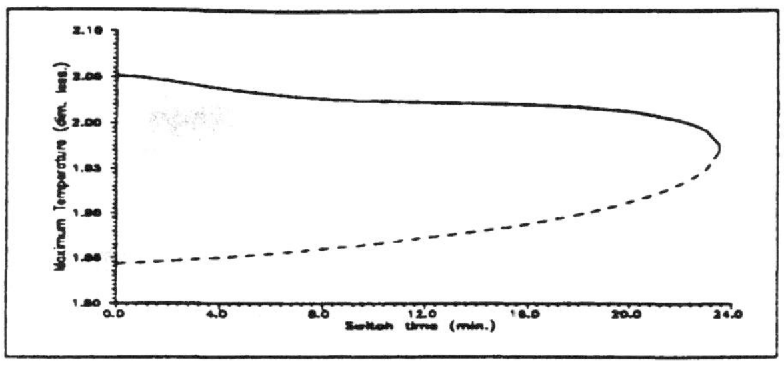

Fig. 7. Maximum temperature aginst the switch time. The dottet part of the curve correspond to the unstable steady state cycles

ture profile in comparison to a large venting time and small vent fraction even if the same energy is released, see fig. 5. This means that a long venting and small vent fraction is preferred, since then the total energy accumulation in the reactor is large which again implies that the system is more robust to sudden decreases in the inlet concentration. Now the problem with a long venting is the risk of letting out unconverted materials from the midzone, but this can be avoided by designing the reactor system with a sufficient length of the active beds.

Fig. 6 shows the quasi static effect of changes in the switch time on the steady state cycles. The system is almost insensitive to small changes in t_{sw} and rather large changes have to be made to see the effect. The movements of the profiles are (approximately) proportional to the switch time. Because of the symmetry of each half-period within a cycle the length of the flat area of the profiles is smaller the longer the movements. This means that a small switch time results in a large flat area and consequently a broader temperature profile.

All the steady states cycles discussed so far are stable steady states. Two stable steady state cycles corresponding to a given set of parameters are obviously possible, Those are the ignited cycle and the extinguish cycle in which no reaction occur at all. Also unsteady states may be found in the reactor system (obviously these can only be found by the Newton-Rapson procedure, and not by an integration in time). For a first order reaction with only one reacting component more than two steady state cycles are unlikely to be found in this kind of system. But an unstable steady state cycle arises in the reactor, and it is *in some sence* found in between the two stable cycles. Fig. 7 shows the maximum solid phase temperature of stable and unstable steady state cycles plotted against the switch time (the rest of the parameters are unchanged from the reference set). Actually the solid phase temperature profile is not completely flat on the top (due to axial dispersion) and for this reason the maximum temperature must be found in the middle of a steady states half-period in the middle of the reactor (this is due to symmetry). For any switch time is it seen that the maximum temperature of the unstable cycle lies below that to the ignited stable cycle, and above that of the extinguished, in which case the maximum temperature equals the inlet temperature. Unsteady states are characterized by incomplete conversion of the reactant.

The steady state analysis of the reactor shows some very important aspects of the changes in the steady state end-of-cycle temperature profiles which make the system hard to control.

4. DYNAMIC BEHAVIOUR

In the following the dynamic behaviour is investigated by *measuring* the gas phase temperature in the locations c_1 and c_2 at the end of a cycle, of which the flow is in the forward direction during the last half-period (see fig. 1).

Fig. 8 through 11 illustrates the extremely slow response when the system is subjected to step changes of various *inputs* When the step changes are induced the reactor is operating at a steady state cycle corresponding to the paramter values of the reference point. The response of a linearized model corresponding to this *state* is plotted on the figures as well (these responses are only evaluated 30 cycles ahead, whereas the nonlinear responses are simulated for 100 cycles). On all four plots the two upper and two lower curves correspond to the gas phase temperature at location c_1 and c_2 respectively.

From the plots it is clear that the linearized model gives a satisfactory description of the process to be used in an MPC algoritm.

An interresting phenomenon is that the switch time exhibit inverse response. The reason is that the average of the solid temperature profile within a cycle is no longer symmetric but shifted in the direction of the flow at the particular half-period when the switch-time was increased.

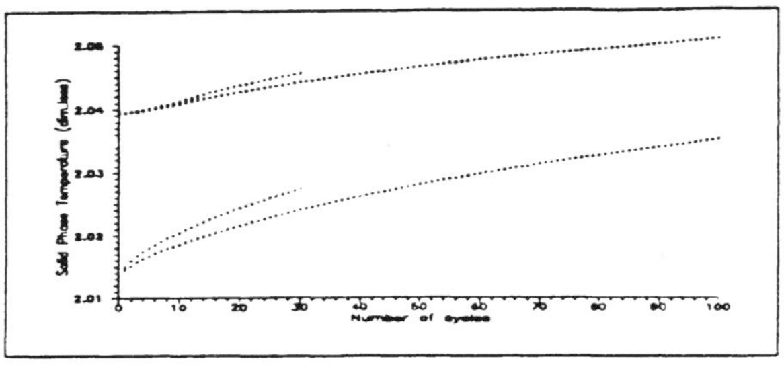

Fig. 8. Response to +5% in the inlet concentration.

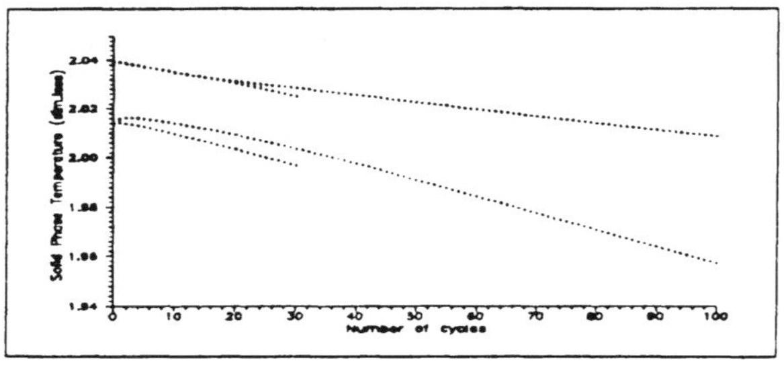

Fig. 9. The purge fraction is raised from 0 to 0.01 while the purge time is 2.5 min (half the switch-time)

5. CONTROL PROBLEM

The general REGENOX unit is designed to run under autothermal operation but due changes in the inlet volatiles concentration control must be applied to avoid extinction of the reactor or overheating causing damage of the catalyst particles.

Three actuators are available for control:

1. the purge of a gas fraction from the midzone
2. heating of the gas stream in midzone
3. alteration of the switchtime

The first and the second actuator should obviously never be used within the same period. The fact that the changes in the switch time gives an inverse response means that one must be very carefully when applying this actuator. In a single loop control design this actuatior will severely limit the obtainable performance. Any controller will have to be tuned to a very sluggish response. In a multivariable control design all three actuators may be used. Here the first and second actuators will take care of fast transinets whereas the switchtime may be used to accomodate the operation to longer lasting disturbances.

The steady state profiles show that both purging and heating changes the shape of the solid phase temperature profiles. This means that upon a step change in the inlet conditions the system can not reach the original steady state profile by control. But the objectives of the control is not to get back into the original steady state cycle, more as it is to keep the reactor from overheating and extinction and still assure full conversion of the reactant.

The linear model must be updated at every cycle, due to the fact that the system can not be expected to return to the original steady state cycle.

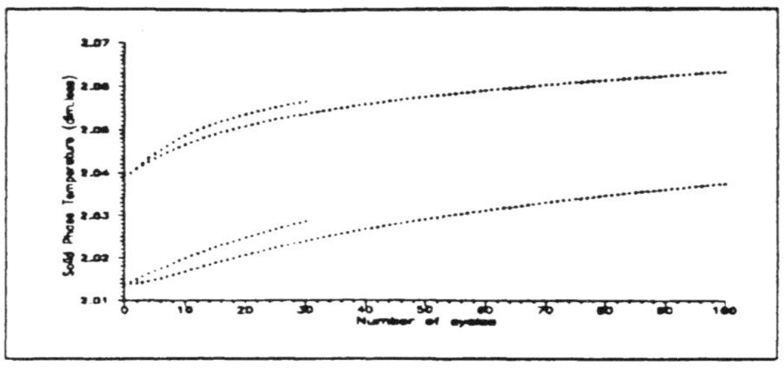

Fig. 10. The dimensionless temperature rise of the gas past the midzone is changed from 0 to 0.01 ($\approx 3^0C$)

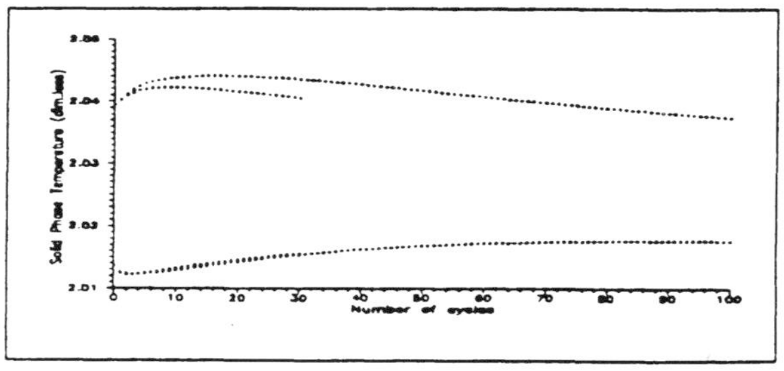

Fig. 11. The switch-time is increased from 5 *min* to 6 *min*

Due to its ability to handle the contraints on the actuators MPC is a very attractive control algoritm. The conntrolled variables are selected as the end-of-cycle temperature at the locations c_1 and c_2 on fig. 1. These temperature should be kept around some specified setpoints. Four meassurements of gas phase temperature at the locations c_1, c_2, c_3 and c_4 are used as sensors.

This performance of this control design will be compared to that of a multiloop PI controller in the presentation.

6. CONCLUSIONS

The paper demonstrates quite complex quasi static and dynamic behaviour of a forced cycled reactor. It is especiallly shown that using the cycle time as an actuator gives rise to a slow nonminimum pahse behaviour, which will severly limit a single loop control design with this actuator. At most a very sluggish response may be obtained.

A control structure for a model based control design is pproposed, where several temperature sensors and midzone heating and venting a fraction of gas from the midzone also are considered as actuators.

7. LITTERATURE

Matros, Yu.Sh. (1985) "Unsteady Processes in Catalytic Reactors", Elsevier

Matros, Yu.Sh. (1989) "Catalytic Processes under Unsteady Conditions", Elsevier

FUZZY LOGIC CONTROLLER DESIGN FOR PH-CONTROL IN A CSTR

Thomas Heckenthaler and Sebastian Engell

Process Control Group
Department of Chemical Engineering
University of Dortmund
44221 Dortmund, Germany
email: ast@astaire.chemietechnik.uni-dortmund.de

Abstract: In this paper two different approaches for the design of fuzzy logic controllers for the control of pH in a laboratory-scale CSTR are shown and compared. First a fuzzy controller was designed based on pure heuristics; the design process was fairly demanding due to the complexity of the system. Second, a fuzzy controller was identified based on a time-optimal controller derived from a coarse mathematical model. After addition of some heuristic rules to the rulebase which increase the robustness, this fuzzy controller shows an excellent performance.

Keywords: pH-control, fuzzy control, fuzzy modelling, time-optimal control.

1. INTRODUCTION

In the chemical industry one often encounters processes with considerable nonlinearity and unknown interactions. An example of such a process is the titration in a continuous stirred tank reactor (CSTR) which has strong gain nonlinearity. Hence common linear control laws (PID) are not able to achieve acceptable control performance. Fuzzy controllers, on the other hand, are known for their inherent nonlinearity which might enable them to handle nonlinear systems better.

The starting point of our research was an existing CSTR in laboratory scale. Our goal in the controller design was to get short regulation times for set point changes as well as good performance in rejecting disturbances. A coarse mathematical model was used to simulate the behaviour of the system. For the design of the fuzzy logic controllers we compared two different approaches. First we used our engineering common sense in order to develop a fuzzy rule base heuristically. Because of the complex system dynamics, this approach - which is the most commonly used in the design of fuzzy logic controllers - turned out to be more difficult than expected. Fuzzy controllers offer large flexibility to build nonlinear relationships, but clear systematics how to compose a rulebase based purely on heuristics are not available. Therefore the performance of such a controller is strongly dependent on the capability of the designer and the CAD-tool.

Our second approach was to make use of a coarse mathematical model of the plant. In order to build a fuzzy logic controller automatically, we computed a time-optimal controller first which is used as a reference to identify the kernel of the rulebase. By adding further rules to the automatically generated basic fuzzy controller in order to cope with modelling errors and measurement noise, it was possible to obtain excellent controller performance in simulations as well as at the actual plant.

This paper is divided into six sections. In section 2 we will describe the plant and its model. Then in section 3 we illustrate the problems encountered in the design of a fuzzy controller based on heuristics. Section 4 will briefly describe the identification of fuzzy controllers approximating a given optimal control law. The model based approach which includes the time-optimal controller is then discussed for the pH-control problem in section 5. Conclusions are given in section 6.

2. DESCRIPTION AND MODELLING OF THE PLANT

The CSTR which we want to control consists of a tank with two inlets and one outlet as shown in Fig. 1. The physical parameters are listed in Table 1.

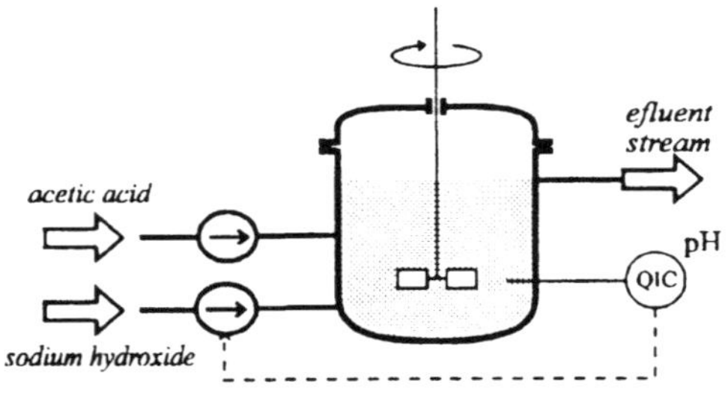

Fig 1: A CSTR for pH-control

Table 1: Physical parameters of the CSTR

Variable	Meaning	Setting
V	volume of tank	5.7 *l*
F_1	flow of acetic acid	20 *l*/h
F_2	flow of sodium hydroxide	[0, 45] *l*/h
$c_{1,e}$	concentration of acid in F_1	0.007 mol/*l*
$c_{2,e}$	concentration of hydroxide in F_2	0.01 mol/*l*
K_a	acid equilibrium constant	1.8×10^{-5}
K_w	water equilibrium constant	10^{-14}

The base flow rate F_2 is used to control the pH value of the solution. The acid flow is held constant or acts as a disturbance. We use diaphragm pumps to obtain the necessary precision of the flow rates. All control actions are performed by a commercial PC-based process control system which has a minimal sampling period of 1s. The sodium hydroxide flow is controlled by modulating an impulse frequency, which leads to a quantisation of the control amplitude because the frequency can assume only certain discrete values.

The mathematical equations of the CSTR can be derived as follows (McAvoy, 1972):

$$\frac{d(c_1)}{dt} = \frac{1}{V_R}\left[F_1 \cdot c_{1,e} - (F_1 + F_2)\cdot c_1\right] \tag{1}$$

$$\frac{d(c_2)}{dt} = \frac{1}{V_R}\left[F_2 \cdot c_{2,e} - (F_1 + F_2)\cdot c_2\right] \tag{2}$$

$$c_{H^+}^3 + c_{H^+}^2 \cdot (K_{HAc} + c_2) + c_{H^+} \cdot \left[K_{HAc}\cdot(c_2 - c_1) - K_W\right] - K_{HAc}\cdot K_W = 0 \tag{3}$$

$$pH = -\log[c_{H^+}] \; . \tag{4}$$

Due to the incomplete dissociation of acetic acid in water and the equilibrium reaction with sodium acetate, the system behaves like a buffer solution between pH 4 and pH 6.5. Consequently, the process gain varies greatly over the range of pH-values which can be controlled (see Fig. 2).

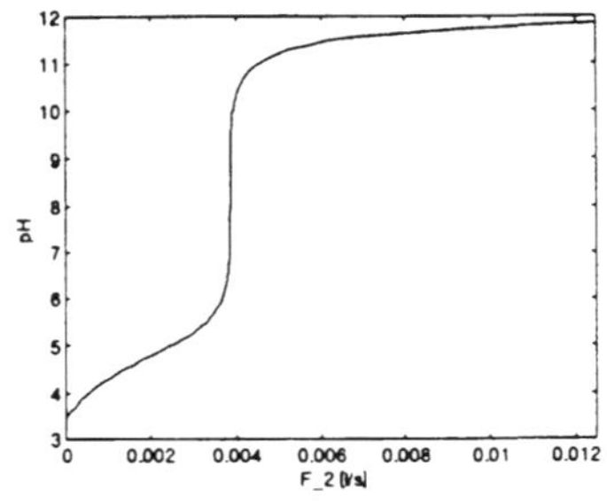

Fig. 2: The steady-state, nonlinear relationship of pH versus base flow rate.

The electrode which measures the pH in the tank exhibits a time delay with a time constant T_{el}=1.4s:

$$\frac{d(pH_{EL})}{dt} = \frac{-\log(c_{H^+}) - pH_{EL}}{T_{EL}} \tag{5}$$

Experimental step responses of the base flow rate showed immediately that the mixing in the tank is far from ideal. Even though the blade mixer is fairly powerful, different mixing zones in the tank exist which result in additional time delays and dead times. We decided not to try to build an accurate model of this very complex behaviour, but rather to lump the dynamical effects from the non-ideal mixing with the time delay of the electrode into one time constant, which was roughly estimated as $T_{El}^{*} = 10s$.

The difficulty of the control problem can be seen from the transients obtained with a fixed PI-controller tuned at pH 7, as shown in Fig. 3. This poor controller performance motivated us to design nonlinear fuzzy logic controllers.

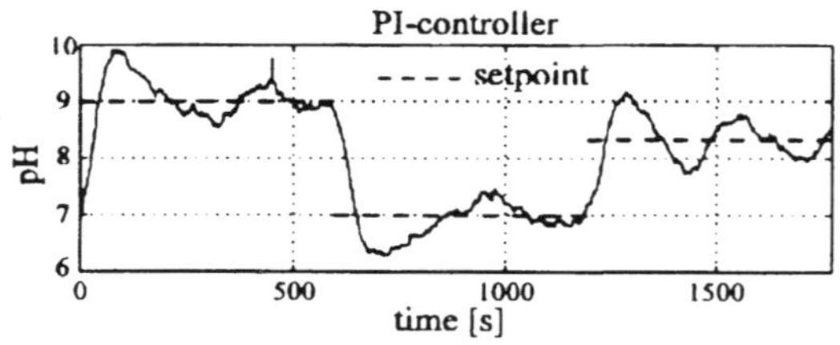

Fig. 3: Performance of PI-control (experiment)

3. FUZZY CONTROLLER DESIGN BASED ON HEURISTICS ALONE

The key to successful fuzzy controller design is the existence of heuristic knowledge which has been acquired by operators over an extensive period of manual control; this knowledge can then be modeled more or less easily by a fuzzy rulebase. For our plant we lacked such a source of knowledge for the design process. Therefore it was reasonable to implement the mathematical model in a fuzzy logic controller design tool to develop a fuzzy controller based on simulations instead of time-consuming on-line training.

First the regulation error, its derivative (here entitled dpH) and the change in the base flow rate were chosen as the controller inputs and outputs respectively. In the next step their range has to be partitioned into qualitative values which is known as fuzzification (see Fig. 4).

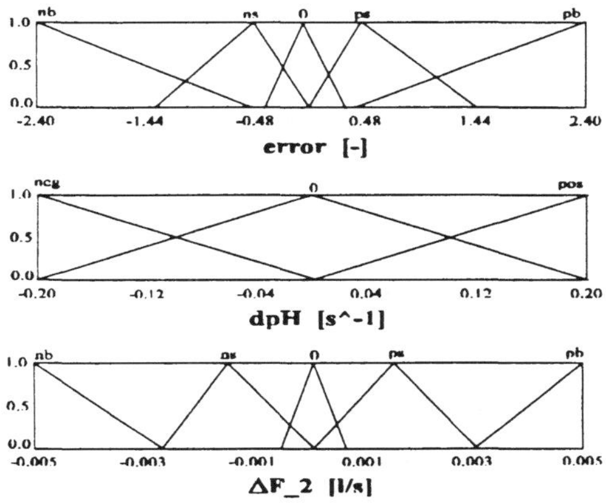

Fig. 4 Fuzzification of controller inputs and output

After an initial set of rules has been formulated, the behaviour of the system can be evaluated in simulations. Now the designer is challenged to change the rules and/or the fuzzification so that the system shows the desired behaviour. This procedure is characterized by a trial and error approach. Due to the numerous degrees of freedom in the fuzzy rulebase, this process can be very time-consuming. Fig. 5 shows typical behaviour of the system for a set of transients, illustrating the difficulties with such an approach for a complex system; even though the rules have been chosen symmetrically (see Fig. 6) the controller may show an excellent response for an initial set point change while for other set point changes the controller performs poorly. The designer has to change the fuzzification and the rules iteratively in order to obtain an acceptable behaviour over the whole range of operation. The success of the controller design is strongly dependent on the capability of the designer and the support provided by the tool which he is using. For example, the possibility to analyze which rules are triggered in every time step of the simulation is indispensable.

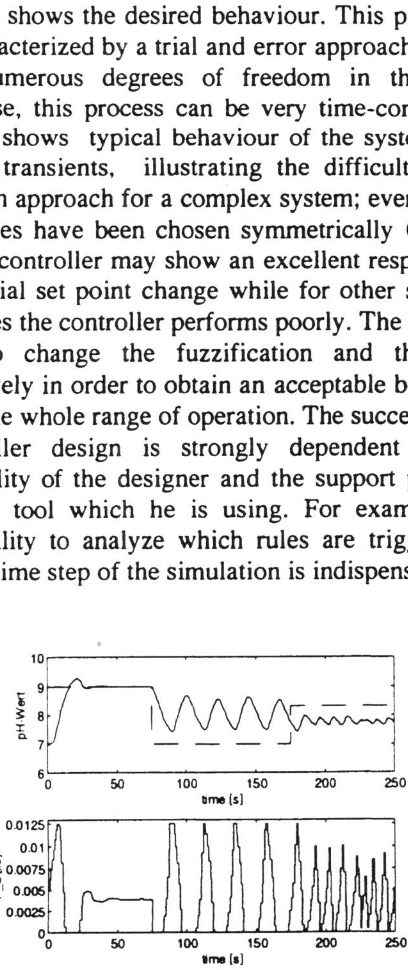

Fig. 5: Responses with fuzzy controller with 9 rules (simulation)

If *pH-Wert any* and ... then *ΔF_2*

↑ *error*	neg	0	pos ← *dpH*
nb	*ns*		*nb*
ns	*pb*		*ns*
0		*0*	
ps	*ps*		*nb*
pb	*pb*		*ps*

Fig. 6: Rulebase for fuzzy controller (9 rules)

To improve the transient behaviour of the fuzzy controller, it is necessary to use the pH value itself as an input as well. Three fuzzy values (*acidic*, *neutral* and *basic*) were chosen to characterize high and low gain regions of the plant. After a long time spent manipulating the rules and the fuzzification, a rulebase resulted with 29 rules which shows acceptable responses for set point changes (Chen, 1994) (see Fig. 7).

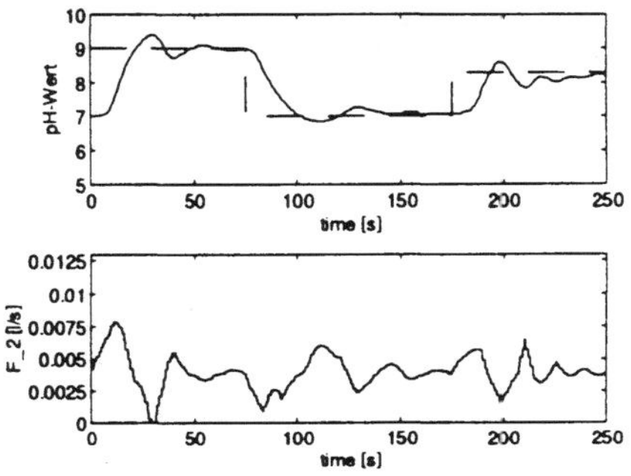

Fig 7: Responses with fuzzy controller with 29 rules (simulation)

4. APPROXIMATION OF OPTIMAL CONTROLLERS BY A FUZZY SYSTEM

The above section has shown that the purely heuristic approach is restricted to plants where either extensive operator knowledge is available or the dynamics of the system are not too complex. Since a coarse mathematical model is often available, the question arises how this information can be used directly for fuzzy logic controller design. The main idea is to compute a model-based optimal controller which can be used to identify the kernel of the fuzzy rulebase. In order to efficiently approximate an analytical control law, it is advantageous to choose a special format of the fuzzy rules which was first introduced by Takagi and Sugeno (1985):

Rule i: IF x_1 is A_1^i, x_2 is A_2^i, ..., x_m is A_m^i

THEN $y_j = p_0^i + p_1^i \cdot x_1 + p_2^i \cdot x_2 + \ldots + p_m^i \cdot x_m$

Here the output is determined by a linear function of the controller inputs x_j and not expressed by fuzzy values. If the fuzzy sets of the input variables have sufficient overlap, a soft interpolation of linear control laws results. The final output y^* is computed easily as the weighted mean of the individual outputs y_i of the rules:

$$y^* = \frac{\sum_{i=1}^{n} \min_j \mu A_j^i(x_j) \cdot y_i}{\sum_{i=1}^{n} \min_j \mu A_j^i(x_j)} \quad . \tag{6}$$

$\mu A_m^i(x_m)$ is the degree of membership of the (crisp) input x_m to the linguistic variable A_m^i. The parameters p_j^i, which determine the consequents of the rules, can be optimized by a least squares approximation of a priori given optimal trajectories. The interested reader is referred to (Heckenthaler and Engell, 1994) for details. The goal of the optimization is to approximate the model based controller characteristic with a minimum number of rules with sufficient accuracy. Therefore it is helpful to optimize the fuzzification as well, i.e. the coordinates of the membership functions. This can be accomplished using multidimensional minimization strategies like the downhill simplex method. To avoid getting trapped in local minima it is advisable to connect them with genetic algorithms.

5. MODEL BASED FUZZY CONTROLLER DESIGN

First we had to choose an optimal controller for this system. Observing the control actions of an experienced operator controlling a plant manually, it is often found that he or she is using the available range of the control variables to obtain quick responses for large deviations. For small ones, on the other hand, he operates more carefully to avoid "bang-bang"-type control. Thus the time-optimal controller seemed to be the right choice for large deviations, and a robust linear controller was used for small deviations.

To compute the time-optimal controller we took the following approach: since two state variables (c_1 and c_2) are not observable, the concentration c_2 is assumed to have a fixed value (5mmol/l) at the beginning of the sampling time. Equation (3) then yields along with the measured pH value an estimate of c_1. Now the state equations (1), (2) and (4) can be integrated for the next sampling interval with a maximal control action F_2, so that the pH value in the tank reaches the set point as fast as possible without overshoot (see Fig. 8). This controller is of course only time-optimal with respect to the assumed pH value in the tank and not with respect to the measured one.

Using this algorithm, 660 trajectories were computed such that the ranges of controller input space (pH, error, dpH) were covered well. Now a fuzzy controller was identified which has the format described in section 4.

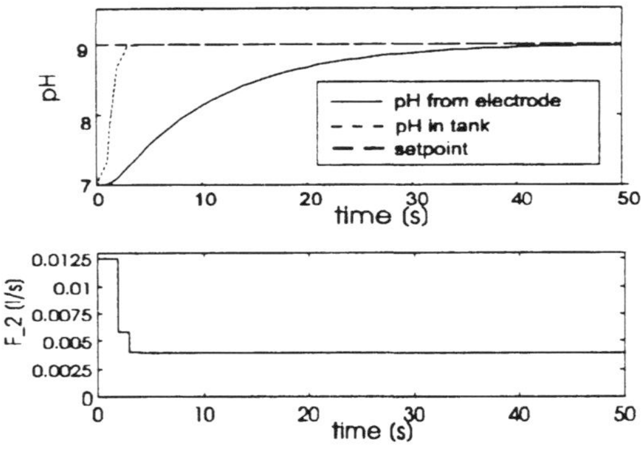

Fig. 8: Time-optimal controller for nominal model

An example of a rule is:

Rule i: If pH is *acidic* and error is *pb* and dpH is *ns*, then $F_2 = p_0^i + p_1^i \cdot pH + p_2^i \cdot error + p_3^i \cdot dpH$

The optimization of the rulebase showed that a fuzzy controller with 16 rules already yields a sufficient approximation of the control law (see Fig 9). A more exact approximation is possible with an increase in the number of rules, but this is not recommended because the model itself is inaccurate. In Fig. 10 the resulting fuzzy values are depicted with solid lines.

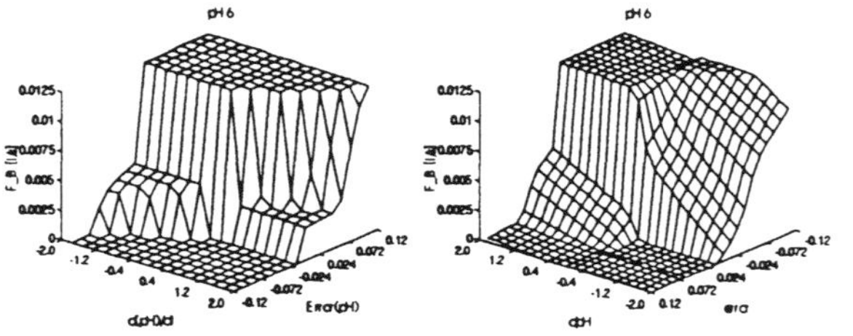

Fig. 9: Time optimal control law (left) and its approximation by a fuzzy controller (right) with 16 rules.

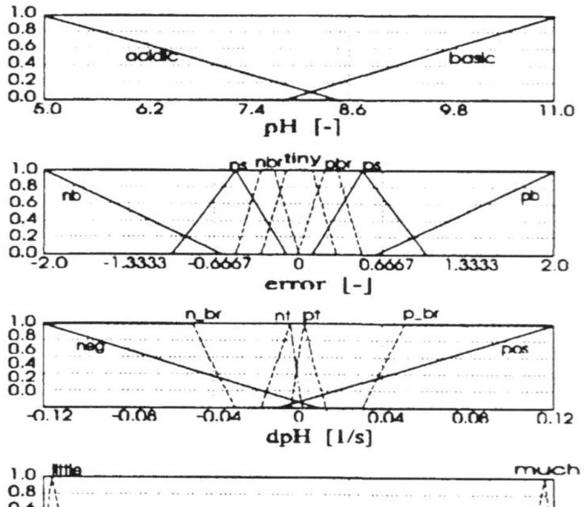

Fig. 10: Fuzzification of controller inputs and outputs (— fuzzy values for modelbased rules, -- fuzzy values for heuristic rules).

Fig. 11 shows the resulting performance for a set point change. As expected, the regulation time is very short without overshoot. The fuzzy controller shows the same deficiencies as its origin, the time-optimal controller, in correcting small deviations with hectic, ineffective control actions. It is therefore necessary to use a more cautious, linear controller for small control errors. This can be done conveniently by adding one rule to the fuzzy rulebase:

Rule 17: If pH is *any* and error is *tiny* and
dpH is *any*, then F_2=a+b·pH +c·error +d·dpH

The parameters a and b were found by approximating the titration curve (Fig. 2) in the manner described in section 4. Actually at least 2 rules for pH values *acidic* and *basic* are necessary to cope with the nonlinearity. The parameters c and d are the gains for the proportional and derivative action which were found experimentally. Fig. 12 shows the effect of these 2 additional rules. The controller actions for small deviations are now much smoother and a steady state is reached.

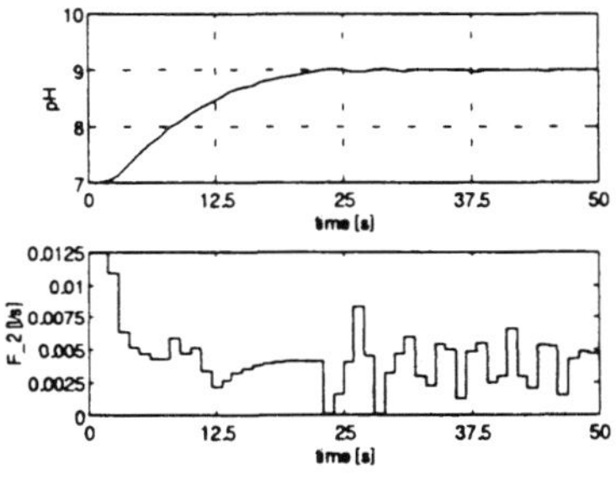

Fig. 11: System behaviour with fuzzy controller with 16 rules (simulation).

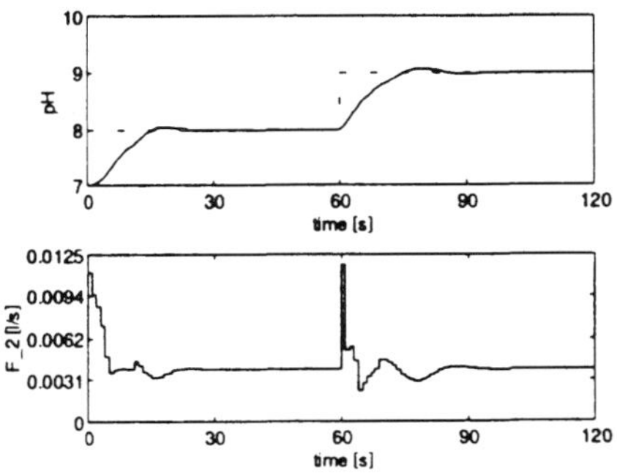

Fig 12: Step point changes with fuzzy controller with 18 rules (simulation)

After the fuzzy controller has been tested in simulations, experiments on the actual plant show how much the modelling errors influence the robustness of the fuzzy controller. It was found that the performance at the real reactor was very good, too. A small overshoot occurred for large set point changes. This observation led to the formulation of two more rules, which are triggered when the pH approaches the set point too quickly:

Rule 19: If pH *any* and error *p_br* and dpH *p_br*,
then F_2 *little*
Rule 20: If pH *any* and error *n_br* and dpH *n_br*,
then F_2 *much.*

Fig. 13 shows the performance at the real plant. The chattering of the controller output at steady state is caused by the measurement noise and the quantisation of the controller output.

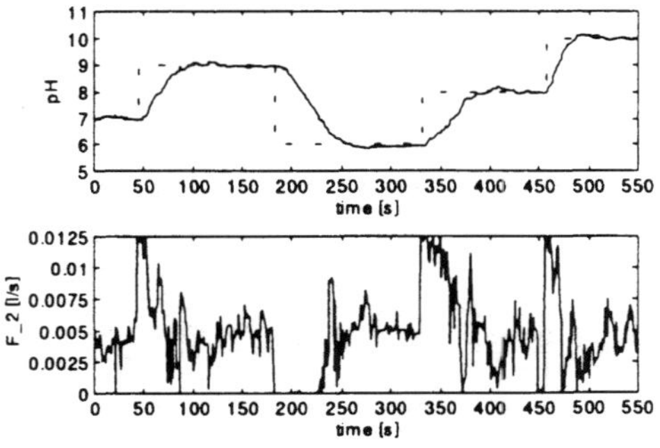

Fig 13: Step point changes with fuzzy controller with 20 rules (experiment)

The rejection of disturbances is very important in pH control. Therefore the fuzzy controller was tested while the concentration of the acetic acid was reduced to 50% from 0.007 to 0.0035 mol/*l* at t=0s (see Fig. 14). Compared to the strong disturbance, only a relatively small remaining regulating error occurs. This is because the time-optimal control characteristic is very steep.

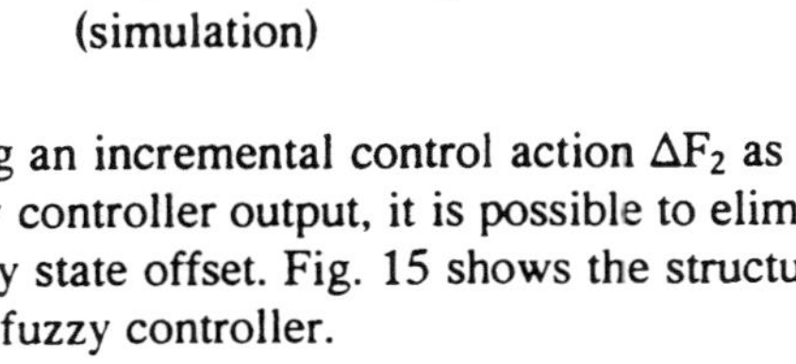

Fig. 14: Response to disturbance c_1(t=0s)=0.0035 mol/*l* with fuzzy controller (20 rules) (simulation)

Using an incremental control action ΔF_2 as a second fuzzy controller output, it is possible to eliminate the steady state offset. Fig. 15 shows the structure of the final fuzzy controller.

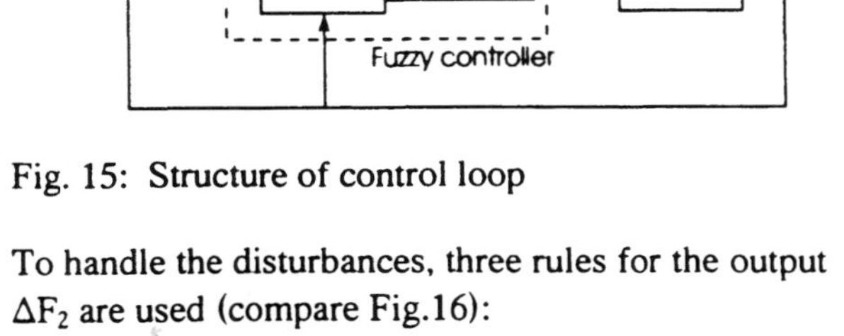

Fig. 15: Structure of control loop

To handle the disturbances, three rules for the output ΔF_2 are used (compare Fig.16):

Rule 21: If pH *any* and error *any* and dpH *any*, then ΔF_2 *zero*

Rule 22: If pH *any* and error *p_br* and dpH *nt*, then ΔF_2 *more*

Rule 23: If pH *any* and error *n_br* and dpH *pt*, then ΔF_2 *less.*

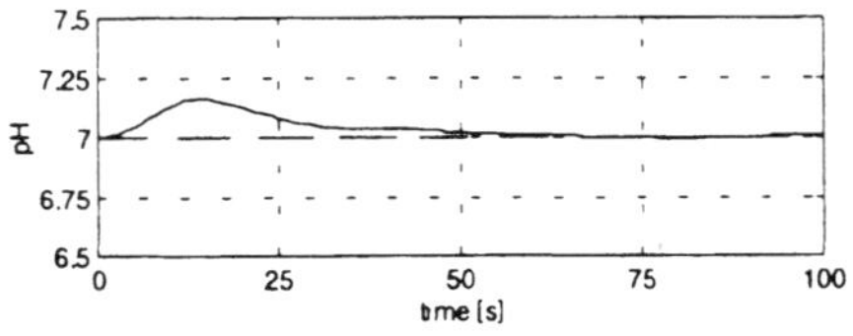

Fig. 16 Response to disturbance c_1(t=0s)=0.0035 mol/*l* with fuzzy controller (23 rules) (simulation)

The effectiveness of these rules can be evaluated from the transients depicted in Fig. 17, where the same disturbance occurred at t=0s. While the first set point change causes a small overshoot, the controller adapts to the disturbance in a comparatively short time.

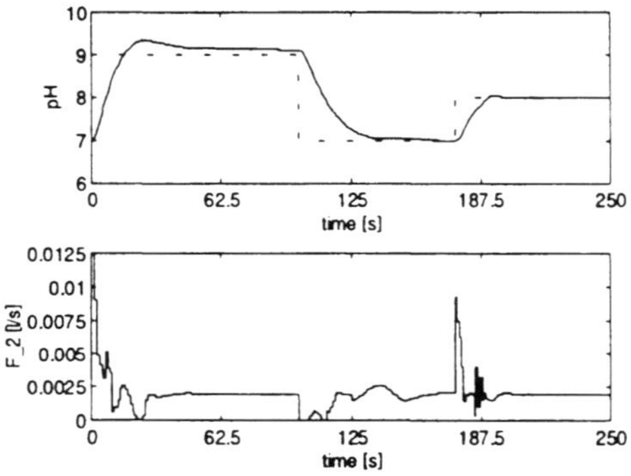

Fig. 17: Response with fuzzy controller (23 rules) with permanent disturbance c_1(t=0s)=0.0035 mol/*l* (simulation)

6. CONCLUSIONS

The design of two different fuzzy controllers has shown that fuzzy technology offers a powerful tool to design nonlinear controllers. The complexity of the neutralization plant showed the limits of a purely heuristic approach, which is strongly dependent on existing operator knowledge and the skill of the designer. Heuristic knowledge for complex systems takes a long time to acquire, and it is difficult to evaluate the quality of the control.

The automatic design of the fuzzy controller based on model information on the other hand offers a systematic design procedure. First the user has to formulate a coarse mathematical model and then an optimal controller is computed for this model. We decided to use time-optimal control which was in our case not very hard to compute. Often the computation of a time-optimal control law for nonlinear systems will be difficult to determine. A promising general approach to compute time-optimal trajectories for nonlinear systems is the cell state space method introduced by Hsu (1985). This optimization procedure requires large memory and computational effort, but the resulting identified time-optimal cell trajectories can be used directly to identify the kernel of the rulebase off-line. The on-line computational effort to determine the output of the fuzzy rulebase, on the other hand, is fairly small, so that on-line control is possible. We are currently investigating and refining this approach for nolinear systems in order to implement it into our design tool (FFUN-tool).

The pH-control of a laboratory CSTR with the model based fuzzy controller has shown that it is advantageous to use a coarse mathematical model as a starting point for the design of the rulebase. Adding heuristic knowledge to a set of model based fuzzy rules is according to our experience simple and effective to correct the deficiencies of the approximated time-optimal behaviour. This increases the robustness of the fast fuzzy controller significantly. The fuzzy rulebase provides a flexible platform to realize different control strategies, which can be integrated efficiently and smoothly.

This research was supported by the Deutsche Forschungsgemeinschaft under grant DFG En 152/8-2. This support is very gratefully acknowledged.

REFERENCES

Chen, M. (1994) Fuzzy control of a neutralization reactor, Student project, Dept. of Chemical Engeneering, University of Dortmund.

FFUN-Tool (Fast-FUzzy-coNtrol) (1994). Design tool for fast fuzzy controllers. Process Control Group, Department of Chemical Engineering, University of Dortmund.

Heckenthaler, T. and Engell, S. (1994). Robust approximately time-optimal fuzzy control of a two-tank system, *IEEE Control System Magazine*. Vol. 14, No. 3, pp. 24-30.

Hsu, C.S. (1985): A Discrete Method of Optimal Control Based on the Cell State Space Concept. Journal of Optimization Theory and Applications, vol 46, no. 4, pp. 547-569.

McAvoy, T. (1972). Dynamics of pH in Controlled Stirred Tank Reactor. *Ind. Eng. Chem. Process Des. Develop.*, Vol. 11, No. 1, pp. 68-70.

Takagi, T. and Sugeno, M. (1985). Fuzzy identification of systems and its application to modelling and control, *IEEE Trans. Syst. Man, Cybern.*, Vol. SMC-15, No. 1, pp. 116-132.

RELIABLE AND EFFICIENT OPTIMIZATION STRATEGIES FOR NONLINEAR MODEL PREDICTIVE CONTROL

Lino O. Santos*, Nuno M.C. de Oliveira* and Lorenz T. Biegler**

**Departamento de Engenharia Química, Universidade de Coimbra, Largo Marquês de Pombal, 3000 Coimbra, Portugal. E-mail: {lino,nuno}@eq.uc.pt.*

***Department of Chemical Engineering, Carnegie Mellon University, Pittsburgh PA 15213, USA. E-mail: biegler@cmu.edu.*

Abstract. Previous experience with predictive control algorithms has shown that the way the optimization problems are formulated and solved has a big impact in the success of the control strategy. Here a multiple shooting formulation is proposed, where a process model is integrated separately inside each sampling interval, and the corresponding equality constraints are added directly to the optimization problem. It is shown that the resulting formulation provides a more reliable framework for the solution of predictive control problems, both in the linear and nonlinear cases. This strategy is compared with the original nonlinear Newton-type (state space) algorithm, on a number of process models with challenging features, including the reactor model from the Tennessee Eastman problem.

Key Words. Nonlinear control systems, Predictive control, Multiple Shooting, State-space methods, Process control.

1. INTRODUCTION

In previous studies, nonlinear predictive controllers have been shown to be advantageous on a variety of process models. Examples where these are essential include:

- model predictive control with sign changes in the gain matrix over desired regions of operation.
- nonsteady state processes where an optimal operating profile is required.
- systems that have complex interactions with constraints and nonlinear phenomena.

These controllers have been developed from a variety of perspectives and can be viewed as nonlinear extensions of widely used linear predictive formulations such as QDMC, or GPC (García et al., 1989). In particular the Newton-type nonlinear controller (Li and Biegler, 1988; Oliveira and Biegler, 1994a; Oliveira and Biegler, 1994b; Oliveira and Biegler, 1995) is able to deal with common difficulties in nonlinear control, such as open-loop instabilities, stability problems introduced by the presence of active hard-constraints, and some amount of ill-conditioning in the predictive model. These features are not only characteristics of the desired steady-states, but can occur over different regions of operation, which makes them, at least in some cases, difficult to anticipate and avoid.

One aspect of the problem that deserves a further look is the issue of ill-conditioning in the predictive model. From past experience, this can occur quite easily with highly nonlinear and open-loop unstable processes. For example, this behavior can show up as numerical instability during the integration of the model equations, as a consequence of large roundoff errors induced by the unstable modes. It can also appear as ill-conditioning of the dynamic matrix, indicating a high sensitivity to certain types of input changes. This is then reflected in a high condition number of the Hessian, which can lead to instability in the matrix factorizations for the quadratic program (QP) or in the line search phase, when stepsizes that guarantee a descent direction cannot be found. Both of these situations lead inevitably to an important deterioration in the performance of the control algorithm, and should be avoided as much as possible.

To overcome this type of difficulties a boundary value approach is adopted, to enhance the stability and problem conditioning (Ascher et al., 1988). These techniques apply a multiple shooting concept to solve the model equations. Here the state trajectory is integrated separately in each sampling interval using the initial state conditions as decision variables. Matching the states across intervals leads to a set of nonlinear equality constraints which, in addition to the input and output constraints, is incorporated in the SQP-based formulation. At each sampling-time new nominal state and input trajectories are thus obtained by solving the quadratic objective in order to simultaneously get optimal controller moves and satisfy the constraints.

While this approach has a number of advantages over current nonlinear MPC controllers, attention must be paid to the optimization algorithm in order to ensure efficient performance. In particular, it is essential to incorporate large-scale features and retain reliable convergence properties. These are preserved, for example, through the use of a reduced space SQP decomposition and trust region strategies. In addition, several authors (Keerthi and Gilbert, 1988; Michalska and Mayne, 1993; Rawlings and Muske, 1993) refer to the use of terminal state constraints to provide closed-loop stability of MPC formulations. In this study, final time constraints on the states or the output variables are also exploited by the numerical procedure as an additional mode of improving the problem conditioning and therefore the algorithmic performance of the method.

Finally, the performance of the multiple shooting control formulation developed in this paper is compared with the previous Newton-type formulation of Oliveira and Biegler (1994a, 1995) and Oliveira (1994), on a number of examples. This include relatively simple linear models as well as the reactor of the Tennessee Eastman plant-wide process control problem (Downs and Vogel, 1993), which provides an example of a highly nonlinear and open-loop unstable process.

2. A MULTIPLE SHOOTING FORMULATION FOR PROCESS CONTROL

The process model considered is of the form

$$\dot{x} = f(x, u, d; \theta) \quad (1)$$
$$y = g(x; \theta), \quad (2)$$

with f and g twice continuously differentiable, where $x \in \mathbf{R}^{n_s}$, $u \in \mathbf{R}^{n_i}$, and $y \in \mathbf{R}^{n_o}$. It is assumed to be a perfect model and also that true values of the state vector are available. For a constant sampling time T it is convenient to define the discrete time operator corresponding to (1) as

$$x_{k+1} = \chi(T; x_k, u_k). \quad (3)$$

The derivatives of the transition function χ and g are defined as

$$\Phi_k = \frac{\partial \chi(T; x_k, u_k)}{\partial x_k}$$
$$\Gamma_k = \frac{\partial \chi(T; x_k, u_k)}{\partial u_k}$$
$$C_k = \left.\frac{\partial g(x; \theta)}{\partial x}\right|_{x=x_k}.$$

Efficient methods for the computation of these sensitivity coefficients are readily available (Caracotsios and Stewart, 1985; Li and Biegler, 1988).

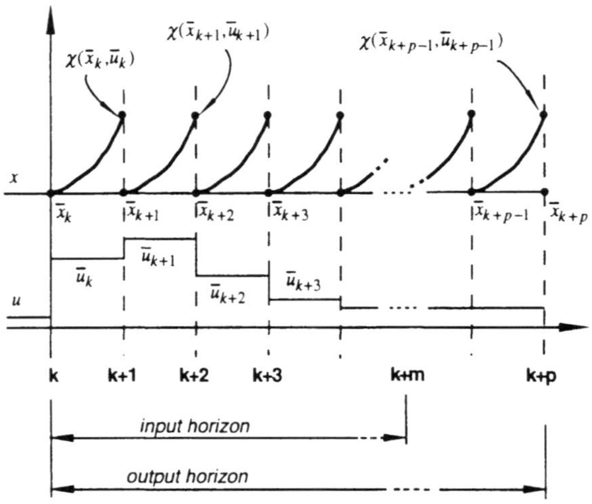

Fig. 1. Profile solution using a multiple shooting approach.

Using a multiple shooting formulation, where the state equations are integrated inside a sampling interval (Figure 1), leads to the following control problem formulation:

$$\min_{u_i, x_i, y_i} J_2 = \sum_{i=1}^{p} O_{k+i} + M_{k+i-1} \quad (4)$$

$$\text{s.t. } \bar{x}_{k+i} = \chi(\bar{x}_{k+i-1}, \bar{u}_{k+i-1}),\ i = 1, \ldots, p \quad (5)$$
$$y = g(x; \theta) \quad (6)$$
$$u(t) = u(t_{k-1} + t_{ih}),\ t \in [t_k + t_{ih}, t_{oh}] \quad (7)$$
$$u_l \le u_k \le u_u \quad (8)$$
$$x_l \le x_k \le x_u \quad (9)$$
$$y_l \le y_k \le y_k, \quad (10)$$

where

$$O_{k+i} = (y_{\mathrm{sp},k+i} - y_{k+i})^{\mathrm{T}} Q_{yi} (y_{\mathrm{sp},k+i} - y_{k+i}),$$
$$M_{k+i-1} = (u_{k+i-1} - u_{r,k+i-1})^{\mathrm{T}} Q_{ui} (u_{k+i-1} - u_{r,k+i-1}).$$

Note that $\chi(\bar{x}_{k+i-1}, \bar{u}_{k+i-1})$ is obtained through the integration of (1) inside the sampling interval $t \in [t_{k+i-1}, t_{k+i}]$ only. To simplify the notation, ***augmented*** vectors U, X and Y containing all values of the correspondent variables inside a predictive horizon are introduced

$$U \in \mathbf{R}^{n_i m} \equiv [\ u_k^{\mathrm{T}} \quad u_{k+1}^{\mathrm{T}} \quad \cdots \quad u_{k+m-1}^{\mathrm{T}}\ ]^{\mathrm{T}}$$

$$X \in \mathbf{R}^{n_s p} \equiv \left[\begin{array}{cccc} x_{k+1}^{\mathrm{T}} & x_{k+2}^{\mathrm{T}} & \cdots & x_{k+p}^{\mathrm{T}} \end{array} \right]^{\mathrm{T}}$$

$$Y \in \mathbf{R}^{n_o p} \equiv \left[\begin{array}{cccc} y_{k+1}^{\mathrm{T}} & y_{k+2}^{\mathrm{T}} & \cdots & y_{k+p}^{\mathrm{T}} \end{array} \right]^{\mathrm{T}}.$$

In previous formulations, the linearization of the model (3) around a nominal trajectory $\bar{U}$ led to the first order model

$$\hat{Y} \approx \bar{Y} + \left. \frac{\partial Y}{\partial U} \right|_{U=\bar{U}} \Delta U = \bar{Y} + \mathcal{S}_m \Delta U,$$

where $\mathcal{S}_m$ represents the *dynamic matrix* of the process, and $\Delta U = U - \bar{U}$. However, the numerical roundoff error in the integration of the state variables and their sensitivities, is given approximately in this case by $\epsilon \exp(LpT)$, where ϵ is the machine precision, pT is the length of the output horizon, T is the sampling interval, and L is the eigenvalue corresponding to the largest unstable mode. On the other hand, in the multiple shooting case the state variables are integrated separately inside each sampling interval, allowing the roundoff error to be drastically reduced to $\epsilon \exp(LT)$.

Since both X and U are treated as decision variables in the multiple shooting formulation, it is necessary to stipulate two independent reference trajectories $\bar{X}$ and $\bar{U}$, for the solution of the problem. Therefore, only $\bar{Y}$ needs to be specified to be compatible with $\bar{X}$, i.e., with $\bar{y}_{k+i} = g(\bar{x}_{k+i}; \theta)$, $i = 1, \ldots, p$. Linearizing the constraints (5) around the nominal trajectories $(\bar{U}, \bar{X})$, leads to

$$\bar{x}_{k+1} - \chi(\bar{x}_k, \bar{u}_k) + I_{n_s} \Delta x_{k+1} - \Gamma_k \Delta u_k = 0$$

$$\bar{x}_{k+j} - \chi(\bar{x}_{k+j-1}, \bar{u}_{k+j-1}) + I_{n_s} \Delta x_{k+j} - \Phi_{k+j-1} \Delta x_{k+j-1} - \Gamma_{k+j-1} \Delta u_{k+j-1} = 0, \quad j = 2, \ldots, p.$$

This set of constraints can be represented in matrix form as $\bar{h} = P\Delta X + G\Delta U$, where

$$\bar{h} = [-\bar{x}_{k+1} + \chi(\bar{x}_k, \bar{u}_k) \ \cdots \ -\bar{x}_{k+p} + \chi(\bar{x}_{k+p-1}, \bar{u}_{k+p-1})]^{\mathrm{T}},$$

$$G = \begin{bmatrix} -\Gamma_k & 0 & 0 & \cdots & 0 \\ 0 & -\Gamma_{k+1} & 0 & \cdots & 0 \\ 0 & 0 & -\Gamma_{k+2} & \cdots & 0 \\ \vdots & \vdots & \vdots & \ddots & \vdots \\ 0 & 0 & 0 & \cdots & -\Gamma_{k+m-1} \\ 0 & 0 & 0 & \cdots & -\Gamma_{k+m} \\ \vdots & \vdots & \vdots & \cdots & \vdots \\ 0 & 0 & 0 & \cdots & -\Gamma_{k+p-1} \end{bmatrix}.$$

$$P = \begin{bmatrix} I_{n_s} & 0 & 0 & \cdots & 0 \\ -\Phi_{k+1} & I_{n_s} & 0 & \cdots & 0 \\ 0 & -\Phi_{k+2} & I_{n_s} & \cdots & 0 \\ \vdots & \vdots & \ddots & \ddots & \cdots \\ 0 & 0 & \cdots & -\Phi_{k+p-1} & I_{n_s} \end{bmatrix},$$

The linearization of constraints (6) is straightforward, leading to the set of equations $\Delta Y = \mathcal{C}\Delta X$, with

$$\mathcal{C} = \mathrm{diag}\{C_{k+i}\}, \qquad i = 1, \ldots, p.$$

It can be easily shown that the dynamic matrix of the original formulation is related to the quantities just defined by

$$\boxed{\mathcal{S}_m = -\mathcal{C}P^{-1}G}. \tag{11}$$

Since in the present case both ΔU, ΔX and ΔY are decision variables, it is convenient to define an augmented decision vector

$$W = [U^{\mathrm{T}} \ \ X^{\mathrm{T}} \ \ Y^{\mathrm{T}}]^{\mathrm{T}} \in \mathbf{R}^{n_i m + (n_s + n_o)p}$$

Therefore, using a Gauss-Newton formulation for the solution of the nonlinear control problem, a quadratic program of the form

$$\min_{\Delta W} \quad J_2 = (\Delta W - W')^{\mathrm{T}} Q (\Delta W - W') \tag{12}$$

$$\text{s.t.} \quad A_c \Delta W = d_c \tag{13}$$

$$W_{ld} \leq \Delta W \leq W_{ud} \tag{14}$$

has to be solved at each iteration, where

$$\begin{aligned} \Delta W &= W - \bar{W} \\ W' &= W_r - \bar{W} \\ W_r &= [U_r^{\mathrm{T}} \ 0 \ Y_{\mathrm{sp}}^{\mathrm{T}}]^{\mathrm{T}} \\ Q &= \begin{bmatrix} Q_2 & 0 & 0 \\ 0 & 0 & 0 \\ 0 & 0 & Q_1 \end{bmatrix} \\ A_c &= \begin{bmatrix} G & P & 0 \\ 0 & \mathcal{C} & -I_{n_o} \end{bmatrix} \\ d_c &= [\bar{h}^{\mathrm{T}} \ 0 \ 0]^{\mathrm{T}} \\ W_{ld} &= W_l - \bar{W} \\ W_{ud} &= W_u - \bar{W} \\ W_l &= [U_l^{\mathrm{T}} \ X_l^{\mathrm{T}} \ Y_l^{\mathrm{T}}]^{\mathrm{T}} \\ W_u &= [U_u^{\mathrm{T}} \ X_u^{\mathrm{T}} \ Y_u^{\mathrm{T}}]^{\mathrm{T}}. \end{aligned}$$

This corresponds therefore to a larger problem than the original QP solved originally, which involved only the ΔU variables. This formulation can also easily consider terminal state constraints of the form

$$x_{k+p} = x_{\mathrm{ref}}, \tag{15}$$

in order to enhance the stability properties of the multiple shooting formulation. Using the augmented quantities defined previously, these can be expressed as $L_p X = x_{\text{ref}}$, with

$$L_p = [0 \quad \cdots 0 \quad I_{n_s}] \in \mathbf{R}^{n_s \times n_s p}.$$

Therefore constraints (13) need only to be substituted by an equivalent set $A_{cf} \Delta W = d_{cf}$, with

$$\begin{aligned} A_{cf} &= \begin{bmatrix} G & P & 0 \\ 0 & \mathcal{C} & -I_{n_o} \\ 0 & L_p & 0 \end{bmatrix} \\ d_{cf} &= \begin{bmatrix} \bar{h}^{\mathrm{T}} & 0 & x_{\text{ref}} - L_p \bar{X} \end{bmatrix}. \end{aligned}$$

These constraints can also be enforced instead as bounds on the terminal state, using a procedure similar to the previous one. In either case, it should be noted that their inclusion in the formulation might introduce some difficulties in the solution of the corresponding problem. This is a consequence of the fact that the final reference state to be used can only be computed rigorously by knowing all of the disturbances affecting the process. In alternative, terminal output constraints of the form $y_{k+p} = y_{\text{ref}}$ can be considered, especially if these variables have unstable modes directly associated with them. However the stability properties associated with the first approach might be lost in this case.

To see the influence of output constraints in the conditioning of the predictive problem, it is noted that the matrix P alone cannot be guaranteed to be well-conditioned. In fact, it can be showed (Ascher et al., 1988; Tanartkit and Biegler, 1995) that the condition number for P is proportional to $\exp(LpT)$ (in the multiple shooting case). On the other hand, augmenting this system with endpoint conditions associated with the unstable modes makes the condition number proportion to LpT instead. Thus, the addition of terminal constraints leads in general to a better conditioning of the problem, with a significant improvement in the performance of any solution method.

3. EXAMPLES

Initially the effects of the proposed approach were examined with linear models. Tables 1 and 2 illustrate the numerical condition of the dynamic and Hessian matrices obtained with simple third order SISO models. As can be observed, the BVP approach alone is able to provide, in most cases, a considerable reduction in the condition numbers (κ) of the Hessian and constraint matrices. This reduction is especially significant when coupled with a terminal state constraint. The effect of the tuning parameters is also visible. For example the relatively small reduction in $\kappa(H_z)$ obtained with the BVP approach in the first example (due essentially to very small terms in the Hessian matrix) can be corrected by using an input weight greater than zero. This constitutes a further indication of the importance of the choice of the tuning parameters for good performance of predictive algorithms.

	MPC		BVP	BTC
Q_u	$\kappa(\mathcal{S}_m)$	$\kappa(H)$	$\kappa(H_z)$	$\kappa(H_z)$
0	1.2×10^5	1.4×10^{10}	3.9×10^9	381
1	same	1.89	2.02	2.00

Table 1 Numerical condition for the system $G(s) = \frac{1}{(s+1)^3}$, with horizons of length $m = p = 15$, and $Q_y = 1$ (MPC – previous predictive formulation; BVP – boundary value formulation; BTC – BVP with terminal constraint).

	MPC		BVP	BTC
Q_u	$\kappa(\mathcal{S}_m)$	$\kappa(H)$	$\kappa(H_z)$	$\kappa(H_z)$
0	1.2×10^{13}	singular	1.9×10^{10}	32
1	same	singular	157	1.6

Table 2 Numerical condition for the system $G(s) = \frac{1}{(s+1)(s-2)(s+3)}$, with horizons of length $m = p = 10$, and $Q_y = 1$.

Consider now the Tennessee Eastman problem, described in detail by Downs and Vogel (1993). This system is composed by a reactor/separator/recycle arrangement involving the manufacture of two liquid products through a network of irreversible and exothermic reactions, with a total of eight chemical species involved (Figure 2). The model to simulate the reactor is taken from the demonstrating application program provided by Downs and Vogel (1993). The model comprises ten state variables (related to the balances to each component and energy flows inside the vessel and in the cooling system), and three output variables (reactor pressure, level and temperature). For manipulation are available the reactor feed rate, the agitator speed and the flow of cooling water. To simplify the analysis, the noise associated with the measurements, the dynamics of the agitator speed and the flow of cooling water are not considered here. The same model is used for both plant simulation and control, although there might be some disturbances present in the system that are obviously not known by the control algorithm. Complete and accurate state information is assumed in this phase.

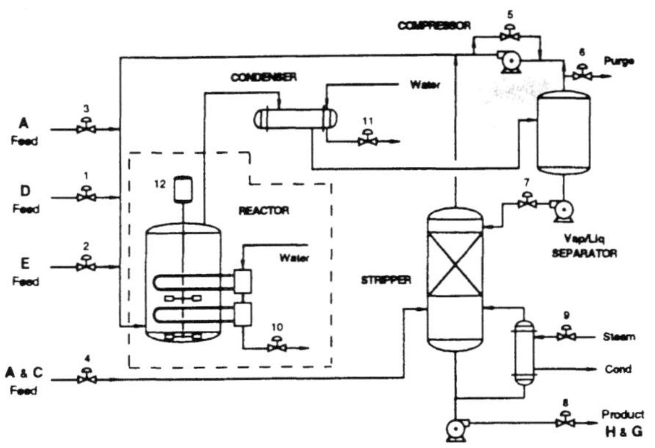

Fig. 2. Diagram of the Tennessee Eastman process.

The process has different modes of operation. Here the conditions of the reactor operation corresponding to the base case with a G/H product mass ratio of 50/50 are used. The problem is simulated using a sampling-time $T = 0.05$h. The control objectives are to keep the reactor operating at the steady-state point and to reject disturbances without violating the operation limits and constraints on the process variables. Figure 3 shows the open-loop and closed-loop responses of the normalized output variables. Although the composition measurement of the reactor outlet stream is not available the numerical solution for the evolution of the G/H mass ratio is also showed in Figure 4. Two different situations are described: the response without any input disturbance and the response to a step of +5°C in the reactor cooling water inlet temperature introduced at time $t = 0.5$h. Weighting matrices for the controller were $Q_y = I$ and $Q_u = 10^{-4} \times I$. As can be observed from this figure, the process is clearly open-loop unstable (as indicated by curves 2 and 3, with and without temperature disturbances, respectively). Curves 4 and 5 show the successful closed-loop results, also with and without disturbances. A slight offset is observed in curve 4 since the integral action was turned off during this study.

Since the nonlinear control algorithm performs an open-loop optimization, its convergence properties are sensitive to any ill-conditioning present in the model. Some experiments performed indicated that with the settings used, the condition number of the Hessian at $t = 0$ is of the order of 10^8, for predictive horizons of length $p = 10$, and $m = 5$. If the output horizon is increased to $p = 20$, this condition number grows to approximately 10^{12}. Using the original Newton-type control algorithm with this system leads to successful results with $p = 10, m = 5$ (curves 4 and 5), but at a large computational cost. An average of 13 SQP iterations were required for each horizon problem. On the other hand, for a longer output horizon $p = 20$ the ODE solver fails at $t = 0.05$h and the

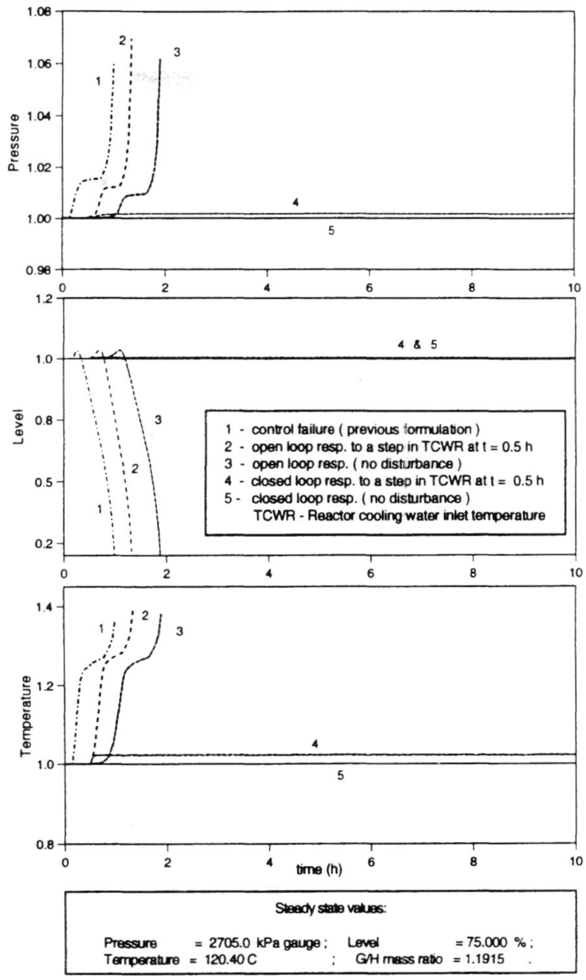

Fig. 3. Normalized output variables.

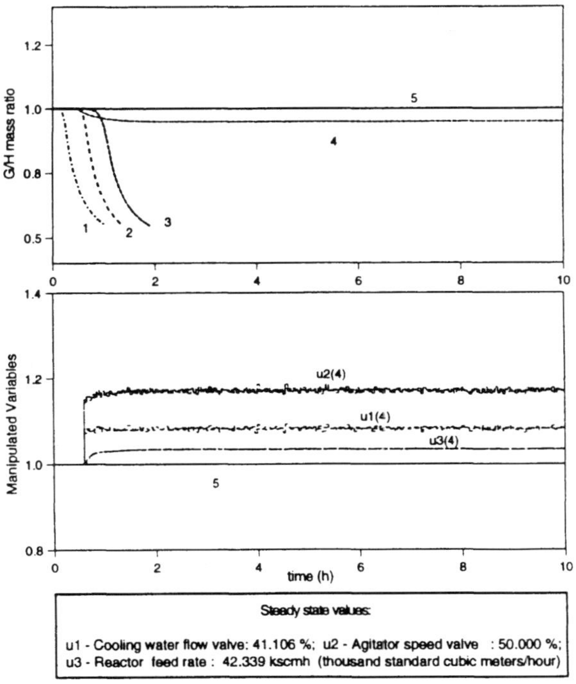

Fig. 4. Normalized G/H mass ratio and manipulated variables.

controller stops. The trajectory of the output predictions for this case is represented in Figure 3 by curve 1. This case illustrates therefore the limitations induced by large roundoff errors on the previous formulation.

These difficulties are however overcome with the nonlinear MPC based on the multiple shooting formulation. Curves 4 and 5 were obtained for horizons of $m = 5$, $p = 10$ regardless of whether endpoint constraints on the outputs were used or not. Moreover, the multiple shooting formulation was two times faster than the previous one and required an average of only 6 SQP iterations for each horizon problem. For a larger output horizon with $p = 20$, the disturbance case (curves 4) were also obtained successfully with and without endpoints. However, for this problem, ill-conditioning was still flagged often by the QP solver for the latter formulation. Finally, for the case of no disturbances (curves 5), only the endpoint formulation was able to solve this problem. Without endpoint constraints, the QP solver fails at time $t = 7$h and the controller terminates. This failure is due to the ill-conditioning of P when endpoint constraints are absent and this difficulty also increases with longer predictive horizons. For instance, curves 4 and 5 are still obtained for a simulation of 20h using the formulation with terminal output constraints with $p = 30$ and $m = 5$.

4. CONCLUSIONS

Multiple shooting formulations for nonlinear MPC are demonstrated to lead to more efficient and reliable performance for challenging process examples. In particular, they are effective in dealing with modeling unstable modes that otherwise lead to failure of previous nonlinear MPC controllers.

With highly unstable models or models that present finite escape time it is also possible to subdivide further the sampling interval or consider smaller intervals for the integration of the state and sensitivity equations. Finally, the addition of endpoint constraints greatly improves the performance of these methods by improving the conditioning of QP subproblem. Both of these factors ensure the applicability of the methodology for more challenging nonlinear control problems.

ACKNOWLEDGMENTS

Partial financial support from the US Department of Energy under Grant DE-FG02-85ER13396 and the Computer Aided Design Lab at Carnegie Mellon University is gratefully acknowledged. The first author is also grateful for financial support provided by JNICT under Programa Ciência.

5. REFERENCES

Ascher, U.M., R.M.M. Mattheij and R.D. Russell (1988). "Numerical Solution of Boundary Value Problems for Ordinary Differential Equations", Prentice Hall, Englewood Cliffs, NJ.

Caracotsios, M., W.E. Stewart (1985). "Sensitivity Analysis of Initial Value Problems with Mixed ODEs and Algebraic Equations", *Computers & Chemical Engineering*, **9**(4), 359–365.

Downs, J.J., E.F. Vogel (1993). "A Plant-Wide Industrial Process Control Problem" *Computers & Chemical Engineering*, **17**(3), 245–255.

Economou, C.G. (1985). *An Operator Theory Approach to Nonlinear Controller Design*, Ph.D. Thesis, California Institute of Technology, Pasadena, CA.

García, C.E., D.M. Prett, M. Morari (1989). "Model Predictive Control: Theory and Practice — a Survey", *Automatica*, **25**(3), 335–348.

Keerthi, S.S., E.G. Gilbert (1988). "Optimal Infinite-Horizon Feedback Laws for General Class of Constrained Discrete-Time Systems: Stability and Moving-Horizon Approximations", *IEEE Trans. Auto. Cont.*, **57**(2), 265–293.

Li, W.C., L.T. Biegler (1988). "Process Control Strategies for Constrained Nonlinear Systems", *Ind. Eng. Chem. Res.*, **27**, 1421–1433.

Michalska, H., D.Q. Mayne (1993). "Robust Receding Horizon Control of Constrained Nonlinear Systems", *IEEE Trans. Auto. Cont.*, **38**(11), 1623–1633.

Oliveira, N.M.C. (1994). *Newton-type Algorithms for Nonlinear Constrained Chemical Process Control*, Ph.D. Thesis, Carnegie Mellon University, Pittsburgh, PA.

Oliveira, N.M.C, L.T. Biegler (1994a). "Constraint Handling and Stability Properties of Model Predictive Control", *AIChE J.*, **40**(7), 1138–1155.

Oliveira, N.M.C, L.T. Biegler (1994b). "A Trust Region Strategy for Newton-type Process Control", *IFAC Symposium on Advanced Control of Chemical Processes, ADCHEM'94*, 283–288, Kyoto, Japan.

Oliveira, N.M.C., L.T. Biegler (1995). "Newton-type Algorithms for Nonlinear Process Control. Algorithm and Stability Results", *Automatica*, in press.

Rawlings, J.B., K.R. Muske (1993). "The Stability of Constrained Receding Horizon Control", *IEEE Trans. Auto. Cont.*, **38**(10), 1512–1516.

Tanartkit, P., L.T. Biegler (1995). "Stable Decomposition for Dynamic Optimization", *Ind. & Eng. Chem. Research*, submitted for publication.

OPTIMAL ESTIMATION OF STATES IN DIFFERENTIAL-ALGEBRAIC EQUATION SYSTEMS

Y. S. Cheng, T. Mongkhonsi and L. S. Kershenbaum

Centre for Process Systems Engineering, Imperial College of Science, Technology and Medicine, London SW7 2BY, UK

Abstract. Based on a minimum least squares criterion and without any assumption on process model error and measuring device noise, an optimal estimation algorithm has been developed for nonlinear differential-algebraic equation systems. The optimal estimator has been applied to an industrial scale fixed-bed reactor with decaying catalysts where phthalic anhydride is produced by the partial oxidation of o-xylene. Although a simplified reactor model was utilised by the estimator, using three real-time temperature measurements, the catalyst activity profile was accurately inferred and the dynamic behaviour of the reactor was excellently tracked. Computational experience indicated that the estimator could be implemented on-line in analogous way.

Keywords. Catalyst deactivation; differential-algebraic equations systems; fixed-bed reactor; optimal estimation

1. INTRODUCTION

The estimation of states in noisy nonlinear dynamic systems plays an important role in identification, optimisation and control. While this problem has been extensively studied since the basic theory for linear discrete systems was introduced by Kalman, most of estimation algorithms are focussed on systems of ordinary differential equations (ODE). However, dynamic modelling of most chemical processes results in a system of differential-algebraic equations (DAE), arising usually from mass and energy balances, which yield the ordinary differential equations, and thermodynamic or kinetic relations, which yield the algebraic constraints between state variables (Pantelides and Barton 1993). Therefore, the problem of state estimation for DAE systems in the presence of input disturbances and measurement errors is very important.

There are many possible criteria which could be used in defining optimal state estimation, such as minimal least squares, maximum likelihood and minimum maximum error. In this study, the minimum least squares criterion, i.e., the minimisation of the integral of the weighted square residual errors in the process model and the measuring device, is employed. This criterion has the advantage of avoiding the nature of the unknown inputs to the system and the measurement errors on the output. The optimal estimation problem can be reformulated as a deterministic optimal control problem, a "formal" approach already used in ODE systems (Detchmendy and Sridhar 1966). The calculus of variations (Jonckheere 1988) is used to obtain an optimal estimation algorithm for DAE systems.

As a case study, the optimal estimation algorithm is applied to an industrial scale fixed-bed reactor where phthalic anhydride is produced by the partial oxidation of o-xylene. The reactor is characterised by having catalysts which deactive both reversibly and irreversibly as a function of time as well as space. The catalyst activity is clearly not directly measurable. However, it may be possible to infer the catalyst activity from the measurements of a small number of temperatures and/or concentrations within the reactor. The estimation of catalyst activity would be extremely useful in deciding an optimum schedule for catalyst reactivation or for disposal and reloading with fresh catalyst.

2. OPTIMAL ESTIMATION FOR DAE SYSTEMS

Consider the class of DAE systems described by

$$\dot{x}_1(t) = f_1(x_1, x_2, t) + \zeta_1(t) \quad (1)$$

$$0 = f_2(x_1, x_2, t) + \zeta_2(t) \quad (2)$$

with initial condition

$$x_1(0) = x_{1,0} + \zeta_1(0) \quad (3)$$

where the n vector state $x(t)$ can be split into an n_1 vector differential state $x_1(t)$, whose time derivative exists explicitly, and an n_2 vector algebraic state $x_2(t)$. $\zeta(t)$ is zero mean random process with unknown statistical characteristics. By reference to $x_1(t)$ and $x_2(t)$, $\zeta(t)$ can be decomposed into an n_1 vector $\zeta_1(t)$ and an n_2 vector $\zeta_2(t)$. Observations of the systems are available at discrete sampling instants $t_k, k = 1, 2, \cdots, M$, and consist of an m vector $y(t_k)$ which is related to the state by

$$y(t_k) = h(x_1, x_2, t_k) + \eta(t_k) \quad (4)$$

where $\eta(t_k)$ is an m vector of unknown measurement error.

Based on the measurements $y(t_k)$ in the time interval $0 \le t_k \le T$, $k = 1, 2, \cdots, M$, it is required to estimate $x_1(t)$ and $x_2(t)$ at some time t. For any admissible estimates $x_1(t)$ and $x_2(t)$, the criterion of estimation is defined by following least square functional

$$I = \frac{1}{2}\zeta_1^T(0)P_{11}^{-1}(0)\zeta_1(0) + \frac{1}{2}\int_0^T \begin{bmatrix} \zeta_1(t) \\ \zeta_2(t) \end{bmatrix}^T R(t) \begin{bmatrix} \zeta_1(t) \\ \zeta_2(t) \end{bmatrix} dt + \frac{1}{2}\int_0^T \eta^T(t)Q(t)\eta(t)dt \quad (5)$$

where the first term minimises the square error of initial condition estimates, the second term minimises the integral squared model error and the third term minimises the integral squared measurement noise. The initial condition estimation error weighting matrix $P_{11}^{-1}(0)$, the model error weighting matrix $R(t)$ and the measurement noise weighting matrix $Q(t)$ are symmetric and positive definite. $Q(t)$ can be expressed by

$$Q(t) = \sum_{k=1}^{M} Q_k\delta(t - t_k)\Delta_k \quad (6)$$

where $\delta(t - t_k)$ is Dirac delta function and Δ_k is the sampling interval at sample k. The optimal estimation problem is to determine $x_1(t)$ and $x_2(t)$ such that the objective defined by Equation 5 is minimised.

If $\zeta_1(t)$ and $\zeta_2(t)$ are considered as control variables, the optimal estimation problem defined by Equation 5 becomes a deterministic optimal control problem, i.e., select the control variables $\zeta_1(t)$ and $\zeta_2(t)$ such that Equation 5 is minimised subject to the constraints described by Equation 1 and Equation 2. With the use of the calculus of variations for DAE systems (Jonckheere 1988) an optimal estimation algorithm is derived. In order to conserve space only final estimation equations are given here. The theoretical derivation may be found elsewhere (Cheng and Kershenbaum 1994).

For convenience, the superscript + and − are used to indicate times immediately after and before the sampling instant respectively, and the estimation error weighting matrix P and the model error weighting matrix R^{-1} are decomposed into the following form

$$P = \begin{bmatrix} P_{11} & P_{12} \\ P_{21} & P_{22} \end{bmatrix} \quad R^{-1} = \begin{bmatrix} R_{11}^{-1} & R_{12}^{-1} \\ R_{21}^{-1} & R_{22}^{-1} \end{bmatrix} \quad (7)$$

Between two successive data points, $t_{k-1} < t < t_k$, there are no measurement data, the following prediction equations are integrated to generate the predicted values $\hat{x}_1(t_k^-)$, $\hat{x}_2(t_k^-)$, $P_{11}(t_k^-)$, $P_{12}(t_k^-)$ and $P_{22}(t_k^-)$.

$$\dot{\hat{x}}_1(t) = \hat{f}_1(\hat{x}_1, \hat{x}_2, t) \quad (8)$$

$$0 = \hat{f}_2(\hat{x}_1, \hat{x}_2, t) \quad (9)$$

$$\dot{P}_{11}(t) = \frac{\partial \hat{f}_1}{\partial \hat{x}_1}P_{11} + \frac{\partial \hat{f}_1}{\partial \hat{x}_2}P_{12}^T + P_{11}\frac{\partial \hat{f}_1^T}{\partial \hat{x}_1} + P_{12}\frac{\partial \hat{f}_1^T}{\partial \hat{x}_2} + R_{11}^{-1}(t) \quad (10)$$

$$0 = \frac{\partial \hat{f}_1}{\partial \hat{x}_2}P_{22} + P_{11}\frac{\partial \hat{f}_2^T}{\partial \hat{x}_1} + P_{12}\frac{\partial \hat{f}_2^T}{\partial \hat{x}_2} + R_{12}^{-1}(t) \quad (11)$$

$$0 = \frac{\partial \hat{f}_2}{\partial \hat{x}_2}P_{22} + P_{22}\frac{\partial \hat{f}_2^T}{\partial \hat{x}_2} + R_{22}^{-1}(t) \quad (12)$$

with the initial conditions

$$\hat{x}_1(0) = x_{1,0} \quad (13)$$

$$P_{11}(0) = P_{11,0} \quad (14)$$

At sampling point t_k, measurement data arrive, the following updating expressions can be used to calculate the updated values $\hat{x}_1(t_k^+)$, $\hat{x}_2(t_k^+)$, $P_{11}(t_k^+)$, $P_{12}(t_k^+)$ and $P_{22}(t_k^+)$.

$$\hat{x}_1(t_k^+) = \hat{x}_1(t_k^-) + \{P_{11}(t_k^+)\frac{\partial \hat{h}^T}{\partial \hat{x}_1} + P_{12}(t_k^+)\frac{\partial \hat{h}^T}{\partial \hat{x}_2}\}Q_k[y(t_k) - \hat{h}] - \{P_{11}(t_k^-)\frac{\partial \hat{h}^T}{\partial \hat{x}_1} + P_{12}(t_k^-)\frac{\partial \hat{h}^T}{\partial \hat{x}_2}\}Q_k \{\frac{\partial \hat{h}}{\partial \hat{x}_1}\delta\hat{x}_1(t_k) + \frac{\partial \hat{h}}{\partial \hat{x}_2}\delta\hat{x}_2(t_k)\} \quad (15)$$

$$0 = P_{22}(t_k^+)\frac{\partial \hat{h}^T}{\partial \hat{x}_2}Q_k[y(t_k) - \hat{h}] - P_{22}(t_k^-)\frac{\partial \hat{h}^T}{\partial \hat{x}_2}Q_k\{\frac{\partial \hat{h}}{\partial \hat{x}_1}\delta\hat{x}_1(t_k) + \frac{\partial \hat{h}}{\partial \hat{x}_2}\delta\hat{x}_2(t_k)\} \quad (16)$$

$$P_{11}(t_k^+) = P_{11}(t_k^-) - \{P_{11}(t_k^+)\frac{\partial \hat{h}^T}{\partial \hat{x}_1} + P_{12}(t_k^+)\frac{\partial \hat{h}^T}{\partial \hat{x}_2}\}Q_k\{\frac{\partial \hat{h}}{\partial \hat{x}_1}P_{11}(t_k^-) + \frac{\partial \hat{h}}{\partial \hat{x}_2}P_{12}^T(t_k^-)\} - \{P_{11}(t_k^-)\frac{\partial \hat{h}^T}{\partial \hat{x}_1} + P_{12}(t_k^-)\frac{\partial \hat{h}^T}{\partial \hat{x}_2}\}Q_k\{\frac{\partial \hat{h}}{\partial x_1}\delta P_{11}(t_k) + \frac{\partial \hat{h}}{\partial \hat{x}_2}\delta P_{12}^T(t_k)\} \quad (17)$$

$$0 = -\{P_{11}(t_k^+)\frac{\partial \hat{h}^T}{\partial \hat{x}_1} + P_{12}(t_k^+)\frac{\partial \hat{h}^T}{\partial \hat{x}_2}\} Q_k\frac{\partial \hat{h}}{\partial \hat{x}_2}P_{22}(t_k^-) - \{P_{11}(t_k^-)\frac{\partial \hat{h}^T}{\partial \hat{x}_1} + P_{12}(t_k^-)\frac{\partial \hat{h}^T}{\partial \hat{x}_2}\}Q_k\frac{\partial \hat{h}}{\partial x_2}\delta P_{22}(t_k) \quad (18)$$

$$0 = -P_{22}(t_k^+)\frac{\partial \hat{h}^T}{\partial \hat{x}_2}Q_k\frac{\partial \hat{h}}{\partial x_2}P_{22}(t_k^-) - P_{22}(t_k^-)\frac{\partial \hat{h}^T}{\partial \hat{x}_2}Q_k\frac{\partial \hat{h}}{\partial \hat{x}_2}\delta P_{22}(t_k) \quad (19)$$

where δ quantity is difference between the updated value and predicated value

3. APPLICATION TO A FIXED-BED REACTOR

The optimal estimator developed for DAE systems is applied to simultaneously estimate the catalyst activity, temperature and concentration profiles in a fixed-bed reactor. The reactor studied here is an industrial scale, jacked cooled, tubular reactor which is designed to carried out the partial oxidation of o-xylene to phthalic anhydride using V_2O_5/TiO_2 catalyst.

It is well known that the computation time for solving prediction equations is crucial for the on-line application of the optimal estimator. Because the reaction network of the partial oxidation of o-xylene to phthalic anhydride is quite complex (Mongkhonsi 1993), the computation of the optimal estimator with a complete kinetic scheme will require excessive CPU time which is impossible to achieve on-line. Therefore, the reaction network is lumped into a simple one in which only the disappearance of o-xylene is involved. The deactivation of the V_2O_5/TiO_2 catalyst can be irreversible and reversible (Mongkhonsi 1993, Lopez-Isunza and Kershenbaum 1992). In order to achieve rapid solution of the optimal estimator, only the reversible deactivation caused by some strongly absorbed compounds which are formed from the non-selective oxidation of o-xylene is considered. These absorbed compounds could be removed by increasing oxygen partial pressure. In order to handle the irreversible deactivation, a maximal available activity is introduced. If no irreversible deactivation occurs and all the deposited compounds are removed, the maximal available activity will equal to one; otherwise, it will be less than one. Although these simplifications would bring errors into the reactor modelling, the optimal estimation algorithm, which is able to handling uncertainty, could effectively compensate for the inaccurate model. The dynamic behaviour of the reactor can be described by

$$\frac{\partial \alpha}{\partial t} = -k_{d_{1,0}} exp(-\frac{E_{d_1}}{\theta}) c\alpha + k_{d_{2,0}} exp(-\frac{E_{d_2}}{\theta}) p_{o_2}(\alpha_m - \alpha) \tag{20}$$

$$\tau\frac{\partial \theta}{\partial t} = -\frac{\partial \theta}{\partial z} + \frac{1}{P_{hr}}(\frac{\partial^2 \theta}{\partial r^2} + \frac{1}{r}\frac{\partial \theta}{\partial r}) + \alpha Da \sum_{s=1}^{3} B_s R_s(\theta, c) \tag{21}$$

$$0 = -\frac{\partial c}{\partial z} + \frac{1}{P_{mr}}(\frac{\partial^2 c}{\partial r^2} + \frac{1}{r}\frac{\partial c}{\partial r}) + \alpha Da \sum_{s=1}^{3} R_s(\theta, c) \tag{22}$$

where α is catalyst activity; θ dimensionless temperature; c dimensionless o-xylene concentration; t dimensionless time; z the dimensionless axial co-ordinate and r the dimensionless radial co-ordinate. The standard boundary and initial conditions are used. Approximate values of all model parameters and, specifically, the kinetics for catalyst deactivation/reactivation have been identified experimentally and given in detail elsewhere (Mongkhonsi 1993). Because the time scales for catalyst deactivation and temperature transient are much longer than that for concentration transient, the concentration is assumed as quasi-steady state.

Equation 20 to Equation 22 is a distributed parameter system and has to be lumped before the application of the optimal estimation algorithm developed for DAE systems. Because the temperature profile of axial direction in the reactor is very steep, and the location of hot spot may vary significantly with small disturbances in operation conditions, orthogonal collocation on finite elements is used for discretization in the axial direction. For the radial direction, due to the symmetry and smooth of the profile, only one collocation point is used. The reactor model Equation 20 to Equation 22 is lumped into the differential-algebraic system in the vector form

$$\frac{d\alpha}{dt} = f_\alpha(\alpha, \theta, c) + \zeta_\alpha \tag{23}$$

$$\frac{d\theta}{dt} = f_\theta(\alpha, \theta, c) + \zeta_\theta \tag{24}$$

$$0 = f_c(\alpha, \theta, c) + \zeta_c \tag{25}$$

where nonlinear functions f_α, f_θ and f_c correspond to catalyst deactivation, energy balance and material balance respectively. ζ_α, ζ_θ and ζ_c represent corresponding modelling errors. The number of finite elements and interior collocation points in each element will be determined by a compromise between accuracy and computational load. In our work, it has been found that using five elements and two interior collocation points within each element can provide satisfactory accuracy in all situations. From Equation 23 to Equation 25 it can be seen that the differential variables are activity and temperature and the algebraic variable is concentration. The following decomposition of state variables and functions is introduced.

$$x_1 = \begin{bmatrix} \alpha & \theta \end{bmatrix}^T \quad x_2 = c \tag{26}$$

$$f_1 = \begin{bmatrix} f_\alpha & f_\theta \end{bmatrix}^T \quad f_2 = f_c \tag{27}$$

With reference to Equation 26 and Equation 27, the estimation error weighting matrix P can be split into the form

$$P_{11} = \begin{bmatrix} P_{\alpha\alpha} & P_{\alpha\theta} \\ P_{\theta\alpha} & P_{\theta\theta} \end{bmatrix} \quad P_{12} = \begin{bmatrix} P_{\alpha c} \\ P_{\theta c} \end{bmatrix}$$

$$P_{22} = P_{cc} \tag{28}$$

The model error weighting matrix can be partitioned similarly. Because catalyst activity can not be directly measured and time delay exists in obtaining concentration measurements from a gas chromatograph, only temperature measurements at discrete axial position are considered.

$$y(t_k) = h(\theta(t_k)) + \eta(t_k) \tag{29}$$

where h is the function which interpolates temperature at any axial position from the values of temperature at the collocation points and η is measurement noise.

With the use of Equation 26 to Equation 29 and on the substitution of Equation 23 to Equation 25 into the optimal estimation algorithm developed in the last section, the following optimal reactor estimator can be obtained. Between two success sampling points, $t_{k-1} < t < t_k$, the prediction equations become

$$\frac{d\hat{\alpha}}{dt} = \hat{f}_\alpha(\hat{\alpha}, \hat{\theta}, \hat{c}) \tag{30}$$

$$\frac{d\hat{\theta}}{dt} = \hat{f}_\theta(\hat{\alpha}, \hat{\theta}, \hat{c}) \tag{31}$$

$$0 = \hat{f}_c(\hat{\alpha}, \hat{\theta}, \hat{c}) \tag{32}$$

$$\frac{dP_{\alpha\alpha}}{dt} = \frac{\partial \hat{f}_\alpha}{\partial \hat{\alpha}} P_{\alpha\alpha} + \frac{\partial \hat{f}_\alpha}{\partial \hat{\theta}} P_{\alpha\theta}^T + \frac{\partial \hat{f}_\alpha}{\partial \hat{c}} P_{\alpha c}^T + P_{\alpha\alpha}\frac{\partial \hat{f}_\alpha^T}{\partial \hat{\alpha}} + P_{\alpha\theta}\frac{\partial \hat{f}_\alpha^T}{\partial \hat{\theta}} + P_{\alpha c}\frac{\partial \hat{f}_\alpha^T}{\partial \hat{c}} + R_{\alpha\alpha}^{-1} \tag{33}$$

$$\frac{dP_{\alpha\theta}}{dt} = \frac{\partial \hat{f}_\alpha}{\partial \hat{\alpha}} P_{\alpha\theta} + \frac{\partial \hat{f}_\alpha}{\partial \hat{\theta}} P_{\theta\theta} + \frac{\partial \hat{f}_\alpha}{\partial \hat{c}} P_{\theta c}^T + P_{\alpha\alpha}\frac{\partial \hat{f}_\theta^T}{\partial \hat{\alpha}} + P_{\alpha\theta}\frac{\partial \hat{f}_\theta^T}{\partial \hat{\theta}} + P_{\alpha c}\frac{\partial \hat{f}_\theta^T}{\partial \hat{c}} + R_{\alpha\theta}^{-1} \tag{34}$$

$$\frac{dP_{\theta\theta}}{dt} = \frac{\partial \hat{f}_\theta}{\partial \hat{\alpha}} P_{\alpha\theta} + \frac{\partial \hat{f}_\theta}{\partial \hat{\theta}} P_{\theta\theta} + \frac{\partial \hat{f}_\theta}{\partial \hat{c}} P_{\theta c}^T + P_{\alpha\theta}^T\frac{\partial \hat{f}_\theta^T}{\partial \hat{\alpha}} + P_{\theta\theta}\frac{\partial \hat{f}_\theta^T}{\partial \hat{\theta}} + P_{\theta c}\frac{\partial \hat{f}_\theta^T}{\partial \hat{c}} + R_{\theta\theta}^{-1} \tag{35}$$

$$0 = \frac{\partial \hat{f}_\alpha}{\partial \hat{c}} P_{cc} + P_{\alpha\alpha}\frac{\partial \hat{f}_c^T}{\partial \hat{\alpha}} + P_{\alpha\theta}\frac{\partial \hat{f}_c^T}{\partial \hat{\theta}} + P_{\alpha c}\frac{\partial \hat{f}_c^T}{\partial \hat{c}} + R_{\alpha c}^{-1} \tag{36}$$

$$0 = \frac{\partial \hat{f}_\theta}{\partial \hat{c}} P_{cc} + P^T_{\alpha c} \frac{\partial \hat{f}^T_c}{\partial \hat{\alpha}} + P_{\theta\theta} \frac{\partial \hat{f}^T_c}{\partial \hat{\theta}} + P_{\theta c} \frac{\partial \hat{f}^T_c}{\partial \hat{c}} + R^{-1}_{\theta c} \quad (37)$$

$$0 = \frac{\partial \hat{f}_c}{\partial \hat{c}} P_{cc} + P_{cc} \frac{\partial \hat{f}^T_c}{\partial \hat{c}} + R^{-1}_{cc} \quad (38)$$

with the initial conditions

$$\hat{\alpha}(0) = \hat{\alpha}_0 \quad (39)$$

$$\hat{\theta}(0) = \hat{\theta}_0 \quad (40)$$

$$P_{\alpha\alpha}(0) = P_{\alpha\alpha,0} \quad (41)$$

$$P_{\alpha\theta}(0) = P_{\alpha\theta,0} \quad (42)$$

$$P_{\theta\theta}(0) = P_{\theta\theta,0} \quad (43)$$

At the sampling point t_k, the updating equations are

$$\hat{\alpha}(t^+_k) = \hat{\alpha}(t^-_k) + P_{\alpha\theta}(t^+_k) \frac{\partial \hat{h}^T}{\partial \hat{\theta}} Q_k [y(t_k) - \hat{h}] - P_{\alpha\theta}(t^-_k) \frac{\partial \hat{h}^T}{\partial \hat{\theta}} Q_k \frac{\partial \hat{h}}{\partial \hat{\theta}} [\hat{\theta}(t^+_k) - \hat{\theta}(t^-_k)] \quad (44)$$

$$\hat{\theta}(t^+_k) = \hat{\theta}(t^-_k) + P_{\theta\theta}(t^+_k) \frac{\partial \hat{h}^T}{\partial \hat{\theta}} Q_k [y(t_k) - \hat{h}] - P_{\theta\theta}(t^-_k) \frac{\partial \hat{h}^T}{\partial \hat{\theta}} Q_k \frac{\partial \hat{h}}{\partial \hat{\theta}} [\hat{\theta}(t^+_k) - \hat{\theta}(t^-_k)] \quad (45)$$

$$P_{\alpha\alpha}(t^+_k) = P_{\alpha\alpha}(t^-_k) - P_{\alpha\theta}(t^+_k) \frac{\partial \hat{h}^T}{\partial \hat{\theta}} Q_k \frac{\partial \hat{h}}{\partial \hat{\theta}} P^T_{\alpha\theta}(t^-_k) - P_{\alpha\theta}(t^-_k) \frac{\partial \hat{h}^T}{\partial \hat{\theta}} Q_k \frac{\partial \hat{h}}{\partial \hat{\theta}} [P_{\alpha\theta}(t^+_k) - P_{\alpha\theta}(t^-_k)]^T \quad (46)$$

$$P_{\alpha\theta}(t^+_k) = P_{\alpha\theta}(t^-_k) - P_{\alpha\theta}(t^+_k) \frac{\partial \hat{h}^T}{\partial \hat{\theta}} Q_k \frac{\partial \hat{h}}{\partial \hat{\theta}} P_{\theta\theta}(t^-_k) - P_{\alpha\theta}(t^-_k) \frac{\partial \hat{h}^T}{\partial \hat{\theta}} Q_k \frac{\partial \hat{h}}{\partial \hat{\theta}} [P_{\theta\theta}(t^+_k) - P_{\theta\theta}(t^-_k)] \quad (47)$$

$$P_{\theta\theta}(t^+_k) = P_{\theta\theta}(t^-_k) - P_{\theta\theta}(t^+_k) \frac{\partial \hat{h}^T}{\partial \hat{\theta}} Q_k \frac{\partial \hat{h}}{\partial \hat{\theta}} P_{\theta\theta}(t^-_k) - P_{\theta\theta}(t^-_k) \frac{\partial \hat{h}^T}{\partial \hat{\theta}} Q_k \frac{\partial \hat{h}}{\partial \hat{\theta}} [P_{\theta\theta}(t^+_k) - P_{\theta\theta}(t^-_k)] \quad (48)$$

When solving the reactor estimation problem, three kinds of parameters must be selected in advance. These are the initial conditions estimation error weighting matrices $P_{\alpha\alpha,0}$, $P_{\alpha\theta,0}$ and $P_{\theta\theta,0}$; the measurement noise weighting matrix Q_k and the model error weighting matrix $R^{-1}_{\alpha\alpha}$, $R^{-1}_{\alpha\theta}$, $R^{-1}_{\alpha c}$, $R^{-1}_{\theta\theta}$, $R^{-1}_{\theta c}$ and R^{-1}_{cc}. The initial conditions estimation error weighting matrices reflect the magnitude of the errors in initial conditions, and thus should be large when little is known about the initial states and small when better initial conditions are available. Generally, these values are not critical. The value of Q_k should be small, when measurement noise is large, and large in correspondence with small measurement noise. In this study, it was assumed that all temperature measurements were independent, therefore, Q_k was a diagonal matrix with all elements having the value of 4.5×10^4. The selection of model error weighting matrix reflects the degree of confidence in the model. Usually, these value greatly influence the performance of the estimator, and thus they must be chosen with caution. In this paper, these values were selected as following.

$$R^{-1}_{\alpha\alpha}(i,j) = 3 \times 10^{-14} S(i,j) \quad (49)$$

$$R^{-1}_{\alpha\theta}(i,j) = 3 \times 10^{-12} S(i,j) \quad (50)$$

$$R^{-1}_{\alpha c}(i,j) = 1 \times 10^{-16} S(i,j) \quad (51)$$

$$R^{-1}_{\theta\theta}(i,j) = 8 \times 10^{-08} S(i,j) \quad (52)$$

$$R^{-1}_{\theta c}(i,j) = 1 \times 10^{-12} S(i,j) \quad (53)$$

$$R^{-1}_{cc}(i,j) = 1 \times 10^{-14} S(i,j) \quad (54)$$

where $S(i,j)$ was chosen to be symmetric and decreasing with distance between z_i and z_j.

$$S(i,j) = 1 - (z_i - zj)^2 \quad (55)$$

In carrying the implementation of the optimal reactor estimator, Equation 30 to Equation 38 are integrated from time t_0 with the initial conditions Equation 39 to Equation 43 to the first sampling instant t_1 to obtain the prediction values, then, the updating values are calculated from Equation 44 to Equation 48 with the use of the measurement $y(t_1)$, these two steps are repeated within any succeeding time interval (t_{k-1}, t_k).

4. PERFORMANCE OF THE REACTOR ESTIMATOR

The performance of the reactor estimator has been examined using dynamic experimental data from an pilot-scale reactor system (Kershenbaum and Lopez-Isunza 1986). Since at this stage of the work no signals are fed back to the reactor, the reactor estimator is tested through off-line implementation. The results are equivalent for on-line estimation, except on-line implementation is constrained by computing time and requires software to collect and process the data in real-time. Only temperature within the reactor are used as measurements. 26 temperatures at different axial location of the reactor are measured at 12 seconds intervals, but only three of those are utilised in the estimator. The remainder are used to represent the actual temperature profile in the reactor. The effectiveness of the estimator can be demonstrated by comparing the actual temperature profile to the estimated one generated by the estimator. Many dynamic experiments have been carried out to investigate the performance of the reactor estimator. Due to the space limitation only the experiment of ramp change in salt bath temperature θ_w is given below. This experiment involved beginning at steady-state operation and performing ramp change in θ_w. After operating of the reactor at steady-state for 30 minutes, ramp increase in θ_w from 383^0C to 403^0C at a rate of $10^0C/min$ was made. The reactor was operated at $\theta_w = 403^0C$ for 70 minutes, then θ_w was decreased from 403^0C to 383^0C at a rate of $0.22^0C/min$. The disturbance pattern for θ_w is plotted in Fig. 1. Three temperature measurements located at 0.4 m, 0.8 m and 1.0 m respectively from the entrance of the reactor were used by the estimator. The estimator assumed steady-state profiles as initial values. Due to the errors in the reactor model, the initial guess of the temperature profile was quite different from the experimental data. In addition to test the dynamic response of the estimator to the ramp change in θ_w, this experiment would also examine the efficiency of

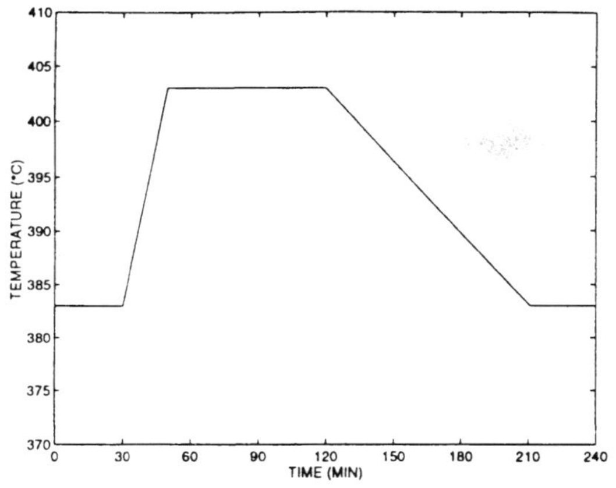

Fig. 1. Ramp change in salt bath temperature.

the estimator in the case of incorrect initial conditions.

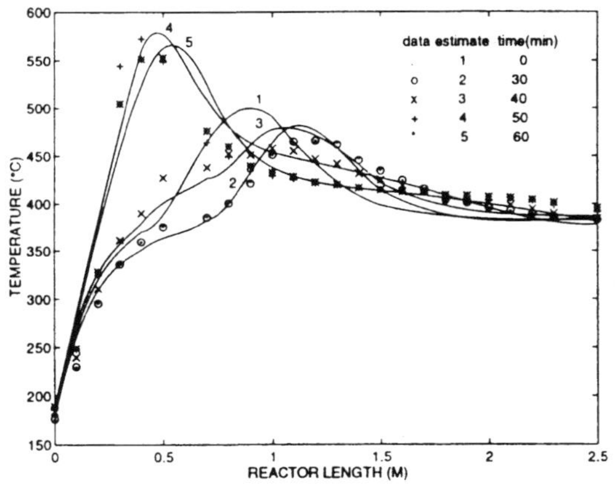

Fig. 2. Comparison of estimated temperature with experimental temperature during ramp increase of salt bath temperature.

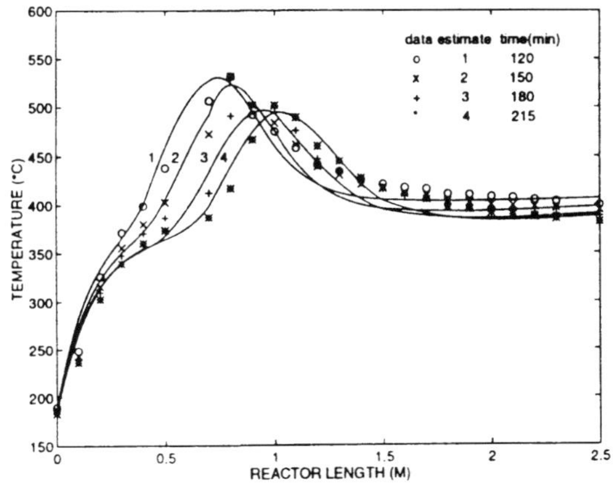

Fig. 3. Comparison of estimated temperature with experimental temperature during ramp decrease of salt bath temperature.

The estimated and experimental temperature profiles are shown in Fig. 2 for the period of the ramp increase in θ_w and in Fig. 3 for the period of the ramp decrease in θ_w. Although the initial temperature profile was wrong, the estimated temperature profile converged to the true data very quickly. The estimated steady-state temperature profile accurately predicted the presence of the shoulder in the region where the reactor temperature was almost equal to θ_w. During the ramp increase in θ_w, for the first several minutes, the hot spot gradually flattened and harmonised with the rest of the temperature. Then, the hot spot increased very fast, developed a peak again and moved upstream. All of these dynamic features were excellently traced by the estimated temperature profiles shown in Fig. 2. For the ramp decrease in θ_w, the hot spot moved very smoothly compared with that in the ramp increase in θ_w. The hot spot decreased and moved downstream. At the end of the experiment, the magnitude and location of the hot spot were different from those before the ramp change in θ_w. The estimator did a nice job of determining shifting temperature profiles.

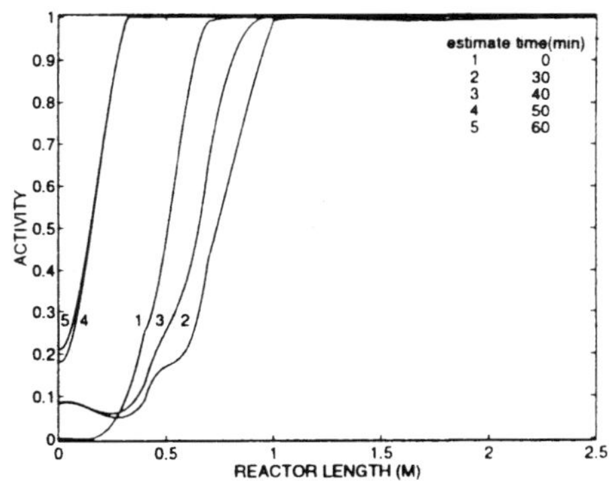

Fig. 4. Estimation of catalyst activity profile during ramp increase of salt bath temperature.

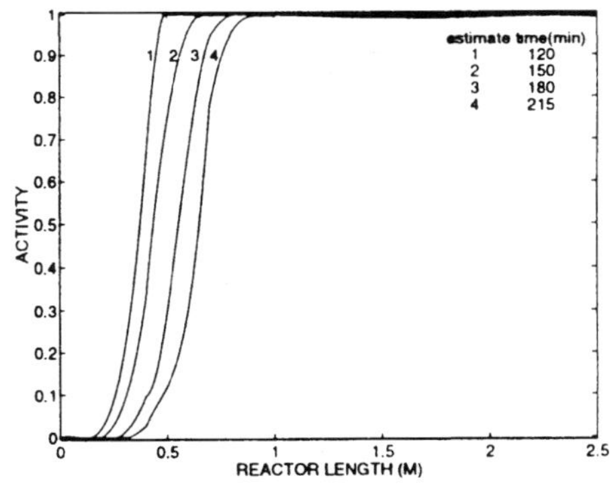

Fig. 5. Estimation of catalyst activity profile during ramp decrease of salt bath temperature.

The estimation of activity profiles during the ramp change in θ_w is given in Fig. 4 and Fig. 5. The difference between the initial guess of the activity profile and the steady-state activity profile before the ramp increase in θ_w is due to the initial guess of the temperature profile being different from the true steady-state temperature profile. During the ramp increase in θ_w, the activity of the catalyst in the front of the hot spot was increased due to the reactivation of the catalyst and the activity profile moved upstream. This caused the upstream movement of the hot spot. For the ramp decrease in θ_w, the activity profile moved downstream very slowly compared with that in the ramp increase of θ_w. At the end of the experiment, a new steady-state activity profile, which is different from that before the ramp change in θ_w, was achieved. The behaviour could be attributed to the irreversible and reversible deactivation and resulted in the different steady-state temperature profiles.

The estimated concentration profiles are shown in Fig. 6 and Fig. 7. The concentration profile moved upstream or downstream with the ramp increase or decrease in θ_w. Two different steady-state concentration profiles were obtained before and after the ramp change in θ_w. These phenomena were consistence with the variation in temperature and activity profiles.

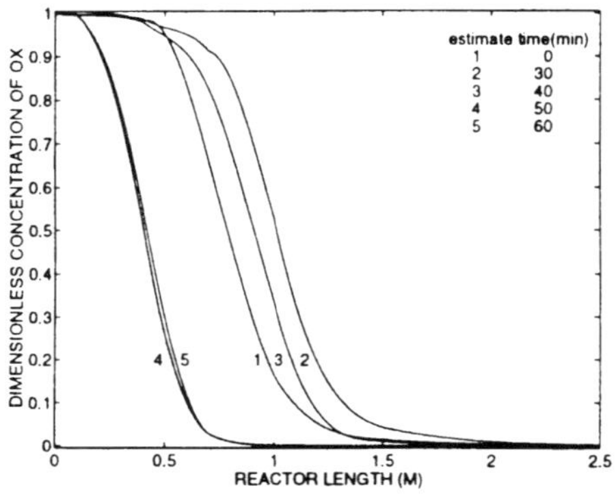

Fig. 6. Estimation of concentration profile during ramp increase of salt bath temperature.

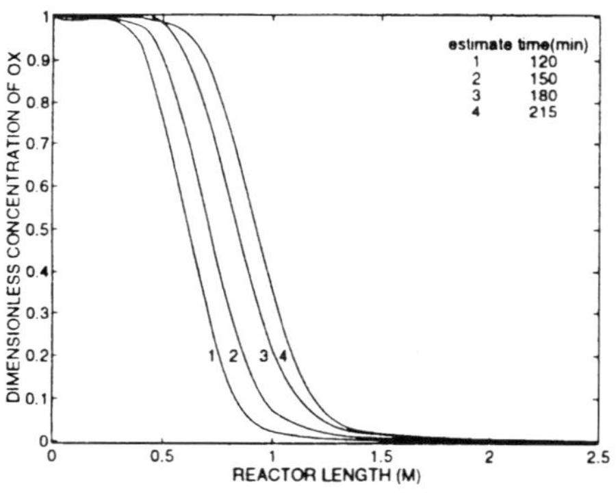

Fig. 7. Estimation of concentration profile during ramp decrease of salt bath temperature.

5. CONCLUSIONS

An optimal estimation algorithm has been derived for nonlinear differential-algebraic equation systems. By choosing the minimum least squares criterion, the optimal estimation problem is converted to an optimal control problem without any assumption of statistical nature of system modelling and measurement error. The optimal sequential estimator is obtained with the use of the calculus of variations.

The optimal estimator has been applied to an industrial scale fixed-bed reactor with decaying catalysts where phthalic anhydride is produced by the partial oxidation of o-xylene. Considering the computation limit for the on-line application of the reactor estimator, a lumped reaction network and a simple catalyst deactivation scheme were used. The orthogonal collocation method is utilised to discretise the distributed parameter reactor model into a differential-algebraic equation system which the optimal estimation algorithm could be applied to.

Using three properly located temperature measurements, excellent estimation has been obtained. The estimator rapidly converged when given a poor initial guess. The dynamic behaviour of the reactor has been successfully tracked. The catalyst activity profile was accurate inferred.

The optimal reactor estimator can been easily applied on-line in analogous way. The computational feasibility of applying the estimator in real-time is demonstrated by the fact that the required computing time is one-seventh of real operation time on our SUN SPARC 10 workstation.

REFERENCES

Cheng, Y. S. and L. S. Kershenbaum (1994). 'Optimal filtering for nonlinear differential and algebraic systems—i. theoretical development'. *Submitted to Comput. Chem. Engng.*

Detchmendy, D. M. and R. Sridhar (1966). 'Sequential estimation of state and parameters in noisy nonlinear dynamic systems'. *J. Basic Engng* **88**, 362–368.

Jonckheere, E. (1988). 'Variational calculus for descriptor problems'. *IEEE Trans. Autom. Control* **33**, 491–495.

Kershenbaum, L. S. and F. Lopez-Isunza (1986). 'Measurement and estimation in catalytic reactors for partial oxidation'. *Trans. Inst. Msmt. Control* **8**, 137–143.

Lopez-Isunza, F. and L. S. Kershenbaum (1992). 'The role of reversible changes in catalyst activity in the observed multiple steady states during partial oxidation dynamics'. *Chem. Engng Sci* **47**, 2817–1364.

Mongkhonsi, T. (1993). Dynamic Behaviour of a Fixed-Bed Catalytic Reactor with Catalyst Deactivation. PhD thesis. University of London.

Pantelides, C. C. and P. I. Barton (1993). 'Equation-oriented dynamic simulation current status and future perspectives'. *Comput. Chem. Engng* **17**, 263–285.

USE OF ITERATIVE DYNAMIC PROGRAMMING IN OPTIMIZATION OF REACTOR SYSTEMS

Rein Luus and Vincent Dong

Department of Chemical Engineering, University of Toronto, Toronto, Ontario, Canada M5S 1A4

Abstract: A batch reactor involving five chemical reactions was used to examine the viability of using iterative dynamic programming (IDP) for highly nonlinear systems which are sometimes encountered by chemical engineers. The effects of the number of grid points, of the number of allowable values for control, and of the region contraction factor on the convergence of IDP were examined. The global optimum could be obtained with reasonable computational effort, in spite of the numerous local optima. The use of stages of different lengths allowed a refined solution to be obtained with a reasonably small number of stages.

Keywords: Dynamic programming, Optimal control, Nonlinear systems

1. INTRODUCTION

Computational difficulties in the optimization of batch reactors arise mainly because of the singular nature of the optimal control policy. Even some two dimensional systems with scalar control can be very difficult to optimize by methods based on Pontryagin's maximum principle, as was shown by Reddy and Husain (1981). An alternative approach is to use iterative dynamic programming (IDP), as developed by Luus (1989). Even with IDP difficulties sometimes arise when there are numerous local optima as encountered in the optimal control of fed-batch reactors (Luus, 1993a, 1993b), and of batch reactors (Luus, 1994).

The purpose of this paper is to examine the application of IDP for solving a very difficult batch reactor optimization problem based on the reaction scheme of Denbigh (1958).

2. OPTIMAL CONTROL PROBLEM

The optimal control of a batch reactor to be considered is based on the reaction scheme of Denbigh (1958) and used for optimal control studies also by Aris (1960), Rangaiah (1985), and Luus (1994). Here, however, an additional reaction is added so that the optimal control policy will have two singular arcs: one leading up to the allowed maximum temperature and the other leading away from the maximum temperature. Five chemical reactions are taking place in the reactor:

$$\begin{aligned} A + B &\rightarrow X \\ A + B &\rightarrow P \\ X &\rightarrow Y \\ X &\rightarrow Q \\ Y &\rightarrow Z \end{aligned} \qquad (1)$$

where X is an intermediate, Y is the desired product, and P, Q and Z are waste products. It is desired to maximize the yield of Y. Although there are seven chemical species, we are concerned only with the concentrations of A, B, and Y , and therefore choose as the state variables their concentrations expressed as mole fractions. The equations describing the system are

$$\frac{dx_1}{dt} = - k_1 x_1 - k_2 x_1 \qquad (2)$$

$$\frac{dx_2}{dt} = k_1 x_1 - (k_3 + k_4) x_2 \quad (3)$$

$$\frac{dx_3}{dt} = k_3 x_2 - k_5 x_3 \quad (4)$$

with the initial condition

$$x^T(0) = [1 \quad 0 \quad 0] \quad (5)$$

The rate constants are given by

$$k_i = k_{i0} \exp\left\{\frac{-E_i}{RT}\right\} \quad , \quad i = 1, 2, \ldots, 5 \quad (6)$$

where k_{i0} and E_i /R are given in Table 1.

Table1. Kinetic parameters for the reactor

i	k_{i0}, s^{-1}	E_i /R
1	10^3	3000
2	10^7	6000
3	10	3000
4	10^{-4}	0
5	1	3100

The constraints on the temperature are

$$273 \leq T \leq 415 \quad (7)$$

The performance index to be maximized is

$$I = x_3(t_f) \quad (8)$$

where the batch time t_f is specified as 500 s. The optimal control problem is therefore to find T in the time interval $0 \leq t < 500$, so that the performance index given in Eq.(8) is maximized.

3. IDP ALGORITHM

The procedure as given by Luus(1989) consists of seven steps:

1) Divide the time interval $[0,t_f]$ into P stages, each of length L,over which the control is kept constant.

2) Choose the number of grid points N at each time stage and the number of allowable values M for the control u. Choose the initial region size $r^{(0)}$ and an initial control policy $u^{(0)}$ for each time stage.

3) By choosing N values of control evenly distributed inside the allowable region r, integrate Eq. (2)-(4) from t = 0 to t = t_f with the given initial condition, to generate the **x**-grid for each time stage. Thus we have N grid points at each time stage, except the first one. For the first time stage there is a single grid point consisting of the given initial condition.

4) Starting at the beginning of the last time stage P, corresponding to time t_f - L, for each **x**-grid point integrate Eq. (2)-(4) from t_f - L to t_f once with each of the M allowable values of control. Choose the control that gives the maximum value of the performance index and store the corresponding value of control for use in step 5) .

5) Proceed to stage P - 1, corresponding to time t_f - 2L. For each **x**-grid point integrate Eq. (2)-(4) once with each of the M allowable values of control from t_f - 2L to t_f - L. To continue integration from t_f - L to t_f choose the control from step 4) that corresponds to the grid point that is closest to the resulting **x** at t_f - L. Compare the M values of the performance index and store the value of control that gives the maximum value.

6) Repeat this procedure for stages P - 2, P - 3, etc., until stage 1, corresponding to the initial time t = 0, is reached. Store the control policy that gives the maximum value for the performance index.

7) Reduce the region r by a factor γ ; i.e.,

$$r^{(j+1)} = \gamma\, r^{(j)} \quad (9)$$

where j is the iteration index. Use the best control policy from step 6) as the midpoint for the allowable values for control and go to step 3). Repeat the procedure for a specified number of iterations, such as 20, and analyze the result.

Although in the above algorithm, all the stages are taken to be of the same length L for simplicity, this does not prevent one from taking stages of different length. This aspect is illustrated in the example.

To obtain adequate convergence for small values of N and M, it may be necessary to repeat the iterations. Such a multipass method is especially useful when dealing with a very high dimensional problem (Luus, 1993c). From computational point of view we would like to choose the smallest values for N and M that will yield convergence to the global optimum. As can be seen with this numerical example, the choice is not always simple.

4. NUMERICAL RESULTS

All computations were performed in double precision on a 486/66 personal computer, using the IBM's OS/2 version 2.1 operating system and the WATCOM FORTRAN 77 version 9.5 compiler. For integration the subroutine DVERK was used.

As the initial control policy, the value of T is chosen to be half-way between the bounds, as specified by Eq.(7), namely $T^{(0)} = 344$, and, as the initial region size, half of the upper boundary minus the lower boundary on T is taken; i.e., $r^{(0)} = 71$.

4.1 Initial runs with constant stage lengths with P = 40

Some preliminary runs showed that at least 9 grid points were required to obtain convergence to the global optimum when 40 time stages of equal length were used. With 9 grid points the convergence was rapid and consistent, as is seen in Fig.1, where a reduction factor $\gamma = 0.80$ was used. The use of 15 allowable values for control instead of 7, yields slighlty faster convergence from the initial value of I = 0.16582 to 0.54820 in 12 iterations instead of 15 iterations required with M = 7. The computation time on the 486/66 personal computer for 20 iterations was 2848 s with M = 7, and 6091 s with M = 15.

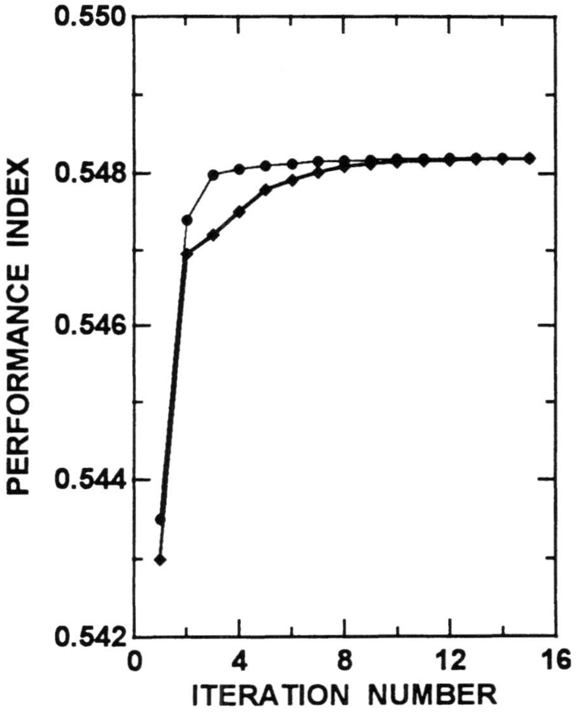

Fig. 1. Convergence profile to I = 0.54820 with N = 9, $\gamma = 0.80$, P = 40. ♦ , M = 7; ●, M = 15

The effect of the reduction factor was not found to be significant with respect to the reliability of getting the global optimum as is shown in Table 2. With the exception of the $\gamma = 0.75$ case, the global optimum was obtained when seven or greater number of allowable values for control were used. Therefore, the use of a very large number of allowable values for control does not ensure getting the global optimum. The reduction factor of 0.80 leads to rapid convergence without premature collapse of the search region.

Table 2. Performance index obtained using $u^{(0)} = 344$, $r^{(0)} = 71$, P = 40, N = 9.

	Reduction factor		
M	$\gamma = 0.75$	$\gamma = 0.80$	$\gamma = 0.85$
3	0.54219	0.54219	0.54521
5	0.54526	0.54526	0.54521
7	0.54820	0.54820	0.54820
9	0.54820	0.54820	0.54820
11	0.54820	0.54820	0.54820
13	0.54820	0.54820	0.54820
15	0.54735	0.54820	0.54820

The optimal control policy with the use of 40 stages each of length 12.5 s is given in Fig. 2. It is noted that the temperature is constant for a long time at the upper bound before being reduced in a stepwise fashion. Therefore, the performance index can be improved by using stages of different length, without the necessity of requiring too many stages.

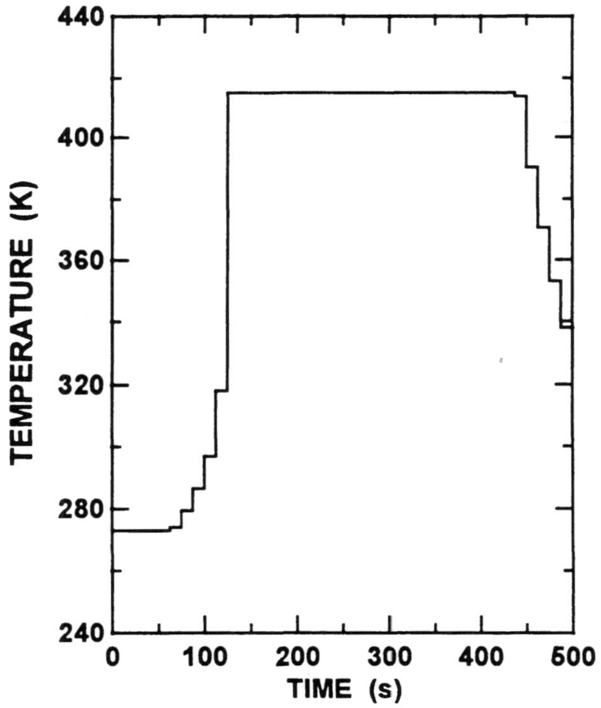

Fig. 2. Optimal control policy with the use of 40 stages, each of length 12.5 s

4.2 Use of stages of different lengths

With reference to Fig. 2, to refine the optimal control policy, the first stage was taken to be 64 s long and the

Table 3. Values of the performance index using 35 stages of uneven lengths as a function of N and M.

Number of allowable values for control, M	Number of grid points							
	N = 7	N = 9	N = 11	N = 13	N = 15	N = 17	N = 21	N = 31
3	0.54829	0.54872	0.54829	0.54872	0.54902	0.54872	0.54830	0.54872
5	0.54830	0.54896	0.54830	0.54830	0.54830	0.54830	0.54973	**0.54917**
7	**0.54917**	0.54830	0.54830	0.54830	0.54830	**0.54917**	0.54902	**0.54917**
9	0.54830	0.54856	0.54830	0.54830	0.54873	**0.54917**	0.54873	**0.54917**
11	0.54790	0.54896	**0.54917**	0.54896	**0.54917**	0.54830	0.54873	**0.54917**
13	**0.54917**	**0.54915**	0.54856	**0.54917**	0.54873	**0.54917**	0.54873	**0.54917**
15	0.54830	0.54856	**0.54917**	**0.54917**	0.54896	**0.54917**	0.54873	**0.54917**
17	0.54830	0.54856	0.54856	**0.54917**	0.54896	**0.54917**	0.54873	**0.54917**
19	**0.54917**	0.54856	0.54856	**0.54917**	0.54896	0.54896	0.54873	**0.54917**
21	0.54830	0.54856	0.54856	**0.54917**	0.54873	0.54896	0.54902	**0.54917**

18th stage was taken to be 304 s long. The remaining time interval was divided into 33 stages, each 4 s long. Numerous local optima were found for this case, as is shown in Table 3. Of the 80 runs, the global optimum of I = 0.54917 was obtained 26 times. Taking a large number of grid points (N = 21) did not guarantee getting the global optimum. However, taking only 7 grid points yielded the global optimum three times. Using stages of different size improved the performance index by 0.2% over the previous case.

The optimal control policy is given in Fig. 3. Although only 35 stages were used, the piecewise constant control policy appears almost continuous, because of the unequal lengths.

To obtain further refinement, a larger number of stages was taken in the sections where the control policy was changing. The first stage was taken to be 64 s long, and the 33rd stage was taken to be 314 s long. The remaining 61 stages were each taken to be 2 s long.

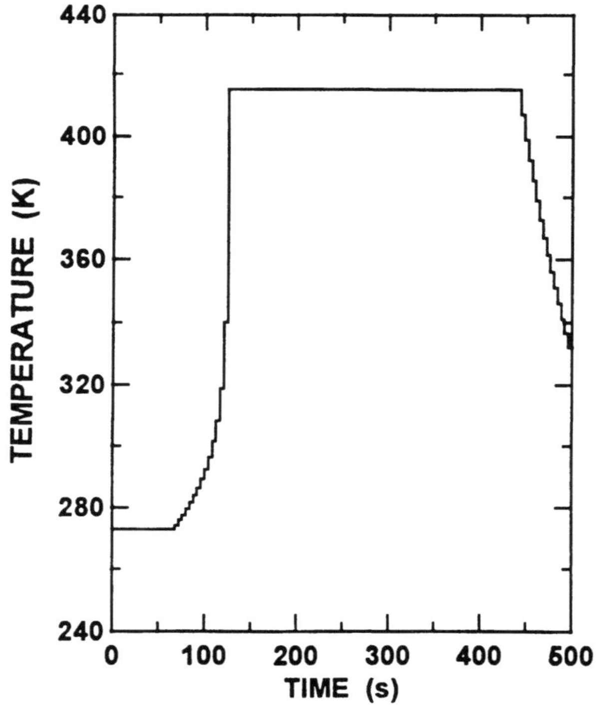

Fig 3. Optimal control policy with the use of 35 stages of different lengths.

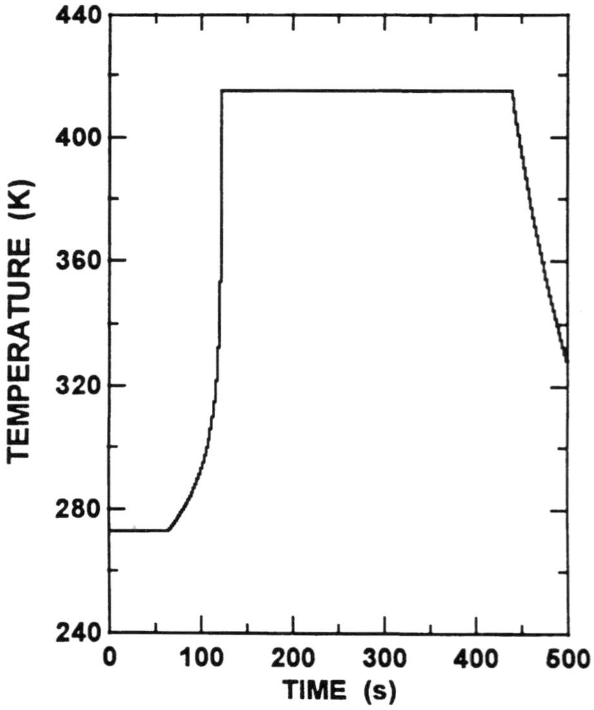

Fig. 4. Optimal control policy with the use of 63 stages of different lengths; I = 0.54937.

The control policy in Fig. 3 was used as the initial control policy and the initial region size of 5.0 was used. The use of 63 stages led to an optimal yield of I = 0.54937, which is a very slight improvement over the previous case. This refined optimal control policy is given in Fig.4.

5. CONCLUSIONS

Iterative dynamic programming has been successfully applied to establish optimal control of a difficult batch reactor optimization problem. To obtain a refined answer without taking an unreasonably large number of stages, stages of different lengths were used. This allowed accurate determination of the optimal control policy.

Computations show that IDP is a useful means of establishing the optimal control policy for very difficult problems.The greatest advantage in its use is the ease of implementation, since no derivatives are required and no auxiliary variables have to be introduced. For systems that exhibit local optima, IDP is especially attractive because of its high reliability in getting the global optimum.

Although some preliminary runs may have to be made to determine the nature of the problem, once the insight is gained, the results are easily obtained. The computations are easily done on personal computers, which makes IDP an attractive method to use.

6. ACKNOWLEDGEMENTS

Financial support from the Natural Sciences and Engineering Research Council of Canada is gratefully acknowledged.

REFERENCES

Aris, R. (1960). On Denbigh's Optimum Temperature Sequence. *Chem. Eng. Sci.* **12**, 56-64.

Denbigh, K.G. (1958). Optimum Temperature Sequence in Reactors. *Chem. Eng. Sci.* **8**, 125-132.

Luus, R. (1989). Optimal Control by Dynamic Programming using Accessible Grid Points and Region Reduction. *Hungarian J. Ind. Chem.* **17**, 523-543.

Luus, R. (1993a).Application of Dyamic Programming to Differential-Algebraic Process Systems. *Computers Chem. Engng.* **17**, 373-377.

Luus, R. (1993b). Optimization of Fed-Batch Fermentors by Iterative Dynamic Programming. *Biotechnol. Bioeng.* **47**, 599-602.

Luus, R. (1993c). Application of Iterative Dynamic Programming to very High-Dimensional Systems. *Hungarian J. Ind. Chem.* **21**, 243-250.

Luus, R. (1994). Optimal Control of Batch Reactors by Iterative Dynamic Programming. *J. Proc. Cont.* **4**, 218-226.

Rangaiah, G. P. (1985). Studies in Constrained Optimization of Chemical Process Problems. *Comput. Chem. Eng.* **9**, 395-404.

Reddy, K.V., and Husain, A. (1981). Computation of Optimal Control with Singular Subarc. *Can. J. Chem. Engng.* **59**, 557-559.

HARD CONSTRAINTS IN CONTROL AND STATE VARIABLES OF MULTIVARIABLE NONLINEAR PROCESSES RESOLVED BY ELEMENTARY NONLINEAR DECOUPLING

Jens G. Balchen* and Bjarne Sandrib*

*The Norwegian Institute of Technology, Department of Engineering Cybernetics,
7034 Trondheim, Norway

Abstract. A model based control strategy for nonlinear multivariable processes having hard constraints on both control and state variables, is derived based upon the concept of Elementary Nonlinear Decoupling (END). In END a *property* space is defined which has the same dimension as the control space and which permits *inverse* control. An optimization procedure is used to design a property transformation that gives the best solution in terms of performance, stability, robustness etc. This control strategy is shown to be convenient when control and state variables are subject to hard constraints. When constraint limits are reached, new property variables are introduced which approximate the variables to be constrained and thereby achieving fast and accurate control on constraint boundaries.

Keywords. Multivariable control, nonlinear control, nonlinear decoupling, hard constraints.

1. INTRODUCTION

When designing a control strategy for a nonlinear multivariable system it is essential that the problem of possible hard constraints in control and state variables can be handled. In most procedures this problem is solved by employing an optimization technique where the consequences of hard constraints are weighed against measures of performance when the optimal control actions are derived (Polak 1971, Mayne and Polak 1993).

In this contribution another approach is taken.

The unconstrained END algorithm (Balchen 1991, Balchen and Sandrib 1994a) is applied as long as non of the control and state variables ($\boldsymbol{u}$ and $\boldsymbol{x}$ respectively) have reached their constraints. When, say, one of the state variables (x_i) reaches a constraint (high or low) a new END strategy is introduced which replaces the state variables (x_i) with appropriate property variables ($\tilde{z}_j$). Thereby an END control scheme is developed which keeps the control and state variables within their boundaries until the system again returns to the unconstrained operating conditions. Obviously if any of the control and state variables stay permanently at their constraints, the system is not properly designed and must be reevaluated from the design phase.

2. SHORT REVIEW OF THE END ALGORITHM

Consider a general nonlinear multivariable process described by

$$\dot{\boldsymbol{x}} = \boldsymbol{f}(\boldsymbol{x}, \boldsymbol{u}, \boldsymbol{v}) \tag{1}$$

where $\boldsymbol{x} \in R^n$: state vector, $\boldsymbol{u} \in R^r$: control vector, $\boldsymbol{v} \in R^s$: disturbance vector, $\boldsymbol{f} \in R^n$: rate function vector

In addition to the model (1), a model must be defined describing the measurements that are extracted from the process forming the basis for estimating the process states

$$\boldsymbol{y} = \boldsymbol{h}(\boldsymbol{x}) + \boldsymbol{w} \tag{2}$$

where $\boldsymbol{y} \in R^m$: measurement vector, $\boldsymbol{h} \in R^m$: measurement functions, $\boldsymbol{w} \in R^m$: measurement uncertainty (noise).

For the system modeled above a property vector ($\boldsymbol{z}$) is defined which describes the quantities (properties) one wants to control to certain values (setpoints). The number of such properties is related to the number of degrees of freedom of control namely r. A mathematical model of the properties is

$$\boldsymbol{z} = \boldsymbol{d}(\boldsymbol{x}) \tag{3}$$

where $\boldsymbol{z} \in R^m$: property vector, $\boldsymbol{d} \in R^m$: property function vector.

Often the property transformation is linear so that

$$\boldsymbol{z} = D\boldsymbol{x} \tag{4}$$

Nonlinear decoupling algorithms have the purpose of finding the control vector ($\boldsymbol{u}$) which drives the system in such a way that $\dot{\boldsymbol{z}} = \dot{\boldsymbol{z}}_d$ is a specified property rate of change trajectory. One solution to this problem is known as input-output linearization employing a differential geometry method to make the system invertible (Isidori 1989). In the END algorithm the invertibility problem is resolved by designing the $\boldsymbol{d}(\cdot)$ or D to achieve ivertibility, proper dynamic behavior of the inner system and robustness.

In the following (4) will be assumed describing the property transformation. Then from (1) we have

$$\dot{\boldsymbol{z}} = \dot{\boldsymbol{z}}_d = D\boldsymbol{f}(\boldsymbol{x}, \boldsymbol{u}, \boldsymbol{v}) \tag{5}$$

which can be solved by an iterative equation solver of the form

$$\dot{u} = K_u(\cdot)\,(\dot{z}_d - Df(\cdot)) \qquad (6)$$

The matrix $K_u(\cdot)$ of (6) is there to assure convergence of (6) and can be found in many ways. One way is to specify a linearized version of (6) to have constant eigenvalues (Λ) such that

$$\delta\dot{u} = -K_u(\cdot)\,D\frac{\partial f(\cdot)}{\partial u}\delta u = \Lambda\delta u \qquad (7)$$

leading to

$$K_u(\cdot) = -\Lambda\left(D\frac{\partial f(\cdot)}{\partial u}\right)^{-1} \qquad (8)$$

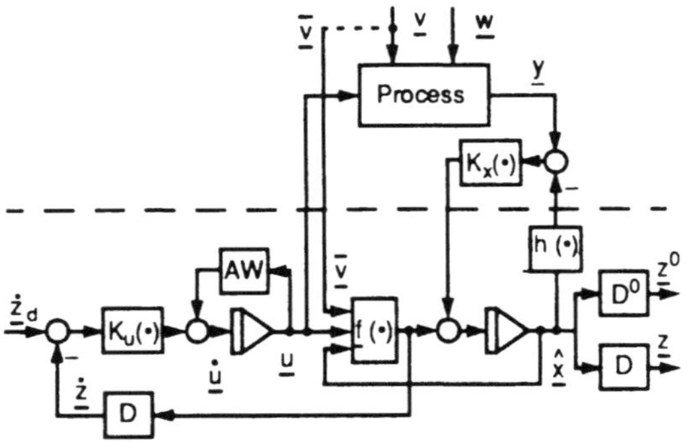

Fig. 1. Blockdiagram of END algorithm.

A block diagram illustrating the END algorithm is shown in Fig. 1. It includes a nonlinear estimator which has an updating matrix denoted $K_x(\cdot)$ of unspecified form since a number of algorithms may be used. It is seen that an estimate of $\dot{z}$ is derived from the process model in the estimator. The feedback loop containing $K_u(\cdot)$ generates the proper control action (u) which satisfies $\dot{z} = \dot{z}_d$. It is also noted that in Fig. 1 the integrator generating u from $\dot{u}$ is equipped with an *antiwindup* facility (*AW*) to prevent this integrator from overcharging when the control variables go into saturation. Furthermore it is noted that the vector v which represents the measured disturbances, gives a direct feedforward into the model. If all dominant disturbances are measured and the model is nearly perfect, the updating signal from the estimator gain matrix $(K_x(\cdot))$ will be small and white. The fewer disturbances that are measured, the more power will be contained in this updating signal.

Assuming for the sake of simplicity that (1) is replaced by its linearized version

$$\delta\dot{x} = A\delta x + B\delta u + C\delta v \qquad (9)$$

and assuming that the eigenvalues Λ of the equation solver are very large, it can be shown that the resulting system after the application of END will be described by

$$\dot{x} = (I - B(DB)^{-1}D)(Ax + Cv) + B(DB)^{-1}\dot{z}_d \qquad (10)$$

It is obvious from both (8) and (10) that for a solution to exist the matrix $D\frac{\partial f(\cdot)}{\partial u} = DB$ must be nonsingular. The least requirement for this to be true is that $dim\ z = dim\ u$.

Furthemore it must be required that the dynamic system described by the state equation of (10) has acceptable behavior i.e. eigenvalues properly located in the left half of the complex plane.

The dynamics of this system is determined by the eigenvalues of the matrix

$$(I - B(DB)^{-1}D)A \qquad (11)$$

From (10) the main system transfer matrix is given by

$$x(s) = H(s)\,\dot{z}_d(s) \qquad (12)$$

where

$$H(s) = (sI - (I - B(DB)^{-1}D)A)^{-1}B(DB)^{-1} \qquad (13)$$

The transfer matrix of (13) can be characterized by the eigenvalues of (11) or when converted into the frequency response matrix $H(j\omega)$ by the maximal and minimal singular values $\bar{\sigma}_H(j\omega)$ and $\underline{\sigma}_H(j\omega)$. Thereby different methods for designing the dynamic behavior can be derived by choosing the elements of the D matrix so that the system becomes invertible with acceptable dynamic properties.

When choosing the properties to be controlled without concern for invertibility one could arrive at the transformation $z = D^0x$ expressing the *desirable* but *nonrealizable* property vector. The reason for the nonrealizability is that the matrix D^0B most often will be singular or the resulting dynamic properties of (10) are not acceptable.

An alternative way of designing D is derived by defining the difference

$$\Delta z = z - z^0 = (D - D^0)x = \Delta Dx \qquad (14)$$

and from that an objective functional

$$J = \int_0^\infty (\Delta z^T Q \Delta z + u^T P u)dt = \int_0^\infty (\underline{x}^T \Delta D^T Q \Delta D \underline{x} + \underline{u}^T P \underline{u})dt \qquad (15)$$

where the matrices Q and P are chosen to give proper weight on the different quantities. An optimization procedure based upon gradient calculations is developed to find $min_{\Delta D}J$ where ΔD is represented by its elements $\Delta D = \{\Delta d_{ij}\}$ converted into a vector. By minimizing (15) the dynamic behavior of (10) is included.

The consequence of the END algorithm of Fig. 1 is that the nonlinear multivariable system is replaced by a set of r integrators from $\dot{z}_d$ to z. This very simple process can be controlled by diagonal feedback as shown in Fig. 2. The inner loop (I) contains the control matrix G and has a reference vector z_0. When there is no saturation in any of the control variables, this loop will have nearly perfect response with very high bandwidth. The outer feedback loop (II) contains a control matrix G^0, a reference vector z_0^0 and an output z^0. This outer loop will be slightly degraded relative to loop (I). In some cases the outer loop (II) can be omitted yielding a higher bandwidth performance.

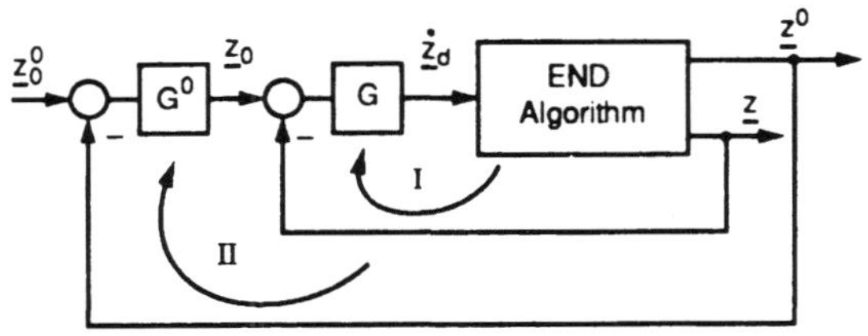

Fig. 2. The feedback control around the END algorithm

3. THE END ALGORITHM WITH CONSTRAINED STATE VARIABLES

The problem arising when control variables ($\boldsymbol{u}$) are constrained (saturated) has been dealt with in Balchen and Sandrib (1994b). In short the solution is that for each combination of saturating control variables $\boldsymbol{u}^{ij\cdots}$, where the superscripts *i* and *j* indicate which control variables are saturating, specific property transformations $D^{ij\cdots}$ are defined expressing the priorities to be employed when the process is to be controlled with fewer degrees of freedom.

Assuming for the sake of simplicity that the control saturation is taken care of by this method, a similar principle is employed to handle *state variable* constraints.

In addition to (1) and (2) the state constrained case has the extra condition

$$\boldsymbol{c}(\boldsymbol{x}) \leq \boldsymbol{0} \tag{16}$$

The simplest realization of (16) is

$$x_{i\ min} \leq x_i \leq x_{i\ max} \tag{17}$$

where x_i is one of the components of $\boldsymbol{x}$.

The proposed strategy is to replace the constraints in $\boldsymbol{x}$ by representative constraints in z and labeling these $\tilde{z}$. In doing this it is observed that the number of state constraints that can be handled at any time is equal to or less than the number of unconstrained (unsaturated) control variables (degrees of freedom). In other words, if there are only two unconstrained control variables available, only two state variables can be controlled to stay on the constraints (maximal or minimal). Thus all the other state variables have to stay inside acceptable operating ranges. The way to achieve this is to design the *D* matrix with this in mind.

The property transformation to be applied when state variables i and j have reached their constraint values is thus

$$\tilde{z} = \tilde{D}^{ij}\boldsymbol{x} \tag{18}$$

(18) indicates that two of the property variables have to be allotted to the constrained states x_i and x_j.

Thus, as an example, if the number of control variables is $r = 3$ and the number of states is $n = 8$ one could have a case with

$$D^0 = \begin{bmatrix} 1 & 0 & 0 & 0 & 0 & 0 & 0 & 0 \\ 0 & 1 & 0 & 0 & 0 & 0 & 0 & 0 \\ 0 & 0 & 1 & 0 & 0 & 0 & 0 & 0 \end{bmatrix} \tag{19}$$

indicating that x_1, x_2 and x_3 are to be controlled. The designed property transformation for this case will be of the form

$$D = \begin{bmatrix} 1 & d_{12} & d_{13} & d_{14} & d_{15} & d_{16} & d_{17} & d_{18} \\ d_{21} & 1 & d_{23} & d_{24} & d_{25} & d_{26} & d_{27} & d_{28} \\ d_{31} & d_{32} & 1 & d_{34} & d_{35} & d_{36} & d_{37} & d_{38} \end{bmatrix} \tag{20}$$

Assuming that states x_4 and x_6 are reaching their constraint values simultaneously, the appropriate $\tilde{D}$ matrix could be

$$\tilde{D}^{46} = \begin{bmatrix} 1 & \tilde{d}_{12} & \tilde{d}_{13} & \tilde{d}_{14} & \tilde{d}_{15} & \tilde{d}_{16} & \tilde{d}_{17} & \tilde{d}_{18} \\ \tilde{d}_{21} & \tilde{d}_{22} & \tilde{d}_{23} & 1 & \tilde{d}_{25} & \tilde{d}_{26} & \tilde{d}_{27} & \tilde{d}_{28} \\ \tilde{d}_{31} & \tilde{d}_{32} & \tilde{d}_{33} & \tilde{d}_{34} & \tilde{d}_{35} & 1 & \tilde{d}_{37} & \tilde{d}_{38} \end{bmatrix} \tag{21}$$

This $\tilde{D}^{46}$ matrix indicates that the original emphasis on x_1 has been kept while x_2 and x_3 have been deemphasized in order to be able to handle the constraints in x_4 and x_6. The elements $\tilde{d}_{ij}$ introduced in $\tilde{D}^{46}$ are mostly very small. Some of them will probably be so small that they can be neglected and made equal to zero. When non of the state variables reach their constraint values, the property transformation will return to that of (20) which has been optimized for the unconstrained case. The design of the matrix $\tilde{D}^{ij\cdots}$ is done offline.

Further discussion of the algorithm is conveniently done in relation to a concrete example.

4. END CONTROL OF A STIRRED TANK REACTOR WITH TWO CONSECUTIVE REACTIONS

Consider a chemical reactor with two consecutive and parallel reactions

$$A \xrightarrow{k_1} B \xrightarrow{k_2} C, \qquad B \xrightarrow{k_3} D$$

where k_i are the reaction rate constants for the three reactions which have reaction rates

$$r_B = k_1 c_A,\ r_C = k_2 c_B \text{ and } r_D = k_3 c_B$$

The reaction rate constants are temperature dependent

$$k_i = k_{i0}\exp\left(-\frac{E_i}{R\theta}\right) \tag{22}$$

where k_{i0}, E_i and R are constants and θ : reactor temperature.

Four differential equations describe the mass balances in terms of the concentrations $(c_A \ldots c_D)$ of each component:

$$\dot{c}_A = (-c_A + c_{A0})\frac{q}{V} - r_B \tag{23}$$

$$\dot{c}_B = (-c_B + c_{B0})\frac{q}{V} + r_B - r_C - r_D \qquad (24)$$

$$\dot{c}_C = (-c_C + c_{C0})\frac{q}{V} + r_C \qquad (25)$$

$$\dot{c}_D = (-c_D + c_{D0})\frac{q}{V} + r_D \qquad (26)$$

The differential equation describing the energy balance is

$$\dot{\theta} = (-\theta + \theta_0)\frac{q}{V} + \frac{1}{\rho c V}(-\Delta H_1 V r_B - \Delta H_2 V r_C - \Delta H_3 V r_D + Q) \qquad (27)$$

where ρ, c, V, ΔH_i are constants, q : feed flow, Q: heat flow.

In summary the state, control and disturbance vectors are:

$$x = [c_A, c_B, c_C, c_D, \theta]^T \qquad (28)$$

$$u = [q, Q]^T \qquad (29)$$

$$v = [c_{A0}, c_{B0}, c_{C0}, c_{D0}, \theta_0]^T \qquad (30)$$

The following numerical values of model parameters are used:

$k_{10} = 1.0 \cdot 10^{15}$, $k_{20} = 4.5 \cdot 10^{15}$, $k_{30} = 1.0 \cdot 10^{14}$,
$E_1 = E_2 = E_3 = 1.3 \cdot 10^5$, $R = 8.3$, $\rho = 1$,
$c = 4.2 \cdot 10^3$, $V = 10^3$, $\Delta H_1 = -5 \cdot 10^4$,
$\Delta H_2 = -5.5 \cdot 10^4$, $\Delta H_3 = -5 \cdot 10^3$
$u_{nom} = [0.513, 70.188 \cdot 10^3]^T$,
$v_{nom} = [1, 0, 0, 0, 350]^T$.

Since the process has two degrees of freedom, we define a desirable property transformation

$$D^0 = \begin{bmatrix} 0 & 1 & 0 & 0 & 0 \\ 0 & 0 & 0 & 0 & 1 \end{bmatrix} \qquad (31)$$

which means that we want to control the concentration of the desirable product B and the temperature θ. Differentiation of (23)-(26) yields

$$B = \frac{\partial f(\cdot)}{\partial u} = \begin{bmatrix} 1/V(-x_1 + v_1), & 0 \\ 1/V(-x_2 + v_2), & 0 \\ 1/V(-x_3 + v_3), & 0 \\ 1/V(-x_4 + v_4), & 0 \\ 1/V(-x_5 + v_5), & 1/\rho c V \end{bmatrix}$$

Even though the matrix $\left(D^0 \frac{\partial f(\cdot)}{\partial u}\right)$ is nonsingular another matrix D must be designed because the dynamics of the inner system (10) becomes unacceptable

An adequate matrix D which secures stability of the inner system will be

$$D = \begin{bmatrix} d_{11} & 1 & 0 & 0 & 0 \\ 0 & 0 & 0 & 0 & 1 \end{bmatrix} \qquad (32)$$

where $d_{11} = 0.5$. An optimization of the other elements of D could have given some improvement of the performance, but that is not important here.

In addition to the END algorithm of Fig. 1 the system is controlled by an inner loop (loop I as shown in Fig. 2 with $G = \begin{bmatrix} 0.05 & 0 \\ 0 & 0.05 \end{bmatrix}$).

The responses of this system are shown in Fig. 3 a-j where one of the system parameters k_{30} is changed from $1.0 \cdot 10^{14}$ to $1.5 \cdot 10^{14}$ and back again.

As can be seen from Fig. 3 g the concentration c_D is first increasing and then decreasing again. If the component D is an undesirablee byproduct and we want c_D to be less than 0.45 we may design a new property transformation.

$$\tilde{D}^4 = \begin{bmatrix} \tilde{d}_{11} & 0 & 0 & 1 & 0 \\ 0 & 0 & 0 & 0 & 1 \end{bmatrix} \qquad (33)$$

which replaces that of (32) when $c_D > 0.45$. At selected intervals short tests are made to see if the system can be returned to the unconstrained transformation (32). In fig. 4g that happens after the fourth test when c_D again becomes less than 0.45. The results of this strategy are shown in Fig. 4 It is clearly seen that c_D is controlled on its constraint with only small overshoots.

5. CONCLUSION

Elementary Nonlinear Decoupling (END) can conveniently be modified to handle hard constraints in states by replacing these states by new properties to be controlled.

6. ACKNOWLEDGEMENTS

This research has been supported by SINTEF (JGB) and STATOIL (BS).

7. REFERENCES

Balchen, J.G. (1993): Design of the property transformation in elementary nonlinear decoupling of multivariable processes. *Modeling, Identification and Control*, **14**, 219-227.

Balchen, J..G., Sandrib, B. (1994a). Elementary nonlinear decoupling control of composition in binary distillation columns. *IFAC Symposium ADCHEM'94, Advanced Control of Chemical Processes*, May 25-27, 1994, Kyoto, Japan.

Balchen, J.G., Sandrib, B. (1994b). Input saturation in nonlinear multivariable processes resolved by nonlinear decoupling, *3rd IEEE Conference on Control Applications*, August 24-26, 1994, Glasgow, Scotland, UK.

Isidori, A. (1989): *Nonlinear Control Systems* (2nd ed.). (Springer Verlag, Berlin).

Mayne, D.Q. and E. Polak (1993). Optimization based design and control *12th IFAC World Congress*, Sydney, Australia, pp 129-138.

Polak, E. (1971). Computational Methods in Optimization. (Academic Press, New York)

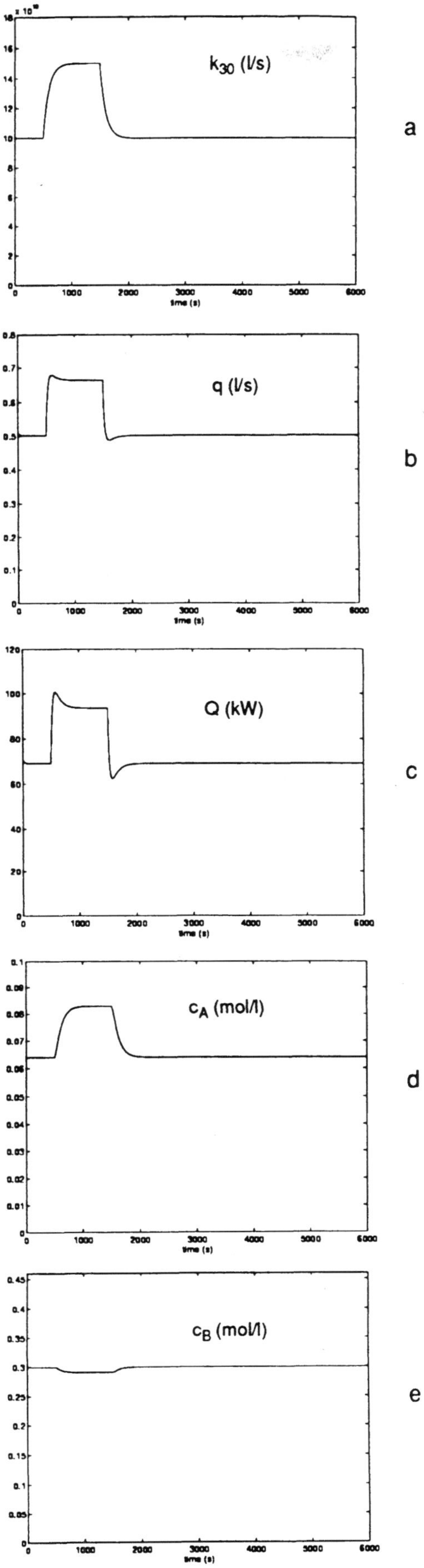

Fig. 3 a-e. Responses of unconstrained END controlled system when k_{30} has 50%change.

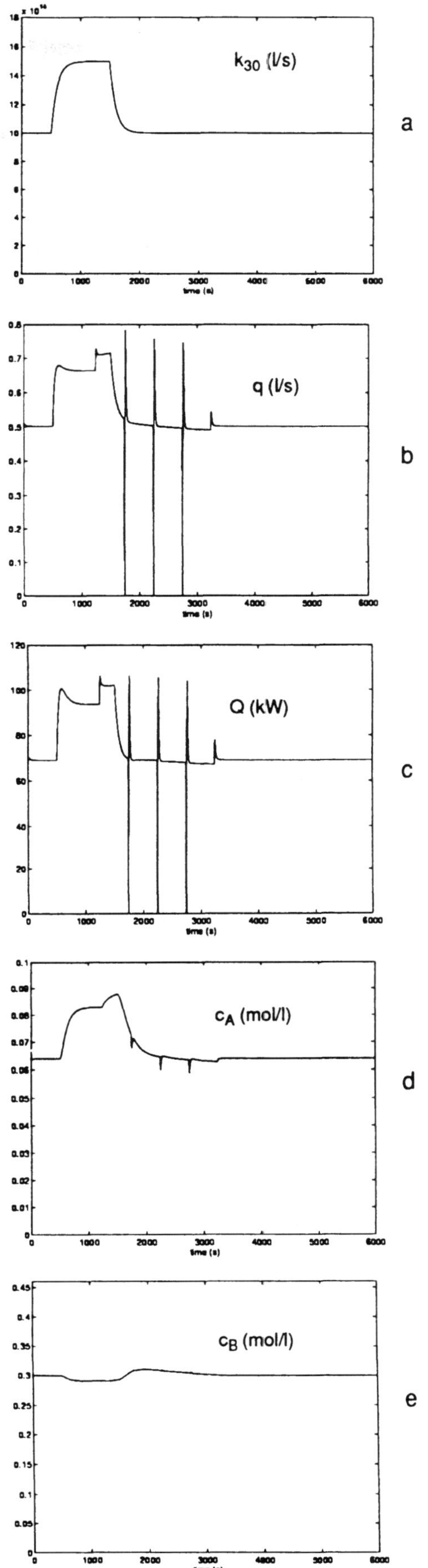

Fig. 4 a-e. Responses of state constrained END controlled system when k_{30} has 50% change.

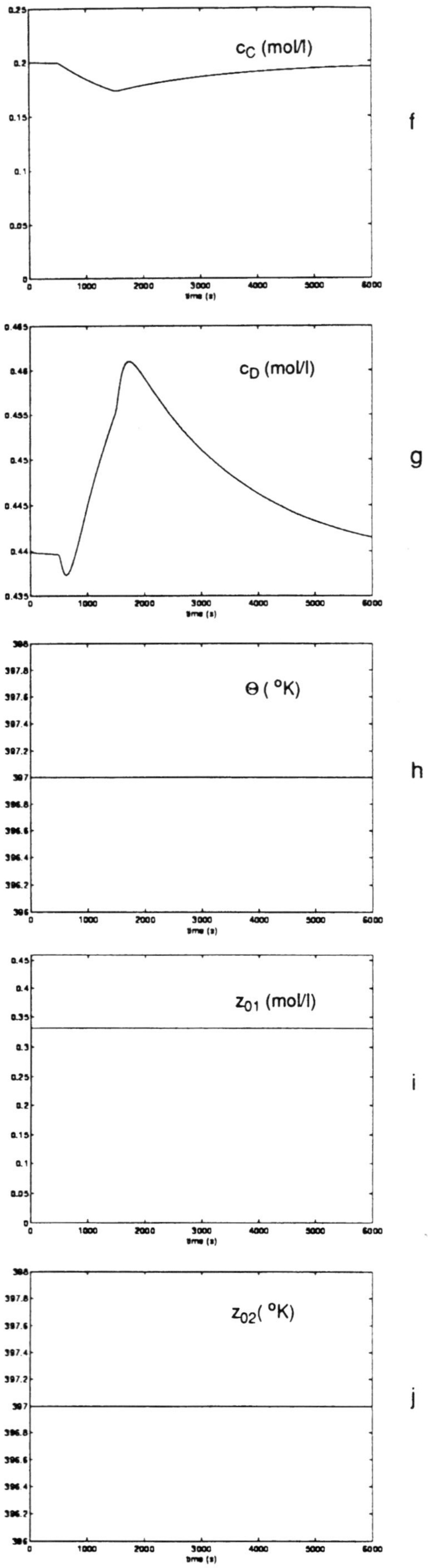

Fig. 3 f-j. Responses of unconstrained END controlled system when k_{30} has 50%change

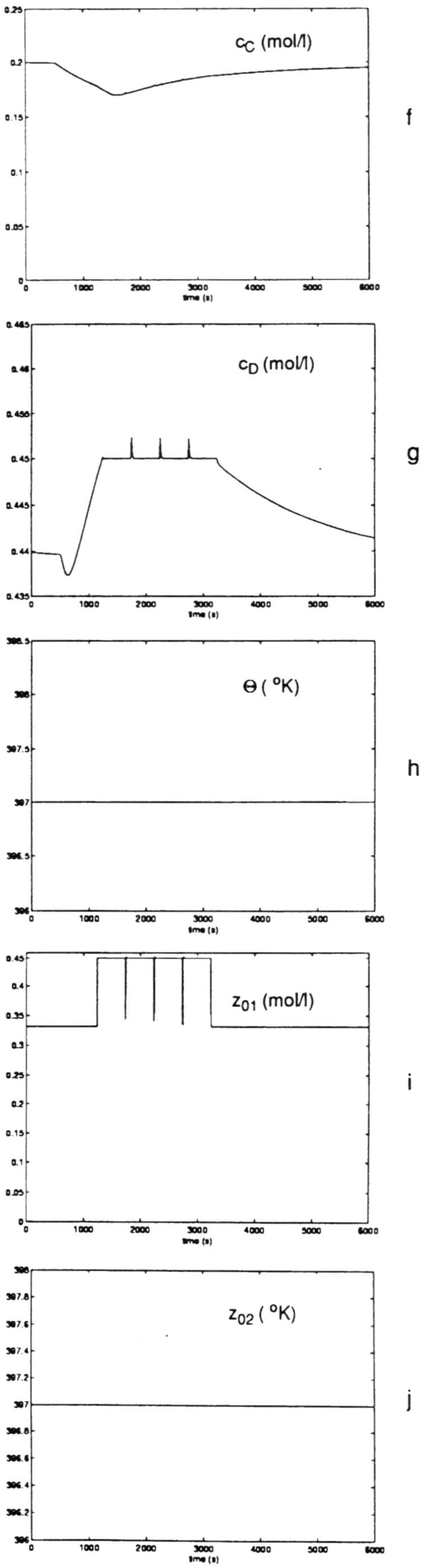

Fig. 4 f-j. Responses of state constrained END controlled system when k_{30} has 50% change.

TIME-VARIABLE MODELS FOR MIXING PROCESSES UNDER UNSTEADY FLOW AND VOLUME

K. Zenger*

**Helsinki University of Technology, Control Engineering Laboratory, Otakaari 5 A, FIN-02150 Espoo, Finland*

Abstract. The classical models describing material transport in a flow system are extended to include cases where the flow and volume of fluid are variable. Time-varying residence time distributions, internal age distributions, weighting functions, and intensity functions are discussed and their mutual dependencies are derived. The results are simplified by presenting the distributions and functions under a volumetric scale. The results extend the traditional theory, in which a scaled time variable has been used to model a system with (different) constant flow rates. Finally, the theory is applied to the results of a pilot plant test, where residence time distributions measured under the conditions of unsteady flow are discussed.

Key Words. Process models; continuous systems; time-varying systems; impulse responses; weighting functions.

1. INTRODUCTION

The traditional theory characterizes the flow of material in e.g. a chemical reactor by using concepts like residence time distribution, internal age distribution, impulse response (C-curve), step response (F-curve), and intensity function, (Levenspiel, 1962; Seinfeld and Lapidus, 1974; Himmelblau and Bischoff, 1968). The connections between these concepts have been shown to possess a particularly simple form, when the functions are presented with respect to the so called *scaled time variable* (usually abbreviated θ), which actually means a transformation of the time scale. The use of the variable is beneficial, because the process model (residence time distribution) becomes invariant even though the flow rate through the process has different constant values.

The idea of a scaled time variable has been extended to include cases, where the flow rate is continuously varying (see e.g. Niemi, 1988; Niemi, 1991; Zenger 1993). The method has introduced a new transformation of the time scale, which is dependent on the integral of the flow rate and on the liquid volume in the process. The *volumetric time scale* coincides with the classical scaled time variable, if the flow rate and volume are constant. The volumetric time scale has been used so far in the applications of control engineering (e.g. Tian *et all.*, 1992; Zenger, 1992), because the possibility of using an invariant process model makes it possible to use classical design methods developed for time invariant models.

Although the original idea of using the new volumetric scale was derived by using the concept of residence time distribution (Niemi, 1977), the theory of the classical models describing material transport in a mixing process has not been discussed systematically in the time-varying case. For certain classes of models like those expressible by differential equations possibly including delay terms (systems consisting of perfect mixers, plug flow vessels, and recycle flows) this is not even necessary, because results can be proved by using the mathematical structure of the equations (Zenger, 1992). For a more general setting, though, the construction of the theory including the classical models in the time varying case would be beneficial. That is the purpose of the current paper.

The outline of the text can briefly be described as follows. Assuming that the flow rate and liquid volume in the system are variable the time-varying residence time distribution and internal age distribution are defined analogously to the classical definitions. The concepts are then related to the weighting function, and a method for determining the residence time distribution under the conditions of variable flow and volume is presented. The volumetric scale is then used to show that interconnections between the models are similar to those of classical theory, in which the scaled time variable has been used in the time-invariant case. However, it will turn out that the analogy is true in the case of variable flow only; the situation is much more difficult if the liquid volume is also changing. A practical application is then discussed, in which time-varying residence time distributions were measured and analysed for two different processes by using tracer tests.

2. TIME-VARYING AGE DISTRIBUTIONS

Consider the flow of fluid through a vessel. The liquid is incompressible, and the process is assumed to be linear with respect to concentrations. The system is closed in the sense that all material is transported with the flow and leaves the system eventually. Furthermore, there is no dispersion at the inlet or outlet of the vessel. Despite variations in the flow rate and in the liquid volume the flow pattern is assumed to be invariant, which means that the statistical distribution of particles passing through the vessel remains the same. Roughly, the invariability of the flow pattern can be interpreted to mean that the model structure of the system does not change.

The standard equations characterizing the system are

$$\dot{V}(t) = Q_i(t) - Q_o(t) \tag{1}$$

and

$$c_o(t) = \int_{t_0}^{t} p(t,\tau)c_i(\tau)d\tau \tag{2}$$

($c_o(t_0) = 0$), where $Q_i(\cdot)$, $Q_o(\cdot)$ $[m^3/s]$ and $V(\cdot)$ $[m^3]$ are flow rates of the incoming and outgoing flow and the liquid volume in the vessel. The concentrations of the dissolved solute at the inlet and outlet are $c_i(\cdot)$ and $c_o(\cdot)$ $[mol/m^3]$, respectively, and $p(\cdot,\cdot)$ $[1/s]$ is the time-varying weighting function.

Let t' be the time, which has passed from the time τ at which a particle entered the vessel. In other words, at the absolute time instant $t \geq \tau$ it holds that $t' = t - \tau$. Define the time-varying residence time distribution (RTD) as follows: $E^*(t',\tau)dt'$ is the part of material in the effluent stream, the age of which belongs to the interval $[t', t'+dt']$. In real time, a particle which has entered at time τ leaves the vessel during the time interval $[\tau + t', \tau + t' + dt'] = [t, t + dt]$. In terms of concentration in the effluent stream the RTD is

$$\begin{aligned} E^*(t',\tau)dt' &= E^*(t-\tau,\tau)dt \\ &= \frac{Q_o(t)c_o(t)dt}{\int_\tau^\infty Q_o(t)c_o(t)dt} \end{aligned} \tag{3}$$

($t \geq \tau$). The denominator of the equation expresses the total amount of solute, which has entered after time τ, because all material is assumed to leave the system eventually. It is interesting to note that the time-varying nature of RTD can be interpreted as a density function, which is obtained for a fixed τ and which depends on the realization of flow and volume in the system.

The equation (3) shows that the liquid volume in the vessel $V(t)$ affects the RTD only implicitly through the term $Q_o(t)$. In the stationary case the RTD is known to be equal to the unit impulse response, and the classical formula

$$E^*(t-\tau,\tau)dt = \frac{c_o(t)dt}{\int_\tau^\infty c_o(t)dt} \tag{4}$$

is obtained, in which the time that the impulse enters is usually set to zero, $\tau = 0$.

In the function $E^*(t',\tau) = E^*(t-\tau,\tau)$ the first argument describes the time that has passed from τ, so that $E^*(t-\tau,\tau) = 0$, as $t < \tau$. It is natural to use the slightly modified expression $E^*(t-\tau,\tau) = E(t,\tau)$, in which the real time is the first argument directly and $E(t,\tau) = 0$, as $t < \tau$.

Taking the relationship $t' = t - \tau$ into account and using the normalization condition for the RTD it follows that

$$\begin{aligned} \int_0^\infty E^*(t',\tau)dt' &= \int_\tau^\infty E^*(t-\tau,\tau)dt \\ &= \int_\tau^\infty E(t,\tau)dt = 1 \end{aligned} \tag{5}$$

It is well known that for linear systems the unit impulse response coincides with the weighting function (for $t \geq \tau$). Thus, in the stationary case it holds that

$$E(t,\tau) = p(t,\tau) \tag{6}$$

when $t \geq \tau$.

In a similar manner as in the case of the time-varying RTD (or exit age distribution) it is possible to define the time- varying internal age distribution. Let $I^*(t',\tau)dt'$ be the part of material in the vessel, the age of which is in the interval $[t', t'+dt']$. Again it is reasonable to use the notation $I^*(t',\tau)dt' = I^*(t-\tau,\tau)dt = I(t,\tau)dt$. The normalization condition (5) is naturally valid for the internal age distribution, too.

3. MODELS AND THEIR INTERCONNECTIONS

3.1. *RTD and weighting function*

Let M moles of tracer be injected into the vessel at time τ. Since all material is assumed to leave eventually, it follows that

$$M = \int_{t_0}^{\infty} Q_i(t)c_i(t)dt = \int_{t_0}^{\infty} Q_i(t)N\delta_\tau(t)dt = NQ_i(\tau) \quad (7)$$

where N is the 'strength' of the concentration impulse. Consequently, $N = M/Q_i(\tau)$, which in the case of a unit impulse ($N = a = 1mol \cdot s/m^3$) leads to

$$M = Q_i(\tau)a \quad (8)$$

Taking the result from the equation (3) into account and interpreting the unit impulse response as the weighting function (more precisely: $c_o(t) = p(t,\tau)a$) the formula

$$E(t,\tau) = \frac{Q_o(t)c_o(t)}{M} = \frac{Q_o(t)}{Q_i(\tau)}p(t,\tau) \quad (9)$$

($t \geq \tau$) is obtained, which states the general dependence between the time-varying RTD and the weighting function.

The result gives rise to a few comments. Again, the liquid volume appears only implicitly through the output flow rate. In the stationary case the RTD coincides with the weighting function, as known. The flow rate in the denominator is calculated at the time that the impulse has entered, while the numerator term means the output flow rate as a function of time. In the case of varying volume it is not possible to determine the RTD by an impulse test, if only the input flow measurement is available. In the case of constant volume but varying flow rate the above equation expresses the relationship, which has been known earlier, see e.g. (Tian *et all.*, 1992).

Next, consider a function $z : t \mapsto z(t)$, which for all t is assumed to be positive, continuous and monotonous. With respect to z the equation (2) can be written as

$$\bar{c}_o(z) = \int_{z_0}^{z} \bar{p}(z,\xi)\bar{c}_i(\xi)d\xi \quad (10)$$

in which it holds that for all t, τ and the corresponding z, ξ

$$\bar{p}(z,\xi) = p(t,\tau)/\dot{z}(\tau) \quad (11)$$

In a similar manner it is possible to change the time variable in equation (5) to get

$$\bar{E}(z,\xi) = E(t,\tau)/\dot{z}(t) \quad (12)$$

which fulfils the normalization condition.

The change of the time variable is particularly useful, if

$$\bar{p}(z,\xi) = \bar{p}(z-\xi, 0) \quad (13)$$

in which case the system is called *z-invariant*. The difficulty caused by the time-varying nature of the process is then removed in the system model by the transformation, because in z-domain the model has a 'time-invariant' form.

Writing the equation (9) in z-domain and taking (11) and (12) into account the equation

$$\bar{E}(z,\xi) = \frac{Q_o(t)}{Q_i(\tau)}\frac{\dot{z}(\tau)}{\dot{z}(t)}\bar{p}(z,\xi) \quad (14)$$

follows. By choosing the transformation

$$z(t) = \int_{t_0}^{t} \frac{Q_i(\nu)}{V(\nu)}d\nu \quad (15)$$

the equation (14) obtains the form

$$\bar{E}(z,\xi) = \frac{Q_o(t)}{Q_i(t)}\frac{V(t)}{V(\tau)}\bar{p}(z,\xi) \quad (16)$$

The result shows that the previously known relationship

$$\bar{E}(z,\xi) = \bar{p}(z,\xi) \quad (17)$$

is valid in the case of varying flow rate and constant volume. Moreover, if the system is z-invariant, the RTD is invariant with respect to z even though the flow rate is changing. It has been shown (Zenger, 1992) that under mild assumptions this holds for all models consisting of perfect mixers, plug flow vessels, and recirculation flows.

As an example of the above theory a method for determining the RTD by using a tracer test is presented, assuming that the flow rates and the liquid volume may be changing during the test. The result is $E^*(t-\tau,\tau) = E(t,\tau)$, where the fixed τ is the time instant at which the tracer impulse enters the system. In practice, the RTD is determined for one particular realization of flow rates and volume.

The method is based on equation (9). The measured impulse response is a multiple of the theoretical unit impulse response $c_o(t) = kp(t,\tau)$, (k is an unknown constant). On the other hand, $Q_i(\tau)$ is also a constant, so that by scaling the measured response as $E'(t,\tau) = Q_o(t)c_o(t)$ and by normalizing it to unity the RTD $E(t,\tau)$ is obtained. It is important to note that the values of the measured response must be multiplied by the output flow rates. In the stationary case this is not necessary, because scaling is then performed automatically by the normalization procedure.

3.2. *RTD, internal age distribution, and step response*

Consider a flow process at time t. The amount of new liquid in the vessel i.e. the material which has entered after the time τ is

$$\begin{aligned} V(t) & \int_0^{t'} I^*(\nu',\tau)d\nu' \\ = & \quad V(t)\int_\tau^t I(\nu,\tau)d\nu \qquad (18) \end{aligned}$$

On the other hand, the amount of old material (liquid in the vessel at time τ) which has left the process by the time t is

$$\begin{aligned} & \int_\tau^t Q_o(\nu')[\int_{\nu'}^\infty E^*(\nu''-\tau,\tau)d\nu'']d\nu' \\ = & \int_\tau^t Q_o(\nu')[\int_{\nu'}^\infty E(\nu'',\tau)d\nu'']d\nu' \qquad (19) \end{aligned}$$

Taking the two previous equations into account the volume of liquid in the vessel at time t is easily calculated to be

$$\begin{aligned} V(t) & = V(\tau)+V(t)\int_\tau^t I(\nu,\tau)d\nu \\ - & \int_\tau^t Q_o(\nu')(\int_{\nu'}^\infty E(\nu'',\tau)d\nu'')d\nu' \qquad (20) \end{aligned}$$

Taking the derivative, using (1) and the relationship

$$\int_t^\infty E(\nu,\tau)d\nu = 1-\int_\tau^t E(\nu,\tau)d\nu \qquad (21)$$

the result

$$\begin{aligned} & \int_\tau^t E(\nu,\tau)d\nu = \frac{Q_i(t)}{Q_o(t)}(1-\int_\tau^t I(\nu,\tau)d\nu) \\ + & \int_\tau^t I(\nu,\tau)d\nu - \frac{V(t)I(t,\tau)}{Q_o(t)} \qquad (22) \end{aligned}$$

is obtained, which gives the dependence between the time-varying RTD and the internal age distribution. The relationship is complicated, and it does not seem to be possible to simplify it generally. However, in the case of constant volume $Q_i(t) = Q_o(t) = Q(t)$, $V(t) = V$, it has the form

$$\int_\tau^t E(\nu,\tau)d\nu = 1-\frac{VI(t,\tau)}{Q(t)} \qquad (23)$$

By changing the time variable according to (15) and by considering the relationship analogous to (12)

$$\bar{I}(z,\xi) = \frac{I(t,\tau)}{\dot{z}(t)} \qquad (24)$$

it is easy to derive

$$\bar{E}(z,\xi) = -\frac{d\bar{I}(z,\xi)}{dz} \qquad (25)$$

The result is similar to the corresponding relationship known in classical theory, in which the scaled time variable θ has been used in the time-invariant case. The z-variable theory can thus be seen as a generalization, which allows continuous changes in the flow rate; however, the analogy seems to break, if the liquid volume is also variable.

For the sake of completeness it should be mentioned that the expected residence time (mean of RTD) can be calculated from

$$\begin{aligned} \bar{t} & = \int_0^\infty \nu' E^*(\nu',\tau)d\nu' \\ & = \int_\tau^\infty (t-\tau)E^*(t-\tau,\tau)dt \\ & = \int_\tau^\infty tE(t,\tau)dt - \tau \qquad (26) \end{aligned}$$

In the stationary case the mean residence time is naturally V/Q, provided that the basic assumptions (see Sec. 2) hold (Guizerix, 1990). In the time-varying case a simple expression is, of course, not possible. If only the flow rate is varying, the mean residence time has the value $z = 1$.

Next, consider the step response (or F-curve) of the system. Let the input concentration be a unit step at time $t = \tau$. According to (2) the response is

$$F(t,\tau) = \int_\tau^t p(t,\tau')d\tau' \qquad (27)$$

which can be written in z-domain as

$$\bar{F}(z,\xi) = \int_\xi^z \bar{p}(z,\xi')d\xi' \qquad (28)$$

Let us further assume that the system is z-invariant. In the case of constant liquid volume the equation (17) is valid. Consequently, for all t, τ and the corresponding z, ξ

$$\begin{aligned} \bar{F}(z,\xi) & = \int_\xi^z \bar{p}(z-\xi',0)d\xi' \\ & = \int_\xi^z \bar{E}(z-\xi',0)d\xi' \\ & = \int_0^{z-\xi} \bar{E}(z',0)dz' \qquad (29) \end{aligned}$$

Consider the equation (23). By using the variable

$$z(t) = \frac{1}{V}\int_{t_0}^t Q_i(\nu)d\nu \qquad (30)$$

it can be written in the form

$$\begin{aligned}\int_{\xi}^{z} \bar{E}(\nu,\xi)d\nu &= \int_{\xi}^{z} \bar{E}(\nu-\xi,0)d\nu \\ &= \int_{0}^{z-\xi} \bar{E}(z',0)dz' \\ &= 1-\bar{I}(z,\xi) \end{aligned} \tag{31}$$

Comparing to (29) the result

$$\bar{F}(z,\xi)+\bar{I}(z,\xi)=1 \tag{32}$$

follows, which, written with respect to the scaled time variable in the stationary case, is known in the classical theory. It is important to note that in addition to the assumption of constant volume the weighting function was also assumed to be z-invariant. That assumption was not necessary in the derivation of equation (25).

The concept of intensity function, which can be used to detect possible bypassing and dead space in the process, can be extended to the time-varying case analogously to the preceding discussion. To this end, let $\Lambda^*(t',\tau)dt'$ be the fraction of particles, which has stayed in the process for the time t', but which will leave on the next time instant. The residence time of those particles in the process belongs thus to the interval $[t',t'+dt']$. Define
$\Lambda^*(t',\tau)dt' = \Lambda^*(t-\tau,\tau)dt = \Lambda(t,\tau)dt$, and notice that the function is not normalized as in the case of the age functions.

The amount of material leaving the system on the interval $[t,t+dt]$ and consisting of particles of ages $[t',t'+dt']$ is $Q_o(t)dtE(t,\tau)dt$. On the other hand, the volume of material in the vessel at time t, the age of which belongs to the interval $[t',t'+dt']$, is $V(t)I(t,\tau)dt$. The fraction of these particles, which leaves on the interval $[t,t+dt]$, is $\Lambda(t,\tau)dt$. Combining this information gives

$$Q_o(t)dtE(t,\tau)dt = V(t)I(t,\tau)dt\Lambda(t,\tau)dt \tag{33}$$

which can be solved to give

$$\Lambda(t,\tau)=\frac{Q_o(t)}{V(t)}\frac{E(t,\tau)}{I(t,\tau)} \tag{34}$$

Because the relationship between the functions $E(\cdot,\cdot)$ and $I(\cdot,\cdot)$ is complicated (see (22)), it does not seem to be possible to simplify the previous equation. However, by using the transformation (15) and by defining

$$\bar{\Lambda}(z,\xi)=\Lambda(t,\tau)/\dot{z}(t) \tag{35}$$

the equation (34) changes into the form

$$\bar{\Lambda}(z,\xi)=\frac{Q_o(t)}{Q_i(t)}\frac{\bar{E}(z,\xi)}{\bar{I}(z,\xi)} \tag{36}$$

(c.f. (16)). The relationship is again complicated except in the case of constant volume, in which case it takes the form

$$\begin{aligned}\bar{\Lambda}(z,\xi) &= \frac{\bar{E}(z,\xi)}{\bar{I}(z,\xi)} = -\frac{\frac{d\bar{I}(z,\xi)}{dz}}{\bar{I}(z,\xi)} \\ &= -\frac{d}{dz}[\ln(\bar{I}(z,\xi))] \end{aligned} \tag{37}$$

That is again in accordance with the classical theory.

4. PRACTICAL TESTS

In order to test the method in the determination of time-varying RTD a laboratory-scale pilot plant was used. Two different processes were used; a 10 meter long U-tube made of rubber to achieve a parabolic velocity profile, and a cylindrical vessel with a good mixer, in order to reach the behaviour of an ideally mixed vessel as well as possible.

The flow rate through the U-tube was calculated to be laminar in all but one test, in which the flow rate was large enough so that turbulent behaviour was possible. The process flow was ion-changed water, into which 1 ml of sodium chloride aqueous solution was injected impulsewise through a premixer. The amount of salt in the solution was changed in some of the tests, because it was not known, what amount would give the most accurate measurement without distorting the velocity profile (a large amount of salt may sink down at the bottom of the tube). The output concentration was determined by using a conductivity meter, which was possible because the relationship between conductivity and concentration is linear for a wide range.

Different constant flow rates and also a sinusoidal flow rate were used, the liquid volume naturally being a constant in the tube. The measured output concentrations were analysed as described in Sec. 3.1. The results, which can be seen in Fig. 1 and in Fig. 2, show that the different RTDs match reasonably well in z-domain except for one particular test. In that test a considerably higher concentration of salt was used in the impulse, and it is obvious that this has changed the velocity profile. Otherwise, the difference between the curves in Fig. 2 may be due to the facts that the system is perhaps not accurately z-invariant and the velocity profile changes a certain amount in different tests.

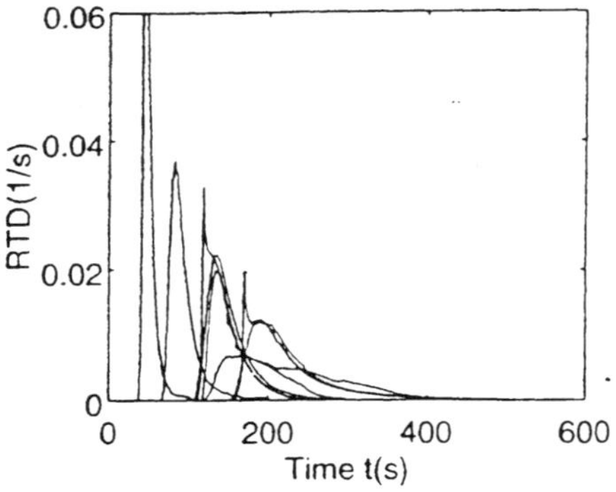

Fig. 1. Residence time distributions of the U-tube process

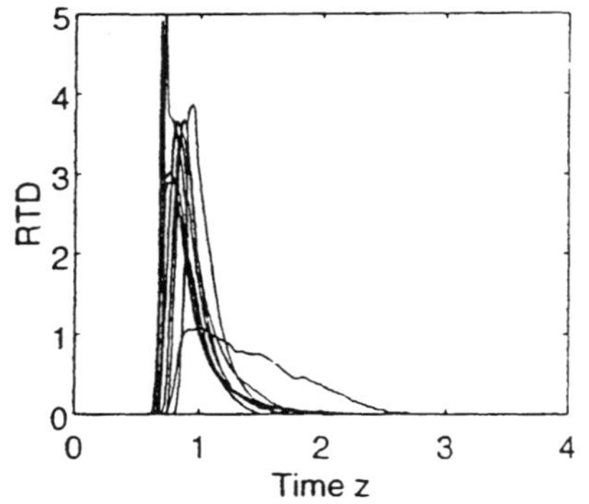

Fig. 2. Residence time distributions with respect to the volumetric scale

The test of the 'perfect mixer' type process was made by using a radioactive tracer (the test was actually carried out earlier than the one discussed above; the pilot plant did not have conductivity meters then). 12 tests were made by using different constant and also sinusoidally varying flow rates and volumes. The results can be seen in Fig. 3, which shows that in z-domain the mixing dynamics of the process is in good agreement with that of the theoretical perfect mixer.

5. CONCLUSION

In the paper it was shown that the classical models of the residence time distribution theory can be extended to the case, in which the flow rate and liquid volume are varying. Under the volumetric scale the interconnections between the models become particularly simple, if the flow rate, but not volume, is changing. The most important question is, whether the weighting function is invariant with respect to the volumetric scale. If that holds even approximately, the difficult time-varying nature of the process model can be simplified, which makes both analysis and design much simpler.

6. REFERENCES

Guizerix, J. (Ed.), (1990). *Guidebook on Radioisotope Tracers in Industry.* International Atomic Energy Agency, Vienna.

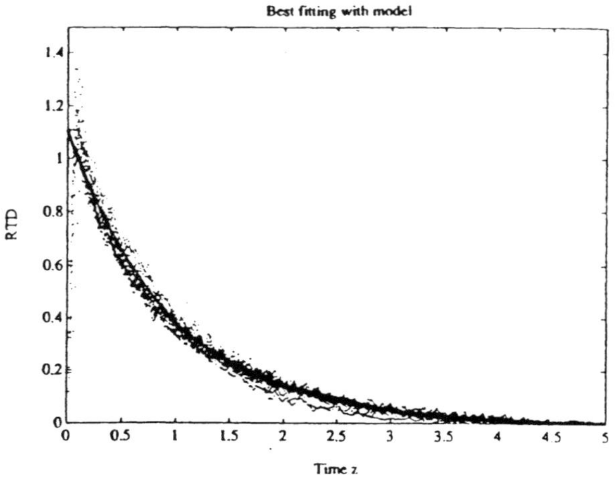

Fig. 3. Residence time distributions of the 'perfect mixer' type process, volumetric scale

Himmelblau, D., M. and K., M. Bischoff. (1968). *Process Analysis and Simulation, Deterministic Systems.* John Wiley & Sons, Inc, USA.

Levenspiel, O. (1962). *Chemical Reaction Engineering.* John Wiley & Sons, Inc, USA.

Niemi, A. J. (1977). Residence time distributions of variable flow processes. *International Journal of Applied Radiation and Isotopes*, Vol. 28, 855–860.

Niemi, A. J. (1988). Variable parameter model of the continuous flow vessel. *Math Comput Modell*, Vol. 11, 32-37.

Niemi, A. J. (1991). Process control under variable flow and volume. Papers of *Int. Symp. on Intelligent Tuning and Adaptive Control* (ITAC 91), Singapore.

Seinfeld, J., H., and L. Lapidus. (1974). *Mathematical Methods in Chemical Engineering,* Vol. 3. Process Modeling, Estimation, and Identification. Prentice-Hall, Inc, USA.

Tian, L., A.J. Niemi and R. Ylinen. (1992). Recursive process identification under variable flow and volume. Selected Papers from the *3rd IFAC Symposium on Dynamics and Control of Chemical Reactors, Distillation Columns and Batch Processes,* Maryland, USA.

Zenger, K. (1992). Analysis and control design of a class of time-varying systems. *Helsinki University of Technology, Control Engineering Laboratory,* Report 88.

Zenger, K. (1993). Analysis and control of time-varying flow processes by using the state-space approach. Proceedings from the *12th World Congress of IFAC, Sydney, Australia,*43-48.

GENERAL METHODS OF KINETIC MODEL REDUCTION

KEITH EDWARDS and THOMAS F. EDGAR

University of Texas at Austin, Department of Chemical Engineering, Austin, TX 78712-1062, USA

Abstract. Two novel approaches to simplification of a kinetic mechanism are presented. The first combines macroscopic reactor balances with linear variable selection techniques from the statistics literature. The second approach formulates the network reduction problem as a combinatorial optimization problem that is solved using genetic algorithms. Both approaches are tested for an 18 reaction, 10 species network in a batch reactor. Results are compared with global solutions found using complete enumeration.

Key Words. Networks; reduction; modeling errors; least squares; genetic algorithms

1. INTRODUCTION

First principles modeling of processes with complex kinetic networks poses challenges when computation time is restricted. This situation arises in a variety of chemical processes such as petroleum refining (Sapre, 1991), atmospheric chemistry (Leone and Seinfeld, 1985), and microelectronics processing (Allred, 1991). In particular, the large number of reactions and species poses problems for first-principles modeling with applications to real-time model-based control. It is desirable to have an accurate kinetic model that can be solved in a reasonable amount of time on the available computer resources, but reaction network size (number of species and reactions) can be a limitation in achieving this goal.

A variety of model reduction techniques have been developed to handle the problem of reaction network size. Many have been employed successfully, but they suffer to some extent to lack of control over accuracy, i.e., the reduced model can be unsatisfactory in some cases. There are other unanswered questions. Is the reduced model the best available? If not, can we find a better one? What is the smallest reduced model that satisfactorily represents the process kinetics? We seek to address these questions by presenting two novel approaches to kinetic model reduction.

The first approach involves the use of variable selection(VS) techniques from the statistics literature. For linear regression models (Rawlings, 1988), a variety of techniques have been developed to tackle the issue of variable selection. All of these techniques seek to answer the following question: given n_p candidate variables, which ones should be included in the regression model? We seek to answer a similar question, though we are working with n_r candidate reactions. Because a first-principles kinetic model is linear in the reaction rates, it is possible to simplify it using linear regression VS techniques.

In particular, the Leaps and Bounds algorithm (Furnival and Wilson, 1974) can be applied to determine a family of reduced kinetic models. Each model can in turn be evaluated to determine if it meets accuracy requirements. The VS approach to kinetic model reduction in principle can be used for any type of reactor transport dynamics provided the reactor model equations are written in terms of macroscopic balances (Bird *et al.*, 1960).

The second approach treats the model reduction problem as a combinatorial optimization problem. The solution space is a finite but exponentially increasing set of configurations consisting of all possible subsets of reactions. A loose bound of 2^{n_r} configurations exist for this problem, where n_r is the number of reactions in the full model.

The formulation resulting from the second approach can be solved using genetic algorithms(GA) (Goldberg, 1989; Davis, 1991), which provide a means of obtaining globally optimal or near optimal solutions to combinatorial optimization problems. The value of the objective function for a particular configuration is a measure of the total number of reactions in the reduced model and the violation of the species error constraints.

The VS and GA approaches to kinetic model reduction have been tested on subsets of a plasma-etching reaction set for a constant-volume batch reactor model. Both techniques were applied to eliminate reactions from an 18 reaction, 10 species network involving monomolecular and bimolecular reactions. Solutions have been compared with global solutions found by complete enumeration.

2. REACTOR MODELING

For all problems tested, a batch reactor model with mass-action kinetics is assumed. Given a full network with n_r reactions and n_s species, the rate of reaction i at time t, $r_i(t)$, is given by

$$r_i(t) = k_i \prod_{j=1}^{n_s} C_j^{\alpha_{ij}}(t) \qquad i = 1, n_r \qquad (1)$$

where k_i is a known rate constant, $C_j(t)$ is the molar concentration of species j at time t, and α_{ij} is the reactant stoichiometry of species j in reaction i. Then the overall reaction rate for species j at time t is given by

$$R_j(t) = \sum_{i=1}^{n_r} (\beta_{ij} - \alpha_{ij}) r_i(t) \qquad j = 1, n_s \quad (2)$$

where β_{ij} is the product stoichiometry of species j in reaction i and $R_j(t)$ is the overall reaction rate of species j.

Setting any number of the rate constants k_i in (1) to zero results in a reduced reaction network having $n'_r \leq n_r$ reactions and $n'_s \leq n_s$ species. The quality of the reduced model may be evaluated over the $[t_0, t_f]$ time domain by integrating both the full and reduced models from t_0 to t_f and comparing the concentration profiles at specific points in time. The model error at time t is measured according to

$$e_j(t) = \left| \log_{10} \left(\frac{C_j^{\text{full}}(t)}{C_j^{\text{red.}}(t)} \right) \right| \qquad j = 1, n_k \qquad (3)$$

where $C_j^{\text{full}}(t)$ and $C_j^{\text{red.}}(t)$ denote full and reduced-model concentrations, respectively, and n_k is the number of species with error bounds. This error measure is chosen over relative error because in our experience it results in improved convergence for the GA approach and elimination of several local optima. The average error along the concentration profiles for the bounded species is given by

$$e_j^{\text{avg}} = \frac{1}{n_p} \sum_{k=1}^{n_p} \left| \log_{10} \left(\frac{C_j^{\text{full}}(t_k)}{C_j^{\text{red.}}(t_k)} \right) \right| \qquad j = 1, n_k \quad (4)$$

where n_p is the number of discrete time points on $[t_0, t_f]$ and t_k is the time at point k. For both model reduction techniques, reduced models are sought that satisfy

$$e_j^{\text{avg}} \leq \text{B}_j \qquad j = 1, n_k \qquad (5)$$

where B_j is a preset bound on the average error for species j.

3. VS FORMULATION FOR NETWORK REDUCTION

For any reaction system, it is possible to develop a differential equation model that is linear in the elementary reaction rates by utilizing macroscopic material balances. Elementary reaction rates as a function of position in a reactor can be integrated over the reactor spatial coordinates to obtain $r_i(t)$, the total moles of reaction per unit reactor volume.

Consider a reactor having n_r elementary reactions and n_s species. If the overall reaction rates $R_j(t)$ at time t are defined as in (2), then the macroscopic balances on the reactor are given by

$$\frac{dM_j(t)}{dt} = M_j^{\text{in}}(t) - M_j^{\text{out}}(t) + R_j(t) \qquad (6)$$

where $M_j(t)$ is the instantaneous moles of species j in the system, $M_j^{\text{in}}(t)$ is the molar rate of species j entering the system, and $M_j^{\text{out}}(t)$ is the molar rate of species j leaving the system.

The overall reaction rates for species in the reactor will always depend linearly on reaction rates $r_i(t)$. By taking advantage of this fact, the macroscopic balance equations can be formulated as a linear regression model

$$\mathbf{Y} = \mathbf{X}\theta + \epsilon \qquad (7)$$

where $\mathbf{Y}$ is a vector of n_d dependent variable values, θ is a vector of n_p parameter values, $\mathbf{X}$ is a $n_d \times n_p$ matrix of independent variable values, and ϵ is a vector of length n_d corresponding to the error between the linear regression model and the true data $\mathbf{Y}$. Linear regression based on the above model form returns a system of the form

$$\hat{\mathbf{Y}} = \mathbf{X}\hat{\theta} \qquad (8)$$

Given the above linear regression model and criteria to evaluate reduced models, VS techniques seek to determine one or a number of subsets of the above model. In short, we seek a model of the form

$$\hat{\mathbf{Y}} = \mathbf{X}_r\hat{\theta}_r \qquad (9)$$

where $\mathbf{X}_r$ is a $n_d \times n_t$ submatrix of $\mathbf{X}$ and $\hat{\theta}_r$ is a n_t parameter subset of n_p.

	Reactions:		Rate:
1)	$A \longrightarrow B + E$		$r_1 = k_1C_1$
2)	$A \longrightarrow C + 2E$		$r_2 = k_2C_1$
3)	$B \longrightarrow C + E$		$r_3 = k_3C_2$
4)	$C \longrightarrow D + E$		$r_4 = k_4C_3$
5)	$F \longrightarrow 2B$		$r_5 = k_5C_6$
6)	$2B \longrightarrow F$		$r_6 = k_6C_2^2$

Fig. 1. Reaction network and reaction rate expressions for batch reactor linear regression formulation. $C_1 - C_6$ are the concentrations of species $A - F$, respectively, and k_i is the rate constant for reaction i.

Example: Batch Reactor Linear Regression Formulation. The linear regression formulation will now be developed for a batch reactor having the reaction network and reaction rates shown in Figure 1.

Since a batch reactor has no feed or product draws, the macroscopic balances are given by (2). Substituting the rate expressions in Figure 1 into (2), the macroscopic reactor balances become

$$\begin{aligned} R_1 &= -k_1C_1 - k_2C_1 \\ R_2 &= k_1C_1 - k_3C_2 + 2k_5C_6 - 2k_6C_2^2 \\ R_3 &= k_2C_1 + k_3C_2 - k_4C_3 \\ R_4 &= k_4C_3 \\ R_5 &= k_1C_1 + 2k_2C_1 + k_3C_2 + k_4C_3 \\ R_6 &= -k_5C_6 + k_6C_2^2 \end{aligned} \qquad (10)$$

Let the binary variable $\lambda_i = \{0, 1\}$ indicate whether reaction i is included in the reduced model. Then (10) can be expressed as

$$\begin{bmatrix} R_1 \\ R_2 \\ R_3 \\ R_4 \\ R_5 \\ R_6 \end{bmatrix} = \begin{bmatrix} -k_1C_1-k_2C_1 & 0 & 0 & 0 & 0 & 0 \\ k_1C_1 & 0 & -k_3C_2 & 0 & 2k_5C_6-2k_6C_2^2 & \\ 0 & k_2C_1 & k_3C_2 & -k_4C_3 & 0 & 0 \\ 0 & 0 & 0 & k_4C_3 & 0 & 0 \\ k_1C_1 & 2k_2C_1 & k_3C_2 & k_4C_3 & 0 & 0 \\ 0 & 0 & 0 & 0 & -k_5C_6 & k_6C_2^2 \end{bmatrix} \begin{bmatrix} \lambda_1 \\ \lambda_2 \\ \lambda_3 \\ \lambda_4 \\ \lambda_5 \\ \lambda_6 \end{bmatrix} \qquad (11)$$

In more compact form,

$$\mathbf{R} = \mathbf{A}\lambda \qquad (12)$$

Note that $\lambda_i = 1, i = 1, n_r$ when the full model concentrations and reaction rates for a specific point in time are substituted into the above expression. But in order to apply VS techniques to determine reaction subsets, we seek a system whereby we can solve for λ_i using least-squares. For this reason, it is necessary to use a linear system that contains reaction rate information for several points along the batch reactor profile. Thus (12) must be modified.

Consider the batch reactor concentrations for n_d different points in time. At each time t,

$$\mathbf{R}^t = \mathbf{A}^t\lambda \qquad (13)$$

Then for n_d different points in time,

$$\begin{bmatrix} \mathbf{R}^1 \\ \vdots \\ \mathbf{R}^{n_d} \end{bmatrix} = \begin{bmatrix} \mathbf{A}^1 \\ \vdots \\ \mathbf{A}^{n_d} \end{bmatrix} \lambda$$

$$\Downarrow$$

$$\overline{\mathbf{R}} = \overline{\mathbf{A}}\lambda \qquad (14)$$

Now the batch reactor dynamics are formulated in terms of a linear regression model. Vector $\overline{\mathbf{R}}$ contains the dependent variable values, matrix $\overline{\mathbf{A}}$ contains the independent variable values, and vector λ contains the model parameters. The linear regression for (14) will return $\lambda_i = 1$ for $i = 1, n_r$ provided that a) full-model concentrations and reaction rates are used in (14), b) appropriate data points are used, and c) the system is properly scaled.

3.1. *Determining Reaction Subsets*

By encoding reaction network dynamic data in the form of (14), the Leaps and Bounds algorithm (Furnival and Wilson, 1974) can be applied to determine a family of subset models, i.e., a family of reaction subsets that approximate the full network kinetics. Leaps and Bounds is an implicit method of finding the best linear-regression subset models without having to evaluate every sub-model. Working from the smallest to the largest reaction subsets, each subset can be evaluated by integrating the reactor dynamics for the particular subset and comparing concentration profiles to those for the full model. The smallest reaction subset meeting predefined error constraints is considered to be an estimate of the global solution. The solution steps are given below.

1. Given a reaction network with n_r reactions and n_s species, choose n_k species that are required to be known accurately in the reduced model. Provide error constraints B_j in (5) for the n_k species that are enforced along the profiles.
2. Using the full-model concentrations and rate constants, generate the $\overline{\mathbf{R}}$ vector and $\overline{\mathbf{A}}$ matrix for the n_k species at appropriate points along the concentration profiles.
3. Treating λ as model parameters to be determined, apply Leaps and Bounds to find n_r best subset models.
4. Integrate the model equations for each subset and evaluate the error between the full and reduced models. The model having the fewest reactions and satisfying the relative error constraints is an approximation to the global so-

lution.

4. GA FORMULATION FOR NETWORK REDUCTION

4.1. *Parameterization*

To aid in parameterizing the network reduction problem it is necessary to introduce a systematic means of representing reduced networks. Let λ_i^r be a binary variable (i.e., 0 or 1) affiliated with reaction i. If $\lambda_i = 1$ in a reduced network, then reaction i is present in the reduced network. Otherwise, reaction i is absent. Similarly, let λ_j^s be a binary variable that denotes the involvement of species j in a reduced network. If $\lambda_j^s = 1$, then at least one reaction involving species j is present in the reduced network. If $\lambda_j^s = 0$, then no reactions involving species j are present.

To compute the number of reactions n_r' and species n_s' in a reduced network, simply sum the binary variables as follows:

$$\begin{aligned} n_r' &= \sum_{i=1}^{n_r} \lambda_i^r \\ n_s' &= \sum_{s=1}^{n_s} \lambda_j^s \end{aligned} \tag{15}$$

4.2. *Combinatorial Optimization Problem*

In encoding the network reduction problem as a combinatorial optimization problem, the first step is to define an appropriate objective function. Since the primary objective is to minimize n_r' subject to certain accuracy constraints, these variables should appear in the objective function. The error bounds (5) may be incorporated into the optimization problems as inequality constraints or as penalties to the objective function. The latter option is favored because it allows all subnetworks to be feasible, thus greatly facilitating the application of genetic algorithms (GA).

In choosing penalty functions, it is important that they are well-scaled and result in a smooth objective function. The following is a suitable choice for incorporating accuracy constraints:

$$P = \frac{1}{n_k} \sum_{j=1}^{n_k} \frac{c_j e_j^{\text{avg.}}}{B_j} \tag{16}$$

where c_j are user-defined penalty coefficients. Other choices for penalty functions exist, but (16) proved suitable for the test problem in the next section.

The network reduction problem may now be encoded as a combinatorial optimization problem:

$$\begin{aligned} \min \quad & \sum_{i=1}^{n_r} \lambda_i^r + \frac{1}{n_k} \sum_{j=1}^{n_k} \frac{c_j e_j^{\text{avg.}}}{B_j} \\ \text{s.t.} \quad & \\ e_j^{\text{avg.}} &= \frac{1}{n_t} \sum_{l=1}^{n_t} \left| \log_{10} \left(\frac{C_j^{\text{full}}(t_l)+\epsilon}{C_j^{\text{red.}}(t_l)+\epsilon} \right) \right| \\ r_i(t_l) &= k_i \prod_{j=1}^{n_s} C_j^{\alpha_{ij}}(t_l) \\ R_j(t_l) &= \sum_{i=1}^{n_r} (\beta_{ij} - \alpha_{ij}) r_i(t_l) \\ k_i^r &= \lambda_i^r k_i \\ n_r' &= \sum_{i=1}^{n_r} \lambda_i^r \\ \lambda_i^r &= 0 \text{ or } 1 \\ i &= 1, n_r \\ j &= 1, n_s \\ l &= 1, n_t \end{aligned} \tag{17}$$

The parameters α_{ij}, ν_{ij}, k_i, and t_l in (17) are presumed to be known, which in turn means $C_j^{\text{full}}(t_l)$ can be calculated. Similarly B_j, c_j, and ϵ are specified. This leaves λ_i^r, k_i^r, $C_j^{\text{red.}}(t_l)$, and n_r' as unknowns. Once the λ_i^r are specified, all the unknown variables and the objective function may be calculated. Since there are n_r binary variables λ_i^r, the optimization problem has n_r degrees of freedom. For a more detailed discussion of GA, see Goldberg (1989) or Davis (1991).

4.3. *GA Encoding*

To parameterize (17) for the GA, a chromosome of length n_r with 0-1 bits is used, where the ith bit in the chromosome determines the involvement of the ith reaction for a subnetwork. In other words, the ith bit is equivalent to λ_i^r. Because the GA is set up to maximize "fitness", The objective functions must be altered by a minus sign and the encodings changed from minimization to maximization.

5. TEST PROBLEM

The network of reactions used in the test problem is shown in Figure 4. The reactions in this network were reduced using both the VS and GA techniques. Reduced models were sought that accurately model CF_2 and F concentrations, and error bounds for these species were chosen to be $B_j = 0.02$, which approximates relative error bounds of 0.05. The time domain for model reduction is $[t_0, t_f] = [0, 0.035]$ where time is in seconds. The scaled species concentration profiles over this time domain are shown in Figure 2, while the scaled species reaction rates are shown in Figure 3. The discrete points in Figures 2 and

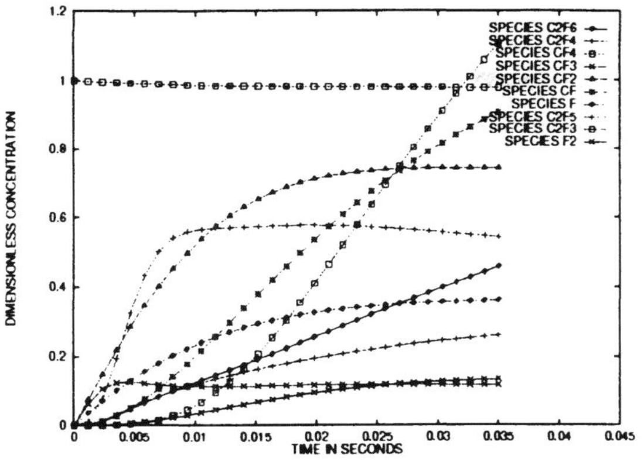

Fig. 2. Dimensionless concentration profiles for 18 reaction network.

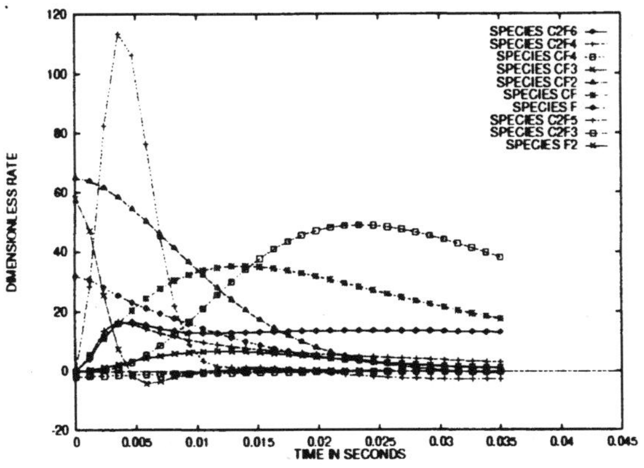

Fig. 3. Dimensionless reaction-rate profiles for 18 reaction network.

3 were used in the reduction techniques, but the concentration and rate profiles were scaled so as to minimize the condition number of $\overline{\mathbf{A}}$ defined in (14). The VS problem was solved using Octave (Eaton, 1994) on an IBM RS6000 running AIX Release 3.2.5.

In performing VS, the coefficient of determination, R^2, was used to rank order the reduced models (Rawlings, 1988). The regression model having all 18 parameters (i.e, data from all 18 reactions) has an $R^2 = 1$, while the regression model having no parameters has an $R^2 = 0$. All other regression models have $0 < R^2 < 1$. The VS procedure written in Octave was run so that the best two models for each model size were generated. Since there are 18 possible parameters (i.e., reactions), a total of 36 candidate models (i.e., reduced networks) were generated.

Table 1 details the GA parameters chosen. Five runs were made for the network reduction problem, with each run corresponding to a different random-number generator seed. Note that tournament selection and one-point crossover were utilized. Also note that "Fraction Function Evaluations" in Table 1 is the total number of individuals over all generations divided by 2^{n_r}. The GA code

Reactions:

1)	CF_4	$\longrightarrow$	$CF_3 + F$
2)	CF_4	$\longrightarrow$	$CF_2 + 2F$
3)	CF_3	$\longrightarrow$	$CF_2 + F$
4)	CF_2	$\longrightarrow$	$CF + F$
5)	C_2F_6	$\longrightarrow$	$2CF_3$
6)	$2CF_3$	$\longrightarrow$	C_2F_6
7)	$CF_3 + F$	$\longrightarrow$	CF_4
8)	$CF_2 + F$	$\longrightarrow$	CF_3
9)	$CF + F$	$\longrightarrow$	CF_2
10)	$2CF_2$	$\longrightarrow$	C_2F_4
11)	C_2F_4	$\longrightarrow$	$2CF_2$
12)	$C_2F_4 + F$	$\longrightarrow$	$CF_3 + CF_2$
13)	$CF_3 + CF_2$	$\longrightarrow$	C_2F_5
14)	$C_2F_5 + F$	$\longrightarrow$	$2CF_3$
15)	$CF + CF_2$	$\longrightarrow$	C_2F_3
16)	$C_2F_3 + F$	$\longrightarrow$	C_2F_4
17)	F_2	$\longrightarrow$	$2F$
18)	$F + F$	$\longrightarrow$	F_2

Rate Constants:

k_1	$= 0.58$	k_2	$= 1.3$
k_3	$= 3.0$	k_4	$= 3.0$
k_5	$= 3.0$	k_6	$= 4.8 \times 10^{12}$
k_7	$= 9.0 \times 10^{12}$	k_8	$= 4.7 \times 10^{11}$
k_9	$= 6.0 \times 10^{9}$	k_{10}	$= 3.0 \times 10^{10}$
k_{11}	$= 3.0$	k_{12}	$= 2.4 \times 10^{13}$
k_{13}	$= 5.4 \times 10^{11}$	k_{14}	$= 6.0 \times 10^{12}$
k_{15}	$= 6.0 \times 10^{11}$	k_{16}	$= 6.0 \times 10^{11}$
k_{17}	$= 2.9 \times 10^{2}$	k_{18}	$= 2.0 \times 10^{7}$

Fig. 4. Reaction network and reaction rate expressions for GA network reduction scheme. Reaction rates are in gmol/cm^3/sec, and species concentrations are in gmol/cm^3.

was written in ANSI C and executed on an IBM RS6000 running AIX Release 3.2.5.

6. RESULTS

Model reduction results found using the VS method are presented in Table 2. Five reduced networks most closely meeting the accuracy requirements for CF_2 and F are presented. Observe that none of the reduced networks meet the accuracy requirements. In general, the error constraints are violated by an order of magnitude or

Table 1 GA parameters chosen for reaction and species reduction problems.

Population Size	30
Chromosome Length	18
Number of Generations	15
Fraction Function Evaluations	0.0017
Crossover Probability	0.6
Mutation Probability	0.01
Tournament Size	5

Table 2 Global and VS solutions for network reduction problem. The subnetworks are the five smallest that meet CF_2 and F error bounds.

solutions	chromosome bits	n'_r	CF_2 error	F error
Global	010100111101001100	9	1.376e-04	2.637e-04
Run 1	101101110000111000	9	2.221e-00	7.530e-01
Run 2	101101110010111000	10	2.221e-00	7.530e-01
Run 3	101101110011111000	11	2.221e-00	7.530e-01
Run 4	101101110101111100	12	2.221e-00	7.530e-01
Run 5	101111110101111100	13	2.219e-00	7.555e-01

Table 3 Global and GA solutions for network reduction problem. Each run corresponds to a different random-number-generator seed.

solutions	chromosome bits	n'_r	CF_2 error	F error
Global	010100111101001100	9	1.376e-04	2.637e-04
Run 1	110001110100100011	9	2.930e-02	6.684e-03
Run 2	110001110100100000	7	2.923e-02	6.930e-03
Run 3	010110111001101100	10	-1.153e+01	5.907e-04
Run 4	110001110100100000	7	2.923e-02	6.930e-03
Run 5	101101101101101000	10	1.365e-01	5.021e-02

more. The reason for this is that all ten species are given equal weight in the VS procedure, so that species errors for reduced networks are all approximately the same order of magnitude. Essentially, the VS procedure is searching for reduced networks that accurately predict all species profiles accurately, not just CF_2 and F. The remedy for this problem would be to use weighted least-squares, where the reaction-rate profiles for CF_2 and F are more heavily weighted than those of the other species.

Model reduction results found using GA are presented in Table 3. The five solutions presented were generated using five different random-number-generator seeds. Note that subnetwork sizes are comparable to those of the global solution, and, except for Run 4, the order of magnitude of the CF_2 and F species errors is less than or equal to that of the error bounds B_j. The best feasible subnetwork (i.e., error bounds are satisfied) is returned in Run 3 ($n'_r = 10$) of Table 3. These solutions compare favorably with the global solutions of Objectives 1 and 2 for which $n'_r = 9$.

The best infeasible solution found using the GA are for Run 2 or 4 ($n'_r = 7$) of Table 3. Note that the chromosomes (i.e., subnetworks) are the same for both solutions. Both solutions satisfy the F error bound and only slightly violate the CF_2 bound ($e^{\mathrm{avg.}}_{CF_2} = 0.029$, $B_{CF_2} = 0.02$).

7. CONCLUSION

Two novel approaches to reaction network simplification have been presented. The GA approach has been shown to return nearly optimal solutions to an 18 reaction, 10 species network reduction problem. Although the VS approach returned sub-optimal solutions, the use of weighted least-squares should improve performance. The VS approach has the advantage that differential equations for the reactor model are integrated once to generate the data. An additional advantage of the approach is that an entire family of reduced models is generated, with each model having varying degrees of accuracy for the bounded species.

The GA approach utilizes an intelligent, randomized search to obtain solutions to the network reduction problem. Globally optimal or near-optimal solutions can be found in a fraction of the time required for complete enumeration. The GA approach has the disadvantage of requiring the equations for several subnetworks to be integrated for each generation, which may result in excessive computational expense for very complicated reactor models.

8. REFERENCES

Allred, Kevin Lyle (1991). Modeling, Optimization, and Control of the Selective Plasma Etching of Silicon and Silicon Dioxide. PhD thesis. The University of Texas at Austin. Austin, Texas.

Bird, R. Byron, Warren E. Stewart and Edwin N. Lightfoot (1960). *Transport Phenomena*. John Wiley and Sons.

Davis, Lawrence (1991). *Handbook of Genetic Algorithms*. Van Nostrand Reinhold.

Eaton, John W. (1994). Octave: A High-level Interactive Language for Numerical Computations. Department of Chemical Engineering, The University of Texas, Austin, Texas.

Furnival, G. M. and R. B. Wilson (1974). Regressions by leaps and bounds. *Technometrics* **13**, 499–511.

Goldberg, David E. (1989). *Genetic Algorithms in Search, Optimization, and Machine Learning*. Addison-Wesley Publishing Company, Inc.

Leone, Joseph A. and John H. Seinfeld (1985). Comparative analysis of chemical reaction mechanisms for photochemical smog. *Atmospheric Environment* **19**(3), 437–464.

Rawlings, John O. (1988). *Applied Regression Analysis: A Research Tool*. Wadsworth, Inc.

Sapre, A. V. (1991). Kinetic modeling at Mobil: An historical perspective. In: *Chemical Reactions in Complex Mixtures: The Mobil Workshop*. pp. 222–253. Van Nostrand Reinhold.

ROBUST DECENTRALIZED CONTROL OF A CSTR WITH COMPLEX REACTION SCHEME

Ralf Müller and Jorge O. Trierweiler

Process Control Group, Department of Chemical Engineering
Universität Dortmund, D-44221 Dortmund, Germany
e-mail: ralf@ast.chemietechnik.uni-dortmund.de

Abstract. A design method for robust controllers with decentralized structure is presented. The frequency response of a full controller obtained by any standard method is approximated by a decentralized controller. As the robustness measure of the approximated controller can be compared to that of the full controller, the price which has to be paid for decentralized control becomes evident.

Key Words. Process control, decentralized control, multivariable control systems, robust control, reactor control, structured singular value (μ), order reduction.

1. INTRODUCTION

The detailed model of an exothermal reaction system with consecutive and side reactions in a continuous stirred tank reactor (CSTR) which was presented by Engell and Klatt (1993) is considered. As the plant is highly nonlinear, robustness plays an essential role in designing a linear controller for the relevant range of operating points. In this paper we will compare the robust performance of a full-feedback controller with that of decentralized controllers.

Continuous stirred tank reactors are central components of many plants in the chemical industry. These systems may exhibit highly nonlinear dynamics, especially when consecutive and side reactions are present. Normally the reaction rates of the side and consecutive reactions are made small compared to that of the desired product by the choice of catalyst and appropriate reaction conditions. If this is not possible to a sufficient degree, the product concentration may be controlled by the inflow to the reactor and the reaction temperature.

Restrictions on the feedback compensator structure are often encountered in chemical plants, when several control stations are provided only with local measurements. Such decentralized information structures result in block-diagonal compensator matrices. Decentralized controllers are also attractive because the information about the feedback is concentrated in the diagonal blocks. This means they are easier to understand and to put into operation and more easily made failure tolerant than general multivariable control systems. The design of robust decentralized controllers remains a demanding problem; standard methods for robust design can not be used for structured compensators.

In our approach the decentralized control design is formulated as a problem of approximating a full-feedback controller. In the first step, a standard design method for centralized robust controllers is applied. In this paper we use D-K iteration to calculate μ-optimal controllers. But any other algorithm may be used to determine a μ- or H_∞-optimal controller. Then the frequency response of this controller is approximated by the desired decentralized structure, such as block-diagonal or one-way decoupling, with elements of specified orders. By changing the structure and orders of the controller, the trade-off between controller complexity and achievable performance can be explored in a systematic manner. Comparing the robustness measure of the approximated controller to that of the centralized controller reveals the price which has to be paid for decentralized control.

If the decentralized approximation of the full-feedback controller can not be improved by further augmenting the controller orders this might be due to the decentralized structure or to the conflict between minimizing the approximation error and maintaining closed-loop stability. To verify if the difficulties are inherent in the decentralized structure we improve the results using the parametrization of all stabilizing decentralized controllers proposed by Date and Chow (1994).

The outline of the paper is as follows: In the second section the method of controller reduction by frequency response approximation is presented. In the following section the procedure using the parametrization of all stabilizing decentralized controllers is introduced. In section 4 the method is demonstrated for the detailed nonlinear model of a CSTR. Some remarks will conclude this report.

2. CONTROLLER DESIGN BY FREQUENCY RESPONSE APPROXIMATION

In this section the approximation of a high-order controller resulting from any design method is considered. In the first step each controller column is approximated independently using an iterative least squares technique. The most important aspect of this procedure is the use of a frequency-dependent weighting function which takes the sensitivity of the closed-loop system to changes of the compensator into account. The controller resulting from the column-by-column approximation is improved by minimizing a mixed L_2/L_∞-criterion for the complete controller. The presented design method was implemented with a graphical user interface using MATLAB (Müller, 1994).

2.1. Column-by-column optimization

For a unity feedback control system with the plant $\underline{P}$ and controller $\underline{C}$, the reference-to-output frequency response is

$$\underline{T} = (\underline{I} + \underline{P}\underline{C})^{-1}\underline{P}\underline{C} \quad . \tag{1}$$

The aim of the approximation is to minimize $\Delta\underline{T}$, the deviation of the closed-loop system frequency response from its desired value. $\Delta\underline{T}$ can be brought to the simple form

$$\Delta\underline{T} = \underline{S}\underline{P}\Delta\underline{C}\underline{S}_0 \, , \tag{2}$$

where $\underline{S}_0$ and S are the nominal and the actual sensitivity functions, respectively. Replacing $\underline{S}$ by $\underline{S}_0$ yields a linear relation of $\Delta\underline{T}$ and $\Delta\underline{C}$:

$$\Delta\underline{T} \approx \underline{S}_0\underline{P}\Delta\underline{C}\underline{S}_0 \quad . \tag{3}$$

As $\underline{S}_0$ is approximately diagonal, each element of $\Delta\underline{C}$ affects only the corresponding column of $\Delta\underline{T}$ and the controller columns can be approximated independently. The nonlinear problem of optimizing the numerator and the denominator polynomials simultaneously is solved iteratively in a least squares sense. As the orders of the numerator polynomials can be different and each coefficient can be set to zero, arbitrary structures as e.g. blockdiagonal controllers can be optimized (for more details see Engell and Müller, 1993).

2.2. The overall optimization

If $\underline{C}_0$ is approximated by a relatively simple transfer matrix, the achieved sensitivity function differs significantly from $\underline{S}_0$. Hence the solution of the column-by-column optimization can be improved by a nonlinear minimization of the overall quadratic error

$$J = \sum_{k=1}^{r}\sum_{i=1}^{r}\sum_{\ell=1}^{N}\left|\frac{\Delta t_{ik}(j\omega_\ell)}{\omega_\ell}\right|^2 , \tag{4}$$

where r and N are the numbers of reference inputs and frequency points, respectively. $\Delta\underline{T}$ is divided by ω to approximate the step responses rather than the impulse responses.

In order to avoid unstable approximations of stable controllers, the denominators are parametrized with positive values of damping and natural frequency. Stability of the closed-loop system follows from the small gain theorem if

$$\left\|\underline{S}_0\underline{P}\,\Delta\underline{C}\right\|_\infty < 1 \tag{5}$$

is satisfied. To emphasize values greater than one (5) is taken to the square which leads to the criterion

$$J = \lambda\sum_{k=1}^{r}\sum_{i=1}^{r}\sum_{\ell=1}^{N}\left|\Delta t_{ik}(j\omega_\ell)\frac{\omega^-}{\omega_\ell}\right|^2 + \left\|\underline{S}_0\underline{P}\Delta\underline{C}\right\|_\infty^2 \; . \tag{6}$$

The normalization with ω^-, the smallest frequency value used for the approximation, is necessary to make the first term, and thus the minimizing argument, independent of the considered frequency range.

The trade-off between closed-loop stability and small approximation errors can be influenced by λ. A greater value of λ reduces the L_2-error but increases the risc of instability. If the approximaton with the default value of 1000 is not satisfactory, λ can be decreased to achieve stability or increased to get a smaller L_2-error. Especially for the decentralized approximation of full-feedback controllers λ is a valuable knob to maintain closed-loop stability.

To minimize the cost functional, a BFGS Quasi-Newton algorithm is used, starting from the solution of the independent optimization of the controller columns. Solving the overall optimization without this starting point is not feasible. Furthermore, the column-by-column approximation is an effective and computationally cheap indicator whether the solution is feasible with the chosen controller structure and the specified orders.

3. THE SET OF ALL STABILIZING DECENTRALIZED CONTROLLERS

The minimization of the L_2/L_∞-criterion for high-order decentralized controllers can give unstable closed-loops for high values of λ. Although (5) is a very good indicator of closed-loop stability, it may for some cases be very conservative. To avoid this problem, the parametrization of all decentralized stabilizing controllers is used to improve the result of the procedure described in last section.

The extension of the stable factor (SF) approach to the design of decentralized controllers given by Date and Chow (1994) is a natural extension of the familiar SF used in the design of full feedback control. The prime motivation for developing the decentralized stable factor (DSF) is that it allows a parametrization of all decentralized stabilizing controllers in the form of a Youla parametrization which has a particular structure and must satisfy an interaction constraint.

3.1. Algorithm for the synthesis of a DSF

Decentralized stable factorization is a special coprime factorization of P that satisfies the decentralized Bezout identity. The following algorithm shows how a DSF can be calculated.

Algorithm 1:

Step 1: Obtain the right coprime factors (RCF) and left coprime factors (LCF) of P

$$P = N_{P0}D_{P0}^{-1} = \tilde{D}_{P0}^{-1}\tilde{N}_{P0} \quad . \tag{7}$$

Step 2: For a stabilizing decentralized controller K_d obtain the corresponding LCF and RCF

$$K_d^{(i)} = \tilde{V}^{-1}\tilde{U} = U V^{-1} \quad . \tag{8}$$

Step 3: Calculate the units

$$\begin{aligned} Z_1 &= \tilde{V}D_{P0} + \tilde{U}N_{P0} \\ Z_2 &= \tilde{D}_{P0}V + \tilde{N}_{P0}U \quad . \end{aligned} \tag{9}$$

Step 4: The right and left DSF of P are

$$\begin{cases} D_P = D_{P0}Z_1^{-1} \\ N_P = N_{P0}Z_1^{-1} \end{cases} \text{and} \begin{cases} \tilde{D}_P = Z_2^{-1}\tilde{D}_{P0} \\ \tilde{N}_P = Z_2^{-1}\tilde{N}_{P0} \end{cases} \tag{10}$$

Step 5: Calculate the interaction matrices

$$\begin{aligned} W_{12} &= -\tilde{N}_{P11}D_{P12} + \tilde{D}_{P11}N_{P12} = \tilde{N}_{P12}D_{P22} - \tilde{D}_{P12}N_{P22} \\ W_{21} &= -\tilde{N}_{P22}D_{P21} + \tilde{D}_{P22}N_{P21} = \tilde{N}_{P21}D_{P11} - \tilde{D}_{P21}N_{P11} \end{aligned} \tag{11}$$

Using the DSF obtained by the above algorithm together with the interaction matrices, the parametrization of all decentralized stabilizing controllers can be developed.

Parametrization (Date and Chow, 1994): The set of all decentralized stabilizing controllers is

$$S(P,K_d) = \begin{Bmatrix} K_d = \left(\tilde{V} - Q\tilde{N}_p\right)^{-1}\left(\tilde{U} + Q\tilde{D}_p\right) \\ = \left(U + D_pQ\right)\left(V - N_pQ\right)^{-1} \\ /Q \in \mathcal{Q}\ ,\det\left(\tilde{V} - Q\tilde{N}_p\right) \neq 0 \end{Bmatrix} \tag{12}$$

where

$$\mathcal{Q} = \left\{ Q \ \ / \ \ Q = Q_w^{-1}\begin{bmatrix} Q_1 & 0 \\ 0 & Q_2 \end{bmatrix} \right\},$$

$$\begin{aligned} Q_w &= \begin{bmatrix} Q_{11} & Q_1W_{12} \\ Q_2W_{21} & Q_{22} \end{bmatrix} \in R_u \text{ unimodular set} \\ Q_w^{-1} &= \frac{1}{Q_{22}Q_{11} - Q_2W_{21}Q_1W_{12}}\begin{bmatrix} Q_{22} & -Q_1W_{12} \\ -Q_2W_{21} & Q_{11} \end{bmatrix} \end{aligned} \tag{13}$$

and $Q_{11}, Q_{22}, Q_1, Q_2 \in R_u$ (set of matrices with elements in the principal ring of stable rational functions).

Remark 1: The parametrization can be used to improve a given decentralized controller K_d using the parameter Q. It is possible to redesign the controllers for each channel independently. For example, to redesign the controller in channel 1, we can choose $Q_2=0$, $Q_{11}=Q_{22}=I$. Then Q_w will be a unit for any stable, proper Q_1. Furthermore, the controller for the second channel will remain unchanged.

Remark 2: The main constraint is that Q_w must be an unimodular matrix. To assure this we choose:

$$\begin{aligned} Q_{11} &= I \\ Q_{22} &= I + Q_2W_{21}Q_1W_{12} \quad . \end{aligned} \tag{14}$$

Substituting in (13) yields

$$\mathcal{Q} = \left\{ Q \,/\, Q = \begin{bmatrix} I + Q_2W_{21}Q_1W_{12} & -Q_1W_{12} \\ -Q_2W_{21} & I \end{bmatrix}\begin{bmatrix} Q_1 & 0 \\ 0 & Q_2 \end{bmatrix} \right\} \tag{15}$$

with the only restriction $Q_1, Q_2 \in R_u$.

3.2. Fequency response approximation with DSF

The low-order controller obtained by the method described in section 2 can be used to calculate the DSF applying the algorithm 1. For a chosen frequency interval calculate the frequency response of all matrices involved and use the following strategy for further optimization (for more details see Trierweiler and Müller, 1995):

Step 1: Initialization: $Q_{01}=I$ and $Q_{02}=I$

Step 2: Optimize $\underset{Q_{1opt},Q_{2opt}}{\text{Min}} \left\| \frac{\|T - T_d\|_F}{\omega^n} \right\|_p$ (16)

where the frequency response $T_d = PK_d(I + PK_d)^{-1}$ is calculated for a given $Q_1=Q_{01}\,Q_{1opt}$ and $Q_2=Q_{02}\,Q_{2opt}$ from (12) and (15).

Step 3: Set $Q_{01}=Q_1$ and $Q_{02}=Q_2$. If the approximation is still not satisfatory, return to step 1.

Step 4: Approximate the decentralized controller frequency responses applying the method from section 2.

Two factors limit the performance which can be achieved by decentralized control. One is associated with the controller order and the other is inherent in the decentralized structure. The above algorithm allows a separate analysis of both restrictions. By steps 2 and 3 the attainable performance with a decentralized control structure can be explored. In step 4 the effect of the controller order can be investigated systematically.

4. APPLICATION TO A CSTR

4.1. Description of the problem

The plant considered here is a CSTR in which cyclopentenol (B) is produced from cyclopentadiene (A). The by-products cyclopentanediol (C) and dicyclopentadiene (D) are produced in unwanted side and consecutive reactions:

$$\begin{aligned} A &\xrightarrow{k_1} B \xrightarrow{k_2} C \\ 2A &\xrightarrow{k_3} D \end{aligned} \quad . \tag{17}$$

The dynamics of the system are described by a system of four nonlinear equations derived from mass and energy balances (Engell and Klatt, 1993).

The product concentration c_B of the reactor is to be controlled in the range $0.7\text{mol/l} \le c_B \le 0.95$ mol/l. The inverse of the residence time, F, and the amount of heat removed by the coolant, $\dot{Q}_K$, are available as manipulated variables. They can only vary in the following ranges:

$$\begin{aligned} 5h^{-1} \le F = \frac{\dot{V}}{V_R} \le 35h^{-1} \\ -8500\frac{kJ}{h} \le \dot{Q}_K \le 0\frac{kJ}{h} \quad . \end{aligned} \tag{18}$$

The reactant cyclopentadiene (A) is fed into the reactor with concentration c_{A0} and temperature T_0. c_{A0} and T_0 are considered as the main disturbances which may vary in the range

$$4.5\frac{\text{mol}}{\text{l}} \le c_{A0} \le 5.7\frac{\text{mol}}{\text{l}}$$
$$125\,°\text{C} \le T_0 \le 135\,°\text{C} \quad . \tag{19}$$

For the nominal model the values of c_{A0}, c_B and T_0 were chosen in the middle of the considered range and T was fixed at 135°C. The linearized nominal plant is given in Table 1.

Table 1 Nominal linearized plant

$$\begin{cases}\dot{\underline{x}} = \underline{A}\,\underline{x} + \underline{B}\,\underline{u} \\ \underline{y} = \underline{C}\,\underline{x} + \underline{D}\,\underline{u}\end{cases} \text{, where}$$

$$\underline{u}_0^T = [\,F \quad \dot{Q}_K\,] = [\,16.2943\text{h}^{-1} \quad -3555\frac{\text{kJ}}{\text{h}}\,],$$

$$T_0 = 130\,°\text{C} \text{ and } c_{A0} = 5.1\frac{\text{mol}}{\text{l}}$$

$$\underline{x}_0^T = [\,c_A \quad c_B \quad T \quad T_K\,]$$
$$= [1.078\frac{\text{mol}}{\text{l}} \quad 0.825\frac{\text{mol}}{\text{l}} \quad 135.0°\text{C} \quad 130.9°\text{C}]$$

$$\underline{A} = \begin{bmatrix} -84.7158 & 0 & -3.7806 & 0 \\ 53.2354 & -69.5297 & 0.7875 & 0 \\ 146.5003 & 208.2508 & -35.8206 & 30.8285 \\ 0 & 0 & 86.6880 & -86.6880 \end{bmatrix}$$

$$\underline{B} = \begin{bmatrix} 4.0225 & 0 \\ -0.8250 & 0 \\ -5.0000 & 0 \\ 0 & 0.1 \end{bmatrix}, \underline{C} = \begin{bmatrix} 0 & 1 & 0 & 0 \\ 0 & 0 & 1 & 0 \end{bmatrix}, \underline{D} = \underline{0} \;.$$

4.2. Uncertainty model

The system uncertainty consists of two major parts. The first one can be modeled by a family of linearized plants corresponding to the different steady-state points. The other part is due to uncertain chemo-physical parameters (e.g.: rate coefficients, activation energies, reaction enthalpies) which are known only within bounds, with a relative uncertainty ranging from 1.3% to 56%.

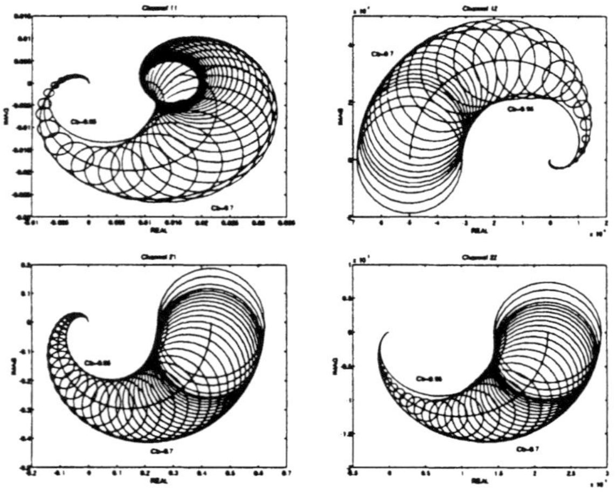

Fig. 1. Nyquist diagram of each channel with uncertainty bounds

Due to the large number of uncertain parameters it is most convenient to model the complete uncertainty as unstructured bands for each element of the transfer matrix. Linearization of the CSTR model for many different operating points showed that most of the possible variations can be included in the uncertainty bands obtained by linearizing the model for the extreme values of c_B. The nyquist diagrams of the nominal plant are shown in Fig. 1 with the uncertainty plotted as circles around the nominal values. The radii of the uncertainties in each channel were approximated by 3rd order transfer functions which are summarized in Table 2.

Table 2 Uncertainty transfer functions

$$l_{g11}(s) = \frac{0.004805\;(0.001001s+1)(0.009398s+1)(0.3586s+1)}{(0.001001s^2+0.03855s+1)(0.158s+1)}$$

$$l_{g12}(s) = \frac{0.00001877\;(0.0003883s+1)(0.00361s+1)(0.04372s+1)}{(0.004281s^2+0.0547s+1)(0.1484s+1)}$$

$$l_{g21}(s) = \frac{0.1887\;(1.06e-006s^2+0.001471s+1)(0.8382s+1)}{(0.01169s+1)(0.1176s+1)(0.8222s+1)}$$

$$l_{g22}(s) = \frac{0.0007439\;(1.888e-006s^2+0.001023s+1)(0.2156s+1)}{(0.008591s^2+0.1264s+1)(0.2895s+1)}$$

Fig. 2 shows the block-structure of the nominal plant with uncertainties that was used for the design of the μ-optimal controller.

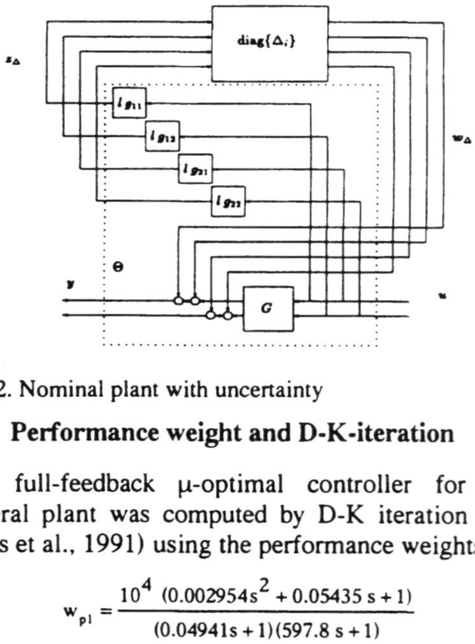

Fig. 2. Nominal plant with uncertainty

4.3. Performance weight and D-K-iteration

The full-feedback μ-optimal controller for the general plant was computed by D-K iteration (see Balas et al., 1991) using the performance weights:

$$w_{p1} = \frac{10^4\;(0.002954s^2+0.05435\,s+1)}{(0.04941s+1)(597.8\,s+1)}$$

$$w_{p2} = \frac{10^4\;(0.00189s^2+0.04348s+1)}{(0.04348s+1)(434.7s+1)} \tag{20}$$

$$\underline{W}_p = \text{diag}\{w_{p1} \quad w_{p2}\}$$

With the second order performance weights in (20) the resulting responses were faster than with a first order weight for the same minimal μ-value.

After five D-K iterations we obtained a controller with 78 states and a μ-norm of 1.23 (see Fig. 4).

The simulation results of product concentration reference steps which are shown in the sequel are calculated for the educt concentrations c_{A0} = 4.5 mol/l, c_{A0} = 5.1 mol/l and c_{A0} = 5.7 mol/l. Furthermore a deviation of +2% in the activation energy of k_1 and k_2 in (17) from its nominal value is simulated. To make the results comparable with the responses given by Klatt and Engell (1995) all simulations are computed for the operating conditions T_k=128.95°C, T_0=130°C and T_{set}=134.14°C.

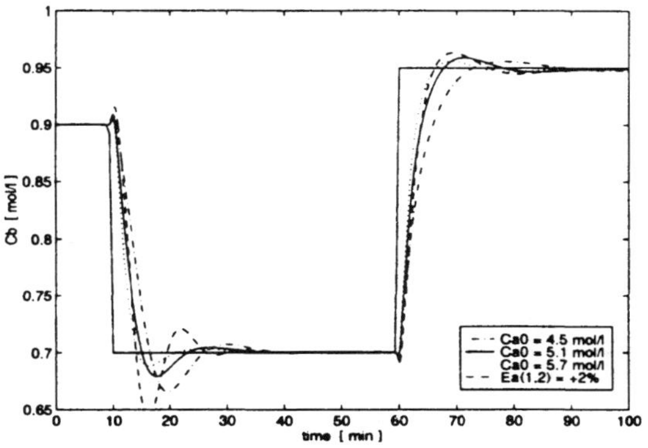

Fig. 3. Simulation of the 78 state controller

Fig. 3 shows the responses with the 78 states controller obtained by D-K-iteration. The diffence between the step responses for the considered educt concentrations are small. The simulation with the augmented activation energy which causes 40% greater reaction rates shows that the controller is insensitive to this deviation, even though it is not included in the uncertainty model.

4.4. Controller reduction by frequency response approximation

To check which μ-value can be achieved with low-order controllers first the frequency response approximation of section 2 is applied to get full-feedback controllers. Optimization of a PI-controller yields

$$\underline{C}_1 = \frac{1}{s}\begin{pmatrix} 11.9\,s + 926 & -15.5\,s + 571 \\ 2652\,s - 213000 & 10900\,s + 198000 \end{pmatrix}.$$

An increase of the controller compexity to 4 states, yields no improvement. Approximation of each column with order 3 gives

$$\underline{C}_2 = \begin{pmatrix} \dfrac{0.00069\,s^3 + 0.497\,s^2 + 31.8\,s + 956}{(0.000048\,s^2 + 0.026\,s + 1)\,s} & \dfrac{-0.00029\,s^3 + 0.37\,s^2 + 31.2\,s + 1110}{(0.0000023\,s^2 + 0.14\,s + 1)\,s} \\ \dfrac{1.14\,s^3 + 18.8\,s^2 - 1500\,s - 194000}{(0.000048\,s^2 + 0.026\,s + 1)\,s} & \dfrac{4.24\,s^3 + 983\,s^2 + 32000\,s + 292000}{(0.0000023\,s^2 + 0.14\,s + 1)\,s} \end{pmatrix}.$$

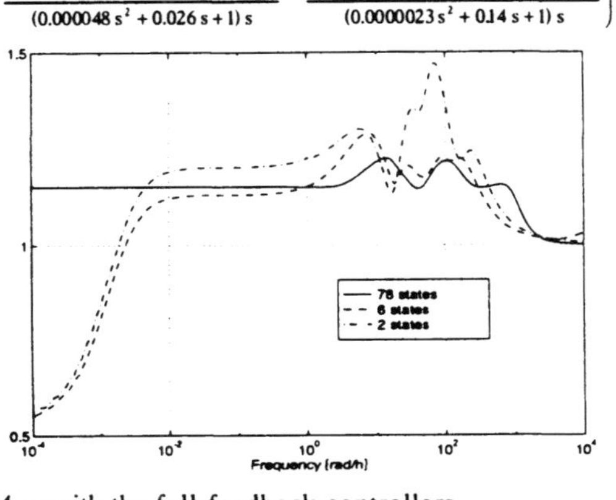

Fig. 4. μ with the full-feedback controllers

The μ-curves of the approximated full-feedback controllers are compared in Fig. 4 with the controller resulting from D-K-iteration. It can be seen that even with PI-controllers the approximation is quite good and with order 6 a μ of 1.28 is achieved. The significantly smaller μ-values of the approximated controllers at low frequencies result from the integral part of the controller, which could not be enforced exactly with μ-synthesis.

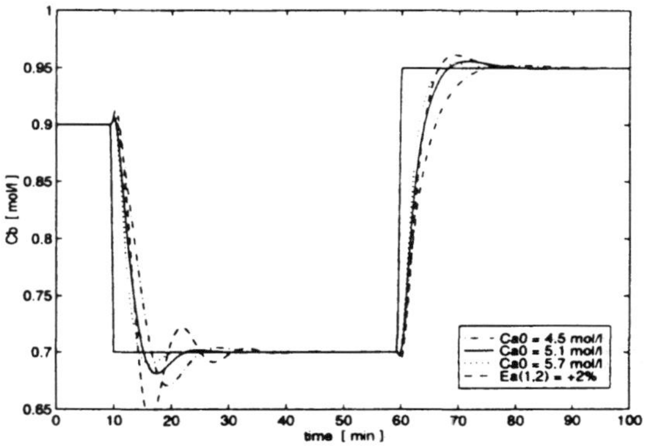

Fig. 5. Simulation of the full-feedback controller $\underline{C}_2$

The simulations with the 6 state full-feedback controller are practically the same as with the original one.

Application of the approximation method described in section 2 to optimize a decentralized PI-controller yields

$$\underline{C}_3 = \frac{1}{s}\begin{pmatrix} 12.5\,s + 786 & 0 \\ 0 & 12500\,s + 142000 \end{pmatrix}$$

The result for two 2nd order controllers is

$$\underline{C}_4 = \begin{pmatrix} \dfrac{0.23\,s^2 + 11.9\,s + 761}{(0.0043\,s + 1)\,s} & 0 \\ 0 & \dfrac{42.2\,s^2 + 5700\,s + 230000}{(0.002\,s + 1)\,s} \end{pmatrix}.$$

The achieved μ of the decentralized controllers and a one way decoupling controller consisting of two PI-controllers in the first column and a single PID-controller in column 2 are shown in Fig. 6.

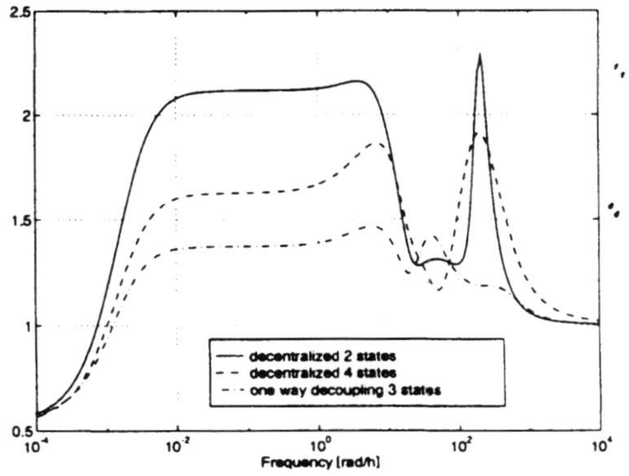

Fig. 6. μ with the decentralized controllers

The decentralized PI-controller achieves already a μ-norm of 2.3; for the decentralized controller having 4 states a μ–norm of 1.89 is obtained. Table 3 summarizes the μ-values for all designed controllers.

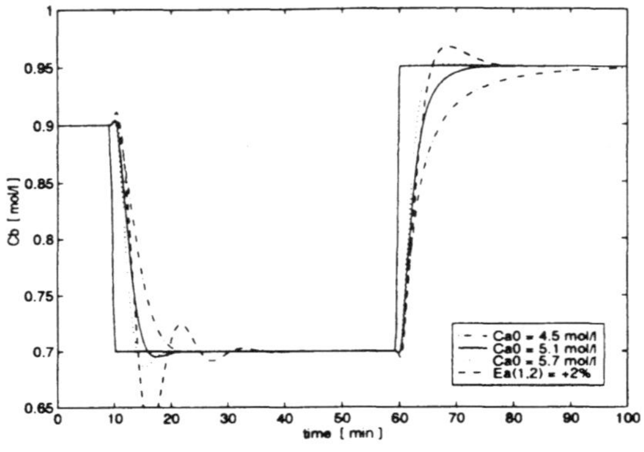

Fig. 7. Simulation of the decentralized PI-controller

The simulation results with the decentralized PI-controller are shown in Fig. 7. The controller is very robust. Only for c_{A0}=4.5 mol/l the responses are remarkably slower than with the centralized controllers. The step responses with $\underline{C}_4$ are only slightly better and therefore omitted.

4.5. Controller design using DSF

With the 4 state decentralized controller a μ-value of 1.89 was achieved. A further increase of the controller order yields no improvement of μ. To improve the results we applied the method based on DSF presented in section 3. In Fig. 8 the μ–curve with the best controller which achieved a μ-norm of 1.86 is shown.

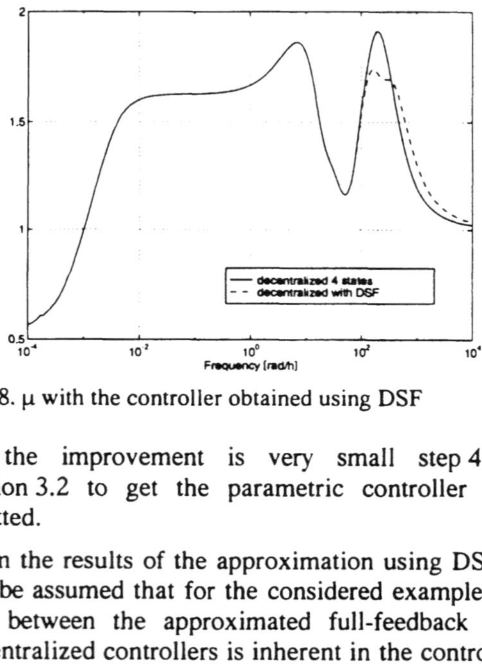

Fig. 8. μ with the controller obtained using DSF

As the improvement is very small step 4 in section 3.2 to get the parametric controller was omitted.

From the results of the approximation using DSF it can be assumed that for the considered example the gap between the approximated full-feedback and decentralized controllers is inherent in the controller structure and can not be compensated by increasing the controller orders.

Table 3 Comparison of the achieved μ-values

	full-feedback			decentralized		
	D-K	$\underline{C}_1$	$\underline{C}_2$	$\underline{C}_3$	$\underline{C}_4$	DSF
states	78	2	6	2	4	-
μ	1.23	1.47	1.28	2.3	1.89	1.86

5. CONCLUSIONS

A systematic method for the robust design of decentralized controllers was presented. The method is not restricted to μ-optimal controllers; any robust design method can be used to calculate the full feedback compensator which is approximated by the decentralized controller.

Two factors limit the performance which can be achieved by decentralized control. One is associated with the controller order and the other is inherent in the decentralized structure. Our method allows a separate analysis of both restrictions.

A shortcoming of the presented design procedure is that the decentralized controller approximates the behaviour of a full feedback controller. This straightforward approach may lead to suboptimal results, as the best centralized controller might not be the best goal for decentralized control. In our present work we try to optimize the decentralized controller directly without this intermediate state.

ACKNOWLEDGEMENTS

The support of this work by the Deutsche Forschungsgemeinschaft under grant En152/3 is gratefully acknowledged.

REFERENCES

Balas, G.J., J.C. Doyle, K. Glover, A. Packard, and R. Smith (1991). μ-Analysis and Synthesis Toolbox. MUSYN and MathWorks, Natick, MA.

Date, R. A. and J. H. Chow (1994). Decentralized Stable Factors and an Parametrization of Decentralized Controllers. *IEEE TAC*,**39**, 347-351.

Engell, S. and K.-U. Klatt (1993). Nonlinear Control of a Non-Minimum-Phase CSTR, Proc. American Control Conference, 2041-2045.

Engell, S. and R. Müller (1993). Multivariable Controller Design by Frequency Response Approximation. Proc. of the 2nd European Control Conference, Vol. 3, 1715-1720.

Grace, A. (1990). Optimization Toolbox for use with MATLAB. The MathWorks, Natick, MA.

Klatt, K.-U. and S. Engell (1995). Gain Scheduling Trajectory Control of a Continuous Stirred Tank Reactor with Complex Reaction Scheme. Submitted to DYCORD+ '95.

Müller, R. (1994). FASTER: Frequency response approximation for the synthesis of regulators, User's Guide. Process Control Group, Universität Dortmund.

Trierweiler, J. O. and R. Müller (1995). Decentralized Controller Design Using Decentralized Stable Factors. Submitted to the 1995 European Control Conference.

Copyright © IFAC Dynamics and Control of Chemical
Reactors (DYCORD+'95), Copenhagen, Denmark, 1995

INPUT MULTIPLICITY AND RIGHT HALF PLANE ZEROS IN IDEAL TWO-PRODUCT DISTILLATION

ELLING W. JACOBSEN

S3 - Automatic Control
Royal Institute of Technology - KTH, 100 44 Stockholm, Sweden

Abstract. The paper addresses the existence of input multiplicity in binary distillation columns. Input multiplicity denotes a situation where the specification of one or more of the plant outputs (e.g., a product composition) results in several possible steady state solutions in terms of the plant inputs (e.g., reflux). We first consider the case where one product composition is specified. In control terms this corresponds to one-point control and is a common arrangement in industry. For such specifications we show that input multiplicity may exist and that the multiplicity implies the existence of RHP zeros at some operating points. At the end of the paper we discuss the case where two compositions are specified and show that input multiplicity is unlikely if they are both product compositions. However, if at least one of the specifications involves a composition or a temperature on a plate inside the column, then input multiplicity may exist and will then imply the existence of RHP transmission zeros at some operating points.

Keywords. Distillation, modelling, dynamics, control, multiple steady states.

1. INTRODUCTION

There has recently been a growing interest in the existence of multiple steady state solutions in distillation columns. Most of the work has concentrated on separations involving complex mixtures forming azeotropes and potential liquid-liquid split. However, as shown in Jacobsen and Skogestad (1991), it is not only in the distillation of multicomponent mixtures with complex thermodynamics that multiple steady states may be experienced. By analyzing models where the internal flow rates of the column depend on the compositions, either through the energy-balance or because some flows are fixed on a mass or volume basis, they show that even ideal binary distillation columns may display multiple steady state solutions.

With multiple steady states is usually understood *output multiplicity*, that is, the existence of more than one solution in terms of the outputs, e.g., product compositions y_D and x_B, when the inputs, e.g., reflux L and boilup V, are specified. However, specifying the inputs is not the only possibility in distillation. With fixed feed conditions, a two-product distillation column generally has two degrees of freedom at steady state which may be used in the following ways

- Specify two inputs, e.g., L and V.
  (open-loop/manual operation)
- Specify one input and one output, e.g., V and y_D.
  (one-point control)
- Specify two outputs, e.g., y_D and x_B.
  (two-point control)

The existence of output multiplicity refers only to the first case. In this paper we address the last two cases, that is, we consider whether the specification of one or more outputs may yield several solutions in terms of the inputs. We denote this type of multiplicity *input multiplicity*. In a recent article Lucia and Li (1992) discuss all three cases referred to above and show that in each case there exist specifications that admit unique solutions. The only multiplicity they find, however, is for the specification of L and V, i.e., output multiplicity, which has previously been shown in Jacobsen and Skogestad (1991). In this paper we show that distillation columns, even in the ideal binary case, may display input multiplicity both with one and two outputs specified.

We start the paper with an example of a column which displays input multiplicity when one product composition is specified (one-point control). We provide an explanation for the observed multiplicity and discuss for which control configurations the multiplicity may exist. It is shown that the input multiplicity is related to a transfer-function zero crossing the imaginary axis, and thus the existence of RHP zeros at some operating points. At the end of the paper we discuss the case where two compositions are specified, and show that input multiplicity is unlikely if they are both product compositions. However, input multiplicity may exist if at least one of the specifications involves a composition or a temperature on a tray inside the column.

2. INTRODUCTORY EXAMPLE

Example 1: We consider a column separating a mixture of methanol and water. The column has 12 trays and the feed (saturated liquid, $F = 1.0$ kmol/min, $z_F = 0.75$) enters at tray 8. The model includes an energy balance and the thermodynamic properties are calculated using the Margules equation. Assuming feed conditions fixed leaves two degrees of freedom at steady state, and we here use the distillate flow

D and boilup V as the independent inputs. In control terms this corresponds to the DV-configuration, in which reflux L and bottoms flow B are used to control the condenser and reboiler levels, respectively. In the following we assume that the distillate flow D is used to manipulate the top composition y_D while the boilup V is kept fixed. Controlling only one product composition while keeping one input in manual is probably the most common arrangement in industrial practice. The nominal operating point we consider has product compositions $y_D = 0.99$ and $x_B = 0.02$, for which we find $D = 0.7525$ kmol/min and $V = 1.5534$ kmol/min. However, as seen from Table 1 and Figure 1, with boilup V fixed at this value the nominal purity $y_D = 0.99$ may be obtained with three different values of the distillate flow D. Thus, the column displays input multiplicity. The reason for the input multiplicity is that the gain $(\partial y_D/\partial D)_V$ changes sign through zero at some operating point.

We also note from Table 1 that x_B decreases with increasing D at all operating points, and in some region of operation the effect of increasing D with V constant is therefore to increase the purity of both products. This is a property which usually is attributed to increases in internal flows, i.e., increasing V with D constant.

Table 1. Steady-state solutions for methanol-water column in Example 1.

V kmol/min	y_D	D kmol/min	L kmol/min	x_B
1.5534	0.9900	0.7044	0.9687	0.1782
1.5534	0.9900	0.7525	1.0351	0.0200
1.5534	0.9900	0.7534	1.0377	0.0169

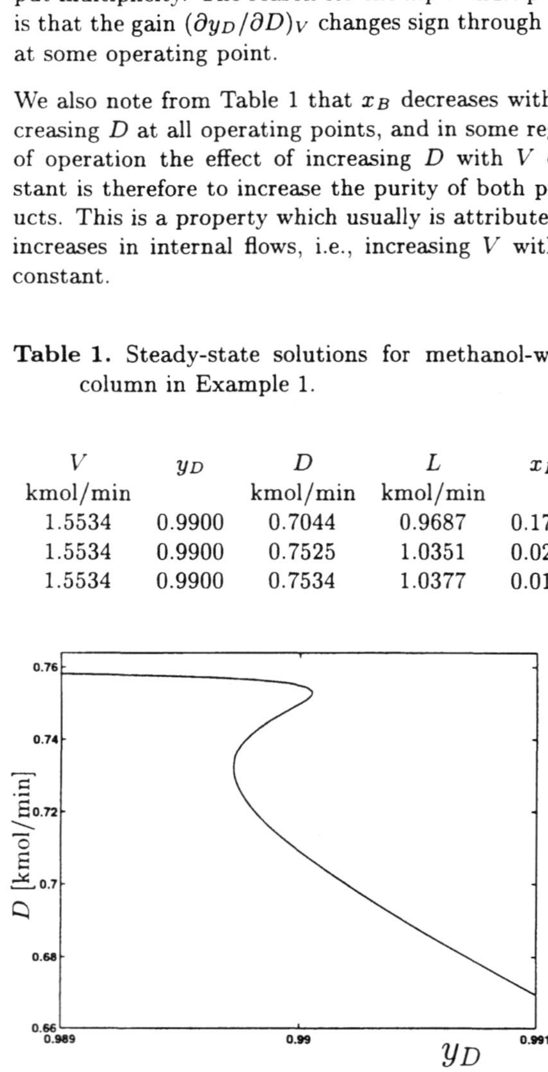

Fig. 1. Distillate flow D as a function of top composition y_D for methanol-water column in Example 1. Boilup $V = 1.5543$.

3. DV-CONFIGURATION

As seen in the introductory example, the DV-configuration may display input multiplicity. The multiplicity is brought about by the gain $(\partial y_D/\partial D)_V$ changing sign through zero at some operating point. This gain is usually assumed to be negative, and before we attempt to understand the source of the multiplicity we need to understand the reason behind this assumption.

The product compositions of a given distillation column (assuming fixed feed conditions) are determined by the relative liquid and vapor flow rates inside the column. In general, increasing the ratio of liquid to vapor flow rates tends to increase the purity in the top of the column, while decreasing this ratio tends to increase the purity in the bottom of the column (recall McCabe-Thiele). It is only in reflecting the change in D on the internal flow rates, through the level control, that the compositions are affected. With the common assumption of constant molar flows (neglected energy balance) we get $(\partial L_i/\partial D)_V = -1$ and $(\partial V_i/\partial D)_V = 0$ for all i, where L_i and V_i denote the liquid and vapor flows from tray i. Thus, increasing the distillate flow D will decrease the ratio of liquid to vapor flow rates throughout the column, thereby decreasing the amount of light component on all trays. The gain $(\partial y_D/\partial D)_V$ will therefore always be negative with the common assumption of constant molar flows, and we may conclude that the type of input multiplicity found in the introductory example cannot be observed in a model with neglected energy balance.

When the energy balance is included in the model the effect of a change in D on the internal flows rates becomes strongly dependent on the operating conditions. Jacobsen and Skogestad (1991) consider the effect of the energy balance on the internal flow rates and show that under certain operating conditions an increase in D with V constant may even increase the liquid flow rates in the top part of the column, i.e., $(\partial L_T/\partial D)_V > 0$ where L_T is the liquid flow from the top tray. As they show, this will correspond to an unstable operating point with the LV-configuration. However, note that an increase in L_T in this case is accompanied by an even larger increase in the vapor flow from the tray below since $dV_{T-1} = dL_T + dD$. Thus, having $(\partial L_T/\partial D)_V > 0$ does not necessarily imply that we increase the *ratio* of liquid to vapor flow rates in the top part of the column. The effect of a change in D on the ratio L_T/V_{T-1} may be written

$$\left(\frac{\partial(L_T/V_{T-1})}{\partial D}\right)_V = \frac{1}{V_{T-1}^2}\left[D\left(\frac{\partial L_T}{\partial D}\right)_V - L_T\right] \quad (1)$$

and we see that in order to have an increase in L_T/V_{T-1} we must require $(\partial L_T/\partial D)_V > L_T/D$. If this is the case, an increase in D results in an improvement of the operating lines in the top of the column and we might expect an increase in the top composition, i.e., $(\partial y_D/\partial D)_V > 0$.

For the methanol-water column $(\partial y_D/\partial D)_V > 0$ for D between 0.7317 and 0.7526 kmol/min, and $(\partial L_T/\partial D)_V > L_T/D$ for D between 0.7326 and 0.7541 kmol/min. Thus, there is a close correspondence between $(\partial y_D/\partial D)_V > 0$ and $(\partial L_T/\partial D)_V > L_T/D$.

From Table 1 we see that, while the gain $(\partial y_D/\partial D)_V$ changes sign with operating conditions, the gain $(\partial x_B/\partial D)_V$ is always negative. The reason for this may easily be seen by differentiating the overall material balance for the light component which yields

$$\left(\frac{\partial x_B}{\partial D}\right)_V = \frac{x_B - y_D - D\left(\frac{\partial y_D}{\partial D}\right)_V}{B} \tag{2}$$

From (2) we see that the gain $(\partial x_B/\partial D)_V$ must be negative whenever $(\partial y_D/\partial D)_V > 0$. In fact, there is no reason to expect $(\partial x_B/\partial D)_V$ to become positive at any operating point. To see this consider the effect of changes in D on the ratio of liquid to vapor flow rates in the bottom of the column. Differentiation of the material balance around the reboiler, $L_2 = V + B$, yields $(\partial L_2/\partial D)_V = -1$, and an increase in D with V constant will therefore always decrease the liquid to vapor flow ratio in the bottom part of the column. In general, at operating points where $(\partial L/\partial D)_V > L_T/D$, an increase in D will cause the ratio of liquid to vapor flow rates to increase in the top section and decrease in the bottom section. Thus, we expect the input multiplicity to exist only for compositions in the top part of the column. For the methanol-water column we find input multiplicity for compositions on trays 8 (the feed tray) to 13. Also note that there are cases where the input multiplicity is observed only on trays inside the column and not in y_D.

4. OTHER CONFIGURATIONS

4.1. LV-Configuration

In order to see if the input multiplicity found with the DV-configuration is conserved when using the LV-configuration consider the transformation

$$\left(\frac{\partial y_D}{\partial L}\right)_V = \left(\frac{\partial y_D}{\partial D}\right)_V \left[\left(\frac{\partial L}{\partial D}\right)_V\right]^{-1} \tag{3}$$

Above we found that an approximate criterion for $(\partial y_D/\partial D)_V$ to change sign was $(\partial L_T/\partial D)_V > L_T/D$ which is almost equivalent to $(\partial L/\partial D)_V > L/D$. Thus, $(\partial L/\partial D)_V$ is likely to be positive whenever $(\partial y_D/\partial D)_V$ changes sign and in that case we see from (3) that the gain $(\partial y_D/\partial L)_V$ changes sign through zero simultaneously. This implies that the input multiplicity found with the DV-configuration exist also with the LV-configuration.

Jacobsen and Skogestad (1991) find that the LV-configuration is the only configuration likely to display output multiplicity, and that the multiplicity is related to the fact that some operating points have $(\partial L/\partial D)_V > 0$. We here find that $(\partial L/\partial D)_V > 0$ is a prerequisite for input multiplicity, and the two types of multiplicities are therefore likely to exist in the same region of operation. Furthermore, for the same reason, all operating points in the region of input multiplicity are likely to be unstable with the LV-configuration (Jacobsen and Skogestad, 1994).

Figure 2 shows reflux L as a function of top composition y_D for the methanol-water column with boilup V fixed. We see that the column with the LV-configuration displays both input and output multiplicity.

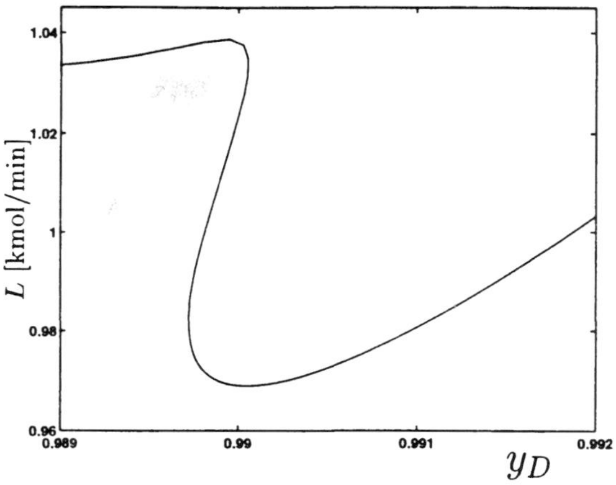

Fig. 2. Reflux flow L as a function of top composition y_D for methanol-water column in Example 1. Boilup $V = 1.5543$.

4.2. (L/D)V-Configuration

The use of flow ratios as independent inputs has been suggested in the literature as a possible improvement over the use of single flows to manipulate compositions. One such configuration is the $(L/D)V$-configuration, also known as "Ryskamps scheme". The transformation between this configuration and the DV-configuration may be written

$$\left(\frac{\partial y_D}{\partial (L/D)}\right)_V = \left(\frac{\partial y_D}{\partial D}\right)_V \frac{D}{\left(\frac{\partial L}{\partial D}\right)_V - \frac{L}{D}} \tag{4}$$

Because $(\partial y_D/\partial D)_V$ is likely to change sign when $(\partial L/\partial D)_V$ is close to L/D, we see from (4) that the gain $(\partial y_D/\partial(L/D))_V$ is likely to change sign as well as go through a singularity when $(\partial L/\partial D)_V$ is close to L/D. A change in the sign of the gain results in input multiplicity while a singularity results in output multiplicity. Thus, when the DV-configuration displays input multiplicity, the $(L/D)V$-configuration is likely to display input multiplicity as well as output multiplicity. Furthermore, we see from (4) that since $(\partial L/\partial D)_V$ becomes larger than L/D in the vicinity of where the gain $(\partial y_D/\partial D)_V$ becomes positive, most operating points are likely to have a positive slope between y_D and L/D.

Figure 3 shows L/D as a function of y_D for the methanol-water column with V fixed. From the figure we see that, as expected, the column displays input as well as output multiplicity. We also note that the slope between y_D and L/D is positive at almost all operating points.

4.3. LB-Configuration

The properties of the LB-configuration are in many respects similar to those of the DV-configuration, and they are both denoted "material balance configurations". One might therefore expect that the input multiplicity found with the DV-configuration exists also with the LB-configuration. However, we found that the reason for the input multiplicity with the DV-configuration was closely related to the fact that

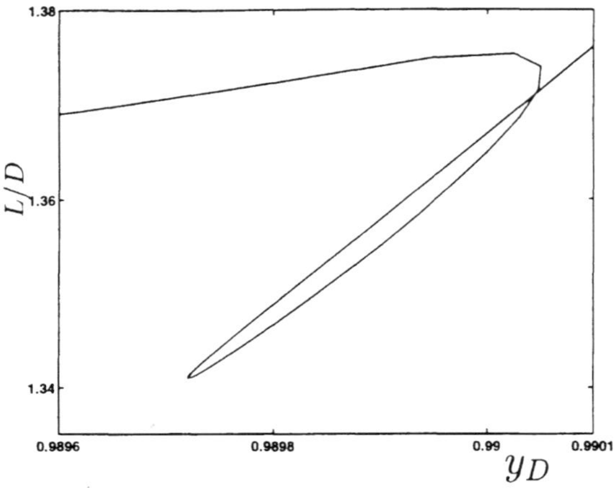

Fig. 3. Reflux ratio L/D as a function of top composition y_D for methanol-water column in Example 1. Boilup $V = 1.5543$.

$(\partial(L/D)/\partial D)_V$ may become positive at certain operating points. With the LB-configuration, however, $(\partial(L/D)/\partial B)_L = L/D^2$ at any operating point, and we therefore expect $(\partial y_D/\partial B)_L$ to always be positive. Thus, input multiplicity between B and y_D is unlikely with L fixed.

However, with the LB-configuration, B is most likely to be used for manipulation of x_B. Following the arguments of improvement in operating lines with the DV-configuration we might expect $(\partial x_B/\partial B)_L$ to change sign when $(\partial V/\partial B)_L$ is close to V/B. It is easily shown that $(\partial L/\partial D)_V > 0$ corresponds to $(\partial V/\partial B)_L > 0$. However, for most separations the heat of vaporization decreases with fraction of light component and $(\partial V/\partial B)_L > 0$ requires $(\partial x_B/\partial B)_L > 0$ (Jacobsen and Skogestad, 1991). Based on this we conclude that $(\partial x_B/\partial B)_L$ is unlikely to become negative for most separations. The only exception is for mixtures where the heat of vaporization increases with fraction of light component, e.g., n-propanol and acetic acid.

5. EFFECT OF OPERATING CONDITIONS

We have seen that the input multiplicity we find is closely related to the fact that $(\partial L/\partial D)_V > L/D$ at some operating points. We here discuss under which operating conditions this gain is most likely to be positive and large.

Internal flows: Increasing the internal flows, i.e., increasing V with D constant, will in general increase the gain $(\partial L/\partial D)_V$ (Jacobsen and Skogestad, 1991). Output multiplicity, which only requires this gain to become positive at some operating point, is therefore favored by large internal flows. However, we find that input multiplicity requires $(\partial L/\partial D)_V$ to be in the order of magnitude of L/D, and we may therefore not make any conclusions on the effect of increasing internal flows on input multiplicity.

Product purities: $(\partial L/\partial D)_V$ is most likely to be positive in columns where $(\partial y_D/\partial D)_V$ is small relative to $(\partial x_B/\partial D)_V$. This is the case when the top product is pure relative to the bottom product.

Column design: With given product purities, the introduction of extra trays in the bottom section of the column will tend to increase $(\partial x_B/\partial D)_V$ and decrease $(\partial y_D/\partial D)_V$ accordingly. Thus, a poor feed tray location and/or overdesign will usually increase the probability of input multiplicity.

The methanol-water column studied has $N_T = 13$ and $N_F = 8$. With the given number of trays and the nominal purities of $y_D = 0.99$ and $x_B = 0.02$, the optimal feed location is close to $N_F = 6$. However, the column displays input multiplicity around the nominal operating point even with this feed tray location. The fact that input multiplicity is found also with an optimal feed tray location is most likely explained by the overdesign of the column. A close to optimal design for the nominal separation is $N_T = 10$ and $N_F = 4$, and with this design there is no input multiplicity.

Input multiplicity is probably unlikely in columns which are optimally designed for their nominal separation. However, although most columns studied in academia probably are optimal in this sense, industrial columns will often be operated in a non-optimal fashion due to changing feed stock as well as product demands and because a certain overdesign is common to allow for flexibility in the operation.

6. DYNAMICS

Jacobsen (1994) show that input multiplicity will imply that some operating points will have a RHP zero in the transfer-function from the input to the output provided the output is dynamically variant at the operating point of zero steady state gain. For the methanol-water column we find that the transfer-function $(\partial y_D/\partial D)_V(s)$ has a RHP zero whenever the steady state gain $(\partial y_D/\partial D)_V(0)$ is positive. This may also be understood from physical arguments. Consider the dynamic effect of an increase in D on y_D with V constant. Initially, the compositions are unchanged and the level control causes the reflux to decrease. The effect of this decrease in reflux is to decrease the top composition. Thus, the initial gain $(\partial y_D/\partial D)_V(\infty)$ is always negative. However, as the compositions throughout the column start to change, the energy balance causes the internal flows to change. At some operating points this will counteract the initial decrease in y_D and we get an overshoot or, when this secondary effect dominates, an inverse response in y_D. An inverse response corresponds to a RHP zero. Because $(\partial y_D/\partial D)_V(\infty) < 0$ there will be a RHP zero whenever $(\partial y_D/\partial D)_V(0) > 0$. For the nominal operating point of the methanol-water column we find that there is a RHP zero located at $0.0049\ min^{-1}$ (assuming 0.5 kmol holdup on all trays).

For the LV-configuration $(\partial L/\partial D)_V$ is likely to be positive when there is a change in the sign of $(\partial y_D/\partial L)_V$, and we therefore expect this configuration to have a RHP pole (Jacobsen and Skogestad, 1994) at the operating points were it has a RHP zero. At the nominal operating point the methanol-water column with the LV-configuration has a RHP-zero at the same location as with the DV-configuration and a RHP-pole at $0.070 min^{-1}$. Similarly, with the $(L/D)V$-configuration the column has the same RHP-zero and a RHP-pole at $0.0175 min^{-1}$.

7. INPUT MULTIPLICITY WITH MASS INPUTS

Jacobsen and Skogestad (1991a) show that fixing one or more of the inputs on a mass or volume basis makes the flows in the column depend on the compositions in a similar manner as with the energy balance, and that output multiplicity therefore also may result from such specifications. We here show that fixing inputs on a mass- or volume basis also may result in input multiplicity. In this part we assume constant molar flows in order to study the effect of using mass flows isolated.

Consider first the D_wV and $(L_w/D_w)V$-configurations (we assume V fixed on a molar basis). Here subscript w denotes mass flows, i.e., D_w is distillate flow in kg/min. Following the arguments used earlier, a likely requirement for input multiplicity with these configurations is that $(\partial L/\partial D_w)_V$ should be in the order of magnitude of L/D at some operating point. However, assuming constant molar flows implies that $(\partial L/\partial D)_V = -1$ at any operating point, and $(\partial L/\partial D_w)_V$ is therefore always negative if we assume $(\partial D_w/\partial D)_V > 0$, which is a reasonable assumption. Input multiplicity with the D_wV-configuration is therefore unlikely with the assumption of constant molar flows. Note also that the $(L_w/D_w)V$-configuration is identical to the $(L/D)V$-configuration if there is a total condenser and D_w and L_w are measured in the same units.

Now consider the L_wB-configuration. We here assume B to be fixed on a molar basis in order to keep the expressions we derive reasonably simple. However, the results obtained are valid also when B_w is used as a manipulated input if we assume $(\partial B_w/\partial B)_{L_w} > 0$. With this configuration the effect of a change in B on the boilup V becomes

$$\left(\frac{\partial V}{\partial B}\right)_{L_w} = -1 + \frac{L\left(\frac{\partial y_D}{\partial B}\right)_L (M_2 - M_1)}{M - L\left(\frac{\partial y_D}{\partial L}\right)_B (M_2 - M_1)} \quad (5)$$

Here M_1 and M_2 denotes the molecular weights of light and heavy component respectively and $M = M_1 y_D + (1-y_D)M_2$ is the molecular weight of the reflux flow. The gains $(\partial y_D/\partial B)_L$ and $(\partial y_D/\partial L)_B$ are always positive (assuming constant molar flows) and we see from (5) that, in the usual case where $M_2 > M_1$, the gain $(\partial V/\partial B)_{L_w}$ will always be larger than -1, and that it may become almost arbitrarily large. Following previous arguments we might expect the gain $(\partial x_B/\partial B)_{L_w}$ to become negative when $(\partial V/\partial B)_{L_w}$ is in the order of magnitude of V/B. However, note that the increase in boilup in this case is accompanied by a similar increase in reflux L, and thus a relatively large increase in L/D which counteracts the effect of the increase in V/B on x_B. How large $(\partial V/\partial B)_{L_w}$ must be in order to have $(\partial x_B/\partial B)_{L_w} < 0$ will therefore depend on the relative effect of changes in V/B and L/D on x_B, respectively. In columns with a pinch in the composition profile the effect of L/D on x_B is small relative to the effect of V/B, and for such columns $(\partial V/\partial B)_{L_w} > V/B$ should be an approximate criterion for input multiplicity with the L_wB_w-configuration.

Example 2: Consider a methanol-n-propanol column with 9 trays, feed at tray 3 ($F = 1$ kmol/min, $z_F = 0.25$). We assume constant molar flows and constant relative volatility $\alpha = 3.55$. The nominal operating point we consider has $y_D = 0.95$ and $x_B = 0.005$, for which we find $L_w = 85.54$ kg/min and $B_w = 44.42$ kg/min. However, as seen from Figure 4 and Table 2, with L_w fixed at this value, the nominal value of $x_B = 0.005$ may be obtained by three different values of B_w. The input multiplicity is also found for the compositions on trays 2 to 5. Note that the input multiplicity found here would persist if the energy balance is included in the model.

Table 2. Steady-state solutions for methanol- n-propanol column in Example 2.

L_w kg/min	x_B	B_w kg/min	V kmol/min	y_D
85.54	0.005	44.42	2.812	0.950
85.54	0.005	42.27	2.630	0.835
85.54	0.005	30.46	2.352	0.503

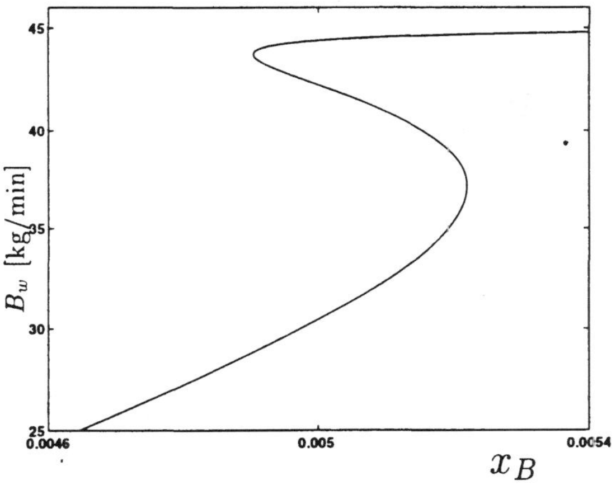

Fig. 4. Bottoms flow B_w as a function of bottom composition x_B for methanol-propanol column in Example 2. Reflux $L_w = 85.54$.

Due to limited space we simply state that input multiplicity with the L_wB_w-configuration implies input multiplicity also with the L_wV- and $L_w(V/B_w)$-configurations. For the methanol-n-propanol column in Example 2 we find that both these configurations displays input as well as output multiplicity.

8. SPECIFICATION OF TWO OUTPUTS

So far we have only considered the case where a single composition is specified while one input is kept constant. The question that naturally arises is whether similar multiplicity may exist when two compositions are specified. Input multiplicity would in this case imply that some operating points would have RHP transmission zeros (Jacobsen, 1994).

Consider first the case where both product compositions, i.e., y_D and x_B, are specified. Lucia and Li (1992) have shown that input multiplicity may not occur with this specification. However, their approach involves rather complicated expressions, and we therefore present simpler arguments here. From differentiation of the overall material balances we obtain

$$\left(\frac{\partial y_D}{\partial D}\right)_{x_B} = \frac{x_B - y_D}{D}; \quad \left(\frac{\partial x_B}{\partial D}\right)_{y_D} = \frac{x_B - y_D}{B} \quad (6)$$

The fact that both gains in (6) always are negative implies that the product flows D and B are uniquely determined whenever y_D and x_B are specified. The only remaining possibility for input multiplicity is then that the internal flows L and V may take different values for given product compositions. This will only happen if, at some operating point, the effect of increasing L and V with D constant is to decrease y_D and increase x_B, that is, the internal flows have a negative impact on the separation. This is unlikely unless there is a significant decrease in tray efficiencies with increased vapor and liquid loads. Patwardhan and Edgar (1991) observed that increasing the internal flows made the separation worse at some operating points of a packed column due to decreased efficiency of the packing with increased flow load. This implies that the column they studied would display input multiplicity if both y_D and x_B were specified.

In practice, composition measurements are often not available or severely delayed, and temperature measurements on plates inside the column are therefore often used for inferring the product compositions. We here demonstrate through an example that in such cases input multiplicity may exist with two temperatures specified.

Example 3: Consider a methanol-water column with $N_T = 15$, $N_F = 7$, $F = 1$ kmol/min (liquid feed) and $z_F = 0.50$. We assume constant molar flows as the multiplicity we find is independent of the energy balance. The nominal operating point we consider has $y_D = 0.99$ and $x_B = 0.05$. For these specifications we find $D = 0.4787$ kmol/min and $V = 1.090$ kmol/min and the solution is unique. However, assume that we decide to control the temperatures on trays 4 and 12, i.e., we specify x_4 and x_{12} rather than y_D and x_B. The nominal operating point has $x_4 = 0.5139$ and $x_{12} = 0.9164$. As seen from Figure 5 this set of specifications does not yield a unique solution in terms of D and V. In addition to the nominal operating point there is a solution with $D = 0.0837$ kmol/min and $V = 0.1983$ kmol/min. At this operating point the product compositions are $y_D = 0.9904$ and $x_B = 0.4552$. Thus, specifying two compositions inside the column does not guarantee uniqueness with respect to the flows and product compositions. Note that the turning point in Figure 5 corresponds to a singular gain matrix and that the plant gain, in terms of the determinant of the gain matrix, changes sign at this operating point. The change in the sign of the plant gain implies that a transmission zero crosses the imaginary axis (Jacobsen, 1994). We find that the second operating point has a RHP transmission zero at $13.1 min^{-1}$. Note that if we instead choose to specify x_6 and x_{11}, then the nominal operating point has a RHP transmission zero at $0.092 min^{-1}$.

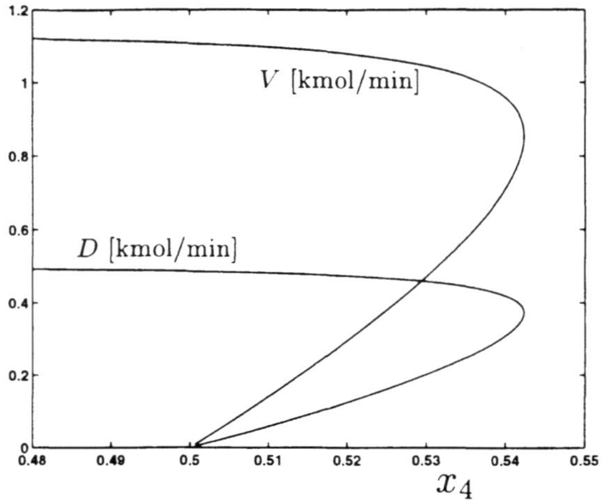

Fig. 5. Distillate flow D and boilup V as functions of composition on tray 4 for column in Example 3. $x_{12} = 0.9164$. Constant molar flows.

We do not provide any detailed explanation for the observed multiplicity here. However, we observe that the multiplicity is related to the existence of an operating point where the gain matrix becomes singular. With y_D and x_B as outputs the gain matrix can not become singular due to the material balance restriction. However, when specifying two compositions inside the column there exists no such restriction. Similar arguments apply for the case where one product composition and one temperature is specified and we find that input multiplicity may exist also in this case.

9. CONCLUSIONS

• Binary distillation columns may display input multiplicity with one product composition and one input specified. The multiplicity implies that certain operating points will have a RHP zero in the transfer-function.

• Input multiplicity is unlikely if both product compositions are specified. However, input multiplicity may exist if at least one of the specifications involve a composition or temperature on a tray inside the column. In this case some operating points will have a RHP transmission zero in transfer-matrix.

REFERENCES

Jacobsen, E.W. (1994). Dynamics of Systems with Input Multiplicity, *AIChE Annual Meeting*, S.F.

Jacobsen, E.W. and S. Skogestad (1991). Multiple Steady States in Ideal Two-Product Distillation, *AIChE J.*, **37**, 4.

Jacobsen, E.W. and S. Skogestad (1994). Instability of Distillation Columns, *AIChE J.*, **40**, 9.

Lucia, A. and H. Li (1992). Constrained Separations and the Analysis of Binary Homogeneous Separators, *Ind. Eng. Chem. Res.*, **31**, 11.

Patwardhan, A. and T. Edgar (1991). Nonlinear Predictive Control of a Packed Distillation Column, *Proc. American Control Conference*, Boston, Ma.

HETEROGENEOUS AZEOTROPIC DISTILLATION INVOLVING AN EMBEDDED TWO-LIQUID PHASE REGION

Hassan S. Mumtaz, Grafton Corbett, Matthew Reagan, Warren D. Seider,

Department of Chemical Engineering, University of Pennsylvania, Philadelphia, PA 19104/6393

Leonard A. Fabiano

Consultant, Wayne, PA

Soemantri Widagdo

Polymer Processing Institute and Department of Chemistry and Chemical Engineering, Stevens Institute of Technology, Hoboken NJ 07030

Abstract: An azeotropic distillation tower involving a ternary system has a built-in decanter tray with an aqueous side-draw. A steady-state model is tuned by varying the side-draw to match an experimental temperature profile. The position of the embedded two-liquid phase region is extremely sensitive to the side-draw. Dynamic simulations show sensitive movements of the interfaces in response to typical disturbances. PI controllers are utilized to expand and contract the two-liquid phase region while rejecting changes in the water composition of the feed.

Keywords: Distillation columns, PID control, dynamic behavior

1. INTRODUCTION

Although heterogeneous azeotropic towers have been studied extensively (Widagdo and Seider, 1994), an industrial dehydration tower exhibits a phenomena that, to our knowledge, has not been reported earlier. The tower involves an embedded section of trays having two-liquid phases that is bounded by trays having a single-liquid phase. Consequently, it becomes important to track the movement of the two interfaces between stages having two and one-liquid phase(s) and to understand how they respond to typical disturbances.

The ternary system of compounds separated by the tower consists of A, water, and C, where A and C are organic compounds (not disclosed due to proprietary restrictions). The heterogeneity of the ternary mixture is partially attributed to water and C which also exhibit a binary azeotrope. Under normal operation, A is concentrated in the distillate, water in the sidedraw, and C in the bottoms product. From time-to-time, the tower encounters operating problems which motivate the study reported herein.

Steady-state simulations are carried out using MAX^{TM}, a flowsheet simulator by Aspen Technology, Inc. To model vapor-liquid-liquid equilibrium, the Non-random Two-liquid (NRTL) model, or Renon equation, is used in conjunction with the Redlich-Kwong equation of state. The flow rate of the sidedraw is adjusted to obtain agreement with the temperature data. In response, very sensitive movements of the embedded two-liquid phase region are predicted, suggesting the need for dynamic simulation to improve the operating strategy for the tower as well as its control systems.

A model of the tower was created using the SPEEDUP™ simulator by Aspen Tech. The usage of MAX and SPEEDUP, with the same data bank and property estimation programs, assured consistency between the steady-state and dynamic simulations.

2. PHASE EQUILIBRIA AND RESIDUE CURVES

The physical and thermodynamic properties of the system, including the binary-interaction coefficients for the NRTL equation, were determined based upon experimental data. Care was taken to obtain interaction coefficients that represent both the VLE and LLE data with acceptable accuracy. To understand the interactions within the A-water-C system, the binary pairs are considered first. Figure 1 illustrates the bubble and dew point curves for the three binary pairs at near atmospheric pressure, computed using MAX. The water-C system exhibits a minimum-boiling azeotrope at 75 mole percent water and 97.2°C, and a large immiscibility gap, while the A-water and A-C systems exhibit no azeotropes. However, there is a marked difference in the volatilities of A and water. A is the lowest-boiling compound (most volatile at 40°C) while C is the highest-boiling compound (least volatile at 144°C).

Next, the ternary system is considered. The residue curves and the binodal curve at 32.2°C is shown in Figure 2. There are four nodes, three at the vertices and the other at the water-C azeotrope. Because the minimum-boiling azeotrope is at a higher temperature than the boiling point of A, there exists a distillation boundary (which connects vertex A with the C-water azeotrope) that cannot be crossed by a residue curve. Above the aqueous sidedraw, A is separated from water, to the right of the distillation boundary, whereas A is separated from C, below the sidedraw, to the left of the distillation boundary. Hence, to carry out the desired separation, the distillation boundary must be crossed.

3. DISTILLATION TOWER

The tower consists of 23 theoretical stages with a liquid feed to stage 12, as shown in Table 1, and a sidedraw from stage 9. Depending on the operating conditions, some of the stages contain both aqueous and organic phases. The sidedraw removes all of the aqueous phase flowing from stage 9, while the organic phase flows to the stage below. This is accomplished using a built-in, "donut" decanter tray that collects the liquid flowing from the tray above and permits phase separation, as illustrated in Figure 3. The cavity in the center of the tray carries both the vapor, rising to the tray above, and the overflowing organic phase. An interfacial level controller maintains the position of the interface between the aqueous and organic phases and prevents the organic phase from escaping in the sidedraw and the aqueous phase from overflowing to the tray below.

4. STEADY-STATE SIMULATION

The RADFRAC™ subroutine in MAX was used to simulate the distillation column. RADFRAC has the ability to associate a decanter with any stage of the tower. To model the decanter tray, specifications are made for the fraction of each liquid phase, L1 and L2 (for the organic and aqueous phases, respectively), to be returned to the tower, with the remainder exiting in the sidedraw. Since all of the aqueous phase is removed from stage 9, in normal operation, the specification for L2 is expected to be zero. Furthermore, since the temperature measurements on the stages are the most plentiful and reliable of the data collected, the temperature profile is the initial criterion for tuning the model. As illustrated in Figure 4, when L2 is specified at zero, the temperature profile of the model does not match the experimental data in the rectifying section.

Due to the critical role of the aqueous phase, the effect of varying the amount of water in the column is examined. In the simulations, this can be accomplished by recycling some of the aqueous phase withdrawn in the sidedraw to the tray below the decanter stage. As the L2 specification is varied, from zero to unity, the L1 specification is kept constant at unity, indicating that all of organic phase is returned to the tower. For $L2 < 0.47$, significant differences, between the experimental and the theoretical temperature profiles, exist in the rectifying section, whereas for $L2 > 0.49$, the differences are primarily in the stripping section. Only at $L2 = 0.48$ ($\pm$ 0.005) is there good agreement everywhere, as illustrated in Figure 5.

In normal operation, none of the aqueous phase is recycled. However, since good agreement is obtained when the recycle fraction is specified at 0.48, the water flow rates in the tower are higher than predicted. One possible explanation for this is that the level controller of the decanter tray may not be functioning properly, and consequently, the entire aqueous phase may not be removed, resulting in the buildup and overflow of the aqueous phase. Another explanation would be the failure of the two liquid phases to separate completely due to an insufficient tray volume. This would result in the entrainment of water in the organic phase. By recycling a fraction of the aqueous phase below the decanter tray, these effects appear to be simulated adequately.

To further clarify the heterogeneous operation in the tower, the average and individual liquid phase compositions are plotted on a triangular diagram in Figure 6. Note the steep tie-lines that connect the phases in liquid-liquid equilibrium and the distillation boundary (not shown exactly), discussed above. Normally, distillation boundaries cannot be crossed by the operating lines in a distillation tower, except when the boundaries exhibit sizable curvature. However, due to liquid phase splitting, the boundary is crossed between stages 9 and 10, the former being the decanter tray. Removal of the aqueous phase in the sidedraw enables the operating line to cross the distillation boundary. In this way, the partial miscibility of the system is used to advantage. A related observation is that water concentrates in the center of the tower because it is less volatile than A (the distillate), and more volatile than C (the bottoms product). It is removed effectively in the sidedraw thanks to the partial miscibility of the system.

Figure 6 also shows that, in the rectifying section, where A is dehydrated, most of the separation occurs between trays 4 and 2 as the liquid phases coalesce. On tray 2, the second liquid phase is no longer present. This is consistent with the temperature profile in Figure 5, which has two plateaus. In the rectifying section, a plateau encompasses stages 5 through 9, with the overall (three phase) compositions close to the distillation boundary. The water/C ratio approaches that at the water-C azeotrope and, since the mole fraction of A is small, the temperature remains nearly constant. Further up the tower, as A is dehydrated, the temperature drops and two liquid phases coalesce.

Below stage 9, as the liquid composition moves toward the C vertex, the other temperature plateau appears near the bottom of the column. Due to the removal of water in the sidedraw, the second liquid phase disappears between stages 10 and 11. Between stages 10 and 15, the temperatures increase rapidly, as A is stripped from C. Below stage 15, the temperature plateau corresponds to the small concentration changes as highly-pure C is produced.

4.1 Sensitivity Analysis

As mentioned above, the column responds very sensitively to the recycle fraction of the aqueous sidedraw. This can be seen by focusing on the responses of the embedded two-liquid phase region. First, Figure 7 shows that the two-liquid phase region expands as the flow rates of water in the column increase. At zero recycle, two-liquid phases appear in the decanter tray only. At the other extreme, when the aqueous phase is completely recycled, two liquid phases appear on every stage, except the reboiler. Note that the upper interface, between stages having two and one-liquid phase(s), is especially sensitive when the recycle fraction varies between 0.4 and 0.5. Also, when increasing the fraction of the aqueous phase recycled, the second-liquid phase does not advance into the stripping section until all of the stages in the rectifying section contain two-liquid phases.

Similarly, the temperature and composition profiles are extremely sensitive to the fraction of the aqueous phase recycled. When none of the aqueous phase is recycled, the bulk of the separation occurs between stages 8 and 15, where the temperature gradient is constant as shown in Figure 4. First, A is separated from water at the bottom of the rectifying section, and then from C at the top of the stripping section, leaving nearly constant temperatures and compositions between stages 1 and 7 and 16 and 23. These profiles are not affected significantly as the recycle fraction is increased to 0.4. Although the composition of water increases substantially, only one additional tray has two-liquid phases, beyond the decanter tray. However, as the recycle fraction increases toward 0.45, the separation between A and water is shifted up the tower. A region, dilute in A, involving water and C phases spreads into the rectifying section, creating temperature and composition plateaus above the decanter tray, as shown in Figure 8. From 0.45 to 0.48, the separation of A from C begins to move down in the stripping section. This movement is drastic from 0.48 to 0.49, as shown in Figure 9. It is accompanied by a sharp increase in the concentration of A in the stripping section, as shown in Figure 10. Further increases of the recycle fraction further concentrate A and spread the temperature composition plateaus.

On the basis of the steady-state simulations, it seems clear that an embedded two-liquid phase region is desirable. Sufficient water is needed to allow the embedded region to contract, without coalescing, when the water concentration in the feed decreases. However, as the region expands (especially for L2 > 0.48), the losses of A increase in the sidedraw and the bottoms product. To select a desirable operating strategy, dynamic simulations, with and without controllers, are needed as described next.

5. PROCESS DYNAMICS

Dynamic simulations are carried out with the SPEEDUP system. The trays in the rectifying section have a residence time of clear liquid between 23 and 29 seconds, whereas in the stripping section the residence time is much smaller, between approximately 2.5 and 3.5 seconds. Note that the trays are modelled with the Francis Weir Formula. The MESH (Material balance, Equilibrium, Summation of mole fraction, and Heat balance) equations include terms that represent the enthalpy and flow rates of the species in a second-liquid (aqueous) phase. Note that when the aqueous phase does not exist, the flow rate variables are zero. Furthermore, during each time step, the stability of the phase distribution on each tray is checked using tangent-plane-distance analysis (Michelsen, 1982). When the distribution is unstable, a liquid phase is added or deleted as necessary. For this tower, this stability analysis is critical in the dynamic simulations to permit the two interfaces between trays having two- and one-liquid phase(s) to be tracked properly in response to typical disturbances. This is established for a similar tower, involving the dehydration of secbutanol, by Widagdo et al. (1992). In the results that follow, this movement seems reasonable, although confirmation using well-tested and reliable algorithms and, better yet, experimental data showing the movements of the interfaces, has not been accomplished. Finally, the decanter tray, reflux accumulator, and reboiler have residence times of approximately one-third hour, between 2.7 and 5.4 minutes, and approximately 9.6 min., respectively.

For these studies, which are limited in scope due to space limitations, the tower is simulated beginning at steady states with the recycle fraction of the aqueous side-draw (L2) at 0.48 and 0.49. At L2 = 0.48, two liquid phases exist on the decanter tray, but not further down in the stripping section. At L2 = 0.49, the aqueous phase moves far down into the stripping section. First, open-loop simulation results are presented to show the movement of the interface below the decanter tray, beginning with L2 = 0.48, after an increase in the weight fraction of water in the feed from 0.0153 to 0.0191 (holding the weight fraction of A fixed). Figure 11 shows the movement of the temperature front, which corresponds to the interface between stages having two- and one-liquid phase(s), to a new steady state in about two hours. The figure also shows that the interface moves up in the rectifying section.

These studies were repeated with a PI controller which measures the temperature on Tray 15 and attempts to maintain the set point (150°C at the L2 = 0.48 steady state) by adjusting the reboiler heat duty. The responses, with a controller gain and reset time equal to 2 and 54.27, respectively, are shown in Figure 12. Observe that the controller maintains the temperatures and compositions below the decanter tray with rather small changes in the boilup rate. However, the movements in the rectifying section are substantial as the composition and temperature fronts move closer to the condenser. To avoid a break-through, a temperature measurement in the rectifying section, coupled with another manipulated variable, such as the reflux rate, may be desirable. Alternatively, it may be possible to prevent the movement in the rectifying section with a model-predictive controller (MPC) that incorporates the SPEEDUP model within an optimization algorithm.

Figures 13 and 14 repeat these simulations, beginning with the steady state at L2 = 0.49, and a decrease in the weight fraction of water in the feed from 0.0153 to 0.0141. Figure 13 shows that, without the controller, the interfaces move upward in the stripping section and downward in the rectifying section. With the controller, as shown in Figure 14, the movement in the stripping section is negligible and in the rectifying section is substantial, but away from the condenser. Note that, in this case, the temperature setpoint on tray 15 is 111.74°C (at the L2 = 0.49 steady state) and the time to reach the new steady state is approximately four hours.

More comprehensive studies, involving a broad range of disturbances and control system designs, are needed to arrive at an improved operating strategy. With additional temperature measurements and manipulated variables, as well as model-predictive control, it seems clear that a strategy can be obtained that retains a sufficient, but not too large, two-liquid phase region, so as to adequately reject the disturbances that arise during normal operation. These studies are underway and will be reported in a more complete manuscript.

6. CONCLUSION

In conclusion, the overall effect of changing the aqueous recycle fraction of the sidedraw is to change the positions of the two separations, A from water, and A from C. When the recycle fraction is zero, both separations occur near the center of the tower, with temperature and composition plateaus located at the top and bottom of the tower. At the other limit, when the recycle fraction is unity, the temperature and composition plateaus are located in the middle of the tower and the separations occur at the extremes of the tower.

At an aqueous recycle fraction of 0.48, the computed temperature profile closely approximates the temperature measurements. Under these conditions, the aqueous phase extends from the decanter tray upward into the rectifying section, but not below the decanter tray. Whereas at an aqueous recycle fraction of 0.49, the aqueous phase extends downward below the decanter tray. This sensitivity suggests difficult operating problems.

To reject typical disturbances, such as changes in the concentration of water in the feed stream, it is desirable to operate with two liquid phases in the rectifying section and below the decanter tray.

The dynamic simulations show that a single PI control loop can reject these disturbances nicely, but not perfectly. Additional measurements and manipulated variables, together with model-predictive control show promise for even better operation in this sensitive region.

Finally, the SPEEDUP simulator appears to be tracking the interfaces between trays having two- and one-liquid phase(s) properly. Confirmation with well-tested procedures for phase-stability analysis and experimental data would be desirable.

7. ACKNOWLEDGMENTS

Dr. Ian P. Willets, of Aspen Tech., prepared the SPEEDUP model, with our specifications, and performed the initial dynamic simulations. His efforts enabled us to carry out the simulation and control studies, many of which were completed by Mr. David Kolesar. Mr. Robert A. Knudsen prepared the binary interaction coefficients for the NRTL solution model and the initial simulations using ASPEN PLUS. Finally, Dr. Oliver Wahnschaft, of Aspen Tech., helped us to prepare the residue-curve map. The contributions of these individuals are acknowledged with appreciation.

REFERENCES

Michelsen, M. L. (1982). The isothermal flash problem, I: Stability. *Fluid Phase Equil.*, **9**, 21.

Widagdo, S., W. D. Seider, and D. H. Sebastian (1992). Nonlinear analysis in process design. *AIChE J.*, **37**, 1, 1-38.

Widagdo, S., and W. D. Seider (1994). Azeotropic distillation - a review, submitted to *AIChE J.*

Table 1. Feed conditions

		Weight fraction
T = 126.7°C		
P = near atmospheric	A	0.1043
Saturated liquid	H_2O	0.0153
	C	0.8804

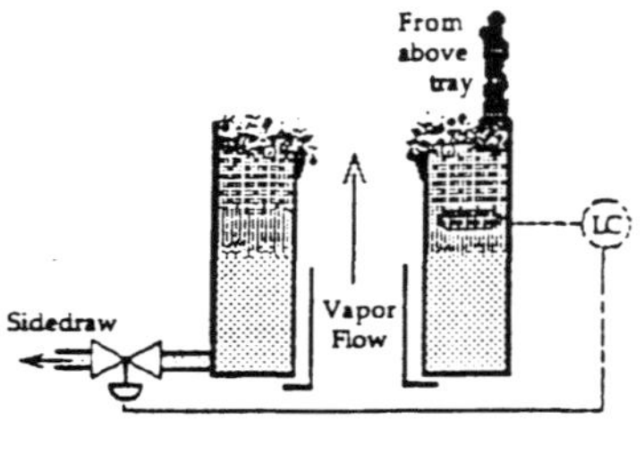

Fig. 3. Decantor tray with sidedraw.

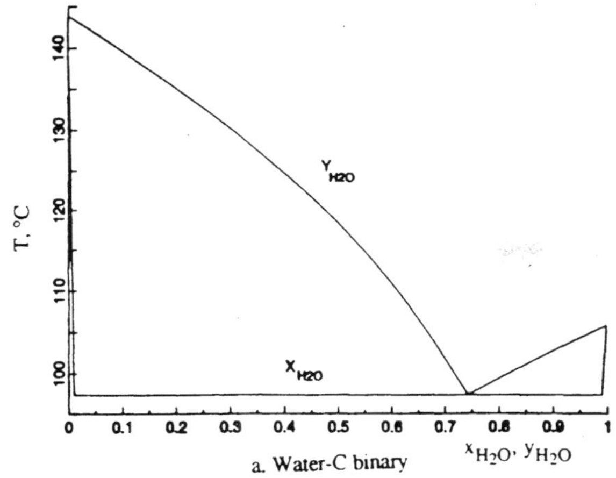

a. Water-C binary

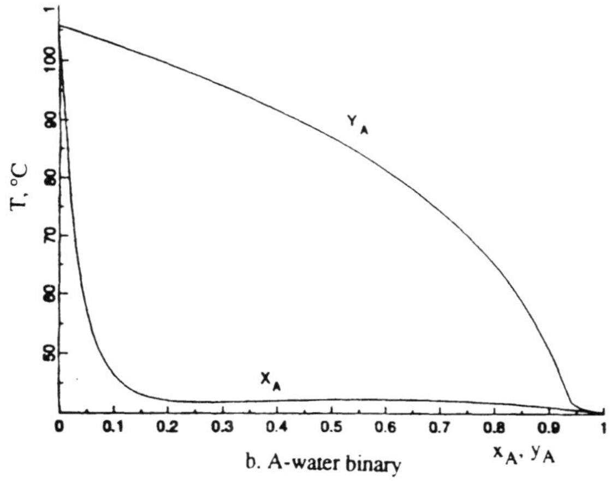

b. A-water binary

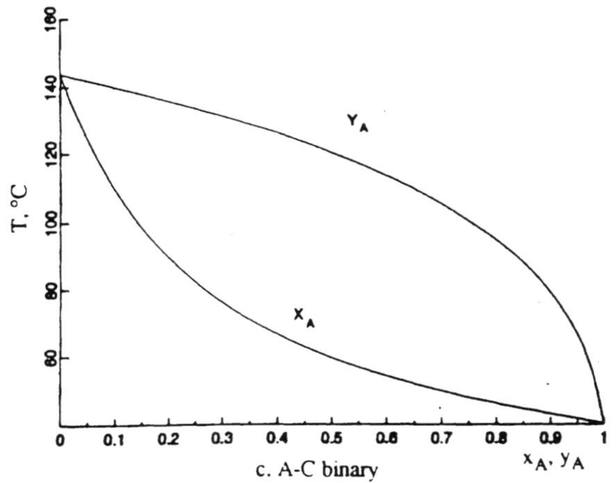

c. A-C binary

Fig. 1. Bubble and dew point temperatures for binary pairs at near-atmospheric pressure.

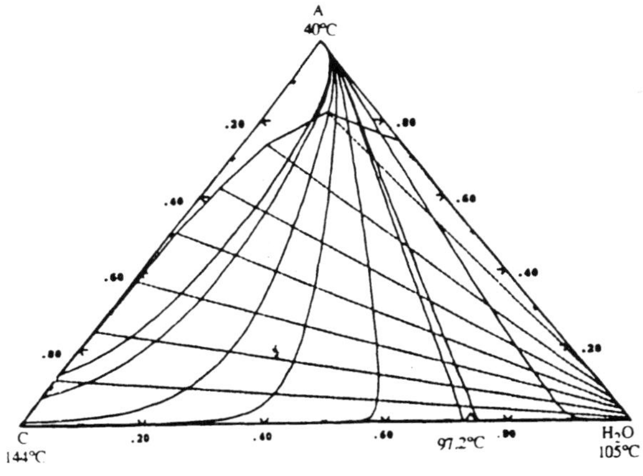

Fig. 2. Residue-curve map for the A-H_2O-C system.

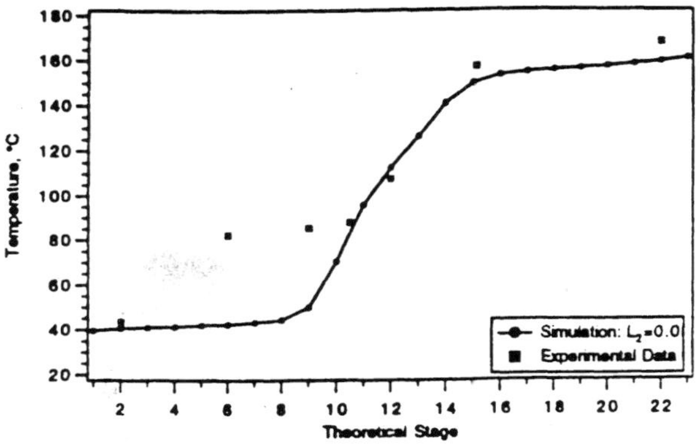

Fig. 4. Temperature profiles. Simulation results for L2 = 0.

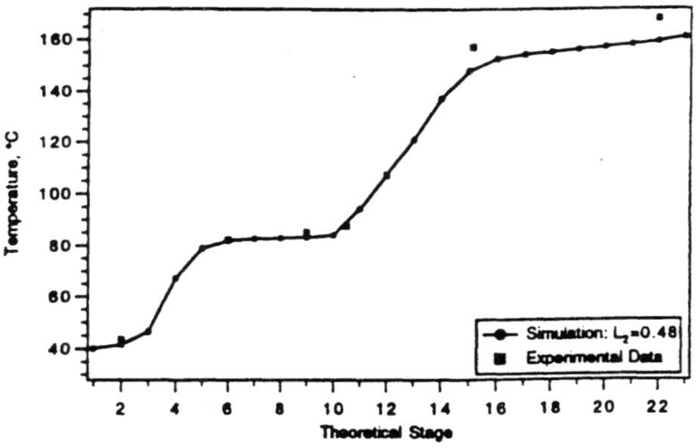

Fig. 5. Temperature profiles. Simulation results for L2 = 0.48.

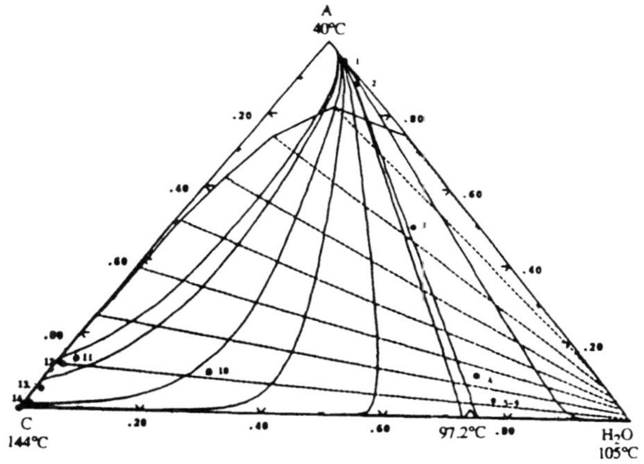

Fig. 6. Operating points for the simulated tower at L2 = 0.48; average liquid mole fractions. • are calculated by RADFRAC.

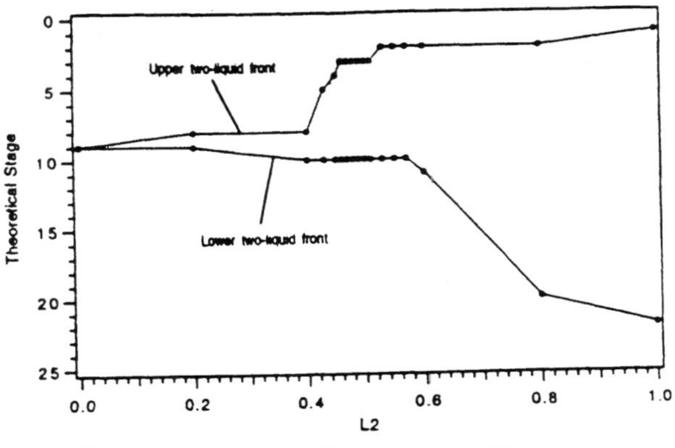

Fig. 7. Trays having two liquid phases (within dashed envelopes) as L2 varies.

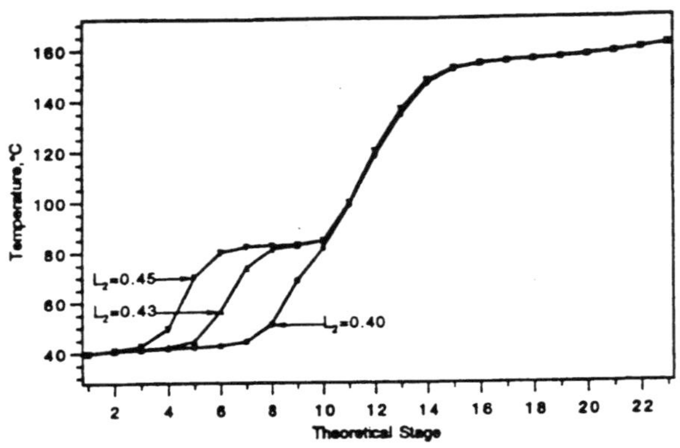

Fig. 8. Temperature profiles. Simulation results for L2 = 0.4, 0.43, and 0.45.

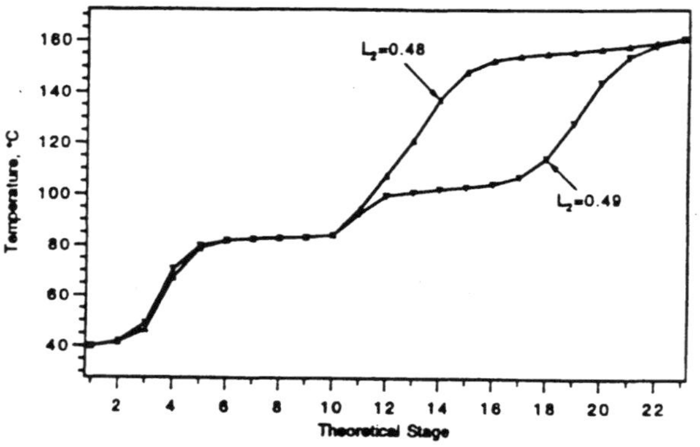

Fig. 9. Temperature profiles. Simulation results for L2 = 0.48 and 0.49.

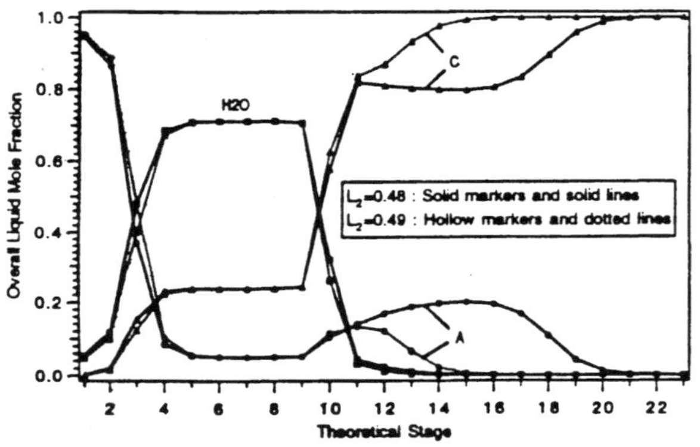

Fig. 10. Overall mole fractions in the liquid phases. Simulation results for L2 = 0.48 and 0.49.

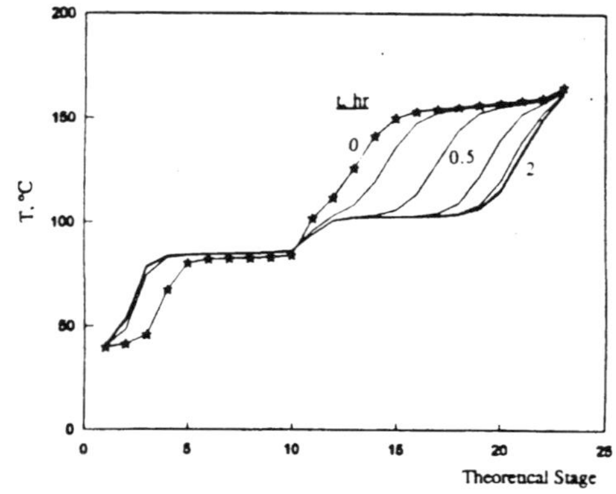

Fig. 11. Temperature response to a step increase in the weight fraction of water in the feed (0.0153 to 0.0191). L2 = 0.48 at the initial steady-state.

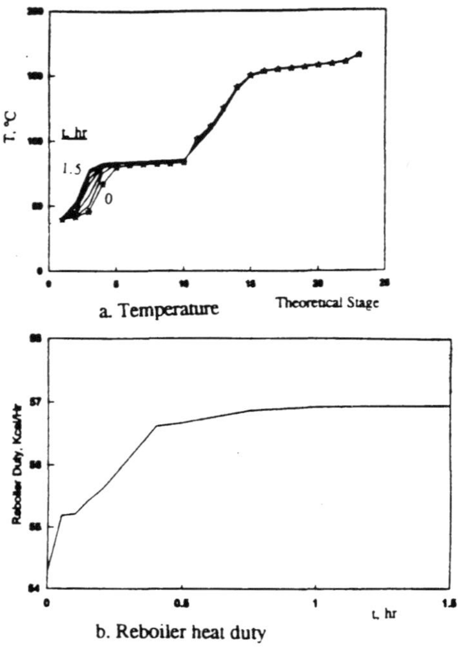

Fig. 12. Responses to a step increase in the weight fraction of water in the feed (0.0153 to 0.0191) with PI control. L2 = 0.48.

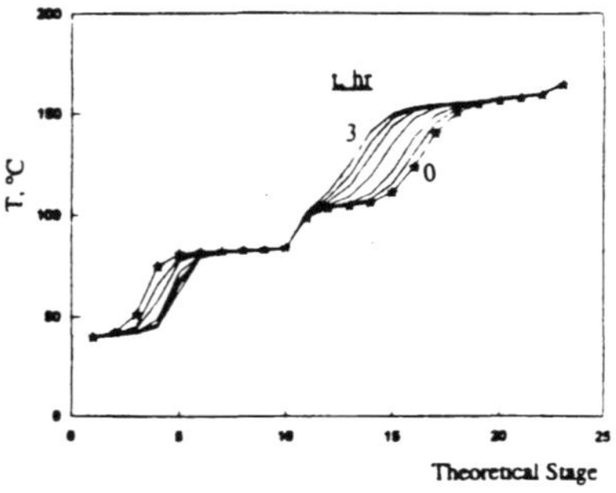

Fig. 13. Temperature response to a step decrease in the weight fraction of water in the feed (0.0153 to 0.0141). L2 = 0.49 at the initial steady-state.

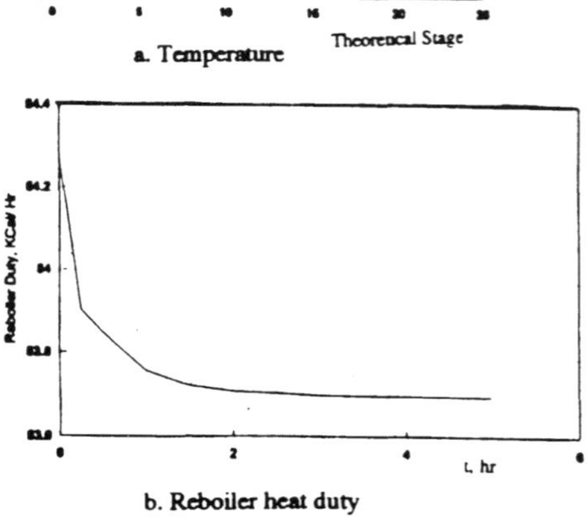

Fig. 14. Responses to a step decrease in the weight fraction of water in the feed (0.0153 to 0.0141) with PI control. L2 = 0.49.

HIERARCHICAL SUB-OPTIMAL CONTROL OF BATCH DISTILLATION WITH SLOP RECYCLING

B.H.L. Betlem, H.C. Krijnsen, Th.H. Timmer

Department of Chemical Engineering, University of Twente
7500 AE Enschede, The Netherlands
E-Mail: B.H.L. Betlem@CT.UTwente.NL

Abstract: Slop recycle increases the production capacity of batch distillation. It avoids exhaustion of the mixture or decreases the difficulty of separation. A hierarchical control approach is proposed. At scheduling level predictions of the required component production times and the ratio between product and slop cut quantities are made. These predictions are based on the degree of difficulty of separation and the mean feed quality and quantity. At conditioning level sub-optimal operation under constant quality control is applied. The production capacity approximates the capacity under dynamic optimal control, and is up to 20% higher than under piecewise constant reflux control.

Keywords: Batch control, Distillation columns, Optimal control, Quality control, Hierarchical control.

INTRODUCTION

In chemical industry the interest in manufacturing small-volume, high-value, specialty chemicals is increasing. For multipurpose flexible plants, batch distillation is an important unit operation. A batch distillation column can handle a wide range of feed compositions and degrees of difficulty of separation. If the distillation step is the bottle-neck operation in the plant, optimization at plant scheduling level can be very profitable. As batch distillation is very energy-intensive, capacity maximization (at constant energy load) reduces energy consumption.

As indicated in literature, slop recycling is a policy to increase production capacity of difficult separations. Different strategies are possible, for example *separate slop recycle*. After each product cut a slop cut that meet no product specification is produced. This cut is stored separately and is added for reprocessing to the next batch just before a similar separation starts. Slop recycling causes mutual dependency of consecutive runs. Fig.1 shows a simple binary version. In this article a coordination hierarchy is proposed to control sub-optimally batch distillation with slop recycle.

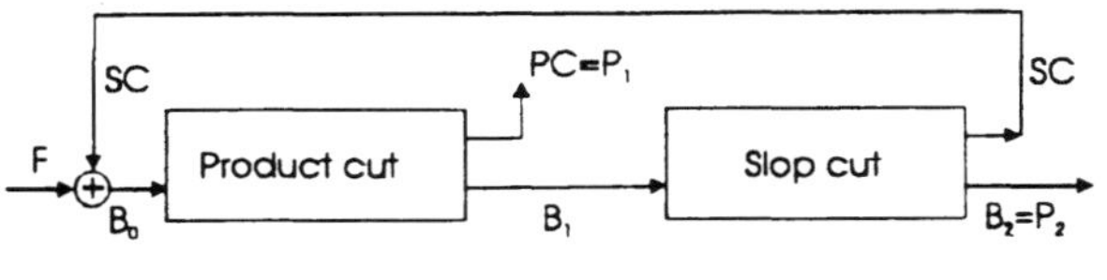

Fig. 1. Slop recycling for binary batch distillation.

After a presentation of previous work on this topic, a comparison is made between the distillation capacities under dynamic optimal, constant quality, and constant reflux control. When the exhaustion or the degree of difficulty of separation is limited constant quality control proves to be as good as dynamic optimal control. In case of slop recycling these limitations are fulfilled. Therefore constant quality control is taken as the starting point for the coordination hierarchy, consisting of: scheduler, conditioner, and controller. The scheduler in this hierarchy needs predictions of the run periods and production performances. A simplified scheduler model is developed to predict the required run time and the product to slop cut ratio. This model is based on quasi-steady state production simulations under constant quality control. In this scheduling model feed quality variations are permitted.

The following basic model assumptions and variables have been used for column simulations:

. binary mixture with constant relative volatility between 1.6 and 3.0,

Table 1 Literature overview concerning slop recycle strategies

	Problem Specifications				Tray hold up	Mixt.	Reflux Strategy / Math. Method	Time Advantage
	X_{P1}	X_{P2}	N	α				
Mayur, May, Jackson (1970)	.85	.90	5	2	-	bin.	DO Pontryagin + PCR	3% over no slop DO over PCR: 4%
Christensen, Jørgensen (1987)	.75 - .95	.50 - .94	2 - 9	1.65 - 2.25	-	bin.	DO / orthogonal collocation	0 - 25% over no slop in one case: 92%
Quintero, Luyben (1990)	.95 + .99	.95 + .99	40	4/2/1 9/3/1	yes	tern.	6 PCR strategies	up to 35% over total slop recycle
Farhat, Czernicki, Pibouleau, Domenech (1990)	.95	.95	22	1 mix Antoi ne	-	tern.	linear and expo-nential reflux ratio profiles / NLP	*fixed final time:* up to 10% over PCR
Mujtaba, Macchietto (1993)	.95	.95	6	3	yes	bin. multi.	PCR 2 level optimization	*goal:* optimization framework
Bonny, Domenech, Floquet, Pibouleau (1994)	.80 - .97	.80 - .85	5 - 30	2 mix Antoi ne	-	tern.	PCR optimization of slop-cut addition / NLP	1.5% and 5.5% over separate distillation of slops

- 22 trays with tray vapour efficiency of 53%,
- no tray or accumulator hold-up,
- top and bottom product fraction have to meet requirements (standard .99 and .01),
- standard feed fraction is 0.5.

The relative volatility is used as independent variable to vary the difficulty of separation.

LITERATURE OVERVIEW

In a review paper, Rippin (1983) reports a number of batch distillation optimization studies. These studies took place in the period from 1963 until 1980. In most cases the objective was minimization of the time required to produce a certain amount of product of a requested composition. They report a time advantage of dynamic optimal control up to 13% over constant reflux control and up to 6% over constant quality control. The studies are generally binary and the effect of hold-up is mostly not treated, although it is acknowledged to be important. The hold-up affects the initial reflux to bring the trays into equilibrium composition.

Kerkhof and Vissers (1978) introduce a degree of difficulty of separation. For a rather low degree constant quality control is as profitable as dynamic optimal control. For higher degrees, dynamic optimal control is always preferable. In this study no requirement for the bottom residue was stated. As optimization objective they use the profit per unit time, taking into account the product and feed values and recharging time. These factors determine the optimum batch time and product amount. Their conclusions apply for that particular batch time, where the profit per unit time under dynamic optimal control is maximal. The exhaustion of the mixture (D/D_{max}) at this point lays between 74% and 95%. When the bottom is exhausted, optimal control gives substantial higher profit.

From 1980 until 1987 only few is published. After this period an increasing interest is perceptible, especially in operating strategies for batch distillation with slop recycle. Table 1 shows an overview of the articles published about this subject. The first article concerning slop recycling was already published by Mayur *et al.* in 1970. They describe one quasi-steady state example.

Christensen and Jørgensen (1987) investigate under which circumstances slop recycling is worthwhile. 31 cases of binary batch distillation in tray columns are presented. They introduce another degree of difficulty of separation (q) based on distillation without slop recycle. The degree of Kerkhof and Vissers appears to be not useful. They indicate three areas:

$q < 60\%$ no time saving is obtained,
$60\% < q < 75\%$ a small improvement is possible,
$q > 75\%$ time savings up to 30 % are gained.

The other articles are more recent and have a different character. They investigate other distillation policies to improve the capacity or look for faster calculation methods. Quintero and Luyben (1990) compare six different strategies of batch distillation. The objective is to maximize a profit function taking into account a fixed recharging time.

Farhat *et al.* (1990) and Bonny *et al.* (1994) do not use the quasi-steady state approach. Farhat *et al.* claim that from an industrial point of view, it is desirable to maximize the productions for a fixed final time. Bonny *et al.* present a general model which allows the calculation of the proportion of each slop cut to be added to each initial load and to each new batch. They illustrate their approach by distilling a fixed quantity with a limited reboiler capacity. This requires multiple runs. Partial slop addition to the next runs show to be more profitable than other strategies. They also find that a fixed reflux ratio appears to be as good as other variable reflux policies (2.3% difference). However, mixing slop cut with fresh feed to reduce the degree of difficulty of the separation is the basic principle of slop recycle. This principle does not agree with fixed time or fixed quantity problems, as future fresh feed is not taken into account.

Mujtaba and Macchietto (1993) developed a calculation structure permitting the calculation of optimal policies for each alternative operation with consideration of multi-period objectives and constraints. A two-level optimal control formulation is proposed. The multi-period operation is decomposed into a sequence of independent optimal control problems. The outer level maximizes the profit function, as the inner loop solves the minimum time problem. The constrained nonlinear optimization problem is solved using a successive quadratic programming technique. Their approach is very powerful.

Literature clearly indicates that slop recycle can be profitable. Different recycle strategies and reflux control policies are used. However no insight was given, *why* a particular control policy is applicable in a particular context.

SINGLE-RUN POLICIES

To describe the difficulty of separation three degrees can be defined:

1. Initial degree of difficulty of separation.

$$\sigma = \frac{(D_{max})_{CQ}}{D_{max}} = \frac{x_D - x_B}{x_B \cdot (1 - x_D)} \cdot \frac{1}{\alpha^{\varepsilon N + 1} - 1}$$

This degree of difficulty is the ratio between the maximal product amount under CQ ($(D_{max})_{CQ}$) and the absolute maximal value (D_{max}). Fig.2 indicates both values for α=2. The degree is introduced by Kerkhof and Vissers (1978).

2. Overall degree of difficulty of separation.

$$\sigma_{int} = \int_{x_{B_0}}^{x_{B_1}} \frac{x_D - x_B}{x_B \cdot (1 - x_D)} \cdot \frac{1}{\alpha^{\varepsilon N + 1} - 1} \cdot \frac{dx_B}{x_{B_1} - x_{B_0}}$$

This is σ modified for the lifetime of the product cut.

3. Degree of exhaustion:

$$D/(D_{max})_{DO}$$

It indicates the exhaustion relative to the maximal exhaustion under optimal control.

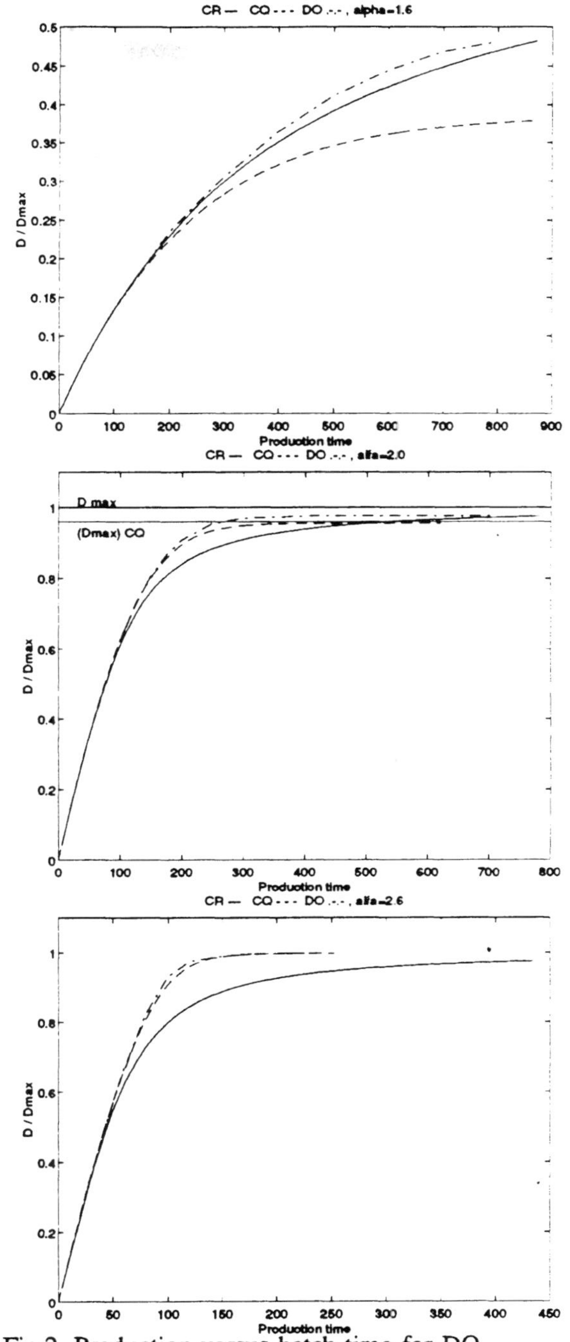

Fig.2. Production versus batch time for DO, CR, CQ control at α = 1.6, α = 2.0 and α = 2.6.

Fig.2 shows a comparison between DO, CR, and CQ control. DO is calculated according to the Pontryagin Principle. DO always performs better than the other two control policies, but the improvement depends on σ and $D/(D_{max})_{DO}$. DO compensates periods of difficult separation and low production with periods of relative simple separation. However, if σ<1% or $D/(D_{max})_{DO}$<40%, the capacity improvement with respect to CQ is <2%. For larger exhaustion CQ is limited by its maximal production. CR is not able to compensate periods of different σ. Only, if no compensation is required or if production time goes to infinity, then it can compete with DO.

SLOP RECYCLE POLICIES

Using slop recycling causes mutual dependency of consecutive runs. A simple strategy to deal with this implication is to assume reiteratively the same feed is offered. This leads to a quasi-steady state operation. Binary distillation with slop recycle is performed under (piecewise)CR and CQ. First, the CR results will be presented and next those of CQ. Finally, both policies are compared.

For CR the two degrees of freedom are the reflux ratios during product cut (R_{PC}) and slop cut (R_{SC}). Thus, the slop quality and quantity are dependent variables. Fig. 3 shows the CR results for α=2.0. The maximum appears to be much more sensitive to R_{PC} than to R_{SC}. This also applies for other α's (not shown). Fig.4 reports the optimal relative slop amount and its fraction. SC and x_{SC} are constant for α>2. When α<2 they strongly increase. It appears that slop recycling is always profitable.

For CQ the degrees of freedom are the slop quality and quantity. Fig.5 includes the CQ results for different α's. The capacity is plotted against one of the degrees of freedom. Above a certain α no slop production is required. Then the optimum has a clear boundary maximum. Fig.6 shows the optimal x_{SC} versus α. The combination of Fig.5 and Fig.6 indicates that for high α's (easy distillation) the optimum is sensible to SC and insensible to x_{SC}. However, for relative low α's the opposite applies. The upper and lower boundary indicate a 2% capacity decrease.

In Fig.7 a comparison between the CR and CQ policy is made. The results for CQ are about 20% better than for CR. If CR is applied, also for high α's, slop recycle improves the capacity substantially. If CQ is applied, only below a particular difficulty of separation slop recycle is advisable (here α<2.6).

Fig. 8 shows the production progress at the highest degree of difficulty α=1.6. The capacity under CQ is higher, despite the large exhaustion during the slop cut. The improvement is evenly partitioned over the product and the slop cut. In both periods the difference in time is about 17% .

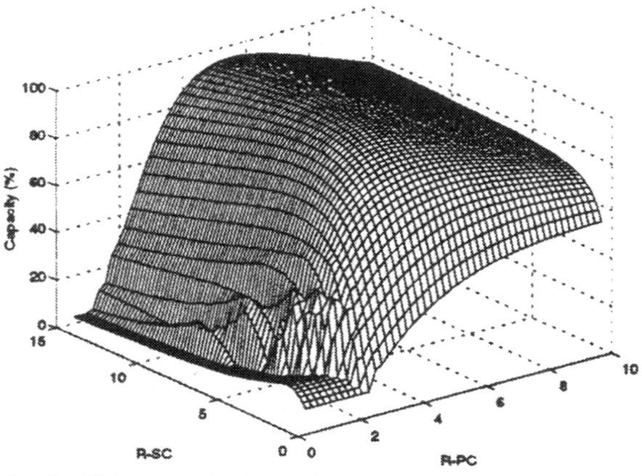

Fig.3. CR-control: Capacity versus R_{PC} and R_{SC}.

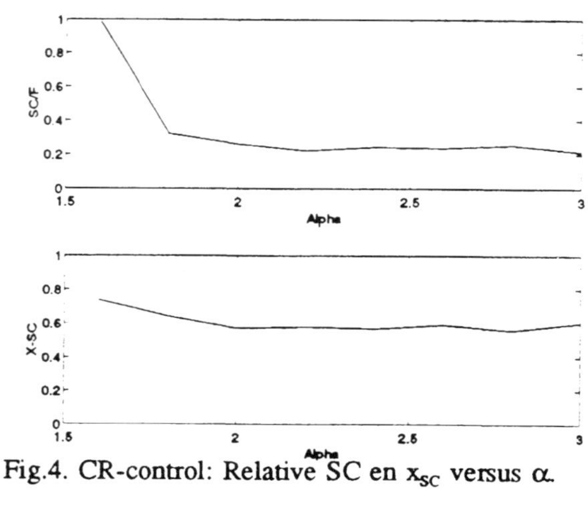

Fig.4. CR-control: Relative SC en x_{SC} versus α.

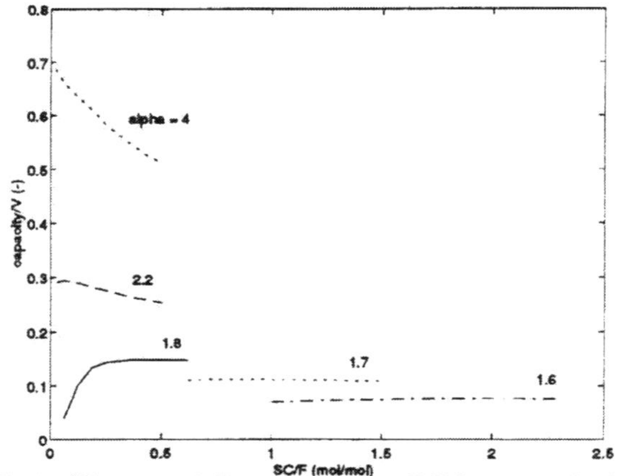

Fig.5. CQ-control:Capacity versus SC for several α's.

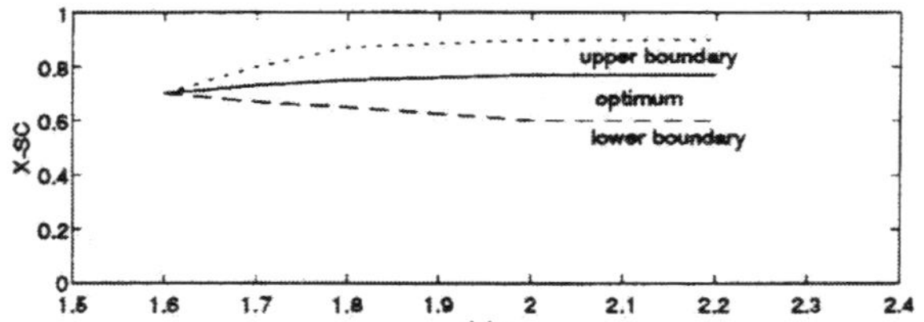

Fig.6. CQ-control: x_{SC} versus α.

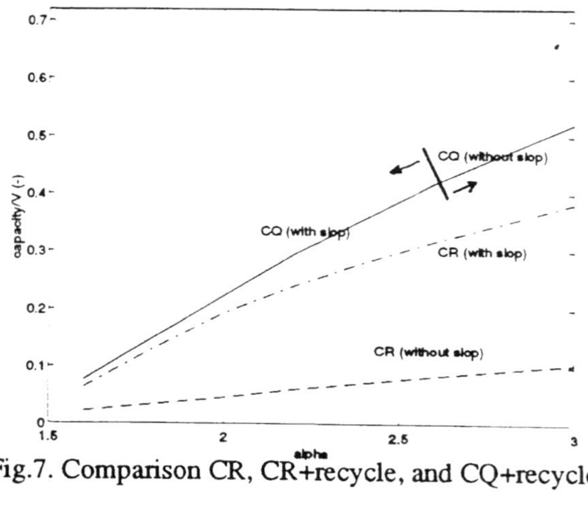

Fig.7. Comparison CR, CR+recycle, and CQ+recycle.

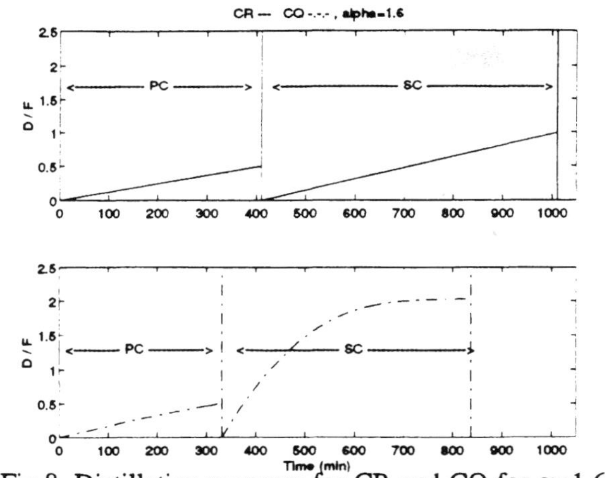

Fig.8. Distillation progress for CR and CQ for α=1.6

The results agree with the single run conclusions. In the worst case α=1.6, for the product cut $D/(D_{max})_{DO}>D/D_{max}\approx20\%$, while during the slop cut σ=0.4%. Under these circumstances, CQ is superior to CR control. As the conditions $\sigma<1\%$ and $D/(D_{max})DO<40\%$ are fulfilled DO will give much the same capacity.

COORDINATOR HIERARCHY

The decision complexity of a (batch) plant requires a hierarchy of coordination functions. Fig.9 proposes such a possible hierarchy for a batch distillation column. (All side-branches of other equipment have been omitted). A multi-level control hierarchy consists of a hierarchy of coordinators each controlling a group of equipment and instruments. The scheduler coordinates the plant as the conditioner is limited to the column. Also the time horizon for the scheduler is much larger. High levels have a wider scope and a more abstract view. At a particular level the view is related to the extent of the process inclusive of all subordinate coordinators which control that particular process. The scheduler

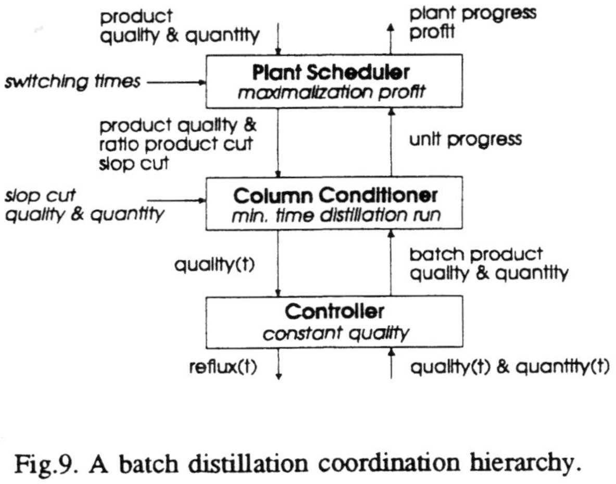

Fig.9. A batch distillation coordination hierarchy.
horizontal →degree of freedom,
downwards →: control,
upwards →: report.

considers conditioned plant units. Therefore, the internal coordinator models should be defined at the right level of abstraction.

The *plant scheduler* optimizes the capacity or profit. Therefor it requires models of every unit describing production time as a function of their performance. and load. Generally, the degrees of freedom are the switching moments. The controls are the switching and operating conditions. In case of the batch distillation it concerns the product quantity and the product to slop ratio. If the objective function contains more economic weights (e.g. utility costs), then more degrees of freedom and controls are available. The *column conditioner* maximizes the capacity with the limited scope of a combined production and slop cut. The controls are the optimal process conditions (set points). In case of batch distillation the qualities are these set points. The *controller* realizes the required quality by manipulating the reflux ratio.

The models at each level should be different. At the plant scheduling level we use a simplified model of algebraic equations describing the capacity as function of σ. The model can be adapted to the local situations. At column level a more elaborate model is required considering multi-component interactions and hold-up.

SCHEDULER RELATIONS

The scheduler needs algorithms to calculate the sum of the total production time and the product to slop ratio. Quasi-steady state simulations under CQ form the basis to determine these relations. When the product requirements are fixed, these times and ratios are a function of the column separation performance. More then ten proposals for degrees of difficulty were tested. The number of Christensen and Jørgensen (1987) is not applicable, because it is based on distillation without slop recycling. D/D_{max} was neither a good measure. Fig.10 indicates the ratio

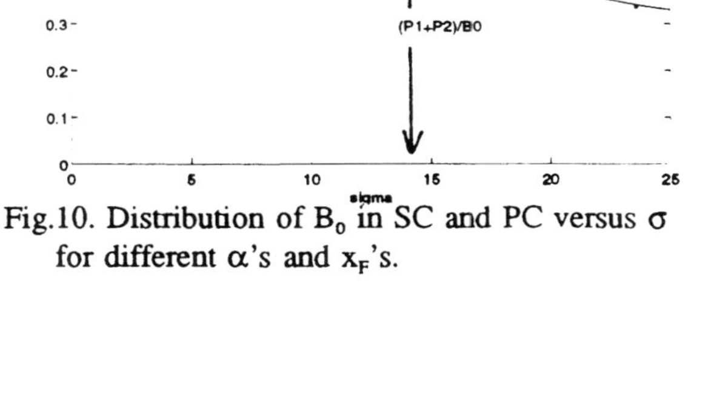

Fig.10. Distribution of B_0 in SC and PC versus σ for different α's and x_F's.

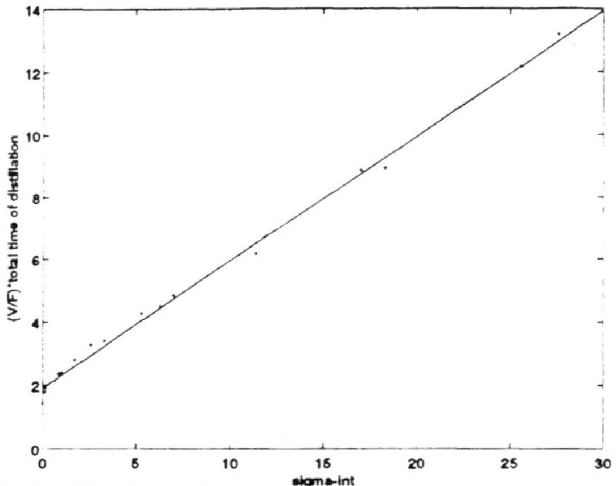

Fig.11. Total production time versus σ_{int} for different α's and x_F's.

between SC and $(P_1+P_2)=P_{total}$ versus σ. This degree has been varied by changing α and x_F. The result is a quadratic relation. Then P_1 can be calculated from the quasi-steady state mass and component balance.

$$P_1 = P_{total} \cdot \frac{x_f - x_{P_2}}{x_{P_1} - x_{P_2}}$$

For x_F the mean value can be taken or the value can be varied according to the current situation. Next x_{SC} and x_{B1} follow from the mass and component balance for the particular run. Fig. 11 shows the total production time versus σ_{int}. This degree can be determined from the data calculated. The result is a linear function.

The parameters of these relations are still a functions of the x_{P1}, x_{P2}, and εN. The parameters are product and column dependent. They allow adaptation for a particular column and different products with defined requirements . If on-line adaptation according to the reported progress is made (with forget factor), every run the scheduler is updated to the current column performance.

CONCLUSIONS

- If constant quality control (or dynamic optimal control) is applied, for batch distillations above a particular degree of difficulty of separation slop recycle is profitable.
- If piecewise constant reflux control is applied, slop recycling is always profitable.
- As the degree of difficulty is low (<1%) or the degree of exhaustion is limited ($(D/D_{max})_{DO}$<40%), constant quality control achieves nearly the same capacity as dynamic optimal control.
- For distillation with slop recycle, constant quality control performs better than constant reflux control. The improvement during the production phase and the slop phase are evenly divided and comes to 20%.
- The required ratio between the slop and product cut can be determined based on the degree of difficulty and the expected future mean feed quality. The required production time can be derived by the degree of difficulty and the amount to be distilled. The parameters of the relation are column and product dependent and allows adaptation in accordance with the reported distillation progress.
- Hierarchical control is necessary to simplify the complexity at scheduling level. The proposed control structure is:
  . scheduler estimates the product to slop cut ratio and the distillation time,
  . conditioner determines minimum distillation time using the slop cut fraction as degree of freedom,
  . at controller level a constant quality controller is required. The quality controller takes care of model incorrectness.

SYMBOLS

B	Bottom	PC	Product cut
CQ	Constant Quality	R	Reflux ratio
CR	Constant Reflux	SC	Slop cut
D	Distillate	V	Vapour rate
DO	Dynamic Optimal	x	Molefraction
F	Feed	α	relative volatility
N	Number of trays	ε	tray efficiency
P	Product	σ	degree of difficulty of separation

REFERENCES

Bonny, L., S.Dominech, P. Floquet and L. Pibouleau (1994). Strategies for Slop Cut Recycling in Multicomponent Batch Distillation. *Chem. Engng. Processing* **33**, pp. 23-31.

Christensen, F.M. and S.B.Jørgensen (1987). Optimal Control of Binary Distillation with Recycled Waste Cut. *Chem.Engng. J.* **34**, pp. 57-64.

Farhat, S., M.Czernicki, L.Pibouleau, S.Dominech (1990). Optimization of Multiple-Fraction Batch Distillation by Nonlinear Programming. *AIChE Journal* **36-9**, pp. 1349-1360.

Kerkhof, L.H.J. and H.J.M.Vissers (1978). On the Profit of Optimum Control of Batch Distillation. *Chem. Engng. Science* **33**, pp. 961-970.

Mayur, D.N., R.A.May and R.Jackson (1970). The Time Optimal Problem in Binary Batch Distillation with a Recycled Waste-Cut. *Chem. Engng. J.* **1**, pp. 15-21.

Mujtaba, M.I. and S.Macchietto (1993). Optimal Operation of Multicomponent Batch Distillation - Multiperiod Formulation and Solution. *Computers chem. Engng.* **17-12**, pp. 1191-1207.

Quintero-Marmol, E. and W.L.Luyben (1991). 2. Comparison of Alternative Slop Handling Operating Strategies. *Ind. Eng. Chem. Res.* **29**, pp. 1915-1921.

Rippin D.W.T. (1983). Simulation of Single- and Multiproduct Batch Chemical Plants for Optimal Design and Operation. *Computers chem. Engng.* **7-3**, pp. 137-156.

HIGH PERFORMANCE DISTILLATION COLUMN CONTROL THROUGH NOVEL CONTROL CONFIGURATIONS

J. ZHANG, M. T. THAM, and A. J. MORRIS

Department of Chemical & Process Engineering
University of Newcastle upon Tyne
Newcastle upon Tyne, NE1 7RU, U. K.

Abstract. Novel distillation control configurations are studied in this paper. In these configurations, composition manipulators are functions of the reflux flow rate, the vapour boil up rate, and the top and bottom product flow rates. Parameters in these functions are selected so as to maximally reduce disturbance sensitivities and control loop interactions, and maximally improve robustness of the control system. Structured singular value analysis is used to assess control system robustness. The proposed technique has been successfully applied to a methanol-water separation column.

Keywords. distillation control, distillation control configuration, robust control, structured singular value analysis

1 INTRODUCTION

Distillation control systems are generally decentralised control systems and there exist a large number of possible control configurations. For binary distillation columns, some frequently used control configurations include the LV configuration, the DV configuration, the LB configuration, and some ratio configurations. Different control configurations will have different degrees of control loop interactions, disturbance sensitivities, and robustness. A desirable control configuration should have less degrees of control loop interactions, be insensitive to disturbances, and be robust. The proper selection of control configurations has been one of the major research topics in distillation control (Skogestad and Morari, 1987a; Skogestad *et al*, 1990). The task of control configuration selection is to compare a set of standard control configurations and select the best one with the above desired properties. Quite often, compromise should be made in situations where no configuration possesses all those desired properties.

To design a high performance distillation control system, it would be more appropriate to design some new control configurations rather than to select from a set of standard ones. As is claimed by Luyben (1988), there usually exists a control configuration which yields a system that is naturally self-regulating for load disturbances and self-optimising and he referred this configuration as an "eigenstructure". The purpose of designing new control configuration is to find out such an "eigenstructure". Several new distillation control configurations have been reported recently. Haggblom and Waller (1990) describe a disturbance rejection and decoupling (DRD) control structure where the composition manipulators are linear combinations of L, V, D, and B. In this configuration, the steady state process model will be a diagonal matrix and therefore achieve steady state decoupling, the steady state disturbance gains with respect to feed composition and feed flow rate disturbances will be zero. The performance of this configuration is much better than the standard ones and the roles of feedback controllers in this configuration are less important than those in other configurations. Some further experimental results of the DRD configuration are presented in (Sandelin *et al*, 1991a, 1991b). Yang *et al* (1991) propose a combined balance control structure which combines the advantages of energy balance and material balance control structures.

This paper aims to provide a better understanding of these new control structures and a systematic way of designing new control configurations. Development of the DRD control configuration to more general cases is presented. Its condition of existence is revealed and its robustness is assessed through the structured singular value analysis. Robust performance is suggested in this paper as a criterion for designing new distillation control structures.

2 MODEL TRANSFORMATION

There exist certain relations between different control configurations and, by using these relations, it is possible to obtained the model of one configuration from that of another configuration. Several researchers have investigated the transformation of distillation control configuration models (Skogestad and Morari, 1987b; Haggblom and Waller, 1988; Yang *et al*, 1990). In this section, a more general approach which considers any number of disturbances are presented.

Let the base configuration be the LV configuration with the following transfer function models:

$$\begin{aligned}\begin{pmatrix}\Delta y\\ \Delta x\end{pmatrix} &= G\begin{pmatrix}\Delta L\\ \Delta V\end{pmatrix} + G_d\begin{pmatrix}\Delta d_1\\ \vdots\\ \Delta d_n\end{pmatrix}\\ &= \begin{pmatrix}g_{11} & g_{12}\\ g_{21} & g_{22}\end{pmatrix}\begin{pmatrix}\Delta L\\ \Delta V\end{pmatrix} + \\ &\begin{pmatrix}g_{d11} & \cdots & g_{d1n}\\ g_{d21} & \cdots & g_{d2n}\end{pmatrix}\begin{pmatrix}\Delta d_1\\ \vdots\\ \Delta d_n\end{pmatrix} \qquad (1)\end{aligned}$$

where G and G_d are process model and disturbance model respectively, x and y are bottom and top product compositions respectively, $d_1, \cdots, d_n$ are disturbances.

In the LV configuration, the condenser level and the reboiler level are controlled by manipulating the top and bottom product flows respectively. There exists a relation between the inventory manipulators, composition manipulators, and disturbances. This relation can be expressed as follows:

$$\begin{pmatrix} \Delta D \\ \Delta B \end{pmatrix} = M \begin{pmatrix} \Delta L \\ \Delta V \end{pmatrix} + M' \begin{pmatrix} \Delta d_1 \\ \vdots \\ \Delta d_n \end{pmatrix}$$
$$= \begin{pmatrix} m_{11} & m_{12} \\ m_{21} & m_{22} \end{pmatrix} \begin{pmatrix} \Delta L \\ \Delta V \end{pmatrix} +$$
$$\begin{pmatrix} m'_{11} & \cdots & m'_{1n} \\ m'_{21} & \cdots & m'_{2n} \end{pmatrix} \begin{pmatrix} \Delta d_1 \\ \vdots \\ \Delta d_n \end{pmatrix} \quad (2)$$

where M and M' are 2×2 and $2 \times n$ dynamic matrices respectively. Since the level control loops are much faster than the composition control loops, the dynamics in M and M' can be neglected. Therefore, only the static elements of M and M' should be determined. The elements of M and M' can be calculated from steady state column data and steady state gains in Eq(1). In the following, let the first two disturbances be disturbances in feed flow and feed composition respectively, i.e. $\Delta d_1 = \Delta F$ and $\Delta d_2 = \Delta z$.

The overall material balance leads to

$$D + B = F \quad (3)$$

The material balance of the light component gives

$$Dy + Bx = Fz \quad (4)$$

Differentiating Eq(3) and Eq(4) gives

$$\Delta D + \Delta B = \Delta F \quad (5)$$

and

$$\overline{D}\Delta y + \overline{y}\Delta D + \overline{B}\Delta x + \overline{x}\Delta B = \overline{F}\Delta z + \overline{z}\Delta F \quad (6)$$

Combining Eq(2) and Eq(5) gives

$$\begin{aligned} m_{11}\Delta L + m_{12}\Delta V + m'_{11}\Delta d_1 + m'_{12}\Delta d_2 + \cdots \\ + m'_{1n}\Delta d_n + m_{21}\Delta L + m_{22}\Delta V + m'_{21}\Delta d_1 \\ + m'_{22}\Delta d_2 + \cdots + m'_{2n}\Delta d_n = \Delta F \end{aligned} \quad (7)$$

Since $L, V, d_1, \cdots$, and d_n are independent, Eq(7) implies the followings:

$$m_{11} + m_{21} = 0 \quad (8)$$

$$m_{21} + m_{22} = 0 \quad (9)$$

$$m'_{11} + m'_{21} = 1 \quad (10)$$

$$m'_{1i} + m'_{2i} = 0 \quad (11)$$

$(i = 2, \cdots, n)$

Similarly, the following can be obtained from Equations (1) to (6),

$$\overline{D}g_{11}(0) + \overline{y}m_{11} + \overline{B}g_{21}(0) + \overline{x}m_{21} = 0 \quad (12)$$

$$\overline{D}g_{12}(0) + \overline{y}m_{12} + \overline{B}g_{22}(0) + \overline{x}m_{22} = 0 \quad (13)$$

$$\overline{D}g_{d11}(0) + \overline{y}m'_{11} + \overline{B}g_{d21}(0) + \overline{x}m'_{21} = \overline{z} \quad (14)$$

$$\overline{D}g_{d12}(0) + \overline{y}m'_{12} + \overline{B}g_{d22}(0) + \overline{x}m'_{22} = \overline{F} \quad (15)$$

$$\overline{D}g_{d1i}(0) + \overline{y}m'_{1i} + \overline{B}g_{d2i}(0) + \overline{x}m'_{2i} = 0 \quad (16)$$

$(i = 3, 4, \cdots, n)$

The elements of M and M' are readily obtained by solving Equations (8) to (16) and are as follows

$$m_{11} = -m_{21} = -\frac{\overline{D}g_{11}(0) + \overline{B}g_{21}(0)}{\overline{y} - \overline{x}} \quad (17)$$

$$m_{21} = -m_{22} = -\frac{\overline{D}g_{12}(0) + \overline{B}g_{22}(0)}{\overline{y} - \overline{x}} \quad (18)$$

$$m'_{11} = 1 - m'_{21} = \frac{\overline{z} - \overline{x} - \overline{D}g_{d11}(0) - \overline{B}g_{d21}(0)}{\overline{y} - \overline{x}} \quad (19)$$

$$m'_{12} = -m'_{22} = \frac{\overline{F} - \overline{D}g_{d12}(0) - \overline{B}g_{d22}(0)}{\overline{y} - \overline{x}} \quad (20)$$

$$m'_{1i} = -m'_{2i} = -\frac{\overline{D}g_{d1i}(0) + \overline{B}g_{d2i}(0)}{\overline{y} - \overline{x}} \quad (21)$$

$(i = 3, 4, \cdots, n)$

In the above equations, the overlined variables represent the nominal values of these variables at steady state.

The composition manipulators in the studied configuration, Δu_1 and Δu_2, can be expressed as linear functions of $\Delta L, \Delta V, \Delta D$, and ΔB as follows

$$\begin{pmatrix} \Delta u_1 \\ \Delta u_2 \end{pmatrix} = G_1 \begin{pmatrix} \Delta L \\ \Delta V \end{pmatrix} + G_2 \begin{pmatrix} \Delta D \\ \Delta B \end{pmatrix} \quad (22)$$

where G_1 and G_2 are 2×2 coefficient matrices. If Δu_1 and Δu_2 are nonlinear functions of $\Delta L, \Delta V, \Delta D$, and ΔB, then linearisation should be applied. The DV configuration, for example, can be represented as

$$\begin{pmatrix} \Delta D \\ \Delta V \end{pmatrix} = \begin{pmatrix} 0 & 0 \\ 0 & 1 \end{pmatrix} \begin{pmatrix} \Delta L \\ \Delta V \end{pmatrix} + \begin{pmatrix} 1 & 0 \\ 0 & 0 \end{pmatrix} \begin{pmatrix} \Delta D \\ \Delta B \end{pmatrix} \quad (23)$$

Eliminating ΔD and ΔB from Eq(2) and Eq(22) gives

$$\begin{pmatrix} \Delta L \\ \Delta V \end{pmatrix} = (G_1 + G_2 M)^{-1} \begin{pmatrix} \Delta u_1 \\ \Delta u_2 \end{pmatrix} - (G_1 + G_2 M)^{-1} G_2 M' \begin{pmatrix} \Delta d_1 \\ \vdots \\ \Delta d_n \end{pmatrix} \quad (24)$$

Substituting Eq(24) into Eq(1) gives

$$\begin{pmatrix} \Delta y \\ \Delta x \end{pmatrix} = G(G_1 + G_2 M)^{-1} \begin{pmatrix} \Delta u_1 \\ \Delta u_2 \end{pmatrix} + [G_d - G(G_1 + G_2 M)^{-1} G_2 M'] \begin{pmatrix} \Delta d_1 \\ \vdots \\ \Delta d_n \end{pmatrix} \quad (25)$$

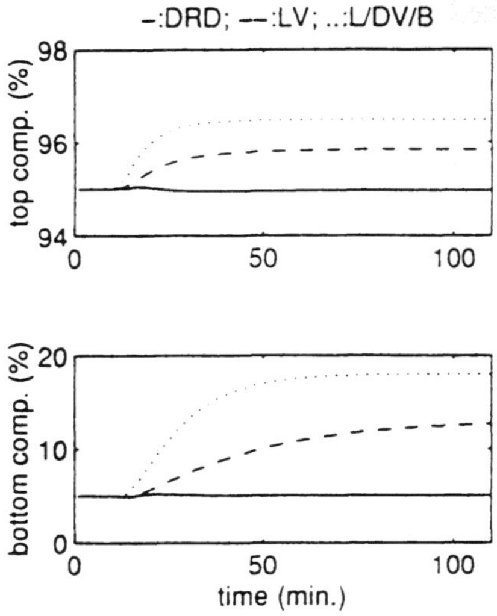

Fig. 1. Open loop responses to a feed composition disturbance

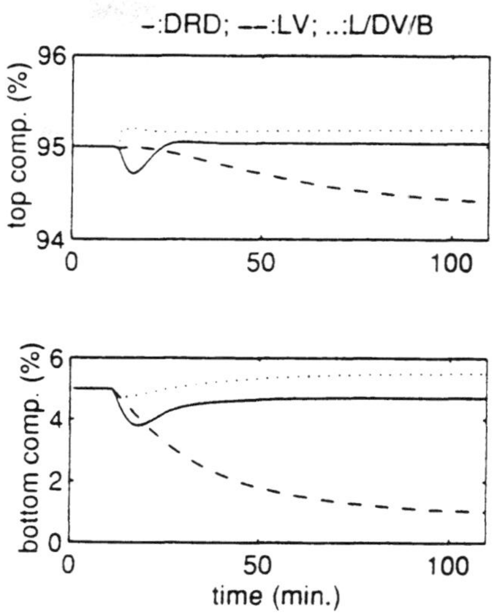

Fig. 2. Open loop responses to a feed rate disturbance

Therefore the process model and the disturbance model for the new configuration are

$$G(G_1 + G_2 M)^{-1}$$

and

$$G_d - G(G_1 + G_2 M)^{-1} G_2 M'$$

respectively.

3 AN EXAMPLE COLUMN

The distillation column studied in this paper is a simulated methanol-water separation column. A nonlinear tray by tray dynamic model has been developed from mass and energy balances (Simonsmier, 1977) and is used in simulation. When designing and analysing control systems, linear approximate models are used. The approximate linear process model and disturbance model of the LV configuration are obtained through step tests and are as follows.

$$G(s) = \begin{pmatrix} \frac{1.09}{1+4.3s} & -\frac{1.27}{1+6.2s} \\ \frac{2.15}{1+14.2s} & -\frac{6.4}{1+15.7s} \end{pmatrix} \tag{26}$$

$$G_d(s) = \begin{pmatrix} \frac{0.17}{1+25.2s} & \frac{0.126}{1+9.8s} \\ \frac{2.9}{1+15.8s} & \frac{0.583e^{-2.5s}}{1+17.6s} \end{pmatrix} \tag{27}$$

4 DESIGN NEW CONTROL CONFIGURATIONS

In all the new configurations studied in this paper, the composition manipulators are linear functions of $\Delta L, \Delta V, \Delta D$, and ΔB, as represented in Eq(22). The levels of condenser and reboiler are controlled by the top product flow, D, and the bottom product flow, B, respectively. The achievable performance of a distillation control system is mainly determined by the process model and the disturbance model. As is indicated in Eq(25), proper selection of G_1 and G_2 can alter the process model and the disturbance model in such a way to enhance the achievable control performance. Methods for properly selecting G_1 and G_2 are presented in the following.

4.1 *Disturbance rejecting and decoupling configuration*

The design objective here is to find suitable G_1 and G_2 to achieve steady state decoupling and disturbance rejection. It is fulfilled if the steady state process model is a diagonal matrix and the steady state disturbance gains are zero. Expressing these mathematically,

$$G(0)(G_1 + G_2 M)^{-1} = G_N(0) \tag{28}$$

$$G_d(0) - G(0)(G_1 + G_2 M)^{-1} G_2 M' = 0 \tag{29}$$

where $G_N(0)$ is a diagonal matrix representing the desired steady state process model.

Combining Eq(28) and Eq(29), the following equation can be obtained,

$$G_2 M' = G_N^{-1}(0) G_d(0) \tag{30}$$

Once G_2 is solved from the above equation, G_1 can be readily obtained from Eq(28) as

$$G_1 = G_N^{-1}(0) G(0) - G_2 M \tag{31}$$

Depending on the number of disturbances considered, Eq(30) could have one solution, many solutions, or no solution.

Transpose the two sides of Eq(30) gives

$$(M')^T G_2^T = G_d^T(0)(G_N^{-1}(0))^T \tag{32}$$

Partition G_2^T and $G_d^T(0)(G_N^{-1}(0))^T$ as follows

$$G_2^T = [A_1 \;\; A_2] \tag{33}$$

$$G_d^T(0)(G_N^{-1}(0))^T = [L_1 \;\; L_2] \tag{34}$$

where $A_1, A_2 \in R^{2\times 1}$ and $L_1, L_2 \in R^{n\times 1}$.

Then the solution of Eq(32) is equivalent to the solution of the following two equations.

$$(M')^T A_1 = L_1 \tag{35}$$

$$(M')^T A_2 = L_2 \tag{36}$$

The existence of solutions can be tested from the ranks of M', $[(M')^T\ L_1]$, and $[(M')^T\ L_2]$. Depending on the number of disturbances, n, there exist the following situations.

(1). $n = 1$.

In this case

$$\rho[(M')^T] = \rho[(M')^T\ L_1] = \rho[(M')^T\ L_2] = 1 < 2$$

where ρ denotes the rank of a matrix, there are infinitely many solutions.

(2). $n = 2$.

In this case, $\rho(M')$ can be 1 or 2. If $\rho(M') = 2$, there is only one solution which is found to be

$$G_1 = G_N^{-1}(0)(G(0) - G_d(0)(M')^{-1}M) \tag{37}$$

$$G_2 = G_N^{-1}(0)G_d(0)(M')^{-1} \tag{38}$$

If $\rho(M') = 1$, there will be no solution if either $\rho[(M')^T\ L_1] = 2$ or $\rho[(M')^T\ L_2] = 2$, otherwise there are infinitely many solutions.

Equations (19) to (21) indicate that if feed flow disturbance is considered, $\rho(M')$ will be 2 and otherwise $\rho(M')$ will be 1. Therefore, there will be only one solution if feed flow disturbance is considered.

(3). $n > 2$.

As indicated in Equations (19) to (21), $\rho(M')$ will be 2 if feed flow disturbance is considered and will be 1 otherwise. However, in this case $\rho[(M')^T\ L_1]$ or $\rho[(M')^T\ L_2]$ will generally be greater than two and therefore, it is not likely that there will be a solution.

In the case of no solution, one can consider only the feed flow disturbance and another important disturbance while ignore other disturbances. In this case, there will be a solution. An alternative is to find G_1 and G_2 through optimisation to minimise the weighted sum of the squared disturbance gains.

4.2 Application of the DRD structure to the example column

In this example column, we consider only the feed composition disturbance and the feed flow disturbance. In this case, there will be a unique solution. In this example, $G_N(0)$ is chosen as the identity matrix. Using Eq(37), Eq(38), and the model of this column presented in Section 3, G_1 and G_2 are found to be

$$G_1 = \begin{pmatrix} 1.4079 & -2.0221 \\ 3.6210 & -9.8801 \end{pmatrix} \tag{39}$$

$$G_2 = \begin{pmatrix} 0.9551 & -0.0142 \\ 6.5325 & 2.0479 \end{pmatrix} \tag{40}$$

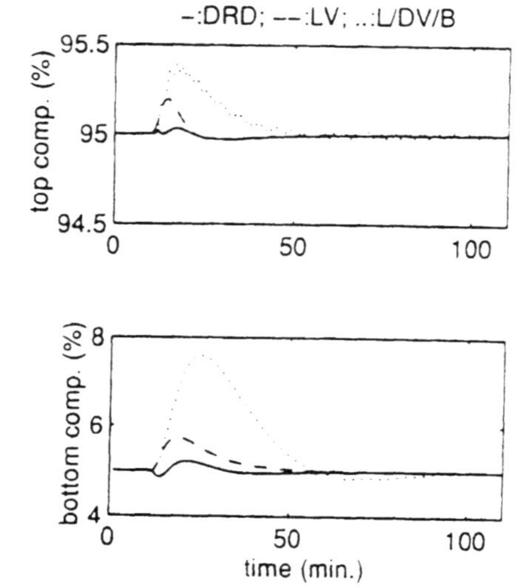

Fig. 3. Closed loop responses to a feed composition disturbance

Open loop and closed loop simulations of this control configuration have been performed. Figure 1 and Figure 2 show that the open loop performance under feed composition and feed flow disturbance respectively. For comparison purpose, the corresponding open loop responses of the LV and $L/DV/B$ configurations are also included.

Figures 1 and 2 show that at steady state the product compositions are hardly affected by disturbances. This suggests that the role of a feedback controller in this DRD configuration is less important than that in other configurations. Due to model parameter uncertainties and the fact that there could also exist other disturbances, a feedback controller is still needed.

Closed loop performance of the DRD, LV, and $L/DV/B$ configurations under feed composition disturbance and feed flow rate disturbance are presented in Figure 3 and Figure 4 respectively. The controllers used are diagonal PI controllers, which are widely used in practice, and they are tuned to achieve maximum robust performance measured by the structured singular value. Figure 3 and Figure 4 clearly demonstrate the good performance of the DRD configuration.

4.3 Robustness of the DRD configuration

The robustness of the DRD configuration can be assessed rigourously using the structured singular value analysis, also known as μ analysis (Doyle, 1982; 1985). Distillation column models are nonlinear and of high order. Using low order linearised models will introduce model parameter uncertainties, such as gain uncertainties and time constant uncertainties, and high frequency dynamic uncertainties. Apart from model uncertainties, there are also input uncertainties due to control valves (Skogestad and Morari, 1988; Skogestad *et al*, 1988). In the analysis here, we only consider model uncertainties and these uncertainties are lumped into a full block output multiplicative uncertainty. A block diagram showing the control system under this uncertainty is presented in Figure 5. The uncertainty weight is determined by identifying several models at different operating conditions and studying the relative model differences. The weight

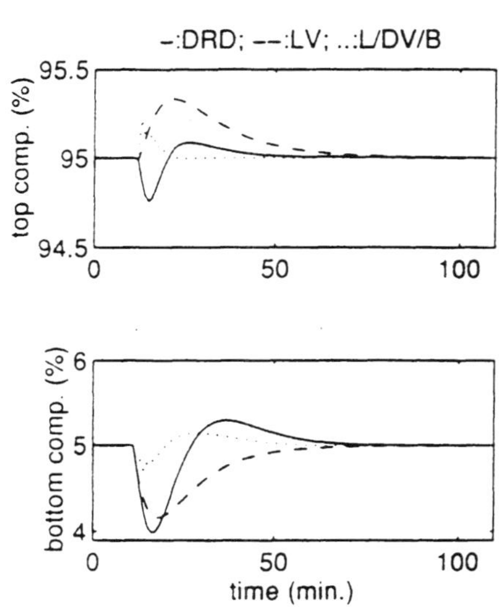

Fig. 4. Closed loop responses to a feed rate disturbance

for model uncertainty is determined as

$$W_o(s) = 0.4\frac{1+4.5s}{1+0.8s} \tag{41}$$

which indicates up to 40% uncertainty at steady state and uncertainty will increase at high frequency. The performance weight, $W_p(s)$, is chosen as

$$W_p(s) = 0.5\frac{1+5s}{5s} \tag{42}$$

which indicates that integral control action is required at low frequency and an amplification of disturbance at high frequency of up to a factor of two is allowed. Such a performance weight is chosen based on the discussions of Skogestad and Lundstrom (1990). The model uncertainty block, Δ_o, is a 2×2 full block perturbation. The disturbances are weighted such that the magnitude of the normalised disturbances are less than one at all frequencies and the resulting normalised errors are also less than one at all frequencies. The disturbance weight, W_d, is chosen as

$$W_d = \mathit{diag}\{1.823\alpha/20, 10\beta/20\}$$

where $\alpha, \beta \in [0,1]$ are used to specify which disturbance occurs more frequently and both of them are chosen as 1 in this example.

Rearranging Figure 5 into the structured singular value analysis framework form (Doyle, 1985), Figure 6, the connection matrix N is found to be

$$N = \begin{pmatrix} -W_oH & -W_oHG'_dW_d \\ -W_pS & -W_pSG'_dW_d \end{pmatrix} \tag{43}$$

where

$$G' = G(G_1 + G_2M)^{-1}$$

$$G'_d = G_d - G(G_1 + G_2M)^{-1}G_2M'$$

$$S = (I + G'C) - 1$$

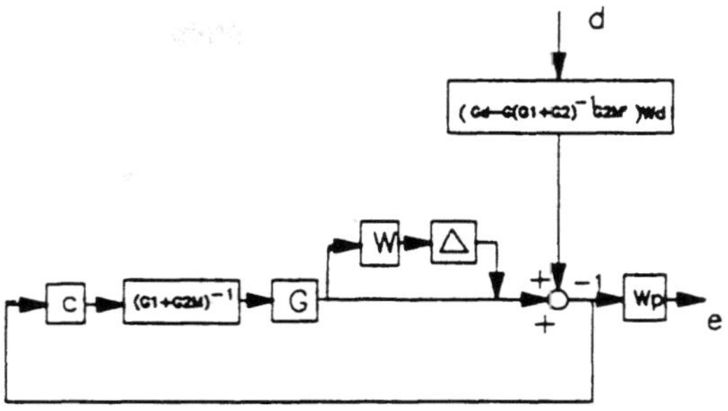

Fig. 5. Block diagram for system under model uncertainty

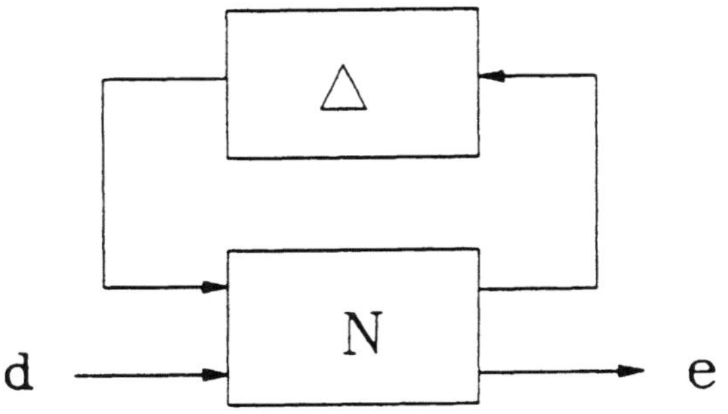

Fig. 6. Framework for μ-analysis

$$H = SG'C$$

Using the results of Doyle (1985), it can be found that robust stability is achieved if and only if

$$\sup_\omega \mu(-W_oH) \leq 1 \tag{44}$$

where $\mu(\cdot)$ is the structured singular value.

Robust performance will be achieved if and only if

$$\sup_\omega \mu(N) \leq 1 \tag{45}$$

where μ is computed with respect to the block diagonal perturbations $\mathit{diag}\{\Delta_o, \Delta_p\}, \Delta_p \in C^{2\times 2}$.

Controller parameters are determined by optimising the achievable robust performance. The robust performance of the DRD, the LV, and the $L/DV/B$ configurations are shown in Figure 7. It can be seen that the DRD configuration can achieve better robust performance than both the LV and the $L/DV/B$ configurations. This demonstrates that despite model uncertainties, the DRD structure can achieve better performance than the LV and the $L/DV/B$ configurations for this particular column.

4.4 μ optimum configurations

The achievable robust performance can be used as a criterion in designing new control configurations. In this approach, the design of control configurations and the tuning of controllers can be carried out in a unique optimisation procedure as follows.

$$\min_{G_1,G_2,C} \sup_{\omega} \mu(N)$$

The optimum G_1 and G_2 are found to be

$$G_1 = \begin{pmatrix} 1.6028 & -2.5335 \\ 1.7338 & -4.791 \end{pmatrix} \tag{46}$$

$$G_2 = \begin{pmatrix} 1.2913 & 0.0938 \\ 3.1807 & 1.0073 \end{pmatrix} \tag{47}$$

The achievable robust performance measured by μ is 0.405 which is slightly better than the DRD configuration. It should be stressed here that for some other columns the improvement can be significant.

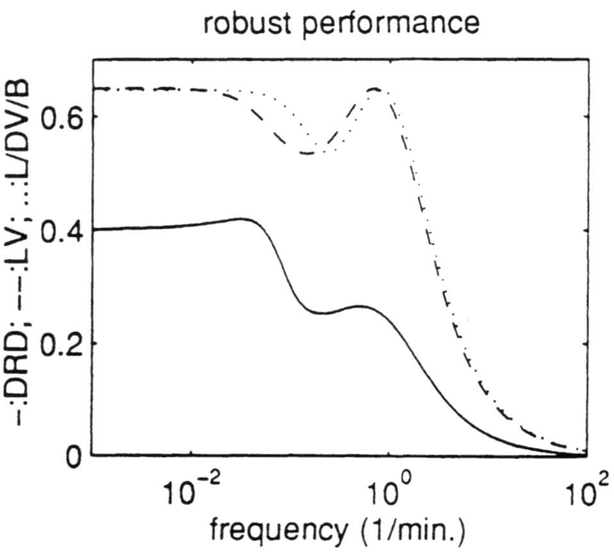

Fig. 7. Robust performance

5 CONCLUSIONS

A systematic method for designing new distillation control configurations is presented in this paper. The DRD control configuration is designed to achieve completely decoupling and disturbance rejection at steady state. The results presented here show that this may not always be possible depending on the disturbances considered. If only one disturbance is considered or two disturbances where one is feed flow disturbance are considered then completely steady state decoupling and disturbance rejection can be achieved. In other cases, this is not likely to be achievable. However, it is possible to design a control configuration in those cases such that disturbance rejection is achieved in a maximally possible way.

Robustness of the designed new control configurations is analysed using the structured singular value analysis. A new criterion, namely robust performance, is suggested in this paper for designing new distillation control configurations. In this approach the design of control configuration and the tuning of controller can be done simultaneously in a unique optimisation procedure. Nonlinear simulation results presented in this paper show that the new control configurations are superior to conventional control structures.

6 REFERENCES

Doyle, J. (1982), Analysis of feedback systems with structured uncertainties, *IEE Proc. Pt.D*, **129**, 242-250.

Doyle, J. (1985), Structured uncertainty in control system design, *Proc. IEEE CDC*, pp260-265.

Haggblom, K. E. and K. V. Waller (1988), Transformations and consistency relations of distillation control structures, *AIChE Journal*, **34**, 1634-1648.

Haggblom, K. E. and K. V. Waller (1990), Control structures for disturbance rejection and decoupling of distillation, *AIChE Journal*, **36**, 1107-1113.

Luyben, W. L. (1988), The concept of "eigenstructure" in process control, *Ind. Eng. Chem. Res.*, **27**, 206-208.

Sandelin, P. M., K. E. Haggblom, and K. V. Waller (1991a), A disturbance sensitivity parameter and its application to distillation control, *Ind. Eng. Chem. Res.*, **30**, 1182-1186.

Sandelin, P. M., K. E. Haggblom, and K. V. Waller (1991b), Disturbance rejection properties of control structures at one-point control of a two-product distillation column, *Ind. Eng. Chem. Res.*, **30**, 1187-1193.

Skogestad, S. and M. Morari (1987a), Control configuration selection for distillation columns, *AIChE Journal*, **33**, 1620-1635.

Skogestad, S. and M. Morari (1987b), A systematic approach to distillation column control, *Distillation and Absorption 87*, Brighton, UK, September 1987, *I. Chem. E. Symposium Series*, No.104, A71-A86.

Skogestad, S., P. Lundstrom, and E. W. Jacobsen (1990), Selecting the best distillation control configuration, *AIChE Journal*, **36**, 753-764.

Skogestad, S. and M. Morari (1988), LV-control of a high-purity distillation column, *Chemical Engineering Science*, **43**, 33-48.

Skogestad, S., M. Morari, and J. Doyle (1988), Robust control of ill-conditioned plants: high-purity distillation, *IEEE Transactions on Automatic Control*, **33**, 1092-1105.

Skogestad, S. and P. Lundstrom (1990), MU-optimal LV-control of distillation columns, *Computers and Chemical Engineering*, **14**, 401-413.

Yang, D. R., K. V. Waller, D. E. Seborg, and D. A. Mellichamp (1990), Dynamic structural transformations for distillation control configurations, *AIChE Journal*, **36**, 1391-1402.

Yang, D. R., D. E. Seborg, and D. A. Mellichamp (1991), Combined balance control for distillation columns, *Ind. Eng. Chem. Res.*, **30**, 2159-2168.

A DAE FRAMEWORK FOR MODELING AND CONTROL OF REACTIVE DISTILLATION COLUMNS

Aditya Kumar* and Prodromos Daoutidis*

**University of Minnesota, Department of Chemical Engineering and Materials Science, Minneapolis, MN 55455, USA*

Abstract. A detailed dynamic model incorporating the vapor dynamics is derived for reactive distillation columns. The modeling equations consist of nonlinear coupled differential and algebraic equations of index two. The open-loop behavior of the DAE model is compared with that of conventional ODE models derived under the assumption of negligible vapor holdup, to illustrate the importance of including the vapor phase dynamics in the combined reaction/separation operation. An output feedback controller is derived on the basis of the DAE model and its superior performance over an analogous controller derived on the basis of the ODE model is demonstrated through simulations.

Key Words. process control, nonlinear control, singular systems

1. INTRODUCTION

Reactive distillation is a favorable alternative to conventional reactor-separator configurations, for chemical reactions that are equilibrium limited or have a high heat of reaction. The simultaneous reaction and separation in a single process unit allows a superior conversion of the reactants, while the inter-phase mass transfer provides an inherent and efficient means for heat removal/addition in the reaction phase (liquid phase). These significant advantages have led to an increasing attention on reactive distillation applications (Saito *et al.*, 1971; Agreda *et al.*, 1990; Doherty and Buzad, 1992).

Existing literature on reactive distillation is abound with techniques for the numerical simulation of steady-state profiles (e.g., Jelínek and Hlavàček, 1976; Chang and Seader, 1988). Recent research efforts have also focused on design aspects. The design and calculation of minimum reflux for reactive distillation with both reaction and phase equilibrium has been addressed in (Barbosa and Doherty, 1988*a*), using a suitable change of variables and residue curve maps (Barbosa and Doherty, 1988*b*). The effect of finite reaction kinetics on residue curve maps has been studied in (Venimadhavan *et al.*, 1994). For processes where the reversible reaction is not at equilibrium, a design technique using fixed-point methods has been proposed (Buzad and Doherty, 1994), while for cases where enthalpy balances are also important, the design problem has been addressed within an optimization framework (Ciric and Gu, 1994).

Despite these advances in steady-state simulation and design, very few results exist on dynamic modeling and simulation (Grosser *et al.*, 1987). On the other hand, the work of (Roat *et al.*, 1986) highlights the limitations of control schemes designed on the basis of steady-state considerations, and the underlying necessity of rigorous nonlinear dynamic models for efficient control. The objective of this work, is to develop a rigorous dynamic model for reactive distillation that captures the fundamental characteristics of the simultaneous reaction/separation operation. The resulting dynamic model, consisting of coupled nonlinear differential and algebraic equations (DAEs), will be used for the design of an output feedback controller, within the general framework for feedback control of nonlinear DAE systems developed in (Kumar and Daoutidis, 1994).

2. DYNAMIC MODELING AND SIMULATION

Figure 1 shows a general reactive distillation column with $N+2$ equilibrium stages (N trays and a partial condenser and reboiler). We will consider a single exothermic reversible reaction in the liquid phase (e.g., esterification) :

$$A + B \rightleftharpoons C + D$$

where the reactants A and B are fed on the column trays at flowrates F_{Ai}, F_{Bi} and temperatures T_A, T_B, respectively. The reaction rate is de-

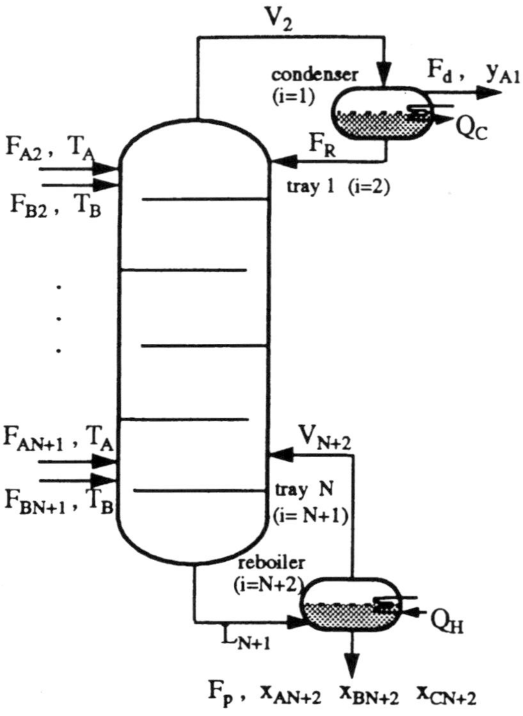

Fig. 1. Reactive distillation column

scribed by :

$$r_{Ai} = M_i^l \rho \left(k_1 \exp(\frac{-E_{a1}}{RT_i}) x_{Ai} x_{Bi} - k_2 \exp(\frac{-E_{a2}}{RT_i}) x_{Ci}(1 - x_{Ai} - x_{Bi} - x_{Ci}) \right)$$

where k_1, k_2 and E_{a1}, E_{a2} are the pre-exponential factors and activation energies of the forward and reverse reactions, respectively. The column is operated with a liquid recycle from the condenser at a flowrate F_R and a vapor recycle from the reboiler at a flowrate V_{N+2}. The product vapor stream from the condenser is withdrawn at a flowrate F_d, while the bottom product is withdrawn from the reboiler at a flowrate F_p.

For ordinary (non-reactive) distillation columns, conventional dynamic models comprised of pure ODEs (e.g., Eckert and Kubíček, 1994), are derived under the assumptions of well-mixed liquid and vapor phases in thermodynamic equilibrium, and a negligible vapor holdup. The assumption of negligible vapor holdup is reasonably valid at low column pressures. However, in reactive distillation columns with exothermic reactions, the temperature and hence the pressure in the individual stages may be significantly high. Moreover, the above assumption implies that the vapor stream from a stage i goes directly to the liquid phase in stage $i-1$, without any inter-phase mass transfer, i.e., both the reactants *and* the products go to the reaction phase (liquid phase). Clearly, this is contrary to the advantageous feature of reactive distillation columns for equilibrium limited reactions, where the product is continuously separated into the (non-reactive) vapor phase to achieve a higher conversion of the reactants. These considerations indicate the need to include the vapor dynamics explicitly in the modeling equations.

In what follows, a detailed dynamic model comprised of coupled differential and algebraic equations will be derived under the standard assumption of well-mixed liquid and vapor phases in thermodynamic equilibrium. For simplicity, it will also be assumed that reactant B and product D are nonvolatile, and the molar densities ρ, heat capacities c_p and latent heat of vaporizations ΔH^v are constant and equal for all the components. For each stage i, the differential equations include the overall and component mole balances in the vapor and liquid phases, and the overall enthalpy balance, while the algebraic equations include the phase-equilibrium relations (Raoult's law, assuming ideal behavior), the ideal gas equation, the Francis weir formula and the pressure-drop correlation. The resulting modeling equations are :

Stage 1 (partial condenser)

$$\begin{aligned}
\dot{M}_1^v &= V_2 - F_d - N_{A1} - N_{C1} \\
\dot{y}_{A1} &= \frac{1}{M_1^v}[V_2(y_{A2} - y_{A1}) - N_{A1}(1 - y_{A1}) + N_{C1} y_{A1}] \\
\dot{M}_1^l &= N_{A1} + N_{C1} - F_R \\
\dot{x}_{A1} &= \frac{1}{M_1^l}[N_{A1}(1 - x_{A1}) - N_{C1} x_{A1}] \\
\dot{T}_1 &= \frac{1}{M_1^l + M_1^v}\left[V_2(T_2 - T_1) + (N_{A1} + N_{C1})\frac{\Delta H^v}{c_p} - \frac{Q_C}{c_p}\right] \\
0 &= P_1 y_{A1} - P_{A1}^s x_{A1} \\
0 &= P_1(1 - y_{A1}) - P_{C1}^s(1 - x_{A1}) \\
0 &= P_1(V_T - \frac{M_1^l}{\rho}) - M_1^v R T_1
\end{aligned} \qquad (1)$$

Stage $i = 2, \ldots, N+1$ (trays $1, \ldots, N$)

$$\begin{aligned}
\dot{M}_i^v &= V_{i+1} - V_i - N_{Ai} - N_{Ci} \\
\dot{y}_{Ai} &= \frac{1}{M_i^v}[V_{i+1}(y_{Ai+1} - yAi) - N_{Ai}(1 - y_{Ai}) + N_{Ci} y_{Ai}] \\
\dot{M}_i^l &= F_{Ai} + F_{Bi} + L_{i-1} - L_i + N_{Ai} + N_{Ci} \\
\dot{x}_{Ai} &= \frac{1}{M_i^l}[F_{Ai}(1 - x_{Ai}) - F_{Bi} x_{Ai} + L_{i-1}(x_{Ai-1} - x_{Ai}) \\
&\quad + N_{Ai}(1 - x_{Ai}) - N_{Ci} x_{Ai} - r_{Ai}] \\
\dot{x}_{Bi} &= \frac{1}{M_i^l}[-F_{Ai} x_{Bi} + F_{Bi}(1 - x_{Bi}) + L_{i-1}(x_{Bi-1} - x_{Bi}) \\
&\quad - N_{Ai} x_{Bi} - N_{Ci} x_{Bi} - r_{Ai}] \\
\dot{x}_{Ci} &= \frac{1}{M_i^l}[-F_{Ai} x_{Ci} - F_{Bi} x_{Ci} + L_{i-1}(x_{Ci-1} - x_{Ci}) \\
&\quad - N_{Ai} x_{Ci} + N_{Ci}(1 - x_{Ci}) + r_{Ai}] \\
\dot{T}_i &= \frac{1}{M_i^l + M_i^v}\Big[F_{Ai}(T_A - T_i) + F_{Bi}(T_B - T_i) \\
&\quad + L_{i-1}(T_{i-1} - T_i) + V_{i+1}(T_{i+1} - T_i) \\
&\quad + (N_{Ai} + N_{Ci})\frac{\Delta H^v}{c_p} + r_{Ai}(\frac{-\Delta H_r}{c_p})\Big] \\
0 &= P_i y_{Ai} - P_{Ai}^s x_{Ai}
\end{aligned}$$

$$0 = P_i(1 - y_{Ai}) - P^s_{Ci} x_{Ci}$$
$$0 = P_i(V_T - \frac{M^l_i}{\rho}) - M^v_i R T_i$$
$$0 = L_i - 1.839 W \rho \left(\frac{M^l_i - \rho V_w}{\rho A} \right)^{1.5}$$
$$0 = P_i - P_{i-1} - \zeta_1 \left(\frac{V_i R T_i}{P_i} \right)^2 - \zeta_2 \frac{M^l_{i-1}}{\rho} \quad (2)$$

Stage N + 2 (partial reboiler)

$$\dot{M}^v_{N+2} = -V_{N+2} - N_{AN+2} - N_{CN+2}$$
$$\dot{y}_{AN+2} = \frac{1}{M^v_{N+2}} [-N_{AN+2}(1 - y_{AN+2}) + N_{CN+2} y_{AN+2})]$$
$$\dot{M}^l_{N+2} = L_{N+1} - F_p + N_{AN+2} + N_{CN+2}$$
$$\dot{x}_{AN+2} = \frac{1}{M^l_{N+2}} [L_{N+1}(x_{AN+1} - x_{AN+2}) + N_{AN+2}(1 - x_{AN+2}) - N_{CN+2} x_{AN+2} - r_{AN+2}]$$
$$\dot{x}_{BN+2} = \frac{1}{M^l_{N+2}} [L_{N+1}(x_{BN+1} - x_{BN+2}) - N_{AN+2} x_{BN+2} - N_{CN+2} x_{BN+2} - r_{AN+2}]$$
$$\dot{x}_{CN+2} = \frac{1}{M^l_{N+2}} [L_{N+1}(x_{CN+1} - x_{CN+2}) - N_{AN+2} x_{CN+2} + N_{CN+2}(1 - x_{CN+2}) + r_{AN+2}]$$
$$\dot{T}_{N+2} = \frac{1}{M^l_{N+2} + M^v_{N+2}} \left[L_{N+1}(T_{N+1} - T_{N+2}) + \frac{Q_H}{c_p} + (N_{AN+2} + N_{CN+2}) \frac{\Delta H^v}{c_p} + r_{AN+2}(\frac{-\Delta H_r}{c_p}) \right]$$
$$0 = P_{N+2} y_{AN+2} - P^s_{AN+2} x_{AN+2}$$
$$0 = P_{N+2}(1 - y_{AN+2}) - P^s_{CN+2} x_{CN+2}$$
$$0 = P_{N+2}(V_T - \frac{M^l_{N+2}}{\rho}) - M^v_{N+2} R T_{N+2}$$
$$0 = P_{N+2} - P_{N+1} - \zeta_1 \left(\frac{V_{N+2} R T_{N+2}}{P_{N+2}} \right)^2 - \zeta_2 \frac{M^l_{N+1}}{\rho} \quad (3)$$

In the modeling equations for the first tray ($i = 2$), L_{i-1} is replaced by F_R and the parameter ζ_2 in the pressure-drop correlation (Eckert and Kubiček, 1994) is set to zero. The above modeling equations (Eq.1-3) comprise a DAE system with the following general representation :

$$\dot{x} = \bar{f}(x, z)$$
$$0 = k(x, z) \quad (4)$$

where the differential variables $x \in \mathbb{R}^n$ include $M^v_i, y_{Ai}, M^l_i, x_{Ai}, x_{Bi}, x_{Ci}, T_i$, $i = 2, \ldots, N+2$, and $M^v_1, y_{A1}, M^l_1, x_{A1}, T_1$, the algebraic variables $z \in \mathbb{R}^p$ consist of N_{Ai}, N_{Ci}, P_i for all stages i, L_i, $i = 2, \ldots, N+1$ and V_i, $i = 2, \ldots, N+2$, and $\bar{f}(x, z), k(x, z)$ are vector fields of dimensions n, p, respectively. Clearly, since the inter-phase mole transfer rates N_{Ai}, N_{Ci} do not appear in any algebraic equation, the DAE system has a high index. It can be verified that the index is two.

The open-loop behavior of a column with $N = 5$ trays was studied through simulations. In the specific column, the reactants A and B were fed in a countercurrent fashion with $F_{Ai}, F_{Bi} = 0$ for all stages except for $F_{A4} = 50$ and $F_{B2} = 50$ mole/s. Moreover, the product C was assumed to be more volatile than the reactant A, with their saturation vapor pressures given by the Antoine relations :

$$P^s_{Ai} = \exp(26 - \frac{3400}{T_i + 20})$$
$$P^s_{Ci} = \exp(27 - \frac{4100}{T_i + 70})$$

A description of the process parameters and variables, and their nominal values is included in Tables 1 and 2. The process is open-loop unstable owing to the unstable dynamics of the liquid holdups in the condenser and the reboiler. Figure 2 shows a representative comparison of the open-loop profiles for T_i, P_i, r_{Ai} for all stages i, as predicted by the rigorous DAE model (Eq.1-3) and a conventional ODE model derived under the additional assumption of negligible vapor holdup, for step changes in the process inputs : $\Delta F_B = 30, \Delta F_R = 10, \Delta F_d = -10, \Delta Q_H = -100000$. Clearly, the ODE model predicts a significantly lower total (over all stages) reaction rate or equivalently, conversion, owing to the limitation imposed by ignoring the vapor holdup as discussed earlier. Moreover, the DAE model, predicts a general increase in the column temperature and pressure owing to the higher reaction rate, while the ODE model predicts a qualitatively opposite behavior.

3. CONTROL PROBLEM

For the reactive distillation column (Fig.1), it is desired to control the liquid holdups M^l_1, M^l_{N+2} in the condenser and the reboiler (open-loop unstable) and the product composition y_{A1}. A higher product purity (lower y_{A1}) can be achieved easily through enhanced cooling in the condenser, albeit at the expense of lower conversion. To ensure that a desired increase in product purity is accompanied by a higher conversion, the temperature in the condenser T_1 will be controlled as well. With these four controlled outputs :

$$y_1 = M^l_1,\ y_2 = y_{A1},\ y_3 = T_1,\ y_4 = M^l_{N+2}\ .$$

and the corresponding manipulated inputs :

$$u_1 = F_R,\ u_2 = F_d,\ u_3 = Q_C,\ u_4 = F_p$$

the resulting DAE system is of the following general form :

$$\dot{x} = f(x, z) + g(x, z)u$$
$$0 = k(x, z)$$
$$y_i = h_i(x),\ i = 1, \ldots, m \quad (5)$$

where $x \in \mathbb{R}^n$, $z \in \mathbb{R}^p$ are the vectors of differential and algebraic variables, respectively,

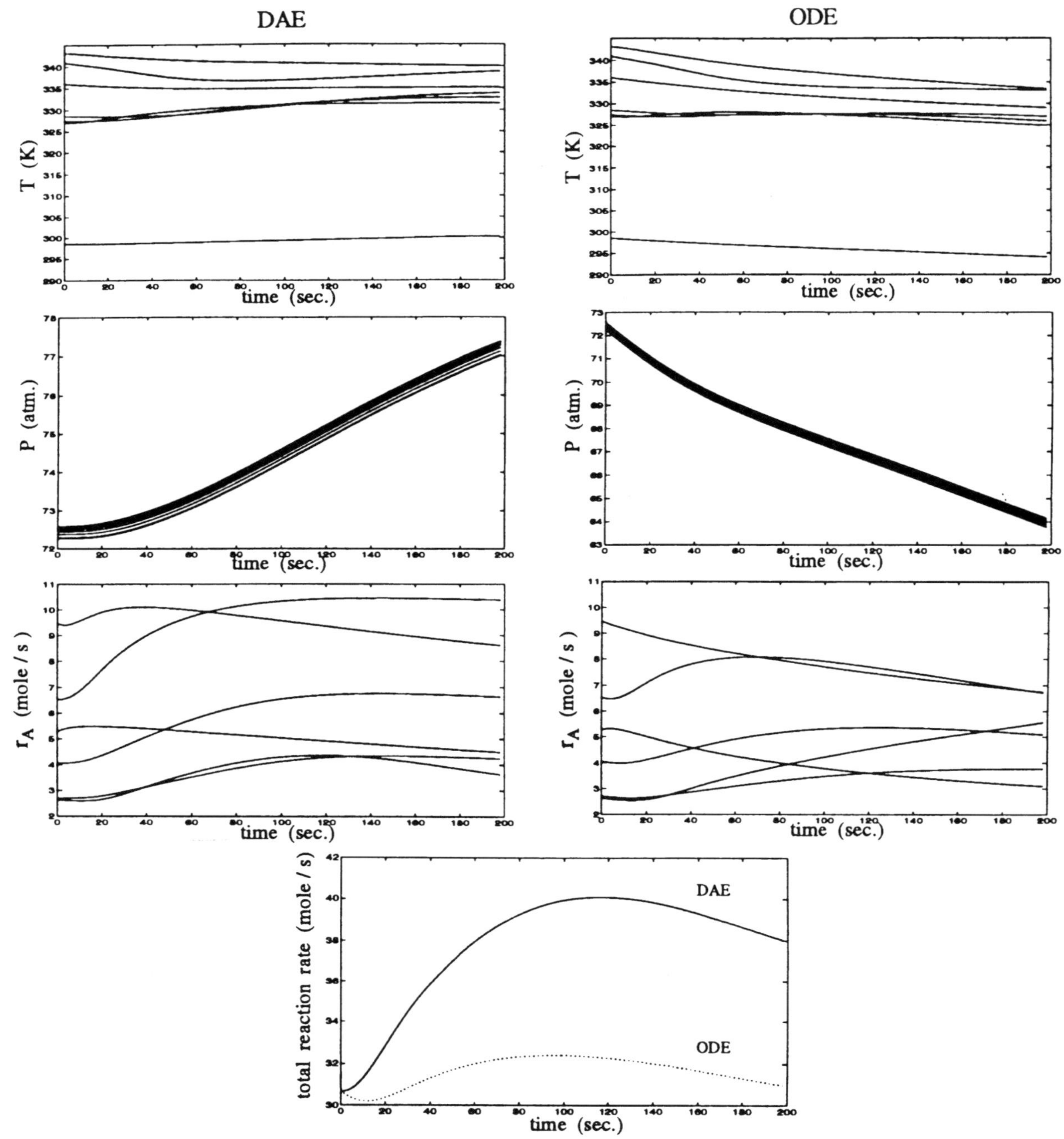

Fig. 2. Comparison of open-loop profiles predicted by DAE and ODE models

$u \in \mathbb{R}^m$ is the vector of manipulated inputs and y_i are the outputs to be controlled. Moreover, $f(x,z), k(x,z)$ are vector fields of dimensions n, p, respectively, $h_i(x)$ are scalar fields and $g(x,z)$ is a matrix of dimension $n \times m$.

In (Kumar and Daoutidis, 1994), a general methodological framework was developed for the feedback control of DAE systems of the form :

$$\begin{aligned} \dot{x} &= f(x) + b(x)z + g(x)u \\ 0 &= k(x) + l(x)z \\ y_i &= h_i(x), \; i = 1, \ldots, m \end{aligned} \qquad (6)$$

The control of systems of the form in Eq.5 can also be addressed within this framework. To this end, consider the dynamic extension where an extended vector of differential variables $\tilde{x} = [x^T \;\; z^T]^T$ and a new vector of algebraic variables $\tilde{z} = \dot{z}$ are defined. Under this dynamic extension, the DAE system of Eq.5 takes the following form :

$$\begin{aligned} \dot{\tilde{x}} &= \begin{bmatrix} f(\tilde{x}) \\ 0 \end{bmatrix} + \begin{bmatrix} 0 \\ I_p \end{bmatrix} \tilde{z} + \begin{bmatrix} g(\tilde{x}) \\ 0 \end{bmatrix} u \\ 0 &= k(\tilde{x}) \\ y_i &= \tilde{h}_i(\tilde{x}), \; i = 1, \ldots, m \end{aligned} \qquad (7)$$

which is affine in the algebraic variables $\tilde{z}$ and hence, in the form of Eq.6. In particular, for the DAE system in Eq.1-3, the dynamic extension is required only for the algebraic variables $P_i, V_i, \; i \geq 2$, i.e. the ones that appear in a nonlinear fashion.

For the extended DAE system in Eq.7, an output feedback controller that induces a well-characterized input/output behavior in the closed-loop system can be derived following the procedure in (Kumar and Daoutidis, 1994). The procedure entails (*i*) the derivation of a state-space realization of the constrained system, and (*ii*) the synthesis of the controller on the basis of the state-space realization. The derivation of state-space realization involves (*a*) the specification of the underlying constraints among $\tilde{x}$ that characterize the constrained state-space, and (*b*) the solution of $\tilde{z}$ in terms of $\tilde{x}, u$, to obtain differential equations for $\tilde{x}$. The output feedback controller is synthesized through a combination of state feedback and state observers.

For the specific 5-tray column considered in Section 2, the phase-equilibrium relations in all stages, and the pressure-drop correlation and ideal gas equation in stages $2, \ldots, 7$, yield altogether $\sigma = 26$ constraints among the extended set of differential variables $\tilde{x}$. The resulting state-space realization described by the differential equations on $\tilde{x}$ has a dimension $\tilde{n} = 59$. A minimal-order state-space realization of dimension $\tilde{n} - \sigma = 33$ was obtained in appropriately transformed coordinates. The relative orders of the four outputs with respect to the manipulated inputs were found to be one each. Thus, an input/output decoupled first-order response was requested for all the outputs. The performance of the output feedback controller was compared with that of an analogous controller derived on the basis of a conventional ODE model.

Figure 3, shows the input and output profiles in a representative simulation run for step changes of -20% and $+3\%$ in the setpoints for y_{A1} and T_1, respectively. Clearly, while the DAE-based controller induces the decoupled first-order responses, the ODE controller shows a significantly poor performance with oscillations and a large overshoot in the temperature. The differences are further apparent in the manipulated input profiles for the two controllers.

4. CONCLUSIONS

This work addressed the derivation of rigorous dynamic models and the design of nonlinear feedback controllers for reactive distillation – a process of increasing importance in the chemical process industry. The inherent advantageous feature of the process, i.e. the simultaneous reaction and separation, motivated the need to include the vapor-phase dynamics explicitly in the modeling equations, thereby yielding an index-two DAE model. The limitations of conventional models that ignore vapor holdup, in comparison with the rigorous DAE model, were illustrated through simulations. The DAE model was then used for the synthesis of an output feedback controller that provides a significant enhancement in closed-loop performance in comparison to an ODE-based controller.

5. ACKNOWLEDGMENT

Financial support for this work from NSF, CTS-9320402, is gratefully acknowledged.

6. REFERENCES

Agreda, V. H., L. R. Partin and W. H. Heise (1990). High-purity methyl acetate via reactive distillation. *Chem. Eng. Prog.* **86**, 40–46.

Barbosa, D. and M. F. Doherty (1988*a*). Design and minimum-reflux calculations for single-feed multicomponent reactive distillation columns. *Chem. Eng. Sci.* **43**, 1523–1537.

Barbosa, D. and M. F. Doherty (1988*b*). The simple distillation of homogeneous reactive mixtures. *Chem. Eng. Sci.* **43**, 541–550.

Buzad, G. and M. F. Doherty (1994). Design of three-component kinetically controlled reactive distillation columns using fixed-point methods. *Chem. Eng. Sci.* **49**, 1947–1963.

Chang, Y. A. and J. D. Seader (1988). Simulation of continuous reactive distillation by a homotopy-continuation method. *Comput. chem. Engng.* **12**, 1243–1255.

Ciric, A. R. and D. Gu (1994). Synthesis of nonequilibrium reactive distillation processes by MINLP optimization. *AIChE J.* **40**, 1479–1487.

Doherty, M. F. and G. Buzad (1992). Reactive distillation by design. *Trans. IChemE* **70**, 448–458.

Eckert, E. and M. Kubíček (1994). Modeling of dynamics for multiple liquid-vapor equilibrium stage. *Chem. Eng. Sci.* **49**, 1783–1788.

Grosser, J. H., M. F. Doherty and M. F. Malone (1987). Modeling of reactive distillation systems. *Ind. Eng. Chem. Res.* **26**, 983–989.

Jelínek, J. and V. Hlaváček (1976). Steady state countercurrent equilibrium stage separation with chemical reaction by relaxation method. *Chem. Eng. Comm.* **2**, 79–85.

Kumar, A. and P. Daoutidis (1994). Feedback control of nonlinear differential-algebraic-equation systems. in press. *AIChE J.*

Roat, S. D., J. J. Downs, E. F. Vogel and J. E. Doss (1986). The integration of rigorous dynamic modeling and control system synthesis for distillation columns: An industrial approach. In: *Chemical Process Control - CPC III* (M. Morari and T. J. McAvoy, Eds.). Elsevier Sci. Pub. Inc., New York. Asilomar, California. pp. 99–138.

Saito, S., T. Michishita and S. Maeda (1971). Separation of meta- and para-xylene mixture by distillation accompanied by chemical reactions. *J. Chem. Eng. Japan* **4**(1), 37–43.

Venimadhavan, G., G. Buzad, M. F. Doherty and M. F. Malone (1994). Effect of kinetics on residue curve maps for reactive distillation. *AIChE J.* **40**, 1814–1824.

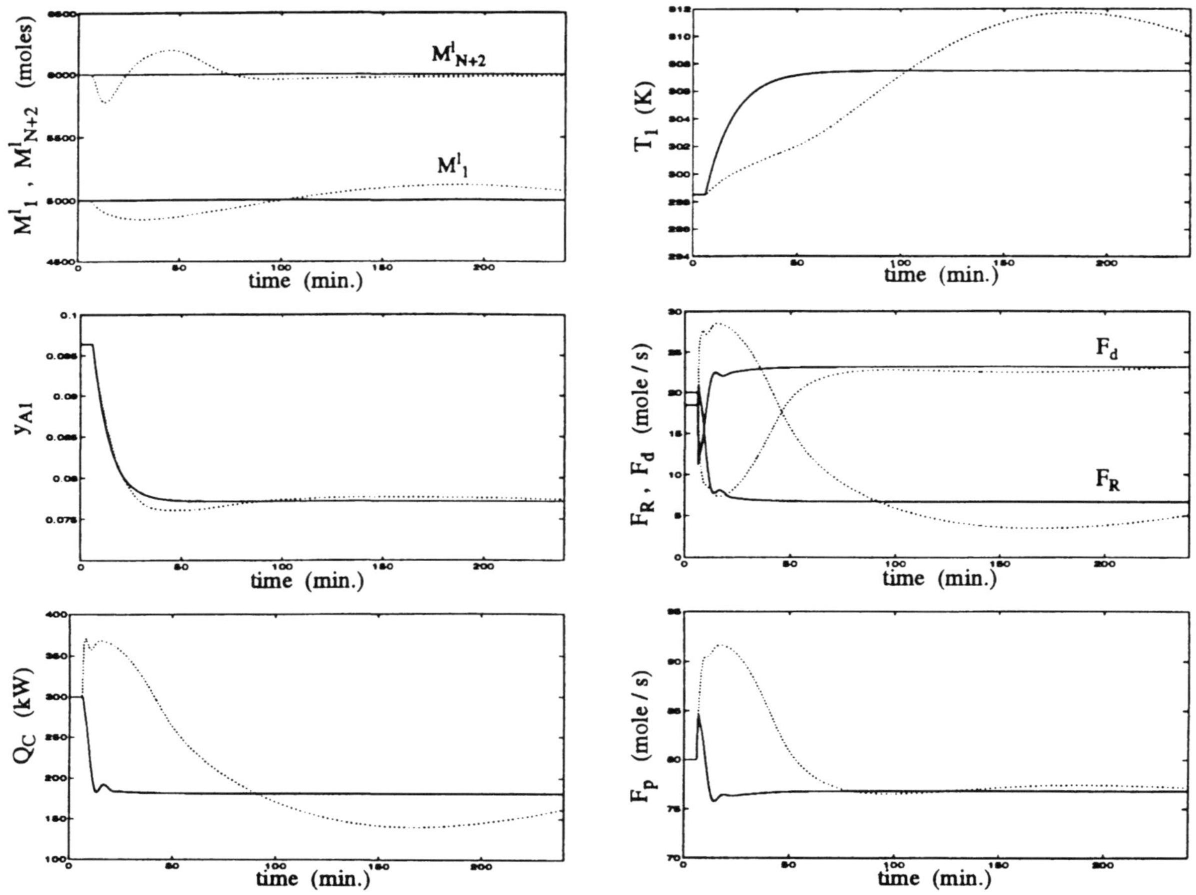

Fig. 3. Closed-loop profiles under DAE (solid) and ODE (dotted) based controllers for a 20% decrease in the setpoint for y_{A1} and a 3% increase in the setpoint for T_1

Table 1 Nominal values of parameters and input and output specifications

Variable	Description	Nominal value
A	cross-sectional area of the absorption column (m^2)	2.0
c_p	molar heat capacity ($J/mole\ K$)	80
E_{a1}, E_{a2}	activation energies ($kJ/mole$)	100, 115
F_d, F_p, F_r	distillate, bottom product and liquid recycle flowrates ($mole/s$)	20.0, 80.0, 18.44
k_1, k_2	pre-exponential factors ($m^3/mole\ s$)	1.0e+10, 1.53e+12
Q_C, Q_H	heat duties in condenser and reboiler (kW)	300, 300
T_A, T_B	inlet temperatures of reactants A,B (K)	300, 300
V_T	total volume for each tray (m^3)	1.0
V_w	volume corresponding to weir height on each tray (m^3)	0.3
W	weir width (m)	0.798
ΔH^v	latent heat of vaporization ($kJ/mole$)	10
ρ	liquid phase molar density ($kmoles/m^3$)	15
ζ_1, ζ_2	parameters in pressure-drop correlation	1.5e+7, 1.0e+4

Table 2 Steady-state values of process variables

Variable	Description	Stage						
		1	2	3	4	5	6	7
L_i	liquid flowrate from stage i ($mole/s$)	-	77.8	73.6	115.1	109.1	103.6	-
M_i^l	liquid holdup ($kmoles$)	5.00	5.20	5.17	5.40	5.37	5.34	6.00
M_i^v	vapor holdup ($kmoles$)	1.97	1.71	1.68	1.73	1.73	1.73	1.56
P_i	pressure (MPa)	7.318	7.321	7.330	7.337	7.343	7.349	7.354
T_i	temperature (K)	298.5	336.1	343.2	327.0	327.4	328.6	341.0
V_i	vapor flowrate from stage i ($mole/s$)	-	38.4	47.8	43.6	35.1	29.1	23.6
x_{Ai}	mole fraction of A in liquid phase	0.156	0.066	0.112	0.385	0.332	0.301	0.218
x_{Bi}	mole fraction of B in liquid phase	-	0.575	0.479	0.250	0.226	0.212	0.242
x_{Ci}	mole fraction of C in liquid phase	0.844	0.292	0.209	0.181	0.209	0.217	0.157
y_{Ai}	mole fraction of A in vapor phase	0.096	0.125	0.257	0.571	0.498	0.465	0.470

IDENTIFICATION OF A PILOT PLANT REACTIVE DISTILLATION PROCESS USING RBF NETWORKS

P. Säynätjoki[*] and L. Hammarström[**]

[*]Department of Chemical Engineering, Helsinki University of Technology, Kemistintie 1, FIN-02150 Espoo, Finland
[**]Neste Engineering, POB 310, FIN-06101 Porvoo, Finland

Abstract. An application of radial basis function networks to identification of nonlinear MIMO models for a pilot plant reactive distillation column is discussed. The studied reactive distillation process is a production of tert-amyl-methyl-ether via the reaction of methanol and isoamylenes. The data for the identifications has been recorded during pilot plant experiments with PRBS changes to the input signals. Three different operation regions of the process have been tested. The process dynamics is taken into account by delayed inputs and outputs in the RBF network. Different data sets have been used for identification and validation. One-step-forward prediction does not give good information about the ability of the model to predict the process behaviour. In this work n-step-forward prediction is used. The validation shows that the models are able to predict all different regions quite well.

Key Words. Process identification; dynamic modelling; nonlinear models; neural network models; radial base function networks

1. INTRODUCTION

Reactive distillation (RD) is a unit operation which combines distillation and reaction in the same vessel. The advantages of RD are high conversion and high selectivity, which means savings of operation and investment costs (Doherty and Buzhad, 1993). Production of TAME has been selected as the subject of this study.

The experimental work was done with a pilot plant of Neste Oy, Finland. The dynamics of the studied process was tested experimentally in different operation regions.

Linear identification tests showed that each of the operation regions needed its own models. This limits or at least makes it more difficult to use the linear models e.g. for control purposes. Thus the nonlinear models and identification are justified.

Neural networks are known to possess flexible function approximation capabilities. The practical use of radial basis function networks for nonlinear system identification has been discussed in many papers (e.g. Chen, *et al.* 1990a, b, Suontausta, *et al.* 1994). Usually RBF networks have been used for single-input single-output systems, but they offer possibilities for larger systems too. A RBF network is usually faster to train than a feedforward perceptron network (Demuth and Beale, 1994).

In this paper the applicability of RBF networks for nonlinear dynamic MIMO-system modelling is studied. Validation of the models is also discussed.

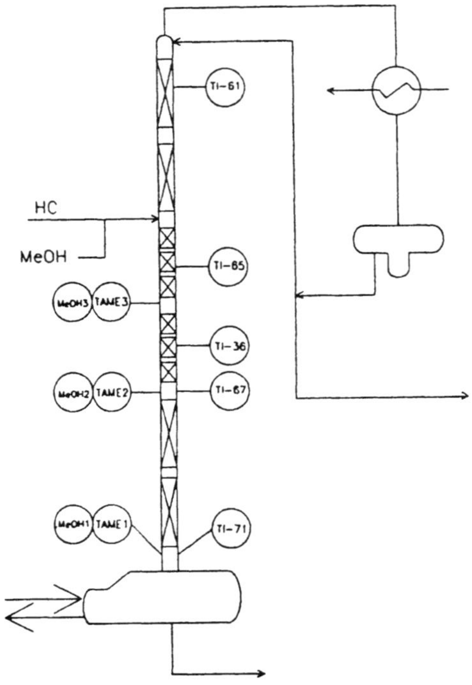

Fig. 1. The pilot reactive distillation column.

2. PROCESS

TAME is produced via reactions of methanol with isoamylenes. The major reactions in TAME production have been described by Bravo *et al.* (1993).

The test runs were done with a pilot RD column at Neste Oy (Pekkanen *et al.* 1994). The distillation column with reaction section was 11 m high and the column inside diameter was 0.15 m. The rectification and stripping sections were packed with IMTP random packing. The reaction section was located in the middle of the column and consisted of six units where both reaction and separation occur. The catalyst was an Amberlyst ion-exchange resin in the form of spheres with a diameter of about 1 mm. A diagram of the column is shown in Fig. 1 where some important measurements are indicated.

3. EXPERIMENTS

The pilot plant RD column was fully instrumented and connected to a digital automation system. Over 60 process variables (temperatures, flows and concentrations) were measured and stored every minute. The dynamic test period for each operation region lasted for 60 hours.

The process dynamics was studied in three different

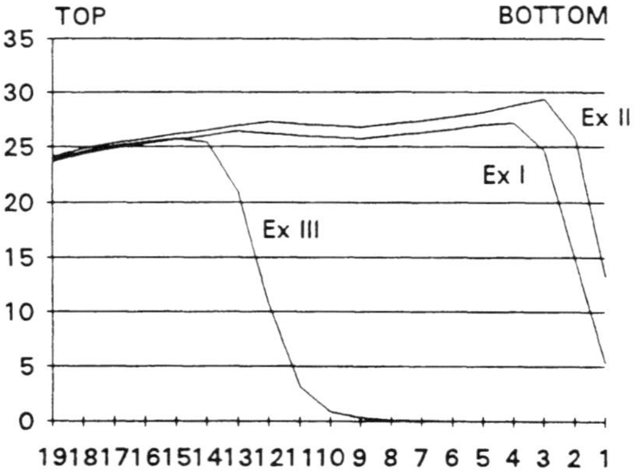

Fig. 2. Methanol concentration profiles for three steady states preceding dynamic runs.

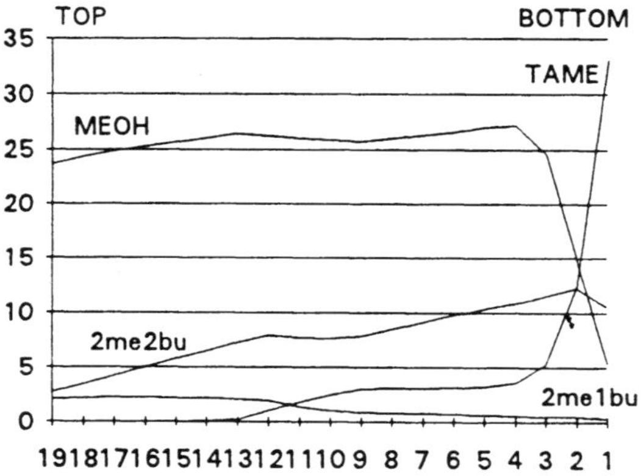

Fig. 3. Concentration profiles of the main reactive components in the first steady state.

operation regions. First the column was brought to and stabilized in a steady state and then PRBS changes in the inputs were made around this point. The steady states differed in the stoichiometric ratio of methanol/isoamylenes. Table 1 contains a summary of the different operation regions. In Figure 2, the methanol concentration profiles for the three steady states preceding the dynamic runs are shown. The amount of methanol in the TAME product was a critical factor and the main specification of the runs. Figure 3 shows the calculated concentration profiles of the main reactive components in the first steady state.

The inputs for PRBS during the dynamic runs were the hydrocarbon feed rate, the methanol feed rate, the reboil rate (steam feed rate to the reboiler) and the reflux flow rate. Maximum, minimum and steady state values of the inputs are shown in Table 1.

Table 1. The minimum, steady state and maximum values for the inputs of the system in operation regions Ex I, Ex II, Ex III

	H	B	R	M		
				Ex I	Ex II	Ex III
Min	10	14.5	40	1.7	2.0	1.3
S-S	15	21.8	60	2.6	3.0	1.7
Max	20	29.0	80	3.3	4.0	2.1

M = MeOH-feed, kg/h
H = Hydrocarbon feed, kg/h
B = Steam feed for reboiling, kg/h
R = Reflux, kg/h

4. NONLINEAR IDENTIFICATION

4.1 Nonlinear System

The system used in this study can be described as a nonlinear functional expansion of lagged inputs and outputs. Chen, *et al.* (1990a) have represented such a systems mathematically

$$y(t)=f_s(y(t-1),...,y(t-n_y),u(t-1),...,u(t-n_u))+e(t) \quad (1)$$

where

$$y(t)=\begin{bmatrix} y_1(t) \\ \cdot \\ \cdot \\ \cdot \\ y_m(t) \end{bmatrix}, \quad u(t)=\begin{bmatrix} u_1(t) \\ \cdot \\ \cdot \\ \cdot \\ u_r(t) \end{bmatrix}, \quad e(t)=\begin{bmatrix} e_1(t) \\ \cdot \\ \cdot \\ \cdot \\ e_m(t) \end{bmatrix} \quad (2)$$

y(t) = system output
u(t) = system input
e(t) = white noise
n_y = max lags in the output
n_u = max lags in the input

A general vector valued nonlinear function $f_s(\cdot)$ may be very complicated and not identifiable in practice. In this study a radial basis function expansion is used to model the input-output relationship of Eq. (1).

4.2 Radial Basis Functions for Modelling

According to Chen, *et al.* (1990a) a radial basis function model with *n* inputs and *m* outputs implements a mapping $f_r: \mathbb{R}^n \rightarrow \mathbb{R}^m$ such that

$$f_r(x)=\Lambda_0+\sum_{j=1}^{M_c}\Lambda_j\phi(\|x-c_j\|) \quad (3)$$

where $x\in\mathbb{R}^n$; $\phi(\cdot)$ is a function from $\mathbb{R}^+$ to $\mathbb{R}$; $\|\cdot\|$ is the euclidean norm; $\Lambda_j\in\mathbb{R}^m$, j = 1,...M_c, are the weight vectors; $\Lambda_0\in\mathbb{R}^m$ is a bias vector and $c_j\in\mathbb{R}^n$, j = 1,...M_c, are the radial basis function centres. If the weight vectors are

$$\Lambda_j=[\lambda_{1j}\ .\ .\ .\ \lambda_{mj}]^T, \quad j=0,1,...M_c \quad (4)$$

the Eq. (3) can be presented in decomposed form

$$f_{ri}(x)=\lambda_{i0}+\sum_{j=1}^{M_c}\lambda_{ij}\phi(\|x-c_j\|), \quad i=1,...m \quad (5)$$

This equation shows that for a MIMO system as many MISO models are identified as the MIMO has outputs, because each element of the system output vector requires its own RBF expansion (Chen, *et al.* 1990a).

The gaussian function is the typical choice for $\phi(\cdot)$

$$\phi(v)=\exp(-v^2/\beta^2) \quad (6)$$

The orthogonal least square learning algorithm (Chen, *et al.* 1991) is available in the Matlab Neural Network toolbox (Demuth and Beale, 1994) and has been used for the selection of the centres of the RBF. The procedure chooses the RBF centres one by one and each selected centre maximizes the increment to the explained variance or energy of the desired output.

4.3 *RBF Parameters*

The number of nodes, the spread constant and the error goal are the design parameters for the RBF network. According to Cembrano and Wells (1992) there is no direct method available for the structure definition of the network. In this work the error goal is defined to be so small that it does not limit the system training. The selection of the number of nodes and the spread constant is based on simulation tests.

4.4 *Dynamics of the Process*

The dynamics of the process is taken into account by delayed input and output values in the network (dynamics structure of the network) when the system is trained. Delayed input values and delayed simulated output values are used when the model is utilized for prediction. An idea of this kind of dynamic models is presented by Narendra and Parthasarathy (1990).

4.5 *Identification in Practice*

Before the training of the RBF network is started, the process variables are chosen with reference to the use of the models. Data treatment is also needed: filtration, evaluation of useful data sets, corrections with initial values for the junction points between data sets in the input matrix, and scaling of the process variables between 0.1 and 0.9. The training of the network is a very memory intensive operation and a computer system with 8 R8000 processors and 2 GB memory has been used.

The difference between the training of and simulation with RBF networks is in the use of old output values. In the training the values are coming from the real measurement data of the outputs but in the simulation the old predicted values are used.

The identification steps are:

1) The determination of the parameters and the dynamics structure for the RBF network,
2) The training of the network,
3) The validation of the model.

In this study the identification was started with a few delayed inputs and outputs (2 delays for each one). The interval between the delays was usually 5 times the interval of the measurements. A good starting value of the spread constant was double the interval between inputs. The first choice for the number of nodes was the number of the inputs in the system. A better choice would have been 0.5 times the number of the inputs, because of the overparametrization tendency of the network.

The best order for finding the final values of the parameters in the identification step 1) was: First the spread constant was optimized and the others were fixed. Then the number of nodes was optimized and the others were fixed. The next step contained the optimizing of the delays and the number of delayed inputs while the others were fixed and the final optimization was for the delays and the number of delayed outputs.

5. MODEL VALIDATION

The validation of the identified models is usually done with one-step-forward prediction. In the studied process the ratio of process time constants to the sampling period is quite large and one-step-forward prediction always gives very good results regardless of the model parameters.

The real prediction ability of the models is studied using n-step-forward prediction. In this study the prediction was done over the whole validation data set. In simulations initial values were given to the model and after that the simulation was performed over the whole data set. Different data sets were used for identifications and for validations. The model prediction ability was compared with sum-squared error (SSE) values.

6. RESULTS

A MIMO model for the whole reactive distillation system was formed by combing 10 different MISO-models. One of the outputs was selected as the main example in this study. The selected output was temperature TI-36 in the middle of the catalytic section of the column. This temperature was selected because it gave good information about the process behaviour.

6.1 *Temperature TI-36*

The aim of the study was to develop a model which was able to predict the behaviour of the temperature TI-36 in the three different operation regions well. Different spread constant values, number of nodes and dynamical structures were tested. The best model for TI-36 was obtained when the network had 10 nodes, a spread constant of 2, input delays of 5 min, 10 min and 15 min and output delays of 5 min, 10 min, 15 min, 20 min, 30 min, 40 min and 50 min.

Fig. 4 shows one-step-forward prediction for TI-36 in operation region I. Figures 5, 6 and 7 show n-step-forward prediction where the n is the length of the whole data set. In the figures the first half of the data set contains the identification data and the second half contains the validation data. Thin lines represent measurements and bold lines predictions.

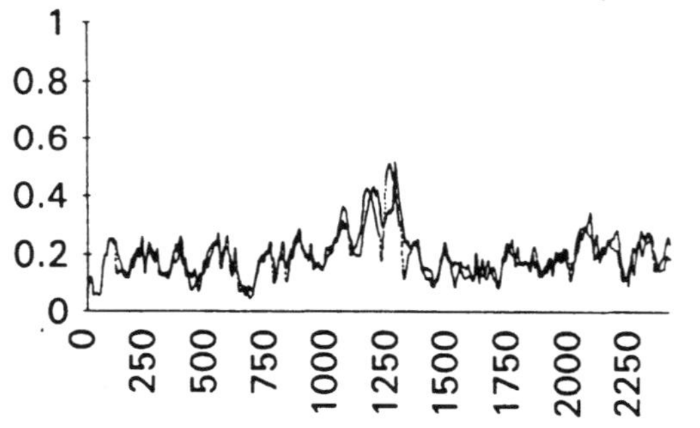

Fig. 4 one-step prediction for TI-36 Ex I.

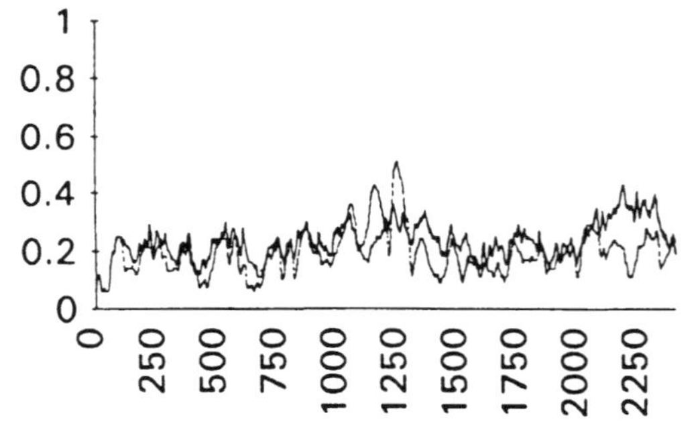

Fig. 5 n-step prediction for TI-36 Ex I.

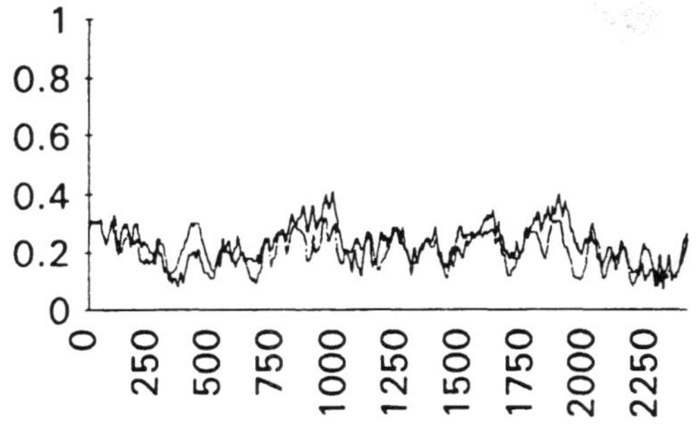

Fig. 6 n-step prediction for TI-36 Ex II.

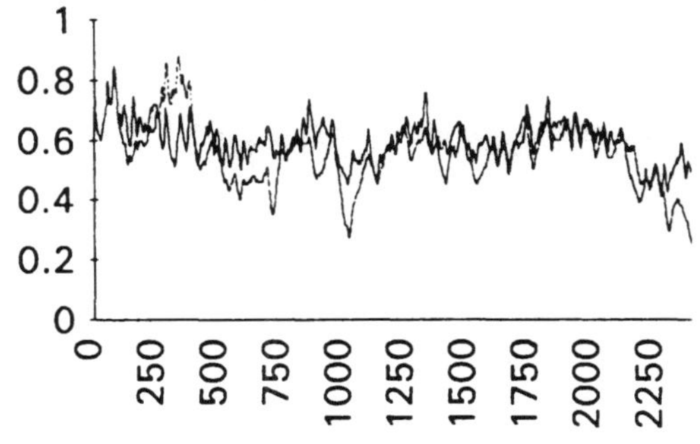

Fig. 7 n-step prediction for TI-36 Ex III.

Table 2 shows SSE values for one-step and n-step predictions for different dynamics structure of the network. The values show the prediction ability of the model for the data set which contains both the identification data and the validation data.

Table 2. SSE-values of TI-36-model predictions for Ex I, II and III in different dynamics structure of the RBF network. Input delays are constant (5 min, 10, min, 15 min), length of simulation period is 2400 min, 10 nodes and spread constant 2 are used.

Output delays min	1-step I	1-step II	1-step III
1, 3, 5, 7, 10, 15, 20	1.6	1.1	2.3
5, 10, 15, 20,30,40,50	4.3	2.8	5.0
15,20,25, 30,35,40,50	7.3	3.8	6.7
Output delays min	n-step I	n-step II	n-step III
1, 3, 5, 7, 10, 15, 20	9.6	4.6	139.4
5, 10, 15, 20,30,40,50	15.3	7.9	19.0
15,20,25, 30,35,40,50	19.7	7.4	30.3

6.2 *Concentration TAME1*

The product concentration in the bottom part (TAME1) was chosen as another example of the ability of the network to predict the system behaviour. The dynamics struture of the network for the best model was the same as for TI-36. The number of nodes was 15 and the spread constant was 2. Because there is a correlation between the number of nodes and the spread constant the effect of the changes of both parameters was tested. Figures 8, 9 and 10 show n-step-forward prediction for TAME1 in the three different operation regions.

The results are not very good and a longer training data set is needed. The bottom part of the column is affected by the big size of the reboiler and its slow dynamics.

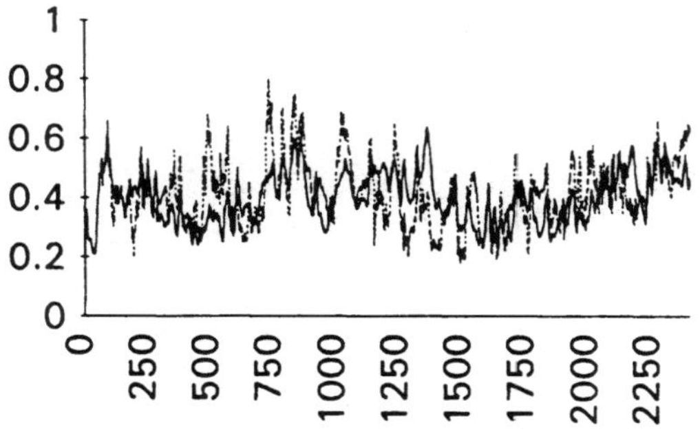

Fig. 8 n-step prediction for TAME1 Ex I.

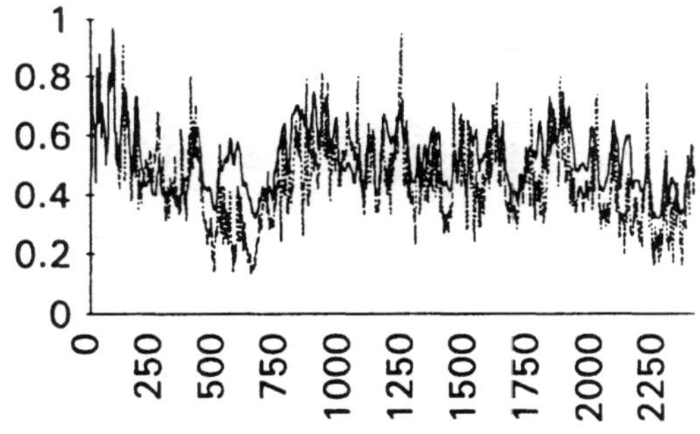

Fig. 9 n-step prediction for TAME1 Ex II.

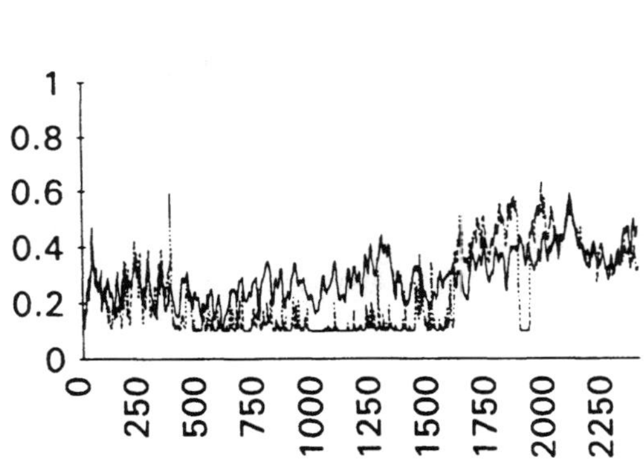

Fig. 10 n-step prediction for TAME1 Ex III.

6.3 *Other Outputs*

During the network training and simulation tests it was noted that the dynamics structure of the network did not need very dramatic changes when different outputs of the RD system were modelled. Some changes in the number of nodes and/or in the spread constant value were usually made.

There were strong correlations between temperatures and some of the concentrations in the column. So the modeling was quite similar for both. Distillate and bottom product flows were easy to model and the results were good.

7. CONCLUSIONS

Neural network identification is based on trial and error. There are no definitive rules how to determine the network parameters, some general recommendations only. The neural network modelling, however, produces useful models for the prediction of process variables.

The dynamics of the process is taken into account by the delayed inputs and outputs in the network when the system is trained. The delays are chosen at regular intervals. The number of nodes, the spread constant, and the number of the delayed inputs and outputs have to be optimized by simulations.

When a good dynamics structure for the RBF network has been found with one output, the whole column model is obtained through optimizing different spread constant values and numbers of nodes for every output separately.

During the first identification attempts, the main problem has been the overparametrization of the RBF network. The initial spread constant and the number of nodes should be quite small. For inputs the delayed values do not have very dramatic effect. Some values are needed and for the RD process three delayed values for every input have been found to be enough. In the delayed outputs quite big delays are needed for good n-step-forward prediction.

Separate validation data for evaluation of the results has been used. The results in Figures 4 - 10 show that there is no difference in the model prediction ability for the identification data set and for the validation data set.

For some outputs identifiability is the worst in the third operation region. (the middle and top part of the column, e.g. TI-36, TI-61) For some outputs the first region is the worst (the bottom part of the column e.g. TI-71, MeOH1). Especially the third region needs long delays for good n-step-prediction. This third region is quite far away from the practical operation conditions of the column. The first region is the most important one and those outputs which are not identified well enough need more training data. This can be taken from the validation data set because in this study 1/2 for identification and 1/2 for validation has been used.

The results show that with RBF networks it is possible to obtain MIMO models which can predict the behaviour of a RD process many hours ahead.

8. REFERENCES

Bravo, J. L., Pyhälahti, A. & Järvelin, H., (1993), Investigations in a Catalytic Distillation Pilot Plant: Vapor/Liquid Equilibrium, Kinetics, and Mass-Transfer Issues, *Ind. Eng. Chem. Res.* **32**, 2220-2225

Cembrano, G. & Wells, G., (1992), Neural Networks for Control, *Application of Artificial Intelligence in Process Control,* Pergamon Press Ltd, Oxford, UK, 388-402

Chen, S., Billings, S.A., Cowan, C.F.N. & Grant, P.M., (1990a), Non-linear Systems Identification Using Radial Basis Functions. *Int. J. Systems Sci.*, **21**, No. 12, 2513-2539

Chen, S., Billings, S.A., Cowan, C.F.N. & Grant, P.M., (1990b), Practical Identification of NARMAX Models Using Radial Basis Functions, *Int. J. Control,* **52**, No. 6, 1327-1350

Chen, S., Cowan, C.F.N. & Grant, P.M., March (1991), Orthogonal Least Squares Learning Algorithm for Radial Basis Function Networks, *IEEE Transactions on Neural Networks,* **2**, No. 2, 302-309

Demuth, H. & Beale, M., (1994), *Neural Network Toolbox for Matlab*, v. 2, The MathWorks Inc. 6-1 - 6-12

Doherty, M. & Buzad, G., (1993), Reactive Distillation by Design, *Trans. Inst. Chem. Eng.* **70**, 448-454

Narendra, K.S. & Parthasarathy, K., 1990, Identification and Control of Dynamical Systems Using Neural Networks, *IEEE Transactions on Neural Networks,* **1**, No. 1, 4-27

Pekkanen, M., Laine, S., Mäkinen, J. & Säynätjoki, P., (1994), Identification and Control Structures of Reactive Distillation, *Literature for the 5th NPC-Workshop (5th Nordic Workshop on Process Control),* January 24-25, 1994, Lyngby, Denmark, 71-77

Suontausta, J., Ruoppila, V.T. & Koivo, H.N., (1994), Modelling of Non-linear Systems Using Radial Basis Function Networks, *10th IFAC Symposium on System Identification, SYSID'94,* Preprints, **2**, Edited by Blanke, M. & Söderström, T., 325-330

REDUCED MODELING OF COMPLEX FLUID MIXTURES BY THE WAVELET-GALERKIN METHOD

R. v. WATZDORF, W. MARQUARDT*

*Lehrstuhl für Prozeßtechnik, RWTH Aachen University of Technology, 52064 Aachen, Germany.
E-mail: (watzdorf,marquardt)@lpt.rwth-aachen.de*

Abstract. A new technique suitable for the reduced, first principle based modeling of complex fluid mixtures applicable to separation and polyreaction systems is presented. The approach brings together the concept of continuous thermodynamics and the promising field of wavelet approximation techniques by means of a Wavelet-Galerkin method. The theory of continuous thermodynamics is reviewed from an approximation point of view and the fundamentals of the Wavelet-Galerkin method are briefly outlined. An implementation of a dynamic process model for a flash evaporator separating a hydrocarbon mixture comprising mass balances and phase equilibria equations is given.

Key Words. Continuous thermodynamics, multicomponent mixtures, wavelets, Galerkin method, separation processes

1. INTRODUCTION

Many fluid mixtures encountered in chemical and hydrocarbon processing industries are ill defined in the sense that they contain far too many components for a detailed compositional analysis and subsequent modeling in terms of pure component mass balances. Common examples of such mixtures are frequently related to processes of high economic relevance and include petroleum (Willmann and Teja, 1986; Cottermann and Prausnitz, 1985a) and reservoir fluids (Pedersen *et al.*, 1984; Whitson *et al.*, 1989) as well as polymer solutions and polyreaction systems (Hutchinson and Luss, 1970).

Nowadays, processing industries involved with complex fluid mixtures have to meet increasingly stringent market demands for products tailored to a more specific chemical composition and produced at competitive costs (Pelham, 1993). Rigorous, first principle based models with high predictive capabilities are essential to effectively support model based techniques both in process design and process control. However, the large number of components renders the computational expenditure prohibitive. Suitable reduction techniques with the following distinctive attributes should be considered: Firstly, a significant reduction of problem size must be obtainable with admissible loss of accuracy. Secondly, the balance between the model's predictive capability and the computational effort should be traded within the reduction scheme by means of an adjustable accuracy.

In the present work, an original approach for the reduced modeling of complex fluid mixtures is developed. The underlying ideas are outlined and some first results are illustrated.

2. MODEL REDUCTION FOR COMPLEX FLUID MIXTURES

The modeling of continuous mixtures is merely a special case of general population balances and can thus be applied in a natural way to both separation processes and polyreaction systems (Aris and Gavalas, 1966). However, as the modeling of multicomponent separation processes imposes more challenging requirements on the reduction technique due to the usually strong scatter of the distributions considered, these processes will be dealt within the following.

2.1. *Pseudocomponent Lumping*

Usually, complex mixtures are characterized by lumping the available compositional information into a certain set of pseudocomponents. Although this technique is well established, the lumping technique is to a large extent arbitrary and has considerable impact on the results obtained (Mansoori and Chorn, 1989). Further, product quality information is only obtainable in terms of pseudocomponent but not in any other species concentration. Consistency of the choice of pseudocomponents in a flowsheet is difficult to obtain.

2.2. *Continuous Thermodynamics*

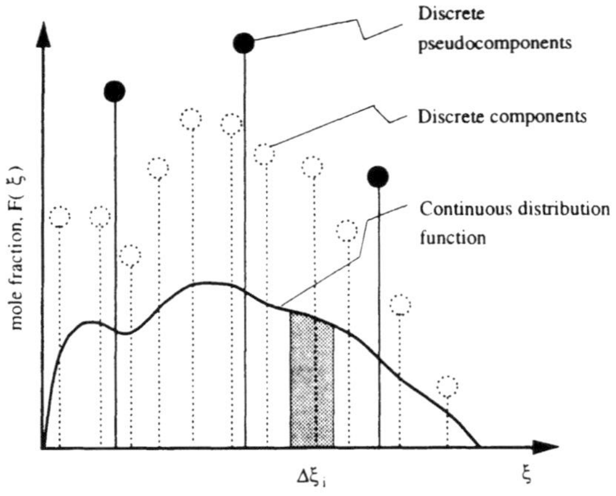

Fig. 1. Concept of continuous thermodynamics: Relation between discrete component mole fractions and distribution function

The framework of continuous thermodynamics, intented to overcome these deficiencies, has received periodic attention throughout the last five decades. The basic idea is to assume an infinite number of components present in the mixture and to describe the compositional information in terms of a continuous distribution function of some independent variables $\xi \in \mathbb{R}$ (e.g. molecular weight, normal boiling point or Single Carbon Number). The relation between the molefraction of any component and the distribution function $F(\xi, t)$ is given by

$$x_i(t) = \int_{\Delta\xi_i} F(\xi, t) d\xi \qquad (1)$$

as depicted by Figure 1. Using definition eq. (1), a complete framework of thermodynamic relations analogous to the discrete case can be derived (Rätsch and Kehlen, 1982; Cottermann and Prausnitz, 1985a).

However, the continuous thermodynamic framework has not gained any relevance for industrial applications. The reason for this is closly related to the choice of the approximation technique for the distribution functions. Before this subject will be addressed, the model equation formulation for continuous mixtures will be briefly reviewed.

2.3. *Flash equations - an example*

The most simple, generic model of an ideal separation stage is a single equilibrium flash unit. Applying the usual assumptions[1] (i.e. negligible vapour holdup, equilibrium, perfect stirred liquid phase, no chemical reactions) and assuming a constant molar holdup in the vessel, the model equations for the general dynamic case relating the distribution functions $F^V(\xi, t)$ and $F^L(\xi, t)$ of the vapor and liquid phase read as follows:

$$\begin{aligned} n\frac{\partial}{\partial t}(F^L) &= PF^P - LF^L - VF^V \ , & (2)\\ 0 &= F^V - K(\xi, T, p)\, F^L \ , & (3)\\ 0 &= 1 - \int_0^\infty F^L(\xi) d\xi \ , & (4)\\ 0 &= 1 - \int_0^\infty F^V(\xi) d\xi \ , & (5)\\ 0 &= P - L - V \ , & (6)\\ 0 &= \alpha P - V \ . & (7) \end{aligned}$$

The value of the phase equilibrium constant $K(\xi, T, p)$ is provided by some suitable thermodynamic property routine. A simple control law on the degree of vaporization eq. (7) is used instead of the energy balance. As in the discrete case, the system of equations is linear dependent and the normalisation constraint eq. (4) is dropped.

3. SOLUTION TECHNIQUES

The set of model equations eq. (2) to (7) can be considered as a PDE-system requiring a suitable numerical approximation technique. Though this step is crucial for the applicability of continuous models it is far less developped than the thermodynamic fundamentals.

3.1. *Method of weighted residuals*

The most natural choice is a method of weighted residuals (MWR), which is conceptually simple, yet mathematically rigorous. An approximation of the unknown distribution functions $F(\xi, t)$ by means of some trial function $\tilde{F}(\xi, \boldsymbol{a}(t))$ comprising a vector of unknown expansion coefficients $\boldsymbol{a} = \{a_1, \cdots, a_m\}$ is constructed giving rise to a residual $R(\xi, \boldsymbol{a}(t))$, which is projected onto a space $\mathbb{W}$ of weight functions w_i:

$$\int_D R\left(\xi, \boldsymbol{a}(t)\right) w_j(\xi) d\xi = 0, \quad \forall j=1,\cdots,m \qquad (8)$$

The optimal approximation is obtained with the Galerkin method using the same functions both for the construction of $\tilde{F}$ and for the weight functions. A detailed discussion of the application of MWR-techniques to engineering problems can be found in Villadsen and Michelsen (1982).

3.2. *Trial functions for continuous distributions*

Most authors use various statistical distribution functions (Radosz *et al.*, 1987; Rätsch and Kehlen, 1982; Peng *et al.*, 1987) for the approximation of $F(\xi, t)$ and determine the unknown parameters

[1] These assumptions are no prerequisites of the Wavelet-Galerkin method. Their only purpose is to furnish a simple illustrative dynamic model of a multicomponent separation process.

by calculating the moments. This is equivalent to a choice of $w_j(\xi) = \xi^j$ in eq. (8). Cotterman and Prausnitz (1985b) introduced a Quadrature method using discrete values of the continuous distribution function at specific quadrature points. A close study reveals that this approach is merely an application of a collocation technique ($w_j(\xi) = \delta(\xi - \xi_j)$) where the collocation points ξ_j are the roots of the Laguerre polynominal of appropriate degree. Cheluget and Vera (1991) use an Extended Spline Fit Technique for the construction of the continuous distribution function. This technique can be considered as a finite element method fitting a cubic polynominal in a least square sense to the data contained in each subinterval.

Due to the small number of adjustable parameters, statistical distribution functions are far not flexible enough to reproduce the usually scattered data. The Quadrature method gives only discrete values of the distribution functions. A more rigorous treatment of the problem shows that Laguerre polynominals have to be used as trial functions in order to obtain an orthogonal collocation scheme. This inherently requires to fit a high order interpolation polynominal giving rise to numerical instabilities (Lohscheidt, 1993). Finite element techniques are flexible but require a large set of more or less arbitrary decisions (e.g. the number and location of intervals, the degree of the local polynominals, the degree of differentiability at the interval boundaries).

The emergence of the wavelet theory has provided an unlimited number of localized approximation functions. Due to their multi-scale analysis capabilities and their rigouros mathematical foundation, the application of wavelets as trial functions within a MWR-scheme for the approximation of the continuous compositional information seems to be highly promising. A thorough treatment of wavelets can be found in (Daubechies, 1992) or (Chui, 1992). The underlying theory of the Wavelet-Galerkin method as required for this application will be briefly sketched in the subsequent section.

4. WAVELET-GALERKIN METHOD

4.1. *Wavelets*

Wavelets are functions of constant shape that are localized both in time and frequency (or space and wave number). A family of wavelets is defined by dilations and translations of a function ψ frequently referred to as 'mother wavelet'. Most common are discrete, dyadic dilations giving rise to a discrete wavelet transform:

$$\psi_{ji}(x) = 2^{j/2}\psi(2^j x - i) \qquad j,i \in \mathbb{Z} \quad . \tag{9}$$

Most wavelets are orthogonal in the sense of

$$\int_{\mathbb{R}} \psi_{ji}(x)\psi_{lm}(x)dx = \delta_{jl}\delta_{im} \quad \forall i,j,l,m \in \mathbb{Z} \quad . \tag{10}$$

The key to the question how to construct an orthogonal wavelet basis and how to perform the calculations involved is the so called scaling function ϕ and the *refinement equation* (11)

$$\phi_{00}(x) = \sum_{i\in\mathbb{Z}} \alpha_i \phi(2x - i) = \sum_{i\in\mathbb{Z}} \alpha_i \phi_{1i}(x) \, . \tag{11}$$

Dilations of ϕ gives rise to a sequence of successive (multiscale) approximation spaces V_j spanned by the translates $\{\phi_{ji}, \forall i \in \mathbb{Z}\}$ of the scaling function. Their particular properties ensure that any square-integrable function f on the real line can be arbitrary closely approximated, e.g. the orthogonal projection operator P_j on the space V_j defined by

$$P_j f = \sum_{i\in\mathbb{Z}} \langle f, \phi_{ji} \rangle \, \phi_{ji} \tag{12}$$

$$= \sum_{i\in\mathbb{Z}} \int_{\mathbb{R}} f(x)\phi_{ji}(x)dx \;\; \phi_{ji} \tag{13}$$

converges $\lim_{j\to\infty} P_j f = f$ for all $f \in L^2(\mathbb{R})$ (Mallat, 1989). The wavelets $\{\psi_{ji}, i \in \mathbb{Z}\}$ are constructed to constitute an orthogonal basis of a space W_j which is the orthogonal complement of V_j in V_{j+1} and by virtue of the multiscale ladder properties of V_j this implies that the whole collection of wavelets $\{\psi_{ij}, j, i \in \mathbb{Z}\}$ is an orthogonal basis of $L^2(\mathbb{R})$, too (Daubechies, 1992).

One of the most interesting aspects of the wavelet basis is that they thus provide direct access to the detail of information lost during the transition from a fine to the next coarser approximation by means of the projection operator P_{j-1}. This detail is reproduced by the appropriately scaled wavelets corresponding to the scaling function:

$$P_{j-1} f = P_j f + \sum_{i\in\mathbb{Z}} \langle f, \psi_{ji} \rangle \, \psi_{ji}(x) \quad . \tag{14}$$

The fact that the space W_j is contained in V_{j+1} gives rise to an analog refinement equation for the wavelet ψ:

$$\psi_{00}(x) = \sum_{i\in\mathbb{Z}} \beta_i \phi(2x - i) = \sum_{i\in\mathbb{Z}} \beta_i \phi_{1,i}(x) \, . \tag{15}$$

Example: The most simple example of an orthogonal wavelet basis is the Haar-basis. The scaling function and the refinement equation appertaining to it are given by

$$\phi^H_{00}(x) = \begin{cases} 1 & : \; 0 \le x < 1 \\ 0 & : \; otherwise \end{cases} \tag{16}$$

$$\phi_{00}^H(x) = \sum_{k=1}^{2} \alpha_k \phi_{1k}^H \quad \alpha_o = \alpha_1 = \frac{1}{\sqrt{2}} \quad . \tag{17}$$

The corresponding wavelet and its refinement equation is given by

$$\psi_{00}^H(x) = \begin{cases} 1 & : \ 0 \le x < 1/2 \\ -1 & : \ 1/2 \le x < 1 \\ 0 & : \ otherwise \end{cases} \tag{18}$$

$$\psi_{00}^H(x) = \sum_{k=1}^{2} \beta_k \phi_{1k}^H \quad \beta_o = -\beta_1 = \frac{1}{\sqrt{2}} \quad . \tag{19}$$

4.2. *Wavelet-Galerkin method*

The Wavelet-Galerkin method exploites orthogonality and local support of the wavelets by using refinable functions both for approximation and weight functions. With regard to eq. (14) either the vectors $\{\phi_{ni}, i \in \mathbb{Z}\}$ or $\{\phi_{00}, \psi_{ji}, 0 \le j \le n, i, j \in \mathbb{Z}\}$ might be used:

$$\tilde{F}(\xi) = a_{00}\phi_{00} + \sum_{j=0}^{n} \sum_{i \in \mathbb{Z}} a_{ji}\psi_{ji} \tag{20}$$

$$= \sum_{i \in \mathbb{Z}} b_{ni}\phi_{ni} \tag{21}$$

Indeed, the two representations are entirely equivalent, and the expansion coefficients $\boldsymbol{a}$ and $\boldsymbol{b}$ are related by a linear operator $\boldsymbol{L}_n$ according to

$$\boldsymbol{a} = \boldsymbol{L}_n \boldsymbol{b} \quad . \tag{22}$$

4.3. *Algorithms for wavelet transformation*

As the solution of the Galerkin equations is obtained in the wavelet domain, algorithms for the decomposition of the inputs and for the reconstruction of the outputs are required. The wavelet transform provides highly efficient recursive algorithms for these tasks (Strang, 1989). Two operators P_j and Q_j can be constructed by virtue of the refinement equations (11) and (15). Given an approximation $f_n \in V_n$ of a square integrable function $f \in L^2(\mathbb{R})$, the operator P_{n-1} projects f onto the next coarse space V_{n-1} while the operator Q_{n-1} extracts the detail information by projection onto the corresponding wavelet space W_{n-1} (Mallat, 1989). These operators are applied recursivly to the vector of coefficients up to any arbitrary level yielding the decomposition of the function in terms of an sum of orthogonal functions according to

$$f_n = \sum_{j=m}^{n-1} \sum_{i \in \mathbb{Z}} a_{ji}\psi_{ji} + f_m \tag{23}$$

$$= \sum_{\lim_{j \to -\infty}}^{n-1} \sum_{i \in \mathbb{Z}} a_{ij}\psi_{ji} \quad . \tag{24}$$

In an anlogous way, the reconstruction operators P^* and Q^* can be constructed and applied (Mallat, 1989).

5. APPLICATION OF WAVELET-GALERKIN METHOD

The Wavelet-Galerkin method is applied to the calculation of a dynamic multicomponent equilibrium flash (eq. (2) to (7)). For computational simplicity, Haar wavelets have been chosen for the construction of the approximation functions. A generalization of the approach to arbitray orthogonal wavelets is presented subsequently.

5.1. *Approximation of the feed distribution*

The feed distribution is given in terms of a possibly large sequence of compositions of the N components. It is easy to construct the continuous distribution function $F^P(\xi) \in V_n$, as the discrete mole fractions x_i already constitute the expansion coefficients b_{ni}^P due to

$$x_i = \int_{\Delta\xi_i} \sum_l b_{nl}^P \phi_{nl}(\xi) d\xi \tag{25}$$

$$= b_{ni}^P \int_{\xi} \phi_{ni}(\xi) d\xi \quad . \tag{26}$$

It is important to note that the distribution constructed in such a manner gives an *exact representation* of the mixture. Using this distribution without tresholding the Galerkin method recovers the full discrete model with the same number of model equations. This is in contrast with approximation techniques using global polynominals where the Weierstraß-Theorem guarantees exact represenation only for the limiting case of an infinite number of coefficients.
However, the problem size can considerably be reduced by tresholding once the distribution function is decomposed into the wavelet domain. In order to achieve a sufficient model reduction, the corresponding equations are eliminated from the system prior to the solution of the Galerkin equations (8). Figure 2 depicts the normalized ℓ^2 error norm calculated by

$$\epsilon_n = \sqrt{\frac{\sum_i (\tilde{x}_i^n - x_i)^2}{\sum_i x_i^2}} \tag{27}$$

using a tresholded representation $\tilde{\boldsymbol{x}}^n$ reduced by n coefficients for three different North Sea Oils (data from (Pedersen *et al.*, 1989)) using DAUB4-Wavelets (Daubechies, 1992).

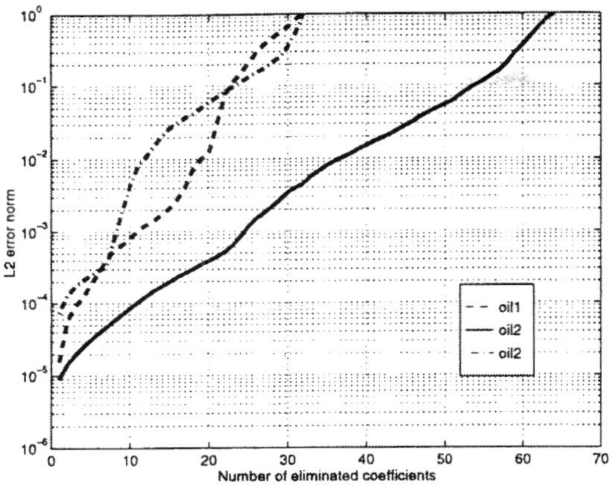

Fig. 2. Normalized ℓ^2-error norm for reduced representations of different North Sea oils

5.2. *Solution of the Wavelet-Galerkin-equations*

Given a feed distribution $F^P \in V_n$ the unknown distribution functions F^L, F^V will be in V_n as no chemical reactions are considered. By virtue of orthonormality the projection eq. (8) of the residuals onto the trial functions $\{\phi, \psi_{ji}\}$ leads to the following simple expressions for the mass balance eq. (2) and summation equation (5)

$$n\frac{da_{k,l}^L}{dt} = Pa_{k,l}^P - La_{k,l}^L - Va_{k,l}^V \qquad (28)$$

$$k\in\{0,\dots n\}, \quad l\in\{0,\dots 2^k-1\} \quad ,$$

$$a_\phi^L = 1 \quad , \qquad (29)$$

$$a_\phi^V = 1 \quad . \qquad (30)$$

The phase equilibria equations (3) give rise to the following quantities:

$$\int_0^N K\psi_{i,j}\phi d\xi, \quad \int_0^N K\psi_{i,j}\psi_{lk} d\xi \quad . \qquad (31)$$

To evaluate these quantities, the phase equilibrium coefficient can be expressed as a piecewise-constant function on N intervals. This *exactly* recovers the discrete model without any deviations due to the continuous model formulation.

However, in order to allow for ill-defined mixtures, a continuous formulation of the phase equilibria coefficient can be applied. For the given implementation, a local VLE-model as suggested by Chimowitz (1993) relating the continuous phase equilibria coefficient to the fugacity of a reference component and the distribution variable ξ was fitted to the rigorous thermodynamic properties:

$$\ln(K(\xi)P) = A_1(\xi)\ln f_{ref}^S + A_2(\xi) \quad . \qquad (32)$$

Using this relation the inner products eq. (31) can be evaluated analyticaly for the case of Haar-wavelets giving rise to a system of differential-algebraic equations (DAE) which can readily be implemented in the dynamic simulation environment DIVA and integrated by standard DAE algorithms. Figure 3 gives a comparison between the steady state solution for the liquid and the vapour phase mole fractions calculated by the Wavelet-Galerkin method without treshold and by a full discrete model. As expected the composition profiles are almost identical, the deviations are due to the interpolation of the physical property routines in the wavelet flash. Note that this is no error but merely a different modeling philosophy compared to the grouping of the multicomponent mixture into single carbon number fractions.

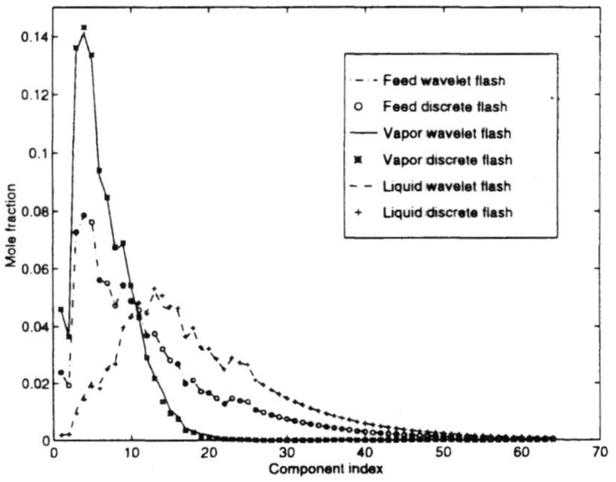

Fig. 3. Composition profiles of equilibrium flash calculated with Wavelet-Galerkin method and full discrete model

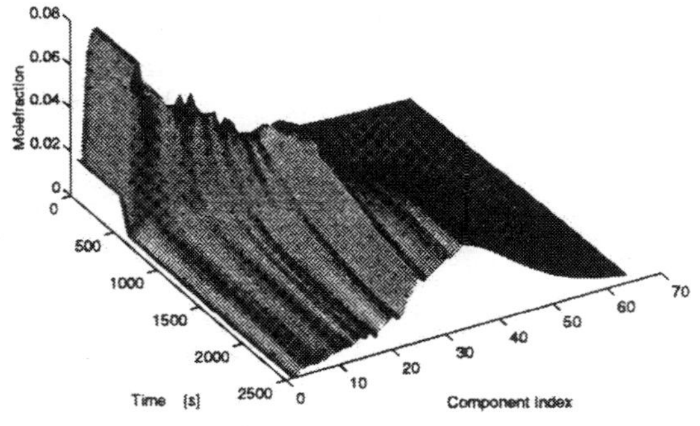

Fig. 4. Step response of liquid phase composition to change in boilup ratio calculated with Wavelet-Galerkin method

The implemented model can be used to perform dynamic simulation runs within the DIVA environment. Figure 4 shows a step response of the liquid phase composition to a change in the evaporization rate α (corresponding to an increase of heat duty) at time $t = 500s$ from $\alpha^- = 0.1$ to $\alpha^+ = 0.9$ calculated by the wavelet flash model. The composition profile is pushed toward the high boiling components and the amount of lights in the liquid phase is strongly reduced.

5.3. *Generalization to other wavelet-basis*

Basically, any wavelet basis can be used to appoximate the unknown distribution functions. If the chosen basis is orthogonal, the resulting set of equations for the mass balances and summation conditions are equal to those given for the Haar example, independent of the particular choice of basis functions. However, the inner products eq. (31) arising from the phase equilibrium equations cannot be evaluated as easy as in the Haar case for a continuous formulation of the phase equilibria. The solution of this problem is due to Dahmen and Micchelli (1993). In our application, it involves the approximation of the phase equilibria coefficients in terms of (possibly different) wavelets and reduces the evaluation of the inner products to the solution of an eigenvector problem. For a thorough description, the reader is referred to the original paper.

6. CONCLUSIONS

MWR techniques can provide a unifying framework for the various approaches related to the reduced modeling of complex fluid mixtures. Even the 'discrete' pseudocomponent lumping can be considered as an application of continuous thermodynamics in conjunction with a collocation technique on finite elements using zero order polynominals for trial functions. However, MWR techniques using a global polynominal approximation are not suitable for the case of hydrocarbon separation problems (Lohscheidt, 1993).

A new ansatz for the reduced modeling of complex multicomponent mixtures has been sketched using a local, wavelet based approximation of the unknown distribution functions. The technique outlined is flexible to accomodate any problem specific wavelet family and is highly computational efficient. It incorporates the full discrete model as a special case, allows for an assessment of the local truncation error due to data compression and requires no involuntary additional assumptions on the physical model.

A first implementation of a dynamic model for a multicomponent separation process highlights the practical applicability of the approach. However, no adaptive model reduction is yet implemented. Further work will address this issue and will generalize the technique for different wavelet families as outlined.

7. REFERENCES

Aris, R. and G.R. Gavalas (1966). On the theory of reactions in continuous mixtures. *Roy. Soc. London Phil. Trans.* **260**, 351 - 393.

Cheluget, E.L. and J.H. Vera (1991). Phase equilibria calculations for ill defined mixtures using the PRSV equation of state. *Can. Jour. Chem. Eng.* **69**, 1374 -1381.

Chimowitz, E.H., S. Macchietto, T.F. Anderson and L.F. Stutzman (1993). Local models for representing phase equilibria in multicomponent, nonideal vapor-liquid and liquid-liquid systems. 1. thermodynamic approximation functions. *Ind. Eng. Chem. Process Des. Dev.* **22**, 217 - 225.

Chui, C.K. (1992). *An introduction to wavelets.* Academic Press. New York.

Cotterman, R.L., R. Bender and J.M. Prausnitz (1985b). Phase equilibria for mixtures containing very many components. development and application of continuous thermodynamics for chemical process design. *Ind. Eng. Chem. Process Des. Dev.* **24**, 194 - 203.

Cottermann, R.L. and J. M. Prausnitz (1985a). Flash calculations for continuos or semicontinuous mixtures using an equation of state. *Ind. Eng. Chem. Process Des. Dev.* **24**, 434 - 443.

Dahmen, W. and C.A. Micchelli (1990). Using the refinement equation for evaluating integrals of wavelets. *SIAM J. Num. Anal.* **30**, 507–537.

Daubechies, I. (1992). *Ten lectures on Wavelets.* SIAM. Philadelphia. Penns.

Hutchinson, P. and D. Luss (1970). Lumping of mixtures with many parallel first order reactions. *Chem. Eng. Jour.* **1**, 129 - 136.

Lohscheidt, M. (1993). Vergleich zur Anwendung von diskreten und kontinuierlichen Gewichtete-Reste Verfahren in der reduzierten Modellierung Komplexer Gemische. Lehrstuhl für Prozeßtechnik. RWTH Aachen.

Mallat, S.G. (1989). A theory for multiresolution signal decomposition: The wavelet representation. *IEEE Trans. Pattern Anal. Machine Intel.* **11**(7), 674–693.

Mansoori, G.A. and L.G. Chorn (1989). Multicomponent fractions characterization: Principles and theories. In: *C_{7+} Fraction characterization* (L. G. Chorn G. A. Mansoori, Ed.). pp. 2-11. Taylor & Francis. New York.

Pedersen, K.S., A. Fredenslund and P. Thomassen (1989). *Properties of oils and natural gases.* Vol. 5. Gulf Pub. Houston,Texas.

Pedersen, K.S., P. Thomassen and A. Fredenslund (1984). Thermodynamics of petroleum mixtures containing heavy hydrocarbons: 1. phase envelope calculations bu use of the Soave-Redlich-Kwong equation of state. *Ind. Eng. Chem. Process Des. Dev.* **23**, 163 -170.

Pelham, R.O. (July 1993). Process optimization in the HPI. *Hydrocarbon Processing* pp. 69 - 73.

Peng, D.-Y., R.S. Wu and J.P. Batycky (1987). Application of continuous thermnodynamics to oil reservoir fluid systems using an equation of state. *AOSTRA Journal of research* **3**, 113 - 122.

Radosz, M., R.L. Cottermann and J.M. Prausnitz (1987). Phase equilibria in supercritical propane systems for separation of continuous oil mixtures. *Ind. Eng. Chem. Res.* **26**, 731 - 737.

Rätsch, M.T. and H. Kehlen (1982). *Kontinuierliche Thermodynamik von Vielstoffgemischen.* pp. 5 - 19. Akademie-Verlag. Berlin.

Strang, G. (1989). Wavelets and dilation equations: A brief introduction. *SIAM Review* **31**(4), 614–627.

Villadsen, J. and M. Michelsen (1982). *Solution of differential equation models by polynominal approximation.* Prentice-Hall. Englewood Cliffs, New Jersey.

Whitson, C.H., T.F. Anderson and I. Søreide (1989). C_{7+} characterization of related equilibrium fluids using the gamma distribution. In: *C_{7+} Fraction characterization* (L. G. Chorn G. A. Mansoori, Ed.). pp. 35–56. Taylor & Francis. New York.

Willmann, B.T. and A. S. Teja (1986). Continuous thermodynamics of phase equilibria using a multivariate distribution function and an equation of state. *AIChE Journal* **32**(12), 2067 -2078.

AN OBJECT-ORIENTED METHOD FOR PROCESS MODELLING

Stein O. Wasbø and Bjarne A. Foss

Department of Engineering Cybernetics, The Norwegian Institute of Technology
The University of Trondheim, N-7034 Trondheim, Norway
fax:+47-735-94399, e-mail: Stein.O.Wasboe@itk.unit.no / Bjarne.Foss@itk.unit.no

Abstract. An object-oriented process modelling method is presented. The mechanisms embedded in object-oriented languages, such as encapsulation and inheritance, are used to build models of a class of chemical processes. The modelling approach has proven to be flexible with regard to model changes, reuse of established models and maintenance. Object-oriented visualizations of the models have been useful in communication with process experts.

Key Words. Process modelling, object-oriented methodology

1. INTRODUCTION

The use of model-based techniques for supervision and control of process units increases rapidly due to improved performance. Hence, the need for efficient methods for model development is required. Object-oriented methodology have been used for developing process models of plants consisting of many units interconnected by mass and energy flows. Each unit model is given as an object. These are interconnected to form a plant model. This has proven to be an efficient way of building models, since the reuse of unit models as well as model changes is well supported by this methodology. Replacing one unit model by another unit model is for instance easily accomplished.

In this paper *we propose a method based on object-orientation for developing models of complex process units* themselves. We derive a set of classes which consists of chemical substances, reactions, zones within the process unit, and transport mechanisms as a basis for the model development.

The proposed methodology is illustrated through two examples, a simple CSTR and a semi-continuous furnace for ferro-manganese production. The CSTR example is included to evaluate the concept for a well-known process unit. The furnace example illustrates the use of the methodology for a complex unit where a large number of chemical substances interact.

This paper is organized as follows: First, a general systems description is given. Then the object-oriented methodology is presented, and the two examples are presented. A discussion on the use of the methodology for process modelling ends the paper.

2. SYSTEM TOPOLOGY AND BEHAVIOUR

The topology of a system is in this context defined as the *physical arrangement* of the building blocks which describe the total system. The topology of most dynamic systems can be described by the same basic components: *devices* aggregating the state variables and *connections* that decide the fluxes between the devices (Marquardt, 1994). As an example, most chemical systems have some kind of reactors connected with pipes and valves which decide the flow.

A process unit may be conceptualized by dividing it into a *topological* structure and a *behavioural* description (Marquardt, 1991). The topological structure describes how elementary building blocks (BBs) interact. The behavioural description describes the behaviour of each building block, i.e. devices and connections. If the topology of a unit and the behaviour of each BB can be regarded as independent of each other, the behavioural description can be placed inside the respective BBs in the topological structure. The unit's complete behaviour will then be defined by the behaviour of each BB in the topological structure of the unit. Thus, each BB needs to know its own behavioural description and the interfaces to interacting BBs. The important implication of this is that changes in the topology can easily be accomplished.

If the topology and the behaviour cannot be regarded as independent, the topology affects the behaviour. Hence, the behaviour of a BB must be revised every time a new topological configuration is made. Independence between topology and behaviour is a strict assumption. Assumptions in the behavioural description of a BB can depend on the unit topology. A topological change might e.g. change the operational conditions of a BB making the existing BB model invalid. This problem is difficult to conquer. Two possible approaches are: 1) make the modelling system flexible, so that the modeller may change the model easily to meet her/his demands. 2) make many models to choose between, with different assumptions incorporated. The first approach requires a higher level of skills of the modeller. The second approach on the other hand, requires a large model library.

In this paper we advocate the following approach, which is similar to 2) above: As argued above, the unit topology and the BB behaviour are in general not independent. It is, however, possible to make the two axes independent inside predefined frameworks. To accomplish this, it is necessary to have well-defined interfaces between the BBs in the system and to clearly define the valid operating range of the BBs. These interfaces may be implemented in different ways. If the interfaces are defined, and they are sufficient to describe the interesting process aspects, it is possible to model system units without being concerned about the dependences between topology and BB behaviour.

The dependency between topology and elementary behaviour will normally be stronger for BBs within a unit than between units in a process plant. This is a major complication when attempting to develop modelling tools for unit processes as compared to plants where unit processes form the building blocks.

An alternative modelling approach, called hybrid phenomena theory (HPT) is presented by Woods (1993). In HPT, the process units are described along three axis: topology, phenomenon and behaviour. HPT has no explicit behavioural description of each BB, but the behavioural description of the whole unit is derived from the topology of the unit and the phenomena taking place.

3. THE OBJECT-ORIENTED MODEL

One main reason why object-orientation is becoming increasingly popular is because it elegantly breaks the world (subject to modelling) into smaller parts that are easier to handle. These parts or objects have well-defined interfaces to the rest of the objects in the world. Each object is described by a set of attributes and a set of methods. Any details about the implementation that is of no interest to the world outside an object may be hidden or *encapsulated*.

Objects with the same attributes and methods are grouped into a *class*. An object is called an instantiation of a class. A method may have *polymorphic* realizations, depending on the use of the method in different applications.

Another important aspect of object-orientation is reuse. If a defined class needs refinements, a new class can be created that *inherits* the attributes and methods of the original class. A class can, by inheritance, be used in several applications. An example of such a class is a cylinder. The cylinder may be used to describe the shape of a tank, but may also describe the shape of an electrode or pellets in a catalyst. The class definition of the cylinder may be reused in all these cases.

These concepts have been used to develop an object-oriented framework for process unit modelling, implemented in C++ (Stroustrup, 1986).

3.1. *Choice of objects*

In general, objects are defined depending on what is modelled and on the usage of the model. The choice of objects is, however, also a matter of the modellers taste. In this paper we define a set of objects that are general in the sense that they can define a large set of process models. The objects are defined in a two-layer hierarchical structure; objects defining BBs, and objects related to matter within BBs.

Building blocks. A process unit is decomposed into *devices* and *connections* (Marquardt, 1994). Each device and connection is an object which carry its own behavioural description, and has defined interfaces toward the surrounding world, implemented as a set of methods. These classes (DEVICE and CONNECTION) are refined to specialized classes, like TANKs and VALVEs. In the following, the term *building block* or BB is used on DEVICEs and CONNECTIONs, and on their specializations; TANKS, VALVEs, etc. The BBs are the main building blocks, where the devices may contain mass, while the connections define the fluxes between the devices it connects. The configuration of BBs define the topology of the process unit.

Inside the building blocks. Each device and connection contains a geometrical description and can also contain a set of *species*, forming one or more phases. We need to model how a set of chemical species interact in a specified environment. It is therefore natural to consider each species as an object. The SPECIES class should contain all the information about the species that is necessary in the modelling. It typically contains methods like .GETHEATCAPACITY(TEMPERATURE) which is a temperature dependent function and .GETDENSITY(). The SPECIES class is further specialized into a GAS, LIQUID and SOLID classes, each including properties special to the three different aggregate states (e.g. GAS.PARTIALPRESSURE(), SOLID.POROSITY(), etc.).

We also need a class to describe the interaction between the species. The class REACTION contains information about activation energy, frequency factor and the rate law of a reaction. The species and reactions are elementary building blocks used inside the devices.

3.2. *Connecting the objects*

Object-oriented methodology should make the modelling more flexible, make it possible to reuse already developed models and make models easier to maintain. It is therefore important to construct the model in a way that supports these demands.

Objects may basically communicate in two ways; by 1) call another object's method, or 2) read or write to another object's attribute. In a truly object-oriented approach, the objects communicate through methods only. In this way, a change in an implementation of an object at a later point can be done without changing other objects, since the implementation is encapsulated in the object.

In our modelling approach, the BBs (DEVICEs and CONNECTIONs) communicate through sets of predefined methods. Another possibility, is to let the units exchange the information through terminals with a predefined format, i.e. like 2) above. This is the policy used in Omola (Mattsson and Andersson, 1992).

3.3. *Simulation*

The behaviour of a unit is described by a set of explicit ordinary differential equations (ODEs). These equations are solved by passing the derivatives of the unit's state variables to a suitable standard equation solver (Euler, Runge-Kutta, etc.). The equation solver returns the updated state variables.

The use of the method is illustrated in the examples in the following sections.

4. MODELLING A CSTR

Consider a simple continuous-stirred tank reactor (CSTR), filled with inert liquid (I) and three chemical species A, B and C. Suppose the exothermic reaction $A + B \rightarrow C$ occurs in the tank. The feed enters the tank from a source. The feed flow is given by the pressure difference over a valve at the inlet. Similarly, the outlet flow is given by the pressure difference between the liquid in the bottom of the CSTR and the drain. The tank has some heat loss to the surroundings. A sketch of the process is given in Fig. 1.

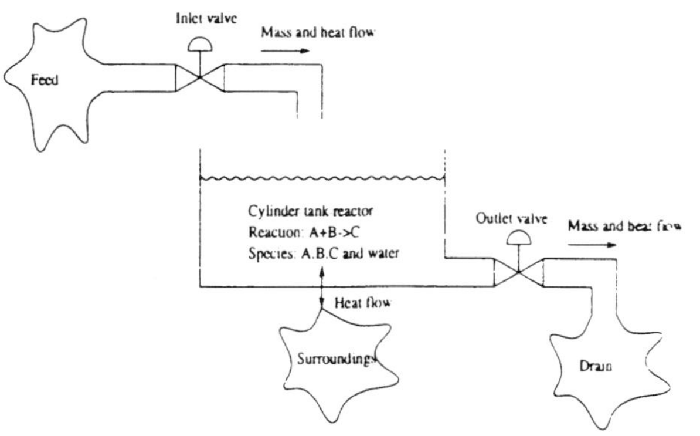

Fig. 1. A simple CSTR-system

The state-space model of the system is given by:

$$\dot{n}_A = F_{A,in} - F_A - rV \quad (1)$$

$$\dot{n}_B = F_{B,in} - F_B - rV \quad (2)$$

$$\dot{n}_C = F_{C,in} - F_C + rV \quad (3)$$

$$\dot{n}_I = F_{I,in} - F_I \quad (4)$$

$$\dot{T} = \frac{1}{C_p}\left(\sum_i (F_{i,in}H_{i,in} - F_iH_i) + Q\right) \quad (5)$$

where

n_i	Mole number of species i
T	Temperature
$F_{i,in}$	Inlet molar flow of species i
F_i	Outlet molar flow of species i
$H_{i;in}$	Enthalpy of species i at inlet
H_i	Enthalpy of species i at outlet
r	Reaction rate per volume unit
V	Reaction volume
C_p	Total heat capacity of the tank
Q	Heat loss to the surroundings

The reaction rate is given by

$$r = f(\cdot)Ae^{\frac{-E}{RT}} \quad (6)$$

where

$f(\cdot)$	The reaction law
A	Frequency factor
E	Activation energy
R	Gas constant

The total heat capacity is defined as

$$C_p = C_{p,tank} + \sum_i C_{p,i}n_i \quad (7)$$

where

$C_{p,tank}$ Heat capacity of the tank
$C_{p,i}$ Heat capacity of species i

The heat conduction between the tank and the surroundings is given by:

$$Q = U(T - T_s) \quad (8)$$

where

U Heat transfer coefficient
T_s Surrounding temperature

The molar flow of each component is given by:

$$F_i = C_i K_v u \sqrt{\Delta p} \quad (9)$$

where

K_v Valve constant
u Valve opening
Δp Pressure difference
C_i Concentration of species i in moles per volume unit

4.1. *Defining objects*

The first modelling step is to define the process topology. A logical choice for the model of interest here, is illustrated in Fig. 2.

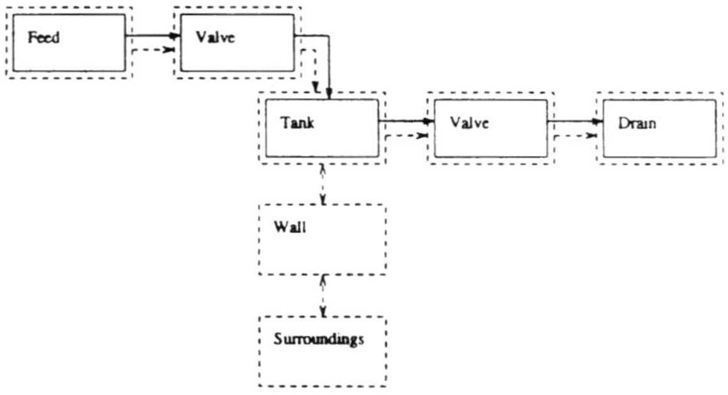

Fig. 2. Process topology. Solid lines: Mass aspects. Dashed lines: Energy aspects

When the topology is defined, specialized connection classes called VALVE and WALL (CONNECTIONS) and TANK (DEVICE) are created. The FEED, DRAIN and SURROUNDINGS may also be considered as specialized DEVICEs.

Inside each device, the internal structure is modelled. The FEED has four species, a liquid pressure and a temperature. The TANK has four species, one reaction, an inlet pressure and an outlet pressure. The reaction in the TANK is connected to the species involved. The DRAIN contains the four species and a liquid pressure, and the SURROUNDINGS has a temperature.

When the internal structure is defined, the devices are connected to the VALVEs and the WALL and the system behaviour is implemented. The behaviour of the TANK is given by equations (1)-(7). Equation (8) describes the behaviour of the WALL, while equation (9) describes the behaviour of the VALVEs. The derivatives of the state variables in the tank are passed to the equation solver and the unit can be simulated. The object-oriented CSTR model is visualized in Fig. 3. Some calculations are omitted in the figure.

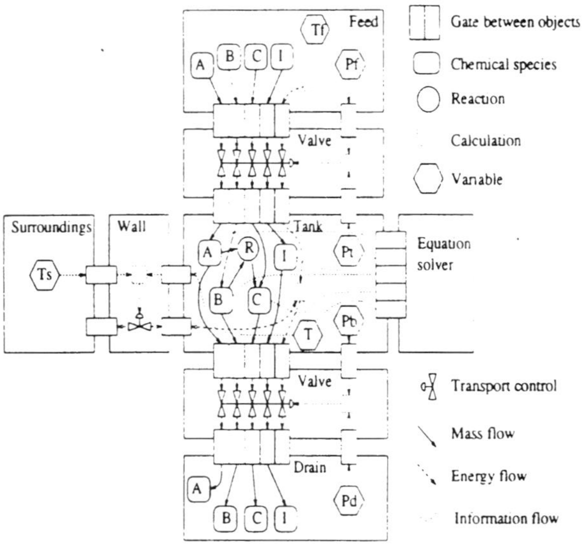

Fig. 3. Object-oriented model of the CSTR.

4.2. *Changing the model*

One of the reasons for choosing an object-oriented approach was that the model should be flexible, and easy to change at a later stage. The chosen approach has both strengths and weaknesses. Some of these will be discusses through some examples.

Case 1: Extra tank. Suppose it was discovered that a model with two CSTRs in series was a better model description of the process. A model adjustment is then necessary. Since the tank has already been modelled in the TANK class, it is sufficient to instantiate the TANK class one more time, and to connect it in the system, instantiating the necessary connection. Then, the equation solver must be made aware of the additional states before the system can be simulated.

Case 2: Extra feed. Consider the CSTR with one inlet and one outlet. Suppose an additional feed pipe is added to the CSTR. No new chemical species are added, and no additional reactions occur in the tank. The changes needed in the model are as follows: A new FEED object and a new VALVE object must be instantiated. In addition, the TANK class must be changed (or a new one created). The new TANK class must include two possible feed flows, and the behavioural method of the class must be changed, to take both flows into consideration. By regarding the two flows as one artificial flow, the behaviour of the tank can be kept unchanged. This, however, implies that a change has to be made somewhere else instead (e.g. in the connection at the inlet of the tank).

Case 3: Reaction invariants. Suppose that the ratio between the concentrations of A and B in the feed is always kept constant. Since the reaction $A+B \rightarrow C$ consumes equal amounts of A and B, the system has one reaction invariant variable (Fjeld *et al.*, 1974). It is sufficient to solve only one of the differential equations describing the behaviour of A and B (cf. equations (1) and (2)). This behavioural change is easily accomplished in the model, by changing the behavioural method in the TANK class.

Case 4: New species and reactions. Suppose that, due to a change in the supply of raw materials, the process changed, and component D is introduced along with a new reaction $C + D \rightarrow B$. The new species and the reaction must be added to the CSTR-class. Methods involving the species in the BBs (TANK, FEED, SINK and VALVE) have to be modified. The behavioural description of the new component must also be added in the BBs. If the modeller suspected such a change, component D could be added in the original classes, without using it in the model until it became necessary.

From the examples above, it can be concluded that changes occuring within each BB are easily accomplished, while changes affecting larger parts of the model will demand larger efforts. Some topological changes are easy to perform without affecting the behavioural description of the BBs. This is the case when the classes can remain unchanged, cf. Case 1. Other types of topological changes will enforce behavioural changes, and demand a modification of one or more classes, or the creation of new classes, cf. Cases 2, 3 and 4. There exist possibilities to get around this. To avoid making a new TANK class when the topology is changed, a 'looser' concept may be applied. This concept could have two separate parts: a topological structure and a behavioural description. A change in the topology would not affect the BB definitions, and the behavioural changes could be done separately when needed. The drawback with this approach is that each unit will no longer contain its own description, and the behavioural description could become difficult to maintain for large systems.

5. HC-FERROMANGANESE FURNACE

The electrical HC-FeMn furnace consists of one large pot. The pot is filled from the top with solid raw materials. The major part of the raw materials are coke and manganese ore. The main reactions taking place in the furnace are reducing manganese- and iron-oxides to a liquid alloy consisting of manganese, iron and carbon. The process is strongly endothermic. The necessary heat is supplied by electricity from electrodes hanging into the pot.

The size of the furnace and the variety of phenomena taking place inside it, makes it unrealistic to model the furnace as one CSTR. In addition, parts of the process is poorly described. This calls for a partitioning of the furnace into subprocesses or zones, giving a topology as shown in Fig. 4.

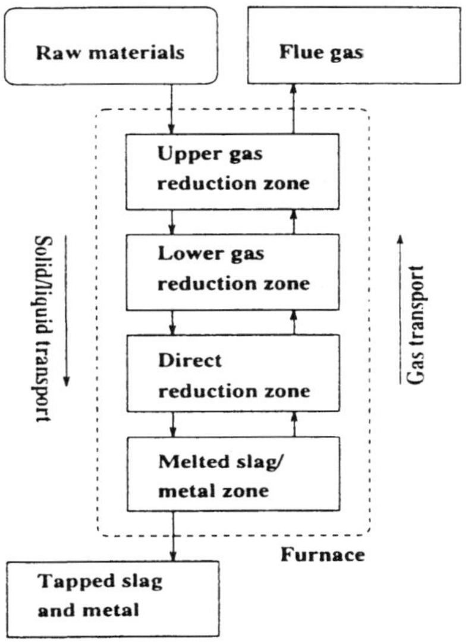

Fig. 4. Furnace zones

The upper parts of the furnace have temperatures in the range $500 - 1000°C$. Here, the oxygen rich solid ores are partly reduced by carbon monoxide gas flowing from the furnace hearth below. In the furnace hearth the partly reduced oxides melt and are further reduced by carbon to metallic manganese and iron. This reduction process produces the gas used in the prereduction in the upper parts of the furnace. The furnace hearth is a carbon-rich region, called coke-bed, filled with melted slag (i.e. different oxides) and metal flowing downstream and gas, being produced, flowing upstream. Below the coke-bed is a slag bath on top of a metal bath. The slag and metal are tapped regularly from tapholes in the bottom of the furnace pot, making the process semi-continuous.

The developed mass-balance model of the furnace (Wasbø and Foss, 1994) consists of four furnace zones and three external zones (raw material, flue gas and tapped slag and metal), i.e. seven devices and a set of connections, as shown in Fig. 4. Each device is modelled as a CSTR. The prereduction part of the furnace is divided into two zones. The coke-bed is one zone, and the slag- and metal baths are placed in one zone. Each zone contains a set of chemical species and a set of reactions between the species. The developed model contains in total 21 different species (in three phases) and 14 reactions . There are about 100 ODEs describing the behaviour of the whole furnace. Between the zones there are objects calculating the mass flows. A simplified illustration of two of the zones is given in Fig. 5.

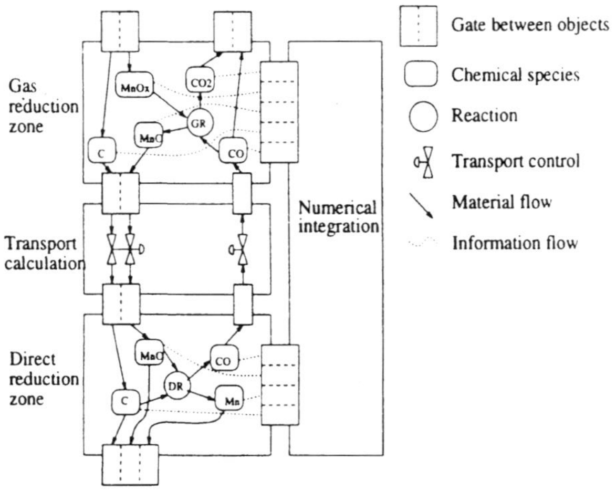

Fig. 5. Simplified illustration of two furnace zones

In a modelling project involving people with different professional background, it is essential to obtain a good communication and mutual understanding of the problem. The HC-FeMn model has been developed in close cooperation with process experts at Elkem a.s. To a large degree, the object-oriented visualization of the model (e.g. Fig. 4 and Fig. 5) has been used as a basis for interdisciplinary discussions. Our experience is that this has aided this interaction significantly.

6. DISCUSSION

The modelling approach used in this paper is an attempt to simplify process modelling of complex chemical unit processes by use of an object-oriented technique. Process units are divided into subsystems with well-defined interfaces. A problem in the modelling of process units is that behaviour and topology are not independent. Hence, a change in topology will, very often, result in a change in behaviour. Within a defined framework, however, it is possible to keep topology and behaviour independent.

The thought of being able to use a BB (being a part of the topology and containing its behavioural description) in several applications is appealing, but difficult. The main problem is that specialized objects (i.e. based on strong assumptions) are more difficult to reuse than generalized objects (i.e. based on weak assumptions).

The developed method is not a general tool, but a tool for structured modelling of chemical systems containing numerous chemical species and reactions, with different flow mechanisms, and with several phases. The model building blocks exchange information through well-defined interface methods where the implementation is encapsulated, and may be changed according to the modellers needs. The internal representation within each building block is hidden to other blocks, only the interface is known. The method has been helpful in the implementation of complex process unit models, since the object-orientation aids organization and reuse of submodels.

7. CONCLUSION

An object-oriented modelling method for process unit modelling has been developed. The method has proven to be useful for efficient modelling of complex unit processes and as a tool in communication with process experts.

Acknowledgments are expressed to professor Sverre E. Olsen for helpful discussions regarding the FeMn-process, and to Dr. Halvard Tveit at Elkem a.s. for his encouragement and cooperation.

8. REFERENCES

Fjeld, M., O. A. Asbjørnsen and K. J. Åström (1974). 'Reaction invariants and their importance in the analysis of eigenvectors, state observability and controllability of the continuous stirred tank reactor'. *Chemical Engineering Science* **29**, 1917–1926.

Marquardt, W. (1991). Dynamic process simulation - recent progress and future challenges. In W. Ray and Y. Arkun (Eds.). 'Chemical Process Control'. CPC-IV, CACHE. pp. 131–180.

Marquardt, W. (1994). Trends in computer-aided process modeling. In 'Proceedings of PSE'94'. pp. 1–24.

Mattsson, S. E. and M. Andersson (1992). The ideas behind omola. In 'Proceedings of the 1992 IEEE Symposium on Computer Aided Control Systems Design'. Napa, California.

Stroustrup, B. (1986). *The C++ Programming Language*. Addison-Wesley.

Wasbø, S. O. and B. A. Foss (1994). Object-oriented model of a ferromanganese furnace. Submitted to INFACON 7, Trondheim 1995.

Woods, E. A. (1993). The Hybrid Phenomena Theory. PhD thesis. The Norwegian Institute of Technology. ISBN 82-7119-520-4.

HYBRID SIMULATION OF FLEXIBLE BATCH PLANTS

**Sebastian Engell, Martin Fritz, Christian Schulz
and Konrad Wöllhaf**

*Lehrstuhl für Anlagensteuerungstechnik (Process Control Group)
Fachbereich Chemietechnik
Universität Dortmund
D-44221 Dortmund
Germany*

Abstract: In this contribution the structure of a simulation system for flexible batch plant operation, which tracks the complete state of the plant and includes both continuous and discrete dynamics, is described. The system consists of two parts, the batch plant and its operational functionality and the recipe, which describes the production process. The object-oriented model of the production process is based on the concept of recipe controlled operation.

Keywords: Batch control, object modelling techniques, discrete-event systems, continuous systems, process simulators, production control.

INTRODUCTION

Dynamic simulation has become a widely used tool in the analysis, optimization, control structure selection, and controller design of continuous flow plants. Such plants can be adequately modeled by systems of algebraic and differential equations. The combination of submodels for the various units of the plant may give rise to very large systems of equations, but effective and reliable software for the solution of the equations is available.

On the other hand, speciality chemicals are in most cases produced in small quantities, and therefore flexible batch plants are most economical for such production processes. These plants are characterized by the lack of a „steady state". Firstly, the operations in the single process units are not stationary but characterized by a trajectory, e.g. the progressing of a polymerization reaction. Secondly, depending on the actual use of the process units, the dynamics and the couplings of the plant are different, not only in the sense of parameter variations but also in their principal structure. In most cases, the sequence of batches in the plant is also variable. Different products are produced in different quantities and under certain constraints (e.g. possible sequences in the variation of product properties due to a non-ideal separation of the batches), with variable urgency or time-pressure. Thus for a complete analysis, orders as a temporal factor (what has to be produced in which timespan and with which quality), and the resulting scheduling of the process units must be included if the operation of a plant is to be simulated.

From a system-theoretic point of view, batch plants have elements of both discrete-event and continuous-time systems (such systems are termed hybrid systems). Simulation of flexible batch plants with (potentially) multifunctional equipment is a task which is different from - and in a sense more demanding than - the simulation of continuous-flow plants, and thus specific solutions have to be developed.

Hybrid systems were first investigated in simulation in (Fahrland, 1970). The program BATCHES (Clark and Joglekar, 1992), implemented in FORTRAN, is commercially available. It is a result of the development of several simulation tools for batch processes at Purdue University (Fruit, *et al.*, 1974).

A survey of the modelling of combined discrete/continuous processes is given in (Barton and Pantelides, 1993).

The simulation program presented in this paper is based on the unit operation concept for multipurpose batch plants. This requires special modelling concepts and simulation strategies, which are not supported by presently available simulation tools. Additionally, the advantages of the object-oriented approach are used here in modelling and implementation.

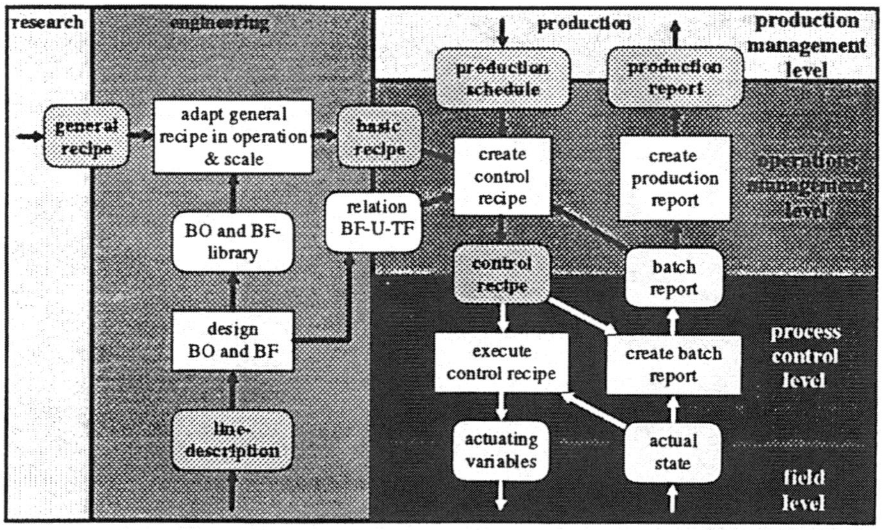

Fig. 1. Block diagram of recipe-based operation according to NAMUR: „Requirements of systems for recipe control“

RECIPE BASED OPERATION

A multipurpose batch plant consists of different vessels (tanks and reactors), auxiliary equipment, and pipes to transport the products. Recipes define the production process of each product, i.e. which substances have to be put together in which quantities and under which conditions. They also contain instructions for the operator or the process control system on temperature and pressure values or curves, variables to be monitored, stopping conditions, etc.

In recent years, rapid progress in the automation of batch processing plants has been made such that in newly equipped plants the task of the operators is not any more the step-by-step execution of the recipes, but the assignment of the equipment, the choice of the production sequences, and the scheduling of the execution of the recipes. For these tasks, it is not sufficient to take the availability of raw materials and other resources and the production orders into account, but the complete state of the batches of substances and the equipment, as well as their future development must be considered. As recipes often contain operations which are performed in parallel, e.g. filling and heating of a vessel and several recipes may be executed in parallel and influence each other, e.g. the effectivity of the heating system may depend on the number of vessels which are heated in parallel, and on external conditions (cooling water may have different temperatures in summer and in winter), the duration and interaction of the process phases in general cannot easily be predicted well.

The concept of automatic recipe driven production has been successfully introduced in the chemical industry in Germany in recent years (NAMUR, 1992; Uhlig, 1987). The basic idea is to describe the chemical and physical operations for the production of a specific product in a basic recipe and to generate the control sequences for the execution of the production automatically from these basic recipes after the decisions on batch sizes and the assignment of equipment have been made.

In figure 1, the standard block diagram of recipe-based operation is shown. Three types of recipes are defined. The general recipe is the general description in terms of unit operations how to produce a certain product. The basic recipe is the translation of the general recipe into a formal description which considers the available operations on typical plants and adapts the quantities from the laboratory to a standard size. The control recipe finally is used for the control of the execution of the recipe on a specific plant. It is in many cases executed in a semi-automatic manner by the DCS. At the generation of the control recipe, the equipment is assigned to the production steps in the recipe, depending on the demand of the production schedule, the operational possibilities, and the actual state of the plant. This means, that actions in the basic recipe which are described by basic functions, activate the technical functions of the plant, as e.g. heating or discharge.

In figure 2, some technical functions of a unit are shown like dosing, inertizing, stirring. The definition of technical functions enables the control of a plant on a high level. During the production there is no need to worry about valves to be opened or closed, because these functions are implemented on the process control level. If a technical function is activated in the control recipe, the actual parameters are loaded down to the controller of the technical function which then is started, and stopped when the task is finished. This may imply, that the decision of the assignment of the equipment is done and that a resource is not available for other production steps.

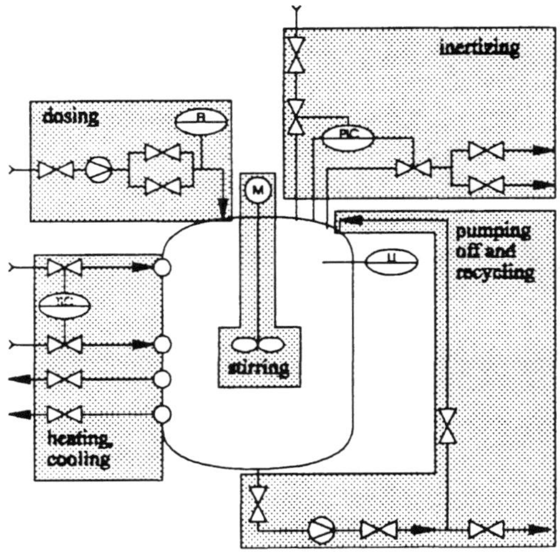

Fig. 2. Technical functions of an unit

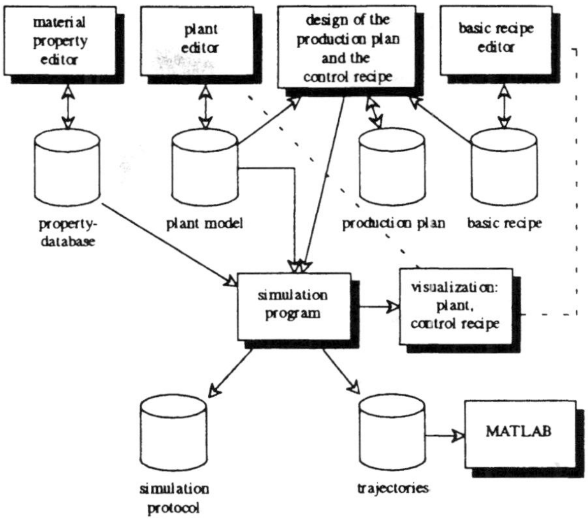

Fig. 3. The components of the simulation program

STRUCTURE OF THE SIMULATION PROGRAM

It is generally agreed that the next step in the automation of batch plants will be the introduction of high-level support for the tactical decisions of the operators. To be effective, this support has to be based on a dynamic simulation of the future development of the plant (Djavdan, 1992). Simulation can be used for different purposes: verification of planned schedules, prediction of delivery dates, short-term reactive rescheduling, bottleneck detection in the planning phase, operator training. Dynamic simulation is at present just being introduced in the manufacturing industries for the same purposes (Moser, 1993). In manufacturing where a very large number of operations of mostly fixed duration is performed, purely discrete models where each operation is represented by the time it needs on a particular machine or station are adequate. This is in general not true in chemical processing operations where most relevant variables change continuously and, as mentioned above, the dynamics cannot be considered as totally fixed. Therefore a simulation system is developed which allows the representation of the complete relevant dynamics, continuous and discrete. The discrete dynamics are generated by the transitions from one step in the recipe to the next, the starting and termination of batch runs, maintenance or cleaning operations etc. Of course, the degree of accuracy of the models of the continuous processes must be chosen with respect to the purpose of a particular simulation and the available information, e.g. on reaction kinetics.

The components of the simulation program are shown in figure 3. Four different model components are needed for simulation. The properties of the materials are stored in a property database and new components can be added by an editor. The plant model is described using a graphical editor. It consists of the equipment and its properties, the plant structure, described by connections and the so-called technical functions (see fig. 2). Independent of a special plant, the basic recipe is defined. A graphical editor is provided for the construction of new recipes using sequential function charts (Schulz, 1993). It offers features to check the recipe for syntactical and semantic errors. The control recipe is defined by the selection of a basic recipe of the product in demand and the choice of a plant equipment. The equipment in the basic recipe is substituted by the equipment of the plant and the basic functions describing the different activities are replaced by technical functions. It is clear that the assignment is only possible if the equipment and technical functions have the properties needed.

When the simulation is started, the plant model and afterwards the production plan and the control recipe are loaded. Several production plans can be simulated sequentially and every production sequence changes the state of the plant model. During the simulation, the execution of the recipes is visualized by showing the active steps in the recipe and the active equipment in the plant. The trajectories of the temperatures, levels etc. are written in a text file. A command script for reading the trajectories in MATLAB (MathWorks, 1990) is automatically generated by the simulation program. Additionally, the events and the activation times of the steps, and the required execution times are protocolled during simulation.

The program, implemented in C++, runs on SUN workstations with Solaris 2.3 and on PCs using MS-Windows. A graphical interface builder XVT (XVT, 1992) is used, that supports the transfer of the program to most important operating systems and hardware platforms. The description of the different models including their graphical description and the simulation results are stored in text files using a specific command language. The models and the data structure, using the object-oriented approach are described in (Engell and Wöllhaf, 1994; Wöllhaf and Engell, 1994).

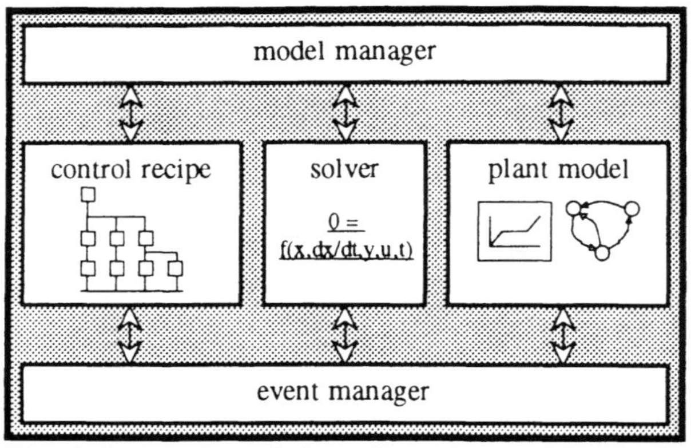

Fig. 4. Interaction of the solver, the model and the event detection.

SIMULATION OF THE HYBRID SYSTEM

The model of multipurpose batch plants where recipes are executed, constitutes a hybrid system which contains discrete and continuous dynamics. As mentioned before, hybrid systems require special simulation techniques.

The model of the actual physical transformations and continuous control units consists of differential or differential- algebraic equations. The switching of input variables or of internal models causes discontinuous trajectories. These discontinuous trajectories cannot be treated adequately by solvers which are designed for continuous functions. If the discontinuous trajectories occur only in the input variables or force a switching of the equations, this problem often is ignored and the automatic step control has to treat the discontinuity by reducing the step size of integration. In this case, the model seems to be a stiff system. This method is not efficient and may even fail because the step size becomes too small. If the switching forces an abrupt change of a state variable, or if the system has discrete state values with memory or the structure of the system changes, this method cannot be applied.

This problem can only be solved, if the discontinuities occur only at the beginning or the end of a simulation interval. The solver is called while the model is continuous. At the end of the interval, the system may change completely and the old state variables are used for the initialization of the new problem. The key question is the choice of the intervals. If the times when switchings appear are known aforehand, the events are called time events. This case can be treated easily. If the switching conditions of the model depend on the values of state variables, they are called state events and the exact time the event occurs has to be determined by iteration. This problem is similar to the numerical solution of an implicit algebraic equation.

State events are caused by the crossing of a border, a threshold of a continuous variable. The definition of a threshold generates a discrete variable which depends on the continuous variable and which can take the values true or false. In the program this is realized by defining a threshold object which compares two continuous values and has the state true or false. The accuracy for event detection related to the value can be chosen differently for every threshold object. The thresholds result from the definition of conditions for transitions which switch the steps in the recipes or by cross values that define the switching of a continuous model. For example if a reactor is empty, the level cannot decrease any more. Actually many thresholds are defined in the models.

The supervision of the thresholds has to be done by a special mechanism. It is important to realize that the actual value of the variables for which a threshold is defined can be wrong because of the step size control of the solver. A step may be rejected. Therefore the handling of model switching cannot be performed on the model level but has to be done by the simulation control. On the model level, the direct use of expressions like „if (A > 10.0) then" is not allowed, a threshold object has to be used instead. The definition of multiple switching conditions must be done by logically connected expressions which use the values of the threshold objects. Furthermore, the limited accuracy of the numerical detection of crossings must be taken into account. Therefore only one threshold object is generated for one cross value to avoid inconsistent states. E.g., if a reactor is empty, the behavior of the model has to change. For the detection of this situation, it is not admissible to define a threshold for the sum of all masses in the reactor, but the logical connection of threshold of all masses to be zero must be used. In this context, it should be mentioned that the evaluation of non physical states like negative masses may make sense for the efficient detection of state events.

The simulation of the hybrid system is performed by the interaction of five objects (see Fig. 4): the control recipe, the plant model, the solver of the continuous system, the model manager, and the event manager. The solver can access the control recipe and the plant model via the model manager.

The threshold objects of the plant model are generated and connected to the event manager during the construction of the model objects. The number of threshold objects and state values of the plant model depends on the actual state of the plant and the number of substances involved. The number is kept as small as possible. This can only be realized by the dynamic generation of state variables and threshold objects during simulation because the objects which are actually needed at a certain instance can only be determined during simulation.

In figure 5, a part of a plant model is shown with several substances and technical functions involved.

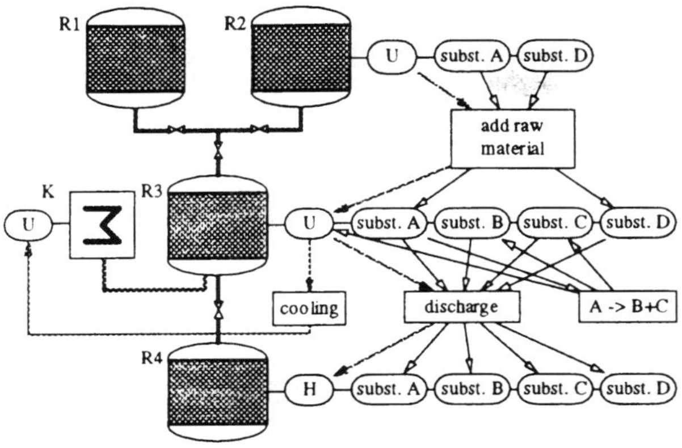

Fig. 5. Dynamic interactions in the plant model

Additionally, in reactor R3 a reaction takes place. Material and energy are transported between the equipment as shown by the arrows. Before the simulation starts, it is not clear, which substance will be in which reactor and which technical functions will be active at the same time. The model must be organized such that any combination of technical functions can be active and any mixture of material can be in the plant. Therefore, during the execution of a recipe, the actual model is generated dynamically.

At the start of the simulation, the plant model and its initial state already exist, the actual state values are stored in the plant model, and are copied into the state vector of the solver. The discrete state is computed from the actual values of the continuous variables. The connections between the plant model and the event manager are updated after every generation or destruction of a model instance in the plant model. The execution of the simulation is started by activating the first step in the control recipe. If a step is activated in the control recipe, the parameters of the actions concerned are downloaded to the technical function of the plant, and the technical function is activated. At the same time, the transition following the active step is activated and threshold objects for the control of the conditions for transitions are created.

In Figure 6, the composition of the global state vector is shown. Each equipment and each technical function have a number of state variables which are stored in a local state vector or in model objects. After an event occurs, a function in the model manager asks the plant model for the actual state. Only the state variables of the active equipment and the active technical functions are considered. Then the global state vector y and a vector of derivatives dy/dt is created, the state variables of the models and the global state vector are linked by a pointer list, and the actual values of the state variables are copied into the global state vector. Up to the next event, the valid state vector is stored in the solver. Before the plant model equations are executed, the actual values are copied down to the model and the solution is copied back from the local state vectors to the solver.

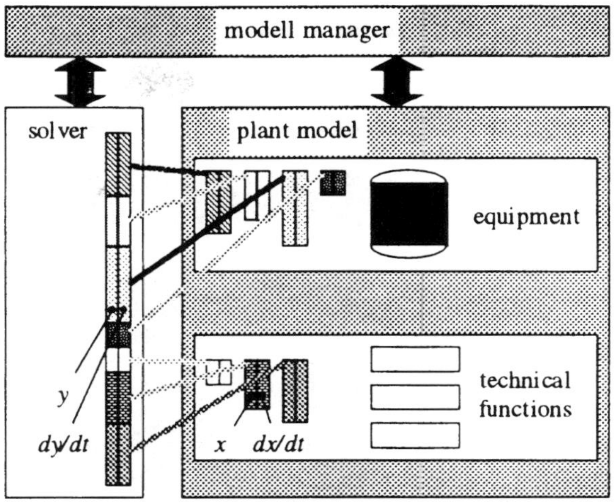

Fig. 6. Composition of the state-vector

If the components of the plant can be described by explicit differential equations, the result is the derivative of the state vector. But this mechanism can also be applied to DAEs. Therefore the residual of the implicit system is stored in an additional vector as an result of the evaluation.

DETECTION OF EVENTS

The detection of time events is not necessary because they are known aforehand, but, as mentioned before, the detection of state events with guaranteed precision is more complicated because an iteration is necessary.

In the simulation program, the detection of state events is realized using a „secant" method. In figure 7, the function $h(t)$ is defined to detect the event if the state of the expression $A(t) > B(t)$ changes. The exact event time is T_e, the estimated event time is T_e^*. If the function $h(t)$ crosses zero, an event is reported to the event manager. If $h(T_i)$ is smaller than the allowed tolerance, the step is valid, and if not, the event manager returns T_e^* as an estimation of the exact event time and the step is repeated in the interval T_{i-1} and T_e^*. There may be more than one event in an interval, therefore the smallest interval must be chosen. After the event is detected precisely enough, the event is accepted to be valid and the model switches into the next state.

Different cases must be taken into consideration. The function $h(t)$ in figure 7 crosses the sections 1, 2, and 3, where section 2 is the tolerance around the threshold. If the function $h(t)$ crosses the threshold $+ \varepsilon$ from positive values, an event is reported and iteration starts if $|h(t)| > \varepsilon$. If the function $h(t)$ crosses from negative values, the event must be reported using the threshold $- \varepsilon$. In case the function $h(t)$ stays inside section 2 for some time steps, a special strategy must be used for event detection. One solution is that the state switches to false after

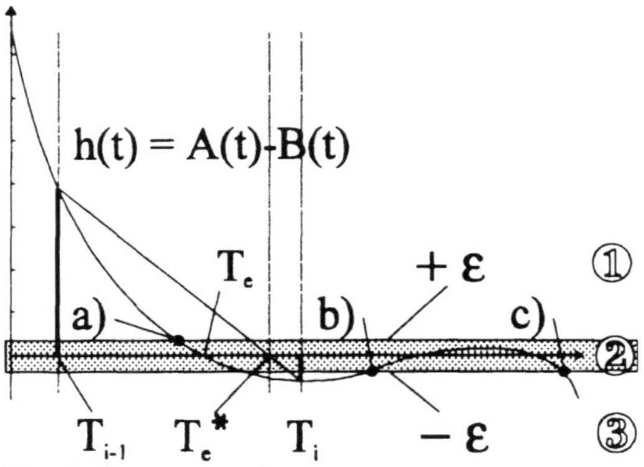

Fig. 7. Detection of state events

crossing the upper tolerance a) line + ε and back to true after crossing the lower tolerance b) - ε and if it returns back to section 3, it switches again to false c). To distinguish the different cases, previous values of *h(t)* have to be stored. This cannot be done directly in the state vector, because to get the dependency of the function *h(t)* from the state vector, the whole system must be evaluated to get e.g. the pressure in a piece of equipment. If the tolerance is chosen with respect to time, the integration interval must be reduced to this tolerance during iteration by bisection. This solution is much easier to implement but more expensive in computation time.

CONCLUSION

In this paper, a simulation program for multipurpose batch plants was presented. The structure of the program allows an independent description of the plant model and of the production steps defined in recipes. During simulation, the complete state of the plant and the recipe are evaluated and the interaction of the discrete and continuous dynamics is taken into consideration. The simulation of the hybrid system requires an adequate simulation technique. The model for the simulation is generated during the simulation run, therefore the tool can handle any kind of combinations of possible actions. This structure allows interactive simulation. The last section briefly described the problems which occur in the computation of the exact event times.

The program described here is a prototype. It contains the implementation of basic models, algorithms for the solution of differential and algebraic equations, a database for physical and chemical properties of the most common substances, a graphical user interface to define the models, and an interface to other programs to evaluate and visualize the results of the simulation. Mechanisms to support automatic step size control, the detection of state events, the switching between models of different structure, and the handling of the global state vector are implemented.

ACKNOWLEDGEMENTS

The research reported here was supported by the Ministerium für Wissenschaft und Forschung of the Province of Nordrhein-Westfalen, and by the Volkswagen-Stiftung. This support is very gratefully acknowledged.

REFERENCES

Barton, P.I. and C.C. Pantelides (1993). The modelling of combined discrete/continuous processes. *AIChE Journal.*

Clark, S. M. and G. S. Joglekar (1992). Simulation software for batch process engineering. *NATO Advanced Study Institute, Antalya,* Turkey.

Djavdan, P. (1992). Design of an on-line scheduling strategy for a combined batch/continuous plant using simulation. *Proc. European symposium on computer aided process engineering-1* (ESCAPE-1), Elsinore, Denmark, pp. 281-288.

Engell, S. and K. Wöllhaf (1994). Dynamic simulation of batch plants. *Computers chem. Engng.* (ESCAPE-3), **Vol.18**, Suppl, pp. 439-444.

Fahrland, D. A. (1970). Combined discrete event continuous systems simulation. *SIMULATION,* **Vol.14**, pp. 61-72.

Fruit, W. M., G. V. Reklaitis and J. M. Woods (1974). Simulation of multiproduct batch chemical processes. *The Chemical Engineering Journal,* **Vol.8**, pp. 199-211.

Moser, M. (1993). *Simulationsverfahren in der Fertigungsleittechnik (Simulation Methods in Shop-Floor Control Systems).* ZwF 88, pp. 565-567, Hanser-Verlag, München.

NAMUR-Empfehlung (1992): Anforderungen an Systeme zur Rezeptfahrweise (Requirments for Batch Control Systems). *NAMUR AK 2.3* Funktionen der Betriebs- und Produktionsleitebene.

Schulz, Chr. (1993): Entwicklung und Implementierung eines Grundrezeptmodels zur Simulation verfahrenstechnischer Mehrproduktanlagen (Development and implementation of a basic recipe model for the simulation of batch plants). Diplomarbeit (Master Thesis). Universität Dortmund, Fachbereich Chemietechnik.

The MathWorks, Inc. (1990): MATLAB User`s Guide. South Natick, MA.

Uhlig, R. J. (1987). Erstellen von Ablaufsteuerungen für Chargenprozesse mit wechselnden Rezepturen. *Automatisierungstechnische Praxis atp,* **Vol.29**, 1. pp. 17- 23.

Wöllhaf, K. and S. Engell (1994). Object-oriented modelling and simulation of batch plants. *Proceedings of IMACS Symposium on Mathematical Modelling,* Vienna, February, **Vol.1**, pp. 72-75.

XVT Software Inc (1992). XVT Reference Manual. Boulder, CO.

NEURAL NETWORK BASED ESTIMATORS FOR A BATCH POLYMERIZATION REACTOR

J. ZHANG†, X. YANG†‡, A. J. MORRIS†, and C. KIPARISSIDES§

†Department of Chemical & Process Engineering
University of Newcastle upon Tyne, Newcastle upon Tyne, NE1 7RU, U. K.
‡Department of Automation, Tsinghua University, Beijing, P. R. China
§Department of Chemical Engineering
University of Thessaloniki, Thessaloniki, Greece

Abstract. Inferential estimation of polymer quality in a batch polymerization reactor using neural networks is described in this paper. Number average molecular weights, weight average molecular weights, polymerization rate, and conversion are estimated from reactor temperature, jacket inlet and outlet temperatures, and the coolant flow rate through the jacket which are easily measurable. Neural networks with mixed types of hidden neurons are used in this study and they are trained by a sequential orthogonal training algorithm. The results reported in this paper demonstrate that difficult-to-measure polymer quality variables can be estimated with acceptable accuracy from easy-to-measure process variables.

Keywords. batch process, polymerization reactor, inferential estimation, neural networks

1 INTRODUCTION

A major difficulty in the control of product quality in industrial polymerization reactors is the lack of suitable on-line polymer quality measurements. Polymer qualities are closely related to such variables as the number average and the weight average molecular weights. The studies of Martin *et al* (1982) indicate that thermal properties, stress-strain properties, impact resistance, strength, and hardness of films of poly(methyl methacrylate) and polystyrene were all improved by narrowing the molecular weight distribution. There is a strong incentive to be able to control the molecular weight distribution accurately during the course of a polymerization.

Although instruments for measuring the number average and the weight average molecular weights are available, these instruments possess substantial measurement delays. These difficult-to-measure variables are related to certain easily measurable variables such as temperatures in the reactor. With this relationship, inferential estimations of these difficult-to-measure variables can be obtained from the measurements of the easily measurable variables. The key in inferential estimation is therefore to find the relationship between the difficult-to-measure variables and the easily measurable variables.

A number of researchers have attempted to find this relationship by studying the polymerization mechanistic and using some state estimation techniques, such as the extended Kalmen filter (EKF), to obtain inferential estimations of polymer quality variables (e.g. Ellis *et al*, 1988; Dimitrators *et al*, 1989; Gagnon and MacGregor, 1991; McAuley and MacGregor, 1991; Kozub and MacGregor, 1992). These approaches, however, require a deep understanding of the polymerization process and the development of a mechanistic model is generally effort demanding.

In this study, we propose to learn the relationship between the difficult-to-measure and easy-to-measure variables in a polymerization reactor from process data using artificial neural networks. Neural networks have been shown to be able to approximate any continuous nonlinear functions (e.g. Cybenko, 1989; Wang *et al*, 1992) and have been applied in chemical process modelling recently (e.g. Bhat *et al*, 1990; Chen *et al*, 1990; Haesloop and Holt, 1990; Narendra and Parthasarathy, 1990; Willis *et al*, 1991). An advantage of neural network based modelling is that a complex nonlinear process model can be discovered from process data only.

Using the learning capability of a neural network, the relationship between polymer quality variables and the measured variables in the reactor, such as reactor and jacket temperatures and coolant flow rate, can be identified from reactor operation data. After training, the neural network can be used to estimate polymer quality variables. Given the measured temperatures and coolant flow rate as network inputs, the network will produce estimated polymer quality variables as network outputs.

This paper is structured as follows. Section 2 describes a batch methyl methacrylate (MMA) polymerization reactor. Section 3 describes the mixed node neural networks used in this study. Neural

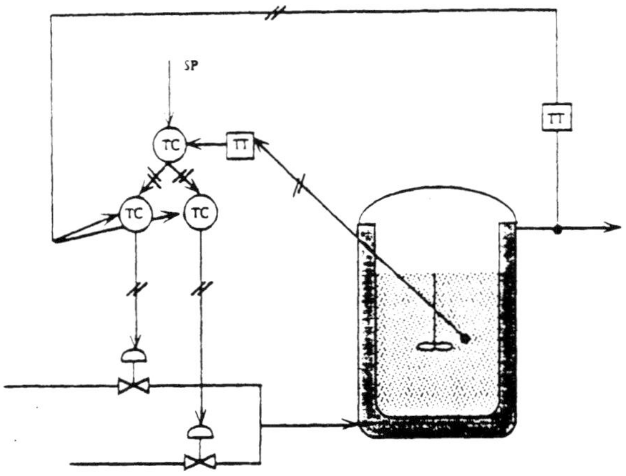

Fig. 1. A batch polymerization reactor

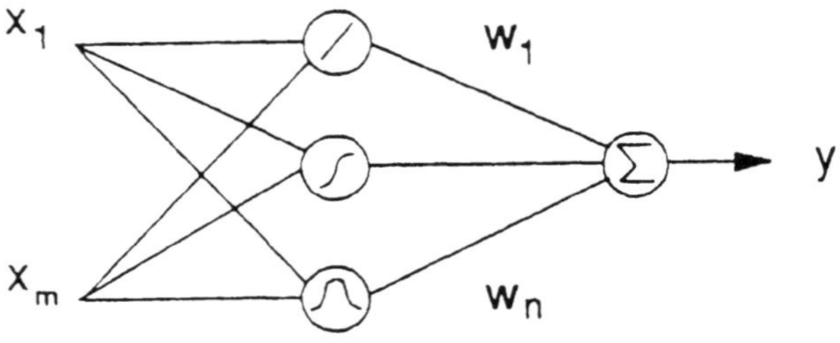

Fig. 2. A neural network with mixed types of hidden neurons

network based inferential polymer quality estimators are presented in Section 4. The last section contains some concluding remarks.

2 A BATCH POLYMERIZATION REACTOR

The batch polymerization reactor studied in this paper is a pilot scale methyl methacrylate (MMA) polymerization reactor developed in the Department of Chemical Engineering, University of Thessaloniki, Greece. The batch polymerization reactor is shown in Figure 1. The jacketed reactor is provided with a stirrer for thorough mixing of the reactants. Heating and cooling of the reaction mixture is achieved by circulating water at appropriate temperature through the reactor jacket. The reactor temperature is controlled by a cascade control system consisting of a primary PID and two secondary PID controllers. The reactor temperature is fed back to the primary controller whose output is taken as the setpoint of the two secondary controllers. The manipulated variables for the two secondary controllers are hot and cold water flow rates. The hot and cold water streams are mixed before entering the reactor jacket and provide heating or cooling for the reactor. The jacket outlet temperature is fed back to the two secondary controllers.

A detailed mathematic model covering reaction kinetics and heat and mass balances has been developed. Based on this model, a rigorous simulation programme was developed and used to generate polymerization data under different batch operating conditions. These data are used to train neural network models.

3 NEURAL NETWORKS WITH MIXED TYPES OF NEURONS

A novel type of neural network using mixed types of hidden neurons is used in this study. The motivation of incorporating neurons with different types of activation functions into a single network is to enhance network representation capability and to reduce the network size. Different types of neurons have different information processing capabilities. Real neural systems usually use various different types of neurons. For example, neurons of the ocular motor system have the sigmoidal characteristics while those located in the visual area have Gaussian characteristics (Ballard, 1988). By incorporating different types of hidden neurons into a single network, the representation capability of the neural network can be enhanced. Studies of Holcomb and Morari (1992) show that neural networks with sigmoidal hidden neurons have difficulties in representing linear mappings. By including one linear neuron the neural network is able to recover any linear characteristics in a much more logical manner.

The type of neural networks used in this paper are single hidden layer networks where output layer neurons have a linear activation function and the hidden layer consists of different types of hidden neurons — linear neurons, sigmoidal neurons, and radial basis function (RBF) neurons. Figure 2 shows such a neural network.

The output of the neural network shown in Figure 2 is

$$\hat{y} = \sum_{i=1}^{n} w_i \phi_i(X, C_i) \quad (1)$$

where w_i is the output layer weight, $\phi_i(X, C_i)$ is the output of the ith hidden neuron in the hidden layer, n is the number of hidden neurons, $X = [x_1\ x_2\ \cdots\ x_m]^T$ is a vector of inputs, and C_i is a vector of parameters associated with the ith hidden neuron.

If the ith hidden neuron is a linear neuron, then

$$\phi_i(X, C_i) = \sum_{j=1}^{m} c_i^j x_j + c_i^{m+1} \quad (2)$$

In the case that the ith hidden neuron is a sigmoid neuron,

$$\phi_i(X, C_i) = \frac{1}{1 + e^{-(\sum_{j=1}^{m} c_i^j x_j + c_i^{m+1})}} \quad (3)$$

When the ith neuron is a RBF neuron with a Gaussian activation function

$$\phi_i(X, C_i) = e^{-\frac{\sum_{j=1}^{m}(x_j - c_i^j)^2}{(c_i^{m+1})^2}} \quad (4)$$

In the above equations

$$C_i = [c_i^1 \; c_i^2 \; \cdots \; c_i^{m+1}]^T$$

When training mixed node neural networks, an initial decision must be made regarding the number of hidden neurons and which neuron is of which type. A technique for developing mixed node neural network using a sequential orthogonal training algorithm has been previously developed by Zhang *et al* (1994). During network training, hidden neurons are added sequentially. The first hidden neuron is used to model the relationship between the input and the output data. The rest of the hidden neurons are used to model the relationships between the input data and the model residuals. When adding a hidden neuron, the new information it introduces is due to that part of its output vector which is orthogonal to the space spanned by the output vectors of the previously obtained hidden neurons. The classical Gram-Schmidt orthogonalisation method is used at each training step to form a set of orthogonal bases for the space spanned by the output vectors of hidden layer neurons. Hidden layer weights associated with each hidden neuron are found through optimisation method such as the gradient descent method. Output layer weights are obtained using the least squares regression. At each training step, all the three types of hidden neurons are tried and the one giving the best fit is selected. The training procedure is terminated when the model error is less than a pre-defined level or the trade off between model accuracy and model complexity, measured by Akaike's Information Criterion (AIC) (Akaike, 1974), reaches its minimum. Details about the algorithm can be found in (Zhang *et al*, 1994).

Through the sequential orthogonal training and by incorporating different types of hidden neurons into a single network, a parsimonious network with enhanced representation capabilities can be developed. A further benefit of sequential training is that it can overcome the problem of local minimum faced by conventional network training methods to certain extent in that the sequential training method utilizes different objective functions at different training stages.

4 NEURAL NETWORK BASED INFERENTIAL ESTIMATORS

In this reactor, the on-line measured process variables are: reactor temperature, jacket inlet temperature, and jacket outlet temperature. The flow rate of water through the reactor jacket is determined by the outputs of the two secondary controllers and is considered here as on-line available information. Polymer quality variables and reactor operation variables include number average molecular weights (Mn), weight average molecular weights (Mw), polydisparsity (Pd), polymerization rate (R), and conversion (X). These variables are not measured and are to be predicted.

In this batch reactor, the polymer quality variables during the course of polymerization is mainly determined by the starting condition, i.e. the reactor temperature setpoint and the initial initiator concentration. Different starting conditions will lead to different polymer growth profiles coupled with different heat generation profiles. Correlation analysis of the reactor operation data indicates that there is a linkage between polymer quality variables and the reactor and jacket temperatures and the coolant flow rate. The three temperature measurements and the water flow rate through the reactor jacket are used here to estimate the polymer quality variables. In this study, the time interval for the estimation polymer quality variables is 0.5 minus.

Neural network based estimators are developed to predict Mn, Mw, and R from the temperature measurements. Polydisparsity, Pd, can be calculated as the ratio of Mn and Mw while conversion can be calculated by integrating R. Three multi-input single output neural network estimators were developed based on the mixed nodes networks.

Inputs to the networks are the time lagged on-line measurements of temperatures and flow while the outputs are polymer quality variables. Studies have shown that the neural network estimators of the following forms give good performance.

$$\begin{aligned} Mn(t) = \; & f(T(t), T(t-1), T(t-2), \\ & T_i(t), T_i(t-1), T_i(t-2), \\ & T_o(t), T_o(t-1), T_o(t-2), \\ & F(t), F(t-1), F(t-2)) \end{aligned} \quad (5)$$

$$\begin{aligned} Mw(t) = \; & f(T(t), T(t-1), T(t-2), \\ & T_i(t), T_i(t-1), T_i(t-2), \\ & T_o(t), T_o(t-1), T_o(t-2), \\ & F(t), F(t-1), F(t-2)) \end{aligned} \quad (6)$$

$$\begin{aligned} R(t) = \; & f(T(t), T(t-1), T(t-2), \\ & T_i(t), T_i(t-1), T_i(t-2), \\ & T_o(t), T_o(t-1), T_o(t-2), \\ & F(t), F(t-1), F(t-2)) \end{aligned} \quad (7)$$

where $T(t)$, $T_i(t)$, $T_o(t)$, and $F(t)$ are reactor temperature, jacket inlet temperature, jacket outlet temperature, and the water flow rate through the reactor jacket, respectively, at time t.

Training and testing data are generated from the simulation of nine batches under different reactor temperature setpoints and initiator concentrations. A portion (about 25%) of the data are randomly selected as training data and the remaining data are used as testing data. Each network is trained sequentially using the sequential

TABLE 1 Network training results

Networks	Hidden neurons	MSE
Mn	10	3.05×10^8
Mw	13	2.83×10^9
R	5	1.6×10^{-6}

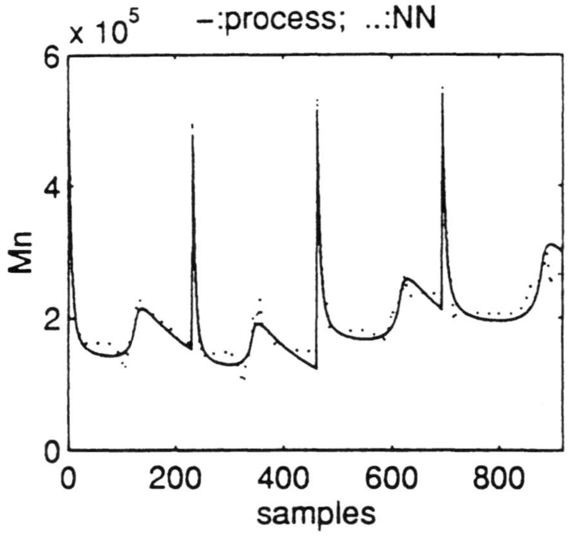

Fig. 3. Neural network estimation of number average molecular weights

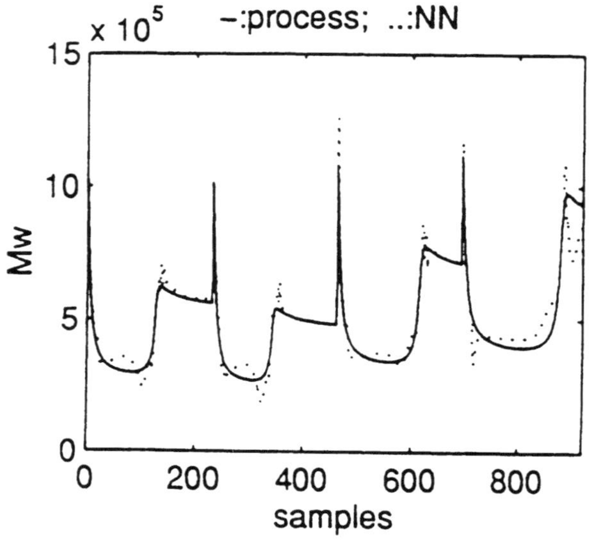

Fig. 4. Neural network estimation of weight average molecular weights

orthogonal training algorithm. After training, the networks are tested on the testing data. The results of network training are summarised in Table 1 where the number of hidden neurons and the mean squared errors (MSE) of the three networks on the testing data are given.

The performance of the neural network based inferential estimators on part of the testing data are shown in Figures 3 to 5 which contain four batches. The estimations of polymerization rates are very accurate as can be seen from Figure 5. Estimations for number average and weight average molecular weights are also very good though they are not as good as those for polymerization rates. During the initial and final stages of each batch, the estimations for molecular weights are very accurate but some errors can be observed in the middle regions of these batches. Noticing the fact that these estimations are based on reactor temperatures and water flow rate only, the results obtained can be regarded as being very good. If additional measurements, such as viscosity measurements, are available, further improved molecular weights estimations could be obtained.

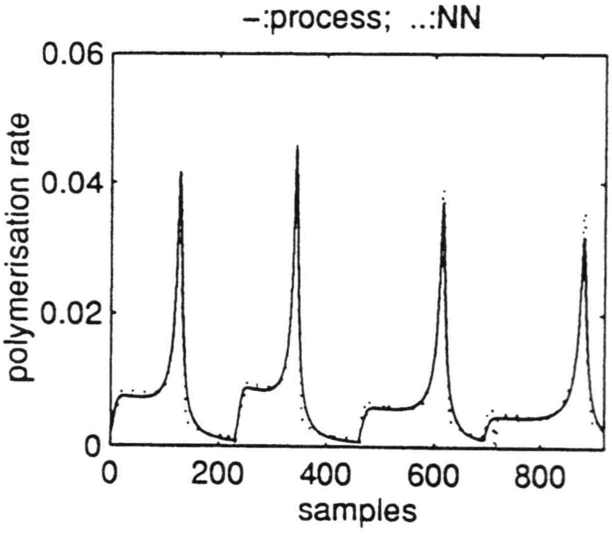

Fig. 5. Neural network estimation of polymerization rate

5 CONCLUSIONS

Studies in this paper have demonstrated that neural network based inferential polymer quality estimators are easy to develop. The only knowledge required is reactor operating data and a detailed mechanistic model is not required. By utilizing the remarkable nonlinear function approximating capability of a neural network, the relationship between the difficult-to-measure and the easy-to-measure variables in a polymerization reactor can be learnt by a neural network using reactor operating data. The mixed node neural network is attractive in that it can enhance network representation capability and provide network size parsimony. Applications to a batch MMA polymerization reactor show that the proposed inferential estimation approach is very successful and attractive.

6 REFERENCES

Akaike, H. (1974). A new look at the statistical model identification, *IEEE Transactions on Automatic Control*, **19**, 716-723.

Ballard, D. H. (1988). Cortical connections and parallel processing: Structure and function, in *Vision, Brain, and Cooperative Computation*, (Eds) M. Arbib and Hamson, MIT Press, pp563-621.

Bhat, N. V. and T., J. McAvoy (1990). Use of neural nets for dynamical modelling and control of chemical process systems, *Computers and Chemical Engineering*, **14**, 573-583.

Chen, S., S. A. Billings, and P. M. Grant (1990). Non-linear system identification using neural networks, *Int. J. Control*, **51**, 1191-1214.

Cybenko, G. (1989). Approximation by superpositions of a sigmoidal function, *Math. Control Signal Systems*, **2**, 303-314.

Dimitrators, J., C. Georgakis, M. S. El-Aasser, and A. Klein (1989). Dynamic modelling and state estimation for an emulsion copolymerization reactor, *Computers and Chemical Engineering*, **13**, 21-33.

Ellis M. F., T. W. Taylor, V. Gonzalez, and K. F. Jensen (1988). Estimation of the molecular weight distribution in batch polymerization, *AIChE Journal*, **34**, 1341-1353.

Gagnon, L. and J. F. MacGregor (1991). State estimation for continuous emulsion polymerization, *The Canadian Journal of Chemical Engineering*, **69**, 648-656.

Haesloop, D. and B. R. Holt (1990). A neural network structure for system identification, *Proc. ACC*, pp2460-2465.

Holcomb, T. and M. Morari (1992). PLS/Neural networks, *Computers and Chemical Engineering*, **16**, 393-411.

Kozub, D. J. and J. F. MacGregor (1992). State estimation for semi-batch polymerization reactors, *Chemical Engineering Science*, **47**, 1047-1062.

McAuley, K. B. and J. F. MacGregor (1991). On-line inference of polymer properties in an industrial polyethylene reactor, *AIChE Journal*, **37**, 825-835.

Martin, J. R., R. W. Nunes, J. F. Johnson (1982). *Polymer Engineering Science*, **22**, 205-228.

Narendra, K. S. and K. Parthasarathy (1990). Identification and control of dynamical systems using neural networks, *IEEE Transactions on Neural Networks*, **1**, 4-27.

Wang, Z., M. T. Tham, and A. J. Morris (1992). Multilayer neural networks: approximated canonical decomposition of nonlinearity, *Int. J. Control*, **56**, 655-672.

Willis, M. J., C. Di Massimo, G. A. Montague, M. T. Tham, and A. J. Morris (1991). On artificial neural networks in process engineering, *Proceedings of IEE, Part D*, **138**, 256-266.

Zhang, J., A. J. Morris, G. A. Montague, and M. T. Tham (1994). Dynamic system modelling using mixed node neural networks, in preprint of *IFAC Symposium ADCHEM'94*, Kyoto, Japan, May 25-27, 1994, pp114-119.

ALAMBIC - A SOFTWARE PACKAGE FOR OPTIMISING DESIGN AND OPERATION OF BATCH DISTILLATION COLUMNS

I.M. Mujtaba*, G. Stuart and S. Macchietto

Centre For Process Systems Engineering
Imperial College, London SW7 2BY, UK.

Abstract: ALAMBIC is a window based, interactive and easy-to-use software package developed for optimising the design and operation of batch distillation columns. Its features are highlighted. The package permits the solution of a wide range of problems encountered in multicomponent and reactive batch distillation, including simulation and optimal operation of a given column, and simultaneous optimisation of design-operation, for single and multiple duties, utilising rigorous models. The problem definition is done interactively by editing simple menus and forms in multiple windows. Results are viewed in graphical form. A case management tool permits the easy solution of multiple problems. The use of the package is demonstrated by two examples from the literature.

Key Words: batch distillation, simulation, optimisation, design, multicomponent, reactive

1. INTRODUCTION

Batch distillation is the most frequent separation method used in batch processes, in particular in the production of fine chemicals and speciality products. The operation is inherently dynamic and has many degrees of freedom. Operation decisions must be made on the structure of the operation (sequence of product and intermediate cuts, recycles, etc.) and the control profiles (reflux ratio, possibly pressure) and times for each cut. Furthermore, the same column is typically used for a variety of separation duties and the problem arises of selecting the best column to meet all requirements. Design and operation choices interact very strongly and may strongly affect the overall economics. Finally, there is often a degree of uncertainty on many process quantities such as the feed composition, relative volatility of species, etc.

Many studies have appeared in the literature (e.g. Farhat et al., 1990; Logsdon et al., 1990; Mujtaba and Macchietto, 1993, 1994a,b) which address individual aspects of the process, for example, i) the dynamic simulation of a given operation for a given column, ii) the development of optimal profiles of key control variables (e.g. reflux ratio) for a fixed design, iii) the determination of optimal design (e.g. no. of plates) and operation (e.g. recoveries, purities, reflux ratio profiles) decision variables, iv) the developement of optimal recycle policies. Much of this work is summarised in a review paper by Macchietto and Mujtaba (1992).

Commercial software packages such as BATCHFRAC (Boston et al., 1981), BASIS (Simulation Sciences Inc., 1989) and ProSimBatch (1992) are available but only for simulating batch distillation operation. To our knowledge,. no commercial package is currently available for optimising the design and operation of general batch columns using rigourous column models and thermophysical properties models.

Diwekar and Madhavan (1991) developed the BATCH-DIST package which includes computer codes for simulation, optimal design and operation of conventional column configurations, for a single separation duty and using short-cut models. The authors state that the package is efficient for rapid analysis of the column behaviour, for rapid simulation of large columns and that it is suitable for preliminary design. Data input to the package is provided through a keyword language in an input file. Problem data must be given in a fixed order (e.g. operation policy, thermodynamic option, column specification, terminating condition etc.). The package however does not deal with

* Current address: Department of Chemical Engineering, University of Bradford, Bradford BD7 1DP, England, UK.

unconventional column configurations, multiple separation duties and multiple case studies. Optimisation is only performed with short-cut models.

Here, a software package, ALAMBIC, is introduced, which is designed to address a much wider class of batch distillation problems, including non conventional column configurations, rigorous column and physical property models with/without chemical reaction, single duty multiperiod operation and multiple duties multiperiod operation. Since many more degrees of freedom are allowed than in other formulations, the design of a suitable problem definition interface and software aspects become more important. Simple keyword orientated languages are no longer sufficient. The overall objective here is to create a user friendly and easy-to-use computer software package which will:

1. Carry out simulation and optimisation of operation and design for general batch distillation problems.
2. Ease problem formulation
3. Simplify input specifications
4. Detect automatically invalid/inconsistent specifications
5. Allow general user supplied thermophysical property and reaction models
6. Facilitate multiple case studies

The package functionality, user interface and architecture are briefly described in section 2, while two illustrative application problems are summarised in section 3.

2. ALAMBIC - OVERVIEW

2.1 Scope

ALAMBIC is a software package being developed at Imperial College for the optimisation of design and operation of batch distillation columns. A unique feature of this package is that it can address a number of novel problems within the same environment. Its main features are outlined in the following and are summarised in Table 1. The general approach followed is to model the batch distillation process in terms of fairly rigorous models, to use standard thermophysical property models through a well defined interface, and to use powerful numerical techniques developed in-house for the solution of numerical integration and dynamic optimisation problems. Table 1 also lists the original references where problem formulations and solution techniques are given in more detail. All equations and numerical aspects are however handled behind the scenes, and all user-program interactions are through high level interfaces.

2.1.1 Batch Column Configurations

Although batch distillation is often considered in the context of a conventional column configuration (with a bottom feed and the products obtained from the top), alternative configurations (collectively called "unconventional columns") have been found, in some cases, to be more advantageous and are indeed used in practice. These include the so called inverted (feed at the top, products from the bottom) and complex (middle feed/product vessel, distillate and bottoms products withdrawn simultaneously) batch distillation columns, which, although known since the mid sixties, have regained some attention lately, in particular, for separating azeotropic mixtures. A comparative study (Mujtaba and Macchietto, 1994a) shows that distinct configurations may be better for different feed conditions and/or product specifications. Thus, it is sometimes essential to consider these alternative configurations. Accordingly, ALAMBIC permits selecting (and optimising) either a conventional, inverted or complex column configuration.

2.1.2 Batch Distillation Model

Available models range from very simple to more detailed (e.g. Mujtaba, 1989, Mujtaba and Macchietto, 1994a). In ALAMBIC a column model consists of a general system of differential and algebraic equations (DAEs) denoted by $f(x, x', t, u, v) = 0$, where x are the state variables, x' their time derivatives, u a vector of time varying control variables and v a vector of constant parameters. The model consists of mass and energy balances and phase equilibrium relations, with thermophysical property variables such as enthalpy, density and K-values calculated from a general purpose thermodynamic package, for multicomponent ideal and nonideal mixtures. Main suboptions include constant molar or volumetric holdup. There is also the option of a simpler DAE model based on equimolar overflow and constant relative volatility assumptions. There are no hydrodynamic models at present, although simple correlations could be readily added. The correct equations are generated automatically for each model option.

2.1.3 Problem type

A batch distillation column can be used for separating several fractions of a single feed mixture (*single duty*) and for processing several mixtures in the same column (*multiple duties*). For each duty, a sequence of product and recycle cuts may be specified, with specifications on product fractions given in engineering terms (amounts, purities and recoveries). For operations problems, it is possible to maximise the recovery of specified species, to minimise energy requirements or to minimise the time required for the entire operation. In fact, it is possible to maximise an economic objective function, with user-given unit costs of energy and of each input, main product and by-product material. It is also possible to optimise an economic function when the column is used for multiple duties, given a production horizon, the fraction of time allocated to each column and the demand for each product. To meet these requirements the optimal column operation (typically reflux ratio, amounts/recoveries and times for each cut) are calculated by solving an optimal control problem. Design problems additionally include the optimisation of just (at the moment) the number of stages, utilising an annualised profit function with user specified cost parameters (Mujtaba and Macchietto 1993, 1994b).

Simulation problems are viewed as special cases of an optimisation problem with no degrees of freedom. At the time of writing, the interfaces are being expanded to incorporate the treatment of uncertain parameters, utilising the solution procedure of Walsh et al. (1995).

Table 1: Features of ALAMBIC

Problem Type	References*
1. Simulation	A (1989)
2. Optimal Recycle Policy-binary mixture	B (1988)
3. Optimal Recycle Policy-multicomponent mixture	B (1992)
4. Optimal operation with reactive distillation	B (1992)
5. Single duty multiperiod operation	C (1992), B (1993)
6. Multiple duties multiperiod design and operation	B (1994b)
7. Optimal operation with unconventional columns	B (1994a)
8. Optimal operation with uncertainty	D (1995)

* A = Mujtaba B = Mujtaba and Macchietto
C = Macchietto and Mujtaba
D = Walsh, Mujtaba and Macchietto

2.2 Numerical Techniques used in ALAMBIC

The underlying numerical techniques used in ALAMBIC have been presented in detail in earlier work and are therefore discussed only briefly here.

Since batch distillation process is inherently a dynamic process, the optimal design and/or operation of such process results in an optimal control problem which can be represented as:

$$\min_{d \text{ (decision variables)}} \quad J \text{ (objective function)}$$

$$\text{s.t.} \quad h = 0 \text{ (equality constraints)}$$
$$g \geq 0 \text{ (inequality constraints)}$$
$$f(x, x', t, u, v) = 0 \text{ (DAE model)}$$

with suitable initial and termination conditions. This optimal control problem is posed as a nonlinear programming problem by approximating the controls (e.g. the reflux ratio profile) by a finite dimensional representation. The time interval of interest, say $[t_0, t_F]$ is divided into a finite number of subintervals, each associated with a set of basis functions involving a finite number of parameters. In particular, a piecewise constant representation of the controls is typically used, as shown in Figure 1, resulting in two parameters per subinterval (a control level and interval end-time). The discretised control parameters, other constant parameters and possibly some design parameters make up the vector d of decision variables to be optimised. Each "function evaluation" in the optimiser requires a full integration of the DAE system and this is achieved by using a Backward Differentiation Formula (BDF) method with special features to handle discontinuities. Solution of the optimization problem requires calculating the derivatives of the objective function and constraints with respect to all the optimisation variables. These are evaluated in an efficient way using adjoint variables. The constrained nonlinear optimisation problem is solved using an efficient and robust Successive Quadratic Programming (SQP) technique (Chen, 1988). Multiperiod and multiple duties problems are solved using a multi-level formulation, thus allowing the easy mix of different models and model options within the same problem.

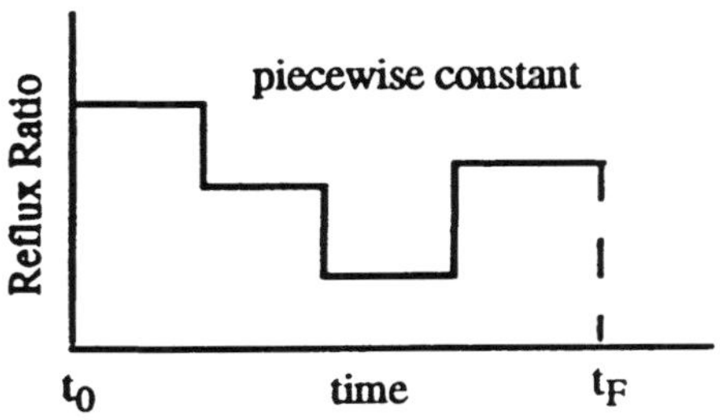

Figure 1. Discretization of Controls

2.3 Software Environment

ALAMBIC is a window based software package consisting of three main components: a problem definition part, a set of solution routines and a graphical results presentation part. For portability and ease of software maintenance reasons the problem definition and results parts are developed using Open Windows Developer's Guide (Devguide), a tool for the design, prototyping and production of user interfaces using the X-Windows standard. Problem definition is performed by utilising simple menus and forms. For example, Figure 2 shows the main menu. By pulling down sub-menus one can easily choose one or more *Case studies*, *Edit* a problem, *Run* the case study and produce *Reports*. The solution routines consist of the computer codes written in standard FORTRAN 77.

Using the ALAMBIC problem definition facilities it is possible to:

1. Edit any menu, sub-menu, form in any order
2. Retrieve, edit, save one or multiple menus and forms so as to generate and run different problems

Case management facilities are thus provided. These together with a *built in check* for invalid and inconsistent specifications give the user a great freedom and make the software easy-to-use.

The *Edit* sub-menu (Fig. 2) allows the user to select (i) the problem type (e.g. simulation or optimisation etc.) (ii) column configuration (conventional, inverted or complex) (iii) model type (simple, based on constant relative volatility, or more detailed based on differential mass and energy balances, with sub-options for constant molar or volume holdup, etc.) (iv) equipment parameters (no. of plates, initial column profiles, etc.) (v) no. of components and their identification, models for physical properties calculations (simple constant relative volatility or rigorous) (vi) operation policy (i.e. sequence of product cuts) (vii) cost functions (column, reboiler and condenser cost, steam cost, feed/product materials unit values etc.) for use in optimisation. Each of these sub-menus can be edited independently.

Some *Edit* sub-menus are simple forms to be filled in to define various activities. For example Figure 3 shows the form to define the sequence of cuts in an operation. Each form has sub-forms and can be independently edited. For example, main-cut 1 of Figure 3 can be used to define the cut in terms of reflux ratio profiles, product specifications (species to be recovered with asssociated recoveries and purities. The initial charge to the column and initialization of the column profile are also specified by filling a form.

Editing menus, sub-menus, forms etc. in the front end creates an input data specification file needed to solve the specific problem defined. Specifications are checked for completeness and (as far as is possible a priori) for consistency. The approach taken is to automatically assume a user will wish to specify all degrees of freedom available. For optimisation problems the user is asked to identify any decision variable to be relaxed and thus optimised and to give lower and upper bounds of these decision variables. For example, in Figure 3 minimum and maximum values of a cyclohexane recovery specification are given. Bounds are used to generate linear constraints for the problem, based on mass balances. These in turn are used during subsequent problem definition steps for simple feasibility checks, as and when further specifications are entered. We found this extremely useful to detect (common) specification mistakes and to avoid at source some infeasible optimisation problems. For example, it is all too easy to specify amounts and purities in a series of product cuts which total more than the amount of a species in the feed charge.

The *Run* sub-menu allows the user to run the job on the same computer or on a remote computer over a network or in the background as a batch job. Once the appropriate option is chosen the run command will (i) retrieve the appropriate computer codes required to solve the problem. A translator generates the required model equations for each problem, together with all (analytic) derivatives and calls to

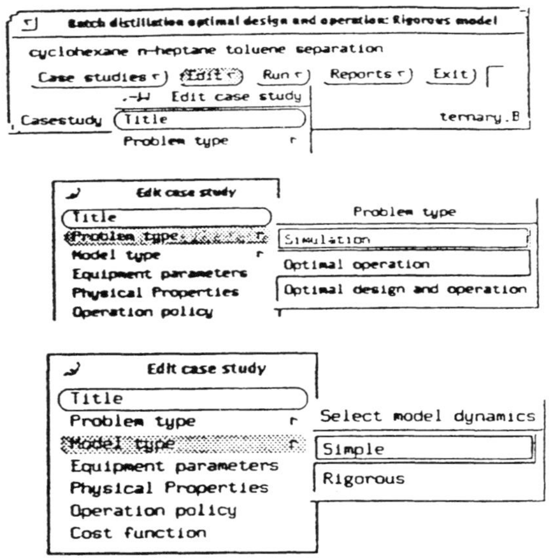

Figure 2: Some ALAMBIC Menus for Problem Definition

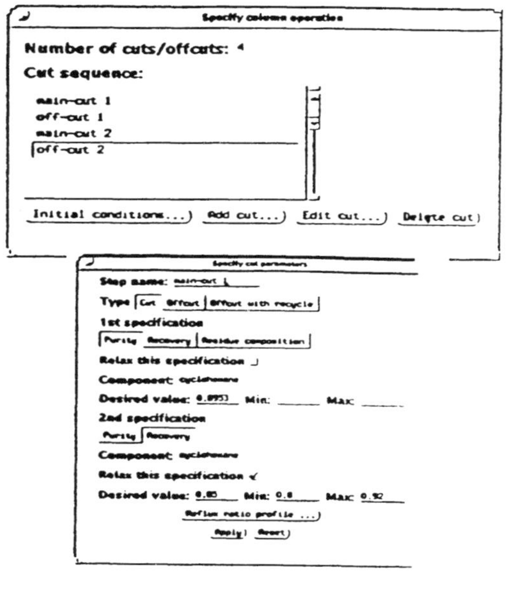

Figure 3: Some ALAMBIC Forms for Problem Definition

external routines (thermophysical properties and reaction models); (ii) compile those codes and (iii) run the problem.

When the numerical solution is completed the *Reports* sub-menu can be used to select the form of the presentation of the results. In the current version all graphics have been implemented using the Matlab package. A sub-menu allows the user to choose different form of plots (e.g. composition, holdup, temperature profiles etc.). A variety of tabular presentations are also easily implemented.

The software architecture is represented in Figure 4.

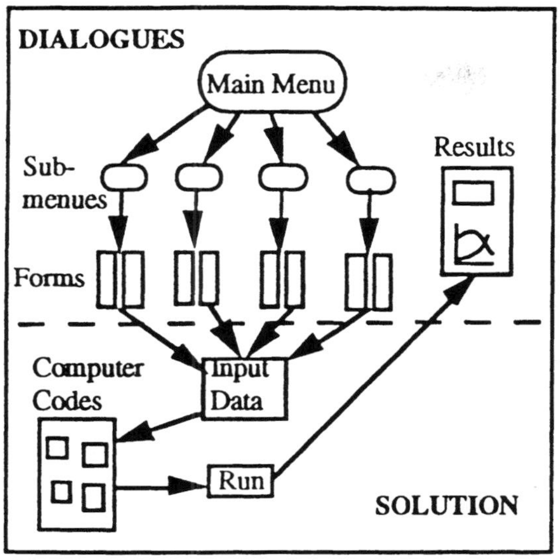

Figure 4: Schematic architecture of ALAMBIC

3. CASE STUDIES

3.1 Simulation

A simple binary batch distillation operation at atmospheric pressure is simulated. Results are compared with the experimental results of Domenech and Enjalbert (1974). The column has 4 internal plates (75% efficiency), a reboiler (constant 3KW input) and total condenser. The feed contains 62% mole Cyclohexane and 38% mole Toluene. The total initial charge is 200 gmols of which 2.5 gmols are in the condenser and 7.5 gmols in the total holdup on plates (equally distributed). The column composition profile is initialised with that of total reflux operation and is operated using a constant internal reflux ratio of 0.75 (external reflux ratio = 3, as used by Domenech and Enjalbert). The rigorous column model and SRK thermodynamic model are used to simulate the operation, with constant molar holdup. Simulation (Figure 5) and experimental results are in good agreement.

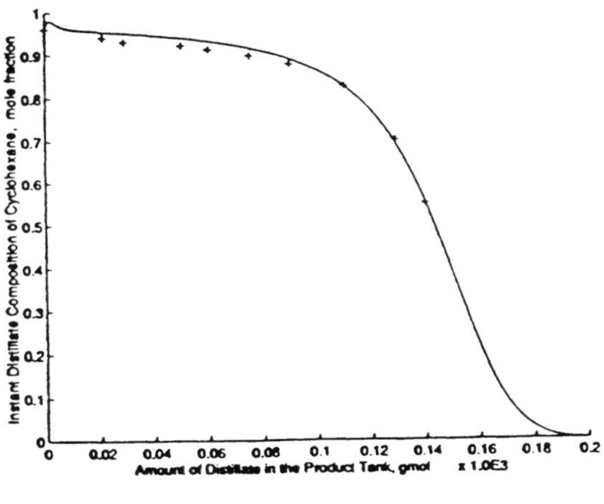

Figure 5: Composition vs Amount of Distillate Simulation and Experimental results

3.2 Optimal Design and Operation

A single duty multiperiod design and operation example (Mujtaba and Macchietto, 1994b) is used to demonstrate the capabilities of the package on more complex problems. As in the base case (Nad and Spiegel, 1987), a conventional column is used. There are 4 distillation tasks in the operation, producing 3 main products {D1, D2, Bf} and 2 off-cut byproducts {R1, R2} from a ternary feed F0 in the sequence F0/D1/R1/D2/R2-Bf. The product specifications are as in the original work. The problem is to find the optimal design of the column (number of internal plates) and the optimal operation policy (reflux ratio) which will maximize the overall profit for this multiperiod operation. The input data, problem specifications and cost coefficients given in Table 2 are defined in ALAMBIC by filling the interactive menus and forms as shown in Figures 2-3. This design and operation problem is formulated and solved as a two level optimisation problem. The main design and operating decision variables are shown in Table 2 with initial values and bounds.

Table 2: Input Data, Product Specifications and Decision Variables. Recoveries $Re^j_{a,b}$ are defined as: amount of component j in state a / amount of component j in state b.

Column:
- Configuration = Conventional
- Still Capacity = Batch size, F0, kmol = 2.93
- Condenser Vapor Load, V, kmol/hr = 2.75
- Condenser Holdup = 1.2% of F0
- Internal Plates Holdup (total) = 1.9% of F0

Operation Sequence: D1/R1/D2/R2-Bf

Feed Mixture:
- Components: 1. Cyclohexane, 2. n-Heptane, 3. Toluene
- Compos., x_{F0}, mole fraction = <0.407, 0.394, 0.199>

Models:
- Column model: Rigourous, const. molar holdup
- Thermodynamic model: SRK equation of state

Specifications:
- Cyclohexane mole fraction in D1, x^1_{D1} = 0.895
- n-Heptane mole fraction in D2, x^2_{D2} = 0.863
- Toluene mole fraction in Bf, x^3_{Bf} = 0.990
- Cyclohexane recovery in R1, $Re^1_{R1,B1}$ = 0.95

Main Decision Variables:	initial value	bounds
Number of Plates, N	= 16	<13 , 21>
Cyclohexane recovery in D1: $Re^1_{D1,F0}$	= 0.85	<0.80, 0.92>
Cyclohexane mole fraction in R1: x^1_{R1}	= 0.40	<0.35, 0.45>
Toluene recovery in D2: $Re^2_{D2,B2}$	= 0.85	<0.60, 0.92>
Toluene mole fraction in R2: x^2_{R2}	= 0.30	<0.30, 0.40>

Costs:
- C_{D1} = 30.0 \$/kmol, C_{D2} = 26.0 \$/kmol,
- $C_{R1} = C_{R2}$ = - 1.0 \$/kmol
- C_{Bf} = 24.0 \$/kmol, C_{F0} = 2.0 \$/kmol, Csu = 0.0

The optimal operation results in Figure 6 show typical profiles of species mole fractions in the distillate product accumulator tank and the optimal (piecewise constant) reflux ratio profile for the whole operation.

The profit with the optimal design and operation is 35% higher than that for the base case coulmn/operation of Nad and Spiegel, calculated using the same cost model. The optimal column has 17 theoretical stages while that of Nad and Spiegel. has 20 stages. Only 6 reflux intervals are required as compared to 20 reflux changes in the base case. There is no need for initial total reflux in the optimal operation as required in the base case (for 2.54 hrs).

Detailed optimisation results are given in Mujtaba and Macchietto (1994b).

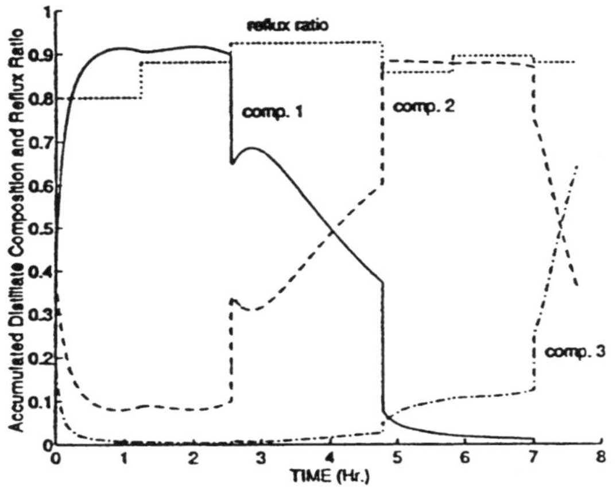

Figure 6: Accumulated Distillate Composition Profile

4. CONCLUSIONS

ALAMBIC offers an interactive, user friendly and easy-to-use software package for simulating and optimising the design and operation of batch distillation columns. The window based front end permits the definition of all problems and specifications in the same environment, through simple menus and forms which are easy to edit and in any sequence. Results are presented in graphical form. ALAMBIC also allows interfacing with user supplied column and thermodynamic models and supports multiple case studies and model management. Many of the problems types handled are not available in currently available softwares.

ACKNOWLEDGMENTS

This work was supported by a grant from EPSRC.

REFERENCES

Boston J.F., H.J. Britt, S. Jirapongphan and V.B. Shah (1981). An advanced system for the simulation of batch distillation operation. *FOCAPD*, 2, 203.

Chen, C.L. (1988). A class of successive quadratic programming methods for flowsheet optimisation. PhD Thesis. Imperial College.

Diwekar, U.M. and K.P. Madhavan (1991). BATCH-DIST: A comprehensive package for simulation, design, optimization and optimal control of multicomponent multifraction batch distillation columns. *Comp. Chem. Engng.*, 15 (12), 833.

Domenech, S. and M. Enjalbert (1974). Modele mathematique d'une colone de rectification discontinue. *Chem. Eng. Sci.*, 29, 1529.

Farhat, S., M. Czernicki, L. Pibouleau and S. Domenech (1990). Optimization of multiple - fraction batch distillation by nonlinear programming. *AIChE Journal*, 36(9), 1349.

Logsdon, J.S., U.M. Diwekar and L.T. Biegler (1990). On the simultaneous optimal design and operation of batch distillation. *Trans IChemE*, 68, Part A, 434.

Macchietto, S. and I.M. Mujtaba (1992). Design of operation policies for batch distillation. *Proceedings of NATO ASI on Batch Processing Engineering*, Antalya, Turkey, May 29-June 7 (1992).

Mujtaba, I.M. (1989). Optimal operational policies in batch distillation. PhD Thesis, Imperial college, London.

Mujtaba, I.M. and S. Macchietto (1992). An optimal recycle policy for multicomponent batch distillation. *Comp. Chem. Engng.*, 16(S), p-S273.

Mujtaba, I.M. and S. Macchietto (1993). Optimal operation of multicomponent batch distillation-Mutiperiod Formulation and Solution. *Comp. Chem. Engng.*, 17(12), 1191.

Mujtaba, I.M. and S. Macchietto (1994a). Optimal design and operation of batch distillation- a comparative study using conventional and nonconventional columns. *Proceedings ADCHEM'94- IFAC Symposium on Advanced Control of Chemical Processes*, p415, Kyoto, Japan, 25 - 27 May.

Mujtaba, I.M. and S. Macchietto (1994b). Optimal design of multicomponent batch distillation columns- single and multiple separation duties. *Proceedings PSE'94- 45h International Symposium on Process Systems Engineering*, Vol 1, p179, Kyongju, Korea, 30 May - 3 June.

Nad, M. and L. Spiegel (1987). Simulation of batch distillation by computer and comparison with experiment. *Proceedings CEF'87*, Taormina, Italy, 737.

Simulation Sciences Inc., (1989). BASIS User Mannual. Fullerton, CA.

ProSimBatch (1992). Manual Utilisateur. Prosim SA. Toulouse.

Walsh, S., I.M. Mujtaba and S. Macchietto (1995). Optimal design of operating procedure for batch distillation column with uncertainty. submitted to ESCAPE-5 conference, Slovenia.

COMPARISON OF INVERTED AND REGULAR BATCH DISTILLATION

E. SØRENSEN* and S. SKOGESTAD**
University of Trondheim- NTH, Department of Chemical Engineering, N-7034 Trondheim, Norway

**Present adress: Imperial College, Centre for Process Systems Engineering, London SW7 2BY, England*

***Author to whom correspondence should be addressed: E-mail: skoge@kjemi.unit.no, phone: +47-73-594154, fax: +47-73-594080*

Abstract. Several alternative strategies have been suggested to increase the productivity of batch distillation columns. One is the use of inverted columns where the feed is charged to the condenser drum instead of to the reboiler. We find that an inverted column is best for separations with small amounts of light component in the feed. The reason is that it takes a long time to move this small amount from the reboiler to the top before removal when a regular column is used. The regular and the inverted column are found not to be mirror images of each other. This is because the inverted column considered in this study is *not* the true inverse of the regular column since the feed and product are in liquid phase. The ideal inverted column with feed in vapor phase is briefly discussed.

Key Words. Batch distillation; optimization; inverted vs. regular coloumn

1. INTRODUCTION

The use of an *inverted* or stripping column where the feed is charged to the condenser instead of the reboiler, was originally proposed by Robinson and Gilliland (1950). They stated that the main advantage with this configuration was that the most volatile components would be collected in the condenser drum in high purity. They briefly discussed the possibility of running the column first in a normal batch fashion to remove the lighter components and then inverted to remove the heavier constituents. They also discussed a combined operation where the feed was added to the middle of the column and the light and heavy components were taken off simultaneously over the top and bottom respectively.

The combined operation was studied by Hasebe at al. (1992) who denoted this column configuration a *complex column*. They also discussed the difference between regular and inverted columns. They claimed that a regular column always has a better separation efficiency than an inverted column if the separation conditions are the same and the relative volatility is constant ("inverted" separations). The conclusion was based on a comparison between the equilibrium curves and operation lines for a regular and an inverted column with all other conditions equal.

Mujtaba and Macchietto (1992) discussed use of a complex column to improved the operation of reactive batch distillation. They found that this column configuration improved conversion and product yield significantly when the reaction products had two extreme boiling points (highest and lowest in the reaction mixture). The use of an inverted column for cases where the reaction product had a higher boiling point than the reactants was suggested but no examples where given. Mujtaba and Macchietto (1994) discussed use of inverted and complex columns for an example with reactive batch distillation. For this example the inverted column gave a lower conversion than the regular column. This was explained in terms of the difference in relative volatility between the heavy components compared to between the light ones.

Chiotti and Iribarren (1991) presented simplified models for binary batch distillation in both regular and inverted columns. They included an intermediate cut which was recycled to the next feed charge. Their models were based on the assumption of pseudo-steady-state in the column. Numerical integration was thereby avoided. They stated that the inverted column was more economical than the regular one for heavy products of high purity. They presented two numerical examples where they optimized the total annual cost for two separations in both an inverted and a regular column. Chiotti *et al.* (1993) extended these models to multicomponent mixtures. They also considered columns which could be used for both rectification (regular column) and stripping (in-

Table 1 Operating parameters.

Column	Regular	Inverted
N	10	10
α	2.0	2.0
V_B $[kmol/hr]$	10.0	10.0
H_F $[kmol]$	10.0	10.0
H_j $[kmol]$	0.001	0.001
H_B $[kmol]$	$H_F - NH_j - H_C$	0.01
H_C $[kmol]$	0.01	$H_F - NH_j - H_B$

verted column) where the products were taken off both at the top and bottom of the column in given sequences. With their very simplified models they found that, for a ternary mixture, the inverted column was optimal when the purity specification for the heavy product was high. It should be noted that both in this paper and in the previous one, the optimal number of trays in the column was in most cases higher for the inverted column than for the regular one.

To the best of our knowledge, no comparison between regular and inverted columns in terms of optimal operation has been given in the literature so far.

In this paper, we will consider the optimal operation of both columns in terms of minimum operating time for the same separations. Intuitively, one may think that an inverted column is best for separations with a large amount of light component in the feed, because one may somehow avoid to "boil off" this large amount. However, as we shall see the opposite conclusion is true: the inverted column is preferable to the regular column, in terms of shorter operating time, when the light component is present in small amounts in the feed mixture. At the end of the paper, we will discuss the operation of the two columns for "inverted" separations (removal of light component in a regular column versus removal of heavy component in an inverted column).

2. OPTIMAL OPERATION

In this section, we consider the optimal operation in both a regular and an inverted column in terms of minimum operating time for given separations. We will show that the inverted column is in fact better than the regular column for separations where the light component is present in a small amount in the feed mixture. The columns are shown in Figure 1. Note that the feed and the product are in liquid phase also in the inverted column.

2.1. *Example studied*

A binary mixture with constant relative volatility is assumed. The operating parameters are given in Table 1. SPEEDUP (1993) is used for the simulations. In the regular column the feed is charged as liquid to the reboiler, and in the inverted column the feed is charged as liquid to the condenser drum. The time and energy input to heat the feed charge to its boiling point is neglected. In any case this time is the same for the two columns since the feed is in liquid phase in both cases. Initially, the column is assumed filled with liquid at its boiling point with a composition equal to that of the feed. This assumption is shown not to be of importance for the results (Sørensen, 1994). It is assumed that the vapor flow V_B can be manipulated directly.

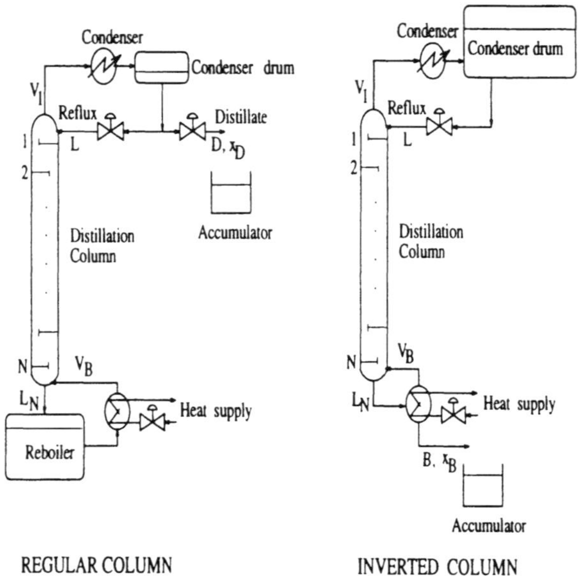

Fig. 1. A regular and an inverted batch column.

2.2. *The minimum time problem*

The optimal operation in terms of minimum operating time can be stated as:

$$\min_{R,R_B} t_f \tag{1}$$

subject to

$$x_A \geq x_A^{spec} \tag{2}$$

$$x_R \geq x_R^{spec} \tag{3}$$

In words, with a given boilup rate, find the optimal reflux ratio R (for the regular column) or reboiler ratio R_B (for the inverted column) which minimizes the operating time t_f subject to constraints on the compositions of accumulated light and heavy product. In the following examples, we have selected to use the same purity specifications for the two products, that is

$$x_A^{spec} = 1 - x_R^{spec} = x^{spec} \tag{4}$$

The optimization parameters R and R_B are assumed piecewise continuously linear over two intervals giving a total of 5 optimization variables ($R(t_0)$, $R(t_1)$, $R(t_f)$, t_1 and t_f). The optimization program DAEOPT developed at Imperial College, London (Vassiliadis, 1993) is used. It should be noted that since more than one optimization vari-

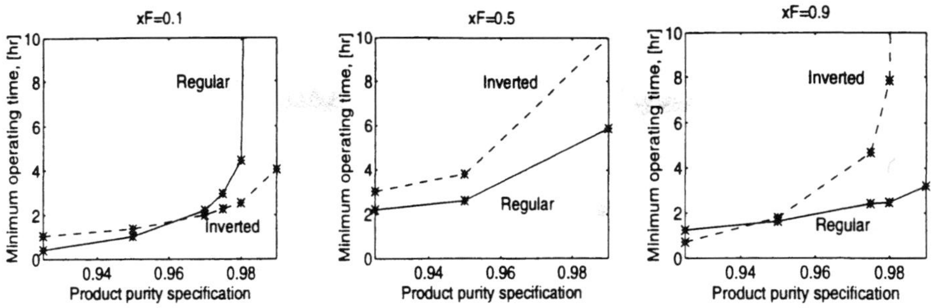

Fig. 2. Minimum operating time t_f as a function of purity specifications x^{spec} for a regular and an inverted column.

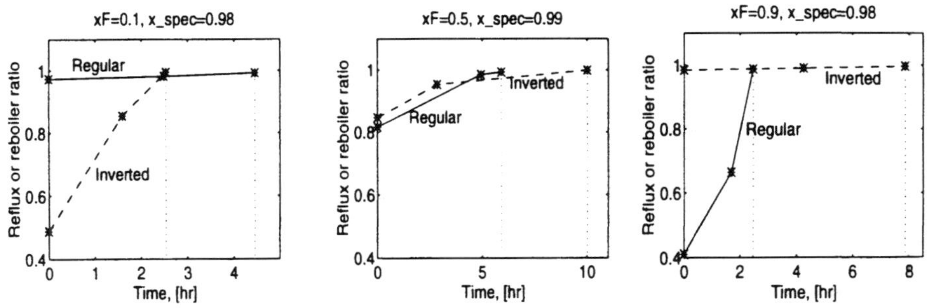

Fig. 3. Optimal reflux ratio R and reboiler ratio R_B.

Table 2 Minimum operating time t_f for varying purity specifications x^{spec} for regular and inverted column (Same purity specification x^{spec} for both products).

Purity specification x^{spec}	Minimum operating time inverted column t_f, *hr*	Minimum operating time regular column t_f, *hr*
Feed composition $x_F = 0.1$		
0.925	1.043	**0.409**
0.950	1.369	**1.024**
0.975	**2.264**	2.940
0.980	**2.532**	4.448
0.990	**4.041**	>100
Feed composition $x_F = 0.5$		
0.925	3.022	**2.200**
0.950	3.807	**2.610**
0.990	10.010	**5.905**
Feed composition $x_F = 0.9$		
0.925	**0.709**	1.248
0.950	1.791	**1.630**
0.975	4.691	**2.410**
0.980	7.856	**2.462**
0.990	>100	**3.192**

able is involved, several local minima may exist.

2.3. *Results*

We consider 13 cases with different values for the feed composition and product specifications (Table 2). Other operating parameters are as given before (Table 1). The minimum operating time is presented for the 13 cases in Table 2 and shown graphically in Figure 2. The corresponding optimal reflux and reboiler ratios are shown for three of the cases in Figure 3 and details on how the compositions change with time are shown in Figure 4 for one case.

From Table 2 and Figure 2 we see that the inverted column has indeed a shorter operating time than the regular column for some separations. We find that the inverted column is better for cases where the products are to be recovered at high purity from a feed *low in light component* ($x_F = 0.1$), whereas the regular column is best for cases where the products are to be recovered at high purity from a feed *rich in light component* ($x_F = 0.9$). The regular column is found to be better than the inverted column for all symmetrical separations, $x_F = 0.5$. However, more importantly, for some specifications the inverted column gives a moderate operating time when the separation is not even possible in the regular column (e.g. $x_F = 0.1$ and $x^{spec} = 0.999$). The inverted column is best when x_F is small because it takes a long time to "move" a small amount of light component from the reboiler to the top of the column when a regular column is used.

The reflux and reboiler ratios in Figure 3 are very close to 1 for separations where a small amount of product is to be withdrawn from the column leaving a large amount of the other product as residue (e.g. $x_F = 0.1$ in the regular column and $x_F = 0.9$ in the inverted column). The compositions of the products in the residual and the accumulator are given in Figure 4 for a case when the inverted column is best ($x_F = 0.1$ and $x^{spec} = 0.98$). For the regular column, the accumulator composition is

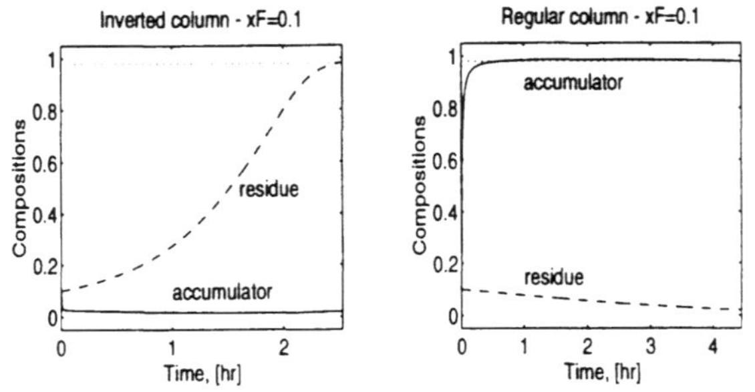

Fig. 4. Composition of product in accumulator or residue ($x_F = 0.1$ and $x^{spec} = 0.98$.)

slightly higher than the specification during most of the run and reaches the specified value at the final time. In the inverted column, the composition in the residue is slowly increasing as the heavy component is removed in the bottom.

3. THE PERFORMANCE OF "INVERTED" SEPARATIONS

In this section we will compare what we denote *"inverted" separations*; the removal of light component from a regular column compared to the removal of heavy component from an inverted column that is, the feed compositions are mirror images of each other x_F(regular column)$=1-x_F$(inverted column).

If we compare the shape of the plots in Figure 2, it can be seen that the separation in the inverted column has a similar operating time profile as the separation in the regular column. The same comparison can be made for the reverse case. Also, the optimal reflux and reboiler ratios have a similar shape and value for "inverted" separations, Figure 3). Nevertheless, the operating time for "inverted" separations have different values, for example:

Inverted column:

$t_f = 2.264\ hr$ for $x_F = 0.1$ and $x^{spec} = 0.975$

Regular column:

$t_f = 2.410\ hr$ for $x_F = 0.9$ and $x^{spec} = 0.975$

In the following we will compare the regular and the inverted column for "inverted" separations in terms of dynamic behavior under total reflux and during production. We will argue that the difference is due to the fact that both the feed and the product are in liquid and not in vapor phase in the inverted column. The inverted column studied so far is therefore *not* the true inverse of the regular column. Consider the two following two cases:

1. A small amount of "withdrawn" component is removed from the feed charge ($x_F = 0.1$ for the regular column and $x_F = 0.9$ for the inverted column).
2. A large amount of "withdrawn" component is removed from the feed charge ($x_F = 0.9$ for the regular column and $x_F = 0.1$ for the inverted column).

The withdrawal and feed end compositions profiles are given in Figure 5 and 6 for the two cases. In both cases an initial total reflux period of 0.05 hr is used to reach steady state before the production period. Constant product flows are assumed. The specification for the withdrawal end composition is for the "withdrawn" component in both cases, $x^{spec}_{D,light} = x^{spec}_{B,heavy}$.

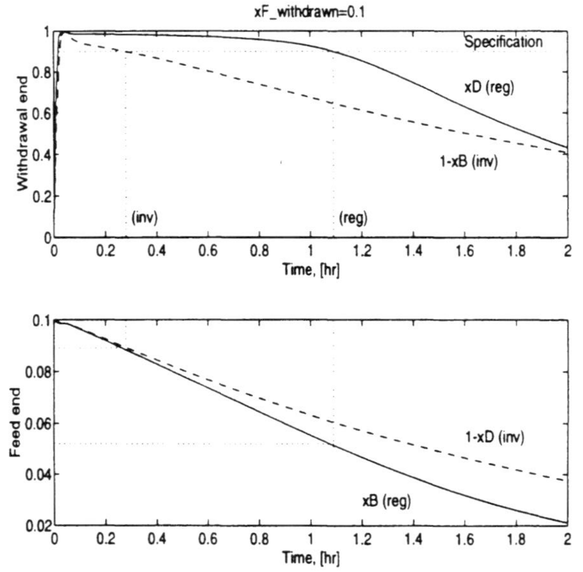

Fig. 5. Withdrawal and feed end compositions profiles for the withdrawn component. ($x_F = 0.1$ for the regular column and $x_F = 0.9$ for the inverted column, $x^{spec}_{D,light} = x^{spec}_{B,heavy} = 0.9$ and $D = B = 0.5\ kmol/hr$.)

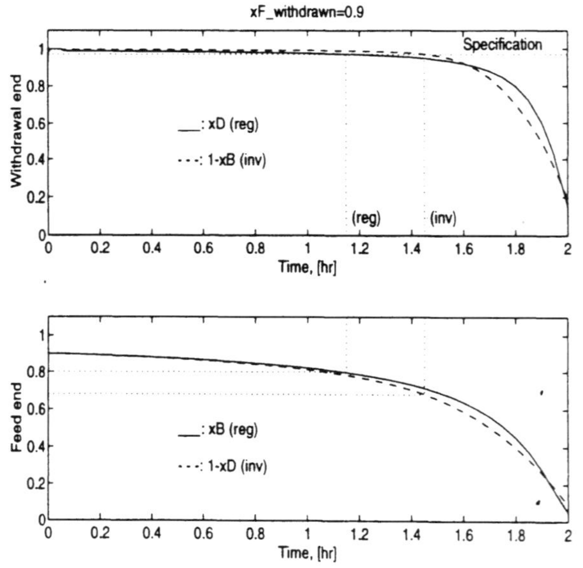

Fig. 6. Withdrawal and feed end compositions for the withdrawn component. ($x_F = 0.9$ for the regular column and $x_F = 0.1$ for the inverted column, $x^{spec}_{D,light} = x^{spec}_{B,heavy} = 0.97$ and $D = B = 5\ kmol/hr$.)

From Figure 5 it can be seen that, for the case with a small amount of "withdrawn component", the product composition for the inverted column drops below the specification $x^{spec}_{B,heavy} = 0.9$ very quickly. For the regular column however, the separation is satisfactory. On the other hand, from Figure 6 we see that for the case with a large amount of "withdrawn" product, the specification for the product composition is held longest for the inverted column.

Conclusion:

- The regular column has a faster approach to equilibrium than the inverted column for "inverted" separations with a *small* amount of "withdrawn" component in the feed (light in the regular column and heavy in the inverted column).
  The regular column is able to maintain a high product composition for a longer period, which again will result in either a larger amount of on-spec product or a shorter operating time for this column.
- For "inverted" separations with a *large* amount of "withdrawn" component in the feed (light in the regular column and heavy in the inverted column), the situation is the opposite. A faster approach to equilibrium and a longer on-spec period is found for the inverted column. The conclusion by Hasebe *et al.* (1992) that the regular column always has a better separation efficiency than the inverted column is thereby found to be incorrect.

4. IDEAL INVERTED COLUMN

The true inverse of a regular batch column is denoted an ideal inverted column in the following. It has the following properties:

- The liquid feed in the reboiler of the regular column is vapor feed in the condenser of the ideal inverted column
- The liquid holdup in the condenser of the regular column is vapor holdup in the reboiler of the ideal inverted column
- The product is taken out as liquid in the top of the regular column but as vapor from the bottom of the ideal inverted column
- The trays are filled with liquid in the regular column but with vapor in the ideal inverted column
- The vapor holdup is neglected in the regular column but the liquid holdup is neglected in the ideal inverted column (with our model assumptions)
- The vapor flow is specified in the regular column but the liquid flow is specified in the ideal inverted column

Finally, for constant relative volatility the separation is equivalent provided the liquid mole fraction of light component in the regular column is the mole fraction of heavy component in the inverted column. This follows since

$$\frac{y_{light}/x_{light}}{y_{heavy}/x_{heavy}} = \alpha = \frac{x_{heavy}/y_{heavy}}{x_{light}/y_{light}} \quad (5)$$

Note that one may directly obtain operational data for the ideal inverted column from data for a regular batch column. One then needs to reverse the data as follows:

$$L^{inv} = V^{reg}, \quad y_F^{inv} = 1 - x_F^{reg},$$
$$y_A^{inv} = 1 - x_A^{reg}, \quad y_R^{inv} = 1 - y_R^{reg} \quad (6)$$

(We use y_F, y_A and y_R to denote that these are vapor phase compositions.)
The data in the right column of Table 2 which are for a regular column with vapor flow $V_B = 10$ $kmol/hr$, will therefore correspond to data for an ideal inverted column with reflux flow $L = 10$ $kmol/hr$. For example, for the case with $x_F = 0.9$ and $x_A^{spec} = 1 - x_R^{spec} = 0.99$ the minimum operating time is 3.19 hr when vapor flow is fixed at $V_B = 10$ $kmol/hr$. Thus, with $y_F = 0.1$ (feed is vapor), $y_A^{spec} = 1 - y_R^{spec} = 0.01$ and *reflux flow* L fixed at 10 $kmol/hr$, the minimum operating time in an ideal inverted column is 3.19 hr. (If we instead fixed the vapor flow at 10 $kmol/hr$ then the operating time would be even shorter.) It is interesting to note from Table 2 that in a practical inverted column this separation needs 4.04 hr which is significantly higher.

5. CONCLUSION

In this paper we have compared a regular and an inverted column in terms of dynamic behavior and optimal operation. The inverted column is found to yield the shortest operating time for separations where the light component is present in a small amount in the feed. This is because in a regular column, the relatively small amount of light component must be moved through the column from the feed end to the withdrawal end which is more time consuming since a very high reflux ratio must be used.
Note that Robinson and Gilliland (1950), who originally proposed the use of inverted columns, stated that the main advantage was that the light component would be collected in the condenser drum in high purity. This is confirmed in this paper.
We have found that the steady state profiles are equal, but the approach to equilibrium is different, for "inverted" binary separations (removal of light component in a regular column and of heavy component in an inverted column). This is because the inverted column is not the true inverse

of the regular column since the feed and product are in liquid and not in vapor phase.
The use of an ideal inverted column is, however, more of academical interest. In practice, one is faced with the decision as to whether to use a regular or a practical inverted column configuration for a given feed mixture. We believe that this paper has at least indicated the answer to this question.

ACKNOWLEDGEMENT

E.K. Hilmen is acknowledged for assistance in preparing the final version of this manuscript.

NOTATION

B	bottom flow for inverted column $[kmol/hr]$
D	distillate flow for regular column $[kmol/hr]$
H_A	liquid holdup in accumulator $[kmol]$
H_B	liquid holdup in reboiler $[kmol]$
H_C	liquid holdup in condenser $[kmol]$
H_F	amount of initial liquid feed $[kmol]$
H_j	liquid holdup on tray j $[kmol]$
H_R	amount in residual $[kmol]$
L	reflux flow $[kmol/hr]$
L_j	liquid flow from tray j $[kmol/hr]$
N	number of trays in the column section (excl. reboiler and condenser)
R	internal reflux ratio $=L/V_1$ for regular column
R_B	reboiler reflux ratio $=V_B/L_N$ for inverted column
t	time $[hr]$
t_0	initial time $[hr]$
t_1	switching time for operating variables $[hr]$
t_f	final operating time $[hr]$
V_j	vapor flow from tray j $[kmol/hr]$
V_B	vapor flow from the reboiler to the column section $[kmol/hr]$
x_A	mole fraction of light component in accumulator
x_B	mole fraction of light component in reboiler
x_D	mole fraction of light component in distillate
x_F	mole fraction of light component in feed
x_j	mole fraction of light component in liquid leaving tray j
x_R	mole fraction of light component in residual
y_B	mole fraction of light component in vapor leaving reboiler
y_D	mole fraction of light component in vapor distillate
y_j	mole fraction of light component in vapor leaving tray j
α	relative volatility

Scripts

inv	inverted column
j	tray number
liq	liquid
norm	normalized
o	initial conditions
reg	regular column
spec	specified value
ss	steady state
vap	vapor
withdrawn	withdrawn component (light in regular and heavy in inverted)

REFERENCES

[1]Chiotti, O.J. and O.A. Iribarren, "Simplified models for binary batch distillation", *Comp. Chem. Eng.*, **15**(1), 1-5, 1991.

[2]Chiotti, O.J., H.E. Salomone and O.A. Iribarren, "Selection of multicomponent batch distillation sequences", *Chem. Eng. Comm.*, **119**, 1-21, 1993.

[3]Hasebe, S., B. Abdul Aziz, I. Hashimoto and T. Watanabe, "Optimal design and operation of complex batch distillation column", *Proc. IFAC Workshop*, London, Sept. 1992.

[4]Mujtaba, I.M. and S. Macchietto, "Optimal operation of reactive batch distillation", *AIChE 1992 Annual Meeting*, Miami Beach, Nov. 1992.

[5]Mujtaba, I.M. and S. Macchietto, "Optimal operation of multicomponent batch distillation. - A comparative study using conventional and unconventional columns", *Proc. ADCHEM'94*, Japan, May 1994.

[6]Robinson, C.S. and E.R. Gilliland, "Elements of Fractional Distillation", *McGraw-Hill Book Company, Inc.*, New York, 4th ed., 1950.

[7]SPEEDUP Release 5.4 User Manual, Prosys Technology Ltd., 1993.

[8]Sørensen, E., "Studies on optimal operation and control of batch distillation columns", Ph.D. Thesis, University of Trondheim- NTH, 1994.

[9]Vassiliadis, V., "Computational solution of dynamic optimization problems with general differential-algebraic constraints", Ph.D. Thesis, University of London, 1993.

APPENDIX

The dynamic models used in this study are valid under the following assumptions:
(1) Staged batch distillation column (trays numbered from the top and down); (2) Perfect mixing on all trays; (3) Negligible vapor holdup; (4) Constant stage pressures and tray efficiencies; (5) Equal vapor flows in the column; (6) Total condensation with no subcooling in the condenser; (7) Constant relative volatility; (8) Constant molar liquid holdup on all trays (liquid dynamics are neglected); (9) Constant molar condenser holdup in the regular column (perfect control); (9) Constant molar reboiler holdup in the inverted column (perfect control).
The reflux ratio R for the regular column is the internal reflux ratio L/V_1, not the usual external ratio L/D. (This is better solved numerically since $R_{internal} \in [0,1]$ whereas $R_{external} \in [0, \infty >$.) The reboiler ratio in the inverted column, R_B, is also defined as an internal ratio V_B/L_N. The external ratio is in this case given as $R_{B,external} = V_B/B$.
It is assumed that the vapor flow V_B can be manipulated directly. The heavy product flow B in the inverted column is assumed to be taken out from the reboiler as liquid with a composition equal to x_B.

EXPERIMENTAL EVALUATION OF A FAULT DETECTION AND IDENTIFICATION SCHEME FOR CHEMICAL PROCESSES

Paulo A.A. Afonso, Jorge L. Ferreira, José A.A.M. Castro and Nuno M.C. Oliveira

[1]*Department of Chemical Engineering, University of Coimbra, Largo Marquês de Pombal, 3000 Coimbra, PORTUGAL. Email: nuno@eq.uc.pt*

Abstract. This paper deals with the experimental implementation of an automatic scheme for fault detection and identification (FDI). It also provides redundant signals for performing an efficient control in case of a faulty situation, improving safety and reliability of plant operations. Two approaches for on-line parameter estimation are tested: a variant of Recursive Least Squares (RLS) algorithm for linear or linearized models and the GREG algorithm (Stewart *et al.*, 1992), suitable for more general nonlinear models. The fault detection method is based on a statistical analysis of the deviation of the process parameters. Simultaneously, an Extended Kalman Filter (EKF) enables the fault identification to be performed, reducing false alarms and providing estimates of process variables for alternative control purposes. Applications take place in a industrial-scale pilot plant. The good performance obtained highlighted the strategy employed when sensor faults were introduced artificially on control loops.

Key Words. Fault detection, Fault identification, Fault tolerant control, Parameter estimation, Process control.

1. INTRODUCTION

The chemical process industry is characterized by a large diversity and complexity of their plants. Traditionally, these have been supervised by human operators at all levels of operation. However, the recent trend of generalized use of automatic supervision systems has introduced the need of improved algorithms for identification and earlier detection of failures and anomalous operation. This is important as the control systems are usually built to use a relatively high degree of confidence in the information received from the instrumentation. Especially in centralized control schemes, a failure in one of the sensors used in a control loop can lead to a totally erratic behavior, or even cause the instability of the system. In parallel with the development of robust (fault tolerant) control strategies, a strong incentive exists therefore for the development of systematic algorithms for fault identification and correction, in chemical processes.

There exist several surveys on the approaches available for the problem of fault detection using process models: Willsky (1976) is the author of a classic paper discussing the most important techniques available at the time, and more recently other authors (Isermann, 1984; Basseville, 1985; Patton, 1989; Frank, 1990) have also focused on this subject. A great variety of fault detection schemes rely on residual generation and testing, such as: parity space approach, dedicated observer schemes and innovation-based approach, fault detection filter approach and parameter identification approach. These fault detection and identification (FDI) strategies were mainly applied in chemical and aerospace industries.

This paper deals with the experimental implementation of such an automated scheme in an industrial-scale pilot plant, used in a natural resin industry. The plant is composed by a stirred tank reactor, equipped with a cooling/heating jacket and an hydraulic stirring system. The plant is fully instrumented, both for measurement (temperature, pressure, level and flow) and actuation (control and solenoid valves, hydraulic pumps). At this stage, our main goal is to continue computing the correct control actions when a failure occurs in a sensor placed in a control loop. This task is accomplished using a dynamic process model, whose parameters may be not very well known and time varying.

The fault detection method described relies on the determination of changes on the process parameters. The normal process model is tracked, including some allowable tolerance. Hence, a change in the estimated parameters may indicate that a fault has occurred. Here, two approaches for on-line parameter estimation were compared: RLS (Recursive Least Squares), suitable for linear or linearized models and the GREG algorithm (Stewart et al., 1992) for non-linear and more fundamental process models. The fault detection step is done through a statistical decision test which analyses the estimated process parameters, and compares their characteristics to the expected properties of the estimates during regular operation. When a failure condition is detected, it is also identified by comparing the estimated values obtained through an EKF with the values from the sensor measurements. In this situation the faulty sensor is automatically disabled, and the process control is carried out using only the corresponding estimated value.

2. PROCESS MODELLING AND CONTROL

The plant where the experimental tests were performed is depicted in Fig. 1. The stirred tank reactor has a capacity of 80 liter and it is fed by two liquid streams coming from two pressurized tanks (Tank 1 and Tank 2). Cold water flows from Tank 1, with flow rate Q1 and temperature T1, while hot water (obtained by steam injection) comes from tank 2 at temperature T2 and flow rate Q2. The cooling fluid has temperature Tc, and inside the reactor, temperature and level are named T and h, respectively. The reactor outflow, governed by gravity and the control valve VC3, drains through a pipe whose outlet into atmospheric pressure is at a level h_0 below the bottom of the tank. The flow rates Q1 and Q2 are manipulated through the control valves VC1 and VC2.

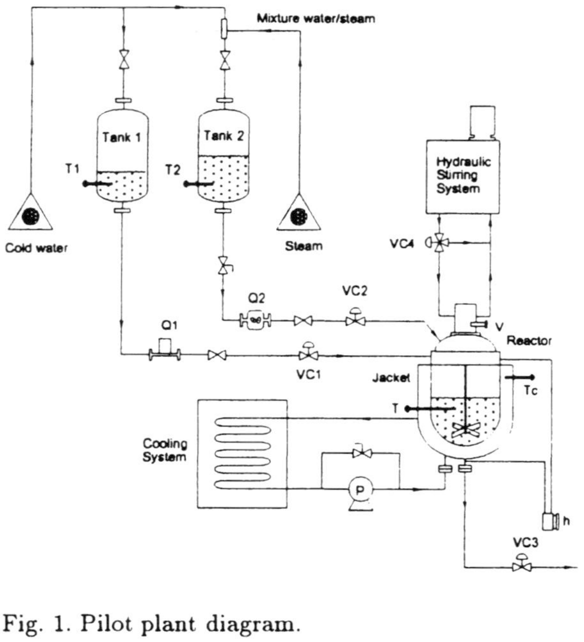

Fig. 1. Pilot plant diagram.

Assuming constant density, the total mass balance to the stirred tank is:

$$\frac{dh}{dt} = \frac{1}{A}(Q_1 + Q_2 - Q_3) \tag{1}$$

The outflow Q3 is a function of both reactor level and VC3 position, and has been approximated by:

$$Q_3 = \beta\sqrt{h + h_0}\,\frac{\alpha \cdot VC_3}{1 + VC_3(\alpha - 1)} \tag{2}$$

The temperature inside the reactor can be obtained through the energy balance, considering perfect mixing and uniform jacket temperature:

$$\frac{dT}{dt} = \frac{Q_2(T_2 - T)}{Ah} + \frac{Q_1(T_1 - T)}{Ah} + \\ + U_g(A + \pi d_r h)\frac{T_c - T}{\rho C_p A h} \tag{3}$$

Figure 2 shows the control loops and the corresponding input and output variables. Controllers C1, C2 and C3 are discrete versions of conventional proportional-integral (PI) controllers. C1 controls the liquid level inside the reactor by acting on control valve VC3. The temperature is controlled by C2 through valve VC1 regulating the flow-rate Q1 from tank 1. On the other hand, VC4 is actuated by C3 in order to keep a constant stirrer rotation speed.

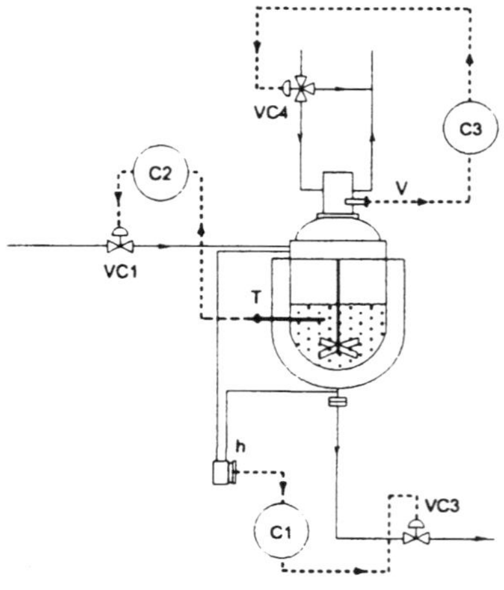

Fig. 2. SISO PI control loops.

3. SENSOR FAULT DETECTION

The performance of the two algorithms for parameter estimation (RLS and GREG) is compared. The set of parameters to be estimated was selected after the evaluation of the model sensibility to all of them. Using just one parameter for each model equation lead to the choice of $p_1 = \beta$ (eq.1) and $p_2 = \frac{U_g}{\rho C_p}$ (eq. 3).

To apply the RLS algorithm, equations (1) and

(3) were previously discretized using an Euler first order approximation.

The present paper employs RLS with constant trace (Fortescue et al., 1981) as described by Goodwin *et al.* (1985). This estimator was chosen for its relative fast convergence speed and numerical stability.

To detect variations on the normal process operation automatically, a statistical test on the estimated parameters based on F-Distribution was used. The experimental tests revealed that these statistical analyses were reliable enough with respect to fault identification. The sensor faults are directly related to the deviations of the estimated parameter, enabling the establishment of fault patterns, the so-called ***fault signatures***, often utilized for fault identification. Simultaneously, the state estimator algorithm (EKF) provides redundant signal for the controlled variables. When the statistical analyses give an alarm indicating that a failure has occurred in one sensor, the state variables given by EKF are compared to sensor measurements. This way it is possible to confirm the sensor fault detection, reducing, therefore, false alarms.

4. FAULT TOLERANT CONTROL STRATEGIES

Despite the great interest on the performance of fault detection and identification schemes, process control in such abnormal conditions is of considerable pratical importance for industrial plants. Bearing this in mind, control strategies were redefined in order to guarantee safe plant operation for this system.

After detecting and identifying a fault in one sensor, that sensor is immediately disabled and the values of the state variables, required for control purposes, are those predicted by the EKF. The quality of the control actions determined by the EKF estimates is directly related to the degree of knowledge of the process model. At this stage, it is important to stress the role of RLS/GREG algorithm that provide an updated model used by EKF, evaluated just before the fault occurrence.

Figure 3 shows a complete scheme of the data flow circuit through the instrumentation and control devices. The solid line represents the process control during normal operation and the dashed lines indicate the strategy used in the case of a faulty situation.

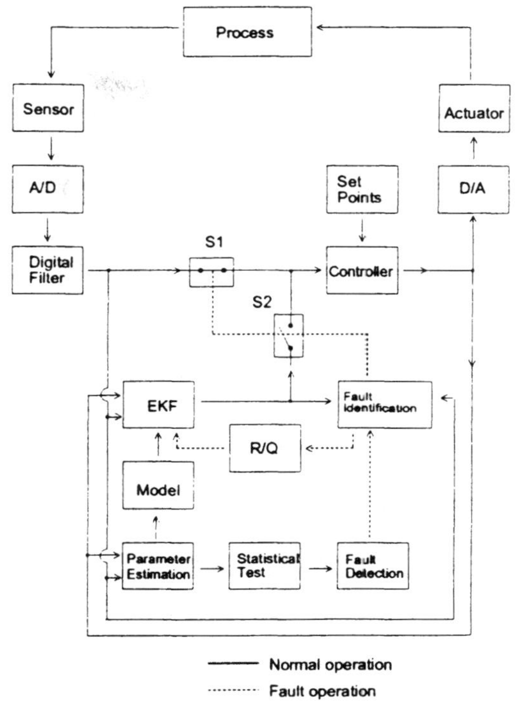

Fig. 3. FDI block diagram.

Table 1 Physical parameters and nominal operating conditions

$h_0 = 10.0\ dm$	$Tc \approx 22^oC$
$A = 16.6\ dm^2$	$T_2 \approx 46^oC$
$\rho = 1\ Kg/dm^3$	$T_1 \approx 15^oC$
$Cp = 4148\ J/KgK$	$Q_1 = 2 \pm 1\ l/m$
$Ug = 755\ J/dm^2.min.K$	$Q_2 = 9 \pm 2\ l/min$
$V = 120\ rpm$	$h = 3 \pm 0.1\ dm$
$\beta = 5.25$	$T = 32 \pm 1\ ^oC$
$\alpha = 2.86$	

5. EXPERIMENTAL RESULTS

This section presents the pratical evaluation of automatic fault detection/identification and control techniques previously described. Table 1 shows the values calculated for the physical parameters of the process and the nominal operating conditions. The sampling time is 0.1 min.

Figures 4 and 5 show the results obtained in the plant, when two faults were artificially introduced in the sensors located in the control loops C_1 (Experiment I) and C_2 (Experiment II).

Both tests had a duration of 120 minutes. During the first 30 minutes, the system was excited by the introduction of disturbances in the set-points of controllers C_1,C_2 and in the flow rate Q_2 (by manipulating VC_2) to improve the initial estimation of the model parameters.

In the first test, a failure occurred in the level sensor, while in the second test the faulty sensor was

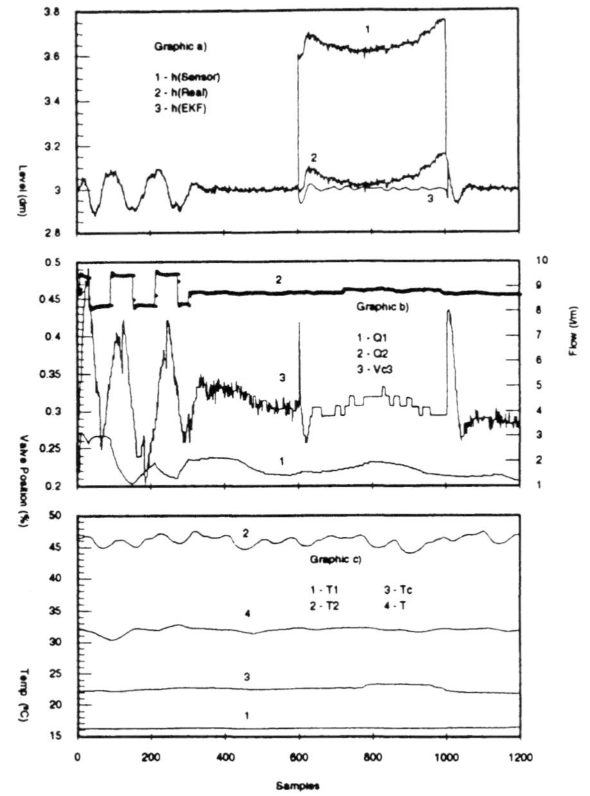

Fig. 4. Evolution of the process variables in Experiment I

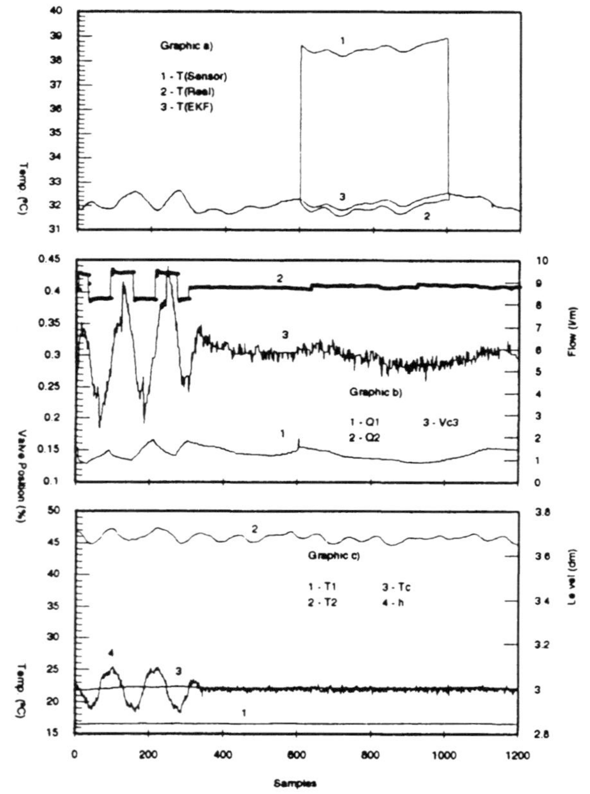

Fig. 5. Evolution of the process variables in Experiment II

the one that measures temperature inside the reactor. The fault introduced is expressed in terms of percentage in relation to the normal operation. In both situations, a failure of 20% occurred between t=60 min and t=100 min.

Figure 6 shows the parameter estimation with RLS and Fig. 7 represents the same parameter estimation with GREG. Figures 8 and 9 show the statistical tests for experiments I and II, with RLS and GREG parameter estimation, respectively.

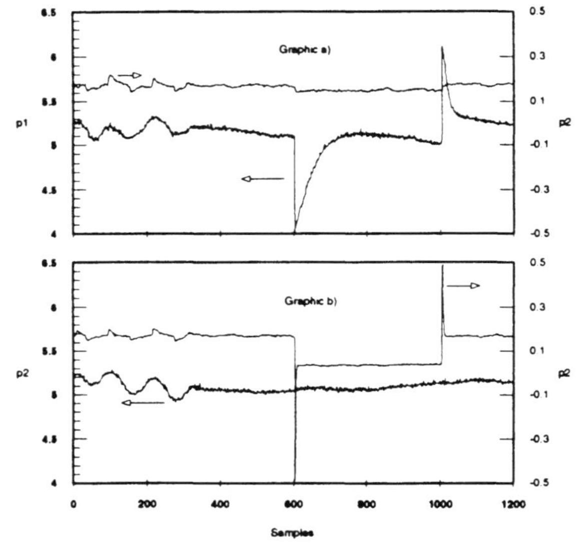

Fig. 6. Parameter estimates obtained through the RLS algorithm: a) Experiment I; b) Experiment II

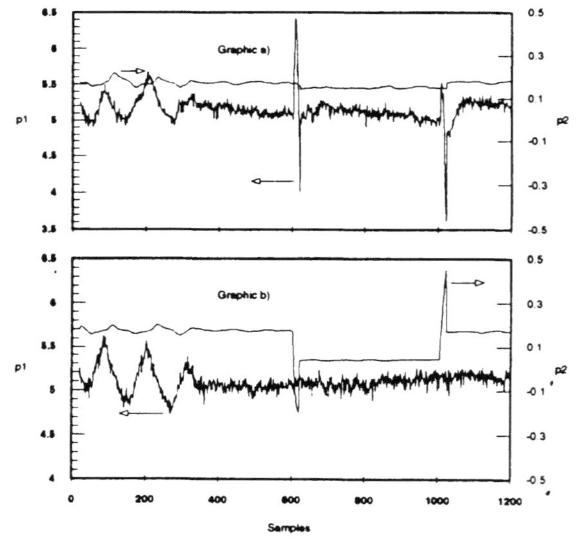

Fig. 7. Parameter estimates obtained through the GREG algorithm: a) Experiment I; b) Experiment II

5.1. *Experiment I*

In Fig. 4(a), h(Sensor) represents the value given by the sensor for the reactor level and h(Real) is the real value of this variable. It is also indicated the EKF estimate - h(EKF), used for control purpose in case of sensor malfunction. Figures 4 (b,c)

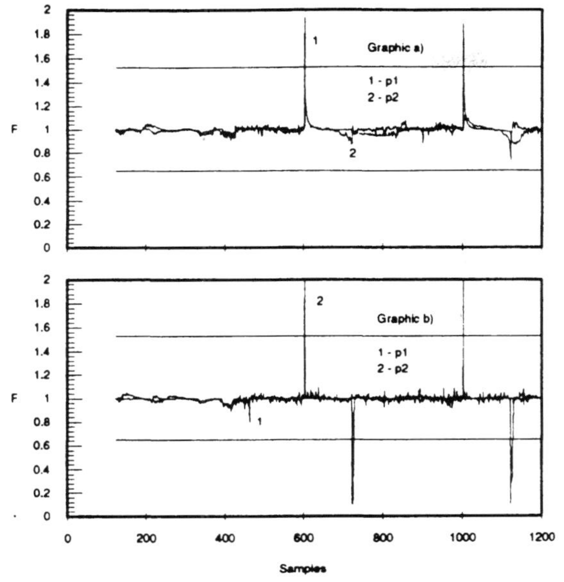

Fig. 8. F - test corresponding to the parameters depicted in Fig. 6

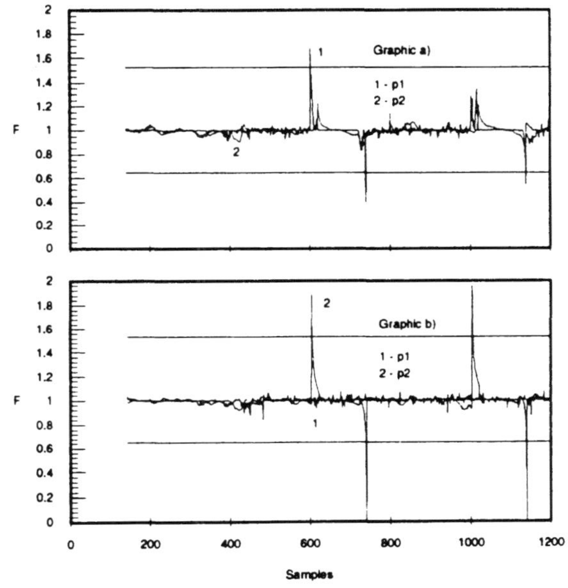

Fig. 9. F - test corresponding to the parameters depicted in Fig. 7

show the behaviour of the other process variables: flowrates Q_1 and Q_2, control valve opening VC_3 and temperatures T, T_1, T_2 and Tc.

Figure 4(a) shows that 60 min after the beginning of the operation, a 20% fault was introduced in the sensor that measures the reactor level. Its effect can be directly observed by the control action of C_1, that manipulates VC_3 -Fig. 4(b), or by the change of the parameter p_1 represented in Figs. 6(a) and 7(a).

To detect failures, a 99% confidence interval was used. Figures 8(a) and 9(a) represent the F-test for RLS and GREG respectively. The limits of the 99% confidence intervals are also represented by the two horizontal lines.

For t=60 min, Figs. 8(a) and 9(a) show an abrupt variation corresponding to parameter p_1. On the contrary, parameter p_2 was not significantly affected, in spite of the influence of the level in the temperature, as expressed in eq.(3).

Thus, it is possible to identify the cause of such changes in these parameters. In order to improve the robustness of fault detection and identification, the estimates given by EKF were compared to the measurements given by the sensors used in the SISO control loops, which are depicted in Fig. 4(a). By comparing h(Sensor) to h(EKF) from the moment the alarm has started (for $t > 60$ min), it was easy to conclude that the level sensor was the faulty one.

At that moment, the fault was detected and identified and the level sensor was immediately disabled and its value replaced by the estimate of EKF, to proceed the level control (through controller C_1), as referred in the scheme of Figure 3. The results can be observed in Fig.4(a), for times greater than 60 min and less than 100 min, where h(Real) is the real reactor level. The small deviation existing between the EKF estimate and the real value of this state variable, is due to the lack of linearity of the sensors, which caused some internal imprecisions to EKF algorithm. However, this difference can be neglected, when compared with the damages that could have happened if the controller was actuated by erratic data from the sensor.

5.2. *Experiment II*

In experiment II, which deals with a fault in the reactor temperature sensor, the general procedure is analogous to experiment I. For t=60 min, a fault of 20% was introduced, as shown in Fig.5(a). The parameter p_2 suffered a sudden change as revealed by Figs. 6(b) and 7(b) and confirmed by the statistical analysis represented in Figs. 8(b) and 9(b). Control actions were determined by the EKF estimations - Fig.5(a) - according to the strategy used in experiment I. Considering the period of time between 60 and 100 min, Fig.5(a) also shows that, in spite of existing a faulty sensor in the control loop, there is no visible deviation on controlled variable T(Real). Once again, if the sensor measuring had not been neglected and replaced by a estimated value, the temperature controller C_2 would induce a quite erratic behaviour.

6. CONCLUSIONS

Several experimental tests have been done in a industrial scale pilot plant to test the scheme proposed for process control under failure conditions, particularly on sensors used in control loops. The success of this FDI scheme is outlined by the results of experiment I and II.

For the model used here, which is quite simple, identical results were obtained for on-line parameter identification with RLS and GREG.

However, with more nonlinear models, the predictive algorithms based on nonlinear estimation techniques such as GREG are expected to offer significant advantages over simpler methodologies like RLS.

This FDI scheme can validate the information that flows through instrumentation and consequently improve plant reliability. Based on the dynamics of the process model, it exhibits simplicity, numerical robustness and little computer effort, which are characteristics needed for industrial applications. Despite of the good results obtained when artificial faults were introduced on variables used on control loops, additional work should be done, considering the fault identification related to sensors and actuators outside the control loops, which will demand different control strategies.

7. NOTATION

A	-	Cross-sectional area of the reactor.
C_1	-	Level controller.
C_2	-	Temperature controller.
C_3	-	Stirring controller.
Cp	-	Fluid heat capacity.
d_r	-	Reactor diameter.
h	-	Height of liquid inside the reactor.
h_0	-	Level offset.
p_i	-	Linear model parameters.
Q_1	-	Hot water inlet flow rate.
Q_2	-	Cold water inlet flow rate.
Q_3	-	Reactor outlet flow rate.
T	-	Fluid temperature in the reactor.
T_1	-	Fluid temperature in tank 1.
T_2	-	Fluid temperature in tank 2.
Tc	-	Coolant temperature.
Ug	-	Global heat transfer coefficient.
V	-	Stirrer rotation speed.
VC_1	-	Valve VC_1 position.
VC_2	-	Valve VC_2 position.
VC_3	-	Valve VC_3 position.
α	-	Constant.
β	-	Constant.
ρ	-	Fluid density.

ACKNOWLEDGEMENTS

Financial support from JNICT BD/612/90-IB is greatfully acknowledged.

8. REFERENCES

Basseville, M., (1985). "Detection of abrupt changes in signals and dynamical systems", *Lectures Notes in Control and Information Sciences*, **77**, Springer-Verlag, Berlin.

Fortescue, T.R., L.S. Kershenbaum, B.E. Ydstie (1981). "Implementation of self-tuning regulators with variable forgetting factors", *Automatica*, **17**, 831–835.

Frank, P.M. (1990). "Fault diagnosis in dynamic systems using analytical and knowledge-based redundancy - A survey and new results", *Automatica*, **26**, 459–474.

Goodwin, G.C., D.J. Hill, M. Palaniswami (1985). "Towards and adaptative robust controller", *Proc. IFAC Conf. Identification and system Parameter Estimation*, York, U.K., 997–1002.

Isermann, R. (1984). "Process fault detection based on modeling and estimation methods - Survey", *Automatica*, **20**, 387–404.

Patton, R.J., P.M. Frank, R.N. Clark (Ed), (1989). "Fault diagnosis in dynamic systems, theory and applications". Prentice-Hall, Englewood Cliffs, NJ.

Stewart, W.E., M. Caracotsios, J.P. Sørensen (1992). "Parameter Estimation from Multiresponse Data" *A.I.Ch.E. J.*, **38**(4), 641–650.

Willsky, A.S. (1979). "A survey of design methods for failure detection in dynamic systems", *Automatica*, **12**, 601–611.

STEADY STATE AND DYNAMIC SIMULATION OF FCC REACTOR UNITS

Rafael Maya Yescas[1,2], David Bogle[2] and Felipe López Isunza[1]

[1] *Dpto. de Ingeniería de Procesos e Hidráulica. Universidad Autónoma Metropolitana - I
Apdo. Postal 55-534, México 09340 D.F. México*
[2] *Dept. of Chemical and Biochemical Engineering. University College London
Torrington Place, London WC1E 7JE. United Kingdom*

Abstract: A dynamic model for an industrial FCC reactor was developed in order to predict steady state operation conditions and responses to perturbations. Dynamic modelling has been performed to predict open loop responses. A new kinetic scheme for the catalytic cracking reaction, consisting of five lumped species, has been proposed. Steady state predictions are in agreement with industrial operating data. Dynamic responses are analysed in order to clarify the very complex and highly interacting nature of the dynamics of the Reactor - Regenerator system of the FCC unit.

Keywords: Digital modelling, Dynamic models, Open loop responses, Fluid-bed catalytic cracking

1. INTRODUCTION

1.1 Background

The production of gasoline by fluid-bed catalytic cracking (FCC) is one of the most common catalytic process in the world, hence, its modeling is an important area of study. In this paper two models are compared in order to predict the behavior of an industrial FCC reactor. The main objectives of the model development are the optimization of operating conditions, in terms of maximizing gasoline yield and the study of the dynamics of the plant.

López-Isunza (1992) developed a dynamic model for the FCC reactor-regenerator system. This model considers the reactor as a cylindrical transported bed unit (riser). Solid and gaseous phases are individually considered. A three lump kinetic scheme and Voorhies' deactivation relationship was used in order to predict the gasoline yield, the coke production and the catalyst activity. The regenerator is considered as a heterogeneous transported bed reactor with internal recycle. The kinetic scheme used provides information about oxygen and coke consumption as well as temperature profiles inside the regenerator. Although both reactors were considered as dynamic units, Lopez-Isunza concluded that the global dynamics are strongly influenced by the regenerator, while the reactor could be considered to be at a *pseudo*-steady state. This observation is in agreement with Ljungquist *et al.* (1992). Dynamic simulations were performed in order to follow the response to disturbances in feed temperature and recirculated amount of catalyst.

In this paper a second model is developed. The reactor is also considered as a heterogeneous riser. However, the model considers that it operates under a *pseudo*-steady state. In this case, the real reactor shape, slightly conical, is considered. A mass balance is used in order to predict the slit velocity at reactor entrance, after the instantaneous vaporization of the gasoil. This calculation is based on the observed void fraction (Schurnmans, 1980) and on the estimation of the gaseous density of gasoil. A five lump kinetic scheme is used. This scheme is able to predict light gasoil, volatile gases, coke and gasoline production. Molecular weights for each entity, with the exception of coke, were estimated and stoichiometric coefficients were assigned. The density of the reactant mixture is predicted along the riser and velocity and void fractions are estimated in order to follow the conservation of mass. Operating data were used in order to estimate the frequency factors and the activation energies. Catalyst deactivation and coke production are modeled using the global conversion of gasoil rather than a residence time function (Maya-Yescas and López-Isunza, 1993).

The regenerator is considered as a heterogeneous CSTR. The kinetic scheme proposed by Krishna and Parkin (1985) is used in order to predict oxygen and coke consumption and CO, CO_2 and H_2O production. Temperature of bubble and emulsion phases is predicted. It is known that there are some temperature gradients along the FCC regenerator. However, in order to accelerate the solution of the dynamic system a CSTR model is used. Furthermore, the global dynamics of the FCC depend on the regenerator inlet and outlet temperatures rather than on temperature profiles (Ljungquist *et al.*, 1992).

1.2 Riser model

The proposed kinetic scheme is a lumped scheme. A good revision of the development of lumped kinetic models can be found in Weekman (1979) and Sapre and Krambeck (1991). In the model presented here the lumps are defined by their boiling points. The following reaction paths are considered. Gasoil (gol, b.p. > 615 K) is supplied. This entity can crack into light gasoil (gli, 490K < b.p. < 615 K), gasoline (ga, b.p. = 490 K), volatile gases (gv) or coke (solid). Also, new gaseous compounds can crack to produce lighter ones or coke (eq. 1).

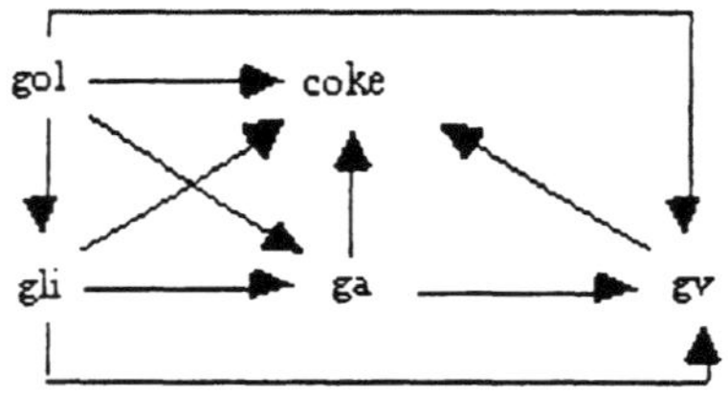

(1)

Second order kinetics have been proposed for the cracking of gasoil and first order kinetics for the other compounds (Decrooq, 1984). Reaction enthalpies were estimated using the stoichiometric relationship in order to follow the law of Hess.

In order to fit the molar generation rates, molecular weights for each lump were estimated and stoichiometric coefficients were assigned (Table 1). The molecular weight of the coke is unknown, hence a coefficient could not be assigned. However, the mass balance equations were used to determine its production rate.

Table 1. Stoichiometric coefficients

	gol	*gli*	*ga*	*gv*	*coke*
gol	1.0	2.2	4.1	10.6	n_{gol}
gli	--	1.0	1.9	4.9	n_{gli}
ga	--	--	1.0	2.6	n_{ga}
gv	--	--	--	1.0	n_{gv}

Catalyst deactivation is modeled as a function of the global gasoil conversion, in order to obtain a relationship which does not explicitly include the residence time of catalyst. However, a crackability factor, "α", is used (Wolf and Alfani, 1982) as shown in eq. 2:

$$\Phi|_z = \exp\left[\frac{\alpha(\mu_{hfo}|_z - 1)}{(\omega_{coq\infty} - \mu_{hfo}|_z)}\right] \qquad (2)$$

The riser model, given in equations 3 to 10, is derived from classical mass and energy balances. Plug flow was assumed and also that only catalytic reactions take place. For component A in the fluid phase:

$$\frac{\partial C_{A_g}}{\partial t} + u_g \frac{\partial C_{A_g}}{\partial z} + \frac{(1-\varepsilon_l)}{\varepsilon_l} k_g a_v (C_{A_g} - C_{A_p}) = 0 \qquad (3)$$

I.C. $C_{Ag}(t=0,z) = C_{Ag}^{O}$ (3.1)

$C_{Ag}(t,z=0) = C_{Ag}i$ (3.2)

and on the catalyst surface:

$$\frac{\partial C_{A_p}}{\partial t} + u_s \frac{\partial C_{A_p}}{\partial z} - k_g a_v (C_{A_g} - C_{A_p}) = \Phi R_A \qquad (4)$$

I.C. $C_{Ap}(t=0,z) = C_{Ap}^{O}$ (4.1)

$C_{Ap}(t,z=0) = C_{Ap}i$ (4.2)

$$m_{coke} = m_{gol}^{O} - \rho_g u_g A_t \varepsilon_l \qquad (5)$$

The energy balances yield, in the gaseous phase:

$$\frac{\partial T_g}{\partial t} + u_g \frac{\partial T_g}{\partial t} + \frac{(1-\varepsilon_l)}{\varepsilon_l} \frac{h_g a_v}{\rho_g C p_g} (T_g - T_p) = 0 \qquad (6)$$

I.C. $T_g(t=0,z) = T_g^{O}$ (6.1)

$T_g(t,z=0) = T_g i$ (6.2)

and in the catalyst particles:

$$\frac{\partial T_p}{\partial t} + u_s \frac{\partial T_p}{\partial t} - \frac{h_g a_v}{\rho_p C p_p} (T_g - T_p) = \Phi \sum_j \frac{(-\Delta H_r)_j R_j}{C p_p} \qquad (7)$$

I.C. $T_p(t=0,z) = T_p^{O}$ (7.1)

$T_p(t,z=0) = T_p i$ (7.2)

The void fraction, reactant mixture density and velocity were also calculated along the riser (eqs. 8-10).

$$\varepsilon_l = \frac{1}{\left(1+\dfrac{m_{cat}}{m_{gas}}\dfrac{\rho_g}{\delta\rho_p}\right)} \tag{8}$$

$$\rho_{gas} = \sum_j C_j PM_j;\ \ j = \text{gol, gli, ga, gv.} \tag{9}$$

$$u_g = \left(\frac{m_{gas}}{\rho_{gas} A_t \varepsilon_l}\right) \tag{10}$$

Kinetic parameters were estimated using data from three steady state operation points. Predicted results were in agreement with observation for an industrial unit (Fig. 1).

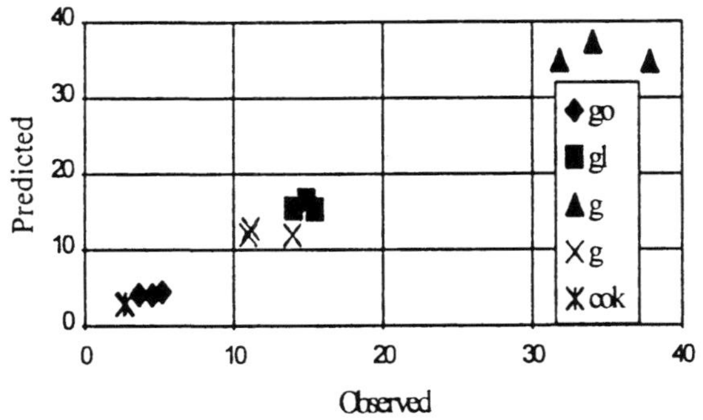

Fig. 1. Mass flows comparison at reactor outlet (kg/s)

1.3 Regenerator Model

The regenerator was considered as a heterogeneous continuous stirred tank reactor. Mass balances were performed for O_2, CO and coke in bubble and emulsion phases. Energy balances were also developed in both phases. The reaction paths proposed by Krishna and Parkin (1985) were used (eq. 11).

$$CH_{v(s)} + \left[\frac{\eta+2}{2(\eta+1)} + \frac{v}{4}\right] O_{2(g)} \rightarrow \frac{\eta}{\eta+1} CO_{(g)} + \frac{1}{\eta+1} CO_{2(g)} + \frac{v}{2} H_2O_{(g)} \tag{11}$$

This approach for the kinetics of coke burning has the advantage that it is possible to consider the ratio $\eta = CO/CO_2$ as a function of temperature. Previous authors give values of activation energies either for reactions or for η. This last activation energy is in agreement with the model proposed by Lee and coworkers (1989). A modification was introduced to the reaction rate expression. Coke concentration was not taken into account in an explicit way. Because of the nature of this reaction, a heterogeneous gas-solid one, when oxygen from air arrives at the solid surface it "finds" the same coke concentration during the entire period that the reaction takes place. Hence, the coke concentration may be assumed to have a constant value when determining the reaction rate. An independent mass balance was performed for coke in order to take into account its consumption. The reaction rate for coke burning was taken as first order in both reactants, coke and oxygen.

2. RESULTS AND DISCUSSION

Model simulations were performed for an industrial FCC unit which processes 40,000 Bbl gasoil/day. Characteristic residence times for gasoil and catalyst in the riser, at steady-state conditions, were 3.6 sec. and 4.1 sec., respectively. The catalyst residence time in the regenerator, also for steady-state conditions, was 10.8 sec. Responses for open loop mode dynamics were obtained for the following disturbances, introduced after steady state is reached, which is considered as zero time.

1) A 3% step increase of catalyst circulation rate after steady state is reached, and re-establishment of original operation conditions after 120 min.

2) A sinusoidal disturbance of feed temperature with an amplitude of 15°C and a period of 20 minutes, and re-establishment of original steady state conditions after 120 min.

3) A 2.5% step decrease in supplied gasoil and re-establishment of original steady state feed conditions after 60 min.

Figure 2 shows the response of the system when a perturbation to catalyst circulation rate is introduced. The yield of gasoline is closely correlated to riser temperature (Fig. 2a) for the kinetic parameters selected. Both curves exhibit a very similar shape. This does not depend on the perturbation since the same behavior is observed in all simulated cases (Figs. 3a and 4a). The yield of gasoline increases because of the increase in temperature. However, the global gasoil conversion decreases due to the more rapid formation of coke (Fig. 2c). Catalyst activity is decreased by coke formation. As well as the faster formation in the riser, the global temperature of the system is increased because of the increase of coke burning in the regenerator.

There is another important interrelation between both reactors. Instantaneous evaporation of the gasoil absorbs energy, which is supplied by the hot catalyst from regenerator. If the mass flow of catalyst increases, then there is more energy available for this process. Consequently, the temperature of the reactant mixture is higher. Also, in this case, the catalyst reaching the gasoil becomes warmer and warmer. It is important to note that recirculated coke from the regenerator to the reactor increases at the beginning of the perturbation. This response is due to increase in recirculated catalyst. However, coke concentration in the catalyst decreases following the effect of temperature. In other words, when the regenerator temperature increases coke burning also increases. Re-establishment of the original conditions shows how quenching of the reaction can be modeled. In this case, a quenching process is exhibited by the system due to the combined effect of a decrease in either riser inlet temperature and coke production. This is a commonly observed phenomenon in chemical reacting systems. The dynamic simulation settles to a new steady state once this effect dominates. Also, it is important to note that manipulation of recirculated amount of catalyst as a control variable is not ideal because it can cause inverse response, as shown in Figs. 2b and c, after 130 minutes (Trickett *et al.*, 1994).

The system response to a perturbation in feed temperature is shown in Fig. 3. The riser outlet temperature follows the perturbation with a delay of one residence time (Fig. 3a). However, the increase in temperature enhances coke production (Fig. 3c) and also increases riser temperature. Hence, the regenerator temperature experiences a 6 °C rise and after this never reaches another steady state, during the simulated time (Fig. 3b). Oscillations continue one hour after the original conditions were re-established. This response influences the global temperature of the system, and riser inlet temperature continues following the regenerator oscillations. In this case, the system again reaches a different steady state after the re-establishment of original conditions. The change in temperature does not produce an inverse response in global temperature nor in the yield of gasoline. However, its increase produces a decreasing effect in gasoil conversion due to the enhancement of coke production. Because the regenerator is not able to burn the coke that is produced, it supplies it to the riser, and conversion can not increase again, as modeled by Eqn. 2.

A response to a decrease in inlet mass flowrate is also shown (Fig. 4). Due to the increase in temperature of the reactant mixture, temperature and gasoil conversion increase (Figs 4a and 4c). The yield of gasoline increases also following the riser temperature increment. An increase in coke production occurs (Fig. 4c) due to the decrease in the gasoil supplied. However, overall coke production decreases. Although the amount of coke in the regenerator is also decreased, the regenerator temperature increases due to the riser temperature increment. After the perturbation is removed, the regenerator continues heating the system and the riser temperature does not decrease (Fig. 4a). Re-establishment of a normal amount of coke production is observed after some hours of simulation. In this particular case, the system reaches the original steady-state after original conditions are established (not shown).

In all of these cases some important characteristics of the FCC system are shown. There is a high degree of interaction between regenerator variables, which control the dynamics of the whole process, and the riser variables, which are the most closely related to commercial targets. Inverse response were also shown to occur.

3. CONCLUSIONS

A dynamic model has been developed to simulate the behavior of an industrial FCC unit. Simultaneously, a five lump kinetic scheme has been proposed for the catalytic cracking of gasoil. Changes in the number of moles, velocity and void fraction were incorporated into the riser reactor model. The catalyst deactivation rate was estimated using the overall conversion of gasoil. In order to take a more fundamental expression for coke burning, coke concentration was taken to be constant during the entire burning reaction. However, an independent mass balance for this solid entity was developed in order to ensure that the mass conservation is satisfied. Despite its inability to predict spatial profiles, a CSTR model for the regenerator is sufficiently good to predict the global behavior of the FCC unit. Nevertheless, it is necessary to use better models for the regenerator of this unit if the knowledge of spatial profiles is required.

Control of this system is difficult because of the high degree of interaction between both reactors. However, good control of the plant is necessary because its central position in a refinery ensures that substantial operational disturbances occur as part of its normal operation. Controllability analysis based on this model is underway in order to develop effective control schemes for the FCC plant.

4. ACKNOWLEDGMENTS

The authors thank CONACYT (National Council of Science and Technology, México) and British Council for the financial support and the Instituto Mexicano del Petróleo (IMP) and Petróleos Mexicanos (PEMEX) for data and useful discussions.

REFERENCES

Decrooq, D. (1984). Catalytic Cracking of Heavy Petroleum Fractions. Editions Technip. Paris.
Krishna, A. S. and E. S. Parkin (1985). Modelling the regenerator in fluid catalytic cracking units. *Chem. Eng. Prog.* April. 57-62.
Lee, L., S. Yu, and C.Cheng (1989). Fluidized-bed catalyst cracking regenerator modelling analysis.

The Chem. Eng. Journal. **40**. 71-82.

Ljungquist, D., S. Strand and J. G. Balchen (1992). Catalytic cracking models developed for predictive control purpouses. in 'Proceedings of DYCORD+ 92', College Park, Maryland, USA. Apr. 26-29.

López Isunza, F. (1992). Dynamic modelling of an industrial fluid catalytic cracking unit. *Comp. Chem. Eng.* **16** Suppl. S139-S148.

Maya-Yescas, R. and F. López-Isunza (1993). Un esquema cinético para el crackeo catalítico de gasóleos en un reactor de lecho transportado. *Avances en Ingeniería Química*. XIV. 71-78.

Sapre A.V. and Krambeck F.J. (1991). Chemical Reactions in Complex Mixtures - The Mobil Workshop. Van Nostrand Reinhold, New York.

Schurnmans, H. (1980). Measurements in a commercial catalytic cracking unit. *Ind. Eng. Chem. PDD*. **19**. 267-271.

Trickett, K. J., I. D. L. Bogle and D. R. Lewin (1994). Towards quantifying inverse response of nonlinear systems. in: Proc. 8th Int. Symp. on Process Systems Engineering. PSE 94. Kyongju, Korea.

Weekman, V. (1977) Lumps, models and kinetics in practice, *AICHE Monog. Series*, **75** (11), 1-28.

Wolf, E., and F. Alfani (1982) Catalyst deactivation by coking, *Cat. Rev.-Sci. Eng*, **24** (3), 329-371.

NOTATION

Latin variables:

A_t.- transversal area (m^2);

m.- mass flow (kg / s);

R_j.- j-th reaction rate (kmol / kg_{cat} s);

u.- velocity (m / s);

Greek Variables:

ε_l.- riser void fraction (m^3_{gas} / m^3_{total});

μ_{hfo}.- gasoil mass flows ratio;

Φ.- catalyst activity factor;

ρ.- density (kg / m^3);

ω.- mass conversion, related to supplied gasoil.

Subscripts:

g.- gaseous current;

p.- catalyst particle.

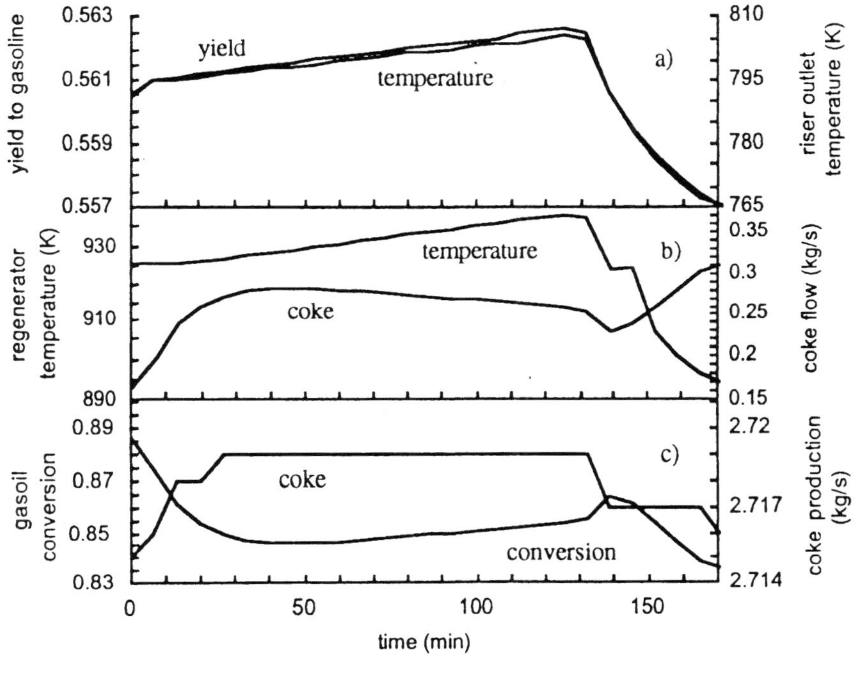

Fig 2. Perturbation in catalyst recirculation rate
a) Yield of gasoil to gasoline and riser temperature
b) Temperature and coke flow at regenerator outlet
c) Production of coke in riser and gasoil conversion

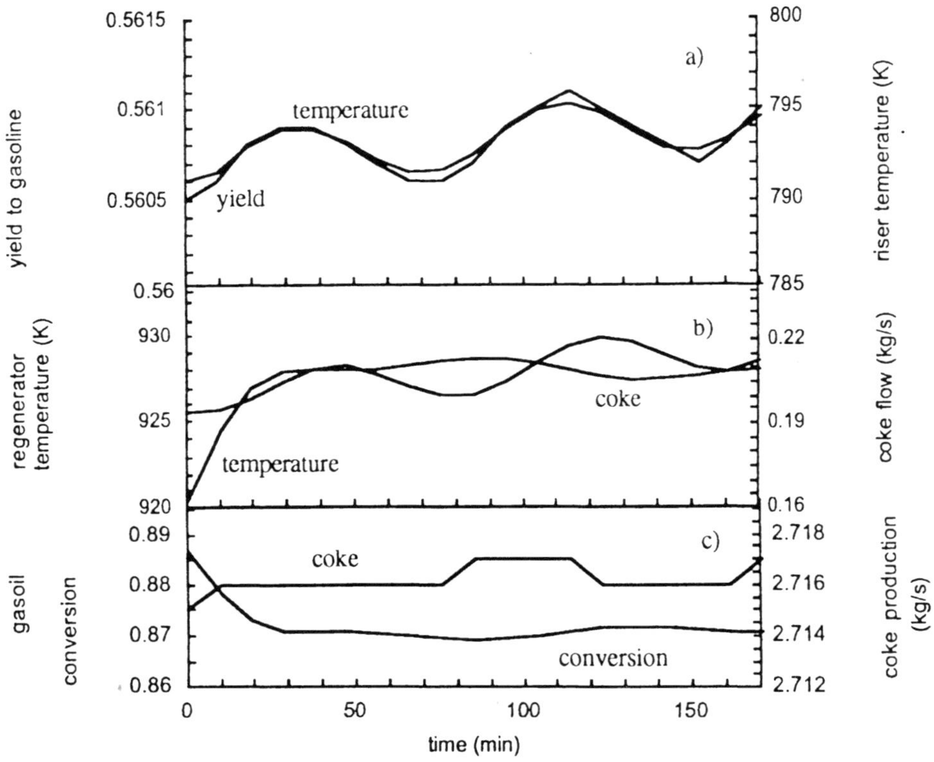

Fig 3. Perturbation in feed temperature
a) Yield of gasoil to gasoline and riser temperature
b) Temperature and coke flow at regenerator outlet
c) Production of coke in riser and gasoil conversion

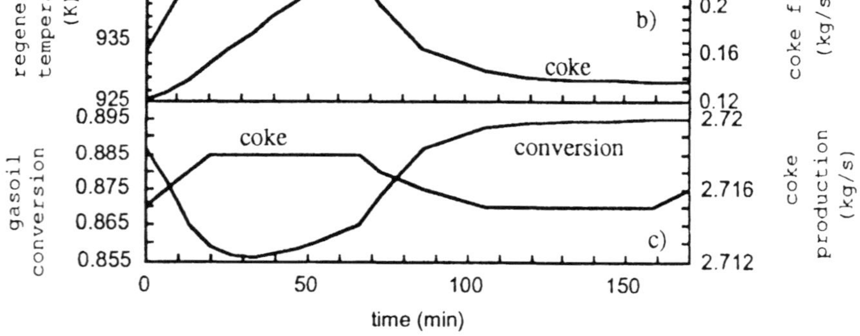

Fig 4. Perturbation in gasoil supply
a) Yield of gasoil to gasoline and riser temperature
b) Temperature and coke flow at regenerator outlet
c) Production of coke in riser and gasoil conversion

DYNAMICS OF THE JAROSITE CONVERSION PROCESS

Gillian Holcroft[1], Lucy Rosato[1], Michel Perrier[2], and Dimitri Berk[3]

[1]*Noranda Technology Centre*
240 Boul. Hymus
Pt. Claire, Qué., Canada

[2]*Depart. of Chem. Eng.*
École Polytechnique
Montréal, Qué., Canada

[3]*Depart. of Chem. Eng.*
McGill University
Montréal, Qué., Canada

Abstract: Noranda's Canadian Electrolytic Zinc plant (CEZinc) in Valleyfield, Québec utilizes the conventional Roast-Leach-Electrowin process to produce zinc metal. Iron removal is carried out in the jarosite conversion circuit which consists of ten continuously stirred tank reactors in series. The objective of this circuit is to maximize zinc ferrite extraction whilst minimizing iron in solution. Several studies have examined the kinetics of the individual reactions occurring in the jarosite conversion process. This work, however, represents the first attempt to model the dynamics of this process. In this study, the first five tanks of the jarosite conversion circuit were piloted and results (which include a total of 300 samples per experiment) were obtained from four process identification tests. The goal of this study was to develop and validate a dynamic model of the jarosite conversion circuit. The dynamic model, of the first two tanks of the jarosite conversion circuit, resulted in an excellent representation of the experimental data. Consequently, this dynamic model can provide a basis for both process control and optimization studies as well as give insight into the development of dynamic models for other heterogenous reaction systems.

Keywords: Dynamic modelling, Simulation, Transient analysis, Process models, Process identification

1. INTRODUCTION TO THE JAROSITE CONVERSION PROCESS

Over 80% of primary zinc is produced via the Roast-Leach-Electrowin (R.L.E.) process (Figure 1). Canadian Electrolytic Zinc (CEZinc) which utilizes the R.L.E. process, is the world's second largest producer of cathodic zinc. During concentrate roasting, zinc and iron sulfides are oxidized at 950°C to produce a calcine comprised of ≈75% zinc oxides (ZnO) and ≈24% zinc ferrites ($ZnO.Fe_2O_3$). Spent electrolyte, which is high in sulphuric acid, leaches the calcine under atmospheric conditions to produce a zinc sulphate solution. This solution is then electrowinned in the cell-house to produce zinc metal.

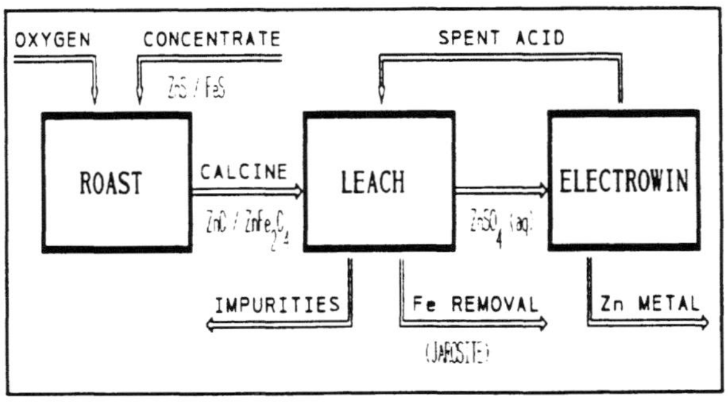

Fig. 1 R.L.E. process

The differences amongst R.L.E. plants lies mainly in the method used to remove iron from the zinc sulphate solutions. Iron removal is an essential step in electrowinning flow-sheets since a large portion of zinc must be extracted from the iron compound ($ZnO.Fe_2O_3$) which formed in the roaster. Furthermore, this extracted iron must be disposed of since it interferes with the zinc sulphate purification process and results in low cell house current efficiencies. The main methods of iron removal are named after the mineral that they generate, namely the Hematite, Goethite and the Jarosite process.

The Jarosite process was developed independently in Europe and Australia in the '60's. Normally the temperature is maintained between 95 and 100°C, a source of alkali is added, and jarosite seed is recycled to provide nucleation sites for jarosite crystal growth. The differences between the conventional jarosite process and the jarosite conversion process are in the addition of calcine in the former for acidity control and that the conversion process results in the simultaneous leaching of zinc ferrite solids and precipitation of jarosite solids. Although many electrolytic zinc plants have adopted the jarosite process, only Canadian Electrolytic Zinc (CEZinc) in Valleyfield, Québec and Outokumpu in Kokkola, Finland employ the jarosite conversion process. CEZinc which produces 230 000 tons per year of electrowinned zinc generates jarosite solids in the leaching section of its plant using ten continuous stirred tank reactors in series. The following are the three main reactions occurring in the conversion circuit, reaction 1 being the leaching of zinc ferrite and reactions 2 and 3 are the precipitation of ammonium and hydronium jarosite respectively.

$$ZnO.Fe_2O_3(s) + 8H^{+} \rightarrow Zn^{2+} + 2Fe^{3+} + 4H_2O(aq) \quad (1)$$

$$3Fe^{3+} + NH_4^{+} + 2SO_4^{2-} + 6H_2O(aq) \rightarrow NH_4[Fe_3(SO_4)_2(OH)_6](s) + 6H^{+} \quad (2)$$

$$3Fe^{3+} + 2SO_4^{2-} + 7H_2O(aq) \rightarrow H_3O[Fe_3(SO_4)_2(OH)_6](s) + 5H^{+} \quad (3)$$

The purpose of the jarosite conversion circuit is to extract the zinc from the zinc ferrite solids and to dispose of the iron as insoluble ammonium jarosite. In the CEZinc conversion circuit, zinc ferrites are leached with concentrated acid and spent electrolyte which bring both the iron and zinc into solution. Ammonia is added to the reactors, further down the train, to precipitate the soluble iron as ammonium jarosite. It is critical to control the exiting iron from this circuit since high iron in solution is detrimental to other areas of the process. The principal objective of this circuit is to maximize zinc ferrite extraction and to minimize the iron concentration in the liquid product. According to Dutrizac (1983), the major control parameters for the conversion process are acid concentration, which is the key variable, temperature, and the molar ratio of acid to iron. If the acid concentration is low, iron will readily precipitate, however, high acid concentrations are advantageous for leaching of zinc ferrites. Consequently, the optimization of the conversion process requires the determination of the most suitable acid level in each stage.

2. CONTROL OBJECTIVES AND PROCESS MODELLING

Control strategy formulation begins by defining the control objectives. In the case of the jarosite conversion circuit, the objective is to consistently maximize zinc extractions and minimize iron in solution. Although process identification studies can be used to generate a parametric model, when this "simplified" approach is used for complex processes, such as jarosite conversion, the closed-loop controller may result in poor performance. The reduction in controller performance occurs since it was tuned under one set of operating conditions without any regard for process non-linearities. Consequently, as the controller goes from one operating point to another the tuning parameters may result in sub-optimal controller performance. In the industry, it is often found that improperly tuned controllers have a destabilizing effect on a process.

An alternative approach to empirical modelling is to develop a mechanistic model of the process. This fundamental approach can sometimes be worth the effort for a variety of reasons. For example, the process model can be used as a basis for controller design. If advanced techniques are being considered, the model can be incorporated directly into the control algorithm. Finally, a process model can be part of a computer simulation for testing alternative control strategies and most importantly provides insight into the fundamental operation of the process.

3. EXPERIMENTAL METHODS

A pilot scale set-up, which simulates the first five tanks of the conversion circuit at CEZinc, was constructed, and used to carry out four process identification experiments. In each of these experiments, the process was operated under "normal" flow conditions for 12 hours. Following this initial steady-state period, the flow-rate of one of the feed streams was increased by a step change and the process response was recorded by taking successive samples from the effluent streams of each of the five reactors. Each sample was subsequently assayed for iron, zinc, lead, and copper. After seven hours of operating under "step change" conditions, the adjusted feed stream flow was returned to its original value, and the process response was once again recorded.

In the pilot set-up, zinc ferrite and jarosite seed slurries, which contain 23% and 33% solids respectively, were independently pumped to the first tank. Two acid streams, spent electrolyte and concentrated acid (raw), which have significantly different sulphuric acid concentrations (182 g/L compared to 1762 g/L for the raw acid) were also pumped to the first reactor. Ammonia was added to the third reactor, as ammonium hydroxide, with the objective of increasing the rate of ammonium jarosite precipitation. Since ammonium is present in the two slurry streams and in the spent acid stream, some jarosite precipitation can occur before the point of ammonium addition.

The reactors, which were constructed in 316L stainless steel, had an operating volume of 2.8 L, contained 4 baffles to enhance mixing, and a riser pipe near the over-flow exit to prevent short-circuiting. A residence time distribution test was carried out to evaluate the mixing characteristics of this tank design. In this study it was found that these reactors, with the risers intact, closely approximated ideal continuous stirred-tank reactors. The temperature was maintained at the desired set-point of 98 °C using temperature controllers.

4. PARAMETERS INVESTIGATED

Experiments were performed with the purpose of studying the transient process response following step changes in the main feed stream flows. This information would then be used to gain a better understanding of the process dynamics as well as help develop and validate the dynamic model. Although data were obtained for all five reactors, only the results from the first two reactors have been used in the modelling work to date.

Step changes of 50%, in the spent electrolyte and raw acid flows, were initiated in order to study the effect of increased acid concentration in the circuit with and without an effect on the reactors residence time (this is due to the fact that the raw acid's nominal flow is much lower compared to the other feed streams).

Although a 50% step change in the jarosite solids concentration results in an increase in the jarosite seeding, it subsequently results in a significant reduction in the reactor's residence time. Only a 25% step change in the zinc ferrite flow was studied in order to evaluate increased loading or production on the jarosite circuit. It was realized that a larger change would lie outside any plausible operating scenario at the zinc plant. In order to simplify the modelling of the jarosite circuit, it was decided not to study the effect of temperature.

5. DETERMINATION OF THE LEACHING REACTION RATE CONSTANT

For a constant volume CSTR with multiple reactions, the change in the number of moles of i with time is equal to zero at steady-state (equation 4). By combining this equation with the definition of the reaction extent we obtain the expression given by equation 5.

$$F_{EXIT_i} - F_{FEED_i} \equiv \sum_{J=1}^{J} \nu_{Ji}\, \varepsilon_J \tag{4}$$

$$r_J = \frac{\varepsilon_J}{V} \tag{5}$$

Although there are three main reactions occurring in the conversion circuit (equations 1,2,3), due to the high concentration of acid present in the first jarosite reactor, we expect a very small amount of jarosite precipitation to occur. With this assumption, the jarosite reaction extent can be neglected and the extent of the first reaction can be calculated from the experimental results by combining equations 1 and 4. The next step is to incorporate an appropriate rate expression into equation 5 and use the experimental reaction extent results to calculate a reaction rate constant.

Following a literature review of various developed rate expressions for zinc ferrite dissolution (Elgersma et al., 1992; Filippou and Demopoulos, 1992; Rastas and Fugleberg, 1979), it was decided to use a rate law of the form suggested by Elgersma (1992). In his study, Elgersma (1992) showed that the global reaction rate for zinc ferrite dissolution depends on the hydrogen ion activity and the ferric ion concentration as follows:

$$r_p^{\bullet} = k \sqrt{\frac{\hat{a}_{H^+}}{\hat{a}_{Fe^{3+}}}} \quad \left(\frac{mass}{time{\cdot}area}\right) \tag{6}$$

In order to simplify equation 6, the ionic activity can be written as the product of its activity coefficient and the ion molality. For electrolytes, the molality is assumed to be equivalent to the molar concentration; therefore the ionic activity can be expressed as: $\hat{a}_i = \gamma\, C_i$. Rewriting equation 6 in terms of molar concentrations, we obtain the following expression for zinc ferrite dissolution:

$$r_p^{\bullet} = k \sqrt{\gamma_{H^+}} \sqrt{\frac{[H^+]}{[Fe^{3+}]}} \quad \left(\frac{mass}{time{\cdot}area}\right) \tag{7}$$

where the species concentrations are given as mol/L. In order to use the above expression, it is necessary to calculate the hydrogen ion activity coefficient. The ionic strength of the solution in the first jarosite tank was estimated to be close to **12.3** giving an ionic activity coefficient of: $\gamma_{H+} = \mathbf{0.16}$. If we assume that the ionic strength of the solution is essentially constant in the first tank, then the activity coefficient can be incorporated into the rate expression by defining a new rate constant as; $k^{\bullet} = k \times (\gamma_{H+})^{1/2}$. Elgersma (1992) found that the rate constant, k, at 95°C, using synthetic zinc ferrite, was in the order of 1.74×10^{-2} g/(min.m^2). However further tests, by this researcher, using industrial zinc ferrite, had shown that the rate constant was half of the above value (8.70×10^{-3} g/min.m^2).

The rate constant per unit volume of the reaction mixture, k^+ can be obtained from:

$$k^+ = \frac{k^{\bullet}\, S\, C_s}{M_A} \tag{8}$$

Given that the specific area of zinc ferrite particles range from 2.00 to 2.67 m^2/g (Filippou and Demopoulos, 1992) and that the average zinc ferrite particle concentration in the pilot reactors was close to 55 g/L, the theoretical rate constant per unit volume for industrial zinc ferrite in these pilot reactors, k^+ ranges from: **1.6×10^{-3} to 2.1×10^{-3} mol/L.min**.

To calculate a rate constant, k^+_1, from the pilot data, we combine equations 5 and 7 to give:

$$k_1^+ = \frac{\varepsilon_1}{V_{TANK_1}} \sqrt{\frac{[Fe^{3+}]}{[H^+]}} \quad \left(\frac{mol}{min{\cdot}L}\right) \tag{9}$$

The four step change experiments each generated three steady state conditions which could be used to determine the average rate constant for zinc ferrite dissolution in the first reactor. This average rate constant was found to be equal to: **1.76×10^{-3}** +/- 0.28 mol/L.min which coincides with Elgersma's (1992) batch reactor findings.

6. DETERMINISTIC DYNAMIC MODEL

A mechanistic dynamic model is obtained from appropriate transient mass, component and energy balances for the system. In this study, since the reactor temperature was maintained constant at 98°C, the energy balance was not considered. Preliminary experiments confirmed that the reactors behave as ideal stirred tanks. Also, since the experiments showed that the slurry and the filtrate densities were found to vary in the 1st reactor by less than 3% during each process identification experiment, it was assumed that they remain constant in the model. Using the above assumptions, the overall mass and component balances can be simplified as follows:

$$Q_{EXIT} = \frac{\sum_{n=1}^{m} \rho_{FEED_n} Q_{FEED_n} - \dot{M}_{EVAP}}{\rho_{EXIT}} \qquad (10)$$

$$\frac{d(C_i)_{EXIT}}{dt} = \frac{\sum_{n=1}^{m} (C_i Q_n)_{FEED} - (C_i Q)_{EXIT}}{V} + \sum_{J=1}^{k} \nu_{iJ} r_J \qquad (11)$$

If we assume that the heterogeneous reactions behave as homogeneous reactions in an ideal stirred tank reactor, then we can write transient component mass balances (mass/L.min) for the solids as follows:

$$\frac{d(C_{s_i})_{EXIT}}{dt} = \frac{\sum_{n=1}^{m} (s_{i_n} s_n \rho_n Q_n)_{FEED} - (C_{s_i} Q)_{EXIT}}{V} + \sum_{j=1}^{k} \nu_{ij} r_j M_i \qquad (12)$$

To simplify the jarosite model, only the main reactions which occur in the conversion circuit were considered. Studies on jarosite precipitation (Dutrizac, 1987; Qian-kun and Rong-jun, 1985) have shown that when operating the jarosite conversion process close to 100°C approximately 25% of the jarosite will be precipitated in the hydronium form and the remainder will be as ammonium jarosite. For modelling purposes, the two jarosite reactions are combined so that there is 75% ammonium and 25% hydronium jarosite formation. This combined jarosite reaction is as follows:

$$\begin{aligned} 15Fe^{3+} + 4NH_4^{+} + 10\ SO_4^{2-} + 31\ H_2O(aq) \rightarrow \\ 29H^{+} + 4NH_4[Fe_3(SO_4)_2(OH)_6](s) \\ + H_3O[Fe_3(SO_4)_2(OH)_6](s) \end{aligned} \qquad (13)$$

Preliminary simulations showed that Elgersma's (1992) expression provided a better representation of the experimental data since it incorporated the "retarding" effect of ferric ions during zinc ferrite dissolution (equation 7). This expression was incorporated into the dynamic model using a rate constant of 1.63 x10^{-3} mol/L.min.

Although two applicable jarosite precipitation kinetic expressions were found in the literature, it was decided to use the rate of iron precipitation as jarosite given by Rastas and Fugleberg (1979) since the form of their equation was easier to work with. When working with industrial CEZinc Jarosite in 1980, they found that the rate of iron removal could be modelled using the following constants for α, β, γ, and δ:

$$-\frac{d[Fe^{3+}]}{dt} = k(T) \times [Fe^{3+}]^{2.406} [H_2SO_4]^{-3.719} [NH_4^{+}]^{1.06} [NH_4\text{-}JAR]^{0.685} \qquad (14)$$

It is important to note that equation 14 is not a rate law but a design equation for a constant volume batch reactor and that the jarosite seed concentration, which refers to the mass of dry jarosite seed per volume of slurry, has been incorporated into the equation. At 98°C the rate constant was found by Rastas and Fugleberg (1979) to be equal to 1.657x10^{-2} (g/L)$^{0.568}$ per min. This rate constant was manipulated, as shown in equation 15, for dynamic modelling purposes. Equation 13 was used as a basis for the ferric iron stoichiometry. In doing this we obtain a corrected rate constant of 1.124 x10^{-4} (mol/L)$^{0.568}$ per min.

$$k = \frac{k^{*}}{-15} \left(\frac{1}{M_{Fe}} \right)^{0.568} \qquad (15)$$

The transient mass balances for the 1st reactor were written using the stream numbering shown in Figure 2. Equation 10 was used to calculate the tank's exit flow. Equation 11 was written for H^{+}, Zn^{2+}, and Fe^{3+} components and equation 12 for zinc ferrite and

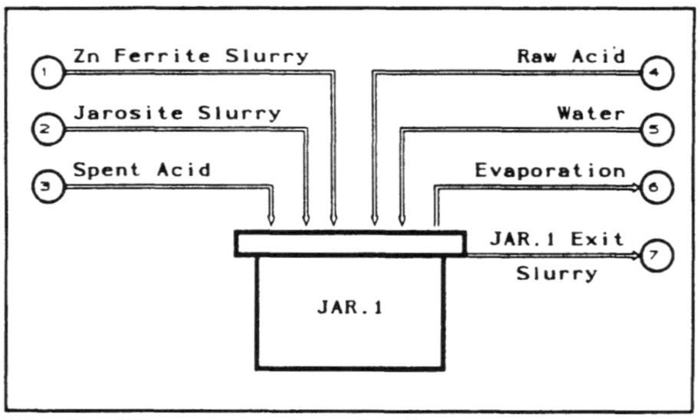

Fig. 2 First conversion circuit reactor

jarosite solids using the stoichiometry of reactions 1, 2 and 3. More details on the equations used in the model are available (Holcroft 1994).

The mass fraction of zinc in the solids was calculated by summing the product of: the solid components multiplied by their known elemental percentage of zinc. This total mass concentration of zinc is then divided by the mass concentration of solids present at the same instant in time. An example of this calculation for the percent zinc in solids is shown below. The same procedure applies for the calculation of the percent iron in the solids.

$$\%Zn\,Solids = \frac{C_{s_{A_7}} \%Zn_A + C_{s_{H_7}} \%Zn_H}{S_7} x\, 100\% \qquad (16)$$

7. SENSITIVITY ANALYSIS

The transient component balance equations for sulphuric acid, zinc and iron in solution and for zinc and iron in the solids were solved using the MATLAB® software package with a 4th and 5th order Runge-Kutta numerical integration technique. In this study, the sensitivity of the rate constants for zinc ferrite dissolution and jarosite precipitation were explored. The purpose of this exercise was to see if an improved representation of the process identification experimental data was possible. In the figures which follow, the vertical lines drawn on each graph represent the step up and the step down times for each experiment.

For the first jarosite tank, it was found that the best model involved using a rate constant of $k_1=1.63\times10^{-3}$ mol/L.min for zinc ferrite leaching and $k_2=1.124\times10^{-4}$ $(mol/L)^{0.568}$ per min for jarosite precipitation. Some of the best representations of the experimental data are seen when we compare the model's predictions with the percent zinc in the solids data. One of these results is shown in Figure 3 which is the jarosite step change experiment. A large drop in the percent zinc following the jarosite slurry step change experiment is due to the diluting effect of the jarosite solids which contain a low percentage of zinc. The percent zinc also drops during the spent and raw acid step change experiments due to increased zinc ferrite leaching which resulted from the larger acid inputs.

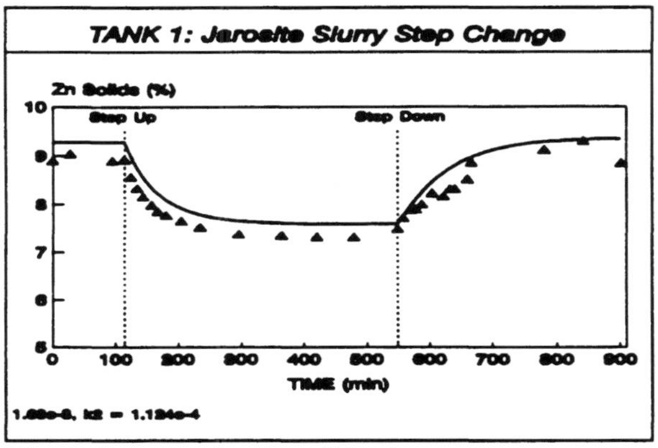

Fig. 3 Percent zinc in solids changes

Figure 4 demonstrates the changes in the ferric iron concentration after a raw acid step change.

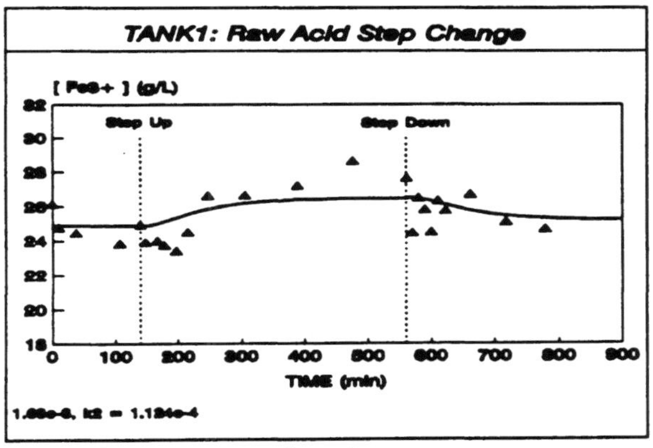

Fig. 4 Iron concentration changes

The degree of jarosite precipitation in the first reactor was studied knowing that the jarosite kinetics are reduced under high acidities. Modelling results have shown that no significance difference resulted in the transient responses when the jarosite precipitation reaction was ignored. This implies that there is negligible jarosite precipitation in the first reactor.

In the second tank, the best modelling results were obtained when the zinc ferrite rate constant was reduced from 1.63×10^{-3} to 0.85×10^{-3} (mol/L.min). This was necessary since the rate constant for the first tank was originally calculated with the assumption that the zinc ferrite solids concentration was 55 g/L. In the second tank, the zinc ferrite solids concentration drops due to continued zinc ferrite dissolution. Unfortunately, the experimental results cannot be used to calculate the zinc ferrite solids concentration in the second reactor since jarosite precipitation is occurring. As the best modelling results were realized for a rate constant of 0.85×10^{-3} (mol/L.min), we can propose that the zinc ferrite solids concentration may have been reduced from 55 g/L to 29 g/L. The fact that the rate constant changes as a function of the solids concentration demonstrates the limitations of modelling heterogeneous reactions as pseudo-homogeneous reactions.

Due to the uncertainty in the estimation of the reaction rate constant in the second reactor, it was decided to examine its sensitivity. Results from this study are shown in Figures 5 and 6. In these figures we see results for cases 1,2 and 3. Case 2 uses the same rate constants in reactor 2 as those employed in the first reactor. Cases 1 and 3 involves changing these nominal values by reducing the zinc ferrite rate constant by 50% (case 1) or by assuming no jarosite precipitation is occurring (case 3). From these figures we see that the best representation of the experimental data occurs using case 1 parameters.

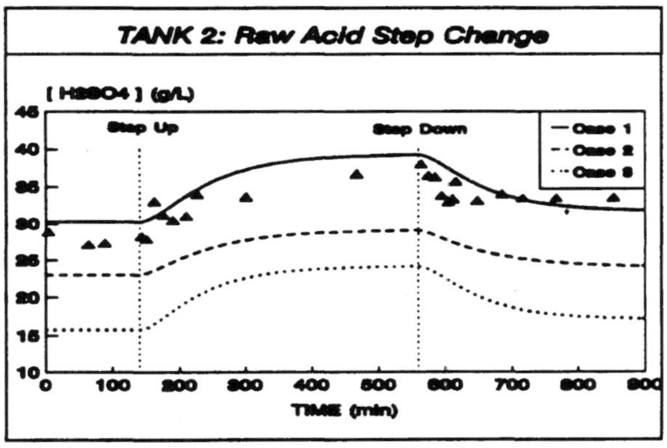

Fig. 5 Acid concentration sensitivity

The importance of the jarosite precipitation reaction in the second tank is obvious from these figures since the model is no longer representative of the data (case 3) when this reaction is omitted.

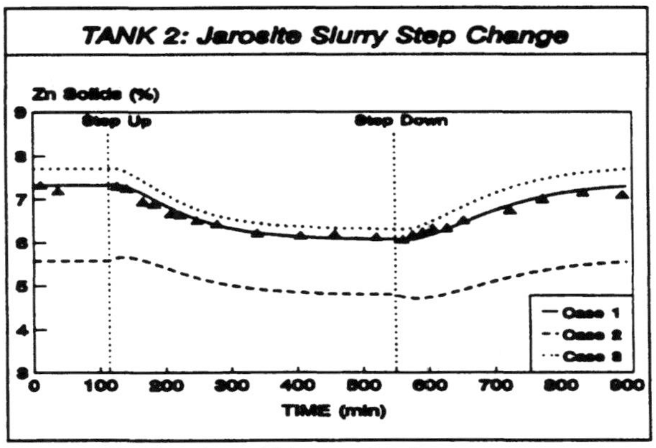

Fig. 6 % Zinc in solids sensitivity

8. CONCLUSIONS

The first five tanks of the jarosite conversion circuit were successfully piloted and results were obtained from four process identification experiments. These data showed the transient effects that the major feed streams had on the reactor performance. Although data was collected from all five reactors, the scope of this paper was limited to analyzing the results from the first two reactors. The data from the subsequent three reactors will be used in the future to extend the dynamic model to incorporate all five reactors.

The data from these pilot experiments gave mass balance closures within ± 12% for both zinc and iron. The transient responses from the process identification experiments were used to validate the dynamic conversion circuit model. Following the development of the dynamic model, case studies confirmed that the available kinetic data could be used to model the jarosite circuit's acid, and iron concentrations in solution as well as the mass fraction of zinc in the solids. Problems with the analysis of iron in the solid phase resulted in model validation difficulties. The zinc ferrite rate constant of 1.76×10^{-3} mol/L.min, which was calculated from the continuous pilot reactor data, was found to agree with the literature value obtained using a batch reactor.

The results from the first tank have shown that the parameter that has the largest impact on the degree of zinc ferrite dissolution is the flow-rate of the raw acid stream. This was observed after changes in this stream's flow did not impact the reactor's residence time. The stream which was found to have the largest effect on the acidity in the first reactor resulted from changes in spent acid flow-rate. Finally the simulation results helped to highlight the initiation of jarosite precipitation in this circuit. This reaction was found to be negligible in the first reactor, where the acidity is relatively high, and was shown to be important in the second tank.

Future work will involve extending the jarosite conversion circuit model to the first five jarosite reactors. Eventually, further work will be undertaken to extend this model to the entire series of jarosite reactors at CEZinc (10 tanks). This dynamic model will be used for optimization studies and will be incorporated into a non-linear control strategy for maintaining the acid and iron at their optimal levels in these reactors.

ACKNOWLEDGEMENTS

The authors would like to thank CEZinc management for supporting this work and for granting us permission to write this paper.

REFERENCES

Dutrizac, J.E., (1983), Jarosite Type Compounds and their Applications in the Metallurgical Industry, Hydrometallurgy Research, Development and Plant Practice, AIME Conference Proceedings.

Dutrizac, J.E., (1987), An Overview of Iron Precipitation in Hydrometallurgy, International Symposium on Crystallization and Precipitation.

Elgersma, F. et al, (1992), Acidic Dissolution of Zinc Ferrite, Hydrometallurgy (29),p173-189.

Filippou, D. and Demopoulos, G., (1992), A Reaction Kinetic Model for the Leaching of Industrial Zinc Ferrite Particulates in Sulphuric Acid Media, Canadian Metallurgical Quarterly, Vol.31, No.1, pp. 41-54.

Holcroft, G., (1994), Dynamics of the Jarosite Conversion Process, M.Eng Thesis, McGill.

Qian-kun, W. and Rong-jun, Ma.,(1985), The Jarosite Process --Kinetic Study, Zinc '85, Chap. 42.

Rastas, J. and Fugleberg, (1979), Kinetics of the Ferrite Leaching and Jarosite Precipitation, Erzmetall, Bd, 32 H.3

NOMENCLATURE

a_i	activity of ionic species i
C_i	molar concentration of species i
C_s	solid component conc. (mass/volume)
F	molar flow rate
k	rate constant
M_i	molecular weight of solid component i
M	mass flow rate
Q	volumetric flow rate
r	reaction rate (mol/time.volume)
r_p	global rate (mol/time.area)
s_i	mass fraction of solid
s_n	mass fraction of solids in stream n
S	Specific area of ZnFerrite particles, m^2/g
t	time
V	reactor volume
$\%Zn_i$	% Zn in species i
M_i	molecular weight of solid component i
υ_i	stoichiometric coefficient of comp i
ρ	density

MODEL-BASED TEMPERATURE CONTROL OF FED-BATCH REACTORS

F. Szeifert, L. Nagy and T. Chovan

Department of Chemical Engineering Cybernetics, University of Veszprem, Veszprem, POB. 158, H-8201 Hungary

Abstract: The temperature control of an autoclave with a cooling jacket is studied. Since the heat removal is limited the temperature is controlled by the feed rate of one of the reactants and the full cooling capacity is applied in the safe region of operation. Approaching the runaway region the reactant feed is stopped and the flow rate of the cooling liquid is manipulated. The algorithm is based on a tendency model derived from a first principle one. It includes the system variables (manipulated, measured, controlled, etc.) and parameters (a priori ones and others estimated from the process measurements).

Keywords: Model-based control; Adaptive algorithms; Predictive control; Parameter estimation; Physical models; Batch control.

1. INTRODUCTION

It is a well-known fact that batch processes are gaining wider ground in chemical industries (Yamatomo and Hashimoto, 1991). This course is motivated first of all by the extraordinary flexibility of batch processes which allows the quick adaptation to market demand and it is made possible by the rapid development of computer process control (Fisher, 1990). Compared to continuos processes the control of batch processes is more difficult (complex sequential control tasks, operating conditions change with time, several different operations are conducted in the same process unit, etc.).

The typical unit of batch systems in polymer, pharmaceutical and fine chemical industries is an autoclave with a heating-cooling jacket. Heating and cooling are generally provided by an alternative heating-cooling system which uses directly cold and hot utility fluids available in the plant at given temperatures. The temperature control is realised by choosing the type of the fluid first and then manipulating the flow rate of this fluid (Rafalimana *et al.*, 1992).

The reaction rate generally shows significant inhomogenities with time in batch chemical reactors (the driving force of spontaneous processes decreases exponentially with time). The heating and cooling capacity is designed for the maximum driving force; therefore it is only partially utilised in most of the processing time. This utilisation can be improved by the application of fed-batch techniques. The basic idea is that the net reaction rate is kept constant by controlling the feed rate of one of the reactants. This way the "operating" point can be set to the maximum of the heating or cooling capacity (leaving an adequate safety region). Consequently the productivity can be increased considerably. At the same time applying this technique difficulties may arise, namely the reactant can be overdosed and reactor runaway may occur which can even cause an explosion.

Different approaches to these problems are discussed by Juba and Hamer (1986) as well as by Rafalimana *et al.* (1992). Modifications of a common PID controller for use in batch systems (e.g. dual mode controller) are presented by Liptak (1986). Davidson (1989) implements the skilled operator's "activities" calculating the integrate term with a modification of Corripio's reset-feedback position algorithm (Intelligent Temperature Controller). The model-predictive adaptive algorithms are in contrast to the intuitive modifications of the usual PID. The robustness and flexibility of the adaptive GPC algorithm (generalised predictive control, Clarke *et al.* (1987)) was demonstrated experimentally for temperature control of batch and fed-batch reactors by Rafalimana *et al.* (1992). On the other hand, Jarupintusophon *et al.* (1994) showed that in spite of the adaptive and predictive features the temperature overshoot cannot be avoided at the end of the heat-up. To eliminate this problem a supervisory control level based on a realistic thermal model was introduced and the relationship between the physical model and the black box model (ARMA) used in the GPC algorithm was established. Then the enthalpy flow in the physical model was estimated using the parameters obtained by the on-line identification of the black box model. In an earlier work Jarupintusophon *et al.* (1991) used an a priori model too to solve the "bain-marie" problem.

The hardware and software solutions of advanced process control systems satisfying different standards (SP88, NAMUR) provide exceptional flexibility for the user. This allows that first principle models, which are used in process design or maybe at the supervisory control level, be applied in the design of lower level controllers. The direct application of first principle models in the control algorithm brings a few problems:

- first principle models are too complex,
- many a priori parameters are involved,
- applying it to a certain problem, the model error can be significant even in case of an adequate model and accurate parameters,
- control algorithms based on the inverse of the process model are very sensitive to model error.

Problems coming from the complexity of first principle models are alleviated by the application of the tendency model concept (Filippi-Bossy *et al.*, 1989). The tendency model is a grey box model based on the detailed first principle model and it takes into account the limitations in sources of information. The idea is that it is not necessary to know everything about the system, however everything what is surely known must be used!

The main points of the strategy are the following (Filippi-Bossy *et al.*, 1989):

- Use all available process information (all knowledge about the process and the accurate process measurements of the previous batches)
- Optimise (and control) the process in an evolutionary approach

* time dependent concentration data for the reactants and products is usually not available,
* only an approximate reaction mechanism and kinetic model are known,
* the initial concentration in the reactor is given,
* the flow rates of the feed reactants are given or measured,
* the relevant final concentrations in the product mixture are measured or estimated,
* the reaction temperature is measured continuously,
* the heat transfer coefficients are given or estimated hence the enthalpy balance can be accurately evaluated.

In the paper the temperature control of a fed-batch reactor based on a tendency model is discussed in details.

2. TENDENCY MODEL OF THE PROCESS

The scheme of the studied fed-batch reactor with an alternative heating-cooling system is shown on Fig. 1. The rigorous model of the system and a study of the process dynamics were presented in another work (Szeifert *et al.*, 1995). The optimal operation of the reactor is a complex problem (Filippi-Bossy *et al.*, 1989). The technique when the temperature profile determined on a higher hierarchy level is maintained by manipulating the feed rate of one of the reactants (valve V1) is interpreted here as a subproblem of the complex one. Apart from a certain safety reserve the whole cooling capacity is utilised by this method. During the fed-batch operation a net chemical reaction is assumed and the following restrictions apply on the rate of reaction:

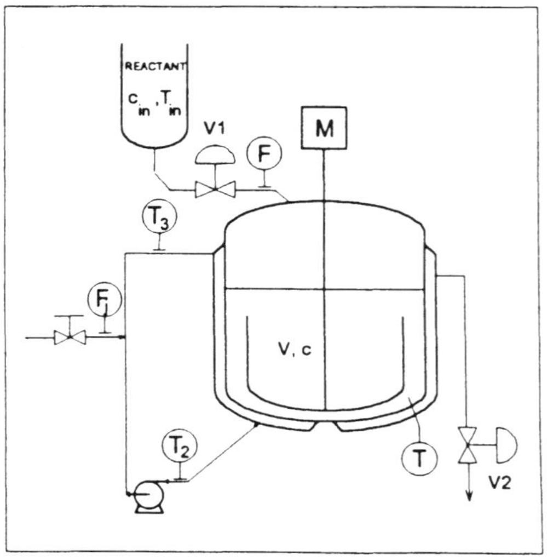

Fig. 1. The scheme of the reactor system

- the rate of reaction increase monotonously with the concentration of the feed component: $\partial r / \partial c > 0$ (it can reach zero only at the end of the feeding)
- an Arrhenius type temperature function is valid in case of exothermic reactions: $\partial r / \partial T = r\, E / (RT^2)$ (quasi irreversible reaction).

The tendency model includes the volume, component and enthalpy balances for the reaction mixture and the enthalpy balance for the jacket:

$$\frac{dV}{dt} = F \tag{1}$$

$$\frac{d(Vc)}{dt} = Fc_{in} - V\,r \tag{2}$$

$$V \rho c_p \frac{dT}{dt} = F \rho c_p (T_{in} - T) + (-\Delta H_r) V r - (UA)(T - T_J) \tag{3}$$

$$\left(V \rho c_p\right)_J \frac{dT_J}{dt} = F_J \left(\rho c_p\right)_J \left(T_{J,in} - T_3\right) + (UA)(T - T_J) \tag{4}$$

where $T_J = T - \dfrac{T_3 - T_2}{\ln\left[(T_1 - T_2)/(T_1 - T_3)\right]}$ is the average temperature in the jacket and t=0, V, c, T, T_J are given (initial values).

The ρc_p, $(\rho c_p)_J$, $(-\Delta Hr)$ thermodynamic properties used in the tendency model (1-4) are given a priori and assumed to be constant in time. The inaccuracy caused by this assumption is compensated by the on-line estimation of two further parameters. Function $V(t)$ can be obtained from the measurement of F by integrating eq. (1). Hence only the value of (UA) is unknown in eq. (4). It can be determined in several ways:

- least square estimation over the $[0,t]$ time horizon (appropriate if (UA) can be regarded as constant; a forgetting factor ($\lambda < 1$) is used if it changes)
- applying a suitable filtering on the measured variables it can be calculated directly from eq. (4).

If the $(UA)(t)$ function is known the $Q_R(t) = (-\Delta H_r) V r$ enthalpy flow can be also estimated on-line using eq. (3) and filtering the measured variables involved in the balance. Then the current reaction rate, $r = Q_R / \left(-\Delta H_r V\right)$ can be calculated and it can be used for the estimation of the reactant concentration by the integration of eq. (2).

Following the above algorithm and using the process measurements shown on Fig. 1 the state variables of the (1-4) tendency model can be calculated at every moment.

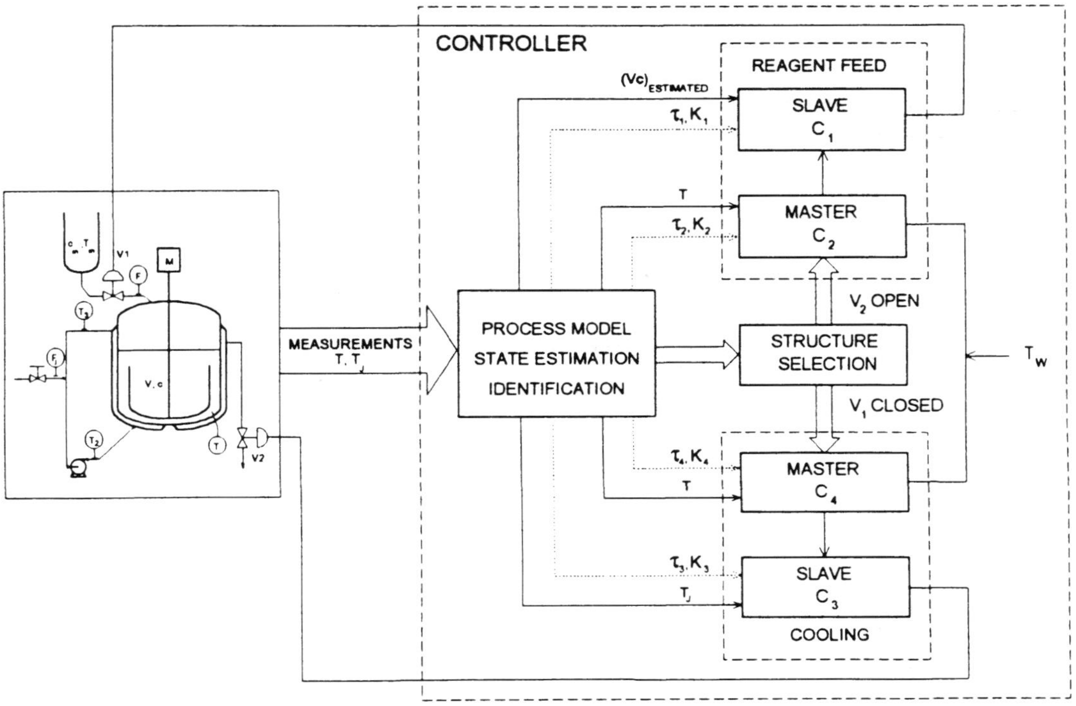

Fig. 2. The Control Scheme

3. MODEL-BASED CONTROL ALGORITHM

The suggested structure of the controller is given on Fig. 2 (see also Szeifert *et al.*, 1995). In case of exothermic reactions the controller is more complex due to the risk of reactor runaway. The relevant criterion serving as a basis for the structure selection is the following (Szeifert *et al.*, 1995):

$$c \le \frac{\dfrac{(UA)}{V\rho c_p} + \left(\dfrac{r}{c}\right)}{\dfrac{(-\Delta H_r)}{\rho c_p} \cdot \left(\dfrac{r}{c}\right) \cdot \dfrac{E}{RT^2}}, \qquad (5)$$

where E/R is an a priori parameter.

If condition (5) is satisfied by the estimated variables then the reactor operates in the "safe region" and the reactant feed is manipulated while using the total cooling capacity. In the opposite case the reactant feed is stopped and the temperature is controlled by manipulating the flow rate of the cooling fluid. The risk of reactor runaway is not present in case of endothermic reactions.

All of the C_1, ... , C_4 controllers used in the scheme are first order, so called predictor-corrector controllers (PCC) (Szeifert and Vass, 1988; Szeifert *et al.*, 1992). For a first order PCC algorithm the process can be given in the following form:

$$\tau \frac{dy}{dt} + y = Ku, \qquad (6)$$

where u and y are the manipulated and the controlled variable respectively while τ and K are parameters or can be functions depending on the process (non linear process).

The τ and K functions can be calculated from the tendency model of the process. In case of constant input, assuming that K and τ do not change (significantly) for a given period of time, the output can be given in analytical form:

$$y(t) = e^{-\frac{t}{\tau}} y(0) + \left(1 - e^{-\frac{t}{\tau}}\right) K u \qquad (7)$$

Applying the discrete time form of eq. (7), where k stands for the discrete time, the model accuracy can be checked, i.e. the model error can be calculated:

$$\varepsilon_M = y_k - y_{k,m} = \\ = y_k - \left[e^{-\frac{T_0}{\tau}} y_{k-1} + \left(1 - e^{-\frac{T_0}{\tau}}\right) K u_{k-1} \right] \qquad (8)$$

where T_0 is the sampling time.

Based on the model error the (7) model can be corrected in the next step (*correction*):

$$y(t) = e^{-\frac{t}{\tau}} y(0) + \left(1 - e^{-\frac{t}{\tau}}\right) K u + \varepsilon_M \qquad (9)$$

This model can be used for the prediction over the $t \ge kT_0$ time horizon. Different u control actions can result in different predicted output functions $\left(y(t), \; / t > kT_0\right)$. The following behaviour of the controller is required (control rule):

the output must reach the setpoint (w) at time $t = \lambda \tau$ taken from the present moment!

where λ is the tuning parameter of the controller. Applying the above requirement the following relationship can be obtained for the manipulated variable:

$$u_k = \frac{w_{k+1} - \varepsilon_M - e^{-\lambda} y_k}{K\left(1 - e^{-\lambda}\right)} \qquad (10)$$

For the design of the C_1 controller the (2) component balance is taken as a basis. Let us assume that a quasi first order net reaction takes place, then eq. (2) can be rearranged as follows:

$$\frac{1}{(r/c)} \cdot \frac{d(Vc)}{dt} + Vc = \frac{c_{in}}{(r/c)} \cdot F \qquad (11)$$

The corresponding substitutions for a first order PCC controller are the following:

$$y = Vc, \; u = F, \; \tau = \frac{c}{r}, \; K = \frac{c_{in}}{c} \cdot \frac{1}{r}. \qquad (12)$$

Giving an adaptive character to algorithm (10) the τ and K functions are determined on-line using eq. (12) and the algorithm presented in Section 2.

The C_2 controller can be designed on the basis of the (3) enthalpy balance and takes the form given below:

$$y = T,$$

$$u = \frac{(-\Delta H_r)\left(\dfrac{r}{c}\right) \cdot (Vc) + F\rho c_p T_{in} + (UA) T_J}{F\rho c_p + (UA)},$$

$$\tau = \frac{V\rho c_p}{F\rho c_p + (UA)}, \quad K = 1 \qquad (13)$$

The design procedures for C_3 and C_4 are similar.

4. SIMULATION AND PHYSICAL STUDY OF THE ALGORITHM

The dynamic behavior of the system shown on Fig. 1 was studied by simulation and its characteristics which are important from control point of view (complex dynamic behavior, risk of runaway, etc.) were explored. Simulation and laboratory experiments were conducted to test the control algorithm presented in Section 3. Parts of the results were presented by Szeifert *et al.* (1995).

In case of slower reactions the difficulties of the control problem increase. In case of fast reactions - ignoring the dynamics of mixing - all methods for controller design provide good results. A comparison between the suggested algorithm and a PID controller is shown on Fig. 3 for the case of a slow reaction. The PID parameters were tuned according to ISE integral criterion and applying the Nelder-Mead search method. Relatively small absolute value for the control error is achieved while the manipulated variable shows some characteristic swings. Whereas the results of the model-based algorithm show favourable control performance. From the point of view of controller operation the initial state of the system is very important (role of initialisation). At the beginning of the process the values of estimated parameters are very "imprecise" and the initial control actions are very unreliable.

The effect of the initial state is demonstrated on Fig. 4. The reactor is started from a temperature below the setpoint and from another one over the setpoint (the rate of reaction is twice as much as on Fig. 3). In the first case while the controller has only an "initial knowledge" it has to act at once since the control error is positive that consequently results in a poor control performance. Therefore it is practical to start the system at a higher temperature or to include an initial "learning" period in the control scheme (and maybe to feed forward the parameters from previous batches).

A priori parameters (ρc_p, $(\rho c_p)_J$, $(-\Delta H_r)$, E) "built" into the controller and used in the above experiments are exactly the same as in the process simulation. (However a model error is inherently present in the previous calculations since the detailed model of the system on Fig. 1 is used for process simulation while tendency model (1-4) is applied in the controller design.) The adaptive characteristics of the controller and the robustness towards the accuracy of a priori parameters are illustrated by Fig. 5. The rate of reaction was increased 10 times while the parameters of the controller were kept unchanged (Fig 5a). On Fig 5b the values of a priori parameters were changed by 50 % by such a way that their effects are superimposed furthermore dead time and white noise were added to the measured signals. The good adaptivity and robustness of the model-based algorithm were justified in further wide-ranging simulation studies. The same results are obtained in physical tests conducted in a laboratory reactor of 5 l volume feeding H_2O_2 into $Na_2S_2O_3$ (oxidation) (Szeifert *et al.*, 1995) as well as feeding HCl into NaOH (acid-base reaction). The results of two experiments of this latter reaction are presented on Fig. 6. Apart from the initial period where the values of the estimated parameters are unreliable the temperature control is excellent. It is also noticeable that applying a tracking setpoint the heat-up to a given temperature can be realized. The industrial application of the method is being prepared.

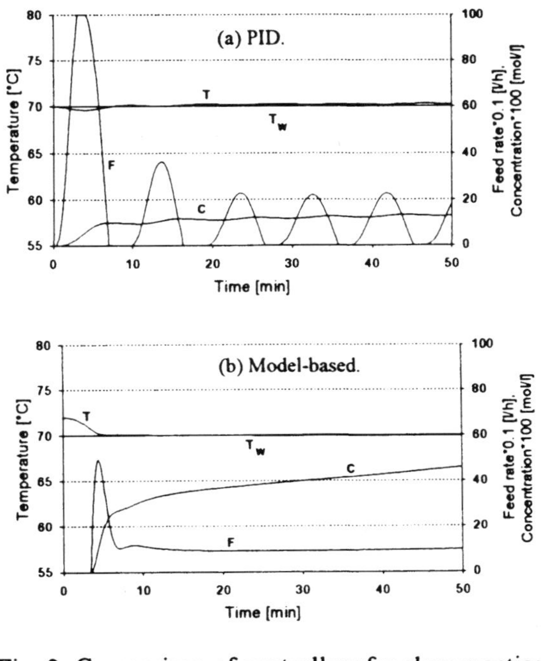

Fig. 3. Comparison of controllers for slow reaction.

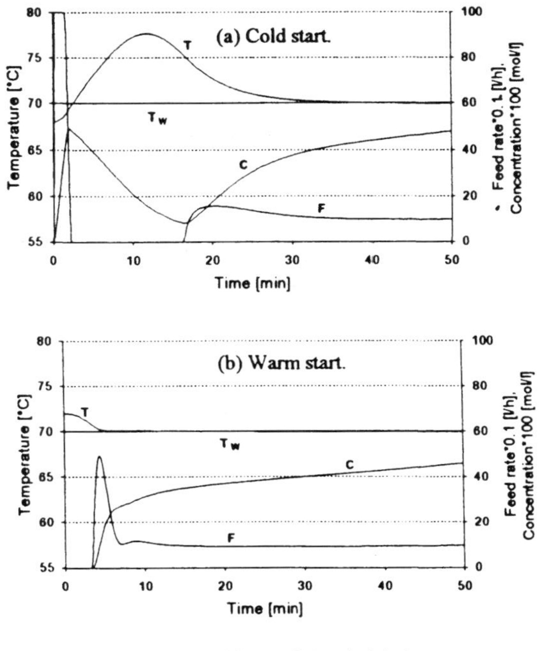

Fig. 4. The effect of the initial state.

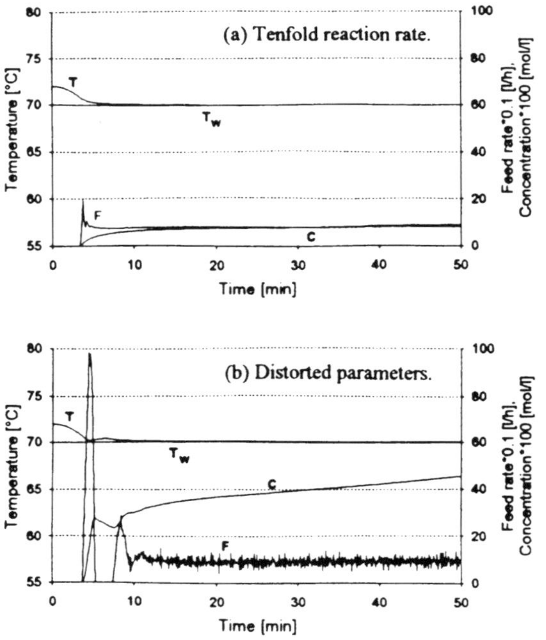

Fig. 5. Robustness and adaptivity of the algorithm.

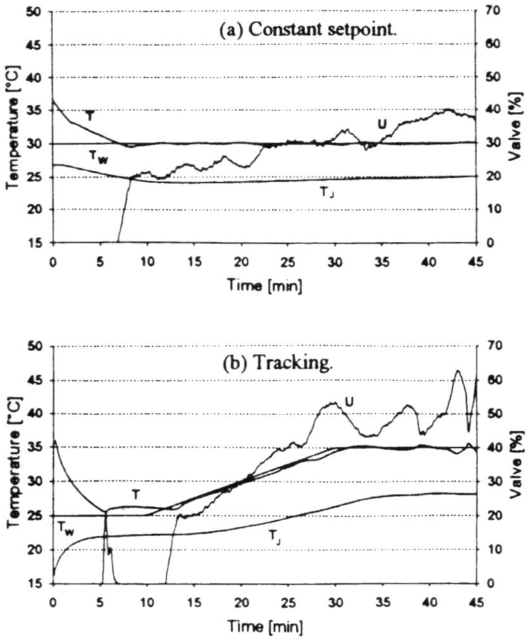

Fig. 6. Laboratory experiments (NaOH + HCl).

5. CONCLUSIONS

A model-based algorithm was developed for the temperature control of a fed-batch reactor. The suggested method evaluates a runaway criterion on-line and manipulates the feed of one of the reactants while using the full cooling capacity in the safe zone of operation. Approaching the critical value the feed is stopped and the temperature is controlled by the flow rate of the cooling liquid. The adaptive properties and the robustness of the algorithm are excellent. The results of wide-ranging simulation and laboratory experiments support the possibility of industrial applications.

ACKNOWLEDGEMENTS

This work was supported by the OTKA 2547 Project and by the 559/92 AMFK Grant (Hungary).

REFERENCES

Clarke, D.W., C. Mohtahi and P.S. Tuffs (1987). Generalized predictive control. *Automatica*, **23**, 126-137.

Davidson, R.S. (1989). Using process information to control a multipurpose batch chemical reactor. *Comp. Chem. Eng*, **13**, 83-85.

Filippi-Bossy, C., J. Bordet, J. Villermaux, S. Marchal-Brassely and C. Georgakis (1989). Batch Reactor Optimization by use of Tendency Models. *Comp. Chem. Eng.*, **13**, 35-47.

Fisher, T.G. (1990). *Batch control systems. Design, application and implementation.* Instrument Society of America, Research Triangle Park.

Jarupintusophon, P., A.M. Gandy, M. Cabassud, M.V. Le Lann and G. Casamatta (1991). Model supported control of a semi-batch reactor. In: *Preprints IV of the 4th World Congress of Chemical Engineering*, 12.3-6. RFA, Karlsruhe.

Jarupintusophon, P., M.V. Le Lann, M. Cabassud and G. Casamatta (1994). Realistic model-based predictive and adaptive control of batch reactor. *Comp. Chem. Eng.* **18**, S445- S449.

Juba, M.R. and J.W. Hamer (1986). Progress and challenges in batch process control. In: *Chemical process control - CPC III* (M. Morari and T. J. McAvoy (Ed.)). Elsevier, Amsterdam.

Liptak, B.G. (1986). Controlling and optimizing chemical reactors. *Chem. Eng.* **26**, 69-81.

Rafalimana, A., M. Cabassud., M.V. Le Lann and G. Casamatta (1992). Adaptive control of a multipurpose and flexible semi-batch pilot plant reactor. *Comp. Chem. Eng.*, **16**, 837-848.

Szeifert, F. and J. Vass (1988). A first order predictor-corrector controller. In: *Proc. of Symp. on Industrial Electronic Measurement and Control, Balatonszéplak.* MATE, Budapest.

Szeifert, F., J. Chovan and L. Nagy (1992). Adaptive optimizing control of a CSTR. *Comp. Chem. Eng.*. **16,** S197-S204.

Szeifert, F., T. Chovan and L. Nagy (1995). Process dynamics and temperature control of fed-batch reactors. Submitted to ESCAPE-5.

Yamatomo, S. and I. Hashimoto (1991). Present status and future needs: The view from Japanese industry, In: *Chemical Process Control - CPC IV* (Y. Arkun and Ray, W.H. (Ed.)), 1-28. CACHE, Austin.

PRACTICAL CONTROL METHODS FOR DISTILLATION COLUMNS USING NEURAL NETWORKS

M. S. Basualdo[1] (1,2), and H. A. Ceccatto(1)

(1) Instituto de Física Rosario (IFIR), CONICET/UNR, Bv. 27 de Febrero 210 Bis, 2000 Rosario.. E-mail: basualdo@ifir.edu.ar. FAX 54-41-821772
(2) Dep. de Electrónica, Fac. de Cs. Exactas, Ingenieria y Agrimensura, UNR.

Abstract -The use of artificial neural networks in SISO IMC structure to control the top column composition is exploited. Here nonlinear process model is replaced by a neural network and its inverse is also incorporated to improve the control action. An IMC multiloop structure is also tested where only top loop includes a neural compensation. The gain variation law of the system is learned by a neural network which serves to adapt this parameter in the internal model and the controller of the IMC scheme. The bottom loop is implemented as a conventional IMC design. This technique can be seen as a feedforward compensation for nonlinearities and gives better performance comparing with the overall linear IMC design. Another control structure using the neural network inverted plant model is combined with a feedback control structure implemented with traditional P, PI and PD controllers is analyzed in this work. This last methodology presents a very good performance depending on which classical controller is used. Concluding the paper a set of validation tests are carried out, via digital simulations, for step variations of the reference and load variables.

Keywords Neural control Nonlinear multiloop control Feedforward compensation Digital Simulations

1-INTRODUCTION

Most of distillation columns show a non linear behavior which is stronger whenever a higher purity products is required. Therefore when attempting to satisfy these requirements it is often difficult to find continuous control laws that are useful in all the relevant regions of the plant. In this context, artificial neural network (ANN) appear to be particularly suitable for controlling chemical processes in a variety of control schemes.

The work reported here involves an investigation into the feasibility of controlling a distillation column using an ANN approach. One of the most studied control strategies is model based control (MBC), where a process model is explicitly used to predict future process behavior, the same process model is essentially inverted to calculate the control action. In this application system identification of the process is done from simulated data for this work. Then the NN is trained with input-output data from the dynamic system to represent the forward dynamics. The same data are switched to generate the inverse model by the NN and its output will drive the system to reach the desired value. The multilayer feedforward architecture is used. The inputs to the network include the reflux, and its corresponding product composition at different operation points. Recurrent network was found to be capable of simulating the dynamic behavior of the top of the column (ANN forward model). Some theoretical aspects of this subject can be found in Bhat and McAvoy (1990), Psichogios and Ungar (1991), Hunt et al (1992). The generation of neural model of the plant and its inverse are used in a SISO Internal Model Control

[1] Author to whom all correspondence should be addressed

structure (Garcia and Morari, 1982). Suggestions for improving performance for multiloop structures are also presented. Control of the two output concentration of the MIMO distillation column where one of the control loop uses a NN compensation is chosen as a test problem. Finally the NN inverse model is implemented as a feedforward compensator of nonlinearities combined with a classical feedback structure employing conventional controllers.

2-DISTILLATION COLUMN MODEL STUDIED.

The column considered here can be seen in figure 1 and its essential operational characteristics are given in Table 1. Non-linear dynamic model is developed under the following assumptions: a multicomponent separation is taking place, in this example a binary mix of Benzene - Toluene is taking place, vapor holdup is negligible compared with liquid holdup, each plate acts as an ideal stage; the system exhibit essentially ideal solution behavior in order to simplify the calculation of enthalpies and vapor-liquid equilibria; perfect mixing occurs in the reboiler, reflux accumulator and on each tray; the column operates adiabatically; negligible energy accumulation on a tray; perfect reboiler and accumulator level control; perfect pressure control. These assumptions result in the following model for each stage and for the reboiler:

Table 1: operational conditions

Number of plates	17
Feed plate	8
Reboiler holdup	3.1 lbmol
Condenser holdup	3.1 lbmol
Reflux flow rate	1.285 lbmol/min
Bottom vapor flow rate	1.99 lbmol/min
Feed flow rate	1.5 lbmol/min
Feed temperature	662°R
Feed composition 1.Benzene 2.Toluene	 0.5 0.5
Pressure drop per plate	0.0085 atm
Condenser pressure	1 atm.

-overall mass balance

$$\frac{dM_n}{dt} = F_n + V_{n-1} + L_{n+1} - L_n - V_n \qquad (1)$$

-component mass balance

$$\frac{d\left(M_n X_{j,n}\right)}{dt} = F_n X_{j,n} + V_{n-1} Y_{j,n-1} + L_{n+1} X_{j,n+1} - L_n X_{j,n} - V_n Y_{j,n} \qquad (2)$$

-energy balance

$$\frac{dM_n h_n^L}{dt} = F_n h_n^F + V_{n-1} h_{n-1}^V + L_{n+1} h_{n+1}^L - L_n h_n^L - V_n h_n^V = 0 \qquad (3)$$

A total condenser is used and Francis weir Formula is employed to relate holdup to liquid flow from a tray. Correlations of physical properties were obtained from the literature and are given as well as the additional details in the Thesis by Basualdo (1990).

It will be assumed here that reflux flow rate (R) is used to control distillate composition XD and bottom vapor flow rate (VB) to bottom composition XB.

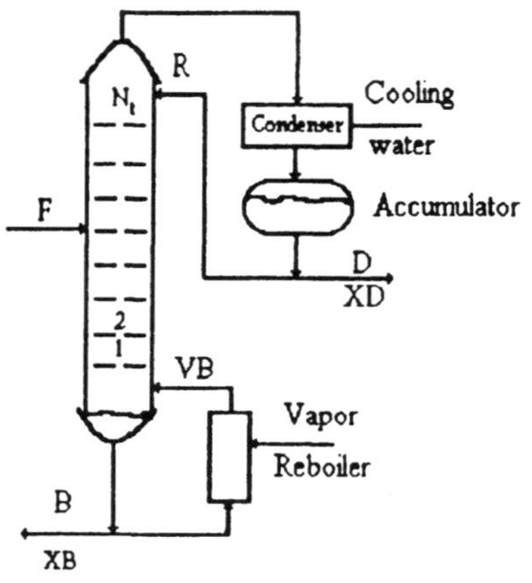

Figure 1: column distillation scheme

3-NEURAL NETWORK ARCHITECTURE

A training dataset for the neural network was obtained by integrating the differential equations describing the physical system with the top manipulated variable (R) reflux rate varying randomly within +20% to -10% of 0.583 Kmol/min in one case and ±2% for other one; a sampling time of 1 min was used. A feedforward network with sigmoidal activation functions was trained using the error back-propagation algorithm (Rumelhart, 1986). Initialization of the network's weights was done by randomly choosing them from a uniform distribution between 0 and 1, and the input-output patterns where also scaled to lie in the same range. The historical input-output data were used to train two neural networks. The first network learns the forward system dynamics, giving the process model. A configuration of ten input neurons, a single hidden layer with ten neurons and an output layer of five output neurons was used for the grater range and the same for input and output neurons but with fifteen neurons for the hidden layer was trained for the ±2% case. The input to the network was the reflux rate, the output for the forward model network was the top composition. A second network was trained to represent the inverse process dynamics. The network configuration was the same as described previously, the input is the composition. The output is the value of the manipulated variable R for the current sampling interval and four predicted values for the next four sampling periods. For this work only the current value is used. In figure 2 the NN predictions of the inverse dynamics are shown.

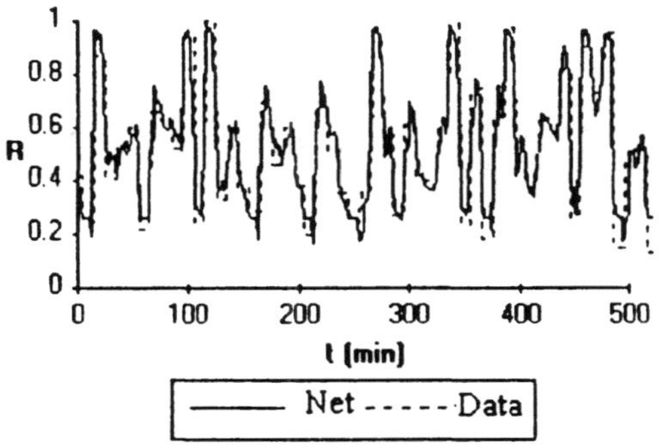

Figure 2: Inverse process model by NN

4- INTERNAL MODEL CONTROL STRATEGY.

4.1 Dual neural network controller. Siso case

The IMC structure shown in figure 3 incorporates models of the plant dynamics and the corresponding inverse represented by neural networks. In this case the model is used in parallel with the plant and the difference takes into account the uncertainty of the model, which is fed back to the system response. The inverse model has been trained with a great steady state information so nearly zero offset can be achieved. Figure 4 shows the dynamic behavior of the top Benzene composition when several reference step changes are required. The same test was made over a linear IMC design. As can be seen both outputs are similar.

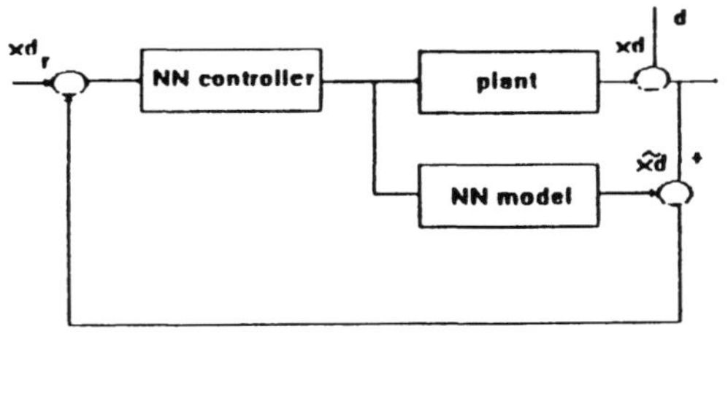

Figure 3: neural network IMC-type

The difference between plant-model for both cases, shown in figure 5, indicates the high capacity of the neural net to mimic a complex system far more accurately than linear model, even when the system is subjecting to a wide range of XD setpoints changes.

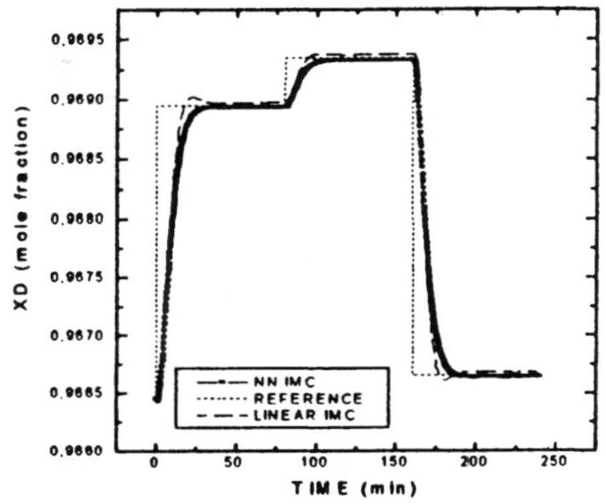

Figure 4: Dynamic response of top composition

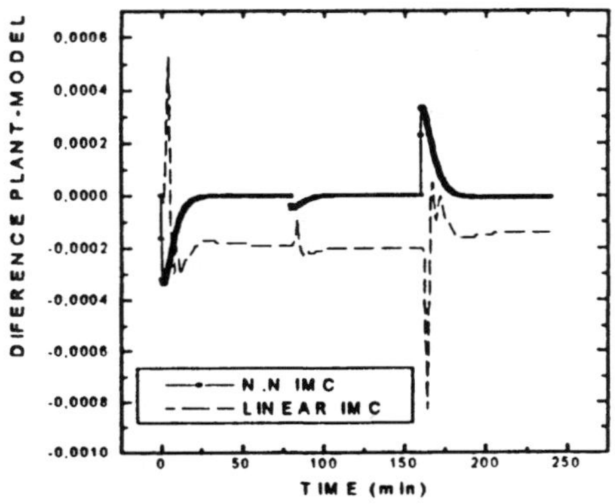

Figure 5: Difference between plant-model for linear and NN IMC-type

Also a better performance of the neural controller can be seen in figure 6.

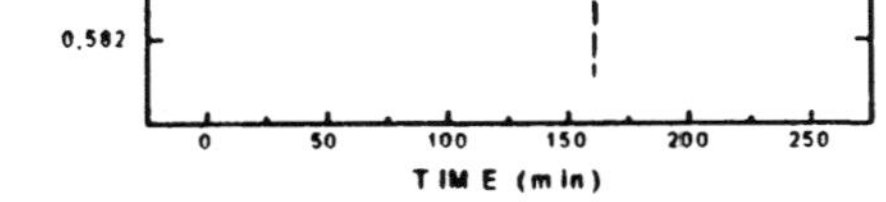

Figure 6: Controllers output for the same step reference changes shown in fig. 4

In Table 2 are given the IAE values for both schemes

Table 2: IAEs for SISO IMC

	IAE
linear IMC	0.568 E-1
NN IMC	0.330 E-1

4.2 Neural network compensation. MIMO case

In this section a slightly different controller architecture is considered, again in the IMC framework. For the column distillation studied here it was found that as XD increases, the process gain decreases and therefore loop gain decreases so the XD transient response become sluggish. Near steady state XD response is faster which is indicative of a higher loop gain. Finally bellow steady state the loop gain is still higher and XD is forced quickly out of this region. As a result of loop gain changes the control system spends more time in the region where XD is higher than its normal operation point (XD = 0.966). Therefore, in this example, the nonlinearity problem is done principally in the process gain. In this context a NN which learns the relation between XD-gain is trained for adapting on-line the plant model as well as the controller gains. Then approximately known forward and inverse model (e.g. linear model) are combined with nonlinear neural network to improve modeling performance (Haesloop and Holt, 1990). The proposed multiloop structure is shown in figure 7.

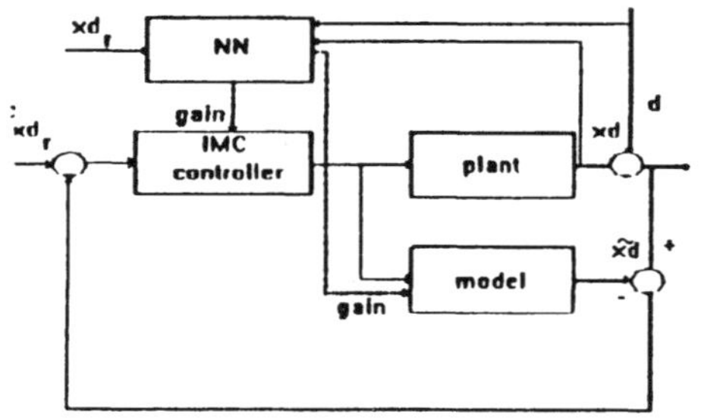

Figure 7: Top loop of the multiloop IMC structure with variable gain using NN

The gain can vary when reference or load changes are required by switching the input to the NN. The bottom control loop is implemented as a conventional linear IMC. The tuning parameters are chosen by using B-M method (Basualdo and Marchetti, 1990). The dynamic responses for step reference changes in top and bottom loops are given in figure 8 and for step change in feed flow rate are given in figure 9. For comparison purposes both IMC (linear and with NN adaptation) structures are tuned with the same filter parameters.

Table 3: IAE values for MIMO IMC

	IAE (linear IMC)	IAE (NN IMC)
Setpoint changes	1.42	0.582
Load changes	0.843	0.401

The real improvement introduced by adapting only the top gain loop is shown in Table 3 where the IAE values are compared with linear IMC multiloop structure with the same tuning parameters.

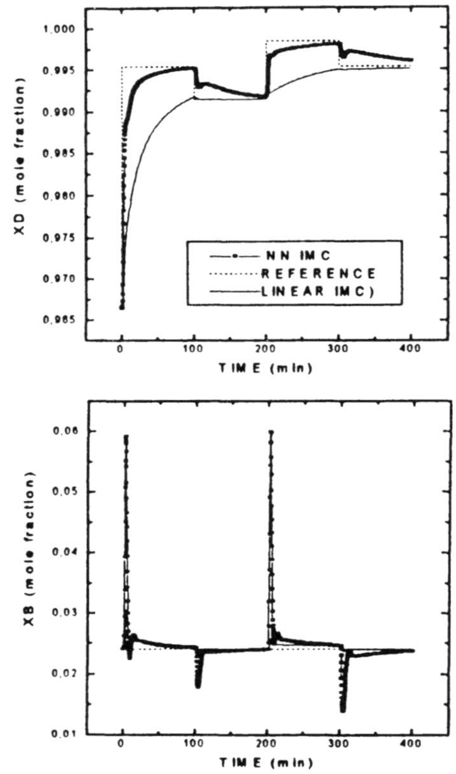

Figure 8: Top and bottom responses for step reference changes using multiloop structure shown in figure 7

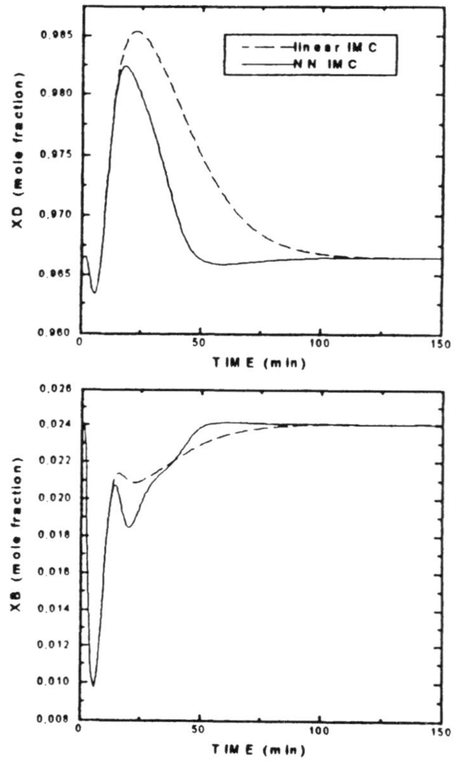

Figure 9: Top and bottom responses for step load change using multiloop structure shown in figure7

5- FEEDFORWARD COMPENSATION USING NN INVERSE PLANT MODEL

For this application the structure shown in figure 10 is employed. Here the inverse NN model obtained previously acts as a feedforward nonlinearity compensator which is combined with the classical feedback structure using conventional controllers. This structure permits the feedback controller to "see" the plant plus the inverse NN model as an overall linear system. Then the tuning problem can be reduced significantly. Therefore depending on the choice of the classical controller and the input-output database employed for the NN the achievable performance can be quite different.

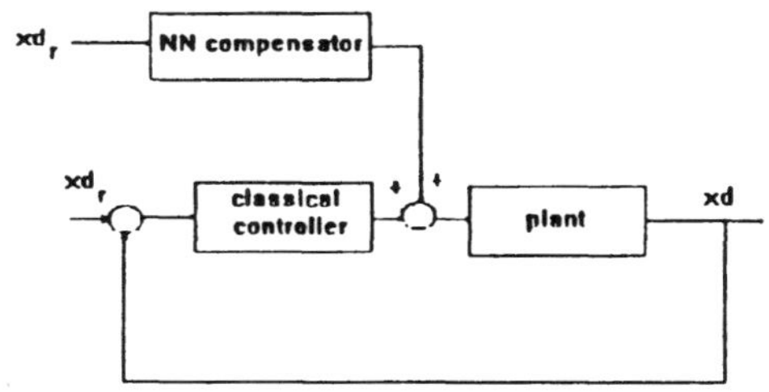

Figure 10: Feedforward NN compensator

As an example with a NN trained principally to learn steady state points its transient behavior can be improved significantly when a derivative action is included. In figure 11 a SISO PD controller is implemented with the feedforward NN compensator. The dynamic response of XD is compared with a feedback PID controller where the same proporcional and derivative constants are used.

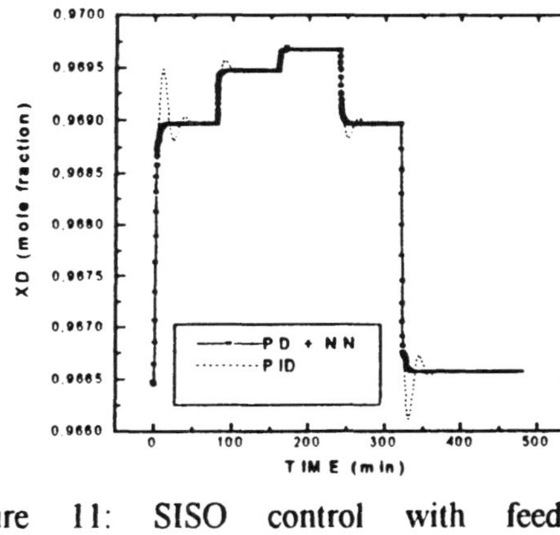

Figure 11: SISO control with feedforward compensator and PD feedback in top loop.

As can be seen from the scheme practically a "perfect" control has been achieved with combined structure. In Table 4 the IAE values for this case are shown.

Table 4 : IAE values for SISO structure

	IAE
NN + PD action	1.96 E-2
LINEAR PID	3.19 E-2

An analogous test was made using a PI controller and the NN feedforward compensator but, in this case, implemented in a multiloop structure. The bottom composition XB is controlled with a conventional PI. The IAE values also indicate the improvement introduced by the inverse neural model. The dynamic behavior of the controlled variables for this case can be seen in figure 12. Here zero offset is provided by the integral action, so a neural net which has learned principally the transient behavior can be proposed. Table 5 presents the top loop IAEs for combined and conventional structures.

Table 5: IAEs for MIMO structure

	IAE for top loop
NN + PI action	0.331
LINEAR PI	0.419

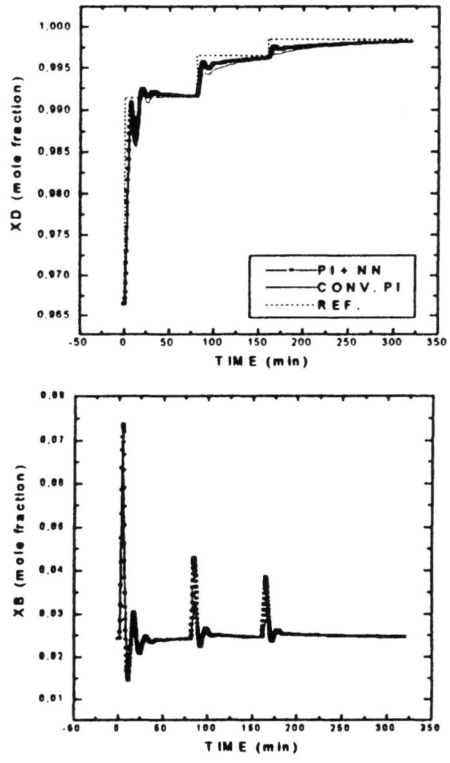

Figure 12: Multiloop structure with feedforward compensator in top loop (XD) with feedback PI. Loop 2 (XB) with conventional feedback PI

6- CONCLUSIONS

Nonlinear control strategies incorporating neural networks were evaluated on the control of SISO and MIMO continuous distillation column. A neural network IMC-type controller where the process model was learned using historical input-output data gives very good performance comparing with the linear IMC design. A simple procedure is suggested which consists on having the IMC controller as well as internal model where the gain is adapted on line by using NN. This methodology permits to improve significantly the transient response of the system comparing with linear multiloop IMC. Another test was made by implementing a feedforward compensation with the NN inverse plant model as a nonlinearity compensator combined with classical feedback structure. Very good results were obtained in setpoints tracking problems when PD action was used in the feedback SISO structure. Also PI action was included in the previous scheme for one of the loop, the other one was implemented as conventional PI. All these control schemes demonstrated to be efficient to handle problems where the process operates with different underlying dynamics. The results seem to indicate that neural nets offer a feasible alternative as system identification and nonlinear compensation tools.

REFERENCES

Basualdo M and Marchetti J, (1990) Tuning Method for Interactive Multiloop IMC, PI and PID Controllers, *Chem. Eng Commun.*

Basualdo M. (1990) *"Dynamic and Control of Distillation Columns"*. Ph.D Thesis . FIQ. U.N.L. (INTEC-CONICET)

Bhat N. y McAvoy T.J. (1990), *Computers Chem Engng,* 14, 4/5, pp. 573-583

Garcia, C.E., Morari M (1982) Internal Model Control 1. A Unifying Review and Some New Results. *Ind. Eng. Chem. Res. Process Des. Dev.* 21, 308-323

Haesloop D., Holt B.R.(1990) A NN. Structure for System Identification. *Proceedings of ACC* IEEE Service Center: Piscataway, NJ, pp. 2454-2459.

Hunt K. J.,Sbarbaro D., Zbikowski R. and Gawthrop P.J. (1992), *Automatica*, 28, 6, pp. 1083-1112

Psichogios D. C., Ungar L. (1991) Direct and Indirect Model Based Control *Ind. Eng. Chem. Res.*, 30, pp. 2564-2573

Rumelhart D.E., (1986) *Parallel Distributed Processing*. MIT Press, Cambridge M

ACKNOWLEDGMENTS

To Rafael Calvo who worked in training the neural networks used in the present paper.

To SECyT of UNR, in part sponsor of this work under PID 5-202/93-026

FINE TUNING OF MULTILOOP PI CONTROLLERS USING FUZZY SET THEORY

Heeman LEE and Sunwon PARK

Department of Chemical Engineering, KAIST, Taejon 305-701, Korea

Abstract: This paper proposes an intelligent way of fine-tuning multiloop PI controllers for distillation processes using linguistic fuzzy sets and fuzzy inference logic. This approach first requires reference dynamic curves which reflect the setpoint tracking and interaction rejection performance requirement by control engineer or operator. Linguistic fuzzy sets which describe the deviation degree from the reference dynamic curve patterns are then defined for each controlled output, and by minimizing the summation of corrected membership values of controlled outputs we can get the proper values of PI parameters which improve the overall performance of the transient and steady state behavior of closed-loop control systems. Initial tuning parameters are set by the conventional multiloop PI controller tuning methods such as the BLT method which guarantees the stability of control system.

Keywords: multiloop PI controllers, fuzzy set theory, reference dynamic curve

1. INTRODUCTION

Chemical processes which are nonlinear, time-delaying, and multivariable systems in their nature, have been considered to be difficult to control with good performance. Today's process control technology thus have been focused on making general algorithm which increases the overall control performance and reduces the additional burden imposed on the process operator because of difficult system-specific control algorithms. Because of this reason most of the chemical plants still rely on multi-loop PI controllers for process control purposes and PI controllers tuning method in multivariable system has long been the major concern among the control engineers and the operators. In multivariable control systems, PI controller tuning procedure is a multiple performance criteria decision making process, and getting the proper controller tuning parameters which give good performance over all the paired loops is considered to be very difficult. Conventional multi-loop tuning methods for multivariable PI controllers such as the BLT method(Luyben, 1986) do not address the control performance directly because they mainly deal with stability of all the controlled outputs by frequency domain stability analysis. For that reason, the real tuning process includes a time-consuming trial and error process to satisfy the required control performance such as overshoot-free, fast response time, no steady state error, good interaction rejection, disturbance rejection, etc. Although an expert system(Jonathan Litt, 1991) may automates this trial and error process, realizing a computer code of this fine-tuning process in the structure of the expert system for multivariable systems is not easy because this tuning process includes a lot of rules and parameters in order to optimize multi-objective functions simultaneously. In this work, a new tuning method using fuzzy set theory and an existing nonlinear optimization technique(GRG algorithm) has been proposed and it gives better results over the conventional BLT tuning method. The proposed method first requires a reference dynamic curve which reflects the performance specification of the control engineer or the operator for each paired loop of multivariable PI control system . This reference curve is then used to evaluate the linguistic performance of controlled output for each sampling time. The basic idea involved in reference dynamic curve is similar with the linguistic performance decision table for fuzzy self organizing controller(Procyk and Mamdani, 1979). The reference dynamic curve evaluates the linguistic performance of controlled output by checking the deviation degree from the reference curve pattern. Deviation degree means the membership value of controlled output toward linguistic performance evaluating fuzzy set, and then the membership value is converted to an appropriate correction factor. The correction factors of all controlled outputs are summed. The goal of optimization is to minimize the summed values over the given time span. In the process of minimizing this single objective function we can significantly enhance the performance of PI controllers compared to that by the conventional BLT tuning method. The simulation result for multivariable distillation system support the advantage of this tuning method.

2. CONVENTIONAL MULTILOOP PI CONTROLLERS TUNING METHOD

Luyben(1986) proposed the BLT method to determine controller setting for the N single-input-single-output controllers in an Nth-order typical industrial multivariable process. This method guarantees the stability for open-loop stable multivariable system, but as Luyben himself indicated in his original paper, performance of control system by this tuning method can be largely deviated from what operators want it to be.

3. FUZZY SET THEORY

3.1 Basic Definition

A classical (crisp) set is normally defined as a collection of elements or objects $x \in X$ which can be finite, countable , or overcountable. Each single element can either belong to or not belong to a set $A, A \subseteq X$. In the former case, the statement " x belongs to A" is true, whereas in the latter case this statement is false. Such a classical set can be described in different ways: one can either enumerate (list) the elements that belong to the set; describe the set analytically, for instance, by stating conditions for membership $(A = \{x | x \leq 5\})$or define the member elements by using the characteristic function, in which 1 indicates membership and 0 nonmembership. For a fuzzy set , the characteristic function allows various degrees of membership for the elements of a given set. If X is a collection of objects denoted generically by x then a *fuzzy set* in $\tilde{A}$ is a set of ordered pairs.

$$\tilde{A} = \left\{(x, \mu_{\tilde{A}}(x)) \middle| x \in X\right\} \tag{1}$$

$\mu_{\tilde{A}}(x)$ is called the membership function or grade of membership (also degree of compatibility or degree of truth) of x in $\tilde{A}$ which maps X to the membership space M. (When M contains only the two points 0 and 1, A is nonfuzzy and $\mu_{\tilde{A}}(x)$ is identical to the characteristic function of a nonfuzzy real numbers whose supremum is finite.) Elements with a zero degree of membership are normally not listed(Zimmerman, 1987).

3.2 Inference logic

As a consequence of the representation of imprecise and uncertain knowledge it is necessary to develop adequate reasoning methods. Since 1973 when Zadeh suggested the compositional rule of inference, a lot of work has been done in the field of fuzzy inference mechanisms . Here in this report we use the maximum minimum inference logic which was developed by Zadeh.

Zadeh's method:
OR(μ_A, μ_B) operation: max(μ_A, μ_B)
AND(μ_A, μ_B) operation: min(μ_A, μ_B) (2)

4. PROPOSED TUNING METHOD USING FUZZY SET THEORY

Tuning multi-loop PI controller is a multi-criteria decision making problem having multiple performance criteria such as

- Overshoot should be small.
- Oscillation should be reduced.
- Response should be fast.
- Offset should be free.
- Damping should be low.
- Disturbance rejection should be good.
- Interaction rejection should be good.

But, conventional tuning methods cannot deal with these performance criteria directly because their main concern is closed-loop system stability based on frequency domain stability analysis. And that is the reason why most of the real tuning processes are much like an expert system . The common rules employed by the operators are as follows(Litt , J. 1991).

- To decrease overshoot, decrease $\mathbf{K}_c$.
- To decrease rise time, increase $\mathbf{K}_c$.
- To decrease damping, decrease.

•To decrease overshoot ratio, increase T_I.

•To decrease damping, increase T_I.

•To increase stability, increase T_I.

•To decrease overshoot, increase T_I.

Computer implementation of the above verbal rules is very difficult since it contains multiple performance criteria which make optimization be difficult to converge. Furthermore, as the number of tuning parameters increases in a multivariable system, it becomes more complicated to decide the weight factors for convergence of each one of the tuning parameters of the controller. On the basis of that idea , we can conclude that it is desirable to diminish the number of performance criteria while still having the aim of satisfying most of the performance criteria. Here to convey all the performance criteria in one objective function, a method employing a reference dynamic curve is proposed. The reference dynamic curve is similar in meaning with the linguistic performance decision table which was used in Fuzzy Self-Organizing Controller(Mamdani, 1979). The performance decision table was used to evaluate performance of the control system at each sampling time judging from the error and change of error of the controlled output, and this linguistic performance decision table suggested the optimal control trajectory if the controlled output just follows the *ZR* element state path. Correction factors inside the above matrix were decided by state values at sampling time and it is utilized to compensate the controller output and rules in the fuzzy controller. These entries ensure a sufficiently fast approach to the set-point, good damping when close to the set-point. The further away from the desired trajectories, the greater the required output corrections one way or another. It is important to realize that these corrections not only take into account the distance from the setpoint but also the rate of approaching to it . Thus a small error and a small change of error is regarded as more satisfactory than a small error but a large change of error because of the over-shoot this is liable to produce . It can also be seen that the table is symmetric about the set-point, as the positive error requires controller output increasing while negative error requires it decreasing. This table is therefore written to express the designer's idea of minimum tolerable response(Mamdani, 1979).

Table 1. Linguistic performance decision table

		CHANGE OF ERROR				
		NB	NS	ZR	PS	PB
E	NB	*NB*	*NM*	*NS*	*NS*	*ZR*
R	NS	*NM*	*NS*	*NS*	*ZR*	*PS*
R	ZR	*NS*	*NS*	*ZR*	*PS*	*PS*
O	PS	*NS*	*ZR*	*PS*	*PS*	*PM*
R	PB	*ZR*	*PS*	*PS*	*PM*	*PB*

Tzafestas and Papanikolopoulos adopted this performance decision table in correcting PID parameters which were initially set by Ziegler-Nichols tuning method. Equation (3) was implemented in this article (S. Tzafestas and N.P.Papanikolopoulos, 1990).

$$
\begin{aligned}
P &= P + CV\{E, CE\} \times k_1 && \text{(Proportional)} \\
I &= I + CV\{E, CE\} \times k_2 && \text{(Integral)} \\
D &= D + CV\{E, CE\} \times k_3 && \text{(Derivative)}
\end{aligned}
\qquad (3)
$$

(E: Error, CE: Change of Error,

$CV\{E, CE\}$: Correction Value in Performance Table)

Parameters k_1, k_2, and k_3 play an important role in the whole procedure since they determine the range of variation of each term. For example, if some tuning method results very small rise time and large overshoot, the integral term should have a large range of variation, whereas the other term can remain unchanged. This range of variation should be matched with the stability interval in order to guarantee stability. Thus, the values of k_1, k_2, and k_3 are determined from both the stability analysis and the particular characteristics of the closed-loop response. In general, parameters k_1, k_2, and k_3 provide large flexibility and can be used in conjunction with all available PID tuning algorithms. The most serious problem faced in Tzafestas and Papanikolopoulos' report was the proper quantization of the performance decision matrix and the selection of the **k** values. All the more Tzafestas' approach is mainly confined to SISO cases so that applying this method to PID controller tuning in multivariable systems requires time consuming search for the values of k_1, k_2, and k_3. In multivariable systems, another disadvantage of using this rectangular performance decision table and as the tuner of PID parameters is that it is not easy to tune horizontal scale, vertical scale, and each element(correction values) of the matrix. This difficulty motivated us to devise a reference dynamic curve. A reference dynamic curve is drawn roughly from previous control outputs when various tuning parameters were used. That is to say , the reference dynamic curve is made to give optimal output trajectory by the sense of the control engineer. Reference dynamic curves are then transformed into a data file in which error and change of error are stored to represent the pattern of the reference dynamic curves. The process of drawing these reference dynamic curves are illustrated in Figures 1 and 2. Figure 1 is for setpoint tracking and Figure 2 for disturbance rejection respectively. After seeing the various controlled outputs by different PI controller parameters, reference dynamic curve is heuristically drawn.

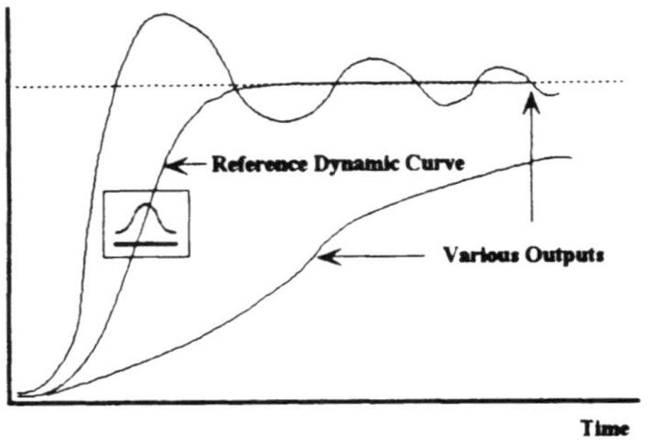

Fig. 1. Drawing reference dynamic curve for setpoint tracking

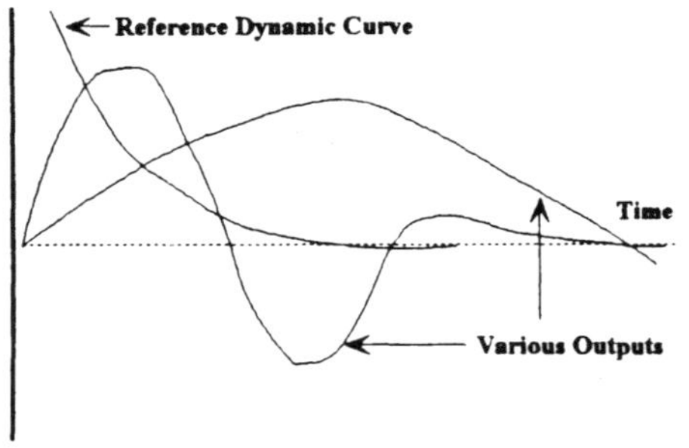

Fig. 2. Drawing reference dynamic curve for disturbance rejection

If there is no previous control output data, first reference dynamic curve is drawn from BLT tuned control output, and in the process of optimization designer can interactively Redraw better reference dynamic curve. The degree of deviation from this reference dynamic curve is used for linguistic performance evaluation of one controlled output at a sampling time. One linguistic fuzzy set for each controlled output is defined in each paired loop of the multi-loop control system. $\tilde{e}$ ={ Performance is good } The membership value of each controlled output for a given error and change of error is decided by the bell shape function in Figure 3.

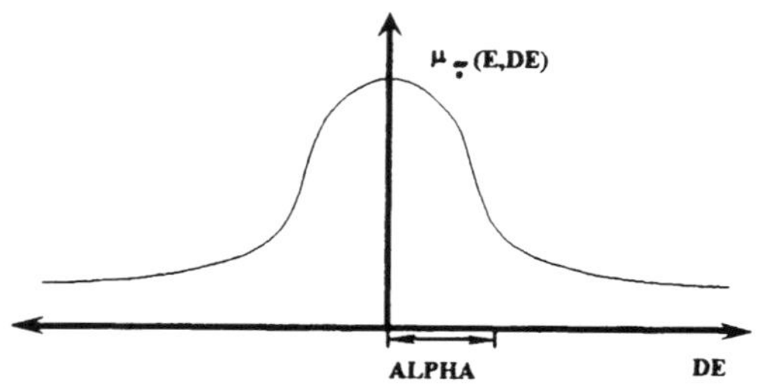

Fig. 3. Linguistic fuzzy set for performance evaluation

$$\mu_{\tilde{e}}(E,CE) = \exp\left(\frac{-(CE - BETA)^2}{ALPHA^2}\right)\Bigg|_E \qquad (4)$$

In the equation (4), **BETA** represents the change of error for reference dynamic curve at given error **E**. **ALPHA** is a scaling factor for each reference dynamic curve and it is usually chosen as the maximum change of error for the reference dynamic curve. For a 4x4 system , 16 fuzzy sets are considered because there exist 4 controlled outputs in four paired loops. Overall performance index of one controller at a given time is inferred by ***t-norm*** operation considering that set-point tracking and regulatory performance are equally important. In this work ***minimum*** operation was used. In Figure 4, μ_{ij} is the membership value judged by the performance of the output variable controlled by the ***i*** th controller in the ***j*** th loop, and this value represents how good performance output value showed at each sampling time.

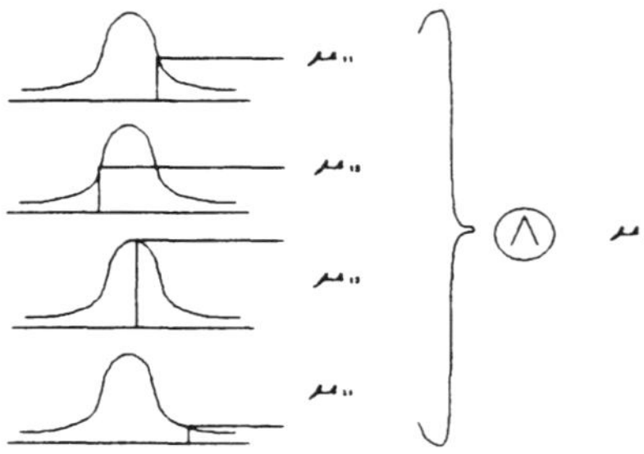

Fig. 4. Fuzzy inference

This membership value which evaluated control performance is then used to make the overall objective function. The goal of optimization here is to maximize the overall goodness of performance of each controller over the concerned time horizon. For easier implementation, the maximization problem is converted to the minimization problem by the following equation which is the typical implication rule of negation.

$$\mu_{BAD} = 1 - \mu_{GOOD} \qquad (5)$$

Real implementation of this membership value includes the ***correction factor*** which is calculated from μ_{BAD} of each controller. The correction factor means the compensation which will be directly used in the optimization function. The correction factor which is similar with the element of linguistic performance decision matrix is obtained easily in the sense that more badness requires extra large correction factor. This common sense is graphically represented in Figure 5. The nonlinear function for this compensation has boundary conditions which represent zero correction factor for membership value zero and infinite correction factor for membership value 1. These boundary conditions are written in Equation (6).

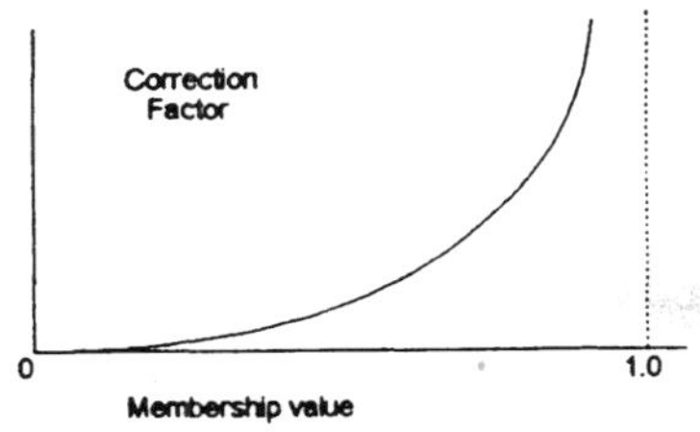

Fig. 5. Graphical representation of correction factor

$$\mathrm{CF}(\mu) \rightarrow \infty \text{ as } \mu \rightarrow 1.0$$
$$\mathrm{CF}(\mu) \rightarrow 0 \text{ as } \mu \rightarrow 0.0 \qquad (6)$$

Each correction factor for each individual control action is summed over the concerned time horizon to make the objective function which should be minimized. The overall scheme of the proposed algorithm is shown in Figure 6.

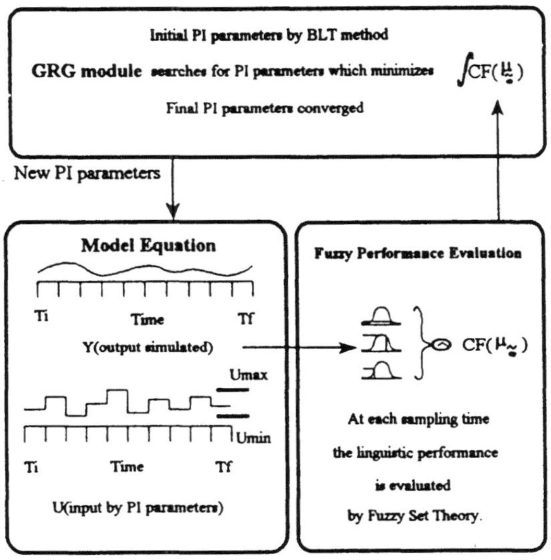

Fig. 6. Overall scheme of proposed algorithm

5. EXAMPLE

Ex) *WOOD-BERRY* column model (Wood and Berry, 1973)

$$\overline{G} = \begin{bmatrix} \dfrac{12.8\exp(-s)}{16.7s+1} & \dfrac{-18.9\exp(-3s)}{21s+1} \\ \dfrac{6.6\exp(-7s)}{10.9s+1} & \dfrac{-19.4\exp(-3s)}{14.4s+1} \end{bmatrix} \qquad (7)$$

Loop pairing is done diagonally according to Bristol's RGA analysis(Bristol,1966). The reference dynamic curves are drawn manually from the dynamic characteristics of output patterns by previous BLT tuning method. Graphically drawn reference dynamic curves are then stored in data file for computer implementation. Reference dynamic curve for the case when error become negative is inverse of reference dynamic curve for the positive error in data file. The optimal control algorithm can also give a good reference for making the optimal reference dynamic curve pattern. Alpha value which decides the degree of fuzziness is chosen by the maximum change of error for reference dynamic curve of each paired loop, and functional relationship between membership value and correction factor used for the simulation is.

$$CF(\mu) = \sqrt{\log(1/(1-\mu))} \qquad (8)$$

The above equation satisfies boundary conditions in Figure 5 and shows nonlinear characteristics described in Figure 5. The simulation results are shown from Figure 7 through Figure 12. The figures were arranged in such a manner that figures showing reference dynamic curves are first displayed and followed by figures showing setpoint tracking performance and interaction rejection performance.

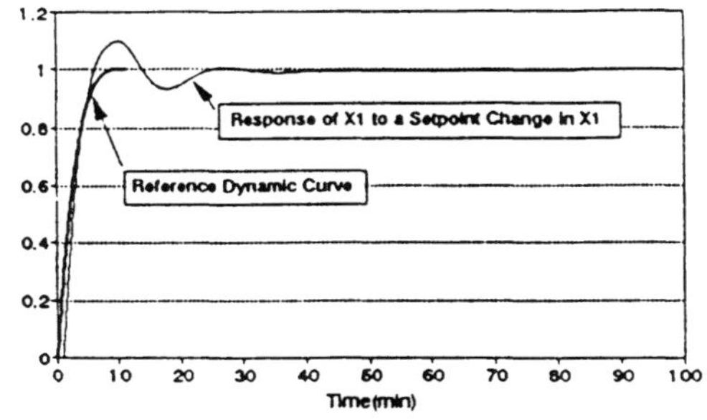

Fig. 7. Reference Dynamic Curve For X1

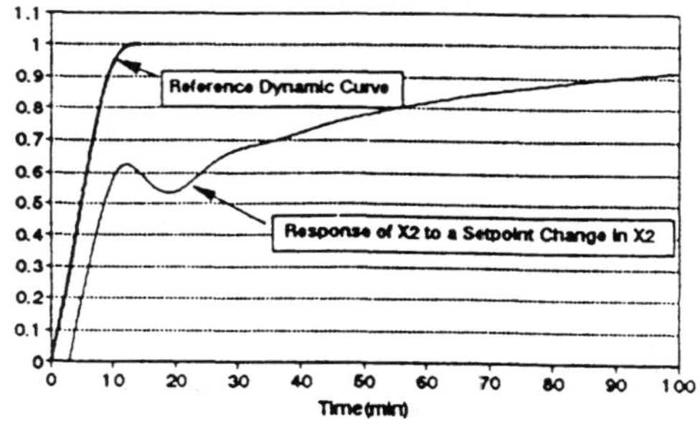

Fig. 8. Reference Dynamic Curve For X2

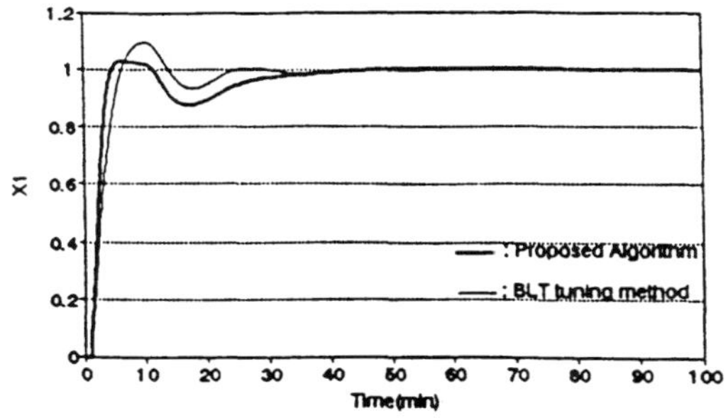

Fig. 9. Response of X1 to Setpoint Change in X1

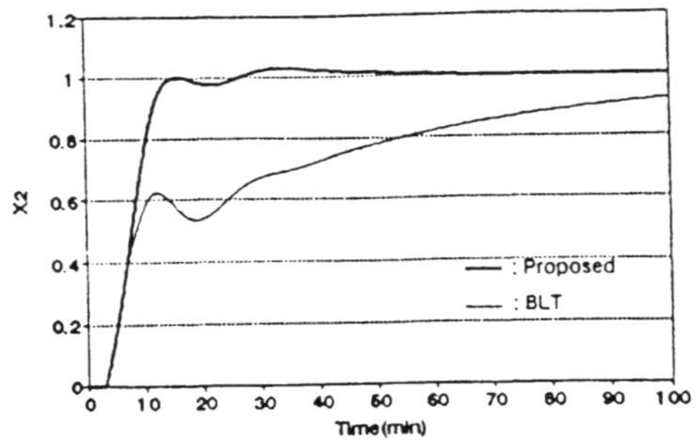

Fig. 10. Response of X2 to Setpoint Change in X2

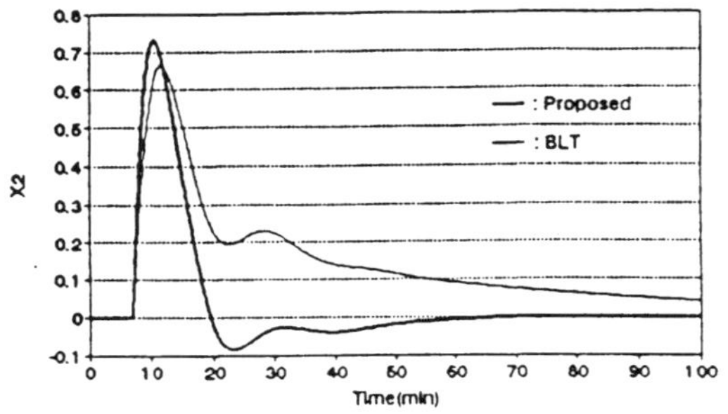

Fig. 11. Response of X2 to Setpoint Change in X2

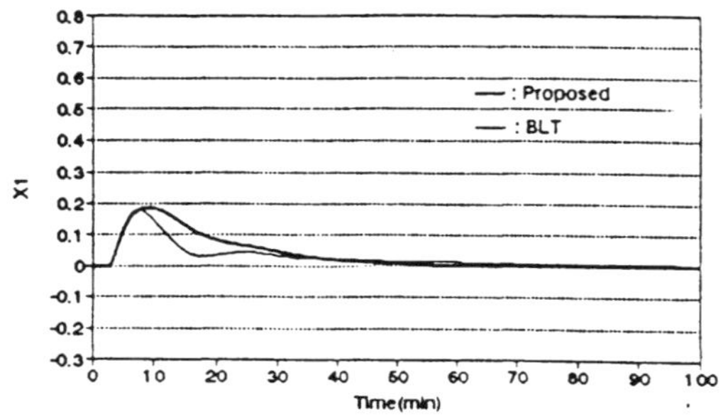

Fig. 12. Response of X1 to Setpoint Change in X2

Figures 7 and 8 illustrate the reference dynamic curves for paired loops 1 and 2 respectively. Each reference dynamic curve was heuristically drawn from the previous output at setpoint change of each loop of which PI controller was tuned by the BLT tuning method. Since this example study belongs to initial design phase, there is no previous output data in addition to those by the BLT tuning method. Figures 9 and 10 compare the setpoint tracking performance between the BLT tuning method and the proposed method. Thick lines represent the output results by the proposed method, while thin lines represent the output results by the BLT tuning method. It is apparent that control output by the proposed method increases the setpoint tracking performance compared to the initial output performance by BLT tuning method. Figures 11 and 12 compare the interaction rejection performance between BLT tuning method and the proposed method. For the evaluation of interaction rejection performance of PI controllers the simulation example shows the figures of controlled output when the setpoint of other controlled output changes. PI controllers fine-tuned by the proposed method show better interaction rejection performance. Table 2 shows the tuning parameters before and after optimization. In every figure, thin lines represent control outputs by BLT tuning and thick lines represent those by the proposed algorithm with initial PI parameters by BLT tuning method.

Table 2. Tuning parameters before and after optimization

		Wood & Berry	
	RGA	**2.01**	**2.01**
BLT	**Kc**	**0.373**	**-0.075**
	Ti	**8.29**	**23.6**
Fuzzy	**Kc**	**0.568**	**-0.0582**
	Ti	**13.3**	**4.24**

6. CONCLUSIONS

A new fine tuning multiloop PI controllers design algorithm for multivariable systems was proposed. This algorithm is based on fuzzy set theory and can attain some design goal which is finding the optimal compromising PI tuning parameters satisfying the multiple performance criteria set by the plant operator. The new idea involved in this work is to draw the reference dynamic curve as an optimal pattern, and in the process of minimizing summation of overall badness judged by the degree of deviation from reference dynamics we can find the good tuning parameters which resulted in good overall performance in transient responses as well as steady state responses. The simulation result shows the advantage of this method.

ACKNOWLEDGMENT

Financial support from Korea Science and Engineering Foundation through the Automation Research Center at POSTECH is gratefully acknowledged.

REFERENCES

Bristol, E. H. (1966) "On a new measure of interaction for multivariable process control", IEEE Trans. Auto. Control , vol. 11, p133

Doukas, N. and Luyben, W. L. (1978), Instrum. Technol., vol. 25, 43.

Ladson, L.S. and Alian D.W. (1982), "GRG2 User's Guide".

Litt, J (1991) , " An Expert System to Perform On-Line Controller Tuning". IEEE Control Systems , p18-p23

Luyben, W. L. (1986) ," Simple Method for Tuning SISO Controllers in Multivariable Systems ". Ind. Eng. Chem. Process Des. Dev. , 25,p654-p660

Procyk, T. J. and Mamdani, E. H. (1979), "A linguistic self-organizing process controller". *Automatica* 15, 15

Tyreus, B. D. (1979) , Ind. Eng. Chem. Process Des. Dev., vol. 18, p177

Tzafestas, S. and Papanikolopoulos, N. P. (1990) , "Incremental Fuzzy Expert PID Control". IEEE Transactions on Industrial Electronics , Vol 37. No 5. p365-p371

Wood, R.K. and Berry, M.W. (1973), "Interaction Control Systems: Steady State and Dynamic Measurement of Interaction". ISA Trans. , vol. 16, pp. 34-41.

Zadeh, L. A. (1973), "Outline of a new approach to the analysis complex systems and decision processes". IEEE Trans. Systems Man and Cybernetics 3(1)

Zimmermann, H. J. (1987), "Fuzzy Set Theory-and Its Applications". 2nd Edition, Kluwer Academic Publishers, p11-p12

MULTIVARIABLE CONTROL OF DISTILLATION COLUMN

Jiří Macháček, Josef Kotyk

University of Pardubice
Department of Automation
532 10 Pardubice, Czech Republic
machacek@hlb.upce.cz

Abstract: Three methods for discrete control of distillation column are compared. The distillation column, separating binary mixture of methanol-water, is regarded as a two inputs - two outputs system. The controller design for all methods is made by the pole placement method, but three different variants are used: The multivariable case, the decoupling with precompensator and the decoupling with precompensator and unit output feedback. The methods are tested on the simulation model of the column.

Keywords: Multivariable systems, distillation columns, multivariable control, decoupling precompensators, polynomial methods, pole assignment.

1. INTRODUCTION

Various methods have been suggested for the control of multivariable systems. The polynomial approach with discrete model of the system and the pole-assignment (placement) method of controller design are used in this paper. The denominator of closed-loop system transfer function is choosen in order to ensure both the stability and the requirement dynamic behaviour. The controller contains also the integral action part ensuring the zero steady-state error under the influence of disturbances.

The decision which method should be used for the control of pilot-plant distillation column, as a typical multivariable system with the interactions between the control loops, is proposed here. Therefore three methods are compared by using the simulation model of column. The first one is the multivariable case of pole-placement method based upon the polynomial theory (e.g. Šebek and Kučera, 1982; Mikleš, 1990). Both second and third methods use the decoupling for the suppression of the undesirable interactions between input and output variables. After the decoupling each input affects only one controlled variable and controllers design can be made for the independent one-variable loops. The decoupling problem with precompensator is solved in works of Waller (1974); Tade *et al.* (1986); Chien *et al.* (1987); Kinnaert *et al.* (1987); Wittenmark *et al.* (1987); Peng (1990); Krishnaswami *et al.* (1991) and others. The comparison of five decoupling methods of this type give Macháček and Kotyk (1994). Decoupling method with precompensator and unit output feedback was proposed by Chen (1984). The main problem of the decoupling methods is the achieving of internal stability, which is ensured when no poles-zeros are canceled (Wang, 1992).

The distillation column is taken as the system with two input and two output signals, with reboiler heating power and reflux flow as action variables.

2. MULTIVARIABLE POLE ASSIGNMENT METHOD

The dynamic behaviour of the column can be described in the neighbourhood of steady state by the discrete linear model

$$\mathbf{A}(q^{-1})\,\mathbf{Y}(q^{-1}) = \mathbf{B}(q^{-1})\,\mathbf{U}(q^{-1})\,, \qquad (1)$$

where $U(q^{-1})$ is the (2x1) input vector, $Y(q^{-1})$ is the (2x1) output vector and $A(q^{-1})$, $B(q^{-1})$ are (2x2) left coprime polynomial matrices. The polynomial matrices have the following structure:

$$\mathbf{A}(q^{-1}) = \mathbf{I} + \mathbf{A}_1 q^{-1} + ... + \mathbf{A}_n q^{-n}$$
$$\mathbf{B}(q^{-1}) = q^{-d} (\mathbf{B}_1 q^{-1} + ... + \mathbf{B}_m q^{-m}) , \qquad (2)$$

where d is the time delay, expressed as a multiple of the sampling period T and n, m are the orders of the polynomials. The initial state of the system output is supposed to be zero. The backward shift operator q^{-1} will be omitted for simplification in all next expressions.

The Eq.(1) can be also expressed as a system transfer function **F**, defined by the left or right matrix fraction

$$\mathbf{F} = \mathbf{A}^{-1} \mathbf{B} = \mathbf{B}_R \mathbf{A}_R^{-1} . \qquad (3)$$

The parameters of the left matrix fraction are obtained directly by identification. Further on it is assumed, without loss of generality, that the matrix **A** is diagonal. This condition is advantageous for some controller design methods.

The block diagram of the system with a general type of controller is shown on Fig.1. The controller consists of three parts: the feedback with the transfer function $\mathbf{P}^{-1}\mathbf{Q}$, the feedforward with the transfer function $\mathbf{P}^{-1}\mathbf{R}$ and the integral action $\mathbf{K}^{-1}$. The matrix **K** is taken as a (2x2) diagonal matrix with the elements $1\text{-}q^{-1}$. The (2x2) controller matrices are defined in following way:

$$\mathbf{P}(q^{-1}) = \mathbf{I} + \mathbf{P}_1 q^{-1} + ... + \mathbf{P}_p q^{-p}$$
$$\mathbf{Q}(q^{-1}) = \mathbf{Q}_1 q^{-1} + ... + \mathbf{Q}_s q^{-s} \qquad (4)$$
$$\mathbf{R}(q^{-1}) = \mathbf{R}_0 + \mathbf{R}_1 q^{-1} + ... + \mathbf{R}_r q^{-r} ,$$

where p, s and r are the orders of polynomials. **W** is (2x1) reference signal vector and **E** is a vector of undefined disturbances, e.g. it includes the inaccurate description of the nonlinear system by linear model, the measurement errors and the influence of other input signals, which contrary to assumptions are not constant.

The equation for the input signals results directly from Fig.1:

$$\mathbf{U} = \mathbf{K}^{-1} (\mathbf{P}^{-1} \mathbf{R} \mathbf{W} - \mathbf{P}^{-1} \mathbf{Q} \mathbf{Y}) . \qquad (5)$$

The solution (1) by use of (5) gives

$$\mathbf{Y} = \mathbf{A}^{-1} \mathbf{B} \mathbf{K}^{-1} (\mathbf{P}^{-1} \mathbf{R} \mathbf{W} - \mathbf{P}^{-1} \mathbf{Q} \mathbf{Y}) . \qquad (6)$$

After substitution of the right matrix fraction (3) and small rearranging Eq.(6) can be rewritten in the form

$$\mathbf{Y} = \mathbf{B}_R (\mathbf{P} \mathbf{K} \mathbf{A}_R + \mathbf{Q} \mathbf{B}_R)^{-1} \mathbf{R} \mathbf{W} . \qquad (7)$$

The controller parameters **P** and **Q** are determined as the minimum-degree solution of the equation

$$\mathbf{P} \mathbf{K} \mathbf{A}_R + \mathbf{Q} \mathbf{B}_R = \mathbf{M} , \qquad (8)$$

where **M** is the (2x2) stable polynomial diagonal matrix, which poles are choosen according to the required dynamic behaviour of the closed loop. The relation between the output and reference signals are then given by (7) and (8)

$$\mathbf{Y} = \mathbf{B}_R \mathbf{M}^{-1} \mathbf{R} \mathbf{W} . \qquad (9)$$

The matrix **R** can be calculated for the step changes of reference signal **W**, as is usual in the technical practice, according to Eq.(10)

$$\mathbf{R} = \mathbf{R}_0 = \mathbf{M}(1) \mathbf{B}_R^{-1}(1) , \qquad (10)$$

which ensures the agreement between **Y** and **W** in the steady state.

3. DECOUPLING METHOD WITH PRECOMPENSATOR

Compensator is placed ahead of the system according to Fig.2.

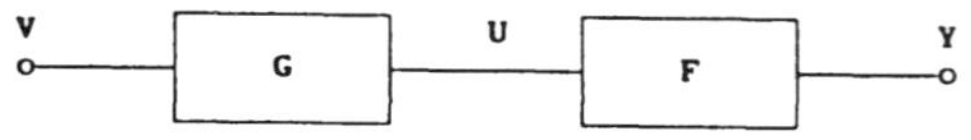

Fig.2. System with compensator

Its transfer function was choosen in the form

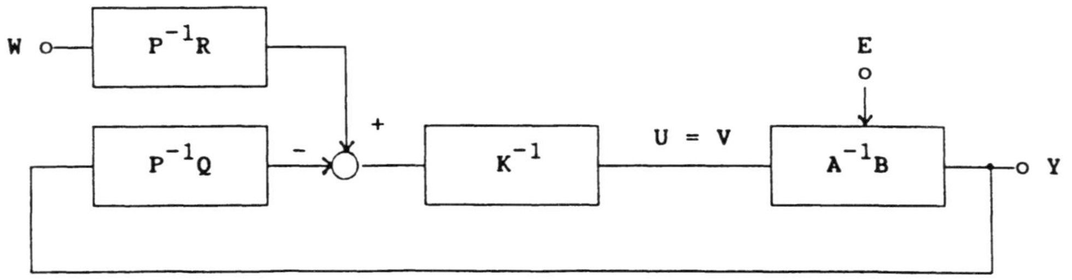

Fig.1. Block diagram of the control system

$$\mathbf{G} = q^{1+d} \text{ adj } \mathbf{B} \ . \qquad (11)$$

The resulting transfer function of system with precompensator

$$\mathbf{F}_1 = \mathbf{F}\,\mathbf{G} \qquad (12)$$

is given by Eqs.(3) and (11)

$$\mathbf{F}_1 = \mathbf{A}^{-1}\,\mathbf{B}\,q^{1+d} \text{ adj } \mathbf{B} = \mathbf{A}^{-1}\,q^{1+d} \det \mathbf{B} \ . \qquad (13)$$

From the comparison of Eqs.(3) and (13) it follows, that the expression $q^{1+d} \det \mathbf{B}$ in transfer function $\mathbf{F}_1$ is taken instead of $\mathbf{B}$ in $\mathbf{F}$. As the matrix $\mathbf{F}_1$ is diagonal, the controller matrices **P**, **Q** and **R** are diagonal too. The right matrix fraction of the system need not to be calculated and Eq.(8) for controller parameters has the simpler form without matrix operations

$$P_{ii}\,K_{ii}\,A_{ii} + Q_{ii}\,q^{1+d} \det \mathbf{B} = M_{ii} \ , \quad i=1,2 \ . \qquad (14)$$

The matrix **R**, according to Eq.(10), is equal to

$$R_{ii} = M_{ii}(1) \ / \det \mathbf{B}\,(1), \quad i=1,2 \ . \qquad (15)$$

4. DECOUPLING METHOD WITH PRECOMPENSATOR AND UNIT FEEDBACK

The block diagram of the system with precompensator and unit output feedback is on the Fig.3.

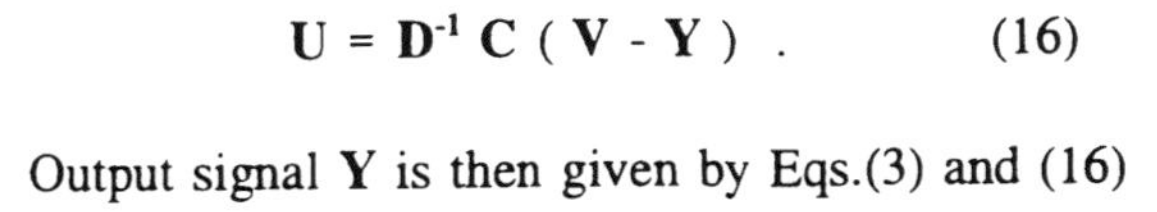

Fig.3. System with compensator and unit feedback

The input variable **U** is given by

$$\mathbf{U} = \mathbf{D}^{-1}\,\mathbf{C}\,(\,\mathbf{V} - \mathbf{Y}\,) \ . \qquad (16)$$

Output signal **Y** is then given by Eqs.(3) and (16)

$$\mathbf{Y} = (\,\mathbf{I} + \mathbf{A}^{-1}\,\mathbf{B}\,\mathbf{D}^{-1}\,\mathbf{C}\,)^{-1}\,\mathbf{A}^{-1}\,\mathbf{B}\,\mathbf{D}^{-1}\,\mathbf{C}\,\mathbf{V} \qquad (17)$$

and using the right matrix fraction (3)

$$\mathbf{Y} = \mathbf{B_R}\,(\,\mathbf{D}\,\mathbf{A_R} + \mathbf{C}\,\mathbf{B_R}\,)^{-1}\,\mathbf{C}\,\mathbf{V} \ . \qquad (18)$$

The poles of the transfer function $\mathbf{F}_2$ of whole circuit from Fig.3 are placed to zero

$$\mathbf{D}\,\mathbf{A_R} + \mathbf{C}\,\mathbf{B_R} = \mathbf{I} \ . \qquad (19)$$

In the same time the decoupling must be ensured, it means that the matrix $\mathbf{F}_2$ must be diagonal. This condition is fulfilled, when the compensator matrices are choosen in the form

$$\mathbf{C} = \mathbf{C}_d\,q^{1+d} \text{ adj } \mathbf{B}_R$$
$$\mathbf{D} = \mathbf{D}_d \text{ adj } \mathbf{A}_R \ , \qquad (20)$$

where $\mathbf{C}_d$ and $\mathbf{D}_d$ are diagonal matrices. The Eq.(19) has then the form of

$$\mathbf{D_d} \text{ adj } \mathbf{A_R}\,\mathbf{A_R} + \mathbf{C}_d\,q^{1+d} \text{ adj } \mathbf{B}_R\,\mathbf{B}_R = \mathbf{I} \qquad (21)$$

and after rearranging

$$\mathbf{D_d} \det \mathbf{A_R} + \mathbf{C}_d\,q^{1+d} \det \mathbf{B}_R = \mathbf{I} \ . \qquad (22)$$

The calculation of matrix $\mathbf{C_d}$ is simple, as all matrices in (22) are diagonal. Besides that the both diagonal terms are equal. The resulting transfer function is given by (18), (19) and (20)

$$\mathbf{F}_2 = q^{1+d} \det \mathbf{B}_R\,\mathbf{C}_d \ . \qquad (23)$$

From comparison with Eq.(3) it is evident that $\mathbf{A} = \mathbf{I}$ and $\mathbf{B} = \mathbf{F}_2$.

The controller parameters are given by modification of Eq.(8)

$$P_{ii}\,K_{ii} + Q_{ii}\,q^{1+d} \det \mathbf{B}_R\,C_{d,ii} = M_{ii} \ , \quad i=1,2 \qquad (24)$$

and the Eq.(10)

$$R_{ii} = M_{ii}(1) \ / \det \mathbf{B}_R(1)\,C_{d,ii}(1) \ , \quad i=1,2 \ . \qquad (25)$$

5. SIMULATION EXPERIMENTS

The experimental distillation column has seven bubble cap trays with inner diameter 150 mm. The two main control loops were closed between the 2nd tray composition (controlled variable y_1) and the reboiler heating power (manipulated variable u_1) and between the 6th tray composition (controlled variable y_2) and the reflux flow (manipulated variable u_2) - see the Fig.4. The other separated auxiliary loops stabilized the flow and the temperature of feed as well as the temperature of reflux. The compositions of the binary mixture methanol-water were measured indirectly by means of the tray temperatures (the recount of temperature into the composition is for known atmospheric pressure unambiguous).

The column was connected with the computer PC/AT by the interface unit. The measuring instruments and control elements were included into the system, so that all input and output values were represented by the electric voltage in range $\pm$ 5 V.

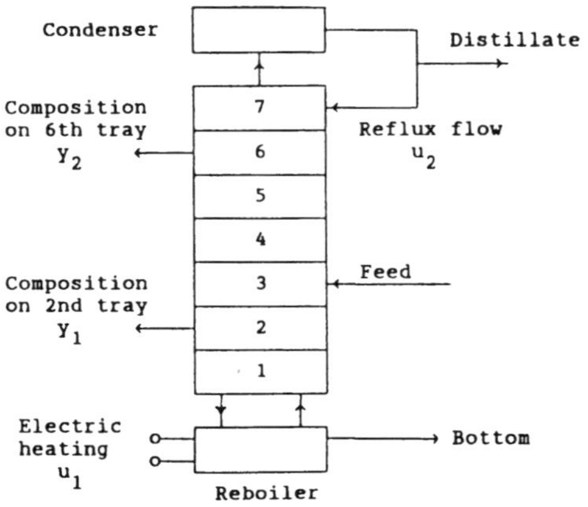

Fig.4. Experimental distillation column

The model of dynamic behaviour of the column was obtained from the identification experiment in the steady state with following parameters:

Feed flow	0.5 l min^{-1}
Mixture compositions	0.25 mol frac. of meth.
Power of. el. heating	7.5 kW

The two independent pseudorandom binary signals were brought at the same time to the both inputs. The column model was choosen in the form of Eq.(1) with 2nd order polynomials (n=2, m=2, d=0) and diagonal matrix **A**. The parameters of the model with sampling interval T=120 s are given by the following matrices (see Eq.(2)).

$$\boldsymbol{A}_1 = \begin{bmatrix} -0.80338 & 0 \\ 0 & -1.06726 \end{bmatrix}$$

$$\boldsymbol{A}_2 = \begin{bmatrix} 0.02616 & 0 \\ 0 & 0.14709 \end{bmatrix}$$

$$\boldsymbol{B}_1 = \begin{bmatrix} 0.03851 & -0.01062 \\ 0.01262 & -0.01990 \end{bmatrix}$$

$$\boldsymbol{B}_2 = \begin{bmatrix} -0.00386 & -0.00192 \\ 0.00316 & 0.00157 \end{bmatrix}$$

The simulation experiments were made using this model. The reference signals w_1 and w_2 were taken in the form of step changes. The output, input and reference signals as a time functions are shown in the Figs.5 to 7 for various types of controllers. First the two one-dimensional controllers were designed without respect to the inner interactions (Fig.5). The

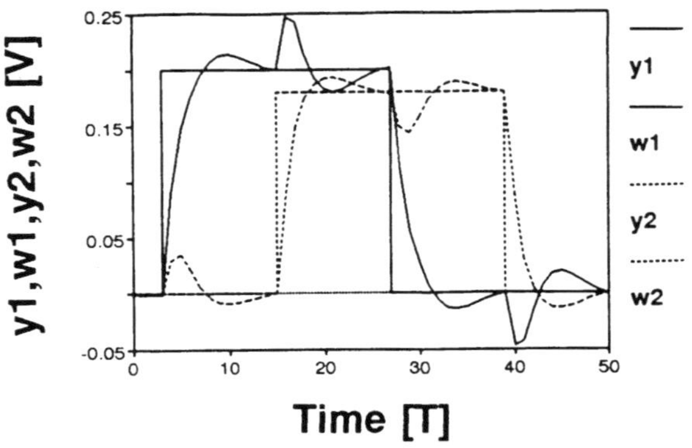

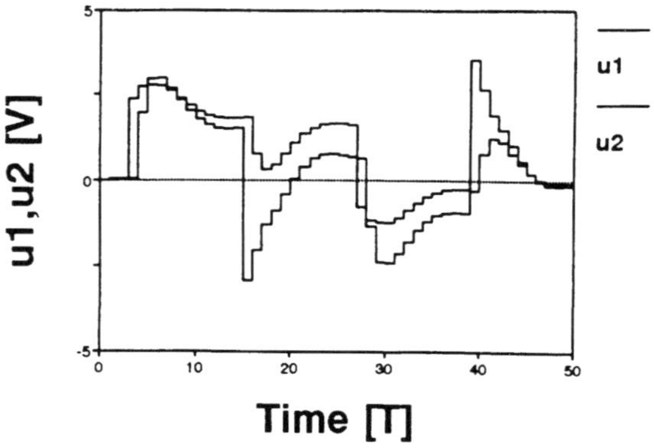

Fig.5. The time responses in the case of the two one-dimensional controllers

coupling of both controllers is evident but the system stability is not broken. The multivariable controller (MULTI) designed according to Eqs.(8) and (10) reduces the interactions but the signals are not absolutely independent, as it can be seen on the Fig.6. The decoupling method DEC 1 using Eqs.(14) and (15) and the one DEC 2 according to Eqs.(24) and (25) have almost the same responses and then only one Fig.7 is presented. Both control loops are in this cases independent. The decoupling methods are the best ones from the point of view of time responses quality. A good results were obtained by use of the investigated methods also in the case of step disturbances on both input and output of system.

The second order polynomials in matrix **M** were designed for all experiments for the 10% overshoot of controlled variables with maximum values in eighth minute (4 T) after the input step change was introduced. As the numerators of closed loop transfer functions differ in the various methods of controller design, the pole-placement must be made individually for every method. The placement of poles in the plane q, which gives the same choosen dynamic behaviour, is given in Tab.1.

The comparisons of several calculation aspects, as polynomial orders of both transfer functions and controller matrices as well as the necessity to calculate right matrix fraction of the system are given in Tab.2.

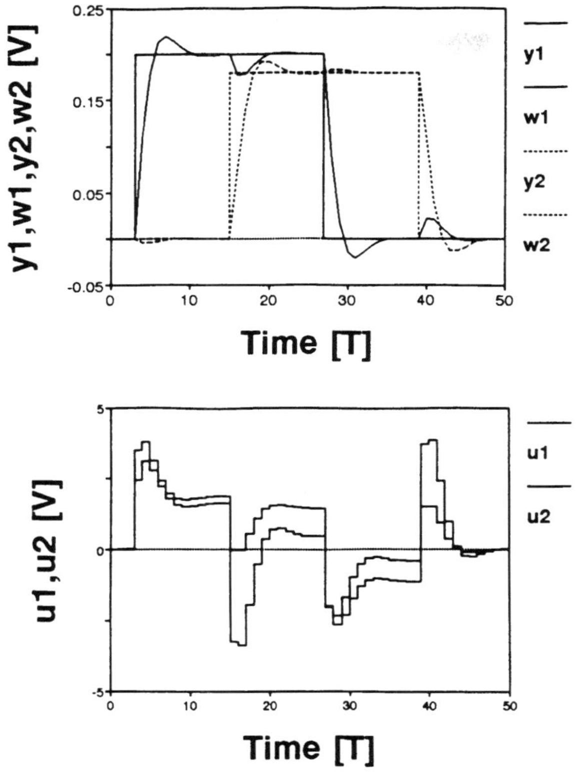

Fig.6. The time responses in the case of the multivariable controller

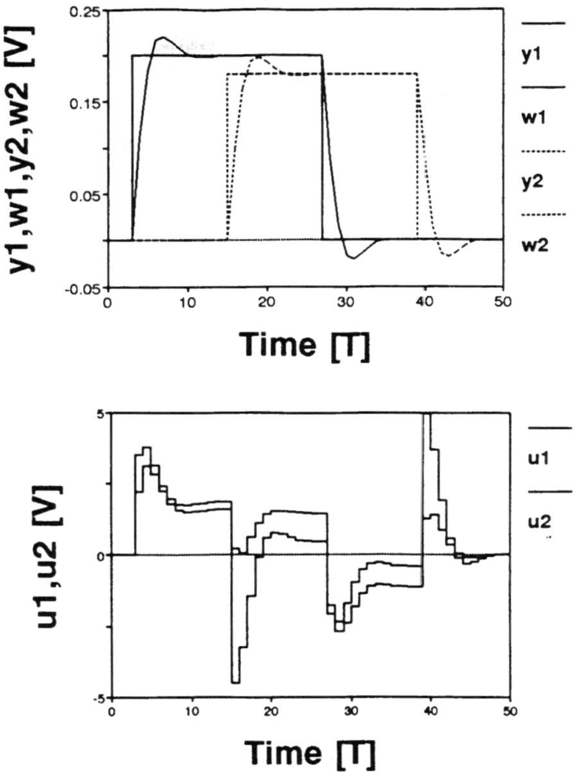

Fig.7. The time responses in the case of the decoupling

Table 1 Pole placement for used methods

Method	Poles
MULTI	0.48 ± i 0.34
DEC 1	0.48 ± i 0.34
DEC 2	0.53 ± i 0.32
2 controllers	0.48 ± i 0.44
	0.37 ± i 0.38

The method DEC 1 has the least demands on calculations and also the best time responses. That is why this method seems to be the most useful one for control of the column. The experimental results for the real column with DEC 1 method are given in the work of Macháček and Kotyk (1994), where the experimental equipment is also described in more details.

Table 2 Calculation aspects of methods

Method	Orders of polynomials: Transfer functions		P	Q	R	Right matrix fraction
MULTI	**F**	2/2	1	2	0	yes
DEC 1	$\mathbf{F}_1$	3/2	2	2	0	no
DEC 2	$\mathbf{F}_2$	6/0	5	0	0	yes

6. CONCLUSIONS

All three methods for multivariable control of the distillation column give good results and can be used for this purpose. The methods differ in extent of calculations. The big time consumption can cause difficulties by adaptive version of those methods, when the parameters of controller are calculated in every time interval.

REFERENCES

Chen, C.T. (1984). Linear System Theory and Design. Englewood Cliffs, NJ: Prentice-Hall.

Chien, I.L., Seborg, D.E. and Mellichamp, D.A. (1987). Self-Tuning Control with Decoupling. *AIChE J.*, **33**, 1079-1088.

Kinnaert, M., Hanus, R. and Henrotte, J.I. (1987). A New Decoupling Precompensator for Indirect Adaptive Control of Multivariable Linear Systems. *IEEE Trans.Aut.Control,* **AC-32**, 455-459.

Krishnaswami, P.R. *et. al.* (1991). Reference System Decoupling for Multivariable Control. *Ind.Eng. Chem.Res.*, **30**, 662-670.

Macháček, J. and Kotyk, J. (1994). Adaptive Decoupling Control of Distillation Column. The 3rd IEEE Conference on Control Application, Glasgow, 263-268.

Mikleš, J. (1990). A Multivariable Self-Tuning Controller Based on Pole-placement Design. *Automatica*, **26**, 293-302.

Peng. Y. (1990). A General Decoupling Precompensator for Linear Multivariable Systems with Application to Adaptive Control. *IEEE Trans.Aut.Control*, **AC-35**, 344-348.

Šebek, M. and Kučera, V. (1982). Polynomial Approach to Quadratic Tracking in Discrete Linear Systems. *IEEE Trans.Aut.Control*, **AC-27**, 1248-1250.

Tade, M.O., Bayoumi, M.M. and Bacon, D.W. (1986). Adaptive Decoupling of a Class of Multivariable Dynamic Systems Using Output Feedback. *IEE Proc.Pt.D*, **133**, 256-275.

Waller, K. (1974). Decoupling in Distillation. *AIChE J.*, **20**, 592-594.

Wang, Q.G. (1992). Decoupling with Internal Stability for Unity Output Feedback Systems. *Automatica*, **28**, 411-415.

Wittenmark, B., Middleton, R.H. and Goodwin, G.C. (1987). Adaptive Decoupling of Multivariable Systems. *Int.J.Control*, **46**, 1993-2009.

SIMULATION OF MULTICOMPONENT BATCH DISTILLATION WITH CHEMICAL REACTIONS PROCESSES

Marta S. Basualdo [1] (*), Carlos A. Ruiz and Nicolás J. Scenna

GIAIQ
Universidad Tecnológica Nacional - Facultad Regional Rosario
(*) Dpto. de Electrónica. FCEIyA. UNR E-mail: basualdo@ifir.edu.ar.
FAX 54-41-821772. Riobamba 326 -2000 Rosario. Provincia de Santa Fe
Argentina

Abstract A well proven multicomponent, reactive, continuous distillation dynamic model has been enhanced to predict batch operations behavior, including simultaneous chemical reactions. This program has been implemented for personal (80486 based) microcomputers in a package called **READYS** (**REA**ctive **D**istillation **DY**namic **S**imulator). Results obtained with the original model applied to multicomponent continuous distillation, startup, shutdown and multiple columns arrangements (distillation trains) were published in several recent papers, demonstrating its good performance. Due to the actual revival in batch operations interest the model has been adapted to handle batch reactive distillation operations. Simulated results for two examples considering both startup and production periods, are presented.

Keywords Reactive batch Distillation Dynamic Simulation Rigorous model

1-INTRODUCTION

Batch processing is the oldest operational mode used for liquid mixtures separation. Actually, it is becoming more popular as the chemical process industries move increasingly toward manufacturing fine and specialty chemicals. Batch distillation columns, operated in transient mode by its own nature, are an important part of this processes. Thus, there is a need for the development of reliable dynamic models for such unsteady state systems.

The concept of reactive distillation is not new. This technique was first applied in the 1920's to esterification processes using homogeneous catalysts and was reviewed by Keyes in 1932. Chemical reactions and multistage distillation can be carried out simultaneously. This combined unit operation is specially suits for those chemical reactions where equilibrium limits the conversion. By continuously separating products from reactants while the reaction is in progress, it can proceed to a much higher level of conversion than otherwise possible. Therefore, exists an increasing attention to reactive distillation systems whose capital and operational costs are cheaper than the equivalent conventional reaction-separation processes. Hence, batch distillation can be conducted, too, with simultaneous reaction, especially when equilibrium must be modified by extracting one or more products from the reaction mixture. Chemical reactions that are characterized by unfavorable reaction equilibrium, high heat of reaction and significant rate of reaction at distillation temperatures are particularly good candidates for reactive distillation, being the production of ethers a particularly promissory area . In this context it is clear that a need exists for a reactive distillation dynamic model capable to simulate such complex processes.

Several dynamic distillation models have been published in the past, however the vast majority contains oversimplifications introduced to overcome numerical troubles or to reduce the problem dimen-

[1] Author to whom all correspondence should be subbmitted

sionality. Therefore poor dynamic prediction can be expected specially during highly discontinuous and transient periods such as startup, shutdown or batch distillation. In the present work, two cases are studied, one taken from literature consisting on the ethyl acetate obtention and the other proposed here. Both, startup and production periods are simulated accounting for the chemical reaction in both stages.

2-DYNAMIC MODEL DESCRIPTION

An enhanced version of the dynamic simulation model described in Ruiz (1986) has been developed. This computer package, implemented in FORTRAN , is actually identified as **READYS**. In previous works, the model has been applied successfully to simulation and control of continuous single and multiple distillation columns (Gani et al., 1985, 1986a, b and c), startup / shutdown of both single and multiple columns (Ruiz et al. 1988), continuous reactive distillation (Ruiz et. al, 1995) and batch distillation without chemical reactions in Basualdo et. al. (1993). This rigorous model, capable to simulate simultaneous chemical reaction, taking into account variations in internal liquid and vapor rates, variable tray liquid holdups including plate hydraulics is presented in this work. Previously, a brief description of the model is done.

In few words, the model consists in a set of ordinary differential equations (ODEs) representing the mass and energy balances related to the system of column/s under study. These ODEs are derived from the mass and energy balances around each plate of the column/s, reboiler, condenser, accumulators and controllers: In equations (1) and (2) given bellow the "differential variable", includes the component and total energy holdups in each tray of all the columns, condensers, reboilers and receivers of the system. It may include the integral of the error, if any PI or PID controller is used. The "procedure variables", include flow rates, pressure drops, plate compositions, temperatures, pressures, chemical reaction evaluations and the input / output specifications. In addition the "design variables " include details of column geometry, column configuration and optional calculations. Time is the independent variable. The dynamic model allows the user to fix and/or control any of the usually manipulated variables such as reboiler heat duty or vapor boilup, distillate rate, reflux rate or ratio, condenser heat duty, etc. Top and bottom dynamics can also be considered and detailed plate hydraulics calculation is available as a user option.

Component mass balance around plate p of column c for component i:

$$\frac{dm^c_{p,i}}{dt} = F^c_p z^c_{p,i} + V^c_{p+1} y^c_{p+1,i} + L^c_{p-1} x_{p-1,i} + EV^c_{p-1} y^c_{p-1,i} + EL^c_{p+1} x^c_{p+1,i} - (PV^c_p + V^c_p + EV^c_p) y^c_{p,i} - (PL^c_p + L^c_p + EL^c_p) x^c_{p,i} + RE^c_{p,i}$$

$$\text{with } c = 1(1)NC;\ p = 1(1)np^c;\ i = 1(1)nc \tag{1}$$

Overall energy balance around plate p of column c:

$$\frac{dE^c_p}{dt} = F^c_p Fh^c_p + V^c_{p+1} H^c_{p+1} + L^c_{p-1} h_{p-1,i} \; EV^c_{p-1} H^c_{p-1} + EL^c_{p+1} h^c_{p+1} - (PV^c_p + V^c_p + EV^c_p) H^c_p - (PL^c_p + L_{p_c} + EL^c_p) h^c_p + Q^c_p + QR^c_p \tag{2}$$

$$\text{with } c = 1(1)NC;\ p = 1(1)np^c$$

The program ***READYS*** can work with different user selectable thermodynamic packages, such as those based on Soave-Redlich-Kwong equation of state, Grayson and Streed correlation or UNIFAC group contribution method. Production or consumption of components due to chemical reaction is computed by a user added routine, allowing to define the plate(s) where reaction occurs, reaction order, stoichiometric coefficients and rate constants (i.e., evaluated via an Arrhenius type equation). Instantaneous, non reversible reactions, can be modeled by fixing appropriately the rate constants.

For the treatment of chemical reaction in distillation systems the traditional approach is to consider equilibrium chemical reactions irrespective of the kinetics, residence time, holdup or other hydraulic considerations. Almost all the previous works for reactive distillation simulation, consider chemical equilibrium. Thus, the calculation of the hydraulic performance is not necessary in those previous models. In this work, special attention will be focused on kinetics because, from a thermodynamic simulation viewpoint, the reaction paths are so important as the final steady states. Condenser, reboiler and tray holdups are considered in this model. The following basic assumptions are made:

(i) Molar vapor holdup is negligible compared to the molar liquid holdup;
(ii) Liquid and vapor leaving each plate are in thermal equilibrium;

(iii) The definition of Murphree plate efficiency applies for each plate irrespective of the chemical reactions in the liquid phase;

(iv) All chemical reactions occur in liquid phase;

(v) There is the possibility of considering

vi) Liquid and vapors are perfectly mixed.

About condition (iii) no methods are reported for efficiency evaluation with simultaneous chemical reactions. Then, the classical Murphree definition is employed. From the "chemical point of view" assumption (vi) is the most restrictive. In fact, this is equivalent to assume each plate as a completely mixed stirred tank reactor. Hence, mixing models for liquid phase are not used. Many different types of trays in the column can be simulated by **READYS**. Chemical reactions can occur elsewhere in the column or in specific locations along it .

2-1 Chemical reaction related equations

The reaction terms $RE^{c}_{p,i}$ (moles reacted or produced per unit of time) are calculated through kinetic expressions, defined in a user added subroutine. The liquid volumes retained on plate p, $lv_p^{\,c}$, are calculated by the detailed hydraulic package included in the model. The kinetic expression used in this work (power law) has the following form:

$$r^{c}_{k,p} = K^{c}_{Dk,p} \prod_{i=1}^{nc} (x^{c}_{p,i}\rho^{c}_{p})^{\alpha_{i,k}} - K^{c}_{Ik,p} \prod_{i=1}^{nc} (x^{c}_{p,i}\rho^{c}_{p})^{\beta_{i,k}} \tag{3}$$

here, ρ is the density of the liquid mixture, $K_{D,k}$ is the constant of the direct reaction, $K_{I,k}$ the constant of the inverse reaction, $\alpha_{i,k}$ the direct reaction exponent and $\beta_{i,k}$ the inverse reaction exponent.

$$RE^{c}_{p,i} = \left(\sum_{k=1}^{nr} r^{c}_{p,k} \; u_{i,k} \right) l^{c}_{v_p} \tag{4}$$

where nr is the number of independent chemical reactions and $u_{i,k}$ are the stoichiometric coefficients.

The reaction constants are expressed as:

$$\begin{aligned} K^{c}_{Dk,p} &= K_{0D_k} \exp\left(\frac{-E_{Dk}}{R\,T^{c}_{p}}\right) \\ K^{c}_{Ik,p} &= K_{0I_k} \exp\left(\frac{-E_{Ik}}{R\,T^{c}_{p}}\right) \end{aligned} \tag{5}$$

Where K^{C}_{0I} and K^{C}_{0D} are the pre-exponencial constants (or frequency factors) for the reaction k. E_{Dk} and E_{Ik} are the activation energies, for the direct and inverse reactions, respectively, R is the gas-law constant and T_p is the plate temperature (in K). It is possible to consider eventual composition functionality for the K^{C}_{Ik} (or K^{C}_{Dk}) values. If the power law kinetics is inadequate, **READYS** allows the user to specify other kinetic reaction equations employing user-supplied subroutines. In the literature many reaction data can be found as the equilibrium constants defined as:

$$K^{C}_{ek,p} = \frac{K^{C}_{D,k,p}}{K^{C}_{Ik,p}} \tag{6}$$

By choosing the appropriate combination of both, direct and inverse reaction kinetic constants, it is possible to simulate the steady state behaviour of such systems. Although, the transient path will be very different, depending on the order of magnitude selected for the individual constants (i.e., among instantaneous and very slow reaction rates).

Instantaneous reactions are simulated by adopting very high values for each kinetics constant but the ratio between them is the same as the equilibrium constant. This technique, applied over several problems, can handle efficiently numerical troubles but the dynamic trajectories will be not realistic. This kind of problem arise when equilibrium based algorithms for non instantaneous chemical reactions are used for dynamic simulation purposes. Obviously, if reversible reactions occur (low kinetics rates) that strategy can not be used.

Heat of reaction is calculated by the following equation:

$$Q\,R^{C}_{p} = \left(\sum_{k=1}^{nr} Q^{C}_{k,p}\, r^{C}_{k,p} \right) l^{C}_{v_p} \tag{7}$$

where each $Q_{k,p}{}^{C}$ is evaluated or given by the user at the reference temperature

3 -SOLUTION OF THE MODEL EQUATIONS

The number of differential equations (ODEs) will depend on the assumption set. The ***generalized*** model is the most detailed and involves all the differential equations described previously. A ***simplified*** form of the model is used whenever instantaneous rate of change for the energy holdup on each plate can be considered. In this case pressure profile is held constant. Hence two different models can be used, depending on the problem complexity:

For the *generalized* model, the problem to be solved is defined as:

"Given the liquid phase composition and enthalpy,

determine the equilibrium vapor-phase composition, temperature and pressure"

For the *simplified* model, the problem to be solved is defined as:

"Given the liquid phase composition and pressure, determine the equilibrium vapor phase composition and temperature"

This set of procedures involves the thermodynamic packages for the equilibrium prediction. Newton-Raphson method is used to update temperatures and pressures on "equilibrium" calculation. The hydraulic routine calculates the liquid and vapor flow rates, checks for flooding, weeping or entrainment conditions and finds out the weeping and entrainment rates. The correlations used for determining the hydraulic variables are mainly empirical. The particular set employed in this work are the typically used in industry. Four different types of plates are considered: sieve plates, two types of valve plates and bubble-cap plates. For special trays, it is possible to incorporate a user subroutine for hydraulic calculations. If the plate pressures are constant, the vapor flow rates are determined algebraically from energy-balance by considering neglected the enthalpy variations. The ODEs are integrated by modified Gear method.

4-RESULTS AND DISCUSSION

To illustrate implementation of READYS package and its performance two examples were run on a PC-80486, which are described bellow:

* separation of two components A, B which simulation is done using the generalized version,

* esterification for the ethyl acetate production using the simplified version.

Case 1, an example specially formulated for this work is presented. In this case, **time varying** column pressure is allowed. The generalized model is applied here, controlling the column top pressure with a Proportional-Integral controller. The following reaction occurs in all plates and still pot:

$$A \rightleftarrows B$$

The reaction term $R^{C}_{p,i}$ is calculated through kinetic expressions given above and the liquid volume retained in plate p is calculated by the hydraulic package included in the program. The pre-exponential terms are evaluated at reference temperature. Table 1 summarizes this case where the adopted values for the chemical reaction are listed in "Notes" Figure 1 shows the evolution of plate pressures considering chemical reaction on the plates and still-pot. Here, the effect of good control action can be evaluated from the scheme.

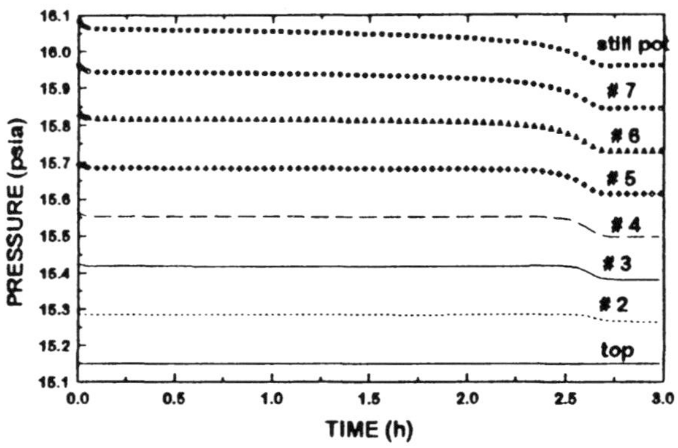

Figure 1: dynamic behavior of trays pressure during production period.

Figure 2a shows the component A evolution on plates and still pot for the production period when top pressure is under PI control. The component B variations in time can be seen in figure 2b. From the scheme it is observed that profiles change abruptly because component B is continually generated by chemical reaction and drown in the destilate until component A is practically consumed. This effect is also, noted at figure 1 when pressure plates decreased at the same time specially in the still pot.

a)

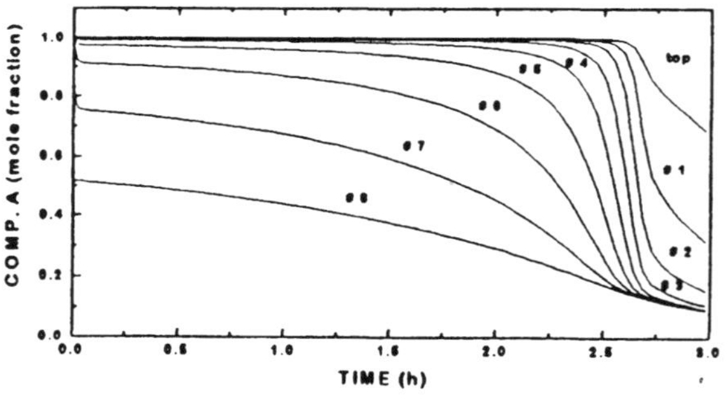

b)

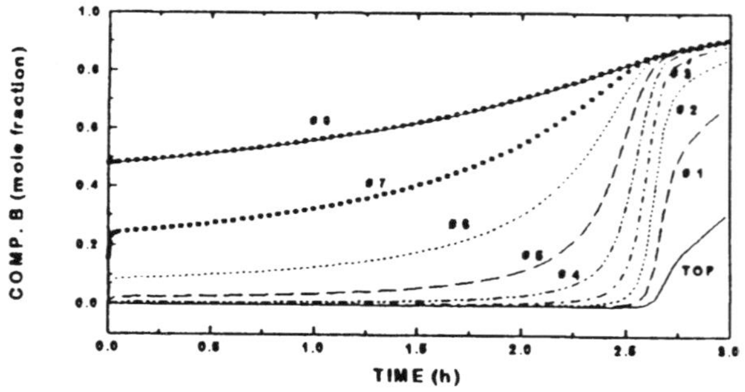

Figure 2: dynamic responses of components A and B during the production period.

Case 2: This problem was taken from Holland (1981) and also proposed to evaluate the **ASPEN** simulator performance in the work of Venkataraman et al. (1990). But, in the previous works a continuous reactive distillation was done. Here the same example is analyzed consisting on the ethyl acetate production by separating the components in a batch reactive distillation form. The details of the system are given in Table 1. The still pot is filled with saturated liquid of acetic acid (A), ethanol (B), and water (C) which compositions are the same given in the cited works. The kinetic reaction is assumed to occur in the liquid phase on all the stages including the still pot, condenser and reboiler. The objective is the production of ethyl acetate (D), according to the following reaction:

$$A + B \rightleftarrows C + D$$

The simplified version of **READYS** is chosen. In this work, holdup varies slightly tray by tray due to the hydraulic calculation (Francis weir formula is used). The initialization is done by considering that chemical reaction occurs in all stages and still pot at the start up stage, working at infinite reflux ratio. The same composition along the column is assumed. Then, for the production period, the reaction continues but the reflux ratio is decreased to ten. Figure 3 shows the components evolution both in top and bottom sections for the two periods.

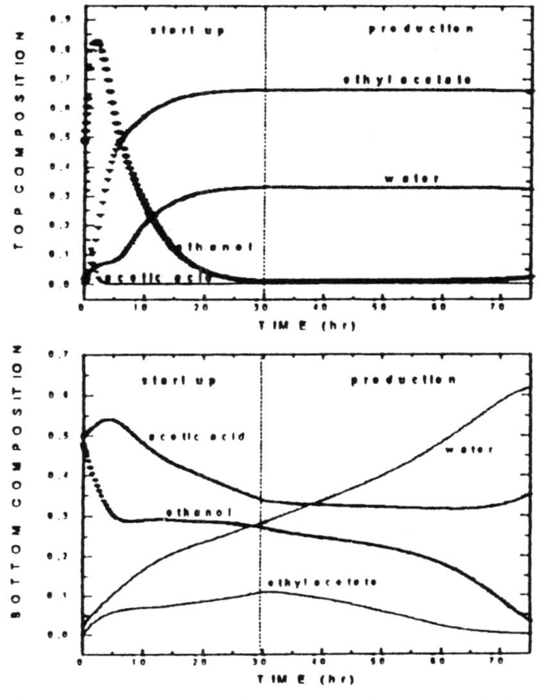

Figure 3: dynamic behavior of top and bottom compositions during start up and production periods.

As can be seen in the scheme top composition consists on ethyl acetate (Etac) and water because the reactants have been consumed at the start up period. Whereas in the bottom, at the start up period, when reaction begins, the Etac composition increased until the production period starts. When the Etac is drown its bottom composition decreased simultaneously with ethanol composition which acts as a limiting reactant. In figure 4 the Etac composition dynamic behavior for many plates can be seen.

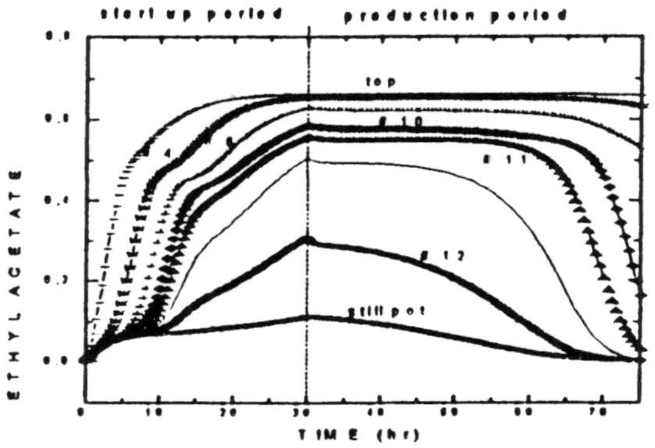

Figure 4: dynamic behavior of Etac plates composition during start up and production periods.

Note the long time required due to the inherently slow reaction rates of this system. In figure 5 the temperature plates evolution for both periods are shown, specially for the still pot the temperature growth becomes evident when Etac is totally drained with distillate. Figure 6 shows internal liquid flow rates for some plates only from production period. Here it can be seen how they decreased when the reflux ratio magnitudes decreased from infinite to ten .

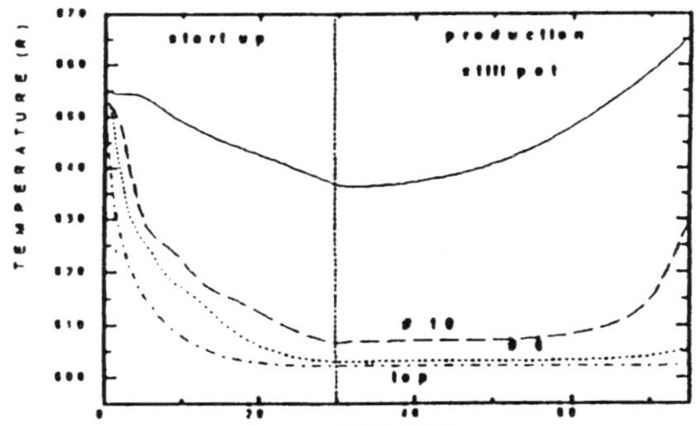

Figure 5: dynamic behavior of plates temperature for start up and production periods.

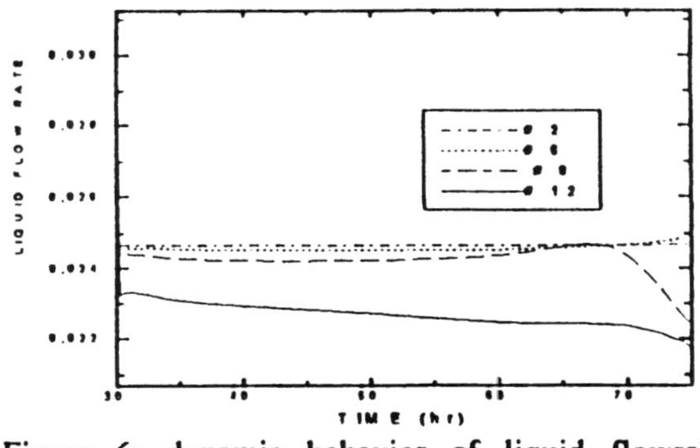

Figure 6: dynamic behavior of liquid flowrates (lbmol/h) for production period only.

Table 1: Description of tests problems

Test #:	I	II
Reference:	This work	Venkataraman al. (1990) modified for this work
General Data:		
Number of Trays	8	13
Number of Components	2	4
Model Option	generalized	simplified
Equilibrium Model:	Chao.Seader	UNIFAC
Total Number of ODEs	24	52
Operational Data:		
Type of Condenser	total	partial
Type of Reboiler	total	partial
Column Top Press. (KPa)	104.2	101.2
Reboiler Heat Duty (KW)	527.4	0.156
Reflux Ratio:		
start up period	∞	∞
production period	2.00	10
Initial charge Composition (mole fraction):		
Comp. A	0.3	0.4963
Comp. B	0.7	0.4808
Comp. C	----	0.0229
Comp. D	----	0.000

Notes

Test Problem I

Reaction Scheme:

K_{0D}	K_{0I}	E_{0D}	E_{0I}
Kmol/m3 sec	Kmol/m3 sec	Joule/gmol	Joule/gmol
100000.0	1.0	50000.0	50000.0

	A	B
Reaction Direct Exponents (α)	1.0	0.0
Reaction Inverse Exponents (ß)	0.0	1.0

Component Data:

	Units	A	B
Critical Temperature	K	507.90	469.78
Critical Pressure	KPa	14.03	15.62
Critical Volume	m3/Kmol	0.368	0.311
Molecular Weight	Kg/Kmol	58.12	58.12
Acentric Factor	dimensionless	0.293	0.239
Hildebrand's Solub. Parameter	(cal/cm3)**0.5	7.266	7.021
Characteristic Molar Volume	m3/Kmol	0.018	0.015

Test Problem II

Reaction Scheme:

k0D	k0I	E0D	E0I
Kmol/m3 sec	Kmol/m3 sec	Joule/gmol	Joule/gmol
483.33	123.00	59445.1	59445.1

	A	B	C	D
Reaction Direct Exponents (α)	1.0	1.0	0.0	0.0
Reaction Inverse Exponents (ß)	0.0	0.0	1.0	1.0

5-CONCLUSIONS

In this work READYS capabilities on handle reactive batch distillation operations are demonstrated. It can simulate properly a wide variation of operational conditions such as the plate hydraulic effects, time varying pressure, conventional control structures and so on. Two academic examples were analyzed here, the first was specially design as hypothetical study case and the other obtained from the literature as a continuous esterification and transformed in a batch example. The results obtained from both seem to indicate proper dynamic responses for generalized and simplified work option.

6-REFERENCES

Basualdo M., Ruiz C. and Campanella E., 1993; "Destilación por lotes ". Chem. Eng. of Spain, Number 294, pp.385-391

Gani, R., C. A. Ruiz and I. T. Cameron, IChemE Symp. Ser., 92, p. 353 (1985)

Gani, R., M. Albuja-Cuadrado, C. A. Ruiz and M. N. Karim, 1986a; "Analysis of the Dynamic Behavior of Distillation Trains", Proc. DYCORD '86 Meeting, IFAC, Bournemouth, England

Gani, R., C. A. Ruiz and M. N. Karim, 1986b; "A Control Strategy for Start-Up Operations of Distillation Columns", Proc. World Congress III of Chem. Eng., Tokyo, Japan

Gani, R., C. A. Ruiz and I. T. Cameron, 1986c; Comp. Chem. Eng., 10, No. 3, p. 181

Holland C.D., 1981; "Fundamentals of Multicomponent Disttillation"McGraw-Hill, New York.

Keyes, D. B., 1932; Ind. Eng. Chem., 24, p. 1096

Ruiz, C. A., 1986; "Desarrollo de una Política de Control para Operaciones de Puesta en Marcha de Columnas de Destilación", Ph.D. Chem. Eng. Thesis, Universidad Nacional del Sur, Bahía Blanca, Argentina.

Ruiz, C. A., I. T. Cameron and R. Gani, 1988; Comp. Chem. Eng., 12, No. 1, p. 1

Ruiz C., Basualdo M. and Scenna N.; 1995, "Reactive distillation. Dynamic Simulation", IChem E, (in press)

Venkatamaran, et al., 1990, Chem. Eng Progr.,pp.45-54

PHYSICAL IMPLICATIONS OF PERIODIC SOLUTIONS IN A FIXED BED REACTOR WITH RECYCLE

Bodil Recke and Sten Bay Jørgensen

Technical University of Denmark

Department of Chemical Engineering

DK2800 Lyngby, Denmark

Abstract: Recycling is often used in industrial processes to reduce cost of raw material and energy supply. The nonlinear effects of introducing recycling on a fixed bed reactor are investigated in the present work. A first order exothermic reaction is chosen as a model reaction. The three cases of recycling investigated are recycling of unconverted reactant, external heat exchange and combined mass and energy. The findings are that bifurcations occur as the operational variables are changed. Investigating the periodic solutions show, that the limit cycle periods in all the investigated cases of recycling, are simple fractions(1, $\frac{1}{2}$, $\frac{1}{3}$) of the thermal residence time. Consequently three periodic solutions appear in various sections of the operational space. A physical interpretation of the periodic solutions is that they correspond to one, two and three ignition points respectively in the reactor. These periodic solutions are further investigated and compared to see if it is the same dynamic effect that causes them to appear in the three recycle cases. The implications of the periodic solutions upon the control structure selection for mass recycle will also be considered.

Keywords: Periodic motion, Partial Differential equations, Instability

1. INTRODUCTION

The concept of recycling can reduce production costs radically, however the introduction of recycling complicates the dynamical behaviour of fixed bed reactors. Consequently it is necessary to fully understand the dynamical effects that are introduced by the recycle. The investigation of multiple steady state solutions have been thoroughly described in the literature, but the physical consequences of periodic solutions have not received such extensive coverage. It is the aim of this article to investigate the effects of periodic solutions, in the case where only one periodic solution exists, as well as in the case of multiple periodic solutions, with or without the existence of additional multiple steady states. The existence of Hopf bifurcations leading to periodic solutions has been reported in several papers, however most of them investigated a system without recycle. The periodic solutions which can occur in a CSTR is described thoroughly in Uppal et al. (1974), this article is also concerned with the effect of variations in the physical parameters. These results where extended by Balakotaiah and Luss (1983). The dynamical behaviour of a fixed bed reactor without recycle is treated in Jensen and Ray (1982), this article is based on a simulation study and include examination of the effect of changes in physical parameters. Similar results have been presented by Puszynski et al. (1981), with the use of a model of a fixed bed reactor, even with high peclét numbers (in the area of the ones used in industry). The fixed bed reactor with internal energy recycling is treated in Bonvin et al. (1980). Lovo and Balakotaiah (1992) investigated uniqueness boundary in the fixed bed reactor both with internal and external heat exchange, and found that multiple solutions could exist in both cases, but they did not report/observe any periodic solutions. Root and Schmitz (1969, 1970) investigated the fixed bed loop reactor and discovered multiple steady states along with periodic solutions but for low peclét numbers. The existence of multiple periodic solutions was first reported by Brabrand (1991), his results were obtained by simulation, and one periodic solution was substantiated by experimental results. The simulation results were systematically described in terms of bifurcation diagrams in Recke and Jørgensen (1994).The purpose of the present paper is further investigation of the similarities of the periodic solutions obtained with different types of recycle, first reported by Recke and Jørgensen (1994). Furthermore it is investigated if the periodic behaviour is similar in the three cases of recycle. Results from application of a controller on the fixed bed reactor with mass recycle are also described, in terms of the changes in the dynamic dependence of the inlet parameters.

2. METHOD

The simulation model used for the reactor is a pseudo homogeneous model consisting of two partial differential equations, describing the temperature profile and the oxygen concentration profile. The reaction is assumed to be first order and have an Arrhenius type temperature dependence. The recycling is described by first order dynamics, resulting in an ordinary differential equation for each type of recycling. Dankwertz' boundary conditions are used, thus assuring zero dispersive flux at the outlet, and continuity at the inlet. The heat exchanger is modelled by a heat exchange efficiency multiplied by the maximum driving force.

The oxygen concentration is assumed to be sufficiently small to have negligible influence on the total concentration. The major assumptions are listed below.

- Finite mass and heat transfer resistance between catalyst pellets and gas phase are accounted for by using $\frac{1}{Pe_i} = \frac{1}{Pe_{gi}} + \frac{1}{NTU_i}$ $i = m \; or \; h$
- The catalyst particles are assumed isothermal.
- Transport parameters are assumed constant throughout the reactor.
- Negligible water adsorption on catalyst pellets.
- Negligible radial temperature and concentration gradients.
- The heat accumulated in the gas phase is negligible compared to the heat accumulated in the catalyst pellets.
- The reaction rate expression is first order in oxygen concentration, with a Arrhenius temperature dependence.
- The mass and energy recycle can each be described by a simple first order system.

The model describing the reactor including both mass and energy recycling is

$$\varepsilon \frac{\partial x_1}{\partial t} = -\kappa \frac{\partial x_1}{\partial z} + \frac{1}{Pe_m} \frac{\partial^2 x_1}{\partial z^2} - R \quad (1)$$

$$\frac{\partial y}{\partial t} = -\kappa \frac{\partial y}{\partial z} + \frac{1}{Pe_h} \frac{\partial^2 y}{\partial z^2} + Bi(y - y_s) + \beta R \quad (2)$$

$$\tau_m \frac{dx_{1,rec}}{dt} = x_{1,out} - x_{1,rec} \quad (3)$$

$$\tau_h \frac{dy_{rec}}{dt} = y_{out} - y_{rec} \quad (4)$$

$$R = Dm x_1 e^{\gamma\left(1-\frac{1}{y}\right)} \quad (5)$$

Boundary Conditions

$$x|_{z=0} = x_{1,feed} + \alpha x_{1,rec} + \frac{1}{\kappa Pe_m} \left.\frac{\partial x_1}{\partial z}\right|_{z \to 0^+} \quad (6)$$

$$y|_{z=0} = y_{feed} + \xi\,(y_{rec} - y_{feed}) + \frac{1}{\kappa Pe_h} \left.\frac{\partial y}{\partial z}\right|_{z \to 0^+} \quad (7)$$

$$\left.\frac{\partial x_1}{\partial z}\right|_{z=1} = 0 \quad (8)$$

$$\left.\frac{\partial y}{\partial z}\right|_{z=1} = 0 \quad (9)$$

For solving this model orthogonal collocation (Villadsen and Michelsen 1978) is used. This procedure approximates each of the profiles by a sum of Lagrange polynomials. This results in N ordinary differential equation for each partial differential equation, where N is the number of internal collocation points. The model used for numerical simulations and locating bifurcation points thus consists of 2N+1 or 2N+2 ordinary differential equations depending on whether one or both types of recycle are used. The following simulation results are obtained with the dimensionless parameters listed at the end of this article, unless otherwise noted. The bifurcation diagrams in this article, are constructed by the use of the routine PATH (Petersen 1989). With this routine it is possible to locate fold, Hopf and period doubling bifurcations. Also it is possible to follow a path of bifurcation points as one or two parameters of the system are varied. The operating variables that have been chosen in this work are the dimensionless feed temperature and the dimensionless feed concentration.

3. RESULTS

Axial profiles for the periodic solutions with a period of approximately 1.1 are shown in figure 1 for combined recycle and in figure 2 for mass recycle. Comparing the time varying profiles in figure 1 to those with mass recycle in figure 2, shows that the dynamics of the mass recycle is quite different from the combined recycle. This difference is most evident when inspecting the location of the point of ignition in the reactor.

The location of an ignition point in the case of only one, can be realized from a drastic decrease in the slope of the concentration profile or by a sudden increase in the slope of the temperature profile. In the case of mass recycle, the ignition point may be located by determining a minimum temperature difference between the highest and lowest attainable temperature at discrete points along the reactor, i.e. an extinction of the thermal wave, as shown in figure 3. The periodic solution with period of approximately half the thermal residence time, shows two points of ignition in the reactor, that is one point of extinction of the thermal wave internally in the reactor, located at z=0.58. The periodic solution with a period of approximately a third of the thermal residence time, that is with three points of ignition, has two extinction points located at z=0.35 and z=0.66, for $y_{feed} = 0.77, x_{feed} = 1.7$. Since it is only possible to see both extinction points with large magnification, graphical illustration is omitted.

The difference between the maximum and minimum temperature in the reactor with different types of recycle can be seen in figure 4, for a cycling period of 1.1. From this figure it is obvious that the dynamic behaviour of the combined and the energy recycling case is very similar. Unfortunately it has not been possible to produce a similar plot for the periodic solutions with periods of approximately 0.54. In the case of mass recycle the thermal inlet conditions are constant and hence development of the cyclic behaviour is associated with the reactant coupling between inlet and outlet. This coupling generates a thermal wave which for a cycling period of 1.1 is amplified passing through the reactor as seen in figure 2 and 3. For a cycling period of 0.54 the thermal wave is first amplified (see figure 3), and then almost extinguished at z=0.58 inside the reactor, whereafter a new wave is generated. The point where the thermal wave becomes extinct represents a nonlinear node point. In the energy and combined recycle cases the thermal wave moves through the reactor and is then coupled back to the inlet temperature with the thermal recycle efficiency. Consequently for these types of recycle the ignition point / points are forced to move during a cycle, by the external energy coupling between outlet and inlet.

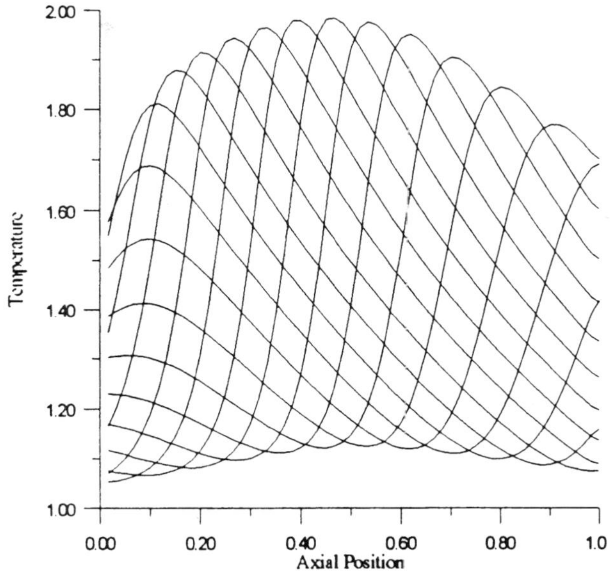

Fig. 1: Time varying temperature profiles for the fixed bed reactro with combined recycle ($\Delta t = 0.1$). $y_{feed} = 0.8, x_{feed} = 1.0, \xi = 0.8$

3.1. Bifurcation diagram with recycle efficiencies

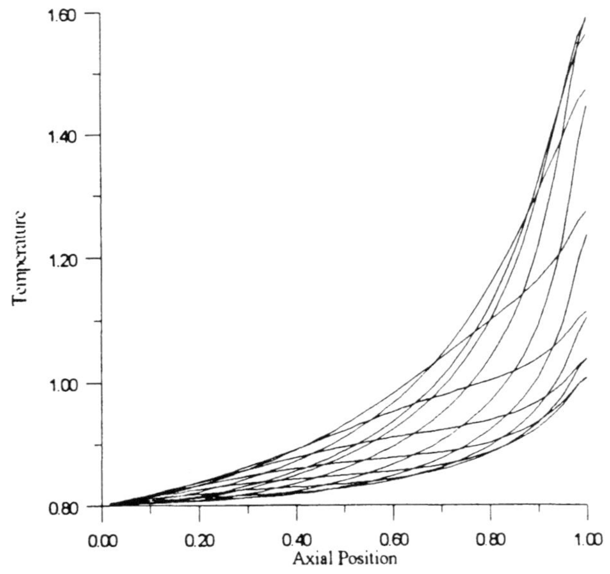

Fig. 2: Time varying temperature profiles for the fixed bed reactor with mass recycle ($\Delta t = 0.1$). $y_{feed} = 0.8, x_{feed} = 1.0$

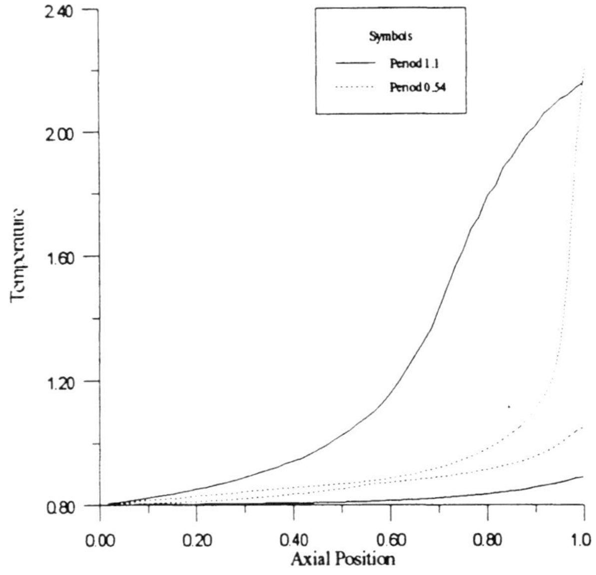

Fig. 3: Maximum and minimum temperature for different periodic solutions in the case of mass recycle. $y_{feed} = 0.8, x_{feed} = 1.25, \gamma = 9.629$

When the two recycle efficiency parameters, the heat exchange efficiency (ξ) and the degree of mass recycling (α), are varied the bifurcation diagram shown in figure 5 results. The feed temperature and concentration are held constant in this diagram, and consequently it would appear different with different values of the feed parameters, but all of the possible dynamical behaviours are believed to be represented in the figure.
The types of solutions which exists in the various areas are shown in table 1.

Table 1 Types of solutions in figure 5

Area	Stable ss	Unstable ss	Per. sol.
I	1	-	-
II	-	1	1
III	-	1	2
IV	-	1	2
V	-	1	3
VI	1	2	2
VII	1	2	2
VIII	1	2	2
IX	2	1	1
X	2	1	2

The difference between the areas with the same number and types of solutions, is the location of the attracting or repelling sets in the phase space, as explained in table 6 of Recke and Jørgensen (1994). However a few vital conclusions are listed here. The area located within the dashed·curves has three steady states. The Hopf bifurcations in these areas can occur either on the upper or the middle steady state. The almost vertical lines in the bifurcation diagram corresponds to the case where the Hopf bifurcation occurs on the upper steady state and the more horisontal ones depicts the locations of the Hopf points on the middle steady state. The very small area denoted III depicts an area with two periodic solutions, but it is only realy possible to se this area in a much larger magnification.

4. CONTROL STRUCTURE CONSIDERATIONS

It seems logical to avoid operation in areas containing unstable limit cycles, because even small disturbances can cause the process states to jump outside the limit cycle and thereby cause reaction runaway at worst. In the case where the unstable limit cycle is enclosed in a stable limit cycle, the consequences would not be as severe, but in a multidimensional state space it is difficult to insure that the unstable limit cycle is enclosed in ALL dimensions. In the case where the opera-

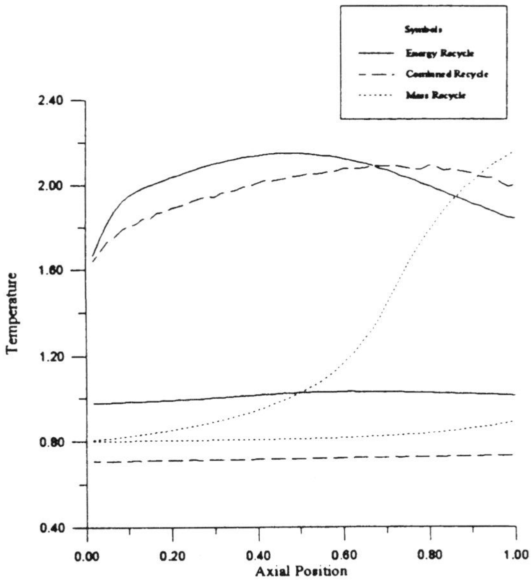

Fig. 4: Comparasion of the maximim and minimum temperatures with different types of recycle, for the periodic solution with period of approximately 1.1. Mass: $y_{feed} = 0.8, x_{feed} = 1.25, \gamma = 9.629$ Energy: $y_{feed} = 0.4, x_{feed} = 2.0, \xi = 0.8$ Combined: $y_{feed} = 0.4, x_{feed} = 0.8, \xi = 0.8$

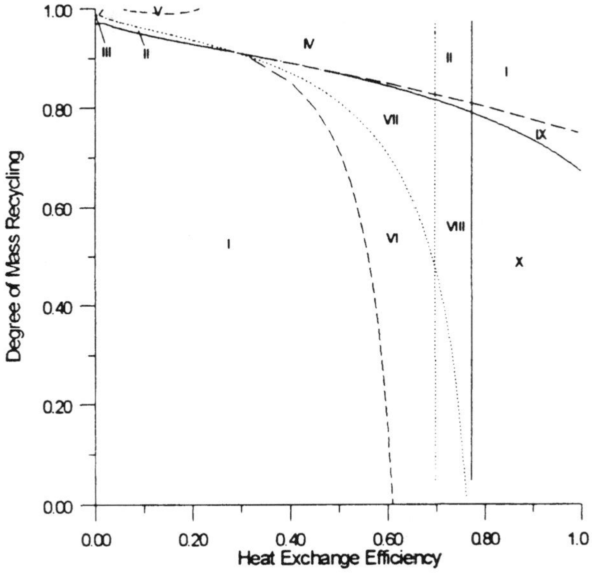

Fig. 5: Bifurcation Diagram for fixed bed reactor with heat exchange efficiency and degree of mass recycling as bifurcation parameters. $y_{feed} = 0.7, x_{feed} = 1.2$, Solid line: Hopf bifurcation to periodic solution with period 1.1, Dotted line: Hopf bifurcation with period 0.54, Dash-dotted line: Hopf bifurcation with period 0.34. Dashed line: fold bifurcation.

tion point is on the upper ignited branch, which is often the case when multiple solutions exists, the distance to the nearby Hopf bifurcation point should be considered, when deciding the operation point. This distance is crucial to the performance required of the controller to be implemented on the process.

4.1. Control of mass recycle reactor

In the case of the fixed bed reactor with mass recycle a simple PI-controller has been investigated. The reason for choosing this type of controller is that previous experimental results, Morud and Skogestad 1994, had shown that such a controller was able to stabilise an unstable steady state surrounded by a stable limit cycle. This work was however done on an ammonia converter, which corresponds to the combined recycling case, described in this article. The PI-controller was introduced to regulate the inlet concentration (combined fresh feed and recycling stream) to the reactor based on a concentration measurement at an arbitrary point in the reactor. This point was chosen in the end half of the reactor. The PI-controller was designed as an optimal controller (LQR) of SISO type, with appropriately chosen weight matrices. The effect on the bifurcation diagrams when the controller was implemented, was that the feed concentration had no effect on the bifurcation diagram, as would be expected since the inlet concentration to the reactor was controlled. When the feed temperature was varied it was seen that a unique stable steady state existed for inlet temperatures below 0.78 and for inlet temperatures above 0.79, regardless of the feed concentration. This shows that by applying a simple controller, the area where undesired dynamical behaviour occur, if periodic or multiple solutions are undesired, can be radically reduced by relatively simple means. It should be stated here that the controller applied in these investigations, had fixed gain and integral time, and it is the belief of the author that by a applying a controller which updates the parameters in a gain scheduling scheme, the area in which uniqueness of the steady state is not guaranteed, will vanish completely, but this remains to be proved.

5. CONCLUSION

The investigation in this paper has shown strong similarities between the periodic solutions occurring when energy or energy combined with mass is recycled. In the case of mass recycle the existence of periodic solutions with periods of approximately one, half and a third of the thermal residence time, corresponding to one, two or three points of ignition respectively, has been shown. Extinction of the thermal wave in the case of multiple ignition points, has also been shown. Furthermore it is shown that introduction of a simple controller can increase the area in the feed parameter space which exhibits uniqueness.

SYMBOLS

Bi	Biot number	0.5
Dm	Damköler number	0.376
Pe	Peclét number	h:118 m:270
r	Ratio of the time constant forrecycle to the thermal residence time	h:0.1 m:1/30
R	Reaction rate	
x	Dimensionless concentration	
y	Dimensionless temperature	
z	Dimensionless axial position	
α	Degree of mass recycle	1.0
β	Ratio between Damköler nr. in mass and energy balance	0.49
ε	Ratio between time constant in mass and energy balance	1/600
γ	Dimensionless activation energy	9.0
ξ	Heat exchange efficiency	
κ	Dimensionless flowrate	1.0

Subscript

1	oxygen
feed	feed conditions
h	heat
m	mass
out	outlet conditions
rec	recycle conditions
s	suuroundings (y_s:0.79)

REFERENCES

Balakotaiah V., D. Luss (1983). Multiplicity Features of Reacting Systems, Dependence of the Steady-States of a CSTR on Residence Time. *Chem. Eng. Sci.* **vol 38**, p.1709-1721

Bonvin D., R.G. Rinker, D. Mellichamp (1980). Dynamic Analysis and Control of a Tubular Autothermal Reactor at an Unstable State. *Chem. Eng. Sci.* **vol 35**, p.603-612

Brabrand H. (1991). *Dynamics, Identification and Control of a Fixed Bed Reactor with Reactant Recycle,* chapter 2. Department of Chemical Engineering, Technical University of Denmark, Lyngby.

Jensen K.F., W.H. Ray (1982). The Bifurcation Behaviour of Tubular Reactors. *Chem. Eng. Sci.* **vol 37**, p.199-222

Lovo M., V. Balakotaiah (1992). Multiplicity Features of Adiabatic Autothermal Reactors. *AIChE Journal* **vol 38**, p.101-115

Puszynski J., D. Snita, V. Hlavacek, H. Hofmann (1981). A Revision of Multiplicity and Parametric Sensitivity Concepts in Nonisothermal Nonadiabatic Packed Bed Chemical Reactors. *Chem. Eng. Sci.* **vol 36**, p.1605-1609

Petersen C.K. (1989). *PATH - User's Guide.* Department of Applied Mathematical Studies & Centre fir Nonlinear Studies, University of Leeds, UK

Recke B., S.B. Jørgensen (1994). Bifurcations in Fixed Bed Reactors as a Consequence of Recycle. *Presented at AIChE Annual Meeting in San Francisco.* (Unpublished).

Root R.B., R.A. Schmitz (1969). An Experimental Study of Steady State Multiplicity in a Loop Reactor. *AIChE Journal* **vol 15**, p.670-679

Root R.B., R.A. Schmitz (1970). An Experimental Study of Unsteady States in a Loop Reactor. *AIChE Journal* **vol 16**, p.356-358

Morud J., S. Skogestad (1994). The Dynamics of Chemical Reactors with Heat Integration. *Presented at 5th NPC-Workshop,* Technical University of Denmark, Lyngby

Uppal A., W.H. Ray, A.B. Poore (1974). On the Dynamic Behaviour of Continuous Stirred Tank Reactors. *Chem. Eng. Sci.* **vol 29**, p.967-985

Villadsen J., M.L. Michelsen (1978). *Solution of Differential Equation Models by Polynomial Approximation. Prentice* Hall, New Jersey

Analysis of Process Dynamics Using Steady State Flowsheeting Tools

Ira Naot and Daniel R. Lewin†

Department of Chemical Engineering, Technion I.I.T., Haifa 32000, Israel
email: dlewin @ techunix.technion.ac.il tel:+972-4-292006 fax:+972-4-230476

Abstract. A short-cut method for generating approximate plant-wide dynamic models using flowsheeting data is used in the diagnosis of potential control problems in a reactor-seperator-recycle process. Simplified analysis enables the positive feedback problem and its causes to be identified, and a systematic testing of alternative control configurations to be carried out, without resorting to rigorous dynamic simulation.

Keywords. Resiliency and controllability diagnosis; chemical process design; flowsheeting; process dynamics; recycle systems

† Author to whom all correspondance should be addressed

1. INTRODUCTION

Continous chemical processes are usually designed with a specific operating point in mind. It is tacitly assumed that this point can be maintained, despite external disturbances, by a suitably designed control system. As some designs can be harder than others to control, some information about the dynamic behavior of a process is therefore required at an early stage of the design. However, at an early stage not much is known about the process and usually only a steady state flowsheet is available. Weitz and Lewin (1994) developed a simple procedure for screening flowsheets for controllability, which relies on the derivation of a simplified dynamic model using steady state flowsheet information. In this paper, we shall extend their results (limited to systems consisting of distillation columns and heat exchangers) to systems also containing reactors and material recycle. First, a review of the simplified dynamic modelling approach is given, including a description of the methods used to model each processing component in the plant, and in particular, the reactor. The case study used to investigate the impact of the method is described next. This is taken from work recently presented by Luyben (1993). Following that, we describe the approximate model generated using the proposed approach, and its closed loop verification using the control system suggested by Luyben. Next, we analyze the approximate model in order to identify the causes for the problematic positive feedback associated with recycle systems (Luyben calls this the "snowball" effect), and systematically screen alternative control schemes.

2. DERIVATION OF AN APPROXIMATE DYNAMIC PLANT MODEL

A procedure for the derivation of an approximate plant-wide dynamic model using flowsheet data was presented by Weitz and Lewin (1994). First, the overall plant is examined conceptually, defining the output variables that need be controlled and the manipulated variables available to carry this out. Next, the flowsheet is decomposed into *component parts*, which are defined as processing units that can be approximated by simple dynamic models. Thus distillation columns, heat exchangers, stirred tank reactors, and surge tanks can all be defined as *component parts*. As a result of this conceptual decomposition, the flowsheet is represented as a connected set of component parts, each expressed in terms of vectors of input and output variables. For example, a component part describing a distillation column may have feed composition, rate and vapor fraction among the input variables, and product rates and compositions among the output variables. The dynamic behavior of the complete plant is determined by the dynamics of its component parts and the connections between them. The following sections describe the methods used for the derivation of component part dynamics. Simulation Science's Pro-II has been used to solve the steady state flowsheet and for generating data for each component part. Once defined, the plant-wide dynamic model is formed by recombining the component parts as dictated by plant topology. For convenience, this step has been carried out using Mathwork's SIMULINK.

2.1 Modeling Distillation Columns.

Distillation columns are distributed parameter systems whose overall dynamic behavior depends on the dynamics of each tray, the reboiler and the reflux drum. As these design details are not yet determined at the time of the analysis, the complete column with the reboiler and condenser is approximated as a first order lag and a matrix of gains and delays. Input variables of the dynamic model are: feed rate, composition and specific enthalpy (temperature or vapor fraction), heating and cooling rates, and reflux and pump around rates. Output variables are product compositions, rates and temperatures. The m-input, n-output transfer function matrix is:

$$\begin{bmatrix} y_1 \\ \vdots \\ y_n \end{bmatrix} = \frac{1}{\tau_c s+1} \begin{bmatrix} k_{1,1}e^{-\theta_{1,1}s} & \cdots & k_{1,m}e^{-\theta_{1,m}s} \\ \vdots & \ddots & \vdots \\ k_{n,1}e^{-\theta_{n,1}s} & \cdots & k_{n,m}e^{-\theta_{n,m}s} \end{bmatrix} \begin{bmatrix} u_1 \\ \vdots \\ u_m \end{bmatrix} \tag{1}$$

The time constant τ_c depends on the residence time in the entire column and is approximated on the basis of liquid flow rate and holdup in the column. The transport delays $\theta_{i,j}$ depend on the flow rate and liquid hold up in the column section between the jth input and the ith output. For example, delay between reboiler boilup rate and bottom product composition is approximated as zero. Static gains of the column component part are determined using steady state simulations of the column with small changes in inputs, assuming

$$k_{i,j} \approx \frac{\Delta y_i}{\Delta u_j} \tag{2}$$

2.2 Modeling Heat Exchangers.

Similarly to distillation column component parts, those partaining to heat exchangers are computed in terms input variables (flow rates, compositions and temperatures) and output variables (exit stream temperatures). The steady state flowsheet simulation provides the required heat exchange area, from which shell side and tube side volumes can be approximated. Shell and tube side flow rates are also provided, and therefore residence times are known. Exchanger delays and time constants are approximated on the basis of the residence times. Static gains are approximated, as in distillation columns, using steady state simulations with small changes in inlet conditions. For more details, see Weitz(1994).

2.3 Modeling Stirred Tank Reactors.

Unlike distillation columns, a CSTR is a lumped parameter system. The dynamic behavior of such a reactor is represented by a set of ordinary differential equations. For a reactor in which m reactions with n reactants take place the state vector includes n reactant concentrations, the reactor temperature, and, in the case of a variable volume reactor, the reactor volume. The state equations are :

a) Material balance for each component:

$$\frac{d(VC_i)}{dt} = F_{in}C_{i,in} - F_{out}C_i + \ldots$$
$$\ldots + V\sum_{j=1}^{m} \nu_{i,j} r_j \qquad j = 1,\ldots,n \tag{3}$$

b) Overall energy balance:

$$\rho \bar{C}_P \frac{d(VT)}{dt} = \rho \bar{C}_P (F_{in}T_{in} - F_{out}T) + \ldots$$
$$\ldots + V\sum_{j=1}^{m} (-\Delta H_j) r_j - UA(T - T_c) \tag{4}$$

c) Overall material balance:

$$\frac{dV}{dt} = F_{in} - F_{out} \tag{5}$$

This set of equations can be linearized and expressed in terms of perturbation variables:

$$V\frac{d\hat{C}_l}{dt} = F_{in}\left(\hat{C}_{l,in} - \hat{C}_l\right) + V\sum_{j=1}^{m}\nu_{l,j}\sum_{i=1}^{n}\frac{\partial r_j}{\partial C_i}\hat{C}_i + \ldots$$
$$+V\sum_{j=1}^{m}\nu_{l,j}\frac{\partial r_j}{\partial T}\hat{T} + \hat{F}_{in}C_{l,in} - \hat{F}_{out}C_l \quad l = 1,\ldots,n \tag{6}$$

$$V\rho\bar{C}_P\frac{d\hat{T}}{dt} = \rho\bar{C}_P F_{in}(\hat{T}_{in} - \hat{T}) + \ldots$$
$$\ldots + V\sum_{j=1}^{m}(-\Delta H_j)\frac{\partial r_j}{\partial T}\hat{T} - UA(\hat{T} - \hat{T}_c) + \ldots$$
$$\ldots + V\sum_{j=1}^{m}\sum_{i=1}^{n}(-\Delta H_j)\frac{\partial r_j}{\partial C_i}\hat{C}_i + \rho\bar{C}_P(\hat{F}_{in}T_{in} - \hat{F}_{out}T) \tag{7}$$

The whole system of equations can be expressed as $\dot{\mathbf{x}} = \mathbf{Ax} + \mathbf{Bu}$, $\mathbf{y} = \mathbf{Cx} + \mathbf{Du}$ where $\mathbf{x} = [\hat{C}_{j,j=1,\ldots,n}, \hat{T}, \hat{V}]^T$, $\mathbf{u} = \left[\hat{C}_{j-in,j=1\cdots n}, \hat{T}_{in}, \hat{T}_c, \hat{F}_{in}, \hat{F}_{out}\right]^T$ and $\mathbf{y}$ is the vector of variables passed as inputs to other parts of the dynamic model. Matrices **A,B,C,D** include constants that are defined by the steady state operating point, and can be computed on the basis of steady state flow-sheeting information. An exception is the derivatives of rate expression terms with respect to product concentration and temperature. These can be computed analytically from the kinetic expressions. The linear dynamic model can be transformed to its transfer function form using:

$$\mathbf{y} = (\mathbf{C}(s\mathbf{I}-\mathbf{A})^{-1}\cdot\mathbf{B}+\mathbf{D})\cdot\mathbf{u} \quad (8)$$

This local model cannot account for potential multiple steady states or chaotic behavior. However, it approximates the dynamic behavior of the reactor close to the selected operating point and can serve as a screening tool, which is the desired objective.

Reactor design using a steady state flowsheeting package is a two step procedure. In the first step, the reactor is designed to operate at a temperature which considers reaction kinetics, side reactions and reactor safety. In the second step, the cooling jacket heat transfer coefficient, UA, is computed on the basis of the specified reactor temperature, available coolant temperature, and the calculated heat generation rate. The design constant UA completes the information required for the linear approximation of the reactor.

3. A PLANT-WIDE CASE STUDY

The test case considered here is an example given by Luyben (1993). The process is one in which two consecutive reactions take place: A→B→C , Where B is the desired product. Since the relative volatility of A is highest, it is possible to use a CSTR with relatively low conversion, separate A by distillation and recycle it, and use a second column for product purification. A flow diagram of the process, defining the process variables, is given in figure 1. The process specifications (control objectives) are to limit the concentration of reactant A in the bottom product of the first column to 1% (thus limiting loss of reactant with the product) and limit the product B concentration in the bottom product of the second column to the same amount (thus limiting loss of product in the waste stream). Steady state design data for the case of a 150 lbmol reactor is given in table 1.

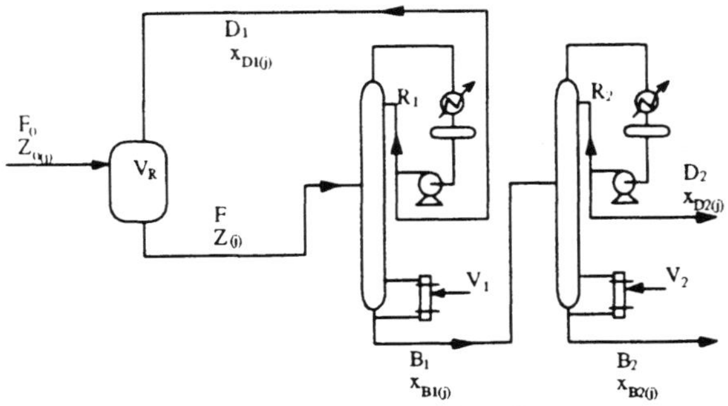

Fig 1: The process flowsheet

The approximation of the process dynamics was based on the following assumptions:

a) The reactor is perfectly isothermal.
b) The reactor, reboilers and reflux drums have perfect level control.
c) Feed and reflux streams to distillation columns are saturated liquid.
d) Sensible heat effects in the distillation columns are negligible.
e) Relative volatilities of A,B,C are constant and in the ratios 4:2:1 respectively.

Table 1 Steady State Design Data

Variable	Value
Reactor:	
Feed rate [lbmol/hr]	100
A fed conc., $Z_{0(1)}$ [mol frac]	1
Reactor Volume V_R [lbmol]	150
Reactor feed F_1 [lbmol/hr]	320
Column 1:	
Number of trays (feed tray)	15(6)
Reflux rate [lbmol/hr]	350
Distillate flow rate D_1 [lbmol/hr]	220
Distillate A conc. $x_{D1(1)}$ [mol frac]	0.95
Distillate B conc. $x_{D1(2)}$ [mol frac]	0.05
Bottom product rate B_1 [lbmol/hr]	100
Bottom A conc. $x_{B1(1)}$ [mol frac]	0.01
Bottom B conc. $x_{B1(2)}$ [mol frac]	0.64
Column 2:	
Number of trays (feed tray)	19(10)
Reflux rate R_2 [lbmol/hr]	170
Distillate rate D_2 [lbmol/hr]	65
Distillate A conc. $x_{D2(1)}$ [mol frac]	0.015
Distillate B conc. $x_{D2(2)}$ [mol frac]	0.975
Bottoms product rate B_2 [lbmol/hr]	35
Bottoms B conc. $x_{B2(2)}$ [mol frac]	0.01

The process was divided to four component parts: feed mixer, reactor and two distillation columns. Figure 2 shows a flow diagram denoting input and output variables of the component parts and the connections between them. It should be noted that in the real plant the multiple feeds enter the reactor directly. The mixer is modeled as a nonlinear gain function so it does not introduce its own dynamic effects. The overall material balance is:

$$\hat{F}_1 = \hat{F}_0 + \hat{D}_1, \quad (9)$$

while the component oncentration balance gives:

$$\hat{C}_{(i)} = \frac{z_{0(i)}F_0 + x_{D1(i)}D_1}{F_0 + D_1} - C^*_{(i)},\ i = 1,2 \quad (10)$$

Note that by definition, $\hat{x}_{D1(2)} = 1 - \hat{x}_{D1(1)}$. Equation 10 is the rigorous form of the component material balance.

In the isothermal, constant volume reactor, the state vector is $x = [z_{(1)}, z_{(2)}]^T$ and the output vector is $y = [z_{(1)}, z_{(2)}, F]^T$. The input vector is $u = [C_{(1)}, C_{(2)}, F_1]^T$. The kinetic expressions reduce to $r_1 = z_{(1)}$ [lbmol/hr ft^3] and $r_2 = z_{(2)}$ [lbmol/hr ft^3], and the reactor transfer function matrix is:

$$\begin{bmatrix} \hat{z}_{(1)} \\ \hat{z}_{(2)} \\ \hat{F} \end{bmatrix} = \begin{bmatrix} \frac{0.71(0.32s+1)}{(0.39s+1)(0.27s+1)} & 0 & \frac{6.7e-4(0.32s+1)}{(0.39s+1)(0.27s+1)} \\ \frac{0.23}{(0.39s+1)(0.27s+1)} & \frac{0.71(0.32s+1)}{(0.39s+1)(0.27s+1)} & \frac{-2.2e-4(0.62s+1)}{(0.39s+1)(0.27s+1)} \\ 0 & 0 & 1 \end{bmatrix} \begin{bmatrix} \hat{C}_{(1)} \\ \hat{C}_{(2)} \\ \hat{F}_1 \end{bmatrix} \quad \text{....(11)}$$

The two columns are modeled using the method described in Weitz and Lewin(1994). The number of trays and feed tray location were taken from Luyben's data and column sizing was performed by the flowsheeting package. The column time constants are based on liquid rate as calculated in the flowsheet, and holdup estimated for a valve tray column operating at atmospheric pressure. Reboiler and reflux drum residence times are five minutes each. The resulting column transfer matrices are :

$$\begin{bmatrix} \hat{x}_{D1(1)} \\ \hat{D}_1 \\ \hat{x}_{B1(1)} \\ \hat{x}_{B1(2)} \\ \hat{B}_1 \end{bmatrix} = \frac{1}{0.278s+1} \begin{bmatrix} 0.5 & -0.12 & 0.0015 & -0.2 & 0.003 \\ 60 & 20 & 0 & 48 & -0.92 \\ 0.2e^{-0.01s} & 0.8e^{-0.01s} & 0.0007e^{-0.01s} & -0.06 & 0.0005e^{-0.02s} \\ e^{-0.01s} & 3e^{-0.01s} & 0.001e^{-0.01s} & -0.11 & 0.001e^{-0.02s} \\ -60e^{-0.01s} & -20e^{-0.01s} & 1e^{-0.01s} & -48 & 0.92e^{-0.02s} \end{bmatrix} \begin{bmatrix} \hat{z}_{(1)} \\ \hat{z}_{(2)} \\ \hat{F} \\ \hat{V}_1 \\ \hat{R}_1 \end{bmatrix} \quad \text{...(12)}$$

$$\begin{bmatrix} \hat{x}_{D2(1)} \\ \hat{x}_{D2(2)} \\ \hat{D}_2 \\ \hat{x}_{B2(1)} \\ \hat{B}_2 \end{bmatrix} = \frac{1}{0.303s+1} \begin{bmatrix} 0.15 & -0.004 & 0.00015 & -0.12 & 0.00021 \\ -0.07 & 0.3 & 0.0046 & -0.1 & 0.002 \\ 0.89 & 17 & 0 & 51 & -0.91 \\ 0.105e^{-0.01s} & 0.3e^{-0.01s} & 0.007e^{-0.01s} & -0.1 & 0.003e^{-0.02s} \\ -0.89e^{-0.01s} & -17e^{-0.01s} & e^{-0.01s} & -51 & 0.91e^{-0.02s} \end{bmatrix} \begin{bmatrix} \hat{x}_{B1(1)} \\ \hat{x}_{B1(2)} \\ \hat{B}_1 \\ \hat{V}_2 \\ \hat{R}_2 \end{bmatrix} \quad \text{...(13)}$$

Time constants are in hours, flow rates are in lbmol/hr, concentrations are in molar fraction, and reboiler rates are in units of 10^6 BTU/hr. The four component parts were connected according to figure 2 using the SIMULINK package to form a dynamic approximation of the entire plant.

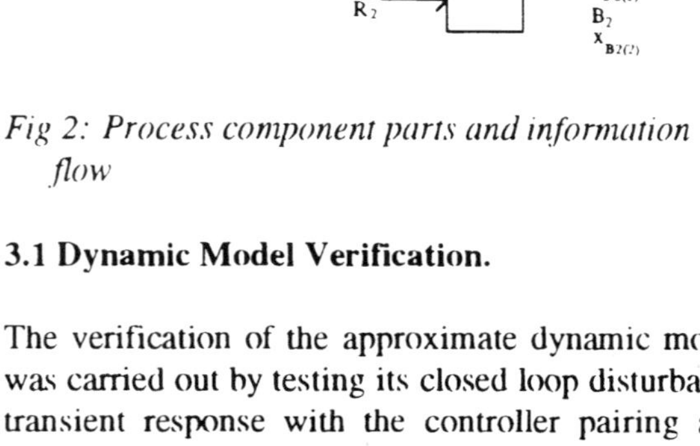

Fig 2: Process component parts and information flow

3.1 Dynamic Model Verification.

The verification of the approximate dynamic model was carried out by testing its closed loop disturbance transient response with the controller pairing and tuning suggested by Luyben. The pairing is: $x_{B1(1)}$-V_1; $x_{B2(2)}$-V_2. Luyben uses bottom product draw as the control variable, but since he assumes perfect level control (using the boilup rate), this amounts to an identical structure. In our verification test, two feedback PI controllers were used, with tuning parameters as given by Luyben. Two step tests were made on the system. In the first the fresh feed composition was changed, switching from pure component A to 90% A and 10% C. In the second test, the feed flow rate was changed.

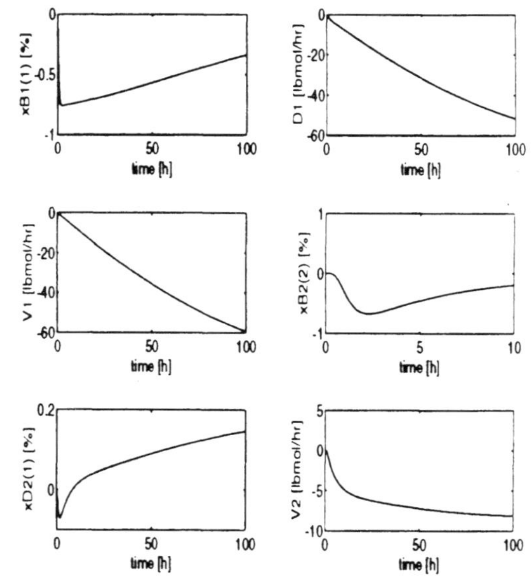

Fig 3: Simulation of a 10% Drop in Fresh Feed A Concentration Using the Linear Dynamic Approximation.

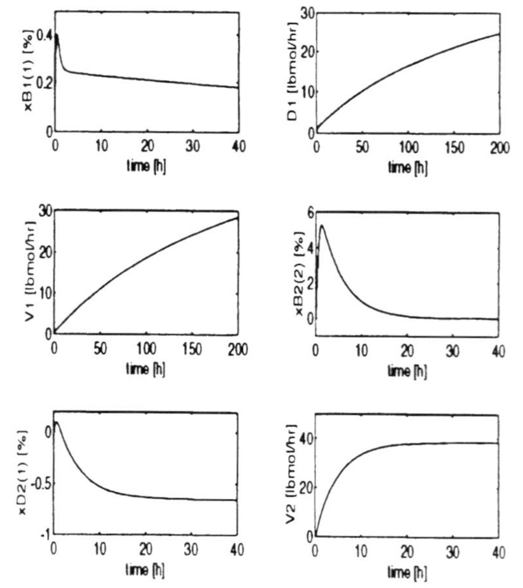

Fig 4: Simulation of a 10 % Increase in Fresh Feed Flow Rate Using the Dynamic Approximation.

The response to these two changes is shown in figures 3 and 4. Many similarities with the results presented by Luyben are observed. Firstly, the positive feedback ('snowball') effect causing large changes in recycle flow rate appears in the two simulations. Secondly, the inverse response of $x_{D2(1)}$ is observed, and thirdly, the maximum deviations of the controlled variables are of a similar magnitudes and occur at roughly the same time (about one hour from the introduction of the disturbance). There is however a large difference in the settling time between the two simulations, but this does not affect the steady state results (magnitude of control action or uncontrolled variables values) or the detection of problems of positive feedback and inverse response.

3.2 Model Sensitivity Analysis.

Following preliminary model validation, sensitivity analysis was carried out in order to find the attributes of the linear approximation which are responsible for the positive feedback effect, and to identify which of the linear gains is particularly sensitive. This information was subsequently used to systematically identify the control structure which will eliminate the positive feedback effect, using the approximate model.

It was found that there are two gains in the dynamic approximation that are responsible for the positive feedback: $k_1 = x_{B1(1)}/V_1$ and $k_2 = x_{B1(1)}/F$. These gains are associated with the required product purity: a high purity product means greater $x_{B1(1)}/V_1$ and lower $x_{B1(1)}/F$. Increasing k_1 by 20% or decreasing k_2 by 20% diminishes the effect, while changes in the opposite direction result in loss of stability in the plant under closed loop integral action. The physical effect of changing these two gains is to change the required control action in V_1 maintain $x_{B1(1)}$ on specification. A change in the boilup rate influences the distillate flow rate directly, thus changing the recycle rate and causing the positive feedback.

Nonlinearities manifest themselves in both the transient behavior of the process and as nonlinear static gains. Our main purpose is provide a screening tool at the early design stage, and here, nonlinearities in the static gains of the component parts are by far the most important consideration. In the process under study, the mixer has been model by a rigorous mass balance, conserving the true nature of the process. The component part approximating the CSTR dynamics has been linearized in the vicinity of the operating point. However, since isothermal operation is assumed, this eliminates the highly nonlinear temperature dependence of the reacton rates. This means that the Jacobian matrices need to approximate only the moderate nonlinearities associated with the concentration and flow dependencies, implying that the linear model should be reliable when dealing with fairly large changes in operating conditions.

The one component part in which the gain nonlinearity is significant is the high purity distillation column. Here, the gains of product rate output variables depend almost entirely on material balances and are therefore almost linear, especially so when constant relative volatility and negligible sensible heat effects are assumed. On the other hand, composition variables have strongly non-linear gains. Consider for example, $x_{B1(1)}$, the concentration of A in the bottoms stream. Its nominal value is 1% and the physical boundaries are between 0 and 100%. This asymmetry means that the gain will depend on the input change direction. For example, whereas a large increase in boilup rate, V_1, will lower $x_{B1(1)}$ by 1% at the most, a small decrease in V_1 will cause a large rise in $x_{B1(1)}$. The same is true for all static gains relating the input variables and output compositions. Thus, the two gains which effect the positive feedback phenomena, k_1 and k_2, are subject to nonlinearities. This will be dealt with by testing the effect of changes in their values on the expected overall performance of the process. Thus, the actual values of k_1 and k_2 to be tested are given by:

$$k_i = k_i^0(1 + \Delta_i \cdot \varepsilon), \quad |\Delta_i| \le 1, \quad i = 1,2 \tag{14}$$

where k_i^0 are the nominal gain values, and an upper bound of 20% uncertainty is assumed on the gain values (ε = 0.2). From the physical arguments described previously, the signs of the perturbations Δ_i must be in opposite directions for the two gains.

3.3 Analysis of the Control Problem.

The positive feedback effect of the recycle system is partly the result of the particular control strategy implemented. Here, the controller pairings were selected on the basis of "common-sense" reasoning for each processing unit separately, rather than considering the process as a whole. A more successful approach is based on the RGA as computed for the entire plant. This enables the systematic analysis of alternative pairings of input variable vectors, selected from the available manipulated variables, with the output variables desired to be controlled. Such an approach can be applied to the case study, were the variables to be controlled are $x_{B1(1)}$ and $x_{B2(2)}$, and in addition D_1, the recycle rate. Two possible sets of manipulated variables will be examined: $[V_1 R_1 V_2]$ and $[V_1 F_0 V_2]$. RGAs for this 3×3 problem were computed for seven combinations of permulations of uncertainty in k_1 and k_2 and are given in Table 2. These results were computed using the approximate linear dynamic model in open loop for step changes in the inputs. Several observation can be made from the data in Table 2:

1) There is only one-way interaction between the overall process and the second column, as

verified by the [0 0 1] on the third row repeated throughout Table 2. This is also apparent when inspecting the flowsheet itself.

2) For the first examined manipulated variable set, the only pairing that will guarantee stable response under integral control is: $x_{B1(1)}$-R_1; D_1-V_1; $x_{B2(2)}$-V_2. However, the large variations in RGA coefficients with uncertainty in the critical gains k_1 and k_2 suggest that strong interactions will be present in the closed loop process. In any case, the "common sense" pairing ($x_{B1(1)}$-V_1; $x_{B2(2)}$-V_2) may lead to unstable behaviour under integral control when the loop D_1-R_1 is closed.

3) Finally, examination of the RGA coefficients for the second set of manipulated variables shows that this configuration is far less insensitive to gain variations. This shows a clear preference of using F_0 to control $x_{B1(1)}$, (rather than $x_{B1(1)}$-R_1). This is apparent if one notes that $\lambda_{2,1}\approx 1$ in the second column of Table 2 for all gain uncertainty settings.

Table 2 RGA for $\mathbf{y} = [x_{B1(1)}\, D_1\, x_{B2(2)}]^T$
With Two Sets of Manipulated Variables

Δ_1	Δ_2	$\mathbf{u}=[V_1\, R_1\, V_2]$	$\mathbf{u}=[V_1 F_0 V_2]$
0	0	$\begin{bmatrix} 0.07 & 0.93 & 0 \\ 0.93 & 0.07 & 0 \\ 0 & 0 & 1 \end{bmatrix}$	$\begin{bmatrix} 0.01 & 0.99 & 0 \\ 0.99 & 0.01 & 0 \\ 0 & 0 & 1 \end{bmatrix}$
+1	0	$\begin{bmatrix} 0.36 & 0.64 & 0 \\ 0.64 & 0.36 & 0 \\ 0 & 0 & 1 \end{bmatrix}$	$\begin{bmatrix} -0.12 & 1.12 & 0 \\ 1.12 & -0.12 & 0 \\ 0 & 0 & 1 \end{bmatrix}$
-1	0	$\begin{bmatrix} -0.75 & 1.75 & 0 \\ 1.75 & -0.75 & 0 \\ 0 & 0 & 1 \end{bmatrix}$	$\begin{bmatrix} -0.09 & 1.09 & 0 \\ 1.09 & -0.09 & 0 \\ 0 & 0 & 1 \end{bmatrix}$
0	+1	$\begin{bmatrix} -0.15 & 1.15 & 0 \\ 1.15 & -0.15 & 0 \\ 0 & 0 & 1 \end{bmatrix}$	$\begin{bmatrix} -0.02 & 1.02 & 0 \\ 1.02 & -0.02 & 0 \\ 0 & 0 & 1 \end{bmatrix}$
0	-1	$\begin{bmatrix} 0.28 & 0.72 & 0 \\ 0.72 & 0.28 & 0 \\ 0 & 0 & 1 \end{bmatrix}$	$\begin{bmatrix} 0.09 & 0.91 & 0 \\ 0.91 & 0.09 & 0 \\ 0 & 0 & 1 \end{bmatrix}$
+1	-1	$\begin{bmatrix} 0.51 & 0.49 & 0 \\ 0.49 & 0.51 & 0 \\ 0 & 0 & 1 \end{bmatrix}$	$\begin{bmatrix} 0.22 & 0.78 & 0 \\ 0.78 & 0.22 & 0 \\ 0 & 0 & 1 \end{bmatrix}$
-1	+1	$\begin{bmatrix} -1.1 & 2.1 & 0 \\ 2.1 & -1.1 & 0 \\ 0 & 0 & 1 \end{bmatrix}$	$\begin{bmatrix} -0.09 & 1.09 & 0 \\ 1.09 & -0.09 & 0 \\ 0 & 0 & 1 \end{bmatrix}$

Thus, the preferable control configuration will be using F_0 to control $x_{B1(1)}$, V_2 to control $x_{B2(2)}$ and keeping V_1 and R_1 (and hence D_1) constant. The use of the fresh feed rate, F_0, as a control variable is problematic, since it should be ideally kept free to control production. In addition, it requires that the overhead composition be regulated by the process feed, which thus acts sluggishly via two large lags (reactor and column). The loss of freedom in production level can be overcome by using the reactor level as a manipulated variable. Increasing the reactor level (and thus its volume) will permit using higher flow rates with the same residence time. The slow speed of response of $x_{B1(1)}$ to F_0 can be overcome by using a combination of feedforward and feedback control. It should also be noted that the recycle rate, D_1, does not have to be kept constant in the proposed scheme. Manipulating D_1 (through changes in V_1) permits control of the selectivity in the reactor. The control structure presented here, using F_0, V_R, and D_1 to control $x_{B1(1)}$, throughput and selectivity respectively was verified using steady state simulations. It was found that such control strategy rejects disturbances in feed composition with little control action and with no positive feedback. Furthermore, this control structure permits control of the reactor selectivity which cannot be achieved if the original control structure is used.

4. CONCLUSIONS

A short-cut method for generating approximate plant-wide dynamic models using flowsheeting data has been presented. Its use in the diagnosis of potential control problems in a reactor-separator-recycle process has been demonstrated. Simplified analysis enabled the positive feedback problem and its causes to be identified, and a systematic testing of alternative control configurations to be carried out, without resorting to rigourous dynamic simulation. Clearly, having such insight into the process at an early stage of design is an advantage, as problematic alternatives can be rejected before investing resources on them.

5. REFERENCES

Luyben, W. L. (1993). "Dynamic and Control of Recycle Systems. 3. Alternative Process Designs in a Ternary System", *Ind. Eng. Chem. Res.* **32**, 1142-1153

Weitz, O. (1994). "Introducing Controllability Measures into Process Design", M. Sc. Thesis, Technion (in Hebrew)

Weitz, O. and D. R. Lewin (1994). "Dynamic Controllability and Resiliency Diagnosis using Steady State Process Flowsheet Data", *Proc. of the IFAC Workshop on the Integration of Process Design and Control*, Baltimore

NEURAL NETWORK IDENTIFICATION OF AN IN-LINE pH PROCESS

S.K. Doherty, J.B. Gomm and D. Williams

Control Systems Research Group, School of Electrical and Electronic Engineering, Liverpool John Moores University, Byrom St., Liverpool L3 3AF, U.K.

Abstract : This study reports the development of an artificial neural network (ANN) representation of a bench scale in-line pH process. The dynamics of the process are characterised by a strong nonlinearity familiar to pH and are further complicated by variable flow which constantly changes the process deadtime. The focus of this paper is on determining a parsimonious ANN model structure which accommodates the variable process deadtime. Three approaches are compared; the first method, which does not require flow data, is to use a delay spread to include all possible values of the delayed variable in the model structure. The other two techniques make use of flowrate; one uses a volume array to represent the deadtime while the other samples at constant intervals of volume thus making the delay a constant number of samples. Results show that the three approaches lead to ANN models with similar prediction accuracy but differ in their computational complexity.

Keywords : Neural network models; Nonlinear systems, Time delays, pH Control.

1. INTRODUCTION

The objective of this research program is to control a pilot scale chemical process using ANN based control strategies and therefore appraise the suitability and limitations of ANNs for process control. The chosen process is an in-line pH neutralisation, which not only fulfils the requirement of process complexity but also offers potential benefits to the waste water and chemical process industries. The control of pH is important in many food and bio-technological industries where inadequate pH regulation can affect both the yield and quality of the product (Shinskey, 1973). Additionally, any chemical industry which discharges effluent must regulate its pH to protect aquatic and human welfare, and to comply with limits imposed by legislation. Hence, there is strong economic and environmental motivation for improved pH control.

The established methods of controlling pH incorporate one or more CSTRs (Continuous Stirred Tank Reactors). The capacity of a CSTR serves to dampen changes in influent concentration giving the control system time to recover from process disturbances and minimises the effect of any errors in estimating the titration curve. However, large tanks also have large mixing deadtimes, which is detrimental to feedback control. Hence the volume of the CSTR is an important design parameter. Often, a series of CSTRs are used, the first for course control and the following larger tanks for progressively finer control (Gulaian and Lane, 1990). The main drawback with using CSTRs in a pH control system is the capital needed to construct the tanks and the stirring equipment.

An alternative approach is to control pH 'in-line' as the influent flows towards its outlet, by injecting neutralizing reagent directly into the pipeline (Rhinehart and Choi, 1988). This is achieved by a combination of feedforward control from a pH measurement upstream of the injection point and feedback control from a pH measurement downstream of the injection point. This offers considerable savings in capital since no tanks or stirring equipment is required and the pipeline will already exist. It also offers potential savings in running costs, since it neutralises small slugs of effluent as they flow in-line and consequently makes more efficient use of neutralization reagents. The main difficulty with either the CSTR or the in-line approach is identifying or estimating the titration curve using available process data (for a practical neutralisation plant it is only possible to measure pH and flowrates - the concentrations and strengths of the influent species would be unknown). The in-line approach introduces further difficulties :-

- The influent is not homogenously mixed, and indeed the extent of mixing is unknown.
- The damping effect due to residence time is minimal, since the in-line pipe volume at the reagent injection point is very small compared with a typical CSTR.
- The deadtime due to the in-line transportation lag will vary with influent flowrate. Hence, pH measurements taken upstream of the neutralisation point for a feedforward control loop must be appropriately delayed to compensate for this.

2. PROCESS DESCRIPTION

The process was designed to serve as a test unit for the experimental evaluation of ANN identification and control strategies. Fig. 1 shows the process configuration for this set of identification experiments. The disturbance flowrate (F_1) serves only to disturb the pH of the fluid entering the tubular reactor. This pH is measured by the pH probe pH_1. The process has two outputs; the in-line effluent pH (pH_2) and the bulk product pH (pH_3), but these experiments are only concerned with the in-line effluent pH. Hence, in this configuration, the process has three inputs (F_{tub}, pH_1 and F_2) and a single output (pH_2). F_2 is the flowrate of neutralising reagent, which would be the controller output in closed loop. For safety reasons the chemical agents used are acetic acid (CH_3COOH) and ammonium hydroxide (N_4OH). Process flowrates are set by peristaltic pumps. The pumps and pH meters are interfaced with a PC for data acquisition.

2.1 *A first principle model*

An approximate first principle model of the in-line process was developed to provide a convenient source of data to preliminary design, initially the ANN identification experiment and subsequently a neuro-controller. The reagent injections at both Tee junctions were modelled as CSTRs (Jacobs, *et al.*,

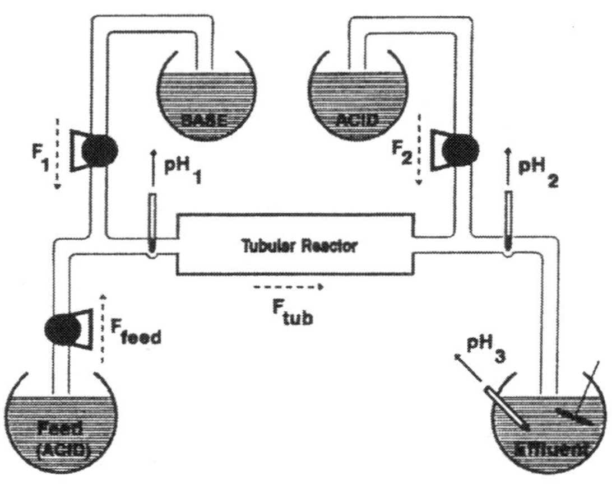

Fig. 1 The in-line pH process

1987) by performing material balances on the relevant chemical reaction invariants (Mc Avoy *et al.*, 1972; Gustaffson and Waller, 1983). Mixing in the tubular reactor was ignored, representing it as a pure transport lag. The dynamics of the pH probes were modelled as first order lags. This mathematical model was implemented in the Advanced Continuous Simulation Language (ACSL) to provide initial data for ANN training and validation.

Unlike the process, this first principle model does not have a deadtime between the disturbance injection point and pH_1. The model also assumes perfect mixing at both injection points with no mixing in the intervening tubular reactor. Given these approximations, the response of the model is similar to that of the process (fig. 2), and was consequently considered suitable for a preliminary ANN modelling investigation. In particular, the model captures the following features of the process:-

- steady-states
- the dependency of the transportation delay in the tubular reactor on the flow rate
- the dependency of the damping on the flowrate; the model exhibits increased damping for the falling edge of the clock pulse as is observed for the process

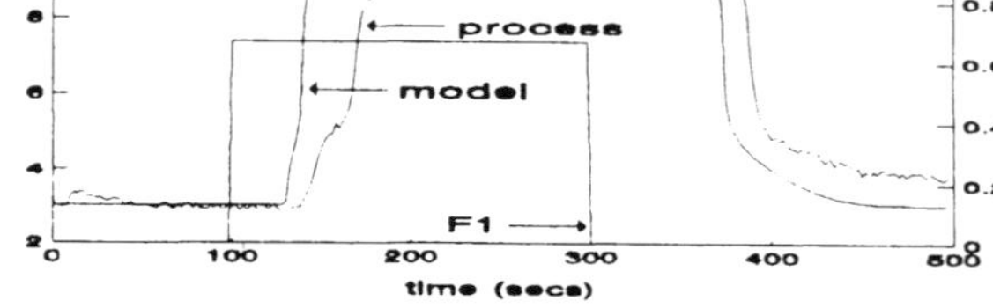

Fig. 2 ACSL Model/Process step responses (F_2=0)

Methods developed in simulation were refined and subsequently applied to the real in-line process. For the bench scale process, a data sample time of 2.0 seconds was chosen on the basis of previous simulation studies of a CSTR pH process (Doherty et. al., 1994) and from the response time of the bench scale process. Because the process uses an acid neutralisation reagent, the flowrate F_1 must maintain pH_1 above a threshold so that in closed loop the control output F_2 is capable of driving pH_2 to any set point below that threshold. For this study, the threshold was chosen as pH 7.0. Fig. 3 shows a typical set of real process data when both F_1 and F_2 are random amplitude signals .

3. RADIAL BASIS FUNCTION ANN MODELS

The radial basis function (RBF) ANN (Broomhead and Lowe, 1988; Moody and Darken, 1989; Chen *et a.*l, 1990) incorporating the thin plate spline function was used to represent the dynamic behaviour of the process. RBF ANNs have attracted considerable attention recently as an alternative network to the more established multi-layered perceptron (MLP) for non-linear modelling. RBF ANNs are linear in their parameters, allowing them to be trained using linear algorithms (e.g. recursive least squares). This results in fast convergence properties and a guarantee of a unique convergence minimum. This is a major advantage over their MLP counterparts, trained using non-linear algorithms (i.e. back propagation), which are non-linear in their parameters and therefore slow to converge and susceptible to local minima impasses.

The I/O mapping for a RBF network with a single output is of the form...

$$y = \sum_{i=1}^{n_h} w_i f(\| c_i - x \|) \qquad (1)$$

where **x** is the input vector, $\mathbf{c}_i$ is the centre vector for

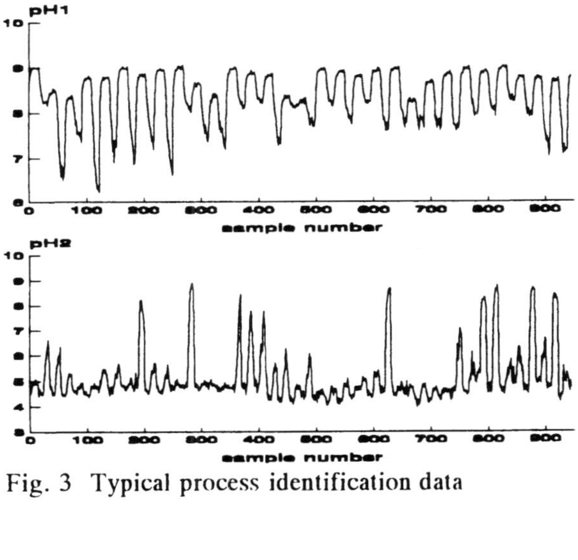

Fig. 3 Typical process identification data

the ith hidden node, w_i is the weight connecting the ith hidden node to the network output, n_h is the number of hidden nodes, and $f(z) = z^2 \log(z)$ is the thin plate spline function. Other functions can be selected for f(z), such as the multi-quadratic function or the gaussian function. The choice of f(z) is not crucial to the approximation capabilities of RBF ANNs (Chen *et al.*, 1990) and consequently the thin plate spline is appealing since, unlike the alternative functions, it does not require the selection of an additional width parameter.

The nonlinear modelling capabilities of an RBF ANN are critically dependant on the positioning of its centre vectors. If the centres do not suitably span the domain of the nonlinear mapping, then poor approximation can result. The k-means clustering algorithm has been proposed to enhance centre positioning by attracting the centres towards clusters of input data. In these investigations it was found that k-means clustering was unsuitable, often causing a deterioration in the model prediction accuracy. Best results were achieved by randomly positioning centres (which remained fixed) throughout the ranges of the network input variables. Model performance was graded using the mean square error (MSE) between the model and process outputs when predicting on unseen test data.

Following standard procedures, delayed values of process inputs and output were used as the ANN inputs so that dynamics could be incorporated in the static network. The NARX (Non-linear Auto Regressive eXogenous) model structure was adopted because of its relative ease of implementation compared with other non-linear I/O model structures (e.g. NARMAX). It assumes that for a single output system the process dynamics can be approximated by a non-linear difference equation of the form...

$$\begin{aligned} y(t) = F\, [&u_1(t-k_1), .., u_1(t-k_1-n_{U1}+1), ... \\ &, u_m(t-k_m), .., u_m(t-k_m-n_{Um}+1) \\ &, y(t-1), .., y(t-n_y)] + e(t) \end{aligned} \qquad (2)$$

where $u_1..u_m$ are the process inputs, y is the process output, e represents process noise, $k_1..k_m$ are the model deadtimes ($k \geq 1$) and *F[]* is a non-linear function to be identified. Hence, the ANN is trained to predict the present process output, y(t), based on past values of u and y.

4. DEADTIME COMPENSATION

The pH_1 signal is certainly a significant model input, since it is in effect a delayed version of the process output (pH_2) prior to being acted upon by F_2, neglecting the mixing effects occurring in the tubular reactor. The delay between pH_1 and pH_2 constantly changes as the reactor throughput varies, providing additional difficulties for successful identification of the process. Three strategies were investigated to compensate for the process deadtime variation in the ANN models.

4.1 Deadtime spread

This method is the easiest to implement and requires no flowrate information. All possible values of the delayed input variable pH_1 are used as network inputs, i.e. $pH_1(t - k_{min})$ .. $pH_1(t - k_{max} - n + 1)$. For this process $k_{min} = 13$, $k_{max} = 21$. This ensures that the relevant lags are presented to the ANN, but additional superfluous lags are also presented, e.g. if the process deadtime is k_{max}, then the lags $pH_1(t - k_{min})$ ... $pH_1(t - k_{max} + 1)$ are irrelevant to the prediction of y(t).

4.2 Transport lag volume array

The deadtime is compensated for by representing the transport of fluid trough the tubular reactor by an array of lagged (pH_1) values. Each array element represents a unit volume. At each sample interval, the number of array elements (N_v) corresponding to the volume of fluid injected into the tubular reactor since the previous sample, is calculated as..

$$N_v = \frac{F_{tub} \times \Delta T}{vol} \tag{3}$$

where F_{tub} = flowrate through the tubular reactor, ΔT = sample time and vol = a unit volume. A feedforward signal (pH_{1FF}) is calculated as the mean value of the last N_v elements of the delay array. All array elements are then shifted forward N_v places, corresponding to the reactor throughput since the last sample. Finally, the current value of the delayed variable ($pH_1(t)$) is copied to the first N_v array elements.

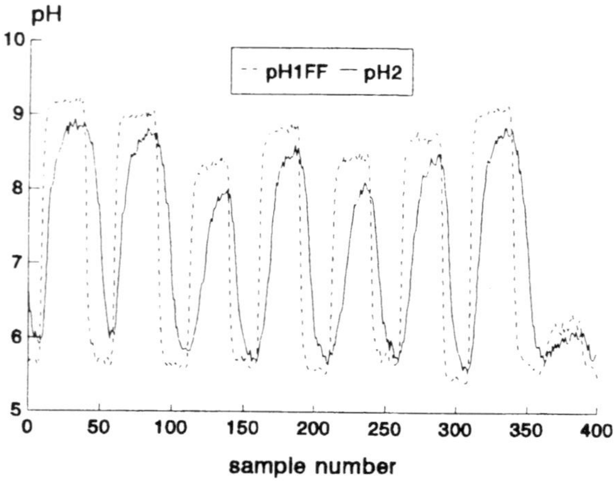

Fig. 4 The feed forward signal pH_{1FF} generated using the transport lag volume array

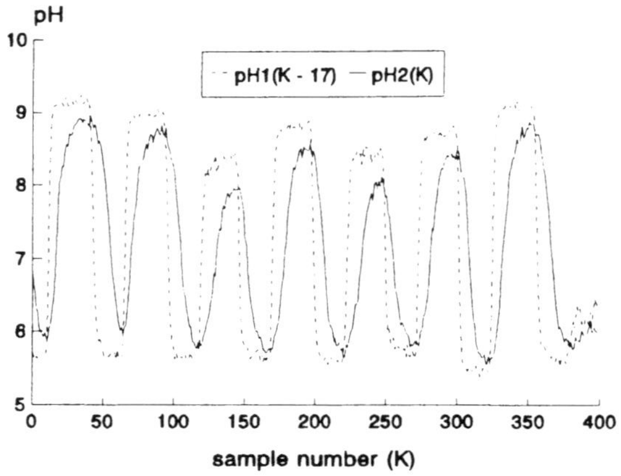

Fig. 5 Sampling in volumetric units, pH_1 is delayed by a constant (17) samples

Fig. 4 compares the feedforward signal and the process output when $F_2 = 0$. pH_2 does not attain the same steady states as pH_{1FF}, and appears as a damped version of pH_{1FF} due to the unknown mixing in the reactor between pH_1 and pH_2. The feed forward signal could be made to approximate pH_2 (with $F_2 = 0$) more closely using a first order filter for its calculation, rather than taking the arithmetic mean. In either case, it is clear that pre-processing the pH_1 signal using a volume array to accommodate the transport lag produces a feedforward signal which does have a dynamic relationship with the process output but, unlike the unprocessed pH_1 signal, the relationship is not now compounded by a time varying deadtime.

4.3 Volumetric sampling

This method also uses flow measurement to try to eliminate the effect of the fluctuating deadtime. At each sample instant, after the process signals are sampled, the time interval (ΔT) to the next sample is calculated as..

$$\Delta T = \frac{\Delta V}{F_{tub}} \tag{4}$$

where ΔV = sample volume. Hence, sampling is performed in equidistant units of volume, rather than in units of time as in the previous two methods. Consequently, the process deadtime is transformed to a constant number of samples as illustrated in fig. 5.

5. RESULTS

The process simulation was to used to initially develop and investigate techniques for the identification of the in-line process. These methods were subsequently applied to the identification of the real bench scale process which are the results presented below.

To identify the process, the RBF networks were configured to receive as inputs delayed values of the process variables F_{tub}, F_2 and pH_2 with a unit deadtime and $n_{u1} \ldots n_{um} = n_y = n$ in eqn.(2), where n is termed the order of the model. The remaining pH_1 inputs were dependent on the method used as follows:

Deadtime Spread: $pH_1(t-k_{min})\ldots pH_1(t-k_{max}-n+1)$
Transport Lag Volume Array: $pH_{1FF}(t-1)\ldots pH_{1FF}(t-n)$
Volumetric Sampling: $pH_1(t_v-k_v)\ldots pH_1(t_v-k_v-n+1)$

with $k_{min}=13$, $k_{max}=21$ and $k_v=17$ (fig.5). The networks were trained to predict the process output $pH_2(t)$, and in each case 30 hidden layer nodes were found to be adequate with little improvement in accuracy for larger networks.

Fig. 6 illustrates the trade-off between model accuracy and NARX order. The goal is to select the smallest model structure with acceptable prediction accuracy. The loss function (LF), which is simply the test MSE, suggests a 2nd order structure, since higher orders do not offer significant reductions in the loss function. Akaike's Final Prediction Error (AFPE) is a model parsimony index which can be used to clarify the ANN model size/accuracy trade-off (Doherty *et al.*, 1994). The AFPE is a weighted function of the LF which penalises small reductions in the LF at the price of introducing extra model parameters. Increasing the NARX model order increases the number of RBF parameters, since it increases the dimension of the centre vectors. Clearly, the AFPE highlights a 2nd order structure as the most parsimonious NARX configuration.

Trained 2nd order networks were tested on a variety of process data collected with random amplitude signals of different frequencies applied to F_2. The predictions of each network on this test data were similar with no network demonstrating a clear superiority in terms of accuracy. All networks were capable of acceptable prediction with typical MSEs of the order of 0.03. Fig.7 shows the predictions of the

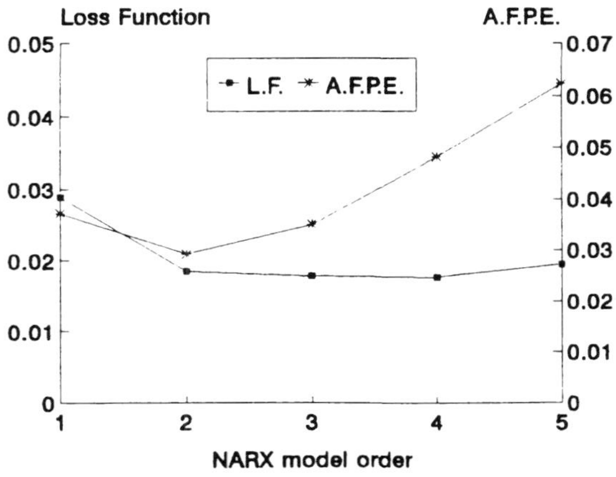

Fig. 6 Selection of ANN NARX model order

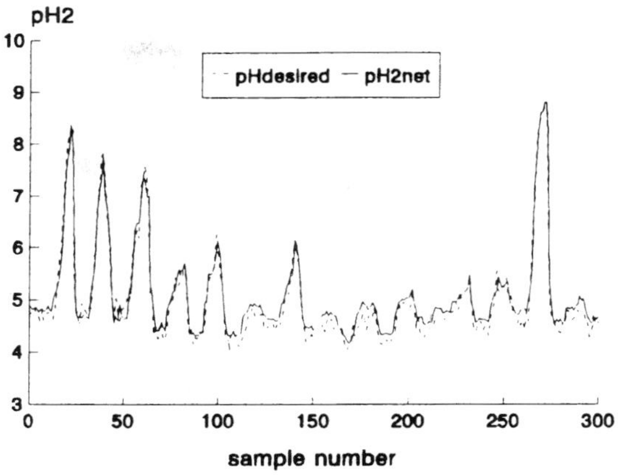

Fig. 7 An ANN model predicting on unseen real process data

transport lag volume array network on test data and is typical of the results obtained.

These results suggest that in this application all three methods for accommodating variable deadtime are valid. However, there are distinctions to be drawn between the implementation complexity and computational effort for the three methods. The delay spread technique is appealing since it does not require the use of flow rate information and is the simplest of the three to implement. However, a significantly larger network was required with this method and the network inherently includes redundancy. Furthermore, the network size increases with the range of the process deadtime and consequently may not be a suitable approach for processes with large deadtime variations.

The remaining two methods both require measurement of the flow rate but benefit by a significant reduction in the ANN model size. The transport lag volume array can be implemented with a conventional constant sample rate but does require additional process knowledge (for the in-line process, this was the volume of the tubular reactor). A disadvantage of this method is the computational effort required in the array manipulation. In comparison, volumetric sampling provides a succinct approach to process identification in the presence of variable deadtime with low computational effort. A possible drawback is the resulting use of a non-conventional variable sample time.

6. CONCLUSIONS

The successful identification of an in-line pH process, dominated by a strong nonlinearity and variable deadtime, was achieved using an RBF neural network. Three methods were investigated for accommodating the variable process deadtime in the identified ANN model. In this application, each resulted in ANN models with comparable prediction accuracies. However, the methods differ in their computational complexity and the amount of process information required for their implementation. This would influence the choice of a method for a particular application.

ACKNOWLEDGEMENTS

This work is funded by Liverpool John Moores University Research Fund and British Nuclear Fuels Ltd. (Sellafield).

REFERENCES

Broomhead, D.S. and D. Lowe (1988). Multivariable functional interpolation and adaptive networks. *Complex Systems*, **2**, 231-355.

Chen, S., S.A. Billings, C.F.N. Cowan and P.M. Grant (1990). Practical identification of NARMAX models using radial basis functions. *Int. J. Control*, **51**, 1191-1214

Doherty, S.K., J.B. Gomm, D. Williams and D.C. Eardley (1994). Design issues in applying neural networks to model highly non-linear processes. *Proc. IEE Control Conf.*, 1478-1483.

Gulaian, M. and J. Lane (1990). Titration curve estimation for adaptive pH control. *Proc. American Control Conf.*, 1414-1419.

Gustaffson, T.K. and K.V. Waller (1983). Dynamic modelling and reaction invariant control of pH. *Chem. Engng. Sci.*, **28**, 389-398

Jacobs, O.L.R., W.A. Badran, C.G. Proudfoot and C. White (1987). On controlling pH. *IEE Proc-D.*, **134** , 196-199.

McAvoy, T.J., E. Hsu and S. Lowenthal (1972). Dynamics of pH in controlled stirred tank reactors. *Ind. Engng. Chem. Process Des. Dev.*, **11**, 68-70.

Moody, J. and C. Darken (1989). Fast-learning in locally-tuned processing units. *Neural Computation*. **1**, 281-294.

Rhinehart, R.R. and J.Y. Choi (1988). Process model-based control of wastewater pH neutralization. *Advances in Instrumentation,* **43**, 351-358.

Shinskey, F.G. (1973). *pH and pION Control in process and waste streams*, Wiley, New York. *Chem. Engng. Sci.*, **38**, 389-398.

SOME ASPECTS OF INTEGRATED PROCESS OPERATION

E.D. GILLES*, G. LAUSCHKE*, A. KIENLE* and M. STORZ*

*Institut für Systemdynamik und Regelungstechnik, Universität Stuttgart, Pfaffenwaldring 9, 70550 Stuttgart, Germany

Abstract. Operation and control of integrated processes is considered. Thereby, the aspect of process integration is examined on two levels. On the lower level, different unit operations, like reaction and heat recovery or reaction and separation, are integrated into multifunctional processing units. On the flowsheet level, different processing units are integrated via mass and energy recycles, like the heat integration of distillation columns. On both levels, complex dynamics including periodic and aperiodic oscillations and multiple attractors can arise as a consequence of strong interaction. For the lower level, nonlinear phenomena are illustrated with two examples. The first example is the so-called circulation loop reactor, which is intended to work as an autonomous periodic system. Second, dynamic operation of a reactive distillation column with multiple steady states is considered. To handle dynamics on the flowsheet level a hierarchical process control concept for plant-wide control is presented. The concept is based on a formal definition of product quality as an intensive state function in the thermodynamic sense. This quality function is evaluated for each product stream in a chemical plant and is employed as a plant-wide controlled variable. Further, such a plant-wide objective function can be used for the integration of process design and control.

Key Words. Chemical industry, nonlinear analysis, travelling waves, limit cycles, chaotic behavior, integrated plant control, quality control, hierarchical control, nonlinear control systems

1. INTRODUCTION

In view of stronger economic competition integrated process operation is becoming an issue of growing importance in processing industries. For an efficient use of mass and energy or for pollution abatement, production processes are integrated to an increasing degree. Modern computer-aided system analysis tools open a wide range of new aspects and possible applications for process integration. This paper focuses on some aspects of process integration on two different levels :

- On the lower level, different unit operations are integrated into multifunctional processing units.
- On the superior level, different processing units are integrated via mass and energy recycles.

Integration of different unit operations within a single multifunctional processing unit has a long tradition in chemical engineering. A classical example is the integration of reaction and separation in a reactive distillation column for the production of esters (Keyes, 1932). The technical benefit of such processing units comes from particularly close interaction between the different operations involved. This interaction is either realized by internal mass and energy recycles or by mutual influence of the basic physical and chemical mechanisms like in reactive distillation. Besides, close interaction can result in complex dynamics including steady state multiplicity, sustained periodic oscillations and even chaos. On the basis of suitable process models, such phenomena can be predicted by rigorous application of nonlinear system analysis. The a-priori knowledge on process dynamics can be utilized in two different ways:

1. In some cases, nonlinear phenomena can be specifically utilized to design novel processes.
2. Suitable process design and operation strategies can be developed to avoid or handle undesired nonlinearities.

The second strategy is common practice today. Usually, undesired nonlinearities are avoided, sometimes at the price of suboptimal design. Examples are given in Seider *et al.* (1990). However, only rare examples are known for the first strategy. In the present paper, both are illustrated exemplarily. First, integration of reaction and heat recovery within a single bifunctional reactor is considered. A novel reactor concept is presented that makes direct use of (nonlinear) travelling reaction zones and is therefore an example for the first strategy, mentioned above. In particular, this reactor is intended to operate as an autonomous periodic system. Second, the integration of reaction and separation is considered. Main emphasis is on reactive distillation. Briefly, dynamic operation of a reactive distillation column with multiple steady states is considered as an example for the second strategy.

On the flowsheet level, similar dynamic phenomena as described above can arise as a consequence of integration via mass and energy recycles between different processing units. Especially if intermediate storage tanks are omitted to provide flexible operation of the plant during load or grade changes. In addition, time constants and sensitivities to feed disturbances can increase dramatically (Morud and Skogestad, 1994; Luyben, 1994). Hence, a control concept as usually employed in the processing industries, which neglects the interactions between different processing units, can be insufficient for highly integrated plants. Further, a plant-wide approach with a single multivariable controller is often not desirable, because of safety, reliability and flexibility. Instead, a hierarchical control concept is proposed (Mesarović *et al.*, 1970; Singh, 1980), where the interactions between different locally controlled units are taken explicitly into account by a coordination control system. In this concept, the local control systems have to simplify the dynamic behavior of the individual processing units by means of nonlinear control techniques, whereas the coordination control system provides optimal operation during load or grade changes. The hierarchical control concept is based on product quality as a plant-wide uniform objective function. In contrast to statistical quality control methods (MacGregor, 1988), process dynamics is explicitly taken into account. In addition, product quality can be used for the integration of process design and control, which is the third main aspect on integrated process operation considered in this paper. These ideas are illustrated with a comparably simple example process. Similar processes have been examined with classical design and control techniques by Luyben (1993) and Lyman and Luyben (1994).

2. INTEGRATION ON PROCESSING UNIT LEVEL

Usually both aspects, integrated design and integrated phenomena, are mutually dependent in a multifunctional processing unit. The aspects of integrated design and integrated phenomena are now discussed in some detail with two examples. Emphasis is on open loop dynamics.

2.1. *Integration of reaction and heat recycling*

In chemical reaction engineering, the most frequent type of integrated design consists of energy or mass recycling. This work is focused on constructively adapted heat recycles and re-

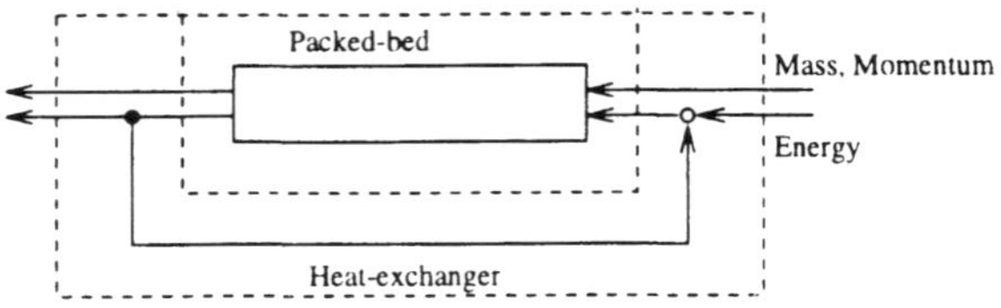

Fig. 1. Principle of heat recycling within a packed-bed reactor

action sections combined in a single integrated reactor. Such a reactor has two design elements, labeled "packed-bed" (PB) and "heat-exchanger" (HE) in Fig. 1. Both elements may contain catalyst. As a systematic modeling approach for the integration of these units a classification for PB and HE is proposed. In this way, work reported in literature and future developments may be considered in a uniform framework. The elements discussed are suitable for gaseous reactants and exothermic reactions. Other conditions may lead to further variants. For the PB elements in Fig. 2 either one dimensional pseudohomogeneous models (PB1-PB4) or the ordinary feedback (PB5) are applied to predict the dynamics sufficiently accurately. The model elements may include a wall energy balance (WEB) in one (PB1) or more sections (PB3). Even if the packed-bed is mostly undivided (PB1) there may exist purposes where more than one section becomes necessary (PB2-PB4), eg. a wall with flanges or other inhomogenities. The HE model elements HE1 and HE2 in Fig. 3 are based on the model PB1 and may be written with two WEBs. HE3 representing cross-current heat-exchange requires a considerable modeling and computational effort if the temperature profiles shall be calculated rigorously. HE4 and HE5 finally represent simple models without spatial information.

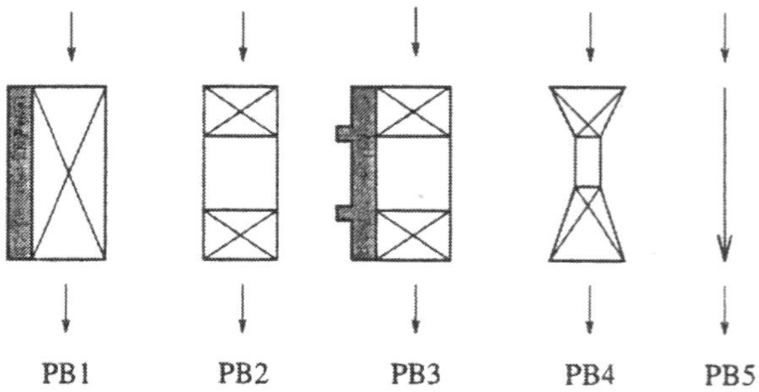

Fig. 2. Packed-bed design variants

As an example of counter-current heat-exchange Bonvin *et al.* (1980) analyzed a configuration of PB5 and HE2 combination without WEB. Adding one WEB in Bonvin *et al.* (1983) leads to similar results. Adomaitis and Cinar (1988), and Adomaitis and Cinar (1991) finally investigated an analogue configuration with two WEBs. In the work cited only one heat-exchanger path is filled

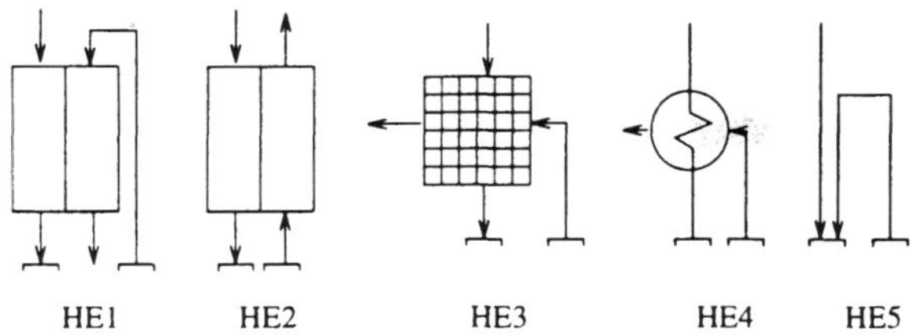

Fig. 3. Heat-exchanger design variants

with catalyst. For these counter-current systems steady state multiplicity has been detected but no sustained oscillations have been found. In contrast Ruppel (1980) presented a reactor with a reduced heat-exchanger model (combination of PB1 and HE4 without WEB). He found also periodic solutions. The stability and bifurcation analysis of Recke and Jørgensen (1994) for the same configuration has revealed steady state multiplicity, Hopf bifurcations, and periodic multiplicity in more detail. Finally, Yakhnin *et al.* (1994) have found Hopf bifurcations also for the minimum of a HE, i.e. periodic boundary conditions (combination PB1 with HE5).

In this work a combination of PB1 and HE1 both with WEB is selected. This reactor configuration, called circulation loop reactor (CLR), is intended to run as an autonomous periodic system based on the phenomenon of self-exciting travelling reaction zones (Balmer and Gilles, 1989). Thus, heat recovery is not for steady state operation but for dynamic triggering of temperature pulses. This is in contrast to an other concept based on travelling reaction zones, the reverse flow reactor (Matros *et al.*, 1988; Matros, 1985), which is operated as a periodically forced system. Like the reverse flow reactor, the CLR is suited for weak exothermic and equilibrium limited reactions.

The basic design shown in Fig. 4 has been developed for the catalytic oxidation of organic compounds (VOCs) in the field of pollution abatement. In the catalytic packed-bed the travelling reaction zones accumulate the reaction heat in a temperature pulse which is transferred through the catalytic co-current HE retaining the shape of the reaction zone. The figure is referred to a laboratory-scale reactor with a double tube as a HE. For an industrial design there exists a wide range of further alternatives for both, reactor loop and HE (detail design for HE1). A single operation cycle of the CLR is explained with Fig. 5 showing some characteristic temperature profiles and the corresponding composition profiles. The axial reactor coordinate z represents the flow path of the reactant mixture. During the start-up the inlet stream has to be heated up to reach ignition temperature in order to start the reaction in the inner tube of the packed-bed HE. Switch-

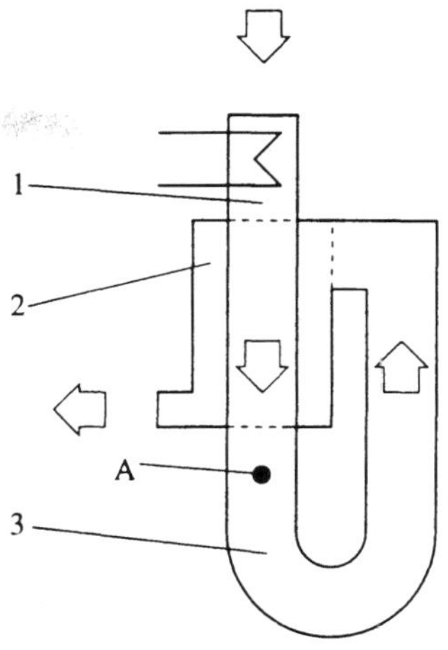

Fig. 4. Circulation loop reactor – 1: heating, 2: co-current heat exchanger, 3: reactor loop (packed-bed), part 2 and 3 are filled with catalyst; A: position of characteristic temperature T_A

ing off the heating, the reaction zone behaves in the sense of wrong-way behavior and moves downstream through the reactor loop section creating a strong temperature rise (phase 1 in Fig. 5). The reaction zone travels with a migration speed considerably lower than the gas velocity into the heat exchanger jacket (phase 2 in Fig. 5). Before leaving the reactor the moving hot spot is heating up the cold inlet stream in the inner tube until ignition conditions. After that re-ignition of a new reaction zone, the old reaction zone annihilates due to a lack of reactant consumed by the new reaction zone (phase 3 in Fig. 5). At this moment one operation cycle is accomplished. The desired, autonomous periodic operation of the reactor is achieved by the described heat recycle of travelling reaction zones. In addition, elevated feed temperatures or reactant concentrations will drive the CLR into a steady-state.

A crucial task for both reactor design and operation, is therefore to determine the existence regions of different attractors depending on operating conditions and design parameters. For this purpose, continuation methods, and dynamic simulation combined with stability analysis have been applied.

Calculating steady state and periodic solutions for the catalytic ethene oxidation on a mixed oxide catalyst, the bifurcation diagram in Fig. 6 is obtained. The characteristic temperature T_A at point A in Fig. 4 is selected as state representation. The stationary solution curve in Fig. 6 is not an ordinary ignition/extinction hysteresis even though a region of three coexisting steady states is identified. The lower branch from heating power $P_h = 0$ to the limit point F_{1L} marks stable extinguished steady states. From this point

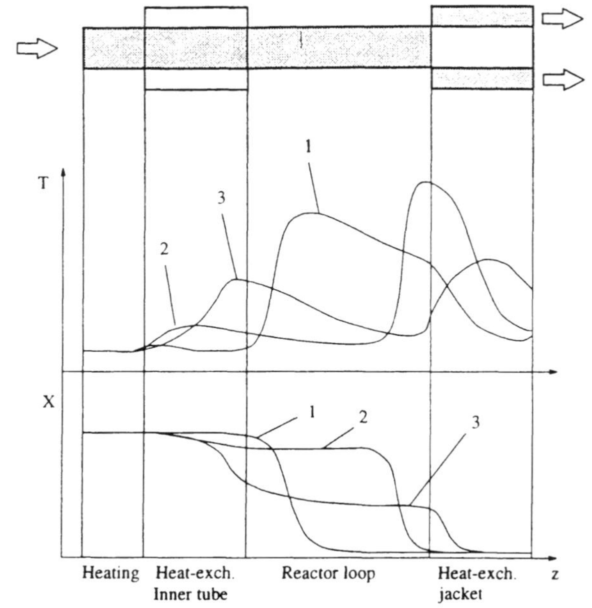

Fig. 5. Principle of the circulation loop reactor – 1: travelling reaction zone in the reactor loop, 2: heat transfer to the inner tube and re-ignition, 3: two co-existing reaction zones in jacket and inner tube during re-ignition

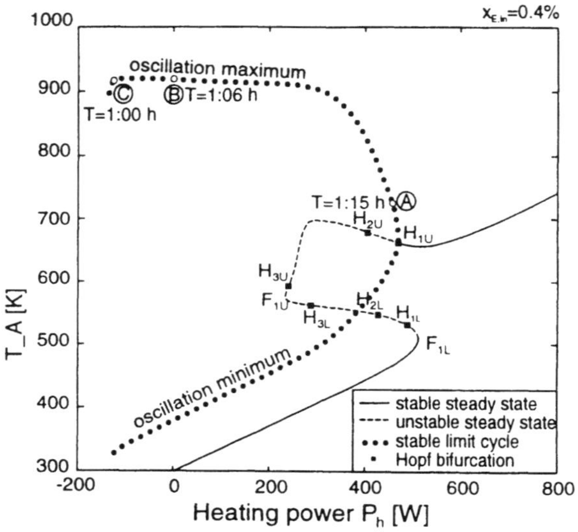

Fig. 6. Calculated bifurcation diagram with steady state and periodic solution branch for molar content $x_{E,in} = 0.004$ ($\tilde{z}_3 = 0.905m$)

to F_{1U} the chemical reaction with incomplete conversion generates unstable steady state solutions. In this region the spatial temperature profile indicates a distinct heat transfer from the reactor jacket to the inner tube. The reaction zone stretches from the inner tube to the middle of the reactor jacket. For this reason a temperature perturbation at the reactor inlet leads to a positive feedback on the reaction zone. At the limit point F_{1U} stability does not change like in a common ignition/extinction hysteresis because the complex eigenvalues which have passed the imaginary axis at the Hopf bifurcation points H_{1L},H_{2L}, and H_{3L} still remain in the right half plane drawing back one after another at higher values of P_h. According to the Hopf theorem (Marsden and McCracken, 1976) all Hopf points except H_{1U} lead to unstable limit cycles. Consequently, the solution curve remains unstable up to the point H_{1U}. Providing complete conversion on the right hand side from F_{1U}, the reaction zone lies entirely inside the inner tube. Between F_{1U} and H_{1U} there exists still positive feedback due to heat recycling. Only above H_{1U} the global heat transfer direction runs from the inner tube to the reactor jacket. This heat transfer direction supplies stability because heat recycling has lost its dominating influence.

By computing a smooth divergence to a limit cycle with small amplitudes, H_{1U} is identified as a supercritical Hopf bifurcation (Guckenheimer and Holmes, 1983). Continuing the limit cycles from H_{1U} the oscillation maximum leads into a saturation at $P_h = 236W$. At this point the periodic behavior is divided into two regimes. On the left hand side circulating reaction zones are observed as depicted in Fig. 5 and on the right hand side there exist so-called swinging reaction zones with variant wave pattern. Autothermal operation is achieved at $P_h = 0W$, but the periodic branch retains stability down to very low feed temperatures corresponding to negative values of the heating power. The limit of the periodic branch is marked by a cyclic fold bifurcation. The period shown at points A - C decreases with decreasing feed temperature. In Fig. 6 all unstable periodic solutions are omitted, those coming from the cyclic limit point as well as H_{1L} to H_{2U}. Further detailed numerical results containing a description of the methods applied and the calculation parameters is given in Kienle *et al.* (1995).

The steady state behavior described above has been verified experimentally as depicted in Fig. 7. The geometrical proportions as well as the catalyst employed in the experiments differ slightly from the properties assumed in the model because not all data were available during the modeling procedure. Nevertheless, the qualitative agreement between experimental and numerical results is excellent. In the experiments as well, a transition to periodic solutions is observed associated with swinging reaction zones in the vicinity of the bifurcation point. These oscillations at point a in Fig. 7 are presented in Fig. 8. Circulating reaction zones (Fig. 9) occur at low feed temperatures (point c in Fig. 7) according to the numerical results. They produce strictly periodic nonlinear oscillations. In contrast to the numerical results in the real process a small region separating circulating and swinging reaction zones has been found. In this region the occurrence of chaotic oscillations

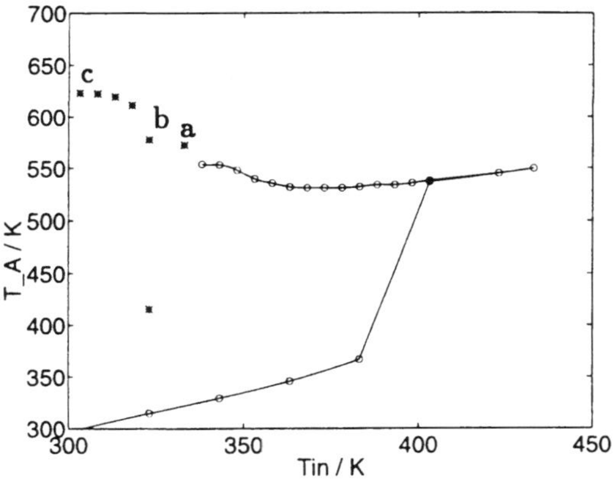

Fig. 7. Experimental bifurcation diagram of ethene oxidation (O_2 in excess) at $0.1\% Pt$ on $\alpha - Al_2O_3$ catalyst, d = 3mm, $X_{E,in} = 0.4\%$, $T_{in} = 295K$, T_A at position A in Fig. 4; a: swinging reaction zones, b: intermittent chaos, c: circulating reaction zones

is observed (point b in Fig. 7). Two asterisks at this point indicate the sets of high and low temperature maxima. The chaotic pattern shown in Fig. 10 is of the intermittent type (Pomeau and Manneville, 1980). At the transitions to chaos both adjacent attractors loose their stability and transform into saddle cycles. The intermittent chaos stems from a competition of these saddle cycles in phase space. Further details on the experimental analysis are given in Lauschke and Gilles (1995).

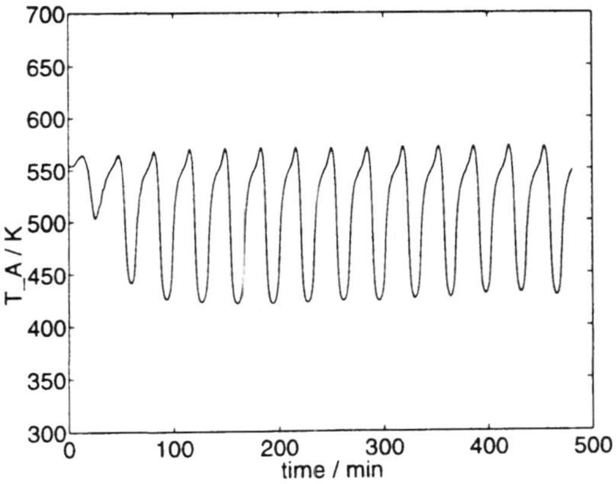

Fig. 8. Experimentally observed oscillations near Hopf-bifurcation (a in Fig. 7) at position A in Fig. 4, regime of swinging reaction zones

In the field of waste air combustion, the CLR has already gained industrial relevance. Applications in further domains of reaction engineering are worked out in current research activities.

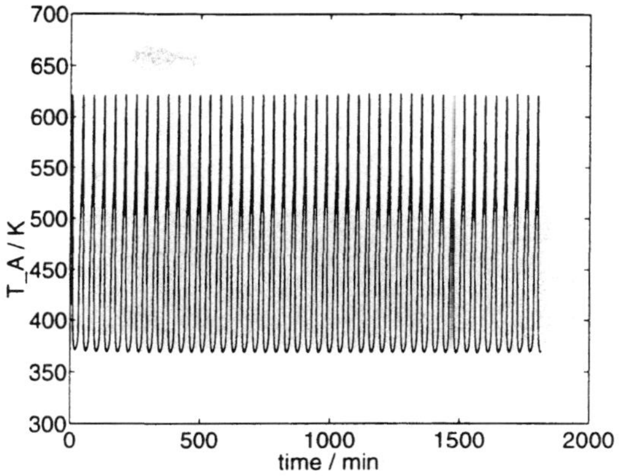

Fig. 9. Experimentally observed oscillations in the regime of circulating reaction zones (c in Fig. 7) at point A in Fig. 4

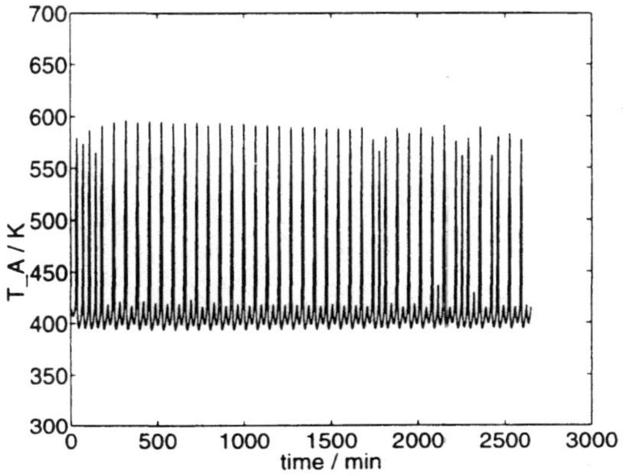

Fig. 10. Experimentally observed chaotic oscillations (b in Fig. 7) at point A in Fig. 4

2.2. *Integration of reaction and separation*

A second class of multifunctional processing units is obtained by combining reaction with separation. The purpose of such an integrated design is either to solve difficult separation problems through selective and often reversible reactions (Gaikar and Sharma, 1989) or to improve a given reaction by overcoming chemical equilibrium limitations through selective separation of reaction products or by suppressing undesired side-reactions. Examples for the latter aspect with emphasis on heterogeneously catalyzed gas phase reactions are discussed in Agar and Ruppel (1988*a*) and Agar and Ruppel (1988*b*). Currently, the most popular type of combined reaction/separation is reactive distillation. This technology is applicable for liquid phase reactions with reasonable reaction rates under moderate temperature and pressure conditions usually encountered in a distillation column. In some cases, reactive distillation is very economical and reduces the total production costs dramatically compared to the conventional technology, where reaction and separation are carried out in different processing units (Agreda *et al.*, 1990; Doherty and Buzad, 1992; DeGarmo *et al.*, 1992). So far, most work in reactive distillation has been reported about steady state design. Only very little work is known on dynamic operation of reactive distillation columns (Doherty and Buzad, 1992). Although, especially in reactive distillation interesting dynamic phenomena can arise due to the interaction of reaction, separation and countercurrent flow as will be illustrated exemplarily in the remainder.

The example to be considered is a reactive distillation column for the production of ethylene glycol as proposed by Ciric and Gu (1994) (Fig. 11).

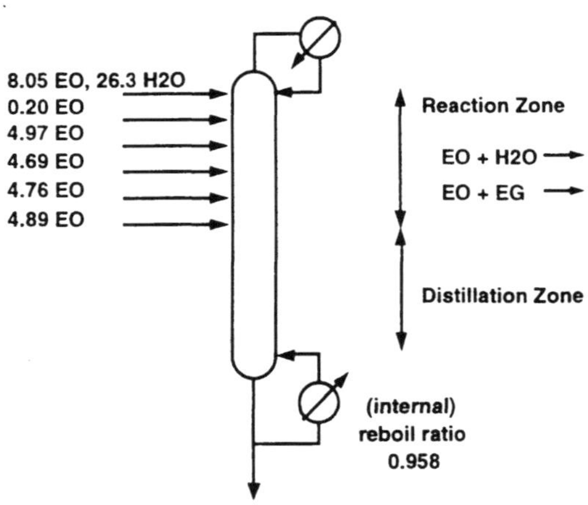

Fig. 11. Optimal distributed feed reactive distillation column for ethylene glycol synthesis (Ciric & Gu, 1994)

Industrially, ethylene glycol (EG) is produced by thermal hydrolysis of ethylene oxide (EO) without catalyst. Further, unconverted ethylene oxide reacts in a side reaction with ethylene glycol to undesired diethylene glycol (DEG) or the corresponding higher homologues tri-, tetra- and polyethylene glycol, respectively, which however are not taken into account in the remainder. Usually, the reaction is carried out in a plug flow reactor. To reduce undesired by-products the reaction is run with a large excess of water which has to be separated from the reaction products. In practice, almost 90 % of ethylene oxide can be converted to (mono-) ethylene glycol (Rebsdat and Mayer, 1990). Instead, almost 95 % yield can be obtained in a reactive distillation column with almost stoichiometric feed composition (Ciric and Gu, 1994). This is attained by running the column at total reflux and high reboil ratios. Thus, unconverted water is recycled within the column and an extremely high ratio of water to ethylene oxide is obtained within the column, which favours the first reaction in Fig. 11.

Ciric and Miao (1994) have shown that multiple steady states occur for the ethylene glycol column. Further, they found that steady state multiplicity of the ethylene glycol column is a consequence of interaction between separation and reaction. In particular, for the optimal design in Fig. 11 three different steady states are possible. The corresponding steady state composition profiles are shown in Fig. 12. There are two stable steady states marked by a solid line in Fig. 12. The first, high conversion steady state has high product composition of desired ethylene glycol at the bottom of the column. Instead, the second undesired one has only low conversion. Inbetween, there is an unstable steady state.

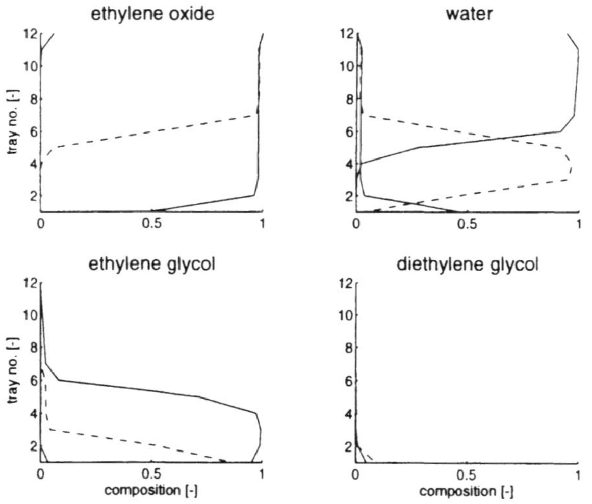

Fig. 12. Multiple steady state composition profiles for the ethylene glycol column. Solid line – stable, dashed line – unstable.

Usually, it should be no problem to operate the column at any steady state no matter if stable or unstable if suitable control is employed. For practical operation, main interest is therefore on column startup. Namely, how to operate the column to get close to the desired steady state before controllers are switched on. Simulation experiments indicate, that some startup strategies may fail, i.e the column moves either towards the undesired steady state with low yields of ethylene glycol or even towards the desired steady state but produces large rates of undesired diethylene glycol during the transient. So far, the best strategy for column startup we found is to operate the column initially with pure water before increasing the ethylene oxide feed to its nominal value (Fig. 13). Thus, the production of desired ethylene glycol is enforced during the whole transient.

Other examples for complex dynamics in reactive distillation are multiple steady states for the industrially highly relevant process for the production of octane enhancer methyl tert-butyl ether (Jacob and Krishna, 1993; Nijhuis *et al.*, 1993). The multiplicities mentioned so far are output

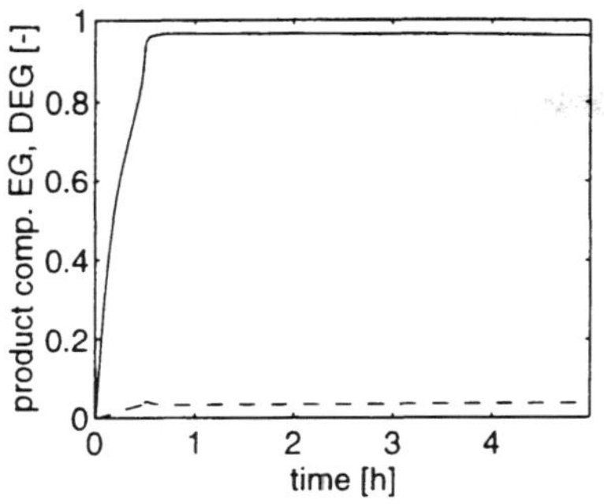

Fig. 13. Startup strategy for the ethylene glycol column: First, column startup is done with pure water. At $t = 0$ ethylene oxide feed is set to its nominal value. Solid line - EG, dashed line - DEG.

multiplicities. Namely, steady state concentration and temperature profiles as well as the product compositions are ambiguous for fixed input variables and fixed design parameters. Instead, input multiplicities for column temperatures as measured variables are known to arise in esterification systems like the methyl acetate process (Roat *et al.*, 1986). Here, the same values for the measured temperatures can be obtained for different sets of input variables. These input multiplicities result in multiple steady states for the closed loop system. Again, in both cases the phenomena described above are a matter of interaction between separation and reaction (Roat *et al.*, 1986; Jacob and Krishna, 1993).

3. INTEGRATION OF PROCESS OPERATION ON THE FLOWSHEET LEVEL

In this section, the plant-wide interactions between different units are considered. A satisfactory integration of process design and control is only possible using a plant-wide uniform objective function as an integrating factor. Here, a formal definition of product quality as an intensive state function in the thermodynamic sense forms the basis to characterize each product stream in a chemical plant with respect to the same production objective.

3.1. *Quality Function for Example Process*

To illustrate the ideas on plant-wide design and control of chemical plants a simple example process is considered having the production objective to form a product B from feedstock A. Further, the undesired byproduct C is obtained according to the consecutive reaction scheme $A \rightarrow B \rightarrow C$. In principle, different plant flowsheets are suitable to meet the desired production objective. One typical flowsheet is shown in Fig. 14. It consists of a continuous stirred tank reactor (CSTR) for reaction, two distillation columns (DC) for separation, and a recycle stream, where unreacted A is fed back into the CSTR to form the desired end product B. In the phase-model (Fig. 15) each product stream is completely characterized by flow rate and thermodynamic state.

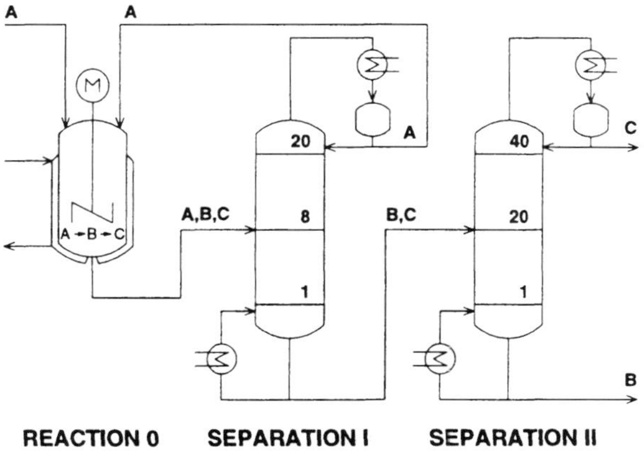

Fig. 14. Simplified flowsheet of example process

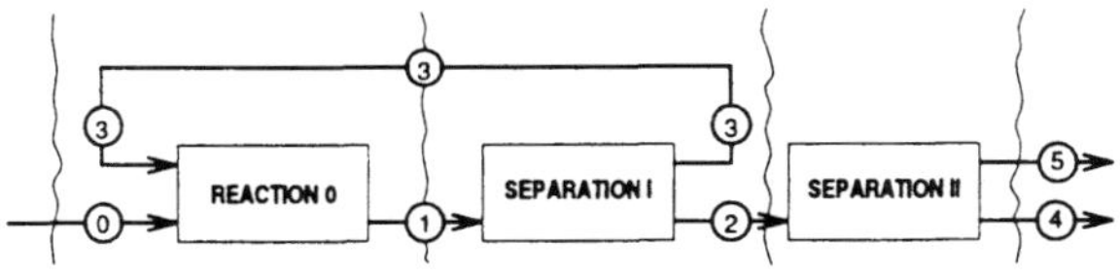

Fig. 15. Phase-model of example process

For plant-wide integration of process design and control an extended quality notion has been defined by Friedrich and Gilles (1991). In this context, it is of interest how end product quality is related to the quality of feedstock as well as intermediate products for given process conditions. Therefore, the following verbal definition of product quality is used :

> Quality of a material stream is the property that enables it to contribute to the production of a specified end product stream.

For so-called "simple" thermodynamic systems (Callen, 1985) this quality is an intensive state function in the coordinate system $\underline{x}^T = [p, T, x_1, \ldots, x_{i_0}]$ of the thermodynamic state variables. Consequently, the partial molar qualities $\bar{w}_i$ are related to the quality of a homogeneous mixture of i_0 components by

$$w(\underline{x}) = \sum_{i=1}^{i_0} \bar{w}_i(\underline{x}) x_i \quad . \tag{1}$$

This quality definition is now illustrated by means of the example process. The objective is to produce pure component B, since only high-purity B shows for example a desired property whereas pure components A and C do not have this property. It is assumed that this technological quality characteristic of the end product is easily measured and a physical relationship between this quality characteristic and the intensive state variables $\underline{x}$ is known. However, this technological quality characteristic does not satisfy the verbal definition of the extended quality given above. This can be shown by contradiction in the following way :

First, it is assumed that the desired property is equivalent to the extended quality we are looking for. For simplicity, it is further assumed that this desired property depends linearly on the composition of the mixture. In Eq. (1) such a linear relationship is obtained for constant partial molar qualities $\bar{w}_i = q_i$. Because A and C do not have the desired property, the quality (1) depends only on the molar fraction x_B. Then, pure B has maximum quality ($q_B = 1$) and the qualities of the pure components A and C both vanish ($q_A = q_C = 0$). This, however, is in contradiction to the verbal definition given above. Since feedstock A can react into product B, substance A is able to contribute to the desired end product quality. Therefore, substance A does have a certain quality ($q_B > q_A > q_C$) with respect to the production of B. But, pure C, as an undesired byproduct, has still minimum or zero quality ($q_C = 0$) with respect to the production of B.

Thus, in quality function (1) each component is weighted with respect to the production objective and not with respect to any technological quality characteristic. This plant-wide uniform quality function can be applied not only to the end product stream of a plant but also to all other material streams. Here, the following fundamental aspects are important :

- The quality function has its maximum value for the desired end product specification, which is expressed in the intensive state variables $\underline{x}_E$. Any technological quality characteristic of the end product like the desired property in the example above is only used to determine these states $\underline{x}_E$. This might be a difficult modeling task which depends on the application and cannot be solved in general. Instead, the abstract quality notion introduced above represents a more general quantity.
- The quality function is interpreted as a special measure of the degree to which the desired production objective has been achieved.
- A compression of information is obtained, since the scalar quality value indicates the "distance" of each product stream from the specified end product stream. Along the contour lines of the surface $w(\underline{x})$ different compositions of a mixture have the same distance to the desired production objective.

The linear quality function with constant partial molar qualities $\bar{w}_i = q_i$ cannot describe any mixing or separation effects, since the molar mixing quality Δw_M vanishes. Additionally, w_{lin} has always its maximum value in the pure components. Therefore, some higher order terms must be considered. One possible expression for the quality function of a homogeneous mixture is given in loose analogy to classical thermodynamic functions, like the entropy or the chemical potential of an ideal mixture :

$$w(\underline{x}) = w_{lin} + \Delta w_M = \sum_{i=1}^{i_0} (q_i x_i + q_i^* x_i \ln x_i). \quad (2)$$

An adequate parameter choice is obtained taking into account with $q_C^* \geq q_A^* \geq q_B^* \geq 0$ that the quality of the mixture is mainly decreased by undesired C but also the partial qualities of A and B in the mixture are decreased ($\Delta w_M \leq 0$). On the other hand, a quality function (2) can be constructed from partial molar qualities $\bar{w}_i$ via the equation of Gibbs-Duhem if the integrability conditions are satisfied. Other possible approaches to construct quality functions are treated in Friedrich (1994).

3.2. *The Qualifying Potential*

Further, it is possible to assign the so-called qualifying potential W to quality w as an extensive state function in the coordinate system $\underline{X}^T = [p, T, n_1, \ldots, n_{i_0}]$ with molar numbers n_i :

$$W = n \cdot w = \sum_{i=1}^{i_0} \bar{w}_i n_i = n \sum_{i=1}^{i_0} \bar{w}_i x_i \quad . \quad (3)$$

For this extensive state function W a balance equation can be formulated for each single processing unit as well as for any part of a plant :

$$\frac{dW}{dt} = J_C[W] + P_L[W] \quad . \quad (4)$$

$P_L[W]$ represents the production of qualifying potential and $J_C[W]$ the convection term, which is expressed in qualifying potential streams $Q_k = L_k w_k$. Generally, the balance equation of W is directly obtained by inserting the differential equations of the state variables into the total differential of W written with differentials in time t :

$$\frac{dW}{dt} = \left(\frac{\partial W}{\partial p}\right)_{T,n_i} \frac{dp}{dt} + \left(\frac{\partial W}{\partial T}\right)_{p,n_i} \frac{dT}{dt} + \sum_{i=1}^{i_0} \bar{w}_i \frac{dn_i}{dt}$$

The qualifying potential W forms the basis to integrate process design and control as will be illustrated in the remainder.

3.3. *Plant-Wide Process Design*

The production of qualifying potential can be used as a plant-wide objective function to evaluate process performance with respect to quality improvement under steady state conditions.

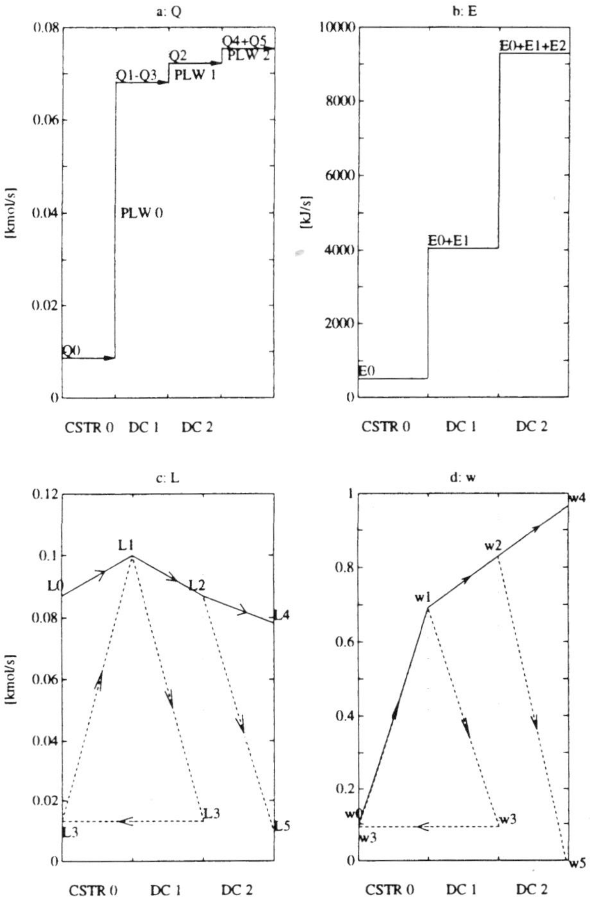

Fig. 16. Total qualifying potential stream (a), heat input (b) as well as phase-model of flow rate (c) and quality (d) with quality function $w = 0.1x_A + x_B + 0.125x_A \ln x_A + 0.1x_B \ln x_B + 0.15x_C \ln x_C$ (explanation in text)

Assigning a qualifying potential stream $Q_k = L_k w_k$ to each molar stream L_k (Fig. 16c) by means of quality function (2), the steady state production terms $P_L[W]_l$ can be visualized for all units (Fig. 16a). The production terms $P_L[W]_l$ are calculated as the differences between the sum of potential streams on the downstream and the upstream side of each section in Fig. 15, where recycle streams are considered with negative sign. During regular steady state operation each unit should contribute to the whole plant production. Therefore, the total stream of qualifying potential (Fig. 16a) and the quality of the main product stream (solid line in Fig. 16d) are continuously increased along the production path. Trivially, it follows directly from the fundamentals of the plant-wide quality definition that during regular steady state operation

- the quality of the end product stream must be increased compared with the quality of the feedstock stream : $w_4 > w_0$.
- the process must produce qualifying potential : $P_L[W] = \sum P_L[W]_l > 0$.

Thereby, each plant fulfills the desired production objective, i.e. the complete conversion of a feedstock stream of pure A into an end product stream of pure B,

$$P_L[W]_{max} = L_0(q_B - q_A) \tag{5}$$

only up to a certain degree η :

$$0 < \eta = \frac{P_L[W]}{P_L[W]_{max}} \leq 1 \quad . \tag{6}$$

In order to assess alternative plants with respect to the same production objective the needed total heat input E of all processing units (Fig. 16b) must be also taken into account. For example, the following plant-wide optimization problem can be formulated with respect to Fig. 16 :

In order to generate a specified total production of qualifying potential for this plant

$$P_L[W] \quad = \quad L_4 w_4 + (L_0 - L_4) w_5 - L_0 w_0 \tag{7}$$

the needed total heat input must be minimized

$$E = \sum E_l \stackrel{!}{=} min. \quad , \tag{8}$$

where the end product stream has been specified by L_4, w_4 as well as the feedstock stream by L_0, w_0. Together with the specified production (7) the byproduct stream L_5, w_5 is also completely determined. Therefore, only the internal variables w_1, w_2, w_3, and L_3 can vary in Fig. 16.

Additionally, the steady state production of qualifying potential can be used as an objective function in process design (Storz *et al.*, 1994) to compare

- alternative processing units, like a CSTR or a continuous tubular reactor as reaction unit for this consecutive reaction scheme,
- alternative plant topologies, like a structure with or without a recycle loop for component A.

Here, it is important to note that alternative plants are compared with respect to the same quality function. Hence, it is only possible to select the most suitable plant with respect to the desired production objective from a given set of alternatives.

3.4. *Plant-Wide Process Control*

In the following, a hierarchical concept for plant-wide quality control is briefly summarized according to Friedrich and Gilles (1991) and Storz *et al.* (1994). It consists of local control systems for each single unit and a coordination control level. The main task of the local controllers is to simplify the quality-oriented I/O-behavior of each unit in a desired way. The coordination control level has to enable information exchange between the idealized units in order to achieve specified quality and desired flow rate of the end product stream (Fig. 17).

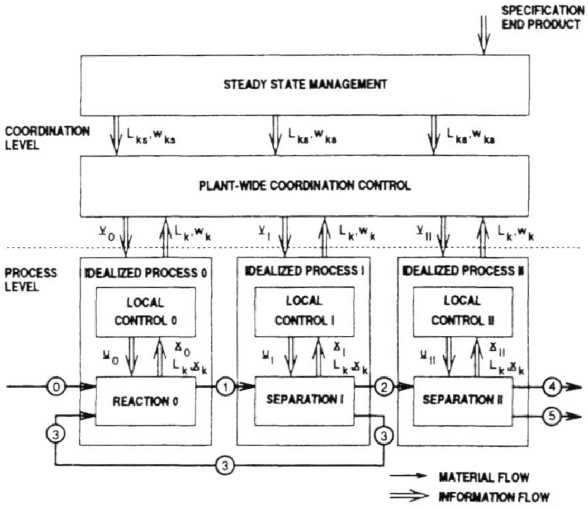

Fig. 17. Hierarchical process control concept.

On the process level, the manipulated variables $\underline{u}$ of each unit are used to match the nonlinear dynamic I/O-behavior between the states $\underline{x}_{in}$ of the inlet streams and the states $\underline{x}_{out}$ of the outlet streams (Fig. 18, left) into an ideally linear I/O-behavior between the qualities w_{in} and w_{out} with new generalized manipulated variables $\underline{v}$ (Fig. 18, right). On the coordination level the material flow is described by qualifying potential streams $Q_k = L_k w_k$. The coordination control task is also simplified, since each unit is represented by an idealized model and the dynamic coupling between the different units is completely determined by inputs Q_{in} and outputs Q_{out}.

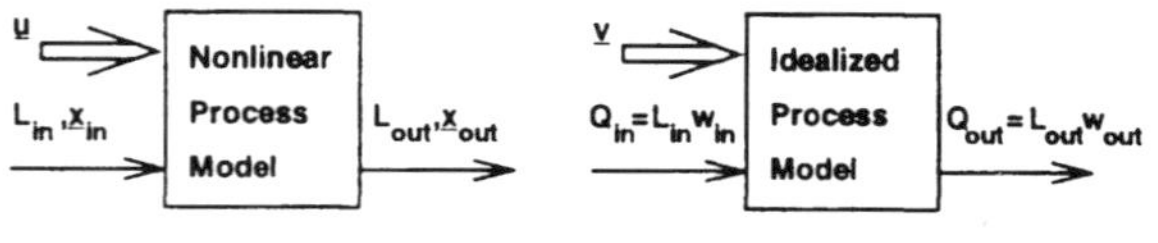

Fig. 18. Model matching problem to give the nonlinear process (left) the I/O-behavior of an idealized quality-oriented reference model (right) .

Local Control on the Process Level. This model matching problem can be solved for example with the method of exact input/output linearization (EIOL) (Kravaris and Chung, 1987; Isidori, 1989). First, the application of EIOL is conceptually outlined assuming that the dynamic behavior of each unit is given by a perfect dynamic model. In this case, all states are available for feedback and any robustness problems are avoided. However, with EIOL only processes with asymptotically stable zero dynamics can be treated. Fig. 19 shows the local control structure of a processing unit with one inlet and one outlet stream.

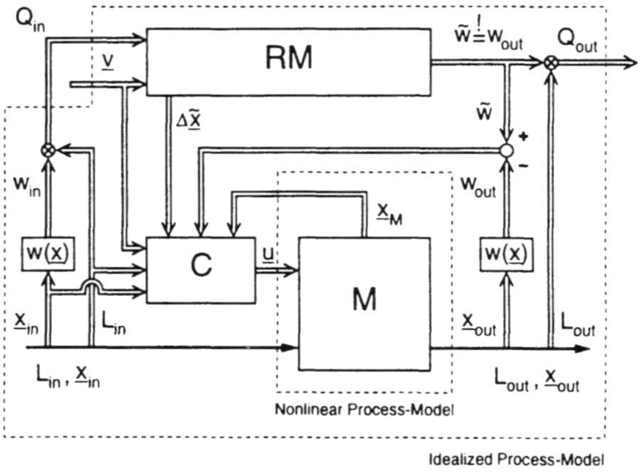

Fig. 19. Local control structure to match the I/O- behavior of nonlinear model (M) into reference model (RM) by nonlinear feedback (C) .

The performance of local nonlinear control systems is briefly demonstrated in a simulation example for the first separation unit. Here, an equilibrium model including the energy balance with constant molar holdup n is assumed. In order to linearize the I/O-behavior of DC1 the outputs $W_2 = n_2 w_2$ in the reboiler and $W_3 = n_3 w_3$ in the condenser are determined by quality function (1). The heating power u_2 as well as the reflux ratio u_3 are used as manipulated variables. For the deviations $\Delta\tilde{W}_k = \tilde{W}_k - W_k^S$ around the equilibrium point the following reference model (RM) with two decoupled first order lags is chosen :

$$\begin{bmatrix}\Delta\dot{\tilde{W}}_3\\ \Delta\dot{\tilde{W}}_2\end{bmatrix} = \tilde{A}_1\begin{bmatrix}\Delta\tilde{W}_3\\ \Delta\tilde{W}_2\end{bmatrix} + \tilde{B}_1\begin{bmatrix}v_3\\ v_2\end{bmatrix} + \tilde{\underline{c}}_1\Delta Q_1 \quad , \quad (9)$$

$$\tilde{A}_1 = \begin{bmatrix}-\frac{1}{\tilde{\tau}_3} & 0\\ 0 & -\frac{1}{\tilde{\tau}_2}\end{bmatrix}, \tilde{B}_1 = \begin{bmatrix}\frac{\tilde{n}_3}{\tilde{\tau}_3} & 0\\ 0 & \frac{\tilde{n}_2}{\tilde{\tau}_2}\end{bmatrix}, \tilde{\underline{c}}_1 = \begin{bmatrix}\tilde{C}_3\\ \tilde{C}_2\end{bmatrix}.$$

The deviations $\Delta Q_k = L_k w_k - L_k^S w_k^S$ are introduced with $\Delta L_k = L_k - L_k^S$ as additional outputs of the reference model on the coordination level :

$$\Delta\tilde{Q}_3 = L_3\Delta\tilde{w}_3 + w_3^S\Delta L_3 \quad , \quad (10)$$
$$\Delta\tilde{Q}_2 = L_2\Delta\tilde{w}_2 + w_2^S\Delta L_2 \quad . \quad (11)$$

Here, the idealized RM (9)-(11) is linear time variant, since molar flow rates L_k and residence times $\tilde{\tau}_k = \frac{\tilde{n}_k}{L_k}$ are varying. Fig. 20 demonstrates the good performance of the controlled unit.

However, in practice it is not possible to measure the complete process state. Then, EIOL must be used together with an observer. This extension is called asymptotically exact I/O-linearization

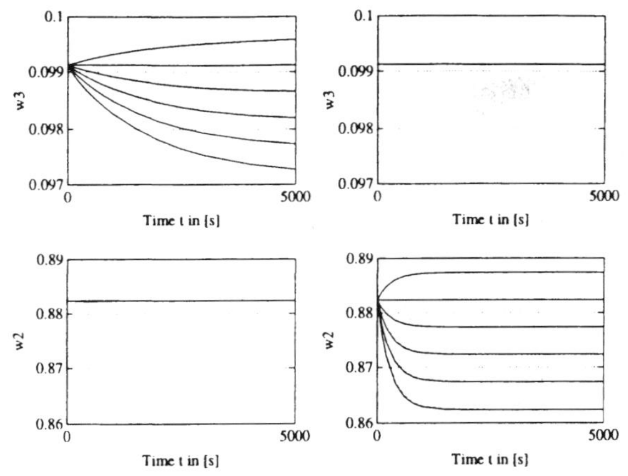

Fig. 20. Linear idealized I/O-behavior of DC1 using quality function $w = 0.1x_A + x_B$ and equal sized step changes in v_3 (left) and v_2 (right).

(AEIOL), since an EIOL of the process is achieved only in an asymptotic sense after the convergence of the state estimator. So far, it is not possible to predict the effect of nonlinear observers on the performance of nonlinear controllers, since no separation theorem holds in the nonlinear case.

Control Concepts of AEIOL. Therefore, two alternative concepts of AEIOL (see Fig. 21 and Fig. 22) have been developed with special reference to output feedback and improvement of robustness properties. The idea is to guarantee separately

- the EIOL of the process model by nonlinear feedback of the model states $\hat{\underline{x}}$
- the convergence between model and process by feedback of the measured output error Δy

The exact linear I/O-behavior between v and model output $\hat{w}$ can be ideally achieved by nonlinear feedback C. The linear I/O-behavior between new input v and process output w is asymptotically obtained, if model M and process P converge. Therefore, the performance of both structures is completely independent of the robustness properties of the I/O-linearization technique.

AEIOL with model correction. This approach (Friedrich *et al.*, 1992) requires that an asymptotically stable nonlinear observer (process model M including correction term E) has been designed for process P. Hence, convergence is achieved. The difference between the classical structure and this new configuration, both given in Fig. 21, is represented by the dashed line. Here, the feedback law C is simply calculated to linearize exactly the observer, i.e. to compensate also for the correction term E. However, the EIOL of the observer model is restricted to the class of processes with relative degree $r = 1$. This method has been successfully applied in experiments to distillation columns (Groebel *et al.*, 1994; Groebel *et al.*, 1995). In some cases, an extension to systems with $r > 1$ is possible.

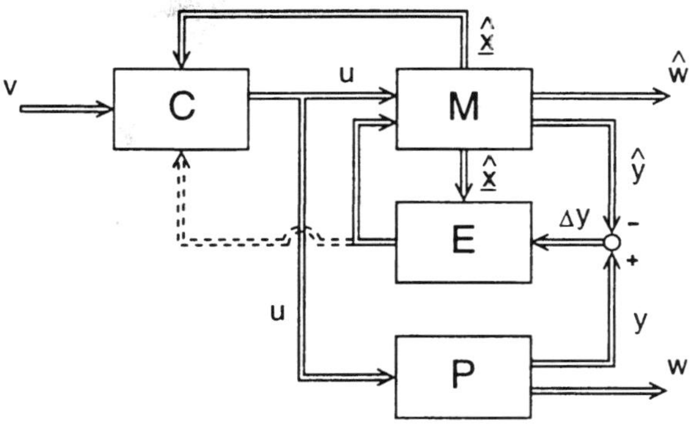

Fig. 21. Simplified block diagram of AEIOL with model correction using a nonlinear observer (solid lines: classical structure; whole diagram: new structure)

AEIOL with additional input. This approach (Friedrich *et al.*, 1992; Gilles *et al.*, 1994) is not restricted to a certain relative degree. First, a suitable process model M is exactly linearized by the feedback law C. Here, the EIOL of the model M has the task to generate a well-defined reference trajectory $(\hat{u}, \hat{\underline{x}})$. To compensate for plant/model mismatch in Fig. 22 an additional feedback Δu is calculated from the measured output error Δy. To design the additional compensation AC an asymptotically stable error dynamics is required using an extended first order linearization around the reference trajectory.

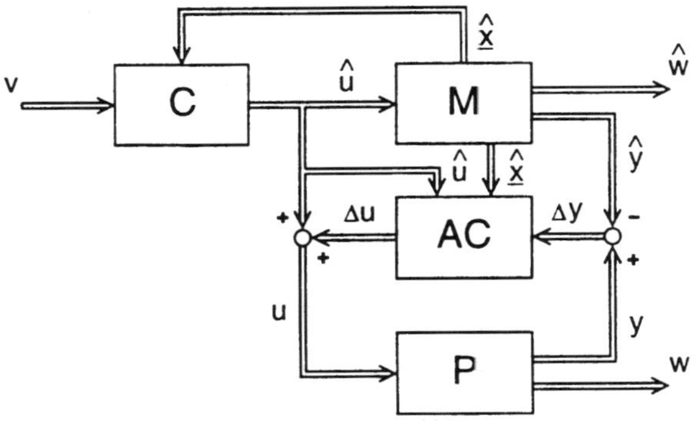

Fig. 22. Simplified block diagram of AEIOL with additional input Δu

Control on the Coordination Level. The coordination control has the task to provide flexible operation of the plant during load or grade changes. Therefore, the coordination control problem is formulated as an optimization problem for changing the operating point of the plant. This new steady state is mainly specified by desired flow rate L_{4S} and quality w_{4S} of the end product stream, but also by the desired production of qualifying potential and the specified inlet stream. The steady state management has the task

- to optimize the internal variables of the plant with respect to a minimal total heat input as outlined in section 3.3

- and to represent this new operating point by the plant-wide set points L_{kS}, w_{kS}.

Since on the process level an ideal model matching is achieved ($\tilde{w}_k \overset{!}{=} w_k$, $\tilde{Q}_k \overset{!}{=} Q_k$), the idealized RMs are connected according to the plant topology. Thus, a complete plant model is obtained similar to RM (9) :

$$\begin{aligned} \Delta\dot{\underline{\tilde{W}}} &= \tilde{A}\Delta\underline{\tilde{W}} + \tilde{B}\underline{v} + \tilde{C}\underline{d} \\ &= \tilde{A}\Delta\underline{\tilde{W}} + \underline{\tilde{v}} \quad . \end{aligned} \tag{12}$$

In order to design the LQR for moving the plant on an optimal trajectory into the new operating point the following quadratic integral criterion is used :

$$I = \frac{1}{2} \int_{t=0}^{t_e=\infty} (\underline{W}^{*T} E^* \underline{W}^* + \underline{v}^{*T} F^* \underline{v}^*)dt \quad . \tag{13}$$

Here, the deviations $\underline{W}^* = \Delta\underline{\tilde{W}} - \Delta\underline{\tilde{W}}_S$ and $\underline{v}^* = \underline{\tilde{v}} - \underline{\tilde{v}}_S$ from the new set point are weighted. For example, an optimal control law is calculated by means of the steady state Riccati-matrix which can be determined off-line for each new set point.

For simplicity, only the recycle loop consisting of CSTR0 and DC1 is considered. To achieve in this case the specified flow rate $L_{2S} = L_{0S}$ of the end product stream the additional manipulated variable $v_0 = \Delta L_0$ is needed. Then, in (12) the deviations $\underline{d}^T = [\Delta L_3, \Delta w_0]$ on the process level must be also taken into account for the coordination.

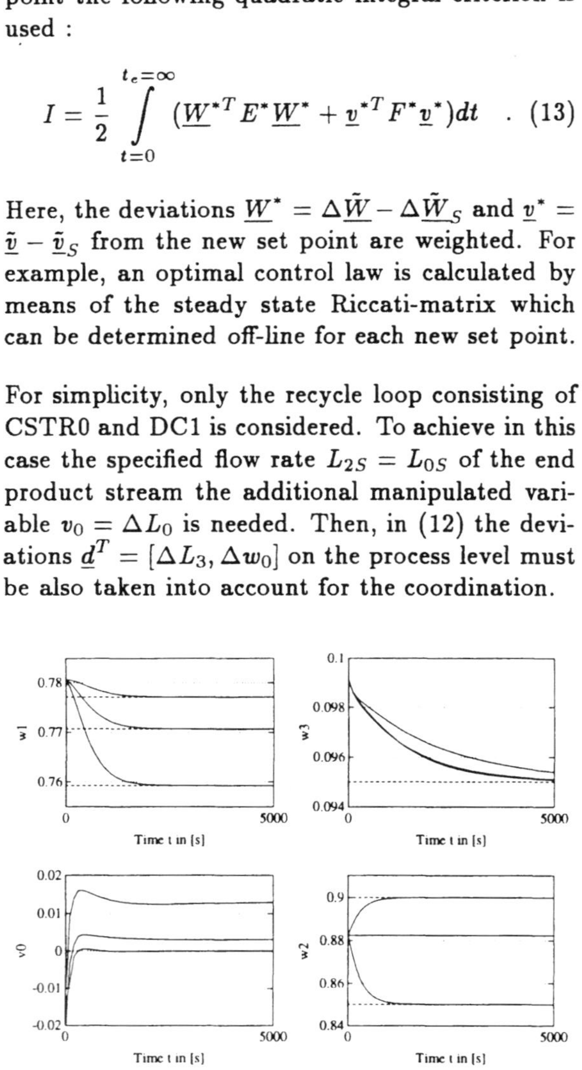

Fig. 23. Optimal quality trajectories (solid lines) into different new operating points (dashed lines) using $w = 0.1x_A + x_B$.

Fig. 23 shows the optimal coordinated control in the case of constant feedstock quality w_0 into different new operating points with specified end product quality w_{2S} and desired productivity L_{2S}. Here, the following situations are considered :

- w_2 as well as L_2 are increased.
- L_2 is increased while constant w_2 is guaranteed for all time t.
- w_2 is decreased to a new operating point, which has the same $L_{2S} = L_2^S$ but needs a smaller total heat input.

The details are given in Storz and Gilles (1995).

4. CONCLUDING REMARKS

Several aspects of process integration have been treated on two different levels. First, integration of different unit operations within a multifunctional processing unit has been considered and illustrated with two examples. Among others, these examples indicate the potential of such techniques. For a systematic utilization of such synergies a systematic classification according to the processes involved is required. Design and operation of such integrated processing units are closely linked. In particular, when these processes are intended to be operated under non steady state conditions like the circulation loop reactor presented in this paper. Development, design and operation of such integrated processes is therefore a challenging task in chemical engineering research.

Second, integration on the flowsheet level has been studied exemplarily with a typical production plant consisting of a reactor followed by a sequence of two distillation columns where unconverted reactand is separated and recycled into the reactor. The outline of a hierarchical control concept has been given. This concept is based on product quality as a plant-wide uniform objective function. It has been briefly shown, that such an objective function can also be used for process design and is therefore one approach to the integration of control and design, which is another challenging task in current chemical engineering research.

In particular, a comparably simple, empirical quality function has been used for the example process. Current and future work is concerned with the construction of more general quality functions on the basis of sound thermodynamics. This is a necessary prerequisite for the application of the quality concept to more complex production processes like polymerization processes. Also, further improvement of the control methods is required. Other control methods for the coordination as well as the local level, like linear as well as nonlinear model predictive control, can be incorporated into the concept and can be compared to the methods already used.

5. ACKNOWLEDGEMENT

The financial support of the BMFT (Bundesminister für Forschung und Technik) for nonlinear dynamics of multifunctional processing units and of the Volkswagen Stiftung for the hierarchical control concept for quality assurance is appreciated.

6. REFERENCES

Adomaitis, R.A. and A. Cinar (1988). The bifurcation behavior of an autothermal packed bed tubular reactor. *Chem. Engng. Sci.* **43**, 887 - 898.

Adomaitis, R.A. and A. Cinar (1991). Numerical singularity analysis. *Chem. Engng. Sci.* **46**, 1055 - 1062.

Agar, D. W. and W. Ruppel (1988*a*). Extended reactor concept for dynamic deNOx design. *Chem. Engng. Sci.* **43**, 2073–2078.

Agar, D.W. and W. Ruppel (1988*b*). Multifunktionale Reaktoren für die heterogene Katalyse. *Chem.-Ing.-Tech.* **60**, 731–741.

Agreda, V.H., L.R. Partin and W.H. Heise (1990). High-purity methyl acetate via reactive distillation. *Chem. Engng. Prog.* **40**, 40–46.

Balmer, J. and E.D. Gilles (1989). Instationäre Prozeßführung zur katalytischen Abgasverbrennung. *Chem.-Ing.-Tech.* **61**, 632–635.

Bonvin, D., R.G. Rinker and D.A. Mellichamp (1980). Dynamic analysis and control of a tubular autothermal reactor at an unstable state. *Chem. Engng. Sci.* **35**, 603–612.

Bonvin, D., R.G. Rinker and D.A. Mellichamp (1983). On controlling an autothermal fixed-bed reactor at an unstable state, I,II,III. *Chem. Engng. Sci.* **38**, 233–244,245–255,607–618.

Callen, H.B. (1985). *Thermodynamics and an Introduction to Thermostatistics.* 2 ed.. J. Wiley. New York.

Ciric, A.R. and D. Gu (1994). Synthesis of nonequilibrium reactive distillation processes by MINLP optimization. *A.I.Ch.E. Journal* **40**, 1479–1487.

Ciric, A.R. and P. Miao (1994). Steady state multiplicities in an ethylene glycol reactive distillation column. submitted for publication.

DeGarmo, J.L., V.N. Parulekar and V. Pinjala (1992). Consider reactive distillation. *Chem. Engng. Prog.* **88**, 43–50.

Doherty, M.F. and G. Buzad (1992). Reactive distillation by design. *Trans. IChemE.* **70**, 448–458.

Friedrich, M. (1994). Qualitätsorientierte Prozeßführung in der verfahrenstechnischen Produktion. PhD thesis. Universität Stuttgart. Düsseldorf 1994.

Friedrich, M. and E.D. Gilles (1991). On a process control framework for quality assurance. In: *ADCHEM '91*. IFAC. Toulouse, France. pp. 235–240.

Friedrich, M., M. Storz and E.D. Gilles (1992). Asymptotically exact linearization of chemical processes. In: *DYCORD '92*. IFAC. College Park, MD. pp. 369–374.

Gaikar, V.G. and M.M. Sharma (1989). Separations through reactions and other novel strategies. *Sep. Purif. Meth.* **18**, 111–176.

Gilles, E.D., M. Storz and M. Friedrich (1994). Asymptotisch exakte Linearisierung des Eingangs-/Ausgangsverhaltens nichtlinearer Systeme. *At* **42**, 135–148.

Groebel, M., F. Allgöwer, M. Storz and E.D. Gilles (1994). Nonlinear control of high purity distillation columns. In: *AIChE Annual Meeting, November 13.-18., paper no. 230c.* San Francisco,USA.

Groebel, M., F. Allgöwer, M. Storz and E.D. Gilles (1995). Asymptotically exact I/O-linearization of an industrial distillation column. In: *ACC'95, June 21.-23.*. Seattle, USA.

Guckenheimer, J. and P. Holmes (1983). *Nonlinear Oscillations, Dynamical Systems, and Bifurcations of Vector Fields.* Springer. New York.

Isidori, A. (1989). *Nonlinear Control Systems.* 2 ed.. Springer Verlag. Berlin, Heidelberg, New York.

Jacob, R. and R. Krishna (1993). Multiple solutions in reactive desillation for methyl tert-butyl ether synthesis. *Ind. Engng. Chem. Res.* **32**, 1706–1709.

Keyes, D.B. (1932). Esterification processes and equipment. *Ind. Engng. Chem.* **24**, 1096–1103.

Kienle, A., G. Lauschke, V. Gehrke and E.D. Gilles (1995). On the dynamics of the circulation loop reactor. Numerical methods and analysis. *Chem. Engng. Sci.* in press.

Kravaris, C. and C.-B. Chung (1987). Nonlinear state feedback synthesis by global input/output linearization. *A.I.Ch.E. Journal* **33**, 592–603.

Lauschke, G. and E.D. Gilles (1994). Circulating reaction zones in a packed-bed loop reactor. *Chem. Engng. Sci.* **49**(24), 5359–5375. Special Issue of ISCRE 13.

Luyben, W. L. (1993). Trade-offs between design and control in chemical reactor systems. *J. Proc. Cont.* **3**, 17–41.

Luyben, W. L. (1994). Snowball effects in reactor/separator processes with recycle. *Ind. Engng. Chem. Res.* **33**, 299–305.

Lyman, P. R. and W. L. Luyben (1994). A method for assessing the effects of design parameters on controllability. In: *IPDC'94,June 27-28* (E. Zafiriou, Ed.). Baltimore, MD, USA. pp. 82–88.

MacGregor, J. F. (1988). On-line statistical process control. *Chem. Engng. Prog.* **84**, 21–31.

Marsden, J.E. and M. McCracken (1976). *The Hopf Bifurcation and its Application.* Springer.

New York.

Matros, Y. Sh., A. S. Noskov, V. A. Chumachenko and O. V. Goldman (1988). Theory and application of unsteady catalytic detoxication of effluent gases from sulfur dioxide, nitrogen oxides and organic compounds. *Chem. Engng. Sci.* **43**, 2061 - 2066.

Matros, Yu. Sh. (1985). *Unsteady Processes in Catalytic Reactors*. Elsevier. Amsterdam, Oxford, New York, Tokyo.

Mesarović, M. D., D. Macko and Y. Takahara (1970). *Theory of Hierarchical, Multilevel, Systems*. Academic Press. New York.

Morud, J. and S. Skogestad (1994). Effects of recycle on dynamics and control of chemical processing plants. *Comp. Chem. Engng.* **18**, 529–534.

Nijhuis, S.A., F.P.J.M. Kerkhof and A.N.S. Mak (1993). Multiple steady states during reactive distillation of methyl tert-butyl ether. *Ind. Engng. Chem. Res.* **32**, 2767–2774.

Pomeau, Y. and P. Manneville (1980). Intermittent transition to turbulence in dissipative systems. *Commun. Math. Phys.* **74**, 189–197.

Rebsdat, S. and D. Mayer (1990). *Ullmanns Encyclopedia of Industrial Chemistry*. Chap. Ethylene Glycol, pp. 101–115. Vol. A 10. VCH Verlagsgesellschaft mbH.

Recke, B. and S.B. Jørgensen (1994). Bifurcations in fixed bed reactors as a consequence of recycle. In: *AIChE Annual meeting San Francisco Nov. 13-18, paper 93f.*

Roat, S.D., J.J. Downs, E.F. Vogel and E.J. Doss (1986). The integration of rigorous dynamic modeling and control system synthesis for distillation columns: an industrial approach. In: *Chemical Process Control-CPC III* (M. Morari and T.J. McAvoy, Eds.). Elsevier. pp. 261–268.

Ruppel, W. (1980). Eine mathematische Beschreibung wandernder Reaktionszonen in Schüttschichten. PhD thesis. Universität Stuttgart.

Seider, W.D., D.M. Brengel, A.M. Provost and S. Widagdo (1990). Nonlinear analysis in process design. Why overdesign to avoid complex nonlinearities ?. *Ind. Engng. Chem. Res.* **29**, 805–818.

Singh, M.G. (1980). *Dynamical Hierarchical Control.* North-Holland. Amsterdam.

Storz, M. and E.D. Gilles (1995). Exact I/O-linearization of chemical processes within a plant-wide quality control framework. In: *ACC'95, June 21.-23.*. Seattle, USA.

Storz, M., M. Friedrich and E.D. Gilles (1994). Process design and control for quality assurance. In: *IPDC'94, June 27-28* (E. Zafiriou, Ed.). Baltimore, MD, USA. pp. 64–69.

Yakhnin, V.Z., A.B. Rovinsky and M. Menzinger (1994). Differential-flow-induced pattern formation in the exothermic A $\rightarrow$ B reaction. *J. Chem. Phys.* **98**, 2116–2119.

OPTIMIZATION IN MODEL BASED CONTROL[1]

D. Q. Mayne*

*Department of Electrical and Computer Engineering, University of California, Davis, California CA 95616

Abstract. This paper examines the design of controllers with emphasis on their ability to handle state and control constraints and nonlinearity. A major motivation is the presence of hard constraints in most applications, not least in the petro-chemical industry where steady state optimization forces the operating point to lie on or near the boundary of the feasible set. Efficient handling of constraints requires nonlinear controllers even if the system being controlled is linear and design of nonlinear systems involves optimization. The interaction between nonlinear control and optimization is explored, particular attention being given to model predictive control of constrained systems, linear and nonlinear.

Key Words. Model based control, predictive control, nonlinear control, constraints, optimization.

1. INTRODUCTION

It is undoubtedly true that most control applications require satisfaction of hard constraints on controls and states; actuators saturate and safe operation requires limitations on states such as velocity, acceleration, temperature and pressure. Efficient handling of constraints requires nonlinear control whether the system being controlled is linear or nonlinear. The closed loop system is, therefore, nonlinear which is the reason for the slow development of an adequate theory. An optimal *feedback* control law for the optimal control problem defined by the system

$$\dot{x} = f(x, u)$$

and cost $\int_0^T \ell(x(t), u(t), t)dt + F(x(T))$ may, in principle, be obtained by solving the Hamilton-Jacobi-Bellman partial differential equation

$$\begin{aligned} -\frac{\partial}{\partial t} V^0(x,t) &= \min_{u \in \Omega} \{\ell(x,u) \\ &+ \frac{\partial}{\partial x} V^0(x,t) f(x,u,t)\} \end{aligned}$$

with boundary condition $V^0(x,T) \equiv F(x)$. However, solution of the Hamilton-Jacobi-Bellman equation is difficult, if not impossible, unless the the system is linear and the cost quadratic or the state dimension n is low. This difficulty has, in general, precluded the development of procedures for designing nonlinear feedback controllers except in specific areas. One such area is adaptive control of linear systems; the adaptive controller is nonlinear but closed loop stability can be established by sophisticated arguments. A second area is model predictive control. Model predictive control avoids solving the Hamilton-Jacobi-Bellman partial differential equation by repeatedly solving online an *open-loop* optimal control problem for the current state, a considerably simpler task, and applying the minimizing control for a short time before repeating the process. In essence, instead of determining the value function $V^0(x)$ for each $x \in X$, it is determined for the sequence of states $\{x[k] := x(k\Delta)\}$ actually encountered. An impossible problem is replaced by a soluble problem; the price is the complexity of the on-line controller (which has to have the capability of solving open-loop optimal control problems) which restricts its use to 'slow' plants.

This paper examines the interplay between optimization and control: how optimization may be employed to determine nonlinear control, and what restrictions this use imposes on achievable control.

2. MODEL PREDICTIVE CONTROL

2.1. *Introduction*

Our purpose here is to explore model predictive control. We are interested in exposing limitations, both inherent and imposed by optimization. We will be concerned with the formulation of finite horizon, open loop, optimal control problem which is both implementable (soluble, at least approximately, by the online controller within ap-

[1] Research supported by National Science Foundation under grant ECS-9312922

propriate time limits) and whose solution ensures satisfaction of the control objectives. We discuss the use of model predictive control for controlling constrained nonlinear systems. To do this, we examine existing MPC strategies, not because they are definitive, but because they reveal underlying problems. We do not pretend to survey the field; for recent surveys see (Garcia *et al.*, 1989), (Bequette, 1991) and (Rawlings *et al.*, 1994).

The control objectives include stability, performance, implementability, constraint satisfaction, robustness, adaptivity, and operator transparency.

Control constraints arise from actuator limitations. State constraints are often imposed for safety. In process control, economic considerations often require the operating point to lie on, or near, the boundary of the feasible region (the subset of X satisfying the state constraints).

Efficient control of a constrained system requires nonlinear control even if the system being controlled is nonlinear. Figure 1 shows the step response of a linear system satisfying an acceleration constraint; this forces the step response to lie below the indicated boundary. A linear zero velocity error controller forces area B to be equal to area A with the result that rapid settling necessitates large overshoot; conversely, low overshoot implies large settling time. Far better performance (rapid response without excessive overshoot) can be achieved with nonlinear control.

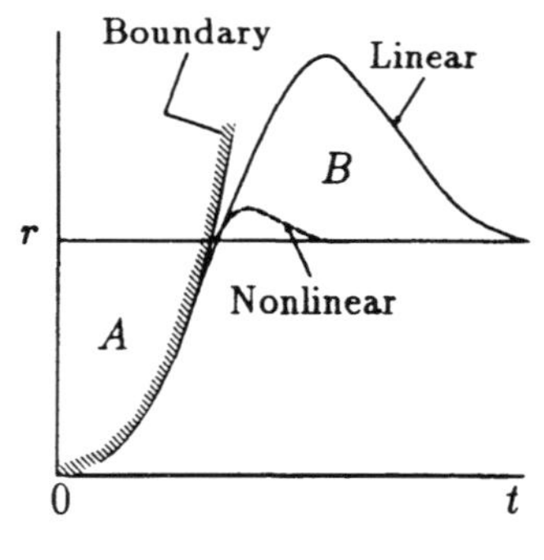

Fig. 1. **Linear vs Nonlinear Control**

2.2. *Problem formulation*

The system to be controlled is assumed to be described by

$$\dot{x} = f(x,u) \quad (1)$$
$$y = h(x) \quad (2)$$

where $f : \mathbb{R}^n \times \mathbb{R}^m \to \mathbb{R}^n$ and $h : \mathbb{R}^n \to \mathbb{R}^r$ are twice continuously differentiable and

$$f(0,0) = 0 \quad (3)$$

and to be subject to the hard control and state constraints

$$u(t) \in \Omega, \quad x(t) \in E \quad (4)$$

where $\Omega \subset \mathbb{R}^m$ is compact and convex and $E \subset \mathbb{R}^n$ is closed and convex; the origin lies in the interior of each of these sets. If the system is linear

$$f(x,u) = Ax + Bu \quad (5)$$

where $A \in \mathbb{R}^{n\times n}$ and $B \in \mathbb{R}^{n\times m}$. In optimal control, $u(\cdot)$ is often assumed to lie in the set $U := \{u(\cdot) \in L_\infty \mid u(t) \in \Omega\}$. To facilitate optimization, and other implementation issues, the control may be finitely parameterized. Among many options, the simplest, and most widely used, is the class of piecewise constant controls with period Δ and values in Ω; let $U_\Delta \subset U$ denote this class of controls. A control $u(\cdot)$ in U_Δ is characterized by the sequence $\{u[k]\}$ where $u[k] := u(k\Delta)$. A control $u(\cdot) \in U_\Delta$ satisfies $u(t) = u[k]$ for all $t \in [k\Delta, (k+1)\Delta)$.

We consider, for simplicity, the regulation problem. The state $x(t) \in \mathbb{R}^n$, the control $u(t) \in \mathbb{R}^m$, and the output $y(t) \in \mathbb{R}^r$ are chosen so that the origin is the set point. We also assume

For all $x \in \mathbb{R}^n$, all $u(\cdot) \in U$ there exists a unique solution $x^u(\cdot)$ to (1) satisfying $x^u(0) = x$.

When f is linear, this assumption is automatically satisfied.

Let $x^u(\cdot; x, s)$ denote the solution of (1), due to control $u(\cdot) \in U$, passing through (x, s) (i.e. satisfying $x^u(s; x, s) = x$). A control $u(\cdot) \in U$ (satisfying, therefore the control constraint $u(t) \in \Omega$) will be called *admissible*; similarly, a state trajectory $x^u(\cdot)$ which satisfies both the control and state constraints will be termed *admissible*.

2.3. *Limitations*

Constraints automatically impose limitations in what can be achieved, even if the system to be controlled is linear. Not every state of a controllable but constrained linear system can be steered to the origin. The set of states which *can* be steered to the origin in time T along admissible trajectories is X_T, defined, for all $T \geq 0$, by

$$X_T := \{x \in E \mid \exists\, u(\cdot) \in U \text{ such that } x^u(t; x, 0) \in E, t \in [0,T]; x^u(T; x, 0) = 0\}$$

Under growth conditions, such as $\|f(x,u)\| \leq M(1+\|x\|)$ for all $x \in \mathbb{R}^n$ all $u \in \Omega$, this set is bounded for all $T \in \mathbb{R}_+$. If the set $f(x,\Omega) := \{f(x,u) \mid u \in \Omega\}$, is convex for all x, then X_T is compact. If $f(\cdot)$ is linear, X_T is compact and convex (even if Ω is not convex). If, in the definition of X_T, U is replaced by $U_\Delta \subset U$, smaller sets will be obtained.

The behaviour of X_T as $T \to \infty$ is of interest. If $f(\cdot)$ is linear, stable and controllable, and $E = \mathbb{R}^n$, then $X_T \to \mathbb{R}^n$ as $T \to \infty$. This result can be extended to include the case when A has simple eigenvalues on the imaginary axis. If A has unstable eigenvalues, then $X_T \to X_\infty$ where X_∞ is a convex, strict subset of $\mathbb{R}^n$. If all the eigenvalues of A are unstable, then X_∞ is compact and convex.

The concept of a *control invariant set*, was used in the analysis of unconstrained linear systems (Wonham, 1974). Gilbert and Tan (Gilbert and Tan, 1991) introduced the concept of an 'output admissible set' to analyse and synthesize linear controllers for constrained linear systems. For our purposes, the extension, by Aubin (Aubin, 1991), of the original definition is useful.

Definition 1 A set $X \subset E$ is said to be control invariant (for the system $(f(\cdot), E, \Omega)$), if, for every $x \in X$, there exists a control $u \in U$ such that $x^u(t;x,0) \in X$ for all $t \geq 0$.

Roughly speaking, if X is control invariant, there exists a nonlinear control law which keeps any trajectory which starts in X in X; this trajectory is admissible. Clearly the set X_T defined above is control invariant; every point $x(t)$ on a trajectory commencing at $x \in X_T$ at time 0 and terminating at the origin at time T can be steered to the origin along an admissible trajectory in a time not exceeding $T - t$; $X_s \subset X_t$ for all s,t satisfying $s \leq t$.

There is thus an inherent limitation on the set of states of a constrained system that can be steered to the origin in finite or infinite time. Any controller must keep the state x in X_∞ if not in X_T; this implied constraint should not be ignored.

2.4. *Stability*

In the process control literature, model predictive control (MPC) is conventionally formulated as solving online a fixed horizon optimal control problem (FHP) of the form

$$\min\{V(x,t,u(\cdot)) \mid u(\cdot) \in U_\Delta; \\ x(s) \in E, s \in [t,t+T]\} \qquad (6)$$

and applying the the minimizing control $u^0(\cdot) \in U_\Delta$, to the plant over the interval $[k\Delta,(k+1)\Delta)$. The procedure is repeated indefinitely. In (6), $V(\cdot)$ denotes the cost function defined by

$$V(x,t,u(\cdot)) := \int_t^{t+T} \ell(x(s),u(s))ds \\ +F(x(t+T)) \qquad (7)$$

The minimizing control $u^0(t;x,t)$ is a time-invariant function $h(x)$ of the current state x, so MPC yields an implicit nonlinear control law even when the system being controlled is linear. When the system is linear, the cost quadratic, the sets Ω and E are specified by linear inequalities, FHP reduces to a quadratic program for which efficient software exists.

But *optimality does not imply stability*; the version of MPC described above does not guarantee closed loop stability. Stability can be achieved, at least when the system is linear and the cost quadratic, by increasing the terminal cost and/or adjusting the horizon T. Tuning (or fiddling) with parameters in this way in order to achieve stability is both undesirable (Bitmead *et al.*, 1990) and unnecessary . Stability can, in fact, be achieved relatively simply (Thomas, 1975; Keerthi and Gilbert, 1988; Mayne and Michalska, 1990) by adding a stability constraint (SC) $x(t+T) = 0$ to the finite horizon control problem $P(x,t)$ as follows:

$$\min\{V(x,t,u(\cdot)) \mid u(\cdot) \in U; \\ x(s) \in E, s \in [0,T]; x(t+T) = 0\} \qquad (8)$$

The terminal cost $F(\cdot)$ becomes irrelevant and can be discarded. For simplicity in presentation, the class of admissible controls is extended to U. With the additional stability constraint $x(t+T) = 0$, the value function $V^0(x)$ (the minimum value of V in (8)) can be shown to be a Lyapunov function for the closed loop system; for suitable $\ell(\cdot)$, $V(x) > 0$ if $x \neq 0$, and $\dot{V}^0(x) \leq -\ell(x,h(x))$ where $h(\cdot)$ is the implicitly defined control law, which establishes stability using a standard Lyapunov argument; see (Mayne and Michalska, 1990) where differentiability of $V^0(\cdot)$ is established (a difficult task because of the terminal equality constraint). A proof, which does not require differentiability of the value function, is given in (Vinter and Michalska, 1991). See also (Michalska and Mayne, 1991).

The addition of the stability constraint $x(t+T) = 0$ does not unduly complicate implementation when the system is linear, the cost quadratic and the sets E and Ω are specified by linear inequalities; FHP is still a quadratic program, (with additional equality constraints) if U is replaced by U_Δ. It is interesting to note that in practice MPC is usually applied to stable plants and T is chosen sufficiently long to ensure the plant has settled to its equilibrium state at time $t+T$; this procedure implicitly ensures satisfaction of the stability constraint.

Subject to some modest conditions, it is possible to achieve closed loop stability by employing an infinite horizon but the resulting optimal control problem cannot be solved in general. But (Keerthi and Gilbert, 1988) show how infinite horizon feedback laws may be approximated by finite horizon laws. An alternative to infinite horizon control laws is the ingenious procedure (Rawlings and Muske, 1993) in which FHP is replaced by a control problem with infinite horizon cost but finite horizon control; the resultant control problem can be solved in the linear quadratic case.

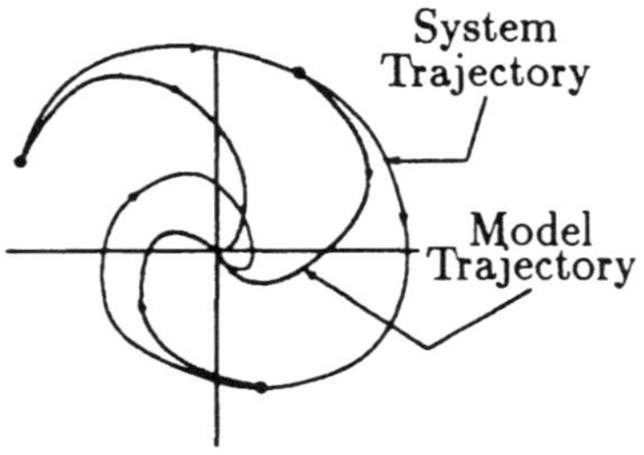

Fig. 2. System and Model Trajectories

It is an interesting fact (Meadows *et al.*, 1995) that receding horizon control can stabilize systems which cannot be asymptotically stabilized with continuous feedback.

With this version of MPC, the trajectory of the controlled system may differ appreciably from the solution to each FHP; an operator cannot necessarily predict future plant behaviour (see Figure 2) and FHP does not 'model' performance accurately.

3. MODEL PREDICTIVE CONTROL OF NONLINEAR SYSTEMS

Since MPC requires the repeated online solution of an optimal control problem, it is important to know whether the problem can be reliably solved. The big divide in nonlinear programming is not between linear and nonlinear but, rather, between convex and nonconvex. If an optimization problem is convex, as FHP is when the system being controlled is linear and $\ell(\cdot)$, $F(\cdot)$ and the state and control constraints are convex, and the feasible set is nonempty, optimization methods exist which ensure convergence to a global minimum (which is unique if the performance criterion is strictly convex). By exploiting duality, lower and upper bounds to the optimal value are easily computed, permitting cessation of the optimization procedure when a predetermined tolerance is satisfied.

Even if $\ell(\cdot)$, Ω and E are convex, but $f(\cdot)$ is not linear, the optimal control problem FHP will not, in general, be convex. A host of difficulties impacting on the implementability of the model predictive controller, ensue.

3.1. *Computation*

Standard descent methods (gradient and conjugate gradient, sequential quadratic programming, interior point methods ...), when applied to a nonconvex problem, yield local, rather than global, minima. Standard stability results, which rely on global optimality, are then inapplicable. Indeed, one cannot even guarantee a feasible solution (a control satisfying the control and state constraints and the stability constraint). In general, the state constraint set E will be characterized by a finite set of inequalities

$$E := \{x \in I\!R^n \mid q(x) \leq 0\} \tag{9}$$

where $q : I\!R^n \rightarrow I\!R^a$ is continuously differentiable. Because the state $x(t)$ is an implicit function of the control $u(\cdot)$, the state constraint (for a given initial state) is implicitly of the form

$$\psi(u(\cdot)) = \max\{q(x^u(s; x_0, t_0)) \mid s \in \mathcal{T}\} \leq 0 \tag{10}$$

where $\mathcal{T}$ is either the interval $[t, t+T]$ or a discrete subset thereof it we wish to avoid semi-infinite optimization. Since $\psi(\cdot)$ is, in general, nonlinear (*even* if q is linear in x), it will have local minima, and any descent algorithm which reduces $\psi(\cdot)$ to find an admissible control may jam at one of these minima. Hence, determination of a feasible control, let alone an optimal control, cannot be guaranteed. Infeasibility, or nonoptimality, of a control $u(\cdot)$ computed by the MPC controller destroys, of course, the theoretical properties of MPC; stability cannot be guaranteed.

Assuming an initial feasible control is available, can feasibility be maintained? This is possible in the absence of disturbances or model error. If the control $u_0(\cdot)$ steers the initial state x to the origin in time T along an admissible trajectory, then (in the absence of disturbances or model error) the control $u_0(\cdot)$, restricted to the interval $[\Delta, T]$, steers $(x(\Delta)$ to the origin in time $T - \Delta$ and so is feasible for the FHP $P(x(\Delta), \Delta)$. However, maintaining both feasibility *and* optimality may be difficult (see below).

The stability constraint is a terminal *equality* constraint which is nonlinear in the decision variable $u(\cdot)$. Satisfaction of nonlinear equality constraints (as contrasted with linear constraints) can only be achieved (if at all) asymptotically (in the sense that accumulation points of sequences generated by the optimization algorithm are feasible). In practice, computation must cease prior to exact satisfaction of the equality constraint. This requires that the MPC strategy be robust to such errors.

An additional, important, factor is the unpredictability of the computation time required to solve nonlinear optimal control problems. Even if the state dimension is relatively low, the dimension of the decision variable and the state constraint can be high (indeed, very high), resulting in complex optimization problems whose solution time can be very variable.

Although implementation of MPC requires the solution of many optimization problems, these problems are not randomly chosen, but strongly correlated. Indeed, the solution to $P(x(t), t)$ provides a feasible, albeit nonoptimal solution, to the successor problem $P(x(t+\Delta), t+\Delta)$. Can this feature be used to simplify online computation by 'tracking' the optimal solution? Indeed it can, but no guarantee against failure can be made. This is illustrated in Figure 3 which shows how a global minimum of $V(\cdot)$ can be transformed into a local minimum or, even, a global maximum, as the parameter x (the initial state) varies (Polak, 1993). Consequently, it is possible, indeed probable, that,

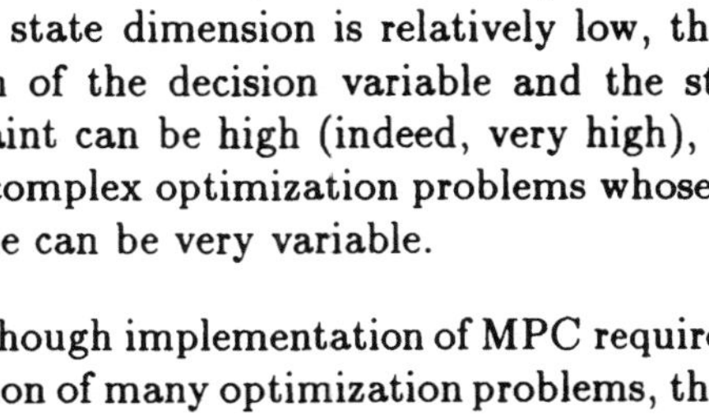

Fig. 3. Transformation of Global Minimum

as time proceeds, an MPC controller, which commences with a global minimum to $P(x(0), 0)$ and tracks the (local) minimum, will eventually generate a nonoptimal solution. The excellent properties of MPC may then be lost.

It seems, therefore, that nonlinear MPC requires the development of alternative approaches. We discuss, in turn, various possibilities (contractive MPC, variable horizon MPC, and hybrid MPC) before turning to an examination of robustness issues.

3.2. *Contractive MPC*

An alternative form of MPC, derived from a procedure (Polak and Mayne, 1981) for the design

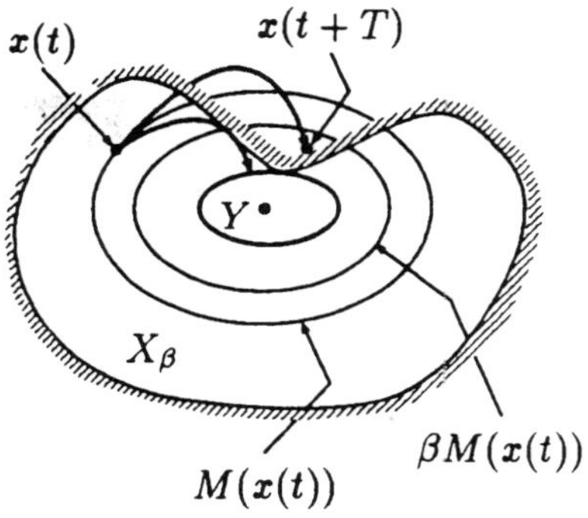

Fig. 4. Contractive MPC

of nonlinear controllers, is proposed in (Yang and Polak, 1993). This is a Lyapunov based approach in which a Lyapunov function $M(\cdot)$, chosen a-priori, is decreased, not continuously, but discretely; it is allowed to increase at other times. In this form of MPC, stability is achieved by replacing the terminal equality stability constraint by an inequality constraint. The resultant finite horizon optimal control problem $P(x, t)$ now takes the form

$$\min\{V(x, t, u(\cdot), T) \mid u(\cdot) \in S;\ T \geq 0;\\ x(s) \in E, s \in [0, T]; x(t+T) \in \beta M(x)\} \quad (11)$$

where $\beta \in (0, 1)$ and, typically,

$$M(x) := \{z \in I\!R^n \mid \langle z, Qz \rangle \leq \langle x, Qx \rangle\} \quad (12)$$

for some positive definite matrix Q; $\beta M(x)$ denotes the set $\{\beta z \mid z \in M(x)\}$; see Figure 4. Note that the decision variables are now the control $u(\cdot)$ and the horizon time T. Suppose that the state is x^k at time t^k and that $(u^k(\cdot), T^k)$ solves $P(x^k, t^k)$. The control $u^k(\cdot)$ is then applied to the system over the interval $[t^k, t^{k+1}]$ where $t^{k+1} := t^k + T^k$; at the end of this interval, the solution to FHP is recomputed. The behaviour of the system is easily understood; it is clear that $x^k \in \beta^k M(x(0))$ so that, under relatively mild conditions, stability ensues *provided* that $P(x^k, t^k)$ can be solved for each k. But solution of $P(x, t)$ cannot be guaranteed for every $x \in X$.

To see why this may arise, let $X_\beta \supset X_\infty$ denote the set of states which may be steered to the *set* $\beta M(x)$ along an admissible trajectory. Whereas X_T and X_∞, defined above, are clearly control invariant sets, it is by no means clear that X_β has this property. Given an arbitrary point x in X_β, the successor point x' generated by 'contractive' MPC will not necessarily lie in X_β as illustrated in Figure 4. Hence, FHP $P(x', t')$ is not necessarily solvable. As in classical MPC, the controller may generate states at which a solution to the finite horizon control problem cannot be found. However (Polak, private communication), there exists an $\alpha \geq 0$ such that $Y = \{x \mid \langle x, Qx \rangle \leq \alpha\} \subset$

$X_\infty \subset X_\beta$. The FHP $P(x,t)$ can be solved for any point in this set. Hence, once the initial state of the system is steered to this set, contractive MPC may be safely employed. Since the initial state should be steered to Y in finite time, a strategy such as finite horizon MPC, discussed next, should be employed.

3.3. *Variable horizon MPC*

The preceding discussion has shown that a satisfactory form of MPC for linear constrained systems is relatively easily obtained by the incorporation into FHP of a stability constraint. When the plant is nonlinear, a host of new difficulties arise. Determining a globally optimal, or even a feasible, solution to FHP is problematic. Even if an initial global solution to FHP is available, it cannot necessarily be tracked as the state x varies. Suppose an MPC controller generates a sequence of states $\{x^i := x(t^i)\}$ where $t^i := i\Delta$. None of the methods discussed briefly above can guarantee that every problem in the sequence $\{P(x^i,t^i)\}$ can be solved. Since the MPC strategies described above rely on optimality, their general use seems to be precluded. It seems that the *most* that can be hoped for is that every member of the problem sequence $\{P(x^i,t^i)\}$ can be solved if the initial problem $P(x^0,t^0)$ can be. MPC has this desirable property if both the problem and control horizons are infinite (and certain stabilizability and detectability assumptions are satisfied). Suppose $P(x^0,t^0)$ is solvable; under the above conditions, the optimal control u^0 for $P(x^0,t^0)$ steers x^0 to the origin. If $P(x^i,t^i)$ is solvable and the optimal control for this problem is $u^i(\cdot)$, then, by Bellman's principle of optimality, the control $u^i(\cdot)$, *restricted* to the interval $[t^{i+1},\infty)$, is optimal for $P(x^{i+1},t^{i+1})$. By induction, every member of the problem sequence $\{P(x^i,t^i)\}$ can be solved, and the minimizing control is, in each case, a restriction of $u^0(\cdot)$, the initial minimizing control. Other desirable properties ensue. The controlled trajectory coincides with the solutions to each $P(x^i,t^i)$, a property not possessed by the controllers previously described.

By permitting horizon time $T \in [0,T_{max}]$ to be a decision variable, as in (Michalska and Mayne, 1993), many difficulties are overcome. In its simplest form, variable horizon MPC has the following ingredients. The cost function (7) includes T as a decision variable:

$$V(x,t,u(\cdot),T) := \int_t^{t+T} \ell(x(s),u(s))ds \qquad (13)$$

so that FHP $P(x,t)$ becomes

$$\min\{V(x,t,u(\cdot),T) \mid u(\cdot)\in U;\ T\in[0,T_{max}];$$

$$x(s)\in E, s\in[t,t+T]; x(t+T)=0\} \qquad (14)$$

The minimizing control is applied to the plant for the period $\min\{\Delta,T\}$ and the process repeated. This version of MPC is similar to the infinite horizon version: if $P(x^i,t^i)$ is solvable and the optimal control for this problem is $u^i(\cdot)$, then, by Bellman's principle of optimality, the control $u^i(\cdot)$, *restricted* to the interval $[t^{i+1},\infty)$, is optimal for $P(x^{i+1},t^{i+1})$. By induction, every member of the problem sequence $\{P(x^i,t^i)\}$ can be solved if $P(x^0,t^0)$ can be solved, and the minimizing control is, in each case, a restriction of $u^0(\cdot)$, the initial minimizing control. *In the absence of disturbances or model error, the solution to every member of the sequence $\{P(x^i,t^i)\}$ of problems is simply obtained as a restriction of the solution to the original problem $P(x^0,t^0)$; only one optimal control problem $P(x^0,t^0)$ need be solved. Moreover, the controlled trajectory coincides with the solution to $P(x^0,t^0)$. Optimal variable horizon MPC is asymptotically stabilizing with a region of attraction $X_{max} := X_{T_{max}}$.* Stability can be established using standard Lyapunov arguments with the Lyapunov function either the value function $V^0(x)$ of (14), or $T^0(x)$, the minimizing value of the horizon T.

Even more can be gained; optimality can be discarded. Consider the following version of variable horizon MPC in which the controller seeks merely to improve the previous control rather than optimize. The finite horizon problem $P(x^i,t^i)$ to be solved by the controller at time t^i is now:

Determine a feasible control $\tilde{u}^i(\cdot)$ in U and horizon $\tilde{T}^i$ in $[0,T^{i-1}-\beta\Delta]$ which satisfies

$$x^{\tilde{u}^i}(t^i+\tilde{T}^i;x^i,t^i)=0 \qquad (15)$$

Determine a feasible control $u^i(\cdot)$ in U and horizon T^i in $[0,\tilde{T}^i]$ which satisfies

$$V(x^i,t^i,u^i(\cdot),T^i) \le V(x^i,t^i,\tilde{u}^i(\cdot),\tilde{T}^i)$$
$$x^{u^i}(t^i+T^i;x^i,t^i)=0 \qquad (16)$$

Apply the control $u^i(\cdot)$ to the system over the interval $[t^i,t^{i+1}]$, $t^{i+1} := t^i+\min\{\Delta,T^i\}$.

This strategy has the following properties:

- If (u^{i-1},T^{i-1}) is feasible for $P(x^{i-1},t^{i-1})$, then $(\tilde{u}^i,\tilde{T}^i)$, where $\tilde{u}^i$ is the restriction of u^{i-1} to $[t^i,t^{i-1}+T^{i-1}]$ and $\tilde{T}^i=T^{i-1}-\Delta$, is feasible for $P(x^i,t^i)$.
- The controller need do no more than accept the feasible control $(\tilde{u}^i,\tilde{T}^i)$.
- If it does, the system and model trajectories coincide.
- The horizon time decreases by $\beta\Delta$ at each iteration (except the last).

If a feasible solution (u^0, T^0) to the initial problem $P(x(0), 0)$ is known, then, by induction, every problem in the sequence $\{P(x^i, t^i)\}$ is (easily) soluble. See Figure 5. Online computation is substantially reduced. A global solution to an optimal control problem is *never* required. In the absence of model error or external disturbances, the solution (u^0, t^0) to the initial FHP provides a solution to all subsequent FHP's; if the original model trajectory provides a good path to the origin (because FHP incorporates all performance and stability objectives), why spoil it with further optimization? Lyapunov analysis is complicated by the fact that the control is not unique. A direct proof of stability is, however, possible. Since the horizon time decreases by $\beta\Delta$ at each iteration, whatever control is used, the controller steers any point x in X_{max} to the origin in a time not exceeding T^0/β along a trajectory which remains in the bounded set X_{max}.

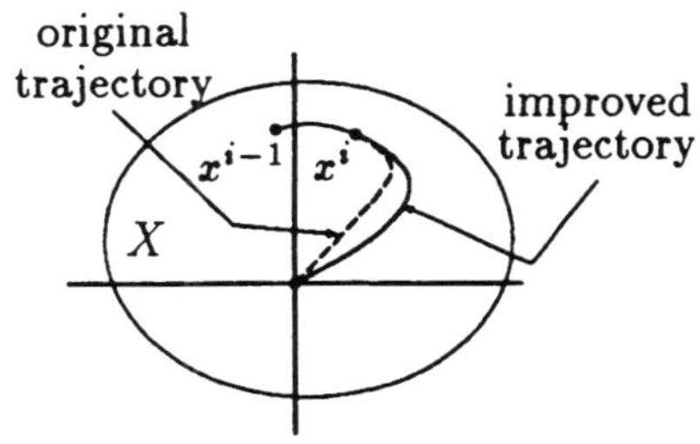

Fig. 5. Variable Horizon MPC

In the absence of model error or disturbance, variable horizon MPC is asymptotically stabilizing with a region of attraction X_{max}.

3.4. *Hybrid MPC*

Hybrid MPC is a variant of variable horizon MPC in which the stability (terminal equality) constraint is replaced by an inequality constraint which is more easily satisfied. In general, an inequality constraint can be satisfied in a finite number of iterations of a suitable algorithm, whereas an equality constraint may be only satisfied asymptotically. To achieve this objective, a local controller stabilizing controller $h(\cdot)$ is employed near the equilibrium point (the origin). Suppose a stabilizing control law $h : W \rightarrow \Omega$, where $W \subset E$ is a positively invariant set for the system $\dot{x} = f(x, h(x))$, is known. The control law $h(\cdot)$ any initial state $x \in W$ to the origin along a trajectory which remains in W and satisfies the control and state constraints. The control law may be linear. The finite horizon problem $P(x^i, t^i)$ to be solved at time t^i becomes:

If $x^i \notin W$, determine a feasible control $\tilde{u}^i(\cdot)$ in U and horizon $\tilde{T}^i$ in $[0, T^{i-1} - \beta\Delta]$ to satisfy

$$x^{\tilde{u}^i}(t^i + \tilde{T}^i; x^i, t^i) \in W \tag{17}$$

Determine a feasible control $u^i(\cdot)$ in U and horizon T^i in $[0, \tilde{T}^i]$ to satisfy

$$\begin{aligned} V(x^i, t^i, u^i(\cdot), T^i) &\leq V(x^i, t^i, \tilde{u}^i(\cdot), \tilde{T}^i) \\ x^{u^i}(t^i + T^i; x^i, t^i) &\in W \end{aligned} \tag{18}$$

Apply the control $u^i(\cdot)$ to the system over the interval $[t^i, t^{i+1}]$.
If $x \in W$, apply the control $h(x)$.

This controller shares many of the properties of its predecessor (16) and steers any initial state to the set W in a time not exceeding T^0/β; the local controller $h(\cdot)$ is employed in W. See Figure 6.

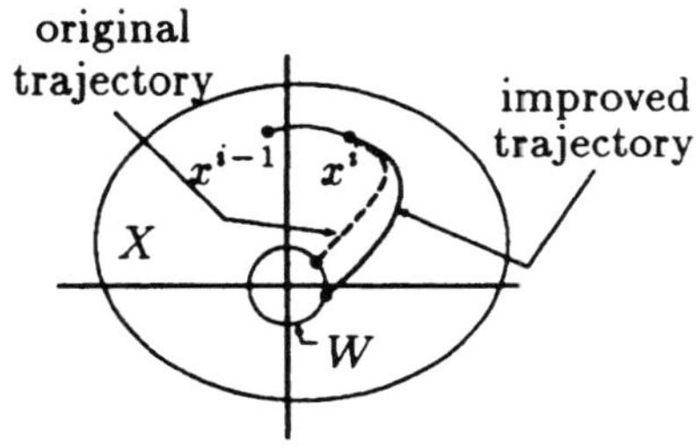

Fig. 6. Hybrid MPC

Hybrid MPC is asymptotically stabilizing with a region of attraction X_{max}.

4. ROBUST MPC

The discussion above assumes that the system being controlled is identical to the model employed by the controller. In reality, $f_R(\cdot) = f(\cdot) + d(\cdot)$, where the dynamics of the real system and the model are described by $\dot{x} = f_R(x, u)$ and $\dot{x} = f(x, u)$ respectively, and $d(\cdot)$ denotes the disturbance or model error. We distinguish three cases:

- $d = d(x, u)$ where $\|d(x, u)\| \leq c\|(x, u)\|$, $c \in (0, \infty)$.
- $d = d(t)$ where $d(t) \rightarrow 0$ as $t \rightarrow 0$.
- $d = d(t)$ where $d(t) \in D$ for all t, D compact, $0 \in \text{interior}(D)$.

and discuss these in turn. It is important to note that even an exponentially decaying disturbance can destabilize a nonlinear system, possibly resulting in finite escape time.

4.1. *Parametric model error*

Suppose parametric model error is present, so that the real system is described by

$$\dot{x} = f_R(x, u) \tag{19}$$

whereas the model is described, as before, by $\dot{x} = f(x, u)$. Assume also that the model error is bounded as follows:

$$\|f_R(x,u) - f(x,u)\| \leq c\|(x,u)\| \tag{20}$$

for all (x, u). The hybrid MPC controller employs, of course, the model in calculating solutions to FHP. The controller can be made robust by employing in $P(x,t)$ conservative constraint sets $E' \subset E$ and $W' \subset W$ in place of E and W respectively. This is illustrated in Figure 7. Consider robust stability. Suppose that, at time t^{i-1}, an admissible control u^{i-1} which steers the *model* from x^{i-1} to $z^i \in W'$ has been computed and that this control is applied to the plant for the period $[(i-1)\Delta,\ i\Delta)$, yielding the next *plant* state $x^i = x(t^i)$. Because of model error, $x^i \neq x^i_m$ which is the model state resulting from the same control being applied to the model. If the model error (i.e. c) is sufficiently small, time t^i can be chosen so that $x^i \in E'$. Because of model error, the control u^{i-1}, restricted to the period $[t^i, t^{i-1} + T^{i-1}]$, steers the model from x^i to the state $\tilde{z}^i \neq z^i$. If the model error is not too large, $\tilde{z}^i \in W$ and the local control law $h(\cdot)$ can steered the model from $\tilde{z}^i$ to W' in a time less than $\Delta/2$, yielding an admissible control-horizon pair $(\tilde{u}^i, \widetilde{T}^i)$ for the subsequent finite horizon control problem $P(x^i, t^i)$; $\tilde{u}^i$ is the restriction of u^{i-1} to the period $[t^i, t^{i-1}+T^{i-1}]$ (this control steers the model from x^i to $\tilde{z}^i$) concatenated with the control which steers the model from $\tilde{z}^i \in W$ to $z^{i+1} \in W'$ and $\widetilde{T}^i \leq T^{i-1} - \Delta/2$. The control $\tilde{u}^i$ thus steers the model from $x^i \in E'$ to W' in time $\widetilde{T}^i \leq T^i - \Delta/2$ so that $(\tilde{u}^i, \widetilde{T}^i)$ is an admissible control-horizon pair for $P(x^i, t^i)$. This control-horizon pair can be improved, if desired. (If model error or disturbance is too large for the above procedure to provide an admissible control-horizon pair, an admissible pair may be determined directly; however, this may require global optimization techniques). Hence again we have the property that, provided a feasible solution (u^0, T^0) to $P(x^0, t^0)$ is available, feasible solutions (u^i, T^i) to all subsequent problems $P(x^i, t^i)$ are easily obtained. Robust stability ensues (Michalska and Mayne, 1993). Let $X'_{max} \subset X_{max}$ denotes the set of states of the model that can be steered in time T_{max} to the set W' along trajectories satisfying the control and conservative state constraint.

Again stability can be established by direct arguments, showing that the horizon time is reduced by at least $\Delta/2$ at each iteration, so that any initial state in X'_{max} is steered to the origin in a time not exceeding $2T^0$.

There exists a $c > 0$ such that Robust Hybrid MPC is asymptotically stabilizing with a region of attraction X'_{max} for every system satisfying (20).

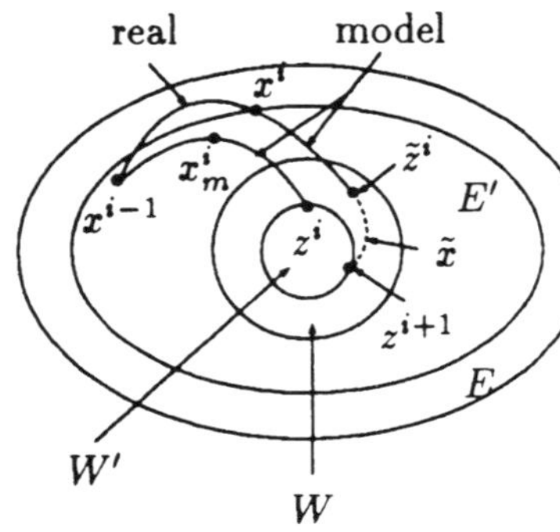

Fig. 7. Robust MPC

Robustness of variable horizon MPC is, perhaps, of more interest. The situation is as depicted in Figure 7 if W' is regarded as a single point, the origin. Control $u^{i-1}(\cdot)$ steers the *model* from x^{i-1} to $z^i = 0$ in time T^{i-1}. Because of model error, the actual state at time t^i is $x^i \neq x^i_m$, and control $u^{i-1}(\cdot)$ (suitably restricted) steers the model from this state to $\tilde{z}^i$. Robustness depends on the ability to find a control which steers the model from $\tilde{z}^i$ to the origin in a time not exceeding $\Delta/2$.

4.2. *Asymptotically decaying disturbance*

This situation typically arises when an observer is used for a linear system (even if nonlinearly controlled) in which case $d(t) = A[x(t) - \hat{x}(t)]$ where $\hat{x}(t)$ is the estimated state. If the controlled system is exponentially stable, converse Lyapunov theorems can be employed to establish robustness (Vidyasagar, 1993; Scokaert and Rawlings, 1995).

4.3. *Bounded disturbance*

Bounded deterministic, or stochastic, disturbances introduce a different problem. It is no longer possible to drive the state to the origin. If the disturbance $d(\cdot)$ is unknown but bounded, then the best that can be hoped for is that the state can be steered to a control invariant set. Consider the discrete time version of (1):

$$x[k+1] = g(x[k], u[k]) + d[k] \tag{21}$$

where $x[k] := x(kT)$ and the disturbance is subject to the constraint $d[k] \in D$ for all k ; the constraint set $D \subset \mathbb{R}^n$ is compact and convex set and contains the origin in its interior.

Definition 2 The set $X \subset E$ is robust control invariant for $(g(\cdot), \Omega, E, D)$ if, for all $x \in X$ there exists a $u \in \Omega$ such that $g(x,u) + d \in X$ for all $d \in D$.

If X is robust control invariant, and the initial state $x[0]$ of (21) is in X, then there exists a *feedback* control $u[k] = h(x[k])$ which maintains the state in X (despite the disturbance).

Because the system (21) cannot be controlled to the origin, MPC must be modified. An obvious possibility is to ignore the disturbance, i.e. to obtain a control by solving FHP for the undisturbed system

$$x[k+1] = g(x[k], u[k]) \tag{22}$$

Because possible future disturbances are neglected, performance can suffer (Rawlings *et al.*, 1994). To overcome this deficiency, the use of

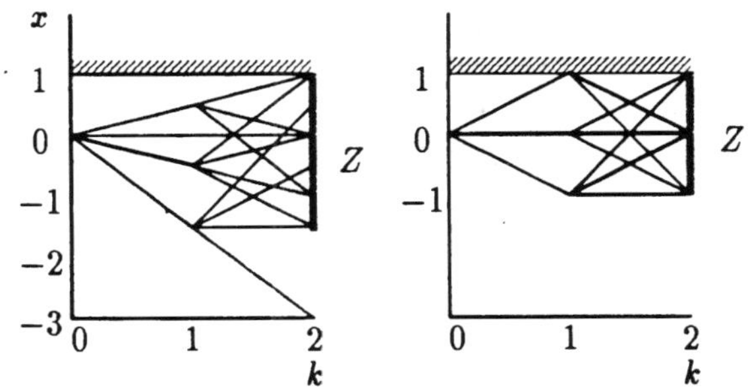

Fig. 8. Min-Max MPC

min-max (differential game) MPC has been proposed; see (Zheng and Morari, 1993). For the sake of later comparison only, consider the following open-loop formulation of the min-max FHP for discrete time systems:

$$\min_{\{u[j]\}} \max_{\{d[j]\}} \{ \sum_{j=k}^{N-1} \ell(x[j], u[j]) + cd(x[j], E)$$
$$+ cd(x(N), E) \mid u[j] \in \Omega;\ w[j] \in W\}$$

where $d(x, E)$ denotes the distance of x from the set E and c is finite but sufficiently large to ensure satisfaction of the constraint $x[j] \in E$ for all possible disturbance sequences. This formulation has two disadvantages. Firstly, it includes no stability constraint (the stability constraint $x(t+T) = 0$ cannot be included since the system cannot be steered to the origin by the control $\{u[j]\}$). Secondly, the worst case scenarios corresponding to a nominal control can be considerably worse than those actually achieved with feedback control.

If the disturbance sequence has a finite number of realizations $\{d^\ell[j]\}$ where $\ell \in \mathcal{L} := \{1, \ldots L\}$ indexes the realization, a 'feedback' formulation of MPC is possible. The feedback FHP $P(x, k)$ is

$$\min_{(\{u^\ell[j]\}, N)} \max_{\ell \in \mathcal{L}} \{ \sum_{j=k}^{N-1} \ell(x^\ell[j], u^\ell[j]) + cd(x^\ell[j], E)$$
$$+ cd(x^\ell[N], Z) \mid u^\ell[j] \in \Omega;\ d^\ell[j] \in D;$$
$$u^{\ell_1}[j] = u^{\ell_2}[j] \text{ if } x^{\ell_1}[j] = x^{\ell_2}[j]\}$$

where $x^\ell[j]$ denotes the solution of (20) due to control and disturbance sequences $\{u^\ell[i]\}_{i=k}^{N-1}$ and $\{d^\ell[i]\}_{i=k}^{N-1}$ and initial condition $x[k] = x$. The terminal constraint $x[N] \in Z$ is the stability constraint for this formulation of MPC; this set is a robust-control-invariant set, preferably the smallest such set, and has to be pre-determined. It appears as if NL control actions $u^\ell[k]$ have to be determined, but the *causality* constraint $u^{\ell_1}[j] = u^{\ell_2}[j]$ for all j, ℓ_1, ℓ_2 such that $x^{\ell_1}[j] = x^{\ell_2}[j]$ enforces a single control action for each state, thereby reducing the number of independent variables. The FHP is a finite dimensional min-max problem for which local solutions can be found by standard algorithms. If the system is linear, the cost quadratic, and the sets Ω, D and Z specified by linear inequalities, FHP reduces to a quadratic program, but of much higher dimension than in the disturbance free case. It is interesting to note that variable horizon, min-max MPC inherits many of the advantages of its disturbance free predecessor; in the absence of errors and disturbances *not* already incorporated in the problem description, *the solution* $\{u^\ell[k] \mid \ell \in \mathcal{L},\ k \in \{0, \ldots, N-1\}\}$ *to* $P(x[0], 0)$ *provides, suitably restricted, a solution to every problem* $P(x[k], k)$ *subsequently encountered.* Optimality can be discarded as before provided the stability constraint $x[N] \in Z$ is retained.

As an illustration, consider the system

$$x[k+1] = x[k] + u[k] + d[k], \quad x[0] = 0$$

with $\Omega = [-1, 1]$, $D = \{-1, 0, 1\}$, $E = (-\infty, 1]$ and $\ell(x, u) = x^2$. Figure 8 illustrates the two strategies. For variable horizon, open loop min-max, the solution to $P(x[0], 0)$ is $\{u^0[k]\} = \{-0.5, -0.5\}$, $\{w^0[k]\} = \{1, 1\}$, yielding the worst case state trajectory $\{0, 0.5, 1\}$. At time $k = 0$, the predicted set of terminal states includes $\{1, -3\}$. At $k = 1$, using horizon $T = 1$, open loop min-max chooses controls $-0.5, 0.5, 1$ if the state is, respectively, $0.5, -0.5, -1.5$. Hence, the actual states at time $k = 2$ lie in the set $Z = [-1.5, 1]$. In fact, Z is a robust control invariant set for this controller and its size is a measure of performance; the set is larger than necessary, as shown below. If the open loop min-max controller employed a fixed horizon of 2, the resultant robust control invariant set would be $[-3, 1]$, even larger.

There are nine possible disturbance sequences $\{-1, -1\}$, $\{-1, 0\}$, $\{-1, 1\}$, $\ldots \{1, 0\}$, $\{1, 1\}$. Let state (x, k) denote state x at time k. Variable horizon, feedback min-max MPC generates control 0 for states $(0, 0)$ and $(0, 1)$, control -1 for $(1, 1)$ and control 1 for $(-1, 1)$. At time $k = 0$, the predicted set of terminal states includes $\{1, -1\}$ which lie in the minimal robust control invariant set Z= $[-1,1]$ for this controller. At $k=1$, the controller applies the precomputed controls $u=-1$ if $x=1$, $u=0$ if $x=0$ and $u=1$ if $x=-1$. The resultant worst case state trajectory is $\{0,1,1\}$. The robust control invariant set $Z=[-1,1]$ for this controller is the minimal achievable.

5. OUTPUT FEEDBACK MPC

The controllers described above assume that the state is perfectly accessible ($y=x$). If this is not the case, state estimation has to be implicitly or explicitly introduced. There are several possibilities for dealing with this case.

(i) The simplest approach, widely used in adaptive control although not without danger, is 'certainty equivalence': estimate the state and use the estimated state $\hat{x}(t)$ in the model predictive controller. The observer can, under reasonable conditions, be designed to ensure that estimation error $\tilde{x}(t) := x(t) - \hat{x}(t) \to 0$ exponentially as $t \to \infty$. Since optimization is employed for the controller, it appears preferable to employ it also for the observer in place of weaker alternatives such as extended Kalman filtering.

(ii) The approach in (i) can lead to instability and even to finite escape time in certain nonlinear systems even if the estimation error is exponentially decaying. It appears necessary to couple the design of the controller and observer to avoid this possiblity. One such approach is described in (Michalska and Mayne, 1995); the sampling frequency is increased as a potential finite escape time is approached so that the observer is no longer independent, as it is in certainty equivalence controllers, of control demands.

5.1. *Observer based MPC*

Since optimization is employed for the controller, it appears preferable to employ it also for the observer in place of weaker alternatives such as extended Kalman filtering. This is what is proposed in (Michalska and Mayne, 1992; Rawlings and Muske, 1993; Robertson *et al.*, 1994; Michalska and Mayne, 1995). Whereas the control looks at the future over the interval $[t, t+T]$, the observer's data is over the *moving horizon* $[t-T_E, t]$ in the immediate past where T_E is the estimation horizon which may differ from the control horizon. The estimation cost function $J(\cdot)$ is defined to be

$$J(x,t) := \int_{t-T_E}^{t} ||h(x^u(s;x,t) - y(s))||^2 ds$$

where $y(s)$ is the actual output $h(x(s))$ of the system, and $h(x^u(s;x,t))$ is the estimated output at time s assuming that the state of the system is x at time t (and the control is $u(\cdot)$). The function $p_t(\cdot) := J(\cdot, t)$ may be regarded as the state of the system at time t. If the system is linear, p_t is quadratic and may be written in the form

$$p_t(x) = (1/2)(x - \hat{x}(t))^T P^{-1}(t)(x - \hat{x}(t)) + a(t)$$

where $P(t)$ is positive definite if the system is observable, and $a(t) = 0$ if there is no process noise or model error. Under these ideal conditions, the estimate $\hat{x}(t) = \arg\min p_t(x)$ is the unique global minimizer of $p_t(x)$ and $\hat{x}(t) = x(t)$.

But, if the system is nonlinear, $p_t(x)$ may have the form shown in Figure 9 if there is no process noise or model error. Again, the estimate $\hat{x}(t) := \arg\min p_t(x)$ is the unique global minimizer of $p_t(x)$ and satisfies $\hat{x}(t) = x(t)$ if the nonlinear system is observable and there is no process noise or model error. Finding $\hat{x}(t)$ is not a simple

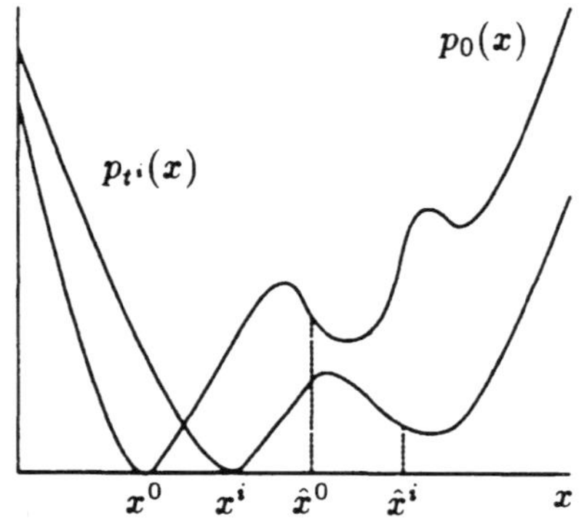

Fig. 9. Observer-Based MPC

matter. Because of nonlinearity, global minimization may be required; see Figure 9.

Under these ideal conditions, if the system being controlled does not have finite escape time, a *conceptual* online observer may be defined as follows. The observer computes a sequence $\{\hat{x}^i\}$ of estimates of $\{x^i := x(t^i)\}$ to satisfy

$$p_{t^{i+1}}(\hat{x}^{i+1}) \leq \beta p_{t^i}(\hat{x}^i)$$

where $\beta \in (0,1)$.

If, among other assumptions, the system is observable, $\hat{x}^i \to x^i$ exponentially as $i \to \infty$ (Michalska and Mayne, 1992; Michalska and Mayne, 1995). These estimators can be harnessed, using the 'certainty equivalence' strategy, to a robust model predictive controller; robustness permits the controller to cope with estimation error. Since even exponentially decaying inputs can destabilize a nonlinear system, fairly strong assumptions have to be made. Results of the following form are obtained for the control constrained control problem (Michalska and Mayne, 1995):

For any compact set Y in the interior of X_{max} and any bound on initial observer error, there exist observer parameters such that the observer-based moving horizon controller is asymptotically stabilizing in Y.

The assumption that $x(0) \in Y \subset \text{interior}(X_{max})$ and that the observer is sufficiently fast, is made to ensure that the state does not escape the set X_{max} of states which can be controlled to the origin; error implicitly reduces the set of states which can be successfully controlled.

The procedure can be extended to deal with systems with finite escape if the the sampling period Δ and the estimation horizon T_E are variable; the system is measured more frequently as a finite escape time is approached.

But results like this do not necessarily invite application. Figure 9 illustrates the type of difficulty which may arise; although the minima of $p_t(\cdot)$ vary with t there is no guarantee that the estimate $\hat{x}(t)$ will escape into the 'correct' valley (whose minimum is $x(t)$) before $x(t)$ moves out of the set X_{max}. The observer described above avoids this problem but only by implicitly assuming that global optimization is available and that there is no process or model error. If the initial estimate $\hat{x}(0)$ is in the 'correct valley', this problem may be avoided. Approximate observers, such as the extended Kalman filter, face this problem more acutely. If model error or disturbance is present, then $x(t)$ cannot be exactly determined, even with the help of global optimization. With noise, $p_t(x) > 0$ for all x; values of x corresponding to low values of p_t are more likely than those corresponding to high values. It is impossible, under these conditions, to control to the origin; control to a control invariant set, as discussed above, is indicated.

6. ADAPTIVE MPC

If the system being controlled has unknown parameters, it may be modelled by

$$\dot{x} = f(x, u, \theta) \tag{23}$$

$$y = h(x, \theta) \tag{24}$$

where $\theta \in I\!R^p$ is a vector of unknown parameters. By adding the dynamics

$$\dot{\theta} = 0 \tag{25}$$

we reduce the problem to one of observer based design in which the observer estimates the extended state (x, θ). But this reduction is not useful because the observability assumption is not appropriate for θ. (If the system is $\dot{x} = Ax + Bu, y = Cx$ and $\theta = (A, B)$, then $y(t) \equiv 0$ if $x(0) = 0$ and $\theta(0)$ is arbitrary.) Adaptive control theory does not, therefore, rely on convergence of $\hat{\theta}$ to θ whereas observer based control usually does requires convergence of $\hat{x}$ to x. While the early literature on generalized predictive control, like that on MPC, did not include a stability constraint, this is no longer the case (Clarke *et al.*, 1991). However, these results rely on linear adaptive control theory precluding consideration of control and state constraints. Since the MPC controller is implicitly nonlinear, a new theory is required. In the ideal case (no model error or disturbance) it is possible, when the system is linear or linearly parameterized, to employ standard algorithms to provide an estimate of θ and to employ this to obtain an estimate of the state. A recurrent problem is that the current estimate $\hat{\theta}(t)$ of θ may not admit a solution to FHP even though $x(t) \in X_T$. This is overcome in (Mayne and Michalska, 1993) by using parameter perturbations which restore controllability without increasing estimation cost. Making use of a property of the innovation process, it is possible to establish, in the ideal case, that the MPC controller is stabilizing. The underlying idea is that if the innovation $e \equiv 0$, then $y \equiv \hat{y}$. Since the MPC controller steers $\hat{y}$ to zero, $y \rightarrow 0$. Using observability we deduce that $x \rightarrow 0$. These results require further development for application but give an indication of the computational effort required for adaptive control.

7. STATE FEEDBACK CONTROLLER

Nonlinear MPC will never achieve the simplicity of linear MPC. Unless a nonlinear MPC controller is exhaustively tested by simulation over the whole range of potential initial states (or observation sequences in the case of inaccessible states) doubt will always remain as to whether or not a state will be encountered for which an acceptable solution to the FHP can be found. It may be necessary to investigate alternative approaches, either to replace, or to be used in conjunction with, MPC to remove some of the disadvantages arising from limitations in what can be achieved from nonlinear optimization.

While there are as yet no visible alternatives to MPC for constrained nonlinear systems, there is considerable activity in the field of nonlinear control of constrained linear systems (Gilbert and Tan, 1991; Gilbert *et al.*, 1994; Mayne and Schroeder, 1994). It is certainly possible that this activity may lead to the new proposals for controllers for nonlinear constrained systems which do not require online optimization.

One approach to nonlinear control of constrained linear systems is described below.

The system to be controlled is described by

$$\dot{x} = Ax + Bu \tag{26}$$
$$y = x \tag{27}$$

where $x(t) \in \mathbb{R}^n$, $u(t) \in \mathbb{R}^m$ and $d(t) \in \mathbb{R}^n$. The control and state are subject to the following hard constraints

$$u(t) \in \Omega, \quad x(t) \in E \tag{28}$$

The design is based on the following simple result (see Figure 10). For any integer j, Σ_j is the convex set defined by:

$$\Sigma_j = \{\mu \in \mathbb{R}^j \mid \mu^i \geq 0, \sum_{i=1}^{j} \mu^i = 1\} \tag{29}$$

Suppose

(i) Control $u_i(\cdot)$ steers state x_i to state z_i in time T along a trajectory that satisfies both state and control constraints, $i = 1 \ldots n+1$,

(ii) $x \in S_1$,

where the simplices S_1 and S_2 are defined by:

$$S_1 := \{\sum_{i=1}^{n+1} \mu^i x_i \mid \mu \in \Sigma_{n+1}\}, \tag{30}$$
$$S_2 := \{\sum_{i=1}^{n+1} \mu^i z_i \mid \mu \in \Sigma_{n+1}\}. \tag{31}$$

Then $x = \sum_{i=1}^{n+1} \mu^i x_i$ for some $\mu \in \Sigma_{n+1}$ and the control $u(\cdot) = \sum_{i=1}^{n+1} \mu^i u_i(\cdot)$ steers the state $x \in S_1$ to the the state $z = \sum_{i=1}^{n+1} \mu^i z_i \in S_2$ in time T along a trajectory that satisfies both state and control constraints.

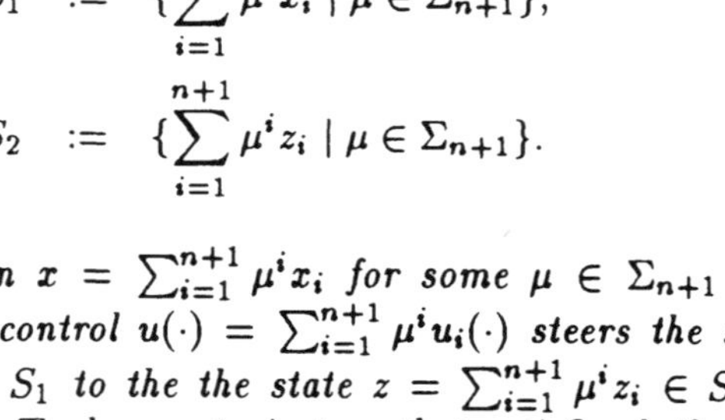

Fig. 10. Simplex control

For simplicity, consider the case when $E = \mathbb{R}^n$ (no state constraints). The discrete-time system corresponding to (26) is

$$x[k+1] = Fx[k] + Gu[k] \tag{32}$$

where $F := \exp(A\Delta)$, $G := \int_0^\Delta \exp AsBds$ and $x[k]$, $u[k]$ denote, respectively, $x(k\Delta)$, $u(k\Delta)$. Assume the sampling period Δ is chosen so that (F, G) is controllable. For every integer j let

$$C_j := -[F^{-j}G,\ F^{-j+1}G, \ldots, F^{-1}G] \tag{33}$$

Construct the nested sequence of sets $\{X_k\}_{k=0}^N$ as follows: $X_k = C_{k+n}\Omega^{k+n}$. Then $X_0 = C_n\Omega^n$ is the set of states that can be controlled to the origin in n steps, and X_k the set that can be controlled to the origin in $n+k$ steps using admissible controls and $X_i \subset X_j$ for all $i, j \in \mathbb{N}, i \leq j$. See Figure 11. For each $k \in 1, \ldots, N$, a set

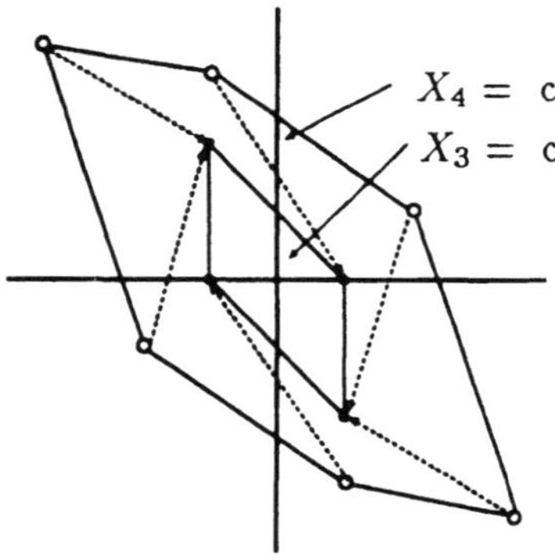

Fig. 11. The Sets X_k

$\{X_{k1}, \ldots, X_{km_k}\}$ of simplices (subsets of X_k), m_k in number, is constructed so that their union covers X_k; see Figure 12. Associate with each vertex x_{kmj} of the simplex X_{km}, a constant control $u_{kmj} \in \Omega$ which steers x_{kmj} into simplex $X_{(k-1)\ell}$ (for some $\ell \in \{1, \ldots, m_{k-1}\}$. Such controls exist by construction. Associate the zero control with the origin. If $x \in X_{km}$ for some $k \in \{1, \ldots, N\}$, $m \in \{1, \ldots, m_k\}$, there exists a $\mu = (\mu^1, \ldots, \mu^{n+1}) \in \Sigma_{n+1}$ such that $x = \sum_{j=1}^{n+1} \mu^j x_{kmj}$. This defines a control $u_{km}(x) = \sum_{j=1}^{n+1} \mu^j u_{kmj}$. Using the result established above, it can be seen that the control law $u_{km}(\cdot) : X_{km} \rightarrow \Omega$ steers any point in $X_{km} \subset X_k$ to a corresponding point in $X_{(k-1)\ell}$ in time Δ. It is easily shown that the control law $u_{km}(\cdot) : X_{km} \rightarrow \Omega$ is affine, i.e. there exist a_{km}, K_{km} such that

$$u_{km}(x) = a_{km} + K_{km}x \tag{34}$$

for all $x \in X_{km}$.

If $x \in X_0$, the control is defined to be

$$u_0(x) = K_0 x \tag{35}$$

where $K_0 = C_n^\dagger$. This control law steers $x \in X_0$ to the origin in n steps, keeping the state in X_0.

The nonlinear control law $h(\cdot) : X_N \rightarrow \Omega$ can now be defined.

$$h(x) := \begin{cases} K_0 x & \text{if } x \in X_0 \\ a_{km} + K_{km} x & \text{else} \end{cases}$$

where (k, m) is such that X_{km} is the *first* set in $\{X_{11}, \ldots, X_{Nm_N}\}$ satisfying $x \in X_{km}$.

All the controller has to do is determine which simplex contains the current state x; if $x \in X_{km}$, $m \geq 1$, then $u = a_{km} + K_{km}x$ and, if $x \in X_0$, $u = K_0 x$ where $\{a_{km}\}$, K_{km}, $m = 1 \ldots N$, and K_0 are precomputed. The number of simplices may be large, but the main demand on the controller is memory. Figure 12 shows how the controller steers a state x in $X_{32} \subset X_3$ to a state z in $X_{21} \subset X_2$. *Normalized* step responses for

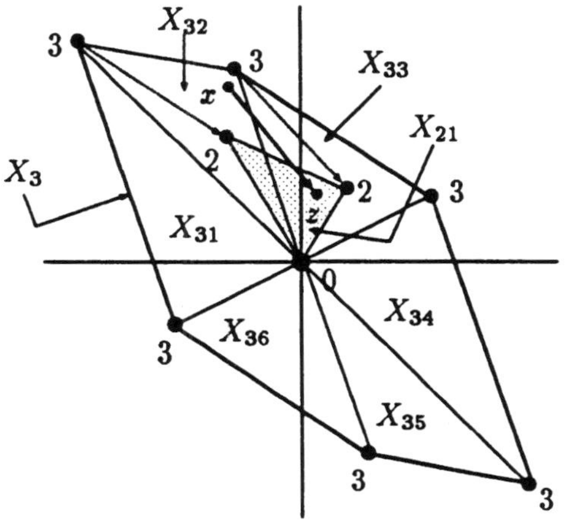

Fig. 12. Nonlinear control

a triple integrator system with control saturation and a six layer control are shown in Figure 13. The faster (slower) responses correspond to smaller (larger) step changes.

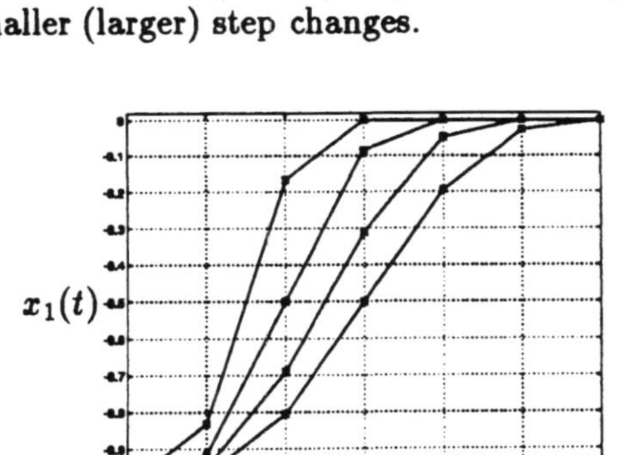

Fig. 13. Step responses for triple integrator

The control law steers any state $x \in X_N$ to the origin along a trajectory which satisfies the control constraints.

The procedure can be extended to handle state constraints, as well as bounded disturbances. The discrete time system is

$$x[k+1] = Fx[k] + Gu[k] + d[k]$$

where $d[k] \in D$. The set X_0 is defined to be a robust control invariant set, and (if $E = \mathbb{R}^n$) the sets $\{X_k\}$ are constructed according to $X_{k+1} = F^{-1}X'_k - F^{-1}G\Omega$ where X'_k is the largest set satisfying $X'_k + D \subset X_k$.

The resultant controller steers any state x in X_N to the invariant set X_0 along a trajectory satisfying the control constraints, and maintains the state in X_0 thereafter.

8. CONCLUSION

Efficient control of constrained systems, whether linear or nonlinear, requires the controller to be nonlinear.

Constraints impose limits on what can be achieved by any controller. The most important of these is the restriction of the set of states X_{max} which can be controlled. For model predictive controllers, this set is generally compact.

Model predictive control of linear systems is relatively easily implemented because FHP (the finite horizon control problem solved online) is convex. Control of nonlinear systems raises formidable problems (global optimization is necessary if FHP and the state estimation problem are not convex); these problems can be considerably alleviated by the use of a variable horizon in FHP.

To ensure stability, FHP should include a stability constraint. If the disturbances decay to zero, this constraint takes the form of an equality constraint $x(t+T) = 0$ on the terminal state. If the disturbances are merely bounded, the stability constraint takes the form $x(t+T) \in Z$ where Z is a robust control invariant set (there exists a controller which can maintain the state in this set despite the disturbances).

Variable horizon model predictive control achieves a region of attraction equal to X_{max} if state feedback is employed. Disturbances, state estimation error, parameter error, all reduce the achievable region of attraction.

Alternative methods of controlling linear systems with constraints are becoming available. These require considerable memory, but much less online computation.

Acknowledgements. I am grateful to Jim Rawlings for his helpful comments.

9. REFERENCES

Aubin, J. P. (1991). *Viability theory*. Systems & control: foundations & applications. Birkhauser. Boston, Basel, Berlin.

Bequette, B. W. (1991). Nonlinear control of chemical processes: a review. *Ind. Eng. Chem. Res.* **30**, 1391–1413.

Bitmead, R. R., M. Gevers and V. Wertz (1990). *Adaptive Optimal Control—The Thinking Man's GPC*. Prentice Hall Int.

Clarke, D. W., D. E. Mosca and R. Scattaloni (1991). Robustness of an adaptive predictive controller. In: *Proceedings of the 30th IEEE Conference on Decision and Control*. Brighton, England. pp. 979–984.

Garcia, C. E., D. M. Prett and M. Morari (1989). Model predictive control: Theory and practice—a survey. *Automatica* **25**(3), 335–348.

Gilbert, E. C., T. Kolmanovsky and K. T. Tan (1994). Nonlinear control of discrete-time linear systems with state and control constraints: a reference governor with global convergence properties. In: *Proceedings of the 33rd IEEE Conference on Decision and Control*. Lake Buena Vista, Florida. pp. 144–149.

Gilbert, E. G. and K. T. Tan (1991). Linear systems with state and control constraints: the theory and application of maximal output admissible sets. *IEEE Transactions on Automatic Control* **AC-36**, 1008–1020.

Keerthi, S.S. and E.G. Gilbert (1988). Optimal, infinite horizon feedback laws for a general class of constrained discrete time systems. *Journal of Optimization Theory and Applications* **57**, 265–293.

Mayne, D. Q. and H. Michalska (1990). Receding horizon control of non-linear systems. *IEEE Transactions on Automatic Control* **35**(5), 814–824.

Mayne, D. Q. and H. Michalska (1993). Adaptive receding horizon control for constrained nonlinear systems. In: *Proceedings 32nd IEEE Conference on Decision and Control*. San Antonio, Texas, USA. pp. 1286–1290.

Mayne, D. Q. and W. R. Schroeder (1994). Nonlinear control of constrained dynamic systems. *International Journal of Control* **60**, 1035–1043.

Meadows, E.S., M.A. Henson, J.W. Eaton and J.B. Rawlings (1995). Receding horizon control and discontinuous state feedback stabilization. *International Journal of Control*. to appear.

Michalska, H. and D. Q. Mayne (1991). Receding horizon control of non-linear systems without differentiability of the optimal value function. *Systems & Control Letters* **16**, 123–130.

Michalska, H. and D. Q. Mayne (1992). Moving horizon observers. In: *Proceedings IFAC Symposium on Nonlinear Control Systems Design*. Bordeaux, France. pp. 576–581.

Michalska, H. and D. Q. Mayne (1993). Robust receding horizon control of constrained nonlinear systems. *IEEE Transactions on Automatic Control* **38**, 1623–1632.

Michalska, H. and D. Q. Mayne (1995). Moving horizon observers and observer-based control. *IEEE Transactions on Automatic Control.*

Polak, E. (1993). On the use of consistent approximations in the solution of semi-infinite optimization and optimal control problems. *Math. Programming* **62**, 385–414.

Polak, E. and D. Q. Mayne (1981). Design of nonlinear feedback controllers. *IEEE Transactions on Automatic Control* **26**(3), 730–733.

Rawlings, J.B. and K.R. Muske (1993). The stability of constrained receding horizon control. *IEEE Transactions on Automatic Control* **AC-38**, 1512–1516.

Rawlings, J.B., E.S. Meadows and K.R. Muske (1994). Nonlinear model predictive control: a tutorial and survey. In: *Proceedings Adchem '94*. Kyoto.

Robertson, D., D. H. Lee and J. B. Rawlings (1994). A moving horizon based approach for least squares state estimation. *AIChE Journal*. Accepted for publication.

Scokaert, P. O. M. and J. Rawlings (1995). Stability of model predictive control under perturbations. In: *Proceedings of the IFAC symposium on nonlinear control systems design*. Lake Tahoe, California.

Thomas, Y. A. (1975). Linear quadratic optimal estimation and control with receding horizon. *Electronics Letters* **11**, 19–21.

Vidyasagar, M. (1993). *Nonlinear system analysis*. second ed.. Prentice Hall. Englewood Cliffs, New Jersey.

Vinter, R. B. and H. Michalska (1991). Receding horizon control for nonlinear time-varying systems. In: *Proceedings 30th IEEE Conference on Decision and Control*. Brighton, England. pp. 75–76.

Wonham, W. M. (1974). *Linear Multivariable Control: A Geometric Approach*. Springer-Verlag. Berlin, Heidelberg, New York.

Yang, T. H. and E. Polak (1993). Moving horizon control of nonlinear systems with input saturation, disturbances and plant uncertainty. *International Journal of Control* **58**, 875–903.

Zheng, Z. Q. and M. Morari (1993). Robust stability of constrained model predictive control. In: *Proceedings of the 1993 American Control Conference*. pp. 379–382.

MULTIPLE STEADY STATES IN AZEOTROPIC DISTILLATION AND THEIR EFFECT ON COLUMN OPERATION AND DESIGN

N. Bekiaris, G. A. Meski and M. Morari*

**Chemical Engineering 210-41, California Institute of Technology, Pasadena, CA 91125, U.S.A.*
E-mail: mm@imc.caltech.edu

Abstract. In this article we study multiple steady states in ternary heterogeneous azeotropic distillation. We show that in the case of infinite reflux and an infinite number of trays one can construct bifurcation diagrams on physical grounds with the distillate flow as the bifurcation parameter. Multiple steady states exist when the distillate flow varies non-monotonically along the continuation path of the bifurcation diagram. We show how the distillate and bottom product paths can be located for tray or packed columns, with or without decanter. We derive a necessary and sufficient condition for the existence of these multiple steady states based on the geometry of the product paths. We also locate in the composition triangle the feed compositions that lead to these multiple steady states. We show that the prediction of the existence of multiple steady states in the case of infinite reflux and an infinite number of trays has relevant implications for columns operating at finite reflux and with a finite number of trays.

Key Words. Distillation columns, design, multiple steady states.

1. INTRODUCTION

Azeotropic distillation is one of the most widely used and most important separation operations in the chemical and the specialty chemical industry. Among their surprising features, it has been discovered that such columns can exhibit multiple steady states. In this article, by multiple steady states we mean what is generally referred to as *output multiplicities* i.e. columns with the same inputs (the same feed, distillate, bottoms, reflux and boilup molar flows, the same feed composition, number of stages and feed location) but different outputs (product compositions) and hence different composition profiles.

The term *heterogeneous azeotropic distillation* covers the general notion of distillation of azeotrope forming mixtures where two liquid phases exist in some region of the composition space. We use the following convention to refer to a given mixture: L (I, H respectively) corresponds to the component which has the lowest (intermediate, highest resp.) boiling point; we also denote the entrainer by E. The locations of the feed, distillate, bottoms, reflux and overhead vapor in the composition triangle are denoted by F, D, B, R and V respectively. The corresponding flowrates are denoted by the same letters in italics.

Two widely used tools for the description of azeotropic distillation are the residue curves (Doherty and Perkins, 1978) and the distillation lines (Stichlmair et al., 1989). Composition profiles of packed (tray resp.) columns operating at infinite reflux coincide with residue curves (distillation lines resp.). Distillation lines do not generally coincide with residue curves. It is assumed that the reader is familiar with these notions and the corresponding maps and boundaries.

As an illustrative example throughout this article we use the mixture ethanol (L) - water (H) - benzene (I-E). Figure 1 depicts the residue curve diagram of this mixture at 1 atm. Ethanol and water form an azeotrope (X) which can be separated using benzene as the entrainer. In this diagram, there are two binary homogeneous azeotropes (X and Y), one binary heterogeneous azeotrope (Z) and one ternary heterogeneous azeotrope (T). There are three interior distillation boundaries in this diagram (TX, TY, TZ) which separate the triangle in three distillation regions. Figure 1 also shows the heterogeneous liquid boiling envelope (line UCW) of this mixture at 1 atm. Liquid compositions located inside the heterogeneous liquid boiling envelope will split in two liquid phases whose compositions will lie on the envelope. Point C is the critical point. We refer to the water-rich phase (line UC) as phase 1 and to the entrainer-rich phase (line CW) as phase 2. The straight lines connecting two liquid phases in equilibrium are the tie lines. The vapor in equilibrium with any liquid in the two-phase region lies on the va-

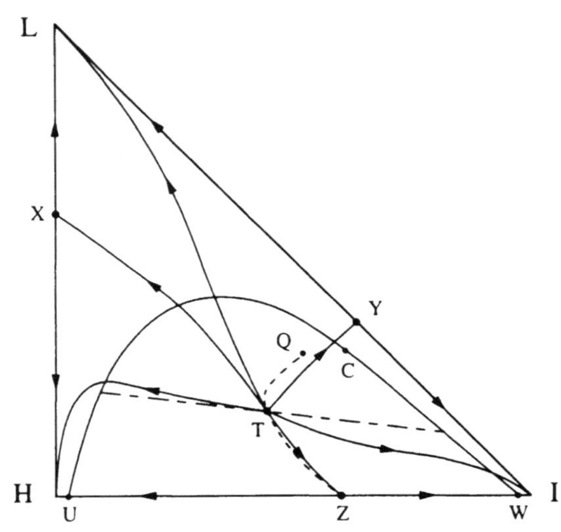

Fig. 1. Residue curve diagram and VLLE of the mixture ethanol (L) - water (H) - benzene (I).

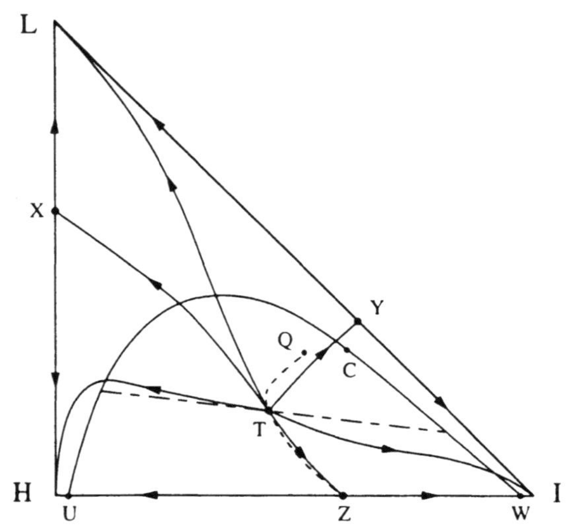

Fig. 2. Distillation line diagram and VLLE of the mixture ethanol (L) - water (H) - benzene (I).

por line ZTQ. Figure 2 illustrates the distillation line diagram for the same mixture at 1 atm. Qualitatively, the distillation line diagram is similar to the residue curve diagram of Figure 1. The locations of the boundaries, however, are somewhat different: In the heterogeneous region, the boundaries TX and TY and the vapor line coincide close to the ternary azeotrope T.

The study presented here is the extension of our previous work (Bekiaris et al., 1993) on homogeneous azeotropic distillation. The basic steps of the analysis of Bekiaris et al. (1993) for the case of infinite reflux and infinite number of trays (or infinitely long columns) (which we hereafter denote as the ∞/∞ case) can be summarized in the following: 1. Determine the characteristics of the ∞/∞ case column profiles: they follow residue curves and they contain a singular point. 2. Using the ∞/∞ case profile characteristics and the material balances determine the feasible product regions. 3. Perform a continuation along the feasible product lines and determine whether multiplicities occur by examining the distillate flowrate along the continuation path by applying the lever rule. Multiple steady states exist when the distillate flow varies non-monotonically along the continuation path and more specifically, for the selected direction of the continuation path, when the distillate flow decreases in a segment along this path.

Using the results of this analysis Bekiaris et al. (1993) derive the necessary and sufficient geometrical condition for the existence of multiple steady states in the ∞/∞ case. This condition checks when D decreases along the continuation path. Finally, Bekiaris et al. (1993) locate the region of feed compositions that lead to multiplicities. The above results have been originally derived for residue curve diagrams. It is obvious that similar results hold for distillation line diagrams. Using the above results, we are able to study the existence of multiple steady states in the ∞/∞ case for mixtures that exhibit two liquid phases. We perform this task in two steps: First for packed and tray columns without decanter and then for columns with decanter.

2. COLUMNS WITHOUT DECANTER

In the homogeneous case, the composition of the distillate is that of the stream drawn from the top of the column. This is also the case when no decanter is used in heterogeneous azeotropic distillation. Hence the techniques developed for homogeneous azeotropic distillation can be directly applied in this case. In particular, the case of heterogeneous mixtures using *packed* columns without decanter is identical to the one studied in our previous article on homogeneous azeotropic distillation (Bekiaris et al., 1993). We study packed columns first.

The only information required in this case is the residue curve boundaries TX, TY, TZ (Figure 1). No information about the VLLE is required. The geometrical condition (Bekiaris et al., 1993) for this particular type of residue curve diagram can be simplified to the following: For the existence of multiple steady states it is required that (1) some line parallel to the LH edge intersects the interior boundary TX more than once, or (2) some line parallel to LI intersects TY more than once, or (3) some line parallel to IH intersects TZ more than once. In Figure 1, the geometrical condition is not satisfied for any interior boundary and therefore no multiplicities exist for packed columns.

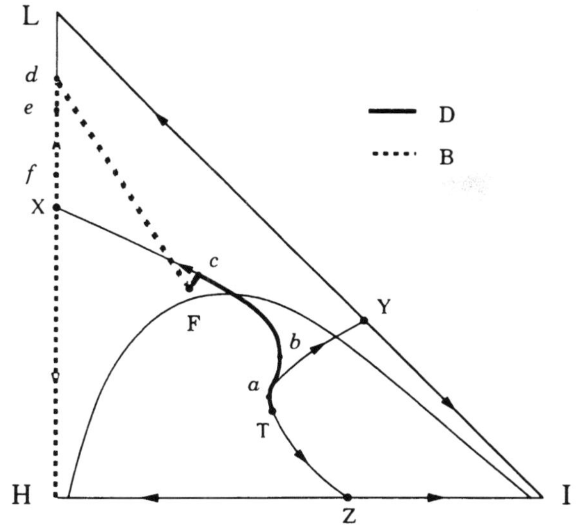

Fig. 3. The distillate and bottoms continuation paths for tray columns without decanter.

Similarly, for tray columns, we only have to check the aforementioned simplified geometrical condition for the interior distillation line boundaries (TX, TY, TZ) which are shown in Figure 3. Points a and b in Figure 3 are defined as the points of TX where the tangent to the boundary is parallel to the LH edge. The condition is not satisfied for the boundaries TY and TZ, it is satisfied, however, for the boundary TX. By applying the geometrical condition in its original form, it is easy to show that the distillate flowrate decreases along the continuation path for distillate compositions located on the ab segment of the boundary TX and bottoms located at any point on the LH edge. Therefore, multiple steady states exist for feed compositions that lie in the abLH region of the composition triangle.

We will next track the whole continuation path (as described above) for a tray column with a homogeneous feed lying in the lower left distillation region where T is the unstable node and H the stable node. The light component mole fraction in the bottoms x_{BL} is recorded along the continuation path. Figure 3 also depicts the feasible distillate and bottoms compositions (D and B) in the composition triangle. Figure 4 shows a sketch of the bifurcation diagram of x_{BL} vs. the bifurcation parameter D. It is apparent that for some distillate flowrate range, three steady states exist. Hence, for the mixture under consideration, there exists a qualitative difference between packed and tray columns due to the differences between residue curves and distillation lines.

Since residue curve boundaries are easier to calculate than distillation line boundaries, we derived the following guidelines on when it is justified to use residue curve boundaries for the study

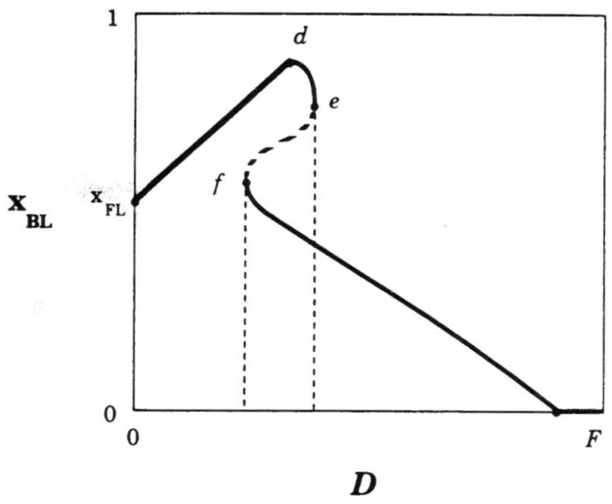

Fig. 4. Bifurcation diagram of the mole fraction of L in the bottoms vs. the distillate flow for tray columns without decanter.

of multiplicities of tray columns: (1) For homogeneous mixtures: residue curve boundaries provide a good approximation except when the boundaries are highly curved (although we are not aware of any mixture where the approximation fails). (2) For heterogeneous mixtures: on one hand, residue curve boundaries inside the two-liquid phase region may provide a very poor approximation of distillation line boundaries. On the other hand, the vapor line gives an indication of the geometry of the distillation line boundaries in the two phase region. Hence, the combination of residue curve boundaries and the vapor line can be used for the qualitative prediction of multiplicities of tray columns. If this prediction is different from the one for packed columns, then it is suggested that the distillation line boundaries should be used for detailed results on multiplicities of tray columns.

3. COLUMNS WITH DECANTER

The existence of two liquid phases in the decanter at the top of a heterogeneous azeotropic distillation column in the ∞/∞ case adds another parameter to the problem because the overall distillate flow is composed of two liquid streams with compositions those of the two liquid phases in equilibrium. The two stream flows, D_1 and D_2, are the independent parameters in this case and $D = D_1 + D_2$. One can reduce the system parameters to one by imposing a "policy" on the distillate. The most common policy used in heterogeneous azeotropic distillation for columns like the one used for the separation of the mixture ethanol - water - benzene, is to recover as distillate a portion of the entrainer-poor phase only (i.e. D_2=0) and recycle a mixture of the two liquid phases.

Since D_2=0, then $D = D_1$ and hence when V lies in the heterogeneous region, the distillate may

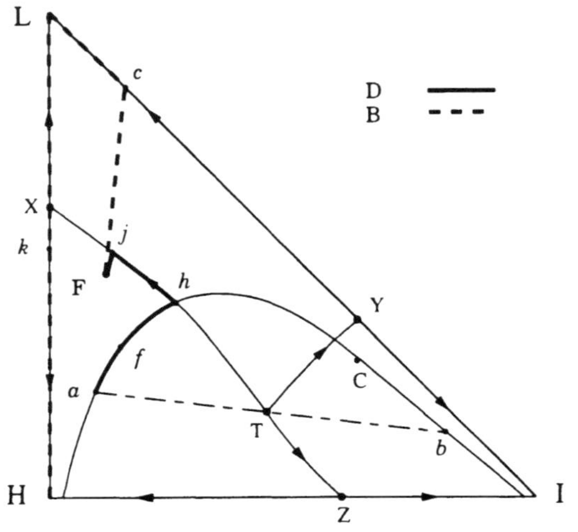

Fig. 5. The distillate and bottoms continuation paths for packed columns with decanter and D_2=0.

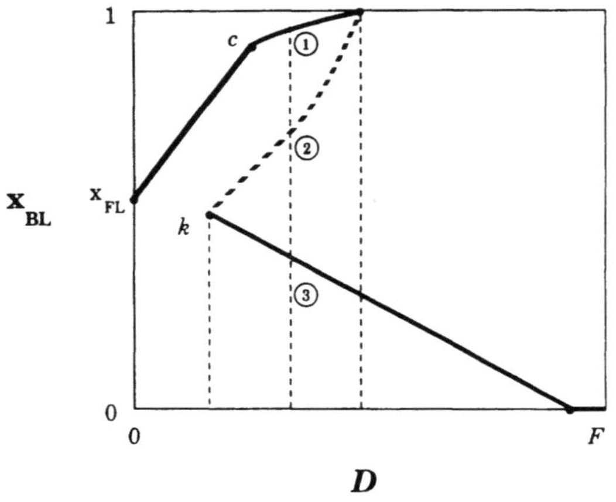

Fig. 6. Bifurcation diagram of the mole fraction of L in the bottoms vs. the distillate flow for packed columns with decanter and D_2=0.

only lie on the part of the heterogeneous liquid boiling envelope on the left of the critical point. Figure 5 shows the distillate and bottoms composition continuation path in the composition triangle when D_2=0 for a packed column. The distillate D follows the line $afhj$F and the bottoms B the line FcLkH. The light component mole fraction in the bottoms x_{BL} is recorded along the continuation path. It is very simple to show using the lever material balance rule that D decreases as D moves from f to h and accordingly B moves from L to k. Figure 6 shows a sketch of the bifurcation diagram of x_{BL} vs. the bifurcation parameter D. It is apparent that for some distillate flowrate range, three steady states exist. Qualitatively similar results are obtained for tray columns using the distillation line boundaries. Some small quantitative differences are expected because of the different boundary locations. The shape of the vapor line does not affect the location of the feasible products.

It is apparent that the key difference between columns with and without decanter is the location of the distillate continuation path. Once this path is located, the geometrical condition for the existence of multiplicities can be directly applied because it only checks if the distillate flowrate decreases along the continuation path. By applying the geometrical condition, it is easy to show that the distillate flowrate decreases along the continuation path for distillate compositions located on the ah segment of the distillate route and bottoms located at any point on the LH edge (Figure 5). Therefore, multiple steady states exist for feed compositions that lie in the ahLH region of the composition triangle.

4. FINITE REFLUX AND FINITE NUMBER OF TRAYS

The ∞/∞ case is the limiting case of high reflux and a large number of trays. Therefore, if the geometrical condition is satisfied for a given mixture then multiplicities will exist for some sufficiently large finite reflux and finite number of trays. However, the inverse is not true. The geometrical condition is only a sufficient condition for the existence of multiplicities when the reflux and the number of trays are finite. In other words, the ∞/∞ case multiplicity analyzed in this article is only one possible cause for multiplicities in finite columns.

In this section, we present steady state bifurcation results for the mixture ethanol (L) - water (H) - benzene (I-E) with the distillate flow D as the bifurcation parameter. These bifurcation results show (1) that the predictions for the existence of multiple steady states in the ∞/∞ case carry over to columns operating at finite reflux and with a finite number of trays and (2) that, although the predictions were made in the ∞/∞ case, it does not mean that multiple steady states do not exist for realistic operating conditions (low reflux and number of trays). Bekiaris et al. (1994) provide the detailed data on the column specifications and the thermodynamic model used. The bifurcation calculations were conducted with AUTO, a software package developed by Doedel (1986). In the following, we present steady state bifurcation results for two tray columns, one without decanter and one with decanter.

4.1. *Column without decanter*

The column characteristics are those of the column studied by Magnussen et al. (1979); 27 ideal stages and R/F =2.4. The location of the

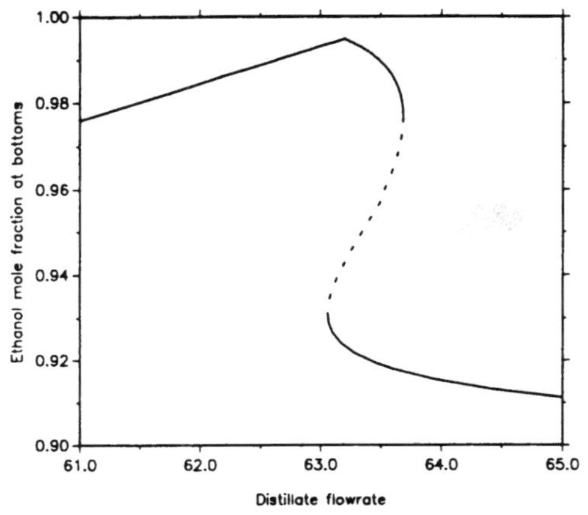

Fig. 7. Computed bifurcation diagram with the distillate flow as the bifurcation parameter for the column without decanter.

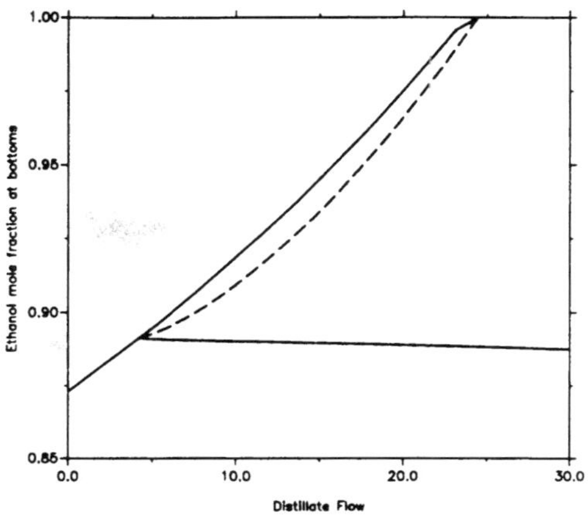

Fig. 8. Computed bifurcation diagram with the distillate flow as the bifurcation parameter for the column with decanter.

overall feed composition in the composition triangle is similar to the one illustrated in Figure 3. Therefore, Figure 4 illustrates the predicted ∞/∞ case bifurcation diagram for this column. Figure 7 shows the computed bifurcation diagram with the distillate flow as the bifurcation parameter. Two stable and one unstable (dashed curve) steady states exist for a narrow distillate flowrate range; a unique stable steady state exists elsewhere. Note the similarity of this bifurcation diagram with that of the ∞/∞ case prediction (Figure 4). Finally, note that, although those multiple steady states were predicted at infinite reflux, they still exist at low reflux values (R/F =2.4).

4.2. *Column with decanter*

The column has 27 ideal stages and R/F =5.0. The location of the overall feed composition in the composition triangle is similar to the one illustrated in Figure 5 and hence, Figure 6 illustrates the predicted ∞/∞ case bifurcation diagram for this column with the exception that the mole fraction of ethanol at point k of Figure 6 is larger than x_{FL}. Figure 8 shows the computed bifurcation diagram with the distillate flow as the bifurcation parameter. Two stable and one unstable (dashed curve) steady states exist for distillate flows between the two limit points; a unique stable steady state exists otherwise. Note that it is purely coincidental that the second (low distillate flowrate) limit point appears to be on the first (low D) stable portion of the continuation path (it does not; the other component mole fractions are different). Again, note the similarity of this bifurcation diagram with that of Figure 6 although some discrepancy from the ∞/∞ case prediction is expected.

5. CONCLUSIONS

In this article we show how the ∞/∞ case multiplicity analysis/prediction method by Bekiaris et al. (1993) can be extended to heterogeneous mixtures. We show how the existence of multiplicities can be predicted and how the region of feed compositions that lead to these multiplicities can be located. Using the material presented in this article, studies on the existence of multiplicities can be easily performed for tray or packed columns with any combinations of reboiler and condenser types (partial, total or subcooled), with or without decanter and for different product policies; it only requires to correctly locate the feasible product regions for the various alternatives. We discuss the differences between packed and tray columns, between residue curve and distillation line diagrams and we derive guidelines on when it is justified to use residue curve boundaries for the study of multiplicities of tray columns.

As an illustrative example throughout this article we use the mixture ethanol - water - benzene. For this mixture, we conclude that multiplicities exist for tray columns without decanter and for (tray or packed) columns with decanter and distillate policy D_2=0. A unique steady state exists, however, for packed columns without decanter. Finally, using numerically constructed bifurcation diagrams, we show that the ∞/∞ case predictions carry over to columns operating at finite reflux and with a finite number of trays.

Acknowledgments: We gratefully acknowledge partial financial support from the Department of Energy and through the Petroleum Research Fund administered by the American Chemical Society.

6. REFERENCES

Bekiaris, N., G. A. Meski, and M. Morari (1994). Multiple Steady States in Heterogeneous Azeotropic Distillation. submitted to *Ind. Eng. Chem. Res.*

Bekiaris, N., G. A. Meski, C. M. Radu, and M. Morari (1993). Multiple Steady States in Homogeneous Azeotropic Distillation. *Ind. Eng. Chem. Res.*, **32**(9), 2023-2038.

Doedel, E. (1986). *AUTO: Software for Continuation and Bifurcation Problems in Ordinary Differential Equations.* Applied Mathematics, Caltech, Pasadena, CA.

Doherty, M. F., and J. D. Perkins (1978). On the Dynamics of Distillation Processes. I. The Simple Distillation of Multicomponent Non-reacting, Homogeneous Liquid Mixtures. *Chem. Eng. Science*, **33**, 569-578.

Magnussen, T., M. L. Michelsen, and Aa. Fredenslund (1979). Azeotropic Distillation Using UNIFAC. *Inst. Chem. Eng. Symp. Ser.*, **56**, 4.2/1-4.2/19.

Stichlmair, J. G., J. R. Fair, and J. L. Bravo (1989). Separation of Azeotropic Mixtures via Enhanced Distillation. *Chem. Eng. Prog.*, (1), 63-69.

COMPARISON OF THE SEPARATION PERFORMANCES OF A MULTI-EFFECT BATCH DISTILLATION SYSTEM AND A CONTINUOUS DISTILLATION SYSTEM

Shinji HASEBE, Taketoshi KUROOKA and Iori HASHIMOTO

Department of Chemical Engineering, Kyoto University, Kyoto 606-01 JAPAN

Abstract: A new type of batch distillation system which separates multiple components simultaneously is proposed. The features of this system are that the steam is supplied to the reboiler of the first column only, and a condenser of one column is heat-integrated with the reboiler of the next column. Therefore, the energy required to separate multiple components is markedly less than that of an ordinary batch distillation system. The energy consumption of the proposed system is compared with that of an ordinary continuous distillation system. Through simulation, it is found that the proposed system has the potential to have a better separation performance than a continuous system if the number of products is increased.

Keywords: Distillation columns, Batch control, Optimization problems, Cascade control, Dynamic systems, Chemical industry

1. INTRODUCTION

Batch distillation is a very flexible separation method in the sense that it can be used even if the number of products and/or feed compositions are changed. Thus, it is a suitable separation method for current chemical industry where the production demand and the lifetime of products are uncertain. However, from the view point of energy consumption, batch distillation has been considered to be an ineffective separation method compared with continuous distillation. Therefore, the energy conservation of a system is one of the key issues in determining whether a flexible separation system consisting of the batch distillation columns is applicable to an actual chemical industry or not.

A batch distillation column with a middle vessel shown in Fig. 1 was proposed by several researchers separately. It differs from the ordinary batch distillation column in the following points:

i) the feed is not supplied to the reboiler but to the middle of the column.
ii) the light and heavy products are withdrawn from the top and the bottom simultaneously. And the material in the feed tank is purified by the recycled stream from the column.

We have shown that the column shown in Fig. 1 can be regarded as a combined system of ordinary and stripping batch distillation columns where the condenser of the stripping column is heat-integrated with the reboiler of the ordinary column. From this viewpoint, it is called a "complex batch distillation column." The complex column is more effective in removing light and heavy impurities from products than the ordinary batch distillation column (Hasebe *et al.*, 1992). The characteristics of this column has been analized by Meski *et al.*, (1993), and the application of this column to distillations with chemical reactions has also been reported (Mujtaba & Macchietto, 1994).

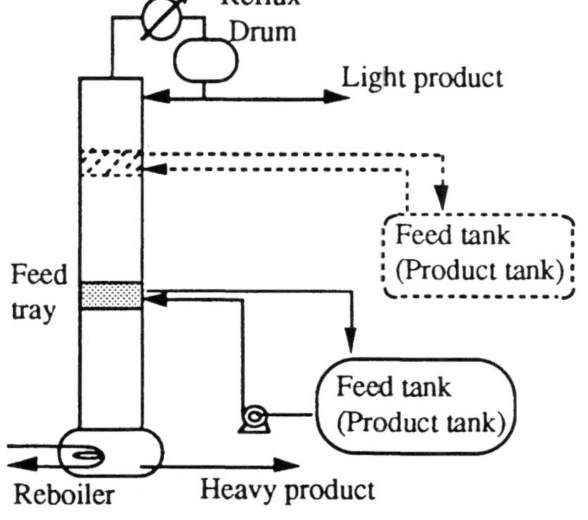

Fig. 1 Structure of the complex batch distillation column

It is possible to separate more than three components at one time by installing another intermediate vessel as shown by broken lines in Fig. 1. However, according to the increase in the number of products to be separated simultaneously, the column becomes taller and the operation of the column becomes more complicated. Through exhaustive simulations involving various conditions, from the viewpoint of energy consumption, it is found that in many cases the operation procedure which has no withdrawal steps of products shows better separation performance than that which has withdrawal steps. The results of the simulations show that every product should be accumulated in a vessel during the whole separation period. It also becomes clear that the flowrate of the recycle flow from the feed tank to the feed tray affects the separation performance and must be large enough to keep the composition of both sections the same.

By taking these facts into account, in this paper we propose a new design for the batch distillation system, and compared the separation performance of the proposed system to that of an ordinary continuous distillation system.

2. MULTIPLE-EFFECT BATCH DISTILLATION SYSTEM

The proposed batch distillation system is shown in Fig. 2. The proposed system consists of a set of batch distillation columns connected sequentially which separate (M+1) components by M columns simultaneously. A feature of this system is that only the first column has a reboiler and only the final column has a condenser. The operation of this system is carried out as follows: Initially, the vessel of every column is filled with the feed, and heat is supplied to the bottom vessel of the first column. During the distillation operation, the vapor from the top of a column is supplied to the bottom of the next column, and the liquid in the bottom vessel is returned to the top of the previous column. By continuing total reflux operation, the ith heaviest component is accumulated in the vessel of column i. The operation is terminated when the compositions of the liquid in all the vessels satisfy the specifications for the final products.

Figure 3 shows a conventional heat-integrated batch distillation system. In this system, the condenser of the first column is heat-integrated with the reboiler of the second column, and the vapor leaving from the top of the first column is used as the heat source of the second column. In the proposed system shown in Fig. 2, the vapor leaving from the top of a column is fed directly to the next column. Therefore, the latent heat of the vapor can be used efficiently without having to consider the temperature difference between the heat source and the heat sink. The heat supplied to the reboiler of the first column is effectively used M times to generate the vapors of M columns. From this viewpoint, the proposed system is hereafter called the "Multiple-Effect BAtch Distillation system, MEBAD."

For the complex column shown in Fig. 1, the starting times of product withdrawals and the reflux and reboil ratios during the product withdrawals are used to control the product compositions. On the other hand, the operation of the MEBAD is simpler because the desired products are obtained only by keeping the holdups of the vessels to their prespecified values during the whole operation period. The large recycle flowrate between the feed tank and the feed tray, which is required for the complex column to keep the composition of both sections the same, is not needed. From these characteristics, the MEBAD proposed here has the possibility to become a flexible and energy-efficient batch distillation system.

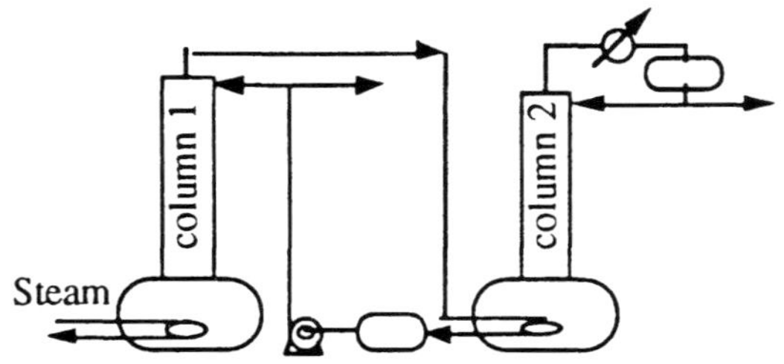

Fig. 3 Heat-integration of batch distillation columns

3. MATHEMATICAL MODEL OF MEBAD

The mathematical model of each column in the MEBAD can be generated by a slight modification of the mathematical model of the ordinary batch distillation column. A fairly simple dynamic model of the MEBAD is used in this paper based on the following assumptions:

i) Constant molar boilup rate at the reboiler
ii) Constant molar liquid and vapor flows
iii) Constant molar holdup on trays
iv) Negligible vapor holdups
v) Constant relative volatilities

The schematic diagram of column i in the MEBAD is shown in Fig. 4. Using the above assumptions, the mass balance equations are derived.

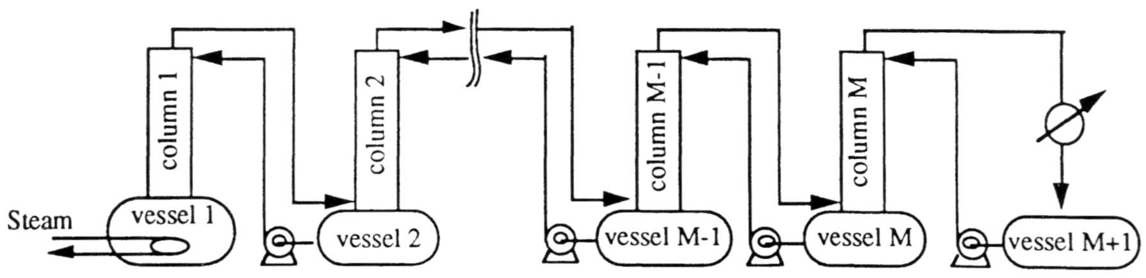

Fig. 2 Multiple-effect batch distillation system

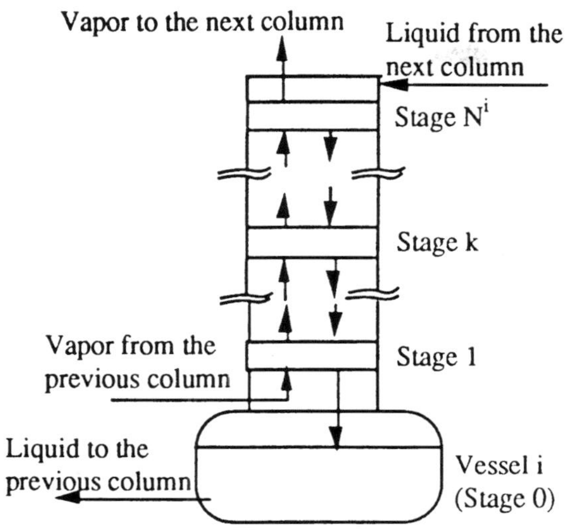

Fig. 4 Structure around column i

In a conventional distillation column, the reflux ratio can be regarded as a manipulated variable used to control the product composition. However, in the MEBAD the reflux ratio cannot be used as a manipulated variable because the system is operated at total reflux during the whole production period. Thus, it is necessary to use other variables to control the product composition.

The specification of each product is given as the lower bound of the composition of the dominant component in the product. In this paper, we treat the case where the relative volatility between two components is not so close to one and the product specification is relatively high. In such a case, it is possible to assume that each product contains only one heavy and one light impurity. The total amount of holdup on the trays is fairly small compared with the holdup in the vessel. Therefore, it is also possible to assume that at the end of the separation operation, only the two components (components i and i+1) are remaining on the trays of column i, and the total amount of each component is the same. Based on the above assumptions, the following mass balance equations of each component can be derived:

$$F z_i^F = z_i^{i-1} P_{i-1} + z_i^i P_i + z_i^{i+1} P_{i+1} + R_i \quad (1)$$
(i = 1, 2, ..., M+1)

$$R_i = (\sum_j H_j^{i-1} + \sum_j H_j^i)/2 \quad (2)$$
(i = 1, 2, ..., M+1)

$$z_{i-1}^i + z_i^i + z_{i+1}^i = 1 \quad (i = 1, 2, ..., M+1) \quad (3)$$

$$z_0^1 = z_1^0 = z_{M+2}^{M+1} = z_{M+1}^{M+2} = P_0 = P_{M+2} = 0 \quad (4)$$

$$H_j^0 = H_j^{M+1} = 0 \quad (5)$$
(j = 1, 2, ..., N^i ; i = 1, 2, ..., M)

where

F : amount of raw material per batch [mol]
H^i_j : liquid holdup of the jth stage of column i [mol]
M : number of columns
N^i : number of stages of column i
P_i : amount of product i per batch [mol] (= holdup of vessel i)
R_i : total amount of component i on the trays [mol]
z^i_j : composition of the jth component in product i
z^F_j : composition of the jth component in raw material

The number of trays for each column, N^i, the holdup on each tray, H^i_j, and the boilup rate, V, are regarded as design variables. Thus, when the design variables mentioned above, and the amount of raw material per batch, F, the composition of raw material $\{z^F_i\}$, and the composition of the dominant component of each product i, z^i_i, are given, there exists (M-1) degrees of freedom.

If the compositions of the light impurity in products 2 through M, (z^i_{i-1}, i = 2, 3, ..., M), are selected as independent variables, the amount of product i per batch and the distillation period per batch are determined by the following steps:

i) Assign proper values to z^i_{i-1}, (i = 2, 3, ..., M).
ii) Calculate P_i (i = 1, 2, ..., M+1) from eqs. (1) through (5).
iii) Execute the dynamic simulation and decide the total batch time. The simulation is terminated when the compositions of all the vessels satisfy the specifications for the final products.

The dynamic simulation must be executed in order to obtain the distillation period even if all of the design and operation variables mentioned above are given. This means that some numerical search method is needed to find the optimal values of $\{z^i_{i-1}\}$ which minimize the distillation period.

4. COMPARISON OF THE SEPARATION PERFORMANCE OF MEBAD AND CONTINUOUS SYSTEM

The MEBAD proposed here can be regarded as a multiple-heat-integrated system in which the condenser of one column is heat-integrated with the reboiler of the next column. Therefore, the energy required to separate multiple components is markedly less than that of an ordinary batch distillation system. In this section, the energy consumption of the MEBAD is compared with that of an ordinary continuous distillation system.

In a continuous distillation system, there are (2M)!/[M!(M+1)!] different structures for separating (M+1) components. In this study, we adopt the most commonly used structure called Direct Sequence, shown in Fig. 5 as the structure for the continuous system.

In the MEBAD, heat is supplied only to the reboiler of the first column. Therefore, qualitatively speaking, the energy consumption per unit amount of product is reduced according to the increase in the number of products which are separated simultaneously. From this viewpoint, the effect of the number of products is examined.

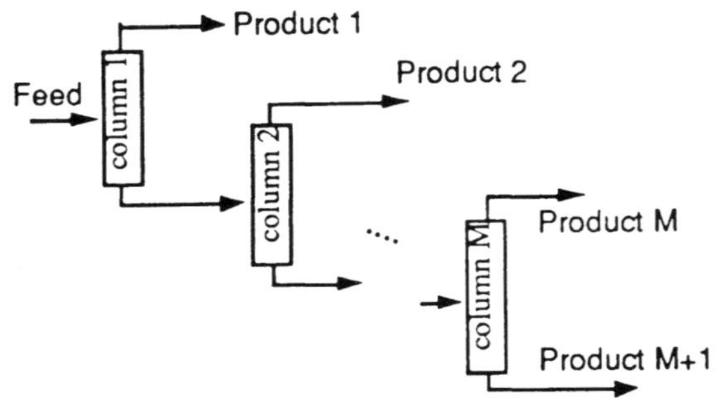

Fig. 5 Continuous distillation system with Direct Sequence

The input data defining the problem is given in Table 1. By changing the number of products from two to five, the separation performances of both systems are compared. In this study, to define separation performance, the following values which show the amount of products per unit boilup rate are used:

<For continuous system>

$$\text{P.I.} = \frac{\text{(Amount of products per unit time)}}{\text{(Sum of boilup rates of all columns)}} \quad (6)$$

<For MEBAD>

$$\text{P.I.} = \frac{\text{(Amount of products per batch)}}{\text{(Boilup rate} \times \text{Total batch time)}} \quad (7)$$

The total batch time in eq. (7) does not include the filling and discharging times of materials to and from the vessels.

In order to make a fair comparison, the same total number of stages are used. Firstly, the minimum reflux ratio of each column is derived for continuous systems using Underwood equations, and the boilup rate of each continuous column is calculated assuming that the reflux ratio is 1.2 times larger than the minimum reflux ratio. Afterwards, the number of stages for each column is determined so as to satisfy the product specification using a steady state simulation system. The result is summarized in Table 2.

Table 1 Input data defining the problems

Number of products (M+1) : 2 through 5
Feed composition : $z^F_i = 1/(M+1)$ $(i = 1,2,...,M+1)$
Relative volatility : $\alpha_{i+1} / \alpha_i = 3.0$ $(i = 1,2,...,M)$
Specification of products: $z^i_i = 0.98$ $(i = 1,2,...,M+1)$
For the MEBAD:
Amount of feed per batch : 50 x (M+1) [kmol]
Holdup on each tray : 50 [mol]

Table 2 Number of trays for continuous system

Number of products	2	3	4	5
column 1	15	16	17	17
column 2	-	17	18	18
column 3	-	-	17	18
column 4	--	--	--	17

(Reflux ratio = minimum reflux ratio x 1.2)

To calculate the performance of the MEBAD, it is assumed that every column has the same number of stages. The total batch time is minimized for each case by optimizing the compositions of light impurities of products 2 through M. The result is summarized in Fig. 6.

According to the increase in the number of products, P.I., calculated by eq. (7) for the MEBAD, increased while P.I., calculated by eq. (6) for the continuous system, decreased. From the view point of energy consumption, batch distillation has been considered to be an ineffective separation method compared with continuous distillation. However, the result in Fig. 6

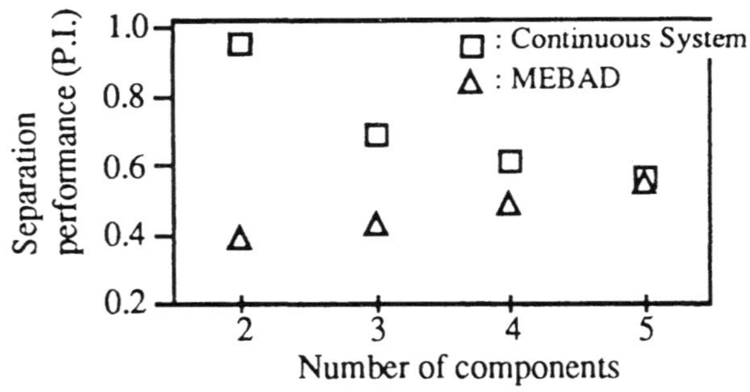

Fig. 6 Comparison of separation performances of MEBAD and continuous system

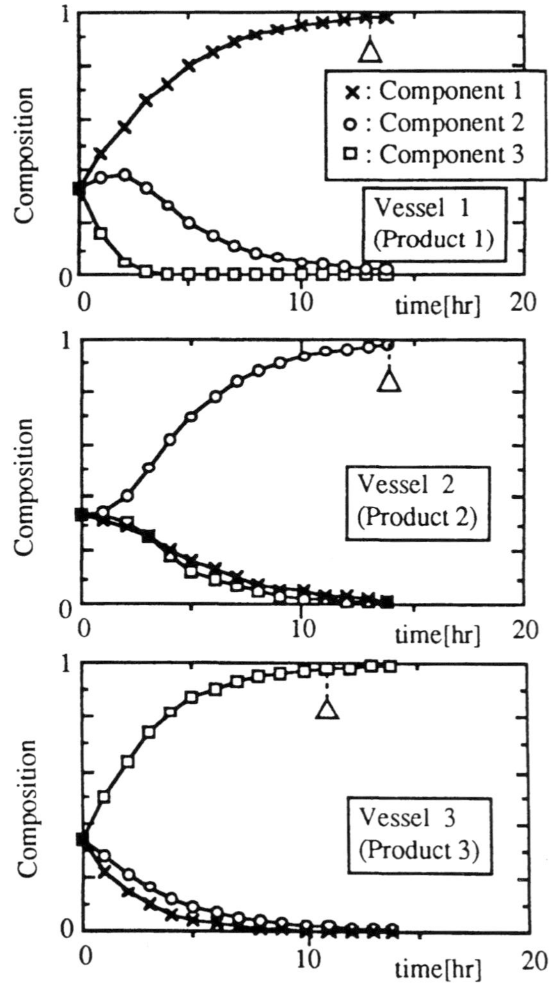

Fig. 7 Composition profiles for MEBAD (Number of products = 3)

shows that the MEBAD has the potential to have a better separation performance than the continuous distillation system if the number of products is increased.

Figure 7 illustrates the changes in the liquid composition, in the vessels, for the three components separated by the MEBAD. Each triangle in the figure illustrates the earliest time in which the product specification (0.98 for the dominant component) is satisfied. In this figure, because product 2 has light and heavy impurities, both of which must be removed from the product, product 2 satisfies its specification last.

5. CONTROL OF MEBAD

In the MEBAD, the holdup of each vessel is calculated in advance by taking into account the amount of feed, feed composition and product specifications. After feeding the prespecified amount of raw material to the vessels, total reflux operation is carried out until the compositions in all vessels satisfy their specifications. During the total reflux operation, by using a level control system, only the holdup of each vessel is kept constant. Therefore, the MEBAD has a much simpler control structure than that of the conventional or complex batch distillation column.

However, the operation policy of keeping the holdup of each vessel constant may not be optimal from the viewpoint of minimizing the total batch time, when the holdups of the vessels are regarded as time-variant variables. The problem of minimizing the total batch time becomes a very interesting optimal control problem. Nevertheless, the problem is beyond the framework of the present paper. We plan to deal with this subject in a future paper.

When the feed composition or the amount of feed is not measured correctly, the products may not satisfy their specifications even if the holdup of each vessel is kept constant at the value calculated by eqs. (1) through (5) for a long period. This is because the MEBAD does not have a feedback control loop to compensate for unexpected errors.

Let us suppose a case where the actual feed composition is given by eq. (8) but the holdup of each vessel is calculated assuming that the concentration of every component is the same:

$$z^F_1 = 0.24,\ z^F_2 = 0.19,\ z^F_3 = 0.18,\ z^F_4 = 0.21,\ z^F_5 = 0.18 \qquad (8)$$

The result of the simulation is shown in Fig. 8. Each line in the figure shows the composition change of the dominant component in the product. It is clear from this figure that products 2 through 5 cannot satisfy their specifications ($z^i_i = 0.98$) even if the total reflux operation is continued for a long time.

In the MEBAD, because the total reflux operation is always carried out, the reflux ratio of each column cannot be used as the manipulated variable to control the composition of the product. Therefore, the holdup of each vessel is selected as a manipulated variable.

From studying the simulations, it becomes clear that an increase in the vessel holdup decreases the final composition of the dominant component. By taking this fact into account, in this research, a single loop cascade control system, shown in Fig. 9, is adopted in order to control the composition of each vessel. If the composition of the dominant component in product i (x_i) does not satisfy its specification (z^i_i) at a certain time, the set point of the holdup of the vessel i (H^{i*}) is reduced.

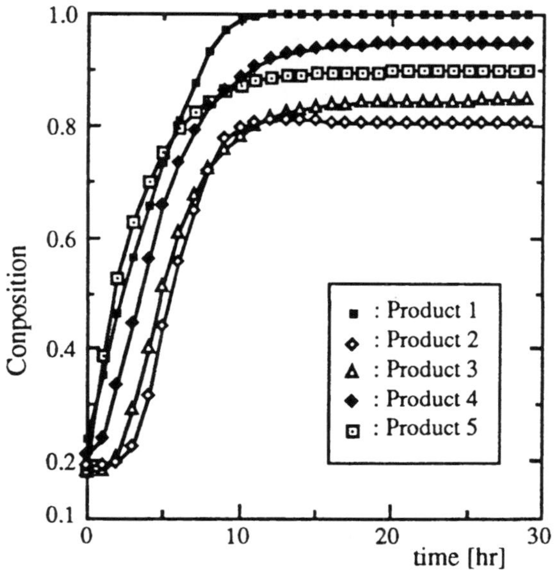

Fig. 8 Composition profiles for the case where the measured data have some errors

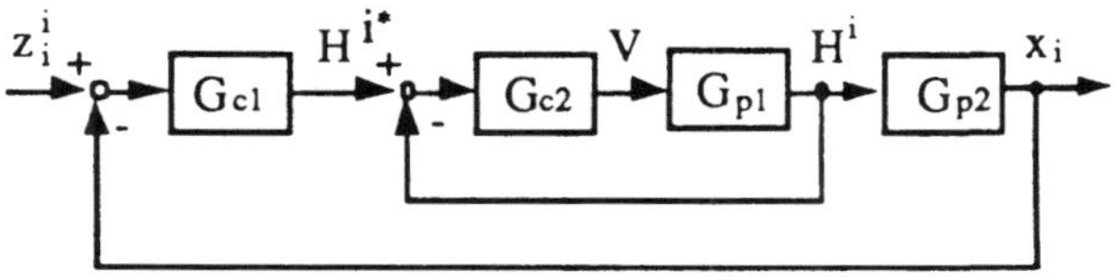

Fig. 9 Cascade control system

We cannot change the holdup of every vessel independently, because the sum of the holdups of vessels is constant. Thus, the number of manipulated variables (holdup of the vessel) is less than the number of controlled variables (composition of the dominant component in the vessel) by one. The remaining controlled variable is used to decide the ending moment of the distillation. That means the operation is terminated when the composition of the remaining controlled variable satisfies the product specification.

When the cascade control scheme shown in Fig. 9 is adopted, the following conditions must be decided to operate the system:

i) Selection of the controlled variables and the pairing between the manipulated variables and controlled variables,
ii) The condition of starting the cascade control,
iii) Structure of the controllers, G_{c1} and G_{c2},
iv) Parameters in G_{c1} and G_{c2}.

When the separation problem is given, these must be decided so as to optimize a certain criteria. However, since the optimization of these conditions is not the subject of this paper and with little space with which to focus on separation performances, a thorough discussion of optimal conditions will have to be taken up in a later paper. We only show that the system can be controlled by deciding these conditions properly.

As an example, the separation problem consisting of five components, explained in the earlier part of this section, is used. In this example, the compositions of products 1, 2, 4, 5 are controlled by using the holdups of the vessels. In order to control the holdup of each vessel, the recycle flowrate from the next column is used at vessels 1 and 2, and the recycle flowrate to the previous column is used at vessels 4 and 5.

The compositions of all vessels are the same as the feed composition at the beginning of the operation. Therefore, executing composition control from the beginning would cause the holdups in the vessels to fluctuate extremely. In this study, cascade control is started when the product composition satisfies one of following conditions:

a) $x_i > z^i{}_i$

b) $dx_i/dt < 0.01$ [1/hr] and $d^2x_i/dt^2 < 0$

where

x_i : composition of component i in product i.

At the beginning of the operation, the composition in the vessel is changed slowly. The second constraint in condition b) is introduced so as not to start the cascade control at the beginning of the operation. The PI controllers are used for both G_{c1} and G_{c2}, and the parameters in the controllers are selected on a trial and error basis.

The result is shown in Figs. 10 and 11. Each line in Fig. 10 shows the composition change of the dominant component in the product, x_i. The holdup profiles are shown in Fig. 11. It is clear from these figures that the composition control system works properly.

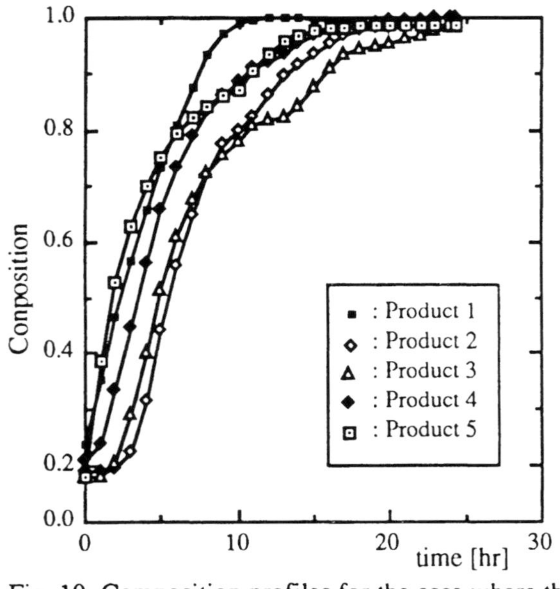

Fig. 10 Composition profiles for the case where the composition control is executed

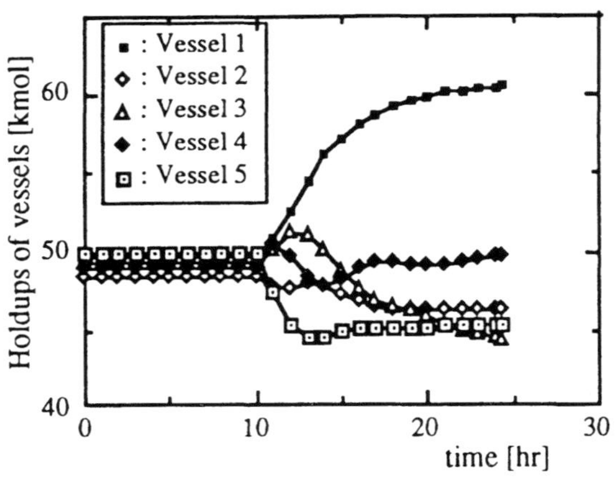

Fig. 11 Holdup profiles

6. CONCLUSION

A new type of separation system called "multi-effect batch distillation system (MEBAD)" is proposed, and the separation performance of this system is compared with that of an ordinary continuous distillation system. Through simulation, it is found that the MEBAD has the potential to have a better separation performance than the continuous distillation system if the number of products is increased. It is also shown that the MEBAD is very easy to control because, even if many types of disturbances occur, only the holdups of the vessels require controlling. By taking these facts into account, we can conclude that the MEBAD has wide applicability as a flexible and energy-efficient separation system.

ACKNOWLEDGMENT

This research was supported by the Grant-in-Aid for Scientific Research from the Japanese Ministry of Education, Science and Culture. We also greatly appreciate the assistance of Mr. Okuyama of our laboratory for computations and figures.

REFERENCES

Hasebe, S., B. B. Abdul Aziz, I. Hashimoto and T. Watanabe, (1992). "Optimal Design and Operation of Complex Batch Distillation Column", *Preprint of the IFAC Workshop on Interaction between Process Design and Process Control*, London, Sept. 7-8, pp. 177-182.

Meski, G. A., M. Morari and A. G. Davidyan, (1993). "Batch Distillation in a Column with a Middle Vessel", *AIChE Annual meeting*, Paper 152a, St. Louis, Nov. 7-12.

Mujtaba, I. M. and S. Macchietto, (1994). "Optimal Operation of Multicomponent Batch Distillation - A Comparative Study using Conventional and Unconventional Columns", *Preprint of ADCHEM'94*, Kyoto, May 25-27, pp. 415-420.

ANALYSIS AND DESIGN OF CASCADE CONTROL SCHEMES FOR DISTILLATION

G. Lorenzi, D. Semino*, C. Scali, A. Brambilla

* *Dipartimento di Ingegneria Chimica, Università degli Studi di Pisa, Via Diotisalvi 2, 56126 Pisa, ITALY. E-mail: semino@sssup1.sssup.it*

Abstract. Different issues related to the introduction of a cascade loop in dual composition control of distillation columns are addressed. In particular it is pointed out how static and dynamic interactions can be reduced by the introduction of the cascade and a measure of this reduction is defined. Two other disturbance-specific parameters are defined in order to address the choice of the loop to which cascade is added and of the pilot tray. Applications to different columns show that the use of the defined parameters can guide advantageously in the analysis and design of cascade control for distillation.

Key Words. Distillation columns, Cascade control, Interaction analysis, Disturbance rejection, Disturbance parameters.

1. INTRODUCTION

The most obvious reason why it is common use to add cascade control, when secondary measurements which are quickly affected by the system disturbances are available, is the possibility of obtaining fast control of primary outputs of slow dynamics. It is not as obvious that, when cascade control is added to one of the loops in a multivariable control scheme, the secondary loop deeply affects the characteristics of the MIMO control problem, particularly as far as the interactions are concerned.

In the paper the problem of composition control in multicomponent distillation is addressed: concentrations of the two key components in the top and bottom products are controlled. Two diagonal controllers with standard algorithms are used and cascade control in one of the two loops is adopted. Decentralized controllers are more robust with respect to multivariable ones, but usually provide inferior performance as they are affected by interactions between the two loops and cannot compensate changes in the less favourable directions of the plant. Therefore the effect of the cascade loop on interactions is a primary concern to understand the benefits of its use. Moreover, as the main interest in industrial distillation columns is about the suppression of disturbances which affect the column through the feed characteristics, the direction of disturbances is what really matters in limiting achievable performance.

Three parameters are introduced for a complete evaluation of benefits of the cascade control and to assist in the different selections about the complete design of the control system. The first parameter, closely related to the Relative Gain of the modified process, evaluates how the closure of the cascade loop reduces the interaction between loops. The second one evaluates how cascade on a particular tray modifies the open-loop gain in the presence of a disturbance and then addresses the choice of the pilot tray. The third parameter evaluates how the offset on a composition is reduced by means of the control of the other composition and then indicates which composition (the *hard* or the *soft* target) has to be associated to the cascade control loop.

The three parameters are computed by referring to steady state conditions; their indications are validated by means of dynamic simulation for two different multicomponent distillation units.

2. INTERACTIONS IN MULTIVARIABLE SYSTEMS WITH CASCADE CONTROL

Fig. 1 shows a 2×2 system in which a series cascade is added to the first loop. When the cascade loop is open, the system transfer function matrix is simply:

$$\mathbf{G} = \begin{bmatrix} g_{11} = g_i g_{11}^* & g_{12} = g_{12}^* g_{11}^* \\ g_{21} & g_{22} \end{bmatrix} \tag{1}$$

and the value of its relative gain:

$$\lambda_G = \frac{g_{11} g_{22}}{g_{11} g_{22} - g_{12} g_{21}}. \tag{2}$$

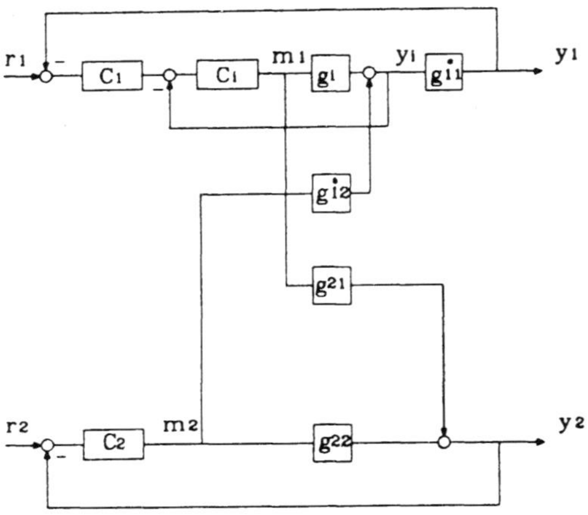

Fig. 1. Series cascade in a 2 × 2 system

Closing the cascade loop changes the transfer function matrix as seen by the primary controllers into:

$$\mathbf{G}' = \frac{1}{1+g_i c_i}\begin{bmatrix} g_{11}g_i & g_{12} \\ g_{21}g_i & g_{22}(1+\frac{g_i c_i}{\lambda_G}) \end{bmatrix} \quad (3)$$

whose relative gain becomes:

$$\lambda_{G'} = 1 + (\lambda_G - 1)\frac{1}{1+g_i c_i}. \quad (4)$$

It is clear that at least at steady-state there is a reduction in the interaction after cascade is added; in particular, if the internal controller has integral action, $\lambda_{G'}(0) = 1$ and interactions between the two loops are eliminated at steady-state.

As far as the value of the interaction parameter at frequency $\omega > 0$ is concerned, one can rewrite equation (4) as:

$$\lambda_{G'} = \frac{g_i c_i}{1+g_i c_i} + \lambda_G \frac{1}{1+g_i c_i} \quad (5)$$

which shows that at frequency at which the complementary sensitivity function is close to one and the sensitivity function is small, the value of the magnitude of the dynamic relative gain stays close to one. Given that the internal controller can be quite aggressive as the result of the fast dynamics, such results are valid for a wide range of frequencies.

The situation in the case of parallel cascade control (Fig. 2) is slightly more intricate. Without cascade control, (1) and (2) still hold. When the cascade loop is closed, the system as seen by the primary controllers becomes:

$$\mathbf{G}'' = \frac{1}{1+g_i c_i} \cdot \begin{bmatrix} g_{11}g_i & g_{12} + c_i(g_{12}g_{i1} - g_{11}g_{i2}) \\ g_{21}g_i & g_{22} + c_i(g_{22}g_{i1} - g_{21}g_{i2}) \end{bmatrix} \quad (6)$$

and the relative gain:

$$\lambda_{G''} = 1 + \frac{\lambda_G - 1}{1+g_{i1}c_i}\left(1 + c_i(g_{i1} - \frac{g_{11}g_{i2}}{g_{12}})\right) \quad (7)$$

whose steady-state value, when integral action in the internal controller is present, is:

$$\lambda_{G''}(0) = 1 + (\lambda_G(0) - 1)\frac{k_{12}k_{i1} - k_{11}k_{i2}}{k_{12}k_{i1}}. \quad (8)$$

It appears that when $k_{12} = k_{11}k_{i2}/k_{i1}$, the series cascade case is reproduced and no steady-state interaction between the two primary loops are present after a secondary controller with integral action has been added. If moreover $g_{12} \approx g_{11}g_{i2}/g_{i1}$ also at frequency $\omega > 0$, the results of the series cascade case for the dynamic RGA extend to this case as well.

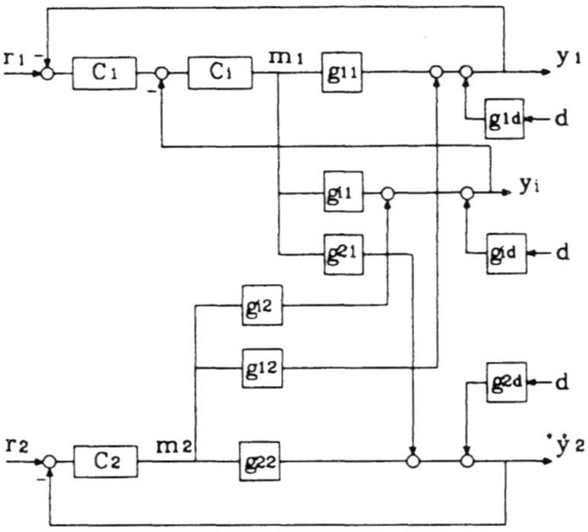

Fig. 2. Parallel cascade in a 2 × 2 system

As far as temperature cascade control in distillation columns is concerned, the parallel cascade control scheme is surely appropriate to describe the statics and the dynamics of the system. According to the values of the system gains, which depend on the choice of the position of the cascade loop and of the pilot tray, interactions between the primary loops can be smoothed or amplified by the introduction of the cascade. The value of the parameter $\lambda_{G''}$ (or equivalently of the terms on which it depends) is therefore significant in order to have a measure of interactions. An appropriate parameter in this respect will be introduced in the next section.

3. PARAMETERS FOR THE SELECTION OF THE CASCADE

When one is interested in a multivariable control problem in which a number of secondary measurements are available, it may not be obvious which are the best choices as far as the introduction of cascade control is concerned. This is typically the case of the control of compositions in distillation columns in which the temperatures on the column trays are available as secondary measurements. It may be also the case of tubular reactors in which temperatures at different spatial positions may be used to infer and control product properties. The more realistic case of parallel cascade is used throughout this section. It is important to point out that all the parameters that will be introduced refer to steady-state conditions.

The effect of cascade control on interactions comes clear from the analysis of the previous section. It may be useful in this respect to introduce the following parameter:

$$\gamma_{y_i} = \left[\frac{(\frac{\partial y_1}{\partial m_2})_{m_1} - (\frac{\partial y_1}{\partial m_2})_{y_i}}{(\frac{\partial y_1}{\partial m_2})_{m_1}} \right] = \frac{k_{11}k_{i2}}{k_{12}k_{i1}} \tag{9}$$

which expresses the fraction of the effect of the manipulated variable m_2 on the output variable y_1, which has been recovered by returning the internal variable y_i at its steady-state value as the result of closing the cascade loop. The value of γ_{y_i} gives therefore the following information:

- $\gamma_{y_i} = 1$ means that after closing the cascade loop there has been a full recover so that the steady-state effect of m_2 on y_1 has been eliminated
- $\gamma_{y_i} = 0$ means that no effect on the interaction is accomplished by the introduction of the cascade loop
- $0 < \gamma_{y_i} < 1$ and $\gamma_{y_i} > 1$ means that a partial or an excessive recover respectively has been obtained as the result of the cascade control
- $\gamma_{y_i} < 0$ corresponds to a higher interaction in the presence of cascade.

Therefore the parameter γ_{y_i} can be used to evaluate to what extent the introduction of the cascade reduces interactions (in particular $\gamma_{y_i} = 1$ corresponds to $\lambda_{G''}(0) = 1$).

Up to this point no attention has been paid on the particular disturbances affecting the system. In the general case the effects of such disturbances on the primary and internal outputs take place through parallel transfer functions as in Fig. 3. It may be appropriate therefore to introduce a parameter (Brambilla *et al.*, 1994) which accounts for the effect on the primary output of the closure of the cascade loop in the presence of the disturbance. The parameter is defined as follows:

$$\gamma_{y_1 y_i} = \left[\frac{(\frac{\partial y_1}{\partial d})_{m_i} - (\frac{\partial y_1}{\partial d})_{y_i, m_2}}{(\frac{\partial y_1}{\partial d})_{m_i}} \right] = \frac{k_{11}k_{id}}{k_{1d}k_{i1}}. \tag{10}$$

It represents the fraction of the effect of the disturbance on the primary output which is recovered by closing the internal loop. The significance of different values of the parameter $\gamma_{y_1 y_i}$ is completely parallel to the one of the parameter γ_{y_i} and is not repeated here for the sake of brevity. Therefore the parameter $\gamma_{y_1 y_i}$ gives information about the appropriateness of the choice of the secondary measurement and it can guide the designer in this choice.

The parameters introduced so far concentrate on the beneficial effect of the introduction of cascade on the output variable y_1. Had one chosen a control scheme and a tray for which both $\gamma_{y_i} = 1$ and $\gamma_{y_1 y_i} = 1$, the introduction of the cascade would lead y_1 back at its steady-state without possible interaction of the other primary control loop. Only one-way interactions have therefore been taken into account. If the cascade loop has a much faster dynamics as expected, one would be interested in having a beneficial effect from the interactions of manipulated variable m_1 on the primary output y_2. In particular it would be appropriate if in the presence of a disturbance, when the output y_1 is lead back to its steady-state (which is the primary objective of the cascade control), also output y_2 is again at its desired value. This is expressed by a third parameter (Brambilla *et al.*, 1994):

$$\gamma_{y_1 d} = \left[\frac{(\frac{\partial y_2}{\partial d})_{m_i} - (\frac{\partial y_2}{\partial d})_{y_1, m_2}}{(\frac{\partial y_2}{\partial d})_{m_i}} \right] = \frac{k_{21}k_{1d}}{k_{11}k_{2d}} \tag{11}$$

which expresses the fraction of the effect of the disturbance on the output variable y_2 which has been recovered as the result of perfect control on output variable y_1. The significance of different values of the parameter is again analogous to the significance of the previous ones. The parameter $\gamma_{y_1 d}$ gives therefore information about the appropriateness of the choice of the control scheme and in particular of the position of the cascade loop.

It is clear that the second and the third parameters are close relatives of other parameters previously introduced in the literature in the analysis of cascade control for distillation (e.g. the relative disturbance gain in Stanley *et al.*, 1985, the open-loop disturbance gain with other outputs controlled in Häggblom and Waller, 1989, the closed-loop disturbance gain in Skogestad and Hovd, 1990); simple relationships can be found to correlate the proposed parameters to those in the literature. The introduction of the parameters in the present form gives however a general framework in which uniform parameters of analogous meaning and similar interpretation are used. A possible drawback of the use of the last parameter for comparison of different structures and different disturbances comes from the fact that a normalization has been accomplished with respect to the open-loop gain, which may happen to be very different for different structure. If one wants to compare quantitatively the relative closed-loop variations, the absolute value of the difference $(\gamma_{y_1d}-1)$ has to be weighted with weight $\omega = k_{2d}\Delta d/y_2^{s-p}$, which accounts for the open-loop gain, the disturbance amplitude and normalizes the value with respect to the set-point.

It is clear from the present analysis that a value of all three parameters close to one would be desired. This would mean that the control of the internal output would imply perfect steady-state control of both the related primary output y_1 and of the other primary output y_2. Moreover occasional remaining offsets could be quickly controlled by the external controller since almost no detuning is necessary as the result of no one-way interactions.

A comparative analysis of the parameters can give relevant information of the system behaviour also in less favourable cases. Had one chosen a configuration for which $\gamma_{y_i} \approx 1$, if an unexpected disturbance with unfavourable values of the other parameters affects the system, the cascade would maintain a beneficial effect on the system. Despite of the need of relevant variations of the temperature steady-state value, the dynamics would anyway be greatly improved by the lack of one-way interactions. Even more relevant, the use of the proposed parameters can make the control designer reconsider control schemes that the use of usual techniquess (RGA, SVD) discourage. All these considerations will come clear from the applications to cascade control of distillation columns in the next section.

4. APPLICATIONS TO DISTILLATION COLUMNS

The use of the parameters which have been introduced in the previous section for the choice of the control scheme in distillation columns is addressed in this section. The control schemes which are taken into account are the $L-V$ scheme and the $D-V$ scheme with cascade control either on the top composition or on the bottom one. The choices to accomplish are therefore the following:

- the control scheme
- the coupling between controlled and manipulated variables
- the position of the cascade loop
- the position of the pilot tray

The three parameters γ_{y_i}, $\gamma_{y_1y_i}$, γ_{y_1d} will guide the choices.

As far as the tuning of the composition controllers is concerned, the technique by Loh *et al.* (1993), which is based on sequential loop closing and relay tuning, is chosen to find the parameters of PI controllers. It is particularly appropriate for a properly chosen cascade scheme, since after the inner loop is closed, one-way interactions are eliminated so that sequential loop closing does not need any iteration. In all other cases, some iterations may be needed to achieve convergence in the controller parameters. A preliminary tuning of the PI controller in the inner loop is accomplished via an IMC-like tuning technique.

4.1. *GPL Splitter*

The characteristics and operating conditions of the GPL splitter are reported in Tab. 1 together with the compositions and temperatures of feed, distillate and bottom product. Propane and isobutane are the key components in the products.

Table 1 Characteristics of the GPL splitter

ideal trays	27		
feed tray	17		
pressure	$15 atm$		
L $(kmol/hr)$	233.1		
V $(kmol/hr)$	199.8		
	F	D	B
T (°C)	78	35.4	93.1
$C2$ $(kmol/hr)$	2	2	≈ 0
$C3$	25.07	24.61	0.46
$i-C4$	19.9	0.13	19.77
$n-C4$	25.6	10^{-2}	25.59
$i-C5$	0.14	≈ 0	0.14
$n-C5$	$2.5 \cdot 10^{-2}$	≈ 0	$2.5 \cdot 10^{-2}$
$C6+$	$3.6 \cdot 10^{-2}$	≈ 0	$3.6 \cdot 10^{-2}$
total	72.77	26.75	46.02

The gains and the dynamics of the column have been determined using software, which has been developed to this purpose. The system transfer function for the $L-V$ scheme is the following:

$$\mathbf{G} = \begin{bmatrix} \frac{-0.0067}{105.8s+1} & \frac{0.00485}{118.25s+1} \\ \frac{0.00729}{94.8s+1} & \frac{-0.00447}{90.8s+1} \end{bmatrix}. \qquad (12)$$

A multiplicative delay of 15 min is added to all transfer functions due to the composition analysers.

A simple RGA analysis shows that $\lambda_{11} = -5.54$ and suggests either to switch the couplings between manipulated and output variable (i.e. to control top composition with boil-up flowrate and bottom composition with reflux flowrate) or to use the $D-V$ control scheme. Analysis of the parameter γ_{y_i} shows that, if cascade is added and the pilot tray is chosen in the section where cascade is present, the $L-V$ scheme can be used, as its RGA gets very close to one. The user could be concerned about integrity but instability would occur only if the very robust temperature loop failed with the other two working, which is really unfrequent.

The other two parameters are analysed thereafter for four typical disturbances:

- feed temperature
- feed flowrate
- ratio $C3/i-C4$ in the feed composition
- ratio $C3/n-C4$ in the feed composition.

While the analysis of $\gamma_{y_1 y_i}$ shows that the column is not very sensitive to the choice of the pilot tray (values close to one are obtained in almost all cases), the parameter $\gamma_{y_1 d}$ changes largely for different disturbances in the different control schemes. It comes out that the best compromise is achieved is achieved if the $L-V$ scheme with the usual couplings and cascade on the bottom loop is chosen ($\gamma_{y_1 d} = 0.92, 0.97, 0.87, 0.89$ for the four disturbances). As the pilot tray parameters give good results for all trays, a sensitive tray (tray 23, $\gamma_{y_1 y_i} = 1.01, 1.07, 1.02, 0.99$ for the four disturbances) in the same section as the cascade is chosen. This corresponds to an RGA for the new scheme $\lambda_{G''}(0) = 1.05$. The comparison of the performances of the control scheme without cascade and with the cascade chosen according to the proposed parameters in the case of a 8% disturbance on the ratio $C3/i-C4$ in the feed composition, shows the very beneficial effects of the designed cascade (Fig. 3). Also an inverse-based controller (steady-state decoupler+PID) manages to eliminate interactions and guarantees a fast return to steady-state for both variables, but it has a larger overshoot as it does not make use of the fast temperature measurement.

A large number of simulations for all disturbances have shown that the proposed scheme, which would have been discarded by a preliminary RGA analysis, outperforms all other control schemes with cascade. Moreover, given that it eliminates one-way interactions without performing an inversion of the process, it is also more robust to uncertainties than inverse-based controllers. Fig. 4 shows for the same disturbance as before a comparison of the proposed control scheme with the above-mentioned inverse-based controller in the case of multiplicative input uncertainty of 10%. As expected, the scheme with the full matrix controller shows oscillations, which do not appear in the cascade control scheme.

4.2. *Gasoline Stabilizer*

A second column is examined in order to show that a comparative analysis of the proposed parameters can give useful information about the dynamic behaviour of the control scheme also when some of the parameters do not have the desired value.

The column is a gasoline stabilizer which separates light components ($C3$, $i-C4$ and $n-C4$) from a very spread mixture of heavy components. All details of the column are available in Lorenzi (1994) and are not reported here for the sake of brevity. The main disturbance acting on the column is a variation in the ratio $n-C4/i-C5$ in the feed composition. The value of the RGA for the $L-V$ scheme without cascade is 5.78. When cascade is added to the top composition loop, the values of γ_{y_i} if tray 2 or 17 is chosen are 1.16 and 1.07 respectively. The value of the parameters $\gamma_{y_1 y_i}$ for these two trays and the chosen disturbance are -0.39 and 0.96 and $\gamma_{y_1 d}$ is 1.9.

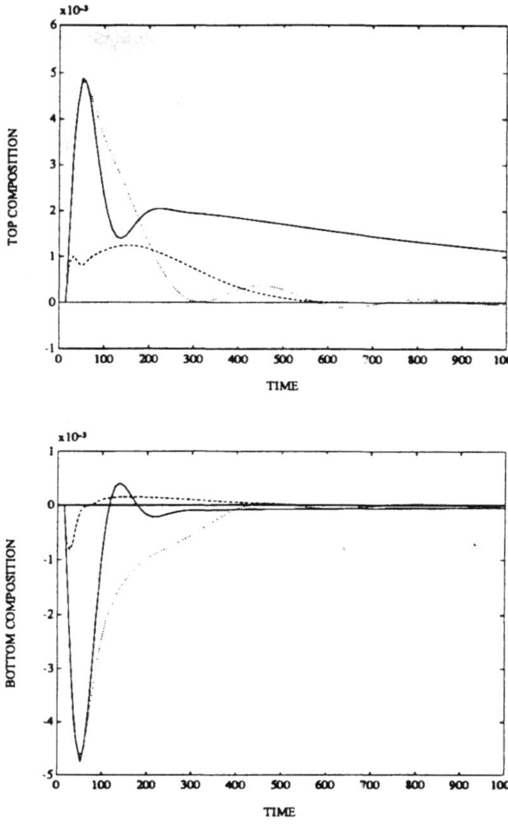

Fig. 3. Top and bottom dynamics for a disturbance in the feed composition. – controller without cascade, - - cascade controller, ··· inverse-based controller

The performances of the system without cascade and with cascade on tray 2 or 17 are compared (Fig. 5) for a 9% disturbance. Without cascade a strong detuning of the controllers is necessary for the delays in the composition analysers and the system interactions; therefore there is a large overshoot and a slow return to steady-state. Good performances are obtained if a correct tray choice is accomplished (tray 17); fast control of the inner loop leads back the top composition to steady-state and the work of the external composition loops is mainly to recover the remaining offset on the bottom composition ($\gamma_{y_1 d} = 1.9$). If tray 2 is chosen, one would expect bad performances, since

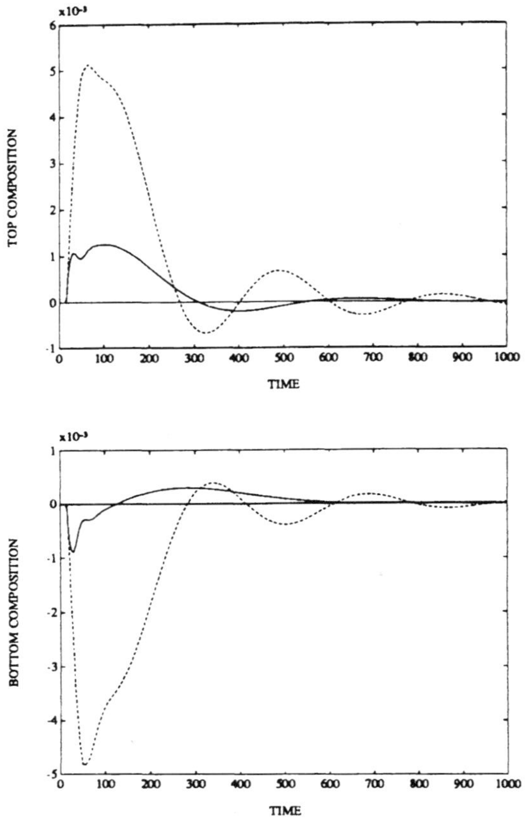

Fig. 4. Top and bottom dynamics for a disturbance in the feed composition in the presence of uncertainty. – cascade controller, - - inverse-based controller

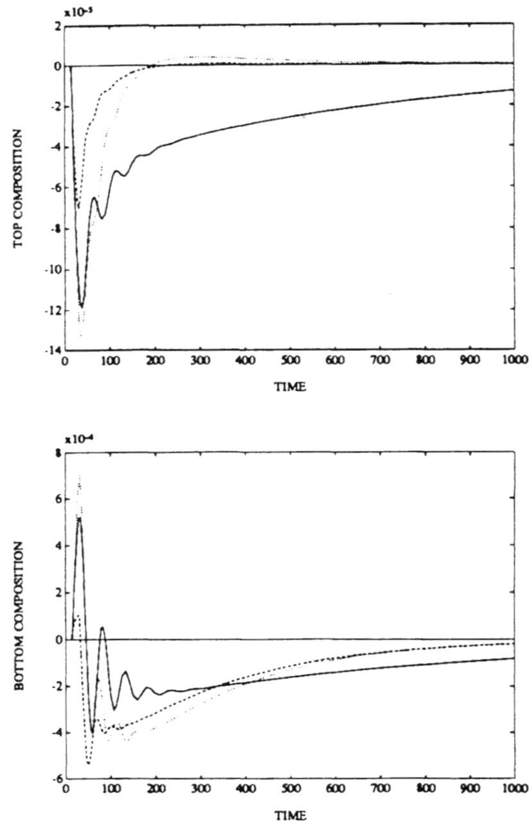

Fig. 5. Top and bottom dynamics for a disturbance in the feed composition. – controller without cascade, - - cascade controller (tray 17), ··· cascade controller (tray 2)

the action of the inner loop amplifies the effect of the disturbance ($\gamma_{y_1 y_i} < 0$). However, the addition of the cascade despite of the unfavourable choice of the pilot tray, decreases the interactions ($\gamma_{y_i} = 1.16$). This is why, if one compares the response to the one of the scheme without cascade, one observes a larger overshoot but a faster return to steady-state as the result of a strong and effective action of the outer loops. All this agrees with what the designer would predict from the analysis of the value of the parameters.

5. CONCLUSION

Three steady-state parameters have been introduced to guide the control engineer in the design of cascade schemes for distillation columns (or generally for multiloop processes). The parameters refer specifically to the interactions after cascade is added, the effect of the addition of cascade to one loop on the opposite one in the presence of a disturbance and the effect of the choice of the secondary measurement. The use of these parameters shows that misleading suggestions, which can be given by classical tools or qualitative analysis, can be avoided and advantageous choices can be made. The application to industrial distillation columns and the analysis of the simulation results shows that the indications given by the parameters are well confirmed by the column dynamics.

6. REFERENCES

Brambilla A., D. Semino, and C. Scali (1994). Design and Control Selections for Cascade Control in Distillation, *Proceedings of IPDC'94*, Baltimore, Maryland, USA, 171-176.

Häggblom K. E., and K. V. Waller (1989). Predicting Properties of Distillation Control Stuctures, *Proc. ACC*, 114-119, Pittsburgh (PA), USA.

Loh A. P., C. C. Hang, C. K. Quek, and V. U. Vasnani (1993). Autotuning of Multiloop Proportional Integral Controllers Using Relay Feedback. *Ind. Eng. Chem. Res.*, **32**, 1102-1107.

Lorenzi G. (1994). Tesi di Laurea, University of Pisa, Italy.

Skogestad S., and M. Hovd (1990). Use of Frequency-Dependent RGA for Control Structure Selections, *Proc. ACC*, 2133-2139, San Diego (CA), USA.

Stanley G., M. Marino-Galarraga, and T. J. MacAvoy (1985). Shortcut Operability Analysis. 1. The Relative Disturbance Gain. *Ind. Engng. Chem. Res.*, **24**, 1181-1188

Evaluation of Dynamic Models of Distillation Columns with Emphasis on the Initial Response

Bernd Wittgens and Sigurd Skogestad*
University of Trondheim, NTH, Laboratory of Chemical Engineering,
7034 Trondheim, Norway

**Author to whom correspondence should be addressed, E-mail: skoge@kjemi.unit.no,*
phone: +47 – 7359 – 4154, fax: +47 – 7359 – 4080

Abstract. The flow dynamics (tray hydraulics) are of key importance for the initial dynamic response of distillation columns. The most important parameters are the liquid holdup, the liquid hydraulic time constant, and the constant representing the effect of a change in vapor flow on liquid flow. In the paper we present methods for determining these parameters experimentally, and compare the results with estimates from available correlations such as the Francis Weir formula.

Key Words. Distillation dynamics; Tray hydraulics

1. Introduction

Simplified models which simplify or neglect flow dynamics (tray hydraulics) and energy balance are often used for studies of distillation column dynamics and control. However, the applicability of such simple models for this purpose is often questioned by practicioners. This critique is indeed reasonable as one knows that the tray hydraulics are crucial in determining the initial dynamic response.

In the paper we consider detailed models of the tray hydraulics and use these to derive expressions for parameters that characterize the flow response. The most important parameters are the liquid holdup M_l, the hydraulic time constant τ_l, the parameter λ (denoted K_2 by many authors) for the initial effect of a change in vapor flow on liquid flow, the fraction of vapor on the tray and the pressure drop. These key parameters are also determined from experiments on our lab-scale column. The final goal of these studies is to see how detailed a dynamic model of a distillation column should be in order to be used for control purposes.

The literature on distillation dynamics is extensive and we do not have space to review it here. In terms of experimental responses for tray columns we only mention the work of Baber et al. (1961, 1962).

2. Tray Modeling

2.1. *Rigorous Tray Model*

In the rigorous model used in the SPEEDUP simulations each stage is divided into two liquid holdups (tray and downcomer) and one vapor holdup. Separated mass and energy balances for tray and downcomer are set up. The holdups are computed from the mass balances, the flows leaving the tray are determined from hydraulic correlations with pressure drop over the stage as driving force. The flash calculation (denoted UV-flash) is performed with given internal energy and a given volume. The tray volume V_{tray} is related to the liquid and vapor molar holdup by:

$$V_{tray} = M_t/\rho_l + M_v/\rho_v + M_{dc}/\rho_l \tag{1}$$

The thermodynamic properties of the components are calculated by the ASPEN PROPERTIES PLUS (1988) package. All rigorous dynamic simulations were performed using SPEEDUP. The main difficulty in these simulations was to find the steady-state operating points.

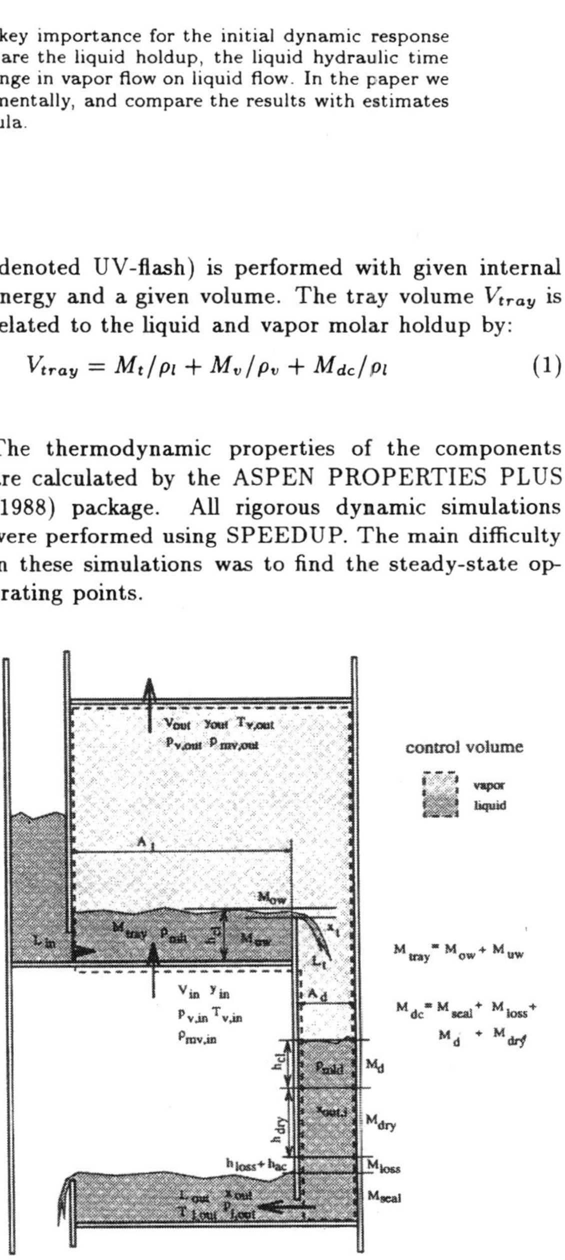

Fig. 1. Typical sieve tray with downcomer

2.2. *Tray hydraulics*

The modeling of the tray hydraulic is based on empirical correlations selected from the literature. Lockett(1986) gives an excellent overview over the different approaches. The chosen correlations are not necessarily the best available, but allow fairly accurate predictions and are easy to implement.

Clear liquid height. Consider the stage shown in Fig. 1. Let M_t denote the holdup on a tray (excluding the downcomer). Then

$$M_t = A_t \rho_l h_{cl} \tag{2}$$

where A_t is the tray area (excluding downcomer) and h_{cl} the clear liquid height. h_{cl} is less than the actual height of liquid on the plate due to bubbles in the liquid which are expressed by the liquid fraction (froth density) ϕ.

$$h_{cl} = \phi h_{weir} + h_{ow} \tag{3}$$

We use the correlation by Bennett (1983)

$$\phi = exp(\pi_1 \; K_s^{\pi_2}); \; K_s = v_t \sqrt{\rho_v/(\rho_l - \rho_v)} \tag{4}$$

where $v_t = V_{in}/(A_t \rho_v)$ is the vapor velocity on the tray. The parameters π_1 and π_2 are empirical constants. Bennett uses $\pi_1 = -12.55$ (in units consistent with the velocity factor K_s in m/s), and $\pi_2 = 0.91$ (dimensionless). We fitted new values to match pressure drop experiments and to get reasonable values for the liquid holdup.

Flow over outlet weir. The liquid flow, L_{out}, over the circular weir from the tray to the downcomer, is computed with a modified Francis weir formula (Perry *et. al.*, 1984). Since the outlet weir is placed off center towards the column wall, the liquid height above the weir h_{ow} will not be constant. There are no existing correlations which deal with the converging flow over outlet weirs in distillation column (Lockett, 1986). Taking the design into consideration, we choose to correct the weir length with a factor of 0.5.

$$h_{ow} = 44300 * \left(\frac{L_{out}}{\rho_l \cdot 0.5 \cdot d_{weir}} \right)^{0.704} \tag{5}$$

We have here assumed that the liquid contains no bubbles as it passes over the weir, that is, h_{ow} is the clear liquid height of liquid over the weir (see Eq. 3); the actual height of fluid on the tray is then h_{ow}/ϕ.

Pressure drop correlation. The "dry" pressure drop through the holes of the tray will contribute to increased liquid holdup in the downcomer. Numerous correlations for the dry pressure drop are available (*e.g.*: Liebson *et. al.* (1957))we have chosen:

$$\Delta p_{dry} = 0.965 \cdot \rho_v v_h^2 \;\; [N/m^2] \tag{6}$$

where $v_h = v_t A_t / A_h$ is the velocity through the holes [m/s].

Summary of Holdup Distribution. The total liquid holdup on a stage is

$$M = M_t + M_{dc} \tag{7}$$

Here M_t is given by Equations (2) - (5). We have

$$M_t = \underbrace{A_t \rho_l \phi h_{weir}}_{M_{uw}} + \underbrace{A_t \rho_l h_{ow}}_{M_{ow}} \tag{8}$$

Note that ϕ depends on the vapor flow V whereas h_{ow} depends on the liquid flow L. We identify four contributions to the liquid holdup in the downcomer.

$$M_{dc} = M_{loss} + M_d + M_{dry} + M_{seal} \tag{9}$$

1. Holdup which corresponds to the height of clear liquid on the tray plus the contribution from the corresponding "wet" pressure drop through the liquid on the tray above:

$$M_d = A_d \rho_l 2 h_{cl} = 2 \frac{A_d}{A_t} M_t \tag{10}$$

2. Holdup corresponding to pressure drop through trays ("dry pressure drop"), which from Eq. 6 is

$$M_{dry} = A_d \Delta p_{dry} / (g \; MW_l) \tag{11}$$

3. Holdup due to pressure drop under the downcomer apron, M_{loss}. This term may usually be neglected. 4. Holdup of liquid in the downcomer seal, M_{seal}, which is independent of vapor and liquid flow.

3. Linear Tray Hydraulics

In order to obtain further insight into the tray hydraulics we shall consider the linearized approximation. Such a simple approximation may also be used in order to simplify dynamic simulations. For simplicity, the vapor holdup is neglected and constant molar flows are assumed. It is assumed that the liquid flow, L_{out}, is a function of the liquid holdup, M and the vapor flow V_{in} (Rademaker *et. al.*(1975), Skogestad *et. al.* (1988)). Taking the total differential of L_{out} then yields

$$dL_{out} = \underbrace{\left(\frac{\delta L_{out}}{\delta V_{in}} \right)_M}_{\lambda} dV_{in} + \underbrace{\left(\frac{\delta L_{out}}{\delta M} \right)_V}_{\frac{1}{\tau_L}} dM \tag{12}$$

Note that this relationship is assumed to hold dynamically. The hydraulic time constant τ_L typically varies from 0.5 to 15 seconds. The vapor constant λ, representing the influence of V_{in} on L_{out}, typically ranges between -5 to + 5 (Rademaker (1975)).

We want to obtain λ and τ_L from correlations for the liquid holdup M as a function of liquid and vapor flow given in (7) to (11). The value of τ_L can be directly obtained from these correlations. To find λ we note that the total differential of M can be written

$$dM = \underbrace{\left(\frac{\delta M}{\delta L_{out}} \right)_V}_{\tau_L} dL_{out} + \underbrace{\left(\frac{\delta M}{\delta V_{in}} \right)_L}_{\tau_V} dV_{in} \tag{13}$$

(this equation holds only at steady state.) τ_L is always positive whereas τ_V may be either positive or negative. Setting $dM = 0$ in Eq. 13 yields

$$\lambda = (\delta L_{out}/\delta V_{in})_M = -(\tau_V/\tau_L) \tag{14}$$

Since both τ and λ depend on the tray loading, their values may be significantly different for the rectifying and the stripping section.

To derive analytic expressions for τ_L, τ_V and λ the holdup expressions in (7) to (11) are linearized:

$$\frac{\delta M_t}{\delta V_{in}} = A_t \rho_l \pi_2 \frac{\phi \ln \phi}{V_{in}} h_{weir} = \frac{M_{uw}}{V_{in}} \pi_2 \ln \phi \quad (15)$$

$$\frac{\delta M_{dry}}{\delta V_{in}} = 2 \frac{M_{dry}}{V_{in}} \quad (16)$$

$$\frac{\delta M_{loss}}{\delta L_{out}} = 2 \frac{M_{loss}}{L_{out}} \quad (17)$$

$$\frac{\delta M_t}{\delta L_{out}} = 0.704 \frac{M_{ow}}{L_{out}} \quad (18)$$

Equations 15 to 18 yield:

$$\tau_V = 2 \frac{M_{dry}}{V_{in}} + \left(1 + 2\frac{A_d}{A_t}\right) \frac{M_{uw}}{V_{in}} \pi_2 \ln \phi \quad (19)$$

$$\tau_L = 2 \frac{M_{loss}}{L_{out}} + \left(1 + 2\frac{A_d}{A_t}\right) 0.704 \frac{M_{ow}}{L_{out}} \quad (20)$$

Note that for our trays which have downcomers which exit above the liquid, we replace $2\frac{A_d}{A_t}$ by $\frac{A_d}{A_t}$. It is important to note that τ_V and thus λ may be either positive or negative. A negative λ means that more liquid is stored inside the column when vapor flow is increased, resulting in a temporary reduction in liquid flow. The main cause for this effect is the increased pressure drop which increases M_{dry}. A positive λ means that *less* liquid is stored inside the column when vapor flow is increased. At first this may seem unlikely. However, in tray columns it is quite common and is caused by a "swelling" (decrease in the froth density ϕ) which pushes liquid off the trays.

Numerical example

We will discuss the experimental data later, but to illustrate the above procedure we shall compute the holdup distribution and hydraulic parameters for the bottom section for experiment no. 6. We use the following data: $Q = 6.45$ kW $\Rightarrow V = 0.156$ mol/s, $L_T = 470$ ml/min $= 0.096$ mol/s, $\Delta p_{dry} = 242$ Pa/tray, $\phi = 0.494$, $h_{weir} = 30$ mm, $h_{ow} = 2.17$ mm, A_d/A_t=0.123. This yields $M_{ow} = 0.243$ mol, $M_{uw} = 1.661$ mol, M_{dry}=0.465 mol, $M_{loss} = 0.0007$ mol (negligible), and we find $\tau_L = 1.11$ s, $\tau_V = -0.77$s, and $\lambda = 0.69$.

Note that the two contributions to τ_V in (19) are +5.96 s (contribution from change in holdup due pressure drop) and -6.73 s (contribution from change in holdup due to from density). It is then clear that τ_V may easily change sign, and so may then $\lambda = -\tau_V/\tau_L$. In fact, for experiment 6 we find for the top part of the column that λ is negative (τ_L=1.32s, $\tau_V = 1.90$s and $\lambda = -1.44$).

Linear flow relationships for column sections

In the following assume that the hydraulic parameters are equal for all trays in the top section (τ_{LT} and λ_T), and for all trays in the bottom section (τ_{LB} and λ_B). Consider a small deviation from steady state. With constant molar flows the material balance on a tray becomes

$$\frac{dM}{dt} = dL_{in} - dL_{out} \quad (21)$$

where from the linearized tray hydraulics (Eq. 12)

$$dL_{out} = \lambda dV + \frac{1}{\tau_L} dM \quad (22)$$

Repeated combination of these equations yields the following expression for the change in liquid flow at the bottom of the column, L_B, in response to changes in reflux, L_T and boilup, V: (similar to Rademaker et. al., 1975), Skogestad et. al., 1988):

$$dL_B = g_L(s) dL_T + g_V(s) dV \quad (23)$$

where

$$g_L(s) = \frac{1}{(1+\tau_{LT}s)^{N_T}} \frac{1}{(1+\tau_{LB}s)^{N_B}} \quad (24)$$

$$g_V(s) = \lambda_B \left(1 - \frac{1}{(\tau_{LB}s+1)^{N_B}}\right) +$$

$$\frac{\lambda_T}{(\tau_{LB}s+1)^{N_B}} \left(1 - \frac{1}{(\tau_{LT}s+1)^{N_T}}\right) \quad (25)$$

The response of L_B to a change in reflux at the top is as expected a cascade of first-order responses, one for each tray. The response of L_B to a change in vapor boilup, V, requires a more detailed derivation. Consider first a column with only one tray. We consider here the effect of a change in V only, so set $dL_{in} = 0$ in (21). Taking Laplace transforms of (21) and (22) and combining yields $\frac{1}{\tau_L} dM = -\frac{\lambda}{\tau_L s+1} dV$, and we get with one tray

$$dL_{out} = dL_1 = \lambda \left(1 - \frac{1}{\tau_L s + 1}\right) dV$$

For two identical trays we get in addition the change in liquid flow from the tray above, dL_{in}, which will be equal to dL_1 given above. dL_{in} will affects dL_{out} through a first order lag, so the total effect of a change in vapor flow with two trays is

$$dL_{out} = dL_1 + \frac{dL_1}{\tau_L s + 1} = \lambda \left(1 - \frac{1}{(\tau_L s + 1)^2}\right) dV$$

For a column with N identical trays

$$dL_{out} = dL_B = \lambda \left(1 - \frac{1}{(\tau_L s + 1)^N}\right) dV$$

For a column with a top and bottom section

$$dL_B = \lambda_B \left(1 - \frac{1}{(\tau_{LB}s+1)^{N_B}}\right) dV + \frac{dL_{T,out}}{(\tau_{LB}s+1)^{N_B}}$$

where $L_{out,T}$ is the liquid entering from the top section. It is itself caused by the increase in vapor, so

$$dL_{T,out} = \lambda_T \left(1 - \frac{1}{(\tau_{LT}s+1)^{N_T}}\right) dV$$

and we derive the desired expression for $g_V(s)$.

Introduce the following approximations (which are good for long column sections)

$$\frac{1}{(1+\tau_{LB}s)^{N_B}} \approx e^{-\theta_B s}; \quad \frac{1}{(1+\tau_{LT}s)^{N_T}} \approx e^{-\theta_T s} \quad (26)$$

where $\theta_B = N_B \tau_{LB}$ and $\theta_T = N_T \tau_{LT}$. Also introduce the total liquid lag from the top to the bottom $\theta =$

$\theta_B + \theta_T$. Then we get from (23) with $dL_T = 0$

$$\begin{aligned} dL_B &\approx \lambda_B(1 - e^{-\theta_B s})dV \\ &+ \lambda_T(e^{-\theta_B s} - e^{-\theta s})dV \end{aligned} \quad (27)$$

In words, a step increase in boilup, dV, will result in an immediate increase in L_B equal to $\lambda_B dV$. This increase will last approximately for the time θ_B (which is the time it takes for a change in liquid flow to propagate through the bottom section), then the increase in L_B will change to the value $\lambda_T dV$ and it will stay at this value for an additional time θ_T. After approximately time $\theta = \theta_B + \theta_T$ the liquid flow L_B will change back to its original value.

4. Obtaining parameters from experiments

4.1. *Liquid holdup*

i) Experimental dumping. The column is emptied by turning of the reflux and the heating. Liquid evaporated during this operation is collected in the accumulator. The combined holdup change of accumulator and reboiler is then the amount of liquid stored on the trays in the column during normal operation.

ii) Theoretical tray calculations using geometric data and correlations, see equations (7) to (11).

iii) Tracer experiments. One may inject a heavy component at the top of the column and measure the time it takes for the wave to move down the column. In general, the time constant for the composition response on a individual tray is $\tau_c = M_i/(L + KV)$, where we can set $K = 0$ for a heavy tracer, and the time constant (apparent delay) for N trays in series is approximately $N\tau_c$.

iv) Experimental temperature responses. Measuring the initial temperature response to a step change in reflux ΔL or vapor flow ΔV enables the estimation of the liquid holdup. Assume

OP1 vapor holdup is negligible
OP2 constant molar liquid holdup M
OP3 equimolar flows (simplified energy balance)
OP4 the local slope of the vapor-liquid equilibrium curve is K_i
OP5 linear relation between liquid composition and temperature, $x = kTk_i$.

The material balance over a stage becomes

$$M\ dx_i/dt = L_i\ (x_{i-1} - x_i) + V_i\ (y_{i+1} - y_i) \quad (28)$$

For a step change in L_i and V_i the internal flows are $L_i = L_i^o + \Delta L$ and $V_i = V_i^o + \Delta V$. Subtract the steady state solution from Eq. 28 and consider the time immediately after the step change when the tray composition still is unchanged. This yields the followinging relationship for the initial slope as a function of ΔL and ΔV:

$$\begin{aligned} M\left(\frac{d\Delta x_i}{dt}\right)_{init} &= \Delta L_i(x_{i-1}^o - x_i^o) \\ &+ \Delta V_i(y_{i+1}^o - y_i^o) \end{aligned} \quad (29)$$

Apply assumptions OP4 and OP5 and rearrange Eq. 29

$$\begin{aligned} M_i\left(\frac{d\Delta T_i}{dt}\right)_{init} &= \Delta L_i(T_{i-1}^o - T_i^o) \\ &+ \Delta V_i K_i(T_{i+1}^o - T_i^o) \end{aligned} \quad (30)$$

From Eq.(30) it is possible to estimate the molar holdup M_i on a stage from observing the initial slope of the tray temperature to changes in reflux or boilup.

4.2. *Liquid hydraulic time constant τ_l*

From Eq. 27 the liquid hydraulic time constant can be determined experimentally by making a change in reflux and observing the delay in liquid response through the column. i) This delay may be observed from the response of the uncontrolled reboiler level, or ii) from the temperature responses inside the column. iii) τ_L may also be estimated theoretically from tray data.

4.3. *Vapor constant λ*

The vapor constant λ represents the initial effect of a change in vapor flow on liquid flow from a stage. Experimentally we can obtain λ by the following means.

i) From temperature responses. For example, for $\lambda \geq 1$ the effect of an increase in V will initially be counteracted by an even larger increase in L and we will, observe an increase in tray temperature in the upper part of the column earlier than in the lower part.

ii) From reboiler level response. Since $dM_B/dt = L_B - V$ we get from Eq. 27 that for a unit step in vapor flow V (keeping the reflux flow constant $\Delta L = 0$), the slope of the response of M_B as a function of time will be approximately $(\lambda_B - 1)$ for the time θ_B, then equal to $(\lambda_T - 1)$ for the next time θ_T, and then remain at -1.

iii) Estimate λ theoretically using hydraulic relationships such as Eq. 19 to Eq. 20.

5. Results

The distillation column is equipped with 11 sieve trays (numbered from the top) of 125 mm diameter. As compared what could be expected in an industrial column, the hole area of the trays is relatively small (A_h is about 5% of A_t), and the holdup in the downcomer is relatively large (about 50% of total holdup). Also, the downcomer design is different from that in Figure 1 in that the downcomer ends above the liquid surface and has a quite large the downcomer seal holdup. The reboiler is a thermosyphon reboiler with a nominal holdup of 3.5 liters and equipped with electrical heating of a maximum effect of 15 kW. The total condenser is connected to an accumulator with a holdup of approximately 1.5 liters. The reflux and feed are fed to the column by metering pumps and enter the column with a temperature approximately 2^oC below the boiling point. The experiments were performed with mixtures of ethanol and butanol. For the theoretical estimation of holdups and time constants we assume constant density $\rho_l = 727.6kg/m^3$ and mole weight $MW_l = 60.1kg/kmol$ corresponding to a 50/50 mixture. We also assume $\Delta h_{vap} = 41320kJ/kmol$. The empirical parameters π_1 and π_2 determined from the pressure drop measurements are presented in Tab. 1.

Table 1 Empirical parameters for Eq.4

	π_1	π_2
rectifier (top) section	-9	0.5
stripping (bottom) section	-12.55	0.8

5.1. *Liquid holdup*

Table 2 lists experimental conditions and some results for experiments 2 and 3. i) The liquid volume determined by dumping the distillation column is denoted $V_{l,dump}$. The corresponding average tray holdup is denoted $M_{i,dump}$. ii) The liquid volume estimated from pressure drop measurements is denoted $V_{l,\Delta p}$. It is found by measuring the differential pressure drop over the column and estimating the liquid holdup based on Eq. 7 to 11. iii) We performed experiments with only ethanol in the column and used butanol as a heavy tracer (exp. 11). With reflux L=538 ml/min the time for the butanol to reach the bottom was about 5.5 min, corresponding to a liquid holdup of approximately 3.0 l. The agreement between these three methods is good and gives a total liquid holdup on the trays of about 3l.

iv) We also used (30) to estimate the holdup from the initial temperature response on trays 3 and 9 to a step response in heat input (Exp.2, Fig. 2) and in reflux (Exp. 3, Fig. 3). The corresponding estimated holdups (denoted M_3 and M_9) show large variations (see M_3) and are also different (mostly larger) from the values found above using dumping. The reason for the variations may be inaccuracy in determining the initial temperature derivative (see e.g. Fig.3). However, the deviation from the dumping experiments may also be because the assumptions for deriving (30) do not hold. This is partly confirmed by the simulated responses (smooth lines in the Figures) which are in good agreement with the experiments. The simulated responses are for the full model (with no assumption about constant molar flow etc.) using holdups which are in agreement with the dumping experiments.

5.2. *Liquid hydraulic time constant* τ_l

For the experimental determination of the hydraulic time constant, the level control of the inventory is placed into manual, such that the positions of the product valves are fixed. The reboiler level will now primarily depend on the liquid flow to the reboiler and the vapor leaving it, although it should be noted that a fixed valve position not necessarily means that the liquid flow through the valve is constant.

Let us first consider experiment 5. i) The hydraulic lag between the increase in reflux until the reboiler level changes is illustrated in Fig. 4. We find $\theta = 31$ s and compute an average hydraulic time constant (Eq. 27) for each stage of $\tau_L = 2.8s$. ii) Using temperature measurements we find that that the liquid delay between the temperature change on tray 3 and 9 is 13 seconds, which give an average hydraulic time constant of 2.2 seconds. iii) Applying Eq. 20 to the set of operational conditions of experiment 5 given in table 3 yields $\tau_{LT} = 1.32$ s and $\tau_{LB} = 1.11$s (which is about half of the experimental value). For experiment 3 we find from the temperature responses in Fig. 3 an average hydraulic time constant of 1.83 s. Applying Eq.(20) we estimate a time constant of 1.41 and 1.20 seconds for the rectifier and stripper, respectively.

Table 2 Data for holdup estimation for experiment 2 and 3

	units	Exp. 2	Exp. 3
F	ml/min	250	250
z_F		0.54	0.56
Δp	$mbar$	22.3	26.1
$Q_{t=0}$	kJ/s	3.60	4.50
$Q_{t=\infty}$	kJ/s	4.05	4.50
$L_{t=0}$	ml/min	250	350
$L_{t=\infty}$	ml/min	250	380
$V_{l,dump}$	l	2.94	3.18
$V_{l,\Delta p}$	l	2.81	2.79
$M_{i,dump}$	mol	3.24	3.51
T_2	^oC	79.11	80.96
T_3	^oC	79.49	84.80
T_4	^oC	80.45	90.07
$d\Delta T_3/dt$	$^oC/s$	0.0018	0.0133
M_3	mol	6.16	2.96
T_8	^oC	90.87	109.79
T_9	^oC	104.68	113.73
T_{10}	^oC	111.15	115.81
$d\Delta T_9/dt$	$^oC/s$	0.0252	0.0045
M_T	mol	8.65	9.06

Table 3 Experiments 3 and 5 for changes in liquid flows. Experimental conditions, measured time delays (θ), and hydraulic time constant (τ_L)

	units	Exp.3	Exp. 5
F	ml/min	250	350
z_F		0.56	0.56
$L_{t=0}$	ml/min	350	470
$L_{t=\infty}$	ml/min	380	557
$Q_{t=0}$	kJ/s	4.50	5.8
Θ_{T3}	s	*	13
Θ_{T9}	s	10	26
$\Theta_{reboiler}$	s	*	31
$\tau_{L,t}$ (Eq.20)	s	1.41	1.32
$\tau_{L,b}$ (Eq.20)	s	1.20	1.11

We note from these and other results that the agreement between experiment and tray calculations is somewhat better for the temperature measurements. The reason may be dynamics in the reflux pump and in the level sensor which do not appear when we consider the difference in temperature response of two trays.

Still, it seems that the estimated values for τ_L from Eq. (20) are too small. However, the rigorous simulations which are based on the same correlations show better agreement as is illustrated by comparing the experimental and simulated reboiler level in Figure 4.

The reason for the difference may be a somewhat more detailed tray model and effects of the energy balance which are also not included in the simple linear analysis leading to (20).

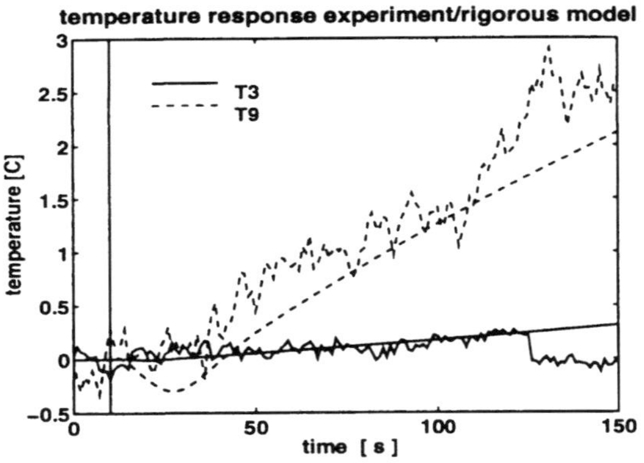

Fig. 2. Experiment 2. Initial response of temperatures on trays 3 and 9 to a step change in heat input at t = 10 s

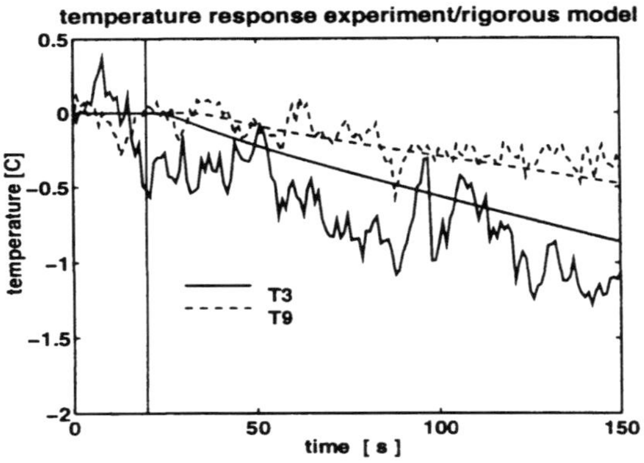

Fig. 3. Experiment 3. Initial response of temperatures on trays 3 and 9 to a step change in reflux flow at t = 20 s

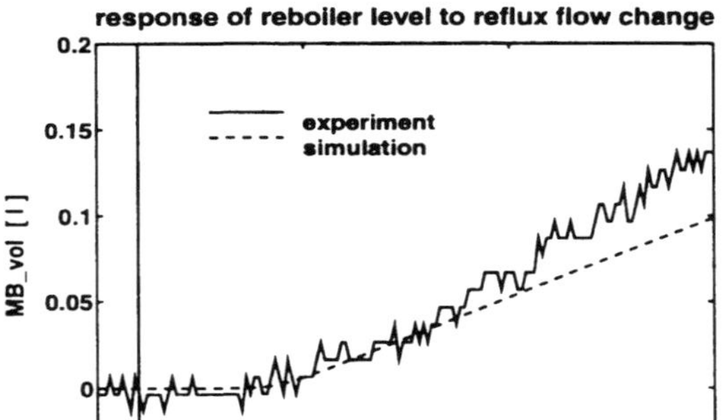

Fig. 4. Experiment 5. Response of reboiler level to a step in reflux flow at t = 20 s

5.3. *Vapor constant* λ

λ may be observed from the initial response to changes in the boilup as changes in the slope of the temperature or level as a function of time. From the temperature measurements for experiment 2 it is difficult to observe any change in the slope so one would expect λ to be close to 0. This does not agree with estimated values from tray calculations which yield λ equal to 2.9 and 5.1 in the top and bottom sections, respectively. As seen from the simulated temperature on tray 9 in Figure 2 (smooth dashed curve) this should yield an inverse response, which should be easy to observe in the bottom section.

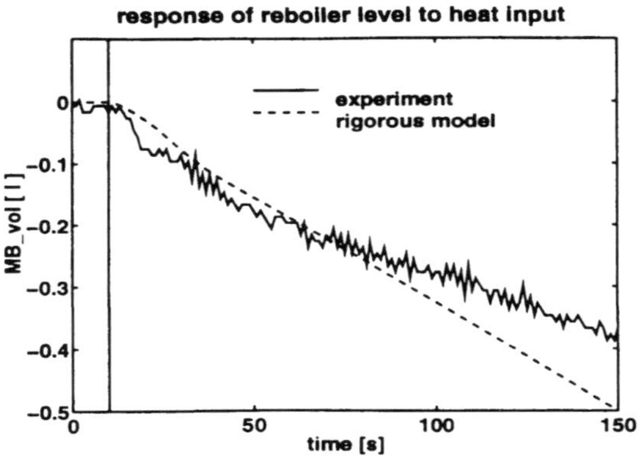

Fig. 5. Experiment 6. Response of reboiler level to a step in heat input at t = 10 s

From the level measurement for experment 6 one can from Figure 2 observe a change in the slope after about 10 s and then another change after about 35 s (from when the change in heat input was applied). This indicates that θ_T is about 25 s and θ_B is about 10 s. This is in reasonable agreement with experiment 5 which has the same initial liquid flow. If we consider the slope of M_B as a function of time then one would from the experimental response in Figure 2 expect λ_T to be about -1 and λ_B to be about -3.

Again, this does not agree very well with the estimated values of about -1.44 and 0.69, and even the relative difference between λ_T and λ_B is in the wrong direction. This deviation may seem very large, but the estimated values are very sensitive to small changes in the data. For example, consider Experiment 6 for which we estimate $\lambda_B = 0.69$. Recall from (19) that λ has two contributions with different signs. Both of these are very sensitive to the value of the vapor flow. For example, the magnitude of the pressure drop term (which yields a negative contribution to λ) is proportional to V, while the froth density term (which yields a positive contribution to λ) may become less important as V is increased (because the froth density ϕ cannot be less than 0). For experiment 6, we obtained in Section 3 τ_{VB} = 5.96 s - 6.73 s = -0.77 s. If we increase the vapor flow by 20% then we find τ_{VB} = 7.15 s - 5.81 s = 1.34 s. τ_L is unchanged at 1.11 s, so we find that a 20% increase in V changes the estimated value of λ_B from 0.69 to -1.21. In fact, an increase in vapor flow by 40% changes the estimated λ_B from 0.69 to -2.95, which is the same as the experimental value of about -3.

6. Discussion and Conclusion

The agreement with the experimental responses and simulations using the detailed model were generally very good, except for the initial effect of changes in vapor flow (e.g., as expressed by the parameter λ). However, this was obtained only after considerable effort (including some parameter fitting) so in general we believe it is difficult to predict accurately the initial response of distillation columns based on only geometric data about the column and thermodynamic data. Two reasons are:

1. The correlations for estimating the tray holdups are empirical and do not seem very reliable. For example, this applies to the formula for estimating the froth density.

2. Even with a good model for estimating the tray holdups, the dynamic response, for example as expressed by λ, is very sensitive to small changes in the parameters. This means that one cannot expect to be able to predict from tray data alone whether one will have inverse response to changes in boilup.

In general, the inverse response is undesirable so one would like to have $\lambda < 0.5$. Fortunately, the models give us insight into how the tray design should be changed to achieve this. The reason for the undesired positive contribution to λ is the swelling due to bubbles. This effect seems difficult to avoid and is probably desirable to improve mass transfer. To make λ more negative one needs to increase the downcomer holdup as vapor flow is increased. This may be achieved by using a larger downcomer area or by increasing the pressure drop over the trays (e.g., by using smaller holes in the plates). Another alternative, which may be more attractive, is to use a packed columns where liquid holdup generally increases with vapor flow, implying that λ is generally negative.

For control purposes one may want τ_L large as this increases the liquid lag from the top to the bottom of the column, and thus tends to decouple the dynamic response in the two column ends. To increase τ_L one may increase M_{ow} by using a shorter outlet weir. Alternatively, one may increase the pressure drop under the downcomer to increase the term M_{loss}. For packed columns all the liquid contributes to the liquid lag, τ_L (as there is no holdup under the weir or in the downcomer), so although the liquid holdup in a packed column is generally smaller than in a trayed column, the decoupling effect from the liquid flow dynamics (as expressed by τ_L) may not be any less.

In conclusion, the dynamic simulations based on a detailed model gave good agreement with experimental responses, while it was found that simpler models based on constant molar flows and linear tray hydraulics gave quite large deviations. Better correlations are needed for predicting the holdup on the trays for the rigorous model. However, even with improved correlations one cannot expect to get accurate predictions of the flow behavior (e.g. λ) because of strong sensitivity to parameter changes. One will therefore in most cases need experimental data to validate the responses and possibly adjust parameters in the model for the tray hydraulics.

REFERENCES

[1]ASPEN PROPERTIES PLUS; User Guide; Release 8 , Aspen Technology, Inc., Cambridge, USA, 1988

[2]Baber, M.F., Edwards, L.L., Harper, W.T., Witte, M.D., Gerster, J.A.; Experimental Transient Response of a Pilot Plant Distillation Column, *Process Dynamics and Control,* 1961, **57**, No. 36, pp. 149-159.

[3]Baber, M.F., Gerster J.A.; Experimental Transient Response of a Pilot Plant Distillation Column: Part II. Response to Liquid and Vapor Rate Perturbations, *AIChE Journal*, 1962, **8**, No.3, pp. 407-415.

[4]Bennett, D.L., Agrawal, R, Cook, P.J., New Pressure Drop Correlations for Sieve Tray Distillation Columns,, *AIChE Journal*, Vol.29, No. 3, pp. 434-443, 1983

[5]Lockett, M.J., Distillation tray fundamentals, *Cambridge University Press*, 1986

[6]iebson, I., Kelly, R.E., Bullington, L.A., How to design perforated trays, *Petrol. ref.*, Vol. 36, No. 127, 1957

[7]Perry, H.P., Green,D., Perry's Chemical Engineers' Handbook, *McGraw-Hill Chemical Engineering Series*, 50th edition, 1984 Chapter 18

[8]Rademaker, O., Rijnsdorp, J.E.,Maarleveld,A., Dynamics and Control of Continuous Distillation Units, *Elsevier, Amsterdam*, 1975

[9]Skogestad, S., Morari, M., Understanding the Dynamic Behavior of Distillation Columns, *Ind. & Eng. Chem. Research*, 1988, No. 10, Vol. 27, pp. 1848-1862

[10]SPEEDUP Release 5.3 User Manual, Prosys Technology Ltd., 1992

Plantwide Control Problem For a Process With Three Distillation Columns and Two Recycle Streams

Michael L. Luyben* and William L. Luyben**

* *DuPont Central Science & Engineering, Experimental Station, P.O. Box 80101, Wilmington, DE 19880-0101 USA (email: luyben@esspt0.dnet.dupont.com)*

** *Department of Chemical Engineering, Lehigh University, Bethlehem, PA 18015-4791 USA (email: wll0@lehigh.edu)*

Abstract. This work presents a plantwide control problem for a complex process typical of many industrial chemical operations. The plant contains two reaction steps, three distillation columns, two recycle streams, and six chemical components. Process details are provided along with disturbances that an effective control strategy must handle.

Key Words. Plantwide Control, Distillation Control, Benchmark Problem

1. INTRODUCTION

A growing amount of evidence shows the importance of considering the effects of the *entire* process in both design and control. Morari and Perkins (1994) recently reviewed some of the area with a specific focus on design for operations. Although real industrial chemical processes typically contain several reaction steps and multiple recycle streams, most process controllability studies in the literature have considered much simpler systems. This paper presents a plantwide control problem for a simplified version of a real complex industrial process illustrating many of the important characteristics of such systems: significant interactions among units with recycle streams and the presence of byproduct/intermediate components. The problem, stated simply, involves:

> *For the process described below containing three reactors, three distillation columns, and two recycles (18 control valves), synthesize an effective control strategy involving inventory, throughput, and product-quality loops to handle three different types of disturbances.*

This process should provide researchers with a test case for various approaches and techniques in control structure selection, controllability analysis, model identification, and model predictive and nonlinear control. The work by Luyben and Luyben (1994) contains an extensive discussion concerning the design and control of this process.

Figure 1 gives a conceptual picture of the system while Figure 2 presents the detailed flowsheet and defines the notation. The first reaction step

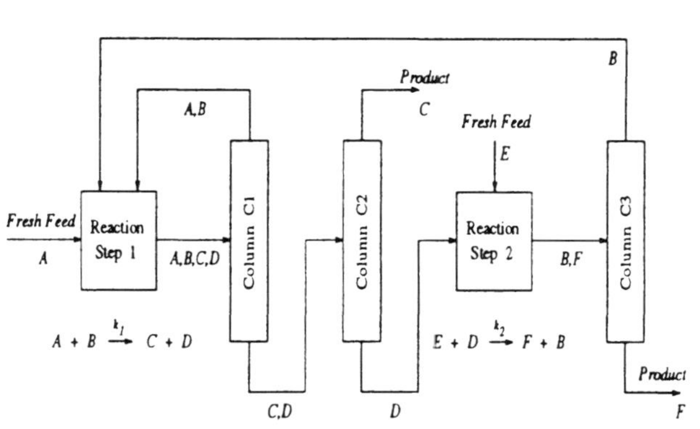

Fig. 1. Conceptual Flowsheet of System

$A + B \rightarrow C + D$ produces desired product C. Component A enters the first CSTR in a fresh feed stream while component B returns in a stream recycled from a downstream unit. The reaction rate (lbmol/hr) is $\mathcal{R}_1 = V_{R1}k_1z_{1A}z_{1B}$. The reactor effluent contains a multicomponent mixture since complete one-pass conversion is not achieved; this effluent is the feed to the first distillation column. The mixtures are assumed to be ideal with volatilities of $\alpha_A > \alpha_B > \alpha_C > \alpha_D > \alpha_E > \alpha_F$. The distillate stream from the first column recycles A and B back to the first reactor. The bottoms product with C and D is fed to the second distillation column. The second column produces a distillate product of C with the desired purity and a bottoms product of D, which is not valuable and so serves as a feed to the second reaction step $D + E \rightarrow F + B$. This reaction produces another desired product F and regenerates reactant B for use in the first reaction step.

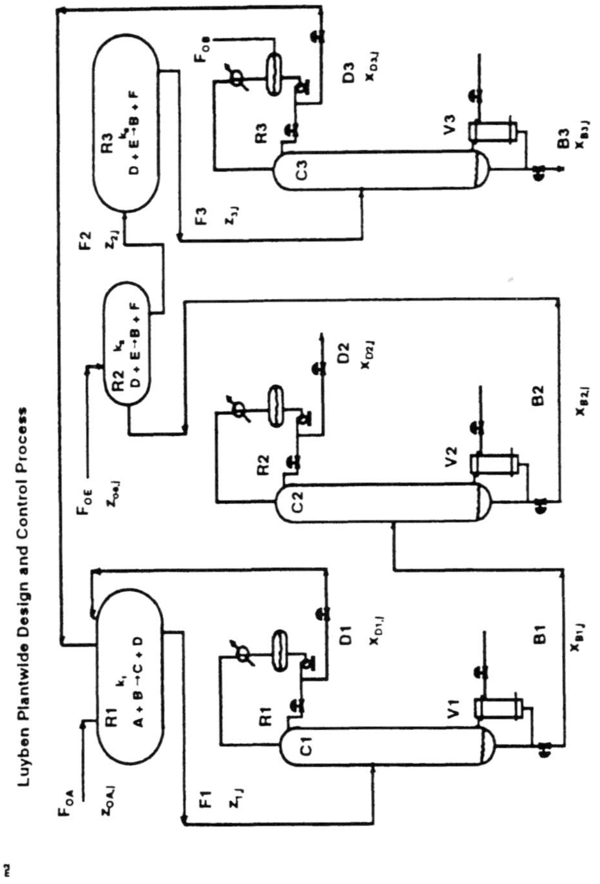

Fig. 2. Detailed Process Flowsheet

A fresh feed stream of E enters the first of two CSTR's in series. Two reactors allow the reaction to proceed to high conversion of component E since we assume that the separation of any unreacted E from product F will be fairly difficult. The reaction rates (lbmol/hr) in the second and third reactors ($i = 2, 3$) are $\mathcal{R}_i = V_{Ri} k_2 z_{iD} z_{iE}$. The effluent from the third reactor is fed into the third distillation column, which produces a bottoms of F with the desired purity and a distillate of B recycled back to the first reactor. From an overall perspective, components B and D are intermediates and exist ideally only within the process itself. However, it is impossible to prevent some loss of these intermediate components, which must be considered both in the steady-state design and in the plantwide control system. We have chosen to add a makeup/purge stream for component B to the reflux drum of the third column (it might also be added to the first reactor). Stream F_{0D} is also used in the dynamic simulations as a makeup or purge for component D; it enters or leaves the process in the base of the second column (not shown in Fig. 2).

2. PLANTWIDE CONTROL PROBLEM

In the paper by Luyben and Luyben (1994), two different design approaches were employed. The designs resulting from the heuristic and nonlinear optimization procedures differed in the purities and flowrates of the recycle streams. We provide here the process data for both designs in Tables 1 through 4 so that control strategies may be tested on both. These data are required to construct a rigorous nonlinear dynamic model of the process. Within the tables and in the model equations, all compositions (z, x_D, x_B) are given in mole fraction, flowrates are in lbmol/hr, and reactor holdups (V_R) are in lbmol.

z_{0A_A}	=	1.0	
z_{0B_B}	=	1.0	
z_{0E_E}	=	1.0	
α_A	=	8	
α_B	=	6	
α_C	=	4	
α_D	=	2	
α_E	=	1.2	
α_F	=	1	
k_1	=	10	hr^{-1}
k_2	=	50	hr^{-1}
ΔH_v	=	250	Btu/lb
c_p	=	0.9	Btu/lb °C
MW	=	50	lb/lbmol
ρ	=	50	lb/ft^3

Table 1 Process Data

		Heuristic	NLP
Holdup	(lbmol)		
C1	M_D	33.7	32.5
	M_R	1.36	1.05
	M_S	1.63	1.58
	M_B	53.8	53.4
C2	M_D	32.1	30.6
	M_R	1.27	1.20
	M_S	1.39	1.33
	M_B	100	100
C3	M_D	100	100
	M_R	0.837	0.505
	M_S	1.04	0.70
	M_B	31.7	24.6
Hydraulic	Time	Constant (s)	
C1	β_R	3.5	5.3
	β_S	2.8	2.7
C2	β_R	3.5	3.5
	β_S	3.1	3.0
C3	β_R	3.6	4.1
	β_S	2.6	2.2

Table 2 Tray Holdups and Hydraulic Time Constants For Distillation Columns

The book by Luyben (1990) contains the differential equations describing the behavior of the reactors and distillation columns. The reactors operate isothermally. The mathematical model for the distillation columns assumes theoretical trays, saturated liquid feeds and refluxes, equimolar overflow, constant relative volatilities, and negligible vapor holdup. Liquid tray hydraulics are included using a linearized version of the Francis weir formula with tray holdups (lbmol) and hydraulic time constants (seconds) in the rectifying (M_R, β_R) and stripping (M_S, β_S) sections. Column base (M_B) and reflux drum (M_D) holdups are set at five minutes based upon total inflow rates, except for locations where purge and makeup streams enter or leave. In the reflux drum of the third column we use 20 minutes of holdup. All composition loops have three minute deadtimes in the composition measurement. Composition transmitter spans are all 0.1 mole fraction in columns and 0.5 mole fraction in reactors. Level transmitter spans are twice the steady-state holdups. The valve spans are all twice the steady-state flowrates, except for F_{0B} where the maximum flowrate is ± 50 lbmol/hr.

Many control structures exist for this multivariable process. The 18 manipulated variables (control valves) include: flowrates of three fresh feeds (F_{0A}, F_{0B}, F_{0E}), flowrates from the three reactors (F_1, F_2, F_3), vapor boilups in each column (V_1, V_2, V_3), bottoms flows from each column (B_1, B_2, B_3), reflux flows to each column (R_1, R_2, R_3), and distillate flows from each column (D_1, D_2, D_3). The obvious controlled variables include the levels in process vessels (reactors, column base and reflux drum) and product stream compositions (D_2 and B_3). Beyond that, however, it is not clear what other variables need to be controlled and so we do not specify them.

The effectiveness of the control strategy is judged on the basis of minimizing the variability in product purities $(x_{D2_B}, x_{D2_D}, x_{B3_D},$ and $x_{B3_E})$ and of not requiring large changes in manipulated variables with particular focus on recycle flowrates (D_1 and D_3). These results are shown with simulations on the rigorous nonlinear model by considering three disturbances on the system:

1. Step change in fresh feed composition z_{0A} from 100% A to 80% A and 20% B.
2. Step change in the specific reaction rate k_1 in the first reactor (+50%).
3. Step change in throughput by a 10% increase in production rate.

These cases provide a good test of upsets such a process might typically face.

It turns out that it is easy to find a control strategy that *will not work* in handling disturbances. Many schemes, which appear on the surface to be logical candidates, do not work. Figure 3 shows a control strategy that might initially seem obvious. Production is set by controlling the flowrate

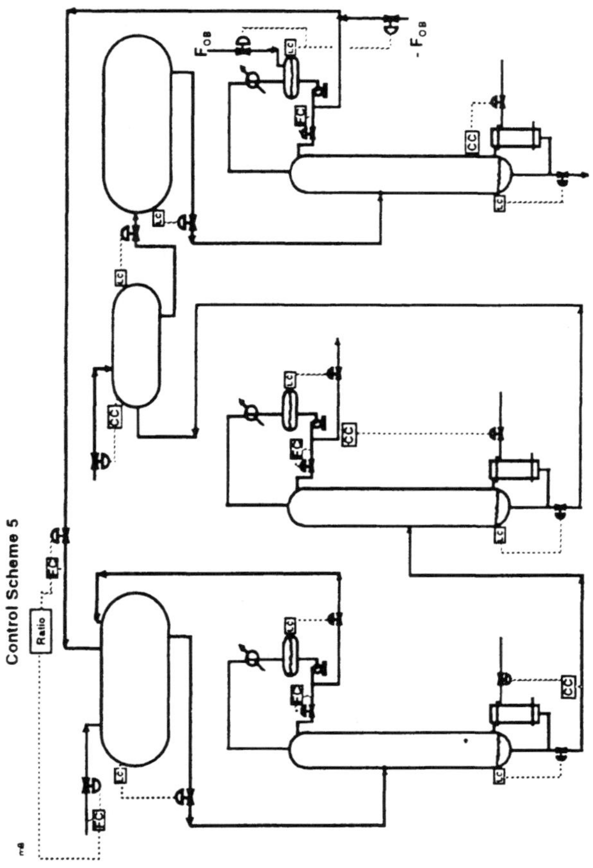

Fig. 3. One Control Strategy

of fresh feed F_{0A}. Component B in the first reactor is maintained by having the flowrate of D_3 as a ratio to F_{0A}. The holdup in the first reactor is controlled by manipulating effluent flowrate F_1. Figure 4 shows the results using this strategy when a disturbance occurs in fresh feed composition z_{0A}. The control system winds up to the point where the first recycle flow D_1 saturates at the upper limit of the control valve, which leads to unacceptable events.

3. CONCLUSIONS

The challenges posed by this plantwide control problem involve the development of effective control strategies. This process contains two reaction steps, two recycle streams, and two intermediate components; such features are typical in industrial processes. This system should serve as a valuable example for examining control structure selection, controllability analysis, model identification, and

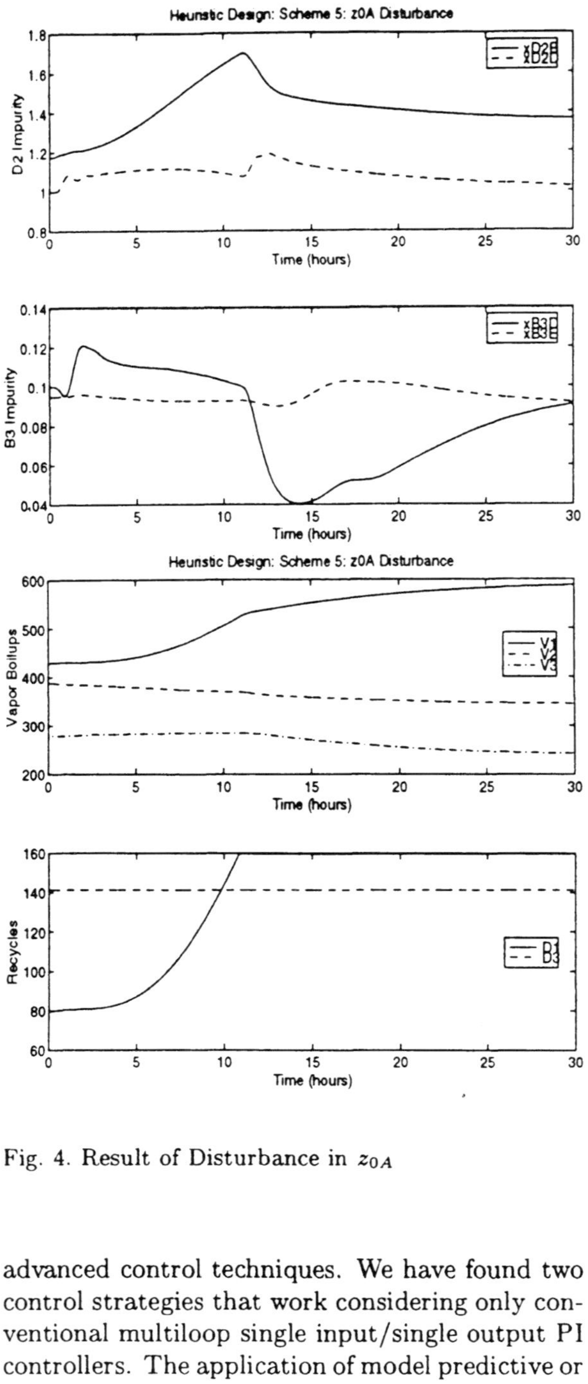

Fig. 4. Result of Disturbance in z_{0A}

advanced control techniques. We have found two control strategies that work considering only conventional multiloop single input/single output PI controllers. The application of model predictive or nonlinear control to this high-dimension process should provide a good test of their practicality on realistically complex processes. Another feature of this problem centers on the issue of assessing and incorporating controllability within the process design. A trade-off naturally exists between economics and controllability and a procedure like the one proposed by Elliot and Luyben (1994) may allow a quantitative analysis.

REFERENCES

[1] Elliott, T. R. and W. L. Luyben, "A Capacity-Based Economic Approach for the Quantitative Assessment of Process Controllability During the Conceptual Design Stage," *Submitted to Ind. Eng. Chem. Res.*, (1994).

[2] Luyben, M. L. and W. L. Luyben, "Design and Control of a Complex Process Involving Two Reaction Steps, Three Distillation Columns, and Two Recycle Streams," *Submitted to Ind. Eng. Chem. Res.*, (1994).

[3] Luyben, W. L., *Process Modeling, Simulation and Control for Chemical Engineers, 2nd ed.*, New York: McGraw-Hill (1990).

[4] Morari, M. and J. D. Perkins, "Design For Operations," *Proceedings of FOCAPD-94*, Snowmass, Jul. (1994).

NOTE

The following supplementary material may be obtained from the authors by sending an email request (addresses given above).

1. Copy of paper by Luyben and Luyben (1994) describing the process in detail.
2. FORTRAN code of nonlinear process model.

B. D. Tyreus has supplied the Antoine coefficients (determined to match the relative volatilities) in Table 5 if column temperature profiles are required in the simulation.

$k_1 =$	10	hr^{-1}	$k_2 =$	50	hr^{-1}	
$F_{0A} =$	100.67	lbmol/hr				
lbmol/hr	F_{0B}	3.172	F_{0D}	-0.843	F_{0E}	98.73
lbmol/hr	F_1	321.5	F_2	236.7	F_3	236.7
lbmol	V_{R1}	993	V_{R2}	870	V_{R3}	870
$i = A$	z_{1i}	0.1879	z_{2i}	0.0000	z_{3i}	0.0000
B		0.0539		0.4046		0.4167
C		0.3247		0.0021		0.0021
D		0.4334		0.1761		0.1640
E		0.0000		0.0125		0.0004
F		0.0001		0.4047		0.4168
lbmol/hr	D_1	79.76	D_2	102.9	D_3	141.1
lbmol/hr	R_1	350	R_2	285	R_3	141
$i = A$	x_{D1_i}	0.7566	x_{D2_i}	0.0005	x_{D3_i}	0.0000
B		0.2021		0.0117		0.7218
C		0.0413		0.9778		0.0035
D		0.0000		0.0100		0.2745
E		0.0000		0.0000		0.0000
F		0.0000		0.0000		0.0002
lbmol/hr	B_1	241.7	B_2	138.8	B_3	98.84
lbmol/hr	V_1	429.8	V_2	387.9	V_3	278.9
$i = A$	x_{B1_i}	0.0002	x_{B2_i}	0.0000	x_{B3_i}	0.0000
B		0.0050		0.0000		0.0000
C		0.4182		0.0036		0.0000
D		0.5765		0.9962		0.0010
E		0.0000		0.0000		0.0009
F		0.0001		0.0002		0.9981
	N_1	30	N_2	30	N_3	40
	N_{f1}	15	N_{f2}	15	N_{f3}	20

Table 3 Initial Conditions For Heuristic Design

$k_1 =$	10	hr^{-1}	$k_2 =$	50	hr^{-1}
$F_{0A} =$	99.67	lbmol/hr			

lbmol/hr	F_{0B}	0.204	F_{0D}	0.824	F_{0E}	100.87
lbmol/hr	F_1	602.2	F_2	251.7	F_3	251.7
lbmol	V_{R1}	596	V_{R2}	319	V_{R3}	495
$i = A$	z_{1_i}	0.2358	z_{2_i}	0.0000	z_{3_i}	0.0000
B		0.0719		0.3586		0.3981
C		0.3969		0.0106		0.0106
D		0.2525		0.1292		0.0898
E		0.0007		0.0438		0.0045
F		0.0433		0.4578		0.4970
lbmol/hr	D_1	351.4	D_2	100.8	D_3	151.0
lbmol/hr	R_1	89.7	R_2	238.2	R_3	73.77
$i = A$	x_{D1_i}	0.4038	x_{D2_i}	0.0001	x_{D3_i}	0.0000
B		0.1197		0.0070		0.6648
C		0.3886		0.9898		0.01768
D		0.0848		0.0032		0.1492
E		0.0001		0.0000		0.0027
F		0.0030		0.0000		0.1656
lbmol/hr	B_1	250.8	B_2	140.0	B_3	100.9
lbmol/hr	V_1	441.1	V_2	339.0	V_3	224.6
$i = A$	x_{B1_i}	0.0000	x_{B2_i}	0.0000	x_{B3_i}	0.0000
B		0.0028		0.0000		0.0000
C		0.4083		0.0178		0.0000
D		0.4876		0.8128		0.0008
E		0.0016		0.0027		0.0070
F		0.0997		0.1667		0.9922
	N_1	26	N_2	36	N_3	26
	N_{f1}	22	N_{f2}	13	N_{f3}	18

Table 4 Initial Conditions For NLP Design

Component	A_j	B_j	C_j	T_{NBP} (°C)
A	14.1523	3500	273.15	32.1
B	13.8646	3500	273.15	40
C	13.4591	3500	273.15	51.8
D	12.7660	3500	273.15	74.1
E	12.2552	3500	273.15	92.7
F	12.0728	3500	273.15	99.8

Table 5 Antoine Equation for Pure Component Vapor Pressures: $\ln P_j^s = A_j + B_j/(C_j + T)$ with P_j^s in psia and T in °C

WAVE-NET BASED ON-LINE QUALITY INFERENCE SYSTEM FOR POLYMERIZATION PROCESSES

Masahiro Ohshima[1], **Alexandros Koulouris**[2], **Sigeyuki Tomita**[1], and **George Stephanopoulos**[2]

1) *Dept. of Comp. Science and Systems Eng., Miyazaki University, Miyazaki 889-1, JAPAN*
2) *Dept. of Chemical Engineering, Massachusetts Institute of Technology, MA 02139, USA*

Abstract: The paper proposes an on-line inferential scheme to predict a polymer property, the Melt Index, from off-line, past measured values of this property and available on-line measured process variables. In the on-line inferential scheme, the L^∞ Wave-Net, which is an artificial neural network with basis functions drawn from the family of wavelets, is utilized by incorporating fundamentals of chemical engineering. The multistep predictor, formulated in a cascaded structure of identical one-step-ahead Wave-Net based predictors, is adaptively learning the functional relationship between the polymer property and process variables. The application of the scheme is illustrated using operating data of an industrial High Density Polyethylene Process. The results of the proposed scheme show that the inferential scheme built upon the Wave-Net can successfully predict the MI and prove its potential as an on-line monitoring scheme to maintain the polymer quality at the highest level.

Keywords: Wave-Net, on-line inference, polymerization process, melt index

1. INTRODUCTION

The changes in customer expectations for product quality have directed the industry's attention to the importance of product consistency. The demand for product consistency is going to become greater and the manufacture of consistently high-quality products using minimum resources and creating little waste are key elements of the competitive edge for industries.

In the industrial polyolefine production processes, the quality improvement has been attracting significant attention. To maintain the polymer quality at a high level, it is required to regulate the polymer properties at the specified level. In the industrial setting, especially at the polyolefine production processes, the specifications of polymer properties are quoted primarily in terms of the polymer density, ρ, and melt index, MI, which is an index of the fluidity of the molten polymer. However, a major difficulty in controlling the polymer properties is the lack of a suitable on-line sensor for these key properties. Due to the lack of the on-line sensor, the polymer properties are occasionally evaluated off-line in quality control laboratories, usually taking two or three hours to obtain the evaluation results. These infrequent measurements with large time-lag hinder the implementation of appropriate counter-measures to take against disturbances that affect polymer properties and often result in a wide variation of product consistency.

One of the recent activities in the polymer production area is to develop an on-line polymer property inferential scheme (Ardel et. al., 1983; McAuley et. al., 1991; Ogunnaike, 1991; Bohm et. al., 1992; Watanabe et. al., 1993). On the basis of either theoretically or empirically derived so-called quality models, which can translate quality requirements into appropriate polymer properties and clarify the quantitative relationship between the polymer properties and the operating conditions, the inferential scheme is developed so as to predict the polymer properties on-line from irregularly time-spaced measurements of polymer properties and available on-line measured process variables. An observer technique (Bohm et. al.,1922), an extended Kalman filter technique (Ogunnaike et. at., 1991) or a recursive parameter estimation technique (McAuely et. al., 1991) can be employed in the inferential scheme to update the model parameters as well as to estimate the state variables.

In these inferential schemes, every time new measurements of polymer properties become available, a nonlinear model is linearized around the present value.

The linearized model is used for updating the parameters and estimating the state variables. Then, the original nonlinear model with updated parameters is used to calculate polymer properties for intersample prediction. However, no theoretical proof of convergence in parameter estimation was guaranteed in the schemes even for the case that the process is time-invariant in nature. Furthermore, the major hindrance in applying these schemes to an industrial process is the difficulty in deriving the quality model. McAuley et. al. (1991) proposed a quality model derived from first principles of polymer kinetics and mass balances for a low density polyethylene fluidized bed reactor. However, the first principles modeling is not easy to derive and it is usually a time-consuming job. The difficulty of the modeling hinders the application of their inferential schemes to other polymerization processes.

In this study, in order to expand the applicability of the inferential scheme to other polymerization processes, the L^{∞} Wave-Net, which was proposed by Koulouris et. al. (1994) and is an artificial neural network with basis functions drawn from family of wavelets, is used to develop the quality model by incorporating the fundamentals of chemical engineering. The Wave-Net part of the quality model is adaptively learning the functional relationship between polymer property (MI) and process variables on the irregularly time-spaced MI measurements and on-line measured process variables. The application of the proposed inferential scheme is illustrated on an industrial High Density Polyethylene process. The results show that MI can be successfully predicted.

Although the inferential scheme presented is discussed in the context of predicting melt index(MI) in a high density polyethylene process, it is applicable to other polyolefine polymerization systems.

2. PROCESS AND DATA DESCRIPTION

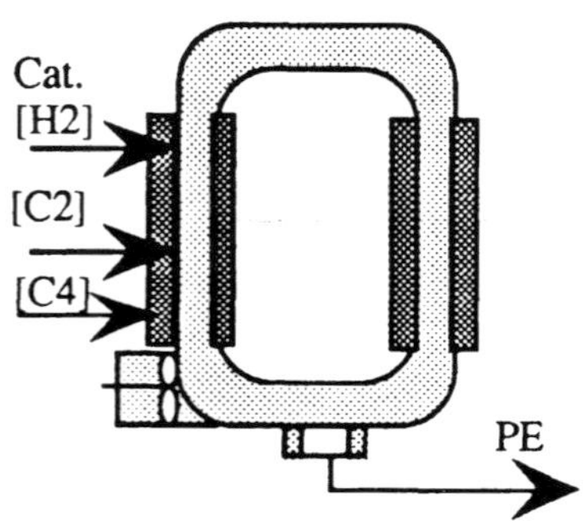

Fig. 1 Outline of HDPE reactor

The outline of the High Density Polyethylene (HDPE) polymerization reactor is illustrated in Fig. 1. The feed is comprised of ethylene, co-monomer, butene, and hydrogen. The catalyst is also fed to the reactor along with solvent. In the HDPE plant, the melt index (MI) is one of the key polymer properties, occasionally sampled by the plant operator and analyzed at the quality lab.

Since the sampling is carried out occasionally, the values of MI measurements are not an equally spaced in time. One typical record of temporal data of measured MI is illustrated in Fig. 2. On the other hand, the hydrogen, monomer, co-monomer and catalyst concentrations are available every 5 minutes. The reactor temperature is measured every second and the polymer residence time at the reactor can be calculated on the basis of the material balance with production rate and polymer concentration. These available on-line measured process variables are illustrated along with the measured MI data in Fig. 3.

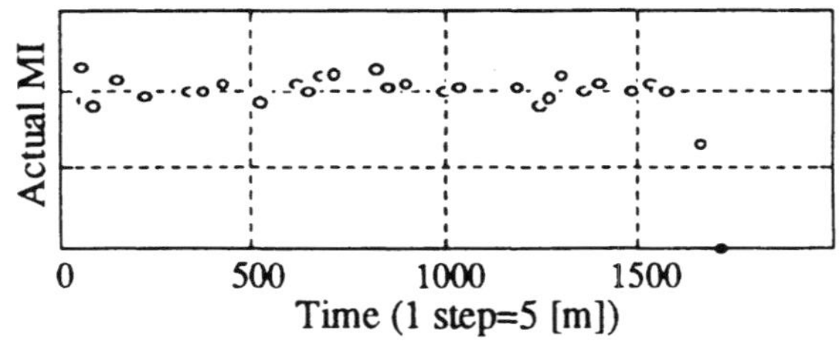

Fig. 2 Measured MI (Data Set A)

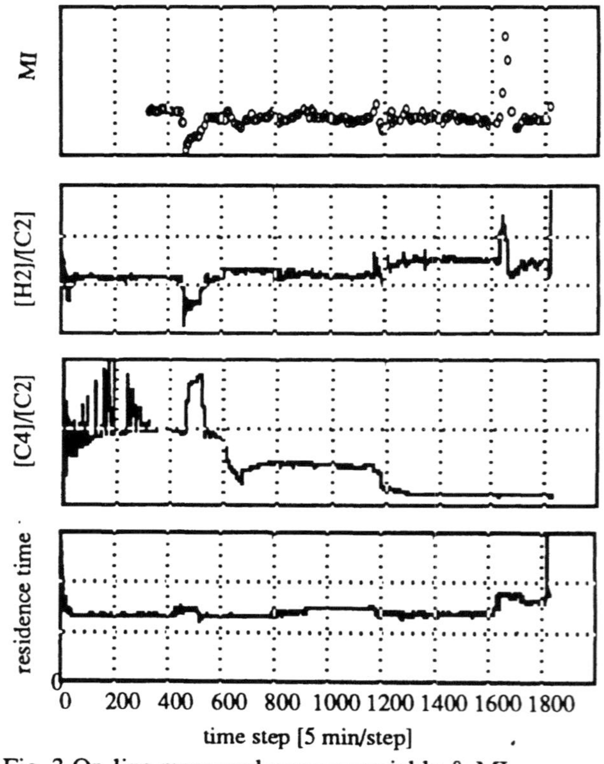

Fig. 3 On-line measured process variable & MI (Data Set B)

3. MODELS FOR MI PREDICTION

Any scheme of inferring the MI from operating data requires a model, which is the so-called quality model, that can describe the dynamic relationship between the MI and on-line measured process variables. In this study, a quality model is developed so as to predict the MI every 5 minutes from the process variables measured every 5 minutes.

The model consists of a Wave-Net (Bakshi et al, 1994; Koulouris at al. 1994) and a mass balance equation. In this section, the structure of the quality model is described with a brief explanation of the fundamental knowledge used for modeling.

3.1 Instantaneously generated polymer and total polymer

The reactions of polyethylene polymerization occur very quickly compared with the residence time of the polymer in the reactor. Thus, it is reasonably assumed that all of the polymer made during a short time interval is similar in structure, which in turn is determined by the operating conditions in the short interval (McAuley et al., 1991). The polymer generated during the short time is called the instantaneously-generated polymer and the polymer properties of the polymer are referred to as instantaneous polymer properties. The properties of the total polymer in the reactor are cumulative properties, i.e. the instantaneously-generated polymer is mixed with the already-generated polymer and constitutes the total polymer. Therefore, the properties of the total polymer are determined by accumulating the instantaneous properties. They are referred to as cumulative properties. In this study, a Wave-Net is developed for describing the relationship between instantaneous MI and process variables, and then a dynamic mass balance on the reactor is utilized to describe the dynamics of the cumulative MI.

3.2 Wave-Net Model for instantaneous MI

McAuley et. al. (1991) developed the following theoretical model for the instantaneous MI of LLDPE (Linear Low Density Polyethylene).

$$MI_i^{-1/3.5}=\sum_{j=1}^{2} w(j)\{k_0+k_1(j)\frac{[H_2]}{[C_2]}+k_2(j)\frac{[C_3]}{[C_2]}+k_3(j)\frac{[C_4]}{[C_2]}+k_4(j)\frac{[R]}{[C_2]}+k_5(j)\frac{[I]}{[C_2]}\}^{-1} \quad (1)$$

where, w(1) and w(2) are the weight fractions of the polymer produced at the two different catalyst sites. MI_i denotes the instantaneous MI and [A] is the concentration of A in the reactor.

Watanabe et. al. (1993) used the following empirical model to describe the relationship between the instantaneous MI and process variables at the HDPE (High Density Polyethylene) process,

$$MI_i=\exp\{k_0+k_1\ln\frac{[H_2]}{[C_2]}+k_2\ln\frac{[C_3]}{[C_2]}+k_3\ln\frac{[C_4]}{[C_2]}+k_4\ln[R]\}^{-1} \quad (2)$$

where, lnA is the logarithm of A.

The complexity of the polymerization reaction mechanism, makes it impossible to derive theoretically an instantaneous MI model for every polyolefine polymerization. In this study, assuming that the MI_i at time t is function of on-line measured process variables at time t, a Wave-Net is employed to learn the functional relationship between the instantaneous MI_i and process variables from the operating data and then, utilized to predict the instantaneous MI_i value from on-line measured process variables. Thus,

$$MI_i^{-1/3.5}=W\_N\{x_1, x_2, x_3, x_4, x_5\} \quad (3)$$

where W_N stands for a Wave-Net expressing the functional relationship between x_i (i=1-5) and MI_i. x_i (i=1-5) represent the process variables.

$$x_1=\frac{[H_2]}{[C_2]},\ x_2=\frac{[C_3]}{[C_2]},\ x_3=\frac{[C_4]}{[C_2]},\ x_4=\frac{[R]}{[C_2]},\ x_5=\frac{[I]}{[C_2]} \quad (4)$$

3.3 Wave-Net in multiple dimensions

The Wave-Net is an artificial neural network with basis functions drawn from the family of wavelets, i.e. the function $W\_N(x_1,x_2,..., x_5)$ is constructed by a weighted sum of a family of wavelets:

$$W\_N(x_1(t),..,x_5(t))=\sum_{s=-\infty}^{\infty}\sum_{u} c_{su}\Psi_{su}(x_1(t),..,x_5(t)) \quad (5)$$

where $\Psi_{su}(\cdot)$ are the multi-dimensional wavelets at scale, s, and translation, u.

Due to the availability of a finite amount of data, the practical expression of Eq.(5) is usually given by:

$$W\_N=\sum_{m=0}^{L}\sum_{k} c_{mk}\Psi_{mk}+\sum_{k} d_{Lk}\Phi_{Lk} \quad (6)$$

where L represents the coarsest scale, with c_{mk} and d_{Lk} being the coefficients of the multi-dimensional wavelets and scaling functions, $\Phi(\cdot)$.

The presence of several input variables, $x_1,..., x_5$, generates multi-dimensional wavelets, which are constructed by taking the tensor product of the single-dimensional wavelet ψ and scaling function ϕ. For two inputs, x_1 and x_3, which are the inputs dealt with in this study, the scaling function is given by,

$$\Phi_{LK}(x_1, x_3)=2^{-L}\phi(2^{-L}x_1-k_1)\phi(2^{-L}x_3-k_3) \quad (7)$$

and the three wavelets by,

$$\Psi_{mK}(x_1,x_3)=\sum_{\ell=1}^{3} 2^{-m}\Psi^{\ell}(2^{-m}x_1-k_1,2^{-m}x_3-k_3) \quad (8)$$

where $\ell=1,2,3$, $K=(k_1,k_3)\in Z^2$ and

$$\Psi^1(x_1,x_3)=\phi(x_1)\psi(x_3)$$

$$\Psi^2(x_1,x_3)=\psi(x_1)\phi(x_3)$$

$$\Psi^3(x_1,x_3)=\psi(x_1)\psi(x_3)$$

3.4 Cumulative MFR Model

Assume that new polymer is instantaneously generated in the stirred tank reactor and mixed with the accumulated polymer and that the amount of the accumulated polymer leaving the reactor is small compared with the polymer hold-up in reactor. Hence, the mixing and residence time behavior can be approximated by that of continuous stirred tank reactor. The dynamics of the cumulative MI can be derived by employing the mass balance equation of CSTR as follows:

Suppose that the polymer reaction rate is Rp(t), the polymer hold-up at the reactor is Hp(t) and production rate (discharged polymer) is Ep(t) at time t. The material balance of the polymer is given by:

$$\frac{dH_p(t)}{dt}=R_p(t)-E_p(t) \tag{9}$$

Denoting the weight averaged molecular weight of the instantaneously-generated polymer by Mwi and the cumulative polymer by Mwc, respectively, the average molecular weight balance is given by:

$$\frac{dH_p(t)M_{wc}(t)}{dt}=R_p(t)M_{wi}(t)-E_p(t)M_{wc}(t) \tag{10}$$

Substituting Eq.(9) into Eq.(10), the resulting equation becomes:

$$\frac{dM_{wc}(t)}{dt}=\frac{1}{\tau(t)}M_{wi}(t)-\frac{1}{\tau(t)}M_{wc}(t) \tag{11}$$

where $\tau(t)$ is $R_p(t)/H_p(t)$, which is equivalent to the polymer residence time.

Employing the empirical relationship between MI and weight averaged molecular weight Mw, i.e., MI $\propto M_w^{-3.5}$ and substituting in Eq.(11), the following equation results

$$\frac{dMI_c^{-1/3.5}(t)}{dt}=\frac{1}{\tau(t)}MI_c^{-1/3.5}(t)-\frac{1}{\tau(t)}MI_i^{-1/3.5}(t) \tag{12}$$

where, MIc and MI_i stand for the cumulative MI and instantaneous MI, respectively.

Compositions in the reactor are measured every five minutes. With respect to the time constant of the reactor, the sampling interval is small enough to assume that the instantaneous MI as well as the residence time remain constant over the interval. Hence, Eq.(12) is equivalent to

$$\begin{aligned}MI_c^{-1/3.5}(t+h)&=\exp(-h/\tau(t))MI_c^{-1/3.5}(t)\\&+(1-\exp(-h/\tau(t))MI_i^{-1/3.5}(t)\end{aligned} \tag{13}$$

where, h is the sampling interval.

Combining Eq.(13) with Eq.(3), the quality model used for predicting MI is formulated and is called the one-step ahead MI predictor: It is given by,

$$\begin{aligned}MI_c^{-1/3.5}(t)&=\alpha(t-h)MI_c^{-1/3.5}(t-h)\\&\beta(t-h)W\_N\{x_1(t-h),...,x_5(t-h)\}\end{aligned} \tag{14}$$

where, $\alpha(t)=\exp(-h/\tau(t))$ and $\beta(t)=1-\alpha(t)$.

4. FORMATION OF ADAPTIVE LEARNING FROM THE INFREQUENTLY MEASURED MI DATA

The Wave-Net defined by Eq.(6) is to be trained on available on-line measured process variables and actual MI data measured at the quality lab. Because the actual MI is measured irregularly, the multiple-step-ahead predictor has to be appropriately formulated to learn the functional relationship between MIi and process variables. In this study, it is formulated in a cascaded structure of identical one-step-ahead MI predictor.

Suppose MI is measured at the time points $\{t, t-n_1h, t-n_2h,..,t-n_jh...\}$. At t-h, MI is predicted by

$$\begin{aligned}MI_c^{-1/3.5}(t-h)&=\alpha(t-h)MI_c^{-1/3.5}(t-2h)\\&\beta(t-h)W\_N\{x_1(t-2h),...,x_5(t-2h)\}\end{aligned} \tag{15}$$

Substitute the predicted $MI_c(t-h)$ into Eq.(14) and continue the same substitution until $MI_c(t-n_1h)$ appears in the equation. The resulting equation becomes,

$$MIc(t)-\left(\prod_{k=1}^{n_1}\alpha(t-k)\right)MIc(t-n_1)=\sum_{j=1}^{n_1}\left(\prod_{k=1}^{j-1}\alpha(t-k)\right)\beta(t-j)W\_N\{x_1(t-j),...,x_5(t-j)\} \tag{16}$$

where, the sampling period h is regarded as unity.

Substituting Eq.(5) into Eq.(16) results in

$$MIc(t)-\left(\prod_{k=1}^{n_1}\alpha(t-k)\right)MIc(t-n_1)=\sum_{su=-\infty}^{\infty}c_{su}\Theta_{su}(t,t-n_1) \tag{17}$$

where, $\Theta_{su}(t,t-n_1)$ is the linear combination of wavelet functions $\{\Psi_{su}(x_1(t-j),..,x_5(t-j))\}$ $(j=1, ..., n_1)$, i.e.,

$$\Theta_{su}(t,t-n_1)=\sum_{j=1}^{n_1}\left(\prod_{k=1}^{j-1}\alpha(t-k)\right)\beta(t-j)\Psi_{su}(x_1(t-j),..,x_5(t-j))$$

By taking the same procedure, Eq.(17) can be extended to describe the relationship between $MIc(t-n_i)$ and $MIc(t-n_{i+1})$:

$$MIc(t-n_i)-\left(\prod_{k=n_i}^{n_{i+1}}\alpha(t-k)\right)MIc(t-n_{i+1})=\sum_{su=-\infty}^{\infty}c_{su}\Theta_{su}(t-n_i,t-n_{i+1}) \quad (18)$$

The structure of the multi-step-ahead predictor is illustrated in Fig. 4 and it is represented by a cascaded structure of identical one-step-ahead MI predictors.

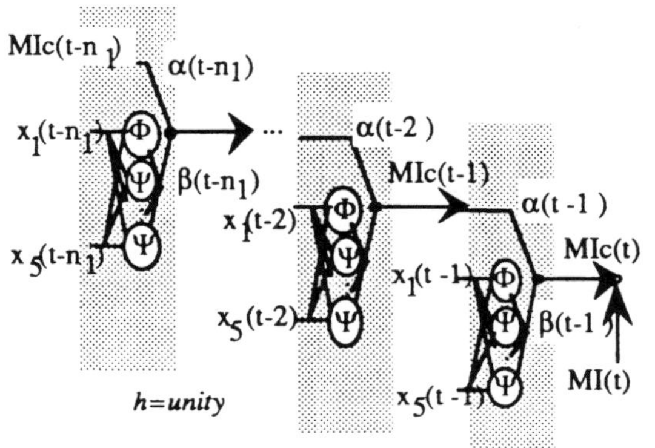

Fig. 4 The structure of the multistep-ahead MI predictor

$\alpha(t)$ and $\beta(t)$ are functions of the residence time of the polymer and are available every, h, 5 minutes. The problem of training the Wave-Net, Eq.(6), is equivalent to that of determining the coefficients c_{su} in Eq.(18), and is formulated by the following min-max problem:

$$\arg\min_{c_{su}}\max_{n_i,\, i\in Z}\left| MIc(t-n_i)-\left(\prod_{k=n_i}^{n_{i+1}}\alpha(t-k)\right)MIc(t-n_{i+1}) - \sum_{su=-\infty}^{\infty}c_{su}\Theta_{su}(t-n_i,t-n_{i+1})\right| \quad (19)$$

Thus, the Wave-Net coefficients are determined by minimizing the l^{∞} norm. Since the third term in the absolute norm is a linear function of coefficients, c_{su}, the min-max problem can be transformed into a Linear Programming problem. As described by Koulouris et. al. (1994), the learning proceeds hierarchically by specifying the type of wavelet functions and the maximum error bound.

Starting with the basis functions at the coarsest approximation, the coefficients of the scaling functions, Φ_{Lk}, are computed first through l^{∞} regression. Subsequent refinement of the input space introduces wavelets of decreasing support. The coefficients of these wavelets are computed through l^{∞} regression on the residual errors. When minimizing the l^{∞} error on the available data, the Wave-Net structure is allowed to evolve. Thus, when the error bound is not satisfied, the new basis functions at finer resolutions are introduced locally so as to make the error within the specified bound.

5. APPLICATION RESULTS

The developed multi-step MI predictor is trained both on Data A and B shown in Figs. 2 and 3, respectively. In the record of Data A, thirty three MI measurements were obtained infrequently during one hundred and sixty hours of operation. In Data B, the MI was measured at one hundred and thirty three points during one hundred fifty hour operation. The concentrations in the reactor were measured by gas chromatograph every five minutes in both operations. The hydrogen to ethylene ratio, x_1, and co-monomer to ethylene ratio, x_3, are taken as the input variables of Wave-Net. The mexican hat wavelets (Daubechies, 1992) were employed for this study.

Figures 5 and 6 show the estimates of the multi-step MI predictor and the actual MI in the data A and B, respectively. The solid line represents the estimates and the asterisk represents the actual MI values. The estimation is carried out by the trained model using the first actual MI measurement as the initial value. As can be seen in the simulation results, the predictor can successfully estimate the MI values for both data. Figure 7 illustrates the functional relationship between MIi and input variables, x_1 and x_3, captured by the Wave-Net. Figure 8 shows the location of the basis functions of the Wave-Net in the normalized input space. The dots represent the input data, x_1 and x_3. The center of each circle corresponds to the center of each Mexican-hat wavelet function and the radius of the circle corresponds to the scale of the wavelet function.

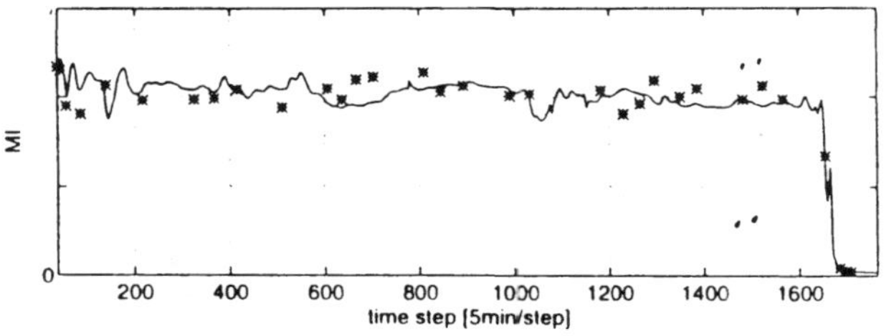

Fig. 5 MI prediction on DATA A

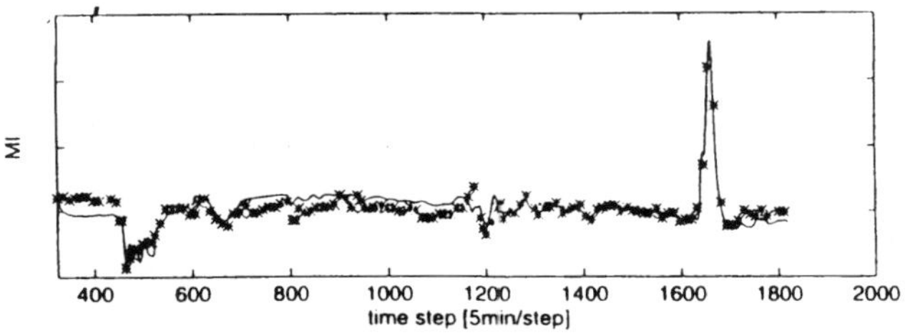

Fig. 6 MI prediction on DATA B

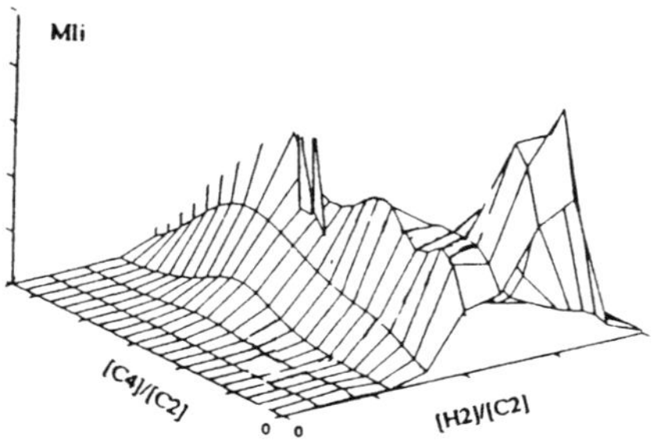

Fig. 7 Functional space approximated by Wave-Net

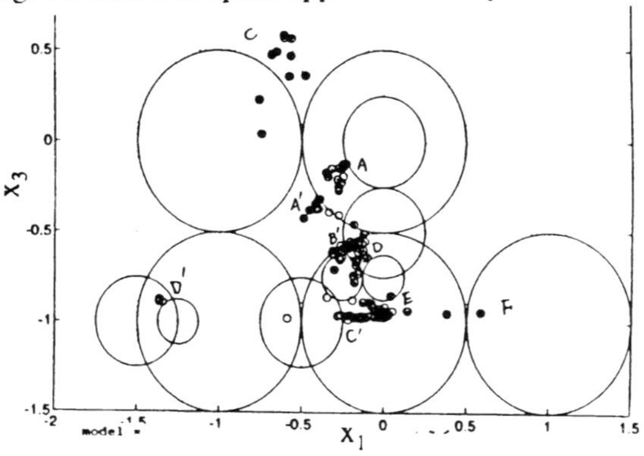

Fig. 8 Location of the basis functions of the Wave-Net in normalized input space

To compare with the predictions of the theoretically derived model, simulation calculations were carried out using Eqs.(1) and (13). The model parameters were determined through nonlinear least squares regression on Data A. Figure 9 illustrates the results of estimation by the model for Data A. It shows good agreement with the measured MI. However, the theoretical model gives overestimates, when used on Data B, as illustrated in Fig. 10.

As illustrated in Figs. 9 and 10, even when the theoretically-derived model is used, the model with one parameter set cannot perfectly describe MI behavior for all operating conditions. Therefore, it is necessary to prepare the appropriate sets of parameter values according to the operating regions or one has to update the parameter values on-line, employing some estimation technique such as Kalman filter or recursive least-squares scheme.

Instead of updating the parameter values, the proposed Wave-Net model can predict the MI behavior successfully on both records, Data Sets A and B, without needing different model structures for different operating conditions.

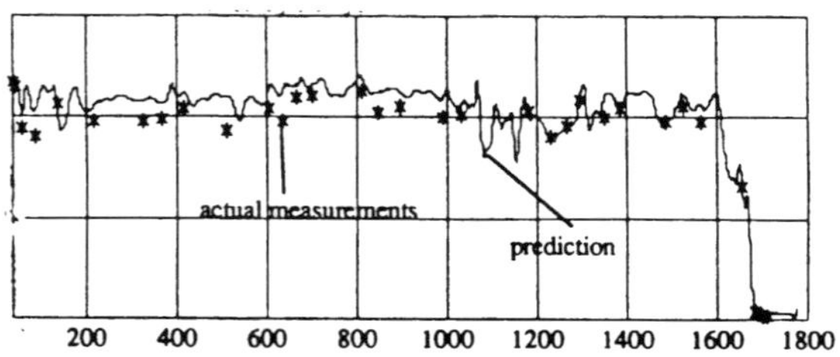

Fig. 9 MI prediction by the theoretically-derived model on Data Set A

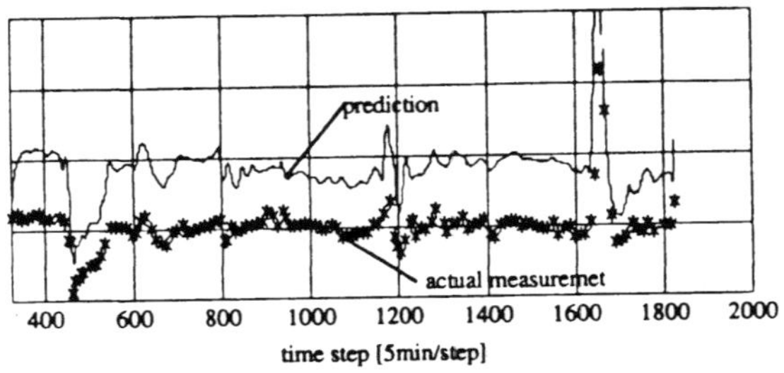

Fig. 10 MI prediction by the theoretically-derived model on Data Set B

6. CONCLUSION

A methodology for the on-line melt index prediction in an industrial high density polymerization process has been presented. The quality model used in the inferential scheme consists of two parts, mass balance and an empirical non-linear map given by the Wave-Net model. The mass balance model relates the melt index of total polymer and that of the instantaneously-generated polymer. The Wave-Net model reflects the functional relationship between the melt index of instantaneously-generated polymer and the process state variables; hydrogen, ethylene, and butane concentrations. By formulating the multiple-step ahead predictor as a cascaded structure of identical Wave-Net models with mass balances, the inferential system can adaptively train the Wave-Net model on the irregularly spaced MI measurements and on-line measured process variables. The Wave-Net coefficients as well as the structure itself can be updated when the new results of laboratory analysis do not satisfy the estimation-error bounds of the predictor. It has been illustrated that the proposed inferential scheme is capable of successfully predicting the MI and has the potential as an industrial on-line monitoring scheme of the polymer quality.

REFERENCES

Bakshi, B. R., A. Koulouris and Geo. Stephanopoulos (1994). *Wavelet Applications in Chemical Engineering*, chap. 5, (R. L. Motard ed.), Kluwer academic publication

Bohm,L.L., (1992). Dechema-Monographen Band 127, p.257

Daubechies, I (1992) *Ten Lectures on Wavelets*, SIAM

Koulouris, A., M. Ohshima, Geo. Stephanopoulos (1994), The L infinite extension of Wave-Net for on-line identification, AIChE annual mtg., San Francisco,

McAuley, K.B. and J.F. MacGregor (1991) On-line Inference of Polymer Properties in an Industrial Polyethylene Reactor, AIChE Journal, vol. 37 No. 6, p. 815,

Ogunnaike, T., et. al. (1991). On-line modeling and control of an industrial terpolymerization reactor, AIChE mtg, LA

Watanabe et. al., (1993). Quality Control System for an Industrial High Density Polyethylene Process, AIChE mtg, St. Louis, # 149m

STATE AND PARAMETER ESTIMATION WITH AUGMENTED KALMAN FILTER APPLIED TO AN INDUSTRIAL POLYMER REACTOR

H. DIDRIKSEN,* M. HILLESTAD** K. S. ANDERSEN*** and J. G. BALCHEN*

*Dept. of Engng. Cybernetics, The Norwegian Institute of Technology, N-7034 Trondheim, Norway

**Statoil Research Centre, Postuttak, N-7005 Trondheim, Norway. Email: mhi@fou.tr.statoil.no

***Borealis a.s, N-3960 Stathelle, Norway

Abstract. An Augmented Kalman Filter is applied to a non-linear model of a gas phase copolymer reactor where 90 state variables and 7 parameters are estimated. The distributed process is represented by a series of volume elements and the model equations are based on the conservation of mass, energy and momentum for the volume elements. The filter estimates the state variables very well. The estimates of the parameters are reasonable even though heat transfer coefficients are affected by mismatches in the energy balance of the model.

Key Words. Gas phase copolymerization, state space modeling, distributed parameter system, state and parameter estimation, Augmented Kalman Filter

1. INTRODUCTION

In model based control and estimation mathematical models of the process are required. The fact that a mathematical model rarely will be an exact representation of the process, imply that model states and parameters have to be updated. Some parameters can be tuned off-line and can be considered as constants, whereas others will vary with disturbances and need to be updated on-line. There are many aspects that make on-line identification of the model difficult; as for example many state variables and non-linearities. Here is an Augmented Kalman Filter (AKF) applied to a model of a gas phase copolymer reactor where these issues are discussed. Control of the reactor is not discussed in this paper.

The copolymer reactor, shown in Figure 6, can be characterized as a *horizontal stirred bed* reactor. The process is a gas phase copolymerization reactor. Polypropene with still active catalyst enters the reactor and a stirrer mixes gas and polymer powder and transports the powder throughout the reactor. Ethene and propene monomer gas are continuously fed to the reactor. The excess gas at the end is recycled. Part of the excess gas is purged to prevent accumulation of inerts and is used to control the pressure in the reactor. The polymerization is exothermic and the heat of the reaction is removed by a heat exchange system.

2. MODEL OF THE COPOLYMER REACTOR

Any automatic control strategy for this process must rely on a model since the objectives are to control unmeasurable or hardly attainable process variables and because of the strong interactions between the variables. An Input-Output model would not be appropriate in this application since estimates of internal process variables are required. In order to design a model based controller, a non-linear state space model of the process is derived from first principles (Strand and Hillestad, 1989). The model presented here is in principle the same as the one described by Strand and Hillestad (1989), however, there are some minor modifications. The on-line model estimator may also be used as an operator support system alone.

The *horizontal stirred bed* copolymer reactor is a distributed parameter system. The flow pattern is modelled by a series of N volume elements in series, with backmixed flows as described by Lapidus and Seinfeld (1974). The model equations are derived from the conservation of mass, energy and momentum for each element. The cooling system, being of a distributed nature, is also modelled by a series of elements. The cooling system is given the same physical discretization as the reactors. Each element is described by the following state variables:

- mass of ethene gas, m_{g1}
- mass of propene gas, m_{g2}
- mass of inert gas, m_{g3}
- mass of ethene in the copolymer, m_{s1}
- mass of propene in the copolymer, m_{s2}
- mass of homopolymer polypropene (produced in the PP-reactor), m_{s3}
- mass of catalyst, m_{s4}
- temperature in the reactor, T
- temperature in the cooling system, T_C

A schematic diagram of a reactor element with cooling system is depicted in Figure 7.

2.1. *Polymerization Reaction*

Only simplified propagation reactions are described here since average molecular weight and molecular weight distribution are not an issue in this application. Ideal copolymerization is assumed and the rate of propagation of monomer j (ethene; $j = 1$, propene; $j = 2$) in element i is modelled:

$$\begin{aligned} r_{ij} &= k_j e^{-E_j/RT_i} m_{s4,i} e^{-k_d t} p_{ij} &(1)\\ t &= i\Delta x / v_s &(2)\end{aligned}$$

where p_{ij} is the partial pressure of monomer j in element i.

2.2. *Conservation of Mass*

The equations below describe the gas-phase ($p = g$) and the solid phase ($p = s$) mass balances:

element no. 1

$$\begin{aligned}\frac{dm_{pj1}}{dt} &= -(1+b_p)w_{pj1} + b_p w_{pj2}\\ &\quad + w_{pj1}^{in} \pm r_{j1} &(3)\end{aligned}$$

element no. $i = 2, \cdots, N-1$:

$$\begin{aligned}\frac{dm_{pji}}{dt} &= (1+b_p)w_{pji-1} - (1+2b_p)w_{pji}\\ &\quad + b_p w_{pji+1} + w_{pji}^{in} \pm r_{ji} &(4)\end{aligned}$$

element no. N:

$$\begin{aligned}\frac{dm_{pjN}}{dt} &= (1+b_p)w_{pjN-1} - b_p w_{pjN}\\ &\quad + w_{pjN}^{in} - w_{pjN}^{out} \pm r_{jN} &(5)\end{aligned}$$

w_p^{in} and w_p^{out} are the inlet and outlet mass flows. The sign of r_{ji} will be "-" for the gas phase and "+" for the solid phase.

The volumetric gas and powder flows are given:

$$\begin{aligned} w_{gi} &= \rho_{gi} A_{gi} v_{gi} &(6)\\ w_{si} &= \rho_s A_{si} v_{si} &(7)\end{aligned}$$

where A_s is the cross section area of the reactor covered with powder and A_g is the area free of powder. The solid velocity will mainly be determined by the stirrer. Since the stirrer is rotating with constant speed, the solid velocity v_s is assumed to be constant.

For the gas phase, a steady state momentum balance is applied:

$$v_{gi} = K_{mg}(p_i - p_{i+1})/\Delta x \quad (8)$$

2.3. *Conservation of Energy*

The gas and solid phase temperatures are assumed to be equal. Hence, it is sufficient to establish one energy balance for each reactor element. The basis for the energy balance is the conservation of internal energy, U.

element no. 1:

$$\begin{aligned}\frac{dU_1}{dt} &= -(1+b_g)h_{g1}w_{g1} - (1+b_s)h_{s1}w_{s1}\\ &\quad + b_g h_{g2} w_{g2} + b_s h_{s2} w_{s2}\\ &\quad + h_{g1}^{in} w_{g1}^{in} + h_{s1}^{in} w_{s1}^{in}\\ &\quad - Q_{C1} + W_{st}/N\\ &\quad + \sum_{j=1}^{2}(-\Delta H_j) r_{1j} &(9)\end{aligned}$$

element no. $i = 2, \cdots, N-1$:

$$\begin{aligned}\frac{dU_i}{dt} &= (1+b_g)h_{gi-1}w_{gi-1}\\ &\quad + (1+b_s)h_{si-1}w_{si-1}\\ &\quad - (1+2b_g)h_{gi}w_{gi} - (1+2b_s)h_{si}w_{si}\\ &\quad + b_g h_{gi+1} w_{gi+1}\\ &\quad + b_s h_{si+1} w_{si+1} + h_{g1}^{in} w_{g1}^{in}\\ &\quad - Q_{Ci} + W_{st}/N\\ &\quad + \sum_{j=1}^{2}(-\Delta H_j) r_{ij} &(10)\end{aligned}$$

element no. N:

$$\frac{dU_N}{dt} = (1+b_g)h_{gN-1}w_{gN-1}$$

$$+(1+b_s)h_{sN-1}w_{sN-1}$$
$$-b_g h_{gN} w_{gN} - b_s h_{sN} w_{sN}$$
$$+h_{gN}^{in} w_{gN}^{in} - h_{gN}^{out} w_{gN}^{out}$$
$$-h_{sN}^{out} w_{sN}^{out}$$
$$-Q_{CN} + W_{st}/N$$
$$+\sum_{j=1}^{2}(-\Delta H_j) r_{Nj} \qquad (11)$$

Q_C denotes the energy removed by the cooling system and W_{st} denotes thermal energy added to the polymer powder by the stirrer. The energy balance over an element for the cooling system is given below:

$$Q_{Ci} = \alpha_C A_C (T_i - T_{Ci}) \qquad (12)$$
$$c_{pC} m_C \frac{dT_{Ci}}{dt} = w_C c_{pC}(T_{Ci-1} - T_{Ci}) + Q_{Ci} \qquad (13)$$

3. MODEL IDENTIFICATION

3.1. *Flow Pattern*

An important model structural "parameter" is the number of volume elements, N. The number of elements has an effect on the model response and the resolution of the axial gradients.

It was found that the number of elements equals $N = 5$ (for each reactor) gives a model structure which describes the process dynamics reasonably good. The choice of N is based upon qualitative knowledge about the axial dispersion in the solid phase, information on the response in temperature to variations in catalyst feed. Increasing N, say to $N = 10$, did not lead to a better performance of the AKF or any significant difference in the estimates. With 5 elements for each reactor, the model will consist of 100 state variables.

3.2. *Time Varying Parameters*

The model contains a large number of parameters. Most of these can be found in chemical handbooks, but some are model specific and must be found from laboratory experiments and from plant data. Some parameters may vary due to unmodeled mechanisms. The copolymerization reaction rate is influenced by many factors. Deviations from the modelled rate, eq. 1, can be due to variations in the concentration of active sites or the size of the polymer particles which alters the diffusion of monomer through the particle to the catalyst. The fundamental mechanisms of catalyst deactivation is not known. There are in other words unmodelled mechanisms and the lumped parameters of the reaction rate will vary with time (or rather process conditions). The heat transfer coefficients can change slightly during a copolymer production campaign. This change may be due to polymer powder which sticks to the the reactor wall. Potential parameters for on-line estimation are:

$$\theta_I = [\, k_e \; k_p \; k_d \; E_e \; E_p \; \alpha_{C1} \; \alpha_{C1} \; w_{flush} \,]$$

It is, however, not possible to estimate θ_I on-line because of the low identifiability of the parameter set. A subset of the parameter vector is selected based on physical knowledge and identifiability analysis. The identifiability of different parameter sets were examined by regarding the corresponding eigenvalue of the covariance matrix of the scaled parameter vector. A parameter combination characterized by an eigenvector $\mathbf{m}$ is poorly identifiable when the eigenvalue is large (Balchen, 1979). The selected parameter vector for on-line estimation is therefor:

$$\theta_{II} = [\, k_e \; k_p \; \alpha_{C1} \; \alpha_{C1} \; w_{flush} \,]$$

3.3. *Observability*

The mass balance of the catalyst is kept out of the model when the AKF algorithm is applied. This is done to obtain an observable linearized model. It is important here to note that the mass of catalyst in the model is not dependent on other of state variables (like temperature, mass of homopolymer etc.). The equations for conservation of catalyst are in other words included when integrating the nonlinear model, eq. 14, and mass of catalyst is not updated by the Kalman filter.

3.4. *Measurements*

The measurements used by the AKF are: temperature at outlet of reactor 1 and 2 (y_1 and y_2), temperatures at outlet of cooling system, molar fraction of ethene and propene in reactor 1 and 2 (y_7 - y_{10}) and gas to flare (y_{11}). The sampling interval of the AKF is $\Delta t = 350s$ which is equal to the sampling frequency of the gas chromatographs. The gas chromatograph analyses the gas in reactor 1 and 2 alternately and the measurement vector alters between:

$$\mathbf{y}_I = [y_1, y_2, y_3, y_4, y_5, y_6, y_7, y_8, y_{11}]^T$$

$$\mathbf{y}_{II} = [y_1, y_2, y_3, y_4, y_5, y_6, y_9, y_{10}, y_{11}]^T$$

3.5. *Augmented Kalman Filter*

The realization of the AKF is as follows:

$$\bar{\underline{x}}(k+1) = \hat{\underline{x}}(k) + \int_{t_k}^{t_{k+1}} f[\hat{\underline{x}}(k), u(k), k]dt \quad (14)$$

$$\bar{\underline{X}}(k+1) = \Phi(k)\hat{\underline{X}}(k)\Phi(k)^T + \underline{V} \quad (15)$$

$$K(k) = \bar{\underline{X}}(k)D^T(k) [D(k)\bar{\underline{X}}(k)D^T(k) + W(k)]^{-1} \quad (16)$$

$$\hat{\underline{x}}(k) = \bar{\underline{x}}(k) + K(k)[y(k) - g(\bar{\underline{x}}(k))] \quad (17)$$

$$\hat{\underline{X}}(k) = [I - K(k)D(k)]\bar{\underline{X}}(k)[I - K(k)D(k)]^T + K(k)W(k)K^T(k) \quad (18)$$

It is important that the parameters are scaled before tuning of the AKF. The parameters are scaled so that a unit change in a parameter causes changes in the measurements of similar magnitude.

The V-matrix is a diagonal matrix and has squared values of 0.1-0.3 % of nominal values of the state variables and 0.2 - 0.4 % for the scaled parameters. The diagonal elements in the W-matrix reflects the inaccuracy and the measurement noise.

3.6. *Simulation Against Plant Data*

The AKF is tested against logged plant data. A selection of the measurements and parameters are shown in Figures 1 through 5. The measurements are normalized. The two most noteworthy situations in the data set takes place at $t \approx 1900$ min and in the period $t \approx 4000 \rightarrow 5400$ min. At $t \approx 1900$ min., there is a fall in catalyst feed in connection with a grade change and altered process conditions in the polypropene reactor. The resulting decrease in polymerization reaction can be seen in a rise in molar fraction of ethene and propene and also a fall in reactor temperature until the flow of cooling water is reduced to keep the temperature. At $t \approx 4000$ min., a fall in polymerization reaction can be traced in a rise in molar fraction of monomer in the gas, particularly for propene. The monomer gas feed is constant around this point of time. The fall in reaction can be related to a decrease in the number of active sites on the catalyst or even more likely to an increase in the productivity. The decrease is most pronounced for the frequency factor for polymerization of propene. The reaction rate increases again at $t \approx 4800$, which is seen in the measured molar fractions and the estimated frequency factors. The change in reaction is again most pronounced for propene.

The heat transfer coefficients vary more than one might expect from a physical explanation. The estimated values will catch up varying mismatches in the energy balance of the model (up to 10 %). The estimated temperatures in the cooling system will also be affected by this mismatch (figure not shown). The AKF is tuned in such a way that most of the mismatch is placed on the estimates of the temperature in the cooling system rather than at the estimates of the temperatures in the reactor. There are many measurements which affect the energy balance and the source(s) of the mismatch can be difficult to place.

Note also in the displayed plant data some unpredictable events at $t \approx 2000$ min. and $t \approx 5200$ min. where a plug in the pipe connecting the two reactors courses some fast responses in the measured temperature and molar fractions and. At $t \approx 4000$ min there is probably a plug in the pipe upstream of the copolymer reactor.

4. CONCLUSION

An Augmented Kalman Filter is applied to a nonlinear model of a gas phase copolymer reactor where 90 state variables and 7 parameters are estimated. The AKF estimates the state variables well. The estimates of the parameters are reasonable even though heat transfer coefficients are affected by mismatches in the energy balance of the model. The catalyst activity is obviously varying and it is therefore important to estimate the frequency factors on-line.

5. REFERENCES

Balchen, J.G. (1979), Stokastiske Metoder og Estimeringsteori (Stochastic methods and Estimation Theory) (in Norwegian), Division of Engineering Cybernetics, NTH, Trondheim.

Lapidus L. and Seinfeld J.H., (1974), Mathematical Methods in Chemical Engineering, Prentice Hall Inc., New Jersey.

Strand K. and M. Hillestad (1989), Modellering for dynamisk simulering av kopolymerreaktor (in Norwegian), available from the authors.

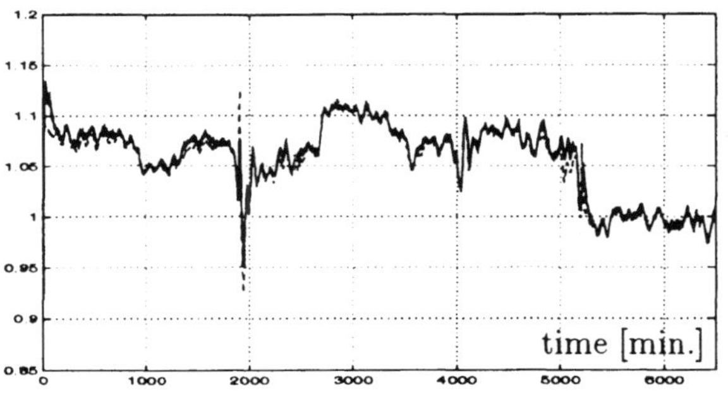

Fig. 1. Temperature at outlet of reactor 1, y_1, measured; dashed line, estimated; solid line.

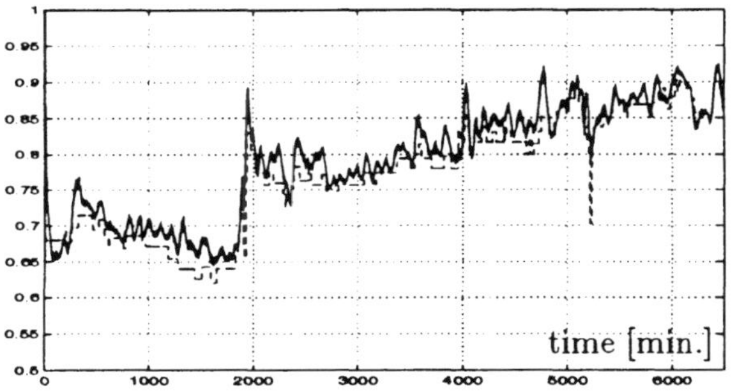

Fig. 2. Molar fraction of ethene in reactor 1, y_7, measured; dashed line, estimated; solid line.

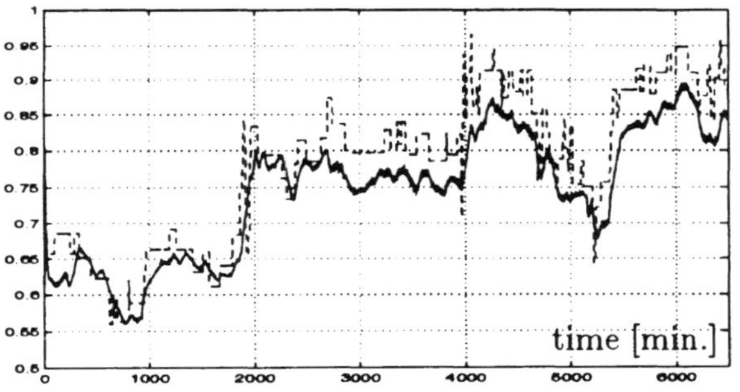

Fig. 3. Molar fraction or propene in reactor 1, y_8, measured; dashed line, estimated; solid line.

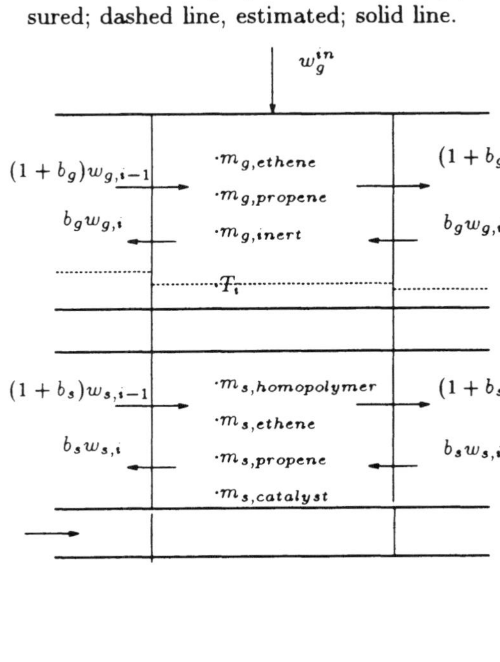

Fig. 7. Reactor element.

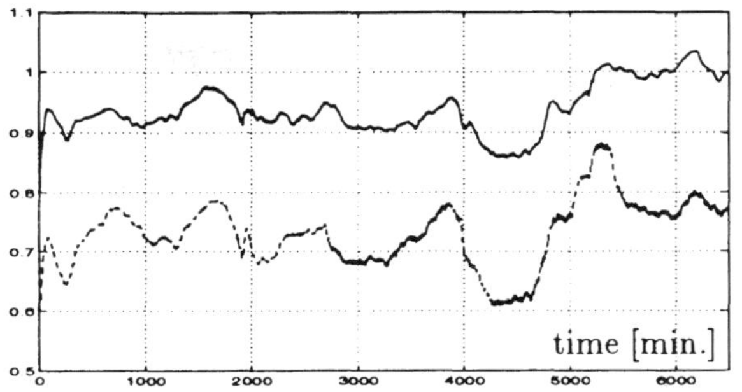

Fig. 4. Estimated frequency factors, k_e; solid line, and k_p; dashed line.

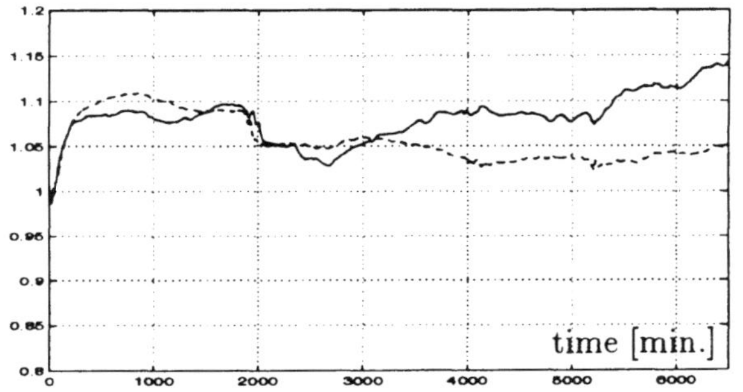

Fig. 5. Estimated heat transfer coefficients for the cooling system.

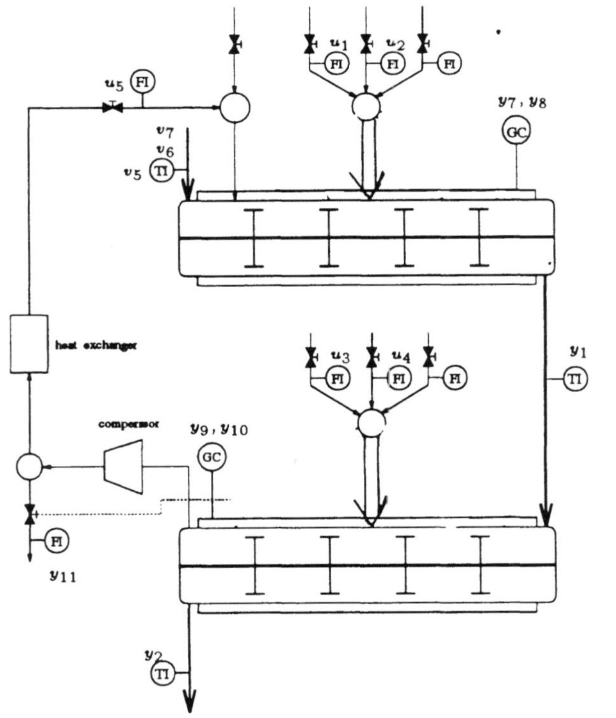

Fig. 6. The copolymer reactor system.

KALMAN FILTER AND NEURAL NETWORK FOR ON-LINE ESTIMATION OF POLYMER CHAIN CHARACTERISTICS

S. Guiochon, G. Defaye, C. Vidal, L. Caralp

Equipe I.A.G.P. - U.R.A. 503 - Ecole Nationale Supérieure de Chimie et de Physique de BORDEAUX - Université BORDEAUX I - 33405 Talence Cedex FRANCE

Abstract: In the past, many efforts have been paid to provide a general way for on-line monitoring of various molecular characteristics while the polymerization reaction proceeds (e.g. average chain length, chain length distribution, etc.). An efficient tool for reaching such characteristics is supplied by the Kalman filtering. The main limitation to its use however, comes from the need of a relevant mathematical model of the process. It was decided to substitute a neural network-based model for the more common kinetic-based model of the average chain lengths evolution.

Keywords: Batch polymerization; On-line state estimation; Extended Kalman filters; Neural-network models; On-line average chain lengths estimation.

1. INTRODUCTION

1.1 Overview

The number of applications using polymer material becomes larger and larger. Each application generally requires specific properties, which are strongly dependent on the structure of the polymer (microscopic or macroscopic). The conditions in which the polymers are synthesized greatly influence their structure, and consequently, their final properties. It is thus important to master this aspect of the synthesis, to estimate and to control in real-time, a certain number of essential characteristics of the polymers. Among the latter group, the number and weight-average lengths (henceforward to be referred to as μ_n and μ_w) and the polydispersity index can be found. Unfortunately, these different magnitudes are not easily accessible due to the absence of physical sensors which permit continuous measurement without a long delay.

Several experiments have been carried out in hopes of surmounting these problems. The tool the most frequently used in order to produce an on-line estimation of the above mentioned magnitudes is the extended Kalman filtering. The principal limitation to its employment lies in the need for a suitable mathematical model of the chemical system. It is possible in certain cases to obtain such a model (e.g. for the radical polymerization of methylmethacrylate and styrene). However, this is no longer possible when more complicated chemical processes are involved. Consequently, the classic Kalman filtering can no longer be used.

In this report, a consistent replacement of the detailed analytic model by a neural network is proposed. The reaction used to test the validity of the approach is the methylmethacrylate polymerization in ethyl acetate as solvent, initiated by AIBN for its ease in experimental operation. Certainly for this type of reaction, the employment of the mathematical model-based Kalman filtering has demonstrated its capabilities. The advantage invested in the use of the neural network can be extrapolated not to be envisaged solely for application within the framework of the poly(methylmethacrylate) synthesis (henceforward to be referred to as MMA).

1.2 Previous treatments of the problem

J.H. Jo and S.G. Bankoff (1975) were probably the first to apply the extended Kalman filtering to estimate the rate of conversion and the number-average chain length of poly(vinyl acetate) produced in a semi-continuous, stirred reactor by free radical polymerization. The state vector was composed of these two magnitudes and the measurement vector consisted of the refractive index of the medium and its viscosity, obtained in continuum without delay. Schuler and Suzhen (1985) proposed a solution to the problem of estimating chain characteristics of the poly(styrene) produced in a semi-continuous reactor. They divided the state variables into two systems, one being observable (it consisted of concentrations of the initiator, monomer, solvent and the temperature of the reaction mixture); the other is non-observable whose values do not influence the magnitudes of the first system (including the moments associated with the distribution functions of the molecular mass, and the concentrations of different chains of polymer). Papadopoulou and Gilles (1986) included the measurements by size exclusion chromatography (SEC) as measurement, thus allowing the possibility of estimation adjustments during the course of the reaction. Several experiments have made beneficial use of this kind of estimations during the course of reactions, in order to control important characteristics of the polymer. For example, Kim et al. (1992) applied the extended Kalman filtering for control in the case of an homopolymerization, Kozub and MacGregor (1992) for a copolymerization.

2. DESCRIPTION OF THE PROCESS

2.1. Chemical process

Table 1. shows the kinetic scheme of the polymerization process. The analytic model that proceeds from it is shown in Table 2. The values of the different constants and physical parameters are indicated in the appendix.

Table 1. Kinetic Scheme of MMA polymerization

- Initiation :

$$I \xrightarrow{k_d} 2R$$

$$R + M \xrightarrow{k_i} P_1$$

- Propagation :

$$P_n + M \xrightarrow{k_p} P_{n+1}$$

- Chain Transfer to solvent and monomer :

$$P_n + S \xrightarrow{k_s} M_n + P_1$$

$$P_n + M \xrightarrow{k_f} M_n + P_1$$

- Termination by disproportionation :

$$P_n + P_m \xrightarrow{k_{td}} M_n + M_m$$

2.2. Reactor configuration

The runs are carried out in a batch reactor with a two liters capacity, around which circulates a heat-transfer fluid permitting the temperature control of the reactive medium. Its temperature is maintained at a constant value of 70°C during the polymerization process due to the use of an internal model control algorithm. In order to avoid the introduction of impurities into the reactor, a continuous release of nitrogen by gas flow is operational. On the other hand, no reactive liquid is distilled - a potential source of impurities (oxygen, water, reaction inhibitor) and hence disturbances for the filter -. The schematic set-up is represented in figure 1.

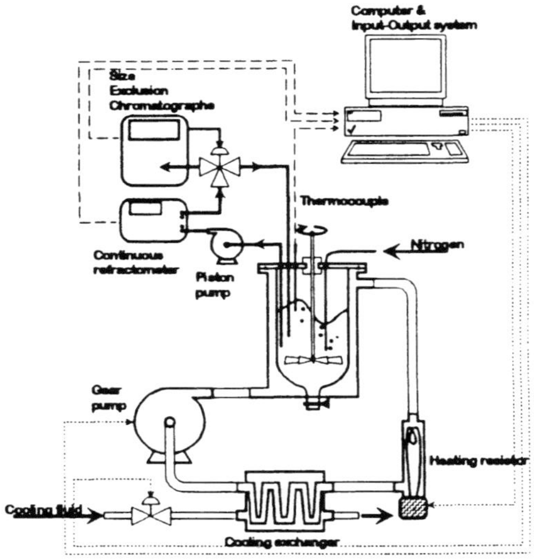

Fig. 1. Diagram of the pilot batch reactor

The available measurement apparatuses are an on-line continuous refractometer and a SEC whose running time (including the elution time and treatment of the results) is approximately 15 minutes. The refractive index variation of the reaction medium is practically linear with the conversion rate, permitting a continuous estimation of the monomer concentration (J.H. Jo and S.G. Bankoff (1975)). As for the SEC, it permits finding the value of the number and weight-average chain lengths.

Table 2. Kinetic equations

$$\dot{[I]} = -k_d.[I] \qquad ; [P] = \sqrt{\frac{2.f.k_d.[I]}{k_{td}}}$$

$$\dot{[S]} = -k_s.[S].[P]$$

$$\dot{[M]} = -k_i.[R].[M] - (k_p + k_f).[P].[M]$$

$$\dot{\lambda}_0 = \left[k_s.[S] + k_f.[M] + k_{td}.[P]\right].\alpha.[P]^2$$

$$\dot{\lambda}_1 = \left[k_s.[S] + k_f.[M] + k_{td}.[P]\right].\alpha.(2-\alpha).\frac{[P]}{1-\alpha}$$

$$\dot{\lambda}_2 = \left[k_s.[S] + k_f.[M] + k_{td}.[P]\right].(\alpha^3 - 3\alpha^2 + 4\alpha).\frac{[P]}{(1-\alpha)^2}$$

$$\alpha = \frac{k_p.[M]}{k_p.[M] + k_s.[S] + k_f.[M] + k_{td}.[P]}$$

The continuous loop of sampling not being operational yet, the different results presented momentarily have been obtained by simulating real-time from analysis results determined off-line.

3. KALMAN FILTERING WITH ANALYTIC MODEL

The filtering method proposed by Kalman (1960) resolves the problem of estimating the state of dynamic linear systems subjected to deterministic inputs with additive Gaussian white noise on the states and outputs. This estimator is consistent and minimum variance. It was extended by Jazwinski (1970) to the case of non-linear systems. The application conditions of this filter demand a particular statistical form of noise and the almost-perfect knowledge of a mathematical model of the process. If the conditions are not fulfilled, it is then only possible to produce a sub-optimal filter.

The form of the chosen filter is identical to that proposed by Schuler and Suzhen (1985). It consists of dividing the state variables into two distinct groups.

The first group is constituted by initiator and monomer concentrations. The measurement vector to which it is associated consists of the sole monomer concentration value, deducted from the refractive index of the medium. This measure is available without delay and a sample period of 10 seconds is attributed to it. The structure of the extended Kalman filter associated with this group of variables is presented in Table 3.

The second group consists of variables not influencing the evolution of the first group. Thus it was chosen to include the 0-th and 2-nd moments of the number chain length distribution. The moment of order 1 being calculated from the monomer concentration, it was not incorporated for the variables of the second group (Kiparissides *et al.* (1990)). Moreover, in order to take into account the possibility of errors in the model (due for instance to the presence of impurities in the core of the reaction medium), it was decided to augment the second state vector by the value of a pre-exponential constant of the kinetic model, directly influencing the value of the different moments, namely Ks. The measurement vector associated with this group of variables is composed of the number and weight-average chain length, issuing from analysis results by SEC. This information is available with a significant delay on the order of 15 minutes. The structure of the Kalman filtering corresponding to the second group will not be described in this paper. The form of the model alone is represented in Table 4.

Table 3. Continuous-discrete extended Kalman filter on first group variables

State vector : $x1 = [n_I \quad n_M]^T$

Measurement vector : $z1 = [n_M]$

System model:

$\dot{x}1(t) = \mathcal{F}_1(x1(t)) + w(t) \qquad w = N(0, Q1)$

$z1_k = C1.x1_k + v_k \qquad v = N(0, R1)$

Estimation prior to measurement :

$\hat{x}1(-)$ and $P1(-)$ are obtained by integration of

$\dot{\hat{x}}1(t) = \mathcal{F}_1(\hat{x}1(t))$

$\dot{P}1(t) = \mathcal{F}_1(\hat{x}1(t)).P1(t) + P1(t).\mathcal{F}_1(\hat{x}1(t))^T + Q1(t)$

Updates:

$K1 = P1(-).C1^T.[C1.P1(-).C1^T + R1]^{-1}$

$P1(+) = [I - K1.C1].P1(-)$

$\hat{x}1(+) = \hat{x}1(-) + K1.[z1 - C1.\hat{x}1(-)]$

Table 4. Model equations for the second group variables

State vector : $x2 = [\lambda_0 \quad \lambda_2 \quad K_s]^T$

Measurement vector : $z2 = [\mu_n \quad \mu_w]^T$

Model equations :

$\dot{x}2 = \mathcal{F}_2(x1(t), x2(t)) + w'(t) \qquad w' = N(0, Q2)$

$z2_k = \mathcal{H}_2(x2_k) + v'_k = [\lambda_1/\lambda_0 \quad \lambda_2/\lambda_0]^T + v'_k$

with $\lambda_1 \approx n_M(t=0) - n_M$ and $v' = N(0, R2)$

Remarks : The temperature of the reaction medium being constantly maintained (fluctuations of less than 0.4°C), it has not been the object of filtering. Furthermore, in order to reduce the risk of errors due to imprecise numerical calculations induced by filtering, all the variables have been normalized.

The results obtained with the analytic model-based filter are presented in figures 2-5. They will be compared to those obtained with the neural models.

4. KALMAN FILTERING WITH NEURAL NETWORK MODELS

4.1. Neural networks

The employment of neural networks as an efficient tool for non-linear modeling is the object of reviewed interest over the past few years. Diverse experiments have demonstrated the advantages that can be drawn from these universal approximators in estimating and controlling complex processes (Thibault et al. (1990), Willis et al. (1991)).

The structure of the chosen network is of a feedforward type and is totally connected (figure 2.). Nerrand et al. (1993) showed that this type of network structure permits representation of a vast grouping of networks (for example, multi-layered networks).

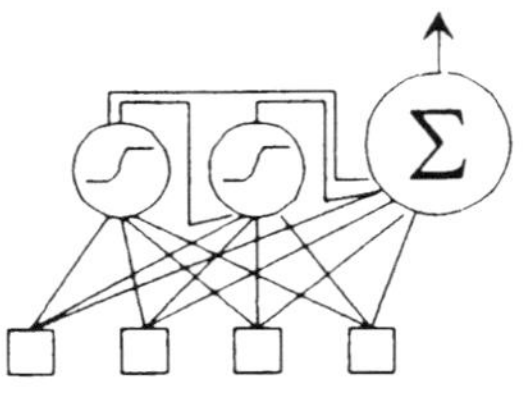

Fig. 2. Network structure

Two types of networks have been tested, permitting the furnishing of an on-line estimate of the number and weight-average chain lengths in continuum during the polymerization. Both use as inputs, the variables that act directly on the value of the average lengths; namely the concentration of monomer (accessible from the measurements by the refractometer), and the initial quantity of the initiator. In all cases, the different inputs and outputs are normalized.

The inputs of thc first network type are 4 in number : initial initiator quantity, instantaneous monomer quantity, time elapsed since on-set of the reaction and a constant value consisting of a bias. Two single-output networks are constructed following this form in order to furnish an estimated value of μ_n and μ_w

The other network form provides as output an estimate of the relative increase of the average lengths in relation to time :

$$\frac{d\mu}{dt} \approx \frac{\Delta\mu}{\Delta t} \text{ with } \Delta t = 1 \text{ seconde.}$$

Its inputs are composed by the initial quantity of initiator, the instantaneous quantity of monomer and a bias.

4.2. Learning phase of the networks

This phase has been realized on the basis of simulations accomplished on the analytic model due to the insufficient number of experimental results. Several reactions have been simulated varying initial quantities of initiator and monomer. The evolution of the values of μ_n and μ_w resulting from these simulations have served in learning the two types of networks. A non-linear optimization algorithm of the Gauss-Newton type has been employed to determine the optimal weights (connections) of both networks.

These two types of neural networks have permitted forming two evolution models of average polymer chain lengths. In the first case, the following model is produced :

$$\begin{cases} x3(t) = \begin{bmatrix} \mu_n & \mu_w \end{bmatrix}^T = \mathcal{N}_1(n_{I0}, n_M, t) + w''(t) \\ z3_k = \mathrm{I}.x3_k + v''_k \end{cases}$$

with $w'' = N(0,\mathrm{Q3})$ and $v'' = N(0,\mathrm{R3})$.

In the second :

$$\begin{cases} \dot{x}3(t) = \mathcal{N}_2(n_{I0}, n_M) + w'''_k \\ z3_k = \mathrm{I}.x3_k + v'''_k \end{cases}$$

with $w''' = N(0,\mathrm{Q4})$ and $v''' = N(0,\mathrm{R4})$.

Their implementation in the extended Kalman filtering is carried out by the following scheme described in Table 5. (only the relative developments of the second type of network are indicated).

Table 5. Kalman filter with neural network-based model

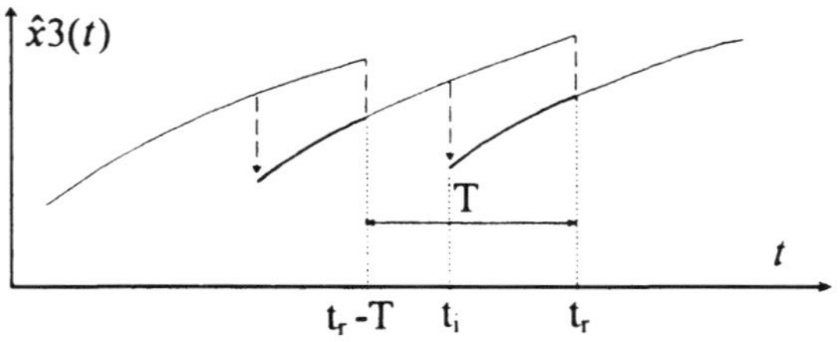

t_i : sampling time of the reacting medium for SEC analysis.

t_r : SEC analysis result outcome time.

T : time interval between two consecutive analyses

For $t_r - T < t \leq t_i$, prediction by neural model :

$\dot{\hat{x}}3(t) = \mathcal{N}_2(n_{I0}, n_M)$

Covariance propagation :

$\dot{P}3(t) = Q3(t)$

At injection time t_i :

$\hat{x}3(-) = \hat{x}3(t_i)$; $P3(-) = P3(t_i)$

$K3(t_i) = P3(-).I.\left[I.P3(-).I^T + R3(t_i)\right]^{-1}$

For $t_i < t \leq t_r$, state estimation :

$\dot{\hat{x}}3(t) = \mathcal{N}_2(n_{I0}, n_M)$

At time t_r, updating takes place :

$\hat{x}3(+) = \hat{x}3(-) + K3(t_i).\left[z3(t_i) - I.\hat{x}3(t_i)\right]$

$P3(+) = \left[I - K3(t_i).I\right].P3(-)$

Updating of the current state estimation and covariance matrix :

$\hat{x}3(t_r) = \hat{x}3(+) + \int_{t=t_i}^{t=t_r} \mathcal{N}_2(n_{I0}, n_M).dt$

$P3(t_r) = P3(+) + \int_{t=t_i}^{t=t_r} Q3(t).dt$

5. EXPERIMENTAL RESULTS

The results obtained with these different filters are resumed by the following figures. The figures 3-6 represent the result of the extended Kalman filtering using the kinetic model (classic approach). They equally show the predictions for the analytic model lacking filtering and their gap with experimental results.

The evolution of the constant Ks testify to the interest in using a state vector augmented by parameters in order to compensate for significant errors in the model. As for figures 7 and 8, they permit the comparison of estimates of the two average chain lengths within the framework of the classic Kalman filtering approach and during the utilization of neural network-based models. The neural model designated as being of the second type leads to quality estimates comparable to those obtained by the classic filtering. This is not the case for the first one, whose chosen network structure is perhaps poorly adapted.

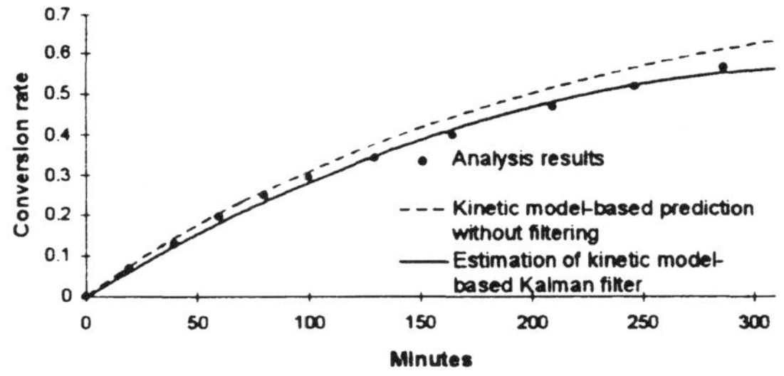

Fig. 3. Estimates of the conversion rate.

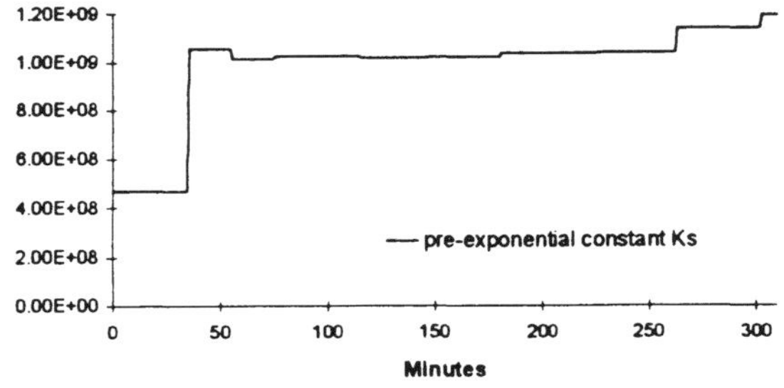

Fig. 4. Estimates of the pre-exponential constant.

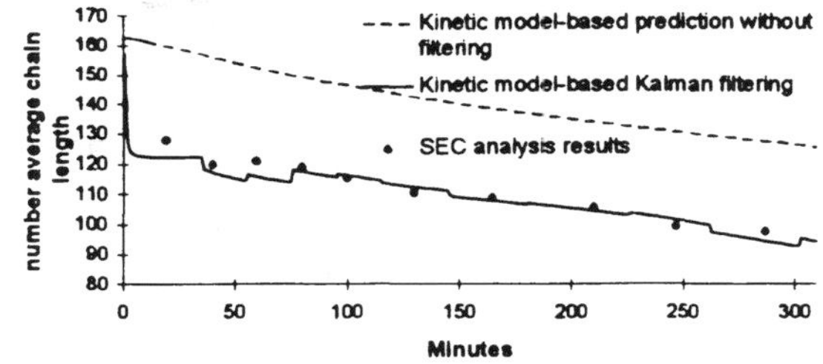

Fig. 5. Estimates of the number-average chain length

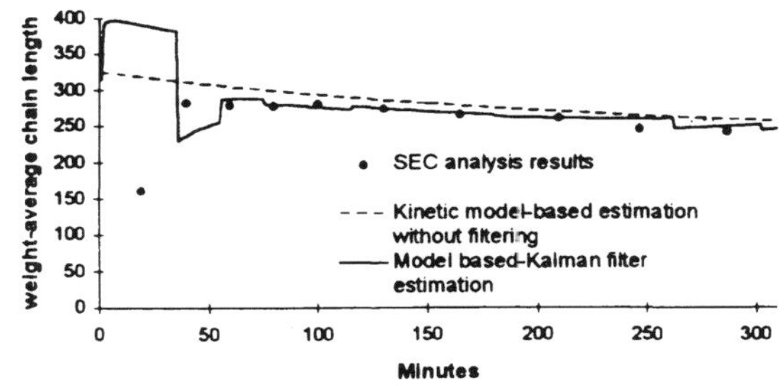

Fig. 6. Estimates of the weight-average chain length

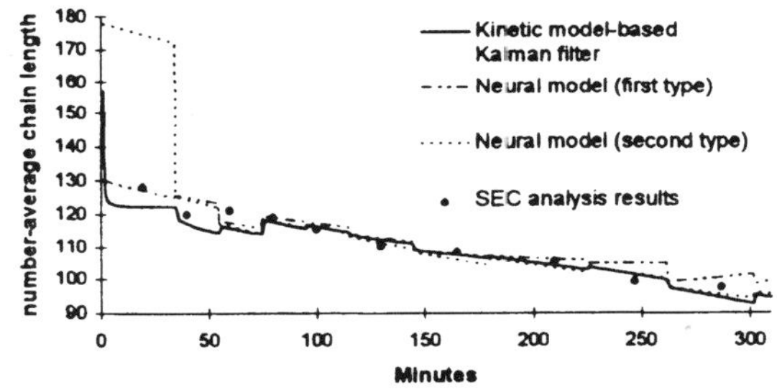

Fig. 7. Neural models versus Kinetic model performances

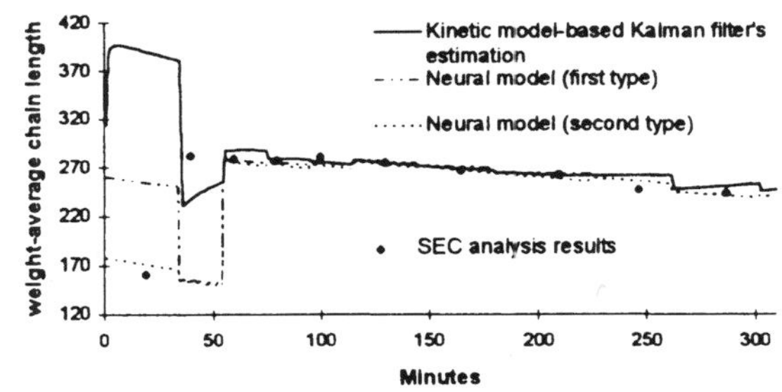

Fig. 8. Neural models versus Kinetic model performances

6. CONCLUSION

The obtained results must be studied more profoundly and subjected to additional validations. Nevertheless, they seem to testify to the advantages possible for the employment of neural networks for on-line estimations of polymer characteristics. The operational simplicity of the approach, compared to the difficulty that can present itself in determining a detailed analytic model, could constitute an incentive justification for its use.

The learning phase of the neural networks was accomplished within the framework simulated results. However, this is not the approach which should be envisaged for an industrial use. Rather, this phase should be preformed on data collected during previous experiments sufficiently rich in information.

The following stage of the procedure is to envisage the control of the μ_n and μ_w in the course of reaction.

7. NOMENCLATURE

f : initiator efficiency.
I : initiator.
R : free radical produced by initiator.
M : monomer.
S : solvent.
P_n : growing polymer chain of length n.
M_n : dead polymer chain of length n.
n_i : normalized mole number for reactant i.
λ_i : i-th moment of the number chain length distribution.
μ_n : number-average chain length.
μ_w : weigth-average chain length.
$w,\dots,w^m$ or $v,\dots,v^m$: gaussian random vectors.
$x=N(0,P)$: indicates that x is a gaussian random process vector with mean 0 and covariance P.

8. REFERENCES

Jazwinski, A.H. (1970). Stochastic processes and filtering theory. Academic Press, New York.

Jo, J.H., S.G. Bankoff (1976), Digital monitoring and estimation of polymerization reactors. *AIChE Journal*, **22**, 361-369.

Nerrand, O., P. Roussel-Ragot, L. Personnaz, G. Dreyfus (1993). Neural networks and non-linear adaptative filtering: unifying concepts and new algorithms. *Neural Comput.*, **5**, 165-199.

Kim, K.J., T. Crowley, K.Y. Choi (1992). On-line estimation and control of polymerization reactors. IFAC DYCORD+'92, Maryland, USA, 161-166.

Kalman, R.E. (1960). A new approach to linear filtering and prediction problems. *ASME Trans. J. Basic. Engng*, **82D**, 35-108.

Kiparissides, C., E. Sidiropoulou, S. Voutetakis, C. Frousakis (1990). Control of molecular weight in a batch polymerization reactor using long-range predictive control methods. *Chem. Eng. Comm.*, **92**, 1-22.

Kozub, D.J., J.F. MacGregor (1992). State estimation for semi-batch polymerization reactors. *Chem. Engng. Sci.*, **47**, 1047-1062.

Papadopoulou, S.A., E.D. Gilles (1986). Continuous estimation of the chain length distribution in a polymerization reactor considering time discrete GPC measurements. Polymer reaction engineering, emulsion polymerization, high conversion polymerization, polycondensation, K.H. Reichert and W. Geiseler, Eds. Huthig&Wepf, Basel-New York.

Schuler, H., Z. Suzhen (1985), Real-time estimation of the chain length distribution in a polymerization reactor. *Chem. Engng. Sci.*, **40**, 1891-1904.

Thibault, J., V. Van Breusegem, A. Chéruy (1990). On-line prediction of fermentation variables using neural networks. *Biotech. and Bioengng.*, **36**, 1041-1048.

Willis, M.J., C. Di Massimmo, G.A. Montague, M.T. Tham, A.J. Morris (1991). Artificial neural networks in process engineering. *IEE Proc.*, **138**, 256-266.

9. APPENDIX

$k_d = K_d.\exp(-E_d / RT)$; $k_p = K_p.\exp(-E_p / RT)$
$k_s = K_S.\exp(-E_S / RT)$; $k_f = K_f.\exp(-E_f / RT)$
$k_{td} = K_{td}.\exp(-E_{td} / RT)$
$K_d = 1.2525E14\ s^{-1}$; $E_d = 29.33$ kcal.mol^{-1}
$K_p = 4.92E5$ l.mol^{-1}.s^{-1} ; $E_p = 4.35$ kcal.mol^{-1}
$K_S = 4.67E8$ l.mol^{-1}.s^{-1} ; $E_S = 15.69$ kcal.mol^{-1}
$K_f = 7.17E9$ l.mol^{-1}.s^{-1} ; $E_f = 17.94$ kcal.mol^{-1}
$K_{td} = 9.8E7$ l.mol^{-1}.s^{-1} ; $E_{td} = .701$ kcal.mol^{-1}
$f = 0.6$

INTERACTIONS BETWEEN PRODUCTION RATE OPTIMIZATION AND TEMPERATURE CONTROL IN GAS PHASE POLYETHYLENE REACTORS

S.A. Dadebo, K.B. McAuley and P.J. McLellan

Department of Chemical Engineering, Queen's University
Kingston, Ontario, CANADA K7L 3N6

Abstract: A nonlinear (feedback) error trajectory temperature controller is used to stabilize a gas phase polyethylene reactor (GPPER) which is open-loop unstable. The controller can saturate when a production rate optimization scheme increases the polymer production rate, and pushes the plant to operate at or near the maximum cooling capacity. The tendency of the reactor system to approach extraneous equilibria is examined using recently developed methods for nonlinear processes. Sufficient conditions for the existence of extraneous equilibria are given. Simulation results are compared with predictions obtained using saturated equilibria equations implied by the temperature output.

Keywords: nonlinear, temperature control, saturation, stability

1. INTRODUCTION

In industrial fluidized bed polyethylene reactors, polymer production rate can be limited by one of several constraints including: monomer availability, capacity of product discharge system, limited market demand, downstream processing limitations, and reactor heat removal capacity. While it is relatively straightforward to optimize the production rate by running the reactor against any of the first four constraints listed above, operation at or near the maximum heat removal capacity can be dangerous. In this article, we examine the potential consequences of running near the heat removal capacity constraint, especially the effect of disturbances and the interaction between the production rate optimization scheme and the reactor temperature controller.

The heat of reaction from this exothermic polymerization process is removed in an external water-cooled heat exchanger on the recycle line. Cooled recycle gas and fresh gas-phase reactants are fed to the reactor to fluidize the bed of polymer particles, to supply reactants for the growing polymer particles and to absorb the heat generated by reaction. The reactor temperature can be controlled by manipulating either the temperature or flow rate of the cooling water to the heat exchanger. Choi and Ray (1985) and McAuley et al. (1994) have shown that fluidized bed polyethylene reactor systems can exhibit open-loop unstable behaviour. Depending on the operating regime, the reactor temperature can exhibit either limit cycle behaviour or run away to an undesirable high temperature steady state. Either situation is extremely undesirable. A temperature excursion beyond the melting point of the polymer leads to polymer particle melting and agglomeration, and results in lost production and costly reactor shutdown and clean-up.

Using a suitable feedback temperature controller, however, the reactor dynamics can be stabilized and it is quite possible to run safely at an open loop unstable steady state. Nevertheless, problems can occur when a production rate optimization scheme increases the polymer production rate, and pushes the plant to operate at or near the maximum cooling capacity. In this situation the temperature controller tends to saturate as the manipulated variable reaches either the minimum possible cooling water temperature, or alternatively, the maximum cooling water flow rate, depending on the control scheme design. Under saturation, feedback temperature control is effectively lost, and the reactor can tend to new undesirable steady states or limit cycle behaviour.

The tendency of the reactor system to approach new equilibria under saturation is examined using recently developed methods for nonlinear processes (Cebuhar and McLellan (1994)). These methods, which are extensions of case specific methods described by Alvarez et al. (1991), simplify the complexity and dimension of the necessary computations by using the controller normal forms implied by the temperature output. When a temperature excursion begins, alternative manipulated variables (e.g. catalyst feed rate, carbon monoxide poisoning) may be able to reduce the heat production rate sufficiently to return the reactor to the desired operating point, thereby preventing a reactor shut-down. Sufficient conditions for the existence of extraneous equilibria are derived. These will form the basis for investigation of the effectiveness of designing safety margins into the production rate optimizer to prevent serious consequences of temperature controller saturation for disturbances of different types and magnitudes.

2. SYSTEM MODEL

The reactor model (eqn.1) is that studied by McAuley et al. (1994). A schematic of a typical GPPER is shown in Fig. 1. The gas exiting at the top of the fluidized bed reactor is recycled and cooled in an external heat exchanger before addition of fresh feed and returned to the reactor. To control ethylene partial pressure in response to changes in the ethylene consumption rate, a PI controller is implemented. When the PI controller is turned off, the monomer feed rate is 198 mol/s. H_{g1} and H_{go} are respectively, the enthalpy of the recycle gas at the heat exchanger outlet and the enthalpy of the gas leaving the reactor. H_f is the enthalpy of the fresh feed and H_p is the enthalpy of the product stream. H_r is the heat generated by reaction. All the remaining model parameters are defined in the glossary.

The heat exchanger model (eqns. 2-5) is similar to the staged model proposed by Alsop and Edgar (1989). They proposed modelling the heat exchanger as a series of small counterflow heat exchangers. Dadebo et al. (1994) have shown that a 2-stage linear temperature driving force model adequately captures the dynamics of a typical heat exchanger. Here a 2-stage linear temperature driving force model is used:

$$V_g\frac{dM}{dt} = F_M - \frac{M}{M+I_n}bt - R_M$$

$$V_g\frac{dI_n}{dt} = F_{I_n} - \frac{I_n}{M+I_n}bt$$

$$\frac{dY}{dt} = F_Y - k_dY - \frac{Op}{BW}Y$$

$$\frac{dT}{dt} = \frac{1}{A_p}(H_f + H_{g1} - H_{go} - H_r - H_p) \qquad (1)$$

$$\frac{d[PIE]}{dt} = P_{sp} - MRT$$

where $A_p = M_rC_{pr} + BWC_{ppol}$

$$F_M = 198 + K_p\left([P_{sp} - MRT] + \frac{1}{\tau_{Ip}}[PIE]\right)$$

$$\frac{dT_{w1}}{dt} = \frac{2F_w}{M_w}(T_{wi} - T_{w1}) - \frac{UA}{M_wC_{pw}}(T_{w1} - T_{g1}) \qquad (2)$$

$$\frac{dT_{g2}}{dt} = \frac{2F_g}{M_g}(T - T_{g2}) + \frac{UA}{M_gC_{pg}}(T_{w2} - T_{g2}) \qquad (3)$$

$$\frac{dT_{w2}}{dt} = \frac{2F_w}{M_w}(T_{w1} - T_{w2}) - \frac{UA}{M_wC_{pw}}(T_{w2} - T_{g2}) \qquad (4)$$

$$\frac{dT_{g1}}{dt} = \frac{2F_g}{M_g}(T_{g2} - T_{g1}) + \frac{UA}{M_gC_{pg}}(T_{w1} - T_{g1}) \qquad (5)$$

The subscripts g and w refer to gas and water respectively. M_w and M_g are the molar holdups in the water and gas sides of the heat exchanger, respectively. Tw_i is the inlet cooling water which is manipulated to control the reactor temperature, T. T_{gj} and T_{wj} (j=1,2) are, respectively, outlet temperature of the gas and cooling water from stage j. The system is control affine (linear in Tw_i) and can be put in the general SISO form:

$$\frac{d\mathbf{x}}{dt} = \mathbf{f}(\mathbf{x}) + u\mathbf{g}(\mathbf{x}) \;;\; y = h(\mathbf{x}) = T \qquad (6)$$

$$\mathbf{x} = [M\ I_n\ Y\ T\ PIE\ T_{w1}\ T_{w2}\ T_{g2}\ T_{g1}]^T$$

The use of a linear temperature difference driving force model for the 2-stage heat exchanger makes it relatively easier to obtain the controller normal forms without sacrificing accuracy. Previous work by Dadebo et al. (1994) has shown that there is little difference between simulation results obtained either by using a linear temperature difference or the standard log mean temperature driving force. The relative order of the system, which may be defined as the number of inherent "integrations" in the process required for a change in the manipulated variable to affect the controlled variable, is 3.

3. FEEDBACK TEMPERATURE CONTROL

3.1 Control Law Development

The error trajectory control (ETC) law is derived from a suitably well-behaved function of the tracking error (McLellan et al. (1990)). An output tracking error, e, is defined as: $e(t) = y_d(t)-y$. y is the output function of the state and y_d is the desired (or setpoint) output. Since the relative order of the system is 3, a third order tracking error function is defined (PI included to account for model inaccuracy):

$$e^{(3)}+\beta_2 e^{(2)}+\beta_1 e^{(1)}+\beta_0 e+K_c[e+\frac{1}{\tau_I}\int_0^t e(\tau)d\tau]=0 \quad (7)$$

Eqn. (7), can be re-arranged for constant $y_d(t)$ to give the control law

$$u=\frac{-L_f^3h-\sum_{i=1}^{2}\beta_i L_f^i h+(\beta_0+K_c)e+\frac{K_c}{\tau_I}\int_0^t e(\tau)d\tau}{L_gL_f^2h} \quad (8)$$

The Lie derivatives, details of which can be found in Isidori (1989), were obtained using Maple®.

3.2 Controller Tuning

The closed loop transfer function for the tracking error function (eqn. 7) is given by:

$$\frac{y(s)}{y_d(s)}=\frac{(\beta_0+K_c)s+\frac{K_c}{\tau_I}}{s^4+\beta_2 s^3+\beta_1 s^2+(\beta_0+K_c)s+\frac{K_c}{\tau_I}} \quad (9)$$

The characteristic equation is then given by:

$$s^4+\beta_2 s^3+\beta_1 s^2+(\beta_0+K_c)s+\frac{K_c}{\tau_I}=0 \quad (10)$$

By placing all poles at $-\omega$, the corresponding equation is $(s+\omega)^4$. Expanding this equation and comparing with eqn. (7) the tuning parameters are:

$$\beta_0=\omega^3,\ \beta_1=6\omega^2,\ \beta_2=4\omega,\ K_c=3\omega^3,\ \tau_I=3/\omega \quad (11)$$

By placing all poles at the same location, there is only one degree of freedom required to tune the controller for desired performance. Details of the above tuning strategy, for a minimum phase system of arbitrary relative order r, can be found in Dadebo et al. (1994). The tuning strategy is much easier than a PID tuning for this open-loop unstable system.

4. HANDLING CONTROL SATURATION

While the production rate is directly influenced by the reactor temperature, care must be taken not to place excessive demands on the cooling water temperature. In practice, operating at the maximum cooling capacity would imply that feedback is essentially lost, and the reactor can tend to new undesirable steady states. Since control saturation induces additional nonlinearity, it is important to examine the potential consequences of operating at or near the maximum cooling capacity in the presence of disturbances.

Chen and Chang (1985) have examined the effects of controller saturation and have developed an algorithm for designing globally stable closed loop systems. Nonetheless, the global stability of saturable systems can be destroyed by limit cycles and higher order attractors as a result of discontinuities introduced by saturation. Recently, Alvarez et al.(1991) have derived necessary and sufficient conditions to ensure structural solvability of the bounded control problem for a class of one-input two-state CSTRs. In this section, the case-specific methods proposed by Alvarez and co-workers are extended to the GPPER problem which is prone to open loop limit cycle behaviour. In order to reduce the dimension of the problem, the controller normal form implied by the output will be used.

4.1 Controller Normal Form

When the relative order r is defined and is less than the system dimension, the nonlinear system can be transformed into a "normal form" (Isidori, 1989):

$$\begin{aligned}\dot z_1&=z_2\\ \dot z_2&=z_3\\ \dot z_3&=a(\mathbf{z},\eta)+b(\mathbf{z},\eta)u\\ \dot\eta&=\mathbf{q}(\mathbf{z},\eta)\\ \text{where}\quad a&=L_f^3h(\mathbf{z},\eta)\\ b&=L_gL_f^2h(\mathbf{z},\eta)\\ \text{with}\quad z_1&=h=T\end{aligned} \quad (12)$$

η is a 6-dimensional vector chosen to ensure that all directional derivatives of η_i (i=1,2,..,6) in the direction of the vector field **g** are zero:

$$L_g\eta_1=L_g\eta_2=....=L_g\eta_6=0 \quad (13)$$

Choosing, for example, $\eta=[M, Y, I_n, T_{g2}, T_{w2}, PIE]^T$ will ensure that the above equation is satisfied. The normal form is obtained via a (local) diffeomorphism:

$$\phi(\mathbf{x})=[z_1\ z_2\ z_3\ \eta_1\ \eta_2\ ....\ \eta_6] \quad (14)$$

4.2 Existence of Extraneous Equilibria

The use of the normal form significantly reduces the amount of computation required and the complexity of the saturation equations. At equilibrium, the reactor temperature is constant and hence its time derivative is zero ($dz_1/dt=z_2=0$). This implies that

$$\dot z_2=z_3=0 \quad ; \quad \dot z_3=0 \quad (15)$$

In general, given an output z_1, the equilibria

equations for a system of arbitrary relative order r, are given by:

$$z_2 = z_3 = \ldots = z_r = 0 \qquad (16)$$

$$\text{and} \qquad \dot{\eta} = \mathbf{0}$$

For the GPPER system this approach allowed us to isolate T_{g1} and T_{w1}, respectively using Maple®. In a seriatim fashion, the other equilibria equations above were solved, leaving only two equations with two unknowns:

$$\dot{z}_3 = a(z_1^*, \eta_2^*) + b(z_1^*, \eta_2^*)u_{sat} = 0 \qquad (17)$$

$$\dot{\eta}_2 = q_2(z_1^*, \eta_2^*) = 0$$

z_1^* is the saturated equilibrium temperature and η_2^* is the corresponding moles of active site concentration of catalyst. It is noted that "a" and "b" in eqn. (17) are functions of z_1 and η_2 only. The above equations were generated in Maple® and an optimized C file was generated and solved using the FSOLVE nonlinear equation solver in Matlab®.

Under saturated conditions the manipulated variable, u_{sat} is either clamped at the maximum or minimum possible value: u^+ or u^-, respectively.

Theorem . Let u^* be the value of the manipulated variable at equilibrium and $z_{1,sp}$ be the temperature setpoint. An extraneous equilibrium exists if

a) $z_1^* < z_{1,sp}$ and $u^* > u^+$.

b) $z_1^* > z_{1,sp}$ and $u^* < u^-$.

Proof. (a) From eqn. (17)

$$u^* = \frac{-a(z_1^*, \eta_2^*)}{b(z_1^*, \eta_2^*)} \qquad (18)$$

Using the control law (eqn. 8), at equilibrium,

$$u^* = -\frac{a(z_1^*, \eta_2^*) - \sum_{i=1}^{3} \beta_{i-1} z_i^* + \beta_0 z_{1,sp}}{b(z_1^*, \eta_2^*)} \qquad (19)$$

In the above relation, the integral mode has not been considered because under saturation, the manipulated variable is clamped at the appropriate bound. This obviates the need for reset windup considerations. Thus whether or not the controller is equipped with anti-reset windup, the tendency of the process to reach an extraneous equilibrium (attractor) cannot be altered when dealing with nonlinear systems, see (Chen and Chang, 1985).

Now $u^* > u^+$ implies that

$$\frac{\beta_0}{b(z_1^*, \eta_2^*)}(z_{1,sp} - z_1^*) > 0 \qquad (20)$$

since $z_2^* = z_3^* = 0$. To ensure stability, the characteristic equation must be a Hurwitz polynomial, implying that all coefficients (β_i; i=0,1,2) should be positive. For the GPPER system, $b(z_1, \eta_2) = L_g L_f^2 h(\mathbf{x}) \equiv 4.08e\text{-}4$. Hence from eqn. (22), $z_1^* < z_{1,sp}$. In a similar fashion, condition (b) can be obtained. QED

The above theorem is expected since, physically, when the extraneous equilibrium is below the setpoint, the manipulated variable (coolant temperature) should be high enough in order to drive the reactor temperature to the setpoint. Therefore, if the coolant temperature is clamped at a value lower than the expected closed loop value, the system will not be able to go to the desired setpoint. A similar analogy can be drawn for condition (b) where a lower coolant temperature is required to drive the system to the reactor temperature setpoint.

5. SIMULATION RESULTS AND DISCUSSION

In spite of the fact that the GPPER is open loop unstable, it is easily stabilized by a feedback controller. As shown by Dadebo et al. (1994), the nonlinear error trajectory controller is much more reliable than a conventional PID in terms of speed, damping and robustness. Fig. 2 shows the reactor temperature response to a set point change in the presence of +20% mismatch in the cooling water flowrate, recycled gas flowrate and the product of the overall heat exchange coefficient and the area available for heat transfer. In this instance the production rate increases from 43 tonnes/h. to 46 tonnes/h. The coolant temperature was clamped between 10 and 60 °C. A typical linear controller would take at least 2 hours to effect such a change. However, the nonlinear controller tuned with ω=0.0195 does this in less than 15 minutes. It is noted that before increasing the production rate and temperature at time t=30 minutes, there was no excessive demand on the controller, in spite of the parametric mismatch. Thus, a change in production rate could cause control problems due to saturation. A similar plot is shown in Fig. 3 where the effect of structural model mismatch in the heat exchanger is considered. In Fig. 3, a controller based on a linear temperature driving force model is used to control a plant simulation using the conventional log mean temperature difference driving force model. In both Figs. 2 and 3, the catalyst flow rate, Fc, was 5.8 kg/h.

The algebraic relations given by eqn. (17) were generated in Maple® and solved in Matlab®. In order to see the effect of cooling water temperature during saturation, the equations were generated with the catalyst flowrate, Fc, and the cooling water temperature as parameters. Fig. 4 shows the effect of cooling water temperature on stability, for a fixed catalyst flow rate Fc=5.8 kg/h if the controller saturates and feedback is lost until an extraneous steady state is reached. It must be emphasized that the melting point of the polymer is about 400 K. For economic reasons, most commercial GPPERs are operated within a temperature range of 353 K to 373 K to allow some safety margin. As shown in Fig. 4, the temperature range of interest is open-loop unstable. An open loop coolant temperature greater than 298 K will only result in undesirable (and physically unrealizable) high temperature steady state. For lower coolant temperatures, steady state multiplicity occurs. Therefore, without feedback control, the reactor temperature will be driven to undesirable steady states. Fig. 5 shows the effect of catalyst flowrate on stability for a fixed coolant temperature of 283 K. It is noted that reducing the catalyst flowrate does not necessarily lower the reactor temperature in the operating region of interest. The current desired operating temperature is 360 K. By comparing Figs. 2, 3, 4 and 5 it can be deduced (using the theorem for predicting the existence of extraneous equilibria) that if the coolant temperatures were clamped permanently at 60 and 10 oC, the reactor temperature would be driven to over 1200 K and 291 K respectively. Open loop simulations have confirmed these observations. However, under feedback control, the controller effects corrective action to ensure that the reactor temperature is driven to the setpoint under favourable conditions.

6. CONCLUSIONS

The nonlinear ETC is very effective in controlling the reactor temperature even in the presence of disturbances. Another advantage of the nonlinear controller is that the tuning is easy. The GPPER is a highly saturable system which exhibits steady state multiplicity depending on, among others, the amount of catalyst used and the coolant temperature and flow. Care must be taken when increasing the production rate. If the plant is pushed to operate at the limit of the cooling capacity for a long time, feedback control is effectively lost and the plant may be driven to extraneous steady states. Sufficient conditions for the existence of extraneous equilibria for the GPPER have been given using recently developed methods for nonlinear systems. Simulation results are in agreement with deductions from nonlinear analysis under saturation. Therefore, it is possible to predict the behaviour of the GPPER using saturation equilibria equations implied by the temperature output.

The assessment of open loop limit cycle behaviour and the use of other manipulated variables like catalyst feed rate and/or carbon monoxide poisoning is the subject of on-going research.

7. REFERENCES

Alsop, W.A. and T.F Edgar (1989), Nonlinear Heat Exchanger Control Through the use of Partially Linearized Control Variables, *Chem. Eng. Comm.*, **75**, 155-170.

Alvarez, J., J. Alvarez and R. Suarez (1991), Nonlinear Bounded Control for a Class of Continuous Agitated Tank Reactors, *Chem. Eng. Sci.*, **46**, 3235-3249.

Cebuhar, W. and P.J. McLellan (1994), A New Method for Assessing Controller Saturation, *Dept. of Chem. Eng. Report*, Queen's Univ., Kingston, Ontario, Canada.

Chen, L-H and H-C. Chang (1985), Global Effects of Controller Saturation on Closed-Loop Dynamics, *Chem. Eng. Sci.*, **40**, 2191-2205.

Choi, K-Y. and W.H. Ray, (1985), The Dynamic Behaviour of Fluidized Bed Reactors for Solid Catalysed Gas Phase Olefin Polymerization, *Chem. Eng. Sci.*, **40**, 2261-2279.

Dadebo, S.A., M.L. Bell, P.J. McLellan and K.B. McAuley (1994), Temperature Control of Industrial Gas Phase Polyethylene Reactors, *Proc. 44th Canadian Chem. Eng. Conf.*, Calgary, AB, Canada, pp. 675-676.

Isidori, A. (1989), *Nonlinear Control Systems*, 2nd Ed., Springer-Verlag, Berlin.

McAuley, K.B., D.A. MacDonald and P.J. McLellan (1994), Effects of Operating Conditions on Stability of GPPERs, *AIChE J.* (in press).

McLellan, P.J., T.J. Harris and D.W. Bacon (1990), Error Trajectory Descriptions of Nonlinear Controller Designs. *Chem. Eng. Sci.*,**45**, 3107-3034.

8. GLOSSARY

AU	product of area and heat exchanger coefficient in heat exchanger (cal/s K)
bt	overhead gas bleed (mol/s)
BW	mass of polymer in the fluidized bed (g)
C_{pg}	gas phase specific heat capacity (cal/mol K)
C_{pw}	specific heat capacity of water (cal/g K)
C_{ppol}	specific heat capacity of polymer (cal/g K)
F_c	flow rate of catalyst (kg/hr)
F_g	flow rate of recycle gas (mol/s)
F_i	flow rate of component i
In	molar concentration of inerts (mol/L)
K_c	PI gain for ETC.
k_d	deactivation rate constant of catalyst site (s^{-1})
K_p	ethylene controller gain (mol/atm s)
M	Monomer gas phase concentration (mol/L)
M_rC_{pr}	product of mass and heat capacity of reactor walls (cal/s K)
Op	polymer outflow rate (g/s)
P_{sp}	setpoint for ethylene controller (atm)
PIE	Integral monomer partial pressure error
R	Molar gas constant (L atm/mol K).
R_M	rate of consumption of monomer (mol/s).
V_g	volume of gas phase in the reactor (L).
Y	molar concentration of catalyst site (mol).
τ_I	integral time constant for ETC (s)
τ_{Ip}	integral time constant for PI controller on ethylene feed (s)

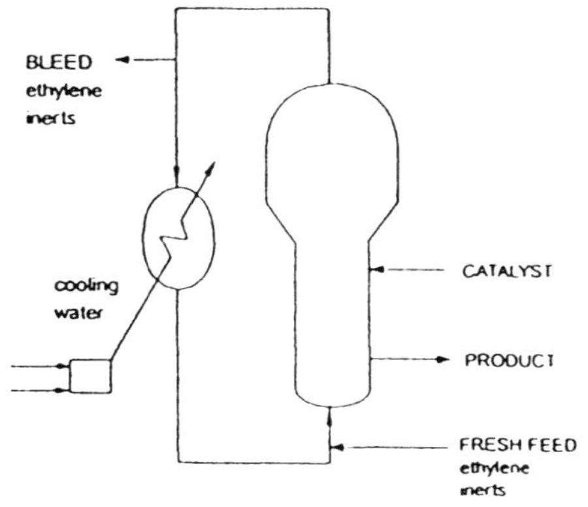

Figure 1. Schematic of a typical GPPER system

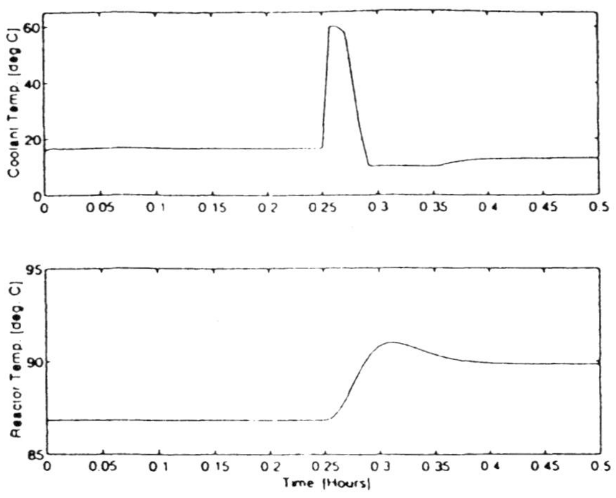

Figure 3. Step response in the presence of structural model mismatch.

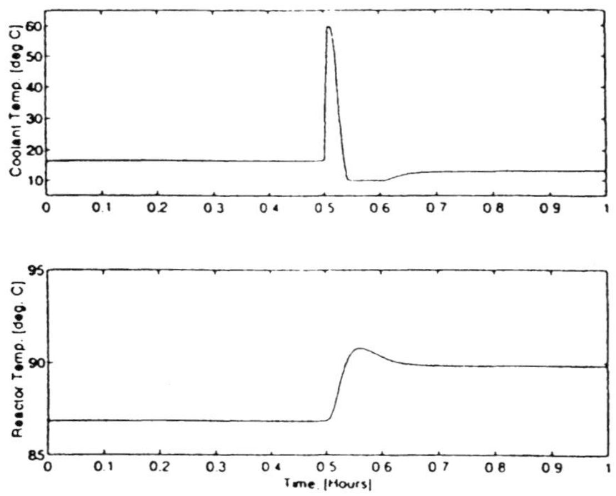

Figure 2. Step response in the presence of +20% mismatch in Fg, Fw, and AU.

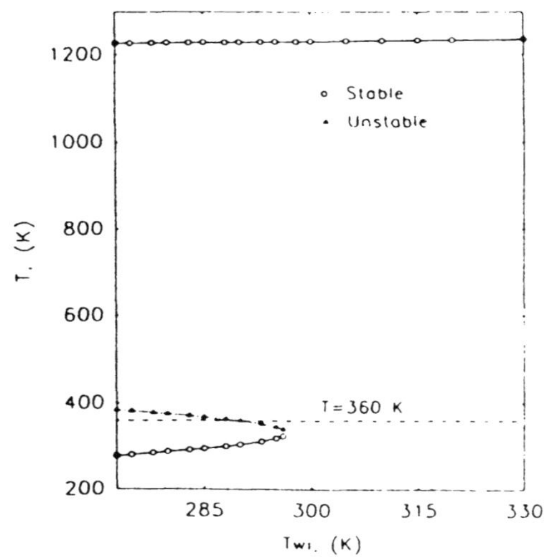

Figure 4. Effect of coolant temperature on stability.

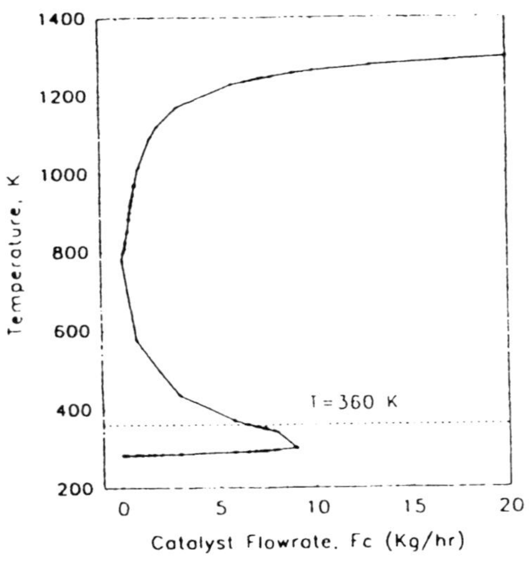

Figure 5. Effect of catalyst flow rate on stability.

CONTROLLABILITY ISSUES CONCERNING PARTICLE SIZE IN EMULSION POLYMERIZATION

Vincenzo Liotta, Christos Georgakis*, and Mohamed El-Aasser**

**Lehigh University, Process Modeling and Control Research Center, Bethlehem, Pennsylvania 18015, USA*

***Lehigh University, Emulsion Polymers Institiute, Bethlehem, Pennsylvania 18015, USA*

Abstract. The utility of the growth mechanism to influence polystyrene particle size distributions is investigated via a competitive growth framework. A dynamic competitive growth model, extrapolated from an experimentally validated unimodal model, is used to simulate the semi-batch growth of two monodisperse particle populations at 50°C. Results show that the greatest relative narrowing between the two populations occurs at the particle monomer concentration ($[M]_P$) saturation limit and decreases as $[M]_P$ is decreased towards a limited reaction lower bound. Simulations run at these limits provide the reachable regions starting from several initial diameter population pairs. The areas of these regions are found to be dependent on the magnitude of the initial diameter ratio and the individual particle sizes.

Key Words. Controllability; reachability; process control; nonlinear systems

1. INTRODUCTION

The product of a conventional emulsion polymerization, referred to as a latex, results from a free radical polymerization of monomer dissolved in aqueous based polymer particles. Industrial applications of these dispersed particles include synthetic rubber, adhesives, paints, and medical products. The particle size distribution (PSD) of these latexes is an important product characteristic due to its influence over the end-use properties including mechanical strength, opacity, and viscosity. Once a certain PSD is linked to a desired property, controlling the PSD towards the appropriate final form becomes an important objective.

Controlling the PSD or any aspect of an emulsion polymerization is, however, a difficult task. Experimental and modeling obstacles arise principally due to the multiphase nature of the process and partially understood physio-chemical phenomena. Therefore, manipulation of the mechanisms which influence the evolution of the PSD such as nucleation, growth and agglomeration is not straightforward. Effort must be placed on a controllability assessment of these candidate 'manipulatable mechanisms' prior to feedback control studies.

In this work, the utility of the growth mechanism to influence the breadth of polystyrene PSDs is investigated via a competitive growth framework. A dynamic competitive growth model is developed using an experimentally validated unimodal model as a basis. Simulations using the monomer feed rate as a manipulated variable to influence the relative narrowing and broadening of two monodisperse populations of particles are presented. In the final section, an assessment of the controllability of the growth mechanism is made.

2. COMPETITIVE GROWTH FRAMEWORK

Assuming growth from a seed distribution is the only operative mechanism, the breadth of a distribution of particles can be altered only if different sized particles grow at different rates. This is the essence of competitive growth. Vanderhoff and coworkers (Vanderhoff and Bradford, 1956; Poehlein and Vanderhoff, 1973) were the first to investigate this phenomenon using two monodisperse populations of particles under batch conditions as a simplified case. There investigation showed a constant relative narrowing between the two particle populations under a variety of experimental conditions. As an extension of that earlier work, this paper explores competitive growth under a semi-batch environment with the objective to improve the general controllability of the growth mechanism for PSD control.

It is imperative at this point to recognize the different competitive growth representations that exist. For instance, the choice to characterize a distribution with either diameter or volume is critical since Chen and Wu (1988) point out that under typical growth conditions, the standard deviation of a distribution decreases with time if diameter is used, and increases with time if volume is used. For a distribution of particles, one must also distinguish between absolute breadth (represented by standard deviation) and relative breadth (represented by standard deviation over mean).

Such distinctions can also be applied to the case where monodisperse populations of particles are of concern instead of a full distribution. In this work, the *relative* growth of two monodisperse diameter populations of particles (of equal number) will be followed by tracking their diameter ratio evolution. If the diameter ratio decreases with time, the populations are relatively narrowing. Conversely, as the diameter ratio increases, the particles are relatively growing away from each other.

3. DYNAMIC COMPETITIVE GROWTH MODEL

A dynamic model has been developed to simulate the homopolymerization of styrene under the following reactor condition assumptions.

- Two monodisperse populations of particles are used as a seed. (Particles in each population are the same size.)
- The number of particles are fixed. (Nucleation and agglomeration are negligible.)
- The reactor is run under semi-batch conditions. (Pure monomer is used as the feed.)
- The reactor is run isothermally at 50°C.
- There exists no monomer droplets. (The monomer concentration $[M]_P$, is not allowed to reach the saturation limit of the particles; $[M]_P \approx 5.4$ mol/l)
- The monomer concentration inside the particles, $[M]_P$, is not a function of diameter. (Gardon (1968b) has argued that this assumption is valid under monomer saturated conditions. Below the saturation limit, this is an even better assumption)
- The solubility of monomer in water is negligible. (This is especially true for styrene.)

Note that all parameters are defined in the nomenclature section along with the literature values of some of the kinetic parameters. In all equations, subscript i denotes population 1 or 2, so that when the notation is used, two equations exist for the one presented.

Equation 1 is the rate of change of the total reactor volume, V_R, or equivalently, the increase in the total monomer swollen particle volume, V_P (population 1 and 2), with time. The next 3 equations are material balances on the mass of polymer in population 1, P_1, total moles of monomer in the reactor, $\{V_P[M]_P\}$, and the total moles of initiator in the reactor, $\{[I]V_W\}$ respectively.

$$\frac{dV_R}{dt} = \frac{dV_P}{dt} = \frac{F_m(MW_m)}{\rho_m} + (R_{p_1} + R_{p_2})V_R(MW_m)(\frac{1}{\rho_p} - \frac{1}{\rho_m}) \quad (1)$$

$$\frac{dP_1}{dt} = R_{p_1}V_R(MW_m) \quad (2)$$

$$\frac{d\{V_P[M]_P\}}{dt} = F_m - (R_{p_1} + R_{p_2})V_R \quad (3)$$

$$\frac{d\{[I]V_W\}}{dt} = -k_d[I]V_W \quad (4)$$

In the above equations, the polymerization rate for a given population can be written as

$$R_{p_i} = \frac{k_p\bar{n}_i[M]_P N_{t_i}}{N_a V_R} \quad (5)$$

In order to solve the above set of equations, the propagation constant, k_p, and the average number of radicals per particle, $\bar{n}$ are required.

Under most conditions, the propagation constant, k_p, does not change with time. However, at low monomer concentrations, $[M]_P$, (low free volume) a diffusion limitation becomes prevalent and must be accounted for. Below a certain critical free volume, v_{fmc}, the propagation constant relation (Sundberg *et al.*, 1981) can be presented as

$$\ln\frac{k_p}{k_{p_o}} = B(\frac{1}{v_{fmc}} - \frac{1}{v_f}) \quad (6)$$

In comparison to the propagation constant, the average number of radicals is much more difficult to obtain since $\bar{n}$ depends on complex phenomena like absorption, desorption, and termination of radicals in the particles. Li & Brooks (1993) have determined that $\bar{n}$ follows a differential equation of the following form.

$$\frac{d\bar{n}_i}{dt} = \sigma_i - k_i\bar{n}_i - \frac{f_i k_t}{V_i N_a}\bar{n}_i^2 \quad (7)$$

The absorption rate of radicals into the particles, σ, can be written as the product of the absorp-

Table 1 Simulated reaction conditions

Initial Conditions	Reaction Conditions
$[I]_o = 0.01$ mol/l	$N_{t_1} = N_{t_2} = 2.5 \times 10^{16}$
$\bar{n}_{1_o} = \bar{n}_{2_o} = 0.0$	$V_W = 1.0$ l
$F_{m_o} = 0.0$ mol/s	

tion rate constant and the aqueous phase radical concentration (obtained via a radical balance not presented here).

$$\sigma_i = k_{a_i}[R]_W \tag{8}$$

Currently some controversy exists on how to obtain the absorption rate constant. In accordance with several other works, (Gardon, 1968*a*; Rawlings, 1985; Kao *et al.*, 1984) a form that relates k_a to the particle surface area known as the collision mechanism is used here.

$$k_{a_i} = \pi k_{mp} D_{s_i}^2 N_a \tag{9}$$

The desorption constant, k, obtained from the work of Asua *et al.* (1989), can be written as

$$k_i = k_{fm}[M]_P \frac{K_{o_i}}{K_{o_i}\beta + k_p[M]_P} \tag{10}$$

while, an empirical form of the termination constant relation was chosen from the work of Hawkett *et al.* (1981).

$$k_t = k_{t_o} exp(-19.0 W_p^{2.1}) \tag{11}$$

3.1. *Model Solution and Validation*

The set of ODE's along with the associated algebraic equations was integrated using LSODE contained in the program ODESSA (Leis and Kramer, 1988). Important initial conditions and other inputs to the program are listed in Table 1.

In this work, the important simulation outputs are the evolution of the monomer concentration inside the particles, $[M]_P$, the average number of radicals per particle, $\bar{n}$, and the dry population diameters, D_1 and D_2. Solution of equation 7 yields $\bar{n}$, while $[M]_P$ can be easily obtained by dividing the solution of equation 3 with the solution of equation 1.

The dry diameter evolution of the monodisperse populations can be obtained by first calculating the total swollen volume of each population. The volume of the first population is obtained by following its polymer mass (solution of equation 2). Subtracting V_{P_1} from V_P (solution of equation 1) readily yields the total swollen volume of the second population. The unswollen particle diameters are tracked in this work instead of the monomer swollen particles primarily because industrial applications of latexes are normally void of monomer. Further arguments for using the dry diameters are given elsewhere (Poehlein and Vanderhoff, 1973).

A unimodal version of this dynamic model has been validated against on-line data obtained in an automated emulsion polymerization control facility (Liotta *et al.*, 1994). All of the model kinetic constants were judiciously chosen from the literature except for the kinetic parameter, k_{mp} (see Equation 9), which had served as the single adjustable parameter. Since all of the kinetic constants in both models are identical, the unimodal validation serves to validate this model. An exception to this may be the form of equation 9 which, due to the controversy surrounding the diameter dependence of radical absorption, should be validated using competitive growth experiments. Never-the-less, the form presented in equation 9 is retained in this study.

4. COMPETITIVE GROWTH SIMULATIONS

Operation of the reactor under semi-batch conditions provides a degree of process flexibility which cannot be achieved with batch reactors. For this work, this flexibility is furnished by the flow rate of monomer into the reactor, F_m. The monomer flow rate is the primary manipulated variable because it influences the amount of monomer accumulated inside the particles ($[M]_P$), which in turn, influences important competitive kinetic processes.

At low $[M]_P$ values, competition between different sized particles for available radicals is altered because of an induced diffusion limitation. This limitation, called the gel effect, influences the growth of larger particles differently than smaller ones. The smaller the value of $[M]_P$, the greater the effect. However, if $[M]_P$ gets too low, the same diffusion limitation causes the reaction rate to drop to an impractical low value. An approximate practical lower limit for $[M]_P$ can be calculated to equal 1.35 mol/l.

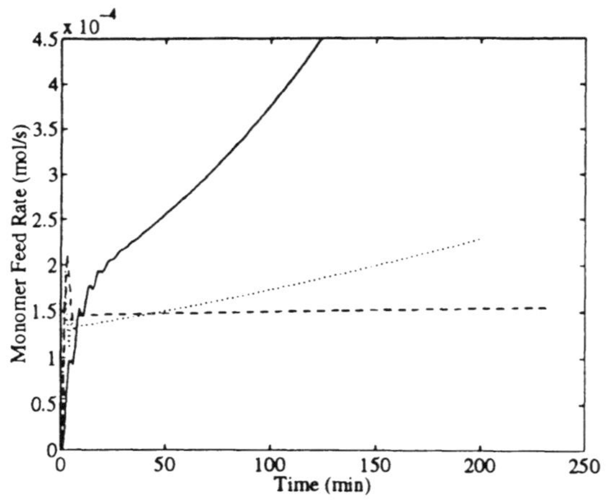

Fig. 1. Monomer feed rate trajectories required to maintain constant particle monomer concentrations at $[M]_P$ = 1.35 mol/l (solid line), $[M]_P$ = 3.2 mol/l (dotted line), $[M]_P$ = 5.4 mol/l (dashed line)

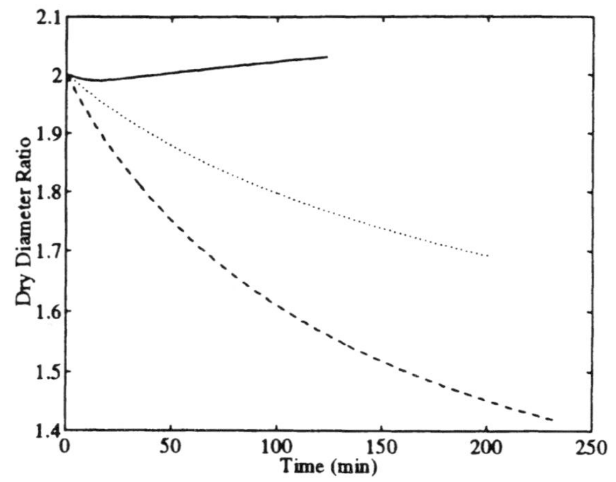

Fig. 2. Competitive growth simulations for D_{1_o} = 100 nm, D_{2_o} = 200 nm at particle monomer concentrations of $[M]_P$ = 1.35 mol/l (solid line), $[M]_P$ = 3.2 mol/l (dotted line), $[M]_P$ = 5.4 mol/l (dashed line).

The impact that $[M]_P$ has on the gel effect, and thus, the competition between different sized particles, requires that it be controlled in order to obtain a particular competitive effect. This objective is realized through the manipulation of the monomer flow rate. Figure 1 shows the monomer feed profiles used to control $[M]_P$ near the high saturation limit, the lower limited reaction limit, and an intermediate value. The solid and dotted curves (corresponding to lower $[M]_P$ values) show the unsteady state behavior typical in semi-batch emulsion polymerization. Although steady states do not exist in semi-batch emulsion polymerization, pseudo-steady states can be induced (Dimitratos *et al.*, 1990) as is demonstrated by the dashed curve (high $[M]_P$) of Figure 1.

PI controllers were applied to the process simulation to maintain the respective $[M]_P$'s at their initial value. Although not easily achievable, it was assumed that the $[M]_P$ values can be measured experimentally. The highly nonlinear behavior of the process was dealt with by tuning the individual controllers at their respective points of operation.

The competitive growth curves for the corresponding monomer concentration cases are presented in Figure 2. For these runs, initial diameters of 100 and 200 nm were grown using a constant amount of monomer (225 g) in order to maintain a fair competitive growth comparison. The results show that for high values of $[M]_P$, the two particle populations grow closer to one another in a realtive sense. As $[M]_P$ decreases, this relative narrowing effect decreases. Finally, at the lower practical bound of $[M]_P$, the diameter ratio begins to increase indicating a slight broadening effect.

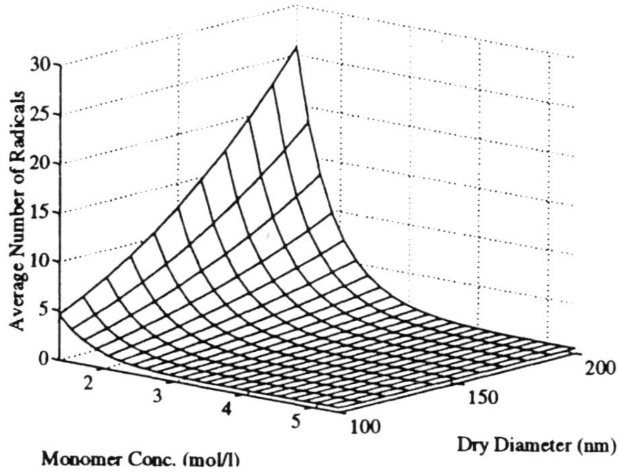

Fig. 3. The effect of particle monomer concentration and particle diameter on the average number of radicals per particle for $N_t = 5 \times 10^{16}$ and $[I]_o = .01$ mol/l

The key to understanding this competition is the realization that different size particles grow at different volume rates (Equation 5) only if they have different average number of radicals. If $\bar{n}_1 = \bar{n}_2$, then the particle ***volume*** growth rates are equal, which makes the smaller particles catch up to the larger particles in ***diameter***. This relative narrowing effect decreases as the difference between $\bar{n}_1$ and $\bar{n}_2$ increases until a point is reached where the particles begin to move away from each other in a relative sense. Figure 3 shows the dependence of $\bar{n}$ on $[M]_P$ and dry particle diameter. By picking two different particle diameters on this plot, one can see the effect that $[M]_P$ has on the difference between $\bar{n}_1$ and $\bar{n}_2$.

For monomer concentrations near the saturation limit (greater than 5 mol/l), $\bar{n}$ remains close to 1/2 throughout the diameter range in Figure 3. As illustrated by the dashed curve in Figure 2,

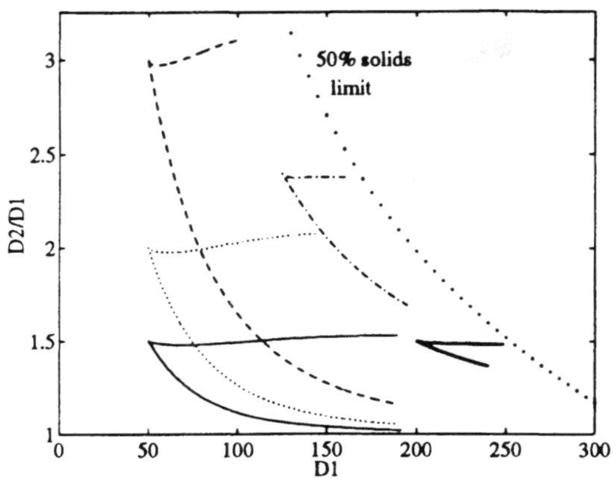

Fig. 4. Reachable regions for several initial diameter pairs. $[M]_P = 1.35$ mol/l at upper bound and $[M]_P = 5.4$ mol/l at lower bound of each region. $D_1 = 50$ nm, $D_2 = 150$ nm (dashed); $D_1 = 50$ nm, $D_2 = 100$ nm (dotted); $D_1 = 50$ nm, $D_2 = 75$ nm (solid); $D_1 = 125$ nm, $D_2 = 300$ nm (dash-dotted); $D_1 = 200$ nm, $D_2 = 300$ nm (bold)

this produces the largest relative narrowing. Alternatively, the slight broadening shown with the lowest value of $[M]_P$ in Figure 2, corresponds to the largest difference in $\bar{n}$ in Figure 3. Thus, the limits of $[M]_P$ correspond to the limits of competitive growth.

5. GROWTH MECHANISM CONTROLLABILITY

The objective in this section is to gauge the utility of the growth mechanism to influence the competitive growth of two monodisperse particle populations under a variety of initial conditions. The resulting conclusions can then be applied to distributions of particles. As in previous cases, the monomer feed rate is manipulated to control the monomer concentration. Although the subsequent simulations are performed for an equal number of particles in each population, equivalent results can be obtained for other basis of competition such as equal initial population volumes.

Figure 4 shows the simulation results of reactions run at the lower and upper $[M]_P$ limits for several initial dry diameter pairs. These simulations were run using the same initial reaction conditions (Table 1) as in previous cases except for the total amount of monomer used. In this figure, all the initial populations were grown with 500 g of monomer. This amount was chosen so that the various cases were kept from exceeding the practical reaction limit of 50% solids.

To relate the curves in Figure 4 to controllability, a couple of definitions must be presented. For nonlinear continuous time systems, Stengel (1986) denotes the difference between reachable and controllable regions. A reachable region is defined as the envelope of $\mathbf{x}(t_f)$ that can be reached from $\mathbf{x}(t_o)$ through the use of control, while a controllable region is defined as the envelope of $\mathbf{x}(t_o)$ from which $\mathbf{x}(t_f)$ can be reached through the use of control.

The bounded regions of Figure 4 represent the practical reachable regions of the respective diameter starting points. Since particles cannot be 'unpolymerized', these regions can only exist to the right of each starting point. In comparison to reachable regions, controllable regions are more difficult to obtain. However, Figure 4 can assist in obtaining an approximate envelope of starting points that can be used to obtain a particular diameter pair. For instance, if populations with $D_1 = 150$ nm and $D_2 = 210$ nm ($D_2/D_1 = 1.4$) were desired, any of the three starting points with $D_1 = 50$ nm could be used to obtain the final pair. The same reason which makes reachable regions expand to the right, forces controllable regions to expand to the left of the final diameter pair.

The effect of changing the initial diameter ratio on the size of the reachable region is demonstrated in Figure 4 by the three diameter pairs using 50 nm particles for the smaller population. As D_2/D_1 increases, the size of the region also increases. This is due to the fact that populations can only effectively narrow (relatively) via the growth mechanism. Thus, the larger the particle diameter ratio, the larger the potential to alter the initial relative size.

Reachable (and controllable) regions are also influenced by the magnitude of the initial diameter pairs. The bold lined region in Figure 4 has the same initial diameter ratio as the solid lined region, yet it has a smaller equivalent area. The reason for this is that the relative narrowing effect diminishes as particles get larger. As explained previously, the greatest relative narrowing (in terms of diameter) occurs at $\bar{n_1} = \bar{n_2}$. As particles become larger (greater than 200 nm), the difference between $\bar{n_1}$ and $\bar{n_2}$ at the saturation limit for two different sized particles becomes greater. In equivalent terms, the slope of the line corresponding to the largest value of $[M]_P$ in Figure 3 will eventually increase as particle diameter increases.

Based on these results, equivalent conclusions can be made for distributions of particles that are bounded by the initial diameter pairs of Figure 4. For this system, the growth mechanism can essentially be used only to relatively narrow diameter distributions. The broader the initial distribution, the more potential there is to alter the initial relative breadth. The potential to narrow a broad distribution also increases as the absolute sizes of the particles decreases.

6. CONCLUSIONS

The monomer concentration has been shown to be a critical parameter influencing the competitive growth between different sized monodisperse particle populations. The extent of this influence is at the practical limits of $[M]_P$. Reachable regions can be specified for initial diameter pairs by running simulations at the $[M]_P$ limits. The area of these regions depend on the magnitude of the initial diameter ratio and the individual particle diameters. Approximate controllable regions can also be obtained from the reachable regions indirectly. Equivalent conclusions drawn for two monodisperse populations can be made for particle distributions bounded by the initial diameter pairs.

7. NOMENCLATURE

D	Unswollen particle diameter [nm]
D_s	Swollen particle diameter [dm]
F_m	Monomer feed rate [mol/s]
$[I]$	Initiator conc. [mol/l]
k	Desorption rate const. [1/s]
k_a	Absorption rate const. [l/mol s]
k_d	Initiator decomp. const. 1.3×10^{-6} [1/s]
k_{fm}	Mon. transfer const. 2.7×10^{-5} [l/mol s]
k_{mp}	Entry rate const. 3.0×10^{-7} [dm/mol s]
k_{p_o}	Propagation const. 253 [l/mol s]
k_{t_o}	Termination const. 6.8×10^{7} [l/mol s]
K_o	Radical diffusion rate [1/s]
$[M]_P$	Monomer conc. in particles [mol/l]
MW_m	Monomer molecular weight [g/mol]
$\bar{n}$	Avg. no. of radicals per particle
N_a	Avagadro's number
N_t	Total number of particles
$[R]_W$	Aq. phase radical conc. [mol/l]
R_p	Polymerization rate [mol/l s]
t	Time [s]
T	Temperature [C]
V	Swollen particle volume [l]
v_f	Particle free volume
v_{fmc}	Critical particle free volume 0.0383
V_P	Total particle volume [l]
V_R	Reactor volume [l]
V_W	Water volume [l]
W_P	Weight fraction polymer
β	Prob. of aq. radical reaction 0.1275
ρ_m	Monomer density 879.25 [g/l]
ρ_p	Polymer density 1042.1 [g/l]
σ	Radical absorption rate [1/s]

8. REFERENCES

Asua, J. M., E. D. Sudol and M. S. El-Aasser (1989). Radical Desorption in Emulsion Polymerization. *J. Polym. Sci.* **27**, 3903–3913.

Chen, S. and K. Wu (1988). Emulsion Polymerization: On the Characterization of the Particle Size Distribution. *J. Polym. Sci.: Part A: Polym. Chem.* **26**, 1143–1155.

Dimitratos, J., M. S. El-Aasser, C. Georgakis and A. Klein (1990). Pseudo-steady States in Semicontinuous Emulsion Copolymerization. *J. Appl. Polym. Sci.* **40**, 1005–1021.

Gardon, J. L. (1968*a*). Emulsion Polymerization. I. Recalculation and Extension of the Smith-Ewart Theory. *J. Polym. Sci: Part A-1* **6**, 623–641.

Gardon, J. L. (1968*b*). Emulsion Polymerization. VI. Concentration of Monomers in Latex Particles. *J. Polym. Sci: Part A-1* **6**, 2859–2879.

Hawkett, B. S., D. H. Napper and R. G. Gilbert (1972). Analysis of Interval III Kinetic Data for Emulsion Polymerizations. *J. Chem Soc. Faraday Trans. 1* **77**, 2395–2404.

Kao, C. I., D. P. Gundlach and R. T. Nelsen (1984). Kinetics of Emulsion Polymerization of Styrene- Simulation Model with Varying Free Radical Capture Efficiency. *J. Polym. Sci.: Polym. Chem.* **22**, 3499–3519.

Leis, J. R. and M. A. Kramer (1988). The Simultaneous Solution and Sensitivity Analysis of Systems Described by Ordinary Differential Equations. *ACM Trans. Math. Soft.* **23**(1), 45–60.

Li, B. and B. W. Brooks (1993). Prediction of the Average Number of Radicals per Particle for Emulsion Polymerization. *J. Polym. Sci., :Part A* **31**, 2397–2402.

Liotta, V., E. D. Sudol, M. S. El-Aasser and C. Georgakis (1994). On-line Monitoring, Modeling, and Model Validation of Semi-Batch Emulsion Polymerization in an Automated Reactor Control Facility. *In Preparation.*

Poehlein, G.W. and J. W. Vanderhoff (1973). Competitive Growth of Polystyrene Latex Particles: Theory and Growth. *J. Polym. Sci.* **11**, 447–452.

Rawlings, J. B. (1985). Simulation and Stability of Continuous Emulsion Polymerization Reactors. Ph.D. Dissertation. University of Wisconsin, Madison.

Stengel, R. (1986). *Stochastic Optimal Control, Theory and Application.* John Wiley & Sons. Princeton, New Jersey.

Sundberg, D. C., J. Y. Hseih, S. K. Soh and R. F. Baldus (1981). . *Am. Chem. Soc. Sym. Ser.* **165**, 315.

Vanderhoff, J. W. and E. B. Bradford (1956). An Investigation of the Kinetics of Emulsion Polymerization by Electron Microscopy. *Tech. Assoc. Pulp Pap. Ind.* **39**(9), 650–655.

MULTIVARIABLE CONTROL OF A CONTINUOUS CRYSTALLIZATION PROCESS USING EXPERIMENTAL INPUT-OUTPUT MODELS.

R.A. EEK* and O.H. BOSGRA*

*Mechanical Engineering Systems and Control Group, Delft University of Technology, Mekelweg 2, 2628 CD Delft, The Netherlands.

Current address of R.A. Eek: Bayer AG, ZF-TST, Building E41, D-51368 Leverkusen, Germany.

Abstract. Results from closed-loop control of a continuous evaporative crystallization process on the basis of identified input-output models are presented. Experimental results show that open-loop identification of a crystallization process may lead to poor models as the process has a tendency to oscillate. Improved experimental conditions are achieved under feedback control, using a simple single-loop PI controller. Identification of low-order models, on the basis of closed-loop data, is studied using linear multivariable input-output models. Controllability analysis reveals that only two output quantities can be controlled independently: the mass production rate and a quantity related to the density of small crystals (fines) in the system. A two input two output model predictive controller (MPC) is designed and tested on the pilot crystallizer. Experimental results show that an improvement of the performance of the closed-loop system is obtained.

Key Words. System identification; Closed-loop identification; Model-based control; Multivariable control; Crystallization

1. INTRODUCTION

Continuous suspension crystallization is widely applied in process industry as a separation and purification process. In volume this unit operation is second to distillation. The crystal size distribution (CSD) is considered as an important process parameter to be optimized, as it strongly determines the performance of downstream solid-liquid separators. The CSD often does not meet the desired product specifications, due to process disturbances, intrinsic physical feedback mechanisms (which can cause cycling), and during plant start-up and shutdown.

In contrast to distillation, results on identification and control of crystallizers are scarce. This is mainly due to the fact that the development of sufficiently accurate models, purely on a first principles basis, is difficult as many complicated phenomena related to hydrodynamics and the birth and growth of crystals play a role. In addition, progress has been hampered due to the absence of robust on-line sensors for the CSD dynamics.

In De Wolf and Van den Hof (1992), system identification techniques have been applied to estimate low order linear models from crystallization process input-output data directly. Miller and Rawlings (1994) present results on parameter estimation within a first principles model, using supersaturation and density measurements from batch crystallization experiments. A review of recent results on identification and control of crystallizers is given by Rawlings *et al.*, (1993).

In earlier work (Eek *et al.*, 1995), results on first principles modelling of crystallizers and the estimation of empirical parameter values are presented. The experimental data sets were obtained from the free start-up responses of a pilot crystallizer using an on-line laser light scattering technique. A main disadvantage using start-up responses is that merely information on the slower dynamics, within a wide (nonlinear) process output range, is obtained. Knowledge on the faster dynamics around the bandwidth of the system is, however, of greater importance, as closed-loop stability is determined in this frequency range. Besides, the estimation of empirical parameters, within the nonlinear model, is a non-trivial problem which may consume a considerable amount of time.

The purpose of this paper is to show that low order models for the control relevant process dynamics can be estimated directly on the basis of experimental input-output data. Secondly, we will show that the desired closed-loop performance can be achieved with a multivariable model predictive control scheme, which employs the identified lin-

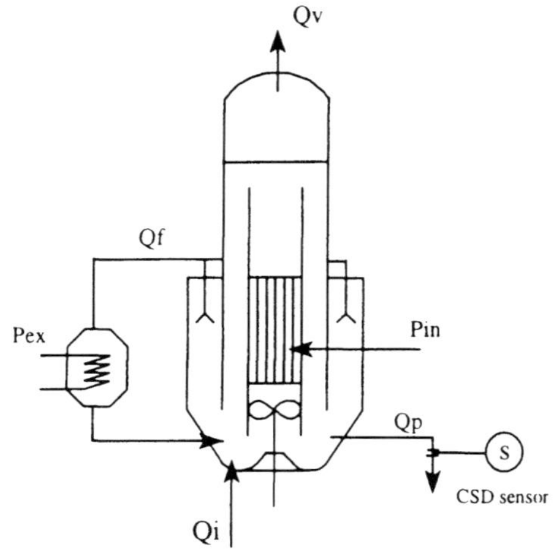

Fig. 1. Schematic drawing of pilot crystallizer, with fines removal system and CSD sensor.

ear input-output model as its internal model

2. PROCESS DESCRIPTION

A schematic drawing of the single stage evaporative suspension crystallizer that is used for identification and controller evaluation is given in Fig. 1. The process is assumed to be isothermal and to have a constant volume of 970 litres. A detailed description of the experimental crystallizer is given in Eek (1995d).

As process actuators, respectively the fines removal rate Q_f, the total heat input defined by: $P_{tot} = P_{in} + P_{ex}$, and the product removal rate Q_p, are considered. Fines are removed selectively with an annular settling zone, in which a slow upward slurry velocity is maintained. Fines are assumed to be dissolved totally by heating (P_{ex}).

Forward laser light scattering is applied as on-line CSD sensing technique. From scattered light intensity patterns, recorded with a Malvern Particle Sizer^{TM} the mean crystal size X_{43} and a parameter related to the number density of fines are estimated. The latter is denoted as y_{r1}. In addition, the slurry density denoted as M_t, measured with an independent density indicator is taken as a third output parameter.

3. PROCESS DYNAMICS

Crystallization processes are in principle distributed parameter processes in which the dynamics are mainly influenced by crystal growth, birth and crystal removal. At first, the process dynamics were considered by applying excitation signals directly onto the process inputs, in an open-loop fashion. As test signals multi-sine signals where chosen for the fines removal rate Q_f, and the product removal rate Q_p, each containing 29 logarithmically spaced frequency components. For every

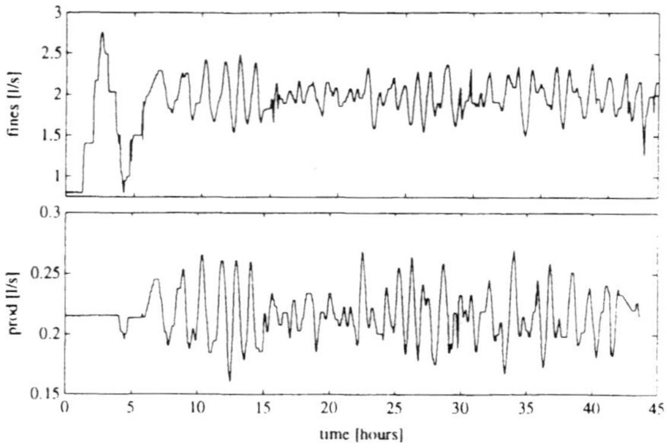

Fig. 2. Sinusoidal excitation signals used for open-loop identification after t>6 hrs.

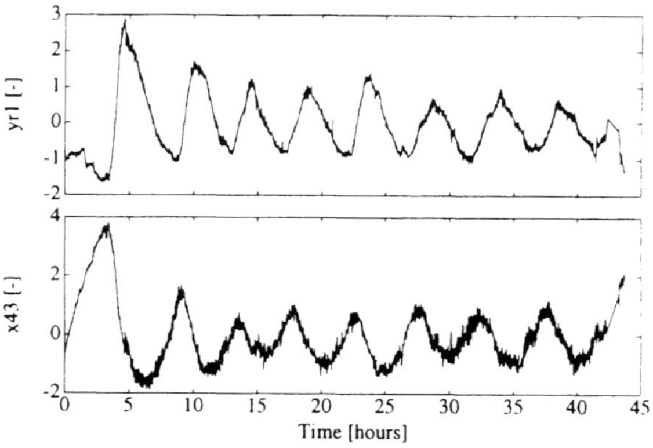

Fig. 3. Open-loop response of y_{r1} and X_{43}, on sinusoidal excitation of the process inputs.

frequency in the first signal the second signal has the first harmonic next to it to obtain uncorrelated signals. In Fig. 2 the designed input signals, are depicted.

A 45 hours lasting experiment, which is referred to as RUN35, is performed with the pilot crystallizer. In Fig. 3, the trends of the signals y_{r1} and X_{43} are depicted.

Visual inspection of the data directly reveals that the process output response is strongly dominated by slow cycling behaviour of the CSD with a time period of approximately 6 hours, which may have been invoked from the non-stationary initial state of the process. This behaviour is also observed after start-up of the crystallizer (Eek *et al.*, 1995a). It is concluded that successful identification results can in principle only be obtained when the process is operated with a closed-loop controller that stabilizes and thus removes the cycling behaviour.

4. Design of a stabilizing controller.

Exercises with the nonlinear process model (Eek *et al.*, 1994b), revealed that stability of the CSD is achieved easily with a single input single output (SISO) PI-control loop, which uses the weighted

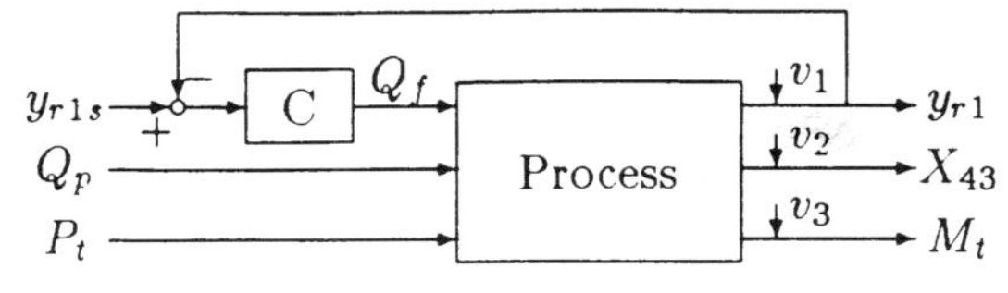

Fig. 4. Block scheme of closed-loop system to be identified.

number of fines y_{r1} as its input, and the fines removal rate Q_f as its output. The resulting closed-loop control configuration is depicted in Fig. 4. Details on the design and the evaluation of the feedback controller are provided in Eek *et al.*, (1995c). The results indicate that this simple closed-loop configuration effectively dampens open-loop cycling of the CSD.

The closed-loop process, which will be considered for identification further, is summarized in Fig. 4. In addition, the controller output Q_f is measured to estimate the open-loop process transfer-functions, following a method that is described in the next section. The signal y_{r1s} is the setpoint for the stabilizing controller, and $v(t)$ is a noise contribution.

5. Closed-loop MIMO identification.

A main and well known problem related to the identification of a process in a closed-loop configuration is that the noise on the outputs is correlated with the process inputs due to the presence of a feedback loop (see, e.g Fig. 1). This correlation generally results in biased estimates for the model, where the bias distribution depends on the characteristics of the noise. To solve this problem, we applied the two-step (TS) method as proposed by Van den Hof and Schrama (1993). In the first step a model for the sensitivity, having $\{y_{r1s}, Q_p, P_t\}$ as inputs and Q_f as output, is estimated, and applied successively to simulate a noise-free controller output: $\hat{Q}_f$. In the second step the inputs $\{\hat{Q}_f, Q_p, P_t\}$ and the three outputs $\{y_{r1}, X_{43}, M_t\}$ are used to estimate the process model in an open-loop manner. Thus, the usual problems of direct closed-loop identification are circumenvented, by subdividing the identification into two successive open-loop identification steps. We refer to Van den Hof and Schrama (1993) for further reading and to Eek *et al.*, (1994b) for details on application to the crystallizer.

5.1. *Design of a closed-loop experiment*

Two closed-loop experiments for identification and successive model validation are performed with the pilot crystallizer. We refer to these experiments as RUN55 and RUN56. All three inputs of the closed-loop process, e.g: y_{r1s}, Q_p, and P_t (see also Fig. 4), have been excited with generalized binary noise signals (Tulleken, 1990), which switch (pseudo-) randomly between two fixed signal levels at fixed time intervals. Choosing a switching probability between $0.5 < p < 1$ will provide more excitation in the low frequency range, conversely, choosing $0 < p < 0.5$ will increase the signal energy in the high frequency range.

As opposed to multi-sine excitation, these signals are much more accepted in processes industry, as they have the advantage that the energy is roughly evenly distributed over the frequency range of interest, hence the maximum signal amplitude is lower.

The real nominal signal values u_{nom} and their amplitudes Δu are given in Table 1. The basic intervals for the GBN signals are chosen 20 minutes. The switching probability was chosen $p = 0.7$. GBN signals used for RUN55 and RUN56 are reasonalby uncorrelated, whereas their frequency contents are comparable. For both experiments, a duration of 50 hours was chosen. This value roughly equals 8 times the dominant time constant in the system.

Several data preprocessing steps were applied to the raw data. After decimation each signal contains 600 data points on a sample interval of 5 minutes. For identification and succesive model validation two data sets, each containing 3 input and 3 output signals, were used.

5.2. *Results two-step identification*

Application of the first step of the two-step identification procedure resulted in a 6th order state-

Table 1 Nominal values and amplitudes of the GBN test signals.

input	u_{nom}	Δu	unit
y_{r1s} (RUN55)	0.5	0.25	[-]
y_{r1s} (RUN56)	0.5	0.20	[-]
Q_p	215	35	[ml/s]
P_t	120	30	[KW]

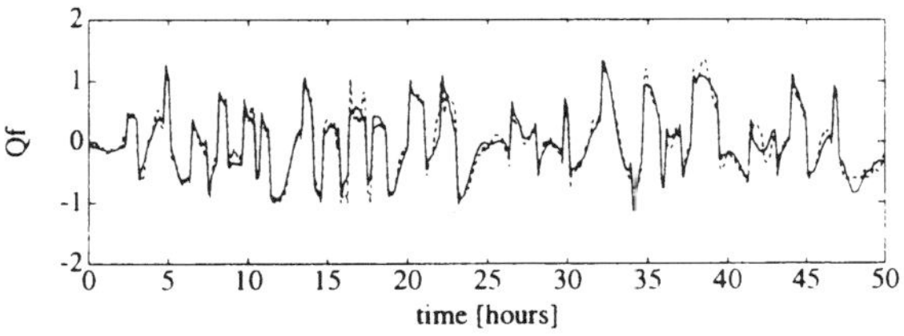

Fig. 5. Measured (dashed, RUN56) and reconstructed (solid) output signal of stabilizing controller estimated on the basis of a 6th order state space model.

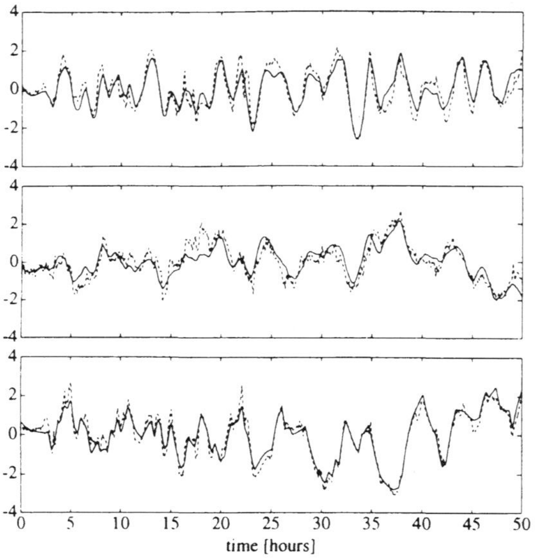

Fig. 6. Fit (solid) on measured output data (dashed, RUN56) of fourth order state space model estimated in second step of the 2 step procedure.

space model in canonical observability form, for the sensitivity model. Note that all three inputs of the process (see Fig. 4) had to be used in this step as all three affect the input Q_f. With the estimated model for the sensitivity a noise free output signal for the stabilizing controller is simulated. The measured and reconstructed output signal of the controller are given in Fig. 5.

In the second step, a 4th order state-space model in canonical observability form is estimated in an open-loop fashion. For the controller output, the signal reconstructed with the model for the sensitivity $\hat{Q}_f$, which is obtained from the first step, is used. An initial state and an offset were also estimated, while the matrix direct coupling matrix in the state space model was kept zero. In Fig. 6 and Fig. 7, the fits on the identification and validation sets are given, respectively.

In Eek *et al.*, (1994b), the data is used for validation of a nonlinear first principles process model as well. In the same contribution several details on the experiment design procedure can be found.

6. Controllability analysis

The input-output model with inputs $\{Q_f, Q_p, P_t\}$ and outputs $\{y_{r1}, X_{43}, M_t\}$ is used to study the controllability of the process. In Fig. 8 the singular values are given. The steady-state condition number is 231, the largest element of the RGA is 33, and the minimal condition number is 136. Over the whole frequency range the condition number is larger than 200, except in a narrow frequency band arround $2{\cdot}10^{-3}$ Hrz. where the condition number is 20.

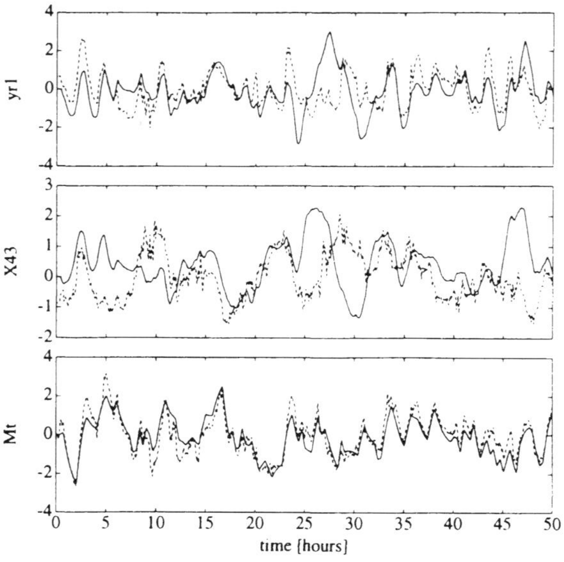

Fig. 7. Validation of fourth order model (solid) using uncorrelated data (RUN55, dashed).

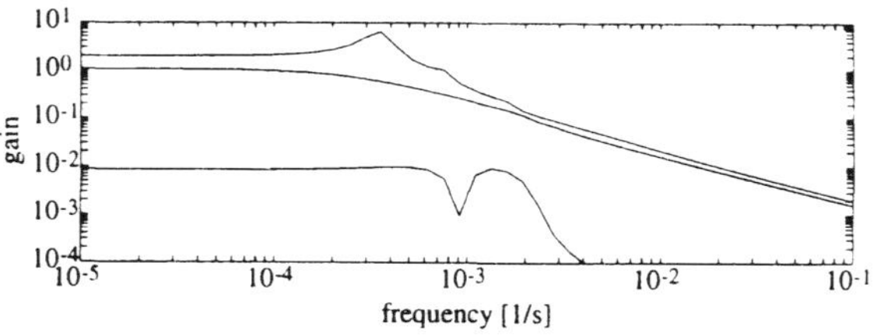

Fig. 8. Singular values of 3 input 3 output transfer function.

These results indicate that the system has a large directionality. From simulated step responses it is found that this is mainly due to a strong dependence of the mean X_{43} on the fines density y_{r1}. The stationary value of the mean crystal size could be written as a direct function of the value for y_{r1}, while in the dynamic case X_{43} only showed a delayed response in comparison to the response of y_{r1}. Consequently, the mean can not be controlled independently and is therefore omitted.

If also the product removal flow is removed as input parameter, which is desirable as this parameter has a large interaction with the process up and down stream, a two input two output system remains. The singular values of this system closely resemble the two dominant singular values in Fig. 8. The condition number of this system is lower then 15 over the whole frequency domain, the RGA is close to an identity matrix, and the largest element, found at any frequency, is 1.8. Additional details on controllability assessment can be found in Eek *et al.*, (1995b).

7. MPC control

Model predictive control is applied to further enhance the closed-loop behaviour of the process. The most striking features of MPC with regard

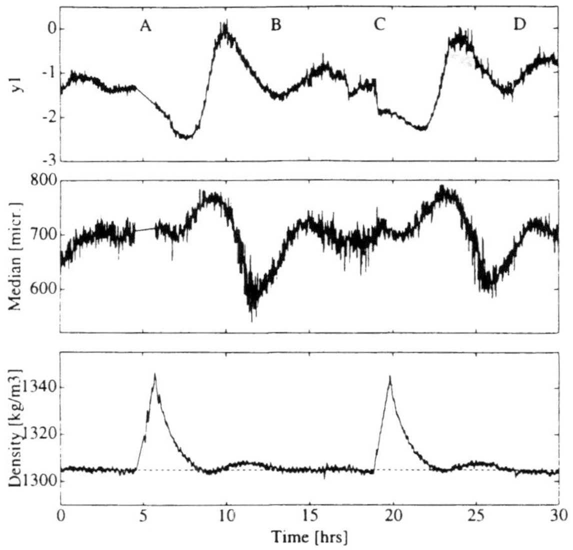

Fig. 9. Experimental responses to temporal batch operation due to simulated blockage of the product removal flow during one hour at A and C; inbetween A and B. and after D the closed loop controller is applied.

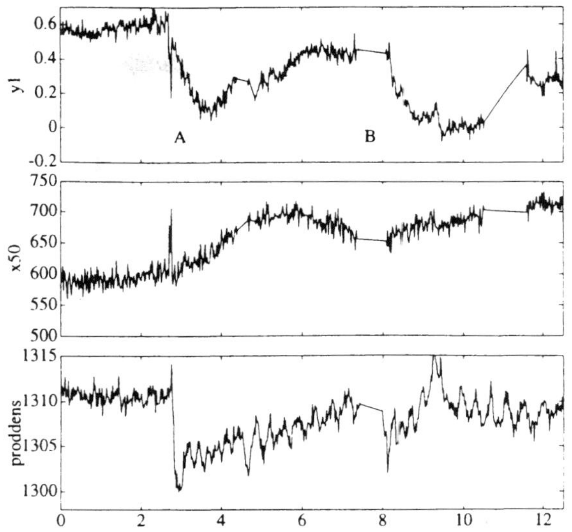

Fig. 10. Open-loop response on injection of 40 litres of water at A; at B the response is corrupted with other undesired process disturbances.

to industrial applications are its ability to handle constraints and the fact that an explicit optimization is made, which enables to account for sensor and actuator failures in an optimal manner. MPC is reviewed by Garcia (1989)

An MPC controller is designed to control the two input two output system described in the previous section. The MPC scheme proposed by Ricker (1990) is applied and the corresponding software toolbox of Morari and Ricker (1994) is used to implement the controller. As internal model the identified input-output model obtained from the previous section is used. Details on the design and implementation aspects are provided in Eek (1995d).

8. Open-loop experimental process responses

Closed loop responses are comparoted to open-loop responses for three different cases:

1. **Temporal batch operation.** In Fig. 9 the measured product density is depicted as a result of temporal batch operation due to a simulated blockage of the product discharge line. After interrupting the product discharge the slurry density in the vessel increases linearly, releasing the valve causes an exponential response to its original value. Also the mean crystal size and the signal y_{r1} are strongly upset.
2. **Water injection.** Fig. 10 depicts the response of different process output parameters after injection of 40 litres of water at A.
3. **Process start-up.** Open-loop start-up responses, as shown in Eek *et al.*, (1995a), reveal that the CSD always shows a second order like badly dampened response.

9. Evaluation of the MPC controller

Using the MPC controller the same aforementioned disturbances are applied after closed-loop start-up of the process. The results are given in Fig. 11 and Fig. 12. After switching on the controller at A oscillations occurred as the controller gain was chosen too large. The gain was lowered gradually in between A and B until the desired response was obtained. After injection of water at B both actuators react. The fines flow is lowered to compensate for the dissolved fines and the heat input is increased to compensate for the dissolved crystal mass. Within two hours, both outputs attain stationary values again, showing an strong improvement in comparison to the open-loop case. Also the oscillation induced after closing the product discharge at C is strongly reduced.

10. CONCLUSIONS

In this paper we have shown that low order linear multivariable input-output models, which are estimated directly from process input-output data, give a sufficiently accurate description of the dynamics of a crystallization process, provided that the process is operated close to its stationary behaviour. The latter condition is imposed with a stabilizing feedback controller that stabilizes the open-loop cycling behaviour of the system. Low order models can be derived via a system identification approach on the basis of measured closed-loop process input-output data. Models are estimated consistently with a two-step identification

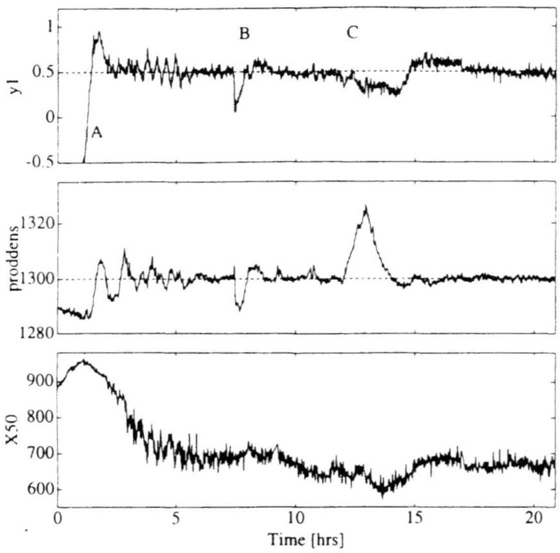

Fig. 11. Process outputs MPC controlled process. Tuning in between A and B; at B 40 litres of water injected and at C the plant is operated batch wise for one hour.

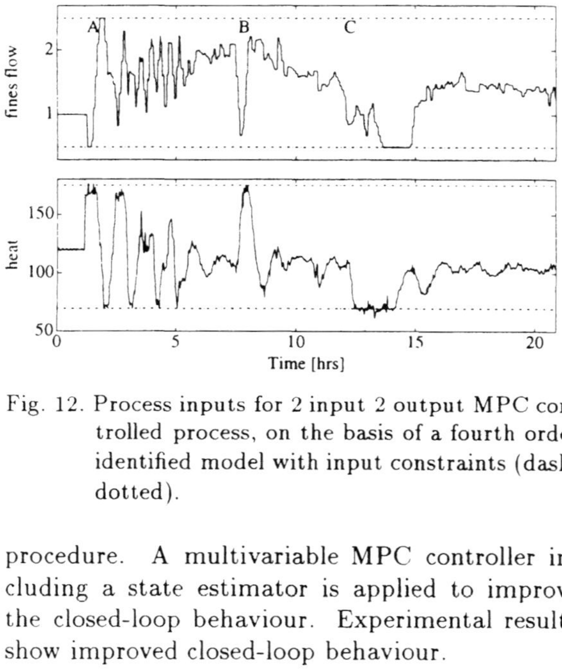

Fig. 12. Process inputs for 2 input 2 output MPC controlled process, on the basis of a fourth order identified model with input constraints (dash-dotted).

procedure. A multivariable MPC controller including a state estimator is applied to improve the closed-loop behaviour. Experimental results show improved closed-loop behaviour.

Acknowledgement

The authors are indebted to the former students Sven Rusticus and Jaap Both who did part of the experimental work, and to the Dutch Foundation of Technology (STW), AKZO, ICI, DOW Chemicals, DSM, E.I. Dupont de Nemours, and Eastman Chemical Company for their financial support of the UNIAK project.

11. REFERENCES

Boxman, A., Merkus, H. G., Verheijen, P. J. T., and Scarlett, B., "Deconvolution of light-scattering patterns by observing intensity fluctuations," *Applied Optics, Vol. 30, No. 33, pp. 4818-4822* (1991).

De Wolf, S., and Hof, P. M. J. Van den, "Identification of a pilot plant crystallization process with output error methods." In: Banyasz C & Keviczky L (Eds.), Identification and System Parameter Estimation; *IFAC Symposia Series* 1992, No. 3, pp. 133-138. Sel. Papers 9th IFAC/IFORS Symp., Budapest, July 8-12, 1991, (1992).

Eek, R. A., Dijkstra, Sj., and Rosmalen, G. M. van , "Dynamic modeling of suspension crystallizers, using experimental process data. ," *AIChE Journal, in press* (1995a).

Eek, R. A., Hoogenboezem, A. J., and Bosgra, O. H., "Design issues related to the control of continuous crystallizers," *Comp. and Chem. Eng., In Press,* (1995b).

Eek, R. A., Rusticus, S. J., and Dijkstra, Sj., "Design of robust stabilizing feedback controllers for a continuous crystallization process," *Submitted to Chem. Eng. Sci,* (1995c).

Eek, R. A., Both, J. A, and Van den Hof, P. M. J. "Closed-loop identification of a continuous crystallization process," *Selected Topics in Ident. Mod. and Control, Delft Univ. Press Vol. 7.,* (1994b).

Eek, R. A., Ph.D. Thesis, Delft University of Technology, (1995d).

Garcia, C. E., Prett, D. M., and Morari, M., "Model predictive control: theory and practice-a survey," *Automatica, Vol. 25, pp. 335-348* (1989).

Ljung, L., "System identification," *Prentice-Hall. Inc., Englewood Cliffs, New Yersey* (1987).

Miller, S. M., and Rawlings, J. B., "Model identification and control strategies for batch cooling crystallizers," *AIChE Journal, Vol. 40, No. 8, pp. 1312-1326* (1994).

Morari, M., and Ricker, N. L., "Model predictive control toolbox, User's Guide," *The Mathworks Partner Series, The Mathworks Inc. Natick, MA* (1994).

Rawlings, J. B., Miller, S. M., and Witkowski, W. B., "Model identification and control of solution crystallization processes: a review," *Ind. Eng. Chem. Res. Vol. 32, pp. 1275-1290* (1993).

Ricker, N. L., "Model predictive control with state estimation," *Ind. Eng. Chem. Res., Vol. 29, No. 3, pp. 374-382* (1990).

Tulleken, H. J. A. F., "Generalized binary noise test-signal concepts for improved identification-experiment design," *Automatica, Vol. 25, No. 1, pp. 37-49* (1990).

Van den Hof, P. M. J. , and Schrama, R. J. P., "An indirect method for transfer function estimation from closed loop data," *Automatica, Vol. 29, No. 6, pp. 1523-1527* (1993).

A DYNAMIC MODEL OF A SUPERFRACTIONATOR: A TEST CASE FOR COMPARING DISTILLATION CONTROL TECHNIQUES

Vikram Gokhale
Scott E. Hurowitz
James B. Riggs

Department of Chemical Engineering
Texas Tech University
Lubbock, TX 79409 USA

ABSTRACT: A detailed dynamic simulator of a propylene/propane (C_3) splitter has been developed to compare dual composition control performance for distillation control techniques. This column represents a large class of industrial distillation control problems: a high reflux ratio, low relative volatility separation with a large numbers of stages (superfractionator). The C_3 splitter was designed to produce polymer grade propylene (99.7% propylene) using 234 trays with a reflux ratio of 12.6. The relative volatility was modeled as a function of liquid composition and pressure. Liquid dynamics were modeled using a hydraulic time constant and equimolal overflow was assumed. The simulator was equipped with loosely tuned PI level controls.

The simulator, which uses the (L,B) configuration, was bench marked against data from an industrial C_3 splitter. A set of standard test scenarios come with the simulator. One of the tests, which involves periodic changes in feed composition, has been shown to provide a more critical evaluation of controller performance than standard step tests. This simulator has been used earlier to compare control results for a diagonal PI controller, a DMC™ controller and a nonlinear model based controller.

INTRODUCTION

Superfractionators are used industrially to separate very close boiling mixtures such as alcohol isomers, C_5 and C_6 isomers, ethyl benzene/styrene and propylene/propane. Because of the low relative volatility, a relatively large number of stages are required with a large reflux ratio, and as a result, these columns are quite sluggish with open loop response times of 5 to 30 hours. Control of this class of columns is important because they directly determine the variability in a number of final products and because these columns are large energy users.

MODEL DESCRIPTION

This distillation control test case is based upon dual composition control of an industrial propylene/ propane (C_3) splitter. C_3 splitters are found in most refineries and many large scale chemical intermediate plants. The C_3 splitter typically produces a polymer grade propylene product (99.5 mole% propylene specification) and a fuel grade propane (approximately 2 mole % propylene). Since the overhead product has an upper limit of 0.5 mole % propane, C_3 splitters are normally operated to produce 99.7% propylene in order to consistantly produce an on specification product. The bottom product purity is based upon the economic trade-off between propylene recovery and energy usage.

Table 1 contains the design conditions for the C_3 splitter test case. The design is based upon producing 99.7 mole % propylene and a bottom product with 2 mole % propylene from a propylene/propane feed with 70 mole % propylene. The column was designed with 234 trays based upon a Murphree tray efficiency of 85% (Kister, 1990) and an operating pressure of 18 atm. The modeling assumptions used in developing the dynamic model of the C_3 splitter are listed in Table 2.

The dynamic column model is based upon dynamic mole balances on propylene for each tray. A hydraulic time constant is used to model the liquid dynamics for the trays with one value of the hydraulic time constant for the entire column. The equimolal overflow assumption is used to calculate the flow rate of vapor leaving each stage.

The vapor/liquid equilibrium was described using a relative volatility which was modeled as an explicit function of pressure and composition (Hill, 1959). As a result, each tray had its own relative volatility. Product composition analyzers and feed composition analyzers (when used) were assumed to have five minute sampling periods.

The test column simulator was equipped with PI level controllers for the accumulator and reboiler. The settings for the level controllers are listed in Table 3. It should be noted that these settings represent loose level controls.

The test column simulator has the (L, B) configuration implemented on it. Gokhale, et al. (1994) evaluated nine possible control configurations for this column and found that the (L, B) configuration yielded the best performance for diagonal PI dual composition control. When the simulation was equipped with perfect level control, the control performance of the (D, B) structure was found to be equivalent to the performance of the (L, B) configuration.

The dynamic model equations were integrated using an Euler integrator (Riggs, 1994) with a step size of 0.3 seconds. A fifty to one ratio of simulated time to CPU was obtained for a 66 MHz 486 PC using Microsoft FORTRAN 5.1.

The dynamic model was bench-marked against dynamic step test data from an industrial C_3 splitter. The industrial data was based upon the (L,B) configuration. The (L,B) configuration is also used industrially (O'Conner, 1993). First, the simulator was found to provide the same general behavior as the industrial plant data (O'Conner, 1993) for openloop step changes in the manipulated variables and the feed rate. Then based upon response times, the hydraulic time constant of each tray was adjusted to match the industrially observed response times as closely as possible. For example, the overhead composition was observed to have an openloop response time of approximately 7 hours for a 0.5% change in the reflux rate. In addition, for a 1% change in the bottom flow rate, the response time for the bottom composition was approximately 25 hours (O'Conner, 1993). A hydraulic time constant of 3 seconds was found to provide the best overall dynamic match. Figures 1a and 1b show the openloop responses of the top and bottom impurities to step changes of +1% in the reflux flow rate and bottoms flow rate, respectively.

TEST SCENARIOS

The following test scenarios were used to test the composition controllers.

1. Setpoint change to 99.85% propylene in the overhead product at t=100 minutes followed by a Setpoint change to 99.5% at t=1000 minutes.
2. Ramp change in pressure from 211 to 226 psia from t=100 minutes to t=160 minutes followed by a step change in feed composition to 65% at t=1000 minutes.
3. Negative and positive 5% step changes in feed composition with changes applied every 250 minutes. At time (t) equal to 250, the feed composition (z) was decreased to 65% propylene. At t = 500 minutes, z was changed to 70%. At t = 750 minutes, z was set to 75%. At t = 1000 minutes, z was set to 70%, at t = 1250 z was set to 65%. Etc.

Test controllers should be tuned for scenario 1 and tested on scenarios 2 and 3. In addition, the period in test scenario 3 (nominally set at 250 minutes) should be varied to assess the sensitivity of the controller to the frequency of the upset.

The control objective in evaluating controller performance should be the variability in the propylene product while keeping the bottom product in the vicinity of 2% propylene.

CONTROLLER TEST RESULTS

The C_3 splitter simulator has already been tested using a diagonal PI controller, a DMC™ controller, and a nonlinear process model based controller (Gokhale, et al., 1995). Figures 2 and 3 illustrate closedloop PI responses for test scenarios 1 and 2, respectively. These closedloop responses represent optimal PI controller tuning.

CONCLUSIONS

The detailed dynamic model of a superfractionator has been developed and bench-marked against industrial data. This simulator represents an important class of distillation control problems and is available for testing distillation control techniques. Contact Dr. James B. Riggs at CPJBR@ttacs.ttu.edu for more information or questions regarding the simulator.

REFERENCES

Gokhale, V., Kulkarni, S., Shukla, N., and J.B. Riggs, (1994), "Analysis of Diagonal PI Dual Composition Controller for Distillation Columns," submitted to I&ECR.

Gokhale, V., Hurowitz, S., and J.B. Riggs, (1995), "A Comparison of Advanced Distillation Control Techniques for a Propylene/Propane Splitter," submitted to I&ECR.

Hill, G.E., (1959), "Propylene-Propane Vapor-Liquid Equilibria" Presented at the AIChE National Meeting, Atlantic City, NJ.

Kister, H., (1993) Distillation Design, McGraw-Hill.

O'Conner, Dan, DMC Corp, Houston, TX, Personnel Communication (1993).

Riggs, J.B., "An Introduction to Numerical Methods for Chemical Engineers," Second Edition, Texas Tech University Press, 1994.

Table 1
Design Specifications for Propylene/Propane Splitter

Number of Trays	232
Feed Tray Location (From Bottom)	64
Feed Flow rate	13.44 kg/sec (106,400#/HR)
Feed Comp. (mole %)	
Light Key	$C_3^=$ - 70
Heavy Key	C_3 - 30
Factor times minimum reflux for design	1.3
Column Diameter	3.96 m (13 ft)
Overhead Pressure	15.0 atm (221 PSIA)
Overhead Product Impurity	C_2 - 0.3 mole%
Bottoms Product Impurity	C_3 - 2.0 mole %
Overhead Flow rate	9.21 kg/sec (73,100#/HR)
Overhead Temperature	34.7°C (94.4 °F)
Bottom Flow rate	4.21 kg/sec (33,400 #HR)
Bottom Temperature	42.3°C (108.1 °F)
Reboiler Vapor flow rate	131.24 kg/sec (1,041,165#/HR)
Reflux Ratio	12.6
Feed Quality	Saturated

Table 2
Modeling Assumptions for Propylene/Propane Splitter

Liquid Dynamic	Hydraulic time constant
Neglible Vapor/Holdup	yes
Value dynamic on all flows	no
Accumulation and reboiler level control	PI
Analyzer delays on product composition	5 minutes
Eqimolal overflow	yes
Residence time in reboiler	5 minutes
Residence time in accumulator	5 minutes
Heat transfer dynamics modeled	no
Saturated liquid feed	yes
Subcooled reflux	no
Pressure dynamics modeled	no
Perfect mixing of liquid on trays	yes
Ideal VLE	no

Table 3
Level Control Tuning Parameters for the [L,B] Structure

	Gain	Reset
Accumulator	0.001	112.5
Reboiler	0.002	225

* Units: Gain: #mol/sec per # mol, Reset: Minutes

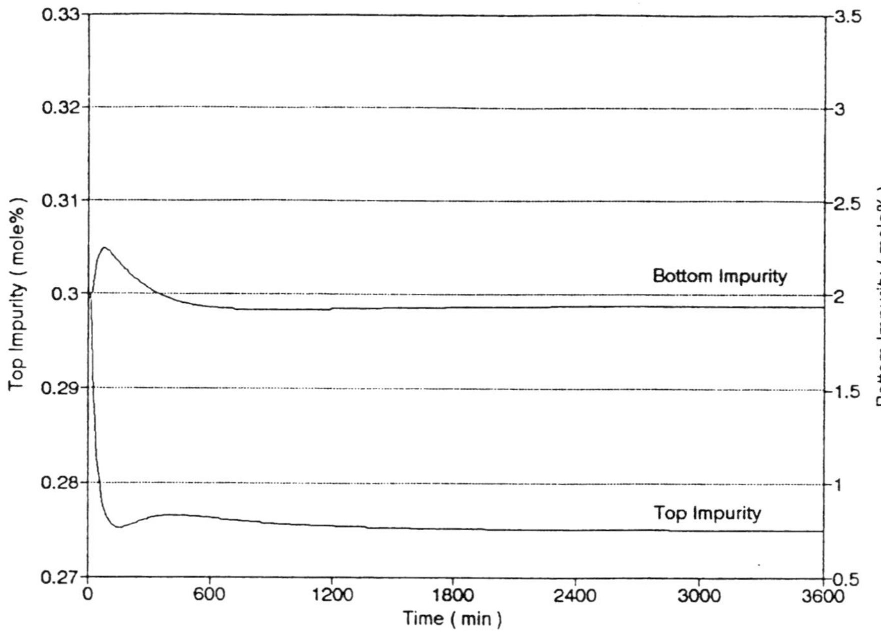

Figure 1 - Openloop Response
(a) +1% Change in Reflux Flowrate

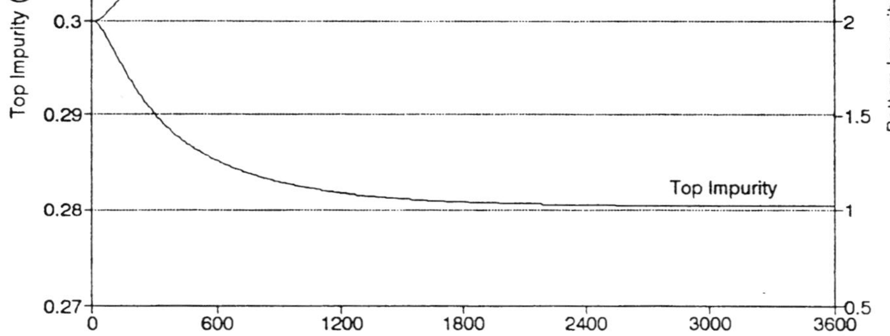

Figure 1 - Openloop Response
(b) +1% Change in Bottoms Flowrate

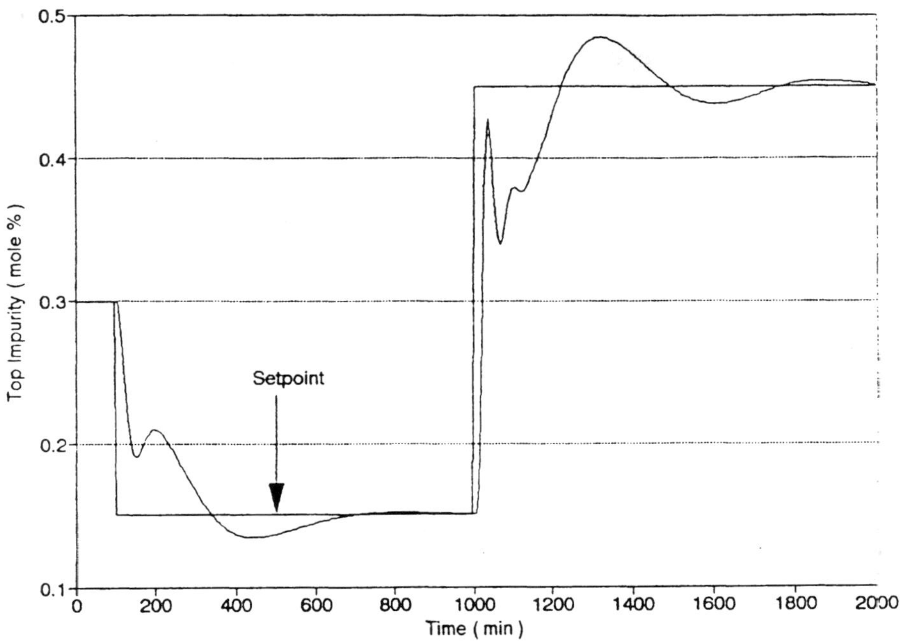

Figure 2 - PI Control for Test Case 1
(a) Top Impurity

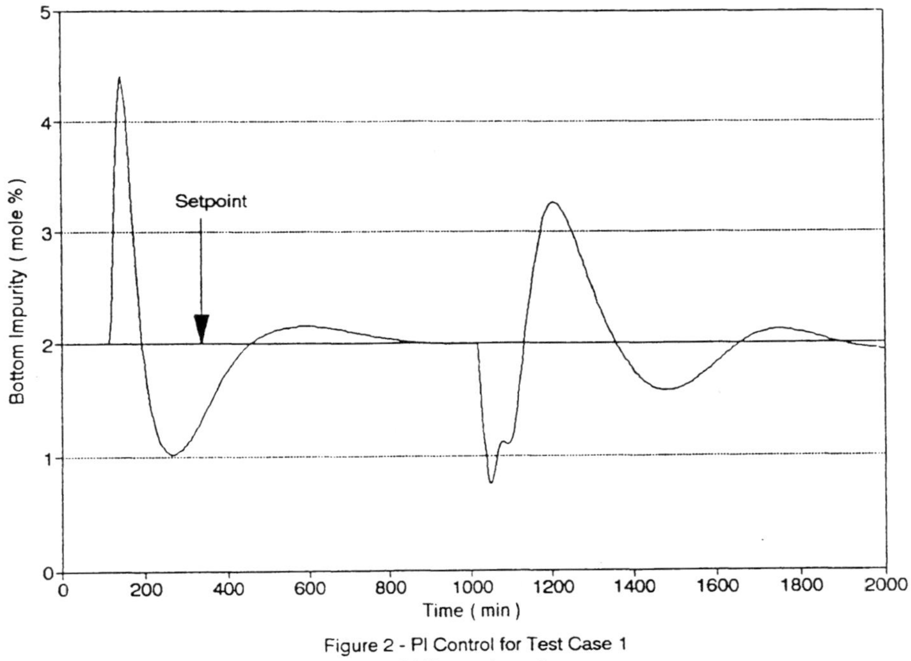

Figure 2 - PI Control for Test Case 1
(b) Bottom Impurity

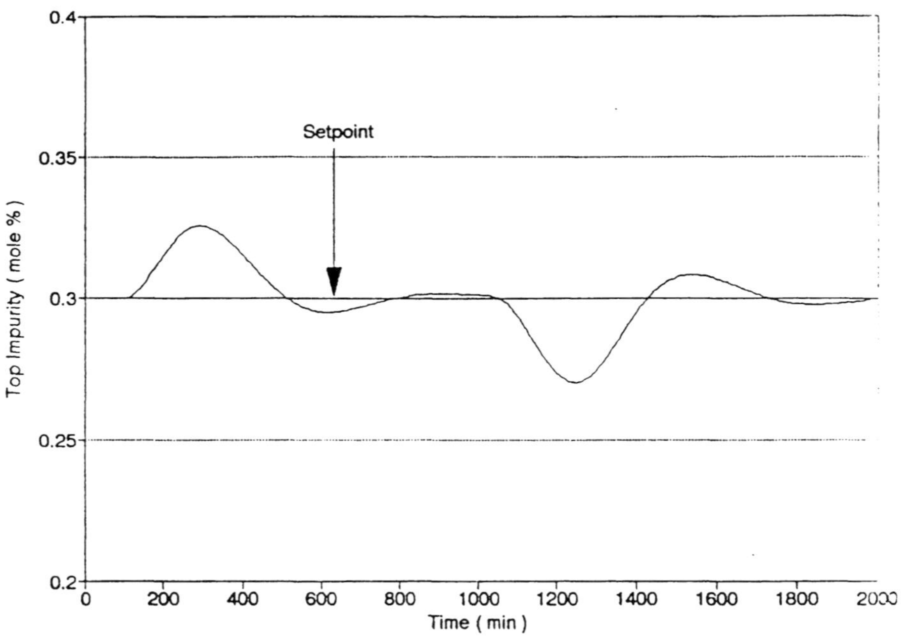

Figure 3 - PI Control for Test Case 2
(a) Top Impurity

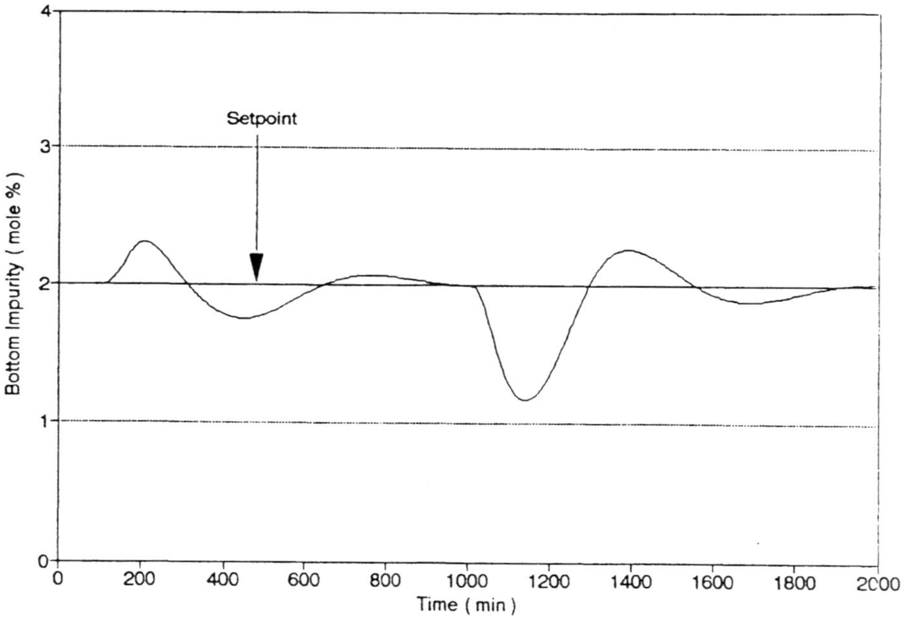

Figure 3 - PI Control for Test Case 2
(b) Bottom Impurity

IDENTIFICATION FOR CONTROL OF HIGH-PURITY DISTILLATION COLUMNS - A BENCHMARK PROBLEM

ELLING W. JACOBSEN*

S3 - Automatic Control
Royal Institute of Technology - KTH, 100 44 Stockholm, Sweden

Abstract. The note presents a high-purity distillation column which is proposed as a benchmark problem for identification of multivariable dynamic models intended for control applications. The column is ill-conditioned, i.e., has strong gain directionality, and is also strongly interactive. This poses a specific problem in identification in that it is difficult to obtain a reasonable model for the low-gain direction of the plant. The low-gain direction is usually hardly visible in open-loop responses but becomes important under feedback control. Furthermore, the column has a single slow pole (time-constant) which easily may become repeated in the identified model, causing a poor prediction of the control behavior of the column, in particular under partial feedback control.
The note includes a detailed description of the example column and the "experimental" conditions available for identification. A specification of the intended use of the identified model is also included and it is stressed that an important control objective is disturbance rejection. Thus, a disturbance model is desired in addition to the usually considered input model. At the end of the note an example is presented which demonstrates that a classical identification method involving independent PRBS signals and ARMAX-type MISO models yields a multivariable model which is poor for feedback control applications.

Keywords. Identification, multivariable, ill-conditioned, interactive, distillation.

1. INTRODUCTION

Tight composition control of distillation columns usually requires some kind of model-based control scheme and thus the availability of a dynamic model, typically linear. For distillation columns operating with reflux L and boilup V as independent inputs, i.e., LV-configuration, the desired linear model is usually written

$$\begin{pmatrix} dy_D \\ dx_B \end{pmatrix} = G^{LV}(s) \begin{pmatrix} dL \\ dV \end{pmatrix} \tag{1}$$

where y_D and x_B denote product purities. In practice, the model $G^{LV}(s)$ must, at least partially, be obtained through experiments on the plant. The traditional approach has been to obtain the individual transfer-matrix elements $g_{ij}^{LV}(s)$ *independently*, e.g., by fitting open-loop responses of the plant to first-order-plus-deadtime transfer-functions. However, several authors (e.g., Skogestad and Morari, 1988) have pointed out that this may be a poor approach if the model is intended for control applications. The reason is that one through this approach usually is unable to capture specific multivariable properties of importance for control.

*E-mail: jacobsen@control.e.kth.se

The main difference between SISO and MIMO plants is the presence of "interactions" or "directions" in the latter case, i.e., the gain of the plant depends on the direction of the input vector. The degree of directionality and interactions is usually evaluated through the Singular Value Decomposition (SVD) and the Relative Gain Array (RGA) (see e.g., Skogestad and Morari, 1988). Plants for which the ratio of the largest to smallest singular value, i.e., the condition number, is large are denoted *ill-conditioned* while those with large values of the RGA elements are denoted *strongly interactive*. Note that strongly interactive plant always are ill-conditioned while the opposite does not necessarily apply.

High-purity distillation columns operating with the LV-configuration are usually strongly interactive. Typically, increasing L and V such that the product flows remain unchanged increases the purity of both products and the corresponding gain is generally small. This is usually denoted changes in the *internal flows* and corresponds to the low-gain direction of the plant. Similarly, changing L and V such that there is a maximum change in the product flows makes one product purer and the other less pure and the corresponding gain is generally large. This is usually denoted changes in the *external flows* and corresponds to the high-gain direction of the plant. A multivariable model intended for control purposes should capture these effects in a reasonable way. However, it is well known that if the interactions are strong then this may be a difficult task. One reason is that the open-loop responses of the plant tend to be dominated by the high-gain direction which renders it difficult to obtain a reasonable model for the low-gain direction. Although often hardly visible in open-loop responses, the low-gain direction becomes important under feedback control and it is therefore crucial to obtain a good model also for this direction of the plant.

We start the note with a brief review of some of the works published on identification of strongly interactive plants in general and high-purity distillation columns in particular. We then present a *linear* distillation column which is proposed as a benchmark problem for which different methods may be tested and compared. The motivation behind using a linear model as the "plant" is that the problem of capturing a plants directionality is present both in the linear and nonlinear plant, and we do not want to confuse this problem with the specific problems caused by nonlinearities. In addition to presenting the model we specify the experimental conditions under which the identification should be performed. The identified model is intended for control applications and it is stressed that an important control objective is disturbance rejection. Thus, a disturbance model is desired in addition to the usually considered input model $G^{LV}(s)$ (1). At the end of the note we apply a standard identification method involving independent PRBS signals and MISO ARMAX models to the benchmark problem. The resulting model is shown to be poor for control applications.

2. A BRIEF REVIEW

Over the years a number of papers discussing the problem of identifying dynamic models for strongly interactive plants in general, and distillation columns in particular, have appeared. We do not intend to provide a complete review of these here, but mention some of the more recent works only. All the papers mentioned either directly address the problem of identifying distillation columns or employ a distillation column as an example[1].

As mentioned above, the traditional approach to identification of distillation columns has been to obtain the transfer-matrix elements independently, i.e., through SISO identification. The main reason for this is undoubtedly that the general literature on system identification has been focused on identification of SISO systems while identification of MIMO systems has been considered a difficult problem. However, Skogestad and Morari (1988) argue that one should be careful about identifying the individual SISO elements of the transfer-matrix of a high-purity distillation column independently as this easily may lead to a poor model for the low-gain direction of the plant. In particular, the identified model will easily have the wrong sign of the determinant at steady state and will in that case be useless for control. Andersen et al. (1989) demonstrate through an example that SISO identification yields a poor model of the plants low-gain direction. Andersen and Kümmel (1991) consider identifying two MISO models and find that the resulting MIMO model also in this case yields a poor prediction of the plants low-gain direction. They also consider using closed-loop identification with a single loop closed combined with MISO identification but find similar results as with open-loop identification. However, by low-pass filtering the closed-loop data prior to fitting the model they obtain a reasonable model for the low-gain direction of the plant.

Jacobsen et al. (1991) propose to use SISO identification and fit the initial responses (high-frequency behavior) only. The motivation behind this suggestion is that it is the initial response which usually is most important for feedback control and that distillation columns generally are significantly less interactive at high and intermediate frequencies than at steady state. To avoid problems with the wrong sign of the determinant of the steady state gain matrix they propose to correct the steady state gains to match an estimated value of the Relative Gain Array. They find that this approach yields models reasonable for multivariable control applications with relatively high bandwidth.

Several authors (e.g., Skogestad, 1988; Alsop and Edgar, 1990; Andersen and Kümmel, 1991) suggest that, for identification, it may be better to use the DV-configuration instead of the LV-configuration. The main motivation behind this is that with the DV-configuration the low-gain direction is closely aligned with changes in the input V only, and this may be utilized to obtain a better model for the low-gain direction. In particular, the determinant of the steady state gain matrix is less likely to have the wrong sign when obtained with the DV-configuration. Koung and MacGregor (1993, 1994) extend this idea to systems where the low-gain direction is not well aligned with a single input, e.g., LV-configuration. The main idea is to utilize a priori knowledge of the plants directionality to obtain better information for the low-gain direction. Koung and MacGregor (1993, 1994) consider the problem of obtaining a model with the correct sign of the determinant of the steady state gain matrix only.

Jacobsen (1994) considers several different approaches to identification of strongly interactive plants and stress that, in order to ensure a model suitable for control applications, both reasonable experimental data for all plant directions as well as a proper multivariable algorithm for fitting the data to a model are required. In cases where knowledge of the plant directionality is available he proposes to use an open-loop experimental design similar to that proposed by Koung and MacGre-

[1]In fact, we have not found any papers addressing this problem that do not mention distillation.

gor (1993). It is demonstrated that this yields satisfactory data for the low-gain direction of the plant, not only at steady state but also at intermediate and high frequencies, even when the knowledge of directionality is somewhat uncertain. In cases where knowledge of the plant directionality is poor or even lacking, Jacobsen (1994) considers using closed-loop identification with a single proportional controller as proposed by Andersen and Kümmel (1991) and shows that any controller gain will yield a closed-loop plant which is less interactive than the open-loop plant. This may be utilized to ensure a sufficient excitation of all plant directions, including the low-gain direction of the open-loop plant. Jacobsen (1994) suggests that the poor results obtained by Andersen and Kümmel (1991) is due to the fact that they fit the obtained data using two MISO models. By fitting the obtained data to a MIMO model Jacobsen (1994) find that a good model is obtained both for the high- and low-gain direction of the plant. A subspace based identification algorithm (e.g., Van Overschee and de Moor, 1994), which requires no a priori knowledge of the model structure, is proposed for fitting the data to a model.

Varga and Jørgensen (1994) apply the experimental design method suggested by Koung and MacGregor (1993, 1994) to a high-purity column. They use the obtained data to fit a MIMO Output Error model with a prespecified structure and find that the low-gain direction is reasonably captured in the resulting model.

Li and Lee (1994) propose to identify both the open-loop plant as well as its inverse and then combine these into a single model. The motivation behind this approach is that the inverse of the plant generally will yield good information about the low-gain direction of the open-loop plant. In order to identify the approximate inverse of the plant they employ four separate closed-loop experiments using four different controllers. Both the open-loop plant and its inverse are fitted using SISO identification methods.

All the papers mentioned above discuss the problem of identifying a reasonable model for the low-gain direction of the plant. A different, although related, problem in identification of strongly interactive plants is discussed in Jacobsen and Skogestad (1994a). A strongly interactive plant will usually contain a single slow pole (time-constant) which tends to dominate all open-loop responses of the plant. Fitting each SISO transfer-function independently, such that they all contain the slow pole, will generally result in a MIMO model with this pole repeated, and hence an inconsistent model. This inconsistency results in a poor prediction of the plants control behavior, in particular under partial control. Jacobsen (1994) find that the only way to avoid this problem is to employ a true MIMO method for fitting the experimental data to a model.

As seen from the above, several different methods for identification of dynamic models for distillation columns have been proposed in the literature. However, most authors demonstrate the goodness of their method through some specific example and it is difficult to make a comparison of different methods. Furthermore, most papers include little or no information on the experimental conditions, e.g., noise levels, used in applying their method for demonstration. Finally, few papers actually demonstrate the quality of their model for feedback control. The purpose of the present note is to propose a benchmark problem for which different proposed methods may be tested and compared. A similar benchmark problem, based on a simplified heat-exchanger, has previously been presented in Jacobsen and Skogestad (1994b). However, the previous problem was less interactive than most high-purity distillation columns and furthermore included predefined PRBS signals to be used in the experiments, thereby not allowing for specific experimental designs to be applied. In this note we present a more "flexible" problem for which experimental designs, closed-loop identification schemes etc. may be tested.

3. PROBLEM DESCRIPTION

The column we consider is separating a binary mixture of ethylene and ethane (C2-splitter), has purities of 99% in both products and is operated with mass reflux L_w [kg/min] and reboiler heat input Q_B [MJ/min] as independent inputs, i.e., L_wQ_B-configuration. This configuration is similar to the LV-configuration. Data for the column are summarized in Table 1. In the nonlinear dynamic model we assume constant relative volatility α, vapor-liquid equilibrium, perfect mixing on each tray, constant pressure, negligible vapor holdups and saturated reflux and feed flows. The liquid flow dynamics are modeled using a linearized Francis Weir formula. A simplified static energy-balance is included in the model. We neglect changes in liquid enthalpy with composition (pure components as saturated liquids at column pressure are chosen as reference of state) and assume a linear relation for the vapor enthalpy. With these assumptions we obtain a model with two states per tray, expressed as fraction of light component and total liquid holdup.

As the "plant" we employ a linear model obtained through linearization of the full nonlinear model around the nominal operating point. The linearized model, which contains a total of 52 states, is available from the author via e-mail. The linear model is written

$$\begin{pmatrix} dy_D \\ dx_B \end{pmatrix} = G(s) \begin{pmatrix} dL_w \\ dQ_B \\ dF_w \end{pmatrix} \tag{2}$$

where $G(s)$ is a 2×3 transfer-matrix expressing the effect of small changes in the manipulated variables L_w and Q_B and disturbance F_w on the product compositions y_D and x_B. The motivation behind employing a linear model as the "plant" is that we mainly are concerned with linear aspects here and do not want to con-

N	N_F	z_F	α	F_w	L_w	Q_B	y_D	x_B
25	13	0.5	1.75	29	69.99	33.23	0.99	0.01

- z_F, y_D and x_B in mole fractions of light component.
- L_w and F_w in [kg/min], Q_B in [MJ/min].
- Reboiler is tray 1.
- Feed is saturated liquid.
- Total condenser with saturated reflux.
- Liquid holdups are $M_i = 0.5$ kmol, including condenser and reboiler.
- Flow dynamics included, hydraulic time constant $\tau_L = 0.067$ min.
- Constant pressure $P = 12$ atm.
- Liquid enthalpy $H_L(x_i) = 0$. Reference state: pure components as saturated liquids at column pressure.
- Vapor enthalpy $H_V(x_i) = 12.45 - 1.39x_i$.

Table 1. Data for example column separating a binary mixture of ethylene and ethane.

fuse matters. Furthermore, most identification methods aim at identifying linear models. Undoubtedly, identifying a linear model from a nonlinear plant will prove harder than from a linear plant, but we believe the latter problem needs to be solved before attacking the former problem. However, for those interested, a nonlinear model is also available from the author upon request.

The singular values and the 1,1-element of the RGA for the input model, i.e., with L_w and Q_B as inputs, are shown as functions of frequency in Figures 2 and 3 (solid lines). At steady state the condition number is 214 and the 1,1-element of the RGA is 37. Thus, the plant is ill-conditioned and also strongly interactive. However, as seen from Figure 2 and 3, the plant is less ill-conditioned and interactive at intermediate and high frequencies than at steady state. This is typical for high-purity distillation columns.

The "experimental" conditions available for identification are:

– The two product compositions y_D and x_B are sampled with a period of 1 min. The total number of samples should not exceed 200.

– The composition measurements contain normally distributed zero mean white noise with standard deviation $\sigma_e = 0.0005$.

– The maximum allowed deviation in both product compositions is 0.01 (the main purpose of this specification is to limit the signal-to-noise-ratio and may, for instance, be satisfied by simply scaling the obtained noise-free outputs and inputs accordingly.).

– The maximum allowed perturbation in the manipulated inputs are 5 kg/min for the reflux L_w and 4 MJ/min for the heat input Q_B

– The manipulated inputs are uncertain such that

$$dL_w = \frac{0.95}{s+1} dL_{wc} \quad dQ_B = \frac{1.05}{s+1} dQ_{Bc}$$

where subscript c denotes command signal. This "uncertainty" applies during identification only, that is, it should not be considered an inherent part of the plant.

– The feed flow F_w may be manipulated during identification and the maximum allowed perturbation is 5.0 kg/min. The disturbance model may be obtained through a separate experiment if desired.

– The distillate flow D_w [kg/min] is measured with uncertainty

$$dD_{wm} = \frac{0.95}{s+1} dD_w$$

where subscript m denotes measured.

– The temperature on tray 20 is measured in ^oC and contains normally distributed zero mean white noise with standard deviation $\sigma_e = 0.15^oC$.

– If feedback control of the product compositions is used during identification then a measurement delay of 5 min. should be assumed.

An example file for generating experimental data with Matlab according to the above specifications is available from the author via e-mail.

The challenge is to identify a model (2) based on the obtained experimental data, possibly combined with the knowledge that the object is a distillation column separating ethane and ethylene at the given pressure. The intended use of the model is for feedback control applications. Setpoint changes as well as feed flow disturbance rejection are considered important. The maximum expected setpoint change is 0.01 for both product compositions and the maximum expected disturbance in the feed flow rate is 5 kg/min. Both product compositions are measured with a delay of 5 minutes. Control of a single product composition (one-point control) as well as control of both product compositions (two-point control) are of interest. In both cases the predicted responses in both product compositions should be considered and compared with those obtained with the "plant".

To allow for a judgement of the ability of a proposed method to identify a reasonable model of the distillation column it is desired that some kind of uncertainty bounds are presented along with the nominal model. The uncertainty bounds may for instance be estimated through Monte Carlo type simulations.

4. MISO-IDENTIFICATION USING AN ARMAX-TYPE MODEL

In this section we employ a fairly standard identification technique to the distillation column presented above. We employ the Matlab System Identification Toolbox (Ljung, 1991) and use MISO-identification with an ARMAX-type model structure. We choose to identify the input model and disturbance model using a single experiment. Independent PRBS signals are applied to the three inputs L_{wc}, Q_{Bc} and F_w with amplitudes 2, 1 and 2, respectively, a minimum switching time of 10 min. and a switching probability of 0.5. The measurements of the product flow D_w and the tray temperature are not utilized. For each output we fit a strictly proper fifth order model which is determined close to optimal based on the Final Prediction Error (e.g., Ljung, 1991).

Figure 1 shows the resulting fit of the magnitude for the transfer-function from L_w to y_D and we see that the experimental conditions allows for a reasonable fit of this transfer-function. Although not shown here, the other five transfer-functions of the 2×3 transfer-matrix (2) are fitted with similar precision.

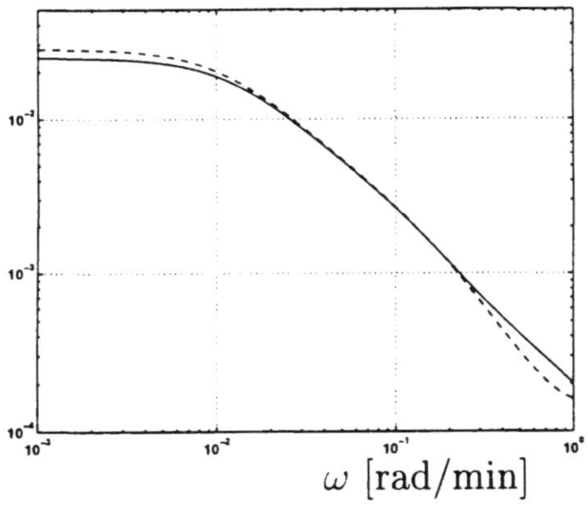

Fig. 1. Magnitude of transfer-function from L_w to y_D for fitted model (dashed) and plant (solid).

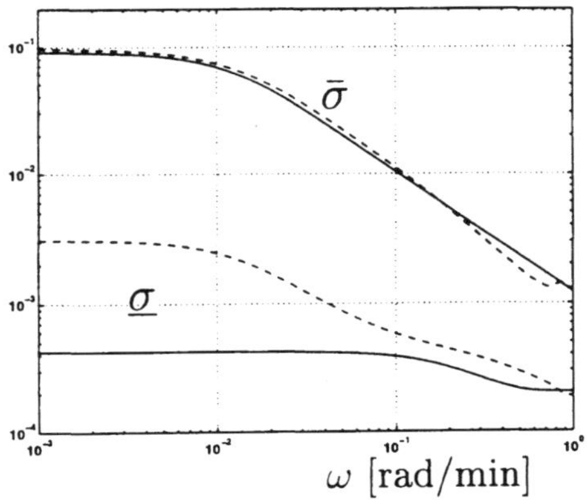

Fig. 2. Singular values for fitted input model (dashed) and plant (solid).

Figure 2 shows the fit of the singular values for the input model (L_w and Q_B as inputs) and we see that, while the high-gain direction is well fitted, the low-gain direction is poorly fitted, in particular at low and intermediate frequencies. This is reflected in the poor fit of the RGA shown in Figure 3. The poor fit of the low-gain direction is at least partially explained by the use of two MISO models in fitting the data (Jacobsen, 1994). From Figure 2 it is also seen that the low-gain direction of the identified model contains a slow pole at approximately 0.01 rad/min, which in the plant belongs exclusively to the high-gain direction. Thus, the identified model is inconsistent in terms of the number of slow poles and will yield a poor prediction of the plant under one-point control (Jacobsen and Skogestad, 1994a).

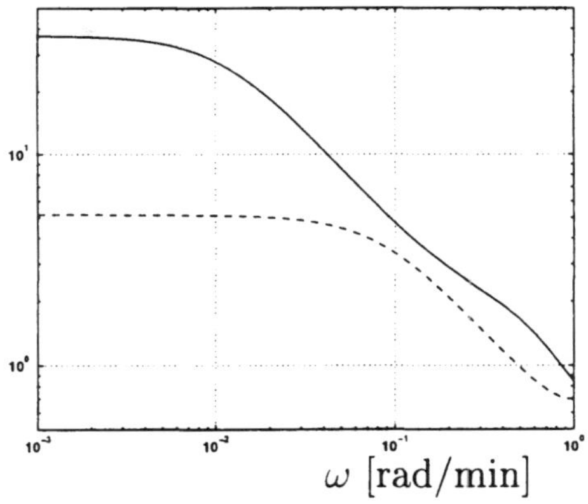

Fig. 3. Magnitude of 1,1-element of RGA for fitted input model (dashed) and plant (solid).

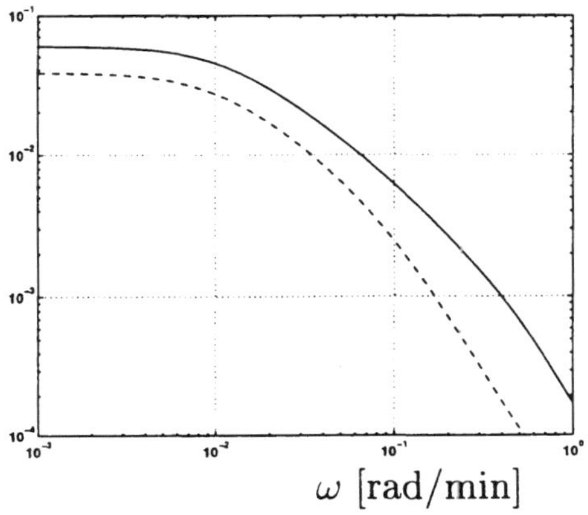

Fig. 4. Magnitude of Closed-Loop Disturbance Gain from F_w to y_D for fitted model (dashed) and plant (solid).

Figure 4 shows the resulting fit of the Closed Loop Disturbance Gain (Hovd and Skogestad, 1994) which takes into account the relative directions of the disturbance model and the input model and we see that we also have a relatively poor multivariable model for the effect of disturbances.

Based on the model analysis presented above we would expect that the model is of poor quality for control applications. Indeed, using the identified model to design a two-point controller using H_∞-optimal design we predict the setpoint response shown by the dashed lines in Figure 5. The solid lines show the corresponding response when applying the same controller to the "plant", i.e., the 52 state full linear model, and we see that the identified model yields a poor prediction of the plant behavior also under two-point control. In particular, the controller when applied to the plant yields a very slow settling towards steady-state. Similar results are found for responses to disturbances in F_w.

Note that we have not presented the requested error bounds, e.g., for the singular values, here. The reason is that even the nominal model is of such a poor quality

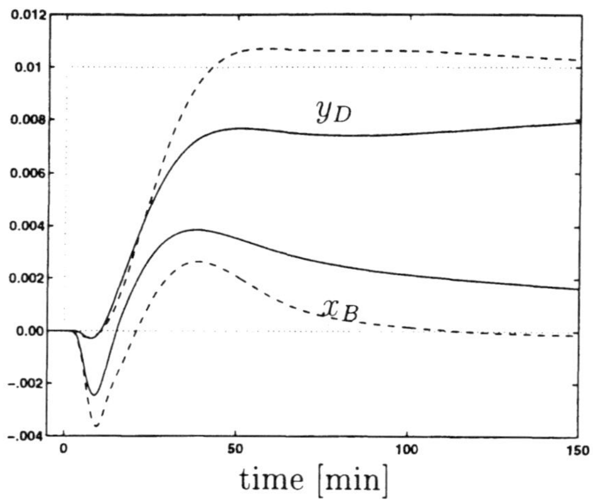

Fig. 5. Closed-loop response for set-point change in y_D with fitted model (dashed) and plant (solid).

that error bounds are of little interest in this case.

5. CONCLUSIONS

- We have presented a high-purity distillation column which is proposed as a challenge problem for identification of multivariable models intended for feedback control applications.
- A detailed description of the experimental conditions is provided along with a model of the column which is available from the author via e-mail.
- A standard identification method involving independent PRBS signals and MISO ARMAX-models were shown to yield a model poor for control applications.

FURTHER INFORMATION

The following is available from the author upon request (e-mail: jacobsen@control.e.kth.se):

- A linear state space model (in ASCII format) for generating noise-free data.
- An example file for generating experimental data according to the specified experimental conditions using the linear model and Matlab.
- A nonlinear model for use with Matlab/Simulink.

REFERENCES

Alsop, A.W. and T.F. Edgar (1990) "Nonlinear Control of a High-Purity Distillation Column by the use of Partially Linearized Control Variables", *Comp.Chem.Eng.*, **14**, 665-678.

Andersen, H.W., M. Kümmel, and S.B. Jørgensen (1989) "Dynamics and Identification of a Binary Distillation Column", *Chem.Eng.Sci.*, **44**, 2571-2581.

Andersen, H.W. and M. Kümmel (1991) "Identifying Gain Directionality of Multivariable Processes", *Proc. of ECC'91*, Grenoble, France, July 1991.

Hovd, M. and S. Skogestad (1994) "Simple Frequency Dependent Tools for Control-System Analysis, Structure Selection and Design", *Automatica*, **28**, 5, 989-99

Jacobsen, E.W. (1994) "Identification for Control of Strongly Interactive Plants", *Preprints NATO ASI on Model Based Control*, Antalya, Turkey, August 1994. Also: paper 226ah, *1994 AIChE Annual Meeting*, San Francisco, November 1994.

Jacobsen, E.W., P. Lundström and S. Skogestad (1991) "Modelling and Identification for Robust Control of Ill-Conditioned Plants - A Distillation Case Study", *Proceedings of 1991 American Control Conference*, Boston, MA, 242-248.

Jacobsen, E.W. and S. Skogestad (1994a) "Inconsistencies in Dynamic Models for Ill-Conditioned Plants: Application to Low-Order Models of Distillation Columns", *Ind.Eng.Chem.Res*, **33**, 3, 631-640.

Jacobsen, E.W. and S. Skogestad (1994b) "Identification of Dynamic Models for Ill-Conditioned Plants - a Benchmark Problem", in *the Modeling of Uncertainty in Control Systems*, Eds. R. Smith and M. Dahleh, Springer-Verlag, London.

Koung, C.W. and J.F. MacGregor (1993) "Design of Identification Experiments for Robust Control. A Geometric Approach for Bivariate Processes", *Ind.Eng.Chem.Res.*, **32**, 1658-1666.

Koung, C.W. and J.F. MacGregor (1994) "Identification for Robust Multivariable Control: the Design of Experiments", *Automatica*, **30**, 10, 1541-1554.

Li, W. and J.H. Lee (1994) "Closed-Loop Identification of Ill-Conditioned Multivariable Systems", Paper 226ag, *1994 AIChE Annual Meeting*, San Francisco, November 1994.

Ljung, L. (1992) *System Identification Toolbox Version 3.0a*, MathWorks Inc.

Skogestad, S (1988) "Disturbance Rejection in Distillation Columns", *Preprints CHEMDATA'88*, Gothenburg, Sweden, 365-371. Published by IVA, Stockholm, Sweden.

Skogestad, S. and M. Morari (1988) "Understanding the Dynamic Behavior of Distillation Columns", *Ind. & Eng. Chem. Res*, **27**, 10, 1848-1862.

Van Overschee, P. and B. de Moor (1994) "N4SID: Subspace Algorithms for the Identification of Combined Deterministic-Stochastic Systems", *Automatica*, **30**, 1, 75-94.

Varga, E.I. and S.B. Jørgensen (1994) "Multivariable Process Identification: Estimating Gain Directions", Paper 229f, *1994 AIChE Annual Meeting*, San Francisco, November 1994.

DISTILLATION COLUMN CONTROL BENCHMARKS: Four Typical Problems

A. Koggersbøl and S. Bay Jørgensen

Department of Chemical Engineering, Technical University of Denmark, DK-2800 Lyngby, Denmark.

Abstract: An industrial size distillation plant is suggested as a benchmark for different distillation control problems. The problems range from simple one point control, via two point control over a sidestream control problem to optimizing control. The plant is fed with a mixture which is almost binary, but with a slight impurity. A range of industrially relevant disturbances are defined, and performance requirements are specified. The choise of model representation and method is left open to the participants. A nonlinear simulation model is made available.

Keywords: Benchmark examples; Computer Simulation; Mathematical models.

1. INTRODUCTION

Several distillation process models have been provided in the litterature in the form of transfer function models, e.g. the SHELL-Control Problem (Prett and Morari, 1987). The emergence of robust control theory have demonstrated that often model uncertainty may be performance limiting in addition to limitations introduced by process dynamics such as right half plane transmission zeros and by disturbance characteristics. Since there have been little consensus thus far with proposing uncertainty models for distillation control problems it is suggested in this paper to adopt an industrial size distillation plant as a benchmark for control studies. In order to investigate the influence of different control problems upon different aspects of solution approaches four different control problems are defined.

An experimentally validated simulation model has been developed (Koggersbøl and Jørgensen, 1995) and used for numerous studies of column operation and column control. Koggersbøl and Jørgensen (1992a+b) investigated dynamics and control configurations for the plant operated both as a two-product binary column and as a ternary sidestream column. Correa and Jørgensen (1993) looked at the column as part of a two-column heterogeneous azeotropic distillation process. They compared performance of different control strategies. Varga et al. (1995) studied model representations for control of the column.

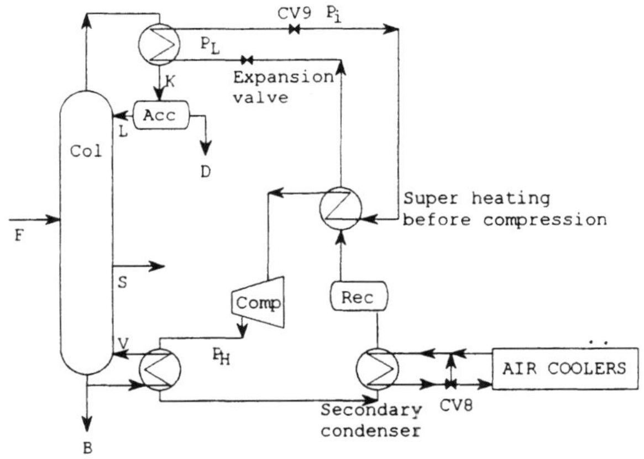

Fig. 1. Distillation plant flowsheet.

The purpose of this paper is to concentrate on description of the four control problems problems, disturbances and control rquirements. First however the plant is briefly introduced with available actuators and measurements.

2. PLANT DESCRIPTION

Figure 1 shows a flowsheet of the process. The column has 19 sieve trays, a thermosiphon reboiler, a total condenser and a reflux drum. It separates a mixture of methanol and isopropanol with a low concentration water impurity. This impurity will mainly escape the column through the bottom product. The heat pump has an expansion valve which throttles high pressure liquid freon (R114) to a lower pressure (P_L) suitable for evaporation in the condenser. After the condenser there is a control valve (CV9) by which the freon vapour flowrate can be manipulated. After superheating the vapour, the compressor elevates the pressure to a high pressure (P_H) suitable for condensation in the reboiler. A small part of the condensation takes place in a secondary condenser which by a cooling water circuit is connected to a set of air-fan coolers. The cooling rate can be manipulated by the control valve CV8, thus controlling P_H. Through a storage tank (Rec) and the pre–compression heat exchanger the freon circuit is closed at the expansion valve.

3. BENCHMARK PROBLEMS

For this process four distillation control problems are proposed:

1. One point control, where the top concentration should be kept. The bottom concentration should satisfy much looser limits.
2. Two point control, where both ends should satisfy equally tight limits.
3. Sidestream control problem, as case 2, but with a sidestream where also performance requirements on the sidestream should be satisfied.
4. Optimizing control of one of the above cases.

The objectives are to:

- Reject disturbances in feed concentration and flowrate. Two process disturbance stress levels are specified.
- Satisfy setpoint tracking requirements. On the level beneath purity control, the requirement is to be able to move the process on a smoth curve between any two points in the operating region (Fig. 2). On the purity control level the requirement is to be able to move smoothly between different purity levels in both products independently.
- The purpose of having a sidestream should be to purge water, and when manipulated, the objective should be to assist in disturbance rejection while minimizing the loss of valuable products through the sidedraw.
- The objective for optimizing control should be to maximize the following function, e.g. with respect to variations in the prize of electrical power around the clock.

$$\begin{aligned} J &= c_D(x_D) \cdot D + c_B(x_B) \cdot B \\ &- c_F(z) \cdot F - c_W \cdot Q_W \end{aligned} \quad (1)$$

where the four contributions are: Value of top product, value of bottom product, cost of feed stream (fuel value), and cost of electrical power. D, B, and F are distillate, bottom and feed flowrates in moles per second, Q_W is electrical energy consumption of the heat pump compressor. x_D, x_B, and z are molefractions of methanol in the distillate, of isopropanol in the bottom product, and of methanol in the feed stream. Prize functions for the four terms are as follows:

$$c_D = 0.31 \cdot x_D - 0.23 \text{ [DKr/mole]} \quad (2)$$

$$c_B = 1.12 \cdot x_B - 0.85 \text{ [DKr/mole]} \quad (3)$$

$$c_F = 0.08 - 0.05 \cdot z \text{ [DKr/mole]} \quad (4)$$

$$c_W = 0.11 \text{ [DKr/MJ]} \quad (5)$$

The control requirements are to:

- Keep the specified concentrations within specified hard limits.
- Ensure zero steady state control error.
- Preferably obtain fast settling.
- Preferably have an over-/undershoot less than 25% of the hard limits.

A mathematical simulation model is available by E-mail to sbj@kt.dtu.dk. The code is in FORTRAN and provides a full simulation, where the controller may be introduced into the CONTROL subroutine, which is executed every 30 – 60 seconds real time on the plant. The mathematical model is under test and preliminary results therefrom are available in Koggersbøl and Jørgensen (1995). Further results will be available, especially on delays in boil-up and reflux flowrates.

The problem statement requests a control configuration and a control design for the selected control problem. The participants are allowed to:

1. Use any design method.
2. Use different methods (or retune) for different stress levels.
3. Filter setpoint changes, as long as the hard limits are observed.
4. Add extra stimulation to the setpoint, as long as the hard limits are observed.
5. Perform identification prior to and/or during control.

Since any mathematical model contains simplifications, it may also be feasible to test a few most promising controllers in practice on the plant.

4. OPERATING CONDITIONS

4.1. *Operating point*

The operating point proposed for purity–control investigations is defined as follows:

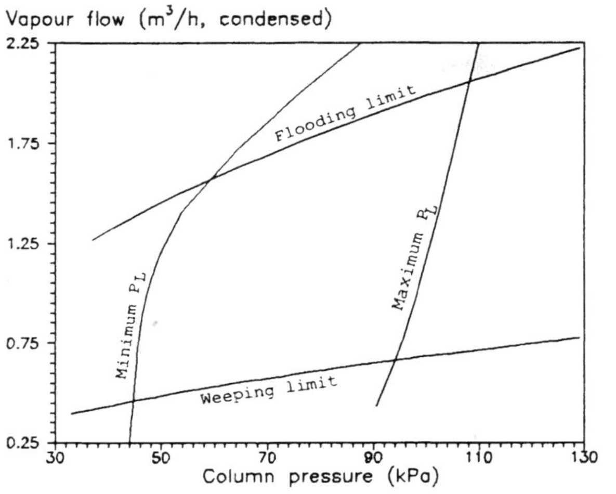

Fig. 2. Operating region of the plant. The region is limited by tray weeping, minimum P_L, tray flooding, and maximum P_L. Column pressure is measured on the top tray.

Feed flowrate	3.0	liters/min
Feed composition:		
Methanol	49.5	mole-%
Isopropanol	49.5	mole-%
Water	1.0	mole-%
Product purities:		
Distillate	97.5	mole-%
Bottom, if controlled	97.5	mole-%
Column top pressure	75.0	kPa

Hard limits on product purities are ±0.5 mole-%. The feed stream is saturated liquid. When used, the sidestream is drawn as saturated liquid.

4.2. *Actuator limits*

Figure 2 shows the limits for boil-up in the possible range of column pressure. In addition there are the following hard limits to be observed.

Heat pump pressures:		
P_L^{max}	600.0	kPa
P_H^{max}	1400.0	kPa
Reflux flowrate, max.	35.0	liters/min
Distillate flowrate, max.	8.0	liters/min
Bottom prod. flowrate, max.	8.0	liters/min

Furthermore, the number of active cylinders in the compressor can be varied between 4 and 16 in steps of 2.

4.3. *Disturbances*

The disturbances to the process are considered to have two frequency contents:

$$\omega_{slow} = 2\pi/(120 \text{ min}) \quad (6)$$

$$\omega_{fast} = 2\pi/(15 \text{ min}) \quad (7)$$

They enter the process simultaneously through feed flowrate and feed composition in the following manner:

$$F(t) = F_0 + A_F \cdot \sin(\omega_{fast} \cdot t) \quad (8)$$

$$z(t) = z_0 + A_z \cdot \sin(\omega_{slow} \cdot t) \quad (9)$$

F(t) is volumetric feed flowrate and z(t) is feed concentration of the light key component. Three situations are considered: The nominal case and two enhanced stress levels. These are as follows.

Nominal case	A_F is low	A_z is low
Stress level 1	A_F is low	A_z is high
Stress level 2	A_F is high	A_z is high

Low and high amplitudes are defined relative to the disturbance free steady state inputs (F_0 and z_0): Low is 5%, and high is 25%.

5. REFERENCES

Correa, R.G.; Jørgensen, S.B (1993); 'Dynamics and Control of Heterogeneous Azeotropic Distillation'; *Proceedings of IFAC World Congress 1993 vol. 1, pp.55–58, Sidney, Australia.*

Koggersbøl, A.; Jørgensen, S.B. (1992a); 'Control Structures for a Sidestream Distillation Column Separating a Ternary Mixture'; *Proceedings of IFAC Symposium DYCORD+92, April 27–29'th, College Park, Maryland, USA, pp.237–242.*

Koggersbøl, A.; Jørgensen, S.B. (1992b); 'Dynamics and Control of a Distillation Column with a Sidestream'; *Distillation and Absorption, IChemE Symposium Series no. 128, pp.A429–A449.*

Koggersbøl, A.; Jørgensen, S.B. (1995); 'Verification of a Distillation Column Simulation Benchmark'; *Submitted for presentation at IFAC Symposium DYCORD+95, June 7–9'th, Helsingør, Denmark.*

Prett, D.M.; Morari, M. (1987); 'Shell Process Control Workshop', *Butterworths, 1987.*

Varga, E.I.; Kovacs, G.Z.; Jørgensen, S.B.; Hangos, K.M. (1995); 'Robust Analysis of Distillation Column Dynamics'; *Submitted for presentation at IFAC Symposium DYCORD+95, June 7–9'th, Helsingør, Denmark.*

THE IMPACT OF PROCESS DIRECTIONALITY ON ROBUST CONTROL IN NON-IDEAL DISTILLATION

M. F. SÅGFORS and K. V. WALLER *

Process Control Laboratory, Department of Chemical Engineering Åbo Akademi, SF-20500 Åbo, Finland E-mail: kwaller@abo.fi

Abstract. The structure of the directionality in non-ideal (real) distillation is examined. The obtained knowledge of the directionality in distillation may be very useful when the two-point control problem is handled. A compensator structure to utilize this information is proposed.

Key Words. Directionality, Non-Ideal Distillation, Multivariable Systems, Robustness

1. INTRODUCTION

The directionality of multivariable processes means that a vector of inputs may be differently amplified according to its direction. Good control performance requires a controller that can compensate for this directionally varying behavior of the process, e.g. a controller based on the inverse of a process model. However, an inverse-based controller may be very sensitive to norm-bounded model uncertainties (Skogestad et al., 1988). Better robustness can be achieved by a diagonal controller, with the expense of performance, which may be strongly dependent on disturbance direction. This trade-off between good nominal performance but high error sensitivity and high robustness but potentially low performance is a central problem in designing controllers for strongly coupled multivariable processes, such as high-purity distillation columns. Good compromises in this respect have been sought for distillation columns through different methods, e.g. μ-synthesis (Arkun and Morgan, 1988; Skogestad et al., 1988; Skogestad and Lundström, 1990), IMC controllers (Semino et al., 1993), SVD-based controllers (Brambilla and D'Elia, 1992) and mixed multimodel H_2/H_∞ controllers (Pensar and Toivonen, 1994).

A pre-requisite for the controller design methods above is an appropriate description of the possible model-plant mismatch. One way to characterize the model error is to consider a region of plants Π in the neighborhood of a nominal model of the plant $\tilde{\mathbf{P}} \in \Pi$ such that also the real plant $\mathbf{P}$ lies in that region ($\mathbf{P} \in \Pi$). Most robust controller design techniques require that the region Π is characterized by magnitude bounds around $\tilde{\mathbf{P}}$. A controller is said to be robust if it satisfies some performance criteria for all models in Π. A crucial step in this approach is to determine the region Π in a "tight" manner. Failing to do this results in a unnecessary cautious controller design, since the region Π then will contain models that cannot occur in practice.

However, the structure of the process-model mismatch in distillation may be such that the modeling of the uncertainty region may be very hard to accomplish. Mapping the region of possible plants by norm-bounded model uncertainty may result in very conservative control (Skogestad and Morari, 1987; McDonald et al., 1988; Koung and McGregor, 1991). Due to the structure of the non-linearity in high-purity distillation, a set of models described by such an uncertainty is likely to contain descriptions of plants that cannot occur in practice. Since the controller design is made for the complete set of modeled plants, the controller is likely to be very cautious.

Model uncertainty due to non-linearity in high-purity distillation has previously been investigated e.g. by McDonald et al. (1988) and Wang and Yu (1993), who study the feasibility of modeling the model-plant mismatch in the frequency domain. Koung and McGregor (1991;1992) follow another approach by establishing several geometrical criteria required for robust control, based on the steady state gain matrix. In this paper, an approach similar to that of Koung and McGregor (1992) is used. The analysis of high-purity distillation in Koung and McGregor (1992) was, however, restricted to a short-cut model for binary distillation (McAvoy, 1983), involving several severe simplifying assumptions. In this paper, we establish structural properties concerning the directionality in non-ideal (real) distillation and discuss the implications for robust controller design. The results are based on an analysis of several experimental and simulated models found in the literature as well as on a rigorous ethanol-water simulator adapted to real steady-state data from the pilot-plant column at Åbo Akademi (Lehtinen, 1994). Based on these results, a feedback compensator design that utilizes the gained information concerning directionality in distillation, is suggested and analyzed.

In practice, model uncertainty due to noisy experiments as well as input- and output uncertainty is also always present. These sources of errors can be modeled fairly well by norm-bounds. This work is, however, concerned with the structure of errors of linear models due to deterministic reasons, such as plant non-linearity, which is likely to be a major source of model-plant mismatch when a linear model is used for the design of controllers for plants with a pronounced non-linear behavior, such as high-purity distillation columns.

Author to whom all correspondence should be addressed.

2. DIRECTIONALITY

Consider a plant $\mathbf{y} = \mathbf{P}(s)\mathbf{m}$. The magnitude and structure of the gain matrix $\mathbf{P}(0)$ can be examined via the singular value decomposition (SVD):

$$\mathbf{P}(0) = \mathbf{U}\ \mathbf{\Sigma}\ \mathbf{V}^T \tag{1}$$

where, for a 2×2 matrix

$$\mathbf{U} = [\bar{\mathbf{u}}\ \underline{\mathbf{u}}] \quad \mathbf{\Sigma} = \mathrm{diag}(\bar{\sigma}, \underline{\sigma}) \quad \mathbf{V} = [\bar{\mathbf{v}}\ \underline{\mathbf{v}}] \tag{2}$$

The singular values, $\bar{\sigma}$ and $\underline{\sigma}$, are the largest and smallest gains, respectively, which the length of an input-vector can be amplified in the outputs by $\mathbf{P}(0)$:

$$\underline{\sigma} \leq \frac{\|\mathbf{P}(0)\mathbf{m}\|_2}{\|\mathbf{m}\|_2} \leq \bar{\sigma} \tag{3}$$

The highest gain $\bar{\sigma}$ is obtained when the input vector is changed parallel with $\bar{\mathbf{v}}$, the output vector is then parallel with $\bar{\mathbf{u}}$ (and vice versa for $\underline{\sigma}$, $\underline{\mathbf{v}}$ and $\underline{\mathbf{u}}$).

The rotation matrices ($\mathbf{U}$ and $\mathbf{V}$) are unitary and orthogonal ($\mathbf{V^T V} = \mathbf{I}$ and $\mathbf{U^T U} = \mathbf{I}$). The SVD can therefore be written in a compact manner, where the rotation matrices are expressed by trigonometric functions (Figure 1)

$$\mathbf{P}(0) = \mathbf{U}\ \mathbf{\Sigma}\ \mathbf{V}^T = \tag{4}$$

$$\begin{bmatrix} \cos\beta & \cos(\beta \pm 90^\circ) \\ \sin\beta & \sin(\beta \pm 90^\circ) \end{bmatrix} \begin{bmatrix} \bar{\sigma} & 0 \\ 0 & \underline{\sigma} \end{bmatrix} \begin{bmatrix} \cos\alpha & \cos(\alpha - 90^\circ) \\ \sin\alpha & \sin(\alpha - 90^\circ) \end{bmatrix}^T$$

The angle α is the positive rotation with respect to the input m_1 of the input direction ($\bar{\mathbf{v}}$) that gives the largest possible gain. Similarly, the angle β is the positive rotation with respect to y_1 of the corresponding output vector, $\bar{\mathbf{u}}$. The fact that the singular values ("gains") are positive by definition is reflected by the $\pm$-signs in the output rotation matrix.

A matrix is said to be ill-conditioned, if there is a large difference between the maximum and minimum singular values. This is expressed by the condition number

$$\gamma = \mathrm{cond}(\mathbf{P}(0)) = \frac{\bar{\sigma}}{\underline{\sigma}} \tag{5}$$

A large condition number means that the gain of the model is strongly dependent on the direction of the input vector. Good control quality regardless of input (or disturbance) direction requests a controller which takes into account the directionally varying process gain by providing large (small) input signals to compensate for weak (strong) directions of the plant. An inverse-based controller could do this, yielding excellent nominal performance, but such a controller may be very sensitive to input uncertainty (Skogestad et al. 1988), failing to provide robust performance. The reason for this sensitivity is that the input uncertainty changes the input directions requiring large (small) control action, which ruins the process-controller fit.

3. ROBUSTNESS

Most robustness analyses do not only involve the plant, but also the controller. One exception is Garcia and Morari (1985), where it was shown that any system is controllable with zero steady-state offset ("stabilizable"), provided that the model-plant mismatch is such that

$$\mathrm{Re}(\lambda_i(\mathbf{P}(0)\tilde{\mathbf{P}}(0))) > 0, \quad \forall i \tag{6}$$

where λ_i is the ith eigenvalue of a matrix. The condition in eq 6 corresponds to the familiar SISO-case showing that a process and its model should have the

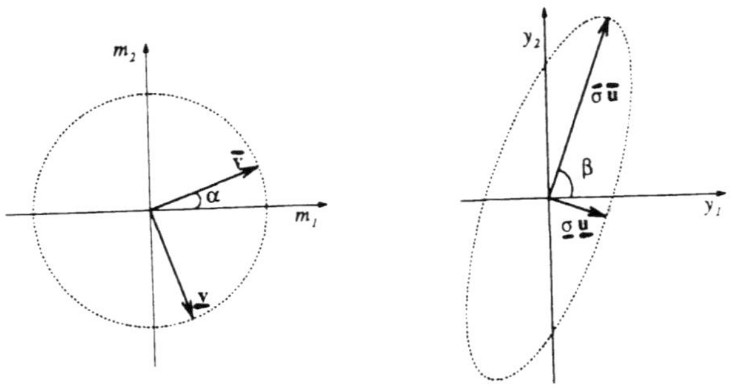

Fig. 1. Geometrical illustration of the singular value decomposition.

same signs in their steady state gains. The implication of the condition is that if eq 6 is satisfied, model based control is feasible provided that the controller is appropriately (de)tuned.

On the basis of this criterion, Koung and McGregor (1992) derived conditions equivalent to that in eq 6 in terms of the rotation matrices in SVD, one of which consists of the following two criteria:

For 2×2 systems, if

(a) $\tilde{\mathbf{V}}^\mathbf{T}\mathbf{V} = \mathbf{I}$

(b) the signs of all the elements of $\mathbf{U}$ remain the same as those of $\tilde{\mathbf{U}}$

then the control system is stabilizable for the locally linearized process.

Criterion (a) implies that we have no mismatch in the input rotation matrix. Criterion (b) implies that the rotation of the outputs, expressed by β in the $\mathbf{U}$ matrix (eq 4) varies in one quadrant only ($n \cdot 90^\circ \leq \beta \leq 90^\circ + n \cdot 90^\circ$, $n \in \{..., -1, 0, 1, ...\}$) in the output space (Figure 1).

The result above tells that an accurate estimate of the input rotation matrix $\mathbf{V}$ is very important provided that precise information concerning the output rotation $\mathbf{U}$ is impossible to obtain.

4. DIRECTIONALITY IN DISTILLATION

Consider the (L, V) structure in distillation, where the gain matrix can be written

$$\begin{bmatrix} \Delta y_D \\ \Delta x_B \end{bmatrix} = \begin{bmatrix} K_{yL} & K_{yV} \\ K_{xL} & K_{xV} \end{bmatrix} \begin{bmatrix} \Delta L \\ \Delta V \end{bmatrix} \tag{7}$$

Previous research considering directionality in distillation (Skogestad and Morari, 1988; Andersen et al., 1989) has given the result that the lowest process gain is connected to changes in the internal flows (L and V) only, and that the high-gain direction is connected to input directions giving maximal change in the external flows (D and B). However, these results built on a number of simplifying assumptions, concerning e.g. the dependence of the external flows to changes in internal flows

$$\begin{bmatrix} \Delta D \\ \Delta B \end{bmatrix} = \begin{bmatrix} -1 & 1 \\ 1 & -1 \end{bmatrix} \begin{bmatrix} \Delta L \\ \Delta V \end{bmatrix} \tag{8}$$

Thus, for the (L, V)-structure, it was found that the lowest process gain is obtained for an equal increase (or decrease) in both inputs. The high-gain is obtained for an increase in one input and a decrease in the other, of the same size. However, the practical usefulness of this result is not directly evident, since the relation in eq 8 is usually not valid. It should generally be written

$$\begin{bmatrix} \Delta D \\ \Delta B \end{bmatrix} = \begin{bmatrix} K_{DL} & K_{DV} \\ K_{BL} & K_{BV} \end{bmatrix} \begin{bmatrix} \Delta L \\ \Delta V \end{bmatrix} \tag{9}$$

The gains in eq 9, which conveniently can be called "flow-gains", can either be measured or calculated by consistency relations based on external material balances (Häggblom and Waller, 1988), giving

$$K_{DL} = -K_{BL} = -\frac{D_0 K_{yL} + B_0 K_{xL}}{y_{D0} - x_{B0}} \tag{10a}$$

$$K_{DV} = -K_{BV} = -\frac{D_0 K_{yV} + B_0 K_{xV}}{y_{D0} - x_{B0}} \tag{10b}$$

In real columns the flow gains are usually very different from the values in eq 8. For experimental columns Häggblom and Waller (1989) report values of K_{DL} ranging from -0.25 to -0.7 and values of K_{DV} from 0.49 to 1.35 thus not even close to the values in eq 8.

The practical usefulness of the observation concerning the relation between directionality in distillation and external and internal flows has been checked in Sågfors et al. (1994b) by comparison of the directions defined by the vectors $\overline{\mathbf{v}}$ and $-[K_{DL}\ K_{DV}]^T$ for a number of published experimental and simulated models. The former vector $\overline{\mathbf{v}}$ is obtained by SVD of the process gain matrix and defines the direction of the largest process gain. The latter vector $-[K_{DL}K_{DV}]^T$ defines the input direction which gives a maximal change in the external flows. The two vectors mentioned were found to be closely aligned for the models. The flow-gains in eq 9 may thus provide a fair estimation of the process directionality in non-ideal distillation. An estimate of the input-rotation matrix can thus be obtained from (Sågfors and Waller, 1994)

$$\hat{\mathbf{V}} = \frac{1}{\|[K_{DL}\ \ K_{DV}]^T\|_2} \begin{bmatrix} -K_{DL} & -K_{DV} \\ -K_{DV} & K_{DL} \end{bmatrix} \tag{11}$$

Based on first principles, Häggblom (1995) derived tight bounds showing the estimate in eq 11 to be very accurate for ill-conditioned distillation columns with large values in the relative gain array (Bristol, 1966). A good estimate of the input rotation matrix may be useful in many control applications, such as in that identification method recently suggested by Koung and McGregor (1993).

In Sågfors and Waller (1994), it was also found that all alcohol-water separation examples had their highest gain for approximately $2\Delta L \approx -\Delta V$. This implies that the ratio K_{DL}/K_{DV} is approximately equal to -2 regardless of actual column or operating point. This may be regarded as a somewhat surprising result, since the flow-gains are known to be strong functions of the operating conditions (Häggblom, 1991). Thus, for alcohol-water separations, an estimate of the input-rotation matrix is given by

$$\hat{\mathbf{V}} = \begin{bmatrix} \cos(-65^\circ) & \cos(-155^\circ) \\ \sin(-65^\circ) & \sin(-155^\circ) \end{bmatrix} = \begin{bmatrix} 0.42 & -0.91 \\ -0.91 & -0.42 \end{bmatrix} \tag{12}$$

prior to any experiments, since $\alpha = \arg(\overline{\mathbf{v}}) \approx \tan^{-1}(K_{DV}/K_{DL})$. It is not very farfetched to expect that other separations may have similar properties concerning the structure of their directionality.

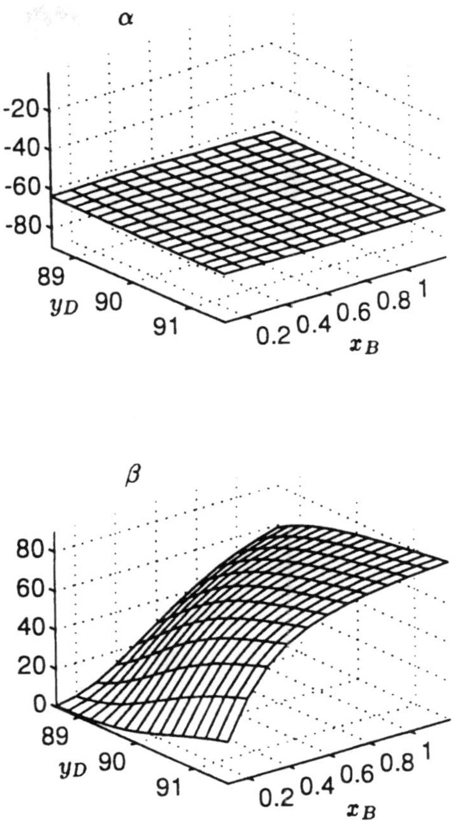

Fig. 2. The angles α and β (eq 4) of the non-linear simulator in the operating space defined by eq 13. Concentrations are expressed in mass %, angles in degrees. The result was obtained by SVD on gain matrices linearized at every grid-line intersection.

4.1 Process Non-linearity and Directionality

A rigorous column simulator (Lehtinen, 1994) adapted to experimental data from the pilot-plant at Åbo Akademi is used to investigate the impact of process non-linearity on process directionality. Good fit between the simulator and data not used in building the model was obtained. The directionality of the process is investigated over an operating region defined by

$$\begin{aligned} 88.6 \text{ mass } \% \le y_D \le 91.6 \text{ mass } \% \\ 0.1 \text{ mass } \% \le x_B \le 1.2 \text{ mass } \% \end{aligned} \tag{13}$$

(It should be noted that y_D is quite pure, though the numerical values could indicate otherwise. The relative volatility of ethanol-water follows a strongly non-linear relationship with an azeotropic point at about 94 mass %.) In Figure 2 the rotation matrices are given in terms of the angles α and β (eq 4) in the operating window studied.

The input rotation matrix is almost constant. This is very good news considering the design of a linear time invariant controller, since it means that the controller should apply large (small) control action in the same direction even if the operating point may change. The angle α is close to $-65°$ in the whole operating space. This is only to be expected, since the simulator in question is describing an alcohol-water separation (cf. eq 12).

The angle β is not nearly constant, but the variations remain in one quadrant indicating that any locally linearized model in the operating space could be used for model-based control without violating the condition in eq 6. By calculation of eq 6 for all combinations of gain matrices linearized in the operating window expressed by eq 13, this has indeed been verified in this study.

The vectors $\overline{\mathbf{u}}$ and $\underline{\mathbf{u}}$ are in close alignment to the single output axes (β close to 0° or 90°) in the output plane when one of the outputs is (relatively) much purer than the other, a fact reflecting that one of the outputs is more sensitive to input changes than the other. The angle β thus tells about the relative sensitivity of the outputs to input perturbations in the units chosen.

The directions of the process directionality are defined by α and β. The actual gains of the process are defined by $\mathbf{\Sigma}$. The condition number for the units chosen is varying between 20.3 and 118.3 in the window, thus reflecting the highly non-linear behavior of the column in the region.

4.2 Norm-Bounded Uncertainty Descriptions

McDonald et al. (1988) showed that modeling of the uncertainty in distillation is a non-trivial task. An attempt to describe the model uncertainty of high purity distillation columns by a magnitude bound in an unconservative manner is made in Skogestad and Morari (1987), where it is assumed that

$$\mathbf{P_{LV}}(0) - \tilde{\mathbf{P}}_{\mathbf{LV}}(0) = \qquad (14)$$

$$\begin{bmatrix} K_{yL} - \tilde{K}_{yL} & K_{yV} - \tilde{K}_{yV} \\ K_{xL} - \tilde{K}_{xL} & K_{xV} - \tilde{K}_{xV} \end{bmatrix} = \begin{bmatrix} d & -d \\ -d & d \end{bmatrix}$$

due to material balance constraints. Using the consistency relations of Häggblom and Waller (1988) (eqs 10) and solving with respect to K_{yL} and K_{yV} result in

$$K_{yL} - \tilde{K}_{yL} = K_{DL}\frac{y_0 - x_0}{D_0} - \tilde{K}_{DL}\frac{\tilde{y}_0 - \tilde{x}_0}{\tilde{D}_0} - \frac{B_0}{D_0}K_{xL} + \frac{\tilde{B}_0}{\tilde{D}_0}\tilde{K}_{xL} \qquad (15a)$$

$$K_{yV} - \tilde{K}_{yV} = K_{DV}\frac{y_0 - x_0}{D_0} - \tilde{K}_{DV}\frac{\tilde{y}_0 - \tilde{x}_0}{\tilde{D}_0} - \frac{B_0}{D_0}K_{xV} + \frac{\tilde{B}_0}{\tilde{D}_0}\tilde{K}_{xV} \qquad (15b)$$

In a narrow region around $\tilde{\mathbf{P}}_{\mathbf{LV}}$, where $(.)_0 \approx (\tilde{.})_0$, and $K_{DL} = \tilde{K}_{DL}$, $K_{DV} = \tilde{K}_{DV}$ and if $B_0 = D_0$, the highly structured magnitude description of model-plant mismatch in eq 14 may be tight, resulting in a fairly unconservative controller design (McDonald et al., 1988). However, in non-ideal (real) distillation, when the flow-gains are functions of the operating point and the compositions and flows may vary, simple descriptions similar to that of eq 14 are not possible to obtain. This was also the result of Wang and Yu (1993) for their particular example. Thus, the problem of how to describe model-plant mismatch in distillation by magnitude bounds in a tight manner still remains unsolved.

5. CONTROLLER DESIGN

Decentralized control can generally not compensate for the directionality in distillation. In practice, during closed loop gain identification of the pilot-plant at

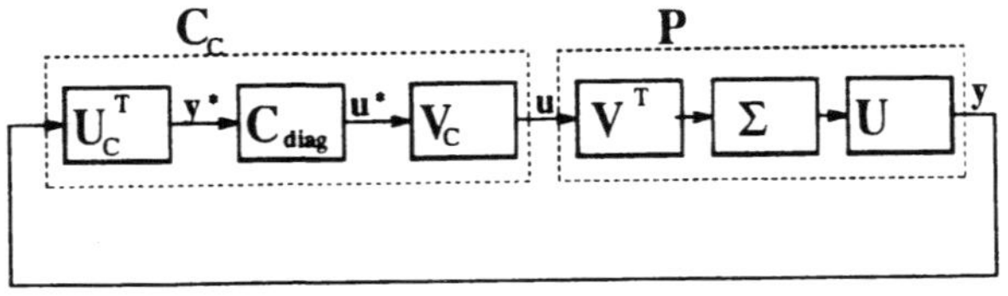

Fig. 3. Block diagram of the process (eq 1) and compensator (eq 16).

Åbo Akademi (Lehtinen, 1994), the operators had to retune the SISO-controllers depending on the direction of the set-point vector, due to the high degree of directionality at the nominal operating point. An obvious controller design objective would be to build in this compensation of directionality in non-ideal distillation into the controller.

In the previous section it was found that the directionality in distillation has a fairly well defined structure. Such information should be utilized in the controller design. Consider the controller (Figure 3)

$$\mathbf{C}(s) = \mathbf{V_c}\mathbf{C_{diag}}(s)\mathbf{U_c}^T \qquad (16)$$

where $\mathbf{V_c}$ and $\mathbf{U_c}$ are real, orthogonal matrices forming weighted sums of the inputs and outputs, respectively, to be used by the diagonal controller $\mathbf{C_{diag}}(s)$. If $\mathbf{V_c} = \mathbf{V}$ and $\mathbf{U_c} = \mathbf{U}$, the pre- and post-compensated system ($\mathbf{u^*} - \mathbf{y^*}$) seen by the controller is decoupled, since $\mathbf{V^T V} = \mathbf{I}$ and $\mathbf{U^T U} = \mathbf{I}$. The principal directions are then independently controlled by the SISO-controllers in $\mathbf{C_{diag}}(s)$.

The compensator structure in eq 16 has previously been studied e.g. by Lau et al. (1985), Bequette and Edgar (1989) and Hovd (1992). In Bequette and Edgar (1989), the (D, V) distillation structure was studied. It was found that $\mathbf{V_c} \approx \mathbf{I}$, i.e. that the input directions requiring maximal and minimal control action, respectively, are aligned with the single inputs (actuators). Since the actuators in the (D, V) structure consist of one internal flow (V) and one external flow (D), this result is in accordance with Section 4 in this paper. A related scheme was studied in Waller and Finnerman (1987), where weighted sums of x_B and y_D were controlled by D and V, respectively. In the (L, V) structure, however, the directions requiring maximal and minimal control action, respectively, are never aligned with the actuators (Sågfors and Waller, 1994).

The matrix $\mathbf{V}$ defines the input directions of the plant to be applied with large and small control action, respectively. The matrix $\mathbf{U_c}$ can be chosen to form weighted sums of the outputs to be controlled by large and small control action, respectively. The actual controller, $\mathbf{C_{diag}}(s)$, determines the degree of compensation of process directionality. (Trivially, the choice $\mathbf{V_c} = \mathbf{U_c} = \mathbf{I}$ would lead to conventional (L, V)-control, which cannot compensate for the process directionality).

5.1 Tuning of the Proposed Compensator

The ideal choices of pre- and post-compensators are

$$\mathbf{V_c} = \mathbf{V} \quad \text{and} \quad \mathbf{U_c} = \mathbf{U} \qquad (17)$$

resulting in a perfectly decoupled system at steady state.

Since the **V** matrix was found to be almost constant regardless of operating point in the example above, it seems that the input compensator $\mathbf{V_c}$ may be chosen from any locally linearized model. The design of the input compensator may also be based on estimate of **V** according to 11, or, if the directions requiring large (small) control action are known *a priori* for the separation at hand, as for alcohol-water separations, on an expression similar to eq 12.

It is very likely that the chosen output-compensator $\mathbf{U_c}$ will not match the matrix **U** as in eq 17, since the output rotation matrix was found to be strongly dependent on the operating point. Thus, if we intend to control the process with a linear controller over a wide nonlinear region, we have to deal with this mismatch.

The diagonal controller may be tuned according to any method for decentralized controllers. A compromise has to be made between two conflicting criteria: Compensation of process directionality (nominal performance) and robustness. Some insight into this is provided by the following example.

Hovd (1992) found that robust μ-controllers designed for some classes of problems have the structure of the compensator in eq 16. One of these classes has the controller design based on a model with scalar dynamics. Assuming scalar dynamics of first order

$$\mathbf{P}(s) = \frac{1}{\tau_1 s + 1}\mathbf{P}(0) \tag{18}$$

which is a fairly common assumption made for distillation models used for controller design (Skogestad et al., 1988), an inverse based IMC first order filter controller (Morari and Zafiriou, 1989) also has the same structure as eq 16

$$\mathbf{C_{IMC}}(s) = \frac{k_1}{s}\mathbf{P}^{-1}(s) = \frac{k_1(\tau_1 s + 1)}{s}\mathbf{P}^{-1}(0) = \mathbf{V}\left[\frac{k_1}{s}(\tau_1 s + 1)\mathbf{\Sigma}^{-1}\right]\mathbf{U^T} \tag{19}$$

where the bracketed expression is a diagonal PI-controller.

A numerical example is provided for the following simple distillation model (Skogestad et al. 1988)

$$\mathbf{P}(0) = \frac{1}{\tau_1 s + 1}\begin{bmatrix} 0.878 & -0.864 \\ 1.082 & -1.096 \end{bmatrix} \tag{20}$$

for which Skogestad et al. (1988) designed both an IMC controller of the kind in eq 19 and a μ-optimal controller. If the IMC controller is realized as in eq 19, it can be written ($k_1 = 0.7$)

$$\mathbf{C_{IMC}}(s) = \mathbf{V}\left[\frac{k_1}{s}(\tau_1 s + 1)\mathbf{\Sigma}^{-1}\right]\mathbf{U}^T = \begin{bmatrix} 0.707 & -0.708 \\ -0.708 & -0.707 \end{bmatrix} \times \frac{(\tau_1 s + 1)}{s}\begin{bmatrix} 0.507k_1 & 0 \\ 0 & 71.87k_1 \end{bmatrix}\begin{bmatrix} 0.625 & 0.781 \\ -0.781 & 0.625 \end{bmatrix} \tag{21}$$

The low-frequency approximation of the μ-controller given by Skogestad et al. (1988) can be decomposed according to

$$\mathbf{C}_\mu(s) = \frac{(\tau_1 s + 1)}{s}\begin{bmatrix} 3.82 & -0.92 \\ 1.73 & -3.52 \end{bmatrix} = \begin{bmatrix} 0.706 & -0.708 \\ -0.708 & -0.706 \end{bmatrix} \times \frac{(\tau_1 s + 1)}{s}\begin{bmatrix} 2.359 & 0 \\ 0 & 5.026 \end{bmatrix}\begin{bmatrix} 0.624 & 0.781 \\ -0.781 & 0.624 \end{bmatrix} \tag{22}$$

Obviously, both the IMC-controller in eq 21, the μ-controller in eq 22 and the controller in eq 16 are similar in the sense that they can be regarded as decouplers with a perfect elimination of interaction on the nominal model, but with a different degree of compensation of directionality. The only difference between the two controllers in eqs 21 and 22 is in the tuning of the diagonal PI-controllers, making the first one potentially non-robust (eq 21) and the second one (eq 22) less sensitive to model errors. Greater robustness is achieved by (de)tuning the compensation of the process directionality by the controller. The IMC-controller has $\mathrm{cond}(\mathbf{C_{IMC}}) = \mathrm{cond}(\mathbf{P_{LV}}) = 141.7$ whereas $\mathrm{cond}(\mathbf{C}_\mu) = 2.13$.

The tuning of the robust μ-controller was performed through established, yet complicated iterative methods. Here we suggest that the controller should be tuned by less complicated methods. A decentralized controller is much easier to tune and understand than optimal state feedback controllers. Further, it has not been possible to capture the actual uncertainty in real distillation by norm-bounds in an unconservative manner, which is a necessity for a successful μ-optimal robust controller design.

5.2 Directionality and Distillation Dynamics

In distillation, it has been found that the second principal direction (low-gain) has a considerably greater bandwidth than the first principal direction (high-gain) (Skogestad and Morari, 1988; Bequette and Edgar, 1989; Sågfors and Waller, 1994). This can be approximately described by the model (Sågfors et al., 1994a)

$$\mathbf{P}(s) = \frac{\overline{\sigma}\,\overline{\mathbf{u}}\,\overline{\mathbf{v}}^T}{\tau_1 s + 1} + \frac{\underline{\sigma}\,\underline{\mathbf{u}}\,\underline{\mathbf{v}}^T}{\tau_2 s + 1} \tag{23}$$

where $\tau_1 > \tau_2$. Using the compensator structure in eq 16, a diagonal PI-controller that takes into account the different bandwidths takes the form of

$$\mathbf{C_{diag}} = \begin{bmatrix} \overline{\sigma}^{-1}\frac{k_1}{s}(\tau_1 s + 1) & 0 \\ 0 & \underline{\sigma}^{-1}\frac{k_2}{s}(\tau_2 s + 1) \end{bmatrix} \tag{24}$$

where the singular values are those of a locally linearized model, the time constants reflect the different bandwidths of the system and the ratio k_1/k_2 is the degree of compensation of process directionality. If $k_1/k_2 = \gamma$, the controller does not provide any directional compensation. If $k_1/k_2 = 1$, the controller is an IMC-controller providing full compensation for the process directionality that ideally results in a closed loop system with a first order response with time constant k_1^{-1} for any direction of set-point changes, provided there is no modeling error in the model in eq 23.

6. CONCLUSIONS

The structure of directionality in non-ideal distillation has been analyzed. Model-plant mismatch due to process non-linearity has been examined in terms of process directionality. Methods of how to generate tight norm-bounded uncertainty descriptions of models of real distillation columns necessary for optimal controller designs are lacking. Norm-bounded uncertainty descriptions are likely to overestimate the error in the description of which input directions require large control action and which input directions require small control action. Process knowledge in distillation can provide information about the directionality.

Based on the structure of the process directionality in distillation, a compensator structure has been suggested. The compensator is not claimed to be "optimal" in any sense, though similarities of the compensator with controllers gained from optimal controller syntheses can be found, such as those with μ-optimal controllers. A compensator structure with diagonal controllers is, however, much easier to generate, tune and understand.

7. ACKNOWLEDGMENTS

The results reported have been obtained during a long-range project on multivariable process control supported by Tekes, the Academy of Finland, Nordisk Industrifond, the Neste Foundation and Neste Oy. This support is gratefully acknowledged.

8. REFERENCES

Andersen, H. W.; Kümmel, M.; Jørgensen, S. B. (1989). Dynamics and identification of a binary distillation column. *Chem. Eng. Sci.* 11, 2571-2581.

Arkun, Y.; Morgan III, C. O. (1988). On the use of the structured singular value for robustness analysis of distillation column control. *Comput. chem. Engng.* 12, 303-306.

Bequette, B. W.; Edgar, T. F. (1989). Non-interacting control system design methods in distillation. *Computers Chem Engng.* 6, 641-650.

Brambilla, A.; D'Elia, L. (1992). Multivariable controller for distillation columns in the presence of strong directionality and model errors. *Ind. Eng. Chem. Res.* 31, 536-543.

Bristol, E. H. (1966). On a new measure of interaction for multivariable process control. *IEEE Trans. Autom. Control.* 11, 133-134.

Garcia, C. E.; Morari, M. (1985). Internal model control. 2. Design procedure for multivariable systems. *Ind. Eng. Chem. Process Des. Dev.* 24, 472-484.

Hovd, M. (1992). Studies on control structure selection and design of robust decentralized and SVD controllers. *Dr. Ing. Thesis,* NTH, Trondheim.

Häggblom, K. E. (1991). Modeling of flow dynamics for control of distillation columns. *Proceedings ACC.*, 785-790, Boston.

Häggblom, K. E. (1995). Static directionality of 2 × 2 systems: A control relevant property with application to distillation. *Submitted to ECC '95*, Roma, Italy.

Häggblom, K. E.; Waller, K. V. (1988). Transformations and consistency relations of distillation control structures. *AIChE J.* 34, 1634-1648.

Häggblom, K. E.; Waller, K. V. (1989). Predicting properties of distillation control structures. *Proceedings ACC.* 114-119, Pittsburgh.

Koung, C-W.; MacGregor, J. F. (1991). Geometric analysis of the global stability of linear inverse-based controllers for bivariate nonlinear processes. *Ind. Eng. Chem. Res.* 30, 1171-1181.

Koung, C-W.; MacGregor, J. F. (1992). Robustness of multivariable linear controllers to process nonlinearities. *Ind. Eng. Chem. Res.* 31, 1085-1096.

Koung, C-W.; MacGregor, J. F. (1993). Design of identification experiments for robust control. A geometric approach for bivariate processes. *Ind. Eng. Chem. Res.* 32, 1658-1666.

Lau, H.; Alvarez, J.; Jensen, K. F. (1985). Synthesis of control structures by singular value analysis: Dynamic measures of sensitivity and interaction. *AIChE J.* 31, 427-439.

Lehtinen, K. (1994). A study of the behavior of a distillation column at different operating conditions. M. Sc. Thesis, Process Control Laboratory, Åbo Akademi, Åbo, Finland (In Swedish).

McAvoy, T. (1983). *Interaction Analysis.* Instrument Society of America, Research Triangle Park.

McDonald, K. M.; Palazoglu, A.; Bequette, B. W. (1988). Impact of model uncertainty descriptions for high purity distillation control. *AIChE J.* 34, 1996-2004.

Morari, M., Zafiriou, E. (1989). *Robust process control.* Prentice-Hall.

Pensar, J. A.; Toivonen, H. T. (1994). A multimodel mixed H_2/H_∞ problem for plants with structured uncertainty. *Proc. ADCHEM '94*, Kyoto, Japan.

Semino, D.; Scali, C.; Brambilla, A. (1993). Design of double filter IMC controller for ill-conditioned distillation columns. *Computers chem. Engng.* 17, 1041-1046.

Skogestad, S.; Morari, M. (1987). Design of resilent processing plants - IX. Effect of model uncertainty on dynamic resilience. *Chem. Eng. Sci.* 42, 1765-1780.

Skogestad, S.; Morari, M. (1988). Understanding the dynamic behavior of distillation columns. *Ind. Eng. Chem. Res.* 27, 1848-1862.

Skogestad, S.; Morari M.; Doyle, J. C. (1988). Robust control of ill-conditioned plants: high purity distillation. *IEEE Trans. Autom. Control.* 12, 1092-1105.

Skogestad, S.; Lundström, P. (1990). Mu-optimal LV-control of distillation columns. *Comput. chem. Engng.* 14, 401-413.

Sågfors, M. F.; Waller, K. V. (1994). Dynamic low-order models for capturing directionality in non-ideal distillation. *Submitted.*

Sågfors, M. F.; Waller, J. B. ;Waller, K. V. (1994a). Modeling and identification of the low-gain direction of ill-conditioned plants. Report 94-1, Process Control Laboratory, Åbo Akademi.

Sågfors, M. F.; Waller, J. B. ;Waller, K. V. (1994b). A model structure for capturing the process directionality in non-ideal distillation. Report 94-2, Process Control Laboratory, Åbo Akademi.

Waller, K. V.; Finnerman, D. H. (1987). On using sums and differences to control distillation. *Chem. Eng. Comm.* 56, 253-268.

Wang, S. J.; Yu, S. W. (1993). Robust feedback design for nonlinear high-purity distillation column control. *Comp. chem. Engng.* 17, 897-908.

ROBUST ANALYSIS OF DISTILLATION COLUMN DYNAMICS

E.I. Varga[1], G.Z. Kovács[1] S.B. Jørgensen[2] and K.M. Hangos[1]

[1] *Systems and Control Laboratory*
Computer and Automation Research Institute

[2] *Department of Chemical Engineering*
The Technical University of Denmark, Lyngby, Denmark

Abstract. The robust stability and performance of a binary distillation column model is analysed using a minimal state space representation. The uncertainty modelling of the column is resulted in an LFT form with a four element diagonal block structure taking into account the dependence between the steady state gains. Robust analysis of the distillation column dynamics is performed and compared using μ- and ℓ_1-analysis tools. The magnitude of possible overshoots is determined by ℓ_1-test for robust performance and the dependence of this magnitude on the dominant time constant (i.e. the purity of the separation) is investigated. Futhermore the effect of the uncertainties in the time constants and the singular values on the robust performance is also analysed using ℓ_1-test. It has been found that the uncertainty in the time constants affects dominantly the robust performance of the open-loop system, i.e. the magnitude of possible overshoots.

Keywords. Distillation column, parametric uncertainty modelling, robust stability, robust performance

1. INTRODUCTION

The success of applying model based control in practice depends strongly upon the accuracy of the underlying dynamic model representation of the plant in terms of control relevant properties. For a multivariable plant the model should be able to capture the interactions and the directionality aspects. It is however well known that this may be difficult to achieve when interactions are strong, as is the case for high purity distillation. In such cases the plant dynamics have a strong directionality dependence which is seen in that the open-loop dynamics are dominated by the high gain, whereas the low gain become important under feedback conditions. Thus it is crucial for control to be able to obtain good estimates of the low gain and its input and output directions.

High-purity distillation columns in the commonly used LV control configuration are ill-conditioned plants where the plant gain is strongly dependent on the input direction since the plant has a high condition number, $\kappa(G) = \frac{\overline{\sigma}(G)}{\underline{\sigma}(G)} \gg 1$, where $\overline{\sigma}(.)$ and $\underline{\sigma}(.)$ are the maximum and minimum singular values. Ill-conditioned plants are known to be difficult to control, especially if there are uncertainties in the plant model. The most important types of uncertainties are as follows:

1. input uncertainties caused by imperfect measurement devices or limited valve resolution
2. disturbance uncertainty caused by unknown disturbances affecting the process
3. uncertain parameters in the linear model (parameters only known approximately or can vary due to changes in operating conditions or nonlinearities)
4. high frequency uncertainty caused by unknown model structure or model order.

Skogestad et al. (1988) concluded that ill-conditioned plants (e.g. high-purity disillation columns) are very sensitive to input uncertainty in the traditional LV-configuration and that time constant and gain uncertainty can be approximated by an equivalent output uncertainty.

As a simple dynamic model of the distillation column the two time constant model (N2) (Skogestad and Morari, 1988) with steady state gains and time constants as uncertain parameters is used and converted to a minimal state space realization (Varga and Jørgensen, 1994). The parameters of this model have been identified in the state space form (Varga and Jørgensen, 1994), however, the minimum singular value predicted by an identified model is very uncertain using traditional identification methods. The precise estimation of the minimum singular value is essential for control purposes in order to fulfill robust stability and performance requirements (Koung and MacGregor, 1993). Using the relation between the uncertainty of the estimated parameters and that of the minimum singular value, the robust stability and performance of a binary distillation column is analysed here.

The results of parametric uncertainty modelling have been used earlier to generate robust model of heat exchanger networks (Varga et al, 1993). Thereafter robust analysis tools (μ and ℓ_1) have been succesfully applied to determine the magnitude of possible large overshoots which can cause operability problems in heat exchanger networks (Varga et al, 1993, 1994). Based on the parametric uncertainty modelling described above the dynamics of the distillation column is analyzed using established robust analysis (μ and ℓ_1) tools (Packard and Doyle, 1993; Khammash, 1993). The proposed method is applied to the the uncertain model of high-purity distillation column in order to determine the maximum magnitude of possible overshoots and the effect of the uncertainty in the minimum singular value on the robust stability.

2. DYNAMIC MODEL OF THE DISTILLATION COLUMN

2.1. *The distillation column model*

A distillation column is modelled by a large number of differential equations (Skogestad and Morari, 1987). For each tray a differential equation can be written for

- the material balance of each component,
- the overall material balance (flow dynamics),
- the energy balance.

However, the dominant part of the dynamics can be approximated by modelling only the composition dynamics. The case study in (Varga and Jørgensen, 1994) considers nonlinear simulation of a binary distillation column separating methanol and isopropanol. The steady-state data of the column are as follows: number of trays, $N = 13$; feed tray number, $N_F = 7$; tray holdup, $H_T = 0.15$ kmole; reflux flow rate, $L = 0.315$ kmole/min; vapor flow rate, $V = 0.350$ kmole/min; feed flow rate, $F = 0.070$ kmole/min; feed concentration $x_F = 0.50$; relative volatility $\alpha = 2.13$; bottom concentration, $x_B = 0.01$; destillate contentration, $x_D = 0.99$. The noise-free simulations are

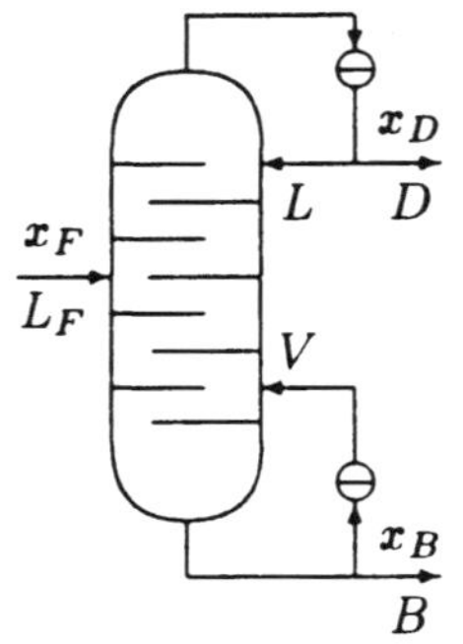

Fig. 1. Binary distillation column

based on the assumptions of constant molar overflows, liquid feed at its bubble point, 100% tray efficiency, constant and perfectly mixed liquid holdups, insignificant vapor holdup and constant relative volatility. Thus only the component balances are needed and the number of states is equal to the number of stages plus two for the condenser and the reboiler. As a reference model the nonlinear model was linearized around the steady state operating point. The manipulated variables are the reflux L and boilup V, i.e. LV-configuration is considered. The outputs are destillate and bottom product compositions of the light component (x_D and x_B).

2.2. *The analytic low-order model*

An approximate simple analytic model for binary high purity distillation without flow dynamics (N2) is given by Skogestad and Morari (1988). The model is based on the properties of the responses for changes in internal and external flows. Changes in internal flows correspond to the high gain and dominant time constant, while changes in external flows correspond to the low gain and small time constant.

$$\begin{pmatrix} X_D \\ X_B \end{pmatrix} = G_{LV}(s) \begin{pmatrix} L \\ V \end{pmatrix} \tag{1}$$

where $G_{LV}(s) = \frac{1}{(1+\tau_1 s)(1+\tau_2 s)} \times$

$$\begin{pmatrix} k_{11}(1+\tau_2 s) & k_{12}(1+\tau_1 s)+k_{11}(\tau_1-\tau_2)s \\ k_{21}(1+\tau_2 s) & k_{22}(1+\tau_1 s)+k_{21}(\tau_1-\tau_2)s \end{pmatrix}$$

where τ_1 and τ_2 are the dominant and small time constants respectively; k_{11}, k_{12}, k_{21} and k_{22} are the steady state gains; X_D, X_B, L and V are the Laplace-transform of the corresponding variables.

From this MIMO transfer function model a minimal state space realization has been obtained in (Varga and Jørgensen, 1994) in diagonal form:

$$A = \begin{pmatrix} -\frac{1}{\tau_1} & 0 \\ 0 & -\frac{1}{\tau_2} \end{pmatrix} \quad B = \begin{pmatrix} 1 & -1 \\ 0 & 1 \end{pmatrix}$$
$$C = \begin{pmatrix} \frac{k_{11}}{\tau_1} & \frac{k_{11}+k_{12}}{\tau_2} \\ \frac{k_{21}}{\tau_1} & \frac{k_{21}+k_{22}}{\tau_2} \end{pmatrix} \quad D = 0 \tag{2}$$

The parameters of the distillation model in state-space form (Eq.2) have been identified in Varga and Jørgensen (1994) using open-loop and closed-loop identification and also SVD-based experiment design method. The resulting parameters have been compared to that of the reference model which is the linearized form of the detailed nonlinear model. The reference parameter values are as follows: $\tau_1 = 318$ min, $\tau_2 = 3.24$ min, $k_{11} = 14.62$, $k_{12} = -14.64$, $k_{21} = 13.38$, $k_{22} = -13.36$. With these parameters the singular values at frequency $\omega = 0$ are resulted as $\overline{\sigma} = 28.0$ and $\underline{\sigma} = 0.0205$. Using closed-loop identification as the best method, the deviation of the estimated values from their reference values are $\Delta\tau_1 = 2.5\%$, $\Delta\tau_2 = 73.8\%$, $\Delta\overline{\sigma} = 3.3\%$, $\Delta\underline{\sigma} = 5.2\%$.

3. UNCERTAINTY MODELING

3.1. *Parametric uncertainty modelling*

Consider a linear time invariant finite dimensional system which has r bounded, real-valued, scalar uncertain parameters in its state space model: $p_i \in [\underline{p}_i, \overline{p}_i] \subset \mathbb{R}$, $i = 1, \ldots, r$. Let $S(p)$ the overall system matrix which is constructed from the system matrices of the state space model, so that the uncertain system can be written as:

$$\begin{pmatrix} \dot{x} \\ y \end{pmatrix} = S(p) \begin{pmatrix} x \\ d \end{pmatrix} = \begin{pmatrix} A(p) & B(p) \\ C(p) & D(p) \end{pmatrix} \begin{pmatrix} x \\ d \end{pmatrix} \quad (3)$$

In order to determine the nominal and uncertain part and to get normalized perturbations one should scale the parameters: $p = p_0 + s\delta$, where $p_0 := (\overline{p}+\underline{p})/2$ is the vector of the nominal value of parameters, $s = (\overline{p} - \underline{p})/2$ is the vector of the scaling factors and δ is the vector of the normalized perturbations ($\delta_i \in [-1, 1]$). Now the uncertain part can be separated from the nominal part:

$$S(p) = S(p_0) + S(s\delta) = S_0 + S_\Delta(\delta) \quad (4)$$

The aim of uncertainty modeling is to transform the uncertain state space model to a Linear Fractional Transformation (LFT) form, where the perturbations are collected in an uncertainty block structure Δ (Steinbuch et al., 1991; Lambrecht et al., 1992). A restricted set of Δ will be considered further: the normalized, square, diagonal, real-valued repeated block structure which is used in parametric uncertainty modeling and analysis:

$$\mathbf{\Delta}_{rr} = \{\mathrm{diag}(\delta_1 I_{k_1}, \ldots, \delta_r I_{k_r}) : \delta_i \in [-1, 1]\}$$

It is assumed that the scaling factors are incorporated in the coefficient matrix M. The uncertain system written in LFT form is:

$$F_\ell(M, \Delta) = S_0 + B_\Delta \left(I - \Delta D_\Delta\right)^{-1} \Delta C_\Delta \quad (5)$$

and the general problem is the following: $S_\Delta(\delta)$ is given, find the constant matrices $(B_\Delta, C_\Delta, D_\Delta)$ and a $\Delta(\delta)$ with dimensions as small as possible such that

$$S_\Delta(\delta) = B_\Delta \left(I - \Delta(\delta) D_\Delta\right)^{-1} \Delta(\delta) C_\Delta \quad (6)$$

Lambrecht et al. (1992) proposed a general method to solve this problem for the case of parametric uncertainty occuring as real rational N dimensional polynomials.

3.2. *Uncertain model of the distillation column*

The dynamic model (2) of the distilation column contains six parameters: two time constants (τ_1,τ_2) and four steady-state gains (k_{11},k_{12},k_{21},k_{22}). The uncertainty in the maximum and minimum singular value ($\overline{\sigma}(\omega)$, $\underline{\sigma}(\omega)$) can be related to the uncertainty in the estimated six parameters. The time constants reflect the frequency dependence of the singular values of the transfer function. The steady-state gains determine the singular values at frequency $\omega = 0$. However, the gains are correlated (Skogestad and Morari, 1987) and there are only two independent parameters and hence two independent source of uncertainty. Using the properties of the low-order model (Skogestad and Morari, 1988), the singular values ($\overline{\sigma}$, $\underline{\sigma}$) at $\omega = 0$ are related to the gains as follows: $k_{11} \approx k_{21} \approx \overline{\sigma}/2$, $-(k_{11} + k_{12}) \approx (k_{21} + k_{22}) \approx \underline{\sigma}$.

The uncertain state-space model of the distillation column can be written as

$$A = \begin{pmatrix} -\frac{1}{\tau_1^0 + s_1\delta_1} & 0 \\ 0 & -\frac{1}{\tau_2^0 + s_2\delta_2} \end{pmatrix}$$
$$C = \begin{pmatrix} \frac{k_{11}^0 + s_3\delta_3}{\tau_1^0 + s_1\delta_1} & \frac{(k_{11}^0 + k_{12}^0) + s_4\delta_4}{\tau_2^0 + s_2\delta_2} \\ \frac{k_{21}^0 + s_3\delta_3}{\tau_1^0 + s_1\delta_1} & \frac{(k_{21}^0 + k_{22}^0) + s_4\delta_4}{\tau_2^0 + s_2\delta_2} \end{pmatrix} \quad (7)$$

The result of the transformation to the LFT form with the uncertainty block structure $\Delta = \{\mathrm{diag}(\delta_1, \delta_2, \delta_3, \delta_4) : \delta_i \in [-1, 1]\}$ is given by

$$D_\Delta = \begin{pmatrix} -s_1/\tau_1 & 0 & 0 & 0 \\ 0 & -s_2/\tau_2 & 0 & 0 \\ -1/\tau_1 & 0 & 0 & 0 \\ 0 & -1/\tau_2 & 0 & 0 \end{pmatrix}$$
$$C_\Delta = \begin{pmatrix} s_1/\tau_1 & 0 & 0 & 0 \\ 0 & s_2/\tau_2 & 0 & 0 \\ 1/\tau_1 & 0 & 0 & 0 \\ 0 & 1/\tau_2 & 0 & 0 \end{pmatrix}$$
$$B_\Delta = \begin{pmatrix} 1/\tau_1 & 0 & 0 & 0 \\ 0 & 1/\tau_2 & 0 & 0 \\ -k_{11}/\tau_1 & -(k_{11}+k_{12})/\tau_2 & s_3 & s_4 \\ -k_{21}/\tau_1 & -(k_{21}+k_{22})/\tau_2 & s_3 & s_4 \end{pmatrix}$$

4. ROBUST ANALYSIS TOOLS

4.1. *Robust stability and performance*

The objective of robust control is to achieve good performance requirements in the presence of uncertainty. Robust analysis tools and robust control design methods have been developed recently dealing with structured uncertainty for the cases when the signals measured in

- the 2-norm (related to energy): μ-analysis (Packard and Doyle, 1993),
- the ∞-norm (maximum amplitude): ℓ_1-analysis (Dahleh and Khammash, 1993).

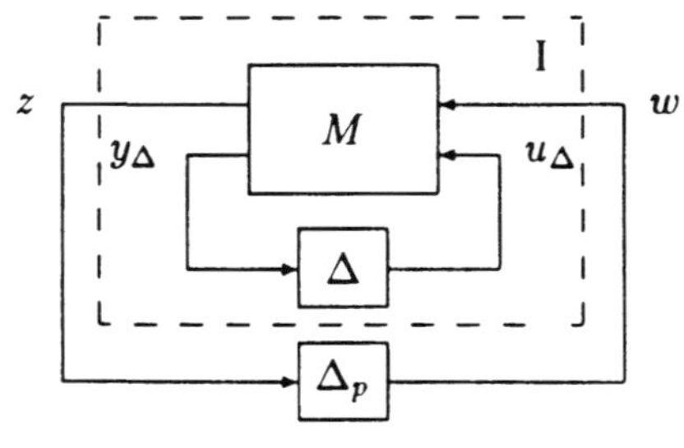

Fig. 2. The robust stability and performance problem formulation

The general problem formulation of robust stability and performance is shown in Fig.2. While Δ models the uncertain part of the system, M is the known part of the system including the nominal plant, the controller, any input and output weighting functions and any weighting functions on the perturbations. M is resticted to be a Linear Time Invariant (LTI) system. The signal w denotes all exogenous inputs, including the control inputs and disturbances, while z denotes the regulated outputs. The map taking w to z is referred as T_{zw}. It is assumed that Δ is structured, i.e. it belongs to the following class:

$$\mathbf{\Delta}(n) := \{\Delta = \text{diag}(\Delta_1, \ldots, \Delta_n) \ : ||\Delta_i|| \leq 1\}$$

Robust stability. The system achieves robust stability iff the system is internally stable for all admissible perturbations (for all $\Delta \in \mathbf{\Delta}(n)$).

Robust Performance. The system achieves robust performance iff

1. the system achieves robust stability, and
2. $\sup_{\Delta \in \mathbf{\Delta}(n)} ||T_{zw}|| < 1$.

The robust performance problem can be transformed to a robust stability problem, which has been shown for both 2-norm and ∞-norm cases (Doyle, 1982; Khammash, 1993). Consider the two systems in Fig.2, where System I corresponds to a performance robustness problem, while System II is formed from System I by connecting z and w through a fictious perturbation, Δ_p, satisfying $||\Delta_p|| \leq 1$.

Theorem 1 Consider the System I having $\mathbf{\Delta}(n)$ as the class of admissible perturbations and System II having $\mathbf{\Delta}(n+1)$ as the class of admissible perturbations. System I achieves robust performance iff System II is robustly stable.

4.2. *μ-analysis*

Measuring the signals in the L_2-norm and assuming time-invariant structured uncertainty, the μ-tests can be used to analyse robust stability and performance.

The structured singular value (SSV, μ) is a matrix function denoted by $\mu_\Delta(.)$ which depends on the underlying structure Δ (a prescribed set of block diagonal matrices) (Doyle et al, 1991).

Definition 1 For $M \in \mathbb{C}^{n \times n}$, $\mu_\Delta(M)$ is defined by $\mu_\Delta(M) := \frac{1}{\min\{\bar{\sigma}(\Delta):\Delta \in \mathbf{\Delta}, \det(I - M\Delta)=0\}}$ unless no $\Delta \in \mathbf{\Delta}$ makes $I - M\Delta$ singular, in which case $\mu_\Delta(M) := 0$ (the symbol $\bar{\sigma}(.)$ denotes the maximum singular value).

The system M is described by the transfer function $G(s) = C(sI - A)^{-1}B + D$ which maps from Laplace-transform of the input signals to that of the output signals. The following Robust Stability (RS) μ-test on $G(s)$ will be used in Frequency Domain (FDμ):

Theorem 2 (Packard and Doyle, 1993) Suppose that $G(s)$ satisfies Nominal Stability, then for all $\Delta \in \mathbf{\Delta}(n)$ the uncertain system is stable iff

$$\sup_\omega \mu_\Delta (G_{22}(j\omega)) < 1 \qquad (8)$$

where $G_{22}(s)$ is the lower right submatrix of $G(s)$.

4.3. *ℓ_1-analysis*

The following notions and notations are applied in robustness analysis. The space L_∞ denotes the set of all real valued functions on $[0, \infty)$ statisfying $\sup_t |x(t)| < \infty$ and the norm of this functions: $||x||_\infty = \sup_t |x(t)|$. L_n^∞ denotes the space of n-tuples of elements of L_∞. The norm of $x = (x_1, \ldots, x_n) \in L_n^\infty$ is $||x||_\infty = \max_i ||x_i||_\infty$.

The impulse response of a linear time invariant causal system given in state space form is

$$\begin{aligned} h(t) &= \mathcal{L}^{-1}[C(sI - A)^{-1}B + D] \\ &= Ce^{At}B + D\delta(t) \end{aligned} \qquad (9)$$

where $\mathcal{L}^{-1}$ denotes the inverse Laplace transformation. The $h(t)$ for $0 \leq t < \infty$ is the kernel representation of the linear time invariant causal system M. The output e_k at time t for any

$u_l \in L^\infty$ is given by $e_k(t) = \int_0^t h_{kl}(t-\tau)u_l(\tau)d\tau$.

Assuming that the nominal system is L_∞ stable, we have $\sup_t \int_0^t |h_{kl}(\tau)|d\tau < \infty$. Therefore h_{kl} belongs to the space $\mathcal{A}$ and the norm is defined as $\|h_{kl}(t)\|_\mathcal{A} = \int_0^t |h_{kl}(\tau)|d\tau$. The induced L_∞ norm of h_{kl} is $\sup_t \|h_{kl}(t)\|_\mathcal{A}$ and the norm of the whole impulse response is:

$$\|h\|_1 = \max_k \sup_t \sum_l \|h_{kl}(t)\|_\mathcal{A} \qquad (10)$$

The system M is described by the impulse response function $h(t) = Ce^{At}B + D\delta(t)$ which maps from the input signals to the output signals in time domain. The following Robust Stability ℓ_1-test on $h(t)$ will be used (RSL1):

Theorem 3 (Khammash, 1993) Given an interconnection of a linear time-invariant stable system h_{22} and n norm bounded perturbation blocks, the system is robustly stable iff $\rho(\hat{h}) < 1$, where matrix $\hat{h}$ consists of the ℓ_1-norm of the subblocks:

$$\hat{h} = \begin{pmatrix} \|[h_{22}]_{11}\|_1 & \cdots & \|[h_{22}]_{1n}\|_1 \\ \vdots & & \vdots \\ \|[h_{22}]_{n1}\|_1 & \cdots & \|[h_{22}]_{nn}\|_1 \end{pmatrix}$$

5. ROBUST ANALYSIS OF THE DISTILLATION COLUMN DYNAMICS

5.1. *Comparing μ and ℓ_1 analysis results*

Measuring the input and output signals in L_2-norm the μ-analysis (Packard and Doyle, 1993), while measuring them in L_∞-norm the ℓ_1-analysis (Khammash, 1993) can be used to analyse the robustness properties of a system. While in the case of the μ-analysis the uncertainty block is assumed to be linear and time invariant, in the case of ℓ_1-analysis it can be nonlinear/time-varying, which makes the ℓ_1-analysis result potentially conservative. However, the calculation of ℓ_1-test is a computationally easy problem (belongs to the class FP) in contrast to that of the μ-test, which is an NP-hard problem (Hangos and Rónyai, 1994). In case of time-varying uncertainty it has been shown in (Shamma, 1994), that for the L_2-norm measured signal the robust stability can be tested by searching for scalings which makes the small gain condition nonconservative:

$$\inf_{D_{\text{admissible}}} \|DMD^{-1}\| \leq 1 \qquad (11)$$

This is an equivalent condition with the L_∞-norm case, which means that assuming time-varying/nonlinear perturbations, the robust stability condition is the same for the two different norm cases. For the time-invariant uncertainty and L_2-norm case the left side of the condition (11) is equal to the upper bound of μ wich also gives a nonconservative condition for uncertainty structures with less than or equal to three blocks.

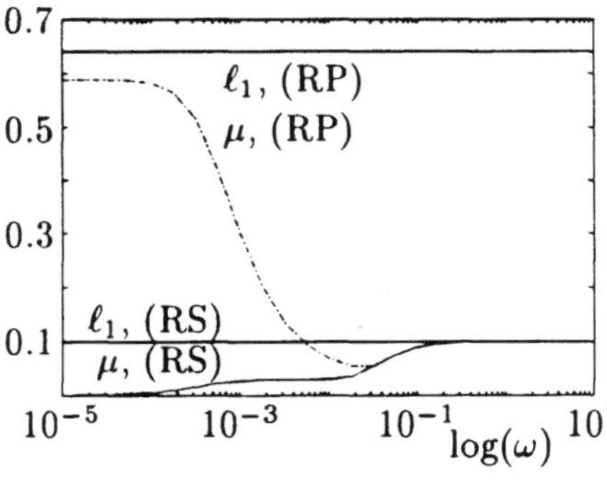

Fig. 3. Comparing the μ and ℓ_1-analysis results

For the distillation column example the result of robust stability and performance analysis is shown in Fig.3. The calculated μ upper and lower bounds have been plotted over the frequencies with horizontal line corresponding to the calculated ℓ_1 condition. While in case of robust stability the results are equivalent, for the robust performance the ℓ_1-analysis is conservative compared to the μ-analysis.

5.2. *Estimating the magnitude of overshoots*

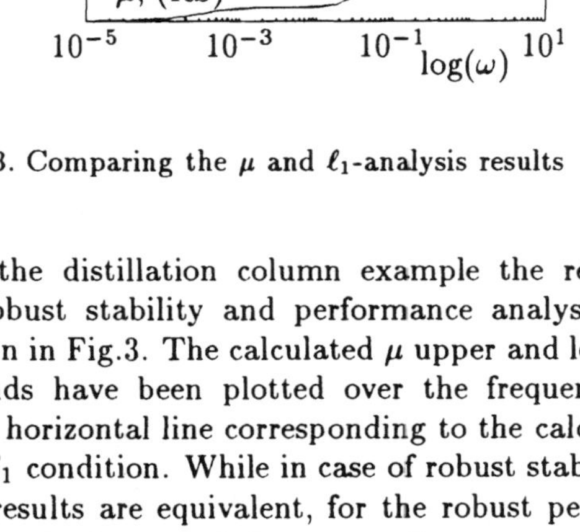

Fig. 4. The magnitude of overshoots vs. τ_1^0

Using the ℓ_1-test for robust performance ($\rho(\hat{h}) < 1$) the tightest bound for the maximum possible overshoot can be determined. Varga et al. (1994) proposed an iterative procedure to solve $\rho(\hat{h}) = 1$ in the magnitude of the output scaling for each output variable. For the distillation column example the estimated magnitude of overshoots have been shown on Fig.4 as a function of the dominant time constant (τ_1) which is related to the purity of the separation. With increasing purity the magnitude of overshoots is decreasing in the presence of uncertainties.

5.3. *The effect of uncertainty sources*

It has been shown in Varga and Jørgensen (1994) that the estimation of the minimum singular value of the distillation column model is very uncertain using traditional identification methods. Using the estimated parameters for control design, the closed-loop system can be unstable depending on the magnitude of the uncertainty in the minimum singular value. The robust stability and performance of the closed-loop system have been analysed using ℓ_1-test as a function of the magnitude of uncertainty from different sources. A model-based IMC controller has been designed to control the system. For the open-loop system the uncertainty of the time constants effets mainly the robust analysis test functions. However, for the closed-loop case (Fig.5) the uncertainty in the maximum singular value has almost the same effect as the time constants and the effect of the minimum singular value is also increased.

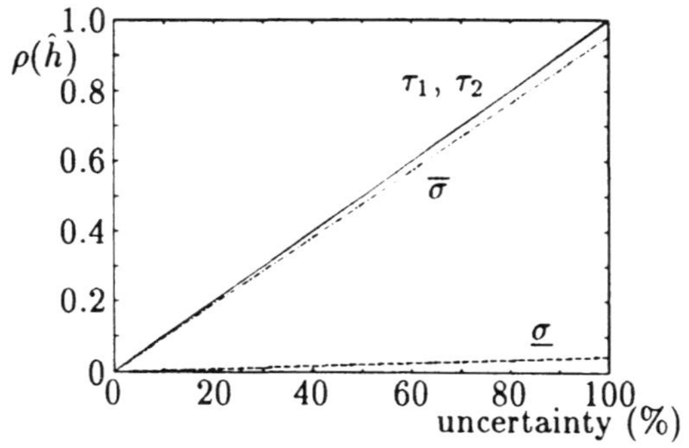

Fig. 5. The effect of different uncertainty sources

6. CONCLUSION

A simple dynamic model of the distillation column, the two time constant model (N2) (Skogestad and Morari, 1988) with steady state gains and time constants as uncertain parameters has been considered in this paper. The uncertainty modelling of the column is resulted in an LFT form with a four element diagonal block structure taking into account the dependence between the steady state gains.

Based on the uncertainty modelling robust analysis of the distillation column dynamics is performed. The results obtained using μ- and ℓ_1-analysis have been compared. The magnitude of possible overshoots of the open-loop system has been determined by ℓ_1-test for robust performance and the dependence of this magnitude on the dominant time constant (i.e. the purity of the separation) has been investigated. Futhermore the effect of the uncertainties in the time constants and the singular values on the robust performance of the open- and closed-loop system has been also analysed using ℓ_1-test. It has been found that the uncertainty in the time constants affects dominantly the robust performance of the open-loop system, i.e. the magnitude of possible overshoots. For the closed-loop system, however, the uncertainty in the minimum singular value also effects the robust stability and performance.

7. REFERENCES

Dahleh, M.A. and M.H. Khammash (1993). Controller Design for Plants with Structured Uncertainty. *Automatica*, **29** (1), 37-56.

Hangos, K.M. and L. Rónyai (1994). The Effect of Model Formulation on the Complexity of Control Problems. *Prep. of IEEE Workshop CMP'94*, 215-227.

Khammash, M.H. (1993). Necessary and Sufficient Conditions for the Robustness of Time-Varying Systems with Applications to Sampled-Data Systems. *IEEE Trans. on Aut. Contr.*, **38** (1), 49-57.

Koung, C.W. and J.F. MacGregor (1993). Design of Identification Experiments for Robust Control. A Geometric Approach for Bivariate Processes. *Ind. Eng. Chem. Res.*, **32**, 1658-1666.

Lambrecht, P., J.C. Terlouw , S. Bennani and M. Steinbuch (1992). Parametric Uncertainty Modeling using LFTs. *Selected Topics in Identification, Modeling and Control*, **5**, 1-10, Delft University Press.

Packard, A. and J.C. Doyle (1993). The Complex Structured Singular Value. *Automatica*, **29**, 71-109.

Shamma J.S. (1994). Robust Stability with Time-Varying Structured Uncertainty. *IEEE Trans. on Autom. Contr.*, **39** (4), 714-724.

Skogestad, S. and M. Morari (1988). Understanding the Dynamic Behavior of Distillation Columns. *Ind. Eng. Chem. Res.*, **27**, 1848-1862.

Skogestad S., M. Morari and J.C. Doyle (1988). Robust Control of Ill-Conditioned Plants: High- Purity Distillation. *IEEE Trans. on Aut. Contr.*, **33**, 1092-1105.

Varga E.I., J. Bokor and K.M. Hangos (1993). Grey Dynamics of Heat Exchanger Networks. *Prep. of IFAC 1993 World Congress*, **2**, 149-156.

Varga E.I., J. Bokor and K.M. Hangos (1994). Robust Analysis of Potential Large Overshoots in Heat Exchanger Networks. *Prep. of ACC'94*, **2**, 1422-1426.

Varga E.I. and S.B. Jørgensen (1994). Multivariable Process Identification – Estimating Gain Directions. *Prep. of Nordic Process Control Workshop 5., NPCW5*, 543-550, Lyngby, Denmark, Jan. 24-25, 1994.

INVESTIGATION OF OPERABILITY AND CONTROLLABILITY PROPERTIES OF A PILOT-SCALE DISTILLATION COLUMN

Kurt E. Häggblom and Kari Lehtinen

Process Control Laboratory, Department of Chemical Engineering,
Åbo Akademi University, FIN-20500 Åbo, Finland

Abstract: The results of an operability and controllability study of a pilot-scale distillation column separating a mixture of ethanol and water are presented. The relative gain for diagonal variable pairing in the LV-structure as well as the condition number are used as measures of controllability. Of particular interest are ill-conditioned operating regions characterized by strong directionality, which is typical of high-purity distillation. Operating constraints imposed by the column and the existing instrumentation are also taken into account. The investigation is carried out by means of a simulator, which has been calibrated with experimental data.

Keywords: Distillation columns, process control, multivariable control systems, controllability, condition numbers, simulators.

1. INTRODUCTION

Distillation is typically an "ill-conditioned" process, especially if the product purities are high. Such a plant has a gain matrix with a high condition number. Another measure of ill-conditioning, which is particularly useful for 2 × 2 systems, is the relative gain (Bristol, 1966) — the higher the relative gain, the more ill-conditioned is the plant. Furthermore, an ill-conditioned plant is characterized by a strong directionality, which means that the (normed) plant gain depends strongly on the input direction, that is, on the interplay between the control variables. Although this is a property of a linear system, the behaviour resembles that of a nonlinear system.

Because of the strong directionality, good set-point tracking in arbitrary directions cannot be obtained for an ill-conditioned plant using decentralized (multi-loop SISO) feedback control, where the controllers operate independently of each other. Good control performance requires a multivariable controller that takes the directionality into account. However, such a controller for an ill-conditioned plant can be very sensitive to model errors and disturbances in the control variables (Skogestad and Morari, 1987a; Skogestad *et al.*, 1988).

It can be shown that a gain matrix with the relative gain λ for a given variable pairing may become singular if the individual gains are perturbed by the relative amount $0.25\lambda^{-1}$. For a plant with $\lambda = 25$, which is not a particularly high relative gain for an ill-conditioned plant, this means that the gain matrix becomes singular if the gains are perturbed by more than 1% in unfavourable directions. Since there exists no controller with integral action that can stabilize two systems whose gain matrices have determinants of different signs (Garcia and Morari, 1985; Skogestad and Morari, 1987b), there is also no such multivariable controller that is certain to stabilize both the plant and the model if the relative uncertainties of the gains are uncorrelated and larger than $0.25\lambda^{-1}$.

If the errors are correlated in a certain way, larger errors can be tolerated. The design of the identification experiment so as to affect the distribution and correlation of errors in a favourable way is therefore important. It is, for example, necessary that the plant is sufficiently excited in all relevant gain directions. This may be achieved by identification in open loop (Koung and MacGregor, 1993) or closed loop (Andersen and Kümmel, 1992).

In this paper, the results of an operability and controllability study of a pilot-scale distillation column are presented. The research was done in order to find an operating region, where identification and control problems typical of ill-conditioned plants such as high-purity distillation columns could be studied. Operating constraints imposed by the column and existing instrumentation were also taken into account. The investigation was carried out by means of a simulator, which was calibrated with experimental data from a few operating points.

The relative gain and the condition number are used as measures of ill-conditioning and controllability. The directionality properties are clearly illustrated and verified experimentally.

2. DISTILLATION COLUMN SIMULATOR

The distillation column under study (Häggblom *et al.*, 1984; Waller *et al.*, 1988; Waller, 1992) separates a mixture of ethanol and water. It has 15 bubble-cap trays and is 0.30 m in diameter. It is equipped with a thermosyphon-type reboiler and a total condenser, and it operates under atmospheric pressure. To facilitate the operability and controllability study, a simulator for the binary distillation column was constructed (Lehtinen, 1994). The simulator was implemented in *Simulink* (MathWorks, Inc.), using a "graphical" block-oriented programming technique.

2.1 Modelling

The simulator was essentially developed from first principles with support of some empirical relationships. Each tray is modelled by a total and a component material balance, an energy balance, Francis weir formula for tray hydraulics, and an empirical relationship for the vapour-liquid equilibrium. The Murphree tray efficiency is used to model the deviation from 100 % tray efficiency. The efficiency of the reboiler is modelled similarly.

It is assumed that the specific enthalpies of liquid and vapour can be expressed as linear functions of their compositions. The dynamic energy balance can then be reduced to an algebraic relationship, where the contribution from the energy balance is contained in a single parameter, which expresses the relative difference between the partial specific enthalpies of vaporization of the two components, ethanol and water (Häggblom, 1991).

The column is equipped with a total condenser producing a subcooled reflux. In the modelling of the condenser it has to be taken into account that neither the heat removal through condensation nor the temperature or the enthalpy of the reflux are constant when the coolant flow rate is constant. The column could not be adequately modelled by assuming any of these quantities to be constant. In accordance with a suggestion by Häggblom (1991), the heat removal was therefore assumed to depend linearly on the flux of enthalpy from the column to the condenser.

2.2 Calibration

In the model, there are a number of uncertain parameters, which were adjusted to fit the behaviour of the simulator to the real column. Since *Simulink* runs under *Matlab* (MathWorks, Inc.), the parameters could be adjusted through numerical optimization, where the simulator was part of the objective function.

The adjusted parameters are the tray efficiencies in the rectifying and stripping sections (assumed constant in each section), the efficiency of the reboiler, the parameter associated with the energy balance, two parameters related with the heat removal in the condenser, and the specific enthalpy of condensation for the steam used to heat the reboiler.

Data from four open-loop experiments with different values for the feed flow rate, the feed composition, the reflux flow, and the steam flow to the reboiler, were used to calibrate the simulator. The product compositions, the temperature on tray 14, the reflux temperature, and the product flow rates were considered in the fit. As a result of the optimization, realistic values for the adjusted parameters were obtained. The behaviour of the simulator was also verified with experimental data not used in the optimization.

The above parameters primarily affect the steady-state behaviour of the simulator. The dynamic behaviour can be adjusted by a number of other parameters (e.g., parameters affecting holdups), but this was not done in the present work.

3. OPERABILITY AND CONTROLLABILITY STUDY

An operability and controllability study of the distillation column was carried out by means of the calibrated simulator. First a feasable operating region was located, after which a more detailed investigation was made. The operability and controllability properties are related both to open-loop and closed-loop specifications.

3.1 Operating Region

The critical operating constraints of the distillation column are: reflux flow rate $L \leq 200$ kg/h, steam flow rate (to reboiler) $V \leq 120$ kg/h. The feed composition z can be varied as desired, whereas the feed flow rate F should not exceed 200 kg/h so as to enable experiments of sufficient duration without switching of feed tanks (and feed composition). The distillate flow rate D should be as large as possible to facilitate control of the reflux drum inventory when using the LV-structure. Furthermore, the bottoms composition x (wt-% ethanol) should not be exceedingly low so as to avoid measurement problems and severe non-linearity. Since ethanol-water forms an azeotropic mixture at 94 wt-% ethanol, the distillate composition y is always below this value, and does not give rise to any particular measurement problems.

Given the above constraints and requirements, it was desired to find an operating region where the column could be classified as "ill-conditioned". By means of simulations, it was found that a satisfactory compromise between the competing requirements was obtained with the feed flow rate $F = 150$ kg/h and the feed composition $z = 25$ wt-% ethanol. This made possible a range of operating points satisfying $D \approx 40$ kg/h, $85\,\text{kg/h} \leq L \leq 160\,\text{kg/h}$, $70\,\text{kg/h} \leq V \leq 105\,\text{kg/h}$, $90.5\,\% \leq y \leq 91.5\,\%$, $0.5\,\% \leq x \leq 1.5\,\%$, $5 \leq \lambda \leq 20$.

3.2 Open-Loop Specifications

A more detailed simulation was performed for the range $L = 120 \ldots 160$ kg/h, $V = 85 \ldots 105$ kg/h. Figures 1 and 2 show the distillate and bottoms composition surfaces vs. L and V. These look relatively well-behaved, although there is a rather sharp change of the slope along the line $L \approx 2.2V - 70$ kg/h. This is revealed more clearly by the gain plots in Figs. 3 - 6. In the region defined by $L = 2.2V - 70$ kg/h, $\Delta V = \pm 1$ kg/h, the gains of y change by a factor of 5 and the gains of x by two orders of a magnitude. The nonlinearity is thus severe in this region. This also means that the compositions and their gains are very sensitive to input uncertainty in the region.

Figure 7 shows the relative gain λ vs. L and V. In the region defined above, the relative gain has a sharp maximum. At lower values of V (and L), λ changes from 5 to 10, and at higher values of V, from 5 to more than 20, when V is perturbed by 1-2 kg/h. The variation of λ is due to the nonlinearities, but their effect on λ is reduced by the fact that the nonlinearities of the gains are strongly correlated. The controllability implied by the relative gain is thus very sensitive to modelling errors and input uncertainty in this region.

At high reflux rates and low boilup rates the compositions approach their upper limits (the azeotropic composition 94 % for y and the feed composition 25 % for x), which naturally results in an increase of the relative gain.

Figure 8 shows the condition number vs. L and V. It is striking that the condition number has a minimum in the region where the relative gain has a maximum, and vice versa. Since the condition number denotes the ratio between the maximum and minimum plant gains, this means that the variability of the normed plant gain is, in fact, smaller in the region where the relative gain is large. This might seem to be in conflict with the previous results and interpretations.

However, the condition number should not be directly related with the relative gain, but with the gains themselves. In the region where the bottoms composition gains are close to zero, the condition number is high since a plant gain close to zero can be obtained by choosing the inputs such that $\Delta y \approx 0$. The fact that the gains for x are close to zero means that it is very difficult to affect x by the inputs L and V. One can also make the interpretation that it is difficult to reduce variations in y to the same level as in x.

It is well-known that the condition number is scaling dependent. If y is scaled by the factor 0.2 and x is scaled by 5, the result shown in Fig. 9 is obtained. Now the condition number is highest in the region where the distillate composition gains are small (note that the scaling has increased the gains related with x and decreased the gains related with y). According to the condition number, the plant is then most difficult to control in this region.

By another scaling (y scaled by 0.6, x scaled by 1.6), large condition numbers are obtained both for high distillate compositions and low bottoms compositions, see Fig. 10. In the region where the relative gain is high, the condition number has a distinct minimum.

Since the scaling of the variables can be related to control specifications in terms of allowed deviations in these variables, the condition number can be taken as a measure of how difficult the specifications are to meet. The relative gain, on the other hand, is more related with the sensitivity to modelling errors and (input) uncertainty, that is, with robustness issues. The relative gain and the condition number thus complement each other.

3.3 Closed-Loop Specifications

The simulations with open-loop specifications cover a large range of product compositions. In practice, the operating region is much more limited. Therefore, a set of simulations in the composition range $y = 90 \ldots 92$ wt-%, $x = 0.05 \ldots 2$ wt-% was performed.

Figure 11 shows contour lines of L and V vs. y and x. The L and V lines are nearly parallell in the whole region, even in the nonlinear part, where the contour lines are curved. This means that the gain matrix is nearly singular.

The figure clearly illustrates the gain directionality of the plant. To move from operating point A ($y = 91$ %, $x = 1$ %) to operating point B ($y = 91.1$ %, $x = 1.5$ %), the required changes in the inputs are $\Delta L = 0.54$ kg/h, $\Delta V = -0.14$ kg/h. A change to operating point C ($y = 91.1$ %, $x = 0.5$ %), on the other hand, requires the input changes $\Delta L = 15.8$ kg/h, $\Delta V = 7.64$ kg/h, which are more than 30 times larger than the previous changes in L and V, although the composition changes are equally large in both cases.

The previous example is not even extreme. A change to operating point D ($y = 91.5$ %, $x = 0.9$ %), which is orthogonal to the change to point B, requires the input changes $\Delta L = 38.5$ kg/h, $\Delta V = 17.5$ kg/h, which are almost two orders of a magnitude larger than the required changes between point A and B. The change from point A to point B is thus in a high-gain direction, whereas the change from point A to point D is in a low-gain direction. The ratio between these gains is (approximately) equal to the condition number.

Note that the principal output directions corresponding to the high- and low-gain directions change significantly when the bottoms composition moves towards zero. The principal input directions, on the other hand, are not much affected, since the angle between the contour lines for L and V remain essentially constant. It can, in fact, be proved that the principal input directions are aligned with changes in the rates of external (product) flows (Häggblom, 1995). These directions can therefore easily be determined from the flow gains.

Figure 7 showed that the relative gain is high only in a narrow region in terms of L and V. Relatively small perturbations in these variables will drop the relative gain to around 5. However, Fig. 12 shows that a large relative gain cannot be

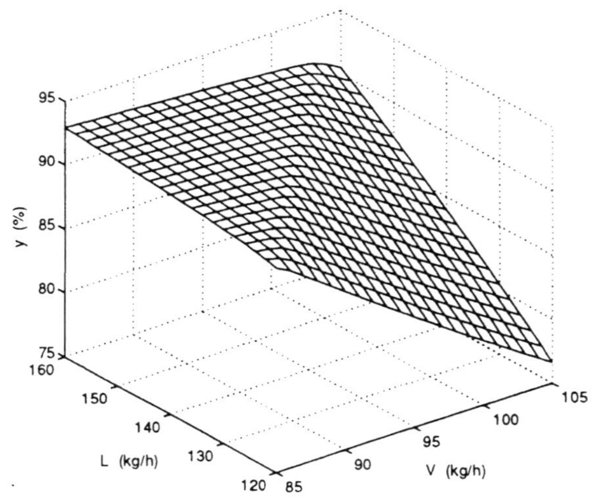

Fig. 1. Distillate composition versus reflux and steam flow.

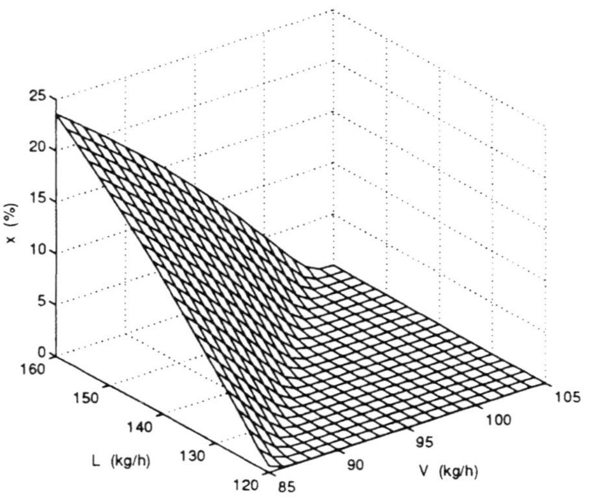

Fig. 2. Bottoms composition versus reflux and steam flow.

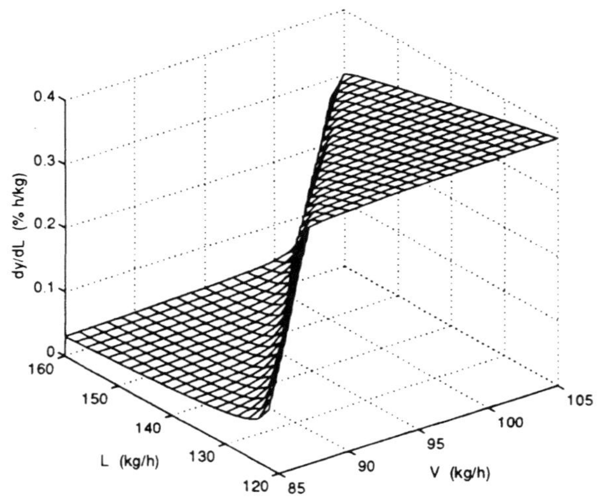

Fig. 3. Gain between distillate composition and reflux.

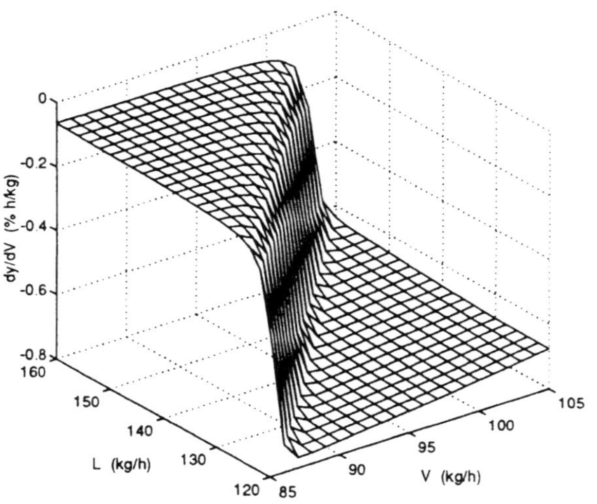

Fig. 4. Gain between distillate composition and steam flow.

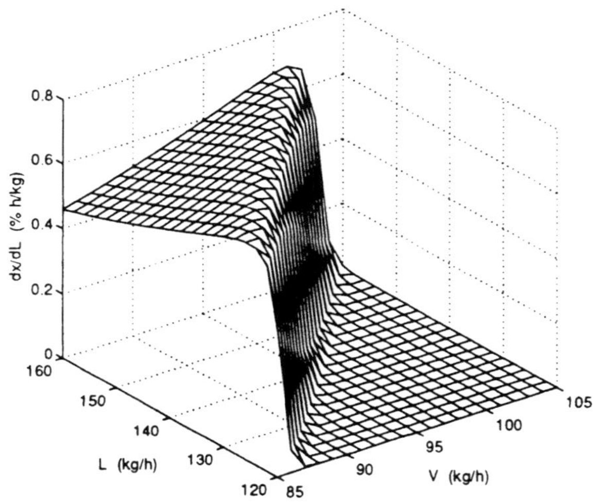

Fig. 5. Gain between bottoms composition and reflux.

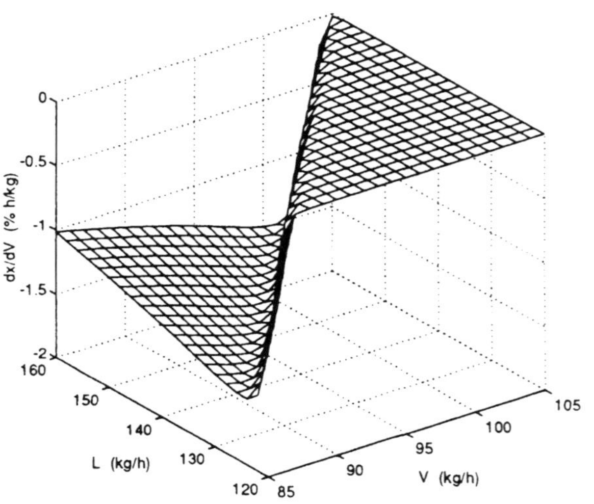

Fig. 6. Gain between bottoms composition and steam flow.

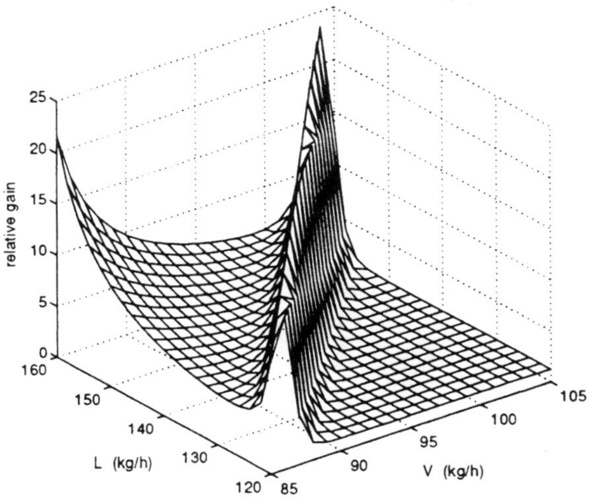

Fig. 7. Relative gain versus reflux and steam flow.

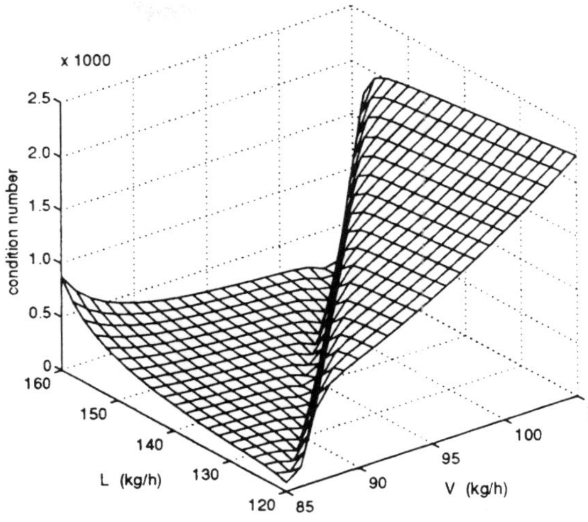

Fig. 8. Condition number versus reflux and steam flow.

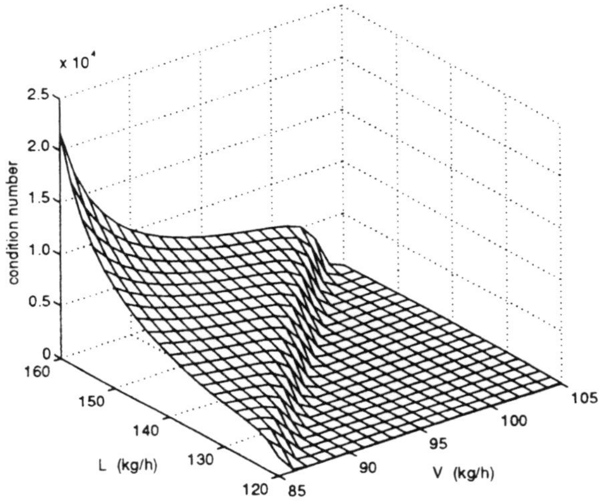

Fig. 9. Condition number versus reflux and steam flow; scaling $0.2y$, $5x$.

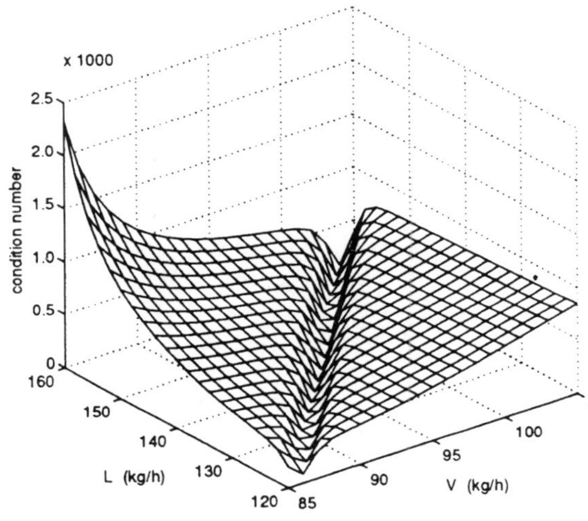

Fig. 10. Condition number versus reflux and steam flow; scaling $0.6y$, $1.6x$.

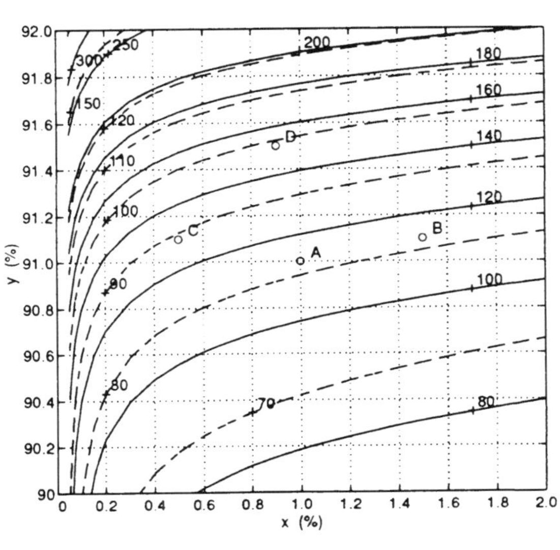

Fig. 11. Contour plot of reflux and steam flow versus product compositions ($L=$ —, $V=$ - -).

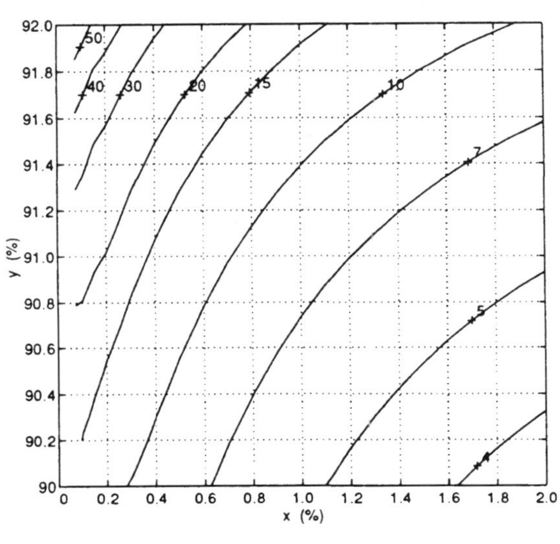

Fig. 12. Contour plot of relative gain versus product compositions.

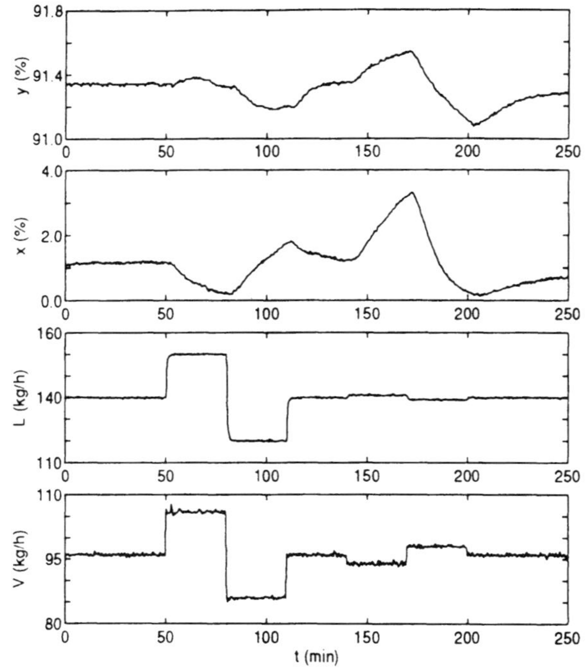

Fig. 13. Experimental illustration of directionality.

avoided if high product purities are desired. This implies that it is in practice difficult to maintain high product purities due to the ever present input uncertainty. The distillation column is, in other words, inherently different to control at high product purities.

4. EXPERIMENTAL VERIFICATION

The simulator was calibrated using data from four steady states, only one of which had the feed flow rate $F = 150$ kg/h and the feed composition $z = 25$ wt-%. Therefore, a number of experimental verifications were made.

One such experiment, which also illustrates the directionality properties, are shown in Fig. 13. The initial steady steady is $y = 91.35$ %, $x \approx 1$ %, $L = 140$ kg/h, $V = 96$ kg/h, which is in good agreement with Fig. 11. At $t = 50$ min, a sequence of three step changes satisfying $\Delta L/\Delta V = 2$ was initiated, and at $t = 140$ min, this was continued with a sequence of step changes with $\Delta L/\Delta V = -0.5$. These ratios were determined from flow gains, which were known from previous experiments. As can be seen, the first sequence of step changes results in a low-gain response, while the second sequence gives a high-gain response. The difference in gains is an order of a magnitude.

5. CONCLUSION

The results of an operability and controllability study of a pilot-scale distillation column separating a mixture of ethanol and water has been presented. In particular, ill-conditioned operating points characterized by strong directionality as well as sensitivity to model errors and input uncertainty were studied. An interesting result of the study is that the relative gain, which is related with robustness issues, and the condition number, which is more related with directionality and control specifications, can be conflicting measures of controllability. This means that they complement each other rather than substitute each other.

ACKNOWLEDGEMENT

Financial support from Tekes and the Academy of Finland is gratefully acknowledged.

REFERENCES

Andersen, H.W. and M. Kümmel (1992). Evaluating estimation of gain directionality. (Part 1 and 2.) *J. Proc. Cont.* **2**, 59–66, 67–86.

Bristol, E.H. (1966). On a new measure of interactions for multivariable process control. *IEEE Trans. Autom. Control* **AC-11**, 133–134.

Garcia, C.E. and M. Morari (1985). Internal model control. 2. Design procedure for multivariable systems. *Ind. Eng. Chem. Process Des. Dev.* **24**, 474–484.

Häggblom, K.E. (1991). Modeling of flow dynamics for control of distillation columns. *Proc. American Control Conference*, pp. 785–790, Boston, USA.

Häggblom, K.E. (1995). Static directionality of 2×2 systems: A control-relevant property with application to distillation. Submitted to the European Control Conference '95, September 5–8, Roma, Italy.

Häggblom, K.E., K.E. Wikman, S.E. Gustafsson and K.V. Waller (1984). Åbo Akademi's pilot-plant distillation column and its control system (in Swedish). *Proc. Finnish Automation Days '84*, pp. II/427–441, Espoo, Finland.

Koung, C.-W. and J.F. MacGregor (1993). Design of identification experiments for robust control. A geometric approach for bivariate processes. *Ind. Eng. Chem. Res.* **32**, 1658–1666.

Lehtinen, K. (1994). Investigation of the behaviour of a distillation column at different operating points (in Swedish). M.Sc. Thesis, Process Control Laboratory, Åbo Akademi University, Åbo, Finland.

Skogestad, S. and M. Morari (1987a). Implications of large RGA elements on control performance. *Ind. Eng. Chem. Res.* **26**, 2323–2330.

Skogestad, S. and M. Morari (1987b). Design of resilient processing plants — IX. Effect of model uncertainty on dynamic resilience. Chem. Eng. Sci. **42**, 1765–1780.

Skogestad, S., M. Morari and J.C. Doyle (1988). Robust control of ill-conditioned plants: high-purity distillation. *IEEE Trans. Autom. Control* **AC-33**, 1092–1105.

Waller, K.V. (1992). Experimental comparison of control structures. In: *Practical Distillation Control* (W.L. Luyben (Ed.)), pp. 313–330. Van Nostrand Reinhold, New York.

Waller, K.V., D.H. Finnerman, P.M. Sandelin, K.E. Häggblom and S.E. Gustafsson (1988). An experimental comparison of four control structures for two-point control of distillation. *Ind. Eng. Chem. Res.* **27**, 624–630.

CONTROL OF CHEMICAL REACTORS USING MULTIPLE-MODEL ADAPTIVE CONTROL (MMAC)

KEVIN D. SCHOTT* and B. WAYNE BEQUETTE*

**Rensselaer Polytechnic Institute, Howard P. Isermann Department of Chemical Engineering, Troy, NY 12180-3590, email: bequeb@rpi.edu*

Abstract. Multiple-Model Adaptive Control (MMAC) is shown to be an effective strategy for controlling nonlinear chemical processes. In this paper, MMAC is applied to two difficult control problems that may be encountered in continuous stirred tank reactors (CSTRs). In the first example, product concentration is controlled in an isothermal CSTR exhibiting input multiplicities and non-minimum phase behavior (Van de Vusse reaction). The MMAC configuration provides disturbance rejection where a fixed-parameter PID controller fails. The second example involves temperature control of a CSTR which has output multiplicities. While a single PID controller can stabilize the plant over a large operating range, its performance can be quite poor since the plant dynamics vary widely. MMAC is shown to give significantly better performance when moving from the open-loop stable regions to the open-loop unstable operating region.

Key Words. Nonlinear Control, model adaptive control, multiplicities.

1. INTRODUCTION

Nonlinear behaviour is a characteristic of chemical processes which presents a challenging control problem. In spite of the wealth of recent research results that have been presented in the literature (see Bequette (1991) for a review), few nonlinear model-based control techniques have been implemented in the process industry. One reason is the lack of nonlinear models to adequately characterize the process (either due to lack of knowledge or the time to develop the models). Another reason is that the nonlinear control methods are not well-understood by many of the engineers that would implement them. The goal of this paper is to present a method, Multiple-Model Adaptive Control (MMAC), which is conceptually straightforward and easily implemented without the requirement of extremely detailed process models.

MMAC is a model-based control strategy which incorporates a set of model/controller pairs rather than relying on a single model and controller to handle all possible operating conditions. It was developed for nonlinear control problems with varying parameters, particularly for situations where the plant was well known (and hence well modeled) in some operating regions, yet poorly known in others. The MMAC strategy was proposed in the early 1970's, yet is only now being applied to chemical processes (Schott and Bequette, 1994; Banerjee et al. 1994) and offers a new technique for managing difficult process control problems. Much of the early MMAC work

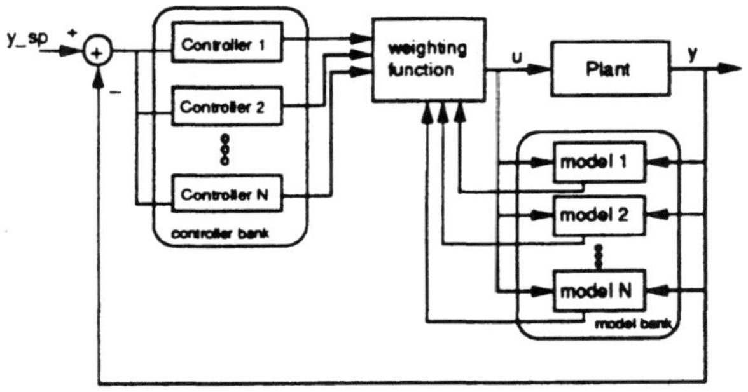

Fig. 1. MMAC system structure

focused on aircraft control (Athans et al. 1977). More recently, He et al. (1986) applied MMAC to blood pressure control and Yu et al. (1992) to simultaneous blood flow/blood pressure control in heart attack patients.

Figure 1 shows a schematic of the MMAC system structure. Instead of using a single, complex model to account for all possible plant operating-conditions, a bank of less complicated, less computationally intensive models are used and the need to explicitly account for all parameter changes in one model is avoided. Each model has a controller whose tuning parameters are based on that model. A weighting function chooses which model or combination of models best represents the current plant input/output behaviour, then produces a weighted-sum of the controller outputs to supply the control action.

This blending of models is a key feature of the MMAC method: if a precise plant model is not available for all operating regions, model-based control techniques may still be used without loss of performance in the regions that are well known. Some systems may operate around one particular setpoint for much of the time and can be described by just one of the models in the model bank. For this case, a single controller is chosen which by itself can adequately control the plant. However, as disturbances enter the system or different setpoints are requested (such as during start-up), the plant dynamics may change and thus require a controller which is tuned differently.

Although the parameters in all the model/controller pairs are fixed, MMAC is adaptive since different combinations of models are used to represent the plant as the plant parameters change. Because all of the models are always available, another MMAC feature is zero start-up time for plant identification. Nearly any control methodology (PID, LQR, NL-MPC, MIMO strategies, etc.) can be used within the MMAC framework. In this work we use PID plus first-order filters for all of the controllers in the controller bank.

2. IDENTIFICATION AND CONTROL WEIGHTING

In this study, the models in the model bank were developed by linearizing and discretizing the plant equations, yielding difference equations

$$\begin{aligned} y_k &= a_1 y_{k-1} + a_2 y_{k-2} + \dots \\ &\quad + b_0 u_k + b_1 u_{k-1} + \dots \end{aligned} \tag{1}$$

The actual plant inputs $u_k, u_{k-1}, \dots$ and outputs $y_k, y_{k-1}, \dots$ were used in calculating the model prediction, so the ith model prediction for time step k is given by

$$\begin{aligned} \hat{y}_{i,k} &= a_1 y_{k-1} + a_2 y_{k-2} + \dots \\ &\quad + b_0 u_k + b_1 u_{k-1} + \dots \end{aligned} \tag{2}$$

The *"weighting box"* of Figure 1 is implemented in several steps. First, the difference between the plant and each model is calculated

$$\epsilon_{i,k} = y_k - \hat{y}_{i,k} \tag{3}$$

A recursive version of Bayes' theorem

$$p_{i,k} = \frac{exp(-\epsilon_{i,k} K \epsilon_{i,k}) p_{i,k-1}}{\sum_{j=1}^{N} [exp(-\epsilon_{j,k} K \epsilon_{j,k}) p_{j,k-1}]} \tag{4}$$

is then used to estimate the probability that the ith model represents the plant at the kth sample. Here K is a parameter controlling the rate of convergence and N is the number of model/controller pairs. Because (4) is recursive, a zero probability for a particular model would force all subsequent probabilities for that model to zero. The probabilities are therefore bounded from reaching zero

$$\begin{aligned} P_{i,k} &= p_{i,k} && for\ p_{i,k} > \delta \\ P_{i,k} &= \delta && p_{i,k} \le 0 \\ &&& and\ \delta \le \frac{1}{N} \end{aligned} \tag{5}$$

This bounding also limits the number of past observations contained in the current estimation of the probability. As K increases, the current difference $\epsilon_{i,k}$ has a greater effect on $p_{i,k}$ than past differences. Finally, these lower-bounded probabilities need to be removed from the final weighting, so the probabilities are renormalized by

$$\begin{aligned} W_{i,k} &= \frac{P_{i,k}}{\sum_{j=1}^{N} P_{j,k}} && for\ P_{i,k} > \delta \\ W_{i,k} &= 0 && P_{i,k} \le \delta \end{aligned} \tag{6}$$

and only $P_{j,k} > \delta$ are included in the summation. The final, incremental control-signal change for the kth sample instant is

$$\Delta u_k = \sum_{i=1}^{N} W_{i,k} \Delta u_{i,k} \tag{7}$$

In both of the examples which follow, discrete velocity-form PID controllers were used, augmented with a first-order filter. Controller tuning parameters were established using the IMC-based PID controller-tuning method with each linear model.

3. EXAMPLE 1. INPUT MULTIPLICITIES

For a Van de Vusse reaction the following parallel and consecutive reactions are taking place

$$A \xrightarrow{k_1} B \xrightarrow{k_2} C \quad \text{and} \quad A + A \xrightarrow{k_3} D$$

The modeling equations are given by

$$\frac{dx_1}{dt} = -k_1 x_1 - k_3 x_1^2 + (x_{1f} - x_1) u \tag{8}$$

$$\frac{dx_2}{dt} = k_1 x_1 - k_2 x_2 - x_2 u \tag{9}$$

where x_1 and x_2 are state variables representing concentrations and k_1, k_2, and k_3 are the rate constants. The concentration x_2 is to be controlled using the dilution rate u as the manipulated variable. For this example, x_2 is the plant output y. Figure 2 shows the steady-state input-output behavior of the system. Parameter values for this example are listed in Table 1.

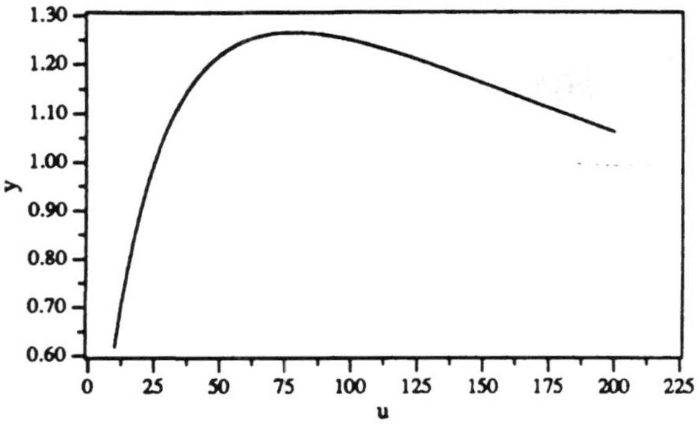

Fig. 2. Steady-state input-output relationship for CSTR with Van de Vusse reaction

Single PID Control. If production of B is to be maximized, then the system must operate as closely to the peak of the input-output curve as possible. However, noise or disturbances may cause the system to shift from one side of the input-output curve peak to the other. The major limitation when using a fixed-parameter PID controller on a system exhibiting input multiplicities is the controller's inability to account for the sign-change of the plant gain should the system cross the peak of the I/O curve. In this particular system, a disturbance in feed concentration shifts the I/O curve lower, forcing what was once a valid setpoint to become unattainable. In such a circumstance, a controller designed on one side of the I/O curve will itself push the system away from its setpoint because the plant and controller have opposite signs. Figure 3 shows the response of a PID controller designed at $u = 65$ as the feed concentration to the reactor is suddenly changed from $x_{1f} = 10$ to $x_{1f} = 9.75$ $\frac{gmol}{l}$ for $t = 0.2$ to 0.37. Once the system moves to the right side of the I/O curve ($u > \sim 77.5$), the single PID drives the system in the wrong direction. Even when the disturbance is removed, the controller cannot recover and the manipulated variable continues to increase until a system constraint is encountered ($u = 200$).

MMAC. Figure 4 shows the response of the MMAC strategy using three models and subjected to the same rectangular-pulse disturbance as in Figure 3. Models were chosen corresponding to dilution rates of 45, 65, and 100 h^{-1}. The MMAC configuration correctly identifies that the system gain has changed sign and then uses the controller with the correct gain to move the system back toward its setpoint. Since the setpoint is still unattainable, oscillations occur as the system gets pushed back and forth between the controllers at the peak. Once the disturbance has ended, the system returns to its original setpoint on the correct side of the input-output curve.

While the oscillations cannot be completely removed, they can be significantly reduced by reducing the interval between models near the peak. The results of a more comprehensive MMAC implementation using 12 models are shown in figure 5. The models are evenly spaced in Δu intervals of 5 from $u = 45$ to $u = 100$. In this case we are attempting to operate directly at the peak of the I/O curve. The first setpoint ($y_{sp} = 1.25$) falls directly on model 4 and the weighting goes primarily to that model. At $t = 2.5$, the setpoint is changed to $y_{sp} = 1.275$, above the I/O curve and unattainable by the system. The plant output responds by moving to the highest attainable output, then starting to oscillate. For the remainder of the run, the setpoint is set just slightly above the peak of the I/O curve, forcing the system to run at its highest achievable productivity. Between $t = 6$ and $t = 8$, a disturbance of $x_{1f} = 9.75$ enters the system. Since the difference between the setpoint and the output is large the system oscillates rapidly between the models closest to the peak. However, because the distance between the models is small, the oscillation amplitude remains small. Although the oscillations may be seen in the manipulated variable, they are imperceivable at the output since the process gain is so small near the peak. Unlike the previous case, here when the disturbance ended the system went to the right side of the I/O curve.

4. EXAMPLE 2. OUTPUT MULTIPLICITIES

The dimensionless modeling equations for a two-state exothermic CSTR for $A \longrightarrow B$ are given by

$$\frac{dx_1}{d\tau} = -\phi x_1 \kappa(x_2) + q(x_{1f} - x_1) \quad (10)$$

$$\frac{dx_2}{d\tau} = \beta\phi x_1 \kappa(x_2) - (q + \delta)x_2 + \ldots$$
$$\delta u + q x_{2f} \quad (11)$$

$$\kappa(x_2) = exp\left(\frac{x_2}{1 + \frac{x_2}{\gamma}}\right) \quad (12)$$

where x_1, and x_2 are dimensionless concentration and reactor temperature and u is the dimensionless jacket temperature. Definitions of the dimensionless parameters and the parameters values used for this example are listed in Tables 2 and 3. Figure 6 shows the steady-state input-output behavior.

Single PID. Although a single PID controller can be tuned to function in both the open-loop stable and OL unstable operating regions, its overall performance may be quite poor. A single PID tuned for operation in an OL stable region can perform well for small setpoint changes close to where the controller is designed to operate, yet the same controller exhibits large overshoot when stepping further away into the OL unstable region (Figure 7a).

MMAC. Four linear models were chosen, corresponding to dimensionless reactor temperatures of 1.30, 2.00, 2.25, and 5.0. Model one is in the lower OL stable branch, models 2 and 3 are in the middle OL unstable region, and model 4 is in the upper, OL stable branch. Figure 7 shows that for the same setpoint change, even this simple implementation of MMAC is able to completely remove the overshoot seen in the single PID case. While such a large setpoint change may not be made routinely, this example shows how MMAC may be used to improve system performance in non-routine situations such as start-up, without sacrificing performance at the system's usual setpoint.

5. SUMMARY AND CURRENT WORK

Presented here are basic results which show that Multiple-Model Adaptive Control is a viable strategy for controlling nonlinear chemical reactors. The use of different models for different operating conditions allows the integration of many control objectives into a single method. Such flexibility is clearly an advantage in cases such as start-up and batch-processing where comprehensive plant models may be difficult to obtain. Currently we are applying MMAC to a batch reaction control problem where the overall heat transfer coefficient changes over the course of the batch and fouling causes additional batch-to-batch variations in the heat transfer coefficient. We are also applying robust performance concepts to gain more insight on estimating the number and spacing of models to achieve a pre-specified performance goal.

REFERENCES

Athans, M., D. Castanon, K-P. Dunn, C.S. Greene, W.H. Lee, N.R. Sandell, and A.S. Willsky (1977). The Stochastic Control of the F-8C Aircraft Using a Multiple-Model Adaptive Control (MMAC) Method - Part I: Equilibrium Flight. *IEEE Trans. Automat. Control*, **AC-22 (5)**, 768-780

Banerjee A., Y. Arkun, B. Ogunnaike, and R. Pearson (1994). Robust Nonlinear Control by Scheduling Multiple Model Based Controllers. 1994 AIChE Annual Meeting, paper 230a, San Francisco, CA

Bequette, B.W. (1991). Nonlinear control of chemical processes: A review. *Ind. Eng. Chem. Res.*, **30**, 1391-1413.

He, W.G., H. Kaufman, and R.J. Roy (1986). Multiple-Model Adaptive Control Procedures for Blood Pressure Control. *IEEE Trans. Biomed. Eng.*, **BME-33 (1)**, 10-19

Schott, K.D. and B.W. Bequette (1994). Control of Nonlinear Chemical Processes Using Multiple-Model Adaptive Control (MMAC). 1994 AIChE Annual Meeting, paper 224j, San Francisco, CA

Yu, C.L., R.J. Roy, H. Kaufman, and B.W. Bequette (1992). Multiple-Model Adaptive Control Predictive Control of Mean Arterial Pressure and Cardiac Output. *IEEE Trans. Biomed. Eng.*, **39 (8)**, 765-778

Table 1. Parameter values for Van de Vusse reaction simulations

$$k_1 = 50.0\ h^{-1}$$
$$k_2 = 100.0\ h^{-1}$$
$$k_3 = 10.0\ \frac{l}{gmol\ h}$$
$$x_{1f} = 10.0\ \frac{gmol}{l}$$

Table 2. Dimensionless variables for the CSTR model

ϕ	Nominal Damkohler number
β	Dimensionless heat of reaction
δ	Dimensionless heat transfer coefficient
x_{2f}	Dimensionless reactor feed temperature
q	Dimensionless reactor flowrate
δ_1	Volume ratio
γ	Dimensionless activation energy

Table 3. Parameter values for two-state exothermic CSTR simulations

$$\phi = 0.072$$
$$\beta = 8.0$$
$$\delta = 0.3$$
$$x_{2f} = 0.0$$
$$q = 1.0$$
$$\delta_1 = 0.3$$
$$\gamma = 20$$

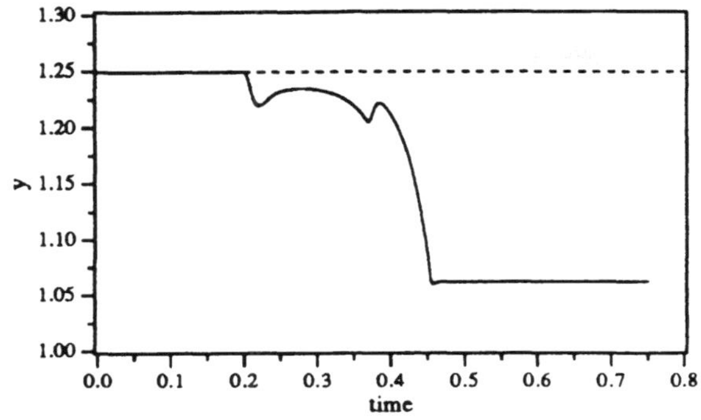

Fig. 3. a. Output

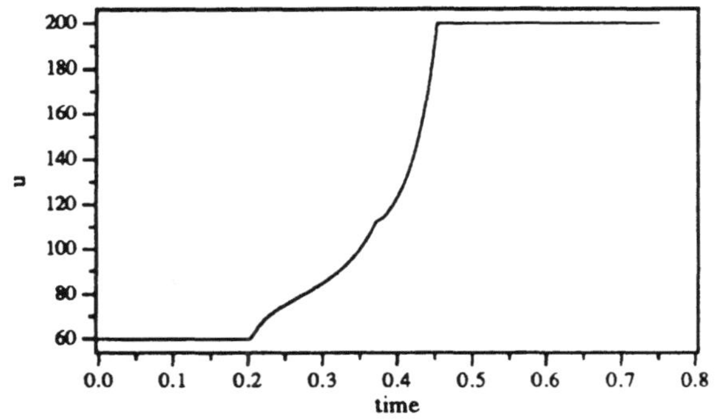

Fig. 3. b. Input; Single PID response to a rectangular pulse disturbance

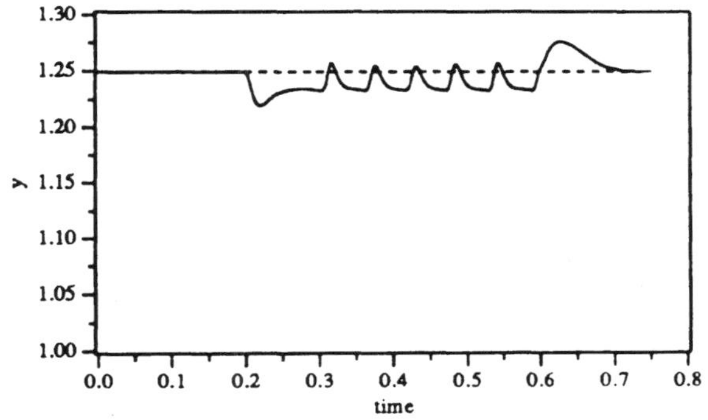

Fig. 4. a. Input

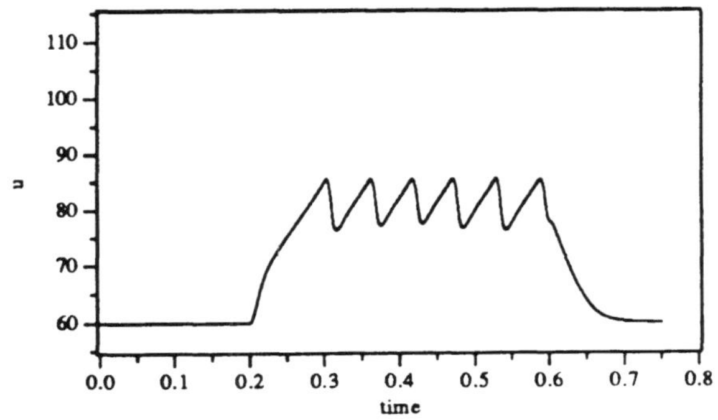

Fig. 4. b. Output

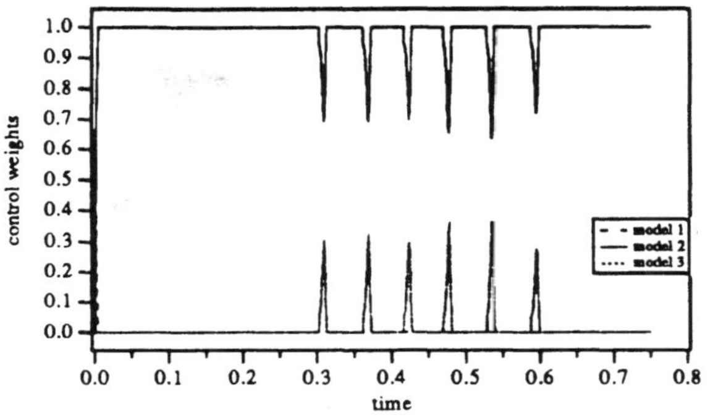

Fig. 4. c. Model weights; MMAC response to the same rectangular pulse as figure 3

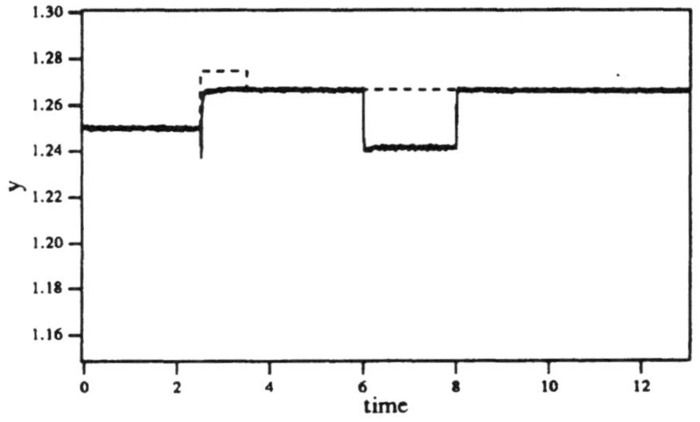

Fig. 5. a. Input

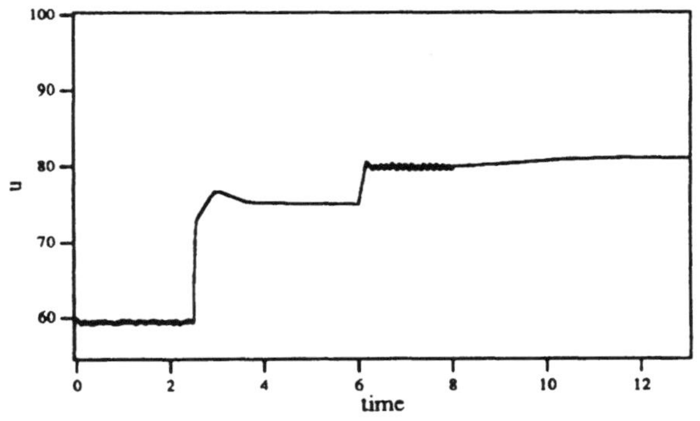

Fig. 5. b. Output

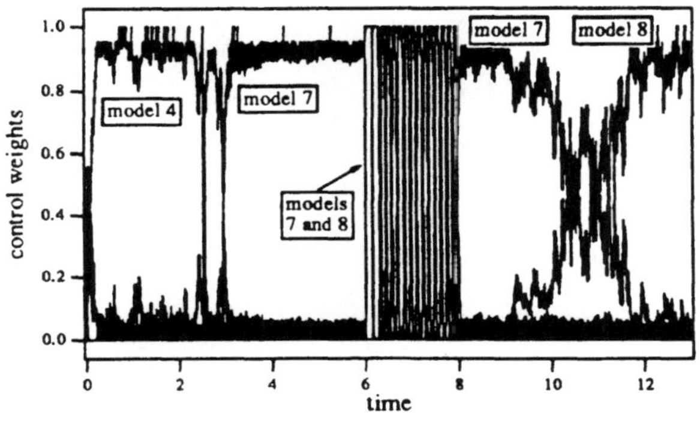

Fig. 5. c. Model weights; MMAC 12-model implementation

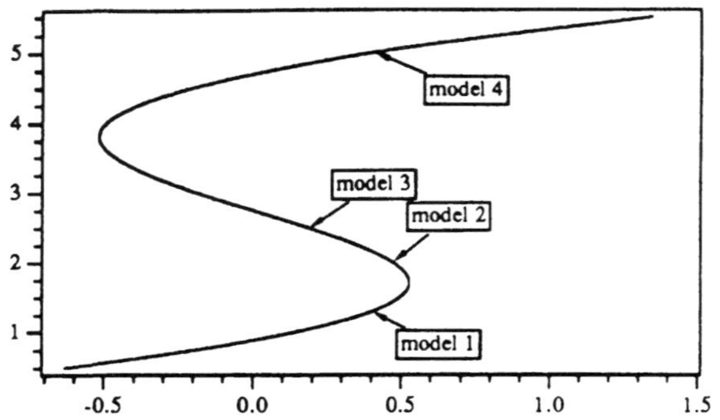

Fig. 6. Steady-state input-output relationship for an exothermic CSTR with output multiplicities. reactor temperature: y ; jacket temperature: u

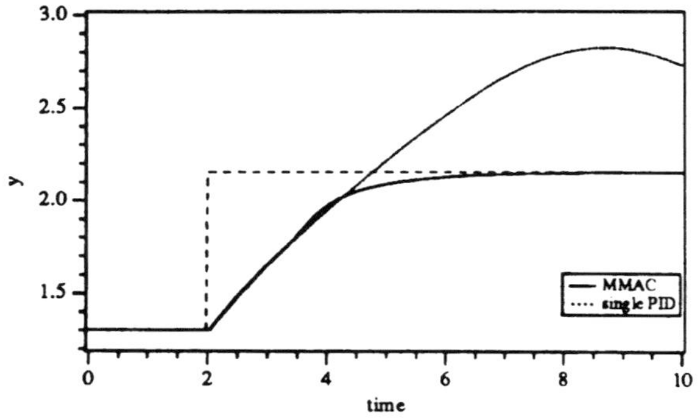

Fig. 7. a. Input

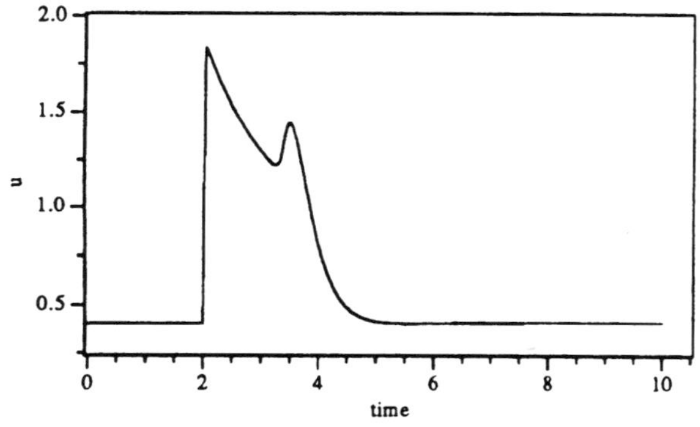

Fig. 7. b. Output

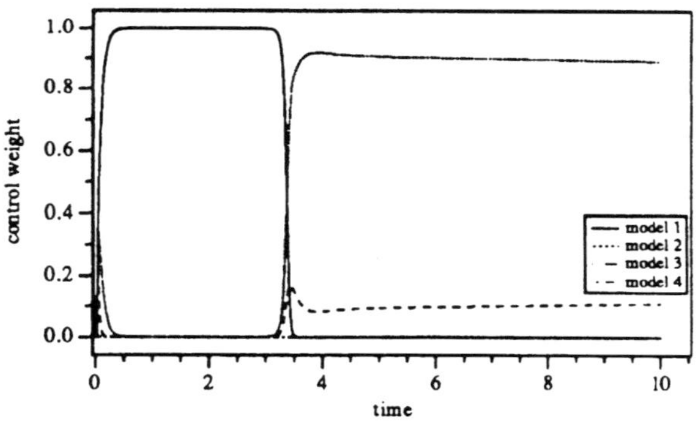

Fig. 7. c. Model weights; Performance comparison of MMAC and a single PID for the same step change in setpoint

FLEXIBILITY ANALYSIS OF DYNAMIC CHEMICAL PROCESSES

V.D. Dimitriadis* and E.N. Pistikopoulos*

* *Centre for Process Systems Engineering, Department of Chemical Engineering, Imperial College, London SW7 2BY, UK*

Abstract. This paper considers the development of a framework for the assessment of *feasibility* and *flexibility* of dynamic chemical processes. The dynamic feasibility and dynamic flexibility analysis problems are formulated as two-stage semi-infinite dynamic optimization problems. It is shown that, when the uncertainty considered is time-invariant or its profile over time is known, both formulations reduce to dynamic optimization problems that can be solved using currently available techniques. The applicability of the approach is illustrated with an example from the area of batch reactors operation.

Key Words. Flexible, Dynamic systems, Chemical industry

1. INTRODUCTION

Flexibility, *i.e.* the ability of a system to maintain feasible operation over a range of conditions, is a key requirement for chemical processes operating in an uncertain environment. This is especially true when dynamics dominate the behaviour of a process, as is the case in batch processes but also in continuous processes during start-up, shut-down or transitions between different operating points. Such processes are inherently dynamic and can be subject to time-invariant as well as time-varying uncertainty.

During the last decade, a lot of research work has focused on providing the necessary theoretical background for the flexibility analysis of chemical processes and incorporating flexibility objectives into the corresponding design procedures (Grossmann and Straub, 1991). Most of this work has concentrated on processes that operate at steady state and can therefore be mathematically described by sets of algebraic equalities and inequalities.

Morari (1983) emphasized the need for a systematic method to assess the dynamic operability characteristics of chemical processes and incorporate such considerations at the design stage. He proposed a general framework for the evaluation of the *dynamic resilience* of process plants that relies on the fact that certain system-related factors pose a limit on the achievable closed-loop behaviour of a process. These factors are (i) non-minimum phase elements (*i.e.* right half plane zeros and time delays), (ii) constraints on the available control action and (iii) mismatch between the actual plant and its mathematical model. However, this applies to process models that are linearized around a specified steady state operating point and can only be used to provide a basis for comparison at the early design stages.

Walsh (1993) studied the effect of uncertainty on the dynamic behaviour of chemical processes as part of an iterative design procedure for waste water neutralization systems. The analysis procedure he proposed determines the ability of a given system to tolerate (i) bounded variations for the parametric uncertainty and (ii) *worst-case* scenarios for the dynamic uncertainty (disturbances).

More recently, dynamic optimization was used to tackle the feasibility problem of dynamic systems in contexts as divert as switchability analysis (White *et al.*, 1994) and safety verification (Dimitriadis *et al.*, 1994). The aforementioned approaches share the assumption of a system model with no degrees of freedom and are thus applicable to systems where the control scheme has already been determined.

This paper presents a framework for the assessment of *feasibility* and *flexibility* of dynamic systems *i.e.* of systems that are described by sets of differential and algebraic equations and are subject to time-varying uncertainty. Based on the steady state flexibility analysis framework (Swaney and Grossmann, 1985), a formulation for the dynamic feasibility problem is presented and a dynamic flexibility index is defined. It is shown that when the uncertainty considered is time-invariant or its profile over time is known (*e.g.* when critical points that limit feasibility lie on the vertices of the time-varying uncertainty space), both problems reduce to dynamic optimization problems that can be solved using currently available techniques. An example from the area of batch reactors operation is used to illustrate the applicability of the proposed approach.

2. DYNAMIC FEASIBILITY/FLEXIBILITY ANALYSIS

In the context of this analysis, the behaviour of a dynamic chemical process is described by a system of differential and algebraic equations, the feasibility of operation being determined by a set of operational inequality constraints which can be either path constraints (constraints that must hold throughout the period of operation) and/or point constraints (constraints that must hold at certain time instances within the period of operation):

$$\mathbf{h}(\mathbf{d}, \dot{\mathbf{x}}(t), \mathbf{x}(t), \mathbf{z}(t), \theta(t), t) = \mathbf{0}$$

$$\mathbf{x}(0) = \mathbf{x}^0$$

$$\mathbf{g}^{path}(\mathbf{d}, \mathbf{x}(t), \mathbf{z}(t), \theta(t), t) \leq \mathbf{0}$$

$$\mathbf{g}_k^{point}(\mathbf{d}, \mathbf{x}(t^k), \mathbf{z}(t^k), \theta(t^k), t^k) \leq \mathbf{0}\ ,\ k = 1..NP$$

where $\mathbf{d}$ is the vector of the design variables that remains constant at the operation stage, $\mathbf{x}(t)$ and $\mathbf{z}(t)$ are the vectors corresponding to the time-dependent state and control variables and $\theta(t)$ are the time-invariant and/or time-dependent uncertain parameters of the process respectively. Time-independent uncertainty is associated with the values of process parameters (*e.g.* heats of reaction, transfer coefficients, *etc.*) whereas time-dependent uncertainty relates to process disturbances that vary continuously throughout the period of operation (*e.g.* flows, temperatures, *etc.*). Control variables correspond to degrees of freedom that can be adjusted dynamically to mitigate against this potential uncertainty. Under these assumptions, the dynamic feasibility problem may be stated as follows:

> Establish if, for *every* possible profile of the uncertain vector, $\theta(t)$, there is *at least one* profile of the control vector, $\mathbf{z}(t)$, such that the process feasibility constraints are satisfied at *any* time $t \in [0, H]$.

This statement is mathematically equivalent to the following max-min-max optimization problem (Halemane and Grossmann, 1983):

[Problem DFT1]

$$\chi(\mathbf{d}) = \max_{\theta(t)\in T(t)} \ \min_{\mathbf{z}(t)\in Z(t)} \ \max_{j\in J, t\in[0,H]} g_j$$

$$\mathbf{h}(\mathbf{d}, \dot{\mathbf{x}}(t), \mathbf{x}(t), \mathbf{z}(t), \theta(t), t) = \mathbf{0}$$

$$\mathbf{x}(0) = \mathbf{x}^0$$

$$T(t) = \{\theta(t) \mid \theta^L(t) \leq \theta(t) \leq \theta^U(t)\}$$

$$Z(t) = \{\mathbf{z}(t) \mid \mathbf{z}^L(t) \leq \mathbf{z}(t) \leq \mathbf{z}^U(t)\}$$

where J is the index set for the feasibility constraints (for notational simplicity, both path and point constraints have been combined into one feasibility constraints vector, $\mathbf{g}$). Problem (DFT1) is the *dynamic feasibility problem* or *dynamic flexibility test problem.* If $\chi(\mathbf{d}) \leq 0$, the proposed design is dynamically feasible in $T(t)$. Otherwise, there exists at least one possible profile, $\theta(t)$, for which no corresponding control action, $\mathbf{z}(t)$, can be found to keep the system feasible at all times within the horizon considered.

The feasibility problem is directly related to the flexibility evaluation problem. The flexibility index, F, defined by Swaney and Grossmann (1985), quantifies the flexibility of a process by establishing the maximum parameter range over which the process can operate feasibly. In the case of dynamic processes, however, the feasible region of operation, and consequently F, change with time. By defining a time horizon, H, the *dynamic flexibility index evaluation problem* can be formulated as follows:

[Problem DFI1]

$$DF(\mathbf{d}) = \max \delta$$

$$\chi(\mathbf{d}) = \max_{\theta(t)\in T(\delta,t)} \ \min_{\mathbf{z}(t)\in Z(t)} \ \max_{j\in J, t\in[0,H]} g_j \leq 0$$

$$\mathbf{h}(\mathbf{d}, \dot{\mathbf{x}}(t), \mathbf{x}(t), \mathbf{z}(t), \theta(t), t) = \mathbf{0}$$

$$\mathbf{x}(0) = \mathbf{x}^0$$

$$\delta \geq 0$$

$$T(\delta, t) = \{\theta(t) \mid \theta^N(t) - \delta\Delta\theta^-(t) \leq \theta(t) \leq$$

$$\leq \theta^N(t) + \delta\Delta\theta^+(t)\}$$

$$Z(t) = \{\mathbf{z}(t) \mid \mathbf{z}^L(t) \leq \mathbf{z}(t) \leq \mathbf{z}^U(t)\}$$

Qualitatively, the *dynamic flexibility index*, DF, represents the largest scaled deviation of the uncertain parameter profile that the design can tolerate while remaining feasible within the horizon considered. Deviations are measured from a nominal profile, $\theta^N(t)$, and are scaled using the expected deviation profiles $\Delta\theta^-(t)$ and $\Delta\theta^+(t)$. Normally:

$$\theta^N(t) - \delta\Delta\theta^-(t) = \theta^L(t)$$

$$\theta^N(t) + \delta\Delta\theta^+(t) = \theta^U(t)$$

3. FEASIBILITY/FLEXIBILITY ANALYSIS WITH GIVEN UNCERTAINTY PROFILE

The dynamic feasibility problem, (DFT1), is greatly simplified when the uncertainty considered is not time-dependent *or* when its variation with time is known (*e.g.* when the critical points that limit feasibility are assumed to lie on the vertices of the time-varying uncertainty space). Then, (DFT1), reduces to a dynamic optimization problem of the following form:

[Problem DFT2]

$$\chi(\mathbf{d}) = \min_{u,\mathbf{z}(t)\in Z(t),t\in[0,H]} u$$

$$\mathbf{h}(\mathbf{d},\dot{\mathbf{x}}(t),\mathbf{x}(t),\mathbf{z}(t),\theta(t),t) = \mathbf{0}$$

$$\mathbf{x}(0) = \mathbf{x}^0$$

$$\mathbf{g}(\mathbf{d},\mathbf{x}(t),\mathbf{z}(t),\theta(t),t) \leq u$$

$$Z(t) = \{\mathbf{z}(t) \mid \mathbf{z}^L(t) \leq \mathbf{z}(t) \leq \mathbf{z}^U(t)\}$$

The dynamic flexibility index evaluation problem can also be transformed into a dynamic optimization problem by assuming that the direction, $\Delta\theta^c(t)$, from the nominal profile, $\theta^N(t)$, on which the critical point that limits flexibility lies, is either known or is assumed to be one of the vertex directions of the time-varying uncertainty space. Then, problem (DFI1) takes the following form:

[Problem DFI2]

$$DF(\mathbf{d}) = \max_{\delta,\mathbf{z}(t)\in Z(t),t\in[0,H]} \delta$$

$$\mathbf{h}(\mathbf{d},\dot{\mathbf{x}}(t),\mathbf{x}(t),\mathbf{z}(t),\theta(t),t) = \mathbf{0}$$

$$\mathbf{x}(0) = \mathbf{x}^0$$

$$\mathbf{g}(\mathbf{d},\mathbf{x}(t),\mathbf{z}(t),\theta(t),t) \leq 0$$

$$\theta(t) = \theta^N(t) + \delta\Delta\theta^c(t)$$

$$\delta \geq 0$$

$$Z(t) = \{\mathbf{z}(t) \mid \mathbf{z}^L(t) \leq \mathbf{z}(t) \leq \mathbf{z}^U(t)\}$$

An example problem, concerned with the operation of a batch reactor in which the heat of reaction is an uncertain parameter, is considered below to illustrate the applicability of formulations

Table 1 Mathematical model and data for batch reactor example

$$\frac{dN_A}{dt} = -kN_A$$

$$\frac{dN_B}{dt} = kN_A$$

$$C_P(N_A + N_B)\frac{dT}{dt} = kN_A(-\Delta H_r) - Q_{CW}$$

$$Q_{CW} = F_{CW}C_{P,CW}\Delta T_{CW}$$

$$k = k_0 e^{-\frac{E}{RT}}$$

$$N_A(0) = 20000\ mol$$

$$N_B(0) = 0$$

$$T(0) = 298\ K$$

$$k_0 = 6\ s^{-1} \quad , \quad E = 28750\ J\ mol^{-1}$$

$$R = 8.314\ J\ mol^{-1}\ K^{-1}$$

$$C_P = 250\ J\ mol^{-1}\ K^{-1}$$

$$C_{P,CW} = 4200\ J\ kg^{-1}\ K^{-1}$$

$$\Delta T_{CW} = 30\ K$$

$$0 \leq F_C W \leq 2\ kg\ s^{-1}$$

(DFT2) and (DFI2).

4. EXAMPLE

A highly exothermic, first-order reaction ($A \rightarrow B$) is carried out in a jacketed batch reactor. The mathematical model that is used to describe the system is given in Table 1. The heat of reaction, ΔH_r, is not known exactly but is considered to vary between $-72500\ J/mol$ and $-82500\ J/mol$ with an expected nominal value of $-77500\ J/mol$. The cooling water flowrate, F_{CW}, can be adjusted during the operation of the reactor so as to satisfy operational constraints and end-time specifications and can therefore be regarded as an available control variable.

The following requirements must be satisfied:

1. The final reactant conversion must exceed

95%, *i.e.*

$$\frac{N_A(t=H)}{N_A(t=0)} \leq 0.05$$

2. Due to safety considerations, the temperature in the reactor must not exceed a maximum of 400 K at any time during operation, *i.e.*

$$T(t) \leq 400\ K \quad , \quad 0 \leq t \leq H$$

3. The temperature of the reacting mixture at the end of the operation must not exceed 308 K, so that discharge and cleaning operations can take place, *i.e.*

$$T(t=H) \leq 308\ K$$

Given that the duration of each batch cycle is 2 hours, it is desirable to determine if the reactor can accommodate the whole range of uncertainty while operating feasibly according to the aforementioned constraints. Note that the uncertainty considered in this example is parametric and hence time-invariant. However, steady state flexibility analysis techniques cannot deal with the problem directly because the process is inherently dynamic.

The dynamic flexibility index evaluation problem, (DFI2), was formulated as a dynamic optimization problem and solved using the *DAEOPT* optimal control code (Vassiliadis *et al.*, 1994*a*; Vassiliadis *et al.*, 1994*b*). *DAEOPT* partitions the time horizon into a specified number of control intervals and utilizes low order polynomial functions to parameterize the control variables within each interval. Equality and inequality path constraints are dealt with efficiently by using point constraints as well as forcing an integral measure of their violation to zero. On the example, the time horizon was divided into six intervals of variable length and the profile of the cooling water flowrate, F_{CW}, was assumed to be piecewise constant over these intervals. Along the positive vertex direction, it is found that the system can handle 78 % of the expected uncertainty range ($DF^+ = 0.7788$, $\Delta H_r^{crit,+} = 81394\ J/mol$). Similarly, along the negative vertex direction, $DF^- = 7.2384$ and $\Delta H_r^{crit,-} = 41308\ J/mol$. The dynamic flexibility index of the system is therefore $DF = 0.7788$ and the critical point of operation corresponds to a value of $\Delta H_r^{crit} = 81394\ J/mol$. The required profile of the cooling water flowrate for the system to operate feasibly at this critical point is shown in Figure 1. In this case the process constraints on final conversion and final reactor temperature are marginally satisfied as shown in Figures 2 and 3.

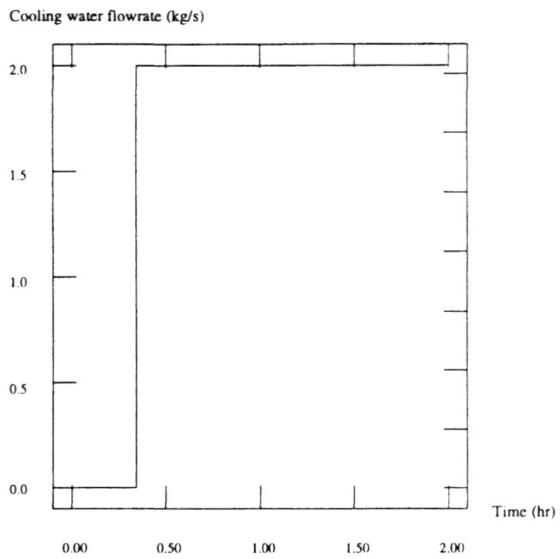

Fig. 1. Cooling water flowrate vs time at critical point for batch reactor example

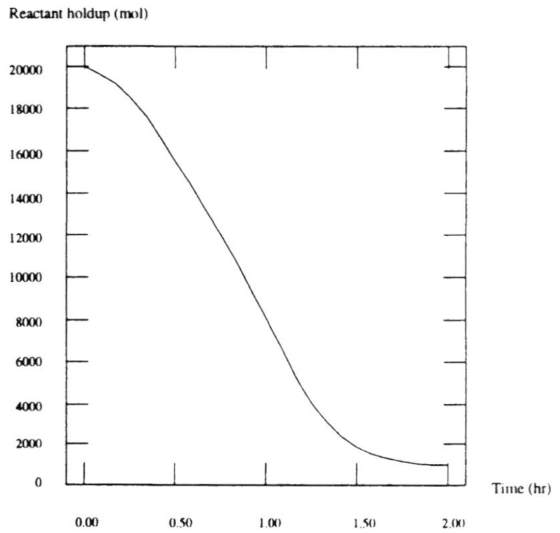

Fig. 2. Reactant holdup vs time at critical point for batch reactor example

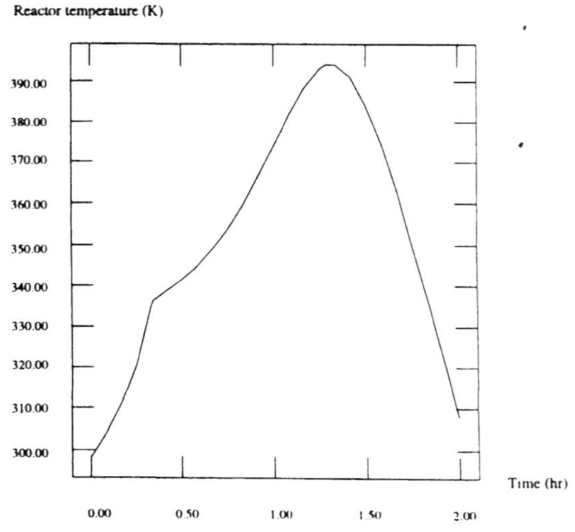

Fig. 3. Temperature vs time at critical point for batch reactor example

5. CONCLUDING REMARKS

A systematic framework has been proposed in this paper for the assessment of feasibility and flexibility of dynamic systems. The dynamic feasibility and flexibility analysis problems were formulated as two-stage, semi-infinite, dynamic optimization problems with an infinite number of decision variables. For time-invariant uncertainty and/or given uncertainty profile over time, both problems reduce to dynamic optimization problems that can be solved using currently available techniques. This was illustrated using an example problem.

When neither of the above assumptions holds, a discretization scheme for the differential equations can be combined with an active constraint strategy to transform both problems into Mixed Integer Non-Linear Programming (MINLP) problems (Dimitriadis and Pistikopoulos, 1994).

The proposed approach extends the steady state flexibility analysis framework to incorporate the time element and has important applications in the areas of operability analysis, control structure selection and controller tuning for dynamic chemical processes (Mohideen *et al.*, 1995).

6. REFERENCES

Dimitriadis, V.D. and E.N. Pistikopoulos (1994). Flexibility Analysis of Dynamic Systems. Submitted for publication to Ind. Eng. Chem. Res.

Dimitriadis, V.D., N. Shah and C.C. Pantelides (1994). Model-Based Safety Verification of Discrete/Continuous Chemical Processes. Presented at AIChE Annual Meeting, San Fransisco.

Grossmann, I.E. and D.A. Straub (1991). Recent Developments in the Evaluation and Optimization of Flexible Chemical Processes. In: *Computer Oriented Process Engineering.* Barcelona.

Halemane, K.P. and I.E. Grossmann (1983). Optimal Process Design under Uncertainty. *AIChE J.* **29**, 425.

Mohideen, M.J., J.D. Perkins and E.N. Pistikopoulos (1995). Incorporation of Flexibility and Controllability in Optimal Design of Distillation Columns. To be presented at the IChemE Annual Meeting, Edinburgh.

Morari, M. (1983). Design of Resilient Processing Plants III - A General Framework for the Assessment of Dynamic Resilience. *Chem. Eng. Sci.* **38**, 1881.

Swaney, R.E. and I.E. Grossmann (1985). An Index for Operational Flexibility in Chemical Process Design. Part I : Formulation and Theory. *AIChE J.* **31**, 621.

Vassiliadis, V.S., R.W.H Sargent and C.C. Pantelides (1994*a*). Solution of a Class of Multistage Dynamic Optimization Problems. 1. Problems without Path Constraints. *Ind. Eng. Chem. Res.* **33**, 2111.

Vassiliadis, V.S., R.W.H Sargent and C.C. Pantelides (1994*b*). Solution of a Class of Multistage Dynamic Optimization Problems. 2. Problems with Path Constraints. *Ind. Eng. Chem. Res.* **33**, 2123.

Walsh, S. (1993). Integrated Design of Chemical Waste Water Treatment Systems. PhD thesis. University of London.

White, V., J.D. Perkins and D.M. Espie (1994). Switchability Analysis. Presented at IFAC Workshop on Integration of Process Design and Control, Baltimore.

CONTROL OF A BATCH REACTOR USING A MULTI-VARIATE STATISTICAL CONTROLLER DESIGN

M. J. Piovoso*, K. S. Dahl* and K. A. Kosanovich**

*E. I. duPont de Nemours & Co., Central Science & Engineering, Wilmington, DE. 19880, USA
**Department of Chemical Engineering, University of South Carolina, Columbia, SC 29208, USA

Abstract. Batch processes play an important role in the chemical, pharmaceutical, and agricultural industries. However, the main characteristics of batch processes, flexibility, finite duration, and nonlinear behavior are associated with both their success and their incompatibility with the usual techniques for control. The batch reactor studied indirectly controls the extent of reaction by using reactor pressure to control temperature which is as an inferential measurement of extent of reaction. The manipulated variable is the flow of the heating or cooling medium through a split range valve. However, just controlling to a reactor temperature profile will not necessarily reduce product non-uniformity and batch-to-batch variability. In this paper, it is proposed to control the reactor via control actions calculated in a reduced representation space of a principal component representation, the score space. This space is determined by the interrelationships among the process variables using Multi-way Principal Component Analysis (MPCA); the control action that will regulate the scores to the nominal score trajectories are calculated. The development of the MPCA representation of the process and the associated controller based on score space data are presented. The performance of this controller formulation in the face of a typical heat transfer disturbance is illustrated.

Key Words. Batch Reactor; Modelling; Multi-way Principal Component Analysis; Multivariate Statistical Controller

1. INTRODUCTION

Batch and semi-batch processes play an important role in the chemical industry due to their low volume - high value products. Examples include reactors, crystallization, distillation, injection molding processes and the manufacture of polymers. Batch processes are characterized by a prescribed processing of materials for a finite duration. Successful operation means tracking a prescribed recipe with high degree of reproducibility from batch to batch. Feed rate, temperature, and pressure profiles are implemented with servo-controllers, and precise sequencing operations are implemented with tools such as programmable logic controllers.

Chemical businesses have a strong desire to increase throughput, uniformity, and quality of products manufactured in batch processes to improve process performance and to maintain critical market share. In order to meet these goals we need to obtain a better understanding of our processes and to identify the potential for improvements. Failure to attain on-aim control of product properties may lead to increased manufacturing costs and loss of market share. The critical properties that define the quality of the product are determined by posterior laboratory analysis. It is not unusual for a large elapsed time (8 hours) to occur between the sampling time and when the results are returned. Timely feedback control based on these results is not possible because of the large delays, although feedback control in a statistical process control sense may be possible.

In an earlier work, Piovoso and Kosanovich (1), showed that an inferential Principal Component Regression (PCR) controller developed from multiple tray temperature measurements could provide better control of the distillate composition in a high purity binary column in contrast to the standard single tray temperature controller. In that study they used Principal Component Analysis (PCA) to generate the scores that are used in the PCR controller formulation. Kosanovich et. al. (2) and Nomikos and MacGregor (3) have illustrated the data analysis capability of Multi-way PCA (MPCA) using single product code taken from an industrial batch process. The analysis was able to point out the major sources of batch-to-batch variability which led to improvements in reactor temperature control and ultimately process performance and product uniformity (4).

In this work, a simulation of a batch reactor (5) is used to develop an MPCA model and to illustrate the score space controller formulation. The paper is organized as follows: a brief overview of MPCA and the score controller formulation are provided.

Second, we describe the batch reactor process and the model development. Third, results of MPCA analysis and the controller performance in the face of a common process disturbance is presented and compared to that of the existing control scheme. Finally, we suggest improvements to incorporate dynamics into the reduced representation and controller formulation.

2. MPCA METHOD

MPCA is an extension of PCA because data from a batch system is three dimensional, that is, in addition to measured variables and time, batch systems have an added dimension of the batch number. To understand MPCA, a brief review of PCA is necessary; the interested reader is referred to the literature references cited for more details (6,7). PCA allows for the decomposition of a single, dependent set of data into a transformed space defined by the eigenvectors of the covariance of the data set. If the data are correlated, their information content can be represented by fewer numbers. PCA performs this reduction by using linear combinations of the variables (scores) that represent in a decreasing order, the maximum amount of variability in the data.

MPCA is implemented by performing a PCA analysis on an unfolded form of the three-dimensional data matrix. The unfolding can be accomplished six different ways. However, the most meaningful unfolding involves taking vertical slices along the time axis corresponding to different points in time and laying the slices side-by-side to produce a matrix $\mathbf{X}$ of size (IxJK), where $i = 1, 2, \cdots, I$ are the batch numbers, $j = 1, 2, \cdots, J$, the variables, and $k = 1, 2, \cdots, K$, sample time. Thus, the first J columns correspond to the all variables and all the batches at time 1; the next J columns represent the same set of variables at the next sampling time, and so on.

As with PCA, MPCA performs a decomposition on the data by finding the direction of maximum variability (loadings) at each point in time and then determining the projections (scores) of the batches onto that direction. Over the K sampling times for each batch, there is a loading matrix $\mathbf{P}_r$ and an associated score vector $\mathbf{t}_r$. The transformation of $\mathbf{X}$ is given as

$$\mathbf{X} = \sum_{r=1}^{R} \mathbf{t}_r \bigotimes \mathbf{P}_r + \mathbf{E} \qquad (1)$$

where R is the number of principal components used to represent the data. The score vectors ($\mathbf{t}$) are orthogonal and the loading vectors ($\mathbf{P}$) are orthonormal. The number of loading matrices and score vectors needed is generally small when there is a high degree of correlation among the data. Each element of the $\mathbf{t}$-vector represents the amount of variability of the corresponding batch in the direction defined by $\mathbf{P}$. Each loading matrix contains information about the relationship among the variables over time within the batch.

3. PROCESS DESCRIPTION

The model used in this work is in the early stages of development and while it does not represent the true industrial batch process in every detail, it captures the basic important operations of the real process and allows us to test different controller formulations that would otherwise not be tried on the real system.

We will begin by first describing the real process and then discuss the model and its limitations.

3.1. *Industrial Process*

A salt solution is boiled in a forced evaporator until the solvent content is reduced to approximately 20% by weight. The evaporator's contents are then discharged to a reactor in which there may be a polymer residue left over from the previous operation. In cycles 1 and 2, the contents are heated at a constant pressure to increase the concentration of the salts. Primarily because the reactor contents are not well mixed, the rate at which the contents at or near the top of the reactor are heated is greater than that at the bottom, as indicated by temperature measurements located along the centerline of the reactor. Consequently, there exists a temperature front with as much as 40°C difference between the top and bottom of material in the reactor. Temperature measurements taken at and near the top of the reactor indicate that no further temperature rise occur until the temperature of the material at the bottom of the reactor reaches the boiling point of the process fluid. Once this temperature is reached, however, the entire mixture is boiling. A measure of the vent rate during these cycles provides an indirect but useful indication of the extent of reaction (average molecular weight).

Once the center temperature has attained a prescribed value, the process is now in cycle 3 of the process. Here, the reactor and heating medium pressure setpoints are reduced. The effect is to lessen the effect of the heat source. Once the reactor pressure reaches the new setpoint, cycle 4 is triggered. The duration of cycle 4 is a priori determined so that the desired viscosity is achieved. The final cycle, cycle 5, signals the end of the batch and the start of expulsion of the reactor contents. The total time for the batch cycle is less than 2 hours.

It is well known that too much heating in the early cycles, will also cause components critical to the yield, to be boiled off with the solvent. Once these components are depleted from the reactor there is no mechanism by which they can be re-added to the existing reactor contents. Additionally, if in cycle 4 the reactor temperature profile has deviated significantly from the reference profile, there is no controller action in place to provide more heating or less cooling to correct for the deviations. Too high a temperature results in the degradation of the product, while too low a temperature implies low yield.

Other known sources that contribute to batch-to-batch variability from within a single product type are variations in the heat content of the heat source, various levels of impurities in the ingredients, and residue buildup over the operating life of the reactor. Any improved control scheme must provide an adequate means of compensation for these disturbances.

3.2. *The Model*

The reactions can be described by a series of first-order consecutive reactions represented as,

$$C_a \longrightarrow C_b \longrightarrow C_c \tag{2}$$

where C_b is the desired product. Too much heating will degrade the polymer to a different grade as represented by C_c, and if the reaction is halted too soon then C_a will not have had a chance to react to produce C_b. Thus the conversion and yield will be low at either extremes. There is an optimum temperature profile at which the process should be operated to minimize batch time and simultaneously produce high yield. From various simulation runs an optimum batch time of 100 minutes and an upper limit on temperature are set to prevent production of C_c, thereby maximizing conversion and yield.

As in the real process, the reaction is controlled by heating or cooling the process. Unlike the real process, the model achieves this by using a split range control valve. That is, heating occurs when the controller output is between 50 and 100% and cooling when the output is between 0 and 50%. In the first cycle, the contents are heated for a prescribed amount of time, (*cf.* real process works off temperature) after which, the heating source is removed and cooling is initiated. Cooling is carried out at a precise rate until the end of the batch. The following energy equations describe the process,

$$\frac{dT}{dt} =$$

$$-\frac{\Delta H_{r,1} k_1 C_a - \Delta H_{r,2} k_1 C_b + h_i A_i (T - T_m)/V}{\rho C_p}$$

$$\frac{dT_m}{dt} = \frac{h_i A_i (T - T_m) - h_s A_o (T_m - T_j)}{\rho_m C_{p,m} V_m} \tag{3}$$

$$\tag{4}$$

$$\frac{d(V_j T_j)}{dt} = F_{w,in} T_{w,in} + \frac{h_w A_o (T_m - T_j)}{\rho_w C_{p,w}} \tag{5}$$

$$\tag{6}$$

$$\frac{dT_j}{dt} = \frac{F_{w,in}(T_{w,in} - T_j) + \frac{h_w A_o (T_m - T_j)}{\rho_s C_{p,s}}}{V_j} \tag{7}$$

where ΔH_i is the heat of reaction i; k_i is the reaction rate of reaction i, h_i is the inside heat transfer coefficient, h_s is the heat transfer coefficient during the heating phase, h_w is the heat transfer coefficient during the cooling phase; T is the reactor temperature, T_m is the reactor metal wall temperature, T_j is the jacket temperature; V and V_j are the volumes of the reactor and jacket respectively; $T_{w,in}$ is the coolant inlet temperature, $F_{w,in}$ is the coolant inlet flow rate; ρ_w, ρ, and ρ_s are the densities of the coolant, reactor contents, and the heat source respectively; A_i and A_o are the areas of the reactor and jacket respectively; and $C_p, C_{p,m}, C_{p,s}$, and $C_{p,w}$ are the heat capacities of the reactor contents, metal wall, heat source, and coolant. The material balances are straightforward to derive, and due to space limitations are not included.

4. Controller Formulation

Batches are characterized by many variables such as temperatures and pressures and variables which are to controlled such as composition and properties. The true variable to be controlled such as composition is often times not measured on-line and lab measurements are delayed too long to be of value for control purposes. Thus, the assumption often made is that if the various variables behave in a predefined fashion, the final batch properties will be within an acceptable range. Control of a batch process is very much a multiple objective controller design problem between the final batch properties and the process variables.

MPCA is a technique which provides a means of addressing these issues. First, the data from a number of batches experiencing different process upsets are generated. From this, MPCA produces both scores and loadings as functions of time. The scores are computed at sample number k based on the time history of the data from the beginning of the batch. The assumption made is that the variables will be at the average value from time k until the end of the batch. Since the data are mean corrected and scaled by the standard deviation, this assumption implies that data from time k onwards will have a normalized value of zero for the purposes of computing the scores. Thus, scores, $\mathbf{t}(k)$, may be computed recursively from the scores, $\mathbf{t}$ (k-1), by,

$$\mathbf{t}(k) = \mathbf{t}(k-1) + \mathbf{x}(k) * \mathbf{P}(k) \qquad (8)$$

where $\mathbf{x}(k)$ is a (1xm) row vector of batch process variables at time k and $\mathbf{P}(k)$ is an (mxr) matrix of loadings for the same time. The parameter r is the number of scores used.

The approach taken in this paper is to define a controller to cause the scores associated with a given batch to match as closely as possible the scores associated with the nominal. Since the routine process data contain variables of differing degrees of importance for the process, we have taken the approach of computing partial scores. The partial scores are generated from the more important variables and their associated loadings via Equation (8). These scores do not contain the manipulated variables information. As shown by MacGregor (7), when the manipulated variables change, the resulting principal component model also changes. The assumption that we are making is that the correlations between the variables used in the computation of the partial scores do not change and these correlations are captured by the original model. Thus, varying the computation of the manipulated variable does not change the existing correlations among the variables used for the partial scores calculation.

The score controller is a proportional only controller. The error between the partial scores for the "ideal" batch and those being generated by the data of the batch to be controlled is used by the score space proportional controller to adjust the bias of the true batch pressure controller, forming a cascaded structure. The score controller is updated once a minute whereas the proportional controller is updated each simulation interval. Thus, the score controller is trimming the standard pressure controller output. Although not used for computing the partial scores, the gain for the score controller is modulated by the sign of the loadings associated with the manipulated variables. This is needed because as the loadings change sign, the sign of the gain of the score space controller must likewise change.

The system being controlled is dynamic and the method discussed here is static. Since the idea is to determine the control action in the score space, the dynamic nature of the process needs to be part of the model. The model will be time varying over the duration of the batch. This is an area of current research for the authors. For the purposes of this work, we use the sign of the manipulated variables' loadings at a point in the future, rather than the present time. This was done to compensate for the natural lags that exist in the system, and experimentation indicate improved control when manipulated in this fashion.

5. SIMULATION RESULTS & DISCUSSION

The parameters of the model are perturbed to represent some of the known causes of variability in the real process. For example, it is well known that variations in the heat transfer will affect the reactor temperature. To represent that effect, the heat transfer coefficient between the reactor vessel and the reactor contents is changed by 16% in both directions. Figure 1 shows the temperature trajectory with these change together with the nominal temperature trajectory. Likewise, to represent a variation in the heat transfer caused by residue left over from a previous batch, the inside area of the reactor is changed in both directions by 11%. Figure 2 illustrates this effect.

It can be observed that in the case of a 16% decrease in the inside heat transfer coefficient, that the current controller cannot cool the reactor contents with the given cooling capacity. For a 11% decrease in the internal reactor area, both the heating and the cooling is slower as compared to the nominal batch. Similar conclusions can be drawn if a lag is introduced to effect measurement delays.

With the PCA controller cascaded to the pressure controller, in the case of an 11% decrease in the inside area, we observe that in the final twenty minutes the temperature trajectory more closely matches the nominal trajectory and the final temperatures are approximately the same. The PCA/pressure controller is applying more cooling in the last twenty minutes but due to the severity of the disturbance and its character, the effect on temperature is small.

If the PCA model incorporated dynamic effects, the PCA/pressure controller would be better able to compensate for this type of disturbance and others within the limitations of the system, i.e.., cooling capacity, total batch time. This is a matter for further ongoing research.

6. REFERENCES

[1]Piovoso, M. J. and K. A. Kosanovich

(1994). "Applications of Multivariable Statistical Methods to Process Monitoring and Controller Design," Intl. J. Contr., **59**, No. 3.

[2]Kosanovich, K. A., M. J. Piovoso, K. S. Dahl, J. F. MacGregor, and P. Nomikos (1994). "Multiway PCA Applied to an Industrial Batch Process", *Proceedings of the American Control Conference*, **2**, Baltimore, MD.

[3]Nomikos, P. and J. F. MacGregor (1993). "Monitoring Batch Processes," *Technometrics*, **IT-14**.

[4]Kosanovich, K. A. and P. D. Schnelle (1994). "Improved Regulation of an Industrial Batch Reactor," to be presented at 1995 AIChE Spring Meeting, Houston, TX.

[5]Luyben, W. L (1990). *Process Modeling, Simulation, and Control for Chemical Engineers.* M^c^Graw-Hill, New York, NY.

[6]Martens, H. and T. Næs, (1989) *Multivariate Calibration*, John Wiley & Sons, New York, NY.

[7]Beebe, K. R. and Kowalski, B. R. (1987) "An Introduction to Multivariate Calibration and Analysis" *Analytical Chemistry*, **I59**.

[8]MacGregor, J. F., T. E. Marlin, and J. V. Kresta (1991) "Some Comments on Neural Networks and other Empirical Modelling Methods," CPC-IV, CACHE Publication, Padre Island, TX.

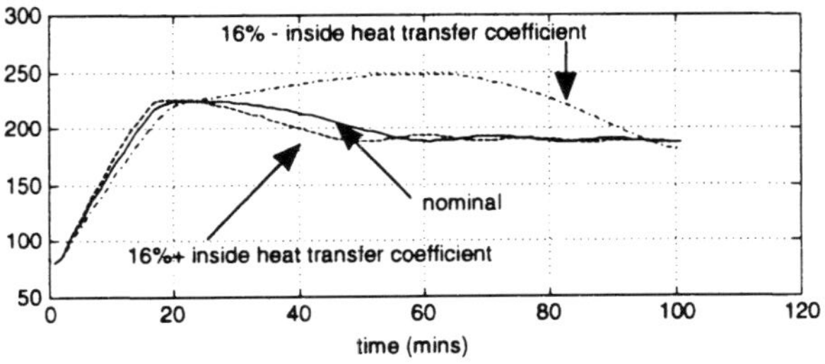

Fig. 1. Variations in the inside heat transfer coefficient.

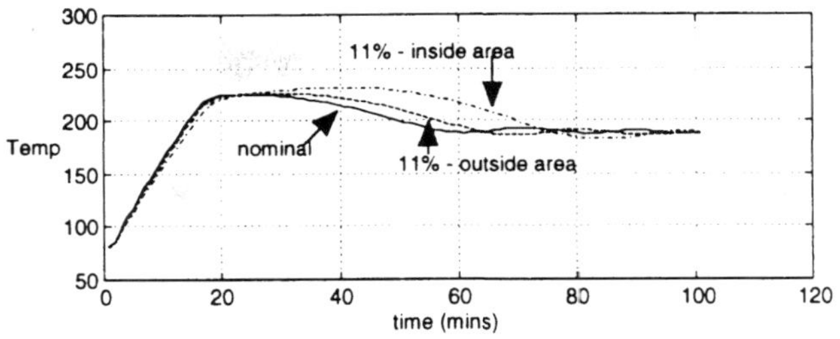

Fig. 2. Variations in the inside heat transfer area.

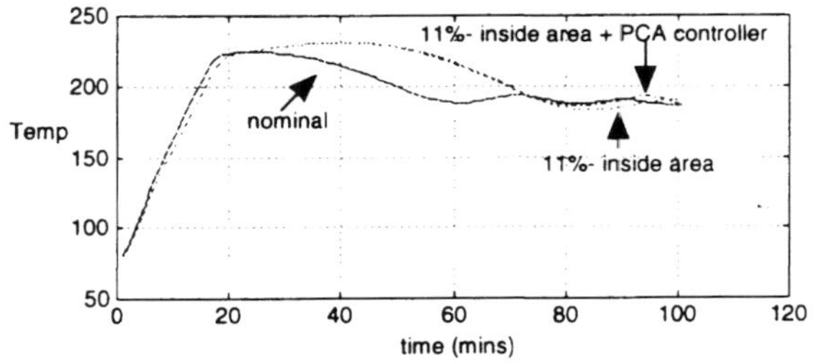

Fig. 3. Temperature performance with PCA/pressure controller.

INTEGRATION OF PROCESS AND CONTROL DESIGNS BY NONLINEAR GEOMETRIC METHODS

Jesús Alvarez, Fernando Zaldo and Salvador Padilla

Centro de Investigación en Polímeros (COMEX)
M. Achar Lobatón 2, 55855 Tepexpan Edo. de México, México

Abstract: For a semibatch process, the problem of simultaneously designing process equipment, operating conditions, and control scheme is casted in terms of the existence of a closed-loop, unique, robustly stable nominal state-trajectory. Using tools from nonlinear dynamics and control, sufficient solvability conditions are given in terms of the output map of the plant and the parameter set associated to the process equipment design. The characterization of the solution yields guidelines and criteria to obtain the process-control design. As an application example, it is considered the batch emulsion polymerization of vinyl acetate in an industrial reactor.

Keywords: Process control, nonlinear control systems, batch control, process equipment.

1. INTRODUCTION

In industrial practice, it is well known that the process equipment design affects, and is affected by, the control design. In general, process designs under stringent efficiency specifications mean operation with enhanced nonlinear behavior, which in batch and semibatch processes, becomes more evident because the state of the plant evolves in ample regions of its state-space. In the area of process synthesis, the process-equipment nonlinear design problem has been addressed with optimization techniques. In fact, these techniques can be seen as elaboration's of the nonlinear optimal control problem. The advantages of these techniques is that they yield quantitative designs, and the disadvantage is that the questions of existence of a solution and of its nature are overlooked. Even optimization process design techniques stem from a nonlinear control theory, in industrial practice the process control design is carried out separately, and mostly with linear-conventional techniques. In the last twenty years, geometric methods have extended to the nonlinear case various classical results in linear systems theory (Isidori, 1989). In chemical engineering, these results have been mostly applied to the construction of nonlinear controllers (Bequette, 1992). To the best of our knowledge, the consideration of a nonlinear geometric approach for the integrated process design problem has not been explored. In this work, as a move towards the integration of process, equipment and control designs, the unified problem is casted within a nonlinear framework that is an adaptation of the one used in the aforementioned nonlinear control methods.

Using tools from nonlinear dynamics and control, sufficient solvability conditions are given in terms of the output map of the plant and the parameter set associated to the process equipment design. The characterization of the solution yields guidelines and criteria to obtain the process-control design.

As an application example, an industrial problem is considered: the semibatch emulsion polymerization of vinyl acetate. The batch must be carried out as fast as possible while maintaining safety and operability margins as well as product quality specifications. The objectives are to yield the basic specifications of: (i) the mixing and heating/cooling equipment, and (ii) a conventional-type control scheme where monomer feedrate is pre-programmed and temperature is handled with a conventional cascaded controller.

2. THE CLASS OF PLANTS AND THE PROCESS-CONTROL PROBLEM

Consider nonlinear plants

$$\dot{x} = f(x, d, u; p)\ ;\ x \in X,\ u \in U,\ p \in P \quad (1a)$$

$$y = h(x, d), \quad y \in Y,\ d \in D \quad (1b)$$

with n states (x), m control inputs (u), s known disturbance inputs (d) and m outputs (y) which take values from the open, bounded and connected sets X, D, U and Y, respectively. The vector p denotes the parameters of the plant. p is partitioned as follows $p=\{p_f, p_d\}$, $p_f \in P_f$, $p_d \in P_d$, with p_f denoting a set of parameters that remain fixed through the design process and p_d denoting a set of free-parameters for process design. Hence, P_d represents a set of admissible process designs. The maps f and h are smooth on their corresponding domains, and the scalar inputs $d_i(t)$ and $u_i(t)$ are piecewise-continuous functions of time.

In general, the process-control design problem consists in simultaneously determining: (i) an admissible nominal plant evolution $\bar{x}(t) \in X$, (ii) its corresponding admissible nominal inputs, $\bar{u}(t) \in U$ and $\bar{d}(t) \in D$, (iii) an admissible plant design $p_d \in P_d$, and (iv) the basic specification of the control scheme.

Consider a nonlinear system (1), for given $\bar{x}_0 \in X_0$, $d(t) \in D$ and $u(t) \in U$ in system (1a), the state trajectory $\bar{x}(t)=\Theta_x[t,t_0,\bar{x}_0,d(t),u(t),\bar{p}]$ of the plant (1) is *nominally stable* if, for any $t_0 > 0$ and $\delta > 0$ there exists a $\delta(\varepsilon, t_0)$ such that

$$\|x_0 - \bar{x}_0\| \le \delta \Rightarrow \|x(t) - \bar{x}(t)\| < \varepsilon,\ \forall t \ge t_0$$

Otherwise, $\bar{x}(t)$ is *unstable*. $\bar{x}(t)$ is *robustly stable* if, for any $t_0 > 0$ and $\varepsilon_r > 0$ there exists a $\delta_0(\varepsilon_r, t_0)$ and $\delta_p(\varepsilon_r, t_0)$ such that

$$\|x - \bar{x}_0\| \le \delta_0, \|p - \bar{p}\| \le \delta_p \Rightarrow \|x(t) - \bar{x}(t)\| \le \varepsilon_r, \forall t \ge t_0$$

Definition 1. The *process-control problem* of plant (1) is solvable if, for given $\bar{x}_0$, D, U and P_d, there exists: (i) a process design $\bar{p} \in P$, (ii) an output map $h(x,d)$, (ii) a state-feedback controller $u=\mu(x,t,\bar{p})$ so that the plant-controller interconnection

$$\dot{x} = f[x, d(t), \mu(x,t,\bar{p}), p],\ x(0) = x_0,\ d(t) \in D,\ x(t) \in X,$$
$$y = h[x, d(t)], \qquad u(t) \in U,\ y(t) \in Y$$

has the following properties:

i) when $p = \bar{p}$, $x_0 = \bar{x}_0$ and $d(t) = \bar{d}(t)$, for a given nominal trajectory $\bar{y}(t)$, there exists a unique nominal state-trajectory $\bar{x}(t)$ and a unique control input trajectory
$\bar{u}(t) = \mu[\bar{x}(t), t, \bar{p}]$

ii) the nominal trajectory $\bar{x}(t)$ is robustly stable,

iii) when $x(0) \ne \bar{x}_0$, $p = \bar{p}$ and $d(t) = \bar{d}(t)$, the dynamics of the output error, $y(t) - \bar{y}(t)$, is linear, noninteractive and has $k \le n$ assignable poles. ■

The time-dependency in the feedback function μ accounts for the feedforward path from the input d(t).

3. SOLVABILITY OF THE PROCESS-CONTROL PROBLEM

3.1 Solvability conditions.

In the next theorem, sufficient conditions are given for the solubility of the process-solvability-control problem.

A map $\alpha(x,p,t)$ is nominally invertible at $(\bar{x}, \bar{p})$ if, in some neighborhood X of $\bar{x}$ $\det[\alpha(x, p, t)] \ne 0$, $x \in X$, $\forall t > 0$, if, in addition, this conditions holds for some neighborhood P of $\bar{p}$, the map a is said to be *robustly invertible* at $(\bar{x}, \bar{p})$. This is, $\det[\alpha(x,p,t)] \ne 0$, $x \in X$, $p \in P$, $t > 0$. For plant (1), the set of m positive integers $\kappa_1 + \ldots + \kappa_m = \kappa \le n$, $\kappa_i > 0$ determine the following maps:

$$\phi_I(x,u,t,p) = [h_1, \ldots, L_f^{\kappa_1 - 1} h_1\ ; \ldots ; h_m, \ldots, L_f^{\kappa_m - 1} h_m]^T \quad (2a)$$

$$\varphi(x;u,t,p) = [L_f^{\kappa_1} h_1, \ldots, L_f^{\kappa_m} h_m]^T \quad (2b)$$

where $\dim \phi_I = n - \kappa$ and $\dim \varphi = m$, and $L_f\alpha$ is the directional derivative of the time-varying scalar field $\alpha[x, t]$ with respect to the vector field $f[x,d(t)]$:

$$L_f\alpha = \sum_{i=1}^{n} (\partial\alpha/\partial x_i) f_i + \partial\alpha/\partial t\ ,\ L_f^{i+1}\alpha = L_f(L_f^i\alpha),\ L_f^0\alpha = \alpha$$

Theorem 1. The *process-control problem* of plant (1) is solvable if there exist m strictly positive integers $\kappa_1, \ldots, \kappa_m$ so that

i) the map $\phi_I(x, u, t, p)$ is independent of u:
$\partial[\phi_I(x, u, t, \bar{p})]/\partial u = 0$, $(x, u, p) \in (X, U, P)$, $\forall t \ge 0$

ii) the map $\varphi(x, u, t, p)$ is robustly invertible for u:
$\det[\varphi_u(x, u, t, p)] \ne 0$, $(x, u, p) \in (X, U, P)$, $\forall t \ge 0$

iii) the restricted internal $(n-\kappa)$-dimensional dynamics

$$\dot{\bar{x}}=f\{\bar{x},\psi[\bar{x},\bar{v}(t);\bar{p})];\bar{p}\},\ \bar{x}(0)=\bar{x}_0,\ \bar{x}\in X_I(t)$$
$$X_I(t)=\{\bar{x}\in X \mid \phi_I(\bar{x},t,\bar{p})=\bar{z}(t)\},\ \dim X_I(t)=n-\kappa$$

is robustly stable, with ψ being the inverse map of φ: $\psi[x,\varphi(x,u,t,p),p,t]=u,\ (x,u,p)\in(X,U,P),\ \forall t\geq 0$

iv) the open sets X and U are the union, over time $t \geq 0$, of neighborhoods X(t) and U(t) of the trajectories $\bar{x}(t)$ and $\bar{u}(t)=\gamma[\bar{x}(t),\bar{v}(t);\bar{p})]$ determined by the preceding dynamics. This is $X=\bigcup_t X(t)$, $U=\bigcup_t U(t)$.

The proof of this theorem and next corollary is a direct consequence of the proof, are given in the appendix. *S* will denote the set of stable matrices.

Corollary 1. Let the solvability conditions of the process-control problem be met. Then,

i) given $\bar{x}_0$, d(t) and $\bar{y}(t)$, the nominal state-control trajectory pair $[\bar{x}(t),\ \bar{u}(t)]$ is given by

$$\bar{v}(t)=[\bar{y}_1^{(\kappa_1)},\ldots,\bar{y}_m^{(\kappa_m)}]^T \quad (3a)$$
$$\bar{z}(t)=[\bar{y}_1,\ldots\bar{y}_1^{(\kappa_1-1)};\ldots;\bar{y}_m,\ldots,\bar{y}_m^{(\kappa_m-1)}]^T \quad (3b)$$
$$\dot{\bar{x}}=f[\bar{x},d(t),\psi[\bar{x},\bar{v}(t)+K_c\{\phi(\bar{x},p,t)-\bar{z}(t)],\bar{p},t\},\bar{p}],$$
$$x(0)=\bar{x}_0(0) \quad (3c)$$
$$\bar{u}=\psi[x,\bar{v}(t)+K_c\{\phi(\bar{x},\bar{p},t)-\bar{z}(t)],\bar{p},t\} \quad (3d)$$

$$A_c=\Gamma+\Pi K_c\in S,\ \Gamma=bd[\Gamma_1,\ldots,\Gamma_m],\ \Pi=bd[\pi_1,\ldots,\pi_m]$$
$$K_c=bd[k_1,\ldots,k_m],\ k_i=bd[k_{i1},\ldots,k_{i\kappa m}],\ \dim k_i=1\times\kappa_i$$

$$\Gamma_i=\begin{bmatrix}0&1&0&\cdots&0\\0&0&1&\cdots&0\\\vdots&&\ddots&&\vdots\\0&0&\cdots&0&0\end{bmatrix},\ \pi_i=\begin{bmatrix}0\\0\\\vdots\\1\end{bmatrix},\ \dim\Gamma=n\times n,$$

$\dim\Pi=n\times m$, $\dim\Gamma_i=\kappa_i\times\kappa_i$, $\dim\pi_i=\kappa_i\times 1$.

ii) given $\bar{x}_0$, d(t) and $\bar{u}(t)$, the open-loop nominal plant dynamics

$$\dot{x}=f[x,\ d(t),\ \bar{u}(t),\ p],\ \bar{x}(0)=\bar{x}_0 \quad (4)$$

is stable/unstable if system (3) is stable/unstable when $K_c = 0$.

iii) the feedback controller is

$$u=\psi[x,\bar{v}(t)+K_c\{\phi(x,\bar{p},t)-\bar{z}(t)\},\bar{p},t] \quad (5)$$

3.2 Implications for process-control design.

The preceding constructive approach to the solvability question suggests the following procedure to design the process and its control:

a) Choose a candidate h(x,d) and test conditions (i) and (ii). If affirmative, build the feedback function (5), otherwise, try a new h(x, d).

b) For a candidate $\bar{y}(t)\in Y$, determine the nominal state and input trajectories, $\bar{x}(t)$ and $\bar{u}(t)$.

c) Test the nominal and robust stability of the nominal state trajectory $\bar{x}(t)$. If $\bar{x}(t)$ is unstable, consider a different nominal output $\bar{y}(t)$. Leave this stage until $\bar{x}(t)$ is robustly stable. If this is not possible, choose a new output map and go back to (a).

d) Adjust $\bar{y}(t)$ so that, for some reasonable X_0 and $d(t) \in D$, the corresponding state and input trajectories are admissible ones.

4. APPLICATION EXAMPLE: A SEMIBATCH EMULSION POLYMERIZATION REACTOR

4.1 The polymerization reactor and its process-control problem

In an emulsion polymerization reactor, monomer is converted into polymer, producing a significant amount of heat (Penlidis *et. al.*, 1985, Hamielec *et. al.*, 1987). Water, emulsifying agent, initiator are loaded into the reactor. First, heat is added to start the reaction. Then, heat must be removed to prevent reactor thermal run-away. Monomer addition and heat exchange policy must be such that the reaction takes place as fast as possible, having adequate safety margins, and yielding a reasonable product variability. The process is described by the following set of equations:

$$\dot{M}=-r_P(M,I,V_w,N,T)+q_M\rho_M$$
$$\dot{I}=-r_I(I,V_w,T)+q_I(C_I-I)/V_w$$
$$\dot{V}_w=q_I$$
$$\dot{V}_e=q_M+q_I-r_P(M,I,V_w,N,T)(1/\rho_M-1/\rho_P)$$
$$\dot{V}^*=q_M$$
$$\dot{N}=r_m(M,I,N)+r_h(M,I,N)+r_f(N)-q_IN/V_w$$
$$\dot{T}=r_P(M,I,V_w,N,T)(-\Delta H_P)(\rho_eC_{pe}V_e)^{-1}$$
$$+q_M\rho_M(C_{pM}T_M-C_{pe}T)(\rho_eC_{pe}V_e)^{-1}$$
$$+q_I\rho_w(C_{pw}T_w-C_{pe}T_e)(\rho_eC_{pe}V_e)^{-1}$$
$$-UA(T-T_j)(\rho_eC_{pe}V_e)^{-1}$$

The states of the reactor are: M (mass of unreacted monomer), I (molar concentration of initiator), V_w (volume of water), V_e (emulsion volume), V^* (volume of added monomer), N (number of latex particles per unit volume of water phase), and T (reactor temperature). The control inputs are: q_M (monomer feedrate) q_I (initiator feedrate), T_j (jacket temperature). The other variables and functions are: T_w (initiator feed temperature), T_M (monomer feed temperature), ρ_P (polymer density), ρ_M (monomer density), ρ_w (water density), U (overall heat transfer coefficient), A (reactor heat transfer area), C_{pM} (monomer heat capacity), C_{pe} (emulsion heat capacity), C_{pw} (water heat capacity), r_P (polymerization rate), r_I (initiator decomposition rate), r_m (micelar nucleation rate), r_h (homogeneous nucleation rate) and r_f (coagulation nucleation rate). With the following notation

$x_1 = M$, $x_2 = I$, $x_3 = V_w$, $x_4 = V_e$, $x_5 = V^*$, $x_6 = N$, $x_7 = T$, $u_1 = q_M$, $u_2 = T_j$.

the reactor has the form (1). The reactor volume (10,000 L) and the monomer-initiator-suspending agent (vinyl acetate-ammonium persulfate-polyvinyl alcohol) are fixed. Thus, the equipment design part of the problem consists specifying the stirrer-mixing system and the cooling jacket (volume, geometry and internal circulation). The volume V_j of the jacket and its internal recirculation q_j must be so that the resistance for heat exchange is basically due to the emulsion-jacket side. Thus, the heat transfer coefficient for a stirred tank is given by (Bondy and Lippa, 1983):

$$h_i = a_b (k/D_T)(D^2N\rho/\mu)^{bh}(\mu c_p/k)^{ch}(\mu/\mu_w)^{dh}$$
$$\ln \mu = \ln \mu_w + 2.5\, \phi_v(1-1/(0.6328\, \phi_v))$$

Where a_b=0.74, bh=2/3, ch=1/3, dh=0.14 and $\phi_v = \phi/(\rho_P(\phi/\rho_P+(1-\phi)/\rho_w))$, $\phi=(\rho x_5-x_1)/(\rho_w x_3+\rho_M x_5)$. These equations account for the dependency of the heat transfer on the emulsion viscosity. Hence the process design parameters are: $p = [a_h, b_h, c_h, q_j, V_j]^T$, $p \in P$. The temperature dependency of the initiator decomposition constant imposes the following restriction on the reactor temperature:

$$T(t) = x_7(t) \le T^* = -E_d/(R\ \ln[.693/A_d t_{1/2}])$$

with E_d, A_d being the activation energy and the exponential factor of the initiator. The water service temperature, the jacket volume, and its recirculation rate q_j, impose the following restrictions on jacket temperature $30°C = T_j^{min} < T_j(t) = u_2(t) < T_j^{max} = 150°C$. The total mass of added monomer has a fixed value M_1, imposing the following restriction on the input $q_M(t) = u_1(t)$: $M_T = \rho_M \int_0^{t_f} u_1(\tau)d\tau = 4{,}500$. The initial nominal condition is: $\bar{x}(0) = [0, 7.9\times10^3, 3890, 3890, 0, 0, 353.15]^T$. Hence our process-control problem consists in finding: the nominal design p, the nominal operation $\bar{x}(t)$, $\bar{u}(t)$ and the basic specification of a control scheme.

4.2 Case I: Temperature and conversion as outputs

Temperature is a rather evident output to be controlled, and conversion is an output commonly used in kinetics and reactor engineering studies. Thus, $y_1 = (V^*\rho_M - M)/M_T$, and $y_2 = T$. The relative degrees associated to this outputs are $\kappa_1 = 2$ and $k_2 = 1$, with condition (ii) of theorem 1 being:

$$\det[\varphi_u(x,u,t,p)] = -(\rho_M/\gamma M_T)(\partial r_P/\partial M) \neq 0 \qquad (5)$$

In this case, the internal dynamics is ($n-\kappa=7-3=4$)-dimensional, where $\gamma = UA / V_e\rho_e C_{pe}$.

From corollary 1, the associated state-feedback controller is given by eq. (5). From standard kinetics, at the earlier polymerization period, $\partial r_P/\partial M \to 0$, meaning that φ may be an ill-condition map, and therefore, the underlying dynamic inversion (*i.e.*, the evaluation of $\bar{x}(t)$ and $\bar{u}(t)$) may be ill-conditioned. In other words, the closed-loop, state-trajectory could be nominally stable and robustly unstable. Physically, this means that in the earlier period, the control inputs do not have a significant effect on the conversion-time curve. Concluding, conversion is not an adequate output.

4.3 Case II: Temperature and unreacted monomer as outputs.

In case I, the ill-condition of $\partial r_P/\partial M$ led to a potentially robustly unstable process. If the output map h_1 had relative degree 1 ($\kappa_1=1$), the presence of $\partial r_P/\partial M$ could be avoided. This is the case if the first output is the amount of unreacted monomer in the reactor: $y_1 = M$, $y_2 = T$. The corresponding condition (ii) (of theorem 1) is: $\det[\varphi_u(x, u, t, p)] = \gamma \neq 0$.

This non-singularity condition is robust in the sense that never tends to zero. From corollary 1, the state-feedback map is given by eq. (5). Application, to the plant, of this controller with K_c=0, yields an internal 5-dimensional dynamics. Its stability (condition (iii) of theorem 1) will be tested with numerical simulations.

4.4 Construction of the nominal operation

From sections 4.2 and 4.3, it is known that case II could have robustness problems. In both cases, I and II, the stability condition of the internal dynamics has to be tested numerically. Application, to case II, of the procedure of section 2.2, assisted by physical insight and equipment design considerations, yields, after three to five trials, the reactor operation of figure 1 (thick lines). It was found that this operation was robustly stable.

To test case I, case II was used to generate the nominal output trajectory $\bar{y}(t)$ (with the output map of case I). Then, the nominal state (thin lines in figure 1), $\bar{x}(t)$ and input $\bar{u}(t)$ was generated with eq. (5). As expected, this state-control pair has a significant deviation from the nominal state-control pair of case II. The closed-loop operation is nominally stable but not robustly stable.

In figure 2, the test of robust stability is shown for an operation (with the output of case II) that is slower than the one of figure 1 (thick lines). As it can be seen, the operation is robustly stable. Only the two states that exhibit the worst deviation are shown. In figure 3, the test of robust stability is shown for an operation that is twice faster than the operation of case II in figure 1 (thick lines). The reactor temperature has an excursion that reaches the maximum temperature imposed by the initiator. As a result, the reactor conversion is not total, meaning that this fast operation is not robustly stable. Concluding, as the semibatch time is reduced, the operation becomes unstable.

4.4 The specification of the control configuration

From case II, one has $\overline{q}_M(t)$ and $\overline{T}(t)$. Set $q_M(t)=\overline{q}_M(t)$ in feedforward fashion and use $\overline{T}(t)$ as a time-varying setpoint in a conventional cascaded temperature controller. Simulations show that this control scheme yields a robustly stable closed-loop operation. This statement was verified with actual plant tests (not shown in this paper).

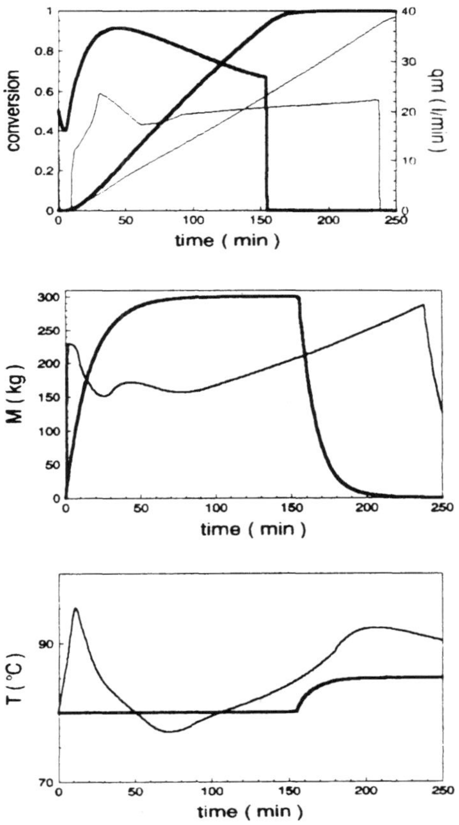

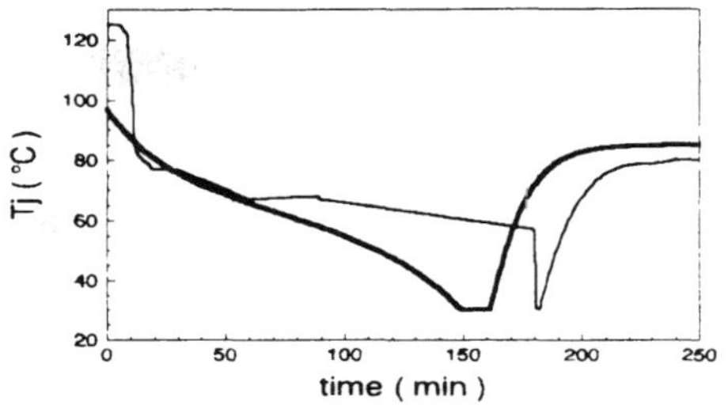

Figure 1. Reactor nominal operation for cases I (thin lines) and II (thick lines).

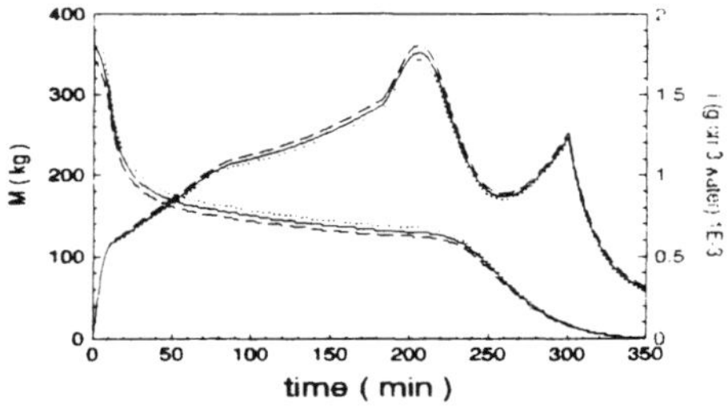

Figure 2. Test of robust stability (varying initial condition and parameters) for case II with a slow operation.

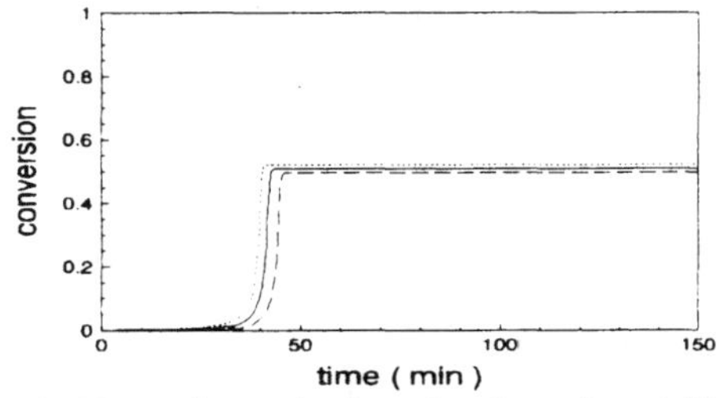

Figure 3. Test of nominal and robust instability for case II with a fast operation.

5. CONCLUSIONS

Sufficient conditions for solvability of the process-equipment-control design problem have been given. The approach to the solvability question was constructive, yielding a procedure to obtain, at the design stage (with specific equipment, operating condition, and control performance specifications), equipment design, the nominal operation, and its basic control scheme. The procedure can be: (i) used to interactively solve the process design problem, or (ii) the point of departure to formulate and solve a well-posed optimization problem. The structural solution to the control problem yielded the essential specification of a control scheme that is valid for a conventional-type implementation or for and advanced one.

REFERENCES

Bequette, B. W., 1991, *Nonlinear control of chemical processes: A review*, Ind. & Eng. Chem. Res., 30, 1391.

Bondy, F. and S. Lippa, (1983), *Heat Transfer in Agitated Vessels*, Chem. Eng., April 4.

Hamielec, A. E., J. F. MacGregor and A. Penlidis, (1987), *Multicomponent free radical Polymerization in Batch, Semi-Batch and Continuous Reactors*, Makromol. Chem., Macromol. Symp. 10/11, 521-570.
Hirsch and Smale (1974), *Differential equations, dynamical systems, and linear algebra*. Academic Press, N.Y.
Hirshorn, R. M., (1979), Invertibility of multivariable nonlinear control systems, IEEE Trans. Automat. Contr.. AC-24, 855-865.
Isidori, A., 1989, *Nonlinear Control Systems*. Springer Verlag, N.Y.
Nijmeijer, H. and Van der Shaft, A., (1990). *Nonlinear dynamical systems*. Springer-Verlag, New York.
Penlidis, A., J. F. MacGregor and A. E. Hamielec, (1985), *A Theoretical and Experimental Investigation of the batch Emulsion Polymerization of Vinyl Acetate*, Polym. Proc. Eng., 3(3), 185-218.
Vidyasagar, M., (1978), *Nonlinear Systems Analysis*, Prentice-Hall, New York.

APPENDIX

Proof of theorem 1. Let $p = \bar{p}$. The following implication is an extension (to time-varying systems) of well-known results (Isidori, 1989; Nijmeijer and van der Schaft, 1990). From conditions (i), (ii) and (iv) follows that (Nijmeijer and van der Schaft, 1990) there exist n-κ scalar maps $\phi_{\kappa+1}(x),\ldots,\phi_n(x)$ so that $\det[\phi(x,t)] \neq 0$, $\phi(x,t)=[\phi_1(x,t),\ldots,\phi_\kappa(x,t),\phi_{\kappa+1}(x),\ldots,\phi_n(x)]^T$. Correspondingly,

$$z=[\phi_1(x,t),\ldots,\phi_\kappa(x,t)]^T,\ w = [\phi_{\kappa+1}(x),\ldots,\phi_n(x)]^T \quad (A1)$$

is an admissible coordinate-change for the state x of the plant. In z-coordinates, plant (1) is given by:

$$\dot{z} = \Gamma z + \Pi\, \varphi[\phi^{-1}(x, t, \bar{p}), u, \bar{p}],\ z(0) = z_o=\phi_I(x_o, 0)$$
$$\dot{w} = \theta(w, z, t, \bar{p}), \qquad w(0) = w_o = \phi_{II}(x_o, 0)$$
$$\theta(w,z,u,t,\bar{p})=\phi_{IIx}[\phi^{-1}(z,w,t,\bar{p}),0)]f\{\phi^{-1}[\phi^{-1}(z,w,t,\bar{p}),t,\bar{p}],u,\bar{p}\}$$

Let v be a new control input. Apply, to plant (1), the feedback controller $u=\psi[x,v,p,t]$ recall property (ii), and obtain

$$\dot{z} = \Gamma z + \Pi v,\ y = \Delta z \quad (A2a)$$
$$\dot{w} = h(w, z, t, \bar{p}) \quad (A2b)$$
$$\eta(w, z, v,t, \bar{p})=\theta\{w, z, y[\phi^{-1}(z,w, t, \bar{p}), v, p,t], t, \bar{p}\}$$

Let $y=\bar{y}(t)$ be a given nominal output. A trivial application of Hirschorn's (1970) structure algorithm to eq. (A2) yields the (n-κ)-dimensional inverse dynamics ($\bar{z}$ and $\bar{v}$ defined in eq. (3)):

$$\dot{\bar{w}} = \eta(\bar{w}, \bar{z}, \bar{v}, t, \bar{p}),\ \bar{w}_0 = \phi_{II}(\bar{x}_0, 0) \quad (A3)$$

Then, for each $\bar{y}(t)$ there is a unique state-input $[(\bar{z}, \bar{w})(t), \bar{v}(t)]$ trajectory pair. Application, at each time t, of the state-control inverse coordinate change (see eq. A1), $\bar{x}(t)=\phi^{-1}[\bar{z}, \bar{w}, t, \bar{p}]$, $\bar{u}(t)=\psi[\bar{x}(t), \bar{v}(t), t, \bar{p}]$, yields a state-control trajectory pair $[\bar{x}(t), \bar{u}(t)]$ uniquely determined by $\bar{y}(t)$. This yields property (i) of definition 1. From the controllability of the pair (Γ, Π) and the block-diagonal structure of matrices Γ and Π, there exists a block-diagonal gain matrix K_c so that the linear controller $v = \bar{v}(t) + K_c[z - \bar{z}(t)]$, applied to system (A2), yields the closed-loop system

$$\dot{e} = A_c e,\ y - \bar{y} = \Delta e,\ e = (z - \bar{z}) \quad (A4a)$$
$$\dot{w} = \eta(w, \bar{z} + e, t, \bar{p}) \quad (A4b)$$

In x-coordinates, the nominal inverse dynamics (2) is given by eq. (A3). From condition (iii), system (A3) is nominally stable. The nominal stability of system (A4) follows from the individual nominal stability of subsystems (A4a) and (A3) (Vidyasagar, 1978). In x-coordinates, the preceding feedbacks yield controller (5). System (A4) yields properties (ii) (the part of nominal stability) and (iii) of definition 1. When $p \neq \bar{p}$, system (A4) becomes:

$$\dot{e} = A_c e + \Pi\, q_e(e, w, p, t),\ y - \bar{y} = \Delta(z - \bar{z}) \quad (A5a)$$
$$\dot{w} = \eta(w, \bar{z} + e, t, \bar{p}) + q_w(e, w, p, t) \quad (A5b)$$
$$q_e(e, w, \bar{p}, t) = 0,\ q_w(e, w, \bar{p}, t) = 0$$

from the smoothness of φ, the piecewise continuity of $[\bar{y}(t), d(t)]$, and the vanishing property of q_e follows that q_e is a bounded disturbance in the following sense:

$$\|q_e(e, \varepsilon, t, p)\| \leq l(E, E, P)\, \|p - \bar{p}\| \quad (A6)$$

where $l(E, E, P) \rightarrow k_e := \sup_{0\leq t\langle\infty} \{\varphi_p [\bar{x}(t), \bar{u}(t), t, \bar{p}]\}$

as $\|p - \bar{p}\|$ and $\|x - \bar{x}(t)\| \rightarrow 0$. The time-integration of eq. (A5a) followed by application of norms and substitution of inequality (A6) yields $\|e(t)\| \leq a \exp(-\lambda t)\ \|e_o\| + a\,l(E,E, P)\lambda^{-1}\|p - \bar{p}\|z\ [1-\exp(-\lambda t)]$. Hence, given ε_e, one can choose $\delta_e = \varepsilon_e/(2a)$ and $\delta_p = (e_e l)/(2ak_e)$ so that $\|e_o\| \leq \delta_e$, $\|p-\bar{p}\| \leq \delta_p$ implies $\|e(t)\| < \varepsilon_e$. The robust stability of system (A5) follows from the robust stability of the internal dynamics (A3). In x-coordinates, this yields property (ii) of definition 1. Condition (iv) guarantees the existence of tube-neighborhoods of $\bar{x}(t)$ and $\bar{u}(t)$ where the properties of definition 1 hold. QED.

DYNAMIC HEAT BALANCE MODEL FOR INDUSTRIAL BATCH REACTORS AND ITS APPLICATIONS

K. S. Lee, C. M. Jeong, S. Yi*, J. S. Son**, and S. C. Yoon**

Dept. Chem. Eng., Sogang Univ., Shinsoodong-1, Mapogu, Seoul, 121-742, Korea
** Samsung Engineering Co., Ltd.*
*** Cheil Industries Co.*

Abstract : A dynamic heat balance model which can be generically applicable to various industrial batch reactors has been proposed. The proposed model is composed by considering the detailed heat balance around the cooling jacket, which has been overlooked in the existing modeling approaches, as well as the heat balance for the reactor contents. In the proposed model, the local heat transfer coefficient in the reactor side is selected as a tuning parameter. The proposed model has been utilized to investigate reactor behaviors as well as to improve control methods for PBL and EPS polymerization reactors.

Keywords: Process model, Process control, Process parameter estimation

1. INTRODUCTION

From the viewpoint of process control, the batch reactor has unique characteristics different from continuous processes. The process runs inherently under unsteady states during the whole reaction period, and as a consequence, it is hard to apply the empirical process modeling technique such as step response analysis. This hampers the controller tuning based on an empirical process model. Moreover, when the reaction is accompanied with large heat of reaction, which is usually observed in polymerization reactions, the process may show significant nonlinearity. Therefore, the conventional fixed gain controllers may not usually perform well over the whole reaction period. Although cascade control is known as a standard control technique for industrial bach reactors (Shinskey, 1988), it may not work satisfactorily unless the controller parameters are appropriately scheduled.

One way to overcome these difficulties is to design a controller based on a physical reactor model. Unfortunately, obtaining a reliable rigorous model, though its usage is much more diverse than the empirical or semiempirical ones, is not generally feasible due to the limited availability of quantitative information on reaction kinetics. In this respect, one of the reasonable compromises is to develop a semiempirical model which considers only the heat balances around the reactor. Although this type of model has only a limited usage compared to the rigorous one, it can play as an adequate model with which a temperature controller is designed. Also the heat balance model can offer additional information on heat transfer characteristics and heat of reaction, which can be utilized in scaleup and/or diagnosis of a reactor (Karlsen and Villadsen, 1987; Bovin *et al.*, 1989).

In this research, we have proposed a generic dynamic heat balance model which can precisely predict the thermal dynamics of a large class of industrial batch reactors. In this modeling, main focus has been given on the heat balance around the reactor jacket. The jacket-side model was composed by taking the temperature distributions along the reactor walls as well as the jacket into account. This model can predict the jacket outlet temperature from the temperature measurements inside the reactor and at the jacket inlet. Using this model, the reactor-side local heat transfer coefficient is estimated such that the prediction error in the jacket outlet temperature is minimized. Once the jacket-side model is set up, heat of reaction is estimated from the reactor-side heat balance equation and then the complete dynamic

heat balance model can be formulated.
The developed model has been utilized in simulating reactor behaviors and/or in upgrading control algorithms for various industrial batch reactors polymeizing PBL (Poly Butadiene Latex), EPS (Expanded Poly Styrene), and ABS (Acrylonitrile Butadiene Styrene). In this paper, results from application to the PBL reactor will be shown in detail and application to the EPS reactor will be discussed only briefly.

2. GENERIC DYNAMIC HEAT BALANCE MODEL

2.1 *Jacket-Side Heat Balance*

For cooling and/or heating, most industrial batch reactors are designed to have a jacket with or without an external coolant recycle loop. When it is necessary to minimize temperature distribution and/or to maximize the heat transfer capability, the recycle loop is installed around the jacket. Otherwise, the recycle loop is not designed. In any case, temperature distribution along the jacket is measurable in most industrial batch reactors.

The jacket-side heat balance model is composed to be able to predict the jacket outlet temperature using temperature measurements at the reactor and jacket inlet. Figure 1 shows the concept of jacket-side modeling. Inside the jacket, bulk flow motion is dominant and a hyperbolic PDE (Partial Differential Equation) should preferably employed in the modeling. The hyperbolic PDE model, however, is not convenient to manipulate numerically, especially when the coolant flow rate varies by control action. Therefore, the jacket was approximated by a series of mixing tanks as in the figure. Likewise, the outer wall as well as the inner wall of the reactor were divided into the same number of separate cells where heat conduction takes place only along the horizontal direction. Using the nomenclatures given in Fig. 1, the resulting heat balance model can be described as follows:

Inner wall :

$$\frac{\partial T_{Iw,k}}{\partial t} = \alpha_{Iw}\frac{\partial^2 T_{Iw,k}}{\partial z^2}, \quad z \in [\,0,\ 1] \tag{1}$$

$$\text{At } z=0, \quad \frac{\partial T_{Iw,k}}{\partial z} = -\frac{h_R}{k_{Iw}}(T_R - T_{Iw,k})$$

$$\text{At } z=1, \quad \frac{\partial T_{Iw,k}}{\partial z} = \frac{h_J}{k_{iw}}(T_{J,k} - T_{Iw,k})$$

Jacket :

$$\frac{M_J}{N}C_{PJ}\frac{dT_{J,k}}{dt} = F_c C_{PJ}(T_{J,k-1} - T_{J,k}) + h_J\frac{A}{N}(T_{Iw,k}(1) - T_{J,k}) + h_a\frac{A}{N}(T_{ow,k}(0) - T_{J,k}) \tag{2}$$

Outer Wall :

$$\frac{\partial T_{Ow,k}}{\partial t} = \alpha_{Ow}\frac{\partial^2 T_{Ow,k}}{\partial z^2}, \quad z \in [\,0,\ 1] \tag{3}$$

$$\text{At } z=0, \quad \frac{\partial T_{Ow,k}}{\partial z} = -\frac{h_J}{k_{ow}}(T_{J,k} - T_{Ow,k})$$

$$\text{At } z=1, \quad \frac{\partial T_{Ow,k}}{\partial z} = \frac{h_a}{k_{Ow}}(T_a - T_{Ow,k})$$

$$k = 1,\ 2,\ ..\ N$$

In eq. (2), T(1) and T(0) denote the temperatures at z=1 and z=0, respectively in the corresponding coordinates. The heat conduction equations for the walls were reduced to a set of ordinary differential equations using one point collocation. The resulting equations can be described by

$$\frac{dT^{o}_{Iw,k}}{dt} = c_1 T_R + c_2 T^{o}_{Iw,k} + c_3 T_{J,k} \tag{4}$$

$$\frac{dT^{o}_{Ow,k}}{dt} = d_1 T_{J,k} + d_2 T^{o}_{Ow,k} + d_3 T_a \tag{5}$$

where the superscript ^{o} denotes the temperature at the collocation point. In the above model, all the parameters except the heat transfer coefficients are easily avaliable. Among three heat transfer coefficients involved, h_a is small compared with other two, so can not affect the jacket temperature sensitively while h_J can be reasonably expressed as a function of coolant flow rate (Perry et. al., 1984) although nonstandard geometry of the coolant path makes it difficult to choose an appropriate correlation. The remaining heat transfer coefficient h_R is, however, difficult to estimate in advance because it depends on physical properties of the reactor contents as wll as agitation speed, of which the functional relationship with h_R is not clearly understood. Based on the above considerations, h_R was rendered to be a parameter which should be estimat- ed from operating data. By rearranging the model equations, we can obtain the following state space model which can predict the jacket outlet temperature from process measurements.

$$\frac{dz}{dt} = \bar{A}z + \bar{b}, \qquad T_{J,N} = \bar{c}^T z \tag{6}$$

where

$$z^T = [\,T_{J,N},\cdots,T^{o}_{Iw,N},\cdots,T^{o}_{Ow,N},\cdots]$$
$$\bar{A} = \bar{A}(F_c;\ h_R)$$
$$\bar{b} = \bar{b}(F_c,\ T_R,\ T_{J,0},\ T_a;\ h_R)$$
$$\bar{c}^T = [\,1,\ 0,\cdots,0]$$

Using this predictor model, h_R can be determined such that the predicted $T_{J,N}$ fits the measured $T_{J,N}$ in an optimal sense.

2.2 *Reactor-Side Heat Balance*

Referring to the notations given in Fig. 1, the reactor-side heat balance can be written as follows.

$$M_R C_{PR}\frac{dT_R}{dt} = h_R \frac{A}{N}\sum_{k=1}^{N}(T_{Iw,k}(0) - T_R) + Q_R + Q_{AUX} \qquad (7)$$

where Q_R and Q_{AUX} denote heat of reaction and auxiliary heat effects caused by agitation, secondary feeding and so forth, respectively. Once h_R is estimated, $Q_R(t)$ can be calculated directly from eq. (7). Combining eq. (7) with eq. (6) produces an augmented state space equation which can be used as a reactor simulator. The resulting state space equation can be expressed as follows.

$$\frac{d\mathbf{x}}{dt} = \mathbf{A}\mathbf{x} + \mathbf{b} \qquad (8)$$
$$\mathbf{y} = \mathbf{c}^T\mathbf{x}$$

where

$\mathbf{x}^T = [\ T_R,\ T_{J,N},\ \cdots,\ T^o_{Iw,N},\ \cdots\ T^o_{Ow,N},\ \cdots\ ]$

$\mathbf{y}^T = [\ T_R,\ T_{J,N}]$

$\mathbf{b} = \mathbf{b}(T_{J,0},\ T_a,\ F_c,\ Q_R,\ Q_{AUX})$

$\mathbf{c}^T = [\ 1,\ 1,\ 0,\ \cdots,\ 0]$

3. APPLICATION TO A PBL REACTOR

3.1 *Process Description*

The concerned process is one operated in Cheil Industries Co., Korea. Figure 2 shows a schematic diagram of the reactor system. Internal volume of the reactor is 50 m^3. The reactor has an external jacket and also internal heat exchanger tubes to provide additional heat transfer area. The coolant recycle loop is not provided in this reactor. The reactor temperature is controlled by directly manipulating either hot water or cold water control valves which are under split range control. During reaction, set point of the reactor temperature is increased stepwise four times from 53 °C to 70 °C. Total reaction time is about 60 hrs but varies depending on batches.

According to the licensor's recipe, the reactor was originally operated under constant-gain PID control with a dead-zone and output limits. Because of inherent nonlinearity by the reaction and also a long response lag by the large volume, the reactor could not be controlled satisfactorily. Figure 3 shows typical profiles of process variables over a selected time span where heat of reaction is large. It can be noticed that the controller exerts an almost bang-bang type output (OP) saturated on the limits. As a result, the reactor temperature oscillates continu- ously. Furthermore, unnecessary heating (OP > 50%) occured sometimes during the reaction by the improperly designed control algorithm.

To remedy the troubled operation, it was decided to develop a reactor simulator first and to evaluate improved control strategies using the simulator before implementation.

3.2 *Reactor Modeling and Parameter Estimation*

The reactor modeling was carried out as was discussed in section 2. Here, the internal heat exchanger was not modeled separately. Instead, heat transfer area of the tubes was added to that of the jacket. In the jacket-side modeling, N=4 was used.

First, h_R was estimated such that the prediction error in the jacket outlet temperature is minimized as follows.

$$\hat{h}_{R,n} = \arg\min\left[\int_{t_n}^{t_n+200}(\hat{T}_{J,N} - T^m_{J,N})^2 dt\right] \qquad (9)$$

where ^ and the superscript ^{m} denote estimate (or prediction) and measurement, respectively. Since h_R is slowly time-varying, separate estimation was carried out over disjoint time intervals each of which has 200 min of span. The prediction error algorithm by Ljung (1983) was used in the calculation. Next, Q_R was calculated from eq. (7) and the complete dynamic heat balance equation was formulated.

In Fig. 4, predicted values of the reactor and jacket outlet temperatures are compared with the measured ones over a time interval. We can see the predicted values agree reasonably well with the measured values. Figure 5 shows the estimates of h_R and Q_R. As might be expected, h_R was found to decrease slowly from a large value at the monomer state to a small value as polymerization proceeds. Estimation of $Q_R(t)$ is very sensitive to measurement noise and also quantization error caused by finite-bit analog to digital conversion since time derivative of the reactor temperature is involved in the calculation. To mitigate the noise amplification, $Q_R(t)$ from eq. (7) was filtered by repeatedly taking the moving average from ten data points (sampled at every minute) until the filtered value fluctuates within a reasonable bound. Actually, the filtering method was arbitrarily chosen since there seem to be no appropriate guidelines how to filter $Q_R(t)$. Nevertherless the overall trend and the cumulative value of $Q_R(t)$ were found to agree well with general understandings on this reaction (Private communication, 1993)

3.3 *Representation of Q_R as a Local Function of Temperature*

As can be seen in Fig. 5, the heat of reaction was estimated to oscillate significantly, which can be considered to reflect the temperature dependency of the heat of reaction. If the oscillation in the reactor temperature is removed by an appropriately designed controller, Q_R will not oscillate, either. In this sense, in order for the reactor model to be a more realistic simulator, it will be desirable to represent Q_R as a function of temperature. Global dependency of Q_R on reaction temperature, however, can not be described unless detailed information on reaction kinetics is available. In this research, an empirical linear relationship between Q_R and T_R was found instead as follows.

First, by inspection we divided $T_R(t)$ and $Q_R(t)$ into the time-varying average profiles and the deviation terms.

$$T_R(t) = \overline{T}_R(t) + \Delta T_R(t) \tag{10}$$

$$Q_R(t) = \overline{Q}_R(t) + \Delta Q_R(t) \tag{11}$$

where the overbar and Δ denote the average profile and deviation term, respectively. In this expressions, it can be reasonably assumed that $\overline{Q}_R(t)$ and $\Delta Q_R(t)$ are contributed by $\overline{T}_R(t)$ and $\Delta T_R(t)$, respectively. Next, we regressed $\Delta Q_R(t)$ by a linear function of $\Delta T_R(t)$ as shown Fig. 6. Through this procedure, we obtained the following local linear relationship between $T_R(t)$ and $Q_R(t)$.

$$Q_R(t) = \overline{Q}_R(t) + a\Delta T_R(t) \tag{12}$$

3.4 *Controller Design*

Two different control problems exist in this process; regulation at the given set points and transition from one set point to another. It was decided to introduce gain-scheduled PID control for the regulation, and bang-bang type open-loop control for the set point change, respectively.

As can be seen from Fig. 2, the process was originally operated almost under relay-feedback during regulation. The oscillation, however, changed slowly as reaction proceeds. The whole reaction time was divided into six disjoint intervals according to the changes in oscillation, and ultimate periods and gains were determined for respective time intervals from the relay-feedback principle (Astrom and Hagglund, 1985). Using the some overshoot strategy (Seborg *et al.*, 1989), the initial PID settings were determined and then adjusted through simulation study.

Action of bang-bang control was designed to be ON, OFF and then switched to a preload value (Shinskey, 1988). Output levels as well as holding time in each action were determined entirely by trial and error using the reactor simulator.

Figure 7 shows performance of the designed controller applied to the PBL reactor. As can be seen, fairly nice set point tracking as well regulation could be attained over the whole reaction period while the on-off movement of the control valve was completely avoided during regulation. Moreover, unnecessary heating which had been observed in the original operation was also eliminated resulting in energy saving as an additional effect.

4. APPLICATION TO AN EPS BATCH POLYMERIZATION REACTOR

The reactor model was also applied to an EPS (Expanded Poly Styrene) batch polymerization reactor to assess the reactor behaviors, especially, the heat transfer characteristics as well as the heat of reaction and also to find an improved control startegy. The EPS reactor system is different from the PBL reactor system in that it has different volume, 100 m^3 and is equipped with an external coolant recycle loop. In addition, the EPS reactor is controlled by a cascade control loop. The structure of the reactor model and the tuning procedure applied to the EPS reactor were essentially same as in the PBL case.

Figure 8 shows performance of the developed model applied to the EPS reactor. It can be observed that the predicted values of the reactor temperature as well as the jacket outlet temperature coincide very closely with the measured values. The estimated heat of reaction is also shown in Fig. 9.

In Fig. 10, predictions of the jacket outlet temperature using two different jacket-side heat balance models are compared: one is without the conduction equation in reactor wall and the other is with the wall dynamic equation. When the the wall dynamics is not considered in the heat balance model, the jacket outlet temperature is predicted to have larger amplitudes than the actual measurements. With the wall dynamics incorporated in the heat balance model, however, the amplitudes are attenuated appropriately while reproducing the actual measurements more closely. This comparison demonstrates the importance of the wall dynamics in precise modeling of the batch reactor although the wall dynamics is overlooked in general.

Redesign of the controllers, especially to reduce the large and frequent control movement is under way using the developed reactor simulator

5. CONCLUSIONS

A generic dynamic heat balance model which is applicable to a large class of industrial batch reactors has been proposed by noticing that the industrial-scale reactors usually have measurable temperarture distributions along the cooling jacket irrespective of its specific design. The proposed model takes into account the temperature distributions along the jacket and the reactor wall in detail and includes one tuning parameter, the reactor-side heat transfer coefficient. The proposed model after tuned with process operating data is able to predict the temperature dynamics, heat transfer characteristics and also the heat of reaction. From practical applications to various industrial polymerization reactors, the foolowing conclusion can be drawn.

- . Detailed distributed modeling of the jacket-side is very important in precise desciption of reactor thermal dynamics although it has been neglected in usual modeling approaches..
- . The reactor-side heat transfer coefficient is a suitable tuining parameter which can adjust the dynamic heat balance model with a sufficient sensitivity.
- . The proposed dynamic heat balance model can paly as a precisely working reactor simulator for industrial batch reactors.

ACKNOWLEDGEMENT

The authors would like to acknowledge the financial support of Cheil Industries, Samsung Engineering Co. Ltd., Automation Research Center in POSTECH, and Korea Science and Research Foundation for financial support.

REFERENCES

Astrom K. J. and T. Hagglund (1985). *Automatic Tuning of simple regulators with specifications on phase and amplitude margins*, Automatica 20, 645-651.

Bovin, D., P. DE Valliere, and D. W. T. Rippin (1989). *Comp. Chem. Eng.*, 13, 1.

Karlsen, L. G. and J. Villadsen (1987). *Chem. Eng. Sci.*, 42, 1153.

L. Ljung and T. Soderstrom (1983). *Theory and Practice of Recursive Identification*, MIT Press, Cambridge.

Perry, R. H., D. W. Green, and J. O. Maloney (ed.) (1984). *Perry's Chemical Engineers' Handbook*, 6th ed., McGraw-Hill, N.Y.

Private Communication with Researchers in Technical Lab. in Cheil Industries Co. (1993).

Seborg, D. E., T. F. Edgar, and D. A. Mellichamp (1989). *Process Dynamics and Control*, John Wiley and Sons, N.Y.,

Shinskey, F. G. (1988). *Process Control Systems: Application Design and Tuning*, 3rd ed., McGraw-Hill N.Y.,

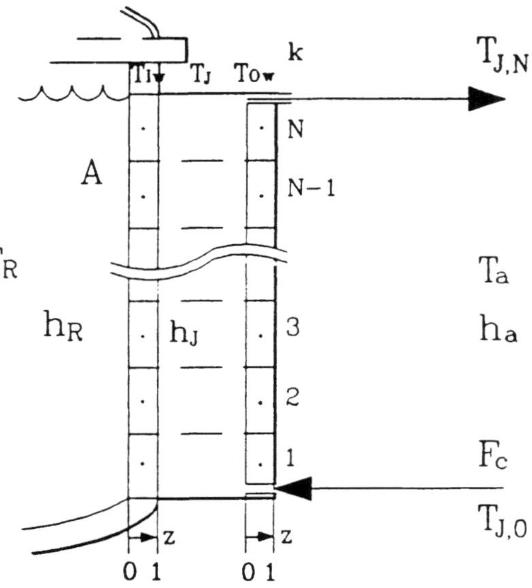

Figure 1. Jacket-side heat balance modeling.

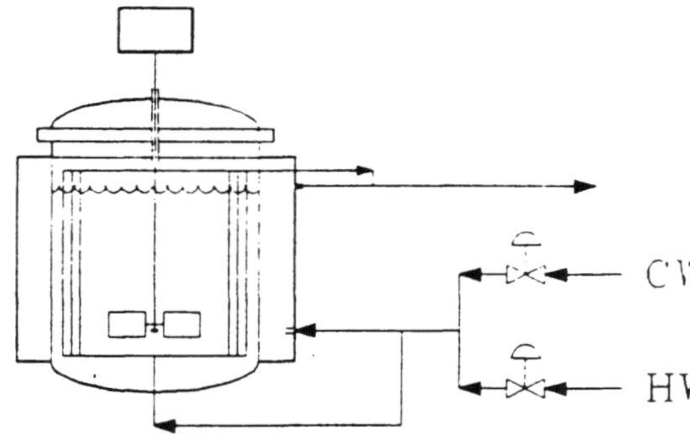

Figure 2. PBL reactor.

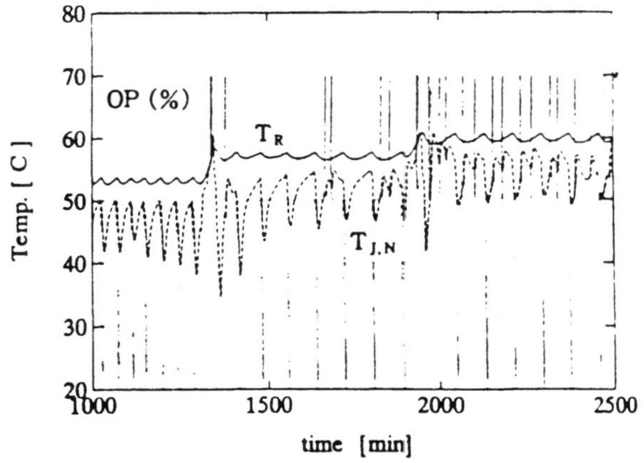

Figure 3. Typical profiles of process variables for PBL reactor.

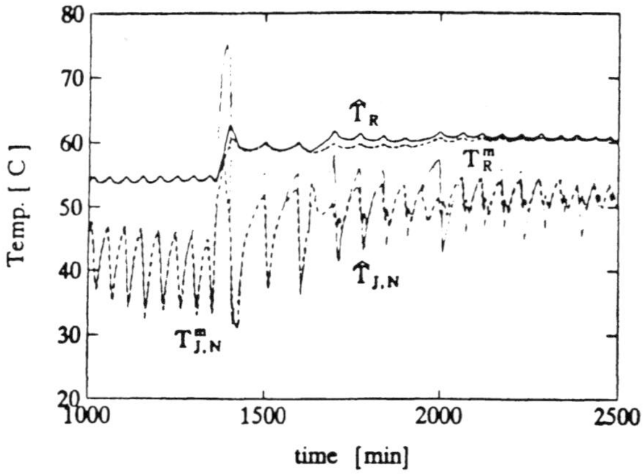

Figure 4. Results of open loop simulation for PBL reactor.

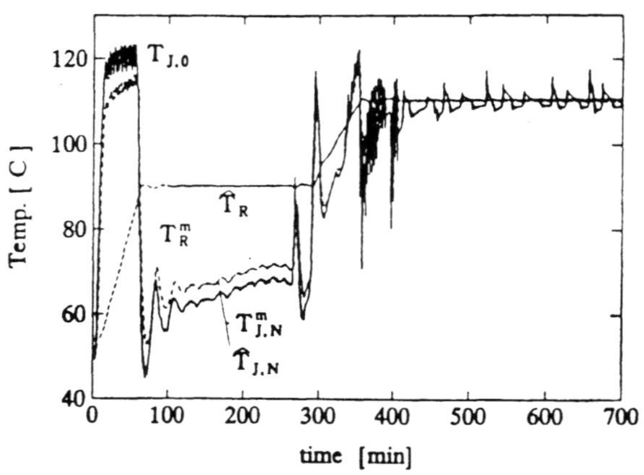

Figure 8. Results of open loop simulation for EPS reactor.

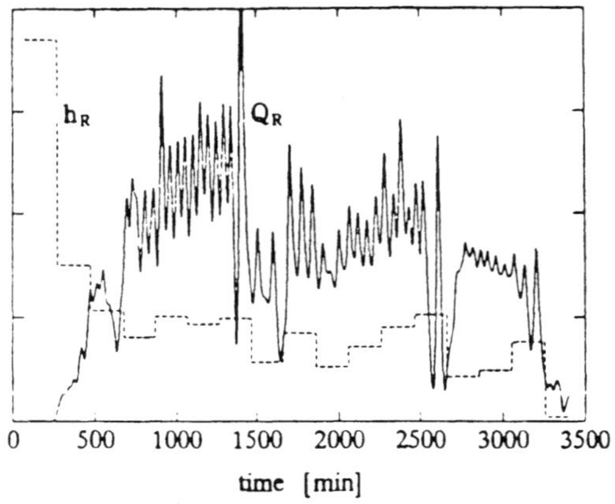

Figure 5. Estimated $h_R(t)$ and $Q_R(t)$ for PBL reactor.

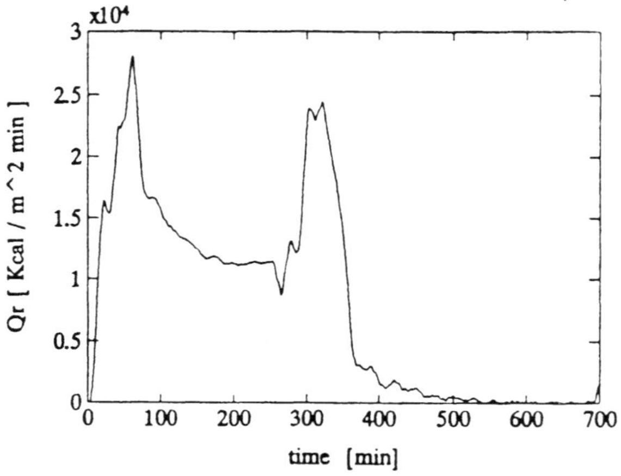

Figure 9. Estimated $Q_R(t)$ for EPS reactor.

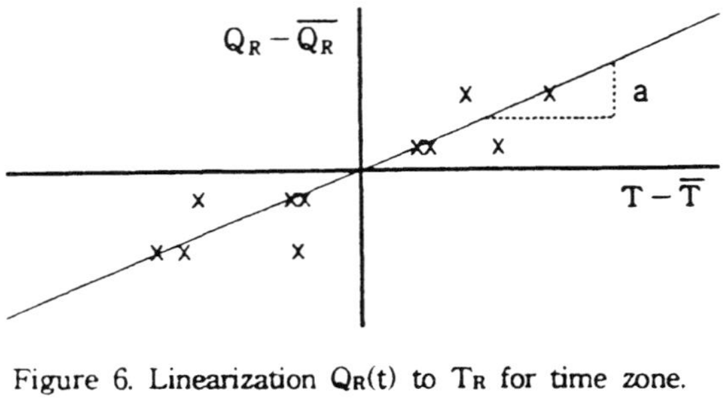

Figure 6. Linearization $Q_R(t)$ to T_R for time zone.

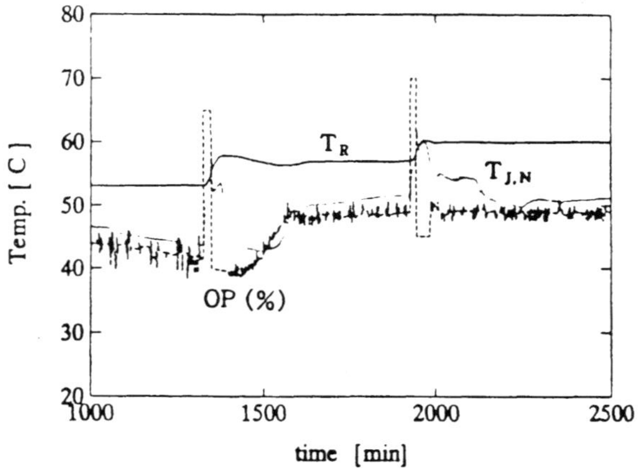

Figure 7. Performance of the improved control method.

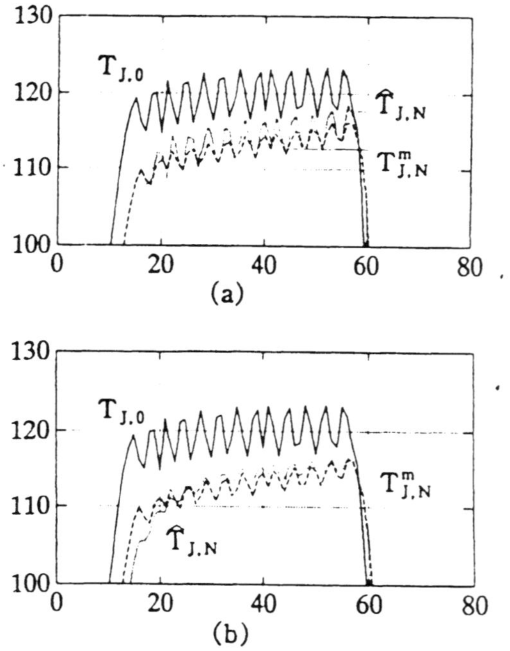

Figure 10. Effects of wall dynamic model
(a) without wall dynamic model.
(b) with wall dynamic model.

METHODS OF DUALITY IN PROBLEMS ABOUT EXPENDIENCY OF USAGE AND OPTIMIZATION OF CYCLIC REGIMES OF TECHNOLOGICAL PROCESSES.

Tsirlin A.

Programm System Institute of Russian Academy of Siences.

Abstract. Here were shown possibiliteis of usage mathematical dual methods for solving quastion about appropriate of usage cyclic regime and for estimate of effictivity of transition to this regime. There are conditions with which optimal cyclic regime can be finded as solution of mathematical programming problem. In this paper there are examples for chemical reactors, heat machines, absorption - adsorption cycles.

Key Words. optimality, cycle length, duality, periodic replecement.

1. INTRODUCTION.

Two kinds of steady-state technological processes are usage in practice. The first one is a static process, when control influences and parameters of the process are constant in time. The second one is a cyclic process, when one or several the process parameters change periodically in time. A length of an apparatus can be used as the time. Then the process parameters change periodically with the position in the apparatus. These processes are widely practiced in chemical technology (regeneration heat-exchange, air separation cycles, cyclic catalytic processes, absorption-adsorption engines etc.)

The following main problems are springing up in a consideration of the cyclic regimes of technological processes.

1. Does the cyclic regime exist to satisfy the problem conditions?
2. What is expediency of a transfer from static to cyclic regime?
3. What is excess of the optimal criterions in the transfer?
4. What is an optimal variation of the controls of the cyclic process?

In this paper some answers the questions are proposed. The solutions were got by duality optimization methods. The same methods were applied to concrete problems.

2. OPTIMIZATION OF CYCLIC PROCESSES AND AVERAGING PROBLEMS OF MATHEMATICAL PROGRAMMING.

Let x be a state variables vector and u be a control vector of a technological process with lumpted parameters. Let τ be a duration of the cycle. Let $f_0(x,y)$ be some characteristic of the process efficiency.

The problem of maximizeing average efficiency per cycle in restrictions of resources and in dynamical connections between x and u is considered. This problem has following form:

$$\overline{f_0} = \frac{1}{\tau}\int_0^{\tau} f_0(x,y)dt \rightarrow \max \qquad (1)$$

in conditions

$$\dot{x}_\nu = f_\nu(x,y); \quad \nu=\overline{1,m}, \quad u \in V \qquad (2)$$

$$\overline{\varphi}_j = \frac{1}{\tau}\int_0^{\tau} \varphi_j(x,y)dt = 0, \quad j=\overline{1,k} \qquad (3)$$

The conditions (3) correspond to restriction on intensities of resources expenses (energy, raw material etc.). The conditions (2),(3) need to complete the cycle conditions $x_\nu(0) = x_\nu(\tau)$ $(\nu = 1,\ldots m)$. Then (2) imply

$$\overline{f_\nu} = \frac{1}{\tau}\int_0^{\tau} f_\nu(x,y)dt = 0, \quad \nu=\overline{1,m} \qquad (4)$$

The problem (1)÷(4) is a problem of optimal

control, if the functions f_0, f_ν, φ_j are continuously on u and continuously differentiable on x. The special form of this problem characterize by free choice of boundary conditions in accord (4) and averaging form both criterion (1) and conditions (3),(4). With the problem (1)÷(4) four auxiliary problems are considered. They are different from (1)÷(4) by absence of the condition (2), which are most difficult to solve:

Problem Qs (quasistatic):

$$\overline{f}_0 = \frac{1}{\tau}\int_0^\tau f_0(x(t), u(t))dt \to \max \quad (5)$$

in conditions (3) and the restrictions

$$f_\nu(x(t), u(t)) = 0, \quad u \in V, \quad \forall t \in [0, \tau] \quad (6)$$

Physically the problem **Qs** corresponds to optimization of the quasistatic regime of stable in neighborhood of every control of $u \in V$.Control changes so slowly that x and u can be taken for each t with conditions (6).

Problem Sm(sliding mode):

$$\overline{f}_0 = \frac{1}{\tau}\int_0^\tau f_0(x, u(t))dt \to \max \quad (7)$$

in the conditions (3) and (4), where u is a time function, which changes so fast that vector x is constant.

Problem Es(estimate):

$$\overline{f}_0 = \frac{1}{\tau}\int_0^\tau f_0(x(t), u(t))dt \to \max \quad (8)$$

in conditions

$$\overline{\varphi_j} = \frac{1}{\tau}\int_0^\tau \varphi_j(x(t), u(t))dt = 0;$$

$$u \in V; \quad j = \overline{1,k} \quad (9)$$

$$\overline{f}_\nu = \frac{1}{\tau}\int_0^\tau f_\nu(x(t), u(t))dt = 0;$$

$$\nu = \overline{1,m} \quad (10)$$

There is not the condition (2) nevertheless $x(t)$ is subject to choice.

Problem S(static):

It is the optimal static regime's problem where x and u are constant.

$$f_0(x, u) \to \max \quad (11)$$

with conditions

$$u_j(x, u) = 0; \quad f_\nu(x, u) = 0;$$
$$u \in \nu; \quad j = \overline{1,k}; \quad \nu = \overline{1,m}$$

It is the problem of mathematical programming.

The inclusions

$$D_{Es} \supset D_C; \quad D_C \supset D_{Sm}; \quad D_C \supset D_{Qs};$$

$$D_{Qs} \supset D_S; \quad D_{Sm} \supset D_S \quad (12)$$

are correct for sets of admissible values of the above five problems.

This inclusions make possible to tell if problem **Es** has no solution, i.e. the conditions (9),(10) and the restriction $u \in \nu_u$ disjoint, then the problem **C** has no solution too. If only one of the problems **Qs** or **Sm** has a solution then the problem **C** has a solution too.

Let I with the corresponding subscript be a maximal criterion value for every discussing problems. The inclusions (12) imply next inequalities

$$I_{Es} \geq I_C \geq I_{Sm} \geq I_S; \quad (13)$$
$$I_{Em} \geq I_C \geq I_{Qs} \geq I_S$$

It makes possible to find a difference Δ between I_C and I_S.

$$\max\left\{ \begin{array}{c} (I_{Sm} - I_S) \\ (I_{Qs} - I_S) \end{array} \right\} \leq \Delta \leq I_{Es} - I_S; \quad (14)$$

The solutions of the auxiliary problem corresponds to the solution of the problem **C** in the case when inequalities (13) become equalities.

For example, if $I_{Es} = I_{Sm}$ or $I_{Es} = I_{Qs}$ the solving of the auxiliary problems **Qs, Sm** and **Es** to draw a important conclusion about cycle regime of the process. All these problems are solved by method of duality, that is their solutions are reduced to minimax principal of Lagrange function.

There are these results:
1.

$$I_{Es} = \min_{\lambda,\mu}\{\max_{x,u}[f_0(x, u) + \sum_{\nu=1}^{m} \lambda_\nu f_\nu(x, u) +$$

$$+ \sum_{j=1}^{k} \mu_j \varphi_j(x, u)]/u \in V\} \quad (15)$$

The condition (15) allows to find x and u simultaneously with the maximal value I_{Es}. Here x and v have not finite numbers of base values $v^i = \{x^i, u^i\}$ more then $(m + k + 1)$. The solution of the problem **Es** is a step function. It takes every base values v^i during γ_i part of the interval τ. The weight coefficients γ_i are determined by conditions (9),(10), which have such forms in this case

$$\left.\begin{array}{ll} \overline{f_\nu} = \sum_{i=0}^{m+k} \gamma_i f_\nu(v^i) = 0; & \nu = \overline{1,m} \\ \overline{\varphi_j} = \sum_{i=0}^{m+k} \gamma_i \varphi_j(v^i) = 0; & j = \overline{1,k} \\ \sum_{i=0}^{m+k} \gamma_i = 1; & \gamma_i \geq 0 \end{array}\right\} \quad (16)$$

2.

$$I_{Q_s} = \min_{\mu_j} \max_{x,u} \{[f_0(x,u) +$$

$$+ \sum_{j=1}^{k} \mu_j \varphi(x,u)] \Big/ \begin{array}{l} f_V(x,u) = 0 \\ u \in V, \ \nu = \overline{1,m} \end{array} \Big\} \quad (17)$$

Here base values of vector $v^i = \{x^i, u^i\}$ are defined too. Amount of base values is not more then $(k+1)$ in this problem. Weight coefficients γ_i are determined by the two last conditions of (16), and the upper index of the sum in (16) is k, not $(m+k)$.

3.

$$I_{Sm} = \min_{\lambda,\mu}\{\max_{u}[f_0(\hat{x},u) + \sum_{\nu=1}^{m} \lambda_\nu f_\nu(\hat{x},u) +$$

$$+ \sum_{j=1}^{k} \mu_j \varphi_j(\hat{x},u)]\} \quad (18)$$

Base values u^i are defined from dual problem (18). Number of u^i are less then or equal to $(m+k+1)$. The vector $\hat{x}$ and weight coefficients γ_i must be found from conditions (16), where $v^i = (\hat{x}, u^i)$, and from the equality

$$\frac{\partial}{\partial x}[\overline{f}_0 + \sum_{\nu=1}^{m} \lambda_\nu \overline{f}_\nu + \sum_{j=1}^{k} \mu_j \overline{\varphi}_j] = 0 \quad (19)$$

It should be noted that all dual problems have requirement of some function maximum and the differentiability of the function have not assumed. So u can be integer-valued in problem **Sm**. Moreover calculation of based value makes independently on estimate of coefficients γ_i in problems **Es** and **Qs**. It makes solving more simple.

3. RESULTS OF USING DUAL METHODS FOR SOME PROCESSES.

3.1. *Cycle rate variation of chemical reactor.*

Let static characteristics of process are known, that is dependence of output flow's concentration C_B and it's cost S_b on rate flow G through the appara tus. One of the problems may be following. t_0 maximize the averaged per cycle cost of product in fixed averaged concentration of one of components (for example the first) in output flow. The rate $G(t)$ is control.

$$\overline{f}_0 = \frac{1}{\tau} \int_0^\tau S(G)G(t)dt \rightarrow \max \quad (20)$$

in condition

$$\frac{1}{\tau} \int_0^\tau (C_{b1}(t) - \overline{C}_{b1})G(t)dt = 0 \quad (21)$$

where $\overline{C}_{b1}$ is the fixed averaged concentration.

$$C_{bi}(t) \geq 0; \qquad \sum_{i=1}^{k} C_{bi}(t) = 1$$

It is sufficient to know dependence $C_b(G)$ to solve problem **Qs** about optimal quasistatic regime. Base values $G^i (i = 1,2)$ defined by solution of dual problem.

$$I_{Q_s} = \min_{\lambda} \Big\{ \max_{G}[S(G) + \lambda(C_{b1}(G) -$$

$$-\overline{C}_{b1}]G \Big/ \begin{array}{l} G \geq 0 \\ C_{bi}(G) \geq 0 \\ \sum_{i=1}^{k} C_{bi} = 1 \end{array} \Big\} \quad (22)$$

If the base value is uniquely ($\gamma_1 = 1; \gamma_2 = 0$) the G_1 is solution of equation

$$C_{b1}(G_1) = \overline{C}_{b1}$$

Otherwise value I_{Q_s} gives lower estimation the advantage of transfer to the cycle regime (here processes are stable).

For example, in ideal displacement reactor with reaction

$$A \xrightarrow{k_1} B \xrightarrow{k_2} C$$

and kinetic equations

$$\dot{C}_A = -C_A; \quad \dot{C}_2 = C_1 - \frac{k_2}{k_1}C_2; \quad \dot{C}_3 = \frac{k_2}{k_1}C_2$$

the output flow's cost is

$$S = 10C_3 - 0.1C_1 - 0.1C_3,$$

where C_1, C_2, C_3 concentrations of components in accordance.

The $C_{bi}(G)$ is given by kinetic equation. Calculations (Tsirlin,1978) have shown that cycle changes of rate are appropriated and imply the advantage in optimality criterion about 7%.

A common appropriate's conditions of cycle rate changes for reversible reactions and two parallel reactions have treat in (Tsirlin, 1978).

3.2. *Extreme efficiency of heat machine with fixed power.*

The heat machine which consists of two heat reservoirs with temperatures T_+ and $T_-(< T_+)$ and working fluid characterized by internal energy E and entropy S is considered. They are change accordingly differential equations

$$\dot{E} = q(T_0, T) - p; \quad E(0) = E(\tau) \tag{23}$$

$$\dot{S} = \frac{q(T_0, T)}{T}; \quad S(0) = S(\tau) \tag{24}$$

where q is heat flow, which depends on the reservoir temperature (T_0 is equal to T_+ or T_-) and the working fluid's temperature T; p is result power.

The efficiency with fixed average power $\overline{p}$ should be maximized. As the maximum efficiency in fixed power is according to minimum getting heat, then the problem has form:

$$\overline{f_0} = \frac{1}{\tau}\int_0^\tau \frac{1}{2}(q(T_0, T) + \\ + \mid q(T_0, T) \mid)dt \rightarrow \min \tag{25}$$

with condition (24) and the request arised form (23)

$$\frac{1}{\tau}\int_0^\tau (q(T_0, T) - \overline{p})dt = 0 \tag{26}$$

The estimation problem Es reduces to the dual problem

$$I_0 = \min_\lambda\{\min_T[q(T_0, T) + \mid q(T_0, T) \mid + \\ + \lambda_1 \frac{q(T_0, T)}{T} + \lambda_2(q(T_0, T) - \overline{p})]\} \tag{27}$$

The function in (27) is convex on T for both possible values T_0 for overwhelming majority of heat-exchange lows. Then we can see that number of base solution is equal to two not to three: $v^1 = (T_+, T_1)$; $v^2 = (T_-, T_2)$.

The parts γ_i of duration τ (the contacts with the hot and the cold reservoirs) define from conditions:

$$\gamma_1 \frac{q(v^1)}{T_1} + \gamma_2 \frac{q(v^2)}{T_2} = 0; \gamma_1 + \gamma_2 = 1; \quad \gamma_i \geq 0 \tag{28}$$

The problem (27) gives us optimal solution of (23)-(26) because of $I_{Es} = I_{Qs}$.

The solution of this problem to linear heat-exchange low

$$q = \alpha(T_0 - T)$$

there is in [4].

There the extreme value of efficiency is

$$\eta^* = \frac{2\overline{p}}{\overline{p} + \frac{\alpha(T_+ - T_-)}{4}} - \\ - \frac{2\overline{p}}{\sqrt{\overline{p}^2 - \frac{\alpha}{2}(T_+ + T_-)\overline{p} - \frac{\alpha^2}{16}(T_+ - T_-)^2}} \tag{29}$$

4. ABSORPTION-ADSORPTION CYCLE.

The problem of extreme productivity of absorption-adsorption scheme of gas separation with thermic adsorption is analogous to problem 3.2. The ratio of the useful product to the expending heat plays the role of efficiency. To maximize the efficiency in fixed amount of useful product the heat expenditure should be minimize. The process have been organized cyclically, working fluid (solution of monoethanolamin) is in contact with separative gas in absorber and with hot steam in adsorber in turn.

Argument t denote the responce time of working fluid element. The fluid purified in absorber is characterized by temperature T_a and concentration of redistribute component Y_a. The steam in desorber have temperature T_d and concentration Y_d. The working fluid is characterized by entropy E and concentration of redistribution component x. They are satisfied to condition

$$\dot{S} = \frac{q}{T} - \frac{\mu n}{T}; \tag{30}$$

$$\dot{E} = q; \tag{31}$$

$$\dot{x} = \lambda n \tag{32}$$

where g and n – heat and mass fluxes, μ - chemical potential of redistribution component in solution ($\mu = \mu^0(T, P) + RT \ln x$), μ^0 – standard chemical potential, R – universal gas constant, λ – constant.

The problem about extreme productivity in average form becomes

$$\overline{n}_g = \mid \overline{n} \mid \rightarrow \max_{\mu, T} \Big/ \begin{array}{l} \overline{\frac{q}{T}} - \overline{\frac{\mu n}{T}} = 0, \\ \overline{q} = 0; \quad \overline{n} = 0. \end{array} \tag{33}$$

with cycle conditions for equations (30)-(32) and with productivity of process N which is defined by output flux from adsorber.

$$N = -\frac{1}{\tau}\int_0^\tau (n - \mid n \mid)dt$$

In the case the processes of heat and mass transfer are characterized by Onzager's equation and the coefficient's matrix in absorber and adsorber are the same, the analytic solution of problem (33) is

$$\overline{n}_g^* = \frac{K\beta\Delta u_1^2}{4(\alpha\beta K\frac{\Delta u_1}{2} + K\Delta u_2 + \sqrt{C})};$$

where

$$C = K^2\Delta u_2^2 + \beta K\Delta u_1^2 + \beta K^2\alpha\Delta u_1\Delta u_2;$$

$$\Delta u_1 = \frac{1}{T_a} - \frac{1}{T_d} \quad \Delta u_2 = \frac{\mu_d}{T_d} - \frac{\mu_a}{T_a}$$

Coefficients α, β and K are expressed throw the coefficients a, b and γ of Onzager matrix. b and γ are the main diagonal elements. a is nondiagonal element.

$$\alpha = \frac{2a}{\gamma b - a^2}; \quad K = \gamma - \frac{a^2}{b}; \quad \beta = b - \frac{a^2}{\gamma}.$$

In the heat and mass transfer laws are different then the solution of the extreme efficiency problem can be found only numerically.

5. CONCLUSION.

Many problems of optimal control for cycle processes can be solved by solution of the minimax dual problem. It facilitates a research essentially.

6. ACKNOWLEDGEMENTS.

This work have been maked up with INTAS support, grant 93-1281.

REFERENCES.

Rozonoer, L.I., A.M. Tsirlin. (1983). Optimal control thermodynamic processes. *Avtomatika & Telemehanika.* **1-3.** Russia.

Tsirlin, A.M. (1978) *Variational methods of chemical apparatus estimation.* Mashinostroenie, Moscow.

Tsirlin, A.M. (1985) *Optimal cycles and cyclic regimes.* Energoatomizdat, Moscow.

Tsirlin, A.M. (1992). Condition of optimality of solving average problem of mathematical programming. *DAN SSSR.*, **1**, 43-48. Russia.

STATUS AND CHALLENGES OF INTELLIGENT PLANT CONTROL

Morten Lind

Institute of Automation
Technical University of Denmark
DK-2800 Lyngby

Abstract: Intelligent plant control is a rapidly expanding research area that is nourished by the industrial interest in improving the efficiency and safety of plant operations, by the general academic interest in developing theories for control of complex systems and by results from AI research. Currently a variety of techniques for building advanced controls has been developed but there is no general consensus on the meaning of basic terms and there seems to be a limited recognition of the importance of common sense engineering knowledge in realising intelligent controls. In the paper it is argued that common sense knowledge of the purpose of plant components and subsystems play a significant role in building dynamic models, in diagnosis and planning and in the design of supervisory displays for the human supervisor. The lack of understanding and formalisation of common sense engineering knowledge within the control domain limits the levels of plant automation that can be achieved and impede the development of methods for design of supervision and control systems. Recent research results of the author on the modelling of plant goals and functions (Multilevel Flow Modelling) show that modelling research that supplement and expands the traditional modelling paradigms of control theory may lead to significant advances in the design of intelligent plant controls.

Keywords: Artificial intelligence, Diagnosis, Intelligent Control, Knowledge representation, Man/machine interfaces, Models.

1. INTRODUCTION

Current research directions within plant control are aiming at the development of so-called intelligent plant control systems based on artificial intelligence and other advanced information processing technology. Designing and operating complex process plants are knowledge intensive tasks and it is the expectation that these new knowledge technologies can make major contributions to the reduction of the risks and costs involved in the design, operation and maintenance of complex production plants. The reason for this optimism is that artificial intelligence offers a range of techniques for representation and processing of knowledge that extends the role of the computer in control applications from a number crunching machine to also being a powerful information processor.

The ultimate acceptance and success of these new information technologies for application in industrial control is however dependent on several factors. On a practical level it should be demonstrated that the cost of the increased complexity of control solutions will lead to economical benefits and more reliable and safe technical solutions to automation problems. On a more theoretical level there should be a clarification of the basic limitations of AI and of the consequences that these limitations have for applications within plant automation. Some of these issues will be addressed in the following.

The overall aim of the present paper is to provide an overview of research in intelligent plant control and to discuss some of the underlying problems and promises of the field. It is not the intention to present a comprehensive review of current research on intelligent plant control.

1.1 Current trends in intelligent plant control.

There are several trends in current research in intelligent plant control. One trend is to improve the performance of existing control solutions. The existing body of established control theory, that is almost exclusively based on the representation of plant knowledge and the formulation of control strategies in mathematical quantitative terms, has been very successful for solving a range of plant control problems. We could mention the use of techniques for identification of plant states and parameters, the synthesis of optimal control strategies and multivariable controls. These solutions are improved by extensions based on symbolic processing techniques derived from AI research (expert control and qualitative control). Within this trend we find also the development of methods for fault diagnosis based on techniques combining quantitative and symbolic processing techniques. Another trend is the development of novel solutions using neural networks and fuzzy set theory. On the systems level you will also find architectures for intelligent systems integrating diagnosis, planning and control concepts. Overall, these trends can be seen as being technology driven, initiated by the advent of various concepts, methods and tools from artificial intelligence research.

Yet another trend is more problem driven and use AI technologies to change the existing levels of plant automation e.g. by the development of automated start-up, shut-down and emergency procedures and computer based systems for supporting plant supervisory personnel in plant monitoring, diagnosis and execution of procedures. We see here that problems that was not tackled by conventional mathematical control theory can be solved by using the wider range of techniques for representation of knowledge offered by AI.

1.2 The concept of intelligence.

The research field of intelligent control suffers from a lack of well established terminology (Antsaklis, 1994). A particular problem is to find generally accepted definitions in the control theoretical literature of the concept of intelligence. Considering the fact that it is also difficult to find widely accepted definitions of human intelligence within psychology, the lack of consensus within the control community may not be so surprising. But even though the topic is difficult to handle it is so important that a few paragraphs will be spent here on definitions. Two complementary ways of defining intelligence will be introduced. The first definition is useful for characterising intelligent control as a technology. The second definition characterise the problems that intelligent control systems are supposed to solve and is accordingly more functional and independent on any particular technology. The discussion of the two alternative definitions will be used to circumscribe and formulate the basic problems of plant modelling that will be discussed in detail later in the paper. These modelling problems comprises main challenges to research in intelligent control.

Intelligence as a capacity. According to Webster(1989) intelligence is 'a capacity for reasoning, understanding and for similar forms of mental activity'. However, reasoning comprise a wide variety of activities such as deduction and induction processes usually associated with logic. The mental activities involved in understanding phenomena comprise various processes of interpretation such as classification and the processes involved in the analysis and synthesis of context. Reasoning is what you do when you play chess by using the rules of the game to reason about possible moves. Reading a book involves problems of context because the interpretation of the text is based on own personal experience and on knowledge of the authors intentions. The variety of these mental processes make it difficult to give a single comprehensive definition of the concept of intelligence as a capacity to reason and understand. This difficulty is illustrated by the availability of a variety of complementary techniques within artificial intelligence research for knowledge *representation* like predicate logic, rule based systems, semantic nets, frame systems and neural networks. In addition, several mechanisms for knowledge *processing* has been developed such as production systems, blackboard architectures and message passing. Each of these techniques has been developed to meet requirements to the simulation of specific mental processes involved in language understanding, vision, diagnosis, robotics and theorem proving. There have been attempts to build general problem solvers like e.g. GPS (Newell and Simon, 1972) and SOAR (Laird et. al., 1986) but the success has been limited. In most cases highly specific knowledge about the task to be solved is required.

Research in Artificial Intelligence has been met with serious criticism from Dreyfus(1979) and Searle(1984). The key point in the critique is that human intelligence is dependent on so-called background knowledge about the world that is only accessible to formalisation indirectly or with difficulty. Background knowledge is therefore difficult to represent and process in a computer. This difficulty would explain the problem of AI systems that their performance degrade in a non graceful manner when pushed to the limit. There are, according to these views, fundamental limits to formalisation of human knowledge. The main problem is to understand the mechanisms involved when humans select problem representation or point of view depending on the situation or context. Furthermore, the context of a given problem is in principle of infinite extension and can therefore only be accounted for partially or in approximation. As will be discussed later in this paper this is not only a problem for research in artificial intelligence it is also a problem for control theory. The problem of context is a bottleneck for understanding the nature of process knowledge. This problem provides therefore also a limit to the level of plant automation that can be achieved.

Intelligence as degree of adaptation. An alternative way to define intelligence is to adopt a functional perspective. According to this view, an organism or system is intelligent if it is adapted to its environment i.e. that it is coping with the demands of e.g. efficiency, autonomy or survival to which it is exposed. This view has been promoted by psychologists with a biological orientation (Piaget, 1966). With this definition it is considered immaterial how the adaptation manifest itself or how that state is obtained i.e. what mental representations or processes that are involved. What counts is the result. Intelligence is in this sense a measure of the degree of fitness between the system and its environment. This fitness can be further defined by the set of relations or constraints that exists between states of the system/organism and its environment as a result of the adaptation. An intelligent system would in this definition have a highly complex relationship with its environment.

The problem of context discussed above is obviously not solved by changing the definition of intelligence. It is still there as a problem of accounting for these relations or constraints that exists between the system/organism and its environment. The only effect of changing the definition is that the problem has got another formulation. However, as will be shown in the next section, this reformulation is a constructive one since it connects the problem of defining intelligent control with problems of plant modelling.

1.3 The concept of intelligent control.

Both definitions of intelligence introduced above can be used to define intelligent control. According to the first definition, where intelligence is defined as a capacity to perform certain types mental of operations, intelligent control means the use of these capacities in the performance of control actions. Examples from current research in intelligent control systems described later can be characterised in this way. According to the second definition, a control system would be intelligent if it is highly adapted to its environment i.e. that it is able to cope with complex plant control requirements. In order to take full advantage of this definition more analysis of the purposes of control actions is required.

The purposes of control actions. The purpose or goal of plant control actions can be to ensure that plant operating goals are achieved and maintained in spite of uncertainties and resource constraints. Other purposes of control actions could be the prevention of safety critical plant states or the mitigation of consequences of failures in plant components or systems. The observation of plant parameters, the identification of plant state, the selection between different possible courses of action and the planning and execution of the control action are all done with one or several of these purposes in view. This *intentionality* i.e. direction towards the *achievement* and *maintenance* of plant goals, the *prevention* and *termination* of critical plant states and the *remediation* of failures are distinctive features of control actions. The realisation of each of these control purposes require different types of knowledge about the plant

This can be illustrated by the following examples; a simple thermostat and a control system for a power plant. These systems differ in the complexity of the constraints they produce and maintain between their inner states and the states of the plant. A thermostat maintains a simple relationship between the setpoint and the temperature of the environment. Its actions are based on knowledge about the setpoint to be obtained and the dynamic causal relations between the means of control and the actual plant temperature. The power plant controller ensure that the many degrees of freedom of the plant are maintained at the desired setpoints and that changes in overall power production requirements are transformed into co-ordinated changes of setpoints and/or into hierarchically organised sequences of actions that changes the configuration of the plant and its subsystems. The actions of the power plant controller are, like the actions of the thermostat, based on knowledge about the plant dynamics. But the actions of the power plant controller are also based on knowledge about relations between the multiple plant goals e.g. priority, interdependence and temporal organisation, and the relations between plant goals and the means or functions provided for their achievement. Knowledge about relations between plant goals and functions play a significant role for the power plant controller but is not important for the thermostat because the *goal structure* of its environment is extremely simple (the causal structure i.e. the dynamics, can however be complex). Knowledge of plant goals and functions are, as shown later, also very important in the remediation of failures.

Evidently, the power plant controller is intelligent (in the second definition) because it maintains a highly complex set of constraints between its own inner state and the state of the plant. But what about the thermostat? The simplicity of the constraint maintained would rank it on the rather the low end of the scale of intelligence. But the attribution of intelligence to the thermostat would also depend on how well it is able to maintain the temperature i.e. on its level of performance. Intelligence in control should therefore be evaluated according to two dimensions: *the complexity of the task* and *the level of performance obtained*. The independence of these dimensions can be illustrated by the fact that the power plant controller could have a very lousy performance but still be able to solve its tasks.

1.4 Autonomy and co-operation.

In the discussions above the social dimension of intelligence has been entirely neglected. In so-called autonomous systems (Maes, 1990) the idea is to build intelligent systems that are characterised by being independent on interaction with other agents. However, it is often the case that the plant to be controlled by an agent comprise other agents whose actions may make demands on common resources or are guided towards conflicting goals. In such cases intelligent action includes the ability to co-operate i.e. to co-ordinate goals and resources with other agents, to communicate, to delegate and take responsibility for tasks. This social dimension of intelligence is explored within the area of distributed artificial intelligence (Bond and Gasser, 1988). Distribution of control tasks between several agents can be required in order to avoid break down in the case of plant or controller failures. In other cases the distribution of tasks is necessary in order to satisfy learning needs of agents. A case in point is the need of keeping operators involved in plant operations even in highly automated plants. Operator involvement is required because otherwise he/she is not able to maintain competence as a supervisor (Bainbridge, 1982). This problem put a limit to the level of automation that can be obtained in plants which require human intervention in critical high risk situations. Those plants can therefore not be controlled without human participation. Developing intelligent control systems for such environments therefore involves the solution of problems at the human-machine interface. Design problems in this area related to plant modelling will be discussed below.

2. A STATUS OF AI APPLICATIONS IN CONTROL

Two main strands of current research in intelligent control can be separated by a distinction between *intelligent control* and *intelligent **plant** control*. In intelligent controls artificial intelligence techniques are used to improve the performance of feedback systems. The purpose of intelligent plant control is to use AI for improving the total operation of the plant and will include high level tasks in addition to the low level loop control such as configuration and fault management, supervision and planning. The main distinction between these two research directions is the difference in the complexity of the task to be solved. This distinction will be used tentatively in the following.

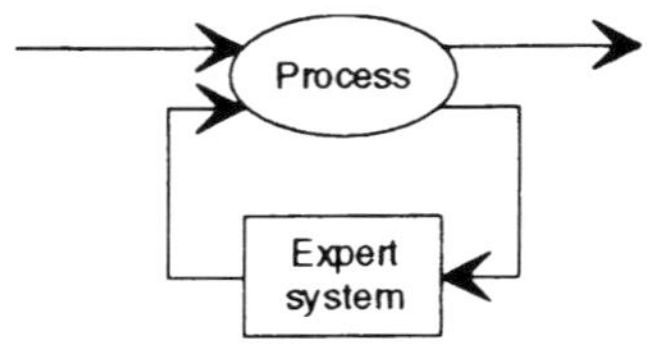

Fig. 1. Qualitative control.

2.1 Intelligent control.

Within intelligent control two approaches can be distinguished namely qualitative control and expert control.

Qualitative control. The basic principle in qualitative control is to use an expert system directly in the feedback loop as shown below. This solution to control problems may be used in cases where it is difficult to establish a mathematical model of the process. In such cases advantage can be taken of the experience of plant operators by using their knowledge as a basis for building a knowledge base for the control system. A notable example of this approach is fuzzy control. Fuzzy control techniques can be used with advantage when the control rules are vague and therefore cannot be described in precise mathematical terms. According to the definitions of intelligence above, these systems are intelligent because they have a capacity for reasoning e.g. by rules or by means of qualitative models of the plant dynamics.

Expert control. In expert control (Åstrøm, et al., 1986) a knowledge based system (e.g. a rule based system) is used to extend the functionality of a conventional numerical controller. This principle can be applied for solving control problems where parameters of the numerical controller depend on various conditions that cannot be fully described in mathematical form. An expert system could be used with rules for setting or adjusting the parameters or algorithms in the numerical controller according to conditions dependent on plant state. In this way knowledge based systems technology could be used to take advantage of the expertise of an human expert controller. This task of this expert system is not to reason about the dynamics of the plant but to reason about the performance of the numerical closed loop controller.

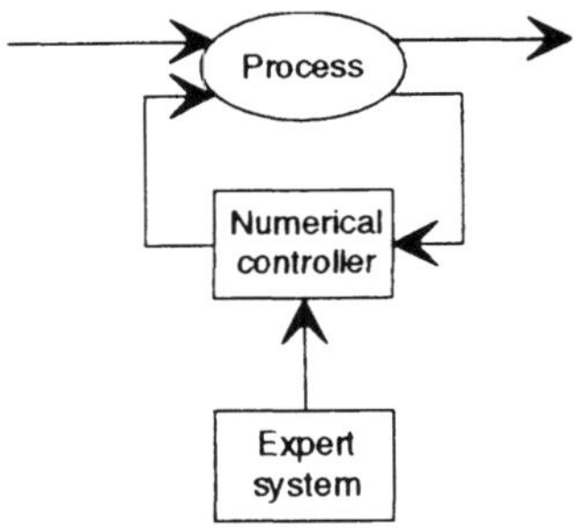

Fig. 2. Expert control.

2.2 Intelligent plant control.

In intelligent plant control AI is used for the global management of plant operations. Applications of AI to be mentioned here includes control, fault diagnosis and remediation, planning and decision support systems for the human operator.

Fault diagnosis. The purpose of fault diagnosis is to identify failures in plant components or subsystems. Usually the objective of diagnosis is to provide guidance to the choice of remedial actions such as e.g. repair of the faulty component or a compensation for the disturbance caused by the failure. Even though diagnosis is one of the classic research areas of AI, diagnostic systems have been around in the process industries years before the development of AI in the form of alarm and disturbance analysis systems. The development within AI has contributed with improved techniques for pattern matching and diagnostic reasoning. For applications within plant control these techniques are combined with methods for signal processing.

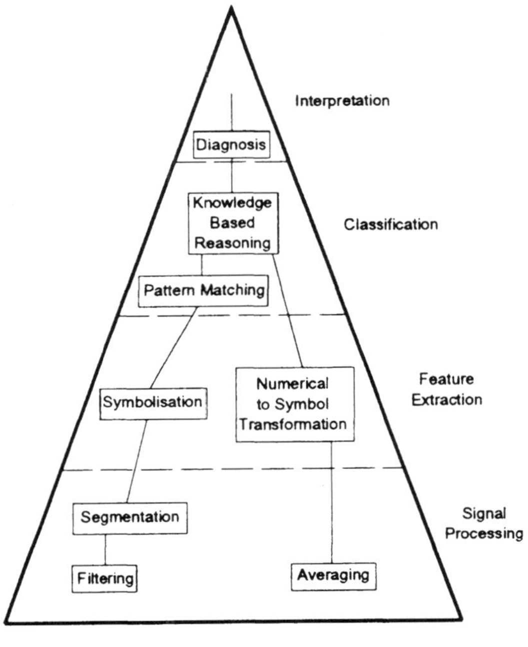

Fig. 3. Levels of state evaluation in diagnosis.

This leads to hierarchical architectures like the one shown in figure 3. The figure show how information about plant failures is derived from sensor signals through several stages of evaluation and transformations including simple filtering, feature extraction, classification and eventual interpretation in terms of e.g. the failed component. A variety of techniques including fuzzy sets, rule based systems, model based reasoning techniques and neural nets can be used to implement classification functions in diagnosis systems.

The field of diagnosis is fairly well developed. General theories of diagnosis has been developed by Reiter(1987) and truth maintenance techniques for the management of the diagnostic reasoning process (deKleer, 1986). Furthermore, commercial toolboxes are available that enable the system developer to build the complex information processing structure required for diagnosis (e.g. in the expert systems shell G2 or the TOPPMUSS toolbox). Areas of active research is the application of model based techniques. In a model based diagnosis system knowledge is represented symbolically in the form of concepts and relations instead of a set of shallow diagnostic rules as used in a rule based system. Model based systems do not suffer from the limitations of rule based systems but are infected with other problems. One of the basic problems is to decide on the concepts to use for describing the plant, also called the ontology in AI parlance. Another closely related problem is how the diagnostic process can be decomposed into sub processes dealing with diagnostic sub problems for plant subsystems. Current systems use descriptions of the physical topology and dynamic behaviour of the components. Other researchers, including the present author, claim that knowledge of plant purposes and functions so-called teleological knowledge is required in diagnosis of complex plant. The fundamental problems at this frontier of research is that it is difficult to formalise the knowledge of plant teleology that humans apply when they diagnose failures in complex systems. Some of these fundamental problems of model based diagnosis are related to the problem of context mentioned above. They will be discussed later in more detail.

Planning. The purpose of planning in intelligent plant control is to enable the system to respond to changing goals of production and/or to be able to perform on-line scheduling of resources or to make plant reconfiguration in situations where a subsystem has failed.

AI planning research has almost exclusively been focused on applications within robotics. However, the basic principles of robot planning systems are also applicable in principle to process control.

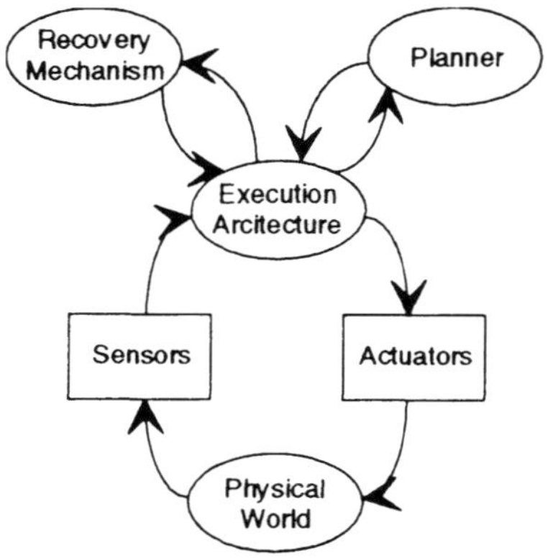

Fig. 4. Basic architecture of a planning system (Zilberstein, 1995).

A typical functional architecture for a planning system is shown in figure 4. The main component of the planning architecture is of course the planner that generate plans of control actions on the basis of some type of knowledge of the plant. A rule based approach could be used or a model based where the action plan is derived from a model of the plant. The planning component is interacting with an execution module that execute the plan. An essential ingredient in this execution module would often be a feedback control system that ensure that the commands in the plan are implemented in the physical world (the plant) in spite of disturbances and other types of uncertainty. Obviously, the plan execution module could in this way also be based on AI techniques.

Another important element in the basic planning structure is the recovery mechanism that is used to handle failures in plan execution. The recovery mechanism is often based on a simple planning mechanism.

The planning research within robotics is fairly advanced. Central topics of research are plan representation, distributed planning, and the development of generic planning systems. A special problem is the design of system architectures that combine planning modules on several levels. These systems incorporate often also diagnostic facilities and comprise comprehensive architectures for intelligent plant control. A prominent example of these architectures will be discussed below.

2.3 Intelligent plant control system architectures.

The basic principle in architectures for intelligent plant control systems is to use hierarchical structures to organise and co-ordinate the complex processing of sensor data and the generation of control actions. But hierarchical structures can be obtained in several ways. In current research two architectures are investigated that are in many ways opposites in their

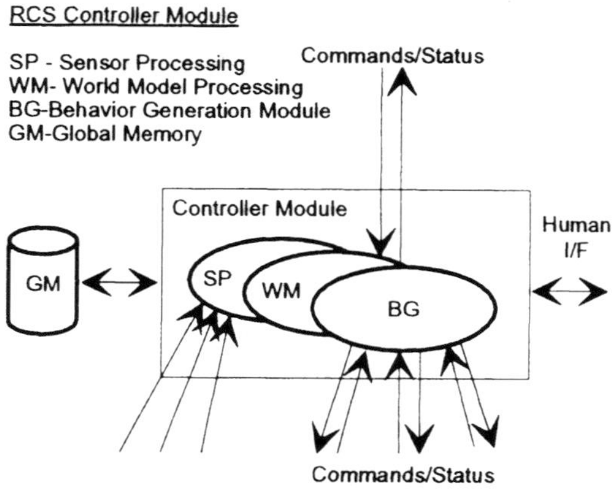

Fig. 5. The basic RCS building block.

basic philosophy. The first architecture discussed below proposed by Albus(1993) and his associates and is based on the principle of deliberation and the other architecture has been developed by Brooks(1990) and is called the subsumption architecture.

The architecture of deliberation. This architecture is based on the principles of deliberate action i.e. the principle that actions are intentional i.e. conditional on goals. Deliberate actions mimic rational reasoning processes and can be organised in hierarchical levels goal decomposition. Albus and his associates has developed a methodology for design of intelligent control system architectures that is based on this principle of deliberation. The basic building block in this so-called RCS architecture is shown in figure 5. This building block includes three modules- sensory processing (SP), world model processing (WM) and behaviour generation (BG). Interfaces are also provided that link the modules into hierarchical structures. The building blocks can be combined into layers and the RCS methodology can in this way be used to build complex architectures.

The NASREM architecture shown in figure 6 is such as multilayer architecture that is build from RCS modules. This architecture is proposed as a standard by NASA/NBS for the design of teleoperated robots. Similar architectures has been developed for submarines and for mining application. The NASREM architecture (as other architectures build by RCS modules) provides an integration of diagnosis (figure 3) and planning functions (figure 4) This integration is obtained via the world model. One of the basic problems in architectures of deliberation is that there is no fundamental principles that can be applied in the design of the contents of the world model. The architecture defines a *processing* structure and not the knowledge structures that are required to support reasoning at the various levels.

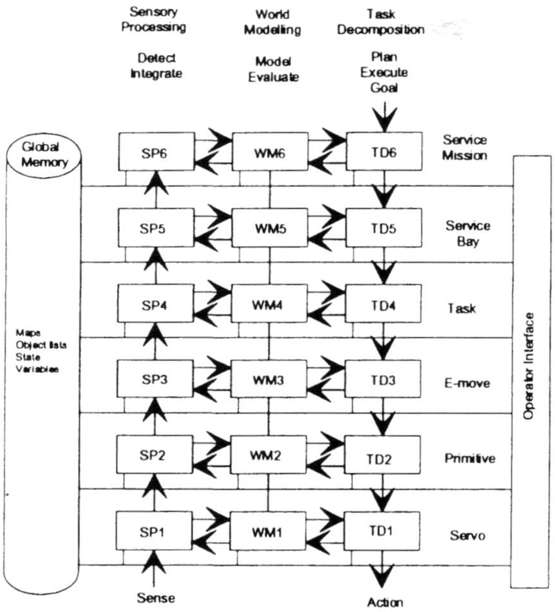

Fig. 6. The NASREM standard reference model.

The number of layers and the contents of the world models are highly dependent on the application. NASREM is proposed as a standard model for telerobotics and is only suitable for this domain. In other application domains the number of layers and the contents of the world models should be adapted to the specific needs. Since the architecture does not provide answers to these questions there is a need for research that investigate the problems of layering and of the design of the world model.

The Subsumption architecture. The basic principle in this architecture is that complex actions can be composed of a set of response patterns that are activated in highly specific situations in the controlled environment (Brooks, 1990). Take for example the architecture proposed for a pipeline surveying vehicle shown in figure 7. In order to perform this complex task the controller need to know how to map the environment and it must know how to find and follow a pipeline. Furthermore the controller must also know how to avoid obstacles and how to move around. The intelligence of the controller is in this way composed of a set of competencies for handling specific situations. Each of these competencies will be activated in specific situations and sometimes conflicts can occur i.e. the controller must choose between alternative responses to the situation. This selection is made on the basis of a priority mechanism.

It is seen that the subsumption architecture does not include an explicit world model as the deliberation architecture does. Furthermore, the generation of control actions is not based on reasoning about goals but on knowledge of action patterns that can be applied to respond to various situations. This means

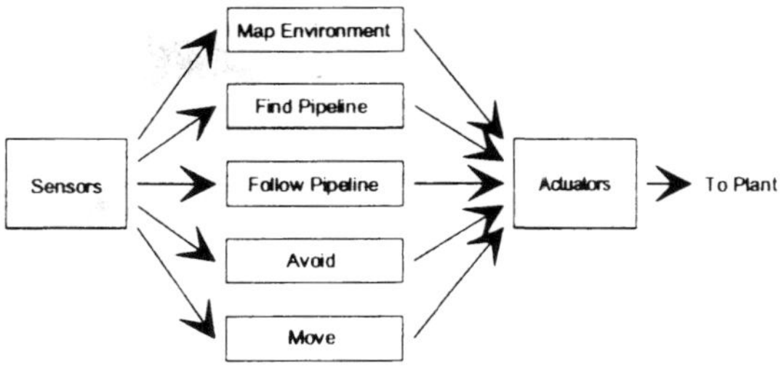

Fig. 7. A subsumption architecture for a pipeline surveying task.

that the controller is highly specialised to environments where the built in responses are adequate for reaching the (implicit) goals for the system. It is therefore much more sensitive to changes in the environment than the deliberation architecture is. The advantage of the subsumption architecture is its simplicity and the ability to produce fast responses.

Models of human cognition. Systems that combine the deliberation and the subsumption architectures have also been proposed. These combined architectures have striking similarities with models of human decision making developed within cognitive psychology. Rasmussen(1986) used these models of human cognition to explain why human operators sometimes make shortcuts in their reasoning. When making shortcuts they shift from rational analysis based on knowledge of plant goals and dynamics to stereotyped responses based on prelearned action patterns. These shifts, that corresponds to jumps between the two architectures mentioned above, is a source of error because the stereotyped responses may not be adequate under unfamiliar critical plant conditions. In such situations a rational analysis would be more appropriate. Models of human cognition are presently used for systems reliability studies or for the design and validation of human-machine interfaces (Hollnagel, 1993) but they may be of interest for applications within intelligent control' in the future.

2.4 The human-machine interface.

The developments within AI and graphics software for visualisation of plant data has generated a renewed interest in the problems of human-machine interaction in control of complex dynamic systems. These new software technologies makes it possible to improve the interaction between the operator and the automated plant. Results from research in natural language are used to design context sensitive dialogues between the operator and the computer. Research in human perception and cognition are used improve the principles for information presentation. Old principles for presentation of plant information based on mimic diagrams are abandoned and substituted by displays developed from the principles of cognitive engineering proposed by e.g. Rasmussen and his co-workers (Rasmussen et. al., 1994). Furthermore, multimedia technology can be used to integrate different modalities of communication such as text, pictures and sound. The technological development in this area facilitate the construction of intelligent human-machine interfaces. The research is centred around the problems of form and content of the interaction. The problem of form relates to the choice of means of presentation such as icons to be used. The problem of content deals with the planning of the information to be presented on the display and is closely connected with problems of plant modelling. The plant modelling problems related to the design of the human machine interface has many commonalties with plant modelling problems in intelligent control and will be discussed later in this paper.

3. AI RESEARCH IN MODELING

The problems of modeling or knowledge representation is one of the main topics of artificial intelligence research. Due to the symbolic nature of computer languages the main interest is here in the development of concepts and structures that can be used to describe our knowledge of the world. One strand of research that is of particular relevance to control applications is the area of naive physics and qualitative reasoning. The aim of these theories is to build systems that can represent knowledge about physical phenomena and that can simulate reasoning processes based on qualitative knowledge (Kuipers, 1994). These models are not meant to be substitutes for quantitative models like differential equations but rather to be used for purposes where the precision of quantitative models is not required for the problem or where the precision is inappropriate because of uncertainty or vagueness of the basic modeling assumptions. AI research addresses in this way modeling problems that are intractable by current mathematical control theories. AI has therefore created high expectations to future developments within control. Leitch (1992) analyze the characteristics of various types of models used in AI and in classical control systems design. This analysis provides as shown in figure 8. a unifying view of different modeling approaches. But controversies has also been generated because proponents of mathematical (i.e. purely quantitative) approaches to control systems claim that the types of control problems presently tackled by AI eventually and in principle can be solved by the formal tools of mathematics. There is however a third alternative position emphasizing the common sense basis of plant models, including mathematical models. This alternative may, as discussed in the remainder of the present paper, provide a reconciliation of these two

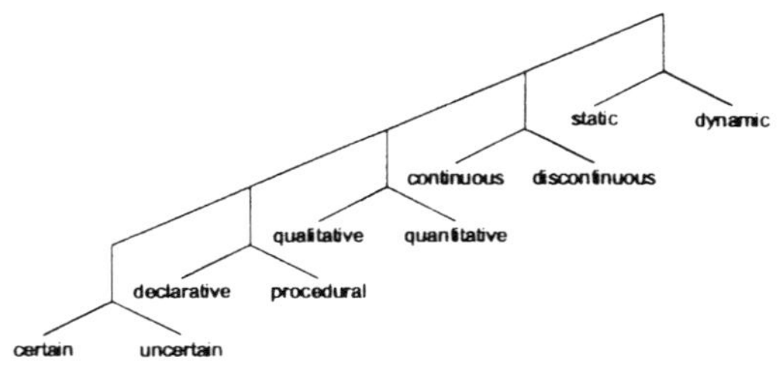

Fig. 8. Characteristic dimensions of plant models (Leitch,1992).

opposing views on the role of AI and mathematics in control. The formalisation of common sense knowledge is one of the main research directions in AI within knowledge representation (Hobbs and Moore, 1985). Common sense knowledge is difficult to formalise because it often is part of the background context for our understanding of the world and is therefore taken for granted or not attended to when building plant models. Thus, common sense knowledge play an important role in the formulation of differential equations but it is hidden in implicit assumptions that tends to be neglected or only addressed in the early conceptual phases of the modelling process. In order to demonstrate the importance of common sense knowledge for control problems three modelling problems in the design of plant supervision and control systems will be analysed in the next section.

4. MODELING PROBLEMS IN PLANT CONTROL

The purpose of this section is to analyse the plant knowledge required in the design of control algorithms, in fault diagnosis and planning and in the design of the human machine interface. Each of the analyses demonstrates the importance of knowing the purposes and functions of plant systems and components. This type of knowledge play an important role for our common sense understanding of physical artifacts like tools and more complex man made objects like process plants but has not until fairly recently been investigated as an object of formalized representation.

4.1 Design of algorithms for regulation.

The basis for design of regulation algorithms are models of plant dynamics. These models include only physical phenomena that are relevant to the design problem which is to meet requirements to closed loop responses or disturbance suppression. The problem is here to account for the criteria of relevance. Usually experts in building dynamic models would refer to their process knowledge or experience as the basis for the selection of phenomena to include in the model. But process knowledge comprises both so-called deep knowledge about the physical mechanisms involved in the plant processes and knowledge about the purpose of plant components and subsystems. This can be illustrated by an analysis of the engine cooling system example shown in figure 9.

Consider the problem of designing a flow regulator for the cooling loop including the control valve V1, the pump P1, the engine and one of the coils of the heat exchanger. Standard regulator design procedures would apply a dynamic model of the energy conversion process in the pump and fluid dynamics. Such a model can be based on detailed theoretical knowledge of the physics involved in the pumping or it could be derived from empirical data on the dynamic performance of the feedwater system using model identification techniques. A physics model is based on the assumption that the physical phenomena going on in the pump are known so that the dynamic behaviour relevant for the task at hand (in this case the moving of the fluid) can be described. As the model is used to design the flow regulator algorithms, phenomena describing how the cooling water is moved by the pump impeller would be relevant to include in the model. Other phenomena not relevant for the task such as the ability of the pump shaft to bend under certain conditions would be ignored. Similarly, the heating of the water caused by the pumping process would also be ignored if the pump is not used for that purpose or if the heating has no influence on the transportation of the water. But if the pump is used to heat up the water (pumps are sometimes used for this purpose), the heat generation and transfer phenomena of the pumping process should be included in the model. The selection of phenomena to include is therefore based on knowledge of the function of the pump- which is to circulate the water. To be more precise, the circulation of water is the *design function* of the pump. This qualification is required because the pump may be used by the plant operator to heat up the water. In such a case the pump have the *use function* of heating the water and a different model would be required to describe the dynamics. It is realised that even though the pump and its surrounding components have a multitude of behavioural properties, only those relevant for the actual function of the coolant circuit are selected. This knowledge about the function or purpose of the pump is taken into account in the modelling process and in the selection of parameters to measure and correlate in an identification procedure. It can therefore be concluded that knowledge about purposes of the plant and its components and subsystems is important for the building of dynamic models. This knowledge serve as a means for the selection of phenomena of relevance but is not explicitly represented in the resulting models. Knowledge of purposes is therefore part of the background or context for the modelling process.

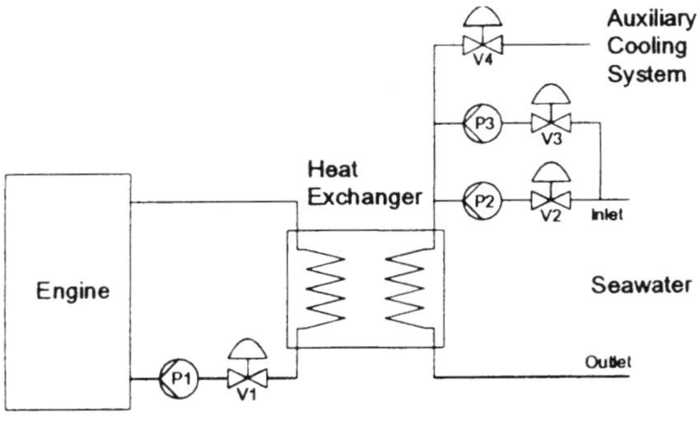

Fig. 9. An engine cooling system.

The dependence of dynamic models on knowledge of plant purposes is reflected in established procedures for control systems design. The typical steps in such a procedure are the following

- Establish operating objectives of process
- Define control-system objectives
- Determine constraints on the process
- Identify characteristics of disturbances
- Determine dynamic characteristics of process
- Develop control algorithm

The first two steps establish knowledge about the goals or purposes of the plant and the control system. The development of the dynamic models for the plant is highly dependent on this analysis because changes in the plant and control systems goals would require changes in the parameters and causal relations that should be represented in the model. However, even though goals are taken into account more systematic procedures are not used to model plant goals structures and their relations with control goals and plant dynamics. It is realised that this lack of formalisation of the background knowledge for modelling plant dynamics hinders the building of control systems that can cope with changes in plant goals or objectives. The ability to cope with such changes in context is a characteristic of human intelligence and would be relevant in the management of multipurpose plant and in situations that require ‘creativity’ such as the remediation of fault situations and the handling of emergencies.

Advances in the building of dynamic models could therefore be obtained if knowledge about purposes could be represented explicitly and systematically taken into account in the modelling process. Eventually, such an approach to model building could also be used in an intelligent controller. In terms of the example above, the controller would be able to adjust its algorithms when the function of the pump is changed from being ‘water circulation’ to being ‘heat generation’.

4.2 *Design of fault diagnosis systems.*

The purpose of systems for plant diagnosis is to detect and interpret the significance of deviations in measured plant parameters from their normal expected values. However, several interpretations are possible depending on the objective of the diagnosis.

Usually the objective is to relate the symptoms of plant malfunction with a possible cause of plant failure. But in situations of high risk and uncertainty it can be dangerous to act according to this objective. Under such circumstances it may be more adequate (i.e. intelligent) to associate the observed symptoms of failure directly with a protective action without knowing the failure cause. Taking into account the uncertainty and the possible consequences this may be the best strategy. Human operators often select the objective of their diagnostic activities in accordance with the actual plant operating situation Rasmussen(1986) found that plant operators used many strategies in diagnosis depending on the situation. The different strategies correspond to different cognitive load. Intelligence in diagnosis thus seems to include the ability to adjust objectives in accordance with the actual situation. In order to simplify the discussion in the following, it will be assumed that the objective of diagnosis is to identify the cause of failure.

One possible and widely used principle for diagnosis is to use knowledge about the relations between plant faults and their possible consequences as a basis for the interpretation of plant events. Several forms of representation of faults are available (fault trees, cause consequence diagrams etc.). However, the modelling of failure causes and consequences is far from trivial. In order to describe this modelling problem the coolant system example in figure 9. will be reused.

Assume that the stem in valve V2 breaks e.g. due to some material weakness. This fault may have a variety of consequences due to the physical interactions in the plant as shown in the fault tree figure 10. below. The fault tree is simplified as it only suggests that there may be several possible causes for each event in the tree. The fact that each event may have several consequences is also ignored. However, these simplifications are not essential for the following discussion of the basic problems of choosing a proper level of abstraction in the model.

The selection of the initiating event and its consequences is however not easy because it is not possible to give simple criteria for their selection. As an example; why is 'the stem of valve V2 is broken' selected as the initial event (cause) and why is 'valve V2 stuck closed' selected as its consequence?. The initial event could also have been described as a change in the metallic substructures of the stem. This modelling problem can, as in the case regulator design, be seen as a problem of relevance that can be resolved if knowledge of plant purposes or functions are taken into account. The following analysis of the objectives of diagnosis will show how this is done. The objective of diagnosis is to give directions for action. There must therefore be a correspondence

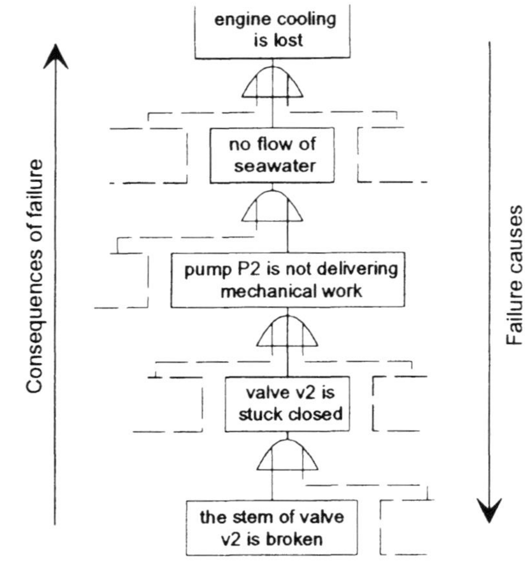

Fig. 10. Fault tree for the engine cooling system.

between the description of failures and the possible courses of action. Failure causes and their consequences should therefore be described so that it is possible to:

- predict the consequences of failure for the plant state and derive the cause
- evaluate the consequence against the objectives of plant operation on a systems, subsystem or equipment level.
- suggest possible courses of action.

The list of events proposed above satisfy these requirements. It describes the temporal development of a disturbance and can accordingly be used for prediction of consequences of failures and explanation their causes.

Table 1 The significance of plant events depends on the plant operating goal in view. Each interpretation of the event is associated with an action that is selected if that view has preference.

EVENT	GOAL	ACTION
Engine cooling is lost	Maintain engine temperature within limits	Reduce energy production
No flow of seawater	Maintain condition for cooling	Use auxiliary cooling system
Pump P2 is not doing mechanical work	Provide pumping capacity	Use standby pump P3
Valve V2 is stuck closed	Maintain manoeuvrability of actuators	Loosen stem of valve V2
The stem of valve V2 is broken	Maintain com-ponents in working order	Repair valve V2

Furthermore, as shown in table 1, each event description is also providing an evaluation of the situation i.e. the signification of the failure situation in view of a specific plant operating objective.

Note that each of the events in the list actually comprise a category. Thus there may be many situations that would have the same consequence 'no flow of seawater'. Likewise we can imagine a variety of situations corresponding to the situation 'engine cooling is lost' both in terms of the degree of loss and in the manner of which the means of cooling has failed. Note also that the event descriptions becomes more abstract towards the top of the table in the sense that they relate to more global aspects of the plant state.

It is realised that it is possible to provide meaningful actions to all event descriptions and that the actions relate to different objectives of plant operation and therefore to *different levels of management* of plant resources. The uppermost level relates to an overall control task dealing with balancing of energy production against energy losses. The next levels relate to the management of redundant cooling capacity. The third level relates to the management of alternative pumping resources. The fourth level relates to control of the valve position and the final level relates to the maintenance of system components.

As can be seen from the table, there would be five ways to respond to the failure depending on the goal to be pursued. The selection between actions would be dependent on multiple interpretations of the plant state. Thus the decision to use the stand-by pump is based on the fact that 'valve V2 is stuck closed' and 'the stem of valve V2 is broken'. If the stem was not broken. the loosening of the stem would be sufficient to remedy the failure. In this way goal priorities are decided on the basis of an analysis of the total situation i.e. all event interpretations. All levels are therefore required in order to make a qualified diagnosis. If levels are ignored, it will not be possible to consider alternative courses of actions i.e. the reliability of plant operations will be reduced (more shutdowns). Furthermore, if levels are ignored the risk of making wrong decisions will be increased. Take as an illustration a case where only the event 'Valve V2 is stuck closed' is given as a diagnosis of 'Engine cooling is lost'. This diagnosis will, if isolated from the context of the other goals and modes of action, result in a stereotyped response (loosen stem of valve V2) to a range of situations which call for much more discrimination.

It can therefore concluded that diagnostic plant models should include use several levels of abstraction. Furthermore, the levels should represent states of the *functions* of plant components or subsystems in view of the operating objectives and the possible means of control. The fault tree in figure 10. mentions three plant functions; the function of the pump as a source of mechanical energy, the function of the coolant circuit as a mass transport device and the function of the coolant circuit as a heat transfer device. The modelling of fault propagation in the plant is therefore based on knowledge of the purposes of plant subsystems and of the objectives of plant operation. Without this background knowledge the description of plant failures would be an ill defined problem. The systematic identification of levels of abstraction in plant models for diagnosis is therefore a major challenge to modelling research.

4.3 Planning of start-up and shut-down sequences.

Current methods for the design of action sequences for e.g. plant start-up or shut downs are based on the use of petri nets or similar formalisms for the representation of sequences. However, these formalisms cannot be used without having background knowledge about the interdependencies between plant goals and sub goals and the conditions for availability of plant functions. It is therefore realised that the conclusions gained from the analysis of modelling problems in diagnosis are also applicable to planning problems. Using the engine cooling system as an example again it can be seen that knowledge about plant goals and functions is directly applicable for the planning of start-ups. Thus, the sequence of goals to be obtained in the start up of the cooling system can be read from table 1 by reading the "goal column" from bottom to top. This conclusion is supported by a more complex example discussed later.

4.4 Problems at the human-machine interface.

Operators supervising industrial plant are faced with a complex task during diagnosis of disturbances and planning of remediating control actions. The difficulties are caused by the inability to directly associate the perceived stream of ongoing events with possible causes of malfunction and with possible courses of action. This separation of the world as perceived from the world as an object of action and satisfaction of needs is called the gulf of perception and action (Hutchins et. al., 1986). If this gulf is not eliminated or diminished humans need to spend cognitive resources on the interpretation of plant sensor data and may easily perform errors. Proper

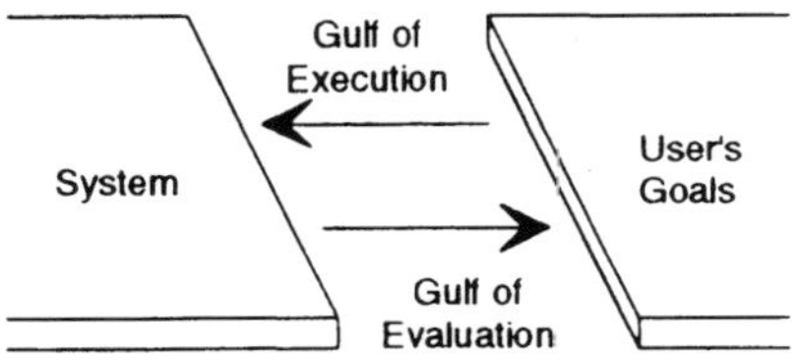

Fig. 11. The gulf of perception and action.

interpretation and evaluation of events and situations may be based on various sources of deep knowledge including knowledge about plant purposes, the causal dynamic interactions between plant parameters and the physical topology of the plant. But the interpretation may also be based on operational experience or on prescribed rules for action as express in written operating procedures. This variety of plant knowledge and therefore varieties of possible interpretations makes the operators conception of complex situations vulnerable to errors and mistakes. A major problem for the operator is to derive information about plant overall performance and about the states of plant functions in dynamic situations from knowledge about plant structure and the behaviour of plant components. Another problem is to infer the plant supervisory control system designers intentions with the operating procedures provided for remediation of plant disturbances. These intentions are important to know for the operator in order to be able to act intelligently in situations where courses of actions specified by the procedures are not available. Access to this type of high level semantic information is essential for the operator in order to be able to understand and respond appropriately in the event of major plant disturbances. However, minor disturbances that occur frequently during normal operation can often be handled by the operator on the basis of recognition i.e. without recourse to complex reasoning. This difference in the cognitive activity required during normal operation and serious malfunctions is a potential source of human error.

5. MODELLING PLANT GOALS AND FUNCTIONS

In the previous sections it was demonstrated that common sense engineering knowledge about plant goals and functions is necessary for building dynamic models, for the diagnosis of faults, for planning of control actions and for planning the information content of human-machine interfaces. Common sense knowledge is usually taken for granted and not considered to be an object of systematic study and explicit formalisation within theories of control. This lack of formalisation of plant purposes is a major bottleneck for further development of control systems engineering in general and to the design of intelligent plant control systems in particular.

The main aim of the present section is to demonstrate the potential value of formalised representations of plant purposes. The author and his students have done research in this area for more than a decade (Multilevel Flow Modelling, MFM). Recent results of these studies show that MFM models can be used to solve various problems in the engineering of supervision and control systems.

5.1 Multilevel Flow Modelling (MFM).

The aim of MFM is to offer a systematic basis for using means-end and whole-part decomposition's in the modelling of complex industrial plant (Lind, 1994). The means-ends distinction is used to represent a plant by the goals of its designer or user, by its functions and its physical components. The descriptions of plant goals, functions and components can at the same time be given on different levels of part-whole decomposition. The two decomposition's lead to plant models with multiple levels of means-end and whole-part abstractions.

The water circulation circuit of a central heating system shown in figure 12 gives examples of descriptions along the two dimensions of the MFM representations. Even though this system is fairly simple it illustrates well general aspects of MFM especially the use of function concepts. It is seen that the physical parts (the components) of the circulation circuit *realise* different functions and that the functions contribute to the *achievement* of plant goals. Note that the functions are each ascribed to the system with a specific goal in view. Thus, if the goal of 'maintain water level within safe limits' is selected then the relevant function of the circulation circuit is to transport the water from the water supply line to the expansion tank. If we, on the other hand, select the goal of 'keep the room temperature within limits,' then the function of the

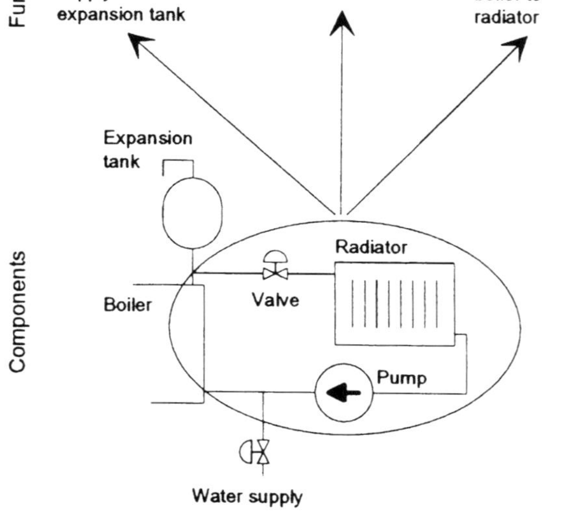

Fig. 12. Goals and functions of the water circulation circuit in a central heating system.

circulation circuit is to transport the energy from the boiler to the radiator. As the functions here describe how the behaviours of the components are useful for the achievement of various purposes or goals, we cannot meaningfully separate the ascription of the functions from the selection of goals. The functional ascription's are dependent on a predefined context of one or several goals.

The principles of the whole-part decomposition of goals, functions and components can also be illustrated. For example, the three goals can be aggregated (the inverse of decomposition) into a super ordinate goal, a whole, specifying that 'the central heating system should operate properly.' Furthermore, as another example, the function of transporting energy from the boiler to the radiator can be decomposed into a chain of two transport functions each describing the functions of the heat transfer between the burning air-gas mixture and the piping and between the piping and the water in the boiler. Finally the circulation circuit can be decomposed into the pump, the boiler, the valve, the radiator and the piping. Even though these three types of decomposition's (and aggregations) in principle are independent, they become strongly coupled when the descriptions of plant components, functions and goals relate as means to some ends. As an example illustrating this dependency, it would not be meaningful to associate the individual components of the circulation circuit with any of the functions mentioned in the example. The functions are the result of the behavioural *interactions* between the components and relate to a specific goal to be achieved. However, in spite of these constraints of *realisation* and *achievement* between the descriptions of goals, functions and components it will often be possible to choose between alternative levels of detail in the decomposition. Eventually, the criterion for choosing between alternatives depends on the application of the model, i.e., if the model is used for diagnosis or for planning.

Often there will be many alternative realisations of the same function and alternative ways of achieving the same goal. Thus, the water circulation function can be realised by different behaviours of the components in the circulation circuit in figure 12. The water circulation can be caused by the revolution of the pump impeller or, under certain conditions, by a difference in temperature between the hot water leaving the boiler and the cold water entering from below (natural circulation).

From this simple example some general observations of the nature of MFM models can be made. Firstly, it is seen that means-end and whole-part decomposition's lead to a multiple of interdependent representations of the same system linked by many-to-many relations of realisation and achievement. Secondly, each plant goal or subgoal will in MFM be a starting point for application of the two types of decomposition. This is because functional ascription's are made within a context of goals. The full implication of the latter aspect of MFM does not show up in the example. We will need the machinery of the formalised MFM concepts introduced later for discussing this important and unique feature of MFM.

Most industrial plants have the features described above and the nature of the many-to-many relations of realisation and achievement between components, functions and goals are fundamental to understanding their complexity. MFM provides formalised concepts to describe these relations.

Goals, functions, behaviour and structure. Before the basic concepts of MFM are introduced and illustrated by an example, a distinction between the concepts of goal and function and the concepts of structure and behaviour will be made. Concepts of behaviour and function are often confused and it is therefore necessary to explain how MFM make the distinction. The distinctions between structure, behaviour, function and goals can be made by a slight elaboration of the means-end aspect of MFM shown in figure 13 where the component level in figure 12 is represented by a combination of a structural and a behavioural level. By this we mean that knowledge of plant components comprises a structural aspect (the distinction of the separate physical parts and their interconnection in the physical topology) and a behavioural aspect (the physical mechanisms or phenomena responsible for the interactions between component variables). These two levels capture the causality of the system as shown in figure 13. As a contrast to these two levels it is also indicated that in MFM the levels of functions and goals are representing the designers' or the users intentions with the system.

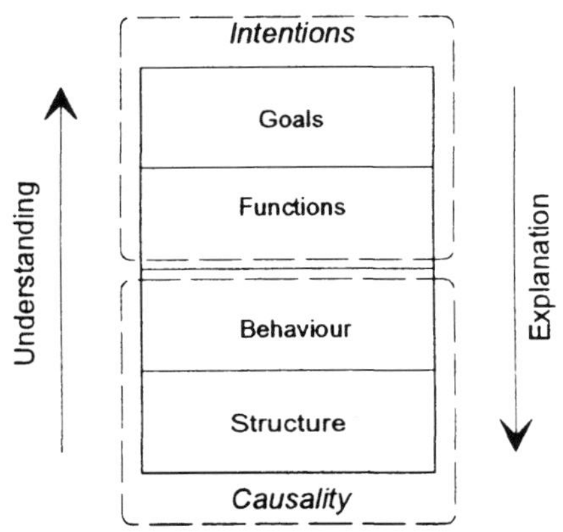

Fig. 13. Plant goals and functions describe intentions. Structure and behaviour are related to causality.

The two views, the causal and the intentional, are connected with two different scientific domains, the natural sciences (causes) and the social sciences (intentions) and support the two different types of reasoning processes- understanding and explanation. The engineering professions combine the two views in the design process. The connection to different scientific traditions has caused disagreements about the definition of the concept of function. Some define function to be equivalent to behaviour whereas others (including MFM) define functions as purely intentional. It is important not to confuse the description of behaviour with descriptions of function. The distinction is necessary in order to characterise a system as an artefact i.e., to make a distinction between systems designed for a human purpose and a purely natural phenomenon.

Multilevel Flow Modelling concepts. The functional elements used for building MFM models are shown in figure 14. MFM is at the present stage of development only dealing with the component level in a very superficial way. Ongoing research on MFM is developing this aspect and is also developing a general basis for the choice of functional concepts. More fundamental research on the foundation of functional models like MFM is presented in (Lind, 1992). The concepts of action in figure 14 are used to represent functions of control systems. Examples explaining the use of these concepts are described by Lind(1994) and will not considered here.

When building MFM models, the initial step is to identify the goals for the system because functional descriptions have no meaning without a previously defined goal context. The example described below

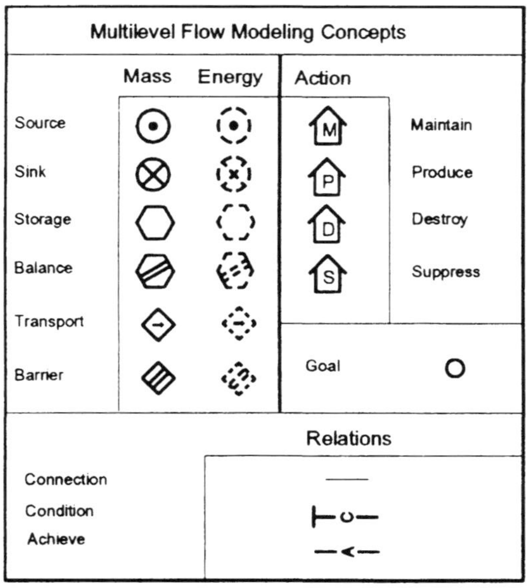

Fig. 14. MFM concepts and symbology.

is in correspondence with this general principle for building MFM models. But the explanation of the model does not reflect the course of the model building process. The MFM model building process is highly iterative.

5.2 A power plant example.

The general nature of MFM models can be illustrated by a model of a conventional power plant developed by Larsen(1993). The model was used for the generation of start-up plans as described below. Only a brief description of the model will be given here and the reader is referred to op. cit. or Lind(1994) for more details. The power plant boiler used in the example is described in (West72) and is shown in figure 15. The corresponding MFM model is presented in figure 16.

First, the plant goals G0,..., G10 will be described and it will be explained how they relate to the plant functions that in MFM are represented as networks of interconnected flow functions -so-called flow structures. Second, the three flow structures in the MFM model that are labelled Structure1, Structure2 and Structure3 will be explained.

The goals and the related flow structures. The main goal G0 of the power plant boiler is 'to generate electricity' and it is achieved by the functions represented by an energy flow structure Structure1. The goal node G0 and Structure 1 are therefore connected by an achieve relation with Structure1 in figure 16. The functions included in this energy flow structure are realized by the burner, boiler, turbines and generator. Structure2 describes the mass flow functions involved in the fuel combustion process, and Structure3 describes the mass flow functions provided by the plant components in the water and steam circuit. The functions of Structure2 achieves three goals G1, G2 and G3. The satisfaction of these goals provides the conditions for delivery and transport of energy in the left part of Structure1. G1, G2 and G3 are therefore related to the respective functions (So1, Tr1 and Tr3) in Structure1 by condition relations. G1 is the goal of providing a proper ratio between the air and fuel flows. G2 is the goal of having a flame and G3 is describing the need of a flow of combustion gases to carry the heat into the economizer. G2 is related to G9, which is a goal of having a chemical reaction between the fuel and the oxygen. This goal is achieved by Structure1, specifically by means of the energy transport Tr1 that again is conditioned by the goal G2. The two goals G2 and G9 therefore support each other. In physical terms this means, that when a flame is first established, it is self-supporting.

This circularity in the MFM model illustrates that the functional levels in an MFM model can combine into cyclic structures of means and ends. MFM models are therefore in general *not* tree structured or hierarchical. These cycles are the sources of start-up problems. In the specific example, the burner flame is needed to ensure transport of energy in the boiler and this transport is a necessary means to maintain the combustion processes in the flame. The combustion process should therefore be initiated by other means than the flame. This is the function of the burner ignition system (not included in the model).

The function described by Structure3 achieve five goals G4, G5, G6, G7 and G8 that all condition the transport of energy from the boiler to the turbine. G4 expresses the need to maintain a minimum level of water in the boiler and G5 is the goal of having a steam flow from the boiler to the turbine. The energy conversion in the turbine only works properly if the vacuum system is operating. This requirement is expressed by G6 that is the goal of 'having the vacuum system in operation'. Cooling is of course also needed and G7 express the goal of 'having a proper flow of cooling water,' while G8 is the goal of 'having a feed water flow'. One of the functions of Structure3 is also conditioned by a goal (G10). The function is Tr14 and represents the transport of mass that is provided by the steam piping connecting the boiler with the turbine. This transport function is conditioned by the goal G10 describing the need to maintain a pressure in the boiler. This goal is therefore achieved by Structure1. This part of the model includes therefore also a cyclic component.

The flow structures. The plant functions provided for goal achievement are represented by the flow functions in Structure1, Structure2 and Structure3. Below we will, without going into too much detail, describe the flow functions in Structure 1 and explain their physical realization. Structure2 and Structure3 will only be described in outline.

Structure1 is composed of the following functions: The source So1 is the energy source provided by the combustion of fuel and air. The energy is transported (Tr1) to the balance function B1. Tr1 is realised by (is a function of) the flame, and is, as mentioned above, conditioned by the goal G2. B1 models the balance between the incoming energy flow (Tr1), the energy transported into the water in the boiler (Tr2) and the energy flow removed by the exhaust gases (Tr3). The balance B2 models the energy balance in the economiser, where the heat from the exhaust gases is used to pre-heat the feed water, which provides an energy transport function, modelled by Tr4. The transport of heat from the economiser to the chimney is modelled by Tr5. This energy is absorbed by the atmosphere that has the function of a sink (Si1). The energy transported by Tr2 ends up in the energy storage St1 provided by the water in the boiler. From this energy storage the energy is transported (Tr6) into a balance (B3) realised by the turbine. From B3 the energy is transported by two flows, one represent the work done on the generator shaft (Tr8) and the other represent the residual heat in the steam on the low-pressure side of the turbine (Tr7). These flows ends in two energy sink functions (Si2 and Si3). Si2 models the energy sink provided by the generator and Si3 models the heat sink provided by the environment through sea water or the atmosphere.

The functions of Structure2 represents the mass flow functions involved in the management of the fuel and air flows in the burner and the fuel supply system. Structure3 comprises the functions that are involved in the management of the flows of water and steam around the boiler and the condenser.

The means-end relations. The plant functions modelled above by the mass and energy flow structures are combined into the multilevel flow model as shown in figure 16. Each of the functional levels Structure1, Structure2 and Structure3 comprises a level of abstraction in the modeling of the system. The functional levels are connected with the goals by achieve and condition relations also called means-end relations.

The meanings of the achieve and condition relations can be explained by following the vertical connections between the flow structures emanating from the node labelled Tr6 in the model in figure 16. The energy transport function Tr6 will only be available if there is proper circulation of mass in the feed water and steam circulation circuit. It is therefore related by a condition relation with G5 that express that condition. But G5 is a goal to be achieved by the functions of the water and steam circuit. G5 is accordingly related by an achieve relation with the flow structure Structure3 describing these functions. In this way we get the condition and the achieve relations shown in figure 16.

5.3 Planning of power plant start-up.

The condition and achieve relations, which interrelate the levels of the MFM model, represent plant knowledge that is efficient in planning start-up control actions (Larsen, 1993). The chains of means-end relations in the model provide efficient heuristic knowledge for searching through the system on many part-whole levels. This type of search would be difficult to formulate (and implement) without a plant representation like MFM. Larsen(1993) has shown that a start-up plan can be generated from an MFM model by defining generic goal hierarchies for each type of flow function in the model. Due to logical interdependencies of the flow functions

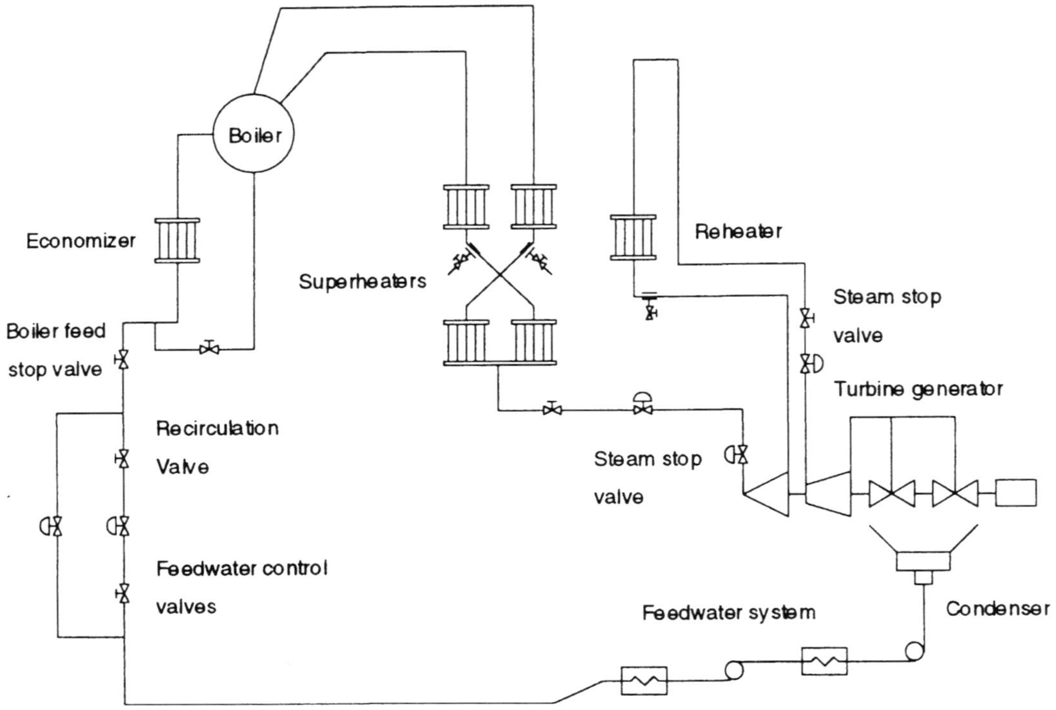

Fig. 15. The power plant boiler (Larsen, 1993).

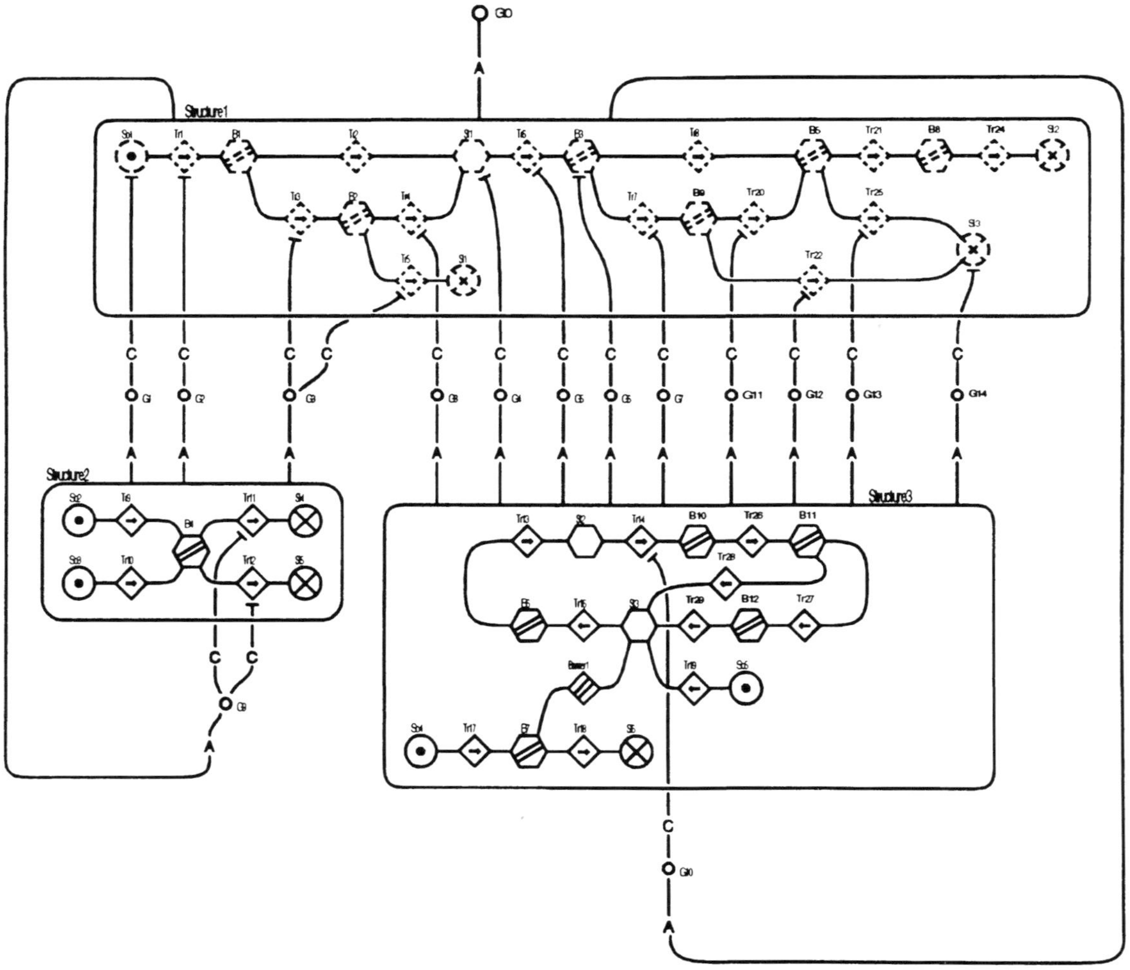

Fig. 16. MFM of the power plant boiler (Larsen, 1993).

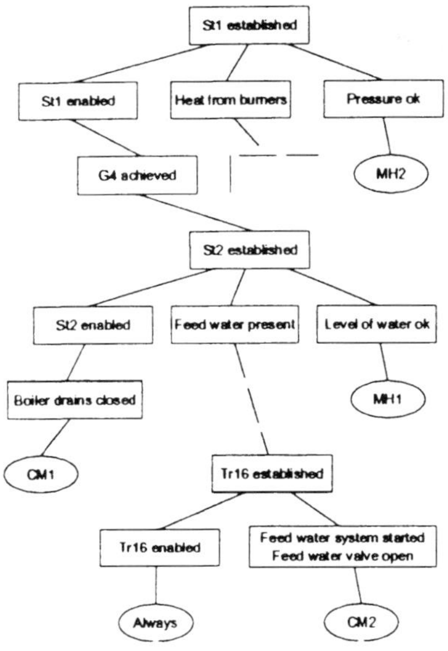

Fig. 17. Goal decomposition hierarchy for a power plant boiler start-up derived from the MFM model. The sequence of actions can be derived from the hierarchy by reading the leaf nodes from left to right (Larsen, 1993).

within each flow structure and the linkage of levels by the condition and achieve relations, these elementary goal hierarchies can be combined into a goal hierarchy for the total plant. An example of a start-up plan generated in this way is shown in figure 17. The top goal of the hierarchy is to establish St1 i.e. the energy storage function of the boiler.

The plan generated does not take into account temporal constraints due to the dynamics of the plant. These constraints can be included by combining the MFM based planning system with a scheduling component based on a dynamic plant model i.e. based on a model of plant causality. The four layers of plant representation presented in figure 13 could in this way be combined into an advanced planning system. The role of MFM would then be to provide an efficient decomposition of the planning problem. The integration of causal and intentional knowledge in planning has not yet been investigated. The study by Gofuku and Lind (1994) described below indicate a promising direction for such research.

5.4 Deriving plant dynamics from goals and functions.

Gofuku and Lind(1994) has investigated the relations between knowledge about plant intentions and causality in a study on plant design methodology. The idea was to exploit the plant knowledge expressed in a functional model like MFM in the generation of a dynamic model. The dynamic model was used to test decisions made in a top down function oriented approach to plant design. One of the main problems is here to provide a linkage between knowledge of plant purposes and causality. This was obtained by combining MFM with the HPT theory developed by Woods(1990, 1991). HPT is related to qualitative process theory QPT developed by Forbus(1984). QPT is a technique for making qualitative simulation based on knowledge of plant structure and behavior. The purpose of HPT is not to generate behaviors but to generate symbolic representations of state space equations (that of course may be used for simulation). The central concept of HPT is the concept of phenomena, that is used to classify the fundamental physical plant mechanisms. The phenomena are components of state space equations and their inclusion and composition in the equations is conditional on the plant state. HPT provides in this way a linkage between plant structure and behavior. The main problem in the linking of MFM and HPT is to combine knowledge of purposes or intentions in MFM with knowledge of causality captured by HPT. This bridge was obtained by taking advantage of design rules mapping plant functions into structure. In order to exemplify such a mapping from function to structure consider the engine cooling systems discussed before. A design rule for this system could say that the heat transport (i.e. the function) between the engine and the heat exchanger is realized by a water circulation circuit comprising the pipes and the pump etc. (the structures). This type of design rules are used to provide the linkage between MFM and HPT (see figure 18).

The combination of MFM and HPT can be utilized in a function oriented approach to plant design as shown in figure 19 (Gofuku and Lind, 1994). The initial design step is to develop an MFM model of the plant. Functions in the MFM model are then mapped by design rules into a component configuration that is a possible candidate solution to the design problem. The candidate solution is then evaluated by the generation of a dynamic model (using HPT) that is used to test the satisfaction of the design specifications. Gofuku and Lind developed the linkage of MFM and HPT and demonstrated the principles on a central heating system.

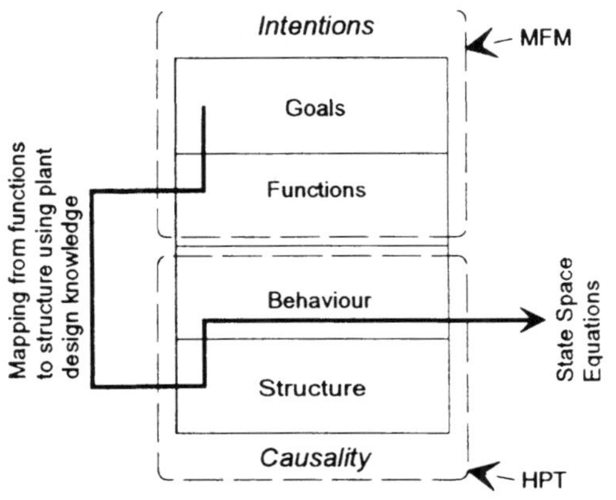

Fig. 18. Plant design knowledge can be used to provide a link between intentional and causal knowledge.

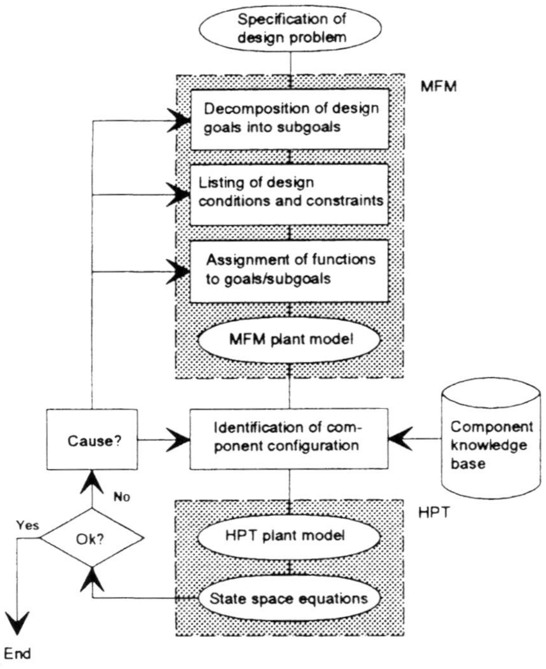

Fig. 19. Flowchart showing how state-space models are generated by combining HPT and MFM.

A software prototype (Smalltalk-80) was developed and state space equations were generated for the central heating system. However, further developments of HPT and MFM and their interconnections are required in order to make full use of these principles.

5.5 Other applications of MFM.

MFM has been also been used for diagnosis and design of human-machine interfaces. A study of MFM based diagnosis is described by Fang and Lind(1995). An application of MFM for design of diagnostic displays is reported by Duncan and Prætorius(1989). The role of MFM in a cognitive engineering approach to information system design is indicated by Rasmussen et. al. (1994).

6. CONCLUSIONS

At the present stage of development artificial intelligence has already made significant impacts on the research in plant control. One of the most important effects of the meetings of control theory and artificial intelligence is an increased awareness of the importance of research in advanced concepts, methods and tools for plant modelling. This comprises both better methods to improve the quality of quantitative models and the development of new types of models that capture other types of plant knowledge than can be expressed in differential equations, such as qualitative knowledge and knowledge about plant purposes. It is becoming realised that the bottleneck in the development of intelligent plant control systems is not to master AI technology per se but to achieve a better understanding of the varieties of plant knowledge used in control and supervision of a dynamic system. It is shown in the paper that common sense engineering knowledge of plant goals and functions, as e.g. represented in Multilevel Flow Models, play an important role in the building of dynamic models of plant, in diagnosis of failure, in the planning of start-ups and in the design of the human-machine interface. It is also argued that the current lack of understanding of the common sense basis of control actions is a major impediment to the further development of the field of intelligent control. The development of concepts, methods and tools that integrates different types of plant models is therefore one of the most important challenges for research in intelligent plant control. In this development the main influence will not only come from artificial intelligence but also from the related field of cognitive science. Research in this direction has been described in the paper.

7. REFERENCES

Albus, J. S. (1993). A Reference Model Architecture for Intelligent Systems Design. In: *An Introduction to Intelligent and Autonomous Control* (Antsaklis, P. J. and Passino, K. M. (Eds.)). Kluwer,Boston.

Antsaklis, P. (1994). *Defining Intelligent Control: Report of the Task Force on Intelligent Control.* IEEE Control Systems Magazine, June.

Bainbridge, L. (1982). Ironies of Automation. In: *Preprints of the IFAC/IFIP/IFORS/IEA Conference on Analysis, Design and Evaluation of Man-Machine Systems.* Baden-Baden, Germany.

Bond, A. H. and Gasser, L (Eds). (1988). *Readings on Distributed Artificial Intelligence.* Morgan Kaufmann Publishers Inc. San Mateo, California.

Brooks, R. A. (1990). Elephants Don't Play Chess. In: *Designing Autonomous Agents.* (Maes, P. (Ed)). The MIT Press, Cambridge, Massachusetts.

Åstrøm, K. J., Anton, J. J. and Årzen, K. E. (1986). Expert Control. *Automatica*, **22**, pp. 277-286.

deKleer, J. (1986). An Assumption-Based TMS. *Artificial Intelligence*, **28**, pp 127-162.

Dreyfus, H. L. (1979). *What Computers Can't Do - The Limits of Artificial Intelligence.* Harper Colophon Books, New York.

Duncan, K.D. and Prætorius, N. (1989). *Flow Displays Representing Complex Plant for Diagnosis and Process Control.* Proc. 2nd. European Meeting on Cognitive Science

Approaches to Process Control, Siena, Italy, October 24-28.

Gofuku, A. and Lind, M. (1994). *Knowledge-based support for efficient design of engineering Systems*. Proc. 2nd. International Conference on Intelligent Systems Engineering. Hamburg Germany, Sept. 5-9.

Fang, M and Lind, M. (1995). *Model-Based Reasoning Using MFM*. Proc. Pacific-Asian Conference on Expert Systems PACES'95, May 15-18. Huangshan, China.

Forbus, K. D. (1984). Qualitative Process Theory. Artificial Intelligence, **24**, pp. 85-168.

Hobbs, J. R. and Moore, R. C. (1985). *Formal Theories of the Commonsense World*. Ablex, New Jersey.

Hollnagel, E. 1993. *Human Reliability Analysis: Context and Control*. Academic Press, London.

Hutchins, E. L., Holland, J. D. and Norman, D. A. (1986). Direct Manipulation Interfaces. In: *User Centered Systems Design* (Norman, D. A. and Draper, S. W. (Eds.)). Lawrence Earlbaum, Hillsdale, New Jersey.

Kuipers, B. (1994). *Qualitative Reasoning*. The MIT Press, Cambridge.

Laird, J., Rosenbloom, P. and Newell, A. (1986). *Universal Subgoaling and Chunking*. Kluwer, Boston.

Larsen, M. N. (1993). *Deriving Action Sequences for Start-Ups using Multilevel Flow Models*. PhD Thesis, Institute of Automation, Technical University of Denmark.

Leitch, R. (1992). Artificial intelligence in control: some myths, some fears but plenty of prospects. *Computing & Control Engineering Journal*, July.

Lind, M. (1992). *A Categorisation of Models and its Application in the Analysis of Planning Knowledge*. Proceedings of Post ANP92 Conference Seminar on Man-Machine systems and Co-operative Activity in Technological Systems, Nov 3-5, Kyoto Japan.

Lind, M. (1994). Modelling Goals and Functions of Complex Industrial Plant. *Journal of Applied Artificial Intelligence*, **8**, pp.259-283.

Maes, P. (1990). *Designing Autonomous Agents*. The MIT Press, Cambridge, Massachusetts.

Newell, A and Simon, H. A. (1972). *Human Problem Solving*. Prentice Hall, New Jersey.

Piaget, J. (1966). *La Psychologie de L'Intelligence*. 9th. Ed, Librarie Armand Colin, Paris.

Rasmussen, J. (1986). *Information Processing and Human-Machine Interaction*. North Holland, New York.

Rasmussen, J., Pejtersen, A. M. and Goodstein, L. P. (1994). *Cognitive Systems Engineering*. Wiley, New York.

Reiter, R. (1987). A Theory of Diagnosis from First Principles. *Artificial Intelligence*, **32**, pp. 57-95.

Searle, J. (1984). *Minds, Brains and Science*. Harvard University Press.

Webster. (1989). *Webster's Encyclopaedic Unabridged Dictionary of the English Language*. Gramercy Books, New York.

West, K.L. et. al. (1972). *Minimum Recommended Protection, Interlocking and Control for Fossil Fuel Unit-Connected Steam Station*. IEEE Winter Power Group Meeting, New York.

Woods, E. A., Telnes, K and Balchen, J. G. (1990). Formalising Energy and Mass-Transfer by Means of the Hybrid Phenomena Theory. *Proceedings of Nordic CACE Symposium*, November 15-16, Technical University of Denmark, Lyngby Denmark

Woods, E. A. (1991). The Hybrid Phenomena Theory. *Proceedings of. International Joint Conference on Artificial Intelligence JCAI 91*, Sydney, August 24-30.

Zilberstein, S. (1995). The Utility of Planning. *ACM SIGART Bulletin*, **6**, 1, pp.42-46.

Risk-Conscious Operation of Batch Processes

Peter Terwiesch* & Dag Ravemark & David W.T. Rippin†

Systems Engineering Group, TCL, ETH Zürich, 8092 Zürich, Switzerland

**Corresponding author. New address: ABB Corporate Research, CHCRC.C1, 5405 Baden, Switzerland.*

†Deceased 26 June 1994.

Abstract. This contribution suggests an approach to the dynamic optimization of batch process operations that takes into account model uncertainty and inter-run variations. As opposed to the standard nominal optimization, a model with a probabilistic uncertainty description rather than an exact nominal model of the process is used. A probabilistic measure of success, such as the probability of making in-spec product, is optimized. Two simulation examples illustrate the application of the proposed technique and demonstrate the increased robustness to variations of the obtained operating profiles.

Key Words. batch, end point, optimization, quality, safety, risk

1. Introduction

Batch or semi-batch process operation is traditionally optimized assuming an exact deterministic model is available. This nominal process model is obtained through optimization of a measure of fit between model predictions and actual plant data. However, due to the innovative nature, the high value, and the comparatively small quantity of many batch products, typically only few modeling runs are conducted. Also, both inter- and intra-run variations are frequent and end-point reproducibility even in the laboratory is often not better than 5% (Terwiesch *et al.*, 1994). Consequently, in many situations use of a single deterministic model for process optimization may represent an over-simplification.

A possible alternative approach is to identify a probabilistic process model, yielding a probability distribution for the parameter space rather than only one nominal parameter vector. In replacement of the optimization for the nominal parameter vector, the optimization can then be performed so as to determine an open-loop policy that accounts for possible uncertainties or variations found in the identification step (Terwiesch, 1994).

In such a probabilistic framework, the objective is a probabilistic measure of success. Such probabilistic measures are often either of the *expected value* type or of the *risk* type. Expected-value objectives express the desire to achieve the best average result, e.g., maximization of expected yield, or to reduce variations in final product composition, corresponding to, e.g., minimization of end-point variance. Possible risk objectives include maximization of probability for making acceptable product, maximization of that profit that one can be 90% sure of achieving, or determining the most rigorous product specifications that can be met in 95% of all batches. Typical objectives are discussed in more detail in section 2.2.

2. Proposed Approach

2.1. *Types of Constraints*

In industrial operation of batch reactors three types of constraints can be distinguished:

Product Specification: definition of final product quality, e.g., upper bounds for by-products or unreacted feed

Safety Regulations: definition of safe operating regime, e.g., an upper bound for accumulation of hazardous intermediates

Manipulated Variable Bounds: range that the operator can technically choose the manipulated variables from, e.g., upper and lower bounds for quantities such as jacket temperature or feed addition rate

Most safety regulations need to be considered as *hard* constraints, i.e., the risk of violating them should be zero theoretically. In practice the parameter space could for example be bounded by plus/minus three standard deviations from the mean. The risk of violating the product specifications may be non-zero. Batches that violate the product specification for some realization values of the parameter vector may be produced, as long as the operating policy is optimal with respect to some probabilistic measure of success.

2.2. *Typical Objectives*

In a probabilistic parametric uncertainty notation, typical objective functions for industrial problems can be written in the following form:

$$\max_{u} \int_{\theta \in \Theta} J(u,\theta) \quad p(\theta)\, d\theta \tag{1}$$

where u, θ, Θ, J, and p represent operating policy, parameter vector, parameter space, profit associated with a particular outcome, and probability, respectively.

Several different types of objectives commonly arise:

Threshold Risk: The profit function J is of the binary type, i.e., $J = 1$ for admissible product and $J = 0$ for off-spec product. The objective then is to maximize the probability of making acceptable product.

α%-Risk Threshold: The probability or relative frequency α of acceptable batches is fixed while some function of the operating policy is optimized. This can be formulated as

$$\max_{u,s} \quad J_1 \quad (u,s,\theta)$$

$$s.t. \quad \int_{\theta\in\Theta} J_2(u,s,\theta)\,p(\theta)\,d\theta \geq \alpha$$

where J_2 is of the binary type, J_1 is some function of the operating policy, and s defines the product specifications. Note that the product specification itself can be subject to optimization. ("How tight can we make the specifications while still making acceptable product with a probability of at least 90%?".)

Expected Value: Find an operating policy that maximizes the expected profit of operation. Examples may include penalties for violation of soft constraints, for example the cost of reworking or disposing of unacceptable batches, or simply J equal to concentration or total amount of produced product

Minimum Variance: Find an operating policy that reproduces a target product specification as tightly as possible despite variations and uncertainty in the system parameters

While in some cases sufficient measurement information for online correction or re-optimization may be available, in most practical situations this is not so and open-loop policy optimization is necessary (see discussion in (Terwiesch *et al.*, 1994)).

2.3. *Computational Approach*

We use an algorithm similar to the one suggested in (Straub and Grossmann, 1993) in the context of plant design optimization. It is illustrated here for the case of two uncertain parameters θ_1 and θ_2:

1. Determine the maximum interval $\Theta_1 = [\theta_1^L, \theta_1^U]$ where a θ_2 exists so that the constraints g_i are met.
2. Collocate Θ_1 with n_1 points θ_1^i and collocation weights w_1^i.
3. For each of these points θ_1^i, find the largest interval $\Theta_2^i = [\theta_2^{L,i}, \theta_2^{U,i}]$ where the constraints g_i are met.
4. Collocate each Θ_2^i with n_2 points $\theta_2^{i,j}$ and collocation weights w_2^j.
5. Compute the probability of a successful batch run given a strategy $\mathbf{u}$ as

$$I = P = \frac{\theta_1^U - \theta_1^L}{2} \sum_{i=1}^{n_1} w_1^i \frac{\theta_2^{U,i} - \theta_2^{L,i}}{2}$$

$$\sum_{j=1}^{n_2} J(\theta_1^i, \theta_2^{i,j}) \quad w_2^j \quad p(\theta_1^i, \theta_2^{i,j})$$

This integration of the probabilistic measure of success is performed within an outer loop for the optimization of the operating policy. $J(\theta_1^i, \theta_2^{i,j})$ represents a weighting of the probability, e.g., profit.

3. A Simple Demonstration Example

Consider a chemical reaction system

$$A + B \xrightarrow{k_1} C$$

$$B + B \xrightarrow{k_2} D$$

conducted in an isothermal semi-batch reactor. While simple, this example characterizes an important class of industrial problems. Since it is also a common benchmark for optimal control approaches, we will assume the same numerical values as given by Young (Young, 1980). The aim is, through addition of reactant B, to convert as much as possible of reactant A to the desired product, C in a specified time. To keep the effect of the undesired reaction to D small, the system is operated in semi-batch mode and B is added in a feed stream with concentration $b_{feed} = 0.2$. The rate equations for this system can be expressed as

$$\dot{[A]} = -k_1[A][B] - [A]\frac{u}{V}$$

$$\dot{[B]} = -k_1[A][B] - 2k_2[B]^2 + \frac{b_{feed} - [B]}{V}u$$

$$\dot{V} = u$$

where the rate constants have the nominal value $k_1 = 0.5$, $k_2 = 0.5$, and where the symbols $[A]$, $[B]$, V, and u denote concentration of A and B, current relative volume, and feed rate, respectively. Initially, the reactor contains $[A](0) = 0.2$ moles per liter of A, no B ($[B](0) = 0$) and is filled to 50% ($V(0) = 0.5$). In this first example, we assume a fixed end-time $t_f = 120$ minutes. The feed rate is constrained to $0 \leq u(t) \leq 0.01$.

While more complex parametrizations, e.g., dynamic profiles in the manipulated variables, can also be treated in a similar manner (Terwiesch, 1994), we assume that the operating policy is parameterized with only 3 degrees of freedom: a constant feed level, u_1, the duration of the feed addition phase, t_1, and the total batch time, t_2. This class of feed addition profiles is probably most frequently used in industry and allows the optimization problem to be kept simple for our demonstration purpose.

Let us assume that the rate constant of the side reaction, k_2, is not as well determined as that of the desired reaction. Its uncertainty can be described by a normal distribution around the nominal value $\mu_{k2} = 0.5$ with standard deviation $\sigma_{k2} = 0{,}08$. Such an extent of uncertainty also represents a typical magnitude for many practical problems.

3.1. *Nominal Optimum*

Using traditional nominal optimization over the described class of staircase feed profiles, one obtains $u_1 = 0.01$, $t_1 = 50$ and $t_2 = 120$, i.e., maximum feed addition until the reactor is filled, then wait as long as possible to let the remaining B react. Even though the example was originally used to demonstrate the benefits of more sophisticated feeding profiles, it reaches more than 97% of the singular optimum (Young, 1980; Terwiesch, 1994). The final product is acceptable when the following specifications are met:

- final concentration of unreacted feed B no greater than 0.0035 moles per liter
- final concentration of undesired side product D no greater than 0.018 moles per liter
- final concentration of desired product C at least 0.058 moles per liter

Introducing a shorthand notation, a batch $\mathbf{b} = ([A]_f, [B]_f, [C]_f, [D]_f)$ is acceptable if $\mathbf{b} \in \mathcal{A}$, where $\mathcal{A}$ is the set of product specifications

$$\mathcal{A} = \{[B]_f \leq [B]_{f,\max} \wedge [C]_f \geq [C]_{f,\min} \wedge [D]_f \leq [D]_{f,\max}\}$$

with numerical values $[B]_{f,\max} = 0.0035$, $[C]_{f,\min} = 0.058$, and $[D]_{f,\max} = 0.018$. An operating policy $\mathbf{u} = (u_1, t_1, t_2)$ is admissible if it belongs to the set $\mathcal{U}$ satisfying

$$\mathcal{U} = \{0 \leq u_1 \leq 0.01 \wedge u_1 t_1 \leq 0.5 \wedge 0 \leq t_1 \leq t_2 \leq 120\}$$

From a practical point of view, since the nominally optimal policy satisfies these constraints even with a certain margin, one could run the reaction with this policy despite the model uncertainty.

However, as Monte Carlo simulations show, the nominal policy (section 3.1) yields acceptable product in only 34.7% of all runs. This value is undesirably small, so that normally more modeling runs would need to be conducted to reduce the model uncertainty, which would be expensive with respect to labour, laboratory, and material cost and prolongated time to market.

3.2. *Risk Minimization*

As an alternative that does not require this additional effort, a minimization of the risk of making unacceptable product is conducted. This optimization problem can be formulated as

$$\max_{\mathbf{u} \in \mathcal{U}} \quad p(\mathbf{b} \in \mathcal{A})$$

Using a Nelder-Mead simplex algorithm, a fast integrator (with relative error $\leq 10^{-5}$), and seven collocation points for each component, the problem is solved within 2 minutes on a DEC μVAX. The optimization yields $u_1 = 0.0062$, $t_1 = 59.9$, and $t_2 = 120$. Note that with the given objective there is no sense in reducing the batch time.

Figure 1 shows the final concentration of C as a function of the uncertain k_2 for both policies together with the constraint $[C]_{f,\min}$. Figure 2 does the same for the other two constraints.

The acceptable region of k_2, i.e., the interval for which a given policy produces a final product that meets the specifications, is shown in figure 3 together with the PDF for $p(k_2)$. One can clearly see (figures 1 and 2) that the conservative risk-minimizing strategy (solid) produces a slightly smaller concentration $[C]$ for the nominal k_2, but extends the acceptable product region drastically (98.26%, solid, compared to 34.70%, dashed).

3.3. *Risk Threshold Optimization*

The optimization problem is now reformulated to illustrate the concept of α%-risk threshold optimization. In the previous section, only product quality issues were addressed, while of course in realistic problems product quantity is also an important issue. One could take that into account either by solving a series of problems with variable lower bounds on product quantity or, as is demonstrated here, by optimizing the product *quantity* that can be reached given a specified risk. Mathematically, the optimization problem then becomes

$$\max_{\mathbf{u} \in \mathcal{U}} \quad V_f$$
$$\text{s.t.} \quad p(\mathbf{b} \in \mathcal{A}) \geq \alpha$$

Figure 4 shows the results of this optimization: the final volume V_f as a function of the risk threshold α. The dashed line indicates the location of the risk minimization optimum from the previous section, which is the upper limit on α for which a solution exists. It fills the reactor only to $V_f = 0.8762$. The nominal solution (filling the reactor with $t_1 = 50, u_1 = 0.01$) results as a limiting case on the other side, as for $\alpha = 34.70\%$ the reactor is filled to $V_f = 1$. The insight that can be obtained from this trade-off curve is that the risk of making unacceptable product can be reduced to about 5% requiring only small reduction in final volume.

When the objective is to maximize the actual quantity of C produced while taking into account that part of the batches need to be rejected, the optimum is located at $\alpha = 0.93$ on the trade-off curve and the reactor is filled to $V_f = 0.9613$.

3.4. *Quality Variance Minimization*

When reproducibility of final product is the main goal, the minimum variance objective is used. Let us assume that reproducibility of $[C]_f$ is the desired goal. Obviously, if the solution space is not constrained by any further specifications, $u_1 = 0$ is a trivial solution that produces no C with no variance. One obtains more meaningful results if the following, more realistic specifications, are made:

$$\min_{\mathbf{u} \in \mathcal{U}} \quad ([C]_f - E\{[C]_f\})^2$$
$$\text{s.t.} \quad p(\mathbf{b} \in \mathcal{A}) \geq \alpha$$

This means that the minimization of the composition variance is only added to the existing product specifications and constraints. For a comparison of the resulting product distributions, see figure 5. Note that the smaller variance with this formulation comes at the cost of a reduction in the mean, and that the final volume is only $V_f = 0.8170$ since no lower bound on volume is enforced.

An interesting overview of the simple investigated problem is given in figure 6, where contour lines of equal probability of making acceptable product are plotted over the (u_1, t_1) plane. Note that the accessible region is bounded by hard constraints $u_1 \leq 0.01$ to the right and by the hyperbola $t_1 \leq 0.5/u_1$ to the upper right. Since k_2 is uncertain, the overall effect of product quality constraints is visualized through constant-risk contour lines (outer 10%, inner 95%). To the lower left, the amount of fed B is insufficient to produce enough C. Near the top right, the feed is stopped too late, so that too much B remains in the final product – a damage that could be dealt with by increasing batch time (if not forbidden due to plant cycle time constraint). Also to the right, near where the hard feed rate constraint becomes active, the amount of final D violates the $D_{f,\min}$ constraint as feed is added too quickly, thus temporarily increasing the concentration of B and favoring the polymerization to D.

4. A Structural Mismatch Example

From a theoretical point of view, it is not clear to what extent structural model-plant mismatch can adequately be described as parametric uncertainty. In practice, on the other hand, even in the most simple batch kinetics studies, verification of residuals between model prediction and observation typically shows clear trends rather than white noise. It is thus the scope of this section to analyze the behaviour of our probabilistic approach when faced with structural mismatch. Since the optimization is conducted for two uncertain parameters, it also demonstrates the integration in the multi-variable case. Consider the following reaction mechanism

$$
\begin{array}{lclcl}
A + B & \underset{k_{2t}}{\overset{k_{1t}}{\rightleftharpoons}} & I & \overset{k_{3t}}{\longrightarrow} & C \\
B + B & \overset{k_{4t}}{\longrightarrow} & D & & \text{(order 2.5)}
\end{array}
$$

This mechanism is subsequently simulated for the *true* kinetics. The reaction is carried out in an isothermal semi-batch reactor and the dynamic model is

$$
\begin{aligned}
[\dot{A}] &= -k_{1t}[A][B] + k_{2t}[I] - [A]\frac{u}{V} \\
[\dot{B}] &= -k_{1t}[A][B] - 2k_{2t}[B]^{2.5} + k_{2t}[I] \\
&\qquad + \frac{b_{\text{feed}} - [B]}{V} \\
[\dot{I}] &= k_{1t}[A][B] - k_{2t}[I] - [I]\frac{u}{V} \\
\dot{V} &= u
\end{aligned}
$$

At time $t = 0$, only A is present with concentration $[A]_0 = 0.2$. The true values of the rate constants are $k_{1t} = 1.1355$, $k_{2t} = 5.0$, $k_{3t} = 4.2539$, and $k_{4t} = 3.5880$. Two modeling runs are conducted. During the run, A and C are measured at 10 points. The final product is analyzed for the four components A, B, C, and D. The short-lived intermediate I cannot be observed. A mechanism is postulated that agrees relatively well with the observations, i.e., produces a good fit in A and C concentration errors.

$$
\begin{array}{lcl}
A + B & \overset{k_1}{\longrightarrow} & C \\
B + B & \overset{k_2}{\longrightarrow} & D
\end{array}
$$

This mechanism is the same as in the previous section and is one that is frequently identified in fed-batch chemical reactions; in our case, it fits the observations well. As in most practical examples, only part of the residuals appear to be due to noise and the remaining part contains a slight trend (not shown).

Using the nominal rather than the true kinetic model, standard regression techniques (Bard, 1974) yield the following estimates and uncertainties*: $\hat{k}_1 = 0.5051 \pm 0.035$, $\hat{k}_2 = 0.5088 \pm 0.094$, and correlation coefficient $\rho = 0.95$. Since this yields the same nominal model as in the previous section, we can keep the quality constraints suggested there.

From a practical point of view, the obtained model structure and parameters are simple and explain the observations. There is no need for further refinement or increased model complexity. Even the violation of the mass balance, where the quantities of A and C do not exactly add up to a constant over time, is very small, since the intermediate I is only short-lived and thus exists in concentrations so small that the difference is easily masked by measurement noise.

Conceptually, the optimization could now be conducted either by full collocation with both parameters or by reduced collocation, e.g., with only one degree of freedom along the main axis of the confidence ellipse. Such a reduction to only one parameter can yield results similar to full collocation whenever both the parameter estimates are strongly correlated and the computing time is a dominant issue. We choose full collocation since a complete objective function evaluation including collocation and integration over the PDF is only a matter of 20 to 30 seconds (on a DEC ALPHA AXP).

Based on the estimated parameter values, nominal optimization, risk minimization, and risk-threshold optimization are compared. The nominal policy is the same as in the previous section. For the other two, uncertainties in parameters k_1 and k_2 are anticipated in the policy optimization by obtaining the probability of making feasible product by integrating over their joint probability distribution.

Figure 7 shows the obtained regions of acceptable product in the (k_1, k_2) plane for each of the three strategies. The axes are chosen corresponding to nominal value ± 3 standard deviations. A full ellipse, as is almost reached in the risk minimizing case (bottom), would correspond to the case in which acceptable product is made for all parameter values within a region bounded by $\Delta\chi^2 \leq 9.0$. This is a region that corresponds roughly to $\pm 3\sigma$ in the scalar case, and is used as a collocation bound here. It is clearly advantageous over $\pm 3\sigma$ collocation in each parameter as long as the true distribution can be approximated as normal since parameter estimates in batch processes are often highly correlated. In all parts of figure 7, the constraint that bounds the acceptable region to the lower left is the upper limit on $[B]_f$, which is reached because k_2 is too small to consume B. It slightly declines to the right as an increase in k_1

* given as expected value ± standard deviation

	Nominal	α–Risk	Robust
u	0.01	0.01	0.0083
t_1	50	46.5	45.5
V_f	1.000	0.965	0.878
α	41.2%	90%	99.2%
$\delta k_1/k_1$	$+\infty$ -0.0438	$+\infty$ -0.1262	$+\infty$ -0.1226
$\delta k_2/k_2$	+0.0847 -1	+0.2670 -1	+0.2585 -1
$\delta k_3/k_3$	$+\infty$ -0.0780	$+\infty$ -0.2102	$+\infty$ -0.2049
$\delta k_4/k_4$	+0.0838 -0.0590	+0.3398 -0.2888	+0.3494 -0.3505

Table 1 Summary of results.

can compensate by consuming some of the B to C. In fig. 7 (top), an additional constraint appears at the upper left. This means that the nominal feeding policy, since it involves a high feedrate, favours the second-order side reaction to an extent at which the by-product D overshoots its upper quality limit. The larger k_1, the more B is also used to make C, such that larger values of k_2 can be tolerated.

As is it is sometimes held against probabilistic approaches that robustness to probabilistic uncertainty might only in part be significant for performance on the true process, deterministic robustness margins are also determined. This is achieved by keeping three of the four parameters of the true model constant while finding a lower and an upper bound for the fourth. These bounds signify to what extent a single parameter can vary without leading to unacceptable product.

Table 1 summarizes the results obtained with the three different operating policies. First, a characterization in terms of the operating variables u_1 and t_1 is given, leading to final volume V_f. Then, the probabilistic measure of performance α, i.e., the probability of making acceptable product is given. In the final part of the table, a deterministic robustness measure, the relative changes in each parameter that are allowed while still making acceptable product, is stated. For example k_1 variations are not bounded above since an increase in k_1 with any strategy produces more and better product. The lower bound on k_1 in the nominal case is given by a 4% decrease, while the other two operating policies are robust up to a 13% decrease. Similar results are found on all other parameters, where the policies obtained through probabilistic optimization based on a model with structural mismatch are considerably more robust to parameter changes than the nominal policy. The -1 as lower bound on $\delta k_2/k_2$ is not a consequence of a quality constraint but simply the bound of physical meaning for this parameter.

It is clear that these results do not have the quality of a mathematical proof. An operating strategy optimized for robustness to changes in two parameters with probabilistic uncertainty must not always yield robustness to deterministic variations in any four parameters of any structurally different model. However, one can adapt the point of view that in most batch processes the true process structure and its full complexity remain unknown, and that the identified model is simply a good local approximation to a more complex physical process. Assuming conceptually that the model parameters depend on the true parameters through a smooth mapping, incorporation of probabilistic uncertainty information at least permits to avoid extreme sensitivity in a neighborhood of what appear to be the best approximate parameter values.

5. Discussion

Admittedly, the presented approach is limited to optimization for uncertainty in only a few model parameters, since computational complexity grows exponentially with the number of uncertain parameters. Also, with the given algorithm only comparatively simple policy parametrizations can be optimized. However, it is the authors' opinion that in most batch optimization problems parametric uncertainty can be lumped into a small number of parameters.

Looking at current industrial operating strategies and equipment (Terwiesch *et al.*, 1994; Ruppen, 1994), the limitation to simple strategies also appears to be no severe drawback and the benefits that may result from application of the proposed technique to industrial problems can be significant.

An open question is whether or how infeasible path optimization methods for simultaneous integration of the model, collocation of the probability density, and optimization of the operating profile could be used to improve computational efficiency.

Acknowledgment: First author acknowledges financial support from the Swiss National Science Foundation.

6. REFERENCES

Bard, Y. (1974). *Nonlinear Parameter Estimation.* Academic Press.

Ruppen, D. (1994). A theoretical and experimental contribution to the implementation of adaptive optimal operation for discontinuous chemical reactors. PhD thesis. Eidgenössische Technische Hochschule, Zürich. Diss. ETH Nr. 10634.

Straub, D.A. and I.E. Grossmann (1993). Design optimization of stochastic flexibility. *Comp. Chem. Eng.* **17**, 339–354.

Terwiesch, P. (1994). Dynamic Optimization of Batch Process Operations with Imperfect Modeling. PhD thesis. Eidgenössische Technische Hochschule, Zürich. Diss. ETH Nr. 10857.

Terwiesch, P., M. Agarwal and D.W.T. Rippin (1994). Batch unit optimization with imperfect modeling: A survey. *J. Proc. Control* **4**, 238–258.

Young, M.J. (1980). Semi-batch reactions offer optimal yields. *Processing* pp. 27–33.

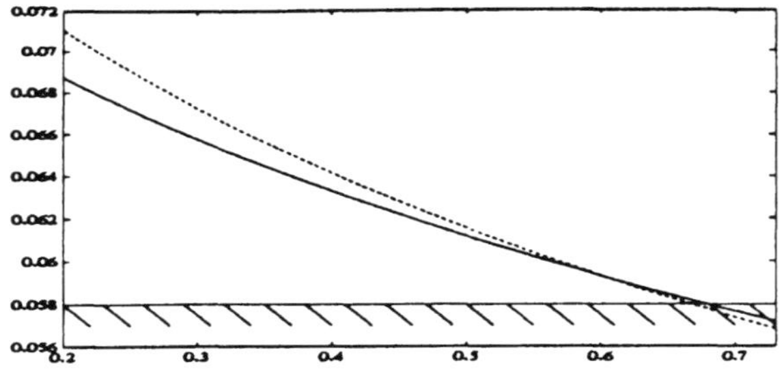

Figure 1: $[C]_f$ as a function of k_2 for risk-minimizing (——) and nominal (- - - -) feed addition profile. Constraint $[C]_{f,\min}$.

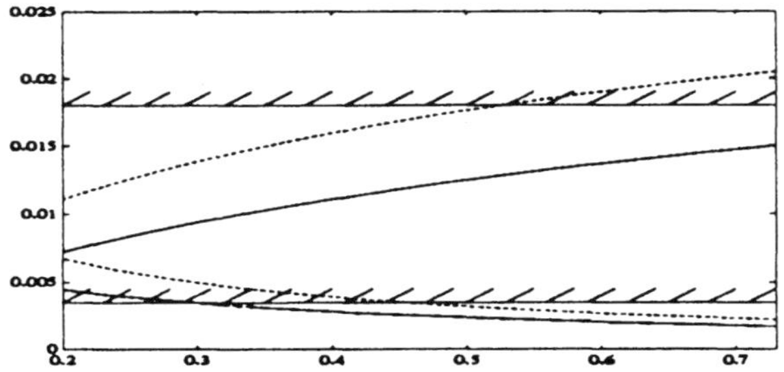

Figure 2: $[B]_f$ and $[D]_f$ as a function of k_2 for risk-minimizing (——) and nominal (- - - -) feed profile. Constraints $[B]_{f,\max}$, $[D]_{f,\max}$.

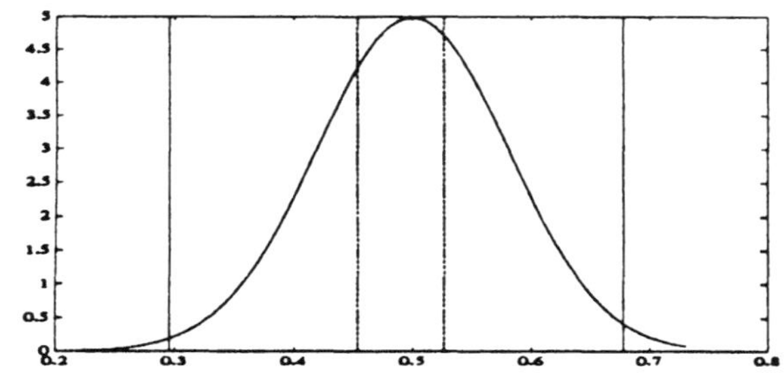

Figure 3: Comparison of risk-minimizing (——) and nominal (- - - -) region of acceptable product; PDF $p(k_2)$.

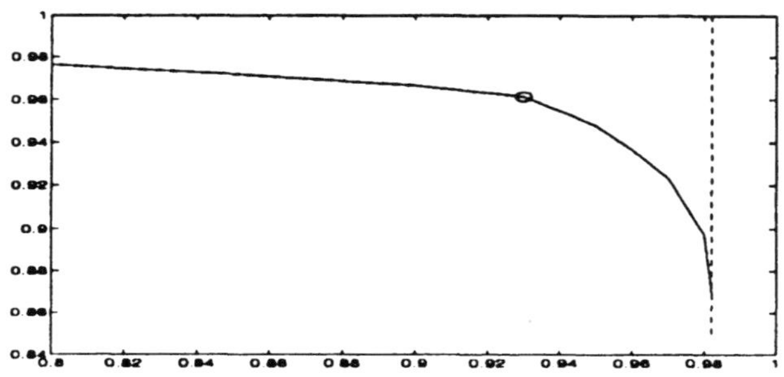

Figure 4: Final volume V_f (——) as a function of risk threshold α. (- - - -) limiting case of risk minimization.

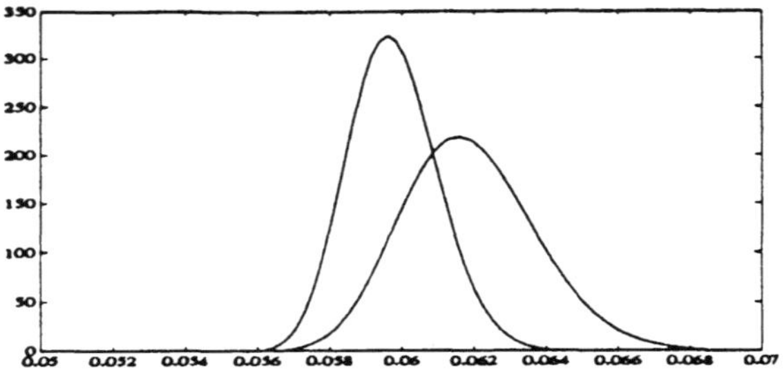

Figure 5: Probability distribution for final product concentration $[C]_f$. left: minimum variance result, right: nominal solution

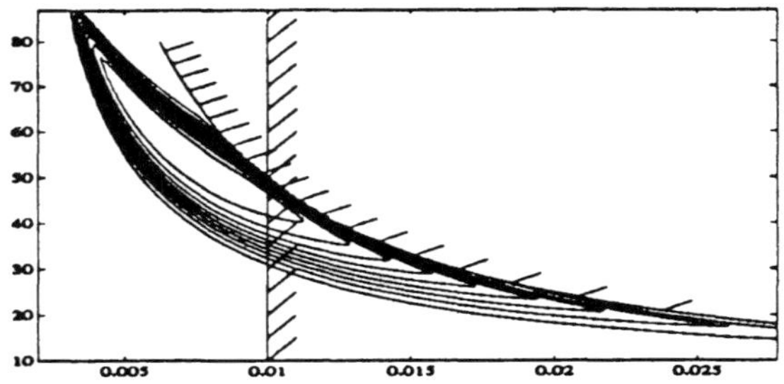

Figure 6: Decision plane (u, t_1) with lines of constant risk.

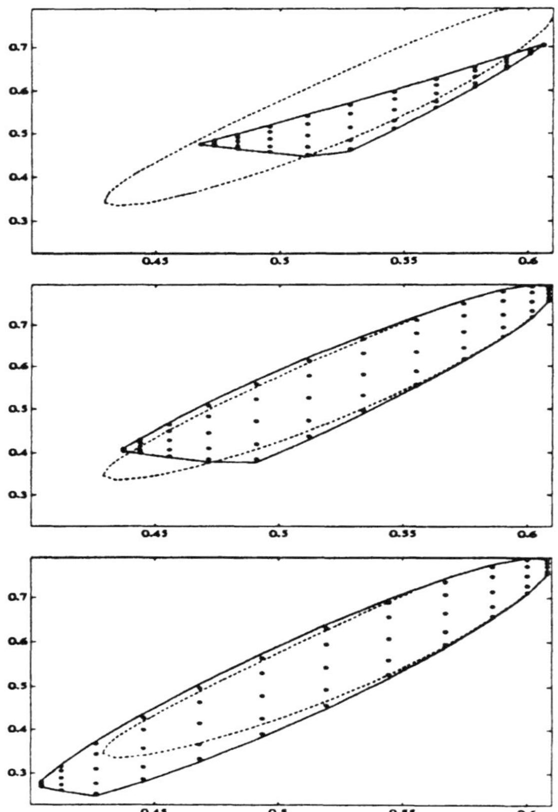

Figure 7: Region of acceptable product, for nominal operating strategy (top), for risk threshold operating strategy (middle) and for risk minimizing operating strategy (bottom), in the (k_1, k_2) plane for ± 3 standard deviations. —— constraints, o: collocation points.

SYNTHESIS OF PROCEDURAL CONTROLLERS FOR BATCH CHEMICAL PROCESSES

Arturo Sanchez, Guillermo E. Rotstein and Sandro Macchietto
Centre for Process Systems Engineering
Imperial College. London SW7 2BY, UK

Abstract. Procedural controllers are employed for the automation of event-driven operations in process systems. In this paper a formal definition for these controllers is proposed, using concepts developed from Supervisory Control Theory (SCT). Some of the limitations of the original SCT for the characterisation of the proposed controllers are identified and suitable theoretical modifications are proposed. An example is used to illustrate the ideas presented. Finally, a controller synthesis method is outlined.

Key Words. discrete event systems, sequential control, procedural control, process automation

1. INTRODUCTION

Batch and semi-continuous process are characterised by a large amount of event-driven activities, such as sequence operations, handling of alarms, emergency procedures and equipment interlocking. In automated plants, these activities are normally executed by procedural controllers. Their synthesis and analysis is a key stage in process automation because these controllers must guarantee both a correct normal operation of the process and a safe handling of abnormal operations (accounting for expected safety and environmental hazards). Although procedural controllers are heavily used in practice, limited theoretical frameworks exist to support their formal design and analysis. The need for such a framework as well as formal methods and tools for synthesis, verification and validation has been stressed by academics and industrialists (Schuler *et al.*, 1991). This is particularly relevant in flexible production environments, characterised by frequent changes in the product recipes, production modes and equipment configuration (and therefore in the required control code).

Work on the synthesis of procedural controllers in process systems has been very limited (Sanchez, 1994). A related formalism that has received some attention in the last ten years is the Supervisory Control Theory (SCT) (Wonham, 1988). Although in its early stages of development, SCT gives some general and rigorous foundations for the study of control issues in event-driven systems. It provides sufficient conditions for the existence of supervisory control structures and means for their synthesis. The *process* is modelled as a state-transition structure, a Finite State Machine (*FSM*), in which transitions are labelled as controllable (those that can be disabled by external intervention) or uncontrollable (those that cannot be prevented from occurring). Control action is exerted upon the process in feedback mode by another *FSM* termed *supervisor* that disables controllable transitions. It is important to emphasize that SCT does not consider mechanisms for enforcing controlled actions. Fundamental concepts (e.g. controllability) are defined starting from this control paradigm and lead to a very limited usability of the theory in process systems (Sanchez, 1994).

The objective of this paper is to illustrate these limitations and propose theoretical modifications for the synthesis of control devices capable of enforcing the execution of transitions rather than only disable them. In section 2 basic definitions and nomenclature are introduced. Section 3 presents a general overview of the original SCT framework. In section 4 an example illustrates its use in a batch process system and how it can fail to give useful results. In the next section, modifications to SCT are proposed, laying the foundations to the introduction of the notion of forced actions and therefore, of a *procedural controller* (section 5). In section 6 the example of section 4 is revisited, in the light of the new concepts, leading in section 7 to the outline of a synthesis method for procedural controllers.

2. PRELIMINARY DEFINITIONS AND NOTATION

In SCT, the process is modelled as an *FSM* :

$$M = \{Q, \Sigma, \delta, q_0, Q_m\}$$

where Q is the set of *states*, $q \in Q$; $\Sigma = \Sigma_u \cup \Sigma_c$ represents the set of transitions, $\sigma \in \Sigma$; δ is the state transition function defined as a partial function $\delta : \Sigma \times Q \rightarrow Q$; q_0 denotes the initial state and Q_m, the set of *marked* states. Each transition is labelled as either controllable, $\sigma \in \Sigma_c$ or uncontrollable, $\sigma \in \Sigma_u$. A set of finite strings of transitions (i.e. state–transition trajectory) generated by an *FSM* is called a *regular language* L, that is

$$L : \{\sigma \in \Sigma^* / \delta(\sigma, q_0)!\}$$

Symbol ! is used here to denote "is defined" and Σ^* is the string set. In other words, if the model of a system is given by an *FSM*, then its behaviour (feasible trajectories from the initial state) is described by the language generated by that *FSM*. The *prefix closure*, $\overline{L}$, of a language $L \subseteq \Sigma^*$, is the language which possesses all the prefixes of the strings of L.

$$\overline{L} : \{\sigma \in \Sigma^* \; / \; \exists t \in \Sigma^*, \; \sigma t \in L\}$$

The prefix closure of a language L realised by a given *FSM* is the set of those partial strings that can be constructed from the initial state of the *FSM* and that can be extended to a complete string which is part of L. It is important to note that in SCT, for every *FSM* M, $L(M)$ is assumed to be closed. That is $L(M) = \overline{L}(M)$ (Ramadge and Wonham, 1987).

The *marked language* of an *FSM* is composed of all strings in which the final *state* is a member of the set of marked *states*, Q_m.

$$L_m(M) : \{\sigma \in L(M) \; / \; \delta(\sigma, q_0) \in Q_m\}$$

It represents all the trajectories finishing at *states* considered relevant for the process, such as the completion of normal or emergency operation steps.

Given $L(M_1)$ and $L(M_2)$, the languages of *FSMs* M_1 and M_2, the language intersection, $L(R) = L(M_1) \cap L(M_2)$, is given by (Hopcroft and Ullman, 1977)

$$L(R) : \{\sigma \in \Sigma^* \; / \; \sigma \in L(M_1) \text{ and } \sigma \in L(M_2)\}$$

Sometimes, it is important to guarantee that all possible trajectories can be extended to complete trajectories. In other words, the language $L(M)$ of M representing the system behaviour must reach the marked *states* of M. If this is true, then it is said that M is *nonblocking*. That is, $L(M) = \overline{L}_m(M)$ (Wonham, 1988).

3. SUPERVISORY CONTROL THEORY

In SCT, control is exerted upon the process by another *FSM*, termed supervisor, which is an *FSM* $S = \{X, \Sigma, \gamma, x_0, X_m\}$, where X is the set of supervisor *states*, $x \in X$; Σ represents the set of transitions, $\sigma \in \Sigma$, labelled as controllable or uncontrollable; γ is the state transition function defined as a partial function $\gamma : \Sigma \times X \rightarrow X$; x_0 denotes the initial *state* and X_m, the set of marked *states*.

The *supervisor* acts in feedback mode disabling controllable transitions in such a way that it restricts the process behaviour to a subset of all the possible trajectories. If the *FSM* represents an input–output model of a process P, then the supervisor S executes synchronously with the process and in this way "controls" it. S/P represents the synchronisation of both *FSMs*. The closed–loop system is given by the *synchronisation* of the process and the supervisor (Kumar *et al.*, 1991; Sanchez, 1994). The closed–loop behaviour is given by the language intersection of the process P and the supervisor S, $L(S/P) = L(P) \cap L(S)$ (Ramadge and Wonham, 1987). Generally speaking, the supervisor can be understood as a device that restricts the assignment of resources among competing activities within the process by disabling controllable transitions only. In general, it is required that if a trajectory s is generated by a supervisor, then any extension of s with one of the uncontrollable transitions realizable by the process must be also generated by the supervisor (i.e. a *complete* supervisor in the sense of SCT).

SCT considers a desired behaviour as *controllable* if it is guaranteed that this behaviour can be maintained within a predefined subset of trajectories. Assume that there exists a language $L(P)$ describing the open–loop behaviour of a process P and a language $K \subseteq L(P)$ which represents the desired behaviour to be imposed upon the process P. K is controllable with respect to $L(P)$ if the uncontrollable transitions that can occur in $L(P)$ never cause a string to exit the space of desired behaviour. In other words, the only transitions that can take the behaviour out of K are controllable and therefore can be disabled by a supervisor.

A supervisor can sometimes block marked *states* that are feasible in the process. A supervisor that avoids this problem is a *proper* supervisor (Wonham, 1988). The existence of a proper supervisor S which, together with a process P, generates a language K that is controllable, is guaranteed in the following theorem.

Theorem 3.1 **Existence of a proper supervisor** (Wonham, 1988). If $K \subseteq L_m(P)$ is a controllable language and S/P is nonblocking then S is a proper supervisor such that $L_m(S/P) = K$.

Controllability in the SCT sense is closed under arbitrary unions (i.e. the union of two controllable languages is controllable) (Ramadge and Wonham, 1987). Thus, the supremal controllable language, $K^{\uparrow}$, is well defined and unique and contains all the controllable sublanguages on a given language L. All the trajectories not included in $K^{\uparrow}$ may drive the system to undesirable behaviour.

4. EXAMPLE

Consider the simple buffer system shown in figure 1. It consists of a tank with a level sensor indicating three discrete positions: minimum, normal and high. Liquid is fed into the tank using the on/off valve FV and is discharged through the on/off valve DV at the bottom of the tank. In the initial state of the system, both valves are closed and the tank is empty. The objective of the exercise is to synthesise a control device to supervise the start–up of the tank, by filling it to normal level and leave it ready to pass on the level control to a continuous regulatory feedback system. All the states in which the level is high must be avoided. These states are identified as *forbidden*. Using SCT the supremal controllable sublanguage, if any, that avoids the forbidden states must be found. This language must include the logic to be executed for the opening/closing of valves as a function of the process responses (level positions).

An input-output model of the process is constructed from the elementary process models shown in figure 1 and described in table 1, using Language Theory operators (Sanchez, 1994). Each model *state* describes the status of the elementary process components by using *state-variables* (Sanchez, 1994), each characterised by a discrete domain. Three *state–variables* in each *state* represent the status of each elementary process component in the following order: (level sensors, FV position, DV position). For the FV and DV valves, *state-variable* value 0 means closed while 1 means open. For the level indicator, value 0 means empty, 1 normal and 2 high. The resulting model is shown in figure 2. It represents all the *states* that can occur in the process and their sequencing using the appropriate transitions, which are numbered according to table 1. The input arrow in *state* 1 indicates the initial process *state*. Marked *states* are the initial and goal *states*.

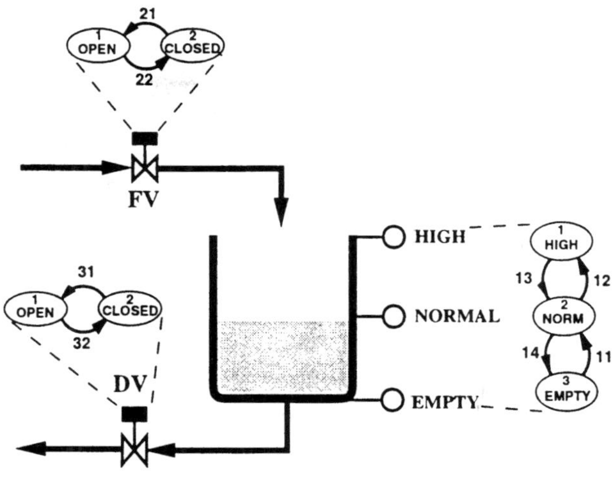

Fig. 1. Process equipment and elementary FSMs.

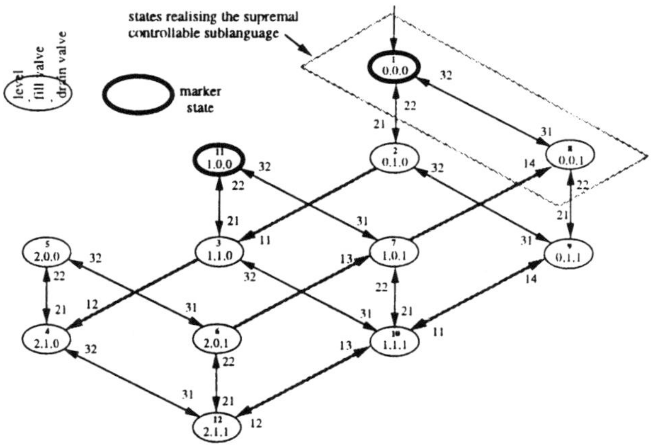

Fig. 2. Overall process model

The operation policy (desired behaviour) to be imposed in the process is shown in figure 3. It is described as a sequence of *states* starting from the initial systems *state* (indicated by an arrow). In this *FSM* the *state–variables* for each *state* are in the same order as in figure 2. The *FSM* indicates that FV must be opened first (controllable transition 21). Then the normal level must be detected (uncontrollable transition 11) and finally FV must be closed (controllable transition 22). The question is then whether it is possible to *guarantee* that this policy can be implemented while keeping the system within a specified "legal" behaviour.

The "legal" behaviour of a system is obtained by eliminating all its forbidden *states* (here, those *states* in which the high level is detected). This means the elimination of *states* 4, 5, 6 and 12 from

elementary component	*state–variable* value	*state* label	transition label	transition description
level indicator	0: empty	1	11*	empty to normal
	1: normal	2	12*	normal to high
	2: high	3	13*	high to normal
			14*	normal to empty
fill valve FV	0: closed	1	21	opening
	1: open	2	22	closing
drain valve DV	0: closed	1	31	opening
	1: open	2	32	closing

Table 1 List of the elementary *FSM* (* = uncontrollable).

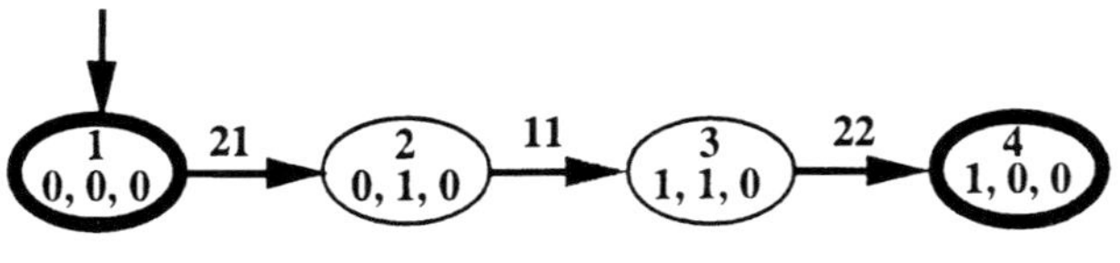

Fig. 3. Dynamic behaviour to be imposed.

figure 2. The *FSM* realising the supremal controllable sublanguage is formed of *states* 1 and 8 as indicated in figure 2 by the shaded line. All the other *states* are connected by uncontrollable transitions to forbidden *states*, or are not reachable (within the superstructure) from q_0. Thus, the "filling up" operation is not controllable. The desired operational sequence described by figure 3 specifies the opening of FV as the first step when the level is empty and DV is closed. Such a transition was eliminated from the *FSM* realising the supremal controllable sublanguage because there is an uncontrollable transition leading the system to a *state* at which high level can be detected. Obviously, one way of avoiding the high level is to never open FV, but then the filling up operation cannot be implemented. Moreover, following the desired operational sequence, once the normal level is detected in *state* 3 of figure 2, the detection of level high (transition 12) may occur spontaneously. Again, the only way of avoiding this situation by disabling transitions is to forbid in the first place the opening of FV. Thus, it is not possible to guarantee that the system can be operated within the proposed specification. Furthermore, it is easy to see that the supremal controllable sublanguage with respect to this specification is empty. Therefore, according to SCT, it is not possible to synthesise a complete supervisor capable of implementing in the process the "filling up" of the tank.

The results obtained raise a question mark on the applicability of SCT, because it is obvious that a "filling up" operation can be safely implemented in the proposed system. The failure of SCT to give useful results even in this very simple case is because control upon a process is understood as the restriction of resources among competing activities (e.g. interlocking). This may well be suitable for manufacturing processes, for instance, in which activities can be easily prevented from occurring (e.g. stop the supply of parts to a production cell, restrict the transit area of an AGV). In process systems, state changes will typically occur spontaneously, as a consequence of internal process dynamics. Control decisions must then be taken, not only to *disable* controllable activities, but actually to *enforce* the occurrence of actions (e.g. if the level reaches the desired value, the feed valve must close at once).

5. NOTIONS OF PROCEDURAL CONTROL

The above discussion shows that the calculated supremal controllable sublanguage is of little use, due to the inability of the supervisor to anticipate the occurrence of uncontrollable transitions. In the previous example, it would be useful to have a description of the closed–loop behaviour which the process can safely undertake and from which different operation policies can be derived (i.e. filling up or emptying the tank). This goes hand–in–hand with the idea of forcing actions to occur and is closer to what is commonly understood as a control action in process control. A *procedural controller* is then defined as a device driving the execution of controllable transitions in a given process. The main difference between a *supervisor* and a *procedural controller* arises at the implementation stage, and is the possibility of *forcing* the execution of transitions in the controlled system. In order to formalise the concept the following assumption is stated first.

Assumption 5.1 Controllable transitions can always be executed *before* uncontrollable transitions.

Even though this assumption may appear to be strong, it is largely a modelling issue. That is, even if a controllable transition has, in principle, very slow dynamics, it can always be divided into two transitions: a fast controllable one, representing the execution decision, and an uncontrollable transition, representing the slow system response to that decision. Assumption 5.1 opens up the possibility of disabling some uncontrollable events by anticipating them with the execution of controllable transitions. Given this assumption, the notion of procedural controller is introduced, together with relaxed SCT concepts that account for the possibility of forcing controllable transitions. The structure of a procedural controller is that of a supervisor but with the possibility of either controllable (at most one) or uncontrollable transitions occurring at each *state*.

Definition 5.1 **Procedural Controller.** A procedural controller is a supervisor, $C = \{X, \Sigma, \gamma, x_0, X_m\}$, in which for each $x \in X$ such that $\gamma(\sigma, x)!$, one of the following is true:
1. $\sigma \in \Sigma_u \wedge \forall \sigma_c \in \Sigma_c \Rightarrow \gamma(\sigma_c, x)$ is undefined.
2. $\forall \sigma' \in \Sigma, \sigma' \neq \sigma \Rightarrow \gamma(\sigma', x)$ is undefined.

In other words, a procedural controller can either be in: 1) a state in which it is waiting for one of a set of uncontrollable transitions to take place (i.e. a "response" from the controlled process) or 2) a state in which it immediately executes the only controllable transition enabled by the controller, anticipating the occurrence of any of the uncontrollable transition that can occur from that process state. The possible existence of behaviour generated by uncontrollable transitions, which is excluded from the closed-loop behaviour, makes it necessary to revise the SCT concepts of completeness and controllability previously mentioned.

Definition 5.2 **Completeness.** A supervisor S is complete with respect to a given process P if for $s \in \Sigma^*$ and $\sigma_u \in \Sigma_u$, the conditions $\gamma(s, x_0)!$ and $\delta(s\sigma_u, q_0)!$ imply that one of the following assertions is true:
1. $\gamma(s\sigma_u, x_0)!$
2. $\exists \sigma_c \in \Sigma_c$ s.t. $\delta(s\sigma_c, q_0)! \wedge \gamma(s\sigma_c, x_0)!$

That is, either the supervisor is complete in the SCT sense, or there exist controllable transitions that can anticipate the occurrence of those uncontrollable transitions that may generate behaviours not included in the supervisor.

Definition 5.3 **Controllability.** A language K is said to be controllable with respect to a language L if it can be partitioned into two languages K_1, K_2, where $K_1 \cup K_2 = K$ such that:
1. $K_1\Sigma_u \cap L \subseteq \overline{K}$
2. $K_2\Sigma_c \cap L \subseteq \overline{K}$

That is, for every *state* in an *FSM* generating a controllable language K, either: 1) all the uncontrollable transitions end in one of the *states* of the *FSM* (and this is the controllability concept in SCT) or, 2) there is at least one controllable transition leaving the *state* and ending in another *state* of the *FSM*. This means that even if there are uncontrollable transitions leading outside the *FSM*, it is possible to synthesise a controller that can anticipate them by proper control actions and keep the process in the desired state space. Clearly, the aim is to synthesise procedural controllers that are complete, and when run in synchronisation with the process to be controlled, generate controllable languages. In fact, both properties are related, as shown in the following proposition:

Proposition 5.1 Given a process P and a supervisor S, then $L(P/S)$ is controllable with respect to $L(P)$ iff S is complete.

At each state there may exist several controllable actions to execute and therefore several complete procedural controllers with the corresponding controllable languages. Furthermore, as in SCT, it can be shown that there is a unique and well defined supremal controllable sublanguage.

Definition 5.4 **Supremal controllable sublanguage.** The supremal controllable sublanguage $K^{\Uparrow}$ of a given language L is given by

$$K^{\Uparrow} = \cup\{K : K \subset L \text{ and } K \text{ is controllable}\}$$

Definition 5.5 **Superstructure of procedural controllers.** An *FSM* modelling a supervisor S, that when running in synchronisation with the process P, generates a controllable language with respect to $L(P)$.

This superstructure encompasses a set of candidate structures for the procedural control of a process. Moreover, using proposition 5.1 it is possible to characterise a superstructure that contains *all* the possible complete procedural controllers.

Definition 5.6 **Maximal Superstructure of procedural controllers.** A superstructure of complete procedural controllers is maximal for a specification language K, if its associated supervisor, when run in synchronisation with the process P, generates the supremal controllable sublanguage of K with respect to $L(P)$.

6. THE EXAMPLE REVISITED

Consider again the example of section 4. Proceeding along the same lines as before, the supremal controllable sublanguage for the "legal behaviour" (i.e. no states at which the level is high) obtained with the new framework, is the *FSM* resulting from eliminating *states* 4, 5, 6 and 12. This realises a controllable behaviour. Moreover, the same *FSM* represents as well the maximal superstructure of procedural controllers for the given process model. It is thus possible to generate from it the procedural controllers for the different operations required, such as the filling up or the emptying of the tank. In particular, the *FSM* in figure 3 is indeed complete with respect to P, and can be used as the procedural controller implementing the "filling up " operation. The uncontrollable transition 12, that exits the state space of desired behaviour, is anticipated (according to assumption 5.1) by the execution of controllable transition 22, that closes the feed valve. The language generated by 3 is thus found, as expected, to be controllable.

7. SYNTHESIS PROCEDURE

The maximal superstructure of procedural controllers for a specification language K and a process P contains all those controllers that generate closed-loop controllable languages (i.e. sets of trajectories) included in the desired behaviour specification K. Once the process model and the maximal superstructure for a given process and specification have been obtained, the next synthesis step is to select from the superstructure the procedural controller considered more convenient. This can be done either 1) by successive specification of the desired behaviour (Sanchez and Macchietto, 1995) (i.e. adding "constraints") thus restricting the superstructure until only one controller remains or 2) by optimisation, selecting trajectories that satisfy a given objective function (e.g. minimum completion time, control cost, hazards probability, etc.).

The synthesis method is summarised in the following steps:

1. Construct the process model (Sanchez, 1994).
2. Construct the model for the "legal" behaviour of the process (Sanchez, 1994).
3. Find the maximal superstructure of procedural controllers in the resulting legal behaviour model.
4. If the superstructure is empty or does not contain the marked strings (i.e. trajectories from the initial to the marked states), then no feasible controller exist. Either there was a modelling error, or there is a design problem. Go back to step 1. Otherwise, continue.
5. Select from the superstructure, a procedural controller satisfying further user specifications or a proposed optimisation criteria.

8. CONCLUSION

The formal definition of procedural controllers has been considered, and a method for their synthesis has been outlined. Given an *FSM* model of the process, and a specification on its legal behaviour, necessary conditions for the existence of procedural controllers were obtained. Furthermore, a maximal superstructure of procedural controllers was characterised. This superstructure included all the existing procedural control structures that can restrict the feasible controlled process trajectories to the specified legal behaviour.

Recently, forced action have been considered within a timed automata framework (Brandin and Wonham, 1994). Accounting for forced actions in a state-transition structure within a qualitative time framework, reduces the modelling and computational complexity of the synthesis problem. The implications and limitations of this new approach are currently under investigation.

9. REFERENCES

Brandin, B. A. and W. M. Wonham (1994). Supervisory control of timed discrete-event systems. *IEEE Transactions on Automatic Control* **39**(2), 329–531.

Hopcroft, J. E. and J. D. Ullman (1977). *Introduction to Automata Theory, Languages, and Computation*. Addison-Wesley. Reading, MA, USA.

Kumar, R., V. Garg and S. Marcus (1991). Using predicate transformers for supervisory control. *Proceedings of the 30th Conference on Decision and Control, Brighton, UK* pp. 98–103.

Ramadge, P. J. and W. M. Wonham (1987). Supervisory control of a class of discrete-event processes. *SIAM Journal of Control and Optimization* **25**(1), 206–230.

Sanchez, A. (1994). Formal Specification and Synthesis of Sequential/Logic Controllers for Process Systems. PhD thesis. University of London.

Sanchez, A. and S. Macchietto (1995). Design of sequential/logic controllers for Chemical Processes. *Submmited to ESCAPE'95.*

Schuler, H., F. Allgower and E. D. Gilles (1991). Chemical Process Control: Present status and future needs. The view from the European industry. In: *CPC IV* (Y. Arkun and W. H. Ray, Eds.). pp. 29–52. Elsevier.

Wonham, W. M. (1988). A control theory for discrete-event systems. In: *Advanced Computing Concepts and Techniques in Control Engineering* (M.J. Denham and A.J. Laub, Eds.). pp. 129–169. Springer Verlag.

TEMPERATURE CONTROL OF A BATCH ESTERIFICATION REACTOR OVERCOMING NON-LINEARITIES USING ADAPTIVE GAIN

P. Djavdan

DSM Engineering & Stamicarbon, Centre for Process Control
P.O. Box 10, 6160 MC Geleen, The Netherlands

Abstract: The design of a Master-Slave control structure for temperature control of a batch esterification reactor is described. The control structure is implemented for the control of a number of reactors in three different sites and is able to guarantee reproducible temperature profiles. The slave controller comprises a gain scheduling strategy to overcome non-linearities in the process gain. In the paper, the non-linear gain of the heating and cooling process which is a result of a wide operation range of reactor temperature is derived using a dynamic model of the reactor and heating and cooling circuit. Simulation is used to clarify the effect of gain scheduling. Implementation results are discussed.

Keywords: Control nonlinearities, Temperature control, Master-slave system, PID, Dynamic model, Adaptive gain, Gain scheduling.

1. PROBLEM STATEMENT

DSM is a chemical company of which the plants are widely spread all over the world. For the quality control of some products it is necessary to have a reproducible temperature profile for each product type. This means that the temperature control should be independent of the physical properties of the plant equipment.

In the case of esterification processes the produced water has to be evaporated. Therefore it is necessary to heat up the reactor. This is often realized by a heating circuit comprising a heating coil or jacket. The heating medium is steam or hot oil depending on the desired temperature in the reactor.

Fig. 1 shows a reactor with internal coils for heating. The heating medium is hot oil. In this configuration a secondary oil circuit is used for each reactor. One primary oil circuit is used to heat up the oil to high temperatures (±280°C). A control valve is installed to control the supply of hot oil into the secondary oil circuit. Using this control valve it is possible to control the oil temperature in the secondary oil circuit.

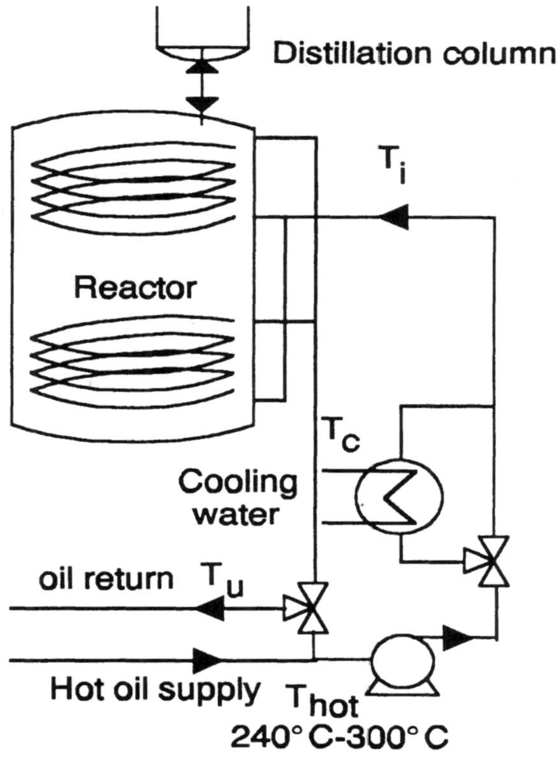

Fig. 1 Layout of the reactor and the oil circuit

A heat exchanger is used for the cooling down

procedures during the batch. The cooling is realized by cooling water. Using these heating and cooling facilities, it is possible to control the oil temperature in the secondary circuit any where between 20 and 280°C.

1.1 Control Requirements

With respect to temperature control, the following operations are required:

Heating up raw materials: The reactor is heated up to reaction temperature by means of hot oil supply into the secondary oil circuit.

Exotherm: During this operation no external heating is required. The reactor contents is heated up by the reaction heat.

Esterification: It is necessary to evaporate the produced water. In order to maintain a sufficient and constant reaction rate the heat input to the reactor is increased.

Constant temperature: In this operation the reactor temperature is kept at a constant level (e.g. 200°C).

Cooling down: After the desired viscosity and acid number is achieved, the reactor contents is cooled down in order to stop the reaction as soon as possible.

The control strategy should be able to realize all these operations.

Although the hot oil temperature is controlled, it can vary between 240 and 300°C depending on the heat requirement in the plant. Also the total flow in the secondary circuit is not constant. There variations effect the oil temperature in the secondary circuit and the heat exchange with the reactor. This means that the temperature control strategy should be able to cope with these disturbances in the oil circuit.

During the distillation operation a varying amount of reflux is returned into the reactor. This changes the heat requirements in the reactor. The viscosity and therefore the heat exchange coefficient in the reactor changes during the batch. Also the volume of the contents of the reactor and therefore the heat exchange area changes during the batch. The control strategy should be able to compensate for these changes.

2. DIRECT CONTROLLER

It is possible to design a direct controller which controls the reactor temperature directly by adjusting the hot oil supply into the secondary oil circuit. This direct control structure has a number of disadvantages:

First, the disturbances in the oil circuit are not measured before having influenced the reactor temperature. It is much better to compensate these disturbances before they have the chance to influence the reactor temperature.

Second, the dynamics of the reactor and the oil circuit are very different. The reactor dynamics are often much slower than the dynamics of the oil circuit. The direct controller must be tuned for reactor dynamics as well as for the dynamics of the oil circuit to be able to cope with both types of disturbances adequately. This is very difficult if not impossible.

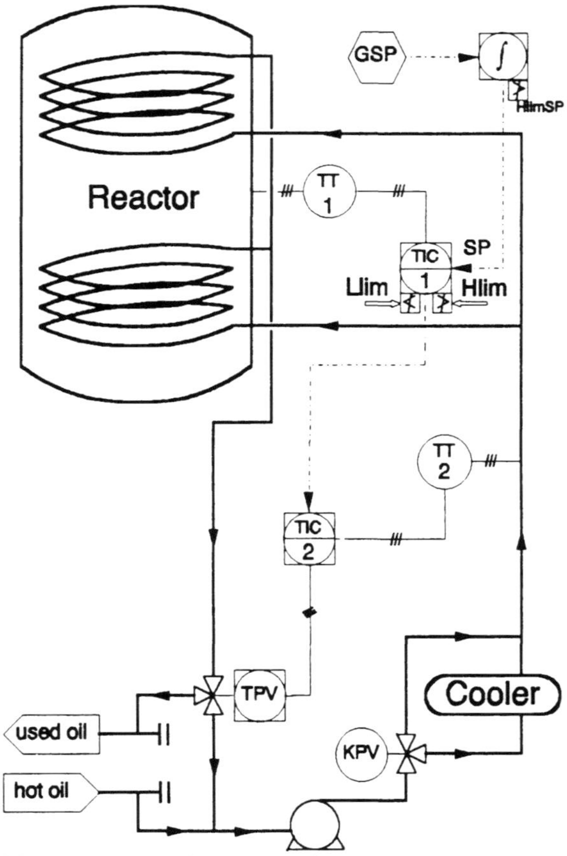

Fig. 2 master-slave control structure

3. MASTER-SLAVE CONTROL STRUCTURE

As described in (Djavdan, 1992) it is better to use a master-slave control structure to control the temperature of a batch reactor.

This control structure is shown in Fig. 2. In this structure the slave controls the temperature of the oil entering the reactor coil, by adjusting the hot oil flow into the secondary oil circuit. The master provides the required setpoint for the inlet oil temperature in order to fulfil the heat requirement in the reactor. The desired reactor temperature is the setpoint of the master controller. The slave will adapt to a suited inlet oil temperature.

This control structure has the advantage that each controller (master and slave) can be tuned separately and therefore properly and disturbances in the oil circuit are suppressed before the reactor temperature is influenced.

The sequence control which is implemented in the Distributed Control System (DCS) indicates the suitable temperature profile for the particular product type. This temperature profile is the setpoint trajectory of the master controller.

3.1 Details of the controllers

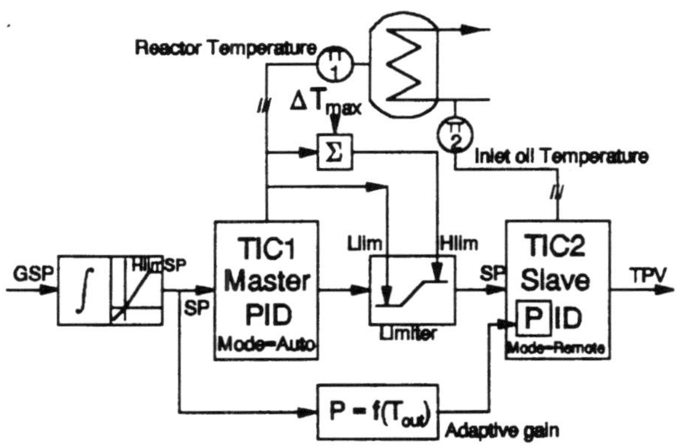

Fig. 3 Detailed scheme

Figure 3 shows a detailed scheme of the master-slave controller. The master is a PID controller. During the heating up procedures, the output of the master is limited between T_r (reactor temperature) and $T_r + \Delta T_{max}$ (maximum allowable oil temperature). An anti reset-windup mechanism is implemented in the controller in order to prevent reset-windup when hitting the above constraints.

The slave is a PI controller with an adaptive gain. The adaptive gain is necessary because of the non-linear gain of the secondary oil circuit. This non-linearity is described in the following paragraph. The output of the slave is limited between 0 and 100%.

4. GAIN SCHEDULING

In the process industry very often the processes are non-linear. Most processes can be characterized by a non-linear process gain and low order dynamics (first or second order dynamics) system. For the control of these systems very often a PID controller is sufficient. In the case that the static non-linearities are significant in the operation range of the controller, it is necessary to adapt the controller gain to the changes in the process gain. This is called gain scheduling. Gain scheduling is a well established method for non-linear control in the process industry. Many Distributed Control Systems (DCS) offer the possibility to implement gain scheduling strategies.

4.1 Reactor model

Using both the material and heat balances as well as a simple first order kinetics (incl. Arrhenius), it is possible to construct a model which can be used for checking the control structure as well as the tuning of the controllers for the esterification reactors. A simple model consists of the following equations:

$$d(M_R.T_r.cp_R)/dt = 0.5.U.A.(T_{oil}-T_r + T_u-T_r) + Q_{reac} - \Phi_{ref}.cp_w.(T_r-T_{ref}) - \Phi_{RawM}.cp_{RM}.(T_r-T_{RM}) - \Phi_{evap}.QV$$

$$d(0.5.M_{oil}.T_{oil}.cp_{oil})/dt = \Phi_{oil}.(T_i-T_{oil}).cp_{oil} - 0.5.U.A.(T_{oil}-T_r)$$

$$d(0.5.M_{oil}.T_u.cp_{oil})/dt = \Phi_{oil}.(T_{oil}-T_u).cp_{oil} - 0.5.U.A.(T_u-T_r)$$

$$d(M_i.T_i.cp_{oil})/dt = (1-x).\Phi_{oil}.T_u.cp_{oil} + x.\Phi_{oil}.T_h.cp_{oil} - \Phi_{oil}.T_i.cp_{oil}$$

$$d(M_R)/dt = \Phi_{RawM} + \Phi_{ref} - \Phi_{evap}$$

$$d([B])/dt = (1-[B]).k_\infty.\exp(-E/RT_r)$$

$$Q_{reac} = M_R.QR.d([B])/dt$$

- T_i = resulting temperature of mixing hot oil and used oil
- T_{oil} = oil temperature in the middle of internal coil
- T_u = outlet oil temperature (used oil)
- T_h = hot oil temperature
- T_r = reactor temperature
- T_{ref} = reflux temperature
- T_{RM} = Raw Material temperature
- M_{oil} = mass of oil in the coil
- M_i = mass of oil in the mixing point
- M_R = mass of the contents of the reactor
- cp_{oil} = cp of oil
- cp_w = cp of water
- cp_R = cp of the contents of the reactor
- cp_{RM} = cp of Raw Material
- Φ_{oil} = flow of the circulating oil in the coil
- Φ_{ref} = flow of the reflux
- Φ_{evap} = flow of the evaporation
- Φ_{RawM} = flow of the raw material into the reactor
- x = Valve opening of the hot oil supply
- Q_{reac} = Reaction heat flux
- [B] = exothermic product concentration
- k_∞ = pre-exponential factor in Arrhenius
- E = Activation energy

R = Gas Constant
QR = Reaction heat [J/kg]
QV = Evaporation heat [J/kg]
U = heat transfer coefficient between reactor contents and circulating oil
A = heat transfer area between reactor contents and circulating oil

4.2 Deriving the adaptive gain

The oil temperature is controlled by adjusting the valve opening that determines the hot oil flow into the secondary circuit. This hot oil is mixed with the outlet oil coming from the coil in the reactor. The effect of the hot oil supply depends on the outlet oil temperature (T_u). When this temperature is almost 50°C (T_{hot} is almost 280°C), the supply has a large effect on the inlet oil temperature. At the time that the outlet oil temperature is almost 240°C, the effect of the hot oil is much less. This means that the gain of the oil circuit (process) is variable and depends on the outlet oil temperature. The gain of the controller should be adjusted to compensate for this change in the process gain. The heat balance of the mixing point can be used to derive the necessary adjustment:

$$d(M_i.T_i.cp_{oil})/dt = (1\text{-}x).\Phi_{oil}.T_u.cp_{oil} + x.\Phi_{oil}.T_h.cp_{oil} - \Phi_{oil}.T_i.cp_{oil}$$

This equation is only valid if the following assumptions are justified:

- The valve is almost a linear valve.
- The total flow of the circulating oil does not change much.
- The mixing of hot oil and used oil in the mixing point is almost ideal.

In order to calculate the necessary gain scheduling we need only the static part of this dynamic equation:

$$0 = (1\text{-}x).\Phi_{oil}.T_u.cp_{oil} + x.\Phi_{oil}.T_h.cp_{oil} - \Phi_{oil}.T_i.cp_{oil}$$

therefore:

$$T_i = (1\text{-}x).T_u + x.T_h = T_u + x.(T_h\text{-}T_u)$$

The static gain of the mixing process in the oil circuit is then proportional to:

$$T_h - T_u$$

This means that in the beginning of the batch where the (T_h-T_u) is almost 200°C, the static gain of the mixing process is 4 times higher than at the end of the batch where (T_h-T_u) is almost 50°C. For a good slave controller we then need to use a controller gain which depends on this factor:

$$K = K_0/(T_h\text{-}T_u)$$

where K_0 is a constant gain and K is the adaptive gain of the slave controller.

In the case that both measurements are available it is possible to use both T_h and T_u measurements for the calculation of the gain scheduling. The changes in T_h however do not have a significant effect on the gain scheduling, therefore the adaptive gain can be:

$$K = K_0/(300\text{-}T_u)$$

The reason that we choose 300 for T_h is that the factor 300-T_u should never become very small (not to mention zero or negative).

In some plants there is no measurement of T_u. In this case it is possible to use an approximation of T_u:

$$T_u = T_r + e^{-\lambda}.(T_i\text{-}T_r)$$

where:

$$\lambda = U.A/(\Phi_{oil}.cp_{oil})$$

For the implementation it is sufficient to use an average of λ during the batch. T_u can then be calculated as a weighted average of T_r and T_i.

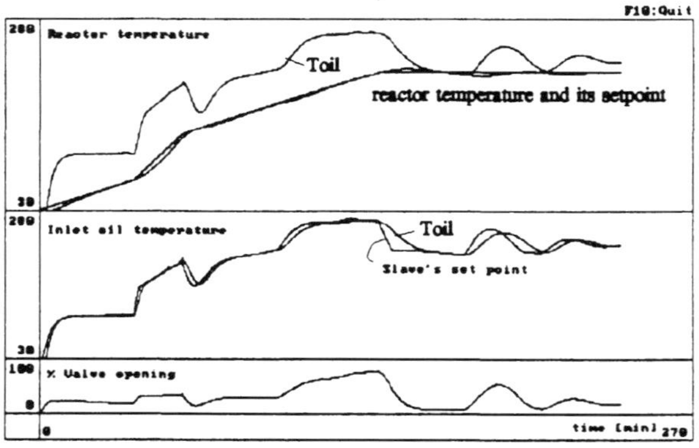

Fig. 4 No gain scheduling low controller gain

4.3 Simulations

The reasoning in last paragraph showed that the process gain of the oil circuit changes dramatically during the batch. In this paragraph some simulations are used to clarify the reason of oscillations when no adaptive gain is used.

Figure 4 shows the performance of the master-slave controller without gain scheduling and with optimal tuning of the slave at the lower oil temperatures. This results in a relatively low controller gain. The effect of this is that the controller gain is too low for the higher temperature values. This on its turn will cause the slave controller to be too slow compared to the master and results in an oscillating process caused by the fact that the slaves PV stays behind its SP (which is dictated by the master). To prevent this the settings of the master controller could be decreased, but this will give a bad performance in realizing the required temperature profiles.

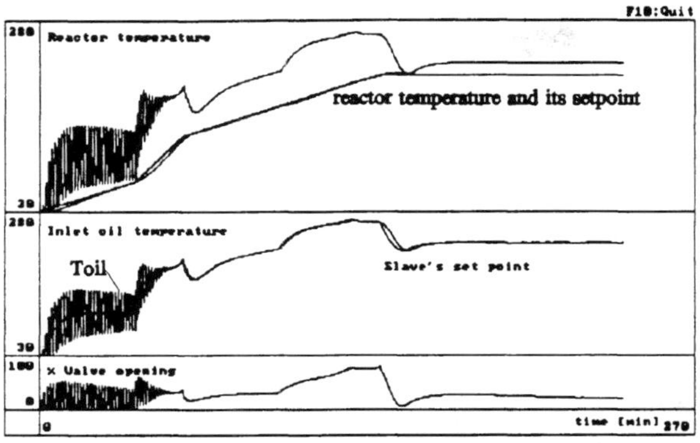

Fig. 5 No gain scheduling with high gain

Figure 5 shows the performance of the master-slave controller without gain scheduling and with optimal tuning of the slave at the higher oil temperatures. This results in a relatively high controller gain. The effect of this is that the controller gain is too high for the lower temperature values and the oil temperature will start oscillating with fast frequencies (oscillation period of almost 1 minute). This oscillation is despite of the stable SP of the slave (as dictated by the master)

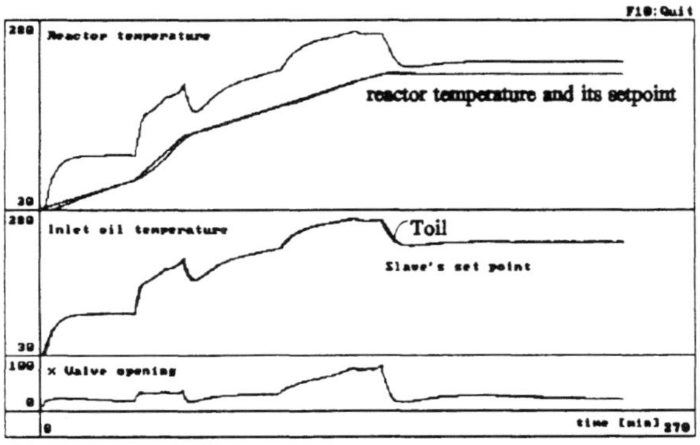

Fig. 6 Gain scheduling

Figure 6 finally shows the result of an adaptive gain as described in this paragraph. The gain of the controller is increased depending on the outlet oil temperature. The overall performance of the both controllers is satisfactory.

5. SPLIT RANGE CONTROL

In some plants it is necessary to use both heating and cooling in order to control the reactor temperature in some operations. In these plants a Split-Range control structure is used as described in (Stephanopoulos, 1984). In this case the control signal is split into two parts, each affecting one of the control valves (cooling and heating). This is shown in Fig. 7. As described in paragraph 3, the control signal is between 0 and 100%. When the control signal is 0%, the cooling valve is completely open and the heating valve is completely closed. When the control signal changes from 0% to 50%, it affects only the cooling valve which is closed at 50%. In the range of 50% to 100% the heating valve is gradually opened..

5.1 Adjusting the gain scheduling

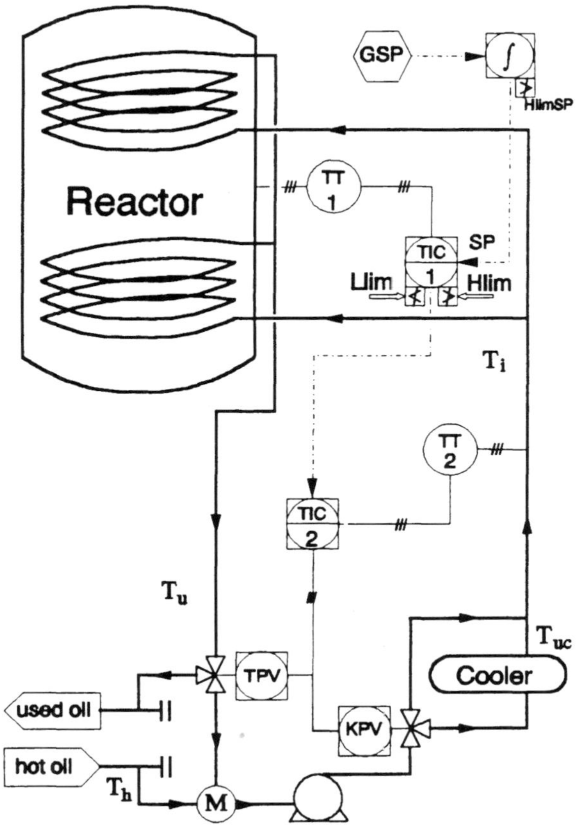

Fig. 7 Split-Range control for the slave

The effect of cooling is completely different from the effect of hot oil supply. As the same controller is used for both heating and cooling in the Split-Range configuration, it is necessary to adjust the controller gain not only depending on the outlet oil temperature T_u, but also depending on the cooling and heating modes.

From the steady-state heat balance of the cooler one can derive:

$$T_{uc} = T_c + e^{-\lambda/x_c}.(T_u - T_c)$$

where:

$\lambda/x_c \quad = U_c.A_c/(\Phi_{oil}.x_c.cp_{oil})$

From the heat balance after mixing:

$$T_i = (1-x_c).T_u + x_c.T_{uc}$$
$$= T_u - x_c.(1-e^{-\lambda/x_c}).(T_u - T_c)$$

where:

U_c = heat exchange coefficient of the cooler
A_c = heat exchange area of the cooler
$\Phi_{oil}.x_c$ = oil flow through the cooler
T_{uc} = outlet cooler oil temperature
x_c = cooler valve position

From this formula, during cooling the gain of the controller must be adjusted according to: $T_u - T_c$.
This means that the gain scheduling is different in the case of cooling. We therefore need an indicator to make distinction between cooling and heating. A good indicator for this is the control signal. A small control signal means cooling. A large control signal means heating. Therefore the formula for gain scheduling is adjusted as follows:

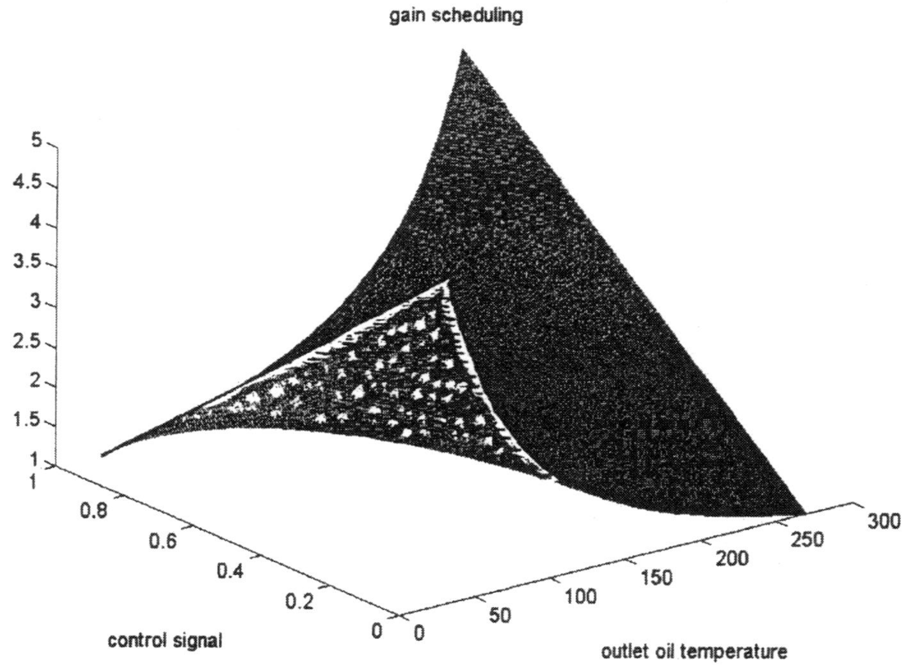

Fig. 8 Gain scheduling for Split-Range

$$K = K_0.(x/(1.1-\tau) + (1-x)/(\tau+0.1))$$

where: $\tau = T_u/300$

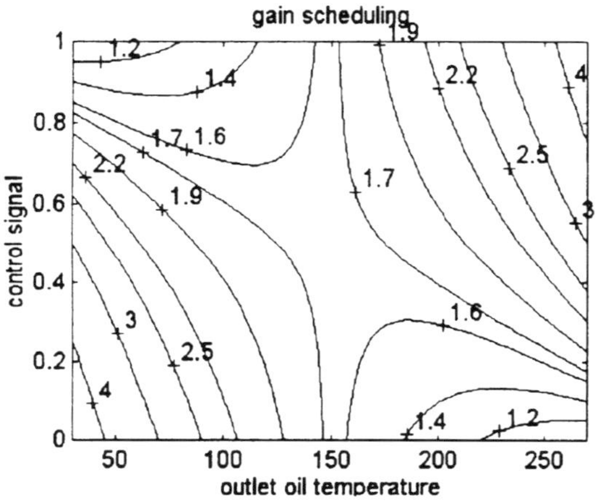

Fig. 9 Gain scheduling Split-Range

This formula is plotted in figures 8 and 9. Fig. 8 shows a three dimensional plot of k depending on T_u and x. Fig. 9 shows the contour lines of the gain.

6. IMPLEMENTATION RESULTS

This control structure with adaptive gain is implemented in different plants which are equipped with different (Distributed) Control Systems.

Every control system needed a slightly different approach. In some plants the adaptive gain had to be implemented based on the reactor temperature. In other plants based on the outlet oil temperature. One of the systems seemed to work internally with a gain, even when the external representation is a Proportional Gain. This caused some problems for the implementation of adaptive gain.
Each controllers had to be tuned separately. For each reactor first the slave controller is tuned and then the master controller. The place of the measurement device in the oil circuit seemed to be very important for the performance of the slave controller. The inlet oil temperaure measurement should be after the cooling and the distance between the measurement and the mixing point should be as short as possible.

After overcoming these problems the controller was able to realize the desired temperature profiles in all different plants.

7. REFERENCES

Djavdan, P. (1992). Batch time optimisation using both reactant dosing and external cooling for tight temperature control. In: *Proceedings DYCORD+'92*. p. 135-140.
Shinskey, F.G. (1988). *Process Control Systems, Application, Design and Tuning*.
Stephanopoulos, G. (1984). *Chemical Process Control, an introduction to theory and practice*. Prentice-Hall, Inc., Englewood Cliffs, New Jersey.

TEMPERATURE CONTROL OF A BATCH REACTOR BY ON-LINE IDENTIFICATION OF A VAPOR PRESSURE LAW

J. Cezerac, V. Garcia, M. Cabassud, M.V. Le Lann and G. Casamatta

Laboratoire de Génie Chimique - URA CNRS 192
Ecole Nationale Supérieure d'Ingénieurs de Génie Chimique
18 Chemin de la loge - 31078 TOULOUSE Cedex - FRANCE

Abstract : In the fine chemical industry the batch or fed-batch reactor functions as the heart of the transformation process. Due to the complexity of chemical synthesis, control of these reactors remains a problem of temperature control commonly performed indirectly via the jacket of the reactor. In this paper, an original approach of thermal control through pressure is proposed by taking advantage of the fast dynamics of the pressure-temperature dependence at the boiling point. The pressure that should be applied to the reactor to track a set-point temperature is computed via an on-line identified vapor pressure law.

Keywords : Batch reactor - Temperature control - Identification algorithms

1. INTRODUCTION

The thermal control of flexible and multipurpose batch reactors is usually performed through heat exchanges with a fluid circulating in the jacket surrounding the reactor. The two common types of heating-cooling systems are the alternative heating-cooling system and the monofluid system (Le Lann, *et al.*, 1994). The first one represents 90% of industrial applications. In these cases, the scale-up of the pilot plant to the large capacity industrial reactor is limited because the control is performed indirectly through the heat transfers between the fluid circulating in the jacket and the reactant mixture. Many problems occur due to the thermal inertia (large time constant) of the jacket wall. To facilitate the operation at industrial scale, reactions are usually carried out in a large amount of solvent at atmospheric pressure. The control of the reaction at constant temperature is ensured by maintaining the reactant mixture at the boiling point, which may be considered constant due to the large excess of solvent. This technique of "thermal control", well-known in the fine chemical and pharmaceutical industry as "refluxing reaction", limits the choice of the solvent to those having a normal boiling point close to the reaction temperature envisaged.

Inspired by the refluxing reactor, we have developed an original approach to the problem of thermal control of batch or fed-batch reactors, described in this paper : thermal control by action on the pressure in the reactor. The principle of this thermal control is to adjust constantly the pressure in the reactor to make up for dynamic changes in the thermodynamic equilibria in order to maintain the temperature at a constant value or to track a desired profile.

This technique of thermal control has many advantages in comparison to classical ones. Thermal control by action on the pressure avoids problems due to poor thermal transfers across a jacket since the thermal exchanges are carried out in the condenser especially designed for this purpose. The capacity of heat exchange is thus significantly improved by a large heat transfer area and by a higher heat transfer coefficient than during classical convective transfer across a jacket.

The refluxing reaction technique with action on the pressure is thus a way to increase the controllability of the batch reactors. Another interest of this technique is the possibility of decreasing the amount of solvent used during the reactions. There are numerous advantages for concentrating the reaction mixture. Decreasing the amount of solvent may increase the batch reactor productivity and lead to a cleaner process that minimizes the consumption, the recycling and the treatment of solvent, and is safer.

This study deals with the theoretical and experimental framework which has been used to develop the technique of refluxing reaction with action on the pressure. The pilot plant reactor and the dynamic model used as simulation are described. The methodology of the thermal control technique based on the on-line estimation of the pressure-temperature relation modelling the boiling mixture is presented. In a first attempt, the availability of a Clausius-Clapeyron law used to describe the relation between pressure and temperature at boiling point for pure liquid has been tested with an homogeneous liquid mixture (constant or composition time-varying). The relevance of thermal control through pressure law is then exemplified in both simulation studies, using a dynamic simulation model taking into account the thermodynamic liquid-vapor equilibria, and experimentally on a pilot plant reactor.

2. PROCESS OPERATION AND MODEL DESCRIPTION

2.1 Pilot plant reactor description

Figure 1 represents the pilot plant developed in the context of this study.

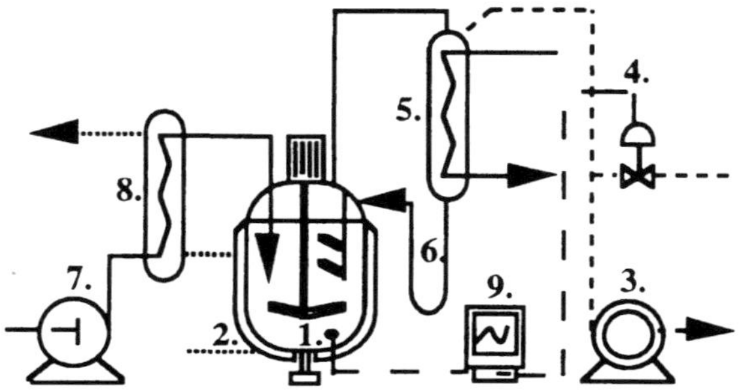

Fig. 1. Pilot plant reactor and its environment

During the thermal control through pressure, the reactant mixture (**1.**) is kept at the boiling point by heat exchange with the hot fluid flowing inside the jacket surrounding the reactor (**2.**). The vapor is condensed (**5.**) and recycled to the reactor (**6.**). The vacuum pump (**3.**) functions continuously to maintain the desired pressure in the system (reactor and condenser). This pressure is computed by the controller implemented in the computer system (**9.**) (data acquisitions, control calculations and supervisory control). In practice, a change of the pressure is attained by varying the control valve opening degree (**4.**) which controls the air flow rate entering the system. The reactor functions in a batch or fed-batch mode (feed pump (**7.**) and preheater (**8.**)).

2.2 Dynamic model description

A dynamic model of the pilot plant has been developed (Cezerac, *et al.*, 1993) in order to minimize the number of experiments on the pilot plant, to study different types of controllers and to perform predictions on the scale-up conditions for industrial plants. Based on mass and energy balances, the following phenomena are modelled :

- the chemical reaction itself (kinetics of a homogeneous liquid phase, feeding of reactants),
- the heat transfers between the thermal fluid flowing in the jacket, the reaction mixture and the ambient surroundings,
- the phase changes due to the variation of the pressure in the reactor (liquid-vapor equilibria).

Heat transfers have been modelled from characteristic data which are easily available on the experimental device (flow rate and temperature of the thermal fluid in the jacket, physical properties of materials used for the reactor and the jacket) and from the computation of heat transfer coefficients via dimensionless physical correlations. Physical and thermodynamic properties of the mixture and liquid-vapor phase equilibria are computed using the computer program, Prophy (Le Lann, *et al.*, 1988). The resulting set of algebraic and differential equations is solved by a Gear method.

3. ON-LINE IDENTIFICATION OF A VAPOR PRESSURE LAW

3.1 Introduction

In 1834, Clapeyron established a relation between pressure and temperature of a pure liquid in equilibrium with its vapor phase, based on the thermodynamic of phase equilibria :

$$\frac{dp}{dT} = \frac{\Delta H_{vap}}{T(V_v - V_l)} \quad (1)$$

where p, T, ΔH_{vap}, V_v and V_l represent respectively the pressure of the vapor phase, the temperature of the pure liquid, the molar enthalpy of vaporization, the molar vapor volume and the molar liquid volume. Some thirty years after Clapeyron introduced his equation, Clausius introduced hypothesis ($V_v >> V_l$ and vapor phase = perfect gas) which led to the formulation of the relation, well-known as the Clausius-Clapeyron law, given by the equation (2) :

$$\frac{d \ln p}{dT} = \frac{\Delta H_{vap}}{RT^2} \quad (2)$$

The integrated form of this equation, with the assumption of a constant value of ΔH_{vap} over the temperature range considered, is :

$$\ln p = Acc - \frac{Bcc}{T} \quad (3)$$

The parameters Acc and Bcc of the Clausius-Clapeyron law are directly related to the characteristics of the pure liquid.

Can the form of the Clausius-Clapeyron law be used to describe the relation between pressure and temperature of homogeneous liquid mixtures at boiling point ? Firstly, to answer this question, the Clausius-Clapeyron law has been used to describe the relation between pressure and temperature of homogeneous liquid mixtures (mixtures of acetone, methanol and water are used in this study) at different given compositions at boiling point. When a chemical reaction is carried out, the mixture composition is time varying. So, secondly, experiments have been performed with a composition time-varying mixture. In each case, it has been possible to identify two parameters Acc and Bcc which provide a good fit of the data (p, T_m) in the range 760-100 Torr (pressure range considered). All the experiments of this part of the study have been carried out in the pilot plant reactor. The Recursive Least Square (RLS) method has been used to identify on-line the parameters of the Clausius-Clapeyron law, based on the minimization of the error between the measured pressure in the reactor, p, at boiling point and the computed pressure with the Clausius-Clapeyron law which is a function of the measured temperature of the mixture, T_m, and the parameters Acc and Bcc. During the experiments the mixture is always at its boiling point.

3.2 On-line identification of the Clausius-Clapeyron law for different mixture compositions

Firstly, for each composition given in Table 1, the experimental relation between the measured pressure and the measured temperature of the boiling mixture (0.5 liter) was established.

Table 1. Volume compositions of acetone/methanol/water mixtures

	acetone	methanol	water
mixture1	0.53	0.29	0.18
mixture2	0.35	0.19	0.46
mixture3	0.26	0.15	0.59

Figure 2 shows that it is possible to fit the experimental points (p_{exp}) with a Clausius-Clapeyron vapor pressure law according to equation (3). Secondly, for each composition, the parameters of the Clausius-Clapeyron law were identified on-line (parameters Acc_{id} and Bcc_{id}) with the RLS method by varying dynamically the pressure in the range 760 Torr to 100 Torr. Figure 2 represents the experimental, p_{exp}, and the identified, p_{id}, vapor pressure as a function of the boiling mixture temperature, T_m. The identified vapor pressure is computed from the measured temperature, T_m, and from the identified parameters Acc_{id} and Bcc_{id} according to the equation (4) :

$$p_{id} = \exp\left(Acc_{id} - \frac{Bcc_{id}}{T_m} \right) \qquad (4)$$

Figure 3 represents the evolution of the parameters Acc_{id} and Bcc_{id} identified on-line by using the RLS method. Figures 2 and 3 show that a Clausius-Clapeyron vapor pressure law, established for pure liquids, can also be used to represent the relation between the pressure and the temperature of a boiling homogeneous liquid mixture. Because the composition of each mixture (mixture1, mixture2, mixture3) does not vary, the identified parameters Acc_{id} and Bcc_{id} are almost constant. For each composition of the mixture, two parameters are associated, which allows one to correctly represent the relation between the pressure and the temperature at the boiling point.

3.3 On-line identification of the Clausius-Clapeyron law for a composition time-varying mixture

Similar studies to the ones developed in section 3.2 have been carried out with a mixture of acetone, methanol and water whose composition was time varying. The initial volume composition of the mixture is 0.53/0.29/0.18 (mixture1 of section 3.2). 0.5 liter of water are fed in the reactor during 300 seconds leading to a final volume composition of 0.26/0.15/0.59 (mixture3 of section 3.2).

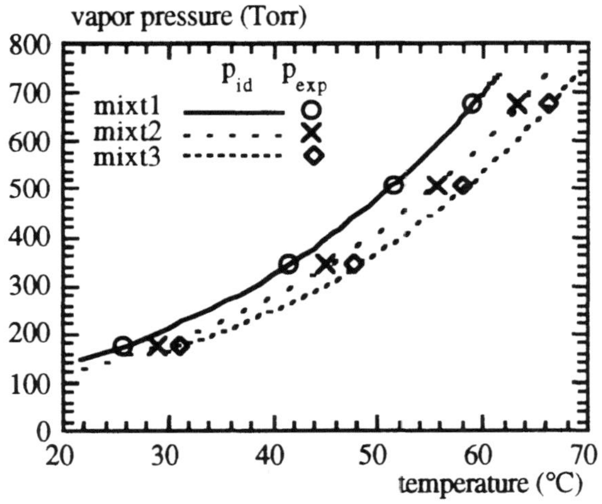

Fig. 2. Vapor pressure law - Identified and experimental data (mixture1, mixture2, mixture3)

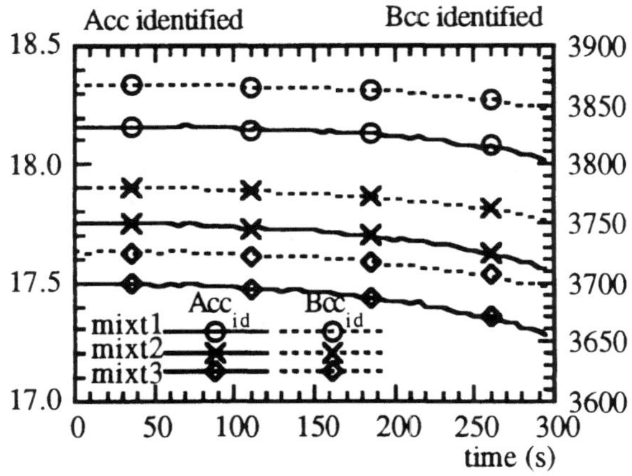

Fig. 3. On-line identified parameters (mixture1, mixture2, mixture3)

Figure 4 represents the experimental, p_{exp}, and the identified, p_{id}, vapor pressure as a function of the boiling mixture temperature, T_m, of the boiling mixture during the feeding of the water. The on-line identified parameters Acc_{id} and Bcc_{id} used to compute p_{id} allow a correctly fit of the experimental points. Figure 5 represents the parameters Acc_{id} and Bcc_{id} identified on-line using the RLS method. In this case, the continuous evolution of the composition mixture implies a continuous evolution of the parameters Acc_{id} and Bcc_{id}. This experiment shows that it is possible to represent the relation between pressure and temperature by a Clausius-Clapeyron law, even if the composition of the homogeneous liquid mixture is time varying (for example when a chemical reaction is carried out). For this case, the Clausius-Clapeyron law is always adequate provided that the parameters Acc_{id} and Bcc_{id} are identified on-line.

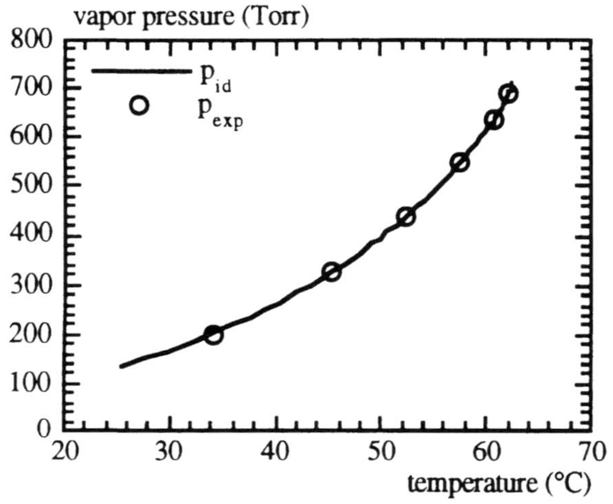

Fig. 4. Vapor pressure law - Identified and experimental data (composition time-varying)

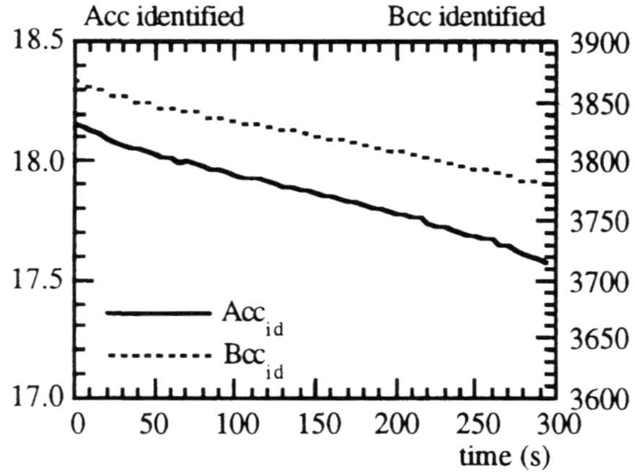

Fig. 5. On-line identified parameters (composition time-varying)

4. THERMAL CONTROL TROUGH PRESSURE BY ON-LINE IDENTIFICATION OF A VAPOR PRESSURE LAW

4.1 Introduction

Thermal control of flexible and multipurpose batch or fed-batch reactors is difficult to achieve with conventional PID controllers. In the fine chemical industry, many different reactions with varying operating conditions are carried out in the same reactor and the parameters of the PID controller require frequent retuning to adjust them to these different operating conditions. The approach developed in this study takes advantage of the instantaneous dynamics of the pressure-temperature loop at the boiling point by acting on the pressure to control the reactor temperature. An immediate environmental benefit of this technique, in comparison with conventional "refluxing reaction", is the possibility to decrease substantially the amount of solvent and consequently to reduce the production of effluents which should be afterwards treated and recycled.

4.2 Methodology for the thermal control through pressure

The methodology for the thermal control through pressure relies on the on-line identification of the relation between pressure and temperature at the boiling point (Clausius-Clapeyron law). This technique can be classified as an adaptive control strategy which adapts the parameters of the control law according to the changes of the process (i.e. changes of the mixture composition due to the chemical reaction). The principle is to compute at each sampling time the value of the pressure which should be applied to the reactor according to the direct pressure-temperature relation. In section 3 it was found that a Clausius-Clapeyron law governed adequately the direct relation between the pressure and the temperature of an homogeneous liquid mixture.

Thus, the control strategy (Figure 6) consists in assuming a Clausius-Clapeyron law for the pressure-temperature law at boiling point and in identifying on-line the parameters Acc_{id} and Bcc_{id} using a RLS method. The pressure value to be applied to the reactor, p_{sp}, is then directly computed as a function of the desired reactor temperature set-point, Tsp (constant temperature or temperature profile) and the parameters Acc_{id} and Bcc_{id} of the Clausius-Clapeyron law according to the equation (5) :

$$p_{sp} = \exp\left(Acc_{id} - \frac{Bcc_{id}}{T_{sp}} \right) \quad (5)$$

A cascade controller has been used to impose the computed pressure from the identified vapor pressure law. The master loop computes the pressure according to the identified parameters of the

Clausius-Clapeyron law. This pressure becomes the set-point of the slave loop constituted by a conventional PID controller. The control signal is then the control valve opening degree which regulates an air in flow rate. Only one set of PID parameters has to be tuned because the relation between the control valve opening and the pressure in the reactor depends only on the plant characteristics and not on the reactant mixture or on the set-point temperature.

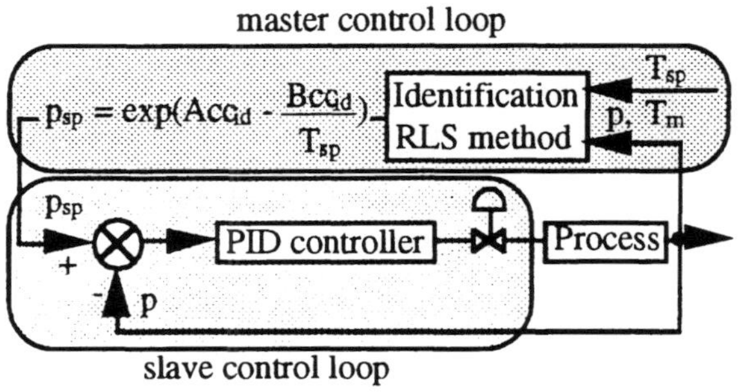

Fig. 6. Strategy of thermal control through pressure

4.3 Simulation and experimental results

The relevance of the control method is exemplified in both simulation studies using the dynamic model (section 2.2) and experimentally on the pilot plant (section 2.1). Two experiments have been simulated. The first one is carried out with a constant composition mixture (experiment1) : 0.5 liter of acetone/methanol/water mixture (0.53/0.29/0.18 volume composition). The second one is carried out with a time-varying mixture composition (experiment2) : the initial mixture is the one used in experiment1 and 0.5 liter of water is fed into the reactor during 800 seconds leading to a final volume composition of 0.26/0.15/0.59.

The methodology of thermal control through pressure by on-line identification of a vapor pressure law, developed in section 4.2, is applied to the tracking of the following set-point temperature profile : T_{sp1} (61.5 °C : normal boiling point of the mixture), T_{sp2} (45 °C) and T_{sp3} (50 °C). This kind of profile can result from an optimization of the operating conditions of synthesis according to criteria involving both concepts of yield, selectivity, productivity, safety or costs of raw and process materials.

Figures 7 and 8 show a satisfactory tracking of the set-point temperature. One can notice a difference between the pressure profiles for experiment1 with a constant composition mixture and experiment2 with a composition time-varying mixture. To ensure the same set-point temperature profile, the pressure in the second experiment has to be lower than that of the first experiment because the mixture is enriched with water and the resulting boiling point temperature is increasing. Figure 9 clearly shows the difference between the constant evolution for experiment1 and the continuous decreasing evolution for experiment2 of the parameters Acc_{id} and Bcc_{id}. These results confirm the remarks made in sections 3.2 and 3.3.

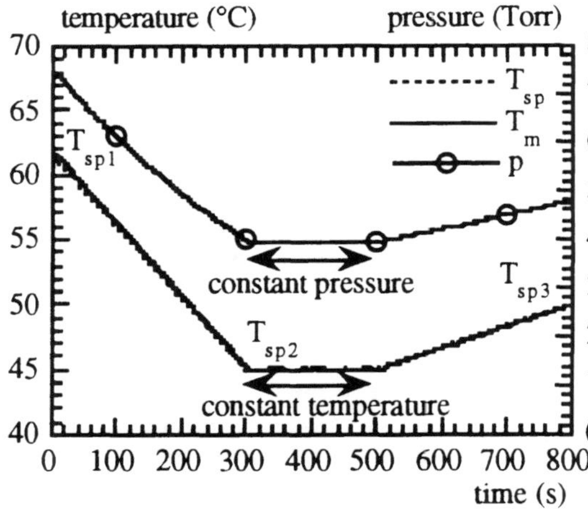

Fig. 7. Thermal control through pressure (experiment1)

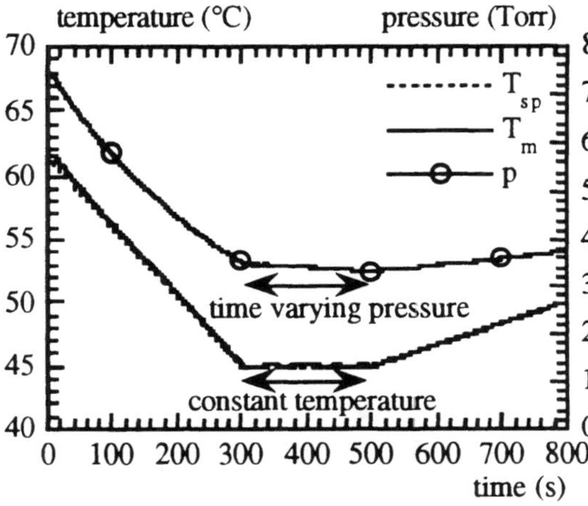

Fig. 8. Thermal control through pressure (experiment2)

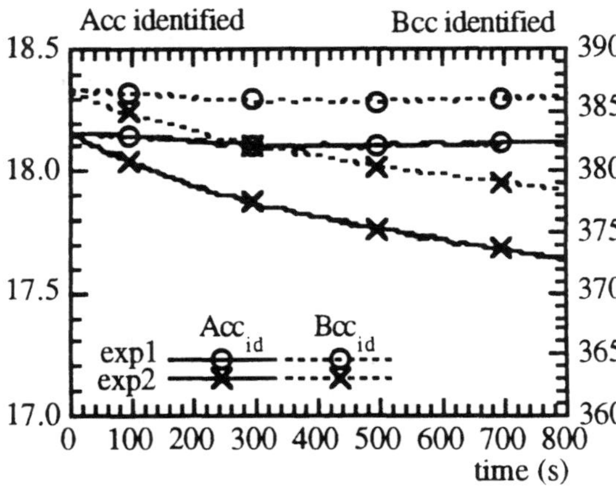

Fig. 9. On-line identified parameters (experiment1, experiment2)

One can notice that the values of the identified parameters Acc_{id} and Bcc_{id} computed during the simulation of the thermal control (Figure 9, experiment2) are the same as those found in the open-loop experiment (Figure 5). Particularly, even if the duration of the two "experiments" is different (800 seconds in the latter one instead of 300 seconds), the final composition is the same (0.26/0.15/0.59) and so, the identified parameters are effectively a function of the composition. The overall time evolution of the identified parameters (Acc_{id} and Bcc_{id}) for these two cases leads to a very small difference which can be explained by the presence of measurement noise (pressure and temperature) in the case of the real-time experiment (Figure 5).

The simulated experiment1 has also been carried out in the pilot plant reactor with the same operating conditions (volume, composition of the acetone/methanol/water mixture, set-point temperature profile). There is a good agreement between the simulated (Figure 7) and the experimental (Figure 10) time evolution of the mixture temperature, T_m, and the pressure in the reactor, p.

In future studies, experiments with a chemical reaction will be carried out in the pilot plant reactor to control the temperature of a pharmaceutical synthesis reaction. The set-point temperature profile will be the result of an optimization of the operating conditions in order to maximize the concentration of a desired product.

5. CONCLUSIONS

In this paper, an original approach of thermal control of batch or fed-batch reactors has been proposed. The thermal control through pressure by on-line identification (RLS method) of a vapor pressure law has been performed with acetone/methanol/water mixtures at constant or time-varying composition. Experiments involving other homogeneous liquid mixtures and/or chemical reactions carried out batch-

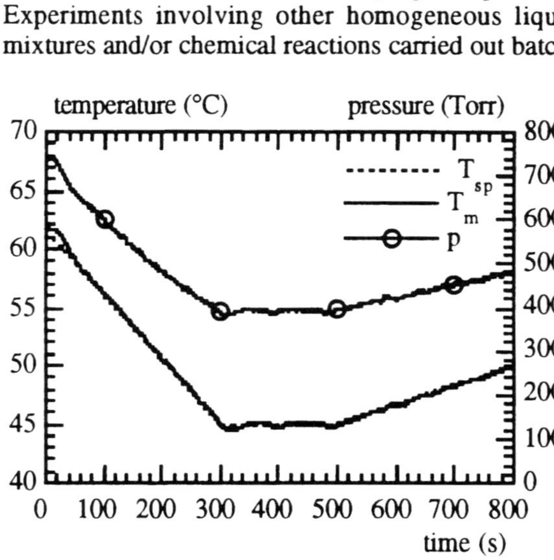

Fig. 10. Thermal control through pressure (pilot plant experiment)

wise will be published elsewhere. For all cases the Clausius-Clapeyron vapor pressure law (two parameters) is adequate to represent the relation between pressure and temperature of homogeneous liquid mixture at the boiling point. When the composition of the mixture is time varying, the parameters of the vapor pressure law have to be modified.

The thermal control through pressure allows one to reconcile two strategies, the minimization of environmental impact and safe operation :

- decreasing the amount of solvent used during reactions in comparison with the classical technique of "refluxing reaction". Thus, the environmental impact of the process is minimized at the source.
- increasing the controllability of batch or fed-batch reactor (especially for exothermic reaction) by thermal exchanges carried out in the condenser.

A common application of thermal control through pressure is a crystallisation process where a decreasing set-point temperature is tracked (Fevotte, 1992). The originality of the study developed in this paper is the generalisation of the thermal control through pressure by on-line identification of parameters of a vapor pressure law to get a direct relation between pressure and temperature (cascade control loop). This kind of thermal control should be applicable to batch or fed batch chemical reactors used in fine chemical industry with arbitrary or predifined temperature profiles.

ACKNOWLEDGEMENTS

The authors are grateful to Elf Aquitaine Company and Sanofi Chimie Company for their scientifical and financial supports which made possible this study.

REFERENCES

Cezerac, J., M. Cabassud, M.V. Le Lann, G. Casamatta, J. Albet and J.M. Le Lann (1993). Développement d'un modèle de simulation dynamique en vue de la conduite automatisée d'un réacteur discontinu par la pression du ciel gazeux. *Récents Progrès en Génie des Procédés*, **7** (29), pp. 1-6, Coord. by C. Jallut, Lavoisier, Paris.

Fevotte, G. (1992). Identification et commande adaptative de cristallisoirs discontinus refroidis par évaporation. PhD Thesis, C. Bernard University, Lyon.

Le Lann, J.M., X. Joulia and B.Koehret (1988). A computer program for the prediction of the thermodynamic properties and phase equilibria. *I.C.E.*, **28** (1), pp. 36-65

Le Lann, M.V., M. Cabassud and G. Casamatta (1994). Adaptative model predictive control. *NATO-ASI Conference on Model predictive Control,* pp. 118-150, 7-17 August, Antalya (Turkey).

EFFECT OF MODEL UNCERTAINTY ON THE TENDENCY MODELING, OPTIMIZATION AND CONTROL OF BATCH REACTORS

Jake Fotopoulos, Christos Georgakis[1] and Harvey G. Stenger, Jr.

Abstract. The optimization of chemical batch reactor processes is strongly affected by the accuracy of the Tendency model used. In this work, the effect of this process-model mismatch on the convergence of the Tendency modeling, optimization and control algorithm is studied. The primary objective is to determine whether an uncertain Tendency model can still be successfully used for process optimization. In the two examples examined, if the magnitude of the uncertainty is limited, it is shown that by operating the next batch under suboptimal but less uncertain conditions, the optimization of the process can be achieved by Tendency models that have some structural mismatch with the process.

Key Words. Tendency Modeling, Batch Reactors, Optimization, Model Uncertainty

1. INTRODUCTION

The successful optimization of a batch reactor process requires the identification of a suitable model which can accurately describe the process. In many instances, especially in the production of fine and specialty chemicals, a detailed model based on fundamental principles and thorough kinetic studies may not be available. An evolutionary modeling and optimization technique known as Tendency Modeling, Optimization and Control (TeMOC) can be used to develop an approximate models for these types of processes (Filippi *et al.*, 1986). This approximate model is used to calculate an optimal operating policy for the next run of the batch reactor. Data from this experiment is used to assess the success of the model and update it. The process is then re-optimized using the updated model. This cycle is repeated until successful optimization is achieved.

Because the models identified in Tendency Modeling are only approximate ones, process-model mismatch exists and may significantly affect the success of optimization. For this reason, it is important not only to quantify the process-model mismatch but also to quantify its effect on the optimization of the process. In this paper, some of the ideas introduced previously (Fotopoulos *et al.*, 1994*b*) are extended to examine how the optimization and model updating steps of the TeMOC cycle are affected by process-model mismatch. First the effect of model uncertainty, represented by the uncertainty of the estimated model parameters, is examined for a two reaction system for which a fixed time optimization problem is solved to determine the optimal temperature profile maximizing the amount of the desired product. If the process-model mismatch, both parametric and structural, is significant, the optimal operating policy calculated through the model may not be optimal with respect to the process. In a second example, the Tendency model features a much smaller number of overall reactions than the simulated process, introducing substantial structural process-model mismatch. The optimization problem is a free-time one where a number of operating conditions are selected to maximize a performance index characterizing the economic productivity of the process.

[1] To whom correspondence should be addressed
Chemical Process Modeling and Control Research Center, Iacocca Hall, Lehigh University,
111 Research Drive, Bethlehem, PA 18015-4732, USA;
Phone: (610) 758-5432; Fax: (610) 758-5297;
E-mail:cg00@Lehigh.edu

2. TENDENCY MODELING, OPTIMIZATION AND CONTROL

A Tendency Model is a low order, nonlinear, dynamic model which approximates the stoichiometry and kinetics of a reactor process using all available process/plant data along with any fundamental knowledge of the process characteristics that is available. The process is modeled in an evolutionary fashion, with the Tendency Model updated as new process data or insight become available.

An initial model, including stoichiometric and kinetic models, is identified from a set of designed experiments, The process is optimized using this model and the experiment is usually run under the calculated optimal operating policy. In many instances, these optimal conditions will lead the process away from the normal operating conditions or from the conditions used in the initial experiments. It is therefore important to assess how accurate the model is and how able the model is to extrapolate to regions of operability away from where it was identified.

2.1. *Uncertainty in Process Modeling*

In this work, the uncertainty in the Tendency Model is quantified directly through the uncertainty in the model's parameters (usually kinetic parameters). This information can then be used to discriminate between different model structures. The uncertainty in the model parameters is found during the parameter estimation step of the TeMOC algorithm.

Biladello, *et al.*, 1993, , describe a general strategy for parameter estimation in differential-algebraic systems by making use of the sensitivities of the process states with respect to the model parameters. These sensitivities can also be used to quantify how good the estimate of the parameters is (Sharma and Arora, 1993) by the construction of a parameter covariance matrix, C_p.

This covariance matrix can be used to place confidence limits on the estimated parameters. It can also be used to determine what effect the process-model mismatch has on the optimization of the process.

2.2. *Process Optimization Under Uncertainty*

In previous work (Fotopoulos *et al.*, 1994*b*), a method to quantify the uncertainty in the optimal operating conditions due to uncertainty in the Tendency Model using a parametric sensitivity analysis of the optimal control problem was developed. Differentiation of the optimal control problem (see Stengel, 1986, for a summary of the optimal control problem) with respect to the model parameters results in the following set of equations:

$$\dot{\mathbf{Z}} = \mathbf{F}(t)\mathbf{Z} + \mathbf{G}(t)\Xi + \mathbf{G_p} \tag{1}$$

with initial conditions:

$$\mathbf{Z}(t_o) = \frac{\partial \mathbf{y}_o}{\partial \mathbf{p}} = \mathbf{0} \tag{2}$$

where $\mathbf{F} = \frac{\partial \mathbf{f}}{\partial \mathbf{y}}$, $\mathbf{G} = \frac{\partial \mathbf{f}}{\partial \mathbf{u}}$, and $\mathbf{G_p} = \frac{\partial \mathbf{f}}{\partial \mathbf{p}}$.

$$\dot{\Omega} = -\sum_{i=1}^{n}\left[\left(\frac{\partial^2 f_i}{\partial \mathbf{y}\partial \mathbf{y}^T}Z + \frac{\partial^2 f_i}{\partial \mathbf{u}\partial \mathbf{y}^T}\Xi + \frac{\partial^2 f_i}{\partial \mathbf{p}\partial \mathbf{y}^T}\right)\lambda_i\right] - \mathbf{F}^T\Omega - \left[\frac{\partial^2 L}{\partial \mathbf{y}\partial \mathbf{y}^T}Z + \frac{\partial^2 L}{\partial \mathbf{u}\partial \mathbf{y}^T}\Xi + \frac{\partial^2 L}{\partial \mathbf{p}\partial \mathbf{y}^T}\right] \tag{3}$$

with terminal condition:

$$\Omega(t_f) = \frac{\partial \phi(t_f)}{\partial \mathbf{p}\partial \mathbf{y}^T} = 0 \tag{4}$$

and a relation for the extremum condition:

$$\sum_{i=1}^{n}\left[\left(\frac{\partial^2 f_i}{\partial \mathbf{y}\partial \mathbf{u}^T}\mathbf{Z} + \frac{\partial^2 f_i}{\partial \mathbf{u}\partial \mathbf{u}^T}\Xi + \frac{\partial^2 f_i}{\partial \mathbf{p}\partial \mathbf{u}^T}\right)\lambda_i\right] + \mathbf{G}^T\Omega + \left[\frac{\partial^2 L}{\partial \mathbf{y}\partial \mathbf{u}^T}Z + \frac{\partial^2 L}{\partial \mathbf{u}\partial \mathbf{u}^T}\Xi + \frac{\partial^2 L}{\partial \mathbf{p}\partial \mathbf{u}^T}\right] = 0 \tag{5}$$

Eq. 5 can be solved for the matrix Ξ, which is inserted into eqs. 1 and 3 to construct a two-point boundary value problem that can be solved for $\mathbf{Z}$ and Ω. The matrix Ξ, which represents the sensitivity of the optimal solution with respect to the parameter vector $\mathbf{p}$, can be determined from seq. 5 once the matrices $\mathbf{Z}$ and Ω are known. Confident limits can then be placed on the optimal operating conditions by relating Ξ to the parameter covariance matrix, $\mathbf{C_p}$.

If the uncertainty in the optimal operating conditions is found to be significant, the next experiment can be run for model improvement rather than process optimization or the next experiment can be run at a more conservative operating policy, assuming the optimization has taken the process away from its normal operating conditions or from the operating conditions in which the Tendency Model was identified at. Mathematically, this operating policy can be expressed as

$$\mathbf{u}_{n+1}(t) = \mathbf{u}_n(t) + \xi\left(\mathbf{u}^*_{n+1}(t) - \mathbf{u}_n(t)\right) \tag{6}$$

where $\mathbf{u}^*_{n+1}(t)$ is the calculated optimum and ξ $(0 \leq \xi \leq 1)$, is the fraction of the distance between this optimum and the previous operating condition, $\mathbf{u}_n(t)$.

The sensitivity of this operating condition to the uncertain parameters can be calculated from:

$$\Xi_{so} = \frac{\partial}{\partial \mathbf{p}}\left[\mathbf{u}_n(t) + \xi\left(\mathbf{u}^*_{n+1}(t) - \mathbf{u}_n(t)\right)\right] = \xi\frac{\partial \mathbf{u}^*_{n+1}}{\partial \mathbf{p}} \tag{7}$$

This new value of Ξ_{so} can now be used to develop confidence limits on the expected profile for the next batch run.

2.3. *Process Model Updating and Re-optimization*

This goal of this paper is to illustrate how knowledge of the uncertainty in the optimization could be used to affect the updating of the Tendency model. Because Tendency Modeling, Optimization and Control is an evolutionary modeling technique, data from the optimization run can be used to update the model and, ideally, decrease the inaccuracy of the model. The reaction stoichiometry is first re-identified and, if different, new kinetics are proposed and the model parameters re-estimated. If the stoichiometry remains the same, the existing model parameters are re-identified, taking the additional data into account. The process is then re-optimized using the updated model.

The same uncertainty analysis can be done on the new optimal operating conditions predicted by the updated model. The entire TeMOC cycle is repeated until convergence is attained.

3. RESULTS

Two examples are presented below. Both are necessary in order to illustrate the methodology but each serves a different, though useful, purpose. The first example, introduced in Fotopoulos, *et al.*, 1994b, is used to illustrate some of the basic concepts and to show how the uncertainty analysis can be useful in the modeling and optimization of batch reactor processes. The second example serves as a test of the methodology on a more complex system. Here, the number of reactions modeled is smaller than the simulated process.

3.1. *Simple Process*

3.1.1. *Identification of Tendency Model.* This example was introduced in Fotopoulos, *et al.*, 1994b. Consider the simple consecutive reaction mechanism:

$$A \rightarrow B \rightarrow C \tag{8}$$

with rate constants k_1 and k_2, respectively. Initial simulations of the system were run and the two possible Tendency Model identified are listed in Table 1. Model I represents the actual reaction

Table 1 Potential Tendency Models identified for Simple Reaction Mechanism

Model I	Model II
A → B B → C	A → B A → C
$r_1 = A_1 exp\left(-\frac{E_1}{RT}\right) C_A$ $r_2 = A_2 exp\left(-\frac{E_2}{RT}\right) C_B$	$r_1 = A_1 exp\left(-\frac{E_1}{RT}\right) C_A$ $r_2 = A_2 exp\left(-\frac{E_2}{RT}\right) C_A$
A_1 = 980.1 E_1 = 15035 cal/mol A_2 = 4709.5 E_2 = 19474 cal/mol	A_1 = 362.2 E_1 = 13979 cal/mol A_2 = 720390 E_2 = 25737 cal/mol

mechanism. Both models were used for process optimization.

3.1.2. *Process Optimization.* The optimization problem for the simple system is to determine the temperature profile to maximize the production of B at a fixed time. The performance index, J, is

$$J = C_b(t_f) \tag{9}$$

Figures 1 and 2 illustrate the optimal temperature profiles using each model, along with the calculated 95% confidence limits based on the uncertainty analysis described above. Full details of these results can be found in Fotopoulos, *et al*, 1994b. The present discussion will investigate the effect of the maximum allowed model uncertainty on the number of updating cycles of the TeMOC algorithm. Model I predicts a performance index of .7344 ± .056 mol/l, at its calculated optimal operating policy (Fig. 1). Using this policy, the simulated process results in a performance index of .7283 mol/l. The predicted performance by model II, under its calculated optimal operating policy (Fig. 2), is .8317 ± .341 mol/l. The simulated process results in a performance of .4472 mol/l using this operating policy, considerably less than the predicted value. The uncertainty in Model II's calculated optimal operating policy would cause one to be hesitant to apply it.

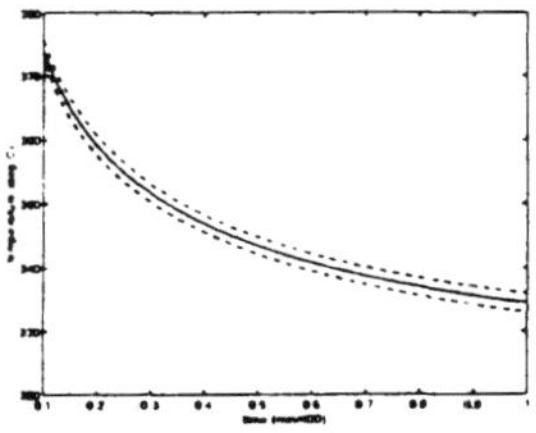

Fig. 1. Uncertainty in Model I's calculated optimal operating policy due to process-model mismatch

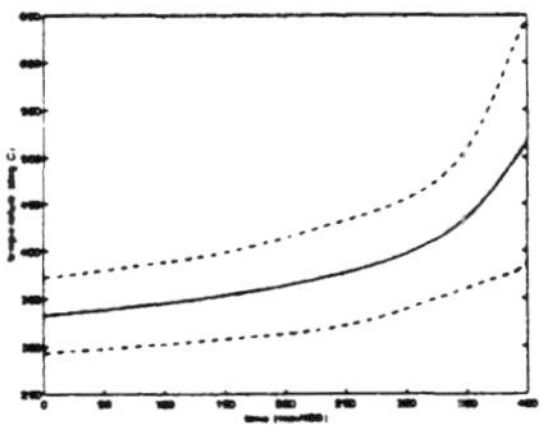

Fig. 2. Uncertainty in Model II's calculated optimal operating policy due to process-model mismatch

If Model II were the only Tendency Model available, its uncertainty may lead one to use the next experiment for further model identification, in order to reduce the uncertainty, or may cause one to operate the process under a more conservative operating policy, closer to the one used in the initial experiments. This more conservative policy can be chosen in order to keep the error in the performance index from exceeding a certain, prespecified value, say 25%.

3.1.3. *Updating of Tendency Model.* The optimization experiment is run under three operating policies:

- Case Ia: Optimal Operating Policy calculated by Model I

- Case IIa: Optimal Operating Policy calculated by Model II
- Case IIb: Suboptimal Policy calculated by Model II such that no more than 25% error exists in performance index

Model updating is performed using the additional data generated by each of these runs.

3.1.4. *Case Ia.* In model I, the uncertainty in the optimal temperature at the final time was ± 1.5 ^oF and with a predicted 7.63% error in the performance index. The new data from the optimization experiment is used to develop the updated model, with new parameters A_1=981.30, E_1=15037, A_2=8321.1 E_2=20129, not much different from the original model listed in Table 1. This updated model is now used for optimization and results in a predicted performance index of .7295 ± .0325 mol/l (4.46% error). The uncertainty analysis indicates that the model should perform very well and indeed it does. The simulated process resulted in a performance index of .7290 mol/l. In each case, the results of the next experiment fell with the calculated uncertainty bounds. The Tendency Modeling algorithm has converged.

Table 2 Parameter estimates for Updated Model II

Parameter	Model II	Model IIa	Model IIb
A_1	362.2	296.54	452.97
E_1	13979	13743	14226
A_2	7.20 x 10^6	1.25 x 10^{21}	5.53 x 10^7
E_2	25737	66276	30178.85
residual	.00211	.00389	.00225

3.1.5. *Case IIa.* The more interesting results come from the updating of Model II, which features the incorrect model structure. At Model II's calculated optimum, the performance index was found to have 40.98 % error. In case IIa, the simulated process is run under this policy and the resulting process performance is much worse than what the model predicted. The results of this optimization run is used to update Model II. The updated model parameters appear in Table II. The additional data did not invalidate the stoichiometric model. The updated model is used to optimize the process. Figure 3 illustrates the new optimal operating policy, along with the calculated confidence limits. The predicted performance index is .6252 mol/l. The simulated process under this operating policy results in a performance of .6746 mol/l. One model updating step has significantly reduced the error in the operating policy as well as allow for a better prediction of the process performance. However, to achieve this, one had to run one experiment in which the model performed very poorly.

Figure 4 demonstrates the uncertainty in the optimal operating policy in a slightly different manner by determining the uncertainty of the time average temperature, T_{av}, as well as the uncertainty of the second moment about this average temperature. As the operating policy is moved further from the conditions under which the model was identified at (corresponding to ξ greater than zero in eq. 6), the larger the uncertainty. The error in the average temperature is ± 13.9 ^oF. For the original model II, this error was 55.2 ^oF. The uncertainty for the updated model is smaller than the one for the original model (Fotopoulos *et al.*, 1994*b*).

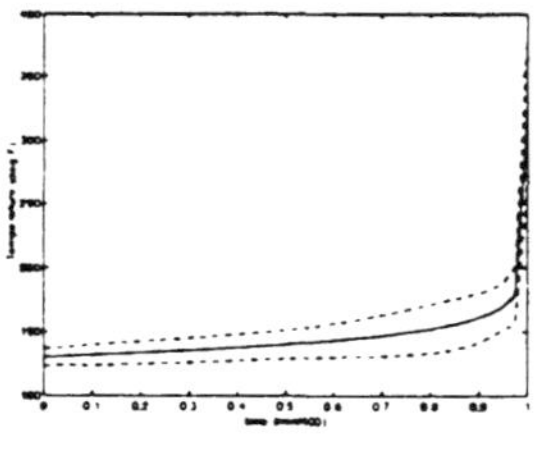

Fig. 3. Optimal operating policy, along with its confidence limits, calculated by Updated Model IIa

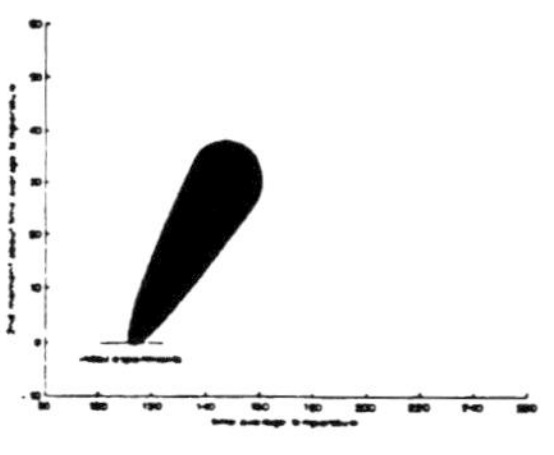

Fig. 4. Uncertainty in the operating policy due to the process-model mismatch of model IIa. Operating conditions range from those used in the initial experiments to the calculated optimal operating conditions of model IIa.

3.1.6. *Case IIb.* Suppose one had decided not to use the next experiment for model development, but process optimization was still the immediate goal. The uncertainty in model II would lead to the use of a suboptimal policy to reduce the predicted error in the performance index. Case IIb illustrates this. An operating policy can be chosen such that the error in the performance index does not exceed some pre-specified amount. For example, at Model II's optimal operating conditions, the performance index was found to have 40.98 % error. By choosing ξ in eq. 6 to be equal to 0.215 and operating under this suboptimal policy, the predicted performance index is .5853 ±

.1463 mol/l (25.00 % error). Using this policy the process results in a performance of .5958 mol/l. The model is able to do a better job in predicting the performance of the process because the new operating conditions are closer to the operating conditions that were used to identify the model. The Tendency model can now be updated with the data from this run. The updated parameters can be found in Table II. Figure 5 shows this updated model's optimal operating policy along with its confidence limits. Figure 6 is the corresponding uncertainty in the time average temperature and the second moment about that temperature for this case.

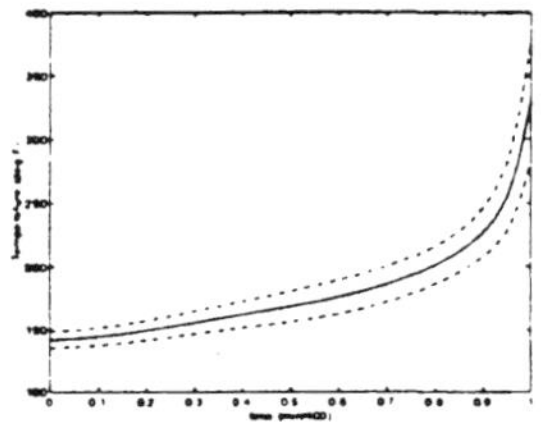

Fig. 5. Optimal operating policy, along with its confidence limits, calculated by Updated Model IIb

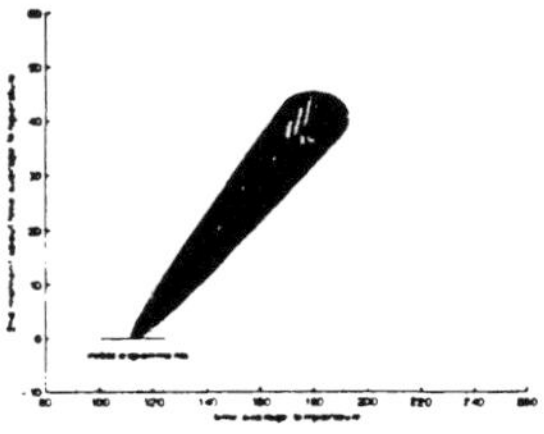

Fig. 6. Uncertainty in the operating policy due to the process-model mismatch of model IIb. Operating conditions range from those used in the initial experiments to the calculated optimal operating conditions of model IIb.

The error in the average temperature has been reduced to 14.89 ^oF. The model predicts a performance of .8137 mol/l where the simulated process results in a performance of .6152 mol/l. Clearly, though the updated model is improving the performance of the process, it is converging at a much slower pace than the updated model in case IIa. At this point, another suboptimal policy would be chosen, keeping the error in J under 25% and re-optimizing. Both updated models are leading the process to the optimum, despite using a model with the incorrect structure. A more conservative approach to optimization, for this simple example, saves one batch from complete failure and slowly leads the process towards an optimum. A daring approach to optimization (case IIa), while leading the process to the optimum more quickly, results in at least one batch with very poor performance.

The uncertainty in the original Model II indicates that the next experiment should be run for model improvement and if this is the case, the updated model can be used for process optimization. This would correspond to a case to IIa, where a good performance for the next experiment is not essential. If it is important to maintain good process performance simultaneously with model development, a strategy similar to case IIb should be implemented.

3.2. *Complex Process*

3.2.1. *Identification of Tendency Model.*

The purpose of the previous example was to first to determine whether the uncertainty analysis developed was a feasible approach to accessing process-model mismatch and its affect on process optimization as well as to examine its effect on the model updating step of the TeMOC algorithm. In order to determine this, a simple example was chosen, in which the number of modeled reactions equaled the actual number of reactions occuring. In a normal TeMOC application, however, the number of reactions modeled will be significantly less than the actual number of reactions occuring, since the technique was developed to model complex reaction networks (Rastogi, 1991). For this reason, a second example is studied to see if the uncertainty analysis is feasible in a more realistic case.

Consider the following 6 reaction, 8 species mechanism, where D_2 is the desired product:

$$A + B \rightarrow C_1 \rightarrow C_2 \tag{10}$$

$$A + B \rightarrow D_1 \rightarrow D_2 \tag{11}$$

$$A + B \rightarrow E_1 \rightarrow E_2 \tag{12}$$

Initial simulated experiments were run using this mechanism for a 1 liter batch reactor using a two level factorial design with the following variables varied: (1) Reactor temperature from 385 to 410 K; (2) Initial amount of reactant A from 360 ml to 600 ml; and (3) Initial amount of reactant B from 200 ml to 300 ml. The batch time for each of the eight initial experiments was 400 minutes. A solvent is necessary to keep the product dissolved, therefore the remaining volume in each experiment was filled with this solvent. The intermediate species, C_1, D_1 and E_1, were assumed to be unmeasurable and, in fact, had completely disappeared by the end of each of the batch runs. In addition, data was only assumed to be available at the end of the run.

The remaining five species were measured, with a measurement error introduced into the simulations that was Gaussian, with mean-zero and a standard deviation of .015 mol/l. The data from the initial experiments were used to develop the Tendency Model. Singular value decomposition of the data matrix (consisting of the change in concentration of each of the measured species for each of the experiments) resulted in singular values of 10.90, 1.35, 0.03, 0.01, 0.01. The number of reactions in the Tendency Model is equal to the number of non-zero or significant singular values. In this case, 2 singular values explain more than 99.5 % of the variance in the data, therefore two reactions are chosen for the model.

Structured Target Factor Analysis (Tsobanakis, 1994; Fotopoulos *et al.*, 1994*a*) was used to identify the following stoichiometric model:

$$A + B \rightarrow 0.5C_2 + 0.5D_2 \qquad (13)$$

$$A + B \rightarrow E_2 \qquad (14)$$

The structure of the model differs significantly from the actual reaction mechanism. Power law kinetic expressions were chosen for the kinetic model and, as before, the estimated parameters were the pre-exponential factors and activation energies for each reaction. The rate orders were assumed to be one. The total number of parameters is therefore four.

3.2.2. *Process Optimization.* The performance index to be maximized for this process is an economic-based performance index that favors the production of the desired product, penalizes the formation of by-products (by use of a cost of separation term, represented by the sum of the all remaining species, other than the desired product, at the end of run), penalizes the cost of the reactants and solvent and attempts to minimize the batch time. For this reason, this problem is a free-time optimization one. The performance index is:

$$J = \alpha \frac{(aD_2 - bA_o - cB_o - dS_o) - e(A + B + C_2 + B_2)}{t_b + t_c} \qquad (15)$$

where A_o and B_o are the initial amounts of reactants used, S_o is the initial charge of solvent, t_b is the batch time and t_c is the time to clean and re-charge the reactor. α is a conversion factor to convert J to \$/day/liter; a,b,c,d,e are the relative costs of the product, reactants, solvent and separation of final product.

The optimization variables are: (1) Initial amount of reactant A; (2) Initial amount of reactant B; (3) Reactor Temperature and (4) Batch time. The initial experiments features values of the performance index varying from -5.6 to 11.2 \$/day/liter. The initial Tendency Model developed calculates the optimal operating policy listed in Table 4.

A comparison of the predicted value of the performance index compared to the simulated process performance under this operating policy is shown

Table 3 Optimal Operating Policy Predicted By Initial Tendency Model

Variable	Optimal Value	Suboptimal ($\xi = 0.5$)
Rxn. Temp. (K)	367.8	382.7
Init. Amt. A (ml)	500.2	460.1
Init. Amt. B (ml)	411.3	330.7
Amt. Solvent (ml)	88.5	209.2
Batch Time (min)	81	240.5
J^{Model} (\$/day/liter)	68.14	31.60
$J^{Process}$ (\$/day/liter)	6.34	23.17

in Figure 5. The value of the process' performance index at the predicted optimal batch time is 6.24 \$/day/liter, considerably less than the predicted value of 68.14 \$/day/liter. If one could determine that the optimal operating policy calculated by the initial Tendency model (Model M0) is not very certain, one could operate the process at a suboptimal operating policy, closer to those conditions which the designed initial experiments were run at. Such a policy (for $\xi = 0.5$) is listed in Table 3. Operating the reactor under this policy, the performance index is found to be 23.17 \$/day/liter, a better performance than under the calculated optimal operating policy.

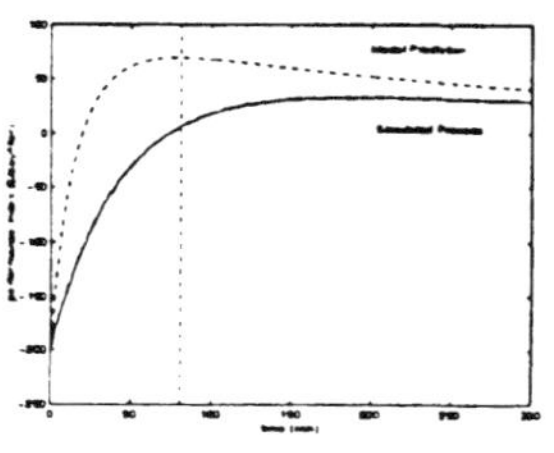

Fig. 7. Comparison of Initial Tendency Model's Predicted Performance Index and Corresponding Simulated Process Performance

The results from each of these optimization runs can be used to update the model. Model M1 is constructed using the results from the experiment run under the calculated optimal operating policy. Model S1 is constructed using the results from the experiment run under the suboptimal conditions. Both updated models are then used for optimization. Table 4 lists the optimal operating policies for the updated models.

Figures 7 and 8 illustrate the process performance using each of these operating policies. Notice that both models are improvements over the initial model's policies, however convergence of the TeMOC cycle is occurring more quickly for the case where the suboptimal operating policy was used (Fig. 8). Because the process-model mismatch of the initial Tendency model is too large, the overall convergence of the TeMOC cycle proceeds more quickly if a more certain, suboptimal

Table 4 Optimal Operating Policy Predicted By the Updated Tendency Models

Variable	Model M1 Optimum	Model S1 Optimum
Rxn. Temp. (K)	347.3	380.1
Init. Amt. A (ml)	495.6	506.0
Init. Amt. B (ml)	409.2	419.6
Amt. Solvent (ml)	95.2	74.4
Batch Time (min)	146	132
J^{Model} ($/day/liter)	38.77	40.29
$J^{Process}$ ($/day/liter)	17.06	31.49

policy is chosen for the next experiment. In the case of Model M1, the process performance index is 17.06 $/day/liter. For Model S1, the performance is 31.49 $/day/liter and the model prediction agrees quite well with the process results. The sensitivity equations described previously can be used to determine how uncertain the initial Tendency model's optimal operating policy is and whether or not a more conservative, suboptimal profile should be implemented. In this case, implementation of an appropriate suboptimal policy would not only provide better immediate performance but would also allow for quicker convergence of the TeMOC algorithm.

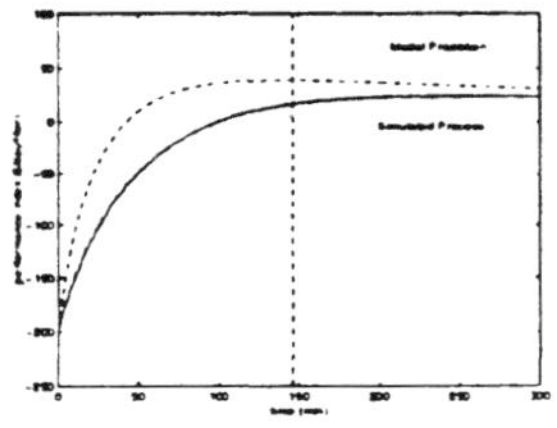

Fig. 8. Comparison of Updated Tendency Model M1's Predicted Performance Index and Corresponding Simulated Process Performance

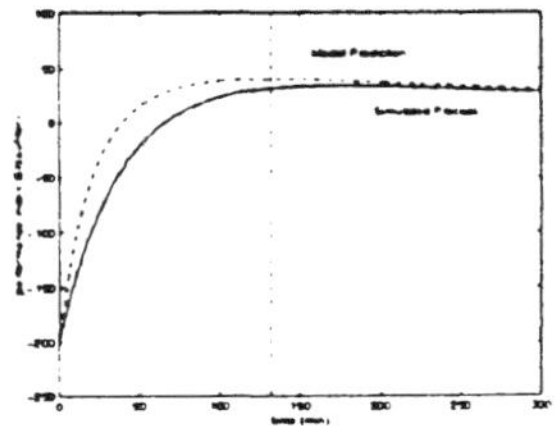

Fig. 9. Comparison of Updated Tendency Model S1's Predicted Performance Index and Corresponding Simulated Process Performance

4. CONCLUSIONS

In this paper, the role of the process-model mismatch on the optimization and model updating steps of the Tendency Modeling, Optimization and Control algorithm has been examined. Specifically two examples have been studied in order to understand how the process-model mismatch's effect on optimization of batch reactor processes can be quantified and how knowledge of this effect can be used to construct a modeling strategy which will efficiently lead the process toward the optimum.

A sensitivity analysis of the optimal control problem can be used to evaluate the effect of the process-model mismatch on the optimization and, if this mismatch is found to be significant, the next experiment can be used for further model identification or it can be run under a more conservative, suboptimal policy, to prevent poor process performance from occuring.

5. REFERENCES

Biladello, P., X. Joulia, J. LeLann, H. Delmas and B. Koehret (1993). A general strategy for parameter estimation in differential-algebraic systems. *Comp. and Chem. Engng.* **17**, 517–525.

Filippi, C., J.L. Graffe, J. Bordet, J. Villermaux, J.L. Barney, P. Bonte and C. Georgakis (1986). Tendency modeling of semibatch reactors for optimization and control. *Chem. Engng. Sci.* **41**, 913.

Fotopoulos, J., C. Georgakis and H.G. Stenger (1994*a*). Structured target factor analysis for the stoichiometric modeling of batch reactors. *Proceedings of the 1994 American Control Conference.*

Fotopoulos, J., C. Georgakis and H.G. Stenger (1994*b*). Uncertainty issues in the modeling and optimization of batch reactors with tendency models. *Chem. Engng. Sci.*

Mezaki, R. and J. Kittrell (1967). Parametric sensitivity in fitting nonlinear kinetic models. *Industrial and Engineering Chemistry.* **59**, 63–69.

Rastogi, A. (1991). Evolutionary optimization of batch processes using tendency models. PhD thesis. Lehigh University.

Rastogi, A., J. Fotopoulos, C. Georgakis and H.G. Stenger (1992). The identification of kinetic expressions and the evolutionary optimization of specialty chemical batch reactors using tendency models. *Chem. Engnn. Sci.* **47**, 2–3.

Sharma, M. S. and N.D. Arora (1993). Optima: A nonlinear model parameter extraction program with statistical confidence region algorithms. *IEEE Trans. on Comp. Aided Des. of Int. Circuits and Systems* **12**, 982–987.

Stengel, R. (1986). *Stochastic Optimal Control.* John Wiley and Sons.

Tsobanakis, P. (1994). Tendency Modeling and Optimization of Fed-Batch Fermentations. PhD thesis. Lehigh University.

Nonlinear Inferential Parallel Cascade Control

Thomas J. Mc Avoy
Nan Ye

Institute for Systems Research
Department of Chemical Engineering
University of Maryland
College Park, MD 20742

Abstract: This paper discusses the use of nonlinear inferential parallel cascade control (NIPCC) to improve control system performance. The goal of NIPCC is to detect and compensate for the effects of unmeasured disturbances faster than the feedback control being employed. NIPCC behaves in a manner that is similar to feedforward control, but NIPCC is feedback in nature. NIPCC can be thought of as a limiting case of developing a soft sensor, and thus, NIPCC draws from both parallel cascade control and inferential sensing. The effectiveness of NIPCC is demonstrated on the Tennessee Eastman test process.

Keywords: Cascade control, inferential, neural networks, process control, soft sensing

INTRODUCTION

Parallel cascade control differs from normal cascade control in that the inner cascade loop is in parallel with the primary loop rather than in series. A classical example of parallel cascade control involves distillation tray temperature control cascaded to a composition loop. Recently, several papers on parallel cascade control have been published . Luyben (1973) appears to have been the first researcher to coin the phrase parallel cascade control. Yu (1988) discussed the design of parallel cascade control for disturbance rejection. Brambilla and Semino (1992) have discussed the use of a nonlinear filter between the outer and cascade loops to improve control system performance. Brambilla, *et al* (1994) recently discussed an approach to the design of parallel cascade controllers for multicomponent distillation columns where top and bottom compositions are controlled. Finally, Pottmann *et al* (1994) discussed a parallel control strategy based on reverse engineering the baroreceptor reflex which is used in animals for short term regulation of arterial blood pressure control.

Inferential sensing, also called soft sensing, involves the use of readily available process measurements to infer variables that are difficult to measure on-line, e.g. composition. Inferential sensing has received a great deal of attention recently, and numerous papers on the subject have appeared. Willis *et al* (1990) have discussed the use of oxygen uptake rate and carbon dioxide evolution to estimate biomass in an industrial fermenter. The use of distillation tray temperatures to estimate and control product compositions has been discussed by several authors (Mejdell and Skogestad (1991); Kresta *et al* (1991)). Potential difficulties in closing feedback loops around soft sensors have been pointed out by Macgregor *et al* (1991). In such applications the closure of a loop can result in a change in the correlation structure of the input inferential variables, with the result that the soft sensor can be inaccurate.

In this paper a somewhat different approach to parallel cascade control is taken. The approach is nonlinear and inferential in nature and it is based on steady state modeling. This approach may be generally applicable to a broad range of process regulation problems where disturbance rejection is the goal. As discussed in the literature, parallel cascade control involves two feedback controllers, each of which needs to be tuned. The approach presented in this paper involves determining an accurate steady state model which can be used to compensate for the effects of unmeasured disturbances. If necessary, linear dynamic compensation can be added to the inferential controller. The steady state model can be multivariable and nonlinear, and in the present application it is implemented using neural networks (Qin and Mc Avoy (1992)). After giving a general discussion of NIPCC, it is applied to the Tennessee Eastman process (Downs and Vogel (1993)) to illustrate its effectiveness. It is shown that NIPCC is capable of achieving a significant reduction in the variance of the product composition fluctuations for the test process.

NONLINEAR INFERENTIAL PARALLEL CASCADE CONTROL (NIPCC)

In order to discuss NIPCC, a distillation column, shown in Figure 1, is used for illustration.

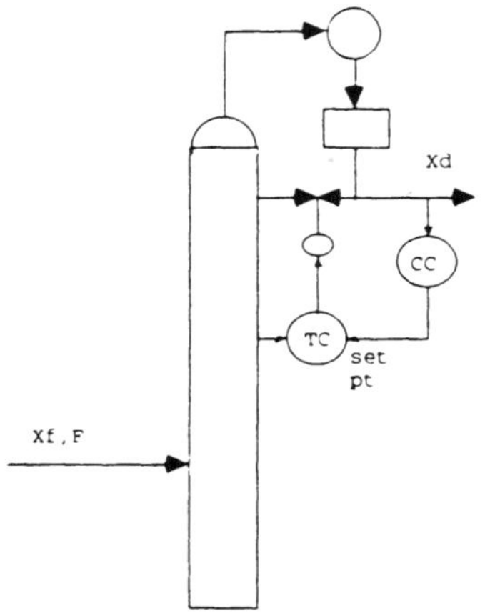

Fig. 1 Distillation Parallel Cascade Control System

The manipulated variable is the reflux flow to the tower. The control problem is to keep the overhead composition as close to set point as possible in spite of fluctuations in the feed flow and composition. Fig. 1 illustrates a standard parallel cascade control system to achieve this objective. A tray temperature controller is cascaded to a composition controller, and a number of authors have discussed how the control tray should be chosen. The rationale behind the parallel cascade approach is that the feed upsets will affect the tray temperature faster than the product composition, and thus compensation for the upset can begin sooner than if composition control alone is used. Fig. 2 gives the block diagram for the control system shown in Fig. 1.

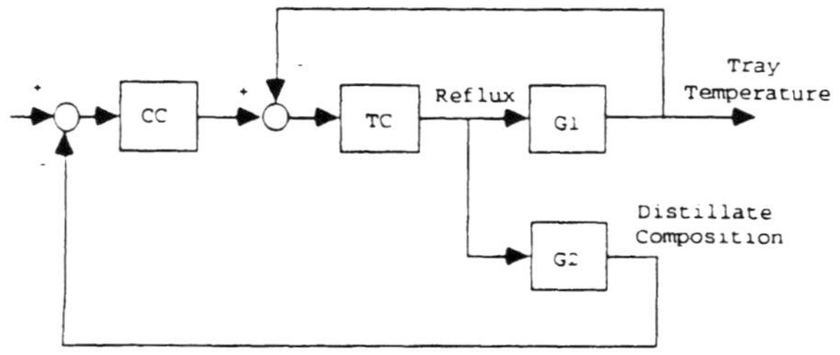

Fig. 2 Block Diagram for Parallel Cascade Controller

As can be seen the manipulated reflux flow affects the tray temperature through block G_1, which is different from its effect on product composition, G_2. These two transfer functions are in parallel and this is where the control approach gets its name. By contrast, the block diagram for a normal cascade control system is shown in Figure 3.

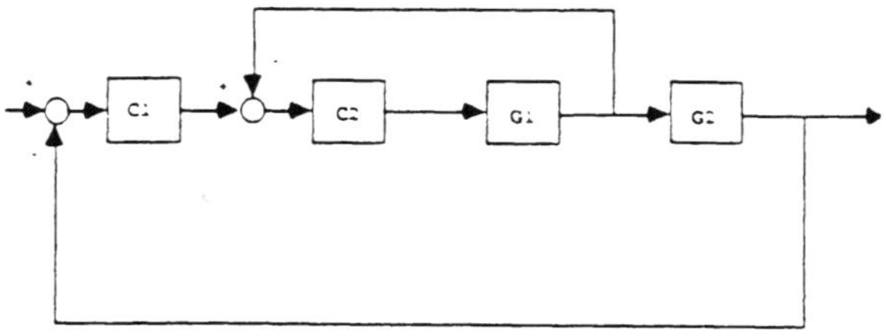

Fig. 3 Block Diagram for Normal Cascade Controller

In this case G_1 and G_2 are in series with one another. If there were a reflux flow controller that got its set point from the tray temperature controller, then the flow and temperature controllers would form a normal, or series cascade control system. Now consider the case where more than one tray temperature is used as part of an inferential scheme to estimate the distillate composition. Such a scheme is illustrated in Figure 4.

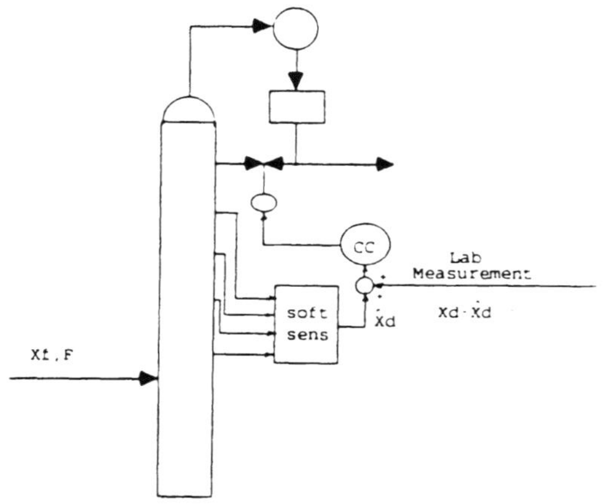

Fig. 4 Soft Sensor Control System

As can be seen the estimated composition is fed to a composition controller which manipulates the reflux valve. When actual composition measurements become available from the lab, then they can be added into the feedback system as discussed by Piovoso and Kosanovich (1991). The control scheme in Figure 4 typically is able to respond to upsets much faster than if the lab values are used for control. As Macgregor *et al* (1991) have pointed out, one potential problem with the scheme shown in Figure 4 is that the closure of the control loop around the inferential sensor may result in an inaccurate model due to a change in the correlation structure of the inferential variables used.

The NIPCC control system considered in this paper uses the same inputs as the soft sensor system shown in Figure 4. However, rather than estimating the controlled variable, the NIPCC scheme estimates the **manipulated variable**. The estimation is steady state and it is carried out subject to the controlled variable(s) being exactly at its setpoint. The NIPCC scheme is illustrated in Fig. 5.

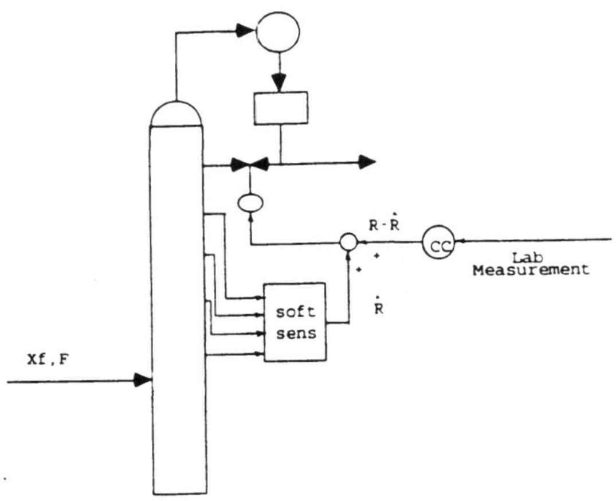

Fig. 5 NIPCC Control System

As with the scheme in Fig. 4, when a lab measurement of composition becomes available it can be use to bias the output of the NIPCC. The composition controller shown in Fig. 5 would be a slow controller that would act to eliminate the effects of plant/model mismatch at steady state. If necessary a dynamic element can be added between the soft sensor and the reflux valve to match the disturbance and manipulated variable dynamics. The block diagram for the NIPCC system is shown in Fig.6.

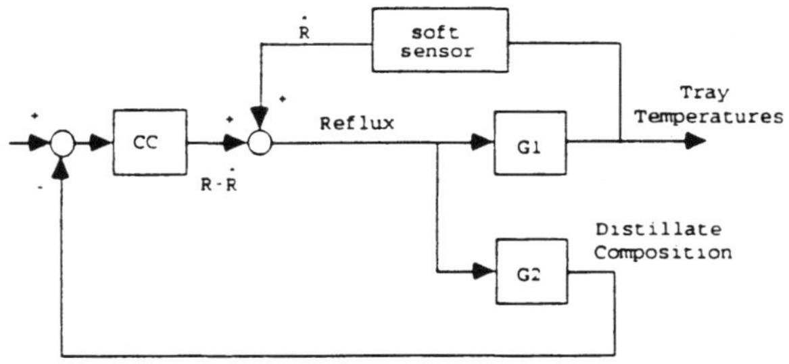

Fig. 6 Block Diagram for NIPCC Controller

As can be seen, Fig. 6 is similar to Fig. 2 in that there are different blocks that affect distillate composition and tray temperatures. In developing data for the NIPCC model, one can either use a process model, or take data on the process while it is operating at its setpoint. In developing the data base it is important that all important disturbances be included. The steady state model between inferential and manipulated variables can be nonlinear, non-square, and multivariable. To develop the model in this case, a nonlinear Partial Least Squares approach (Qin and Mc Avoy (1992)) can be used with the inputs being the inferential variables and the outputs being the manipulated variables. The NIPCC approach avoids the correlation problem, discussed above, since it is developed under precisely the closed loop conditions where the model will be used.

ILLUSTRATION OF NIPCC

To illustrate the inferential parallel cascade approach, the Tennessee Eastman process (Downs and Vogel (1993)), shown in Fig. 7, will be used. The Tenessee Eastman process involves a real plant that has been disguised for proprietary reasons. The process produces two products from four reactants. Also present are an inert and a by-product making a total of eight components: A, B, C, D, E, F, G, and H. The reactions are:

$$A(g) + C(g) + D(g) \rightarrow G(liq) \quad \text{Product 1}$$

$$A(g) + C(g) + E(g) \rightarrow H(liq) \quad \text{Product 2}$$

$$A(g) + E(g) \rightarrow F(liq) \quad \text{By-Product}$$

$$3\,D(g) \rightarrow 2\,F(liq) \quad \text{By-Product}$$

The process has five major unit operations: the reactor, the product condenser, a vapor/liquid separator, a recycle compressor, and a product stripper. The gaseous reactants are fed to the reactor where they react to form liquid products. The gas phase reactions are catalyzed by a non-volatile catalyst dissolved in the liquid phase. The reactor has an internal cooling bundle for removing the heat of reaction. The products leave the reactor as vapors along with the unreacted feeds. The catalyst remains in the reactor.

There are three modes of process operation corresponding to three G/H product ratios. The product mix is normally dictated by product demands. The plant production rate is set by market demand or capacity limitations. The process has 41 measurements and 12 manipulated variables. A prerequisite for most studies on this problem is a process control strategy for operating the plant. The control objectives for this process are typical for a chemical process:

1. Maintain process variables at desired values.
2. Keep process operating conditions with equipment constraints.
3. Minimize variability of product rate and product quality during disturbances (stream 11).
4. Minimize movement of valves which affect other processes (in this case the gas feeds as described below).
5. Recover quickly and smoothly from disturbances, production rate changes, or product mix changes.

The authors of the Tennessee Eastman problem point out that it is an appropriate testbed for a number of topics. These include: plant-wide control strategy design, multivariable control, optimization, predictive control, estimation/adaptive control, nonlinear control, process diagnostics, and education.

Earlier we have designed a base control system for this process (Mc Avoy and Ye (1994)). To smooth out product flow variations optimal averaging level control was employed effectively (Ye and Mc Avoy (1995)) on the separator and the stripper reboiler.

Optimal feedback averaging level control

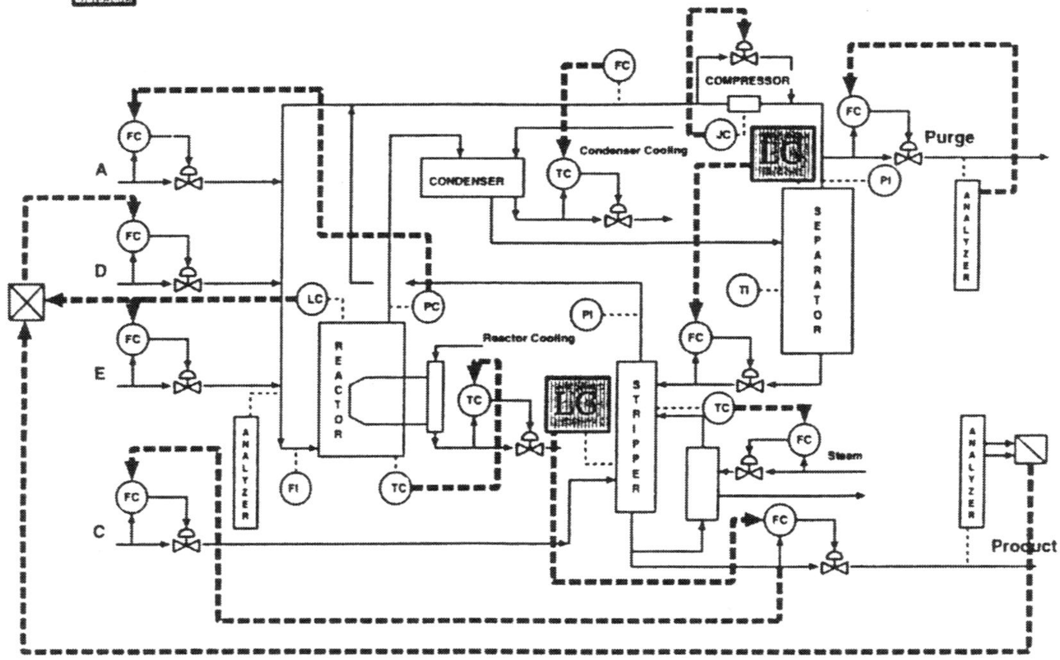

Fig. 7 Schematic of Tennessee Eastman Process and Base Control Scheme

Figs. 8 and 9 show the performance of the base control system, also shown in Fig. 7, utilizing optimal averaging level control for random composition upsets in feed stream C (Disturbance IDV(8)). Operation at three different product ratios, 10/90, 50/50, and 90/10 is shown. The inferential parallel cascade control system is illustrated in Fig. 10. As can be seen this scheme consists of a steady state model followed by first order filters. In order to build the model, one solves the steady state process equations for random changes in disturbance IDV(8), subject to achieving perfect product composition and flow control. The inputs to the steady state model consist of variables not used in the base control system. These include stand alone sensors, set points of inner cascade loops, and valve positions. A total of 13 variables, shown in Table 1, were used as inputs.

Table 1 Input Variables for NIPCC Model Development

INPUT VARIABLES
A Feed Flow Setpoint
E Feed Flow Setpoint
Purge Rate Setpoint
Separator Underflow Setpoint
Stripper Flow Setpoint
Reactor Cooling Temperature Setpoint
Condenser Cooling Temperature Setpoint
Compressor Recycle Valve Position
Total Feed to Reactor
Separator Temperature
Separator Pressure
Stripper Pressure
G/H Ratio Setpoint

The outputs are setpoints for the two controllers used for product composition and product flow control, namely the D feed flow controller, and the C feed flow respectively. A total of 720 data sets for the 13 inputs and 2 outputs were generated. These calculations were carried out for all three operating modes, with 240 point generated for each mode. A nonlinear partial least squares approach, based on neural networks (Qin and Mc Avoy (1992)), was used to model the resulting data. One of the operating constraints that must be met for the Tennessee Eastman plant, involves the fact that some streams are constrained with regard to how fast they can be changed. Both the C and D feeds are so constrained. In order to meet these constraints, first order filters were added to the NIPCC scheme as shown in Fig. 10. The filter for the D feed was 1/(0.56 (hr) s + 1), and for the C feed 1/(0.36 (hr) s + 1).

Figs. 8 and 9 show the improvement that can be achieved using the inferential approach. As can be seen the variance in both the product flow and G composition is reduced dramatically. Table 2 give the percent reductions in variance that is achieved between the base control system, with averaging level control, and the NIPCC system.

Table 2 Percent Reduction in Variance of NIPCC Scheme Over Base Scheme

	G/H=90/10		G/H=50/50		G/H=10/90	
	Flow	G	Flow	G	Flow	G
% Reduction	67%	56%	55%	44%	60%	60%

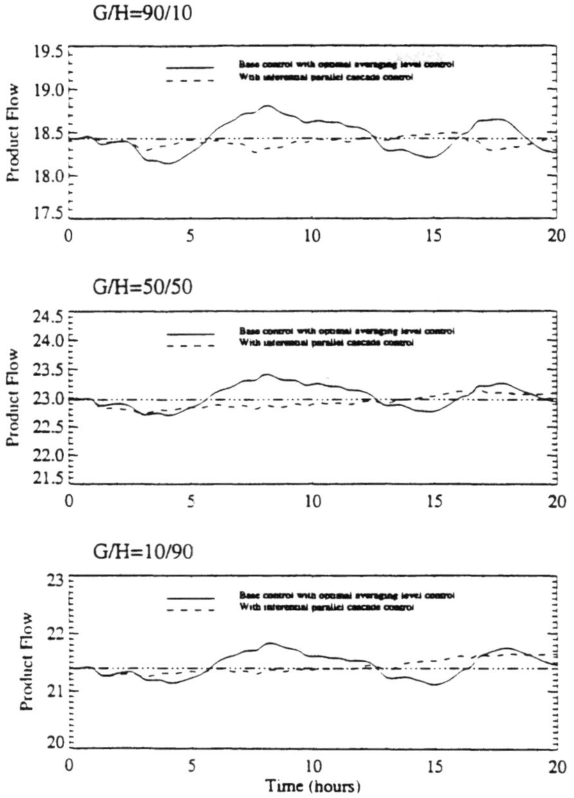

Fig. 8 Response of Product Flow. Base Control and NIPCC.

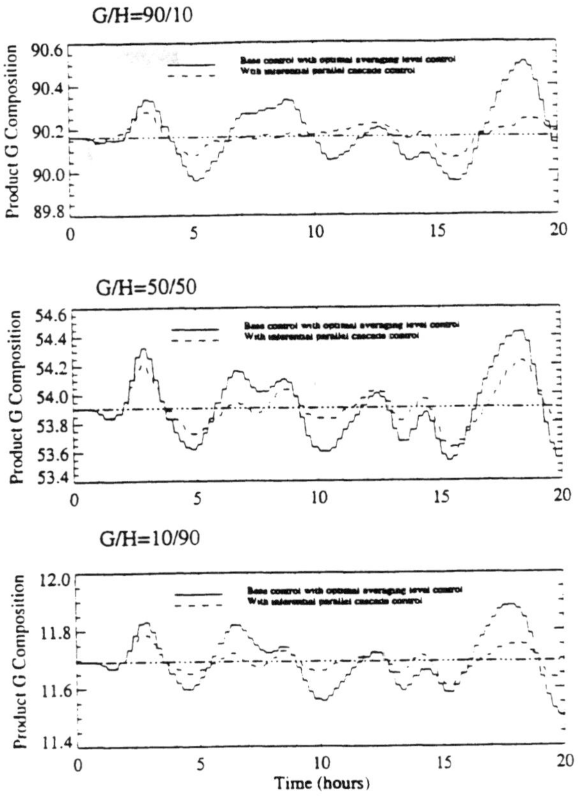

Fig. 9 Response of G Composition. Base Control and NIPCC.

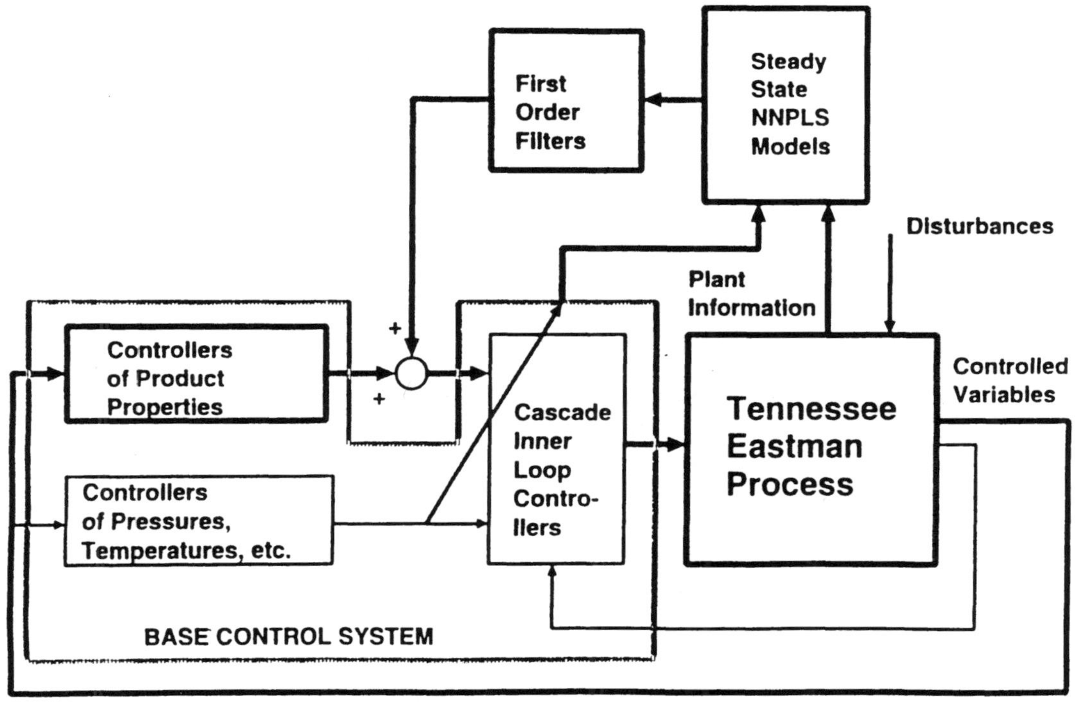

Fig. 10 Nonlinear Parallel Cascade Control Applied to the Tennessee Eastman Test Process

It should be noted that while averaging level control significantly improves the behavior of the product flow over the base control system, it causes the G composition to vary more than in the base case. The NIPCC scheme coupled with the averaging level scheme is able to achieve both good flow filtering and good composition filtering. For this particular application, inferential parallel cascade control is very effective. It should be possible to apply the approach used in this problem to a broad range of processes.

CONCLUSIONS

This paper has discussed the use of nonlinear inferential parallel cascade control (NIPCC) to improve control system performance. NIPCC is a feedback approach that detects unmeasued disturbances through inferential measurements. It then compensates for these disturbances in a manner that is similar to feedforward control. NIPCC can be thought of as a limiting case of developing a soft sensor. The effectiveness of NIPCC has been demonstrated on the Tennessee Eastman test process. It has been shown that for random feed composition fluctuations the variance in product flow and composition can be reduced significantly by using NIPCC, for all of the process operating modes. The NIPCC control approach is a promising method for improving the performance of plant wide control systems, particularly when a steady state model is available. Further work is required to assess stability aspects of the proposed approach, and how dynamic elements in the approach should be tuned, and how they affect overall performance.

REFERENCES

Brambilla A., and Semino, D. (1992). Nonlinear Filter in Cascade Control Schemes. *I&EC Research*, **31**, 2694-2699.

Brambilla, A., Semino, D., and Scali, C. (1994). Design and Control Selection of Cascade Loops in Distillation. In: Preprints of IFAC Workshop on Integration of Process Design & Control, June 27-28, Baltimore, MD, p.171.

Downs, J., and Vogel, E. (1993). A Plant-wide Industrial Process Control Problem. *Computers Chem Engin*, **17**, 245-255.

Kresta, J., Macgregor, J., and Marlin, T. (1991). Multivariate Statistical Monitoring of Process Operating Performance. *Canadian J. Chem. Engin.*, **69**, 35-47.

Luyben, W. (1973). Parallel Cascade Control. *I&EC Fundamentals*, **12**, 463-467.

Macgregor, J., Marlin, T., Kresta, J., and Skagerberg, B. (1991). Multivariate tatistical Methods in Process Analysis and Control. In: Proc Fourth International Conference on Chemical Process Control, South Padre Island, Texas, pp 665-672.

Mc Avoy, T., and Ye, N. (1994). Base Control for the Tennessee Eastman Problem. *Computers Chem Engin*, **18**, 383-413.

Mejdell, T. and Skogestad, S. (1991). Estimation of Distillation Compositions from Multiple Temperature Measurements Using Partial Least Squares. *I & EC Research*, **30**, 2543-2555.

Piovoso, M., and Kosanovich, K, (1994). Applications of Multivariate Statistical Methods to Process Monitoring and Controller Design. *Int. J. Control*, **59**, 743-765.

Pottmann, M., Henson M., Ogunnaike, B., and Schwaber, J. (1994). A Parallel Control Strategy Abstracted From The Baroreceptor Reflex. In: Proc. of 1994 American Control Conf., Baltimore, MD, pp. 97-101.

Qin, J., and Mc Avoy, T. (1992). Nonlinear PLS Modeling Using Neural Networks. *Computers Chem Engin*, **16**, 379-391 (1992).

Willis, M., DiMassimo, C., Montague, G., Tham, M., and Morris, A. (1991). Inferential Measurements via Artificial Neural Networks. In: Proc. Symposium on Intelligent Tuning and Adaptive Control, Singapore.

Ye, N., Mc Avoy, T., Kosanovich, K., and Piovoso, M. (1995). Optimal Averaging Level Control for the Tennessee Eastman Problem. *Canadian J. Chem Engin.*, in press.

Yu, C. (1988). Design of Parallel Cascade Control for Disturbance Rejection. *AIChEJ*, **34**, 1833-1838.

NONLINEAR OBSERVER DESIGN WITH APPLICATION TO CHEMICAL REACTORS

Masoud Soroush*

*Department of Chemical Engineering, Drexel University, Philadelphia, PA 19104. E-mail: Masoud.Soroush@coe.drexel.edu

Abstract. This study concerns the problem of nonlinear state observer design. It presents a nonlinear observer design approach, which allows to adjust the rate of the decay of observer error and is easy to use. Furthermore, in many chemical engineering cases, it is straightforward to prove that the observer-error dynamics is globally asymptotically stable. Two classes of chemical reactors (classical chemical reactors with parallel reactions and free-radical polymerization reactors) are considered to illustrate the application of the observer design approach. The global asymptotic stability of the observer-error dynamics is proved for the two classes of processes.

Key Words. Continuous-time systems; nonlinear systems; observers; process control; reactor control.

1. INTRODUCTION

In practice, the measurements of all the state variables of a process are rarely available, and in such cases, reliable information on the unmeasurable state variables are obtained by using a state observer. It is generally agreed that linear observers are inadequate for nonlinear processes, and therefore there is a need for nonlinear observer design methods. Apart from the use of nonlinear observers in nonlinear controller design, observers can play a key role in process monitoring, wherein an early detection of hazardous conditions is needed for a safe operation.

The lack of a practical and general nonlinear state-observer design method is an important but yet unsolved control and monitoring problem. In the case of linear systems, the observer design problem has been studied extensively and powerful observer design methods are available, whereas in the nonlinear case, because of the difficulties involved in dealing with nonlinear systems, the available results are either problem-dependent or suitable for a very restricted class of nonlinear systems (Dootingh et al., 1992; Kantor, 1989; Krener and Isidori, 1983; Krener and Respondek, 1985).

A variety of observer/state-estimator design methods have been used for nonlinear processes. These include but not limited to:

- the output injection method (Krener and Isidori, 1983; Krener and Respondek, 1985), which can ensure the global asymptotic stability of the error dynamics but can be used for a very restricted class of nonlinear processes such as the reactor example described in (Kantor, 1989).
- "open-loop", full-order, nonlinear state observers (e.g. Daoutidis and Kravaris, 1992). This approach is applicable to every nonlinear process which operates *only* within the domain of attraction of a nominal, asymptotically stable equilibrium (steady-state operating) point. Open-loop observers can be applied to a wider class of processes, if they are reduced-order observers driven by measurable outputs. This approach may be applicable to a nonlinear system operating at an unstable equilibrium point. A disadvantage of these approaches is that the rate of the decay of the observer error is not adjustable. Open-loop, reduced-order observers were used successfully in the real-time polymerization-reactor-control studies (Soroush and Kravaris, 1992a, 1993, 1994).
- the so called extended Kalman filtering, whose design is based on a local linear approximation of process model (linear time-varying process description). This method has been used widely in chemical engineering since the 1970s.
- other "closed-loop" approaches such as extended Luenberger observer, Newton observer, and high gain observer methods (Dootingh et al., 1992; Grizzle and Moraal, 1990; Zeitz, 1987).

In summary, among the available observer design methods, only the method of output injection can guarantee the global stability of observer-error dynamics. This method however is applicable to a very restrictive class of nonlinear processes.

This study concerns a practical method of reconstructing the states of nonlinear processes described by a mathematical model of the form

$$\begin{cases} \dot{x}(t) = f(x(t), u(t)), \quad x(0) = x_0 \\ y(t) = h(x(t)) \end{cases} \tag{1}$$

where $x=[x_1 \cdots x_n]^T \in X$, $u=[u_1 \cdots u_m]^T \in U$, $y=[y_1 \cdots y_q]^T$ $(q \geq 1)$ are the vectors of state variables, measurable inputs, and measurable outputs respectively. Here $X \subset \mathbb{R}^n$ and $U \subset \mathbb{R}^m$ are connected, open sets such that $x_{eq} \in X$ and $u_{eq} \in U$, where (x_{eq}, u_{eq}) is a nominal equilibrium pair: $\dot{x}_{eq}=f(x_{eq}, u_{eq}) \equiv 0$, $\dot{u}_{eq} \equiv 0$. It is assumed that (a) the vector functions $h(x)$ and $f(x,u)$ are analytic on X and $X \times U$ respectively; (b)

$$rank \left\{ \frac{\partial h(x)}{\partial x} \right\} = q, \quad \forall x \in X, \tag{2}$$

that is, the measurable outputs are not redundant; (c) all the state variables can be reconstructed by using the measurable inputs and/or outputs (process is at least locally detectable); and (d) the measurable inputs $u_1(t) \cdots u_m(t)$ are bounded.

The objective is to derive state observers of the general form

$$\begin{cases} \dot{\eta}(t) = F(\eta(t), y(t), u(t)), \quad \eta(0) = \eta_0 \\ \hat{x}(t) = H(\eta(t), y(t)) \end{cases} \tag{3}$$

where $\hat{x}$ is the vector of the state estimates, such that $||\hat{x}(t) - x(t)|| \longrightarrow 0$ as $t \longrightarrow \infty$, for every nonlinear process of the form of (1). The notation $||.||$ represents the Euclidean norm of a vector.

Section 2 includes some mathematical preliminaries such as the definitions of universal equilibrium point, universal global asymptotic stability, observability, and detectability. Section 3 presents the nonlinear observer design approach as well as the corresponding observer-error dynamics. Section 4 illustrates the application and performance of the state-observer approach. In particular, the following two classes of chemical reactors, which are known to exhibit steady-state multiplicities, are considered: (i) non-isothermal continuous stirred tank reactors (CSTRs) with parallel reactions, and (ii) free-radical polymerization CSTRs.

2. PRELIMINARIES

2.1. *Global Asymptotic Stability and Universal Global Asymptotic Stability*

Definition 1: The system of (1) (equilibrium point of the system, x_{eq}) is said to be globally asymptotically stable (GAS), if $||\bar{x}(t) - x_{eq}|| \longrightarrow 0$ as $t \longrightarrow \infty$, for every $x_0 \in \mathbb{R}^n$, where $\bar{x}(t)$, $t \geq 0$, is the solution of the differential equations of (1) for $u(t) \equiv u_{eq}$.

The following definition describes the notion of "universal" equilibrium point, which will be used later to study the observer error dynamics.

Definition 2: The system of (1) is said to have a universal equilibrium point x^*_{eq}, if $\dot{x}^*_{eq} = f(x^*_{eq}, u(t)) \equiv 0$, for every bounded $u(t) \in \mathbb{R}^m$.

Systems with universal equilibrium points are rare and normally do not represent well-defined or real processes. For instance, a linear system possesses a universal equilibrium point, when the matrix transfer function relating its inputs to its states is identically equal to zero. As we will see later, the derived observer-error dynamics has a universal equilibrium point at the origin, which is a very desirable feature of the error dynamics.

Definition 3: The system of (1) is said to be universally globally asymptotically stable (UGAS), if the system has a universal equilibrium point x^*_{eq} such that $||x(t) - x^*_{eq}|| \longrightarrow 0$ as $t \longrightarrow \infty$, for every $x_0 \in \mathbb{R}^n$ and for every bounded $u(t) \in \mathbb{R}^m$.

The following proposition states a *sufficient* condition for the universal, global, asymptotic stability of a system; it is a straightforward application of Lyapunov's direct method. The proposition will be used later to show that the observer-error dynamics of the reactor examples are UGAS.

Proposition 1: A system of the form of (1) with a universal equilibrium point x^*_{eq} is UGAS, if $(x - x^*_{eq})^T f(x,u) < 0$, for every $(x,u) \in (\mathbb{R}^n - \{x^*_{eq}\}) \times \mathbb{R}^m$.

Proof: Using the decrescent, radially unbounded, globally positive definite function $V(x,t) = (x - x^*_{eq})^T (x - x^*_{eq})$, we see that $\dot{V}(x,t) < 0$, for every $(x,u) \in (\mathbb{R}^n - \{x^*_{eq}\}) \times \mathbb{R}^m$ and for every $t \geq 0$. Thus, according to Lyapunov's direct method for non-autonomous systems, the system is UGAS. ■

Remark 1: Every physically meaningful system of the form of (1) normally operates (is defined) *only* within a physically feasible region; that is, for such a process there exists a bounded, connected set $\Omega \subset \mathbb{R}^n \times \mathbb{R}^m$ such that $(x(t), u(t)) \in \Omega$ for every $t \geq 0$. In applying the above definitions and proposition to these processes, the feasible operating set Ω should be considered rather than the entire set $\mathbb{R}^n \times \mathbb{R}^m$.

2.2. *Observability and Detectability*

The notions of observability and detectability for linear and nonlinear systems have been studied extensively and precise criteria for their assessment are available in the systems literature (e.g.,

Chen, 1984; Zeitz, 1987). These notions will play a key role in the theoretical results presented in the subsequent sections.

Definition 4: A system of the form of (1) is said to be observable on $X \times U$ (locally observable), if

$$rank \left\{ \frac{\partial}{\partial x} \begin{bmatrix} y \\ \frac{dy}{dt} \\ \vdots \\ \frac{d^{n-1}y}{dt^{n-1}} \end{bmatrix} \right\} = n, \quad \forall (x,u) \in X \times U. \quad (4)$$

In the case of linear time-invariant processes:

$$\begin{cases} \dot{x}(t) &= Ax(t) + Bu(t) \\ y(t) &= Cx(t) \end{cases}$$

where A, B, and C are constant matrices, the observability rank condition simplifies to

$$rank \left\{ \begin{bmatrix} C \\ CA \\ \vdots \\ CA^{n-1} \end{bmatrix} \right\} = n \quad (5)$$

which is the standard rank condition for observability of linear time-invariant systems (e.g. Chen, 1984) and indicates that observability is a global property in linear time-invariant systems.

Definition 5: A system of the form of (1) is detectable on $X \times U$ (locally detectable), if there exits an $n \times q$ smooth matrix function $\psi(\xi, u, y)$ such that the nominal equilibrium point of the system

$$\dot{\xi} = f(\xi, u) + \psi(\xi, u, y)[y - h(\xi)];$$

that is, $(\xi, u, y) = (x_{eq}, u_{eq}, h(x_{eq}))$, is locally asymptotically stable. Equivalently, the system is locally detectable, if there exits a function $\psi(\xi, u, y)$ such that all the eigenvalues of the $n \times n$ matrix

$$\left. \frac{\partial f(\xi, u)}{\partial \xi} \right|_{(x_{eq}, u_{eq})} - \psi(x_{eq}, u_{eq}, h(x_{eq})) \left. \frac{\partial h(\xi)}{\partial \xi} \right|_{x_{eq}}$$

lie in the left half of the complex plane. Note that in order to be able to reconstruct the states of a process, the process must be at least locally detectable.

3. NONLINEAR OBSERVER DESIGN APPROACH

To reconstruct the states of a nonlinear process, a full-order or a reduced-order nonlinear state observer can be used. For example, the observer of (3) is full-order when $\eta \in \mathbb{R}^n$, and is reduced-order when the vector η has less than n components. In the case of nonlinear processes, apart from the advantage of simulating and implementing a lower-order dynamic system, the use of a reduced-order observer, in many cases, makes (a) the observer design problem simpler and (b) the design of an observer with a globally asymptotically stable observer-error dynamics possible.

Consider a nonlinear process of the form of (1). The condition of (2) ensures the existence of a locally invertible state transformation of the form

$$\begin{bmatrix} \eta_1 \\ \vdots \\ \eta_{n-q} \\ y \end{bmatrix} = \mathcal{T}(x) = \begin{bmatrix} x_1 \\ \vdots \\ x_{n-q} \\ h(x) \end{bmatrix}$$

where $x_1, \cdots, x_{n-q}$ are $(n-q)$ state variables that satisfy the rank condition

$$rank \left\{ \frac{\partial}{\partial x} \begin{bmatrix} x_1 \\ \vdots \\ x_{n-q} \\ h(x) \end{bmatrix} \right\} = n$$

The system of (1), in terms of the new state variables $\eta_1, \cdots, \eta_{n-q}, y$, takes the form

$$\begin{cases} \dot{\eta} &= F_\eta(\eta, y, u) \\ \dot{y} &= F_y(\eta, y, u) \end{cases} \quad (6)$$

where F_η and F_y are vector functions with appropriate dimensions, and $\eta = [\eta_1, \cdots, \eta_{n-q}]^T$.

In terms of the above transformed system, a system of the form of (1) will be detectable on $X \times U$ (locally detectable), if all the eigenvalues of the $q \times q$ Jacobian matrix

$$\left. \frac{\partial}{\partial \eta} F_\eta(\eta, y, u) \right|_{(\eta_{eq}, y_{eq}, u_{eq})} \quad (7)$$

where $(\eta_{eq}, y_{eq}, u_{eq})$ satisfies $F_\eta(\eta_{eq}, y_{eq}, u_{eq}) \equiv 0$ and $F_y(\eta_{eq}, y_{eq}, u_{eq}) \equiv 0$, lie in the left half of the complex plane. This condition is a sufficient condition for local detectability; it will be a necessary and sufficient condition for local detectability, if $\partial F_y / \partial \eta \equiv 0$. Recall that observability implies detectability but the converse is not true.

Once an estimate of η, $\hat{\eta}$, is obtained by using an open-loop or closed-loop, reduced-order observer designed for the first $(n-q)$ differential equations of (6), the state estimate, $\hat{x}$, are calculated by using the relation $\hat{x} = \mathcal{T}^{-1}(\hat{\eta}, y)$.

3.1. *Open-Loop, Reduced-Order Observer*

If all the eigenvalues of the matrix of (7) lie in the left half of the complex plane, then the states can be reconstructed by using an open-loop, reduced-order observer of the form

$$\begin{cases} \dot{\hat{\eta}} &= F_\eta(\hat{\eta}, y, u) \\ \hat{x} &= \mathcal{T}^{-1}(\hat{\eta}, y) \end{cases} \quad (8)$$

The dynamics of observer error, $e = \hat{\eta} - \eta$, is governed by the system

$$\dot{e} = F_\eta(e+\eta, y, u) - F_\eta(\eta, y, u) \quad (9)$$

which has a universal equilibrium point at $e = 0$. This equilibrium point is locally asymptotically stable, because the process was assumed to be locally detectable. Obviously, this approach does not allow to adjust the rate of the decay of the observer error. Although the observer is open-loop, it may be applicable to a process operating at an unstable operating point.

3.2. *Closed-Loop, Reduced-Order Observer*

If the process is locally observable, then (a) the local asymptotic decay of the observer error can be guaranteed, and (b) the rate of the decay of the observer error will be adjustable. These can be achieved by using a closed-loop, reduced-order observer of the form:

$$\begin{cases} \dot{\hat{\eta}} &= F_\eta(\hat{\eta}, y, u) + L[\dot{y} - F_y(\hat{\eta}, y, u)] \\ \hat{x} &= \mathcal{T}^{-1}(\hat{\eta}, y) \end{cases} \quad (10)$$

where the $[(n-q) \times q]$ matrix L is the observer gain. In general, to ensure a local asymptotic decay of the observer error, the observer gain has to be a function of $\hat{\eta}$, y, and u: $L = L(\hat{\eta}, y, u)$. A constant L, as proposed by Wallman (1979), however may be adequate for many cases.

If the observer gain is chosen to be constant, then the observer of (10) is equivalent to

$$\begin{cases} \dot{z} = F_\eta(z + Ly, y, u) - LF_y(z + Ly, y, u) \\ \hat{x} = \mathcal{T}^{-1}(z + Ly, y) \end{cases} \quad (11)$$

whose implementation does not need the time-derivative of the measured outputs. The corresponding observer-error dynamics is governed by

$$\begin{aligned} \dot{e} = {} & F_\eta(e+\eta, y, u) - F_\eta(\eta, y, u) \\ & -L\{F_y(e+\eta, y, u) - F_y(\eta, y, u)\} \end{aligned} \quad (12)$$

which has a universal equilibrium point at $e = 0$. The observer gain, L, should be chosen such that the universal equilibrium point e=0 is at least locally asymptotically stable. The local asymptotic stability of the equilibrium point e=0 ensures that locally $||\hat{x}(t) - x(t)|| \longrightarrow 0$ as $t \longrightarrow \infty$, for every nonlinear process of the form of (1). As expected, the closed-loop observer of (11) with $L = 0$ is exactly that of (8).

4. APPLICATION TO CHEMICAL REACTORS

4.1. *Reactors with Classical Reactions*

Consider the CSTRs, where the N parallel exothermic chemical reactions

$$\begin{array}{ccc} A & \xrightarrow{k_1} & P_1 \\ & \vdots & \\ A & \xrightarrow{k_N} & P_N \end{array}$$

take place. The dependence of the reaction rate constants $k_1, \cdots, k_q$ on temperature is given by $k_i = Z_i \exp(-E_{a_i}/RT)$, $i = 1, \cdots, N$. The reactor temperature and heat input to the reactor are assumed to be the measurable input and output respectively.

A mole balance on A and an energy balance for the reactor (under standard assumptions) give the reactor model:

$$\begin{aligned} \dot{C}_A &= -\sum_{\ell=1}^{N} k_\ell(T) C_A^{n_\ell} + \frac{C_{A_i} - C_A}{\tau}, \qquad C_A(0) > 0 \\ \dot{T} &= \frac{1}{\rho c_p} \sum_{\ell=1}^{N} (-\Delta H_\ell) k_\ell(T) C_A^{n_\ell} + \frac{T_i - T}{\tau} \\ & \quad + \frac{Q}{\rho c_p V}, \qquad T(0) > 0 \end{aligned} \quad (13)$$

where C_{A_i}>0 and C_A represent the inlet and outlet concentrations of the reactant, A, respectively; $(-\Delta H_\ell) > 0$ denotes the heat of the ℓth reaction; n_ℓ ($n_\ell \geq 0$) denotes the order of the ℓth reaction; T_i and T represent the inlet and outlet stream temperatures respectively; τ is the reactor residence time; c_p and ρ are the heat capacity and density of the reacting mixture respectively; Q and V represent the rate of heat input to the reactor and reactor volume respectively. Note the states C_A and T always satisfy $0 < C_A$ and $0 < T$.

Chemical reactors of the above class have been studied extensively and are known to exhibit steady-state multiplicities and other exotic features (e.g., Adebekun, 1992; Soroush and Kravaris, 1992b; Uppal et al., 1976).

Here, the measurable output $y = T$, and the measurable input $u = Q$. The system of (13) is already in the form of (6). The η subsystem is

$$\dot{C}_A = \frac{C_{A_i} - C_A}{\tau} - \sum_{\ell=1}^{N} k_\ell(T) C_A^{n_\ell} \quad (14)$$

which is GAS. These chemical reactors with temperature as the measured output are observable everywhere [using the condition of (4)], and therefore detectable everywhere (see also the following paragraph).

4.1.1. *Open-loop, reduced-order observer.* For this case, the open-loop, reduced-order observer of (8) takes the form

$$\dot{\hat{C}}_A = \frac{C_{A_i} - \hat{C}_A}{\tau} - \sum_{\ell=1}^{N} k_\ell(T) \hat{C}_A^{n_\ell}, \qquad \hat{C}_A(0) > 0$$

and the observer-error dynamics is given by

$$\dot{e} = -\frac{e}{\tau} - \sum_{\ell=1}^{N} k_\ell(T) \left\{ [C_A + e]^{n_\ell} - C_A^{n_\ell} \right\}, \; e(0) = e_0$$

whose universal equilibrium point (e=0), according to Proposition 1 and Remark 1 (recall that

$C_A(t) + e(t) = \hat{C}_A(t) > 0,\ \forall t \geq 0)$, is UGAS.

4.1.2. *Closed-loop, reduced-order observer.* For this class of reactors, the closed-loop observer of (11) becomes

$$\begin{cases} \dot{z} = \dfrac{C_{A_i} - LT_i}{\tau} - \dfrac{z}{\tau} - \displaystyle\sum_{\ell=1}^{N} k_\ell(T)\left[z + LT\right]^{n_\ell} \\ \quad \bullet \left[1 + \dfrac{L(-\Delta H_\ell)}{\rho c_p}\right] - \dfrac{LQ}{\rho c_p V}, \quad z(0) = z_0 \\ \hat{C}_A = z + LT \end{cases}$$

where $L \in \mathbb{R}$ is the observer gain. The corresponding observer-error dynamics is

$$\dot{e} = -\frac{1}{\tau}e - \sum_{\ell=1}^{N}\left[1 + \frac{L(-\Delta H_\ell)}{\rho c_p}\right]$$
$$\bullet k_\ell(T)\left\{\left[e + C_A\right]^{n_\ell} - C_A^{n_\ell}\right\}, \quad e(0) = e_0$$

whose universal equilibrium point (e=0), according to Proposition 1 and Remark 1, is UGAS, if the gain $L > 0$. Thus, $|\hat{C}_A - C_A| \longrightarrow 0$ as $t \longrightarrow 0$, for every possible e_0, for every possible $T(t)$, and for every possible $Q(t)$.

4.2. *Free-Radical Polymerization CSTRs*

Polymerization reactors are highly nonlinear processes characterized by steady-state multiplicities, exotic dynamics, and potential reactor runaway (e.g. Alvarez, et al., 1993; Schmidt et al., 1981). Another feature of the reactors is that there are no reliable sensors with fast sampling rate and small time-delay for measuring many of the process variables, whose measurements are essential for the effective monitoring and control of the reactors.

Consider the polymerization CSTRs, where free-radical polymerization reactions take place. Under a set of standard assumptions such as those listed in (Daoutidis et al., 1990), mole balances on the monomer and initiator and an energy balance for the reactor yield a mathematical model of the process:

$$\dot{C}_m = \frac{C_{m_{in}} - C_m}{\tau} - C_m\sqrt{\frac{2f^* C_i k_i(T)}{k_t(T)}}$$
$$\bullet \left[k_p(T) + k_{f_m}(T)\right], \quad C_m(0) > 0$$
$$\dot{C}_i = -k_i(T)C_i + \frac{F_i C_{i_{in}} - FC_i}{V}, \quad C_i(0) > 0 \qquad (15)$$
$$\dot{T} = \gamma k_p(T) C_m \sqrt{\frac{2f^* C_i k_i(T)}{k_t(T)}} + \frac{Q}{\rho c_p V}$$
$$+ \frac{T_{in} - T}{\tau}, \quad T(0) > 0$$

where the reaction rate constants $k_\ell(T) = Z_\ell exp[-E_{a_\ell}/(RT)]$, $\ell = p, f_m, t, i$ and $\gamma = (-\Delta H_P)/(\rho c_p)$. C_m, C_i, and T represent the monomer and initiator concentrations and reactor temperature respectively. The rest of the notations are the same as those used in (Daoutidis et al., 1990). Note that the states C_m, C_i, and T always satisfy $0 < C_m$, $0 < C_i$, and $0 < T$.

The reactor temperature and the density of the reacting mixture can readily be measured on-line. Monomer conversion, x_m, defined as

$$x_m = \frac{C_{m_{in}} - C_m}{C_{m_{in}}}$$

can be inferred from the density measurements, and thus the monomer concentration can be calculated on-line from:

$$C_m = C_{m_{in}}(1 - x_m)$$

Now, consider the case in which both the reactor temperature and monomer concentration are measured: $y_1 = C_m$ and $y_2 = T$. The system of (15) is already in the form of (6), and the η subsystem is

$$\dot{C}_i = -k_i(T)C_i + \frac{F_i C_{i_{in}} - FC_i}{V}$$

which is GAS. These chemical reactors with temperature and monomer concentration as the measured outputs are, according to the condition of (4), observable everywhere and therefore detectable everywhere (see also the following paragraph).

4.2.1. *Open-loop, reduced-order observer.* For these reactors, the open-loop, reduced-order observer of (8) takes the form

$$\dot{\hat{C}}_i = -k_i(T)\hat{C}_i + \frac{F_i C_{i_{in}} - F\hat{C}_i}{V}, \quad \hat{C}_i(0) > 0$$

and the observer-error dynamics is given by

$$\dot{e} = -\left[\frac{1}{\tau} + k_i(T)\right] e, \quad e(0) = e_0$$

whose universal equilibrium point (e=0), according to Proposition 1, is UGAS.

4.2.2. *Closed-loop, reduced-order observer.* For these reactors, the closed-loop, reduced-order observer of (11) takes the form:

$$\begin{cases} \dot{z} = -\left[k_i(T) + \dfrac{1}{\tau}\right] z - k_i(T)\left[L_1 C_m + L_2 T\right] \\ \quad -C_m \theta \sqrt{\dfrac{k_i(T)}{k_t(T)}}\left[z + L_1 C_m + L_2 T\right] \qquad (16) \\ \quad \bullet \left\{L_2 \gamma k_p(T) - L_1\left[k_p(T) + k_{f_m}(T)\right]\right\} + \alpha_0 \\ \hat{C}_i = z + L_1 C_m + L_2 T \end{cases}$$

where L_1, $L_2 \in \mathbb{R}$,

$$\theta = \sqrt{2f^*},$$

$$\alpha_0 = \frac{F_i C_{i_{in}} - F[L_1 C_{m_{in}} + L_2 T_{in}]}{V} - \frac{L_2 Q}{\rho c_p V}.$$

The corresponding error dynamics is governed by

$$\dot{e} = -\left[k_i + \frac{1}{\tau}\right]e - C_m\theta\sqrt{\frac{k_i(T)}{k_t(T)}}\left[\sqrt{C_i + e} - \sqrt{C_i}\right]$$

$$\bullet \{L_2\gamma k_p(T) - L_1\left[k_p(T) + k_{f_m}(T)\right]\}, \quad e(0) = e_0$$

whose universal equilibrium point ($e = 0$), according to Proposition 1 and Remark 1 (recall that $C_i(t) + e(t) = \hat{C}_i(t) > 0, \ \forall t \geq 0$), is UGAS, if the gain $L = [L_1 \ \ L_2]^T$ is chosen such that

$$L_1\left[1 + \frac{k_{f_m}(T)}{k_p(T)}\right] \leq \gamma L_2 \qquad (17)$$

for every possible T. For example, any values of L_1 and L_2 which satisfy $L_1 < 0 < L_2$, will also satisfy the condition of (17).

5. CONCLUSIONS

The notions of universal equilibrium point and universal, global, asymptotic stability were defined. By using Lyapunov's direct method for non-autonomous systems and the decrescent, radially unbounded, globally positive definite function $||x - x^*_{eq}||^2$, a *sufficient* condition for a system to be UGAS was derived.

A simple and practical method of nonlinear observer design was presented. The method was used to design state observers for two classes of chemical reactors, which are known to exhibit highly nonlinear dynamics with steady-state multiplicities. For these reactors, the global asymptotic stability of the observer-error dynamics was shown by using Lyapunov's direct method for non-autonomous systems. There are many nonlinear processes for which this approach can be used to design nonlinear state-observers with globally asymptotically stable error dynamics. The page limitation did not allow (a) to include simulation results and (b) to consider more classes of nonlinear processes and present their observers as well as the corresponding UGAS error dynamics.

6. ACKNOWLEDGMENT

Acknowledgment is made to the donors of The Petroleum Research Fund, administered by the ACS, for support of this research.

7. REFERENCES

Adebekun, A. K. (1992). The Robust Global Stabilization of a Stirred Tank Reactor. *AIChE J.*, **38**, 651.

Alvarez, J., J. J. Alvarez, M. A. Barron, and R. Suarez (1993). The Global Stabilization of a Two-Input Three-State Polymerization Reactor with Saturated Feedback. *Proc. of ACC*, 2917.

Chen, C. T. (1984). Linear System Theory and Design. *Holt, Rinehart and Winston.*

Daoutidis, P., and C. Kravaris (1992). Dynamic Output Feedback Control of Minimum-Phase Nonlinear Processes. *Chem. Eng. Sci.*, **47**, 837.

Daoutidis, P., M. Soroush, and C. Kravaris (1990). Feedforward/Feedback Control of Multivariable Nonlinear Processes. *AIChE J.*, **36**, 1471.

Dootingh, M. van, F. Viel, D. Rakotopara, J.P. Gauthier, and P. Hobbies (1992). Nonlinear Deterministic Observer for State Estimation: Application to a Continuous Free Radical Polymerization Reactor. *Comput. & Chem. Eng.*, **16**, 777.

Grizzle, J. W., and P. E. Moraal (1990). Newton Observers and Nonlinear Discrete-Time Control. *Proc. of the 29th IEEE Conference on Decision and Control*, Honolulu, Hawaii, 760.

Kantor, J. C. (1989). A Finite Dimensional Nonlinear Observer for an Exothermic Stirred - Tank Reactor. *Chem. Eng. Sci.*, **44**, 1503.

Krener, A. J., and A. Isidori (1983). Linearization by Output Injection and Nonlinear Observers. *Systems Control Lett.*, **3**, 47.

Krener, A. J., and W. Respondek (1985). Nonlinear Observers with Linearizable Error Dynamics. *SIAM J. Control Optimization*, **2**, 197.

Schmidt, A. D., and W. H. Ray (1981). The Dynamic Behavior of Continuous Polymerization Reactors-I. *Chem. Eng. Sci.*, **36**, 1401.

Soroush, M., and C. Kravaris (1994). Nonlinear Control of a Polymerization CSTR with Singular Characteristic Matrix. *AIChE J.*, **40**, 980.

Soroush, M., and C. Kravaris (1993). Multivariable Nonlinear Control of a Continuous Polymerization Reactor: an Experimental Study. *AIChE J.*, **39**, 1920.

Soroush, M., and C. Kravaris (1992a). Nonlinear Control of a Batch Polymerization Reactor: an Experimental Study. *AIChE J.*, **38**, 1429.

Soroush, M., and C. Kravaris (1992b). Discrete-Time Nonlinear Controller Synthesis by Input/Output Linearization. *AIChE J.*, **38**, 1923.

Uppal, A., A. B. Poore, and W. H. Ray (1974). On the Dynamic Behavior of Continuous Stirred Tank Chemical Reactors. *Chem. Eng. Sci.*, **29**, 967.

Wallman, P. H. (1979). Reconstruction of Unmeasured Quantities for Nonlinear Dynamic Processes. *Ind. Eng. Chem. Fundam.*, **18**, 327.

Zeitz, M. (1987). The Extended Luenberger Observer for Nonlinear Systems," *Syst. & Cont. Lett.*, **9**, 149.

pH CONTROL IN THE PRESENCE OF PRECIPITATION EQUILIBRIA

Raymond A. Wright

The Dow Chemical Company
1400 Building
Midland, MI 48667

Costas Kravaris

The University of Michigan
Dept. of Chemical Engineering
Ann Arbor, MI 48109

Abstract: The control of pH to minimize solubility of precipitates is a challenging problem. A first principles model for a precipitation process, including acid/base equilibria and precipitation equilibria is developed. A minimal order realization of the full order model is derived, which consists of linear first-order dynamics and a static nonlinearity in the output map. The control problem becomes linear after it is redefined in terms of an equivalent output. The approach is illustrated in a zinc hydroxide precipitation example.

Keywords: pH control, Nonlinear control, Model-based control, Dynamic models, Process models, Control oriented models, Process control

1. INTRODUCTION

The chemical treatment of a liquid process stream resulting in formation of a solid phase is important in the chemical process industry. Coagulation, flocculation, and precipitation are examples of operations used for product manufacture as well as waste water treatment. In such systems, the solubility of the solid and/or presence of necessary intermediates is often dependent upon the pH. Several specific chemical examples can be found in texts such as Eilbeck and Mattock (1987). As a result, tight control of pH is necessary for the maximum formation of solids and optimal operation of the process.

Standard pH control algorithms are usually developed for systems with fast acid / base equilibria. The objective of the algorithm is thus to compensate for the static nonlinearity related to the titration curve, which will vary with time as the inlet concentrations change. Applications of systems control theory have resulted in a wide range of algorithms; from adaptive linear control (Proudfoot *et al.*, 1983), to nonlinear control (Parrish and Brosilow, 1988; Wright and Kravaris, 1991; Mahuli *et al.*, 1992), to adaptive nonlinear control (Gustafsson and Waller, 1992; Wright and Kravaris, 1993).

The purpose of the present work is to develop an algorithm specifically for the control of pH in processes where precipitation occurs. This paper begins with a discussion and review of precipitation equilibria, including solubility and complex ion formation. A representation of the nonlinear static part of the precipitation process including general acid/base equilibria will then be developed. A dynamic process model is then given and a reduced-order model for control design is derived. The control algorithm for precipitation processes is then designed. Finally, computer simulation results demonstrating the performance of the proposed method for a simple precipitation process will be given.

2. PRECIPITATION EQUILIBRIA

The formation and precipitation of a solid phase from liquid streams can be complicated. General discussion of processes involving the formation of solids can be found in chemistry texts (e.g. Kolthoff, et al, 1969) or books on specific areas, such as waste water treatment (e.g. Eilbeck and Mattock, 1987). While these discussions give extremely useful insights into such processes, they generally are not in the form necessary for control engineers. The mechanisms behind the formation of the solid phase can be different depending on the process and chemical system and the development of a general model would be quite difficult.

An important industrial process is the recovery of heavy metals (Narvarte and Koch, 1985), frequently by the precipitation of the metal hydroxide. In the following development a metal hydroxide with two hydroxyl groups is used. This choice represents a common number for the hydroxides of heavy metals and simplifies the development, as compared to using an unspecified, arbitrary number, for the limited space available. Development for a species with a different number of hydroxyl groups is straightforward.

The equilibria for the dissociation of the hydroxyl groups in the aqueous phase are given by:

$$M(OH)_2 \rightleftarrows MOH^+ + OH^-$$

$$MOH^+ \rightleftarrows M^{2+} + OH^-$$

These are obviously identical to the equilibria for any dihydroxyl base. The corresponding dissociation constants for these reactions are given by:

$$K_{b1} = \frac{[MOH^+][OH^-]}{[M(OH)_2]} \qquad (1)$$

$$K_{b2} = \frac{[M^{2+}][OH^-]}{[MOH^+]} \qquad (2)$$

Note that these constants are larger for stronger dissociations. A strong base (one that is completely dissociated) would have infinite K_b. The precipitation of the metal hydroxide from the aqueous phase is represented by:

$$M(OH)_2\downarrow \rightleftarrows M(OH)_2$$

The constant corresponding to precipitation that is normally reported in reference material is the solubility product constant, which is defined as:

$$K_{sp} = [M^{2+}][OH^-]^2 \qquad (3)$$

Many metal hydroxides will form complex ions in solutions with excess hydroxyl ions. These generally differ depending on the metal (Kolthoff *et al.*, 1969; Charlot, 1974), but a form common to several metals is the addition of a third and fourth hydroxyl group. These equilibria are given by:

$$M(OH)_2\downarrow + OH^- \rightleftarrows M(OH)_3^-$$

$$M(OH)_2\downarrow + 2OH^- \rightleftarrows M(OH)_4^{2-}$$

The constants corresponding to these ion formations are defined as (Kolthoff *et al.*, 1969):

$$K_{c1} = \frac{[M(OH)_3^-]}{[OH^-]} \qquad (4)$$

$$K_{c2} = \frac{[M(OH)_4^{2-}]}{[OH^-]^2} \qquad (5)$$

For metals that form different complex ions, constants may be defined in a similar manner, depending upon the ion. Note that the constants defined in (4) and (5) will be zero if a particular metal does not form that complex ion.

The equilibrium representing the dissociation of water must also be included when considering aqueous systems. The dissociation constant for water is given by:

$$K_w = [H^+][OH^-] \qquad (6)$$

The solubility of the metal, S_M, will be the sum of concentrations of all species existing in the liquid phase (Charlot, 1974). By definition, this is:

$$S_M = [M^{2+}] + [MOH^+] + [M(OH)_2] + [M(OH)_3^-] + [M(OH)_4^{2-}] \qquad (7)$$

An expression for S_M in terms of constants and hydrogen ion concentration may be obtained by combining (1-7). This results in:

$$S_M = \frac{K_{sp}[H^+]^2}{K_w^2} + \frac{K_{sp}[H^+]}{K_w K_{b2}} + \frac{K_{sp}}{K_{b1}K_{b2}} + \frac{K_{c1}K_w}{[H^+]} + \frac{K_{c2}K_w^2}{[H^+]^2} \qquad (8)$$

From (8) it can be seen that a minimum will exist if the complex ions are formed. This is consistent with the equilibria expressions that represent the complex ions forming with the precipitate at higher hydroxyl ion concentrations. Thus, beginning from low hydrogen ion concentrations, where the metal hydroxide is very soluble, as hydroxyl ions are added, the metal hydroxide will precipitate and the overall solubility of the metal decreases. However, at high hydroxyl ion concentrations, the complex ions will begin to form and the overall solubility of the metal increases. Note that if the metal does not form complex ions, its solubility will continue to decrease as the hydroxyl ion concentration increases.

3. PRECIPITATION IN THE PRESENCE OF ACIDS AND BASES

In general species other than the ones involved in precipitation will be present in the process. For example, the metal may be in the form of a salt which dissociates and then forms hydroxide. The anion originating from the salt must be considered, as well as the cation of the base being used to initiate the precipitation. In this section, the equations necessary to compute the pH from analytical concentrations of other species will be developed. The discussion that follows will make use of concentrations of each chemical species. It would be more precise to use activities of these species instead of concentrations, but under the assumption that all species will be present in dilute concentrations, the activity coefficient would approach 1. Additionally, the development will assume that only one species is precipitating. This results in a more readable development and equations in the space permitted. The generalization to more precipitating species is relatively straightforward.

The modeling method used in this section and the next is a method developed for pH processes in McAvoy *et al.* (1972) and generalized in Gustafsson and Waller (1983) and Wright and Kravaris (1991). The static part of the model consists of using an electroneutrality condition and combining this with the definition for the dissociation constants and the analytical (or total ion) concentration. For a pH process this results in (Wright and Kravaris, 1991):

$$\sum_{i=1}^{n} a_i([H^+])x_i + A([H^+]) = 0 \qquad (9)$$

where x_i is the analytical concentration of the anion of an acid or the cation of a base and

$$A([H^+]) = [H^+] - \frac{K_w}{[H^+]} \qquad (10)$$

The definition of analytical concentration and the resulting $a_i([H^+])$ function are given in Table 1 for common acids and bases.

The pH may then be calculated by solving (9) for $[H^+]$ and using the definition of pH:

$$pH = -\log[H^+] \qquad (11)$$

At this point it is important to notice the difference between the dissociation constants, K_b, the complex ion constants, K_c, and the solubility product constant, K_{sp}. The dissociation constants will hold at all pH values. The concentrations they relate may be larger at some pH values than others, but the relation will always hold. The equilibria showing the formation of the complex ions depend on the solid phase (Kolthoff *et al.*, 1969; Charlot, 1974) and therefore are assumed to occur only if precipitation occurs. K_{sp} is like a constraint. It marks the point at which the solid phase will form and is valid only if there is a solid phase present. At pH values or concentration values below where the precipitate will form, the concentration relation shown in equation (3) will not hold. At conditions where there is no precipitate, the system will be described by (9-11) and the metal hydroxide can be treated as any other dihydroxyl base.

Once a solid phase is formed the resulting equations will be different. Without loss of generality, label the metal hydroxide as species 1 and all others as species 2 through n. Beginning with the electroneutrality condition and developing as before results in:

Table 1: Definitions of $a_i([H^+])$ and Analytical Concentration for Common Species

i-th ionic species	Analytical conc.	$a_i([H^+])$
Anion of HA	$[HA] + [A^-]$	$-\dfrac{1}{1 + \dfrac{[H^+]}{K_a}}$
Anion of H_2A	$[H_2A] + [HA^-] + [A^{2-}]$	$-\dfrac{2 + \dfrac{[H^+]}{K_{a2}}}{1 + \dfrac{[H^+]}{K_{a2}} + \dfrac{[H^+]^2}{K_{a1}K_{a2}}}$
Cation of BOH	$[BOH] + [B^+]$	$\dfrac{[H^+]}{[H^+] + \dfrac{K_w}{K_b}}$
Cation of $B(OH)_2$	$[B(OH)_2] + [BOH^+] + [B^{2+}]$	$\dfrac{2[H^+]^2 + \dfrac{K_w}{K_{b2}}[H^+]}{[H^+]^2 + \dfrac{K_w}{K_{b2}}[H^+] + \dfrac{K_w^2}{K_{b1}K_{b2}}}$

$$\sum_{i=2}^{n} a_i([H^+])x_i + A([H^+]) + P_M([H^+]) = 0 \qquad (12)$$

where

$$P_M([H^+]) = \frac{2K_{sp}[H^+]^2}{K_w^2} + \frac{K_{sp}[H^+]}{K_wK_{b2}} - \frac{K_{c1}K_w}{[H^+]} - \frac{2K_{c2}K_w^2}{[H^+]^2} \qquad (13)$$

With the formation of a second phase, one degree of freedom must be lost. This manifests itself by the loss of the analytical concentration from (12). Once a species begins to precipitate, its analytical concentration in the liquid phase is fixed by the solubility (8) and is not necessary to calculate pH. The pH of this system may then be calculated by solving (12) for $[H^+]$ and using (11).

4. A DYNAMIC MODEL FOR PRECIPITATION PROCESSES

In the previous section the static relations for a precipitation process were developed. In this section, the dynamics for a precipitation process will be developed and the overall model will be formulated. The dynamic system under consideration (see Figure 1) consists of a CSTR with a process stream flowing into it. The pH of this stream is to be manipulated in order to minimize solubility by manipulating the flowrate of the titrating stream. It is assumed that solid phase being formed leaves with the effluent stream to a point downstream where it can settle and be separated.

The analytical concentrations are conserved quantities

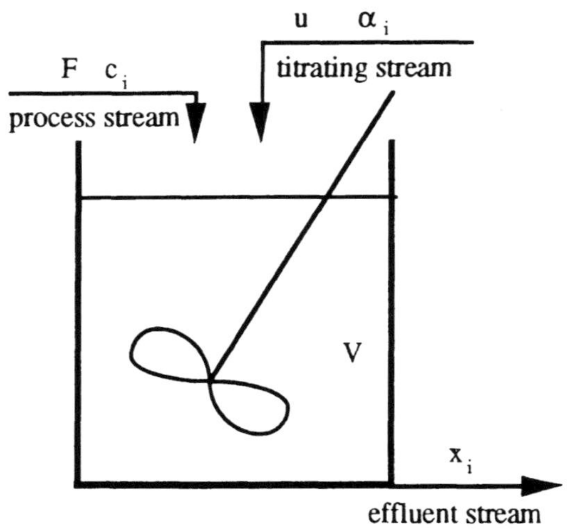

Figure 1: pH process with precipitation

(or reaction invariants). Letting the metal be the first species and all others being 2 through n results in the following set of mass balances for this system:

$$V\frac{dx_M}{dt} = F(c_M - x_M) + u x_M \quad (14)$$

$$V\frac{dx_i}{dt} = F(c_i - x_i) + (\alpha_i - x_i)u\ , \quad i = 2, ..., n \quad (15)$$

It makes no sense in this process to consider the case of the metal entering in the titrating stream. It is also possible to define a rate of precipitate formation as:

$$r_M = \left[x_M - S_M(pH)\right](F + u) \quad (16)$$

where r_M is in gram-equivalents per unit time. Clearly precipitate will only be formed if x_M is greater than S_M. Note that in terms of calculating pH, (14) is redundant.

As with pH processes, the full model with (15) as the state equations and (12) as the output map, is unobservable and uncontrollable. This full model from first principles can be reduced using the same method as used for pH processes (Wright and Kravaris, 1991). By defining:

$$\frac{c_2 - x_2(t)}{c_2 - \alpha_2} = \frac{c_3 - x_3(t)}{c_3 - \alpha_3} = ... = \frac{c_n - x_n(t)}{c_n - \alpha_n} = X \quad (17)$$

The second through n-th state equations collapse into one:

$$V\frac{dX}{dt} = u - (F + u)X \quad (18)$$

Equation (12) can then be written as

$$\frac{A(pH) + P_M(pH) + \sum_{i=2}^{n} a_i(pH)\, c_i}{\sum_{i=2}^{n} a_i(pH)\,(c_i - \alpha_i)} = X \quad (19)$$

Furthermore, the inverse of the titration curve for the precipitation process may be defined as:

$$\mathbf{T}(pH) = -\frac{A(pH) + P_M(pH) + \sum_{i=2}^{n} a_i(pH)\, c_i}{A(pH) + P_M(pH) + \sum_{i=2}^{n} a_i(pH)\, \alpha_i} \quad (20)$$

The state variable, X, in the reduced model is related to $\mathbf{T}$(pH) by:

$$X = \frac{\mathbf{T}(pH)}{1 + \mathbf{T}(pH)} \quad (21)$$

The reduced model is thus (18) with state/output map (21). This model represents a much more macroscopic view of the process than the model given by (15) and (12). The main concern is the nonlinearity as a whole, represented by T(pH), instead of the individual concentrations of species in the system.

The objective of the control structure which will be developed in the next section is to regulate pH at pH_{sp}. It is convenient for control purposes to redefine the state of the reduced model as:

$$\mathbf{Y} = \mathbf{T}(pH_{sp}) - \left[1 + \mathbf{T}(pH_{sp})\right] X \quad (22)$$

The dynamics of the model are then given by:

$$V\frac{d\mathbf{Y}}{dt} = F\mathbf{T}(pH_{sp}) - u - (F + u)\mathbf{Y} \quad (23)$$

Note that $F\mathbf{T}(pH_{sp})$ is the steady-state value of the manipulated input, u, corresponding to pH_{sp}. The state/output map becomes;

$$\mathbf{Y} = \frac{\mathbf{T}(pH_{sp}) - \mathbf{T}(pH)}{1 + \mathbf{T}(pH)} \quad (24)$$

5. CONTROL STRUCTURE

From the minimal order realization given by (23) and (24) it is relatively straightforward to develop the control structure. For most processes, the flowrate of the titrating stream will be much smaller than the flowrate of the process stream, i.e. u << F. In this case the state equation (23) becomes linear and the model has only a static nonlinearity in the output map. From (24) it is clear that $\mathbf{Y}$ will go to zero if and only if pH goes to pH_{sp}. Thus $\mathbf{Y}$ is an equivalent output to pH. As a result, a PI controller in terms of $\mathbf{Y}$ can be used to control the process:

$$u - F\mathbf{T}(pH_{sp}) = K_c\left[\mathbf{Y} + \frac{1}{\tau_I}\int_0^t \mathbf{Y}\, dt\right] \quad (25)$$

The PI controller parameters are related to the z-domain closed-loop poles P_1 and P_2 by

$$K_c = F \frac{e^{-(F/V)T} - P_1 P_2}{1 - e^{-(F/V)T}} \tag{26}$$

$$\tau_I = \frac{T\left(e^{-(F/V)T} - P_1 P_2\right)}{1 - \left(P_1 + P_2\right) + P_1 P_2} \tag{27}$$

These results are completely analogous with those given in Wright and Kravaris, 1993 for pH processes. Only the definition of the inverse of the titration curve must be changed to reflect the presence of precipitates.

The proposed approach can be easily extended to an adaptive control setting, if **T**(pH) needs to be estimated on-line. The identification problem can be formulated by postulating an appropriate representation of the nonlinearity, in the same vein as in Gustafsson and Waller (1983) or Wright and Kravaris (1993) for pH processes without precipitation. The latter allows for varying degrees of chemical knowledge of the process to be incorporated. Once a representation is chosen, a standard identification algorithm, such as Ljung and Soderstrom (1983) or Papadoulis, et al (1987), can be used.

6. COMPUTER SIMULATIONS

The method detailed in the previous sections will now be illustrated in simulations of a simple precipitation process. The system chosen for study is the precipitation of zinc hydroxide. Aqueous zinc chloride enters the system and a controlled amount of sodium hydroxide is added. The parameters used in the simulations are given in Table 2. The equilibrium constants are taken from Charlot (1974). The solubility of zinc hydroxide as a function of pH is shown in Figure 2. The minimum solubility occurs at a pH value of 11.2.

For this particular system, **T**(pH) is given by:

$$\mathbf{T}(pH) = -\frac{A(pH) + P_M(pH) - c_2}{A(pH) + P_M(pH) + \alpha_3} \tag{28}$$

This function is shown in Figure 3 for both initial and final conditions used in the simulations. For comparison purposes, the T(pH) corresponding to the final conditions, but assuming no precipitate is also shown in Figure 3. It is clear from the figure that assuming a solid phase exists is correct above pH values of 6.5.

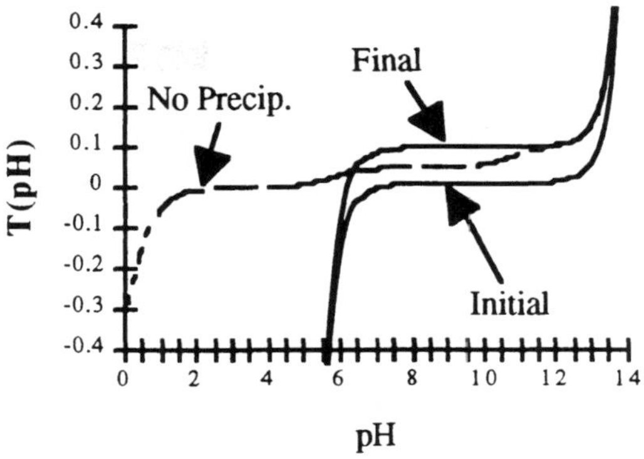

Fig. 3: Titration curves for simulation conditions

Table 2: Parameter Values Used in Simulations

Parameter	Value
V	10 l
F	1 l/min
initial u	10.8 ml/min
T	6 sec
initial Zn: c_M	0.01 N
final Zn: c_M	0.1 N
initial Cl : c_2	0.02 N
final Cl: c_2	0.2 N
α_3	2.0 N
pK_{b1}	3.3
pK_{b2}	8.3
pK_{c1}	3.2
pK_{c2}	2.0
pK_{sp}	17
pH_{sp}	11.2

Equation (28) may be combined with the state/output map of the reduced order model, (24). This results in:

$$Y = \left[A(pH) + P_M(pH)\right] - \left[A(pH_{sp}) - P_M(pH_{sp})\right] \tag{29}$$

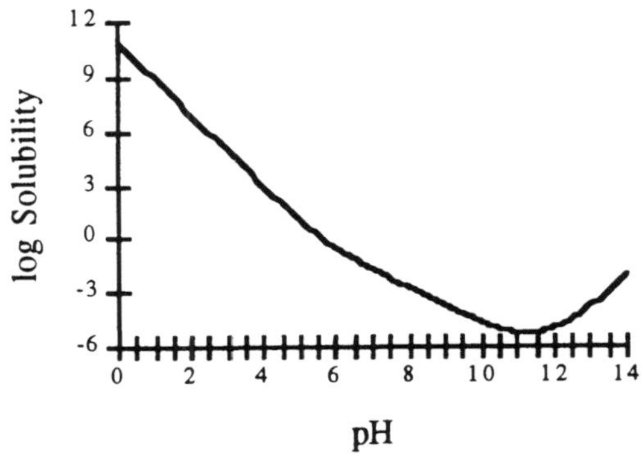

Fig. 2: Log of solubility as a function of pH for the zinc hydroxide system.

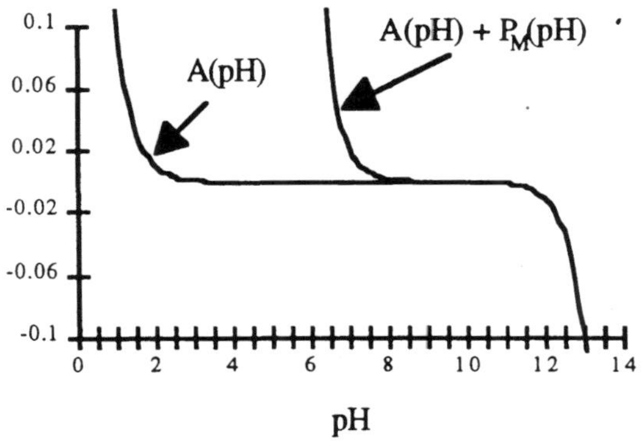

Fig. 4: Nonlinearities for simulated precipitation process and strong acid / strong base pH process.

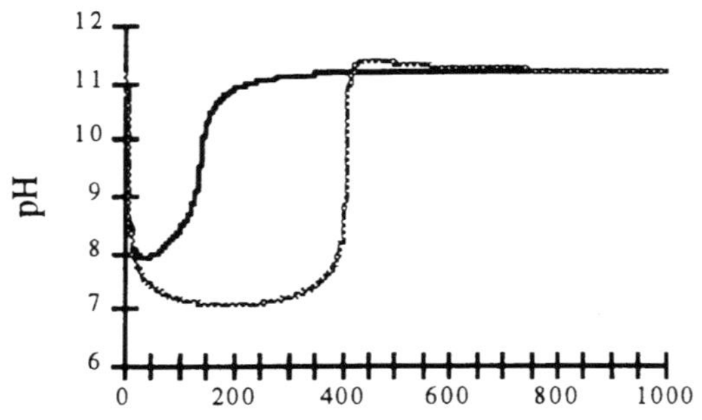

Fig. 5: pH response of simulated system

This is analogous to the result obtained for strong acid/strong base pH processes without precipitation (Wright et al, 1991). The functions A(pH) and A(pH)+PM(pH) are shown in Figure 4. This demonstrates the differences in the nonlinearity for this precipitation system and a strong acid / strong base pH process in the absence of precipitation.

Figures 5 through 7 show the simulations results for the conditions given in Table 2. Each figure has the response using the proposed method (solid line) and the response generated using the control law for a strong acid / strong base system without precipitation (dotted line).

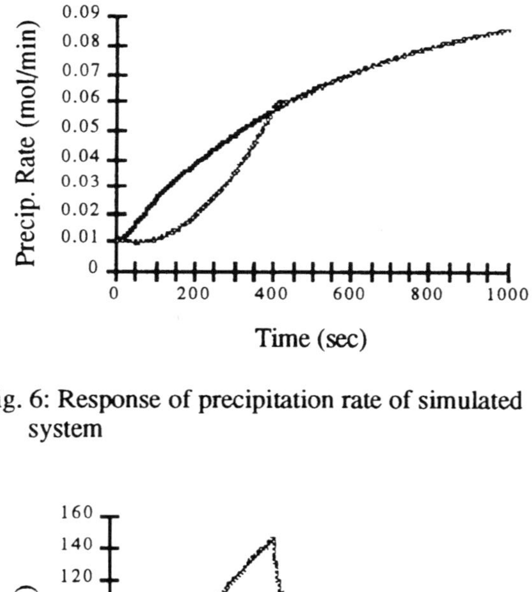

Fig. 6: Response of precipitation rate of simulated system

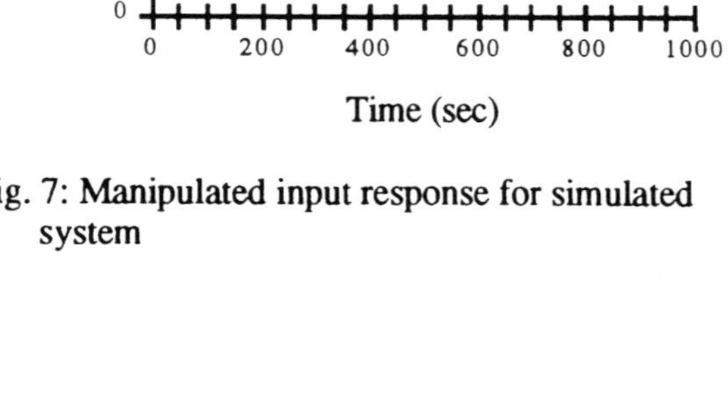

Fig. 7: Manipulated input response for simulated system

The poles for the simulations are set equal at 0.9. Figure 5 shows that the pH using the precipitation equilibria is quickly returned to setpoint, while the control using the strong acid / strong base is much more sluggish. The difference in minimum pH values is less than 1 unit, but the difference is precipitation rates is significant. The solubility actually becomes an order of magnitude greater if the precipitation equilibria are not used. This represents metal lost through the process.

8. REFERENCES

Charlot, G. (1974). *Chimie Analytique Quantitative, Vol. 1*, Masson et Cie, Paris.

Eilbeck, W. J. and G. Mattock (1987). *Chemical Processes in Waste Water Treatment*, Ellis Horwood Limited, Chichester.

Gustafsson, T. K. and K. V. Waller (1983). Dynamic Modeling and Reaction Invariant Control of pH. *Chemical Engineering Science*, **38**, 389-398.

Gustafsson, T. K. and K. V. Waller (1992). Nonlinear and Adaptive Control of pH. *Ind. Eng. Chem. Res.*, **31**, 2681-2693.

Kolthoff, I. M., E. B. Sandell, E. J. Meehan and S. Bruckenstein (1969). *Quantitative Chemical Analysis*. The Macmillan Company, London.

Ljung, L. and T. Soderstrom (1983). *Theory and Practice of Recursive Identification*, MIT Press, Cambridge.

Mahuli, S. K., R. R. Rhinehart and J. B. Riggs (1992). Experimental Demonstration of Non-linear Model-based In-line Control of pH. *J. Proc Cont.*, **2**, 145-153.

McAvoy, T. J., E. Hsu and S. Lowenthal (1972). Dynamics of pH in a Controlled Stirred Tank Reactor. *Ind. Eng. Chem. Process Des. Dev.*, **11**, 68-70.

Narvate, S. and C. Koch (1985). Distributed Control, Multiple pH Loops Optimize Heavy Metal Waste Treatment. *Chem. Processing*, 134-136.

Papadoulis, A. V., C. A. Tsiligiannis and S. A. Svoronos (1987). A Cautious Self-Tuning Controller for Chemical Processes. *AIChE J.*, **33**, 401-409.

Parrish, J. R. and C. B. Brosilow (1988). Nonlinear Inferential Control. *AIChE Journal*, **34**, 633-644.

Proudfoot, C. G., P. J. Gawthrop and O. L. R. Jacobs (1983). Self Tuning PI Control of a pH Neutralization Process. *Proc. IEE. D*, **130**, 267-272.

Wright, R. A. and C. Kravaris (1991). Nonlinear Control of pH Processes Using the Strong Acid Equivalent. *Ind. Eng. Chem. Res.*, **30**, 1561-1572.

Wright, R. A., M. Soroush and C. Kravaris (1991). Strong Acid Equivalent Control of pH Processes: An Experimental Study. *Ind. Eng. Chem. Res.*, **30**, 2437-2444.

Wright, R. A. and C. Kravaris (1993). On-Line Identification and Nonlinear Control of pH Processes. *Proc. of American Control Conference*, San Francisco, 1167-1171.

PARTITIONING-SPACE APPROACH FOR NONLINEAR PROCESS CONTROL

Z. H. Liu and S. Macchietto

Centre for Process Systems Engineering, Imperial College, London SW7 2BY, UK.

Abstract. A partitioning-space approach is presented for the design of nonlinear controllers. Easy to check conditions are presented for partitioning a general nonlinear SISO system into two subsystems, resulting in a cascade control structure which can accommodate standard or nonlinear control techniques in each subsystem. A particularly useful application is to a class of nonlinear systems for which model uncertainties can be restricted to just one subsystem. The resulting cascade structure is robust with respect to measurement errors, sever model mismatch and disturbance rejection. Simulation and experimental results on a batch reactor are presented.

Key Words. Nonlinear control; Partitioning-space; Batch reactor; Experiments

1. INTRODUCTION

The need of tight control to ensure high performance and consistent product quality is well understood. However, the existence of process nonlinearities, frequent changes in processing conditions (batch size, raw materials, products, etc.) in flexible plants present major obstacles to achieve a good control. During the last decade, significant progress has been made in the control of non-linear systems. Globally Linearizing Control (GLC, Kravaris and Chung, 1987) is one of the controller design approaches that are based directly on a nonlinear process model or an exact linearization of a nonlinear model. A number of applications have been reported using these approaches (Henson and Seborg, 1991; Dore, 1993), but very few experimental studies have been presented. This can be attributed to several factors, in particular:

i) the requirement for a perfect model in order to generate a feedback control law which exactly cancels all the nonlinearities between input and output

ii) extensive computation involved for high order systems

iii) implementation difficulties

iv) the applicability only to systems which satisfy disturbance matching conditions and to minimum phase non-linear systems.

Obtaining a good model for an entire process system is typically difficult. On the other hand, portions of the system are often easy to model. In this paper, a partitioning-space approach is developed which is intended to make the best use of the well known part of the system model for the nonlinear controller design, and to resolve some of the other problems above. A theoretical analysis is described first. The approach is then demonstrated by two examples: a simple illustrative simulation of the level control of two noninteracting tanks, and the experimental temperature control of a pilot scale multipurpose batch reactor.

2. SYSTEM PARTITIONING

Consider the following SISO control affine nonlinear system:

$$\dot{x} = f(x) + g(x)u + \gamma(x)d$$
$$y = h(x) \quad (1)$$

where x is an n-dimensional state vector, u is a scalar input, d is a disturbance, y is a scalar output, f(x), g(x) and γ(x) are vector functions, and $h(x)$ is a scalar function. It is assumed that f(x), g(x), γ(x) and $h(x)$ are of class C^∞ (i.e. have continuous derivatives of all orders). The above system can always be transformed into the following form:

$$\dot{x}_I = f_I(x) + g_I(x)u + \gamma_I(x)d$$
$$\dot{x}_{II} = f_{II}(x) + g_{II}(x)u + \gamma_{II}(x)d \quad (2)$$
$$y = h(x)$$

where $x=[x_I \ x_{II}]^T$, $f(x)=[f_I(x) \ f_{II}(x)]^T$, $g(x)=[g_I(x) \ g_{II}(x)]^T$, $\gamma(x)=[\gamma_I(x) \ \gamma_{II}(x)]^T$. The partition is chosen such that the following conditions are satisfied:

$$\frac{\partial y}{\partial x_i} = 0 \quad \forall x_i \in x_I, \quad \frac{\partial y}{\partial x_j} \neq 0 \quad \forall x_j \in x_{II} \qquad (1.c)$$

If either x_I or x_{II} are empty, then a partition does not exist.

The SISO system in Eq (2) is said (Isidori, 1989) to have relative degree r in a region Ω if, ∀x∈Ω

$$L_g L_f^{\,i} h(x) = 0 \qquad \forall i = 0, \ldots, r-2$$
$$L_g L_f^{\,r-1} h(x) \neq 0$$

From this definition, the following new corollary is obtained.

Corollary 1: The relative degree of the system given by Eq (2) is greater than one if

$$g_j(x) = 0, \quad \forall g_j(x) \in g_{II}(x). \qquad (2.c)$$

Proof:

$$L_g h(x) = \frac{\partial h(x)}{\partial x} g(x) = \frac{\partial h(x)}{\partial x_I} g_I(x) + \frac{\partial h(x)}{\partial x_{II}} g_{II}(x)$$

As $\partial h(x)/\partial x_I = 0$, $L_g h(x)=0$ if $g_j(x)=0$, $\forall g_j(x) \in g_{II}(x)$.

Condition (2.c) provides a simple sufficient condition to identify systems which have a relative order greater than one without explicit calculation of each order of Lie derivatives.

In the rest of work presented, the partitioning space approach is only considered for systems in which $g_{II}(x)=0$, since the design of controllers for systems of order one is rather simple. The system in Eq(2) is, therefore, simplified as follows:

$$\dot{x}_I = f_I(x) + g_I(x)u + \gamma_I(x)d$$
$$\dot{x}_{II} = f_{II}(x) + \gamma_{II}(x)d \quad (3)$$
$$y = h(x)$$

Assume the system (3) has a relative degree q with respect to the disturbance d. A necessary and sufficient condition for the solution of the disturbance decoupling problem is (Isidori, 1989):

i) $q>r$ if the disturbance cannot be measured. This is often called disturbance matching condition (Henson and Seborg, 1991)

ii) $q \geq r$ if the disturbance can be measured.

From this definition, the following new corollary is obtained:

Corollary 2: The disturbance matching condition is not satisfied for a system described by Eq (3) if

$$\partial h(x)/\partial x_{II} \, \gamma_{II} \neq 0 \qquad (3.c)$$

Condition (3.c) provides a simple way to identify partitioned nonlinear systems which do not satisfy the disturbance matching condition. For a system which does satisfy condition (3.c), the disturbance cannot be decoupled using the input-output feedback linearization approach, even if the disturbance can be measured.

Nonlinear system partitioning. For the system in Eq (3), if $\exists\, x_c$, $x_c \in x_I$ such that

$$\frac{\partial f_j(x)}{\partial x_c} \neq 0, \quad f_j(x) \in f_{II}(x) \qquad (4.c)$$

$$\frac{\partial h}{\partial x_{II}} \left(\frac{\partial f_{II}(x)}{\partial x_c} + \frac{\partial \gamma_{II}(x)}{\partial x_c} \right) \neq 0 \qquad (5.c)$$

Then, the system can be partitioned into two subsystems SysI and SysII, as in Eq (4)

$$\begin{cases} \text{SysI} \begin{cases} \dot{x}_I = f_I(x) + g_I(x)u + \gamma_I(x)d \\ y_c = x_c \end{cases} \\ \text{SysII} \begin{cases} \dot{x}_{II} = f_{II}(x, x_c) + \gamma_{II}(x)d \\ y = h(x) \end{cases} \end{cases} \qquad (4)$$

The output y is not directly dependent on state variables x_I in SysI, however, y is a function of all state variables x_{II} in SysII. A state variable x_c in SysI is selected to serve as a new output of that subsystem. The selection of x_c is arbitrary, as far as it satisfies the conditions in Eq (4.c) and (5.c). However, in order to implement the control law, x_c should be a physically meaningful process variable, such as temperature or flow rate, that can be measured. The x_c should also be selected in such a way that the closed loop response of the u-x_c loop is faster that of the x_c-y loop. The partitioning space gives an extra dimension for control structure selection. A schematic representation of the partitioning of nonlinear systems is shown in Fig. 1.

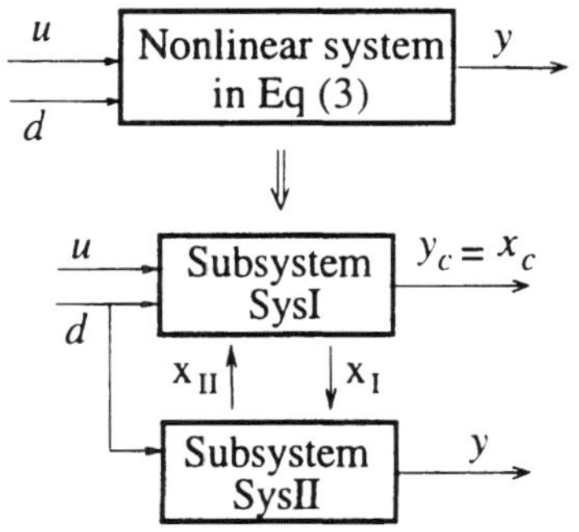

Fig. 1 Partitioning nonlinear systems

Corollary 3: For the system described by Eq (4), the disturbance d can be decoupled for the SysII if the disturbance d can be measured.

Furthermore, if $\partial h(x)/\partial x_{II}\, \gamma_{II} = 0$, the disturbance d can always be decoupled from the output in subsystem SysII even if d cannot be measured. Also, the relative degree of subsystem SysI is reduced by one with respect to both the control input u and disturbance d.

3. CONTROLLER DESIGN

In the linear control theory, the motivations behind cascade control are well understood, such as local rejection of disturbances and suppression of uncertainty/nonlinearities. A classical example is flow control in which the secondary loop is used to compensate for flow rate changes. The uncertainty on the valve inlet and outlet pressure variations, effect of wear on valve, and leakage of the valve are suppressed in the secondary loop. To decouple the design of the secondary loop from the primary requires that the closed loop bandwidth of the secondary be greater than that of the primary. Cascade control structures have been mainly used in control of linear processes or processes with mild nonlinearity around the operating region, using conventional controllers.

For the partitioned nonlinear system described in Eq (4), a cascade control structure can be formed by selecting x_c (an output of SysI) as a new intermediate control input for SysII. The outer loop (x_c-y) tracks the output y using the intermediate input x_c, and the inner loop (u-x_c) achieves servo control of the x_c, using the original control input u. Unlike the exact input-output linearization approach that linearizes the nonlinear system in the whole space, the partitioning-space approach allows the design of a controller for each subsystem separately. Various design approaches can then be applied for each subsystem, depending on the confidence on the process model and characteristics of each subsystem.

For example, let us consider the GLC approach to design a controller for each subsystem. Assuming that the subsystem SysII can be transformed into the control affine form:

$$\dot{x}_{II} = f'(x) + g'(x)x_c$$
$$y = h(x)$$

the GLC procedure developed by Soroush and Kravaris can be applied. Under the following state feedback:

$$x_c = x_c^* = \frac{v_2 - h(x) - \beta L_{f'} h(x)}{\beta L_{g'} h(x)}$$

the subsystem SysII can be transformed into a linear form:

$$\beta \frac{dy}{dt} + y = v_2$$

The linear controller design techniques can be used for the transformed linear system. This results in a control law which calculates the desired manipulated action x^*_c in order to drive the output y so as to follow a defined trajectory. However, if the nonlinear system cannot be transformed into a control affine form, the input-output linearization of general nonlinear processes (Henson and Seborg, 1990) can be considered.

The subsystem SysI is in the control affine form. Therefore, the GLC procedure can be directly applied. The secondary GLC controller functions a servo control to drive x_c as close and as fast to the desired trajectory x^*_c as possible. This results in the cascade control structure shown in Fig. 2 where Ψ_1 and Ψ_2 are state feedback transformations for SysI and SysII, respectively.

Figure 2 provides a general cascade control structure with either linear or nonlinear controllers. By choosing both state-feedback transformations equal to 1 (Ψ_1=1, Ψ_2=1), a conventional cascade structure with linear controllers is obtained. Nonlinear controllers using various design techniques may also be used for control of each subsystem.

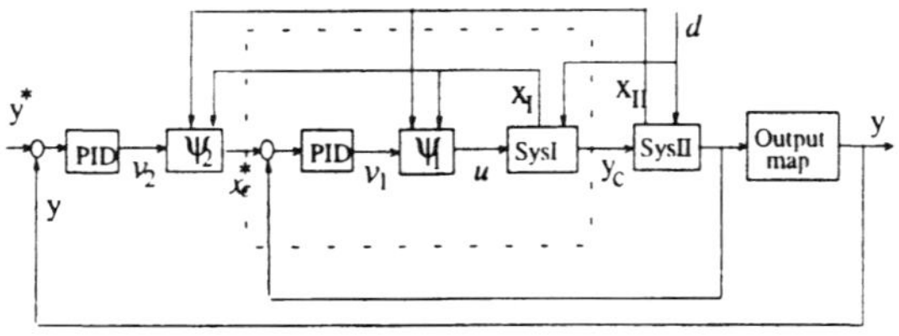

Fig. 2 A cascade control structure for nonlinear process control

4. APPLICATIONS

4.1 Illustrative example

Let consider the level control of two noninteracting tanks in series shown in Fig. 3. The objective is to control the level (x_2) in tank T2 by manipulating the inlet flow (u) of tank T1. The T1 outlet flow is assumed to be proportional to the level (x_1) in T1, while the T2 outlet flow is assumed to be proportional to the square root of x_2. The system can be represented as follows:

$$\begin{aligned} \dot{x}_1 &= -\alpha\gamma_1 x_1 + \alpha u \\ \dot{x}_2 &= \beta\gamma_1 x_1 - \beta\gamma_2\sqrt{x_2} \\ y &= x_2 \end{aligned} \qquad (5)$$

where y is the system output, and α, β, γ_1, γ_2 are constant parameters determined by the cross sections of the tanks and outlet pipes.

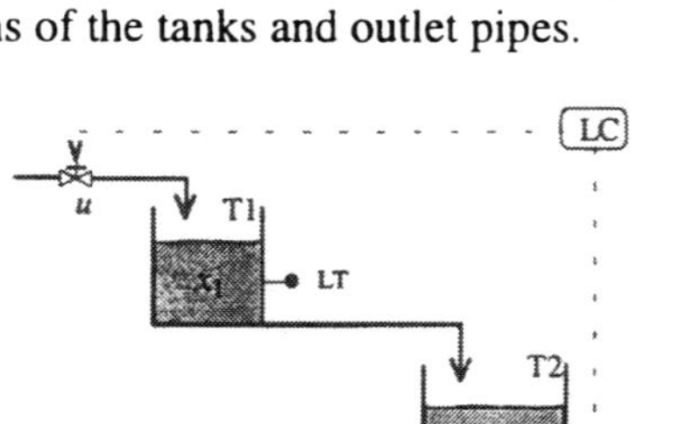

Fig. 3 Two noninteracting tanks in series

The relative degree of the system is r=2 ($L_g h(x)=0$, $L_g L_f h(x)=\alpha\beta\gamma_1\neq 0$) in the full state space $\Re^n$, n=2. Under the following input/output linearizing state feedback for the full system

$$u = \frac{1}{k_2\alpha\beta\gamma_1}(v - x_2 - \beta\gamma_1(k_1 - k_2\alpha\gamma_1)x_1$$
$$+\beta x_2 \frac{1}{\sqrt{x_2}}(k_1 x_2 + 0.5k_2\beta\gamma_1 x_1) - 0.5k_2\beta^2\gamma_2^2 \quad (6)$$

the closed-loop linear system v - y is given by:

$$y + \sum_{i=1}^{r} k_i \frac{d^i y}{dt^i} = v$$

With a step change in the T2 level setpoint from zero (initial T2 level) to 8 meters, and initial T1 level of 10 m, the level responses of the GLC controller are shown in Fig. 4, assuming all model parameters are known exactly (α=0.1, β=0.05, γ_1=0.2625, γ_2=1.05).

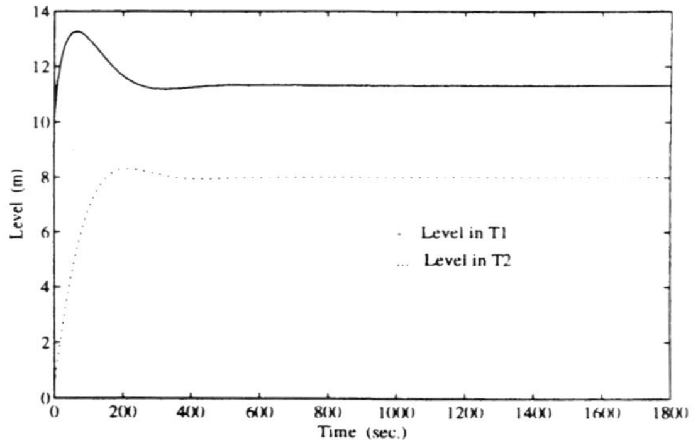

Fig. 4 GLC control with perfect model

However, if the model parameter γ_1 used in the control law Eq (5) is 50% bigger than the true value, under the same control law and controller tuning parameters, the performance of the GLC controller is severely degraded. This is shown in Fig. 5.

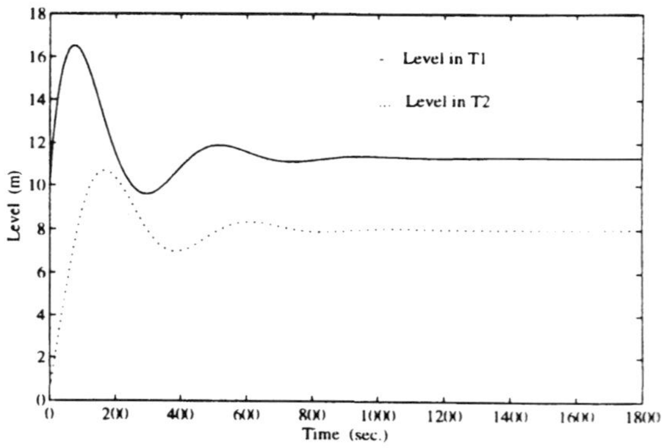

Fig. 5 GLC control with mismatch on a model parameter

The simulation results demonstrate that the performance of the GLC controller in the given example depends on the quality of the model. Model mismatch leads to degraded control performance. As a consequence of the poor control, T1 must be designed with a big safety margin to avoid overfilling the tank which, in turn, increases its cost.

Considering the partitioning space approach, the system in Eq (5) can be partitioned into two subsystems by selecting the level in T1 as an intermediate control input (in fact, it is the only possible variable that can be used). The partitioned system model is shown in Eq (7).

$$\begin{cases} \text{SysI} \begin{cases} \dot{x}_1 = -\alpha\gamma_1 x_1 + \alpha u \\ y_c = x_1 \end{cases} \\ \text{SysII} \begin{cases} \dot{x}_2 = -\beta\gamma_2\sqrt{x_2} + \beta\gamma_1 x_1 \\ y = x_2 \end{cases} \end{cases} \qquad (7)$$

The GLC approach is used for the controller design of subsystem SysII which has a relative degree of one. The GLC control law is given in Eq (8).

$$x_1^* = \frac{1}{k_1\beta\gamma_1}(v - x_2 - \beta\gamma_2\sqrt{x_2}) \qquad (8)$$

Compared to Eq (6), the control law for the partitioned system is much simpler and easier to derive. A PI controller with constants k=2.0, τ_c=0.002 is used for SysI as a secondary controller. The control responses with a perfect model and with the same model parameter mismatch as above, using the GLC controller cascaded to the PI controller are shown in Fig. 6. Compared to Fig. 4 and Fig. 5, the cascaded control structure suppresses the effect of the model mismatch on the control system performance.

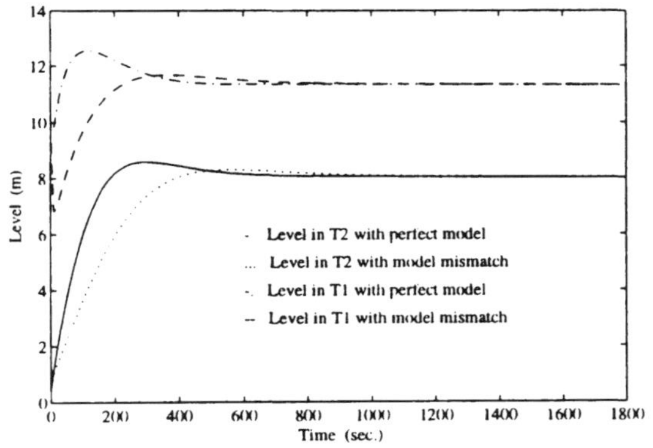

Fig. 6 Level responses under a cascaded GLC and PI control.

4.2 Temperature control of a batch reactor

The temperature control of a 100 l multipurpose jacketed batch reactor is investigated experimentally for a starch liquefaction process catalysed by *α-amylase*. A detailed description and a model of the system are given by Liu and Macchietto (1994). Achieving good control in such a reactor is difficult due to i) nonlinear flow characteristics in the jacket recirculating route; ii) a constraint on the reactor jacket temperature; iii) a wide range of operating conditions, iv) variable control structure.

Conventional control: Figure 7 shows two experimental temperature responses of the reactor using cascaded PI controllers. In the first experiment, the setpoint of the reactor temperature is changed to 65 ^{0}C from initial room temperature (25 ^{0}C). The reactor temperature overshoots by 4 ^{0}C. With the same controllers and tuning parameters as in the first experiment, but the setpoint changed to 85 ^{0}C from an initial reactor temperature of 105 ^{0}C, the reactor temperature overshoots by 17 ^{0}C, and it takes 30 minutes to bring it back to the setpoint.

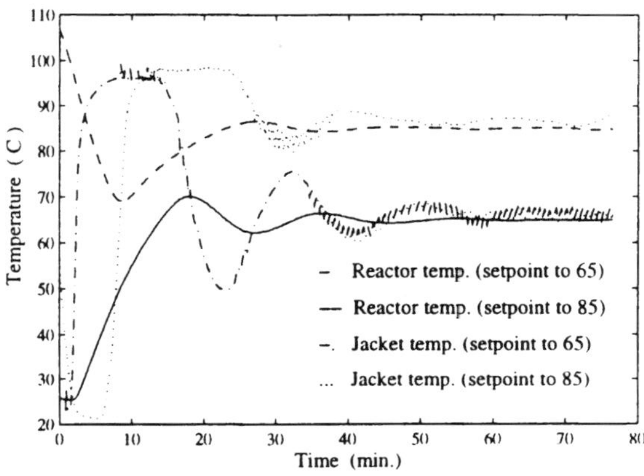

Fig. 7 Reactor temperature responses with cascaded PI controllers - experimental results

These results demonstrate that due to the nonlinearities and variable control structures of the system, PI type controllers perform poorly and must be retuned for different process conditions (e.g. reference profiles, batch sizes, raw materials). The experiments during the plant commissioning have also shown that tuning PI controllers for this type of systems is a difficult and time consuming task.

For this reactor, direct use of a nonlinear design approach (e.g. GLC) is unlikely to be effective since some model parameters in the jacket recirculation route (e.g. heat exchanger coefficient) are subject to a large degree of uncertainty. Even if a suitable state feedback were found, and the resulting direct feedback control structure was able to ensure a good tracking response of the reactor temperature, stability of the jacket control could not be guaranteed.

Partitioning-space approach: The system model is partitioned into two subsystem: SysI representing the dynamics of the reactor, and SysII representing the jacket and its recirculating route. The jacket temperature chosen as intermediate control variable x_c satisfies Eq (4.c) and (5.c). To demonstrate the ability of accommodating various nonlinear control approaches in the controller design, the Generic Model Control (GMC, Lee and Sulivan, 1988) is used to design a nonlinear controller for the primary loop, which is cascaded to a conventional PI controller. All model uncertainties are confined to the secondary loop. For the same setpoint changes of Fig. 7, the new reactor and jacket temperature responses are shown in Fig. 8. In both experiments, the cascade controllers with a single set of tuning parameters performed far better. More detailed experimental results with various process conditions (reference profiles, model mismatch, measurement errors, batch sizes) and all controller parameters are given elsewhere (Liu and Macchietto, 1995). The experimental work has shown that the resulting nonlinear controllers are easy to tune, and that no retuning of controllers is required in order to cover a very wide range of operating conditions.

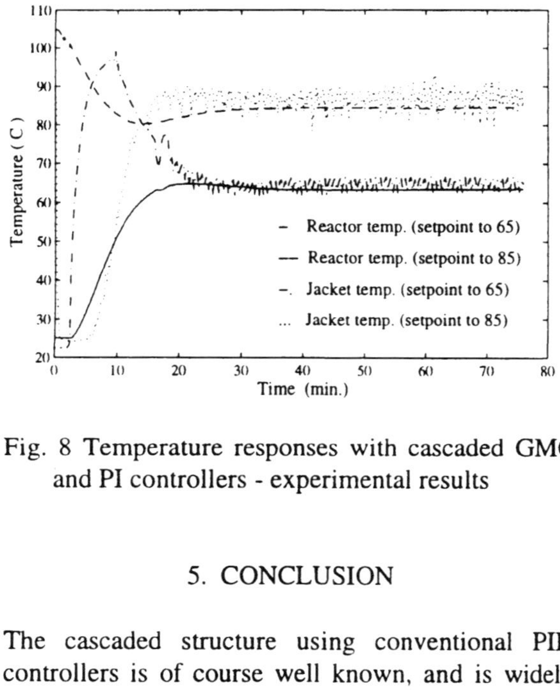

Fig. 8 Temperature responses with cascaded GMC and PI controllers - experimental results

5. CONCLUSION

The cascaded structure using conventional PID controllers is of course well known, and is widely used in the industries. This paper provides theoretical justification to the decomposition used and extends its support to nonlinear controllers. Two criteria have also been presented which permit the easy screening of candidate variables as intermediate outputs.

The partitioning-space approach for nonlinear processes was presented and demonstrated by two examples. By partitioning the nonlinear systems into two subsystems, a cascade control structure is obtained which can accommodate various controller design techniques. The resulting cascade control structure combined with nonlinear model based controllers offer better robustness and disturbance decoupling properties compared to direct feedback. Unlike most model based control techniques that require a good quality model of the entire system, the approach allows the controller design for each subsystem to be carried out individually, permitting the utilisation of nonlinear design techniques only in the well modelled subsystems. This extends the applicability of model based nonlinear techniques to systems with model mismatch in some parameters. The partitioning approach presented also reduces the relative order of the subsystems, resulting in simple control laws.

Acknowledgements: This research was funded by APV Baker and EPSRC/BBSRC.

5. REFERENCES

Dore, S. (1993), Application of Geometric Nonlinear Control in the Process Industries - A Case Study. PhD Thesis, Imperial College of Science, Technology and Medicine, University of London.

Henson, M. H. and D. E. Seborg (1991), Critique of exact linearization strategies for process control. *J. Proc. Cont.* Vol. 1. May, 122-139.

Isidori, A. (1989), *Nonlinear Control Systems: An Introduction.* Springer-Verlag, New York.

Kravaris, C. and C. B. Chung (1987), Nonlinear state feedback synthesis by global input/output linearization, *AIChE Journal*, Vol. 33, No. 4, 592-603.

Lee, P. L. and G. R. Sullivan (1988), Generic Model Control, *Comput. Chem. Engng.*, Vol. 12, No. 6, 573-589.

Liu, Z. H. and S. Macchietto (1994), Model based control of a multipurpose batch reactor with a variable control structure. Proceedings of the *ACoFoP III* Symposium, (J. Bimbenet, E. Dumoulin and G. Trystram, Eds), Elsevier, Paris, France.

Liu, Z. H. and S. Macchietto (1995), Model based control of a multipurpose batch reactor - An experimental study, submitted to *ESCAPE5*, Bled, Slovenia, June 11-14.

LIMITATIONS TO INPUT-OUTPUT ANALYSIS OF CASCADE CONTROL OF UNSTABLE CHEMICAL REACTORS

LOUIS P. RUSSO* and B. WAYNE BEQUETTE*

*Rensselaer Polytechnic Institute, Howard P. Isermann Department of Chemical Engineering, Troy, NY 12180-3590, USA. Email: bequeb@rpi.edu

Abstract. In this paper we show that the traditional methods of analyzing cascade control (based on series or parallel input/output relationships) should not be used when the primary and secondary processes are coupled, and the process is open-loop unstable. The connection between cascade control and traditional state feedback control is studied. In addition, we develop a cascade control system methodology which incorporates a specified inner-loop controller in the outer-loop controller design. This cascade control design procedure is compared to a conventional cascade control design procedure and to a noncascade control strategy.

Key Words. Cascade control; chemical reactors; unstable processes; state-space analysis; input/output analysis

1. INTRODUCTION

Input/output (transfer function) analysis is often used to design feedback control systems for chemical processes. Cascade control systems are usually designed by first tuning a slave control loop based on the transfer function model of the secondary process. If the slave loop can be tuned "fast" enough, then the master loop can be tuned based on the primary process transfer function, independent of the slave loop. If the bandwidth of the slave loop is close to that of the primary loop, then the loops must be iteratively tuned, until acceptable performance of the entire closed-loop process is obtained.

Implicit in the typical controller design process is the assumption that the secondary and primary processes are decoupled. This is illustrated by the typical series cascade control block diagram, shown in Figure 1. The output of the secondary process is the input to the primary process in a series cascade control block diagram. Similarly, a parallel cascade control structure (in which the manipulated input has a direct effect on both of the outputs) is shown in Figure 2.

The motivating problem is cascade control of an exothermic continuous stirred tank reactor (CSTR) operated at an open-loop unstable operating point. Although many new nonlinear control techniques have been developed (Bequette, 1991; McLellan et al., 1990; Kravaris and Kantor, 1990), we use linear control for two reasons:

- Linear feedback and cascade controllers are still the dominant strategies used in industry.
- The linear techniques are well-known and easy to use for comparison purposes.

2. CSTR MODELING EQUATIONS

The standard two-state CSTR model (Uppal et al., 1974) describing an exothermic diabatic irreversible first-order reaction ($A \rightarrow B$) is a set of two nonlinear ordinary differential equations obtained from dynamic material and energy balances (with the assumptions of constant volume, perfect mixing, negligible cooling jacket dynamics, and constant physical parameters).

$$\frac{dC_a}{dt} = \frac{Q}{V}(C_{af} - C_a) - k_0 exp\left(\frac{-E_a}{RT}\right) C_a \quad (1)$$

$$\frac{dT}{dt} = \frac{Q}{V}(T_f - T) - \frac{UA}{V\rho C_p}(T - T_c) + \left(\frac{-\Delta H}{\rho C_p}\right) k_0 exp\left(\frac{-E_a}{RT}\right) C_a \quad (2)$$

where C_a and T are the concentration of component A and the temperature in the reactor, respectively. An additional energy balance around the cooling jacket assuming perfect mixing yields

$$\frac{dT_c}{dt} = \frac{Q_c}{V_c}(T_{cf} - T_c) + \frac{UA}{V_c \rho_c C_{pc}}(T - T_c) \quad (3)$$

where T_c is the cooling jacket temperature. Equations 1 - 3 can be written in dimensionless form

$$\frac{dx_1}{d\tau} = q(x_{1f} - x_1) - \phi x_1 \kappa(x_2) \quad (4)$$

$$\frac{dx_2}{d\tau} = q(x_{2f} - x_2) - \delta(x_2 - x_3) + \beta\phi x_1 \kappa(x_2) \quad (5)$$

$$\frac{dx_3}{d\tau} = \delta_1[q_c(x_{3f} - x_3) + \delta\delta_2(x_2 - x_3)] \quad (6)$$

where x_1, x_2, x_3, and q_c are the dimensionless concentration, reactor temperature, cooling jacket temperature, and cooling jacket flowrate, respectively. These dimensionless variables and parameters (for example: ϕ, q, β, etc.) are defined in Table 1. Representative values of the parameters can be found in Russo and Bequette (1995a, 1995b). Case 1 is open-loop stable over the entire region of operation for the two-state CSTR model but open-loop unstable in a certain operating region for the three-state CSTR model, while Case 2 exhibits ignition/extinction behavior for the two and three-state models.

It is well-known that the exponential relationship of reaction rate with respect to reactor temperature is one of the major nonlinearities of the CSTR. Russo and Bequette (1995a, 1995b) studied both the steady-state nonlinearities (output and input multiplicities, infeasible operation regions) and the dynamic nonlinearities (Hopf bifurcation behavior) of this three-state CSTR model. Sistu and Bequette (1995) determined conditions under which input multiplicity behavior results in the open-loop system having unstable zero dynamics (nonminimum-phase behavior).

3. CONTROL SYSTEM DESIGN

A common control configuration in chemical processes is cascade control. In a cascade control configuration we have one manipulated variable (in this case the dimensionless cooling jacket flowrate q_c) and more than one measurement (the dimensionless reactor temperature x_2 and dimensionless cooling jacket temperature x_3). Cascade control can also be thought of as multiple state feedback. The major benefit of cascade control is that action is taken to reject inner-loop disturbances before they effect the outer-loop. Therefore, the inner-loop of the cascade control strategy is generally tuned very tightly in order to obtain a fast response. In addition, intuitively we expect the performance of a cascade control system to be superior (generally) to a noncascade control system, since more system information (outputs) is being fed back to the control system.

3.1. *Transfer Function Models*

The transfer functions between the deviation variables $x_2'(s)$, $x_3'(s)$, and $q_c'(s)$ are:

$$\begin{aligned} x_2'(s) &= g(s)q_c'(s) \\ &= k\frac{(c_1 s + 1)}{(a_3 s^3 + a_2 s^2 + a_1 s + 1)}q_c'(s) \quad (7) \\ x_2'(s) &= g_1(s)x_3'(s) \\ &= k_1\frac{(c_1 s + 1)}{(b_2 s^2 + b_1 s + 1)}x_3'(s) \quad (8) \\ x_3'(s) &= g_2(s)q_c'(s) \\ &= k_2\frac{(b_2 s^2 + b_1 s + 1)}{(a_3 s^3 + a_2 s^2 + a_1 s + 1)}q_c'(s) \quad (9) \end{aligned}$$

where the gains and coefficients of the transfer functions are defined in Table 2 (the $'$ indicates deviation variable). We can view these equations in block diagram form, as shown in Figure 3.

The poles of the two-state model relationship (relating $x_2'(s)$ to $x_3'(s)$), $g_1(s)$, are the zeros of the $x_3'(s)$ - $q_c'(s)$ relationship in the three-state model, $g_2(s)$. When $g_1(s)$ is unstable at a particular operating point, then $g_2(s)$ will have right-half-plane zeros. We would think that the performance of the "inner-loop" process would be reduced, since right-half-plane zeros cannot be "inverted" for "perfect" control. We find that this is not the case, because the real cascade system is actually a multiple state feedback control. Since the real controller measures two outputs and manipulates a single input, there are no problems with the secondary process right-half-plane-zeros.

3.2. *Series and Parallel Cascade Control*

The two most typical cascade control configurations are the series and parallel structures, which were shown in figures 1 and 2. Luyben (1973) stressed the differences between conventional series cascade control and parallel cascade control. The three-state CSTR transfer function models have a parallel cascade control structure, since the manipulated variable q_c (dimensionless cooling jacket flowrate) has a direct influence on x_2 (dimensionless reactor temperature) and x_3 (dimensionless cooling jacket temperature) through parallel transfer functions.

In this subsection we show that the traditional methods of analyzing cascade control (based on series or parallel input/output relationships) should not be used when the primary and secondary processes are coupled, and the process is open-loop unstable. Consider the series cascade control case, shown in Figure 1. The state-space realization is:

$$A_{series} = \begin{pmatrix} a_{11} & a_{12} & 0 & 0 & 0 \\ a_{21} & a_{22} & a_{23} & 0 & 0 \\ 0 & a_{32} & a_{33} & 0 & 0 \\ 0 & 0 & 0 & a_{11} & a_{12} \\ 0 & 0 & a_{23} & a_{21} & a_{22} \end{pmatrix}$$

$$B_{series} = \begin{pmatrix} 0 \\ 0 \\ b_{31} \\ 0 \\ 0 \end{pmatrix}$$

$$C_{series} = \begin{pmatrix} 0 & 0 & 1 & 0 & 0 \\ 0 & 0 & 0 & 0 & 1 \end{pmatrix}$$

where states $1-3$ represent the secondary process and states $4-5$ represent the primary process transfer functions, respectively. The coefficients of the A_{series} and B_{series} matrices come from the original open-loop A and B matrices. The first ($C_{series_{13}} = 1$) and second outputs ($C_{series25} = 1$) correspond to the dimensionless cooling jacket and reactor temperatures, respectively. The state-space realization of the two open-loop transfer functions in series results in a state-space model with a non-minimal realization, because the primary and secondary processes are actually coupled. This state-space realization is not controllable and when the primary and secondary processes are unstable it is not stabilizable (Russo, 1995).

Consider the case of parallel cascade control (Figure 2) of the jacketed exothermic CSTR. The state-space realization is also not minimal when the parallel form is used. This configuration suffers from internal stability problems (Russo, 1995) when the primary and secondary processes are unstable. Hence, the unstable modes cannot be stabilized and a process so-simulated will be unstable while the actual process may be stable. These non-minimal realizations can be a serious problem if one attempts to use a commercial simulation package such as Simulink. The actual coupled process is best represented by Figure 4, where it is shown clearly that the primary and secondary outputs are coupled through the process model.

3.3. *IMC-based Control System Design*

Internal model control (IMC) (Morari and Zafiriou, 1989) provides a clear framework for controller design. One advantage to the IMC procedure is that it simplifies controller tuning because a single tuning parameter, λ, is used as opposed to the three parameters (k_c, τ_I, τ_D) for a PID controller. Linear controllers are often not valid over a wide range of operating regions due to process nonlinearities. However, since we are concerned here with operation in a small region where the stability characteristics do not change, linear controllers generally provide acceptable performance.

It is well known that for open-loop unstable responses, the IMC controller must be implemented in conventional feedback form since the IMC form is internally unstable (Morari and Zafiriou, 1989).

$$g_c(s) = \frac{q(s)}{1 - \tilde{g}(s)q(s)} \tag{10}$$

Here $g_c(s)$ and $\tilde{g}(s)$ are the feedback controller and process model transfer functions, while $q(s)$ is the IMC controller. The IMC controller is:

$$q(s) = \tilde{g}_{-}^{-1}(s)f(s) \tag{11}$$

where $\tilde{g}_{-}^{-1}(s)$ is the invertible part of the process transfer function and $f(s)$ is a filter transfer function which is added to make the IMC controller proper. Rivera et al. (1986) demonstrate that for a large number of single input single output (SISO) linear systems the IMC design procedure yields PID-type controllers. There are many possible choices for the IMC filter structure; one possible structure for open-loop unstable systems is

$$f(s) = \frac{e_k s^k + e_{k-1}s^{k-1} + \ldots + e_1 s + e_0}{(\lambda s + 1)^n} \tag{12}$$

Here k is the number of distinct poles in the right-half-plane (RHP), n is chosen to make the IMC controller $q(s)$ proper, and λ is the filter time constant. The IMC filter is chosen such that its steady-state gain is unity; for open-loop unstable systems, the filter must also be unity at the k unstable poles (Morari and Zafiriou, 1989)

$$f(s) = 1 \quad \text{at} \quad s = p_i \tag{13}$$

where p_i is the ith unstable pole of $g(s)$. Rotstein and Lewin (1991) showed that the IMC design procedure for some first and second-order unstable transfer functions (without numerator dynamics) results in PID feedback controllers. The choice of the IMC filter $f(s)$ and subsequently controller structure for the inner and outer-loops of the cascade control scheme depends on the number of right-half-plane (RHP) zeros and poles.

3.4. *Cascade Control System Design*

In subsection 3.2 we saw that traditional methods of analyzing cascade control (based on series and parallel input-output relationships) should not be used when the primary and secondary processes are coupled, and the process is open-loop unstable. In this case input-output relationships do not adequately describe the process dynamics, hence we need to incorporate the full state-space model in our control system design procedure.

Let us consider the ubiquitous case when a PI/P controller structure is used for the outer and inner loops, respectively. It is well-known that integral action in the outer-loop can be attained by appending an additional state variable to the model:

$$\dot{x}_4 = -x_2 \tag{14}$$

The state feedback controller is given by:

$$u = -Kx \tag{15}$$

where the feedback gain row vector is:

$$K = \begin{bmatrix} 0 & k_{c2}k_{c1} & k_{c2} & \frac{k_{c2}k_{c1}}{\tau_{I_1}} \end{bmatrix} \tag{16}$$

As one can see, cascade control can be thought of as partial state feedback. Note that the first feedback gain is constrained to be zero with this controller structure (because we are assuming that the 1st state can not be measured or reconstructed with an observer). Hence arbitrary relocation of the "closed-loop" poles is not possible. One potential problem with this type of design procedure is that a stabilizing feedback gain (K) may not exist, i.e. the constrained controller structure may not in general be able to stabilize the process. Understanding the effect of controller structure is the focus of current work. In some industrial applications the control structure is limited to P, PI, or PID.

In order to develop the controller design procedure, we use the following state-space model, which assumes proportional control on the inner-loop.

$$\dot{x} = \bar{A}x + \bar{B}x_{3sp} \tag{17}$$

$$y = \bar{C}x \tag{18}$$

where $\bar{A}$, $\bar{B}$, $\bar{C}$ are given by:

$$\bar{A} = A - \begin{bmatrix} 0 & \dots & 0 & \bar{B} \end{bmatrix} \tag{19}$$

$$\bar{B} = Bk_{c2} \tag{20}$$

$$\bar{C} = \begin{bmatrix} 0 & 1 & 0 \end{bmatrix} \tag{21}$$

where k_{c2} is the proportional gain on the "inner-loop" controller. The design transfer function is:

$$\begin{aligned} y(s) &= \bar{C}\left(sI - \bar{A}\right)^{-1}\bar{B}\,x_{3sp}(s) \\ &= \bar{g}(s)\,x_{3sp}(s) \end{aligned} \tag{22}$$

The IMC-design procedure can be used on $\bar{g}(s)$, which in our case turns out to be minimum-phase and is of the following form:

$$\bar{g}(s) = \bar{k}\,\frac{(\tau_n s + 1)}{(\tau_1 s + 1)(\tau_2 s + 1)(-\tau_u s + 1)} \tag{23}$$

For a perfect linear model and process, the closed-loop response will be given by:

$$x_2(s) = \frac{\theta s + 1}{(\lambda s + 1)^3}x_{2sp}(s) \tag{24}$$

where θ is chosen to satisfy the IMC filter $f(s)$ unity gain condition at the unstable pole.

3.5. *Control Case Studies*

The IMC-based feedback cascade controller is obtained using equation 22 with equations 10 and 12. The resulting "outer-loop" controller is a PI with a double lead-lag. Generally, one of the poles of this controller is fast hence a reduced-order PI with a lead-lag controller can be used with very little performance degradation. We compare the robustness of standard noncascade and cascade feedback control with our new cascade design procedure, applied to the nonlinear process. The "inner-loop" proportional gain is $k_{c2} = -0.5$ and the IMC filter time constant is $\lambda = 0.2$ for both sets of simulations.

The first case study considers the influence of the cascade control system design. The reactor parameters are Case 2 conditions (Russo and Bequette; 1995a, 1995b); the nominal setpoint is $x_2 = 2.7$ (which is open-loop unstable). A setpoint filter ($\tau_F = 0.45 \approx 2\lambda$) is used to remove the overshoot which results from the ISE-optimal IMC design procedure. Figures 5 and 6 demonstrate that our cascade design procedure is superior to the normal cascade control design where the inner-loop controller design is not incorporated in the "outer-loop" design.

The second case study deals with the effect of an unmeasured disturbance on the performance of cascade and noncascade controllers. The reactor parameters are Case 1 conditions (Russo and Bequette; 1995a, 1995b); the nominal setpoint is $x_2 = 2.3$ (which is open-loop unstable). An unmeasured disturbance in the dimensionless cooling jacket feed temperature (x_{3f}) occurs at $\tau = 1$ (x_{3f} changes from -1 to -0.2). Figures 7 and 8 demonstrate that the cascade control scheme does a better job of rejecting the unmeasured disturbance in the dimensionless cooling jacket feed temperature. An interesting point is that if we had analyzed a two-state CSTR model (neglecting cooling jacket dynamics) one would think that it would be possible to arbitrarily detune the system, because Case 1 is open-loop stable for the two-state CSTR model.

4. CONCLUSION

We have shown in this work that traditional methods of analyzing cascade control should not be used when the primary and secondary processes are coupled (through the state-space model) and the process is open-loop unstable. The parallels between traditional full state feedback control and cascade control were demonstrated. The effectiveness of the new cascade control procedure over conventional cascade and noncascade control was studied.

5. REFERENCES

Bequette, B.W. (1991). Nonlinear control of chemical processes: A review. *Ind. Eng. Chem. Res.*, **30**, 1391-1413.

Kravaris, C. and J.C. Kantor (1990). Geometric methods for nonlinear process control. 1-2. *Ind. Eng. Chem. Res.*, **29**, 2295-2323.

Luyben, W.L (1973). Parallel cascade control. *Ind. Eng. Chem. Fund.*, **12**, 463-467.

McLellan, P.J., T.J. Harris, and D.W. Bacon (1990). Error trajectory descriptions of nonlinear controller designs. *Chem. Eng. Sci*, **45**, 3017-3034.

Morari, M. and E. Zafiriou (1989). *Robust process control.* Prentice Hall, Englewood Cliffs, NJ.

Rivera, D.E., M. Morari, and S. Skogestad (1986). Internal model control. 4. PID Controller design. *Ind. Eng. Chem. Proc. Des. Dev.*, **25**, 252-265.

Rotstein, G.E. and D.R. Lewin (1991). Simple PI and PID tuning for open-loop unstable systems. *Ind. Eng. Chem. Res.*, **30**, 1864-1869.

Russo, L.P. and B.W. Bequette (1995a). Impact of process design on the multiplicity behavior of a jacketed exothermic CSTR. *AIChE J.*, in press.

Russo, L.P. and B.W. Bequette (1995b). Effect of process design on the open-loop behavior of a jacketed exothermic CSTR. *Comp. Chem. Eng*, in press.

Russo, L.P. (1995). Ph.D. Thesis. Rensselaer Polytechnic Institute, Troy, NY, in progress.

Sistu, P.B. and B.W. Bequette (1995). Model predictive control of processes with input multiplicities. *Chem. Eng. Sci.*, in press.

Uppal, A., W.H. Ray, and A.B. Poore (1974). On the dynamic behavior of continuous stirred tank reactors. *Chem. Eng. Sci.*, **29**, 967-985.

APPENDIX

Table 1. Dimensionless variables and parameters for the three-state CSTR model.

x_1	=	$\frac{C_a}{C_{af0}}$	x_2	=	$\frac{T-T_{f0}}{T_{f0}}\gamma$
x_3	=	$\frac{T_c-T_{f0}}{T_{f0}}\gamma$	q_c	=	$\frac{Q_c}{Q_0}$
γ	=	$\frac{E_a}{RT_{f0}}$	$\kappa(x_2)$	=	$exp(\frac{x_2}{1+\frac{x_2}{\gamma}})$
β	=	$\frac{(-\Delta H)C_{af0}}{\rho C_p T_{f0}}\gamma$	δ	=	$\frac{UA}{\rho C_p Q_0}$
ϕ	=	$\frac{V}{Q_0}k_0 e^{-\gamma}$	q	=	$\frac{Q}{Q_0}$
τ	=	$\frac{Q_0}{V}t$	δ_1	=	$\frac{V}{V_c}$
δ_2	=	$\frac{\rho C_p}{\rho_c C_{pc}}$	x_{1f}	=	$\frac{C_{af}}{C_{af0}}$
x_{2f}	=	$\frac{T_f-T_{f0}}{T_{f0}}\gamma$	x_{3f}	=	$\frac{T_{cf}-T_{f0}}{T_{f0}}\gamma$

Table 2. Transfer function models.

The transfer functions $g(s)$, $g_1(s)$, and $g_2(s)$ are:

$$g(s) = k\frac{(c_1 s+1)}{(a_3 s^3+a_2 s^2+a_1 s+1)}$$

$$g_1(s) = k_1\frac{(c_1 s+1)}{(b_2 s^2+b_1 s+1)}$$

$$g_2(s) = k_2\frac{(b_2 s^2+b_1 s+1)}{(a_3 s^3+a_2 s^2+a_1 s+1)}$$

where k, k_1, and k_2 and c_1 - a_3 are:

$$k = \delta\delta_1(x_{3f} - x_{3s})\frac{C_1}{A_3}$$
$$k_1 = \delta\frac{C_1}{A_3}$$
$$k_2 = \delta_1(x_{3f} - x_{3s})\frac{B_2}{A_3}$$

c_1	=	$\frac{1}{C_1}$	b_1	=	$\frac{B_1}{B_2}$	b_2	=	$\frac{1}{B_2}$
a_1	=	$\frac{A_2}{A_3}$	a_2	=	$\frac{A_1}{A_3}$	a_3	=	$\frac{1}{A_3}$

C_1 - A_3 are defined in terms of the CSTR parameters.

$$B_1 = \frac{qx_{1f}}{x_{1s}} + (q+\delta) - \beta q\frac{(x_{1f}-x_{1s})}{(1+\frac{x_{2s}}{\gamma})^2}$$

$$B_2 = \frac{(q+\delta)qx_{1f}}{x_{1s}} - \beta q^2\frac{(x_{1f}-x_{1s})}{(1+\frac{x_{2s}}{\gamma})^2}$$

$$A_1 = B_1 + \delta_1(\delta\delta_2 + q_{cs})$$

$$A_2 = B_2 + \delta_1[(\delta\delta_2 + q_{cs})B_1 - \delta^2\delta_2]$$

$$A_3 = \delta_1[(\delta\delta_2 + q_{cs})B_2 - \frac{\delta^2\delta_2 q x_{1f}}{x_{1s}}]$$

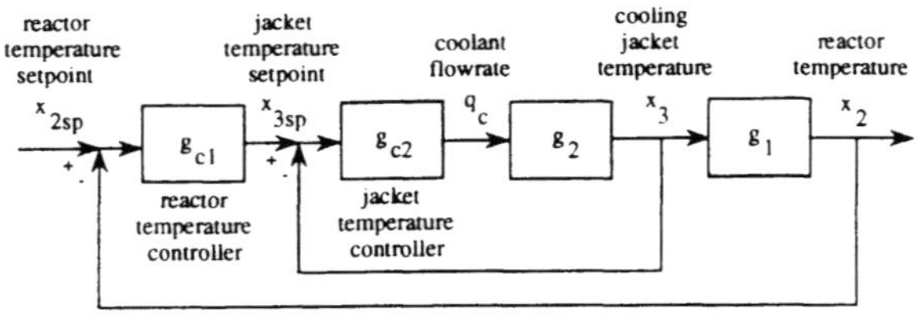

Fig. 1. Series cascade control structure for a CSTR.

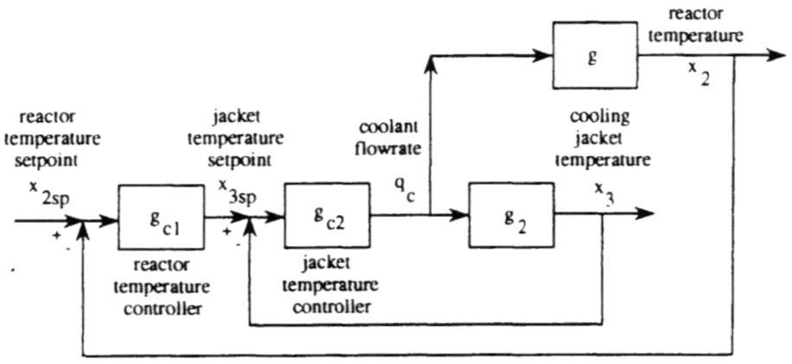

Fig. 2. Parallel cascade control structure for a CSTR.

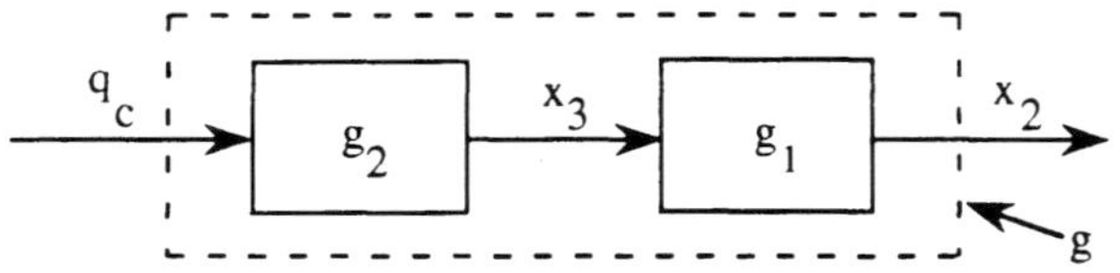

Fig. 3. Relationship between the open-loop transfer functions.

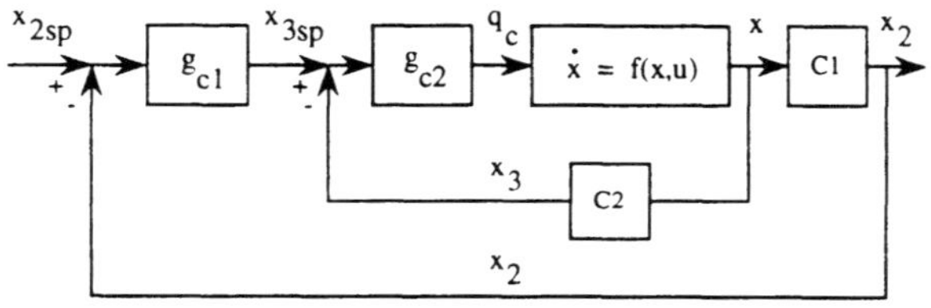

Fig. 4. Coupled Process Formulation using a State Space Representation. The controller design is based on a linear state-space model. Actual implementation is on a nonlinear process.

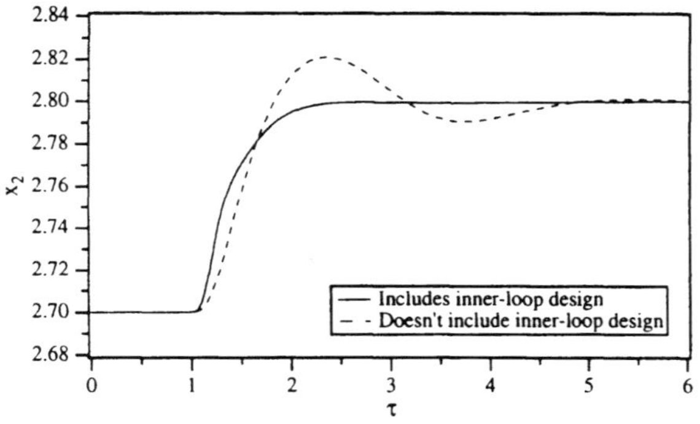

Fig. 5. Comparison of the closed-loop dimensionless reactor temperature performance of the two cascade control strategies.

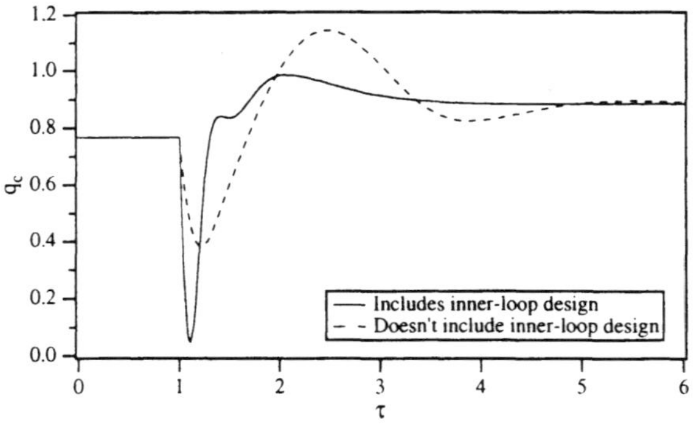

Fig. 6. Comparison of the closed-loop dimensionless cooling jacket flowrate performance of the two cascade control strategies.

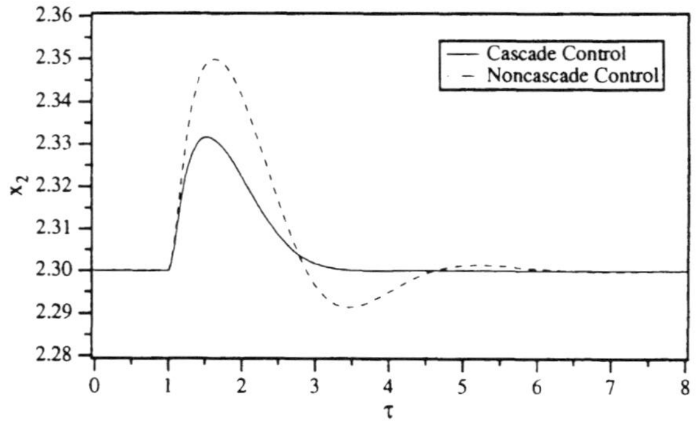

Fig. 7. Comparison of the closed-loop dimensionless reactor temperature performance of the cascade and noncascade strategies to a disturbance in the cooling jacket feed temperature.

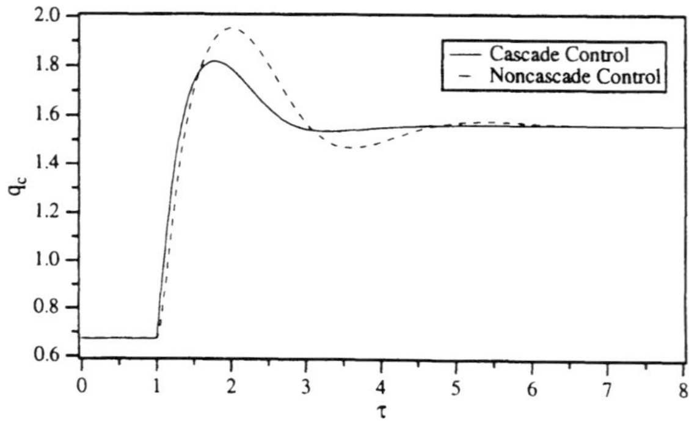

Fig. 8. Comparison of the closed-loop dimensionless cooling jacket flowrate performance of the cascade and noncascade strategies to a disturbance in the cooling jacket feed temperature.

AUTOMATIC CONTROL OF CHEMICAL REACTORS AT UNSTABLE STEADY-STATES

D. Semino* M. Gori

* *Dipartimento di Ingegneria Chimica, Università degli Studi di Pisa, Via Diotisalvi 2, 56126 Pisa, ITALY. E-mail: semino@sssup1.sssup.it*

Abstract. A tuning technique is developed to control unstable processes of unknown dynamics with conventional PID controllers. The tuning is accomplished by making use of the process knowledge which is constructed by closed-loop operation of the process with a relay controller. Performance and robustness tests are made both on linear systems and on nonlinear chemical processes including a styrene polymerization reactor. Limits and extensions of the tuning technique are carefully examined.

Key Words. Process control, Chemical industry, Unstable, Relay control, PID control, Adaptive control.

1. INTRODUCTION

There are a number of cases in which it is appropriate to operate a chemical reactor at an unstable steady-state. A well-known example is the adiabatic reactor in which an exothermic reaction takes place; it may happen that among the three possible steady-states only the middle unstable one has reasonable conversion and temperature. More intricate cases appear in polymerization reactors for which also at constant temperature a wide variety of multiplicities and instabilities may be present. In these cases it may not be easy to have an adequate form of the model in proximity of the steady-state of interest to which available techniques (e.g. Ziegler-Nichols type PID controllers in De Paor and O'Malley, 1989, Internal Model controllers in Morari and Zafiriou, 1989, IMC-like PID controllers in Rotstein and Lewin, 1992) can be applied, given that the model which is needed to describe the physical and chemical phenomena is usually highly nonlinear and mathematically involved and that all kinetic and physical parameters may not be available. Moreover it may happen that, even if stability is guaranteed at the design stage, unforeseen instabilities appear during operation (e.g. as a result of the presence of feed impurities).

It is therefore appropriate to develop a tuning technique which does not require much a-priori knowledge of the process and is able to deal with changes in operating conditions and unmodelled disturbances. To this aim a well-known tuning procedure by Åstrom and Hägglund (1984) has been extended to unstable nonlinear systems. The basics of the tuning procedure have been presented in a previous work (Semino, 1994). Here a deeper insight into how the reconstructable knowledge can be used to tune the controller and into the limits and extensions of the technique is presented together with a comparison for different processes with the performances of other control strategies which require, however, a wider process knowledge.

2. TUNING TECHNIQUE

In this section the basic idea which is used to get information about the process dynamics from closed-loop relay operation (Fig. 1) is firsty recalled; thereafter the tuning technique by which the reconstructed knowledge is used to tune the controller parameters is developed into detail.

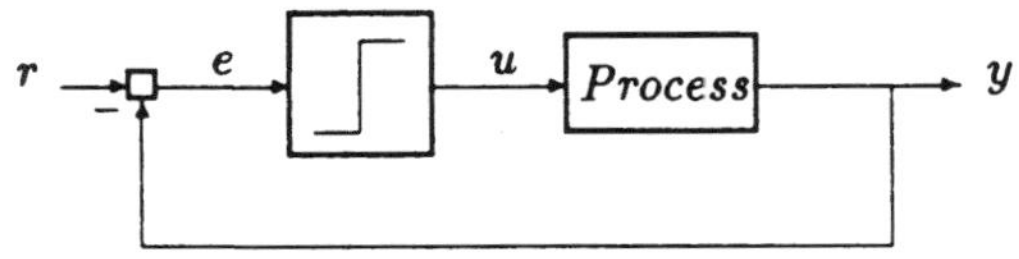

Fig. 1. Feedback loop with a relay.

The behaviour of unstable systems when the feedback loop is closed with a relay controller has been carefully examined in some previous work (Semino, 1994). The application of the describing function analysis has shown that stable limit cycles are present only if an interval of conditional stability exists, and, in case it does, their frequency and amplitude are related to the characteristics of the point correspondent to the maximum gain (point C in Fig. 2) by the following relationships:

$$AR_C = \frac{\pi a}{4d} \qquad \omega_C = \frac{2\pi}{T} \tag{1}$$

where AR_C and ω_C are the amplitude and frequency of point C, a and T are the amplitude and period of the sustained oscillations and d is the relay amplitude.

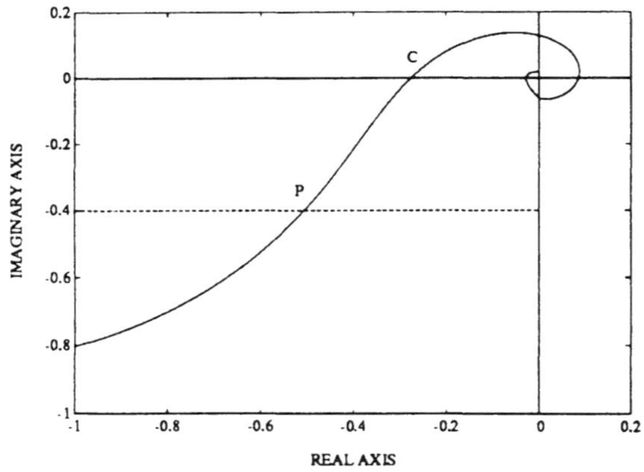

Fig. 2. Describing function analysis for a relay with (– –) and without hysteresis.

Closed-loop operation with a relay with hysteresis provides the characteristics of a point P belonging to the branch of the Nyquist plot in the third quadrant of the complex plane which departs from C (Fig. 2); this point is determined by the intersection of a line parallel to the real axis with y-value $\pi\epsilon/4d$ (ϵ is the relay hysteresis as in Fig. 3), and the Nyquist plot, provided that such an intersection exists. Gain and frequency of point P are given by the following relationships:

$$AR_P = \frac{\pi}{4d}|\sqrt{a^2 - \epsilon^2} + i\epsilon| \qquad \omega_P = \frac{2\pi}{T} \tag{2}$$

where the symbols have the usual meanings.

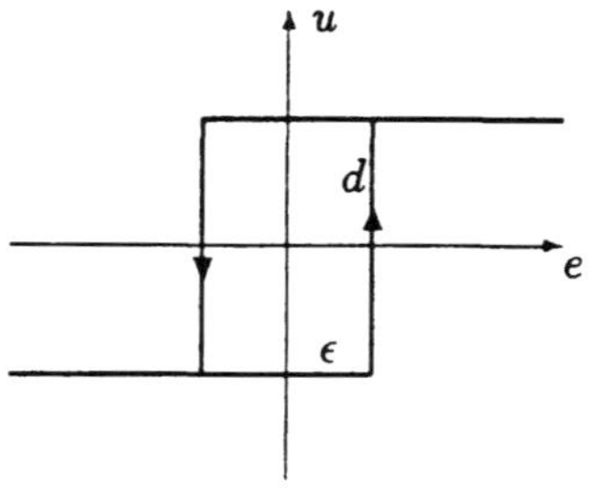

Fig. 3. Characteristic curve for a relay with hysteresis ϵ and amplitude d.

It may be appropriate to have non-zero hysteresis not only to get further knowledge about the process but also to avoid oscillations with very high frequency and small amplitude which occur when point C is very close to the origin in the complex plane.

Few simple tests can therefore be undertaken on the unstable process of unknown dyanmics in order to obtain some points on the Nyquist plot which give information about the interval of conditional stability. How the reconstructed knowledge can be used to choose the controller parameters, is not, however a straightforward matter. Semino (1994) suggested to choose a point among the obtained ones and to move it in the complex plane by a PI action in order to guarantee some phase margin; however both the choice of the point and the one of the phase margin were somewhat arbitrary; moreover an arbitrary phase margin cannot be guaranteed for a generic unstable plant (the closer the plant is to the conditions in which stabilizability is not possible, the smaller is the phase margin that can be obtained).

In the following the tuning matter is developed into detail. Let us consider point P_1 and point P_2, which have the smallest and largest imaginary part among the identified ones (corresponding therefore to the minimum and maximum hysteresis); P_1 coincides with C if the relay with no hysteresis works without trouble. The integral and derivative times of the PID controller are chosen so that the point at frequency ω_{P_1} is not moved from P_1 and the phase of the point at frequency ω_{P_2} changes by a fraction of the angle θ in Fig. 4 through rotation in the clockwise direction (and moves to point P_2').

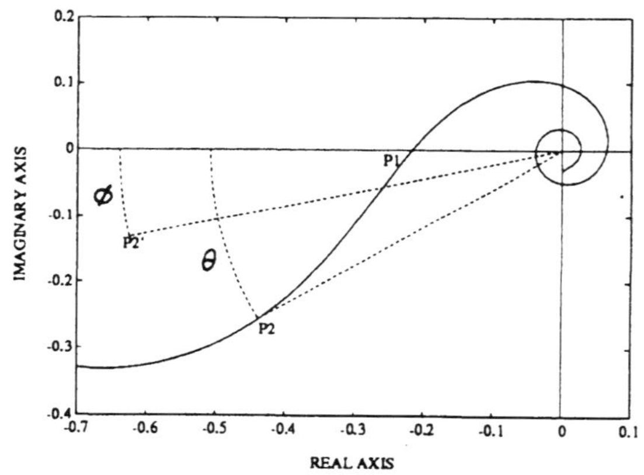

Fig. 4. Grafical representation of the determination of integral and derivative time.

Let us understand the motivations under these choices:

- the presence of the two points still in the third quadrant of the complex plane guarantees the maintenance of an interval of conditional stability still after the introduction of integral and derivative action
- point P_1 close to the real axis gives information about the maximum gain
- as the phase shift is zero at ω_{P_1} a phase lag

is present at lower frequency so that P_2 is to be rotated in the clockwise direction:

  - $\phi = 0$ corresponds to the highest rotation which still guarantees the presence of P_2 in the third quadrant and to the strongest admissible integral and derivative actions
  - $\phi = \theta$ corresponds to no integral and derivative action given that with PID action when the phase shift is zero at two different frequencies, it is zero everywhere

- the absence of phase shif at ω_{P_1} guarantees a contribution to the amplitude of the system which decreases with frequency for $\omega < \omega_{P_1}$ which acts in the sense of widening the interval of conditional stability
- no arbitrary choice is introduced; freedom is still left on the value of the angle ϕ which is chosen according to a criterion given below.

Once the integral and derivative time have been determined, the choice of the controller gain becomes the primary concern. Such a gain is chosen in the interval between the inverse of the real part of P_2' and the inverse of the real part of P_1 in order to be safe about stability. Among the possible gains, one would like to choose the combination of the maximum and minimum ones that guarantees the highest phase margin. This would require however a larger knowledge about the process that is actually available.

To build up a criterion, let us think about the characteristics of the processes that are taken into consideration. Together with the unstable portion of the plant, such processes own some stable part (stable poles, time delays), whose effect on the conditional stability is negative: the higher is the contribution to the frequency behaviour of this stable part, the smaller the interval of conditional stability becomes (Semino, 1994). What is chosen as a reference process is a system with one unstable pole ($1/\tau$) and a time delay (θ) midway in the range of those for which stabilizability is admitted ($\theta = \tau/2$). For this process the value of the coefficient h in the linear combination of the above-mentioned maximum and minimum gains:

$$K_C = hK_{max} + (h-1)K_{min} \tag{3}$$

for which the phase margin is the highest is given as a function of ϕ (Fig. 5). Tests on different linear and nonlinear plants and comparisons with the performances of other tuning techniques (see next section) have shown the validity of such criterion.

The angle ϕ is still left to be chosen: it is clear from the tuning procedure that the aggressiveness of the controller increases as ϕ decreases. Therefore ϕ is chosen according to the robustness which is required for the controlled system. If small errors have been accomplished in the identification

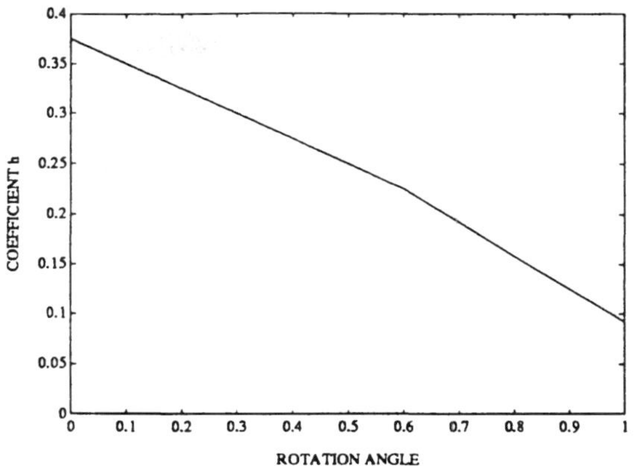

Fig. 5. Coefficient h as a function of ϕ.

phase and no big changes are expected during operation (as the result of strong nonlinearities or time-varying parameers) a value of ϕ close to zero can be chosen; if, on the other hands, robustness is a strong issue, a value of ϕ close to θ is suggested. Large values of ϕ are moreover suggested when strict constraints on the manipulated variable are present.

Let us summarize the tuning procedure for the sake of clarity:

1. few closed-loop tests with a relay with increasing hysteresis are undertaken
2. amplitude and frequency of the oscillations are measured for the smallest and largest hysteresis and they are used to identify two points on the Nyquist curve (P_1 and P_2 in Fig. 2)
3. the integral and derivative time are chosen in order not to move point P_1 and to rotate P_2 of an angle ϕ ($0 < \phi < \theta$ as in Fig. 4); ϕ is chosen according to robustness considerations
4. the controller gain is chosen by combining the inverse of the real parts of point P_1 and P_2' (Fig. 4) according to equation (3) where h is plotted in Fig.5.

3. APPLICATIONS

3.1. *Linear Systems*

As a first example the process $P(s) = e^{-0.3s}/(s-1)$ is considered. The tuning procedure developed in the previous section is used to obtain PID controllers without making use of any knowledge about the process but the sign of its gain. PID parameters for $\phi = 0$ (a), $\phi = 0.4\theta$ (b), $\phi = 0.8\theta$ (c) are listed in Tab. 1. Performances of the three controllers for a step set-point change are compared in Fig. 6 where it comes clear that detuning is accomplished by increasing the angle ϕ.

Table 1 Parameters for the PID controllers

	K_C	τ_I	τ_D
$\phi = 0$ (a)	2.581	2.077	0.026
$\phi = 0.4\theta$ (b)	2.297	3.648	0.0148
$\phi = 0.8\theta$ (c)	1.686	10.972	0.0049

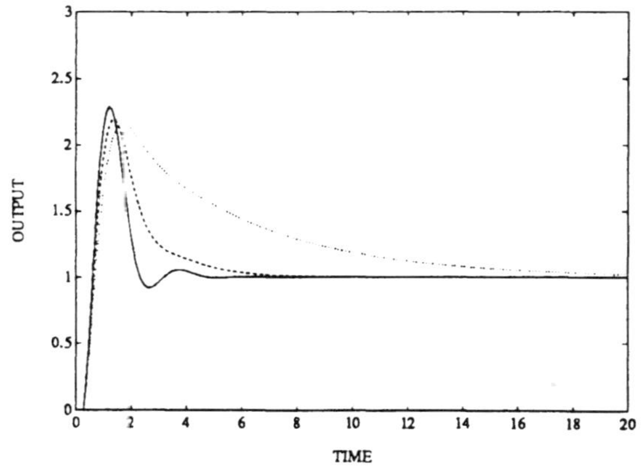

Fig. 6. Performances of three PID controllers. – $\phi = 0$ (a), - - $\phi = 0.4\theta$ (b), ··· $\phi = 0.8\theta$ (c)

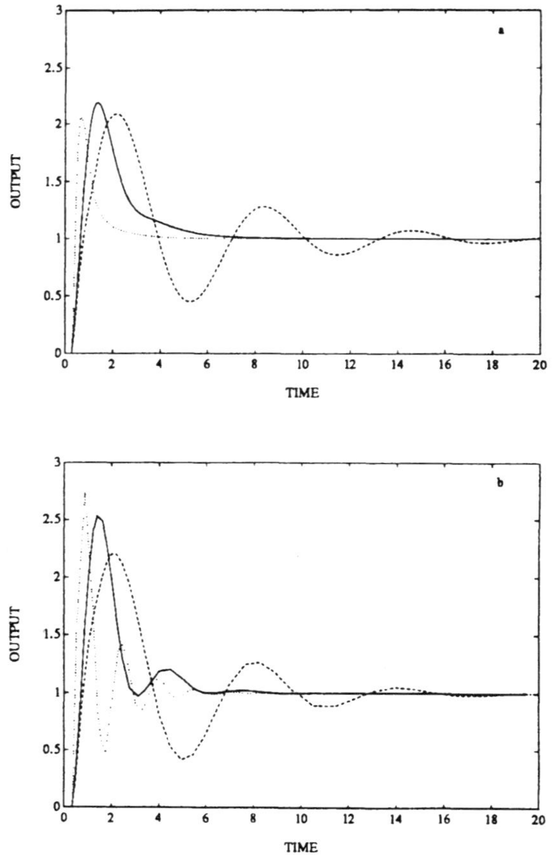

Fig. 7. Comparison of different controllers in the nominal case (a) and in the presence of uncertainty (b). – PID (b), - - De Paor PID, ··· IMC

Two controllers from the literature are taken for comparison: a PID controllers tuned according to De Paor and O'Malley (1989) and an IMC controller in the version for unstale plants (Morari and Zafiriou, 1989). A third order filter with a time constant ($\lambda = 0.15$) designed to guarantee robustness to 20% uncertainty on the time delay is adopted in the IMC controller. It is important to point out that both controllers require for the design the knowledge of the process transfer function.

Performances of these controllers and controller (b) are compared for a set-point change both in the nominal case (Fig. 7a) and in the case when the time delay is 20 % bigger than the nominal one (Fig. 7b). The designed controller outperforms the other conventional one and gives acceptable performances also when compared to the IMC controller.

No other tests on linear processes are reported here for the sake of brevity. However similar results are obtained for different values of the time delay and when stable poles or zeroes are present in the transfer function. As expected, performances tend to deteriorate for processes with a small interval of conditional stability: in this case large overshoots and oscillations can by no means be avoided by making use of conventional controllers.

3.2. *Nonlinear Chemical Reactors*

A chemical reactor where an exothermic first order irreversible reaction takes place refrigerated by a cooling jacket is considered. The material and thermal balances are the following:

$$\frac{dX_A}{dt} = -\frac{q_0 X_A}{V} + k_0 e^{-E/RT}(1 - X_A) \qquad (4)$$

$$\frac{dT}{dt} = \frac{(-\Delta H_r)c_{A0}}{\rho c_P} k_0 e^{-E/RT}(1 - X_A) + \frac{q_0}{V}(T_0 - T) - \frac{UA}{V\rho c_P}(T - T_J) \qquad (5)$$

and the physical, kinetic and operating parameters are listed in Tab. 2.

For this system three steady-states are admissible but only the middle unstable one has reasonable conversion and temperature so that a control system is required to manage to keep conversion and temperature at their steady-state values $X_A = 0.216$, $T = 327K$. Tuning techniques which require a process model would imply the knowledge of all physical and kinetic parameters and a

Table 2 Parameters for the chemical reactor

$V = 0.04m^3$	$\rho = 10^{-3}kg/m^3$
$A = 0.1m^2$	$c_P = 4.184kJ/kgK$
$q_0 = 6 \cdot 10^{-5}m^3/sec$	$k_0 = 4.48 \cdot 10^{-6}sec^{-1}$
$c_{A0} = 3 \cdot 10^{-3}mol/m^3$	$E = 15000cal/mol$
$T_0 = 298K$	$-\Delta H_r = 50000cal/mol$
$T_J = 303K$	
$U = 0.348kW/m^2K$	

local linearization of the system. This would lead to complicated transfer functions for which tuning of conventional controllers is not easily available in the literature unless careful case-specific simplifications are accomplished (Rotstein and Lewin, 1992). The IMC tuning technique (Morari and Zafiriou, 1989) would lead on the other hand to involved controllers of heavy implementation.

Therefore the proposed tuning technique is applied to control reactor conversion with feed flowrate. Due to the strong nonlinearities, which are present in the process, $\phi = 0.9\theta$ is chosen. Closed-loop behaviour of both the controlled and the manipulated variable (on which operating constraints have been imposed) for a step set-point change is shown in Fig. 8. The dynamic behaviour is very good. Similar tests in the case of disturbances have shown the good robustness of the controller as well.

A second chemical reactor for which performances are examined is a solution polymerization reactor which is carefully described in a paper by Hidalgo and Brosilow (1990). The interested reader is asked to refer to this paper for both the system equations and the process parameters. As explicitely stated in the referred paper, the aim of the first level controller for such a system is to stabilize the reactor by controlling its temperature. This is accomplished by a nonlinear model predictive controller which acts both on the monomer flowrate and on the coolant flowrate in a coordinated matter. If the cooling water can maintain the desired process variable trajectory, the monomer flowrate remains at its nominal value. Otherwise, the monomer flowrate changes in order to assist the cooling water in controlling the temperature. In this way, large disturbances and set-point changes are tolerated and the monomer flowrate goes back to its nominal value when the process is at the desired steady-state.

The tuning method developed above is tested for this control problem as well. A primary controller which manipulates monomer flowrate is tuned at first. Subseqently a secondary loop is added in which coolant flowrate is used to control monomer

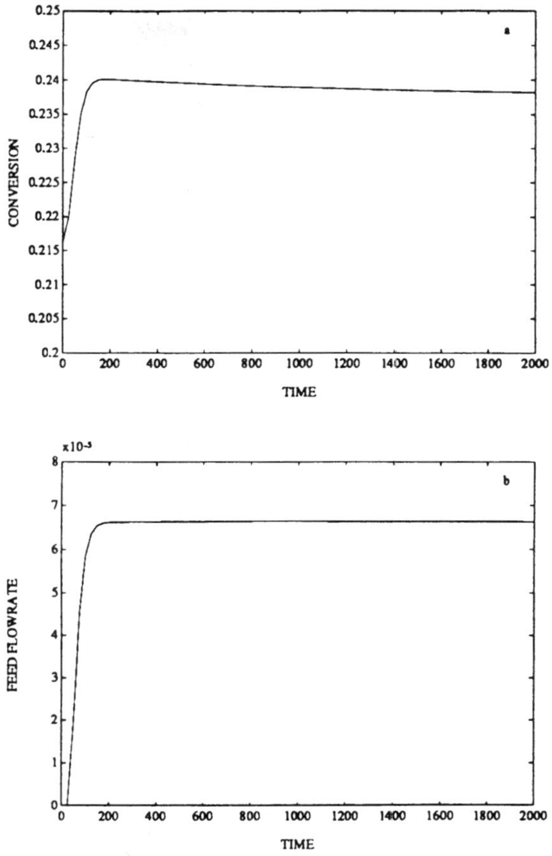

Fig. 8. Set-point change in conversion for the exothermic reactor. (a) conversion, (b) feed flowrate

flowrate, so that a global effect similar to the one of the coordinated controller is obtained. Here again $\phi = 0.9\theta$ for the strong nonlinearities and the strict constraints. Performances for a step set-point change of 10^oC are reported in Fig. 9 for the manipulated and controlled variables. The controller is successful in achieving its aim and performances are not much worse than those of the controller by Hidalgo and Brosilow (1990) which have been obtained by a more involved controller which requires both a nonlinear model and a computationally heavy calculation of the control action.

4. LIMITS AND EXTENSIONS OF THE TUNING PROCEDURE

The first clear limit of the tuning strategy is the impossibility to stabilize a generic unstable process by conventional controllers unless some conditions on the frequency behaviour (related to the process dynamic characteristics) are satisfied. As an example, for the process $P(s) = e^{-\theta s}/(\tau s - 1)$ no stabilizibility by PI controller is achievable if $\theta > \tau$ and by PID controllers if $\theta > 2\tau$. Such a problem is not specific of the developed tuning

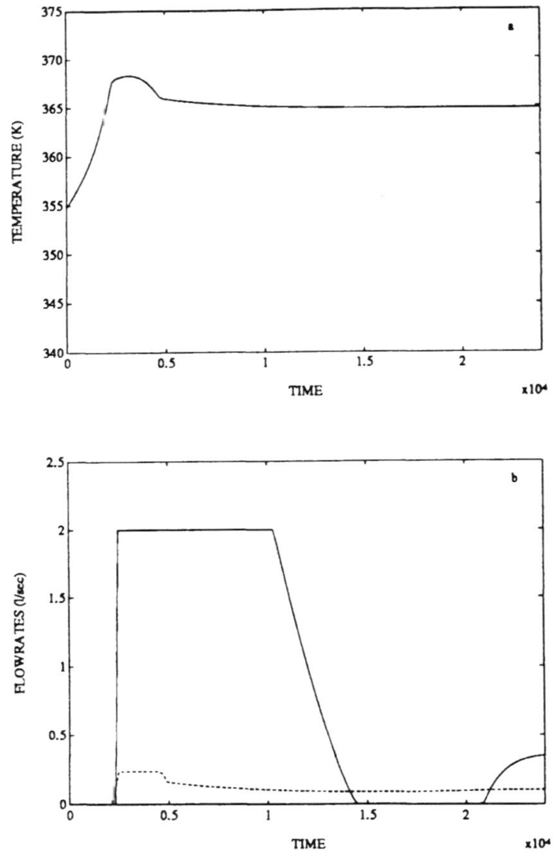

Fig. 9. Set-point change in temperature in the styrene polymerization reactor. (a) temperature, (b) – coolant flowrate, - - monomer flowrate

procedure but concerns the general class of controllers; however, it has to be taken into account in order to explain unsuccessful attempts.

Moreover the use of closed-loop relay tests reduces somewhat more the set of processes for which an interval of conditional stability is detected. For the above mentioned process, an interval of conditional stability by proportional controllers should be identified for $0 \leq \theta < \tau$. Limit cycles are expected to appear in closed-loop relay operation for all the processes satisfying this conditions as the application of the describing function analysis has shown. However this analysis involves a certain number of approximations. A more precise dynamic analysis (accomplished by summing up the responses given by successive steps) shows that sustained oscillations appear only if $0 \leq \theta < \ln 2 \cdot \tau$. Similar limitations, which reduce slightly the class of processes for which the interval of conditional stability is identified, can be obtained for processes described by different transfer functions in an analogous way.

On the other hand, the tuning procedure can be extended to the control of multivariable unstable processes by sequential loop closing in a way analogous to the one developed for stable systems by Loh *et al.* (1993). Improvements in performances can be obtained by iterating the sequential loop closing a few times till convergence in the controller parameters is achieved.

5. CONCLUSION

A tuning procedure for unstable systems which makes use of the process knowledge reconstructed from closed-loop relay tests has been developed. Given that only few tests and few calculations are needed, the controller parameters can be easily updated during operation if required.

Applications both to linear systems and to nonlinear chemical reactors have shown that the achievable performances compare well to those of more advanced controllers which require however a wider process knowledge.

It has been pointed out that the techniqe is applicable only if specific conditions on the process dynamics are respected, which are slightly more restrictive than those imposed by a generic conventional controller. Possible extensions to multivariable systems have been indicated.

6. REFERENCES

Åström, K. J., and T. Hägglund (1984). Automatic Tuning of Simple Regulators with Specifications on Phase and Amplitude Margins. *Automatica*, **20**, 645-651.

De Paor A. M., and M. O'Malley (1989). Controllers of Ziegler-Nichols Type for Unstable Process with Time Delay. *Int. J. Control*, **49**(4), 1273-1284.

Hidalgo P. M., and C. B. Brosilow (1990). Nonlinear Model Predictive Control of Styrene Polymerization at Unstable Operating Points. *Computers Chem. Engng.*, **14**, 481-494.

Loh A. P., C. C. Hang, C. K. Quek, and V. U. Vasnani (1993). Autotuning of Multiloop Proportional Integral Controllers Using Relay Feedback. *Ind. Eng. Chem. Res.*, **32**, 1102-1107.

Rotstein G. E., and D. R. Lewin (1992). Control of an Unstable Batch Chemical Reactor. *Computers Chem. Engng.*, **16**, 27-49.

Morari M., and E. Zafiriou (1989). *Robust Process Control.* Prentice Hall, Englewood Cliffs, New Jersey.

Semino D. (1994) Automatic Tuning of PID Controllers for Unstable Processes. *Proceedings of ADCHEM'94*, Kyoto, Japan, 337-342.

MULTIVARIABLE QUALITY CONTROL OF A CRUDE OIL FRACTIONATOR

M. V. de Oliveira Magalhães[1] and Darci Odloak[2]

1- PETROBRAS, Industrial Department, Rio de Janeiro, Brazil
2- Universidade de São Paulo, Chemical Eng. Department, São Paulo, Brazil

Abstract This paper describes the industrial application of a multivariable predictive controller to a typical crude oil fractionator where jet fuel and diesel fuel are the main specified products. The controller functional specification includes the main targets of the column operation, accounting for 14 controlled variables, 6 manipulated variables and 2 disturbances. The implemented algorithm is a linear DMC, that makes use of a linear programming routine specifically developed to approach the problem of bounded variables, since the system variables are assumed to be controlled by range. The controller was successfully implemented in one of the Petrobrás refineries at Paulinia (Brazil) and some practical results are presented here.

Keywords-Multivariable control, predictive control, distillation columns, model based control, quality control.

INTRODUCTION

The atmospheric fractionator is an important equipment in a crude unit, because of the large feed flow and the amount of energy consumed. The products specifications basically follow the market demands and the economic objective of the crude unit is to keep the products the closest as possible to its quality specifications. To exceed the specifications is less critical than to violate them. Any violation on the product specifications leads to an expensive reprocessing or to degrade the stream to a less valuable pool. The experienced operator keeps a safety margin from the market specifications to prevent any violations during transient conditions. The frequent change of the feed quality and flow is the major incentive for the application of advanced control on crude units. Benefits as the minimization of off-spec products and the energy consumption in the oil heaters can amount to millions of dollars a year.

The application of predictive multivariable controllers to crude fractionators has already been reported in the literature. Muske et al.(1991) describes the application of the IDCOM-M package to a crude tower of the refinery of Sakai in Japan. The proposed control configuration is very sophisticated with the control functions distributed in three different layers. The system structure is very similar to the one considered in this paper. In the control level, the main difference concerns the manipulation of the set-point of the overhead temperature controller instead of the reflux flow as done in this work. This single difference substantially affects the dynamic response of the whole system. The manipulation of the overhead temperature partially decouples the controlled and manipulated variables of the system. As can be seen from the transfer functions given by Muske et al.(1991), in open loop, the yield of a side draw does not practically affect the cut-points of the products above that draw and a triangular structure is obtained. This can be explained, since any change on the flow of the side draws is compensated by the overhead temperature controller that changes the reflux flow and tends to keep the ratio L/V constant. The disadvantage of this control configuration is that the overhead temperature controller can not work at its saturation limit. If for instance the reflux valve is wide open, as can occurs when the unit is operating at its maximum feed flow, the whole set of transfer functions of the system are no longer valid and the multivariable controller may become unstable. In this work the predictive controller manipulates the reflux flow and the overhead temperature is one of the variables of the system that are controlled by range. In particular this variable has to be kept above a minimum value to prevent the condensation of the stripping steam inside the tower. Following this configuration, in open loop, the controlled variables are affected by all the manipulated variables and the triangular structure is lost, but we find a more general approach to manipulate the reflux flow instead of the overhead temperature. Other similar applications of multivariable controllers are reported by Cutler and Finlayson (1988), O'Connor, Grimstad and Mckay (1991) and Hsie (1989).

The Linear Dynamic Matrix Control (LDMC), was developed by Morshedi , Cutler and Skrovanek (1985) as an alternative to the conventional DMC. In the LDMC the control actions are the solution of a linear problem that minimizes the absolute value of a residue function of the error on the controlled variables. The main feature of the LDMC controller is the possibility of rigorously attending the constraints on the manipulated and controlled variables. The usual approach to solve the LP problem is to use the revised simplex algorithm, but for systems with a large number of manipulated variables may turn the solution of the problem prohibitive in terms of computation time. Since most of the controlled and manipulated variables shall be kept inside well defined ranges, we found particularly useful the "bounded variables" approach of the simplex algorithm (Murty, 1976) for the solution of the LP problem.

PROCESS DESCRIPTION

A schematic representation of the atmospheric crude tower is shown in figure 1. The reduced crude from the bottom of the pre-flash tower is heated and partially vaporized in two parallel heaters. The heated feed is then sent to the atmospheric tower where five liquid streams are produced. At the top of the tower heavy naphtha is produced and usually used as a petrochemical feed stock. There are three side draws: jet-fuel, light diesel and heavy diesel, that are sent to three stripping towers where live steam can be injected to control the flash-point of these streams. At the bottom of the crude tower we have the atmospheric resid that is sent to the vacuum section of the crude unit to the recovery of the gasoils. The commercial diesel in Brazil is a mixture of the light diesel and heavy diesel streams and eventually the heavy naphtha is also added to the diesel pool. The studied crude tower has two pumparounds that substantially affect the heat distribution along the tower. The upper pumparound preheats the crude and the lower pumparound exchanges its heat dutyi with the fresh crude and also at the reboiler of the gasoline stabilizer. At the bottom of the crude tower live steam is injected to maximize the recovery of heavy-diesel from the resid stream. To prevent the entrainment of resid to the heavy-diesel section, a minimum liquid flow must be kept in the flash zone of the tower. In the case here studied this liquid flow is guaranteed by a controller of the level the liquid that overflows a dam at the first tray above the flash zone.

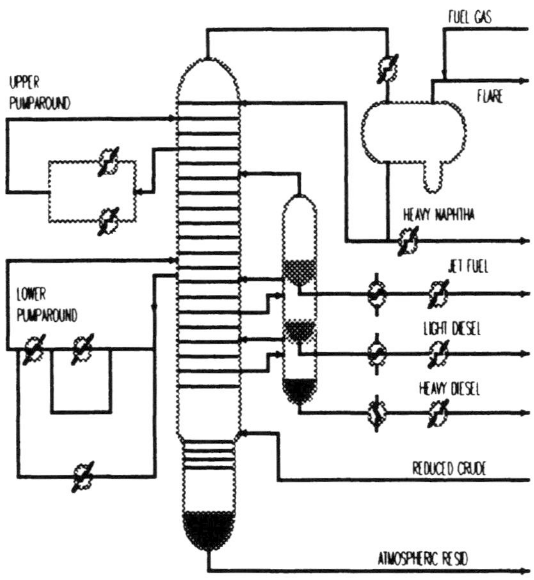

Figure 1 - Schematic of the crude tower

THE CRUDE FRACTIONATOR CONTROL PROBLEM

In the conventional control scheme the heavy naphtha boiling point is controlled by an overhead temperature controller on the DCS, that manipulates the reflux flow. If the tower pressure and the overhead temperature are kept constant, the operators expect that the naphtha composition will also remain constant. This is certainly not true since the naphtha end-point is also affected by other variables as the L/V ratio and the feed composition. The pressure on the overhead drum is controlled by a split-range controller that admits or release gas from this drum. The distillation ranges of the side products, except the heavy diesel are manually controlled by manipulating their flows. The 85% boiling point of the diesel blend is also manually controlled by the liquid overflash above the flash zone or by the outlet temperature of the feed heater. The liquid level on both the jet-fuel and the light diesel strippers are controlled on the DCS by the manipulation of inlet liquid flow to the strippers. The level of the heavy diesel stripper is controlled by the outlet liquid flow. The heat duties of the external pumparounds can also be manipulated by the operator to impose a proper liquid vapor flow in the tower and maximize the heat recovery in the oil preheating system. The major disturbances to the atmospheric tower are related to the feed quality and flow. These disturbances demand frequent correction by the operators on the flows of the product withdraws, overhead temperature, pumparounds heat duties and heater's outlet temperatures. A multivariable predictive controller can certainly perform better than the human operator in these circumstances. Then in general lines the objective of the advanced control is to keep the controlled variables the closest possible to their set-points without violating the constraints on the remaining variables of the system. For the crude fractionator we can consider as controlled variables, the distillation ranges of:

- the overhead heavy naphtha
- the jet-fuel stream
- the light diesel stream
- the heavy diesel stream

These variables are controlled on externally defined set-points that are selected depending on the market conditions. For the particular fractionator considered in this study, an usual demand is to specify the jet-fuel stream and the diesel blend that normally includes the light and heavy diesel and eventually external streams from other units. We consider two possible cases:

Case I. There are four controlled variables with defined set-points related to the product composition specifications. These are the ASTM 90% boiling points of the naphtha, jet-fuel and light diesel cuts and the ASTM 85% of the diesel pool.

Case II. The controlled variables with defined set-points are only the jet-fuel 90% ASTM and the 85% ASTM of the diesel blend.

All the variables related to the compositions or specifications of the products are inferred from the system measured properties (temperatures, pressures and flows). Other variables can also be controlled but not necessarily on a set-point. These variables are controlled by range and the objective is to keep them inside maximum and minimum values. In the system here considered these variables are:

- max. 10% ASTM of the jet-fuel
- max. 50% ASTM of the jet-fuel
- max. freezing point of the jet-fuel
- min. flash-point of the jet-fuel
- max. cloud-point of the diesel pool
- max. flooding % on the jet-fuel section
- max. flooding % on the diesel section
- min. tower overhead temperature
- min. liquid level in the jet-fuel stripper
- min. liquid level in the diesel stripper

The controller can manipulate:

- the reflux flow rate at the top of the tower.
- the jet-fuel flow.
- the light diesel flow.
- the feed heaters temperature controllers.
- the stripping steam to the jet-fuel stripper.
- the lower pumparound heat duty.

The controller also includes as measured disturbances the feed flow and the heat duty of the upper pumparound.

THE CONTROLLER EQUATIONS

The LDMC algorithm (Morshedi , Cutler and Skrovanek, 1985) defines a vector of residues ρ linearly related to the control move Δu and to the predicted error E' on the controlled variables:

$$\rho = (\mathbf{A^T.G.A + R).Du - A^T.G.E'} \qquad (1)$$

where $\rho = [\rho_1\ \rho_2 \cdots \rho_{ML-1}\ \rho_{ML}]^T$

and M is the number of manipulated variables and L is the number of future control actions considered in the controller. The matrix A is the modified dynamic matrix of the system, that contains the step response coefficients of the controlled variables whose corresponding predicted error components are not zero. For those variables controlled by range with no predicted error, or the predicted value of the variable is inside its range, the corresponding rows of matrix A are also zero. The diagonal matrix G weights the importance of the controlled variables, and R is also a diagonal matrix of the moving suppression factors. The control actions are then calculated by the solution of the LP problem

$$\underset{u,x,z}{\text{Minimize}} \quad \phi = \sum_{i=1}^{M.L} w_i.(x_i + z_i) \qquad (2)$$

Subject to

$$\mathbf{x - z = (A^T.G.A + R).Du - A^T.G.E'} \qquad (3)$$

(M.L equalities)

$$\mathbf{u, x, z \geq 0}$$

We also have constraints on the manipulated variables:

- On their absolute values:

$$\mathbf{u_{min} \leq u \leq u_{max}} \quad \text{(2.M.NL inequalties)}$$

- On their moves

$$\mathbf{\Delta u_{min} \leq \Delta u \leq \Delta u_{max}} \quad \text{(2.M.(NL-1) inequalities)}$$

This LP problem has a total of 4.M.NL-2.M +M.L equations, where NL is the number of future instants where the constraints on the manipulated variables are imposed. Usually we assume NL < L to guarantee the feasibility of the solution of the LP. The revised simplex algorithm is usually the first choice to solve this problem, and since the inequalities are transformed to equalities by the inclusion of slack variables the number of variables of the LP problem is substantially increased. A particular approach can be followed if some or all the variables of the problem are constrained inside a limited range. This is the "bounded variables" approach (Murty, 1976) and it gives a more economic solution in terms of computation effort. The LP problem with "bounded variables" can be formulated as:

Minimize $\mathbf{c^T.x}$

Subject to

$$\mathbf{A.x = b} \qquad (4)$$

$$x_j \le U_j \qquad j \in J = \{1,\dots,n_1\} \qquad (5)$$

where n_1 is the number of bounded variables.

and $\quad x_j \ge 0 \qquad$ for any j

The LP problem defined by the LDMC can be formulated with the structure defined by eqs (4) and (5) after a suitable change of variables. In this approach, using the simplex nomenclature a bounded nonbasic variable can be on any of its bounds 0 or U_j, and this turns the selection of the nonbasic variable to enter the basis a little more complicated than in the usual simplex algorithm. The benefit to the solution of the control problem is due to a smaller number of equations and variables. The reduction in the computation time gained with this approach depends on the size of the problem in terms of the number of manipulated variables and control horizon. For the size of problem here considered the "bounded variable" simplex algorithm can be ten times faster than the conventional revised simplex.

THE OPEN-LOOP DYNAMICS OF THE ATMOSPHERIC FRACTIONATOR

To gain insight on the dynamic behavior of the system consider the step responses of the crude fractionator for some of the manipulated variables. These responses correspond to pulse transfer functions obtained by linear regression of the plant data. Figure 2a corresponds to a step change on the flow of jet-fuel that reduces the internal reflux flow below this draw and as a consequence an heavier vapor flow goes to the top. A heavier naphtha is produced since the top reflux flow is kept constant. Hence the jet-fuel 90% ASTM increases, since the higher jet-fuel flow changes the cut-point between the light diesel and the jet-fuel. Part of the diesel components are incorporated by the jet-fuel flow that becomes heavier. Some of the light components of the diesel are lost to the jet-fuel, and the light and heavy diesel cuts also become heavier following a response with almost the same gain as the jet-fuel. We have a cut-point exchange between these two products. A different pattern is followed by the diesel blend (mixture of light diesel, heavy diesel and heavy naphtha), represented by the 85% ASTM boiling point as also shown in fig.2a. The diesel blend boiling point decreases with an underdamped behavior, with a slower dynamics than the jet-fuel and a significative dead time. Apparently the underdamped response is related with the liquid overflash flow control. The reduction on the internal reflux below the jet-fuel draw disturbs the overflash control loop that shows the underdamped response. Other properties of the diesel blend as the cloud point follow the same type of response. The behavior of the fractionator to a step change on the reflux flow can be analogously explained. In this case an increase on the reflux flow results in a higher L/V ratio and the step responses shown in fig. 2b for the jet-fuel and diesel blend boiling points, have opposite signs of the responses obtained for the step change on the jet-fuel flow.

For a step increase on the heat duty of the upper pumparound, the boiling points of all the products are reduced since the L/V ratio along the tower is increased. At the top section of the tower this happens because the increase in the pumparound duty causes a reduction on the vapor flow to the top section of the tower and the top reflux flow is kept constant. Bellow the pumparound, the internal liquid flow is increased by the extra condensation of vapor at the pumparound. The net effect on the diesel blend is then similar to the effect of an increase on the top reflux flow, as shown in fig.2d.

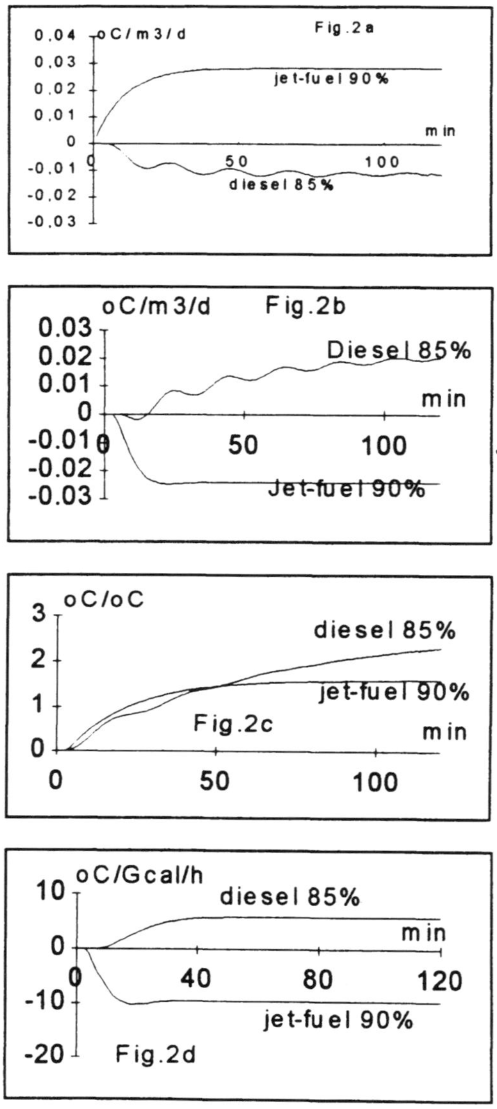

Figure 2. Open loop responses of the crude fractionator for step responses on (a)- Jet-fuel flow; (b)-Reflux flow; (c)-Feed heater temperature and (d)-Upper pumparound heat duty, for Case II.

We also observe that the heat duty control loop seems to smooth out the effect on the internal reflux flow and the underdamped response of the diesel blend properties is not observed. It is also interesting to compare the step responses of the two controlled variables of case II for the feed heater temperature. Obviously the increase in the open loop feed temperature causes an increase in the boiling point of the products since a heavier vapor is injected into the tower while the internal liquid flow remains constant. But observing fig.2c, we note that the dominant time constant of the diesel blend boiling point is larger than the jet-fuel response. This seems to be not coherent with the position of the diesel draws that are closer to the feed tray than the jet-fuel draw. Apparently, the heavier vapor flow initially affects both products with the same intensity, but the delays of the liquid flow inside the tower has an important role on the final response of the system.

PLANT RESULTS

An example of the performance of the developed controller for Case I configuration in the real plant can be seen on figure 3, obtained from the crude unit at the refinery of Paulínia in Brazil. The crude tower was on a steady-state with the controller in operation. At instant 25min the set-point of the level of liquid overflash was slightly reduced and this caused an unmeasured disturbance on the system. The variable most affected by this disturbance is the diesel boiling point as shown in fig.3c. The controller compensates this effect by slightly increasing the heat duty of the lower pumparound (fig.3d) to increase the internal reflux in the diesel section. At instant 70 min the system was disturbed by a small step change in the feed flow and a step reduction on the set-point of the light diesel 90% ASTM boiling point (fig.3a). This intensifies the effects of the first disturbance and the control actions are more clearly observed. We observe that one of the variables controlled by range, the jet-fuel freezing-point (fig.3e) tend to violate its maximum constraint but is brought back to an acceptable range. The remaining variables controlled by range stay inside their range during all the time of observation. After about 50min all the product specifications are at their required set-points and the system stabilizes.

Another example of the practical performance of the controller concerns Case II where only the jet-fuel and the diesel blend are controlled with external set-points while the remaining variables of the system are controlled by range. Figure 4 shows the plant tendencies for about 12 hours of normal operation.
The set-point of the diesel blend (fig.4a) was continuously changed to attend the production scheduling while set-point of the jet-fuel remained constant during all the period considered. We observe that the controller performed quite well even when the system was severily disturbed by the upper pumparound heat duty (fig.4d) that is used to preheat the crude. We note that this heat duty was increased of about 0.7Gcal/h after instant 450min.

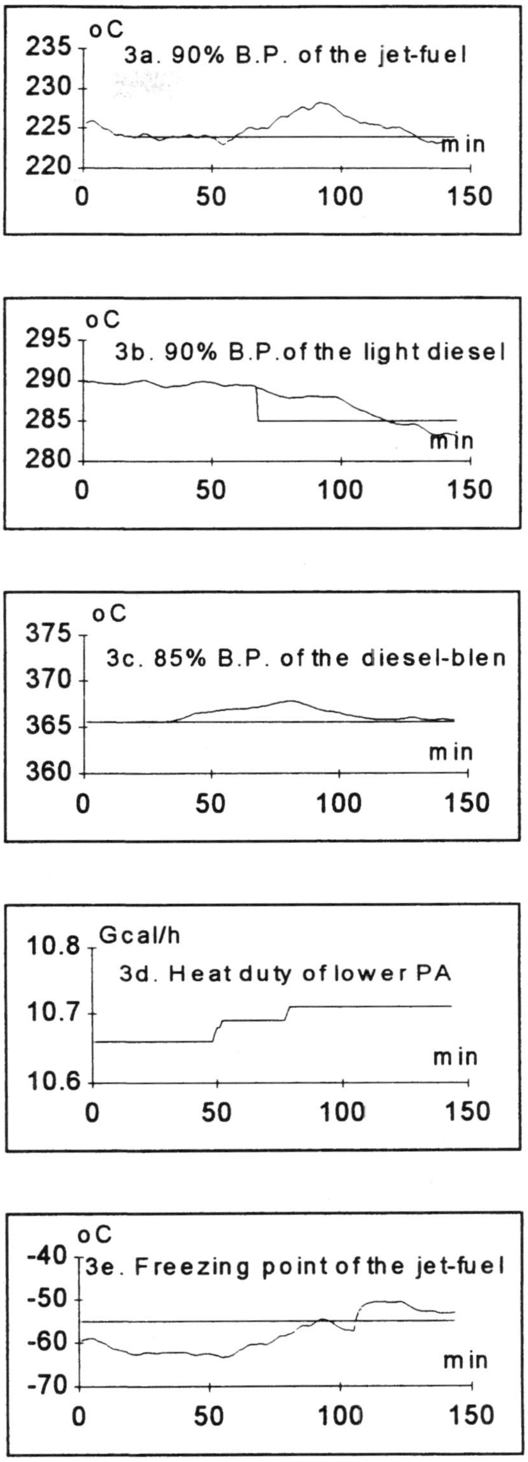

Figure 3. Plant responses for the multivariable controller for Case I configuration.

This is a measured disturbance to the controller and had almost no effect on the controlled variables. To compensate this disturbance the controller reduced the top reflux flow to mantain the internal reflux constant below the pumparound. The jet-fuel flow (fig.4e) was transitorily reduced to compensate the effect of this disturbance on the jet-fuel boiling point and brought back to almost its original level after a few hours.

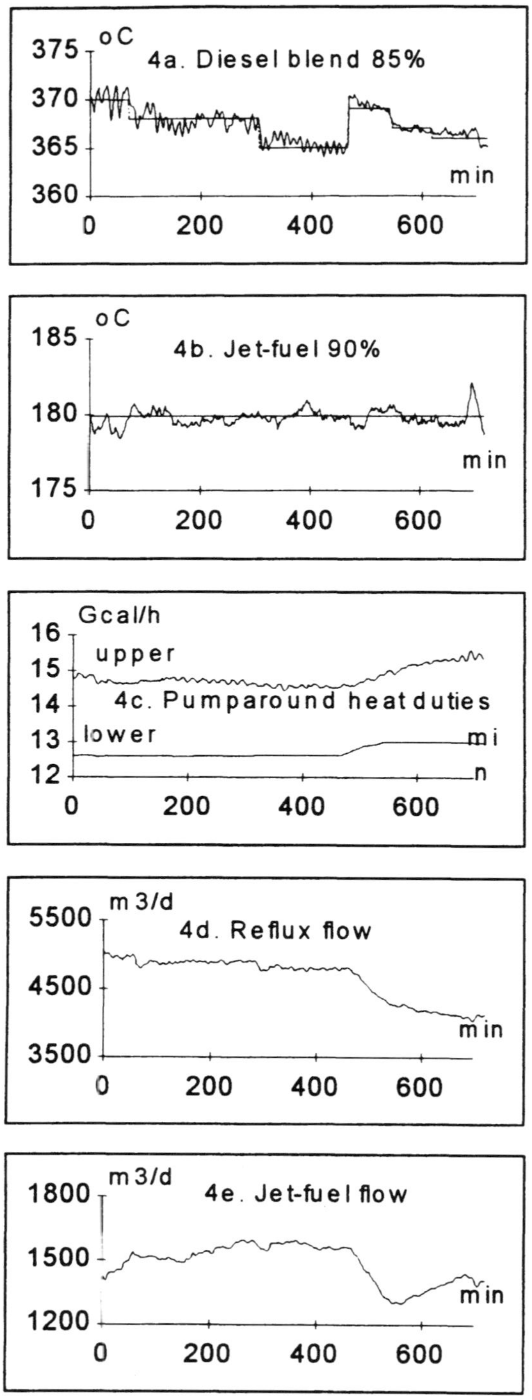

Figure 4. Plant responses with the multivariable controller for Case II.

CONCLUSION

A multivariable controller was successfully applied to the quality control of a large scale industrial crude atmospheric fractionator at the refinery of Paulínia in Brazil. The developed control algorithm shows that the LDMC formulation is adequate to large systems where several or eventually all variables are controlled by range, and the manipulated variables have min/max constraints. In this study it was adopted the "bounded variables" approach of the simplex algorithm that shows to be particularly useful in reducing the computation effort and allowing the application of the controller to systems of larger order while keeping the computation time at an acceptable level. The selected control configuration uses the reflux flow as a manipulated variable instead of the set-point to the overhead temperature controller as usually suggested in the literature.

REFERENCES

Cutler, C. R. and S. G. Finlayson, *Design Considerations for a Hydrocracker Preflash Column Multivariable Constraint Controller*, IFAC Conference, Atlanta, 1988.

Hsie, W. L., *Modeling Simulation and Control of a Crude Tower*, Ph.D. Dissertation, University of Maryland, 1989.

Morshedi, A. M., C. R. Cutler and T. A. Skrovanek, *Optimal Solution of Dynamic Matrix Control with Linear Programming Techniques (LDMC)*, Proc. Am. Control Conf., Boston, pp. 199-208, 1985.

Murty, K. *Linear and Combinatorial Programming*, John Wiley & Sons, New York,1976.

Muske, K., J. Young, P. Grosdidier and S. Tani, *Crude Unit Product Quality Control*, Comp. and Chem. Engng. 15, 9, 629-638, 1991.

O'Connor, D. L., K. Grimstad and J. Mckay, *Application of a Single Multivariable Controller to two Hydrocracker Distillation Columns in Series*, ISA Meeting, Anaheim, 1991.

COMPOSITION ESTIMATORS FOR A DISTILLATION COLUMN: THEORY AND EXPERIMENTS

Roberto Baratti[1], Alberto Bertucco, Alessandro Da Rold, Massimo Morbidelli[2]

Istituto di Impianti Chimici - Universita' di Padova
Via Marzolo, 9 - 35131 Padova Italy

[1]*Dipartimento di Ingegneria Chimica e Materiali - Universita' di Cagliari*
Piazza d'Armi - 09123 Cagliari Italy

[2]*Dipartimento di Chimica Fisica Applicata- Politecnico di Milano*
Via Mancinelli, 7 - 20133 Milano Italy

Abstract: This work is a contribution to the development of non-linear estimators based on the Extended Kalman Filter theory to monitor the dynamics of a continuous distillation column. The performances of the estimators are checked by comparison with the behavior of a pilot distillation unit where fractionations of ethanol-water binary mixtures are carried out. A simplified model of the column is presented, based on the orthogonal collocation approach, in order to extend the applicability of the estimator also to multicomponent distillation columns with high number of stages.

Keywords: Estimators, Extended Kalman filters, Distillation columns, Dynamics, Orthogonal collocation

1. INTRODUCTION

The possibility of a real-time prediction of the dynamic behavior of a chemical plant is of paramount importance for control purposes, especially when one needs to have informations on non-directly accessible variables or when the time delay of the measurements are incompatible with the process time constants. This is the case, for example, of distillation units, where the purity constraints on the products must be achieved in spite of changes in the operating conditions such as, for example, the external reflux flowrate or the feed composition. Time constants in these plants are usually as high as many hours, and the amount of materials processed as large as hundreds of tons per day, so that it is extremely valuable, at least from the economic point of view, to be able to simulate in real-time the behavior of the plant under transient conditions. On the other hand, the on-line measurement of the compositions of top, bottom and side products of a distillation column is a typical problem in industrial practice, since the use of composition analyzers is discouraged by high investment and maintenance costs and is affected by large response delays.

This problem can be overcome by using temperature measurements, together with a suitable estimation procedure, for controlling the product compositions (cf., Kister, 1990). Much interest has been devoted to this topic in the last two decades: starting from the work of Joseph and Brosilow (1978), a number of studies have been recently involved in the development of estimators for either batch (Quintero-Marmol *et al.*, 1991) and continuous distillation columns (among others, Lang and Gilles,

1990 and Mejdell and Skogestad, 1991a, 1991b and 1993). In most of these cases, the performances of the proposed estimators was not checked by comparison with real plant data. Very recently, Baratti *et al.*, 1994, developed a procedure able to infer the product compositions from indirect and incomplete on-line temperature measurements, based on a simplified mathematical model of the process and an Extended Kalman Filter (EKF, cf. Gelb, 1974) as an estimator technique. The proposed approach was tested on a pilot-size distillation unit in the case of binary mixtures, with particular interest to the robustness of the nonlinear estimator to model errors.

In the present work, two problems are discussed: first, the possibility of using the previously proposed estimator in the calculation of the movement of the temperature wave along the column; second, the development of a 'reduced' estimator, where the EKF procedure is coupled with an orthogonal collocation model for the column. This last point is crucial in order to extend the previously proposed approach to multicomponent distillations: in fact, if Nc is the number of components and Np is the number of plates considered in the column, the use of a complete stage-by-stage model leads to a system of $[(Nc-1)Np+1](Nc-1)Np/2$ equations, the solution of which requires significant computing effort (with reference to commonly adopted process computers) even with three components and a number of plates as low as few dozens.

2. THE MODELS

The definition of a suitable process model is an essential step. Various models with increasing accuracy and complexity can be applied for the simulation of the distillation column dynamics (cf., Gani *et al.*, 1986); however, the analysis presented in their previous work (Baratti *et al.*, 1994) led the authors to the following conclusions, at least for the systems considered (see the referred paper for details):

- only the composition dynamics needs to be accounted for;
- vapor and liquid molar flowrates are assumed to be constant in each column section, i.e. energy balances are neglected. The subcooling effects of feed and external reflux are taken into account;
- changes in volumetric hold-ups of condenser/reflux drum, reboiler/column bottom and trays are neglected;
- an accurate description of vapor-liquid equilibria is essential, i.e. the relationship between temperature and composition of liquid and vapor streams must be represented correctly.

According to the equations reported there, only few parameters are needed for the application of that estimator: the set-points of the feed and external reflux flowrates, the feed composition and temperature, the external reflux temperature, the heat duty at the reboiler, the total pressure drop in the column, plus the values of parameters involved in the calculation of geometric hold-ups and of vapor-liquid equilibria.

2.1 Complete model.

The system of equations of the complete model of the column is reported in the work by Baratti *et al.* (1994). It is noteworthy that, unlike most studies in the literature, systems with highly nonideal phase behavior were considered, for which the activity coefficients were calculated through the NRTL thermodynamic model. A Murphree-efficiency correction was also included for the vapor composition calculation.

2.2 Reduced model.

In spite of the simplifying assumptions summarized above, the resulting system of equations can be satisfactory handled within a non-linear estimator scheme for the distillation of binary mixtures, but is too large for practical application to multicomponent mixtures, even in the case, considered in the experimental part of the work, of a 30-stage single distillation unit.

In order to further reduce the model complexity, an orthogonal collocation approach was considered. The application of the orthogonal collocation technique to distillation problems has been set forth by several authors (cf., Stewart *et al.* 1985), both for steady-state and for dynamic simulation (cf., Carta *et al.*, 1994). This kind of model reduction is based on the idea that a distillation column can be represented as a distributed system where the temperature, composition and flow profiles can be represented as continuous variables along the height of the unit. Accordingly, the distillation column is divided into modules and the variables in each one of them are approximated by Lagrange polynomials, fitted to the complete process model at the collocation points; these are designed in order to give a good approximation on the full set of stages, and will be referred to as collocation stages, even though they do not have a physical meaning.

In the present paper, the collocation stages are defined as the zeros s_j of the Hahn polynomials, calculated for $\alpha=\beta=0$, as suggested by Stewart *et al.* (1985); the balance and equilibrium equations are written referring to such points. The column has to be divided into at least two modules: one represents the enrichment section, the other the stripping section; the condenser, reboiler and feed stage can either be treated apart or as boundaries of one module. For each module, n internal grid points are considered, plus the two entry points at the boundaries. For a binary mixture, the material balance of the reference component at collocation stage j (numbering from the bottom upwards) is given by:

$$M_j \frac{dx_j}{dt} = L\left[\sum_{k=1}^{n+2} W_k(s_j-1)\,x_k - x_j\right] + V\left[\sum_{k=1}^{n+2} W_k(s_j+1)\,y_k - y_j\right] \qquad (1)$$

where x_k and y_k are the liquid and vapor mole fractions of the streams leaving collocation stage k, M_j is the molar tray hold-up, L and V the liquid and vapor molar flowrates, and W_k is the weight function, calculated by:

$$W_k(s) = \prod_{k=1,k\neq j}^{n+2} \frac{s-s_k}{s_j-s_k} \qquad (2)$$

The equilibrium equations are expressed by:

$$y_j = K_{A,j}\,x_j \qquad (3)$$

$$(1-y_j) = K_{B,j}(1-x_j) \qquad (3')$$

where $K_{i,j}$ is the equilibrium ratio of component i, given by:

$$K_{i,j} = \frac{\gamma_i(T_j,x_j)\,P_i^s(T_j)}{P_j} \qquad (4)$$

In Eq. (4) γ_i indicates the activity coefficient of i, P_j and T_j the pressure and temperature at stage s_j; T_j is calculated as the equilibrium temperature at composition x_j.

Once the solution is obtained at the collocation stage, the Lagrange polynomials provide values of variables x, y and T at any stage s.

It is noteworthy that all the assumptions made for the complete model, as summarized in Section 2. above, hold also for this reduced model. In particular, L and V are kept constant in each column section.

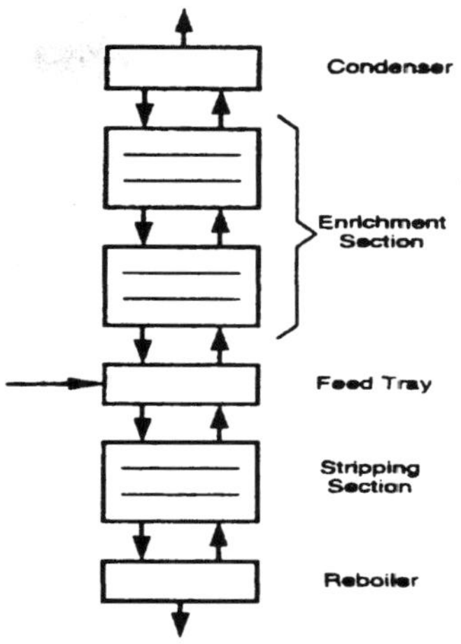

Fig. 1. Schematic of the distillation column as modeled by orthogonal collocations

Unfortunately, the reduced model as presently outlined could not be applied to the distillation problems here considered, since the composition and temperature profiles along the column at the examined operating conditions exhibited steep and flat profiles in the enriching section. Therefore, it was necessary to modify it into a moving boundary finite element collocation method. As a result, the colum was divided according to the sketch in Fig. 1, with the stripping section represented by a single module and the enrichment one splitted into two elements. Since the currently proposed algorithms for steep-and-flat profiles (cf., Srivastava and Joseph, 1987 and Hwang, 1991) do not apply due to the non-ideality of the vapor-liquid equilibria, the boundary between these two elements was changed according to an heuristic method, i.e. when the temperature difference between the two adjoining trays at the boundary became greater than a prefixed value, optimized by a trial-and-error procedure. Note also in Fig. 1 that the condenser, the reboiler and the feed tray were treated as separate stages.

2.3 Nonlinear estimator.

The estimation technique applied to both the models examined is the stochastic non-linear Extended Kalman Filter (Gelb, 1974). According to this method, the state estimate at time t_{l+1} is calculated by suitably updating the values of the state variables obtained by integration of the process model with the estimate at time t_l as starting point. Namely, this optimal estimate is obtained using a set of secondary measurements at time t_{l+1}, through a correction of the model prevision which accounts for both model informations and actual measurements:

$$\mathbf{x}_{l+1/l+1} = \mathbf{x}_{l+1/l} + \mathbf{K}_{l+1}(\mathbf{z}_{l+1} - \mathbf{h}_{l+1/l}) \qquad (5)$$

In Eq (5) $\mathbf{z}_{l+1}$ is the vector of measured values at time t_{l+1}, $\mathbf{h}_{l+1/l}$ the vector of the observations (the temperatures on the selected trays) as predicted by the model from the state values $\mathbf{x}_{l+1/l}$ (in this case, the mole fractions in the liquid phase), and $\mathbf{K}_{l+1}$ is the gain matrix. The numerical integration of the material balance and the Riccati equations was performed by means of the Heun's method (cf., Carnahan *et al.*, 1969); as initial conditions, the values of the reference steady-state were always taken.

The EKF application was devised to predict the changes in time of the distillate and bottom stream compositions and of the composition profiles in the column, as a function of disturbances imposed to the process (see next Section 3.). As secondary measurements the temperatures of two or three suitable trays only were used. The error covariance matrices relative to the model prediction and to the measurement were tuned by fitting to experimental product composition profiles as a function of time in a reference run. Then, the method was validated in several other runs by comparison of the predicted compositions with those measured off-line, which were not used for further tuning of the estimation procedure.

More details on the application of the complete-model EKF to the distillation column and on the regression procedure can be found in Baratti *et al.* (1994).

3. EXPERIMENTAL DATA

As oulined in the referred paper, in order to perform a significant test of the proposed estimator, experiments were carried out on a distillation column of diameter and height large enough to ensure realistic hydraulic and thermal behavior. The pilot plant used has a diameter of 0.3 m and 30 sieve trays, with tray spacing 0.20 m, and is equipped with a shell-and-tube natural circulation reboiler and a water-cooled total condenser, which is open to atmosphere. A more detailed description of this apparatus is reported by Bertucco *et al.* (1985). The feed, containing binary ethanol-water mixtures, is introduced on the 23th tray from top. The plant is equipped with electro-pneumatic regulation valves on the steam, feed, reflux, distillate and bottom lines and can be fully operated by a on-line compatible PC through a standard RS-232 interface.
The procedure adopted in each experimental run can be summarized as follows: first, the operating conditions were selected in order to obtain the desired steady state, using a stedy-state simulation program. Then, the plant was started up; when steady state conditions were reached, the scheduled disturbances in the operating variables were applied to the column and the resulting transient behavior was monitored by sampling distillate and bottom compositions (analyzed off-line with a gaschromatograph) and temperatures of nine internal trays (measured by thermoresistances).

Table 1 Summary of operating conditions of the experimental runs considered. Step changes:
- steam flowrate from 111.1 to 81.1 kg/h in Run 1;
- reflux flowrate from 117.9 to 208.7 l/h in Run 2 .

	Run 1	Run 2
feed flowrate [l/h]	154.5	154.2
feed temperature [°C]	20.9	18.9
ethanol feed comp. [%mol]	14.0	13.6
external reflux flowrate [l/h]	122.4	117.9
steam flowrate [kg/h]	111.1	111.0

In particular, step changes on the reflux volumetric and on the steam by-weight flowrates are considered here; the operating conditions of the two experimental runs discussed below are reported in Tab. I.

4. RESULTS AND DISCUSSION

As already shown by Baratti *et al.*, 1994, the estimator coupled with the complete model proved in all cases to be able to reproduce the experimental transient response of the product compositions by using only two temperature measurements, i.e. those of the column bottom and of tray no. 26 from bottom.

As about the composition profiles along the column, in Fig. 2 an example is reported, concerning Run no. 1. It is shown by Fig. 2a that the estimator maybe unable to ensure the same accuracy as for the product composition profiles (for example, at t=21 min after the disturbance). Note that experimental and calculated temperatures along the column are compared, since temperatures and mole fractions are equivalent for binary mixtures.

The failure in reproducing the movement of the temperature wave in the column can be explained by looking at the shape of these curves: they present a sharp change from the lower to the upper value, and the propagation of the step change in the steam flowrate is like a wave that affects a limited number of trays at any time considered. Therefore, if the two pilot trays are far from one another, as in the case of Run 1, the corresponding temperatures do not change as long as the wave has passed tray no. 26 but has not reached the bottom of the column yet. In this time range the estimator cannot take any advantage on secondary measurements, i.e. the prediction is based on the model only, which is clearly unsufficient. To overcome this problem, a third temperature has been used in the estimator, namely the one on tray no. 16. A remarkable improvement is obtained in this way, as reported in Fig 2b. This result is especially important when, for example, distillation columns with side streams are considered.

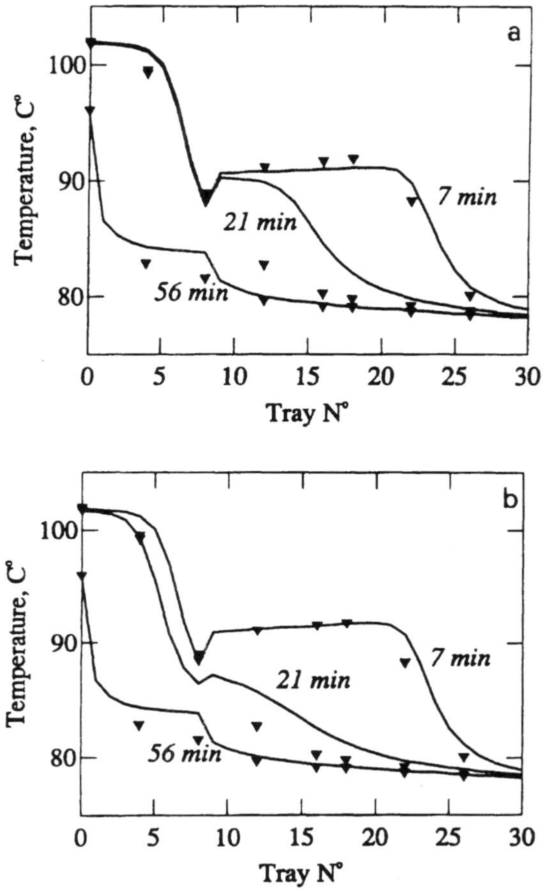

Fig. 2. Temperature profiles along the column as a function of time after the step change in the steam flowrate: experimental data points (Run no. 1) versus curves provided by the complete-model estimator. Secondary measurements on:
(*a*) tray 26 and column bottom;
(*b*) trays 16 and 26 and column bottom.

Figs. 3 and 4 refer to the estimator with the reduced model (in this case only 20 material balance equations were used), and presents a comparison between experimental and calculated composition profiles of top and bottom products in Runs no. 1 and 2, respectively. Again, as secondary measurements, the temperatures on tray no. 26 and in the column bottom were selected. It is clear that the estimator is able to capture the main features of the unit dynamics with quantitative accuracy. Note that data of the same runs had been reproduced also with the complete-model estimator, with results very similar to those presented here. Although these are only preliminary results, they demonstrate that it is possible to simplify drastically the model complexity without reducing appreciably the accuracy in the estimation of product composition profiles.

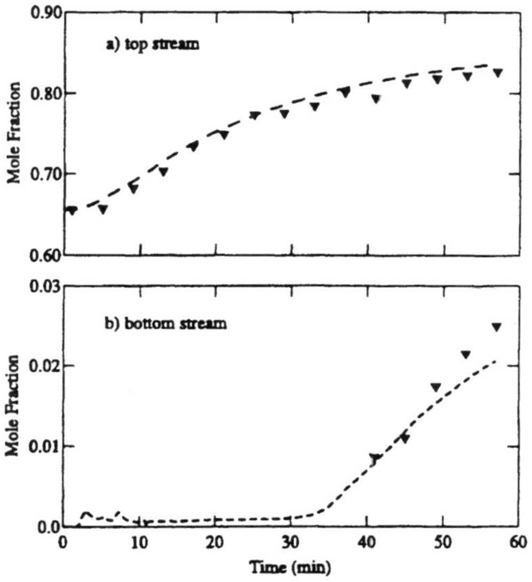

Fig. 3. Product composition profiles for Run no.1: comparison between experimental values and calculated ones by reduced-model estimator.

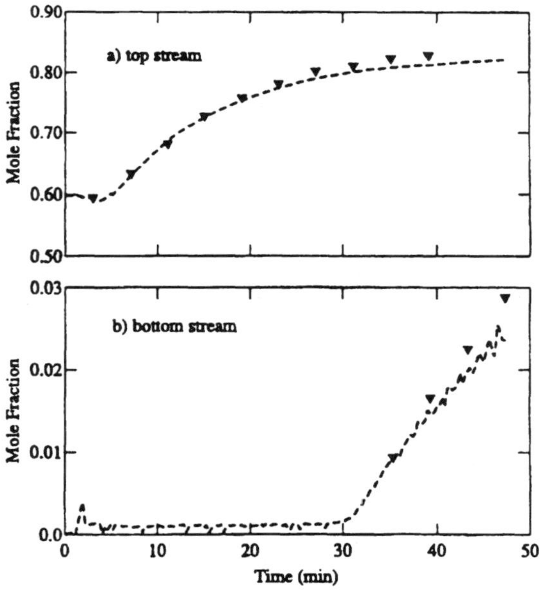

Fig. 4. Product composition profiles for Run no.2: comparison between experimental values and calculated ones by reduced-model estimator.

5. CONCLUSION

Two EKF-based estimators were presented for the simulation of the dynamics of a distillation columns.

Calculations were compared with experimental results of product composition and column temperature profile changes with time for the fractionation of an ethanol-water binary mixture. It was shown that:

- the internal dynamics of the column (tray temperature profiles) is predicted satisfactorily by the complete-model estimator.
- the coupling of the orthogonal collocation method with the EKF is successfull.
- the reduced-model estimator is satisfactory in reproducing the distillate and bottom experimental composition profiles.

More work is in progress to further reduce the complexity of the system of equations in view of applications to multicomponent distillation plants.

ACKNOWLEDGMENTS

The financial support of the Consiglio Nazionale delle Ricerche-Progetto Finalizzato Chimica Fine is gratefully acknowledged.

REFERENCES

Baratti, R., A. Bertucco, A. Da Rold and M. Morbidelli (1994). Development of a Composition estimator for Binary Distillation Columns. Application to a Pilot Plant. *Chem.Eng.Sci.*, accepted for publication

Bertucco, A., S.A. Rienzi, G.B. Guarise and A. Trotta (1985). *Controllo di un Impianto Pilota di Distillazione.* CLEUP Ed., Padova.

Carnahan, B., H.A. Luther and J.O. Wilkes (1969). *Applied Numerical Methods, p. 362.* John Wiley & Sons, New York.

Carta R., Tola G., Servida A. and Morbidelli M. (1994). Performance of collocation models for simulating transients multistage separation units *Computers Chem. Engng,* in press.

Hwang, Y.L. (1991). Nonlinear wave theory for Dynamics of Binary Distillation Columns *AIChE J.* **5**, pp. 705-23.

Gani, R., C.A. Ruiz and I.T. Cameron (1986).A Generalized Model for Distillation Columns-I. *Computers chem.Engng* **10**, pp. 181-198.

Gelb, A. (1974). *Applied Optimal Estimation.* M.I.T. Press, Cambridge.

Joseph, B. and Brosilow, C.B. (1978). Inferential Control of Processes: Parts I, II and III. *AIChE J.* **24**, pp. 485-509.

Kister, H.Z 1990. *Distillation Operation*, p 545. McGraw-Hill, New York.

Lang, L. and D. Gilles (1990). Nonlinear Observer for Distillation Columns. *Computers Chem. Engng* **14**, pp. 1297-1301.

Mejdell, T. and S. Skogestad (1991a). Estimation of Distillation Compositions from Multiple Temperatue measurements using partial Least-Squares regression. *Ind.Eng.Chem. Res.* **30**, pp. 2543-2555.

Mejdell, T. and S. Skogestad (1991b). Composition Estimator in a Pilot-Plant Distillation Column Using Multiple Temperature. *Ind.Eng.Chem. Res.* **30**, pp. 2555-2564.

Mejdell, T. and S. Skogestad (1993). Output Estimation Using Multiple Secondary Mesurements; High-Purity Distillation. *AIChE J.* **39**, pp. 1641-1653.

Quintero-Marmol, E., W.L. Luyben, and C.Georgakis (1991). Application of an Exented Luenberger Observer to the Control of Multicomponent Batch Distillation. *Ind.Eng.Chem. Res* **30**, pp. 180-1880.

Srivastava, R. K. and B. Joseph (1987). Reduced-order models for separation columns-IV-VI. *Comp.Chem.Eng.* **11**, pp.159-164, pp. 165-176.

Stewart, W. E., K. L. Levien, M. Morari (1985). Simulation of fractionation by orthogonal collocation. *Chem.Eng.Sci.* **40**, pp. 409-421.

PREDICTION AND ANALYSIS OF THE EFFECT OF REBOILER AND CONDENSER HOLDUP ON THE DYNAMIC BEHAVIOUR OF DISTILLATION COLUMNS

Michela Montesi and Alessandro Brambilla

Dipartimento di Ingegneria Chimica-Università degli Studi di Pisa
Via Diotisalvi,2 - 56100 PISA (ITALY)

Abstract: A prediction method of the dynamic behaviour of distillation columns based on the steady-state composition profiles is derived. The peculiarity of the method comes from the ability to take into account specifically the fluodynamic model of the "non active" hold-ups (reflux drum and sump). The method has been developed for packed columns because the rôle of the "non active" hold-ups becomes more important for columns with low internal ("active") hold-up but its extension to tray columns is straightforward. The application of the method has also been extended to changes of the internal flows.

Keywords: dynamic behaviour, distillation columns, prediction methods

1. INTRODUCTION

There have been in the last 30 years several efforts to develop a general method for determining the dynamic characteristics of distillation columns from steady-state operating conditions. Despite of the actual possibility to solve quite easily the dynamic model of distillation columns, the availability of a method for a reliable prediction of the column dynamic based on steady-state data is still appealing if it becomes a tool for improving the understanding and if it widens the capability of the previous methods.

The analysis of the contribution to the column dynamics of the "non active" hold-ups of the columns, such as the reflux drum and the sump liquid holdup, falls in these arguments. The motivation of such analysis derives from the following considerations:

- many columns, often as a consequence of a retrofitting, are equipped with high efficiency packings which are characterized by low liquid hold-up in comparison with trays. It follows that the ratio between the "active" internal hold-up and the "non active" hold-up decreases significantly for the former columns and, therefore, the way to consider the "non active" holdups has a greater effect on the global dynamics;
- the fluodynamic behaviour of the "non active" holdup may play an important role in the global column dynamics. For example the different control schemes of the column pressure with their different hardware realizations (mainly in the presence of hot by-pass) lead to different fluodynamic behaviour of the reflux drum (i.e. mixing degree);
- the effect of the "non active" holdups on the global dynamics strongly depends on the operating conditions, in particular on the presence of pinch zones in the column and on the product considered. The paper shows that at some operating conditions, commonly met in practical applications, the amount of the "non active" hold-up at one end of the column (e.g. the reflux drum) does non affect at all the dynamic behaviour of the compositions of the product at the other end of the column (e.g. bottom product).

The paper presents an original method for determining the column dynamic from steady-state data taking into account separately the "non active" holdups with their specific fluodynamic model. An extension of the method allowing to predict the time constant also for changes of the internal flows is presented in the paper.

2. THE RIGOROUS MODEL

In order to verify the accuracy of the results obtained with the simplified method, a rigorous model suitable to represent the dynamics of distillation column with low liquid hold-ups has been developed. The columns present this characteristic when the contact between the two phases is provided by structured packing instead of trays. The development of a specific model for continuous packing is motivated by the use of this internal as a common solution when the retrofitting (revamping) of an existing column is examined.

A rate-based approach has been adopted and a general model has been developed both for binary and multicomponent distillation by Montesi (1992). The model equations are solved by the orthogonal collocation method which allows to considerably reduce the computation time maintaining an adequate accuracy of the results.

In the present paper the dynamic analysis has been carried out for the iC4/nC4 fractionator; the scheme and the main characteristics of this column are shown respectively in Figure 1 and in Table 1.

Table 1. Main characteristics of the column

Feed flowrate (kmol/hr)	680
Feed status (q)	1
Feed composition iC4 mole fraction	0.5
Distillate flowrate (kmol/hr)	340
Distillate composition (iC4) mole frac.	0.993
Bottom composition (iC4) mole frac.	0.007
Bottom Vapour flowrate(kmol/hr)	2380
Pressure (atm)	6.6
Column diameter (m)	3.2
Number of packings	4
Height of each packing (m)	6
Liquid distributor hold-up (kmol)	6.33
Liquid hold-up for each packing (kmol/m)	4.34
Reflux drum hold-up (kmol)	200
Sump hold-up (kmol)	200

3. THE PROPOSED SIMPLIFIED METHOD

The method proposed in this paper for the estimation of the principal dynamic parameters of a distillation column is mainly directed to the following aims:

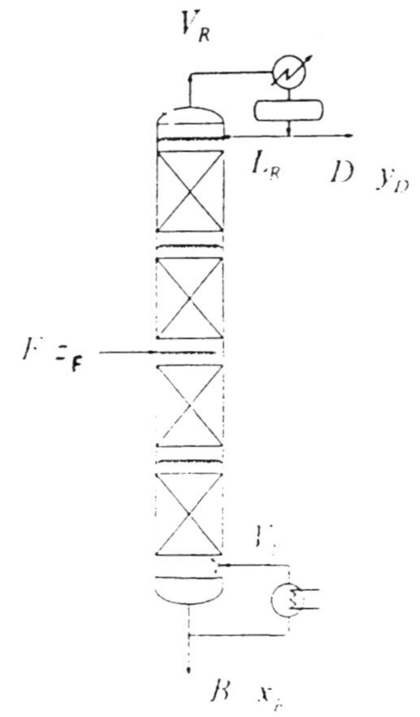

Fig. 1 Scheme of the packed column

- to specifically take into account the contribution to the dynamics of the "non active" hold-ups (reflux drum and sump) in columns with low liquid hold-up;
- to extend the estimation to the cases where only changes in the internal flows occur.

The basic idea, according to Rademaker, *et al.* (1975), seems to date back at the beginning of the century (Lord Raleigh) and it has been adopted in several different ways by the authors who, in the last thirty years, have made an effort to predict the dynamics of distillation columns using steady-state information (Moczek, *et al.*, 1965; Wahl and Harriot, 1970; Weigand, *et al.*, 1972). More recently, Skogestad and Morari (1987) have recalled the efforts of these authors pointing out the misconceptions originated by their work in the chemical engineering community; moreover they have restated the assumptions on which also the method adopted in this paper is based together with the extent of its validity. The only restriction that has been removed in this paper concerns the case of changes of internal flows.

On the basis of the following assumptions:

1. the flow changes have a negligible dynamics with respect to the composition dynamics;
2. the column has the same dynamic response at every height (equivalent to the assumption that all trays have the same dynamic response);
3. the vapour hold-up is neglected

the overall mass balance equation for the generic component can be derived:

$$\frac{d}{dt}\int_0^L m(z)x(z,t)dz = F_f z_F - D_f y_D(t) - B_f x_B(t) \quad (1)$$

where m is the liquid hold-up in the packing , x is the liquid composition and the subscript f represents the new steady-state value.

Taking into account the balance at the steady-state :

$$F_f z_F - D_f y_{Df} - B_f x_{Bf} = 0 \qquad (2)$$

and introducing the differences :

$$\Delta y_D f(t) = y_D(t) - y_{Df}$$
$$\Delta x_B f(t) = x_B(t) - x_{Bf} \qquad (3)$$

eq.(1) becomes:

$$\int_0^L m(z)\Delta x(z)dz \frac{df(t)}{dt} = -D_f \Delta y_D f(t) - B_f \Delta x_B f(t) \qquad (4)$$

The integration of eq.(4) gives a first order response:

$$f(t) = \exp(-t/\tau) \qquad (5)$$

where :

$$\tau = \frac{\int_0^L m(z)\Delta x(z)dz}{D_f \Delta y_D + B_f \Delta x_B} = \frac{\int_0^L m(z)\Delta x(z)dz}{\Delta(Fz_f) - y_{D0}\Delta D - x_{B0}\Delta B} \qquad (6)$$

represents the principal time constant for the column (the subscript 0 represents the reference steady-state value of the variables). The estimation of τ requires the knowledge of the composition profile along the column before and after the change considered and the liquid hold-up profile along the column. Equation (4) has been applied to the column schemes reported in Fig.2. The value of the dominant time constant calculated by Eq. (6) applied to the scheme of Fig.2a is defined as τ_∞ (in fact, the assumption that the composition of the reflux and boil-up are constant corresponds to an infinite hold-up of the reflux drum and sump). τ_0 indicates the dominant time constant obtained applying Eq.(6) to the scheme of Fig.2b where the reflux drum and sump have null hold-up.

4. "DECOUPLED" COLUMNS

In order to investigate the effect of the "non active" liquid hold-ups on the dynamic behaviour of the column, operating conditions have been selected for which the influence of the amount of the "non active" hold-up at one extreme of the column is negligible on the dynamic behaviour of the product composition at the other extreme. These conditions are commonly encountered when both products have high purity as in the case of the column considered in this paper.

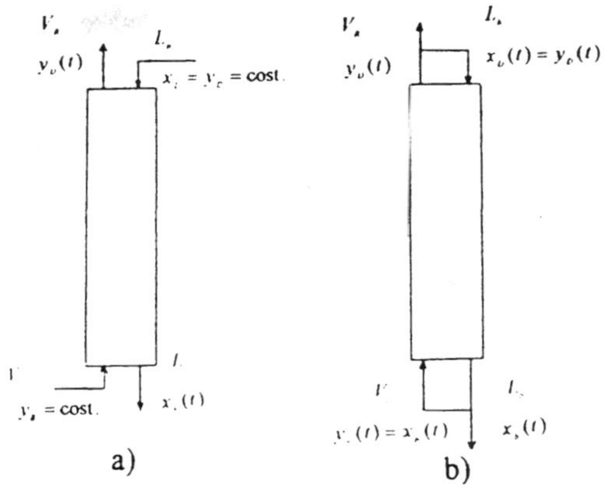

Fig.2 a) Scheme of the column with reflux and boil-up composition constant
b) Scheme of the column with null hold-up of the reflux drum and sump

In fact, when a change of -5% of vapour flow is considered for the two column schemes shown in Fig.3 it can be observed that the corresponding change of the top product composition is independent from the scheme of the bottom. Analogously, as shown in Fig.4, for the same vapour flow change the variation of the bottom product composition is independent from the scheme of the top. From these observations it is possible to assert that, at the operating conditions considered, the dynamic behaviour of the composition of one product is not affected by the amount and the fluodynamic model of the "non active" hold-up at the other extreme of the column; in fact each one affects the dynamics of the product composition relative to its extreme, but does not affect the behaviour of the composition of the other product.

For the scope of the present analysis the column that shows this behaviour is defined as "decoupled". At these conditions the dynamic response of the product compositions to a vapour flow change of -5% (at reflux flow constant) obtained with the rigorous model are shown in Fig.5. The parameter of the curves is the residence time of the "non active" hold-ups $t_D = v_D/V_R$ and $t_B = v_B/L_S$ with v_D and v_B respectively representing the reflux drum and the sump hold-up. The responses clearly confirm the above statements concerning the decoupling of the effects of the two extremes of the column. This particular dynamic behaviour is explained by the composition profile changes along the column as shown in Fig.6 for a -1% change of vapour flow where the two curves inside the initial and final profiles represent the profiles corresponding to the schemes of Fig.3a and 4a. The "decoupled" condition is associated to the presence and formation of pinch zones in the column.

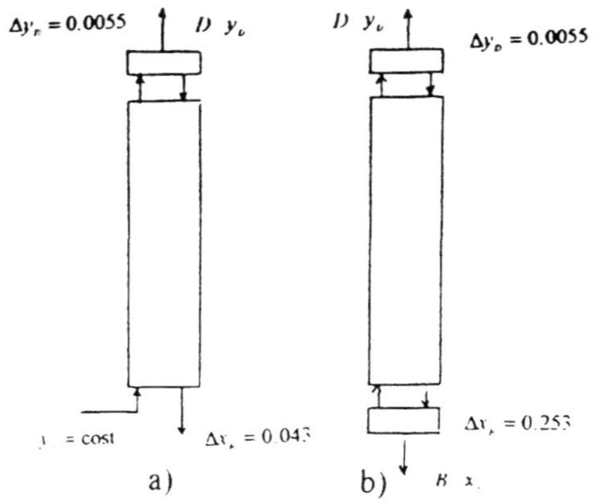

Fig.3 Distillate composition variations for a change of -5% of the vapour flow with constant reflux

An additional confirmation of the "decoupling" between the two extremes of the column for the operating conditions considered in the paper is shown in Fig.7a and b where the time constants of the responses of the two products are reported in function of the vapour flow changes.

Fig.7a and b show that the "decoupling" is independent from the amplitude of the change and make evident the non linearity of the dynamics.

5. INTERNAL FLOW CHANGES

When the overall material balance of the column remains constant and only the internal flows change, the new composition profile of the column (see Fig.8) shows that a displacement of the components from one section of the column to the other prevails maintaining the inventory of the component constant in the column; in opposition to the case of external flow changes where an increment or a decrement of the components in the column is predominant.

In presence of internal flow changes the representation of the column according to the scheme shown in Fig.9 is useful for the extension of the approach described above for the estimation of the principal dynamics. In fact, the application of Eq.(4) to all the column is not possible because the denominator is zero:

$$D_f\Delta y_D + B_f\Delta x_B = \Delta(Fz_F) - y_{D0}\Delta D - x_{B0}\Delta B = 0 \quad (6)$$

According to the scheme of Fig.9, the column is divided in two sections: one above and the other below the height z^* where the composition profile at the new steady-state intersects the initial profile. For the section below, with the same assumptions previously expressed, the new balance for the generic component gives:

$$\frac{d}{dt}\int_0^{z^*} m(z)x(z,t)dz = L_{Sf}x(z^*) - V_{Sf}y(z^*,t) - B_0x_B(t) \quad (7)$$

with $x(z^*)$ time independent.

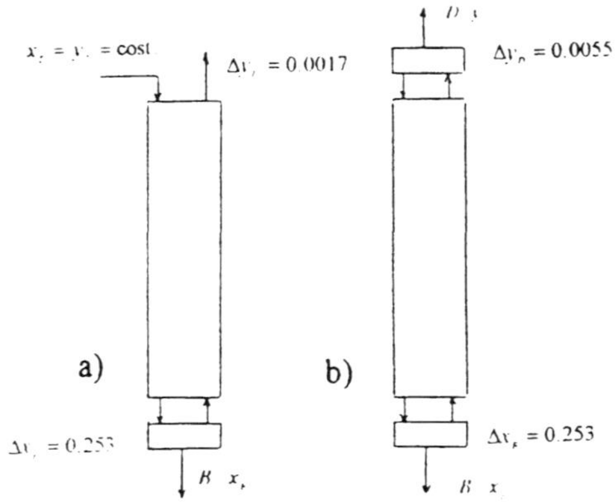

Fig.4 Bottom product composition variations for a change of -5% of the vapour flow with constant reflux

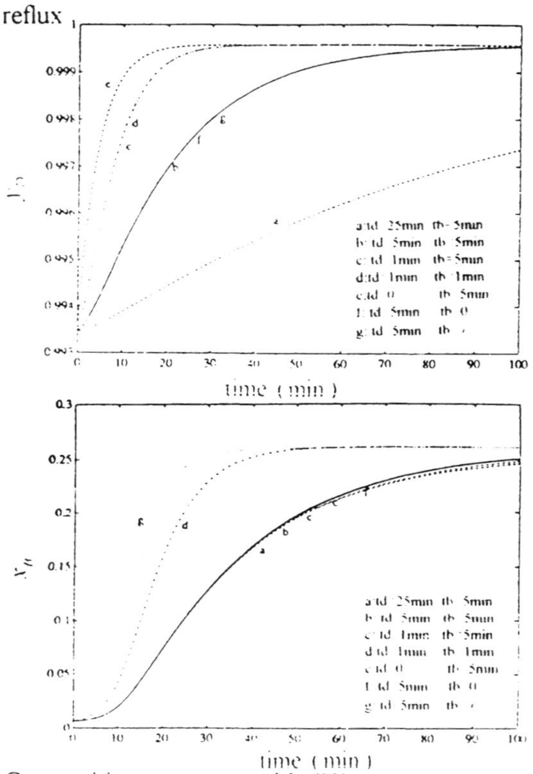

Fig.5 Composition responses with different hold-up of reflux drum and sump.

The mass balance on the same section at the new steady-state is given by:

$$L_{Sf}x(z^*) - V_f y_f(z^*) - B_0 x_{Bf} = 0 \quad (8)$$

The following equation is obtained subtracting Eq.(8) to the second member of Eq.(7) and applying the second assumption previously mentioned :

$$\int_0^{z^*} m(z)\Delta x(z)dz \cdot \frac{df(t)}{dt} = -V_f\Delta y(z^*)f(t) - B_0\Delta x_B f(t) \quad (9)$$

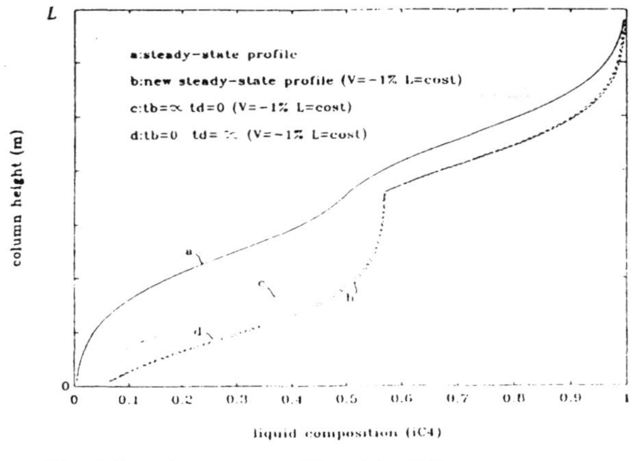

Fig.6 Steady-state profile with different arrangements of column extremities.

Solving Eq.(9) gives a first order response:

$$f(t) = \exp(-t/\tau_S) \qquad (10)$$

with

$$\tau_S = \frac{\int_0^{z^*} m(z)\Delta x(z)dz}{V_f \Delta y(z^*) + B_0 \Delta x_B} \qquad (11)$$

τ_S represents the time constant characterizing the dynamics of the bottom product composition.

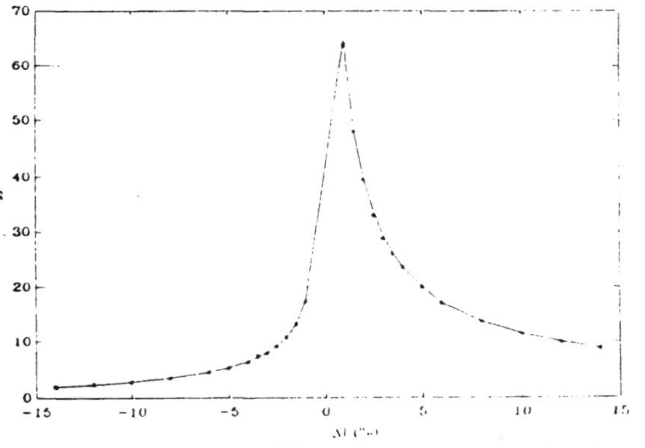

Fig.7 Distillate composition time constant: * scheme of Fig.3a ; — scheme of Fig.3b.

Analogously, applying the same procedure for the section above the height z^*, the time constant for the response of the top product composition can be derived:

$$\tau_R = \frac{\int_{z^*}^{L} m(z)\Delta x(z)dz}{V_f \Delta y(z^*) + D_0 \Delta y_D} = \frac{\int_{z^*}^{L} m(z)\Delta x(z)dz}{V_f \Delta y(z^*) - B_0 \Delta x_B} \qquad (12)$$

When the described procedure is applied to the scheme of Fig.9a the two calculated time constants are defined as $\tau_{S,\infty}$ and $\tau_{R,\infty}$, while the corresponding time constant for the scheme of Fig.9b are defined as $\tau_{S,0}$ and $\tau_{R,0}$ (in analogy with the definition given for the case of external flow changes). It can be shown that also for internal flow changes there is a decoupling between the two extremes of the column in a sense described before for changes of the external flows.

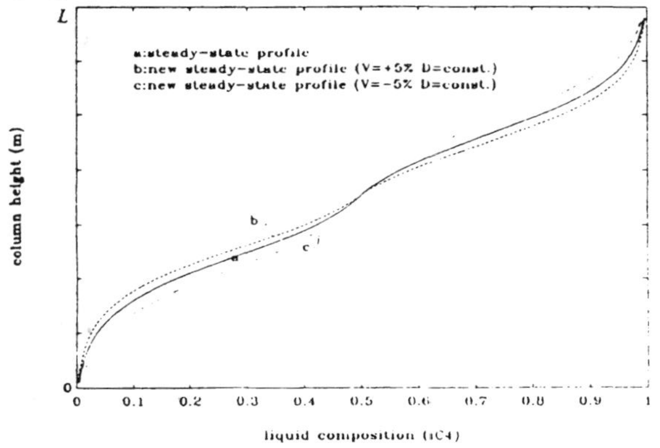

Fig.8 Steady-state profiles for a change of internal flows (ΔD=0).

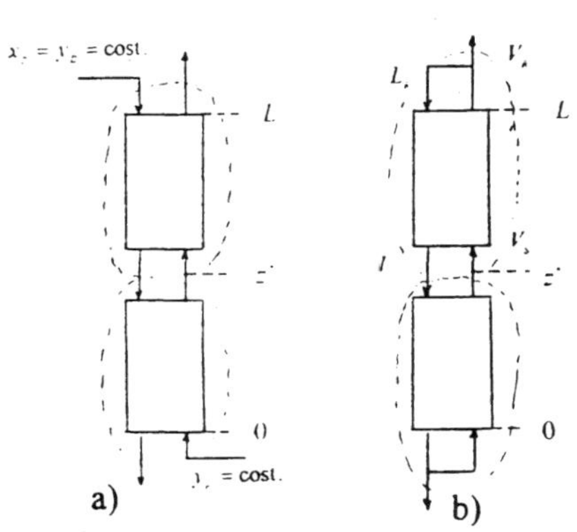

Fig.9 Column scheme adopted for the analysis of the internal flow changes.

6. TIME CONSTANT PREDICTION WITH "NON ACTIVE" HOLD-UPS

In order to analyze the effect of the amount and the fluodynamic model of the "non active" hold-ups located at one extreme of the column, conditions corresponding to a "decoupling" of the two extremes are considered. With this assumption, the response of the composition at one extreme, for example at the top of the column, can be calculated either by adopting the scheme of Fig.3a or 3b. This allows to represent such dynamic response in terms of transfer functions by a block diagram with positive feedback as shown in Fig.10. The effect of the presence of recycle on the dynamics of distillation column with the same structure has been analyzed previously by Kapoor, *et al.* (1986).

The block solution gives:

$$\frac{y_D(s)}{V_S(s)} = \frac{g_1(s)\cdot g_2(s)\cdot g_3(s)}{1 - g_2(s)\cdot g_3(s)\cdot g_4(s)} \qquad (13)$$

where $g_1(s)\cdot g_2(s) = g_\infty(s) = k_\infty/(\tau_\infty s + 1)$ represents the transfer function corresponding to the scheme of

Fig.4a; $g_3(s) = g_c(s)$ represents the transfer function of the reflux drum ; $g_2(s) \cdot g_4(s) = g_R(s)$ is the transfer function of the positive feedback and will be characterized utilizing the estimation of the time constant τ_0 relative to the particular case of null reflux drum hold-up ($g_c(s) = k_C$) with the procedure described below.
Eq. (13) for null reflux drum becomes:

$$\frac{y_{D,0}(s)}{V_S(s)} = \frac{g_\infty(s) \cdot k_C}{1 - g_R(s) \cdot k_C} \qquad (14)$$

Assuming a first order transfer function for $g_R(s)$:

$$g_R(s) = \frac{k_R}{\tau_R s + 1} \qquad (15)$$

Eq.(14) becomes:

$$\frac{y_{D,0}(s)}{V_S(s)} = \frac{k_\infty k_C(\tau_R s + 1)}{(\tau_\infty s + 1)(\tau_R s + 1) - k_R k_C(\tau_\infty s + 1)} \qquad (16)$$

The application of the simplified method described above (Eq.(5)) to the scheme of Fig.2b allows to write the following equation:

$$\frac{y_{D,0}(s)}{V_S(s)} = \frac{k_0}{\tau_0 s + 1} \qquad (17)$$

and to estimate τ_0 from the steady-state profiles (Eq.(6)). The combination of Eq.(16) and (17) gives:

$$\frac{k_0}{\tau_0 s + 1} = \frac{k_\infty k_C(\tau_R s + 1)}{(\tau_\infty s + 1) \cdot (\tau_R s + 1) - k_R k_C(\tau_\infty s + 1)} \qquad (18)$$

The values of k_R and τ_R can be related to $k_\infty, k_C, k_0, \tau_\infty$ and τ_0 using the method of moments. Equating the zeroth and first moment of the impulse response of the first and second member of Eq.(18) gives:

$$k_0 = \frac{k_\infty k_C}{1 - k_R k_C} \qquad (19)$$

$$\tau_0 = \tau_\infty + \frac{k_R k_C \tau_R}{1 - k_R k_C} \qquad (20)$$

It follows that:

$$k_R = \frac{k_0 - k_\infty k_C}{k_0 k_C} \qquad (21)$$

and

$$\tau_R = (\tau_0 - \tau_\infty) \cdot \left(\frac{k_\infty k_C}{k_o - k_\infty k_C} \right) \qquad (22)$$

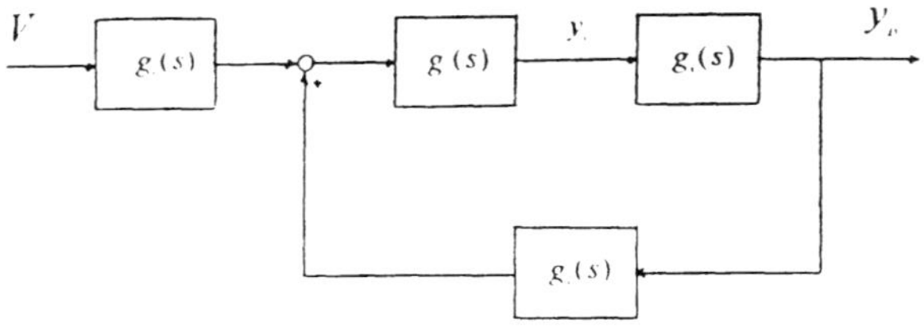

Fig.10 Block diagram

When a specific model for the reflux drum is selected, it can be expressed in terms of its transfer function $g_C(s)$ and its effect on the response of the composition to the change of a vapour flow can be observed antitrasforming the following transfer function:

$$\frac{y_D(s)}{V_S(s)} = \frac{g_\infty(s) \cdot g_C(s)}{1 - g_R(s) \cdot g_C(s)} \qquad (23)$$

If the effect of the reflux drum is analyzed only in terms of the dominant time constant of the response, the procedure of equating the zeroth and first moment of the impulse response can be applied and the following relationship derives:

$$\tau_D = \tau_\infty + \frac{\tau_C + k_R k_C \tau_R}{1 - k_R k_C} \qquad (24)$$

In Eq.(23) it has been assumed a first order model for the reflux drum :

$$g_C(s) = \frac{k_C}{(\tau_C s + 1)} \qquad (25)$$

This procedure can be used to estimate the time constants for the other flow changes and for the bottom product.

7. ANALYSIS OF THE RESULTS OBTAINED WITH THE PROPOSED METHOD

A comparison of the results obtained with the procedure proposed in this paper with the responses obtained by the rigorous simulation is described in this section. A comparison with other methods based on steady-state data is not described because all these methods take always into account both the "non active" hold-ups of the column regardless of the fact that the column could operate in conditions of "decoupling" (as it has been defined in this paper). In such conditions, the amount of the "non active" hold-up of one extreme does not affect the dynamic response of the product composition at the other extreme; the prediction error, then, is guaranteed and it is at least of the order of magnitude of the time constant of the "non active" hold-up. The described method, analogously to the others based on the same approach, is able to predict only the dominant time constant of the column which corresponds to the response time constant of the product whose purity deteriorates.

It is object of further investigations the development of a method for the prediction of the time constant also for the product whose purity increases.

For the column described in Tab.1 the time constants predicted with the proposed method (τ_{calc}) and those obtained by rigorous simulation (τ_{rig}) are shown in Tab.2 and in Fig.11 for changes of V_S. The comparison shows the excellent accuracy of the predicted time constants. The ability of the proposed method to predict the effect of the model of the "non active" hold-up has been verified for two cases:

a) reflux drum completely mixed with different residence times;
b) reflux drum unmixed equivalent to a time delay.

For case a), the time constants of the distillate composition for a +5% change of the vapour flow obtained using Eq.(23) and the rigorous simulation are shown in Tab.3. The time constants are practically the same.

For case b) the response obtained by Eq.(23), shown in Fig12, shows the ability of the proposed method to represent specific models of the reflux drum with a good agreement with the results obtained by rigorous simulation.

Table2 Parameters used in Eq.(24)

	ΔV %	τ_0	τ_∞	k_R	τ_R	k_0	k_∞	τ_{calc}	τ_{rig}
T	+5	20	17	.82	.64	-5	-.89	48	44
O	+2	38	31	.85	1.2	-5.7	-.86	71	67
P	+1	59	45	.87	2.1	-5.9	-.79	97	90
	+ 1	153	64	.92	7.2	-4	-.3	29	216
B	-.1	131	56	.86	12	3.9	.54	163	165
O	-1	43	30	.87	1.9	5.9	.78	76	75
T	-2	27	21	.86	.97	5.8	.81	59	59
T	-5	15	13	.83	.39	5.0	.84	41	38

Table3 Time constants for case a

t_D (min)	$\tau_{D,calc}$ (min)	$\tau_{D,rig}$ (min)
0	20	20
1	27.5	25
5	49	46

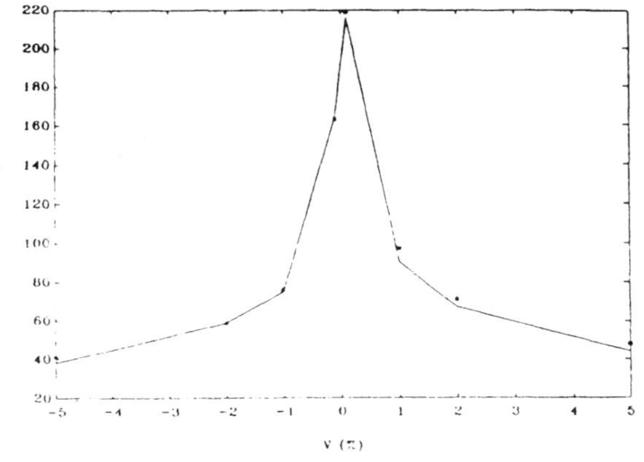

Fig.11 Comparison of time constants : * proposed method; — rigorous simulation.

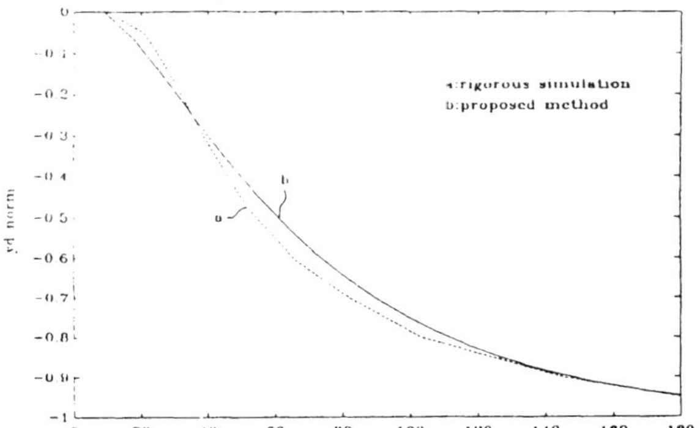

Fig.12 Distillate compositions responses:reflux drum model equivalent to 10 min time delay.

REFERENCES

Kapoor, N., Mc Avoy, T.J., Martin, T.E. (1986). Effect of recycle structure on distillation tower time constants. *AIChE Journal* **32**,411-418

Moczek, J.S., Otto, R.E., Williams, T.J. (1965).Approximation models for the dynamic response of large distillation columns. *Chem.Eng.Progr.Symp.Series* **61**,136-146

Montesi, M. (1992). Thesis in Chemical Engineering. University of Pisa.

Rademarker, O., Rijnsdorp, J.E., Maarleveld, A. (1975). *Dynamics and control of distillation columns*. Elsevier, New York.

Skogestad, S., Morari, M. (1987). The dominant time constant for distillation columns. *Comp.Chem.Eng.* **11**,607-617

Wahl, E.F., Harriot, P. (1970) Understanding and prediction behaviour of distillation columns. *Ind.Eng.Chem.Proc.Des.Dev.***9**,396-407.

Weigand, W.A., Jhawar, A.K., Williams, T.J. (1972) Calculation method for the response time to step inputs for approximate dynamic models of distillation columns. *AIChE Journal* **18**,1243-1252.

Development and Experimental Verification of a Simulation Tool for Column Startup

Ch. Kruse[1)]; G. Fieg[2)]; G. Wozny[3)]; L. Jeromin[2)]

[1)] COGNIS GmbH; Postfach 130164; 40551 Düsseldorf; Germany
[2)] Henkel KGaA; 40191 Düsseldorf; Germany
[3)] Institut für Prozeß- und Anlagentechnik; TU Berlin ; 10623 Berlin; Germany

ABSTRACT A simulation tool for column startup is developed in the presented paper. The approach used is based upon a mathematical model founded on the conservation laws, the equilibrium relationship as well as pressure, holdup and heat transfer correlations. The global Newton-Raphson method is used to solve the described equations.
Previous studies about simulation of column startup can be found in the literature. However only a few papers describe experimental investigations aimed to verify the numerical results obtained. Experimental studies have been made with the help of a laboratory-scale distillation column for the startup operation. It is shown that the simulated results agree closely with these obtained experimentally.

KEYWORDS Distillation columns; dynamic modeling; dynamic behaviour; simulation; temperature and concentration profiles

1. INTRODUCTION

The startup behaviour is interpreted as the change of the column from its initial state to a steady-state when a defined product composition is obtained. The multitude of different products and possible downtime and maintenance time requires even in continous plants frequent startups. The startup procedure represents the most complex processes in industrial practice because of the simultanous changes in many relevant process variables. Because these dynamic transitions are always non-productive procedures the aim of the investigations is to minimize the time needed.
Dynamic simulation has been used increasingly in recent years, especially in connection with distillation columns. In the presented work a simulation tool for column startup is developed using the model of equilibrium stage. This program proves to be a valuable tool in the planning and preparation of experimental work. Based on this the amount of extensive experimental studies can be minimized. The program facilitates an extended view in these complicated process and broadens the necessary understanding. It allows the calculation of temporal changes of temperatur, concentration, pressure, holdup, liquid and vapour flow profiles along the column height.
The dynamic simulation of startup and shutdown operations of distillation columns is interesting from the view point of modelling, design, control and operability. Here attempts are made to answer the question of which valves have to be switched at what time in order to achieve the demanded product purities within the shortest possible time in a safe manner. The development and application of such simulation program have been reported earlier by Gani et al. (1987) and Ruiz et al. (1988).
Practical experiences and notes concerning column startup and shutdown operation have been featured in the open literature by production engineers. Here the concern has ranged from the startup planning, a checklist for control equippment, etc. to practical hints and experience of troubleshooting [Kister 1979, Fulks 1982; Meier 1982].

Only a few papers describe experimental investigations [Yingshang et al. 1986; Fieg et al. 1993] that are aimed to verify the numerical results obtained. Therefore, often only the product concentrations or temperature profiles has been used to proof the developed simulation model. The verification of this tool by means of concentration profiles along the column height seems to be necessary and useful. The aim of the presented paper is to verify the simulation model and to get additional valuable informations of the temporal and local spreading of the relevant process variables with the help of product sampling along the column height during the startup operation. The experimental findings are compared with dynamic simulation results.

2. SIMULATION MODEL

The mathematical model used is based on the well-known conservation laws (energy and mass balance) and equilibrium relationship. Heat losses can also be considered in the model. These equations are formulated for each stage and for the peripheral systems (condenser and reboiler). The stages are numbered from the top of the column (the first stage) to the bottom (the last stage).

Component balance:

$$\frac{d\left(HU_j^L x_{i,j}\right)}{dt} = L_{j-1}x_{i,j-1} - L_j x_{i,j} + F_j x_{i,j}^F + V_{j+1}y_{i,j+1} - V_j y_{i,j} \quad (1)$$

Energy balance:

$$\frac{d\left(HU_j^L h_j^L\right)}{dt} = L_{j-1}h_{j-1}^L - L_j h_j^L + F_j h_j^F + V_{j+1}h_{j+1}^V - V_j h_j^V - Q_j \quad (2)$$

Mass balance:

$$\frac{d\left(HU_j^L\right)}{dt} = L_{j-1} - L_j + F_j + V_{j+1} - V_j \quad (3)$$

Summation equation:

$$\sum_{i=1}^{NK} y_{i,j} = 1 \quad (4)$$

Vapour-liquid equilibrium:

$$y_{i,j}^* = K_{i,j}^{VL}\, x_{i,j} \quad (5)$$

Murphree-tray efficiency:

$$y_{i,j} = M_j\, y_{i,j}^* - y_{i,j+1}\left(M_j - 1\right) \quad (6)$$

Pressure correlation for packed columns (Stichlmair et al. 1989):

$$p_j = f\left(design, x_{i,j}, y_{i,j}, L_j, V_j, T_j\right) \quad (7)$$

Because the studies described here are carried out on a packed column, a Mackowiak holdup correlation is used and implemented into this simulation tool.

Holdup correlation for packed columns (Mackowiak 1991):

$$HU_j = f\left(design, x_{i,j}, y_{i,j}, L_j, V_j, T_j\right) \quad (8)$$

Heat transfer equations:

$$Q_j = f\left(design, x_{i,j}, y_{i,j}, L_j, V_j, T_j\right) \quad (9)$$

with

j=1...NJ (number of theoretical stage) and i=1...N (number of component).

The model equations are based on the following simplified assumptions:

- ideally mixed liquid phase
- negligible storage effect of the gas phase

$$HU_j^V = 0 \quad (10)$$

The system of equations is solved simultanously with the Newton-Raphson algorithm. In each time step the differential equations are transformed with the modified Euler procedure. The difficulty of the development of such suitable simulation tool is that the used equation system possibly varies at each iteration step because of the simultanous changes in the relevant process variables. The separation stages are determined with the help of the data given in the brochure of the Sulzer company.

3. DESCRIPTION OF THE COLUMN

Fig. 1 shows the laboratory packing column with a diameter of 70 mm used in the present paper. It is packed with two types of packings, namely Sulzer BX and Sulzer DX. The column is made of glas. The unit is equipped with numerous temperature control points and a process control system. The control system enables analogues quantitities to be measured and relevant process variables to be regulated and stored. The column is fitted with sampling points at the top and bottom. Three additional samplers are placed at the packing heights 2.0m, 1.0m and 0.5m. This equippment enables the determination of steady-state as well as dynamic concentration profiles along the column height. The distillation column is operated at a vaccum (20 mbar top pressure). More details are given by Fieg et al. (1993).

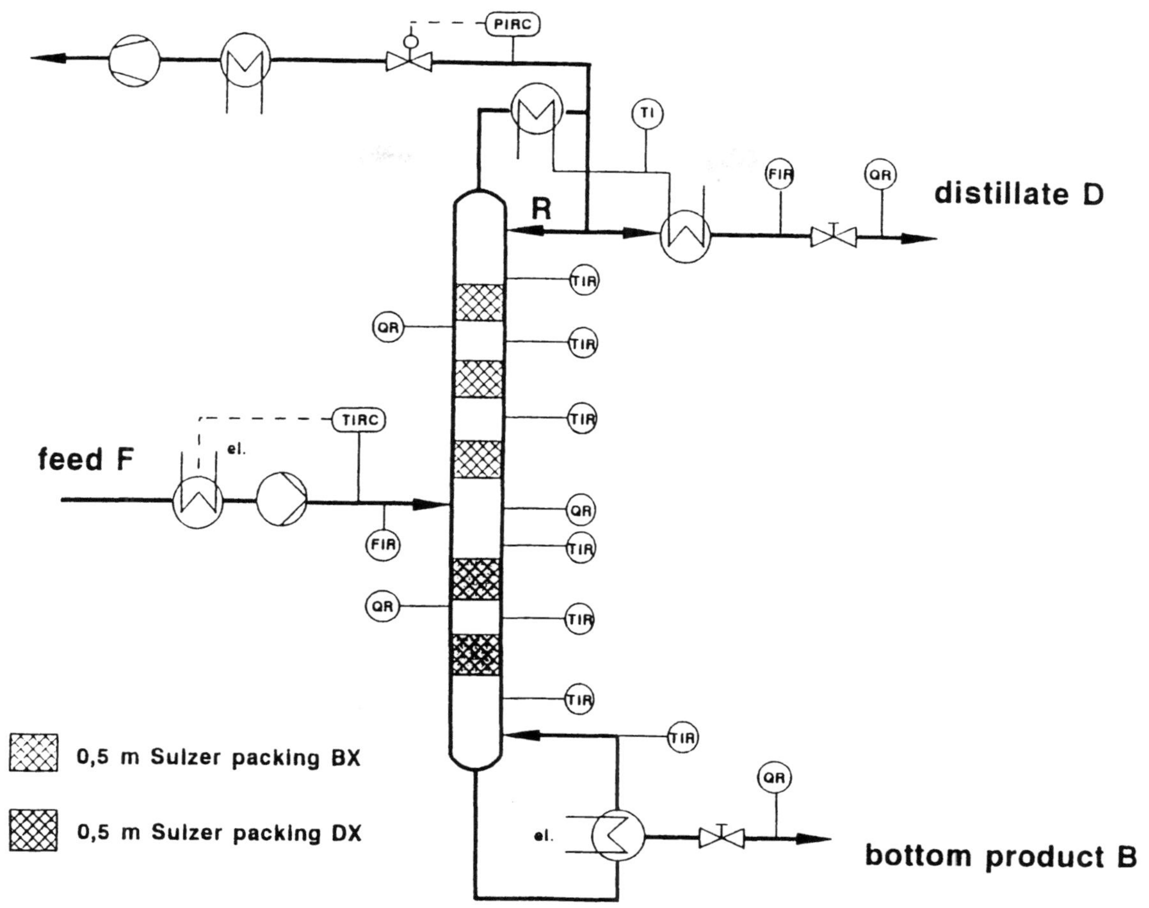

Fig. 1.Laboratory-scale distillation column

4. EXPERIMENTS

Within the framework of the experimental studies the startup behaviour of the column is investigated. For the experimental studies a Palm kernel fatty alcohol with the following composition is used from the production plant:

Table 1: Steady-state concentrations with Palm kernel fatty alcohol

[% by wt.]	C_{10}	C_{12}	C_{14}	C_{16}	C_{18}	C_{20}
feed composition	0.6	49.6	18.3	10.1	21.2	0.2
distillate composition	0.5	74.0	25.0	< 0.5	/	/
bottom composition	/	< 0.5	< 8.0	63.0	28.0	< 0.5

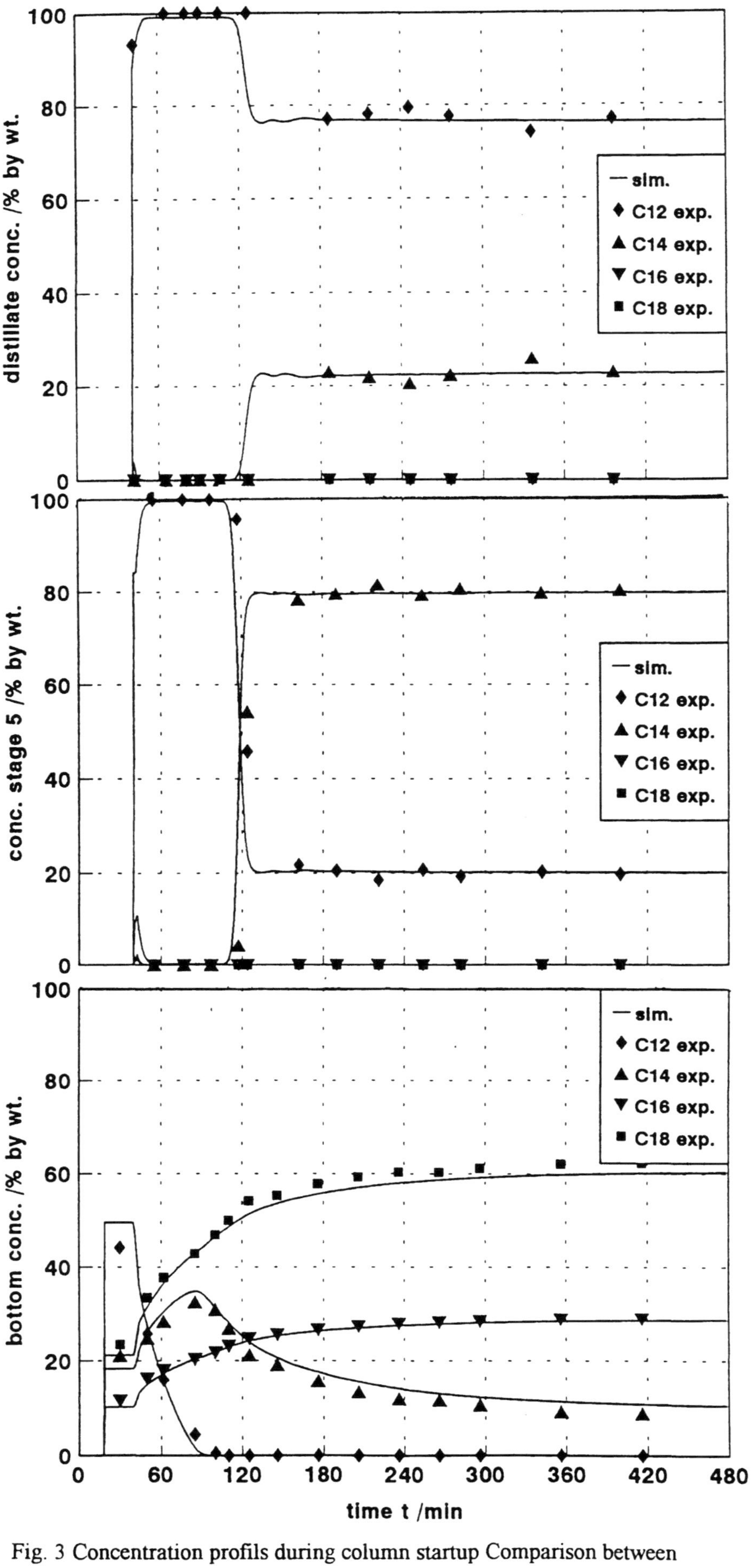

Fig. 3 Concentration profils during column startup Comparison between simulation and experimental results

For the startup study with Palm kernel the desired overhead product is 74% C_{12} and 25% C_{14}. The bottom product should contain no more than 10% C_{14} and 1% C_{12}. At a feed rate of 1.5 kg/h and a reflux ratio of 0.7 a distillate rate of ca. 1 kg/h is removed. For the experimental investigations usually the conventional startup method is used. The remarkable feature of this method is that from initial start of the operation the desired process variables (the constant feed, the distillate, the reflux ratio and the corresponding reboiler heat input) based on practical knowledge are set.

5. RESULTS

As mentioned above, the column was packed with two types of packing. According to literature from the Sulzer company the total column packing height corresponds to 30 theoretical stages including the condenser and reboiler. The measured temperature curve during conventional startup operation is shown as a function of time in Fig. 2. At the beginning of the experiments the column temperature was 20°C. A rise in temperature was observed at first below the feed tray because of feed pre-heating and the reboiler heat input. The temperature above the feed tray started to increase only when the vapour reached this column region. As can be seen, the distillate was obtained at the time point 40 min and was removed at constant reflux ratio thereafter.

Fig. 3 shows the concentration profiles along the column height (distillate, 5th stage and bottom product) as a function of time. At the beginning the C_{12} concentration was about 99% in the distillate and on stage 5. After 120 min the middle-boiling component (C_{14} fatty alcohol) increased to ca. 22%. in the distillate and about 80% on stage 5. The steady-state in this column region was attained after approximately 130 min. As it can be seen in the bottom of the column the start concentration corresponded to the feed composition. During reboiler heat input the C_{12} and C_{14} components decreased to the steady-state while the C_{16} and C_{18} fatty alcohols increased within 6 h to their constant values.

The results of the experimental studies are compared with dynamic simulations by using the mathematical model discussed above. Such simulations provide insights about the time depended behaviour of a distillation column. The results obtained are shown in Fig. 3 at the corresponding measuring points as a function of time. The comparison between the simulated and measured steady-state and dynamic temperature and concentration profiles show good agreement within the margin of accuracy of the measurements.

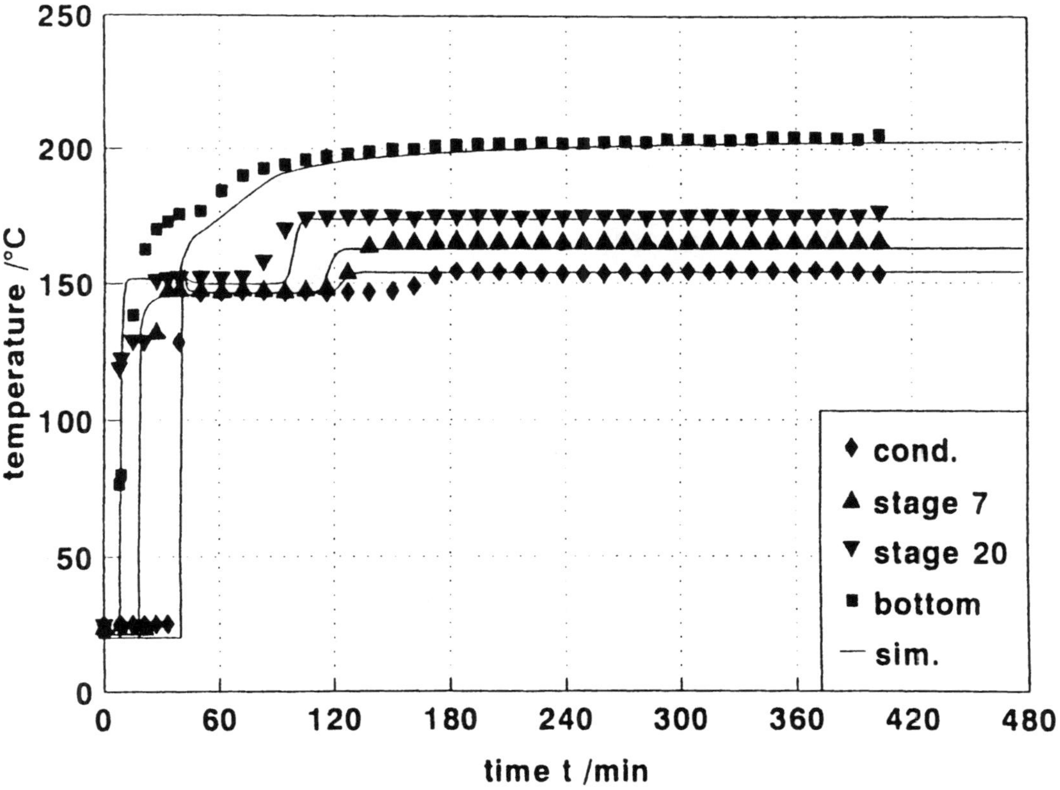

Fig. 2. Measured temperature curves during column startup Comparison between simulation and experimental results

6. CONCLUSION

In the presented paper a simulation tool for column startup is developed. An experimental verification of the program is carried out with many investigations at a laboratory-scale distillation column. The results obtained were very encouraging. It is found that the simulation results agree closely with these obtained experimentally. The good agreement confirms the used mathematical model and the assumptions taken. On this basis the program is suitable for further optimizations of the startup procedure. In previous works a new time-optimal strategy for startup operation was developed and verified experimentally [Kruse et al. 1994]. The remarkable feature of this modified strategy is that in the first step the column is operated at total removal of distillate for a certain short period. At a fixed moment the switchover of the desired reflux ratio is required in the second phase. The correct reflux ratio is well-known with the help of suitable tools or on the basis of wide practical knowledge.

The reached temporal advantages are determined. With the modified mode of startup operation the time required to attain a new steady-state can be reduced even up to 80% in comparison with the conventional method.

In summary, dynamic simulation has proved to be a valuable tool in the planning and preparation of the experimental work. Based on the good agreement between experimental results and simulation the amount of experimental work and hence the costs incurred can be clearly reduced.

ACKNOWLEDGEMENT The authors would like to thank the BMFT (Bundesministerium für Forschung und Technik) and the AIF (Arbeitsgemeinschaft industrieller Forschungs-vereiningung), Germany, for providing financial support for this work.

7. REFERENCES

Fieg G.; G. Wozny; Ch. Kruse (1993). Experimental and theoretical studies of the dynamics of startup and product switchover operations of distillation columns. *Chem. Eng. & Proc.*, 32, 283-290

Fulks B. D. (1982) Planning and organizing for less-troublesome plant startup. *Chem. Eng.*, 6, 96-106

Gani R.; C. A. Ruiz; I. T. Cameron (1987). Studies in the dynamic of startup and shutdown operations of distillation columns. *Proc XVIII th Congr. Use of Computers in Chemical Engineering*, Italy, 541-546

Kister H. Z. (1979). When tower startup has problems. *Hydrocarbons Process*, 58, 89-94

Kruse Ch., G. Fieg; G. Wozny(1994). A new time-optimal strategy for column startup and product changeover. *Proc. 5th intern. Symposium of Process Systems Engineering PSE'94* in Kyongju/Korea, 1217-1223

Mackowiak J. (1991). *Fluiddynamik von Kolonnen mit modernen Füllkörpern und Packungen für Gas/Flüssigkeitssysteme.* Salle + Sauerländer, Arau

Meier F. A. (1982). Is your control system ready to startup? *Chem. Eng.*, 22, 76-87

Ruiz C. A.; I. T. Cameron; R. Gani (1988). A generalized dynamic model for distillation columns. Study of startup operations. *Comput. & Chem. Eng.*, 12, 1-14

Stichlmair J.; J. L. Bravo; J. R. Fair (1989). General model for prediction of pressure drop and capacity of countercurrent gas/liquid packed columns. *Gas Separation & Purification* , 3, 19-28

Sulzer Company. Trennkolonne für Destillation und Absorption. Chemtech brochure

Yingshang D.; Z. Ruwen, J. Lei, W. Shihang, Y. Guocong (1986). Startup dynamics of plate column. *J. of Chem. Ind.. & Eng. (China)*, 3, 306-311

8. NOMENCLATURE

B	bottom rate	[kmol/s]
D	distillate rate	[kmol/s]
F	feed rate	[kmol/s]
HU	holdup	[kmol]
h	enthalpy	[kJ/kmol]
K	equilibrium constant	
L	liquid flow	[kmol/s]
M	Murphree tray efficiency	
N	number of components	
NJ	number of theoretical stages	
p	pressure	[bar]
Q	duty	[W]
T	temperature	[K]
t	time	[s]
V	vapour flow	[kmol/s]
x	liquid concentration	[mol%]
y	vapour concentration	[mol%]

Subscripts

i	component
j	stage number

Superscripts

F	feed
L	liquid phase
V	vapour phase
*	equilibrium

VERIFICATION OF A DISTILLATION COLUMN BENCHMARK

A. Koggersbøl and S. Bay Jørgensen

Department of Chemical Engineering, Technical University of Denmark, DK-2800 Lyngby, Denmark.

Abstract: This paper presents the validation of an elaborate model of a semi industrial scale distillation column equipped with an indirect heat pump. The model is derived for ordinary operation within the designed operating window. The paper describes the basis assumptions of the model. From experimental data model parameters are found and the model is validated against static and dynamic experimental data. The performance of the model is generally good, however, it does not catch some of the more complex behaviour of the process.

Keywords: Benchmark examples; Mathematical models; Process models; Process simulators.

1. INTRODUCTION

With the significantly increased utilisation of model based process design and optimization and especially of dynamic models for process simulation and control design there is a need for validated test cases or benchmarks of important process examples. Such test cases may be applied for development of simulation models and for development of new control structures and of new control designs.

In the chemical industry unit operations are often tightly interconnected to minimize the load on energy and raw materials. Interconnections tend to give rise to complex dynamic behaviour and to more difficult control problems. The purpose of this work is to provide a benchmark for testing e.g. control designs on a plant with the above characteristics. The paper describes development of a dynamic model of an energy integrated distillation column. This is done by first describing and commenting the assumptions which were chosen as the platform of the model, and then presenting validation results. Model equations are not presented here since they are easily derived from the assumptions.

2. PLANT DESCRIPTION

Figure 1 shows a schematic of the process. The column has 19 sieve trays, a thermosiphon reboiler, a total condenser and a reflux drum. It separates a mixture of methanol and isopropanol with a low concentration water impurity. The heat pump has an expansion valve which throttles high pressure liquid freon (R114) to a lower pressure (P_L) suitable for evaporation in the condenser. After the condenser there is a throttling valve (CV9) by which the freon vapour flowrate can be manipulated. After superheating the vapour the compressor elevates the pressure to a high pressure (P_H) suitable for condensation in the reboiler. A small part of the condensation takes place in a secondary condenser which by a cooling water circuit is connected to a set of air-fan coolers. The cooling rate can be manipulated by the control valve CV8, thus controlling P_H. Through a storage tank (Rec) and the pre-compression

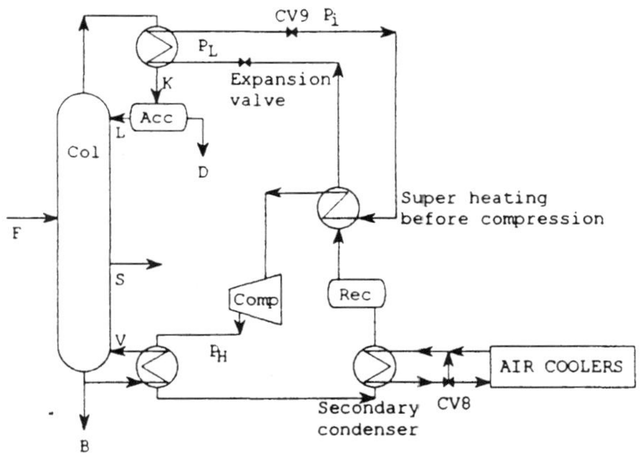

Fig. 1. Distillation plant flowsheet.

heat exchanger the freon circuit is closed at the expansion valve.

3. MODELLING ASSUMPTIONS

A mathematical model is formulated from first engineering principles. The basis for the model development is a set of assumptions motivated by the desire to obtain a model which is suitable for understanding process dynamics with the aim of leading to simplified models for control design. The model should however not become unneccessarily complex. The basic assumptions deal first with the number of phases and balances. Additional assumptions treat the behavioural aspects within each phase and balance, first for the distillation column and then for the heat pump:

A. 1 The number of flowing phases is limited to one liquid and one gas phase

A. 2 Balances are formulated for components and energy.

A. 3 Vapour phase assumptions:
1. Quasistationary and ideal vapour phase.
2. Constant molar vapour flow rate.
3. The tray pressure drop is the summed contributions from liquid weight and the dry hole pressure drop using correlations from Zuiderweg(1982).

A. 4 Liquid phase assumptions:
1. Ideal liquid phase mixing is assumed on the trays, on both sides in the reboiler and on the coolant (heat pump) side of the column condenser and in the accumulator.
2. The column side of the condenser is assumed to consist of a subcooled well mixed liquid and a well mixed vapour region at the bubble point temperature. Thus reflux is subcooled.
3. Tray hydraulics are modeled assuming spray flow regime using correlations from Zuiderweg(1982).
4. The feed is saturated liquid.
5. Liquid flows induced by pumps in volumetric units.
6. Constant heat of vapourisation.

A. 5 Total energy balance assumptions:
1. Quasistationary energy balances around column ends.
2. Heatloss is negligible.
3. Negligible pressure drop due to friction in tubes and heat exchangers.
4. Liquid heat capacities are composition (but not temperature) dependent.
5. Dynamic energy balances in condenser and accumulator.
6. Linear correlations between heat transfer coefficients and freon flowrate and heat pump low pressure.
7. Infinite efficiency of the air-fan coolers.

A. 6 Interphase transport assumptions:
1. The thermodynamic equillibrium is described using UNIFAC.
2. The reboiler is assumed to provide an ideal equillibrium stage.
3. A (constant) Murphree stage efficiency describes the approach to equillibrium.

A. 7 Heat pump vapour phase:
1. The vapour holdup is assumed quasistationary
2. Expansions over the expansion and the throttling valves are assumed isenthalpic.
3. Compression in the piston flow compressors is assumed adiabatic and reversible, i.e. isentropic.
4. Pressure changes are assumed concentrated to three locations:
   - (a) Over the compressors, where pressure increases to P_H.
   - (b) Over the expansion valve, where pressure decreases from P_H to P_L
   - (c) Over the throttling valve CV9 where pressure decreases to the compressor inlet pressure: P_i.

A. 8 Heat pump Liquid phase:
1. Dynamic heat pump fluid hold up in Rec. and on heat pump side of the column condenser. Both volumes are assumed well mixed.
2. The apparent liquid level on the heat pump side of the condenser is controlled by a mechanical float which uses the expansion valve as actuator. The total volumetric hold-up is modelled assuming a bubble rise velocity.
3. Quasistationary temperature profiles in secondary condenser and compressor gas preheater.

A. 9 A Redlich-Kwong equation of state is used for the heat transfer medium.

The main differences between the developed model and standard distillation column model assumptions used in the litterature can be pinpointed to three points:

- This column is closed to environment, therefore pressure dynamics are modelled. The vapour flow rate is determined implicitly from two energy balances around the column ends and the tray pressure drops.
- Realistic nonideal thermodynamics.
- Realistic tray hydraulics, thus flow dynamics are not instantaneous. This assumption reduce the liquid amount on the sieve trays compared to the standard assumption of emulsion flow regime. Thus leading to faster column dynamics both for the slow and fast time constants (Skogestad and Morari, 1988). This assumption also leads to a smaller contribution to the tray pressure drop.

The major limitations of this model may be variations in tray efficiencies, which at present are not incorporated. An additional limitation may be that thermal delays in the heat pump heat exchangers are not included. However both these limitations can clearly quite easily be accomodated if experimental results prove this to be desirable.

The assumption regarding constant molar vapour flowrate is reasonable because the molar heats of vapourization of the two main components differ from the mean value by only some 2% and the temperature dependence is insignificant within the relevant operating range (4 % change on a 20^oC temperature change) (Smith and Van Ness, 1975). Reid, Prausnitz, and Poling (1987) illustrate that the heat capacity of a liquid is almost constant around and below the normal boiling point. Therefore the heat capacity of a mixture is assumed only to be concentration dependent.

The assumptions of isenthalpic expansions and adiabatic compression are basically assumptions of ideal behaviour. With the present instrumentation on the plant it is not possible to validate these assumptions experimentally. However, with respect to the dynamic behaviour of the system these assumptions have little significance. Static gains with respect to the heat pump pressures P_L and P_i may on the other hand be affected significantly by e.g. non-ideal (non-adiabatic) compression, and a far better description may be obtained of the non-linear gain of the expansion valve CV9 which is of great importance for low-pressure control purposes, if the thermodynamics of the compressors were modelled with greater detail.

4. VALIDATION

Three experimental runs comprising 15 steady states were used for estimation of model parameters. As an example the plots on figure 2 show estimation of stage efficiencies by comparing simulated temperature profiles to experimental data covering most of the operating region. The PT100-elements are more accurate and better calibrated than the thermocouple elements so in case of significant differences between the two the former is to be trusted. Clearly, by using a single constant stage efficiency for all trays it is not possible to obtain a perfect temperature-profile fit, however, it seems that the temperature profile is reasonably well estimated assuming a value of 50–60%.

Using the estmated parameters transients from two experiments were simulated: 1) Transients from operation at total reflux. This may reveal all but the slowest time constant of the system. 2) Transients from changing an external flow in the production case. This will reveal the slowest time constant which is related to disturbance of the external material balance.

4.1. *Total reflux*

An experiment with 13 step changes in P_H, P_L, and the number of active compressor cylinders is shown in figure 3. 8 steady state operating points were established. Reboiler, condenser, and accumulator levels and heat pump high and low pressures are under feedback control.

The plots show the comparison between transients from the experiment and the results of a simulation where the same setpoints changes were applied as during the experiment.

On the plots of the heat pump pressures it is seen that in the simulation the transients are generally faster than in the experiment. This indicates that the heat pump model doesn't fully encompass the constraints in the energy inputs and outputs which are present in reality. At times 5.3 and 11.4 hours the cooling valve CV8 controlling P_H actually saturates which is why the decending and climbing rates of P_H suddenly decrease during these large steps. The behaviour is nonlinear since it seems that reducing P_H is easier than increasing it. The model does not catch this characteristic probably due to simplifications in the model of the compression (isentropic) and the model of the air coolers connected to the secondary condenser (infinite efficiency: cooling water return temperature equals temperature of the surroundings). The mismodelling of these characteristics is, however, only important for large stepchanges in the setpoints of P_L and P_H. The case which is usually of interest is the use of P_H and P_L for control purposes and in this case the increments of the pressure setpoints at each sampling instance are never as large as 50 kPa. The mismodelling should therefore have little significance for the control results.

In a number of situations (at time 2.93 hr, 5.22 hr, and 11.36 hr) where P_L was reduced or P_H was increased the CV9 valve of the simulation saturated at fully open in an attempt to hold P_L down to the setpoint. It seems as if the model of the compressors doesn't give them enough suction power compared to the live conditions. This inaccuracy of the model is probably due to the fact that the compression is modelled adiabatically and heatloss from tubes and from the column was set to zero in the simulation. Such heatloss would if all other conditions remained unchanged result in a decrease of P_L. Nonadiabatic operation of the compressors would increase the slope of the com-

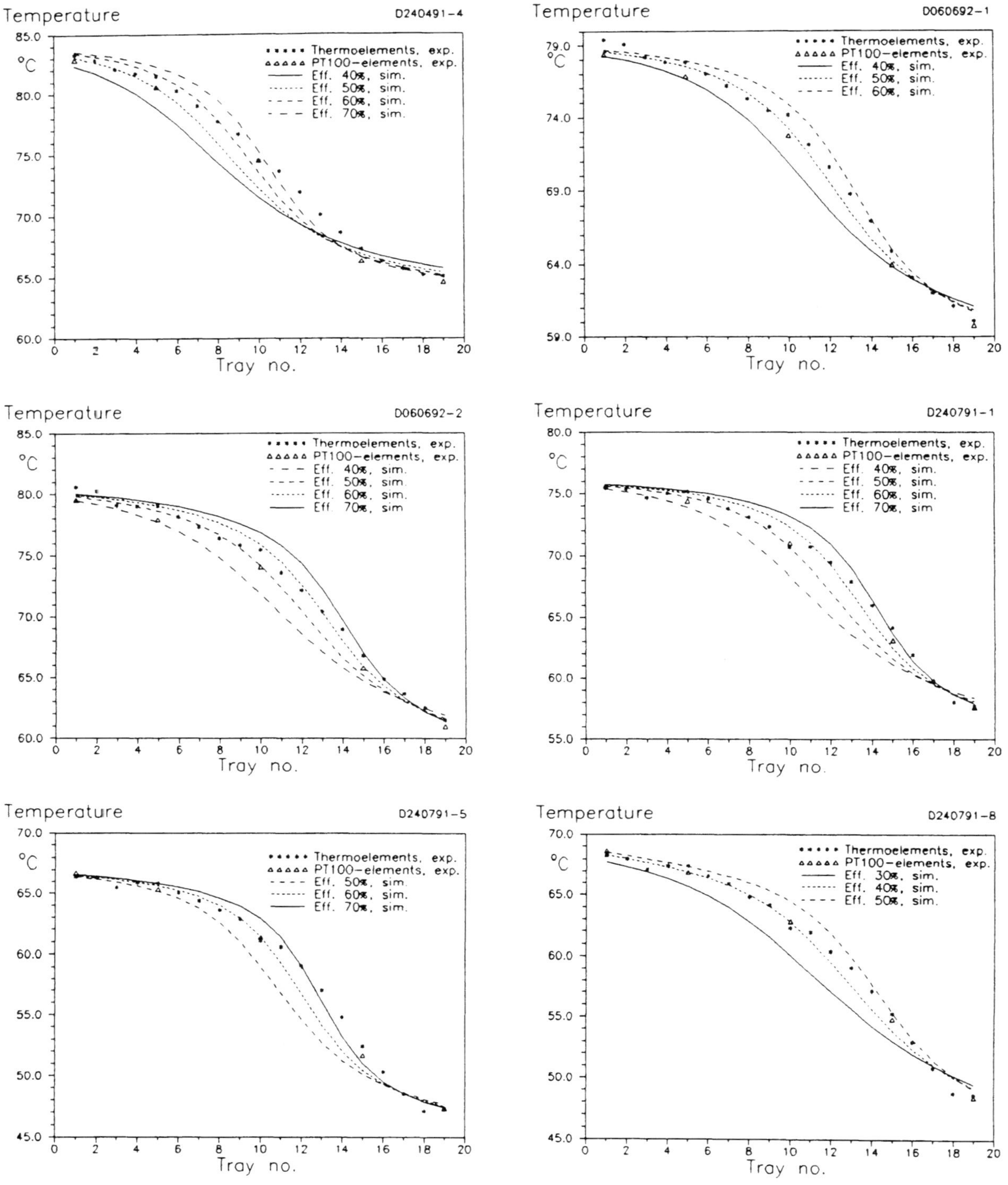

Fig. 2. Estimation of overall tray efficiency: Temperature profiles are fitted to experimental data by adjusting the efficiency.

pression curve in a P,H–diagram with the result that the pressure drop across CV9 should be larger in order to fulfil the enthalpy balance for the compression circle.

Estimation of heat transfer coefficients has proven to be encumbered with relatively large uncertainties. These affect the accuracy of the simulated column pressures and especially the simulated internal flowrates. The latter is clear as the simulated reflux flowrate suffers from quite large offsets.

Apart from these offsets and the overshoots caused by the overshoots of P_H the dynamic behaviour of the experimental reflux rate seems to be reasonably well described by the model. The model seems to have too fast dynamics for the column pressures but the deviations from the experiment can all be related to the fast and overshooting be-

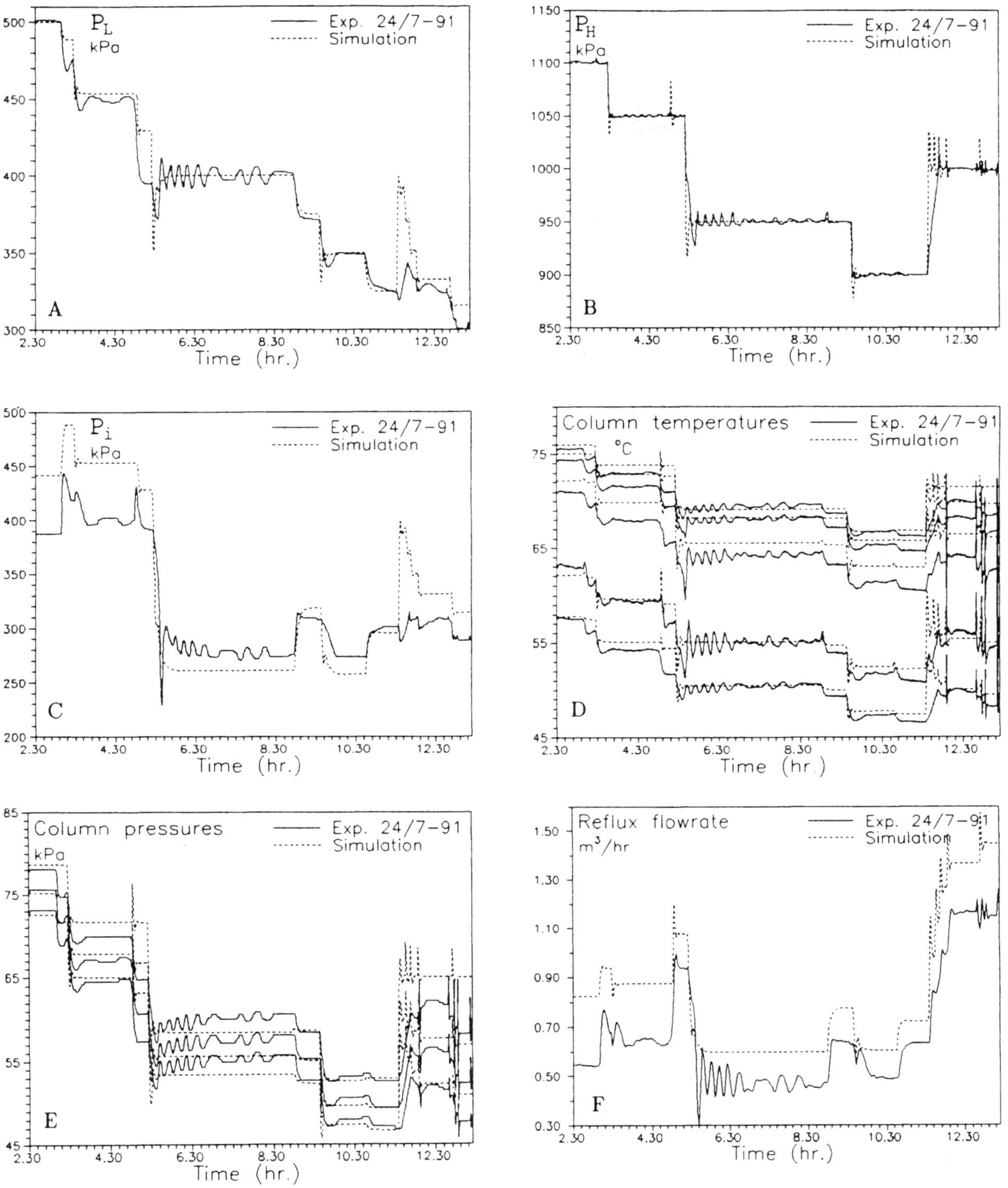

Fig. 3. Simulation of transients following a series of step changes in the setpoints to P_L and P_H. The column is operated at total reflux such that only the faster dynamics are excited. Plot A and B: Heat pump low and high pressures. These plots compare the simulated disturbances to the plant with the actual ones resulting from the setpoint changes. Plot C: Compressor inlet pressure pressure. Plot D: Temperatures on trays 1, 5, 10 ,15 ,and 19. Plot E: Pressures on trays 1, 10, and 18. Plot F: Volumetric reflux flowrate.

haviour of P_H. In situations where only P_L is changed the speed of response of the column pressures does match that of P_L both for the simulation and the experiment. So the model assumption that column pressure has no dynamics of its own seems to be justified.

Combining the reflux rate and the column pressures it seems that in the model the vapour flowrate affects the pressure drop across a tray more than in the experiment. At time 8.76 hours the setpoint for P_L is reduced by 25 kPa with the result that the reflux rate is increased and the pressure on the top tray is decreased. In the experiment the pressure is decreased on all trays but in the simulation the increased vapour flowrate re-

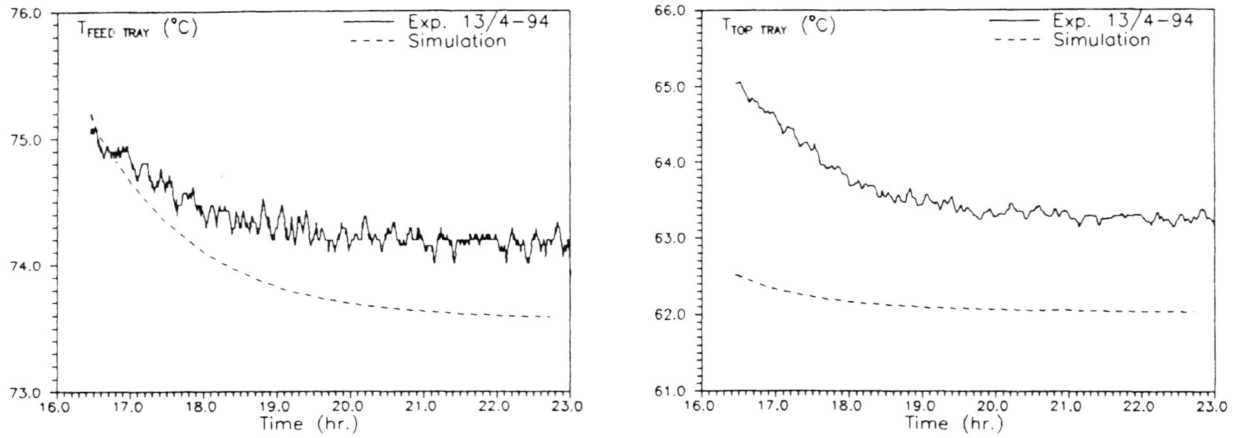

Fig. 4. Simulation of transients following a step change in the topproduct flowrate. The column is operated in the D,V-configuration such that the slow dynamics related to changes in the external flows are visible

sults in an increased pressure drop across the trays such that the pressure in the reboiler is actually increased. It appears that the exponent of Zuiderwegs (1982) equation 10

$$\Delta P_{dry} = \frac{1}{2}\rho_g \left(\frac{u_{g,h}}{C_D}\right)^2$$

should be slitely less than 2 for our tray design (6mm holes in 15mm Δ pitch). In this equation ΔP_{dry} is the pressure drop through the tray holes [Pa], ρ_g is vapour density [kg/m^3], $u_{g,h}$ is hole vapour velocity [m/s], and C_D is a discharge coefficient.

4.2. *Production case*

From a production experiment transients following a single step change in the top-product flowrate from 1.30 to 1.22 l/min are recorded. The feed flowrate is 3 l/min and boil-up rate is 18.8 l/min. Boil-up rate and top tray pressure are controlled in addition to the control loops implemented during the total reflux experiment. Being a disturbance to the external material balance, this excites the slowest time constant of the distillation column which is visible in the temperature plots on figure 4. The plots show that the model does encompass this slow dynamic element since the time constant of the simulated responses resemble that of the experimental ones. However, it is clear that the magnitude of temperature gains in the model are not accurate, and that the inaccuracy depends very much on the conditions on the actual tray. This is very likely to be a result of the "Constant stage efficiency"-assumption which from figure 2 is known to have only a limited validity. If the stage efficiency during this transient is different from the previously estimated value, and even more if it varies from tray to tray, then the shape of the temperature profile will not be estimated correctly resulting in faulty estimates of the temperature gains since these are closely correlated to the slope of the temperature profile.

5. CONCLUSIONS

The main conclusions are that the model in general perform well by predicting the time constants of the actual plant and the approximate process gains. The accuracy of predicted steady state temperature profiles is somewhat limited by the assumption of constant stage efficiency. It has also become evident that for accurate simulation of a distillation column which is heat integrated and closed to environment modelling of heat transfer rates in heat exchangers is crucial. Furthermore, the assumption of ideal compression is seen to restrict the accuracy of steady state gains from the control valves CV8 and CV9 to the heat pump pressures.

6. REFERENCES

Reid, R.C.; Prausnitz, J.M.; Poling, B.E. (1987); *'The Properties of Gases and Liquids'*; Fourth edition, McGraw-Hill Book Company.

Skogestad, S.; M. Morari; (1988); 'Understanding the Dynamic Behaviour of Distillation Columns'; *Ind. Eng. Chem. Res*, **27***(10), p.1848.*

Smith, J.M.; Van Ness,H.C. (1975); *'Introduction to Chemical Engineering Thermodynamics'*; McGraw-Hill Book Company, Third Edition.

Zuiderweg, F.J. (1982); 'Sieve trays: A view on the state of the art'; *Chem. Eng. Sci.*,**37***(10), p.1441.*

AUTHOR INDEX

Afonso, P.A.A. 147
Alvarez, J. 363
Andersen, K.S. 281

Balchen, J.G. 51, 281
Baratti, R. 475
Basualdo, M.S. 171, 191
Bekiaris, N. 243
Bequette, B.W. 345, 457
Berk, D. 159
Bertucco, A. 475
Betlem, B.H.L. 87
Biegler, L.T. 33
Bogle, D. 153
Bosgra, O.H. 305
Brambilla, A. 255, 481

Cabassud, M. 419
Caralp, L. 287
Casamatta, G. 419
Castro, J.A.A.M. 147
Ceccatto, H.A. 171
Cezerac, J. 419
Cheng, Y.S. 39
Chovan, T. 165
Corbett, G. 81

Da Rold, A. 475
Dadebo, S.A. 293
Dahl, K.S. 357
Daoutidis, P. 99
de Oliveira, N.M.C. 33
de Oliveira Magalhães, M.V. 469
Defaye, G. 287
Didriksen, H. 281
Dimitriadis, V.D. 351
Djavdan, P. 413
Doherty, S.K. 209
Dong, V. 45

Edgar, T.F. 63
Edwards, K. 63
Eek, R.A. 305
El-Aasser, M. 299
Engell, S. 27, 123

Fabiano, L.A. 81
Ferreira, J.L. 147
Fieg, G. 489
Foss, B.A. 117
Fotopoulos, J. 425
Fritz, M. 123
Fujiwara, T. 9

Garcia, V. 419
Georgakis, C. 299, 425
Gilles, E.D. 215
Gokhale, V. 311
Gomm, J.B. 209
Gori, M. 463
Guiochon, S. 287

Häggblom, K.E. 339
Hammarström, L. 105
Hangos, K.M. 333
Hansen, J.E. 21
Hasebe, S. 249
Hashimoto, I. 249
Heckenthaler, T. 27
Hillestad, M. 281
Holcroft, G. 159
Hurowitz, S.E. 311

Jacobsen, E.W. 75, 317
Jeong, C.M. 369
Jeromin, L. 489
Jørgensen, S.B. 21, 197, 323, 333, 495

Kershenbaum, L.S. 39
Kienle, A. 215
Kiparissides, C. 129
Koggersbøl, A. 323, 495
Kosanovich, K.A. 357
Kotyk, J. 185
Koulouris, A. 275
Kovács, G.Z. 333
Kravaris, C. 445
Krijnsen, H.C. 87
Kruse, Ch. 489
Kumar, A. 99
Kurooka, T. 249

Lauschke, G. 215
Le Lann, M.V. 419
Lee, K.S. 369
Lee, H. 177
Lehtinen, K. 339
Lewin, D.R. 203
Lind, M. 381
Liotta, V. 299
Liu, Z.H. 451
López Isunza, F. 153
Lorenzi, G. 255
Luus, R. 45
Luyben, M.L. 269
Luyben, W.L. 269

Macchietto, S. 135, 407, 451
Macháček, J. 185
Marquardt, W. 111
Maya-Yescas, R. 153
Mayne, D.Q. 229
McAuley, K.B. 293
McAvoy, T.J. 433
McLellan, P.J. 293
Meski, G.A. 243
Mongkhonsi, T. 39
Montesi, M. 481
Morari, M. 243
Morbidelli, M. 475
Morris, A.J. 93, 129
Mujtaba, I.M. 135
Müller, R. 69
Mumtaz, H.S. 81
Murakami, S. 9

Nagy, L. 165
Nan Ye 433
Naot, I. 203
Nishitani, H. 9

Odloak, D. 469
Ogunnaike, B.A. 1
Ohshima, M. 275
Oliveira, N.M.C. 147

Padilla, S. 363
Park, S. 177
Perrier, M. 159
Piovoso, M.J. 357
Pistikopoulos, E.N. 351
Preisigke, H.-J. 15

Ravemark, D. 401
Reagan, M. 81
Recke, B. 197
Riggs, J.B. 311
Rippin, D.W.T. 401
Rosato, L. 159
Rotstein, G.E. 407
Ruiz, C.A. 191
Russo, L.P. 457

Sågfors, M.F. 327
Sanchez, A. 407
Sandrib, B. 51
Santos, L.O. 33
Säynätjoki, P. 105
Scali, C. 255
Scenna, N.J. 191
Schott, K.D. 345
Schulz, C. 123
Seider, W.D. 81
Semino, D. 255, 463
Skogestad, S. 141, 261
Son, J.S. 369
Sørensen, E. 141
Soroush, M. 439
Stenger, Jr., H.G. 425
Stephanopoulos, G. 275
Storz, M. 215
Stuart, G. 135
Szeifert, F. 165

Terwiesch, P. 401
Tham, M.T. 93
Timmer, Th.H. 87
Tomita, S. 275
Trierweiler, J.O. 69
Tsirlin, A. 375

Varga, E.I. 333
Vidal, C. 287

Waller, K.V. 327
Wasbø, S.O. 117
Watzdorf, R.V. 111
Widagdo, S. 81
Williams, D. 209
Wittgens, B. 261
Wöllhaf, K. 123
Wozny, G. 489
Wright, R.A. 445

Yang, X. 129
Yi, S. 369
Yoon, S.C. 369
Yoshinari, T. 9

Zaldo, F. 363
Zenger, K. 57
Zhang, J. 93, 129

Printed in Dunstable, United Kingdom

91898356R00288